D0606440

15th EDITION
FROBISHER & FUERST'S
MICROBIOLOGY in Health & Disease

ROBERT FUERST, Ph.D.

Professor of Biology
Head of Microbiology Research Laboratory
Texas Woman's University
Denton, Texas

W. B. SAUNDERS COMPANY

Philadelphia London Toronto
Mexico City Rio de Janeiro Sydney Tokyo

W. B. Saunders Company: West Washington Square
Philadelphia, PA 19105

1 St. Anne's Road
Eastbourne, East Sussex BN21 3UN, England

1 Goldthorne Avenue
Toronto, Ontario M8Z 5T9, Canada

Apartado 26370—Cedro 512
Mexico 4, D.F., Mexico

Rua Coronel Cabrita, 8
Sao Cristovao Caixa Postal 21176
Rio de Janeiro, Brazil

9 Waltham Street
Artarmon, N.S.W. 2064, Australia

Ichibancho, Central Bldg., 22-1 Ichibancho
Chiyoda-Ku, Tokyo 102, Japan

Library of Congress Cataloging in Publication Data

Frobisher, Martin, 1896–

Frobisher & Fuerst's Microbiology in health & disease.

Includes bibliographies and index.

1. Medical microbiology. I. Fuerst, Robert, 1921– . II. Title: Frobisher and Fuerst's Microbiology in health & disease. III. Title: Microbiology in health & disease. IV. Title: Microbiology in health and disease.

QR46.F75 1983 616′.01 82–42506

ISBN 0–7216–3944–5

Listed here is the latest translated edition of this book together with the language of the translation and the publisher.

Spanish (*14th edition*)—Nueva Editorial Interamericana, S.A., de C.V., Mexico 4 D.F., Mexico

Spanish (*13th edition*)—Nueva Editorial Interamericana SA de DV, Mexico 4 D.F., Mexico

French (*13th edition*)—Les Editions HRW Ltee., Montreal, Quebec, Canada

The cover photograph is a scanning electron micrograph of *Aspergillus flavus.* (Courtesy of Gary T. Cole, Department of Botany, University of Texas of Austin/BPS.)

Microbiology in Health and Disease ISBN 0–7216–3944–5

Last digit is the print number: 9 8 7 6 5 4 3 2

PREFACE

In preparing the revision for the fifteenth edition of this book, the present author extensively revised, corrected, updated, and rewrote a substantial part of its content for the fourth time. This task was made less difficult throughout earlier editions as the result of the encouragement provided and the extensive experience and strength supplied by Martin Frobisher, Sc.D., the book's original author since its first edition in 1919. Dr. Frobisher is not responsible for the content of the fourteenth and fifteenth editions although his kind letters were still appreciated.

Seemingly the easiest changes to make in revising a textbook of this type would be to enlarge its content to include the extensive new knowledge in the fields of microbiology and community health. Yet in the interests of the student and the reader, and in keeping with the scope of the material to be covered in the course for which this book is written, expansion must be discriminatively avoided wherever possible. Regretfully, tempting lengthy discussions must be abbreviated or some concepts deleted entirely. Great selectivity of newer areas of knowledge has had to be exercised to keep the book within limits. Wherever newer data have been added, obsolete material has been deleted. However, this time more expansion than usual was encouraged, for which the author is thankful.

The results of the revision are a moderate increase of total pages, many of them showing new selected figures and tables that should help to make the text more understandable and pertinent; supplementary readings and other references have been updated to include especially selected publications of the last three years. The format of the book has been completely revised and modernized, which I hope will be pleasing to the reader.

New materials in this edition are: modern concepts in immunology, presented in a completely revised section, including a new chapter, updating of herpesvirus–caused diseases, discussions of developments pertaining to other viruses, discussions of legionnaires' disease, toxic shock syndrome, vaccines and globulins used to prevent rabies, and illustrations of technics used in genetic engineering, as well as new information about hybridomas and anticlonal antibodies. The most recent nomenclature has been employed for all organisms, although old names are also mentioned for the convenience of readers who have become accustomed to older names. Time and effort have been expanded in the selection and preparation of new illustrative and tabular materials; it is believed that these aspects of the book have been greatly improved. For many of the beautiful photographs and diagrams, we are indebted to numerous scientists whose contributions to the book are individually acknowledged in the legends to the illustrations.

Special acknowledgment and appreciation are due for technical advice pertaining to certain hospital-related areas of the book, freely provided, mostly by telephone, by Jan F. Fuerst, M.D., son of the author; and to Dr. Margaret Pittman and Rita C. Parker, who helped to find photographs of women who were pioneers in the field of bacteriology. The author is grateful to the administration of the Texas Woman's University for relieving him of the duties of Chairman ad interim of the Department of Biology, which he performed for three semesters. This enabled him to complete the revision of this book.

A helpful ancillary volume has also been prepared; it is a laboratory manual written especially for this text.

The author is indebted to Gesine A. Franke, R.N., without whom he would not have been associated with the publication of this book and whose true friendship he values.

Words cannot express the kind understanding, hospitality, and helpful assistance provided by Katherine Pitcoff, Nursing Editor of the W. B. Saunders Company, who expertly served as the author's mentor for the fifteenth edition of this book. Other members of the W. B. Saunders Company who were instrumental in the production of this edition were Laura Tarves; Karen McFadden; Karen O'Keefe; Betty Cobbs; Stephany Scott; Walter Verbitski; and all those other experts who worked on this revision with the added expertise and craftsmanship that for decades have been so characteristic of the entire W. B. Saunders organization.

Finally, how can encouragement and understanding be measured? Both were provided generously by the author's wife, Rachel. Love was added from his grandchildren, Bernie, Stephanie, Mandy, Sandra and David.

ROBERT FUERST, PH.D.

CONTENTS

Section One INTRODUCTION TO THE STUDY OF MICROBIOLOGY **2**

1 THE HISTORY OF MICROBIOLOGY 4

2 CLASSIFICATION OF MICROORGANISMS; CELLS AND CELL STRUCTURES 14

3 CHARACTERISTICS OF ALGAE, PROTOZOA, YEASTS, MOLDS, AND HELMINTHS 34

4 CHARACTERISTICS AND TYPES OF BACTERIA 48

5 CHARACTERISTICS OF RICKETTSIAS, CHLAMYDIAS, AND MYCOPLASMAS 68

6 CHARACTERISTICS OF VIRUSES 76

Section Two MICROORGANISMS AND THE WORLD AT LARGE **100**

7 GENERAL BACTERIAL PHYSIOLOGY 102

8 MICROORGANISMS IN THEIR ENVIRONMENT 126

9 LABORATORY STUDY OF MICROORGANISMS 142

10 MICROORGANISMS IN OUR ECOLOGICAL SYSTEM 164

Section Three INHIBITION, REMOVAL, AND DESTRUCTION OF MICROORGANISMS **188**

11 DEFINITIONS AND BASIC PRINCIPLES OF STERILIZATION AND DISINFECTION 190

12 STERILIZATION BY HEAT, RADIATION, CHEMICALS, AND FILTRATION 202

13 DISINFECTION 218

14 ANTIMICROBIAL AGENTS AND CHEMOTHERAPY 230

15 STERILIZATION AND DISINFECTION IN HEALTH CARE 248

Section Four INFECTION, IMMUNITY AND ALLERGY 262

16 THE FUNCTIONS AND COMPONENTS OF NORMAL BLOOD 264

17 INFECTIONS AND ANTIGENS 280

18 NONSPECIFIC RESISTANCE TO INFECTION 288

19 LYMPHOCYTES AND ANTIBODIES 298

20 SPECIFIC RESISTANCE TO INFECTION 312

21 ACTIVE AND PASSIVE IMMUNITY 326

22 ALLERGY AND AUTOIMMUNE DISEASES 340

Section Five PATHOGENIC MICROORGANISMS 352

23 FACTORS IN TRANSMISSION OF COMMUNICABLE DISEASES 354

Part A Pathogens Transmitted from the Intestinal and/or Urinary Tract 360

24 SALMONELLOSIS, SHIGELLOSIS, AND CHOLERA 360

25 BRUCELLOSIS AND LEPTOSPIROSIS 386

26 INTESTINAL PROTOZOA AND HELMINTHS 394

27 ENTERIC VIRAL INFECTIONS 404

28 FOODS AS VECTORS OF MICROBIAL DISEASE; SANITATION IN FOOD HANDLING 418

Part B Pathogens Transmitted from the Respiratory Tract 434

29 INFECTIONS BY PYOGENIC COCCI 434

30 DIPHTHERIA 456

31 LARYNGOTRACHEITIS, CONJUNCTIVITIS, WHOOPING COUGH, AND “TRENCH MOUTH” 468

32 TUBERCULOSIS AND HANSEN'S DISEASE 476

33 RESPIRATORY VIRAL, MYCOPLASMAL, AND CHLAMYDIAL INFECTIONS 494

Part C Bacteria Causing Venereal and Related Infections 516

34 CHLAMYDIAL DISEASES 516

35 THE TREPONEMATOSES 522

36 GONORRHEA, CHANCROID, TRICHOMONIASIS, VENEREAL HERPES, PREVENTION OF VENEREAL DISEASES 536

Part D Pathogens from the Soil 548

37 BACTERIAL PATHOGENS OF THE SOIL: THE ANAEROBES AND THE AEROBES 548

38 THE MYCOSES 560

39 SECONDARY AND ACCIDENTAL INFECTIONS; ARTHROPOD-BORNE BACTERIAL INFECTIONS 574

40 ARTHROPOD-BORNE RICKETTSIAL AND VIRAL INFECTIONS 586

41 ARTHROPOD-BORNE PROTOZOAL AND HELMINTHIC DISEASES 600

42 ALLIED HEALTH PERSONNEL, ASSISTANTS TO THE PHYSICIAN 614

APPENDIX A
Classification and Nomenclature of Bacteria as Given in Bergey's Manual of Determinative Bacteriology (Eighth Edition) 627

APPENDIX B
Sterilization Charts 634

INDEX 643

15th EDITION

FROBISHER & FUERST'S

MICROBIOLOGY

in Health & Disease

SECTION ONE

INTRODUCTION TO THE STUDY OF MICROBIOLOGY

Theobald Smith (1859–1934), an American pathologist, showed that *Babesia bigemina*, a sporozoan protozoan parasite that causes Texas fever in cattle, was transmitted by ticks. This finding helped to solve the riddle of such diseases as trypanosomiasis, malaria, and yellow fever. He was a pioneer in the comparative etiology and immunology of infections and parasitic diseases. (Courtesy of the late Professor Simon Henry Gage.)

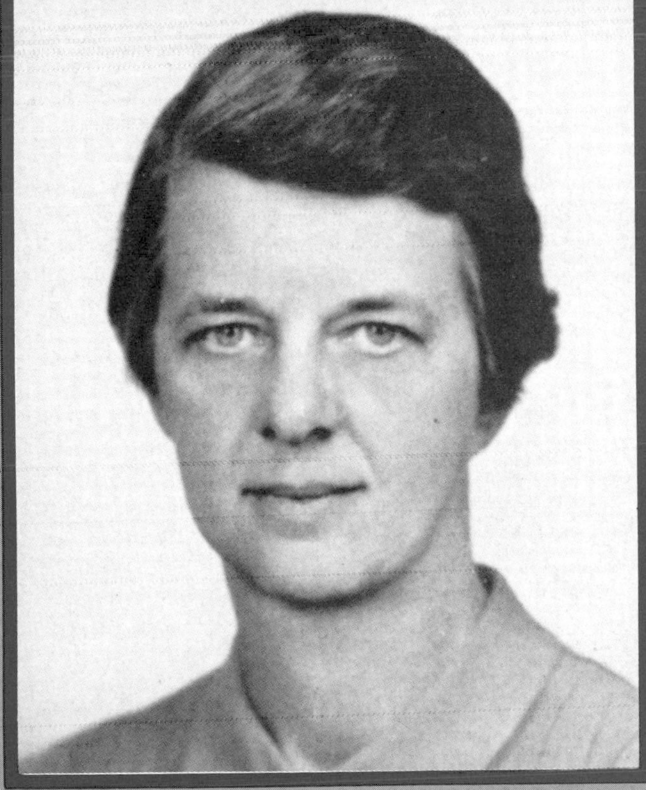

Rebecca Craighill Lancefield (1895–1981) formulated the "Lancefield classification of the beta-hemolytic streptococci," a system most important in classifying and identifying virulent human pathogens belonging to the genus *Streptococcus*. She served as President of the Society of American Microbiologists. (From *Chronicles of the Society of American Bacteriologists 1899–1950* by Barnett Cohen. Copyright © 1950 by the Society of American Bacteriologists.)

THE HISTORY OF MICROBIOLOGY

Microbiology is the study of living organisms that are so small as to be visible only through microscopes. Some of these *microorganisms* require yet greater magnification in order to be seen, which can be obtained only by using a powerful instrument called the electron microscope. Microorganisms comprise algae, fungi, bacteria, viruses, and protozoa.

Microorganisms are the oldest, the most numerous, and the most diversified form of life on earth. They shape our environment, decompose dead animal and plant matter, and keep our soil fertile. Most microorganisms are useful to us and only a very few are true pathogens. These we have to control; the others we can use for the welfare of humanity.

Why Do We Study Microbiology? In this century alone, owing in large part to the application of discoveries in microbiology, your life expectancy has been increased by approximately 50 per cent. Not very long ago, at the beginning of this century, people died of illnesses we no longer even contend with as causes of death except in the most neglected or advanced cases: tuberculosis, rheumatic heart disease, and syphilis. You have inherited a healthier world without fear of epidemics of smallpox, typhus, polio, plague, diphtheria, rubella, measles, or mumps. We know how to protect our children against tuberculosis by giving them more wholesome food and numerous environmental, medical, and other advantages. Because of microbiology we know how to preserve our food better and we eat and live in cleaner, better houses under more sanitary conditions. Due in part to advances in microbiology, you, the people living in the twentieth to the twenty-first centuries, are the tallest, brightest, healthiest, and best looking generation ever to inhabit this planet.

On July 20, 1969, Neil Armstrong, standing on the moon, uttered the historic words: "One small step for man, one giant leap for mankind." Yet space technology is only in its infancy; the universe is vast and is here to be conquered by human beings—let us hope, in peace! What life exists on other planets? What microorganisms will the space traveler meet? What benefit will we derive and what dangers will we encounter? The progress of mankind never stops for long. We must ask what bacteria, protozoa, or viruses we shall encounter as man moves farther and farther out into space.

Yes, microbiology and medicine must be ready to face the challenges of the future, as they have faced them during this turbulent century, which will be soon completed. Will you be ready for the year 2000 and beyond, or even for the proliferation of knowledge today? Are scientists really ready now to redesign living organisms? Some people have questioned that we should. Yet this is already being done, using microorganisms. This most recent and exciting development, called "recombinant DNA technology" began in 1973 and is based on knowledge gained over the last 30 years in microbial genetics. (DNA is deoxyribonucleic acid, the nucleic acid that determines the heredity of an organism.) We are now able to insert genetic material of any living organism into selected bacteria and thus change them so that they can perform special tasks, normally inherent in the donor cell, possibly of mammalian origin. These gene manipulating technics are referred to as *"Genetic Engineering."*[1] This is truly the most exciting new development in microbiology, genetics, and biochemistry in this century.

---

[1]See Chapter 7.

CHAPTER 1

The science of microbiology is essential to members of the health professions and, in fact, to everyone who is interested in maintaining health and preventing disease. Knowledge of the method of transfer of microorganisms from one person to another can reduce the number of colds and incidents of "flu" that you and your family suffer; knowledge of sterilization and disinfection will teach you precautions in bandaging minor cuts and handling food to prevent food poisoning and infection; knowledge of immunity will help you understand the relation of microbiology to health and disease, and the importance of immunization procedures to prevent smallpox, typhoid fever, and influenza.

These are only a few simple examples of the application of microbiology to everyday life. Thanks to health procedures based upon principles of microbiology, health workers can not only assist in the prevention of disease but also help to promote recovery when disease does occur. It is obvious that everyone in the health professions should understand the underlying principles of microbiology, health maintenance, and disease prevention so that they will understand preventive procedures, make intelligent adaptations of them, and teach auxiliary personnel (attendants, orderlies, maids, technicians, and others) how to proceed correctly. Is the study of microbiology difficult? Not excessively so. It is a very logical science and certainly one of the most fascinating.

Evolutionary Succession of Life Forms. All available evidence shows that the various types of organisms existing today slowly developed from earlier forms. For example, some of the present-day species of plants and animals have evolved, through millions of generations, from more primitive ancestors. It may be inferred that the simplest and most lowly creatures alive today are descendants of those that existed in the earliest ages, probably more than four billion years ago. Although we do not actually know what the first plants or animals were like, they may have been something like present-day bacteria or other microorganisms, possibly viruses or Protozoa. Certain components of cells in higher forms of life superficially resemble bacteria and may have descended from free-living bacteria-like organisms by evolutionary processes of endosymbiosis.

A more recent theory suggests that mitochondria and other organelles of the "higher" forms (eukaryotes), instead of being derived from phagocytized "primitive" microorganisms (prokaryotes), evolved from the prokaryotes themselves by intracellular metamorphoses of already existing parts of the prokaryotes, especially the cytoplasmic membrane.

EARLY OBSERVATIONS THROUGH THE MICROSCOPE

Though the lineage of microorganisms is primordial, they were not discovered until fairly recent times, about 1680. Because of their minuteness, a knowledge of their existence had to await the invention of the microscope. Although Zacharias Janssen is credited with having devised the first compound microscope in 1590, the first person to see and also describe microorganisms was *Antony van Leeuwenhoek* (1632–1723), a Dutch draper and haberdasher of Delft who wanted to observe the weave of fine cloth (Fig. 1–1). His microscopes, which he made himself, consisted of a biconvex lens held in a metal frame

Figure 1–1 Antony van Leeuwenhoek (1632–1723). (From Dobell, C.: Antony van Leeuwenhoek and his "Little Animals." New York, Staples Press, 1932.

(sometimes gold and silver!). These lenses magnified up to 270 diameters. With such crude instruments he examined water from pools, the tartar from his teeth, feces from a patient with dysentery, and many other substances and fluids. He was amazed to see in all these substances what he called "animalcules"—tiny organisms, spherical, cigarette-shaped, or spiral in form, some of which were in rapid motion. The drawings that he made are still in existence

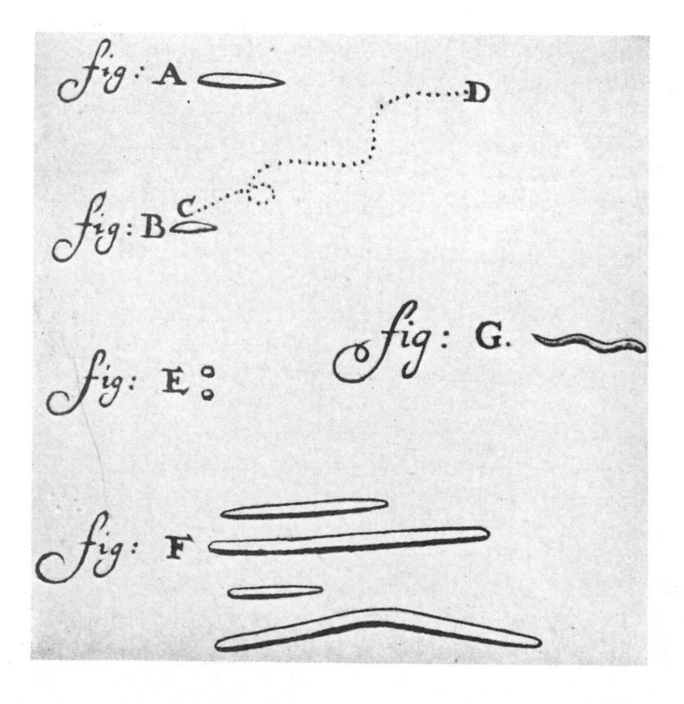

Figure 1–2 Drawings of bacteria made by van Leeuwenhoek in 1684.

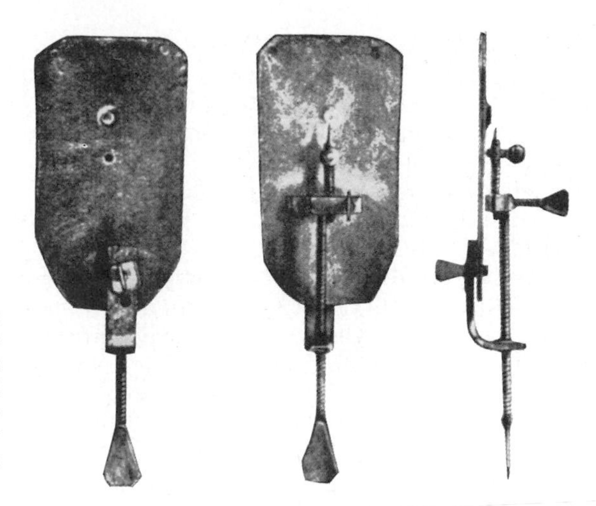

Figure 1–3 One of van Leeuwenhoek's original microscopes is preserved at the University of Utrecht. The single minute lens was mounted in the perforated metal plate and held close to the eye. The two vertical screws surmounted by a fine point were used to hold the object before the lens and adjust its position.

and prove that what he saw actually were bacteria and protozoa and other microorganisms (Fig. 1–2). Over a period of 40 years van Leeuwenhoek wrote 125 letters that were translated into English and reported in the Transactions of the Royal Society of London. In addition, he sent 27 papers to the French Academy of Sciences, which were published in the Academy's Memoires. It is fortunate that he described his findings to the Royal Society of London and left 26 of his 247 microscopes to the Society after his death. Later on most, if not all, of these heirlooms were lost, but not before Baker[2] left us with exact descriptions of all of them. (The University of Utrecht claims to have one of van Leeuwenhoek's original microscopes, shown in Figure 1–3.) Van Leeuwenhoek's discoveries opened up a whole new world to investigators who followed him, a world that can be explored by all who learn to use our present-day microscopes.

[2]Henry Baker, in his work *The Microscope Made Easy* (1742).

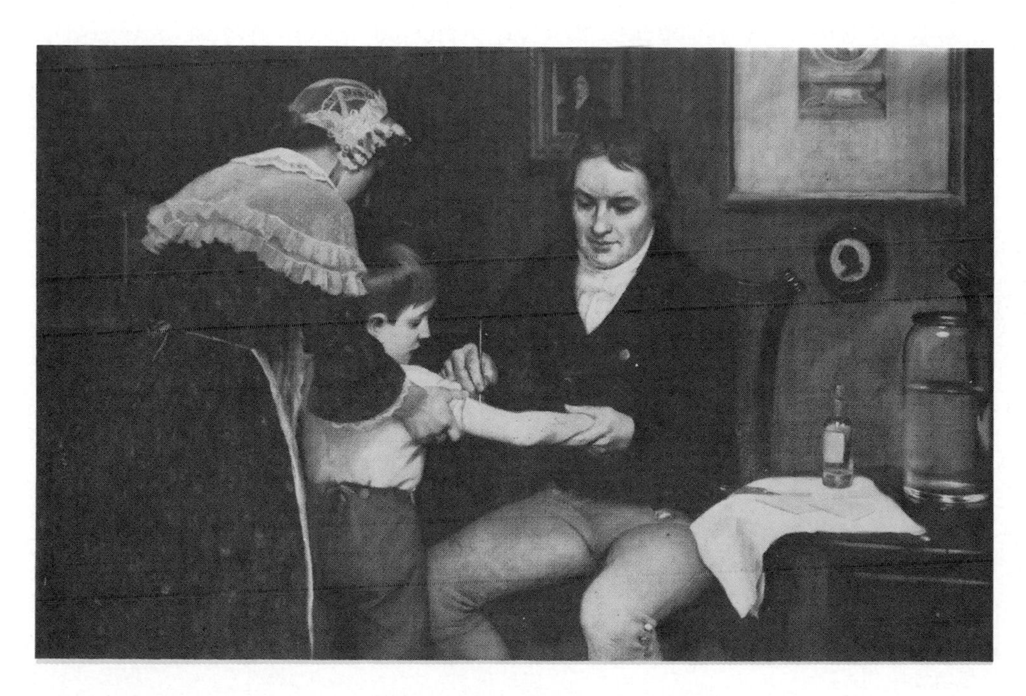

Figure 1–4 Edward Jenner (1749–1823) performing his first vaccination, May 14, 1776. (Courtesy of the Bettmann Archive, Inc., New York.)

Figure 1–5 Ignaz Philipp Semmelweis (1818–1865), pioneer of asepsis in obstetrics. (From Sigerist: The Great Doctors. Courtesy of W. W. Norton and Company, Inc.)

Early History of Disease Prevention. Long before it was known that microorganisms are the cause of communicable diseases, methods to prevent disease were practiced. For example, in 1796 Edward Jenner (Fig. 1–4) showed that smallpox could be prevented by vaccination with cowpox. Semmelweis (Fig. 1–5) later fought the spread of disease in the maternity wards of hospitals with chemical disinfectants. The rules and methods of the new science of preventive medicine, however, are largely based on the life work of two men, Louis Pasteur (1822–1895) and Robert Koch (1843–1910) (Figs. 1–6 and 1–7).

Great Discoveries in the Science of Microbiology. Shortly after the middle of the last century, the French chemist Pasteur, famous for founding the science of polarimetry, proved that fermentation and putrefaction were caused by living microorganisms and noted a similarity between these processes and infectious diseases. He spoke of spoilage of wines and beer as "diseases" of those beverages. It was still believed by many scientists that living forms could arise spontaneously from dead matter. Spallanzani had disproved this "doctrine of spontaneous generation" almost a century earlier; however, it was Pasteur who devised a simple experiment to show that, so far as was known on the basis of data then available, living organisms originated only from living organisms, never from nonliving matter (Fig. 1–8).

Although the question of spontaneous generation was thus apparently resolved in the nineteenth century, it was raised again in the twentieth century with infinitely more sophistication under the heading "chemical evolution." Today microbiologists and biochemists, using simple, naturally occurring substances and electric sparks, gases, and steam under high temperature and pressure, can synthesize many complex organic compounds that formerly had been found only in living cells. In fact it is now possible to build genelike structures as they may have occurred originally some two billion or more years ago. But the fanciful eighteenth century "doctrine of spontaneous generation" is dead: bacteria come only from existing bacteria.

After studying fermentation of wine, Pasteur investigated a communicable disease in silkworms; as a result he formulated the *germ theory* of disease. In dealing with diseases in humans, he insisted that bandages should be cleaned and instruments in the hospital boiled. Joseph Lister (1827–1912), an English surgeon, applied Pasteur's discoveries to surgery even before the actual bacteria that cause surgical infections had been discovered. Lister's work is the basis of present-day aseptic surgical technique (Fig. 1–9).

Figure 1–6 Louis Pasteur, 1822–1895. "Chance favors the prepared mind." (From Carpenter: Microbiology. 4th ed. Philadelphia, W. B. Saunders Company, 1977).

In ingenious experiments Pasteur protected sheep against anthrax by vaccinating them with a preparation of anthrax organisms. The Pasteur "preventive treatment" of rabies was developed to protect against the disease by injections of dried material from the spinal cords of animals that had died from rabies. Joseph Meister, a small boy bitten by a mad dog, was the first human whose life was saved by this method and who did not get the disease hydrophobia (rabies) because he was protected by Pasteur's injections. It is an irony of history that half a century later Joseph Meister, who was then the gatekeeper of the Pasteur Institute, killed himself when ordered by curious German soldiers to open Pasteur's crypt.

At the time of Pasteur's discoveries, Ferdinand Cohn (1828–1898) was one of the foremost microbiologists in Germany. He extended Pasteur's findings in his own laboratory, where he worked with algae and fungi. Cohn became so interested in bacteriology that he wrote some of the earliest books about bacteria. In them, he followed the Dane O. F. Müller's suggestion (1786) and classified bacteria into genera and species. Cohn was of great help to Robert Koch and encouraged him to publish his paper on anthrax.

Two hundred years before the advent of bacteriology, Robert Boyle suggested that some diseases are caused by some action of living microorganisms. M. A. Plenciz, a Viennese physician, published a germ theory of infectious diseases in 1762 based purely on speculations, and in 1839 Schönlein made the admirable discovery of the infectious nature of ringworm. It was the gifted anatomist Henle who proposed that infectious diseases might be caused directly by microorganisms, but his theory was also deficient in experimental data. One of Henle's students was Robert Koch, who became a famous physician and supplied all the evidence needed to prove further the "germ theory of disease," as formulated in the meantime by the great Pasteur and others.

Figure 1–7 Robert Koch, 1843–1910. In Koch's laboratory the techniques of medical bacteriology were developed. (From Clifton, C. E.: Introduction to the Bacteria. New York, McGraw-Hill Book Co., Inc., 1950.)

Pure Culture Study and the Etiology of Infectious Diseases. The development of methods for isolating bacteria in pure culture was among the most important early advances in microbiological techniques and was largely the work of Robert Koch. A pure culture of microorganisms is one in which only one kind of organism is actively growing in test tubes, plates or flasks on substances that are appropriate as food.[3] Under natural conditions many microorganisms of different species usually exist together in the same environment. In the feces of a typhoid patient, for example, there may be typhoid bacilli, but they are mixed with billions of cells of other species of bacteria and other forms of microorganisms. It is part of the work of the medical diagnostic microbiologist to isolate a pure culture of the typhoid bacillus from the feces of the patient.

In 1876 Koch isolated in *"pure culture"* the bacterium causing anthrax from the spleens of infected cattle and was able to infect mice with the culture. Previously, in 1850, C. J. Davaine had observed little threadlike bodies (anthrax bacilli) in the blood of animals that had died of anthrax. A pure culture is the growth of a single species, the progeny of one cell, with all individuals genetically identical.

Having become involved in questions concerning the true etiology of infectious diseases, Koch summarized what he believed to be the evidence necessary to prove an organism to be the cause of a disease. This evidence

[3]Such a nutrient preparation is called a *culture medium*.

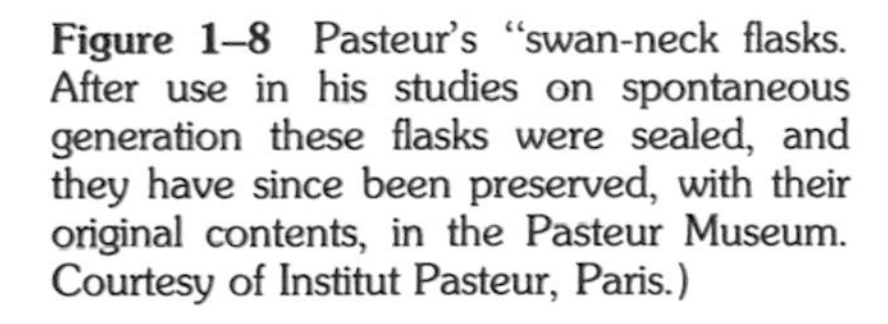

Figure 1–8 Pasteur's "swan-neck flasks. After use in his studies on spontaneous generation these flasks were sealed, and they have since been preserved, with their original contents, in the Pasteur Museum. Courtesy of Institut Pasteur, Paris.)

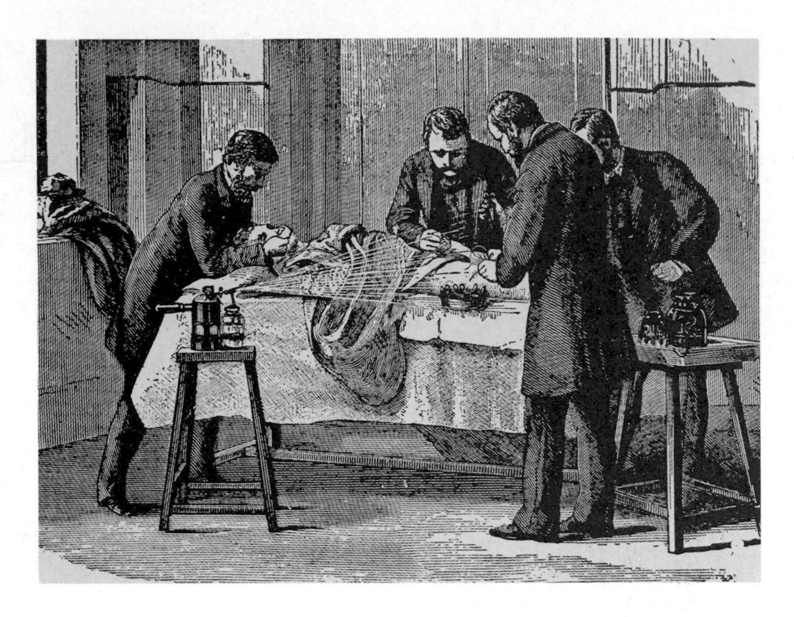

Figure 1–9 Lister operating with carbolic spray. The illustration represents the general arrangement of surgeon, assistants, towels, and spray in an operation performed with (supposed) complete aseptic precautions (1882). Note the carbolic mist spraying over the field of operation. Also note the coats. (W. Watson Cheyne.)

consists of four postulates, generally called *Koch's postulates*. Essentially, they are as follows:

1. The organism causing the disease must be found in all observed cases of a given disease in pathological relationship to its symptoms and lesions.
2. The organism must be isolated from these victims of the disease into *pure culture* for study in the laboratory.
3. When the *pure culture* is inoculated into a susceptible animal, possibly humans, it must reproduce the disease condition or, as later modified, engender specific antibodies in the new host.
4. The organism must be isolated again into *pure culture* from such experimentally caused infections.

Using these rules and techniques, Koch discovered the tubercle bacillus, the cholera bacillus, and modes of transmission of numerous other diseases. Discovery after discovery was made in Koch's laboratory by himself, his pupils (Fig. 1–10), and his assistants, among them Gaffky and Kitasato.

Figure 1–10 One of the groups of famous scientists who studied microbiology under Koch. Standing, left to right: Alphonse Laveran (1845–1922), discoverer of the malarial parasite; Émile Roux (1853–1933), codiscoverer of diphtheria toxin; Edmund Étienne Nocard (1850–1903), French veterinarian and mycologist; George H. F. Nuttall (1862–1937), British microbiologist. Sitting, left to right: Robert A. Koch (1843–1910), discoverer of the tubercle bacillus and pioneer microbiologist; Karl Joseph Eberth (?) (1835–1926), discoverer of the typhoid bacillus; Elie Metchnikoff (1845–1916), Russian zoologist and discoverer of phagocytes and phagocytosis. (Courtery of Wiley A. Penn, Director of Laboratories, Department of Health, Savannah, Georgia.)

The discoveries and the work of Pasteur, Koch, and many other dedicated pioneer investigators were the beginnings of the sciences of bacteriology, mycology, virology, parasitology, and immunology, all areas of study in microbiology. Within the 15 years following 1880 the organisms causing many important infectious diseases were discovered and isolated in pure cultures in which each could be studied as a single, living entity. These discoveries made it possible for preventive medicine and specific therapy to be firmly established.

The Health Professions and Microbiology. Although nursing in some form has existed from the start of human life on earth, the beginning of professional nursing is usually credited to Florence Nightingale (1820–1910), who began her work more than 120 years ago during the war in the Crimea (1854–1856). Microbiology was then in its infancy.

Subsequent discoveries concerning the relationships of microorganisms to disease necessitated the development of nursing procedures far more complex and refined than those used by "The Angel of the Crimea." In her time little was known about the transfer of infections from person to person. The development of aseptic surgery after Lister's discoveries (about 1865) opened the whole field of operating room technique to the surgeon and the professional nurse.

Figure 1–11 Emil von Behring (1854–1917), discovered with Kitasato of tetanus toxin and of tetanus antitoxin. In emergency rooms 250 to 500 units of human hyperimmune tetanus gamma globulin are commonly used for prophylaxis after injury in nonimmunized persons. Hyperimmune human gamma globulin eliminates problems of hypersensitivity. It is retained longer and is more effective than horse antitoxin for prophylaxis or treatment.

The earlier discovery of preventive measures such as administration of smallpox vaccine (Edward Jenner, 1796), and later of diphtheria and tetanus toxoids and antitoxins, or immune sera (von Behring, Fränkel, and Kitasato, 1890) (Fig. 1–11), to say nothing of polio and measles vaccines (Enders, Weller, and Robbins, 1949; Salk, 1954; Sabin and others, 1954–1967) and gamma globulins[4] for prevention of measles, rabies, pertussis, and some other diseases—all of these made necessary the development of new concepts and education in the preparation and administration of these substances. The discovery of diagnostic tests in the field of microbiology has necessitated more advanced microbiologic training of medical technologists, doctors, nurses, and other health professionals in methods for collecting specimens and reporting specific laboratory findings, so that therapy can be started early. The discoveries of specific chemicals (chemotherapeutic agents such as sulfonamides) and antibiotic substances (such as penicillin, streptomycin, the tetracyclines, and chloramphenicol) have altered medical care and presented new problems and new opportunities for the health team.

Discoveries of means by which infections are transmitted led to the development of effective methods to prevent the spread of disease. For example, the transmission of some diseases through the bites of insects and arachnids was demonstrated in 1893 by Theobald Smith in connection with Texas fever, a disease of cattle caused by infectious ticks. In 1895, Sir Ronald Ross (1857–1932), an army physician in India, demonstrated the transmission of malarial protozoa by mosquitoes. The parasite had, however, been seen in human red blood cells as early as 1881 by Laveran (Fig. 1–10), a French army surgeon in Algeria. In 1900, the transmission of the yellow fever virus by a particular species of mosquito *(Aedes aegypti)* was demonstrated in Cuba by Drs. Walter Reed (1851–1902), James Carroll (1854–1907), Aristides Agramonte (1868–1931), and Jesse Lazear (1866–1900) of the United States Army. The discoveries of the modes of transmission of malaria and yellow fever are among the most brilliant in medicine and have been epoch-making in their results.

Modern surgical procedures would be impossible without disinfection and sterilization. Today's dairy, frozen and packaged food, and canning industries are dependent on microbiology. Sanitation of water supplies and reclamation of polluted waters (previously called sewage disposal) are possible only because of knowledge gained from microbiology in the past century. The human life span has been extended partly because of what is now known about prevention

[4]Protective proteins (antibodies, Chapter 18) derived from the blood serum of immune persons or animals.

of deadly transmissible diseases. In many ways every health profession is dependent upon the knowledge, understanding, and utilization of information from microbiology.

Thanks to the efforts of many scientists, the incidence of certain microbial diseases has been decreased. Only constant vigilance on the part of all members of the health professions, however, will continue this downward trend in preventable diseases. The Registered Medical Microbiologist, the practicing physician, and the professional nurse need to apply knowledge from microbiology to the everyday practice of their professions. Although certain techniques and skills could be learned by rote, the truly professional person understands the scientific facts and principles underlying these techniques. The qualified professional also knows how these techniques should be modified for certain kinds of patients (e.g., those who have had cardiac surgery); what techniques must be used in the operating room, the delivery room, and the nursery, and how these can be safely modified in emergency conditions (e.g., war or other disasters; and why patients, families, and the general public should be informed about immunizing agents that are available, such as polio vaccines.

A trained member of the health team who is knowledgeable about microbiology can easily differentiate truth from misinformation and misinterpretation of medical literature written for the laity, and he or she can recognize error in popularized versions of health practices that are widely circulated through the mass media. The health worker will find that there is rarely a duty or act performed in ordinary living where knowledge of microbiology does not apply.

SUMMARY

Microbiology, the study of microscopically small living organisms, has progressed in this century from a relatively simple science to one producing great advances in preventive, diagnostic, and therapeutic medicine. It has improved food preservation and sanitary conditions, and it has increased the human life expectancy by approximately 50 per cent. We are at the threshold of even greater advances involving "recombinant DNA," transferring genetic material of mammalian origin into selected bacteria, which will then produce specific chemical products for medical or industrial uses.

Antony van Leeuwenhoek, a Dutch dry goods merchant, was the first person to see, record, and describe in great detail microorganisms, which he observed with his primitive microscopes. Other early pioneers in microbiology were Jenner, who showed that smallpox can be prevented with cowpox vaccination; Semmelweis, who fought the spread of disease in maternity wards; Lister, who introduced aseptic surgery; and Pasteur, who proved that bacteria come only from existing bacteria. Pasteur also developed the process of pasteurization and vaccinations against anthrax and hydrophobia (rabies). In Germany, Koch established a set of rules, called "Koch's postulates," to determine the etiology (cause) of an infectious disease. Many of the methods used in modern microbiology laboratories were first developed in Koch's laboratory by his coworkers.

Out of the numerous discoveries that followed, the microbiological sciences of bacteriology, mycology, virology, parasitology, and immunology developed. A qualified member of the health team must be knowledgeable in all of these and related areas.

REVIEW QUESTIONS

1. What are microorganisms? Why do we study them?.
2. List some discoveries in microbiology that were beneficial to us. Why?
3. What is "Genetic Engineering"?
4. List the important contributions of van Leeuwenhoek, Jenner, Semmelweis, Pasteur, Lister, and Koch to microbiology and to medicine.

5. How do you determine the etiology of infectious diseases according to Koch's postulate?
6. Why is microbiology considered to be an "applied science"?
7. What should the health worker expect to learn in a course of microbiology?

Supplementary Reading

Ainsworth, G. C.: Introduction to the History of Mycology. New York, Cambridge University Press, 1976.

Asimov, A.: Asimov's Biographical Encyclopedia of Science and Technology. Rev. ed. Garden City, N.Y., Doubleday & Co., Inc., 1972.

Brock, T. D.: Milestones in Microbiology Englewood Cliffs, N.J. Reprint ed. Washington, D.C., American Society for Microbiology, 1975.

Bulloch, W.: The History of Bacteriology. New York, Oxford University Press, 1960.

Collard, P.: The Development of Microbiology. New York, Cambridge University Press, 1976.

Demain, A. L.: Application of the microbe to the benefit of mankind: Challenges and opportunities. *Amer. Soc. Microbiol. News, 38*:237, 1972.

Dobell, C.: Antonj van Leeuwenhoek and his Little Animals. New York, Russell & Russell Publishers, 1938 (reprinted, Dover Publications, 1960).

Dolan, J. A.: Nursing in Society: A Historical Perspective. 14th ed. Philadelphia, W. B. Saunders Company, 1978.

Dols, M. W.: The Black Death in the Middle East. Princeton, Princeton University Press, 1976.

Fox, S. W. (ed.): The Origins of Prebiological Systems and of Their Molecular Matrices. New York, Academic Press, Inc., 1965.

Lechevalier, H. A., and Solotorovsky, M.: Three Centuries of Microbiology. New York, Dover, 1974.

Marquardt, M.: Paul Ehrlich. New York, Henry Schuman, Inc., 1951.

Newerla, G. J.: Medical History in Philately. Milwaukee, American Topical Association, 1964.

Porter, J. R.: Antonj van Leeuwenhoek: Tercentenary of his discovery of bacteria. *Bacteriol. Rev., 40*:260, 1976.

Raff, R. A., and Mahler, H. R.: The non-symbiotic origin of mitochondria. *Science. 177*:575, 1972.

Schierbeek, A., and Swart, J. J.: The Collected Letters of Antonj van Leeuwenhoek. Amsterdam, Swets & Zeitlinger, 1939–1979.

Schwartz, R. M., and Dayhoff, M. O.: Origin of prokaryotes, eukaryotes, mitochondria and chloroplasts, *Science, 199*:395, 1978.

Stewart, I. M., and Austin, A. L.: A History of Nursing From Ancient to Modern Times: A World View. 5th ed. New York, G. P. Putnam's Sons, 1962.

Vallery-Radot, R.: The Life of Pasteur. New York, Doubleday-Doran & Co., 1926 (reprinted, Dover Publications, 1960).

CLASSIFICATION OF MICROORGANISMS; CELLS AND CELL STRUCTURES

GENERAL INTRODUCTION

Any rational, simple, and generally accepted classification of microorganisms is at the present time impossible. It has never been easy to relate "single cell" organisms to each other. It has been possible to organize an animal kingdom, including the protozoa (primitive animals), and a plant kingdom, containing algae, bacteria, and other fungi. The establishment of a third kingdom, named Protista, was first proposed in 1866 by Ernst Heinrich Haeckel (1834–1919),[1] who placed all "lower living forms," including the microorganisms, in this kingdom. Microbiologists in general did not recognize a third kingdom until very recently; now, however, it seems prudent to group the bacteria and the blue-green algae in the kingdom *Procaryotae*.[2]

TWO BASIC CELL TYPES: PROKARYOTES, EUKARYOTES

All living forms (except viruses) are alike in that they contain genetic nuclear material, referred to generally as DNA (deoxyribonucleic acid), in the cells. This nuclear material is located in an area called the *nucleus*. In one type of cells the nucleus (*karyon,* nut, kernel) may be well defined, being enclosed within a nuclear membrane. It is called a true nucleus, or *eukaryon* (Greek *eu*, true, real) and the cell is said to be *eukaryotic*. All higher forms of life, including ourselves, consist of eukaryotic cells.

In the cells of lower (primitive) life forms, the nuclear material (DNA) is not enclosed within a membrane, but rather is distributed in masses throughout the cytoplasm; this primitive type of nucleus is called a prokaryon, and the cells are said to be *prokaryotic* (Greek *pro,* primitive). The present day prokaryotes, the bacteria and blue-green algae, basically are not different from their fossil ancestors, whereas evolution seems to have favored the eukaryotes with great diversity (Fig. 2–1).

The "nucleus" area (Feulgen stain–positive,[3] chromatinic bodies) of typical prokaryotic cells consists of a single, very long, circular, threadlike molecule of DNA folded along the long axis of the bacteria. It is not enclosed in a membrane, is not segregated from the surrounding cytoplasmic material, and, despite contrary reports, never seems to manifest any mitotic or meiotic phenomena. The prokaryotic "nucleus" is, therefore, commonly referred to as nucleoid

---

[1]Haeckel was a field biologist and first used the term *ecology* in his work.

[2]See *Bergey's Manual of Determinative Bacteriology*. 8th ed. Baltimore, Williams & Wilkins Co., 1974. Although Bergey's uses "procaryotic" and "eucaryotic," the more common current usage is "prokaryotic" and "eukaryotic."

[3]The Feulgen reaction is based on the free aldehyde group of acid hydrolyzed DNA that reacts with Schiff's reagent to produce violet color.

CHAPTER 2

(nucleus-like). There may be from one to four nucleoids per bacterial cell, with multiple nucleoids present only in actively multiplying cells. The bacterial chromosome is usually circular, but during conjugation it becomes linear, opening at the sex factor gene, and is in direct contact with the plasma membrane.

There are also numerous striking differences between eukaryotic and prokaryotic cytoplasms. The cytoplasm of eukaryotes contains numerous discrete, membranous, and membrane-enclosed functional bodies *(organelles)*

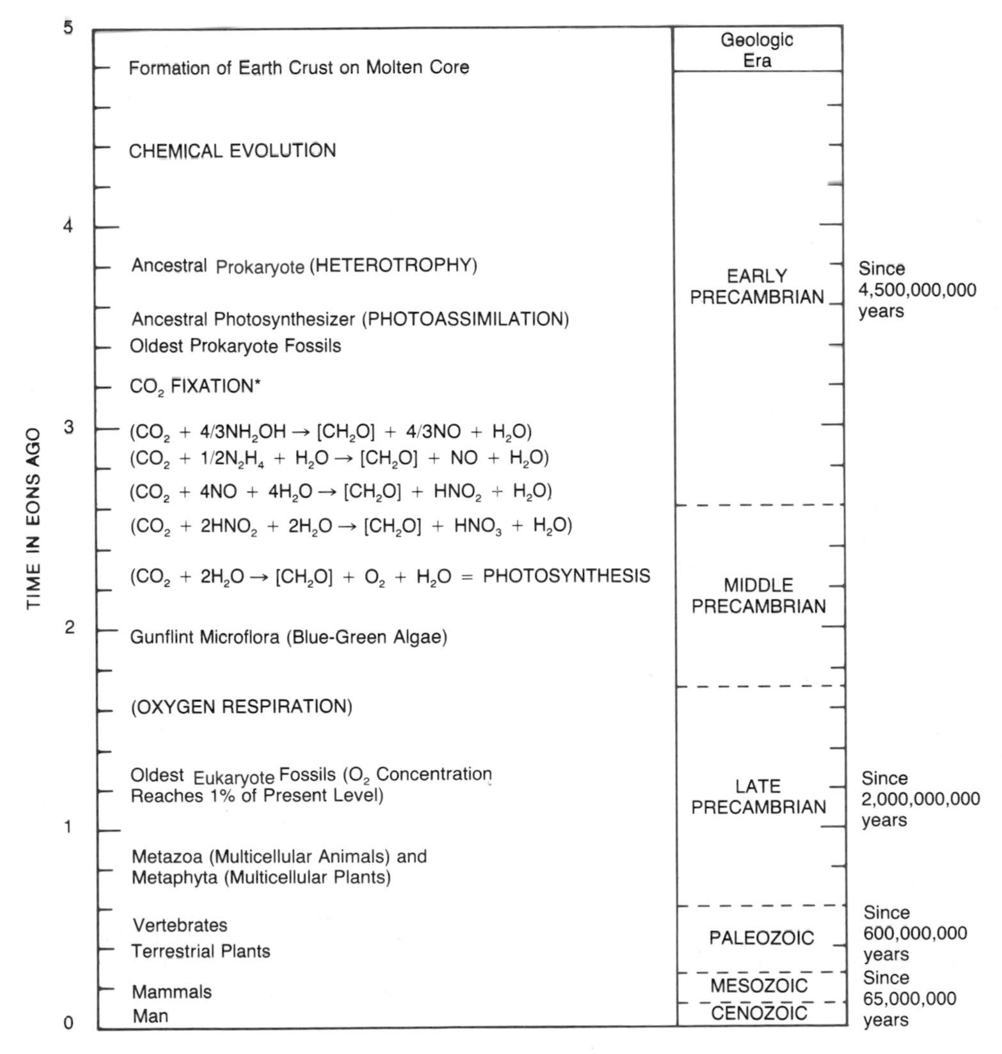

Figure 2–1 Major developments in the evolution of photosynthesis and respiration. From Olson, A., and Cooney, C. L.: Immobilized Enzymes in Food and Microbial Processes. New York, Plenum Publishing Co., 1974.)

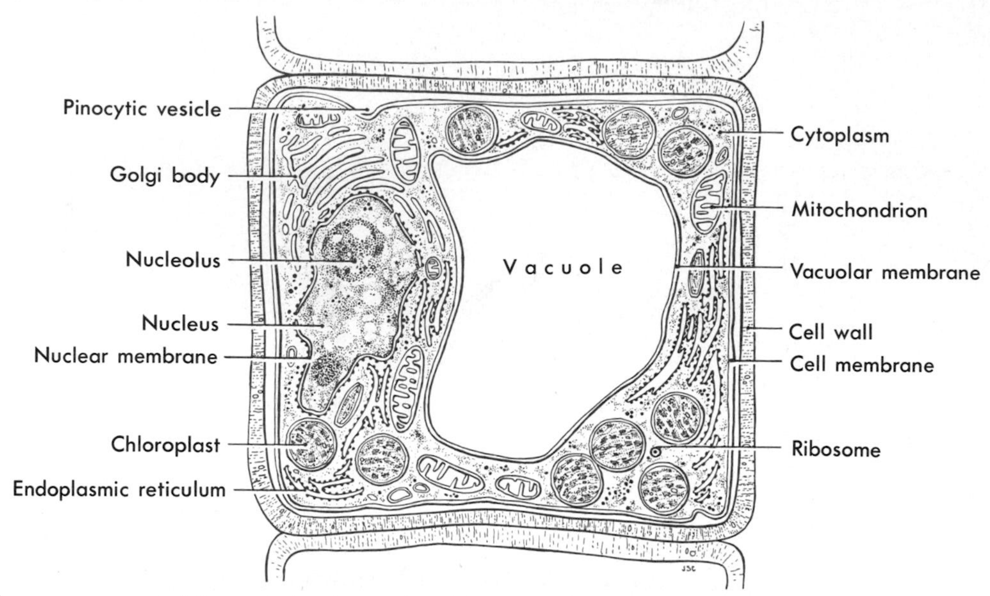

Figure 2–2 Diagram of a eukaryotic *plant* cell based on various cytological studies and use of the electron microscope. The *nucleus* controls hereditary properties and all other vital activities of the cell. Both the nucleus and the *nucleolus* have functions in the synthesis of cell material. The well-defined *nuclear membrane* appears to have pores or openings for communication with the cytoplasmic structures and direction of their synthetic and energy-yielding activities. The *cytoplasm* contains immense numbers of granules called *ribosomes,* concentrated especially along the periphery of a cell-wide labyrinth of connected, narrow sacs called the *endoplasmic reticulum.* These granules are involved in the continuous enzymic reactions that synthesize cell materials under direction from the nucleus. The *mitochondria* are involved in another set of enzymic reactions called *biological oxidation,* which yield the energy for all the cell activities. The *pinocytic vesicle* or invagination is a means of ingesting fluids or extremely minute food particles. *Golgi bodies* comprise an apparatus that may secrete the cellulose that in plants forms the thick, nonliving *cell wall.* This wall supports the plant body and is penetrated by many tiny holes that permit passage of materials between cells. Inside the cell wall is the *cell membrane,* or plasma membrane, a very thin structure that functions as a regulatory apparatus for the nutrients and waste products that pass through it. *Chloroplasts* are the small bodies, typically disk-shaped, that contain the chlorophyll that imparts the color in green plants and participates in photosynthesis. *Vacuoles,* common in cells of plants and lower animals but rare in cells of higher animals, are bubble-like cavities filled with watery liquid and bordered by a thin vacuolar membrane. There are undoubtedly still other structures whose nature and function await elucidation by future cytologic studies. Compare this complex cell with pictures of bacterial cells shown in this chapter and others. (From Villee, C. A.: Biology. 6th ed. Philadelphia, W. B. Saunders Company, 1972.)

(Fig. 2–2).[4] The cytoplasm of prokaryotes contains principally submicroscopic, protein-synthesizing particulates, the ribosomes, which contain rRNA (ribosomal ribonucleic acid). Other functional portions, except certain curious membranes, are missing, obscure, or very primitive (Fig. 2–3). While eukaryotes have larger 80 S[5] endoplasmic reticulum–oriented ribosomes, the ribosomes of prokaryotes are of the 70 S type with subfractions from 50 S to 30 S consisting of about 20 different proteins (see Table 2–1). Aggregates of 70 S also exist: they are called polyribosomes or polysomes and are attached to mRNA (messenger RNA). About 80 per cent of all RNA is rRNA; the rest is mRNA and many different tRNAs (transfer RNA).

On centrifugation at 100,000 G among several peaks, a broad heterogeneous 5 S band may be isolated from the protoplasm. It is the ground substance of the bacterial cell. Most of the cellular DNA, about 90 per cent, is found in the 8 S fraction.

---

[4]Actually, there are two other types of eukaryotic distinctions of cells, based on nuclear structure: the heterokaryons and the homokaryons. The heterokaryons are exemplified by a condition in fungi in which protoplasm and nuclei flow freely from cell to cell through the structure of the organism called the mycelium, an exchange that even occurs in the spores. The result is that if more than one strain is growing in a flask, the different nuclei mix freely and express their genetic influence in all cells. In the homokaryon, only one type of nucleus is present, although in these fungi one may still find five nuclei in one cell, two in another, and perhaps none in the third, at a specific moment.

[5]S for Svedberg unit is a measure of the rate at which a given solute, like a protein, will suspend in a field of ultra-centrifugal force, roughly a measure of its molecular weight. One S has a sedimentation coefficient of 1×10^{-13} sec. A Svedberg unit of 80 S corresponds to a molecular weight of 4×10^6 daltons, and 70 S equals 2.7×10^6 daltons.

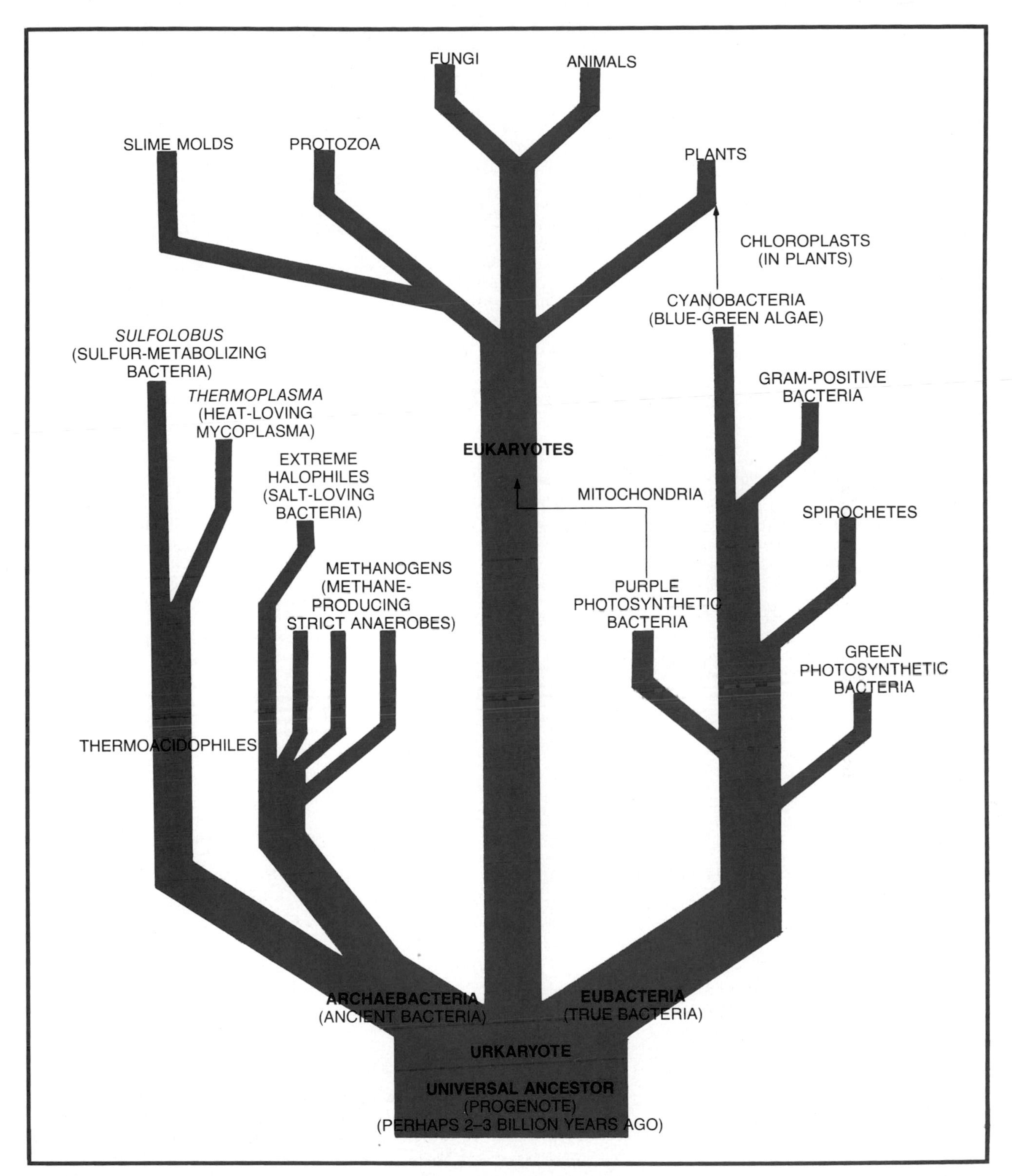

Figure 2–3 This phylogenetic tree is based on the discovery that the archaebacteria are fundamentally different from all other bacteria, which are designated the eubacteria, or true bacteria. Both eubacteria and archaebacteria are alike in being prokaryotic cells: simple cells that lack a nucleus and are very different in their structural properties from eukaryotic cells, which have a nucleus and several other subcellular organelles. Genealogically, however, archaebacteria and eubacteria are no more closely related to each other than either group is to eukaryotes. It is proposed that the archaebacteria, the eubacteria, and the urkaryote—the original eukaryotic cell—stemmed from a common ancestor (the progenote) much simpler than the simplest present-day cells (prokaryotes). Eukaryotes evolved after the urkaryote became a "host" for bacterial endosymbionts that developed into mitochondrion and chloroplast. (From Woese, C. R.: Archaebacteria. Scientific American, June 1981, pp. 98–122.)

Table 2–1. Two Main Classes of Ribosomes*

RIBOSOMES			SUBUNITS	
Liver	81	80S	60	Molecular weight = 4×10^6 daltons
Yeast	80		28	RNA = 45%
Bean	78		40	
Neurospora	77		18	
Mitochondria	73	70S	50	Molecular weight = 2.7×10^6 daltons
Escherichia coli	70		23	RNA = 65%
Chloroplasts	67		30	
Rhodospirillum	66		16	

*Ribosomes are "engines" used by the cell to assemble amino acids into proteins under the direction of tRNA, mRNA, and rRNA.

From DeRobertis, E. D. P., Saez, F. A., and DeRobertis, E. M. F., Jr.: Cell Biology. 6th ed. Philadelphia, W. B. Saunders Company, 1975.

Two Types of Prokaryotes: Eubacteria, Archaebacteria

A recent study disclosed that *archaebacteria*[6] (mostly *Methanobacterium* species, some halophiles, and related species) and *eubacteria* (true bacteria) are as different from each other as they are from the eukaryotes. Differences discovered between archaebacteria, eubacteria, and eukaryotes are based on 16 S ribosomal RNA genetic sequencing and paleontological studies of early microfossils. A proposed phylogenetic tree is shown in Figure 2–3.

The methanogenic bacteria *(Methanobacterium)* live only in oxygen-free environments and generate methane from the reduction of carbon dioxide.

[6] *Archaios*, Greek for ancient.

After DeRobertis, E. D. P., Saez, F. A., and DeRobertis, E. M. F., Jr.: Cell Biology. 6th ed. Philadelphia, W. B. Saunders Company, 1975.

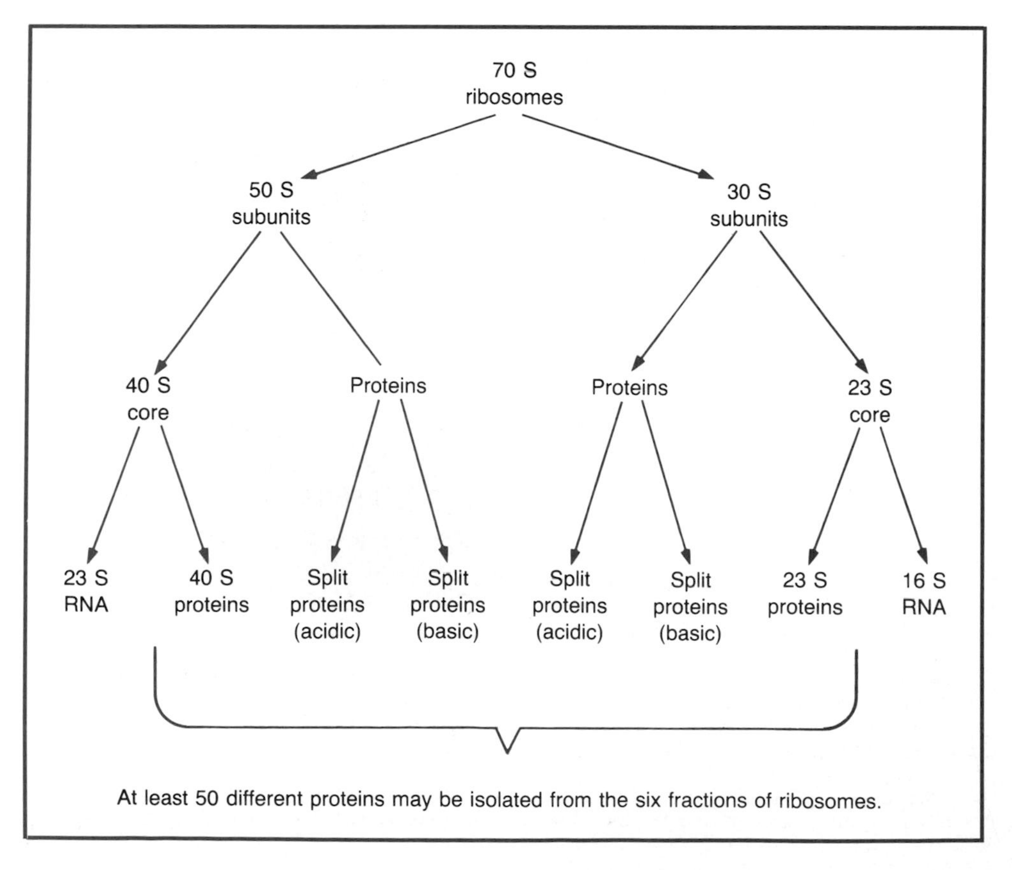

Figure 2–3 also shows how endosymbionts,[7] like mitochondria and chloroplasts, may have developed from free living bacteria into internal essential organelles of eukaryotic cells, as has been proposed by some investigators.

A CLASSIFICATION OF MICROORGANISMS

The principal[8] groups containing *unicellular* organisms may be listed as follows:

I. Eukaryotes that are microscopic animals.
   A. Protozoa[9]: e.g., amebas, malarial parasites
II. Eukaryotes that are microscopic plants
   A. Fungi–true fungi or Eumycetes: e.g., yeasts, molds, "mushrooms."
   B. Eukaryotic true algae: These require sunlight for growth. None is pathogenic[10] for humans. They have been classified into six groups, each with distinct pigments, food reserves, and reproductive cells. Thus they are known as the green algae, the brown algae, the red algae, various seaweeds, and so forth. Some, like the euglenoids, have been classified with the protozoa by some investigators.
   These eukaryotes show some relationship to higher forms of life. The blue-green algae (prokaryotic) are sometimes included as a seventh group.
III. *Procaryotae* (in accordance with *Bergey's Manual of Determinative Bacteriology*, 8th ed., 1974).
   A. Phototrophic prokaryotes.
      1. The Cyanobacteria[11] (Greek *cyan(o)*, blue) also require sunlight for growth. These are the blue-green algae. They are phototrophic and produce oxygen in light. They are much larger than the bacteria, but resemble them in many ways. The blue-green algae are now considered a major group of bacteria.
      2. Phototrophic (photosynthetic) bacteria ("Photobacteria")[12] *(Rhodospirillales)*.
         a. The purple bacteria. These consist of the purple nonsulfur bacteria (*Rhodospirillaceae*-formerly *Athiorhodaceae*) and the purple sulfur bacteria (*Chromatiaceae*-formerly *Thiorhodaceae*).
         b. The green sulfur bacteria *(Chlorobiaceae)*.
   B. Prokaryotes indifferent to light (Scotobacteria").[13]
      1. The common bacteria (formerly called fission fungi or *Schizomycetes*). All the bacteria that used to belong to the *Eubacteriales* (true bacteria) belong here, including *most bacterial pathogens*.
      2. Actinomycetes and related organisms, the group of coryneform bacteria. Here belong diverse organisms like those that produce streptomycin and other antibiotics, the pathogen that causes tuberculosis, and, according to *Bergey's Manual*, the diphtheria organism.
      3. Rickettsias.[14] These are obligate intracellular scotobacteria that live

---

[7]Greek *endon* for within, *symbioun* for to live together.

[8]A few microscopic, multicellular animals of medical importance, e.g., certain parasitic worms (microfilarias), are not listed here (Chapter 41).

[9]*Protos* is a Greek root meaning primitive; *zoa* is from the Greek *zoon*, for animal.

[10]*Pathos* is Greek for sadness or pain; *genic* is from a Greek word meaning to produce or cause. Hence any agent, living or not, that produces pain or disease is said to be pathogenic.

[11]The blue-green algae are also known as *Cyanophyta*, with their photosynthetic pigment distributed through the cytoplasm.

[12]Greek *phot(o)*, light.

[13]Greek *scoto*, darkness.

[14]A group of microorganisms named for their discoverer, James Howard Ricketts (Chapter 39).

Table 2–2. Characteristics of Major Groups of Microorganisms[1]

EUKARYOTES			PROKARYOTES	
PROTOZOA	MOLDS AND YEASTS (EUMYCETES)	GREEN ALGAE[5] (CHLOROPHYTA)	BLUE-GREEN ALGAE[6] (CYANOBACTERIA)	BACTERIA (SCHIZOMYCETES)
Grow in dark; some on lifeless media.	Grow in dark on lifeless media.	Require sunlight; grow on lifeless media.	Require sunlight; grow on lifeless media.	Grow in dark on lifeless media.
Multiply by mitosis and cytokinesis or by quite complex sexual means. The term binary fission[2] should be reserved for the prokaryotes.	Multiply by budding, conidia,[4] and sexual means.	Multiply by fission and sexually.	Multiply by fragmentation or cell division.	Multiply principally by binary fission.
Much larger than bacteria; in volume usually hundreds to thousands of times greater.	Size much greater than bacteria – filaments and other structures often macroscopic.	Much larger than bacteria; may reach 75 meters in length.	From microscopic size to as much as 1 meter in length.	Round forms do not exceed about 5 micrometers,[7] usually not over 2 to 3 μm in diameter; rod forms commonly 2 μm in diameter and 10 μm in length; some curved species may exceed 500 μm in length but not over 1 to 2 μm in diameter.
Do not pass filters.[3]	Do not pass filters.	Do not pass filters.	Do not pass filters.	Generally cannot pass filters.
Internal cell structures well developed; nucleus, vacuoles, "mouth" opening, etc.; no true cell wall.	Cell structure, especially nuclei, visible with ordinary microscopes.	Cell structure like higher plants.	Cell structure resembles bacteria; even forms spores that may be dormant.	No definite organs in the cell visible with ordinary microscopes; granules, spores, capsules visible.[8]
Can ingest solid particles (except a few parasitic forms).	Cannot ingest solid particles.	Cannot ingest solid particles.	Cannot ingest solid particles.	Cannot ingest solid particles.
Cellular.	Cellular.	Cellular.	Cellular.	Cellular.

[1]Not including helminths. Also not included: Slime molds, the symbiotic fungal-algal plant forms called lichens, rusts, smuts, mosses, perhaps even liverworts, and mushrooms.

[2]Multiplication by *binary fission* means that the cell divides into two new individuals (daughter cells), each sharing approximately equally genetic parts of the "parent" cell. Sex is not involved; hence multiplication by binary fission is asexual. In the process of budding, most yeasts divide into two or more unequal parts. This is not binary fission.

[3]*Filters* in this sense mean unglazed porcelain, clay, paper, or other fine-pored materials that permit the passage only of fluid and of microorganisms much smaller than algae, protozoa, bacteria, yeasts, molds, or rickettsias. With a few exceptions (see mycoplasmas), among microorganisms only viruses can pass through such filters.

[4]*Budding* and *conidia* formations are types of asexual multiplication in which daughter cells grow out from the parent cells like a bud or branch from a tree. Many buds may occur on a single parent cell. Each later becomes a complete, independent cell.

[5]The most diverse of all algae. The brown and red algae do not produce land forms.

[6]These have the simplest nutritional requirements of all living things—only N_3, CO_2, minerals, and water. They appear to be the first colonizers of land. Note: Blue-green algae may appear to be almost any color—even red as found in the Red Sea!

[7]Micrometer (μm) is a unit of length commonly used in microscopy. It is 1/25,400 (0.000039) of an inch, 10^{-6} meter, or 0.0001 centimeter. The following line is 1 centimeter (cm) in length: ________(see page 24).

[8]For definitions of these structures see more detailed description of bacteria (Chapter 4).

Table 2–2 *Continued.* Characteristics of Generally Smaller Forms of Microorganisms.

PROKARYOTES			NEITHER EUKARYOTIC NOR PROKARYOTIC
L FORMS OF BACTERIA	MYCOPLASMAS[9]	RICKETTSIAS AND CHLAMYDIAS	VIRUSES
Grow on lifeless media; their development from bacteria is promoted by high salt concentration or toxic agents, especially penicillin: they are pleomorphic forms of bacteria; some revert easily to the bacterial form while others do not.	Grow on lifeless media or on chick embryos or in tissue culture.	Grow only inside living cells.	Grow only inside living cells.
L forms are larger than most mycoplasmas.	Multiply by modified form of bacteria-like fission and subdivision.	Method of multiplication probably binary fission.	Method of multiplication "biosynthetic."
L forms do not form a rigid mucopeptide cell wall, but they do contain DNA and carry on metabolism.	Highly pleomorphic; limp, fragile cell membrane; produce very minute forms that can pass bacteria-retaining filters. They are the smallest "free-living" cells.	Much smaller than bacteria. Rickettsias have rod, spherical, and spiral cells.	Extremely minute. Not visible with ordinary microscopes. By electron microscope some are seen to have a concentric, spherical, helical, cuboidal, or "tadpole" form.[10]
	Internal structure similar to bacteria; no cell wall.	One species of rickettsias can pass bacterial filters; chlamydias pass with difficulty, if at all.	Pass bacterial filters.
	Mode of nutrition includes serum (for plasma proteins) and, for pathogens, a steroid. They resemble bacteria without cell walls.	No definite internal structures visible with ordinary microscopes. Electron microscope reveals bacteria-like internal structure.	No internal structures visible with ordinary microscopes.
		Mode of nutrition like bacteria in some respects; dependent on certain enzymes of host cell.	Depend on genetically directed alterations of synthetic mechanisms, and use of genetic materials, of host cell.
Cellular.	Cellular.	Cellular.	Noncellular.

[9]Also called pleuropneumonia-like organisms (PPLO).
[10]Most of our first knowledge about viruses comes from studies with bacteriophage, special types of viruses pathogenic in bacteria.

in eukaryotic cells as parasites. They are modified, smaller forms of bacteria.

a. Rickettsias cause Rocky Mountain spotted fever, typhus fever, and other so-called rickettsial diseases.
b. Chlamydias[15] are closely related to rickettsias. Some diseases caused by these organisms are trachoma, psittacosis or "parrot fever," and lymphogranuloma venereum, a venereal disease.

4. Mycoplasmas *(Mollicutes*[16]*)*. These are scotobacteria without cell walls. They are prokaryotic organisms bounded by a single triple-layered membrane. They are also known as pleuropneumonia-like[17] organisms, or PPLO and cause respiratory and other diseases.

[15]From the Greek *chlamys*, meaning a cloak or mantle. They were formerly thought of as "mantle viruses."
[16]Latin *mollis* soft, pliable–a class with pliable cell boundaries.
[17]Pleuropneumonia is a disease of cattle caused by one member of this group of microorganisms (Chapter 4).

IV. Viruses[18]–unclassified: neither cells in the general sense, nor alive as life is commonly defined. They cause poliomyelitis, influenza, measles, and other diseases. Viruses are neither eukaryotic nor prokaryotic.

For convenience we shall give a brief general description of each main group and then turn especially to the bacteria for detailed discussion, not because they are most important but because they furnish convenient working material and are good illustrations of general principles underlying the whole science of microbiology. Some general characteristics and relationships of microorganisms are seen in Table 2–2. This listing is continued for the smaller bacteria and viruses. If all microorganisms are listed in order of relative size, the following approximation of a size relationship results:

protozoa > algae > yeasts and molds > bacteria > L forms of bacteria[19] > mycoplasmas > rickettsias > chlamydias > viruses

However the student is warned *not* to regard this listing as an absolute fact. Many algae are much, much longer than some protozoa, and some forms of mycoplasmas may be as small as viruses or as large as bacteria.

UNICELLULAR ORGANISMS

With the exception of viruses, which have only a few of the fundamental properties of living matter, all living things, including humans, consist of living cells. Large animals and plants are composites of billions of microscopic cells. These are differentiated into integrated systems of organs in many-celled animals, or *metazoa*, and into leaves, flowers, and stems in many-celled plants, or *metaphyta*. Each unicellular organism consists of a single microscopic cell. In a few species the cells form loose aggregates, but with a few possible exceptions among primitive animals (the protozoa), each cell is a complete organism and alone can reproduce the entire structure.

The Cell

The cell is the smallest and simplest unit of living matter capable of independent life and self-reproduction. Exceptions are seen in the modified smaller forms of bacteria now classified as chlamydias and rickettsias; these can reproduce only as intracellular parasites. Mycoplasmas are free-living cells. Viruses are smaller units, but they are not cells as currently defined, and so far as is known, they are incapable of independent life or reproduction. Cells vary in size, structure, and physiology according to their species. The human body is made up of billions and billions of different tissue cells; liver cells differ from muscle cells, nerve cells differ from pancreatic cells and from bone cells, and so on. Basically, however, all cells are similar. Whether animal or vegetable, eukaryotic or prokaryotic, they *all* possess numerous structures and properties in common. The reader should not be misled and think of animal or plant cells when talking about bacteria. A drawing of a common bacterial cell is shown in Figure 2–4.

Protoplasm. The word is derived from the Greek *protos*, for first, original, or ancestral, and *plasma*, for substance. Before the advent of high-power

[18] *Virus* is derived from a Latin root meaning "slimy substance" or poison (Chapter 6).

[19] L forms of bacteria (after the Lister Institute in London, where they were first isolated) are bacteria without cell walls that are highly pleomorphic.

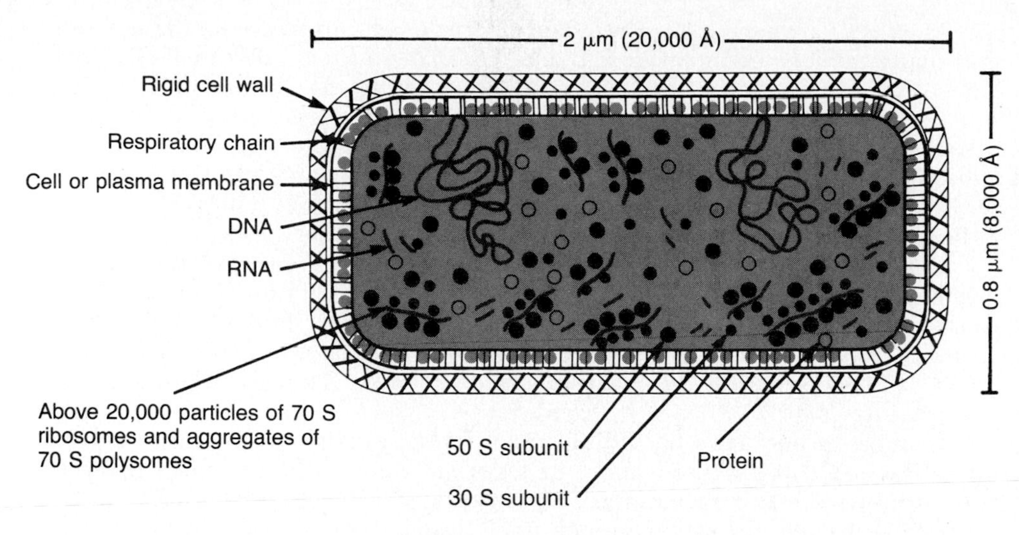

Figure 2–4 Diagram of a prokaryotic cell, representing the bacterial species *Escherichia coli,* containing two chromosomes 1 mm long (10^7 Å), each consisting of a single circular molecule of DNA. Although only one chromosome is normally expected per *E. coli* cell, two (as shown here) may exist just before cell division. With slow cell division and rapid nucleic acid replication, even four chromosomes may temporarily coexist in a cell. The ribosomal subunits of 50 S and 30 S refer to the different molecular weights that can be separated on the basis of their sedimentation coefficient (S). (From DeRobertis, E. D., Saez, F. A. and DeRobertis, E. M. F., Jr.: Cell Biology. 6th ed. Philadelphia, W. B. Saunders Company, 1975.)

microscopes protoplasm was thought to be the first and only known living substance, animal or vegetable. Thomas Huxley (1825–1895) called protoplasm "the physical basis of life." The term protoplasm, as the name of a single substance, is now obsolete, although it is sometimes used to mean, simply, "cell contents." As seen with low-power lenses, protoplasm appears to be watery, transparent, colloidal, usually colorless, and very much like the raw white of an egg. When analyzed chemically, it is found to consist largely of carbon, oxygen, hydrogen, sulfur, phosphorus, and nitrogen. Many other elements (e.g., iron, sodium, chlorine, and magnesium) are usually present in smaller amounts.

Studies using the compound microscope, with ordinary (visible) light, and using the electron microscope, have revealed that, far from being a single "living distinct substance," protoplasm consists of many distinct submicroscopic bodies called *organelles* in a somewhat viscous matrix. In a typical eukaryotic cell (animal or plant), organelles include: nucleus, mitochondria, chloroplasts, endoplasmic reticulum, ribosomes, and numerous other structures (Fig. 2–2).

Cell Structure

The cell is the basic unit of life. In *one-celled organisms,* cells grow to a certain size by taking in materials from the external environment and converting them into structural and functioning molecules within the protoplasm. After a certain amount of growth, the cell divides into two daughter cells, each closely resembling its parent cell.

In order to divide, *eukaryotic cells* undergo *mitosis,* during which the nuclear membrane breaks down and the chromosomes, which have previously duplicated themselves, are divided equally and form two new nuclei; *cytokinesis* follows, with the cytoplasm also dividing equally, each part containing one nucleus. By this process the organism grows and, in multicellular organisms, also replaces worn out or injured tissues.

In all cells the protoplasm forms its own self-containing cell membrane and, in plants and bacterial cells, a rigid cell wall. In prokaryotes the entire structure—cell wall plus cell membrane and contents, and any external appendages present in some bacteria, such as flagella, pili, and capsules—are now generally included in the term cell.

Most cells are extremely minute (of the order of 1 to 20 μm in diameter[20]), and many millions of them are required to form so small an animal as a flea. Cells vary greatly in size, however, and some are several inches in diameter. For example, all eggs are single cells. An ostrich egg is certainly far from microscopic, although it might be objected that much of the egg is reserve food substance and is not properly part of the actual microscopic reproductive cell itself. Some cells, such as nerve cells, may have portions several inches in length, although these filaments may be of microscopic fineness.

Cell Membrane and Cell Wall. The immediate covering of the "protoplasm" of each cell is a *cell* (or *cytoplasmic*) *membrane*, a thin, limp sac. The electron microscope shows it to consist of several layers and various minute granules and molecular aggregations that are absolutely essential to the life of the cell. This constitutes the external surface of most animal cells and also of mycoplasmas.

In most "plant" cells, including algae, yeasts, molds, and bacteria (except mycoplasmas), there is also a more or less rigid outer *cell wall*, which protects and supports the cell membrane and consists of a hard or tough substance formed by the cell. *Cellulose*, common in all familiar plants, is such a substance. A good example of cellulose structure is wood, which consists of a mass of microscopic cellulose chambers, each containing protoplasm. Another hard substance found in cell walls of many animals and a few plants is *chitin*. Examples of other structures made of chitin are the shells of insects, crabs, and lobsters. In bacteria the cell wall contains tough substances called *mucopolysaccharides* or peptidoglycans. These are important in gram-negative bacteria. Other constituents in bacterial cell walls, called mucopeptides, contain N-acetylglucosamine and N-acetylmuramic acid unique to bacteria. There are also teichoic acids, which are phosphate polymers of a rather complex nature, found only in gram-positive bacteria. These are not entirely restricted to the cell wall but occur also between the cell wall and cell membrane (Fig. 2–5). One important difference between gram-positive bacteria and gram-negative ones is that the latter's cell walls contain 6 to 11 times more lipid (11 to 22 per cent) and are more resistant to penicillin, sulfonamides, phenol, detergents, and basic dyes.

Mesosomes. Mesosomes seem to be more important in gram-positive bacteria. They are irregular pockets in the cytoplasmic membrane and may function in secretion.

Nucleus. A typical eukaryotic cell contains a definite organelle, usually situated near the center, which is called the *nucleus*. It is a very important part of the cell: without it the cell soon dies, but the nucleus may live by itself for some time. Indeed, there are some cells (e.g., spermatozoa) that appear to consist almost entirely of nuclear material. The nucleus is the center and controlling agency in all vital functions of the cell, and it takes a leading part in the processes of its reproduction and inheritance.

---

[20] 1 μm = 0.001 millimeter or 1/25,400 or 3.937×10^{-5} inch.

A system of nomenclature for metric units now used by many microbiologists and the American Society for Microbiology follows:

m = *milli* = 0.001 = 10^{-3} meter or gram or liter (thousandth)
μ = *micro* = 0.000,001 = 10^{-6} meter or gram or liter (millionth)
n = *nano* = 0.000,000,001 = 10^{-9} meter or gram or liter (billionth)
p = *pico* 0.000,000,000,001 = 10^{-12} meter or gram or liter (trillionth)
For example:
mm = millimeter = 10^{-3} meter (= 10^7 Å)
μm = micrometer = 10^{-6} meter (formerly, μ = micron)
nm = nanometer = 10^{-9} meter (formerly, millimicron = mμ)
Å = Angstrom = 10^{-10} meter (= 0.1 nm; 10 Å = 1 nm)
pm = picometer = 10^{-12} meter (formerly, micromicron = μμ)
pg = picogram = 10^{-12}g
ppm = μg/ml (cc); μg/g; 10^{-6} g/g or ml (cc)

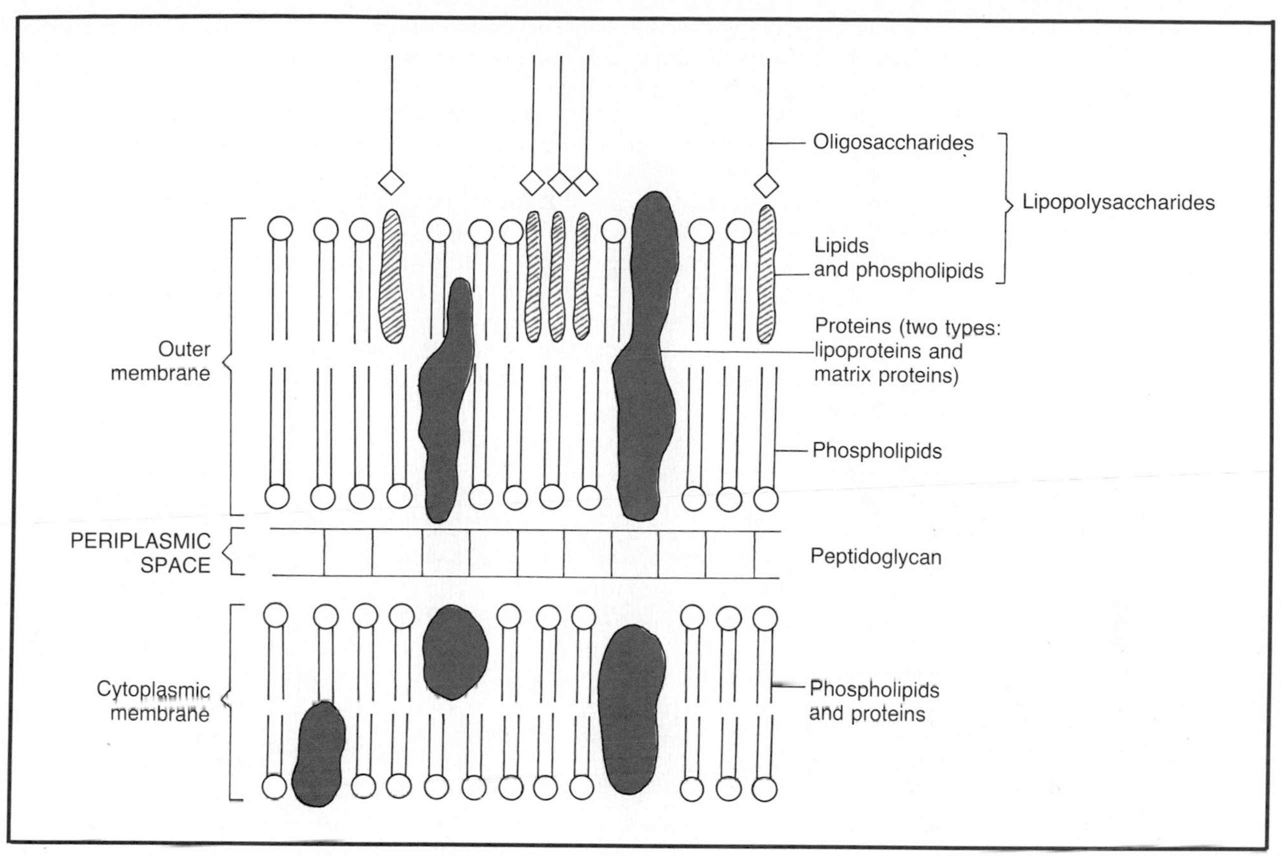

Figure 2–5 The cell wall of gram-negative bacteria. Molecules up to 800 molecular weight can pass through the outer membrane and peptidoglycan layers, but the cytoplasmic membrane excludes most hydrophilic molecules except water. The outer membrane is not very permeable to hydrophobic substances, and as a consequence the gram-negative bacteria are becoming more resistant to antibiotics.

For years it has been thought by many microbiologists that certain eukaryotic organelles may have evolved from phagocytized, free-living, prokaryotic organisms. Like these primitive cells, the mitochondria and the photosynthetic plastids of eukaryotic cells perform metabolic functions, have their own nucleic acids, and synthesize their own protein components. Certainly the influence of the nucleus may not be overlooked, but the Precambrian ancestors (primitive bacteria, perhaps) of these ingested cell organelles must have required at least a billion years to become true endosymbionts.

Deoxyribonucleic Acid (DNA) and Ribonucleic Acid (RNA). These nucleic acids occur in all typical cells. DNA determines the hereditable characters and the entire makeup and functioning of the cell. The chemically complex DNA is formed into long double-spiral strands or helices (Fig. 2–6). These strands are composed of regularly repeated molecular groupings called *nucleotides*, arranged in certain patterns. DNA contains D-2-deoxyribose (called simply deoxyribose), the purines adenine (A) and guanine (G), the pyrimidines cytosine (C) and thymine (T), and also phosphate. The per cent base ratio (GC to AT) of DNA has been found useful in classifying and identifying organisms. Animals and higher plants have an excess of AT (in man 1.40 AT to 1.0 GC), whereas in bacteria and viruses there is much variation from species to species. DNA extracted from cells by chemical means contains approximately 10,000 paired deoxynucleotide units on each strand. The nature and arrangement of these nucleotides determine hereditary characters. In some viruses the strands are single.

The chemical components of RNA nucleotides are D-ribose, the purines adenine (A) and guanine (G), the pyrimidines cytosine (C) and uracil (U), and phosphate, which links them together. In some viruses the sugar D-glucose may replace deoxyribose, and 5-hydroxymethylcytosine may be present instead of cytosine. Cytosine may also be missing in some species that contain 5-methyl-

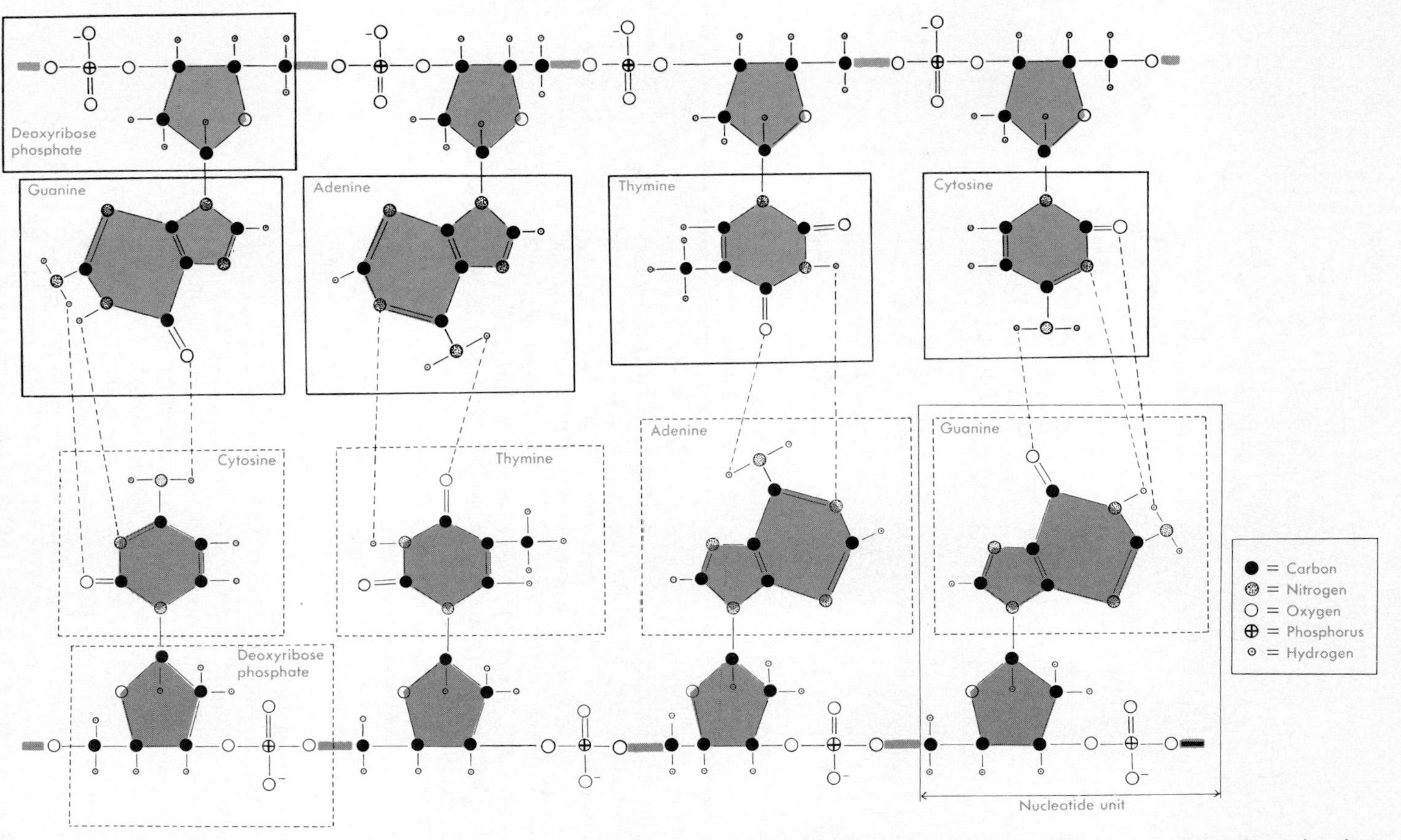

Figure 2–6 Diagram of the repeating nucleotide units making up the structure of DNA. In the lower part of the diagram is a chain of deoxyribose-phosphate groups (enclosed by dashed lines, lower left) linked in a row from left to right by covalent bonds (heavy colored lines) between a phosphate oxygen of one and a carbon atom of the deoxyribose of another. To each deoxyribose-phosphate group there is attached either a purine base (adenine or guanine) or a pyrimidine base (cytosine or thymine). The three groups—deoxyribose, phosphate, and base—together constitute a *nucleotide unit* (enclosed by a solid line, lower right). The nucleotides in DNA are called deoxyadenylic acid, deoxyguanylic acid, deoxycytidylic acid, and deoxythymidylic acid. Thousands of nucleotide units are linked together, in varying sequences, in long twisted filaments of nucleic acid.

In the upper half of the diagram is a chain of nucleotide units corresponding to the lower half of the diagram but inverted. The upper chain is connected to the lower chain by hydrogen bonds (dashed lines) between the purine and pyrimidine bases, as shown. Note that adenine always pairs with thymine, and guanine with cytosine. In DNA the two interconnected chains are commonly twisted together in a double helix. However, single-stranded DNA is known to occur in some viruses. In RNA there is commonly only a single strand. However, under some conditions of cell or virus reproduction double-stranded RNA has been observed.

cytosine. The genetic material of viruses contains either DNA or RNA, never both, unless one is present in minute amounts.

Each hereditary character is associated with a particular segment of the long DNA (or RNA in some viruses) molecule. Such a segment may consist of (rarely) one to (commonly) hundreds of nucleotides. Each such segment constitutes a gene or genetic determinant. Any alteration in a gene produces a hereditary change in the organism of which the gene is a part. This change is called a *genetic mutation*. Such mutations are often produced by x-rays, ultraviolet light, atomic bomb radiation, various chemicals, and possibly by some viruses. The smallest unit of DNA subject to such mutation — a nucleotide base pair — is spoken of as a *muton*.

In the nucleus of most cells, genes are arranged as long, microscopically visible threads called *chromosomes*. The supporting structure of a eukaryotic chromosome consists of basic nuclear proteins (usually protamine or histone) on which the DNA-containing genes are arranged. During asexual reproduction of cells the chromosomes replicate themselves and are divided between the daughter cells. In eukaryons this is accomplished by a process called *mitosis*. When eukaryotic sexual cells are formed, the maternal and paternal chromosomes in each cell combine, replicate, divide, and are redistributed to the sex cells in a complex process of cell division called *meiosis*. True sexual cells (gametes) do not occur in prokaryons, although a primitive form of conjugation between certain cells of some species of bacteria is observed. Often genes are very closely linked on the same chromosome to neighbor genes, each of which produces almost identical phenotypic effects in the organism. For example, in a virus called T4 at least 40 such genes, also called *cistrons*, have been described. Although many microbiologists use the terms gene and cistron interchangeably, actually cistron should be used only to describe genetic material when applied to *complementation analysis*, a genetic method of studying hereditary DNA arrangement on the chromosome.

Nucleotides are adenine deoxyribose phosphate, guanine deoxyribose phosphate, cytosine deoxyribose phosphate, or thymine deoxyribose phosphate (see Fig. 2–6). The deoxyribonucleotide adenine deoxyribose phosphate also exists as deoxyadenosine-3′-monophosphate or as deoxyadenosine-5′-monophosphate, and so on. If any mutational structural change occurs in a single nucleotide pair, out of many thousands of polydeoxyribonucleotides within a single gene, the mutational unit may be referred to as a *muton*. When nucleotide pairs are involved in genetic recombination, each single nucleotide acts as a recombination unit. This smallest unit, not divisible by recombination, is called a *recon*. As will be shown, a virus particle (called a *virion*) consists of a single molecule of nucleic acid coated with protein. Thus, a virus particle in many ways resembles an independently existing chromosome. In some viruses RNA is the genetic material, in place of DNA. The difference is that in DNA viruses one finds deoxyribose and thymine, whereas RNA viruses contain ribose and uridine. The other nitrogen bases are identical in both types of viruses.

The linear sequence of nucleotides in the double-helix DNA (Fig. 2–7)[21] of the genes is exactly copied prior to cell division. The nucleotide DNA order template is also transmitted via the enzyme RNA polymerase to messenger RNA (mRNA). At least three other RNA types have been recognized to exist besides mRNA: nuclear RNA (also part of the chromosomes), soluble or transfer RNA (sRNA or tRNA), and ribosomal RNA (rRNA). By means of a complex mechanism RNA is then involved to transmit the message or *code* to ribosomes and to determine the order of amino acids that make up the enzymes of the cell. Thus the linear sequence of amino acids in the enzymes (proteins) synthesized in the cytoplasm is related directly to the linear sequences of nucleotides in the mRNA as copied from the DNA.

---

[21]The figure shows a right-handed, universally accepted helix. Recent evidence based on x-ray diffraction studies of material from bacteriophage λ proposes left-handed DNA helices as well.

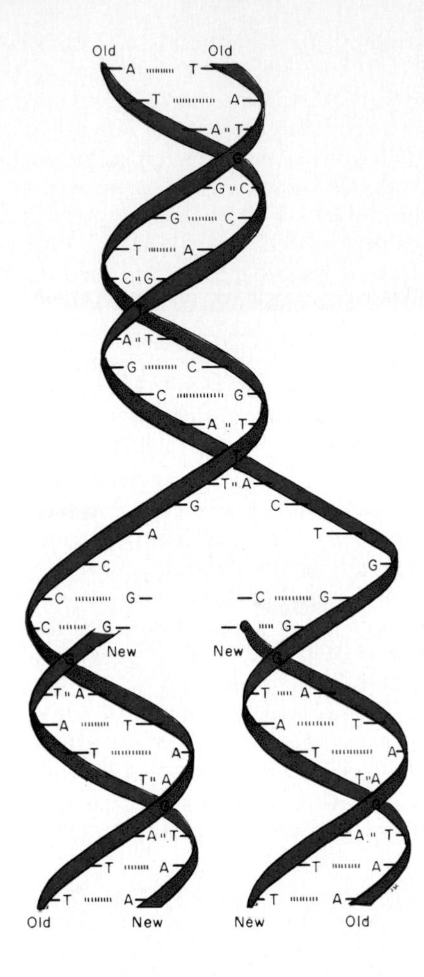

Figure 2–7 Diagram illustrating the semiconservative replication of DNA by a process involving the separation of the original base pairs and the unwinding of the two strands of the double helix. Each strand then serves as a template for the synthesis of a new complementary strand. (From Villee, C. A.: Biology. 5th ed. Philadelphia, W. B. Saunders Company, 1967.)

The triplet code letters in messenger RNA have been determined with some certainty for all amino acids found in proteins. However, a specific amino acid may be linked to another specific amino acid not only by one set of code directions but perhaps by several. For example, Table 2–3 shows that alanine may be attached to a polypeptide under the direction of GCU, GCC, GCA, or GCG.

The sequences of amino acids in the polypeptide repeating units (amino acids) of enzymes, arranged by mRNA according to the nucleotide sequence of the chromosomal DNA, determine whether a protein formed by a ribosome shall be a dehydrogenase, viral, or other protein, or it may perhaps determine the quality of a beefsteak. We may think of some mutations as the result of one or more changes in the DNA structure such that a "nonsense" protein or enzyme is produced that has no biologic function. In some experiments certain nucleotide sequences, (e.g., UAA, UGA, and UAG) have been shown to eliminate certain amino acids, creating a gap rather than adding any known amino acid to a peptide chain. We may thus think of UAA, UGA, or UAG as determining the punctuation of the code rather than being part of the code itself.

Protein synthesis of the enzymes required for metabolic reactions to proceed starts with the message from DNA that determines the codon[22] in mRNA and the amino acid that it specifies. The message is transmitted from mRNA to the anticodon of three consecutive nucleotides in a molecule of tRNA, and then the specific amino acid is selected to be added to the peptide chain.

[22]A codon is a triplet code, like GCU, GCC, GCG, etc.

Table 2–3. Genetic Code*

[The nucleotide compositions of RNA codewords have been obtained by directing amino acids into protein in *Escherichia coli* extracts with randomly ordered polyribonucleotides synthesized with polynucleotide phosphorylase. Nucleotide sequence in codewords is not known; thus the order of bases is arbitrary. The tentative summary of RNA codewords shown in the table is considered to be *potential* codewords; that is, they code for amino acids in cell-free systems, but all codewords may not be applicable to cells in vivo. The genetic code may be regarded as quite universal in the sense that messenger RNA codons, when translated, are usually translated into the same amino acids in all organisms that have been examined. This applies to *E. coli,* human hemoglobin, bacteriophage T4, tobacco mosaic virus, yeast, *Neurospora*, and others.]

AMINO ACID	RNA CODEWORDS*					
1 Alanine	GCU	GCC	GCA	GCG		
2 Arginine	CGC	CGU	CGA	CGG	AGA	AGG
3 Asparagine	AAU	AAC				
4 Aspartic acid	GAU	GAC				
5 Cysteine	UGU	UGC				
6 Glutamic acid	GAA	GAG				
7 Glutamine	CAG	CAA				
8 Glycine	GGU	GGC	GGA	GGG		
9 Histidine	CAC	CAA				
10 Isoleucine	AUU	AUC	AUA			
11 Leucine	UUG	UUA	CUU	CUC	CUA	CUG
12 Lysine	AAA	AAG				
13 Methionine	AUG**					
14 Phenylalanine	UUU	UUC				
15 Proline	CCC	CCU	CCA	CCG		
16 Serine	UCU	UCC	UCG	UCA	AGU	AGC
17 Threonine	ACU	ACC	ACA	ACG		
18 Tryptophan	UGG					
19 Tyrosine	UAU	UAC				
20 Valine	GUU	GUC	GUA	GUG**		
CHAIN TERMINATION	UAA	UAG	UGA			

*The code triplets account for all 64 possible combinations (codons), as based on data presented by Jukes, T. H., and Gatlin, L.: Recent Studies Concerning the Coding Mechanism. Progress in Nucleic Acid Research and Molecular Biology. Vol. 11. New York, Academic Press, 1971.

This amino acid code was almost determined as shown here in Braun, W.: Bacterial Genetics. 2nd ed. Philadelphia, W. B. Saunders Company, 1965.

**Starts the synthesis of a polypeptide chain ("initiator codons").

Figure 2–8 shows how the genome of a microorganism duplicates itself. DNA sends the message by transcription via RNA and translation of the code to protein, which, as the enzyme, carries out the work the microorganism needs for its life processes. More than 50 different tRNAs and at least 20 known aminoacyl-tRNA synthetases are involved in specific selections of amino acids to be added to polypeptide (protein) chains on the ribosomes. The process of cell division is illustrated in Figure 2–9.

The Nucleolus. The nucleolus is a spherical body inside the nucleus of eukaryotes. It appears to play a role in protein synthesis in these cells. It consists only of nucleoprotein containing RNA.

In 1968 another type of RNA was reported. It was thought to represent more than 75 per cent of the RNA synthesis of the cell and was suspected to be the precursor of "true" messenger RNA; it had escaped detection because it is very unstable.[23] It appears to have been a mixture of RNA and precursors to mRNA, named *heterogeneous nuclear* RNA or hnRNA. About 10 per cent of

[23]Bernhard, R.: Verdict of cancer meeting: Information too sketchy. *Sci. Res.*, April 29, 1968, pp. 47–52.

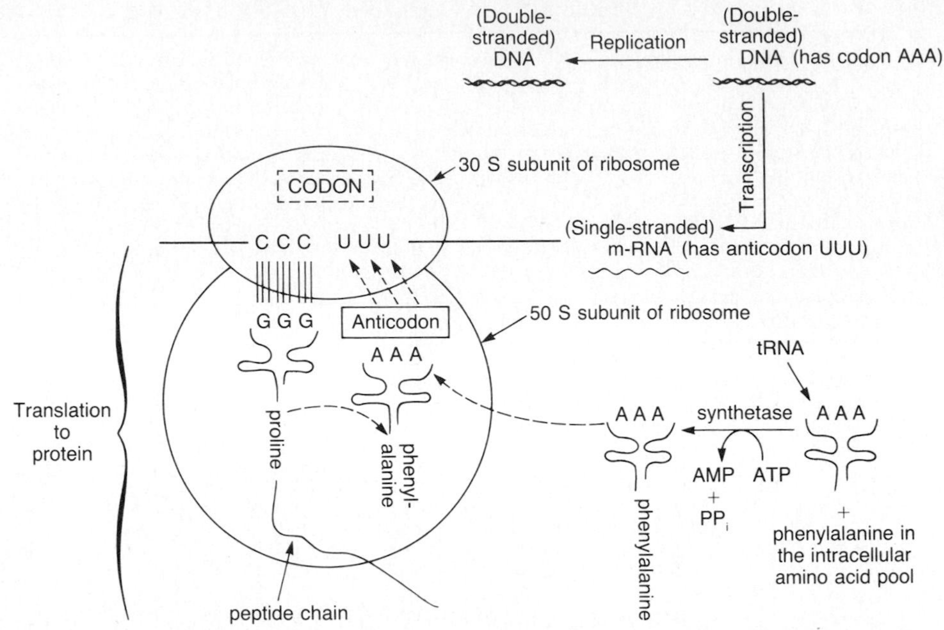

Figure 2–8 Protein synthesis. Because of base pairing, some triplet codons transcribe as shown below:

DNA	mRNA	amino acid
AAA	UUU	phenylalanine
GGG	CCC	proline
TTT	AAA	lysine
TAC	AUG	methionine

When mRNA receives the triplet code message AAA from DNA by transcription, it moves out of the nucleus to the ribosome and becomes the template. Its message is UUU, which is the phenylalanine codon. The 30 S and 50 S subunits of the ribosome unite to form 70 S and attach to the mRNA. The amino acid phenylalanine is being picked up from the free intracellular amino acid pool; this catalyzes peptide formation and leads to linkage with aminoacetyl-tRNA synthetase, an enzyme. The tRNA AAA region recognizes by base pairing the UUU on the mRNA, and a peptide bond is formed by phenylalanine to the adjacent proline, another amino acid on the peptide chain. This chain adds subsequently one amino acid at a time until elongation of the protein is terminated by a chain punctuation message. (After Gottschalk, G.: Bacterial Metabolism. New York, Springer-Verlag, 1979.)

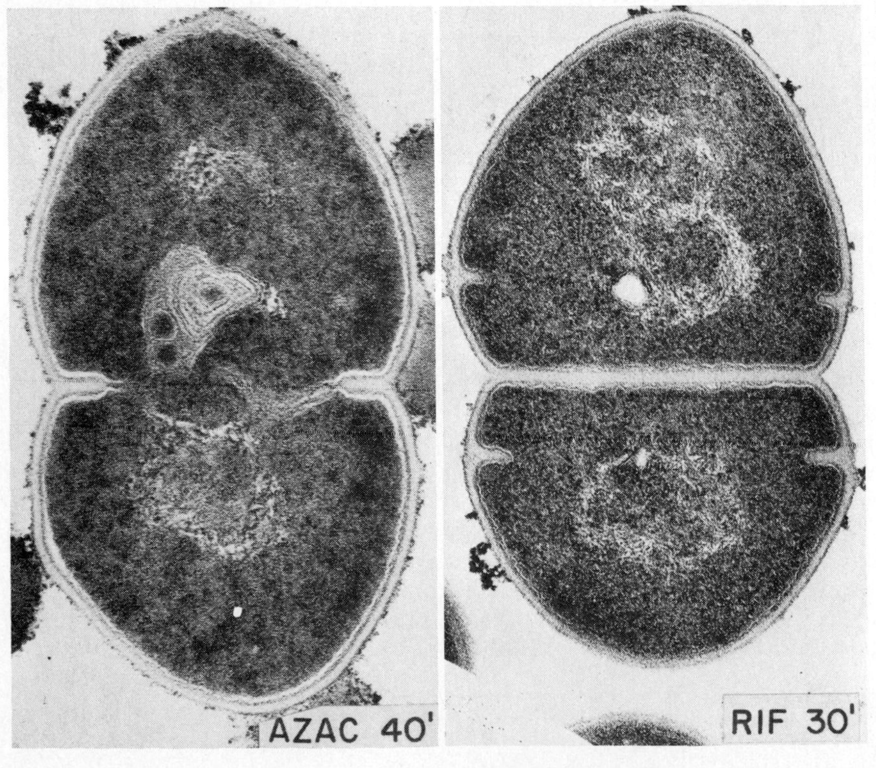

Figure 2–9 Electron micrographs of *Streptococcus faecalis* duplicating itself. *Left,* Nuclear material can be seen in the cell (treated with 5-azacytidine) before the septum has fully formed between the two daughter cells. *Right,* Nuclear material is also seen in the two cells (treated with the antibiotic rifampin) at the beginning of the four-cell stage. (Courtesy of Daneo-Moore, L., and Higgins, M. L.: *J. Bacteriol., 109*:1210, 1972.)

hnRNA survives and becomes mRNA. The functions of the other hnRNA fractions are at present unknown.

Centrosome. This is a small condensed structure containing some DNA, usually located near the nucleus of eukaryotic cells and having a function that is associated with the process of cell division.

Cytoplasm. The remainder of the protoplasm is called the *cytoplasm*. The cytoplasm of a living cell consists of all the portion enclosed within the cell membrane (cytoplasmic membrane) except the nucleus. The cytoplasm may contain many food particles (for example, fat, starch, and sulfur), pigment granules, and special protein and vitamin complexes (coenzymes) with various functions related to digestion and reproduction.

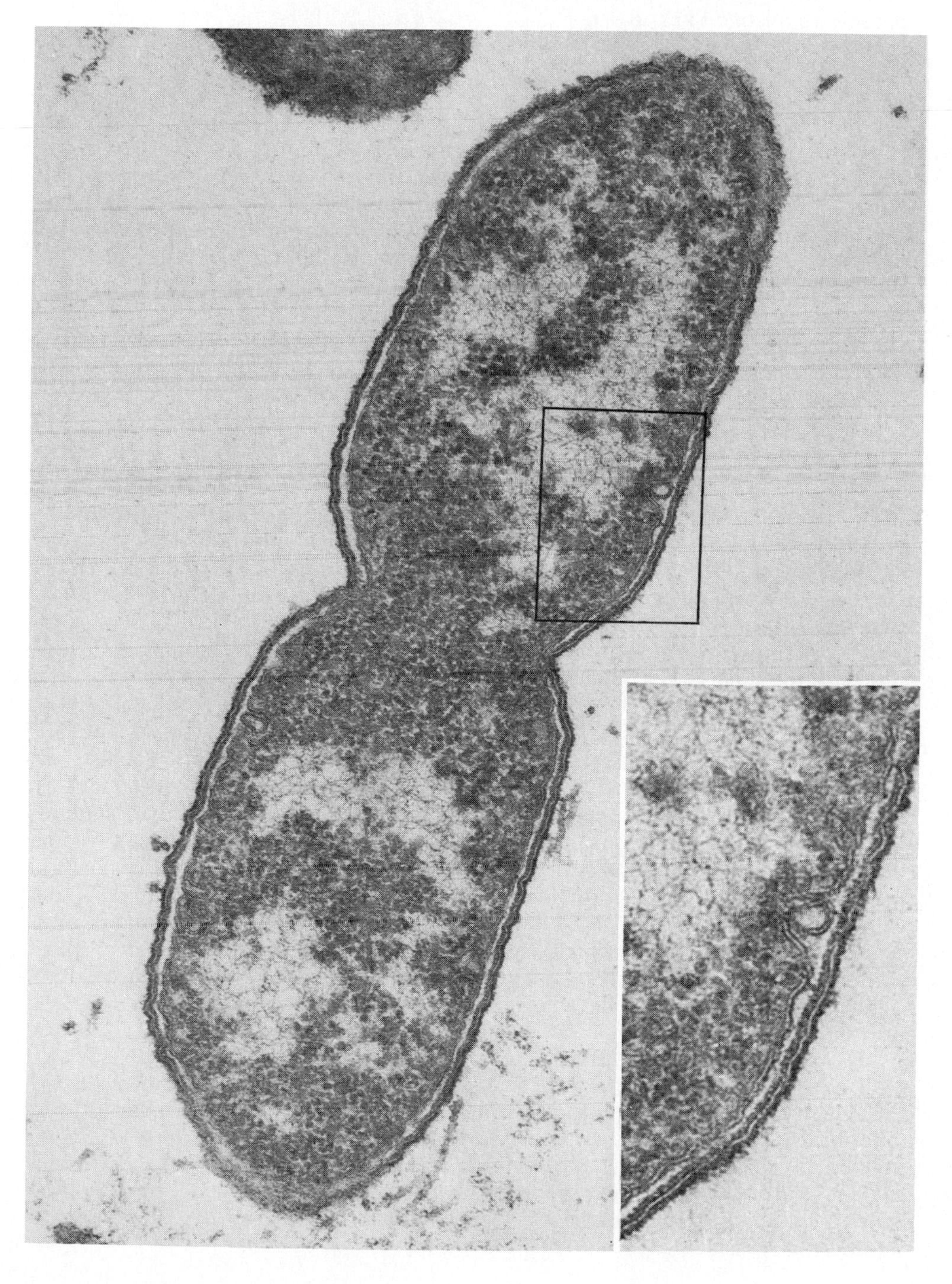

Figure 2–10 Electron micrograph of the bacillus *Erwinia amylovora.* The cell, fixed at room temperature, shows the double-track structure of the cell wall (about 35,000 ×, inset about 95,000 ×), (Courtesy of Huang, P., and Goodman, R. N.: *J. Bacteriol.*, *107*:361, 1971.)

The cytoplasm of *eukaryotic cells* contains such structures as mitochondria, the sites of respiration; endoplasmic reticulum supporting millions of ribosomes (of the larger 80 S ribosome type), the sites of protein synthesis; Golgi bodies, the sites of secretion synthesis; lysosomes, sacs of digestive enzymes; chloroplasts, sites of photosynthesis; and other structures and organelles that may be seen and studied with special microscopic techniques, especially electron microscopes.

As previously mentioned, *prokaryotic* (especially bacterial) cytoplasm contains chiefly ribosomes (of the small 70 S type), possibly attached to membranous materials (Fig. 2–4). In photosynthetic species there are also *chromatophores* that contain the photosynthetic pigments bacteriochlorophyll now designated as bacteriochlorophylls a, b, c, and d. The prokaryotic cytoplasm seems very sparsely furnished (Fig. 2–10).

HABITAT AND MODE OF LIFE

Unicellular, or single-celled, microorganisms of one type or another are found almost everywhere on the surface of the earth. Some live deep in the sea; others inhabit pools of stagnant water; and still others can exist only in the soil or in the blood or feces of various animals. The entire life work of unicellular organisms, whether plant or animal or bacterial, consists of feeding, excreting waste products, and multiplying. Their physiology is, therefore, relatively simple, although, strangely enough, the chemistry of life and reproduction of all cells are remarkably alike and are obviously mere modifications of a single basic design, like different makes of automobiles.

SUMMARY

Modern classification divides living organisms into prokaryotes and eukaryotes. Prokaryotes do not have a well-defined nucleus—their DNA is not enclosed within a membrane—whereas eukaryotes have a "true nucleus." Blue-green algae (cyanobacteria), archaebacteria, and bacteria (eubacteria) are prokaryotes, while animals and plants—including all other algae, the fungi, and the array of microscopic plants—are eukaryotes. Protozoa are animals and thus belong to the eukaryotes. Viruses are neither eukaryotes nor prokaryotes since they are not true cells.

The cell is the smallest and simplest unit of living matter capable of independent life and self-reproduction. While the human body consists of billions of diverse specialized living cells, each bacterium is a single cell that has to carry out all necessary functions required for life. The unit of measurement for microscopic cells is the μm, which equals 1/25,400 of an inch. In a eukaryotic cell (plant or animal) the organelles include the nucleus, mitochondria, chloroplasts in green plants, endoplasmic reticulum, and ribosomes. In all cells the protoplasm forms its own self-containing cell membrane and, in plants and bacteria, a rigid cell wall. In prokaryotes the entire structure (i.e., cell wall plus cell membrane and contents, and any external appendages present in some bacteria, such as flagella, pili, and capsules) are now generally included in the term cell.

Deoxyribonucleic acid (DNA) and ribonucleic acid (mRNA and tRNA) exist in all typical cells. DNA determines the hereditable genetic information of the cell, which is coded into its double spiral (double helix) deoxynucleotide units of adenine (A), guanine (G), cytosine (C), and thymine (T). Subsequent thymine is coded as uracil (U) in RNA. DNA sends a triplet code message by transcription via RNA (mRNA to tRNA), and then translation of the code to the peptide chain forming protein, which, as an enzyme, carries out the life processes of the microorganism.

REVIEW QUESTIONS

1. In what ways are prokaryotes different from eukaryotes?
2. How are the blue-green algae related to other algae and to bacteria?
3. Describe the structure of cells of large animals and plants. How does the cell structure of microorganisms differ from these? How are they alike?
4. What is a cell? List different types of cells.
5. What is protoplasm? What structures may be found in it? How are they related to the functions of living organisms?
6. In relative order of size list different major groups of microorganisms. How are they related to each other? List characteristics used to distinguish them from each other.
7. What are the functions of DNA and RNA? What are genes? What is the genetic code? How does it work?
8. Define the following terms: transcription, translation, protein, ribosomes, cytoplasm, cell wall, cell membrane.
9. How are viruses different from cells?

Supplementary Reading

Boyd, R. R., and Hoerl, B. G.: Basic Medical Microbiology. 2nd ed. Boston, Little, Brown & Co., 1981.

Braun, W.: Bacterial Genetics. 2nd ed. Philadelphia, W. B. Saunders Company, 1965.

Buchanan, R. E., and Gibbons, N. E. (eds.): Bergey's Manual of Determinative Bacteriology. 8th ed. Baltimore, The Williams & Wilkins Co., 1974.

Cantarow, A., and Schepartz, B.: Biochemistry. 4th ed. Philadelphia, W. B. Saunders Company, 1967.

Committee on Form and Style of the Council of Biology Editors: CBE Style Manual. 4th ed. Arlington, Va., Council of Biology Editors, 1978.

Curtis, H.: Biology, 3rd ed. New York. Worth Publishing Co., 1979.

Davidson, J. N., Cohn, W. E., et al. (eds.): Progress in Nucleic Acid Research and Molecular Biology. Vols. 1 to 27. New York, Academic Press, 1963–1982.

Dayhoff, M. D. (ed.): Atlas of Protein Sequence and Structure 1972. Vol. 5. Washington, D.C., National Biomedical Research Foundation, 1972.

DeRobertis, E. D. P., Saez, F. A., and DeRobertis, E. M. F., Jr.: Cell Biology, 6th ed. Philadelphia, W. B. Saunders Company, 1975.

DeWitt, W.: Biology of the Cell: An Evolutionary Approach. Philadelphia, W. B. Saunders Company, 1977.

Fawcett, D. W.: The Cell. 2nd ed. Philadelphia, W. B. Saunders Company, 1981.

Frobisher, M., Hinsdill, R. D., Crabtree, K. T., and Goodheart, C. R.: Fundamentals of Microbiology. 9th ed. Philadelphia, W. B. Saunders Company, 1974.

Gottschalk, G.: Bacterial Metabolism. New York, Springer-Verlag, 1978.

Margulis, L.: Origin of Eukaryotic Cells. New Haven, Yale University Press, 1970.

Poindexter, J. S.: Microbiology. An Introduction to Protists. New York, Macmillan, 1971.

Ragan, M. K., and Chapman, D. J.: A Biochemical Phylogeny of Protists. New York, Academic Press, 1977.

Sharp, J. T.: The Role of Mycoplasmas and L Forms of Bacteria in Disease. Springfield, Ill., Charles C Thomas, 1970.

Spencer, J. H.: The Physics and Chemistry of DNA and RNA. New York, Holt, Rinehart & Winston, 1972.

Stanier, R. Y., and Doudoroff, M.: The Microbial World. 4th ed. Englewood Cliffs, N.J., Prentice-Hall, Inc., 1976.

Villee, C. A.: Biology. 7th ed. New York, Holt, Rinehart & Winston, 1977.

Woese, C. R.: Archaebacteria. *Scientific American*, June 1981, pp. 98–122.

Woese, C. R., Magrum, L. J., and Fox, G. E.: Archaebacteria. *J. Mol. Evol.*, *11*:245, 1978.

CHARACTERISTICS OF ALGAE, PROTOZOA, YEASTS, MOLDS, AND HELMINTHS

ALGAE

Blue-green algae are of interest because of their prokaryotic structure and other similarities to bacteria. Bacteria and blue-green algae, now called cyanobacteria, are *the only known prokaryotic organisms*. (See Chapter 2 and Fig. 3–1.)

Although algae are rarely of pathogenic significance to humans, it is worth noting that some species of eukaryotic algae such as *Chlorella* are "cultivated" on the surface of sewage effluents in large artificial lagoons. They help decompose the organic matter in the sewage.

It has been suggested that eukaryotic algae may play an important role in the space age since, when cultivated under artificial sunlight (ultraviolet light) in space ships and submarines, they not only produce life-giving oxygen like other green plants but also take up poisonous carbon dioxide from the atmosphere and help purify human wastes. They may also provide food, although experiments performed in this direction have been less than successful, since the tough cell wall produces digestive and other difficulties. Still, algae do yield some food products. Eukaryotic algae synthesize several polysaccharide gums of commercial value. Agar, an important ingredient of microbiologic culture media (Chapter 9), is derived from a common seaweed alga, *Gelidium*. Similar gums are algin and carrageen.

PROTOZOA

Protozoans, relatively large in size but still mostly microscopic, are unicellular animals that are very complex in structure and activities. Many of these minute organisms grasp and take solid food particles into their single-celled bodies, swim or creep about, have a definite (though, it may seem to us, simple) sex life, and sometimes form quite complex communities. Most species of protozoa are harmless to humans, living on dead organic matter or on bacteria (which are much smaller). They are found in the sea, lakes, rivers, sewage, and damp soil. A few protozoa are found only in animals and plants, where they cause disease. Examples of pathogenic protozoa are those that cause malaria, amebic dysentery, and African "sleeping sickness" (trypanosomiasis). For purposes of study some can be propagated in test tubes or flasks and their reproduction can be watched under the microscope on glass depression slides.

Structure. The protozoan cell is typically eukaryotic in structure and in general resembles other typical animal cells. Unlike plant cells, which have cell walls, animal cells, including protozoa, characteristically lack morphologically and chemically distinct cell walls. In many species of free-living protozoans, a much thickened and toughened outer membrane or plasmalemma serves most of the functions of a cell wall.

Multiplication. Protozoa multiply by dividing into two. First the nucleus divides by a process called *mitosis*; then the cytoplasm and cell membrane divide by cytokinesis (cytoplasmic division) to surround the two nuclei; and the two *daughter cells* then separate as two new individuals. In addition to cell division,

CHAPTER 3

some species of protozoa exhibit definite differentiation of sexes, such as that seen in malarial parasites. The differentiation of sexes in dioecious species of protozoa is much more highly developed than in bacteria, yeasts, or molds.

Nutrition. The individual cells of most protozoa differ from all bacteria, and from individual cells of virtually all plants, in having the power to ingest solid food particles.[1] Ingestion of solid particles of food is distinctively an animal characteristic and is often called a *phagotrophic* type of nutrition. However, many protozoan species may be nourished, as are yeasts, molds, bacteria, and plant cells, by diffusion of dissolved food matter through the cell membrane.

Life Cycles. Many of the protozoa pass through a definite and readily demonstrable series of stages in their development and thus differ greatly from most bacteria. These *life cycles,* as they are called, are characteristic for each species and often are relatively complicated. A good illustration of a protozoan life cycle is that of the malarial parasite (Chapter 41). Some protozoa, especially the malarial parasite, multiply sexually in one or more sanguivorous[2] arthopod[3] hosts (certain flies, mosquitoes, and other insects). The offspring then become mature inside the arthropod and are ready to infect humans or other susceptible animals when the arthropod bites them.

Trophozoite and Cyst. The actively growing, feeding, and multiplying

[1]An exception is the Venus's-flytrap *(Dionaea muscipula),* a plant that catches and ingests insects that settle on its leaves.

[2]Latin *sanguis,* blood; *vorare,* to eat; blood-eating, as a mosquito or flea.

[3]Greek *arthron,* joint; *podion,* foot; hence jointed-legged creatures such as mosquitoes, ticks, flies, and shrimp.

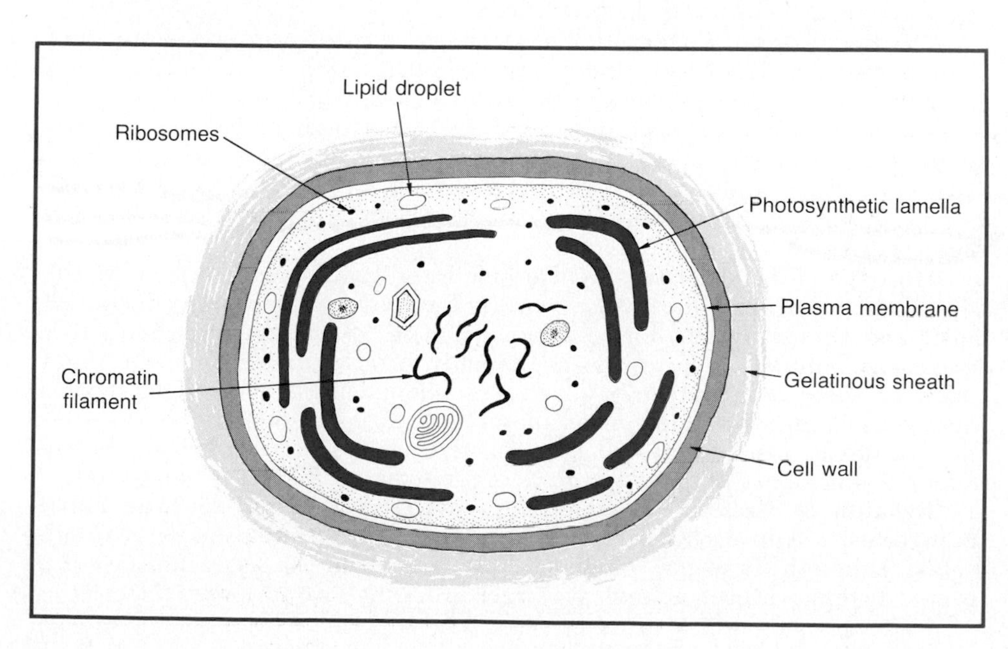

Figure 3–1 Diagram of the major structures found in blue-green algae. Outside the cellulose-containing cell wall is a protective gelatinous sheath. Photosynthetic pigments are bound to simple membranous lamellae. Lipid droplets are common in the cytoplasm. Genetic material is located on chromatin filaments, generally restricted to one region of the cell. (From Clark, M. E.: Contemporary Biology. 2nd ed. Philadelphia, W. B. Saunders Company, 1979).

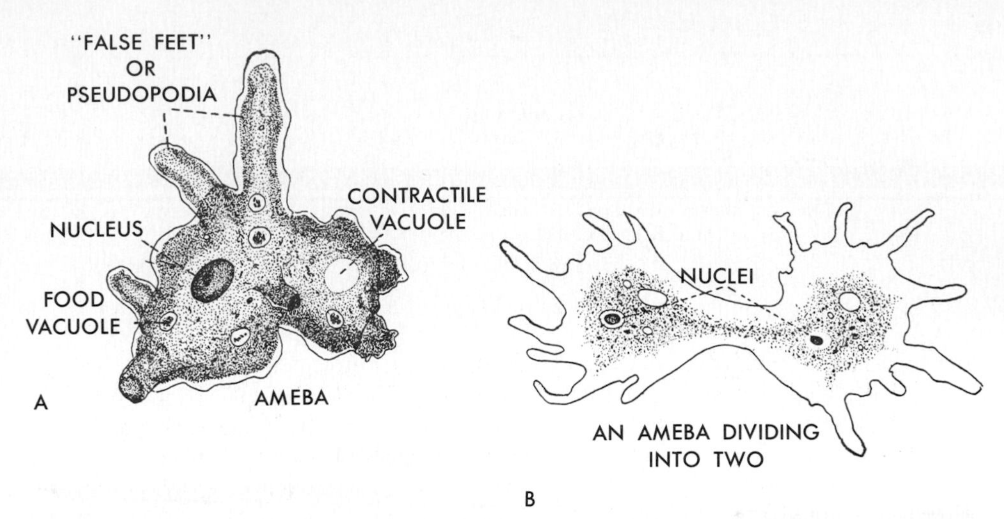

Figure 3–2 A species of harmless ameba from stagnant water: trophozoite stage. Others are not so harmless, for example, *Entamoeba histolytica*. Note the numerous pseudopodia, the nucleus, and the vacuoles inside the cell. This organism, itself microscopic in size (18 to 25 μm), feeds upon bacteria, and is thousands of times larger than bacteria. (MacDougall and Hegner: Biology. New York, McGraw-Hill Book Co., Inc.)

form of protozoa is called the *trophozoite*[4] stage (Fig. 3–2). Many protozoa also exhibit a thick-walled, inactive stage in which they are dormant and resist drying. This is the *cyst stage* (Fig. 3–3). The cyst stage is important in the transmission of the ameba *(Entamoeba histolytica)* that causes amebiasis (dysentery, liver abscess, and so forth).

Classification. All protozoa are motile in at least one stage of their life cycle. The means of motility differs among them and furnishes a basis for classification into four subphyla, as shown in Table 3–1. At least one species of each group is pathogenic to humans. Many others are pathogens of lower animals and plants. The important human pathogens are discussed more fully in Chapters 27 and 41.

TRUE FUNGI (EUMYCETES)

Fungi are plants that lack chlorophyll, the green pigment responsible for photosynthesis. The group of Eumycetes includes large, edible mushrooms, "puff-balls," and microscopic yeasts and molds. Like the protozoans, the cells of Eumycetes are eukaryotic in structure.

Characteristics of Eukaryotic Fungi. Most true fungi grow as branching, filamentous tubes filled with protoplasm. When growing in this manner they are called *molds*; when growing in the form of ovoid, budding cells they are commonly referred to as *yeasts*. In many species, environmental conditions (e.g., temperature, oxygen supply, nutrition) may cause a "yeast" to grow in a filamentous form and a "mold" in a yeast-like form. *The terms yeast and mold have no scientific status*; they are used for convenience in discussing different forms of fungi.

More than 70,000 species of eukaryotic fungi have been described. Mycology is the name given to the science concerned with the study of fungi (including molds and yeasts). Except for some aquatic types (Saprolegniales), most true fungi are nonmotile. The cell walls of many true fungi resemble the outer cuticle of some animal cells in that they contain chitin, the tough, flexible substance (chemically resembling cellulose) in skeletal and shell tissues of insects and crustacea. True fungi and most bacteria are heterotrophic or chemoorganotrophic—that is, they must have organic matter as part of their food.

Relation of Bacteria (Schizomycetes or Fission Fungi) to True Fungi (Eumycetes). Approaches to the classification of microorganisms vary considerably. Although there may be general agreement on the characteristics of a species, a genus, or even a family, as larger groupings are proposed differences

[4]Greek *trophe*, nutrition; *zoon*, animal; an actively feeding form of animal.

Table 3–1. Classification of the Phylum Protozoa[1]

Phylum Protozoa
Subphylum I. *Sarcomastigophora* (flagellates and amebas)
Subclass I. *Mastigophora*[2] (flagellates) (move with long, whiplike lashes called *flagella;* reproduction asexual)
Intestinal parasites
Giardia lamblia—occasionally causes mild, epidemic enteritis (transmission like that of other enteric infections)
Trichomonas hominis—pathogenicity and transmission same as *Giardia lamblia*
Parasites in genitourinary tract
Trichomonas vaginalis—causes vulvovaginitis (transmission by sexual contact, also by fomites[3])
Blood parasites
Trypanosoma rhodesiense and *T. gambiense*—cause African sleeping sickness (transmission by insect vectors[4])
Trypanosoma cruzi—causes South American trypanosomiasis (Chagas' disease) (transmission by insect vectors)
Blood and tissues
Leishmania donovani—causes kala-azar (transmission by insect vectors)
Leishmania tropica—causes oriental sore (transmission by insect vectors)
Leishmania braziliensis—causes espundia (transmission by insect vectors)
Subclass II. *Opalinata* (superficially appear to be ciliates; all are parasitic)
Subclass III. *Sarcodina* (amebas; all are phagotrophs; move with pseudopodia[5]; reproduction asexual)
Entamoeba histolytica—causes amebic dysentery (transmission like that of other enteric infections from feces; Chapters 24, 26, 28).
The "true" slime molds belong here.
Subphylum II. *Sporozoa* (move with pseudopods in immature stages only; male form is flagellate; reproduction alternately sexual and asexual). All are parasites.
Plasmodium (vivax, malariae, falciparum, ovale)—causes different types of malaria (transmission by insect vectors). Others cause diseases of primates, birds, and reptiles.
Subphylum III. *Cnidospora* (sporozoans). All are parasitic, usually in lower vertebrates. Pasteur studied a disease in silkworms caused by an organism in this subphylum.
Subphylum IV. *Ciliophora* (move with cilia; reproduction usually asexual; occasional conjugation)
Balantidium coli—causes enteritis and ulcers of the colon (transmission like that of other enteric infections).

[1]This table is based on the now widely accepted, revised classification proposed by the Society of Protozoologists. It is expected to undergo further revision in the future.
[2]It is thought that *Glaucocystis* may be the missing link to more primitive algae; it has a chloroplast (eye spot) and exhibits photosynthesis.
[3]Plural of fomes; any object that may transfer infectious microorganisms from one person to another, e.g., eating utensils, toothbrushes, handkerchiefs, and so on.
[4]A vector is any agent, living or inanimate, that transmits infectious agents.
[5]"False feet," or having the functions but not the structure of feet (Greek, *pseudo,* false or imitation; *podion,* foot).

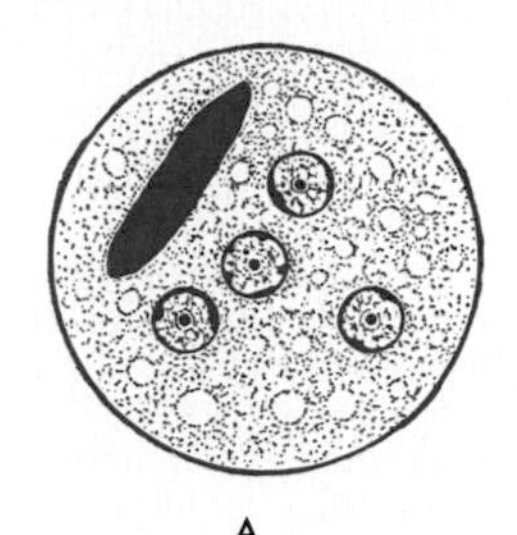

A

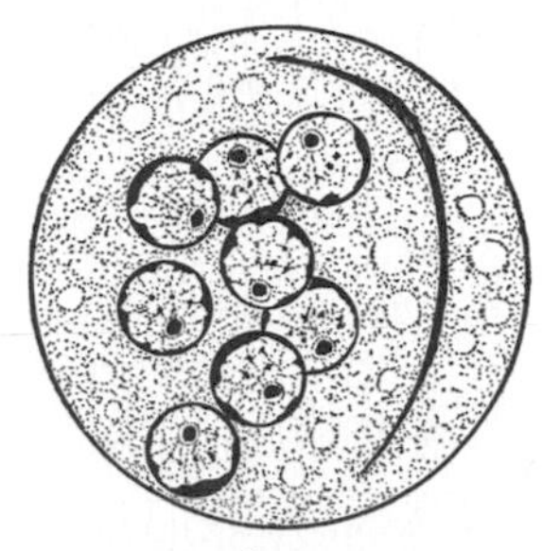

B

Fig. 3–3 *A,* Cyst of *Entamoeba histolytica* (7 to 18 μm in size). *B,* Cyst of *Entamoeba coli* (10 to 25 μm), generally considered harmless. The larger dark circular objects inside the cysts are nuclei formed by nuclear fission within the cell that produces the cyst. The numbers and size of the nuclei are characteristic of species. *E. histolytica* has four mature nuclei. *Entamoeba coli* has eight mature nuclei, *Iodamoeba bütschlii* has only one mature nucleus, and so on. The elongated bodies in the cysts are food or other reserve material. Recognition of the cysts in feces of patients is an important means of diagnosis. (Hegner, Root and Augustine: Animal Parasitology. New York, Appleton-Century-Crofts, Inc.)

of opinion between specialists in *taxonomy*[5] become very great. Table 3–2 illustrates the relative position of "True Fungi" (Phylum III, Subphylum 2) in the Plant Kingdom. They are eukaryotes. Not included are the bacteria (Schizomycetes); these prokaryotes are classified in Table 3–3 and in Chapter 4.

Thallophytes. Among the fungi the *thallophytes* are distinguished from all other plants by having no specially differentiated organs such as roots, leaves, or flowers, as seen in Table 3–2. A single cell may constitute the entire plant or, if the plant is multicellular like a mold or seaweed, a single cell by continued cell fission may reproduce the entire plant asexually.

YEASTS

Yeasts are typically unicellular organisms, although several cells often cling together after fission (Fig. 3–4). Yeast cells are 20 to 100 times as large as most

[5]Taxonomy is the science of the classification of organisms.

Table 3–2. Classification of Fungi in the Plant Kingdom

Kingdom *Plantae*
Phylum I. *Tracheophyta* (vascular plants). Contain chlorophyll; require sunlight.
  Subphylum 1. Pteridopsida (ferns and seed-bearing plants such as lilies, grasses, roses, zinnias, oak trees).
  Subphylum 2. Sphenopsida (horsetails).
  Subphylum 3. Lycopsida (club mosses).
Phylum II. *Bryophyta* (liverworts and true mosses). Contain chlorophyll; require sunlight.
  Class 1. Musci (mosses: e.g., *Sphagnum, Polytrichum*).
  Class 2. Liverworts (e.g., *Marchantia*).
Phylum III. *Thallophyta* (nonvascular). May or may not contain chlorophyll. No differentiated leaves, stems, flowers, or roots. Relatively simple structure. Can generally grow in lifeless media. (Bacteria have some of the characteristics of thallophytes but are prokaryotic in structure.)
  Subphylum 1. Red, brown, and green algae. Contain chlorophyll; require sunlight. (Represented by various seaweeds and green pond-scums.)
  Subphylum 2. Fungi. Do not contain chlorophyll. Do not require sunlight.* With some possible exceptions, can grow in lifeless media.
    Class 1. *Myxomycetes.* Slime molds. These have a motile ameboid stage and closely resemble protozoa. They reproduce by "spores," but are as far removed from the true fungi as are the bacteria. Some bacteriologists regard them as bacteria.
    Class 2. *Basidiomycetes.* Reproduce sexually by basidiospores, and also asexually. Represented by mushrooms, puffballs, and rust and smut diseases on plants.
    Class 3. *Ascomycetes.* Most yeasts and many molds. Reproduce sexually by ascospores, and also asexually. Some are parasitic on plants. Truffles are a type of ascomycete.
    Class 4. *Phycomycetes.* Molds with coenocytic hyphae. Multiply asexually by sporangiospores; also reproduce sexually. Represented by water molds, common bread mold, many saprophytes; some are parasites on plants and insects.
    Class 5. *Deuteromycetes* (Fungi Imperfecti). Reproduce only by asexual spores. Many are very common molds. Most fungi that can cause disease in man belong to the Fungi Imperfecti. In addition there are many that are parasitic on plants. The imperfect fungi include fungi that cannot be properly assigned to either class 3 or 4.

*A few species of harmless, photosynthetic bacteria are exceptions. Organisms in Kingdom Plantae are eukaryotes.

bacteria. Everyone is familiar with bakers' yeast, which is composed of countless numbers of these unicellular plants compressed together, dried, and granulated. Each granule contains millions of yeast cells. Stained smears of the moistened material on glass "slides" must be viewed with the microscope in order to see the yeast cells. A recent study described 39 genera with 349 different species of yeasts.

Yeasts are gram-positive (a type of staining process described later) and may also be stained with simple dyes like methylene blue. Staining makes microscopic cells more readily visible. In general, yeasts are grown on simple media (nutrient materials) much as are bacteria, and their colonies (masses of growth visible to the naked eye, discussed in Chapter 9) resemble those of true bacteria. Sabouraud's agar is a favorite medum for the cultivation of yeasts and molds. It contains 1 per cent neopeptone (partly digested meat), 4 per cent maltose or glucose, and 1.5 to 2.0 per cent agar (to solidify the medium unless broth is desired), and the pH is adjusted to 5.6 (acidic).

Structure of Yeast Cells. Yeast cells are eukaryotic in structure. They are usually oval or egg-shaped, although in some species long filamentous cells, as well as filaments of cells, are formed, resembling those of molds. There is a well-developed and rather large, globular nucleus and a thick cell wall, composed in part of a chitin-like substance. There are also vacuoles containing waste substances, functional organelles, food granules of various kinds (some evidently of glycogen), and fat.

Habitat of Yeasts. Yeasts are widely distributed in nature in much the same situations as are bacteria. They are commonly found on grapes and other fruits and plants, the spores passing the winter in the soil. Wine made from grapes depends to some extent on the varieties of "wild" yeasts that settle upon them naturally. Yeasts are always found in dung, soil, and milk, and are not infrequently observed in throat cultures, river water, and dust.

Table 3–3. Classification of Bacteria in the Kingdom *Procaryotae*

Kingdom *Procaryotae*
Cyanobacteria. Myxophyceae or Cyanophyceae (blue-green algae). Contain chlorophyll; require sunlight. (Some seaweeds, scums on stagnant water, etc.)
Bacteria (Schizomycetes or Fission Fungi)*

Part 1. Phototrophic bacteria Order *Rhodospirillales*	Part 9. Gram-negative anaerobic bacteria
Part 2. The gliding bacteria Order *Myxobacteriales* Order *Cytophagales*	Part 10. Gram-negative cocci and coccobacilli Family *Neisseriaceae*
Part 3. The sheathed bacteria	Part 11. Gram-negative anaerobic cocci
Part 4. Budding and/or appendaged bacteria	Part 12. Gram-negative chemolithotrophic bacteria
Part 5. The spirochetes Order *Spirochaetales*	Part 13. Methane-producing bacteria
Part 6. Spiral and curved bacteria	Part 14. Gram-positive cocci
Part 7. Gram-negative aerobic rods and cocci	Part 15. Endospore-forming rods and cocci
Part 8. Gram-negative facultatively anaerobic rods Family *Enterobacteriaceae* Family *Vibrionaceae*	Part 16. Gram-positive, asporogenous rod-shaped bacteria
	Part 17. Actinomycetes and related organisms
	Part 18. The rickettsias
	Part 19. The mycoplasmas

*Only species of medical interest are described in detail in this book. This is an outline from *Bergey's Manual of Determinative Bacteriology* (1974). See Table 4–3 for species of bacteria of medical importance.

Multiplication of Yeasts

Asexual. Most species of yeasts reproduce asexually by a process called budding, in which, instead of forming two equal cells as the bacteria do, the new cell is at first much smaller than the other and is referred to as a daughter cell. The daughter cells, or *buds*, often cling to the parent cell; clumps or chains of cells are thus formed (Fig. 3–4). A few species of yeasts reproduce by fission like bacteria and protozoa.

Sexual. Ascospores (i.e., spores in sacs, or *asci*) are formed during sexual reproduction by many species of Eumycetes, including yeasts. All sac-forming fungi are called Ascomycetes. In some yeasts (nonfilamentous Ascomycetes), the nucleus of a single cell undergoes meiotic division and the haploid daughter nuclei remain in the cell as ascospores. Later they fuse and form diploid cells

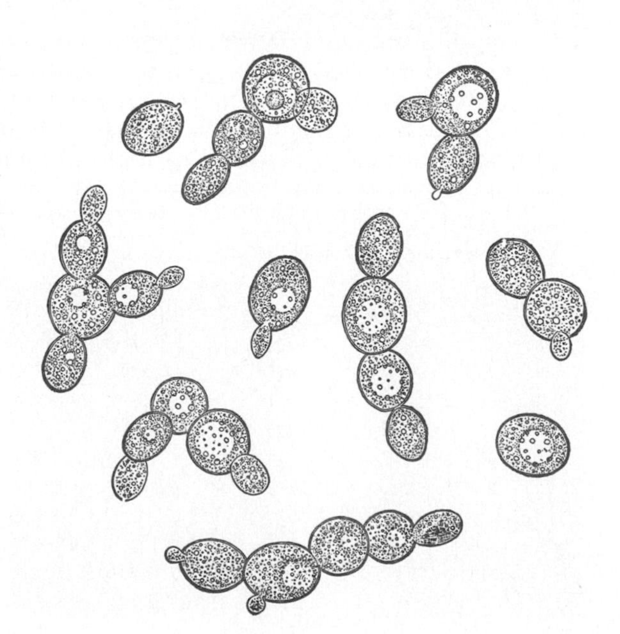

Figure 3–4 Yeast cells. Brewer's yeast actively multiplying by budding. This species is called *Saccharomyces cerevisiae.* The single large internal vacuoles and the numerous small fat drops are shown, as are buds in various stages of development and the cell walls. Nuclei not visible here. Internal granules are ribosomes, mitochondria, and other organelles found in all eukaryotic cells, but *not* in prokaryotic cells. (Highly magnified.) (Sedgwick and Wilson.)

that are released as budding diploid cells. Usually four, six, or eight spores are formed by divisions of the nucleus within a sac, the number tending to be characteristic of the species (Fig. 3–5). Ascospores are less resistant than bacterial spores, and they are killed by a temperature of 62 C in a short time, but they are dormant and resistant to drying.

Often asci are formed following a process in which two adjacent haploid cells send out projections that meet and form a copulation canal, through which there is an intermingling of nuclear material. The fertilized diploid nucleus divides meiotically to form a number of haploid ascospores. These may reproduce indefinitely asexually as haploid cells. Some very interesting studies have been made of the heredity and breeding of yeasts (which are quite analogous to heredity and breeding in dogs, cattle, or humans) and the development of progeny that have special value in fermentation processes (e.g., wine making) and the like.

Activities of Nonpathogenic Yeasts

Yeasts, especially yeasts of the genus *Saccharomyces*, which includes the bakers' and brewers' yeasts, readily use various kinds of sugar as food. They give off enzymes that bring about the chemical changes in sugars, called *fermentation*, during which alcohol and carbon dioxide are formed. Both of these products are found in beer, new wine, and rising dough, all of which are yeast-fermented products. The holes in bread are due to CO_2 gas bubbles. The foam on beer and the effervescence of champagne also result from carbon dioxide formed by yeast.

There are various species of "wild" yeasts and torulas,[6] which produce different by-products during their metabolic processes. Some of these impart undesirable flavors and odors to beer and wines. The use of *pure cultures* of desirable yeasts is, therefore, of great importance in industry, where flavor and other qualities of the product depend on the particular species and strain of yeast employed.

MOLDS

The differentiation between yeasts and molds is not a sharp one, since, as previously mentioned, many yeastlike fungi have properties of growth like molds, and vice versa, depending on environmental circumstances. For purposes of this discussion, only those forms that commonly grow in definitely woolly and filamentous colonies will be considered molds.

[6]Torulas are yeastlike fungi that do not form sexual spores or cause alcoholic fermentation, e.g., Fungi Imperfecti.

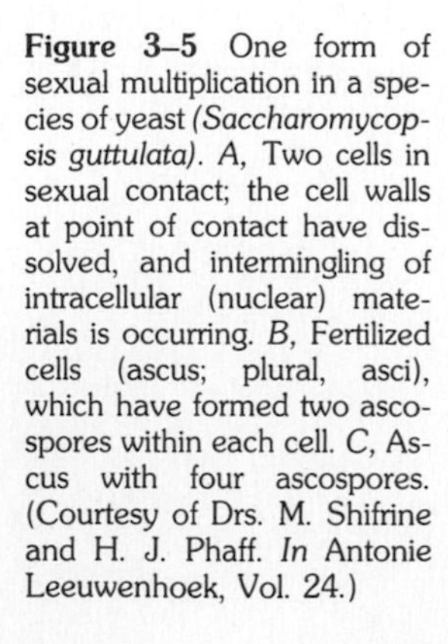

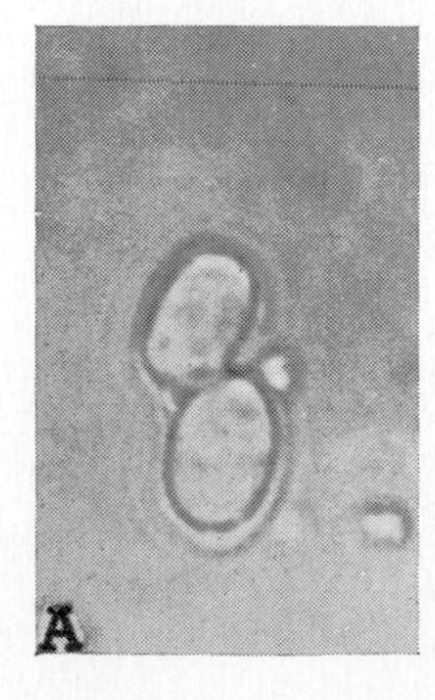

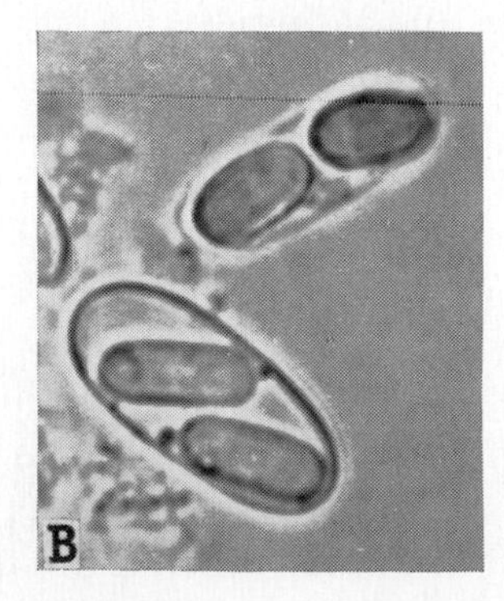

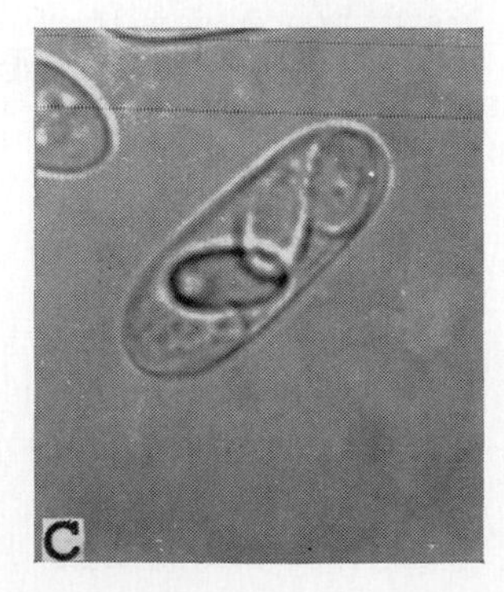

Figure 3–5 One form of sexual multiplication in a species of yeast *(Saccharomycopsis guttulata)*. *A*, Two cells in sexual contact; the cell walls at point of contact have dissolved, and intermingling of intracellular (nuclear) materials is occurring. *B*, Fertilized cells (ascus; plural, asci), which have formed two ascospores within each cell. *C*, Ascus with four ascospores. (Courtesy of Drs. M. Shifrine and H. J. Phaff. *In* Antonie Leeuwenhoek, Vol. 24.)

Molds are as widely distributed in nature as yeasts, most of them being saprophytes. Many live in the soil and take an active part in the decomposition of organic matter. Everyone is familiar with the green, brown, black, white, or yellow molds often seen on stale bread, fruits (especially overripe oranges), and jars of jelly, or on rags, books, or old leather shoes that have lain in a damp place for a long time. If such growths are examined with a magnifying glass, they will be seen to consist of beautiful, almost fairy-like plants, all glistening and transparent. The "fruit" (a mass of spores) is usually deeply colored and grows in great profusion. Each little colored ball or frond is a mass of spores. Molds are cultivated and manipulated in the laboratory in much the same manner as are yeasts and bacteria.

Typical molds are made up of branching tubular filaments, or *hyphae*, forming a woolly growth, the *mycelium*, from which the asexual reproductive cells (*spores* or *conidia*) develop. The mycelial filaments often penetrate into the medium on which the mold is growing, thus acting like roots.

In some molds (Phycomycetes[7]) the filaments throughout the whole plant or mycelium are *coenocytic* (undivided by transverse walls), the nuclei being spaced more or less regularly in the filament. Only rarely are these species pathogenic for humans. In other molds — Ascomycetes, Basidiomycetes (e.g., mushrooms), and Fungi Imperfecti — nuclei and cytoplasm seem to be segregated inside segments ("cells") of the hyphae by means of partitions called *septa*.

[7]From the Greek *phykos*, seaweed; *mykes*, fungus; hence, marine and aquatic fungi. Some species are found on land.

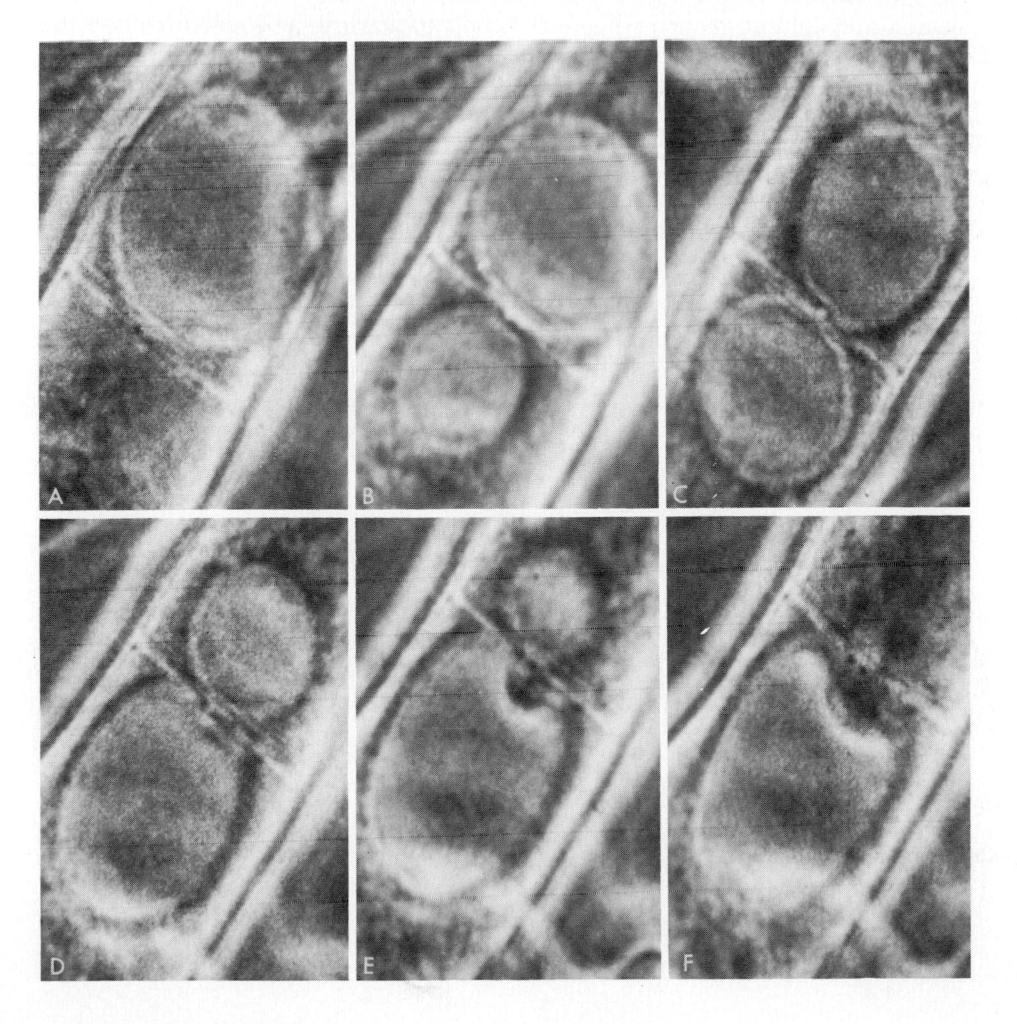

Figure 3–6 Protoplasmic flow through a septum of a hypha of *Neurospora crassa*, from *A* to *F*. The result is complete mixing of nuclei and genetic determinants, and thus of enzymes and their metabolic products. If two strains are inoculated into one flask, the cultures grow like one organism.

Such molds are said to be *septate*. However, the septa in Ascomycetes are not restrictive of protoplasmic flow or the transfer of nuclei. For example, in *Neurospora crassa* (an ascomycete) the septa are perforated; protoplasm and nuclei pass freely through them (Fig. 3–6). Sometimes one may find several nuclei in one cell and in other cells none. The flow of protoplasm carries nuclei and cellular materials not only from cell to cell in the hyphae but also, at certain stages, from conidial spore to spore.

In molds we thus see the beginnings of *division of function* between *vegetative* cells, which provide for nutrition and anchorage, and *reproductive* cells. This is one step higher in organization than bacteria and the simpler species of protozoans, in which all the functions of the organisms are carried on in the single cell.

Reproduction of Molds

Asexual. Molds reproduce *asexually* by one or more of five different methods, as follows:

Oidia[8] ***or Arthrospores.*** These are short fragments of mycelium that become detached by *fragmentation*. This occurs only in septate molds.

Blastospores. These are much like the buds of yeasts, but they develop along mold hyphae.

Chlamydospores. Along some hyphae certain cells develop thick protective walls and appear to go into a resistant, dormant stage. They separate and resume independent growth when conditions of warmth and moisture become favorable again, each forming a new mycelium.

Sporangiospores. Sporangiospores are formed only by the Phycomycetes. In terrestrial species they are minute, rounded, thick-walled bodies resistant to climatic heat and drought. They are produced in large numbers in globular envelopes at the tips of special hyphae. The envelope is called a *sporangium*, hence the term sporangiospores (Fig. 3–7). In most aquatic species, they are liberated from the sporangium as flagellate, free-swimming zoospores.

[8]Oidium is from a Greek root meaning egg, i.e., an ovoid body.

Figure 3–7 *Left*, characteristic structures of *Rhizopus*. The rootlike filaments *(rhizoids)* are seen at *A*, the stolon or spreading filament at *E*, and a *conidiophore* or conidia-bearing hypha at *B*. Details of the sporangium are seen at *C* and *D*, and conidia leaving the *columella* at *F*. Chlamydospores are seen at *G*. This is *asexual* reproduction. (From Conant, N. F., et al.: Manual of Clinical Mycology. 3rd ed. Philadelphia, W. B. Saunders Company, 1971. *Right*, a species of *Rhizopus*, a coenocytic mold. Note the branches resembling roots *(rhizoids)*, the long connecting filament by which the mycelium spreads (*stolon*) and the vertical branches *(sporangiophores)* holding the sacs *(sporangia)* full of *sporangiospores*. (Swingle, D. B.: Plant Life. 2nd ed. D. Van Nostrand Co., Inc.)

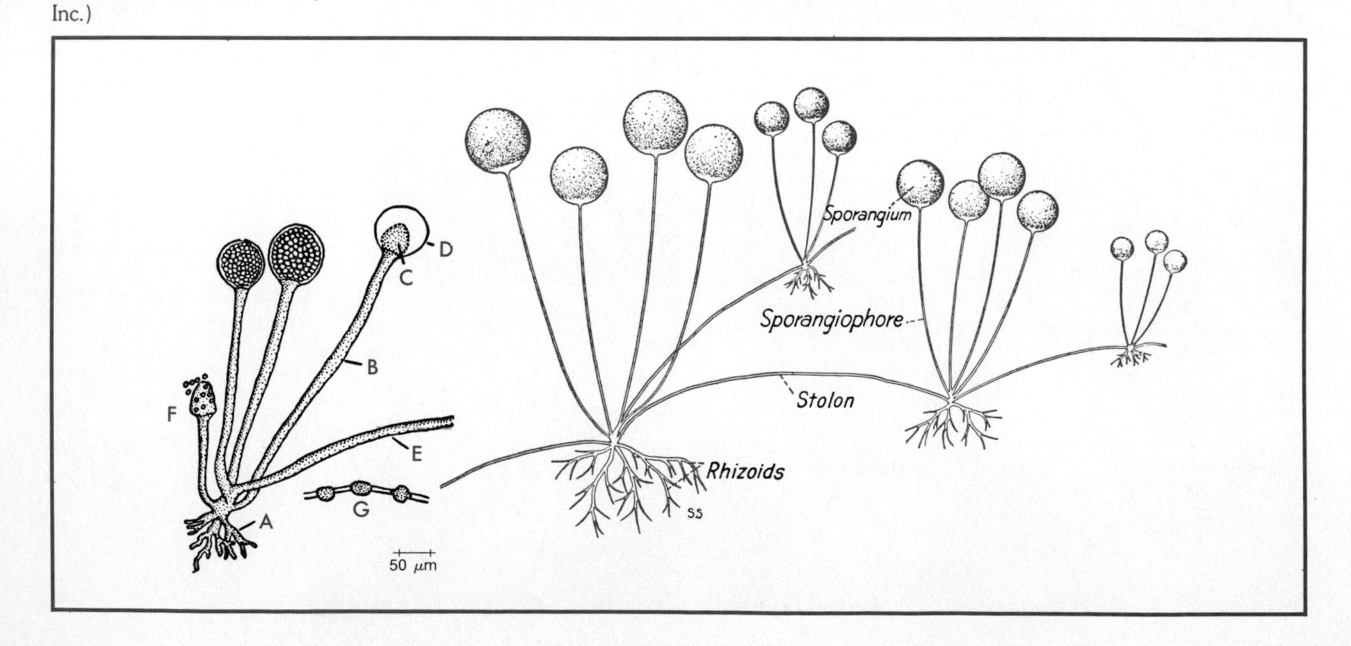

Conidiospores. Conidiospores[9] are produced in long chains, free of an enclosing membrane, by certain Ascomycetes (Fig. 3–8). The forms, arrangements, and colors of sporangiospores and conidia are very useful in the identification of molds. The green or blue powdery material on a spoiled orange is a mass of conidiospores. Since each conidium is able to grow into a new mycelium, it is easy to understand how quickly spoilage can spread.

Sexual. Sexual reproduction occurs in Phycomycetes and also in the filamentous Ascomycetes. The process is very distinctive in each species or genus. For example, in two common genera of Phycomycetes called *Mucor* and *Rhizopus*, two cells of the same culture grow into contact, the separating walls dissolve, and the cell contents mingle as one cell. This cell, called a *zygospore*, acquires a hard, thick outer shell and becomes dormant for a time, much as does a seed. Eventually, finding conditions of moisture, temperature, and nutrition favorable, it bursts or germinates, sending forth a filament (Fig. 3–9). During the resting stage the zygospores resist unfavorable conditions such as drought, food deprivation, and exposure to sunlight, in this respect resembling the spores of yeasts and bacteria. Although more heat-resistant than the spores of yeasts, they are very much less so than the endospores of bacteria (Chapter 8).

In the filamentous Ascomycetes (e.g., *Aspergillus* and *Penicillium*), the sexual spores are formed, as in *Mucor* and *Rhizopus*, by fusion of cells in the mycelium. However, instead of a single zygospore resulting, the fertile cell nucleus divides to form a number of spores that are held together for a time in a multiple sac or complex ascus. The structure holding these sexually produced spore-sacs is called an *ascocarp* and is usually found suspended in the mycelial network.

Fungi Imperfecti

This term is used for a large and rather miscellaneous collection of molds and yeasts that, although closely resembling well-known species, have never been observed to produce sexual spores and are therefore designated imperfect.

[9]Conidiospores are also called conidia (pl.) or conidium (sing.). The Greek word *konis* means dust.

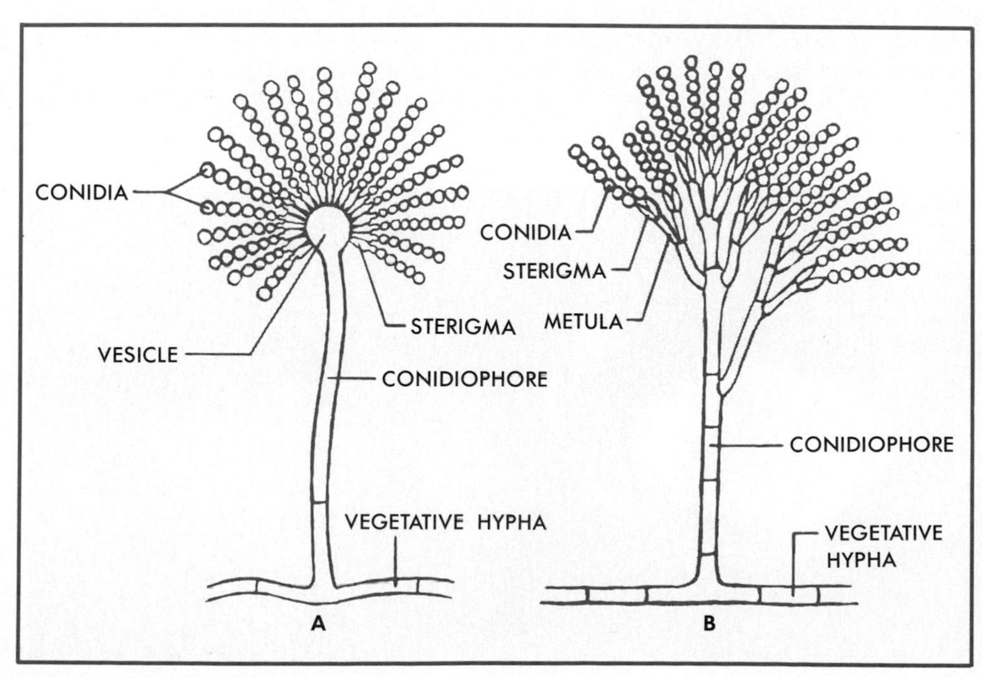

Figure 3–8 Two common species of mold, showing structures. *A, Aspergillus; B, Penicillium.* For discussion, see text. (Carpenter, P. L.: Microbiology. 4th ed. Philadelphia, W. B. Saunders Company, 1977.)

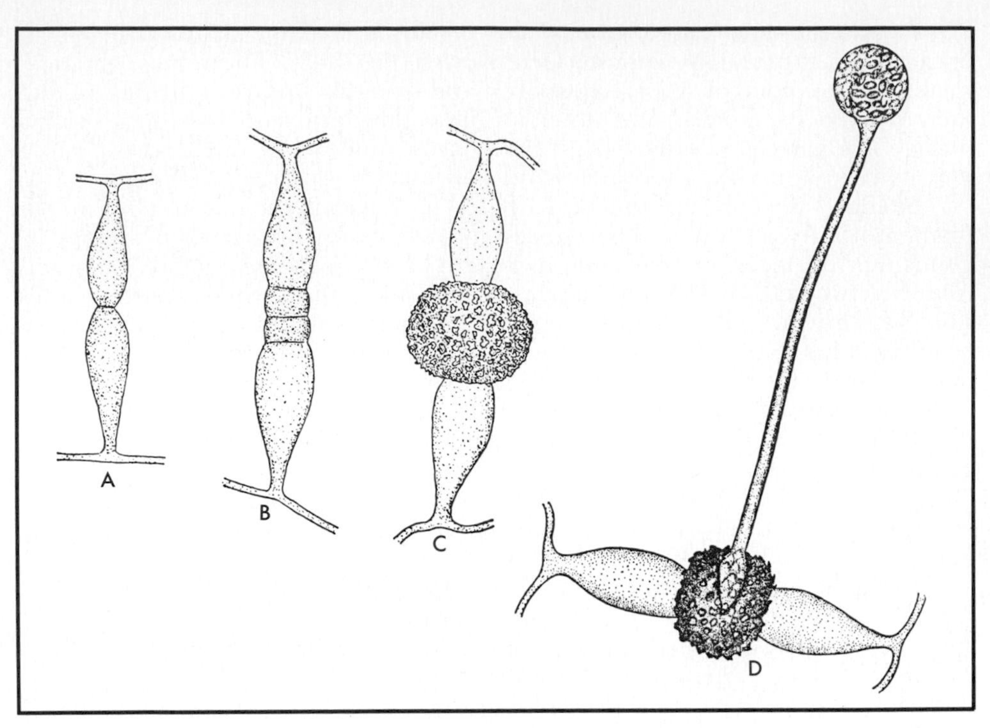

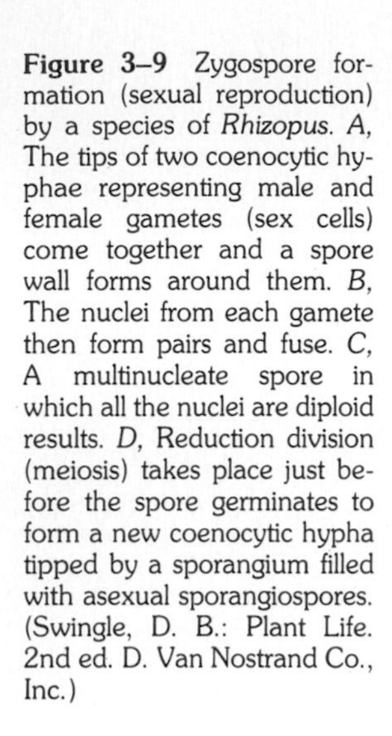

Figure 3–9 Zygospore formation (sexual reproduction) by a species of *Rhizopus*. *A*, The tips of two coenocytic hyphae representing male and female gametes (sex cells) come together and a spore wall forms around them. *B*, The nuclei from each gamete then form pairs and fuse. *C*, A multinucleate spore in which all the nuclei are diploid results. *D*, Reduction division (meiosis) takes place just before the spore germinates to form a new coenocytic hypha tipped by a sporangium filled with asexual sporangiospores. (Swingle, D. B.: Plant Life. 2nd ed. D. Van Nostrand Co., Inc.)

Many of the commonest molds are Deuteromycetes, which are the Fungi Imperfecti. They cease to be designated Fungi Imperfecti when we discover their sex life.

Growth of Molds

Molds may be readily cultivated on ordinary bacterial media; however, Sabouraud's medium is preferred (Fig. 3–10). They can grow where there is relatively little moisture, and can get sufficient water from humid atmospheres. Because of their ability to obtain nutriment from solutions of high osmotic pressure, such as syrups, jams, jellies, and pickling brines, they are often found on the surface of such materials in preserved foods. They are, in general, aerobic plants that are sensitive to sunlight.

Figure 3–10 A giant colony of a species of *Aspergillus* (about one-half life size) that has grown for a week on nutrient agar in a Petri dish. Note the aerial conidiophores (left) and the mass of dark conidia in the central portion of an older colony (right). (Smith and Sasaki: *Appl. Microbiol.*, Vol. 6.)

Common Species of Molds

There are several genera of molds that everyone commonly sees. We may mention *Aspergillus, Penicillium, Mucor,* and *Rhizopus*. The Ascomycetes *Neurospora crassa* (Fig. 3–6) and *Neurospora sitophila*, pink "bread molds," have been extensively studied in the laboratory by microbial geneticists because these species offer certain advantages for investigations of genetic fine structure and biochemical metabolic blocks and pathways. Some common molds produce poisonous metabolic products (*aflatoxins, mycotoxins*) when growing in certain foods, especially in moldy livestock feeds.

Aspergillus. One of the commonest species of mold is *Aspergillus glaucus*, which forms gray-green conidiospores (Fig. 3–8A). It is seen on spoiled food and mildewed clothing, especially shoes, during the summer or in the tropics. A species causing infection in the lungs is called *A. fumigatus. Aspergillus flavus*, common on spoiled stock feeds, such as ground peanuts, produces *aflatoxin*. Just how common this poison and the related mycotoxins may be in human foods is not yet clear.

Penicillium. Like aspergilli, the penicillia are very common in nature and contribute to the spoilage of food and other objects composed of organic matter. Various species of *Penicillium* form the familiar, dusty, blue-green growth seen on decaying oranges and other fruits. The conidiospores are borne at the ends of branched filaments in an arrangement suggestive of a tiny paint brush, from which the name *Penicillium* (from Latin *penicillus*, for paint brush or pencil) is derived (Fig. 3–8B). The drug penicillin is named for *P. notatum*, the fungus from which it was first isolated.

Some species of *Penicillium* are known to give a characteristic flavor to cheese, such as *P. roquefortii* in Roquefort cheese and *P. camemberti* in Camembert cheese. The blue veins in the Roquefort cheese are made up of masses of conidiospores.

Mucor and Rhizopus. *Mucor* is the generic name for one of the many familiar white or grey molds that grow as cottony tufts on damp organic substances such as decaying manure. Some species of these molds occasionally cause infections of the lung called mucormycosis. The disease is often fatal.

The black mold seen on old bread is often a closely related genus called *Rhizopus*. These two genera belong in the group of Phycomycetes that reproduce by asexual *sporangiospores* and sexual *zygospores* (Fig. 3–9). They differ most distinctively in that *Rhizopus* has little rootlike attachments on the hyphae (Fig. 3–7).

HELMINTHS

The term helminth is used to designate worms that are parasitic in humans or animals. Three major groups of helminths are the following: (1) phylum Nemathelminthes, class Nematoda, *nematodes* or roundworms such as hookworms, pinworms, and the trichina worms found in pork; (2) phylum Platyhelminthes, class Cestoidea, *cestodes* or flatworms represented by various tapeworms; and (3) phylum Platyhelminthes, class Trematoda, *trematodes* or flukes, large flattish worms, the most dangerous species of which are found in tropical areas, especially in the Orient.

None of the *adult* helminths found in the United States is a genuine microorganism, although some, such as the pork trichina worm (1.5 mm), are quite small. Diagnosis and study of helminthic diseases require the use of microscopes in order to find and identify the microscopic eggs of the worms and to study the minute identifying structures of the adults. In the filarial worms, the nematodes that cause the tropical disease *filariasis* (one symptom of which is *elephantiasis*), the infective larvae are truly microscopic in size. They are transmitted by biting flies and mosquitoes and are found in the blood of patients.

Since adult helminths are highly complex animals and are not truly microorganisms, they are not discussed here in detail. The student who is interested in them will find numerous excellent textbooks devoted largely to them. In this book representative species are discussed among the organisms transmitted from the intestinal tract (Chapter 27) or from the blood (Chapter 41).

APPLICATION FOR THE HEALTH PROFESSIONS

A basic knowledge of the characteristics of protozoa, yeasts, molds, viruses, and helminths is frequently valuable in understanding symptoms, treatment, and prevention of infectious diseases. It is important to recognize and readily identify with some understanding the scientific names of microorganisms as they appear in professional literature and medical records. Medical aspects of these organisms are discussed later in appropriate chapters.

SUMMARY

Only the bacteria and blue-green algae (cytobacteria) are prokaryotes.

Protozoa are primitive animals, eukaryotic in structure and much larger in size than bacteria. They may also ingest solid food particles such as bacteria. Protozoa exhibit definite differentiation of sexes, and some have a complex life cycle of many morphological forms. Many human diseases are caused by protozoal pathogens, including amebic dysentery, giardiasis, Chagas' disease, kala-azar, African sleeping sickness, ulcers, colon infections, genitourinary tract infections, and malaria.

Fungi are plants that lack chlorophyll. Most, the molds, grow as branching filamentous tubes filled with protoplasm, whereas the yeasts form budding ovoid cells. However, the terms yeast and mold have no scientific status since the morphology of the cells depends on environmental conditions.

Yeast cells are 20 to 100 times as large as most bacteria. Bakers' yeast and brewers' yeast are used in commercial fermentation. In nature yeasts may be found almost everywhere, including in so-called yeast infections. Yeasts may reproduce either asexually or sexually by means of special spores.

Molds reproduce either asexually by forming spores as conidia, blastospores, oidia, chlamydospores, or sporangiospores, depending on the species of mold involved; or by means of sexual spores derived from two sexes of the same strain. Sexual spores are either ascospores or zygospores. Fungi Imperfecti have no sexual partners and thus cannot reproduce sexually. Many of the commonest molds belong to this group.

Molds grow on very ordinary media; in fact they seem to grow too readily on almost everything. Some of the common genera of molds are *Aspergillus, Penicillium, Mucor, Rhizopus*, and *Neurospora*. You have seen them on mildewed shoes, in good cheese, as yeast on grapes, on oranges, on old books, and on top of preserves.

Human and animal parasitic worms are called helminths. The three major groups of helminths are roundworms, flatworms, and flukes. Helminths are rather complex animals and are more common than one may suspect.

REVIEW QUESTIONS

1. What are cyanobacteria?
2. What do you know about the nutrition of algae? Name a prokaryotic and some eukaryotic algae.

3. How do protozoa take in food? How are protozoa related to animals? Name some diseases caused by protozoans.
4. What are trophozoites? What are cysts? What is meant by "life cycle"?
5. What are fungi, eumycetes, yeasts, molds, and schizomycetes?
6. How do yeasts multiply? What practical commercial uses of yeasts can you list?
7. Name some spores (other than bacterial spores). What are their functions?
8. What foods do molds prefer? Name some common molds.
9. What are three major groups of helminths?

Supplementary Reading

Barnett, H. L., and Hunter, B. B.: Illustrated Genera of Imperfect Fungi. 3rd ed. Minneapolis, Burgess Publishing Co., 1972.

Bodily, H. L., Updyke, E. L., and Mason, J. O. (eds.): Bacterial, Mycotic and Parasitic Infections. 5th ed. New York, American Public Health Association, 1970.

Burmeister, H. R., and Hesseltine, C. W.: Biological assays for two mycotoxins produced by *Fusarium tricinctum*. *Appl. Microbiol.*, *20*:437, 1970.

Burnett, J. H.: Mycoenergetics; An Introduction to the General Genetics of Fungi. New York, John Wiley & Sons, 1975.

Center for Disease Control: Waterborne giardiasis outbreaks. *Morbid. Mortal. Rep.*, *26*:169, 1977.

Christensen, C. M.: The Molds and Man. 3rd ed. Minneapolis, University of Minnesota Press, 1965.

Conant, N. F., Smith, D. T., Baker, R. D., and Callaway, J. L.: Manual of Clinical Mycology. 3rd ed. Philadelphia, W. B. Saunders Company, 1971.

Emmons, C. W., Binford, C. H., and Utz, J. P.: Medical Mycology. 3rd ed. Philadelphia, Lea & Febiger, 1977.

Faust, E. C., et al.: Animal Agents and Vectors of Human Disease, 4th ed. Philadelphia, Lea & Febiger, 1975.

Faust, E. C., Russell, P. F., and Jung, R. C.: Craig & Faust's Clinical Parasitology. 8th ed. Philadelphia, Lea & Febiger, 1970.

Grell, K. G.: Protozoology. New York, Springer-Verlag, 1973.

Hegner, R. H.: Big Fleas Have Little Fleas, or Who's Who Among the Protozoa. Baltimore, The Williams & Wilkins Co., 1938.

Hsu, T. C., and Fuerst, R.: The Biology of *Neurospora crassa*. 16 mm. motion picture with sound. Released by the University of Texas, M. D. Anderson Hospital, January 1958.

Lee, W.-S.: Wound infection by *Prototheca wickerhamii*, a saprophytic alga pathogenic for man. *J. Clin. Microbiol.*, *2*:62, 1975.

Lockhart, W. R., and Liston, J. (eds.): Methods for Numerical Taxonomy. Washington, D.C., American Society for Microbiology, 1970.

Lodder, J. (ed.): The Yeasts, a Taxonomic Study. Amsterdam, North Holland, 1970.

Manwell, R. D.: Introduction to Protozoology. 2nd ed. New York, Dover Publications, Inc., 1968.

Meyer, M. C., and Olsen, O. W.: Essentials of Parasitology. 3rd ed. Dubuque, Wm. C. Brown Co., 1980.

Rose, A. H.: Yeasts. *Sci. Amer.*, *202*:136, 1960.

Sleigh, M.: The Biology of Protozoa. Balimore, University Park Press, 1975.

Smith, J. E., and Berry, D. R. (eds.): Industrial Mycology. The Filamentous Fungi Series, vol. 1. New York, Halsted Press, 1975.

Stanier, R. Y., Doudoroff, M., and Adelberg, E. A.: The Microbial World. 4th ed. Englewood Cliffs, N.J., Prentice-Hall. 1976

CHARACTERISTICS AND TYPES OF BACTERIA

Since many laboratory techniques and other health-related procedures center on bacteria, a general survey of these organisms, given in this chapter, will illustrate to the student basic phenomena of microbiology. Later chapters descibe in detail the more important pathogenic species and the modes of bacterial growth, transmission, and control.

Classification. Space prohibits a complete classification of all the microorganisms discussed in this book. Classifications of protozoa are to be found in textbooks on parasitology or on animal agents of disease; classifications of pathogenic yeasts and molds are found in books on clinical mycology and tropical medicine (see Supplementary Reading, Chapters 2, 3, and 38). Appendix A gives an outline of the latest classification and nomenclature of bacteria, including actinomycetes, rickettsias, chlamydias, and mycoplasmas (PPLO). It follows the system presented in the 1974 edition of *Bergey's Manual of Determinative Bacteriology.* Viruses are discussed briefly elsewhere in this book.

Why Bacteria Were Classified as Fungi. In earlier classifications the bacteria were grouped with the thallophytes, and technically could still be considered fungi. All fungi, except a few alga-like bacteria of no medical significance, can grow without sunlight and are characterized by the absence of the green coloring matter chlorophyll, which is necessary for photosynthesis and is so familiar in trees and grass.

THE BACTERIA

Bacteria are prokaryotic microscopic organisms that do not produce oxygen, even those few species that are photosynthetic. Cell multiplication of bacteria involves growth and usually binary division, with occasional unequal division and budding. More than 1,700 species of bacteria are known. Currently they are arranged into 19 Parts, only 7 of which contain human pathogens of consequence.

Nomenclature of Bacteria. The long and apparently meaningless names often borne by bacteria, as well as by higher plants and animals, need not be a source of confusion. They are based on a long-standing binomial (two-name) system: the first name is that of the genus, the second name that of the species.

A *species* of microorganisms is a group of which all the individuals are essentially alike, identifiable, and belonging to that group and not to some other group. (In actual practice, few bacterial species are so clearly distinguishable.) A *genus* is a group of similar species.

A *strain* consists of the progeny of a particular group of individuals of the same species. For example, a species of protozoa (say, *Entamoeba histolytica*) derived from a patient named Albert Jones would be called the "Albert Jones" strain of *E. histolytica.* A *clone,* in microbiology, consists of the progeny of a single cell.

Each species of organism, as has been said, has two names: the first name, that of the genus, is written with the initial letter *capitalized*; the second, the name of the species, is written with the initial letter *not capitalized.* When printed, genus and species names are properly italicized—e.g., *Staphylococcus aureus.*

CHAPTER 4

When written by hand or typed, the genus and the species must be underlined: for example, Staphylococcus aureus. The names are usually of Latin or Greek derivation and are often derived from the names of places or persons associated with the discovery of the organism.

The names of the bacteria are also intended to be descriptive of the cardinal features of the organisms. This saves time and avoids confusion. For example, by having a general agreement as to just what characters are possessed by organisms classified in the genus *Bacillus*, it saves a great deal of time in writing and talking to use that word in place of the long list of properties to which it refers such as "strictly aerobic, spore-forming, rod-shaped bacterium." Similarly, by applying the name *anthracis* to the species of *Bacillus* causing anthrax, we save repetitions of long, detailed pathologic and clinical descriptions. To give another example, the name *Salmonella typhi* indicates, by general agreement among microbiologists, a facultative, nonspore-forming, gram-negative, motile, rod-shaped bacterium, not fermenting lactose, not liquefying gelatin, fermenting glucose without gas formation, and causing typhoid fever. Its generic name, *Salmonella,* is derived from an American microbiologist named Salmon. It is obviousy efficient to use only two distinctive words to express all these facts.

Even this convenient nomenclature, expressing the many properties of an organism with only two names, may be further abbreviated. For example, if *Escherichia coli* is mentioned the first time in a report, its full name is used, but subsequent citations of the organism may be written as *E. coli,* as long as the genus cannot be confused with another genus (e.g., *Escherichia* with *Entamoeba*).

Structure of Bacteria

Size of Bacteria. It is easy to understand why a microscope is necessary for the discovery and study of microorganisms when we realize that even very large bacteria are only about 100 μm long, and *Escherichia coli* is only about 5 to 10 μm in length. Nearly two billion (2,000,000,000) medium-size bacteria may easily be contained in a single drop of water, and about 30 billion would hardly weigh as much as a dime.

The highest power of an ordinary compound microscope (oil-immersion lens) magnifies about 1,000 diameters. An object 0.5 μm in diameter when magnified 1,000 times (× 1,000) appears about the size of a period on this page. Most bacteria are about 3 μm in diameter, although their length may be

Table 4–1.

Some Useful Relationships of Measurements
1 inch = 2.54 cm
1 cm = 10 mm = 1/2.54 inch
1 mm = 1,000 μm (micrometer)
1 μm = 0.001 mm = 0.00003937 or 1/25,400 inch = 1,000 nm (nanometer)
1 nm = 0.001 μm = 10.0 Angstroms (Å)
1Å = 0.0001 μm = 0.0000001 mm = 1/254,000,000 inch

See also system of metric units on page 24.

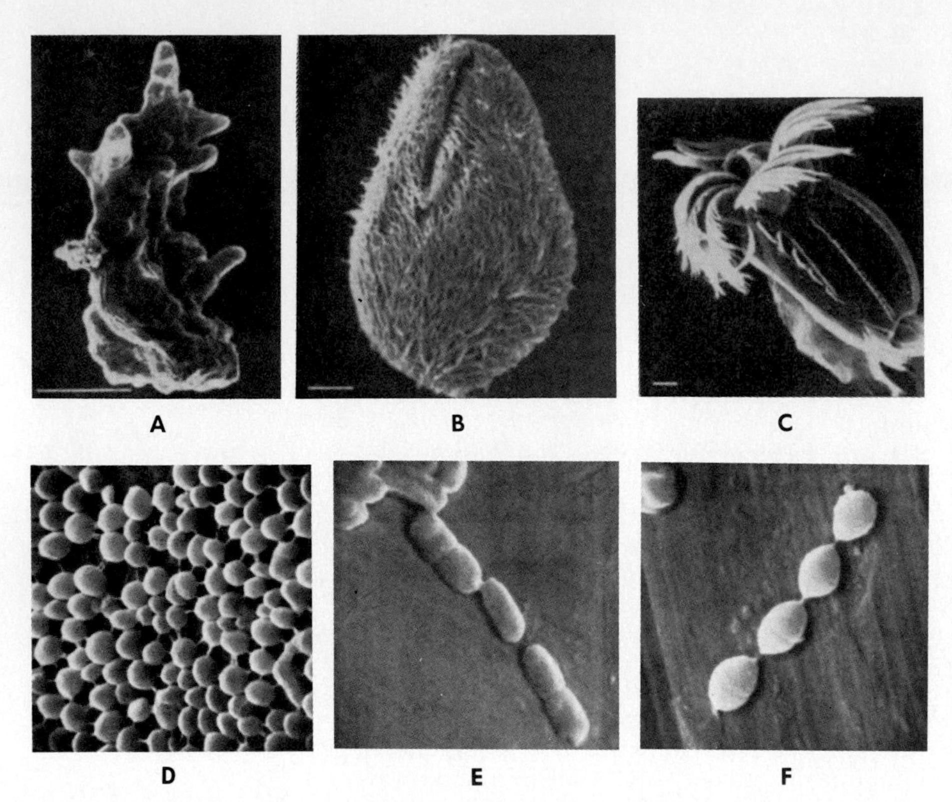

Figure 4–1 *A, Amoeba proteus.* Preservation of lobopods is shown (scale, 100 μm). *B, Nyctotherus ovalis.* Holotrichous ciliate from the hind gut of the roach *Blabarus discoidalis* (scale, 100 μm). (From Small, E. B., and Marszalek, D. S.: Scanning electron microscopy of fixed, frozen, and dried protozoa. *Science, 163:*1064, 1969. Copyright 1969 by the American Association for the Advancement of Science.) *C. Uronychia* sp. (scale, 10 μm). (From Frobisher, M., et al.: Fundamentals of Microbiology. 9th ed. Philadelphia, W. B. Saunders Company, 1974.) *D, Staphylococcus aureus* (× 7,500), demonstrating typical spherical shape, grape-like clustering, variation in size and shape, dividing cells, involutional and atypical forms, and intercellular bridges. *E, Proteus vulgaris* (× 7,500). *F, Streptococcus pyogenes* (× 7,500), demonstrating diplococcus shape, chaining, and intercellular bridges. (From Klainer and Betsch: *Journal of Infectious Diseases, 121:*339–343, 1970. © University of Chicago Press, all rights reserved.)

greater. Most viruses cannot be seen at all with ordinary microscopes. Rickettsias are visible with such microscopes and in general have forms like bacteria but are much smaller.

Another type of magnifying instrument called the *electron microscope* gives clear images of bacterial structures at enlargements up to one million and more diameters. A human hair magnified × 100,000 would have a diameter about twice that of a modern tunnel like those leading into New York City! Pictures made with an electron microscope are called electron micrographs. By means of the electron microscope, viruses can be made visible and their inconceivably minute anatomy studied. (For further details about the electron microscope see Chapter 9.) A complex microscope, an outgrowth of the development of the electron microscope, also uses electrons but gives an astonishing "3-D" perspective. This instrument is called a scanning electron microscope. (See Fig. 4–4.)

Morphology of Bacteria. Van Leeuwenhoek's drawings (Fig. 1–2) show that he observed spherical types of bacteria that today are called *cocci* (singular, coccus; from a Greek word meaning berry); elongated, cylindrical forms, some of which were cigarette-shaped, now classified under the general heading of *bacilli* (singular, bacillus; from a Greek word meaning rod); and others curved, wavy, or *helicoidal,* like a coiled wire spring, now known as *vibrios* and *spirilla.* Some of the helicoidal (spiral) bacteria are capable of flexing, twisting movements. These organisms are grouped in the order Spirochaetales and are known as *spirochetes.* Figures 4–2 and 4–3 present the four basic morphologic types of bacteria: cocci, coccobacilli, rods, and helicoidal forms.

Morphologic classification of bacteria

Spherical coccus (plural, *cocci*):
- *diplococcus* (pairs)
- *streptococcus* (chains)

staphylococcus (irregular clusters)
gaffkya (groups of 4 cocci)
sarcina (cubical packets of 8 cocci)

Cylindrical:
bacillus (plural, *bacilli*) (straight, commonly sausage- or cigarette-shaped rods); many other elongated shapes and forms; outnumber all other bacterial forms

Spherical-elongated coccobacilli (Fig. 4–2)

Helicoidal:
vibrio and *spirillum* (from single curve of *Vibrio* to five and more crockscrew-like turns in the *rigid* genus *Spirillum*)
Spirochetes (curved or spiral; *flexible*; Fig. 4–3)

Classification of Cocci. The division of the spherical and spheroidal bacteria into genera is based primarily on the distinctive grouping of the cells after fission. The cells of many species remain together in pairs after fission. In order to differentiate these pair-forming cocci they are called *diplococci,* the prefix *diplo* being derived from a word meaning *two* or *a pair*. There are two medically important genera of diplococci: *Neisseria* and *Branhamella,* named after Albert Neisser and Sara Branham, and one species, *Streptococcus pneumoniae* (the cause of lobar pneumonia), which previously was known as *Diplococcus pneumoniae.* Cocci that cling together in long chains as they continue to divide are called *streptococci (strepto,* chain), e.g., genus *Streptococcus* Fig. 4–5). Still other cocci form neither regular pairs nor chains, but irregular groups and masses like clusters of grapes. These are called staphylococci (*staphylo,* cluster), e.g., genus *Staphylococcus.* Some cocci of uncertain pathogenicity form tetrad groups of four cells; others (*Sarcina*) produce cubical packets.

Types of Bacilli. Bacilli or rods are quite diverse in shape and size. Two groups that are of medical interest are spore-forming long rods that are gram-positive—i.e., they stain purple with Gram's stain. Others are slender and much

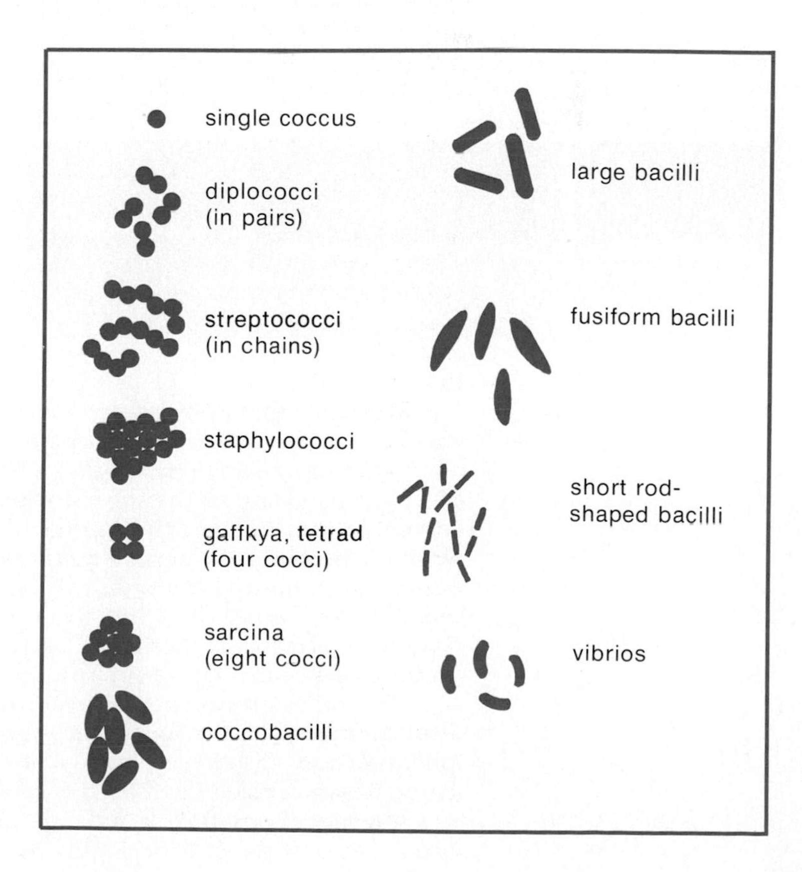

Figure 4–2 Some basic morphologic forms of bacteria, cocci, bacilli, coccobacilli, and vibrios. Spiral forms are not shown here but are presented in Figure 4–3. Note that pleomorphic (abnormal, greatly distorted) forms of all these organisms are sometimes observed, as well as long filaments of very long rods that may then break up.

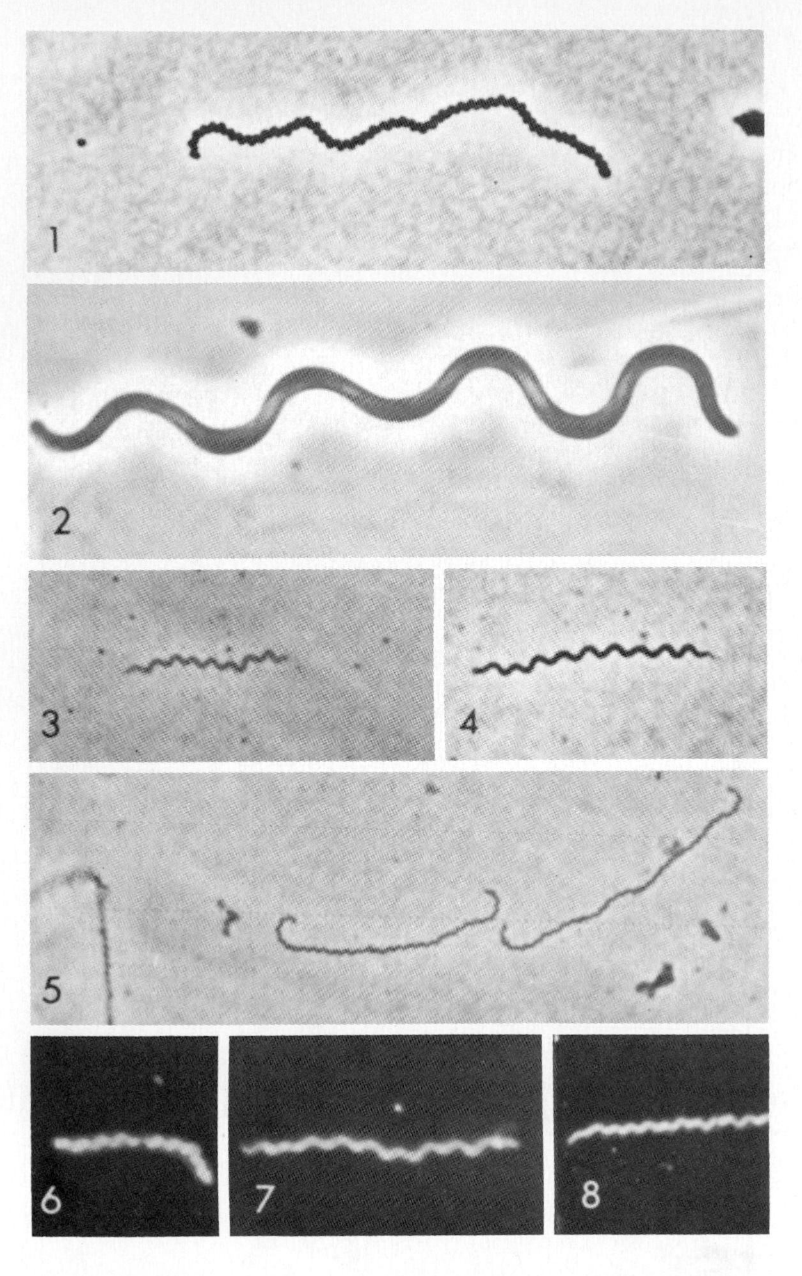

Figure 4–3
1. *Spirochaeta plicatilis* from activated sludge.
2. *Cristispira* sp. from crystalline style of clam, *Cryptomya californica.*
3. *Treponema pallidum* from rabbit testis.
4. *Borrelia anserina* from chicken blood.
5. *Leptospira interrogans* serotype icterohaemorrhagiae from Fletcher's medium.
6. *Treponema phagedenis* Reiter treponeme.
7. *Treponema refringens.*
8. *Treponema denticola.*

1–5. Phase-contrast photomicrographs by Dr. Daisy A. Kuhn, magnification × 2,200. Material for *T reponema supplied by Dr. James N. Miller, for Leptospira* and *Borrelia* by Dr. Ernst L. Biberstein. *6–8.* Dark-field microscopy, magnification × 2,300 by Dr. R. M. Smibert, (From Buchanan, R. E., and Gibbons, N. E. (eds.): Bergey's Manual of Determinative Bacteriology. 8th ed. Baltimore, Williams and Wilkins Company, 1974.)

smaller bacilli that do not form spores and are gram-negative—i.e., stained red by Gram's stain. Many gram-negative species important in medicine belong to the genera *Escherichia, Salmonella, Shigella, Pseudomonas,* and *Proteus.*

To one group of the spore-forming rods just mentioned, which grow only in the presence of air, Cohn, their discoverer (1876), gave the name *Bacillus.* Strictly speaking, the term *Bacillus* is limited to the genus of spore-forming, rod-shaped, mostly gram-positive bacteria that grow only in the presence of air. When so used, the word *Bacillus* is italicized and begins with a capital B. The term bacillus, not italicized and with a small b, is widely used for *any* rod-shaped bacterium—gram-positive or gram-negative, spore-forming or not.

The other group of gram-positive, spore-forming rods is the genus *Clostridium.* Members of this genus grow only in the absence of air. This strange phenomenon, discovered by Pasteur and called *anaerobiosis,* will be discussed later. No bacteria other than the genera *Bacillus* and *Clostridium* (with possibly one or two exceptions of little or no medical significance, e.g., *Sporosarcina ureae,* a coccus form) form highly heat-resistant spores.

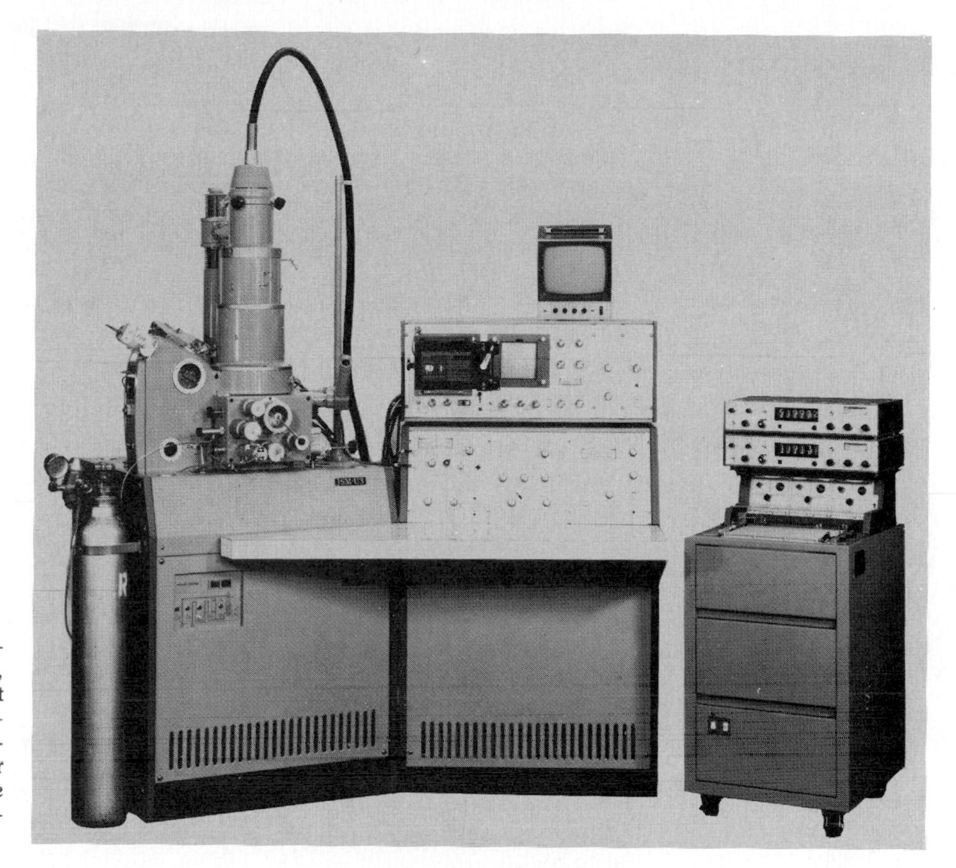

Figure 4–4 A scanning electron microscope, with the electron column, sample port, and vacuum equipment in the left console; controls and cathode ray tubes for viewing in the center console; and a digital counter for various specialized applications to the right. (Courtesy of JEOL USA, Medford, Mass.)

Coccobacilli. An example of a coccobacillus is *Moraxella lacunata,* a very short, plump rod approaching coccus shape. It causes one type of conjunctivitis in humans. Other coccobacilli are *Bordetella pertussis,* a cause of whooping cough, and species of *Brucella,* which may be either coccobacilli or short rods.

Types of Helicoidal Bacteria. There are two types of helicoidal bacteria: *rigid* and *flexible.* The *rigid* helicoidal species occur in several families of Schizomycetes. Some are coiled in several turns; others are merely comma-shaped. Of the former, only one species is pathogenic, *Spirillum minor,* the cause of ratbite fever. Of the many comma-shaped species, only a few are important human pathogens, notably *Vibrio cholerae,* the cause of Asiatic cholera, and the biotype (no species) *Vibrio El Tor,*[1] which produces a cholera-like disease.

ORDER SPIROCHAETALES. The *flexible* helicoidal bacteria of this order are commonly referred to as spirochetes. The pathogenic species are slender, spiral microorganisms about 0.2 to 2 μm in diameter and ranging from 5 to 20 μm in length. About half the known species of spirochetes are harmless saprophytes, living in the soil, in decaying organic matter, and in stagnant pools. Many have a rather complicated structure, like a plastic tube wrapped around a central fibrillar rod or axial filament composed of a kind of elastic protein, with properties suggestive of protozoa. Others, like *Cristispira,* have a crest or ridge of chitin-like substance along one side (Fig. 4–3).

The Anatomy of the Bacterial Cell

Capsules. Capsules are a very important part of bacterial anatomy. Many, probably most, species of bacteria produce a gummy or slimy coating over their

---

[1]*Vibrio El Tor,* named after the El Tor lazaret (Sinai) where it was first isolated from pilgrims returning from Mecca, is now properly designated as *V. cholerae* biotype *eltor,* one of the four biotypes of the species *V. cholerae*: biotype *cholerae,* biotype *proteus,* biotype *albensis,* and biotype *eltor.*

surface, which is called a capsule, soluble specific substance, or SSS. It is sometimes a very thin film and invisible even when the bacteria are examined by special methods. At other times it is thick and may easily be demonstrated by simple methods (Fig. 4–5). Capsular material of many species of bacteria consists of starchlike or gummy substances called *polysaccharides*. The chemical composition of the capsule is often absolutely distinctive of a type or species of bacteria and is said to be *type- or species-specific*. This distinctiveness is exceedingly valuable in medical work, since it affords a reliable means of diagnosis in bacterial infections (see Chapter 29). Capsules generally confer increased virulence on bacteria that possess them, probably because, like a jacket or armor, they insulate the bacteria from the phagocytes, antibodies, and other defensive mechanisms of an infected body and from many other deleterious influences.

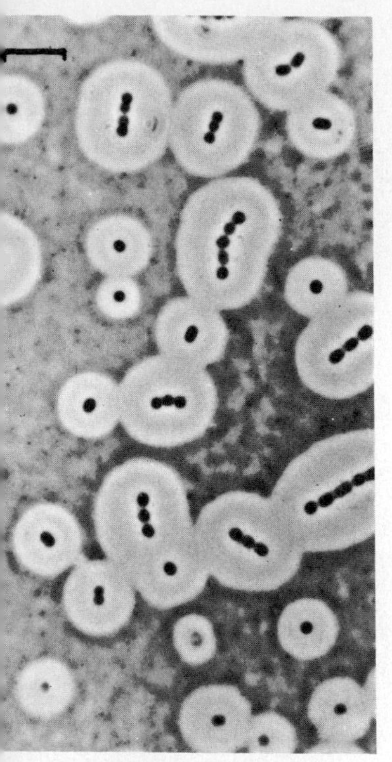

Figure 4–5 Photomicrograph of encapsulated streptococci suspended in India ink for contrast. The small dark spheres are cocci; the surrounding light areas, capsules. These capsules are unusually large. The straight line represents 10 μm at this magnification (about × 1,500). (Taylor, W. H., and Juni, E.: *J. Bacteriol.*, *81*:688, 1961.)

Sheath. In some pathogenic species a very thin, capsule-like surface layer on individual cells is often called a *sheath antigen*, microcapsule, or envelope. *Sphaerotilus natans* and the genus *Leptothrix*, which have iron and manganese encrusted sheaths, are commonly found in polluted streams.

Capsules and Health. Easily identifiable capsules may not always be present on bacterial cells, and they constitute a variable factor in bacterial structure. Usually they are present on bacteria infecting the body. Hence, bacteria in pus, mucus, and feces from an infected person are particularly dangerous because of the increased virulence conferred on them by their capsules. This is a very important fact that should be remembered.

Motility of Bacteria

The Gliding Type. Some bacteria of no medical significance exhibit *gliding motion* when the cells are in contact with a solid surface.

The Spirochete Type. In spirochetes a rotary movement results from structures within the inner core and the outer coat of the cell, called axial fibrils (see Fig. 4–6).

Bacteria with Flagella. Flagella (singular, flagellum, which means "whip") are long, extremely fine, hairlike appendages distributed on the outside of the cell in various ways. They enable flagellate organisms to swim about in fluids (Figs. 4–7 and 4–8). Many types of microorganisms have flagella—for example, protozoa and certain algae. Among bacteria, only certain species of rod-shaped bacteria (bacilli) and species of *Vibrio* and *Spirillum* have them. The spherical bacteria (cocci) generally do not have flagella.[2]

Bacterial flagella consist of a protein called flagellin (mol. wt., 20,000). They extend from the protoplasmic material inside the cell and are attached to both the cell membrane and cell wall by ringlike structures, as a pipe may be fastened to metal plates (see Fig. 4–9). Unless prepared with certain stains, flagella are not readily visible with ordinary microscopes, but the motion produced by them is readily observed by mounting and examining a drop of fluid culture under a microscope, as described in Chapter 9. The average speed of bacteria with flagella has been estimated at between 25 to 55 μm per second, which is rather fast.

Flagella located at one or both *ends* of rods are said to be *polar*; located indiscriminately over the surface of the rod, they are said to be *peritrichous*. The location of flagella is of use in identifying species.

Bacteria with Pili or Fimbriae. Pili, like flagella, are very thin, hairlike, external appendages of a cell, protruding from the cytoplasm through the cell membrane, cell wall, and capsule if present. They are generally peritrichously arranged and number around 150. They consist of a specific protein called *pilin*. (See Fig. 4–10.)

Unlike flagella, pili are rigid, relatively short, straight, and not associated with motion. Their functions are not yet fully known, but certain forms of pili (F pili) appear to play a role in the primitive sexual conjugation processes in certain species of bacteria. They have so far been seen only on gram-negative bacteria.

[2]There are exceptions, e.g., *Nitrococcus mobilis*.

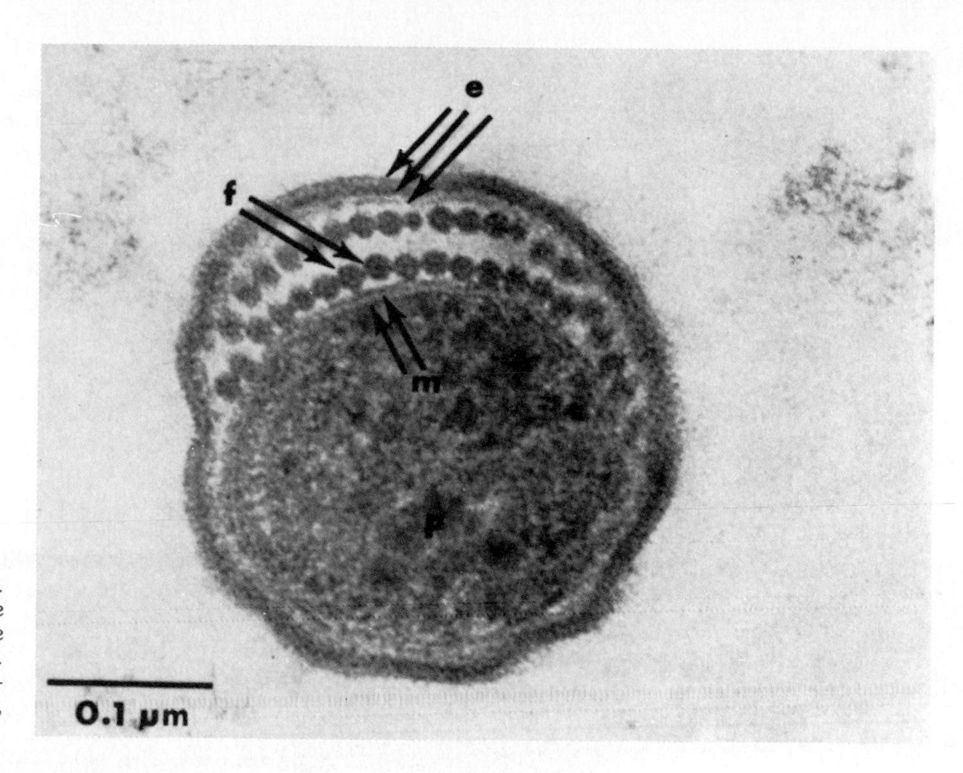

Figure 4–6 Electron micrograph. This cross section of a spirochete shows the axial filament between the cell wall and the cell membrane. (From Listgarten, M. A., and Socransky, S. S.: *J. Bacteriol.*, *88*:1092, 1964.)

Bacteria That Produce Prosthecae. Prosthecae are a type of bacterial appendage that differ from flagella, pili, or fimbriae and are even less well understood. They occur in some "budding" bacteria, but neither organelle depends on, or excludes, the other. No pathogens producing prosthecae are known.

Endospores. Of the 1,700 or more species of bacteria known, only two genera of medical significance (*Bacillus* and *Clostridium*) have the power of forming within each cell a small, glistening, highly heat-resistant, hard, and thick-walled round or oval spore (Fig. 4–11).[3] These are dormant *endospores*, and they contain the essential parts of the protoplasm of the cell in a condensed, probably dehydrated form. Bacterial spores are far more resistant to drying, sunlight, heat, and disinfectants than the ascospores, conidiospores, and sporangiospores of yeasts and molds and the cysts of protozoa, though the term spore is sometimes used for all such bodies. Bacterial *endospores* (Fig. 4–12) can resist drought, sunlight, heat (even boiling), disinfectants, and other unfavorable conditions for long periods. Although the numerous spores produced by eukaryotic fungi are reproductive stages of normal life cycles, sexual or asexual, bacterial spores must be considered to be formed only as means of survival when conditions for growth and metabolism are unfavorable, since only one endospore (rarely two) is produced per cell. When conditions for life become difficult for *Bacillus* or *Clostridium* cells, the formation of a spore permits the organism to survive in a dormant state for many years, if necessary, until a favorable environment prevails again and the bacteria can grow, reproduce, and (alas!) cause us harm.

Spores and Sterilization. Spores are of tremendous significance in sterilization and surgery because they are so very resistant to chemical disinfectants, boiling, and other usually destructive environmental conditions. It is of great importance that anyone working with (or against!) microorganisms learn which organisms produce spores and how to kill both the spores and the organisms.

Spores and Dust. A spore may leave a bacillus, which then becomes nothing

Figure 4–7 Types of flagellation: *A*, monotrichous; *B*, lophotrichous; *C*, amphitrichous; *D*, peritrichous. (From Frobisher, M., et al.: Fundamentals of Microbiology. 9th ed. Philadelphia, W. B. Saunders Company, 1974.)

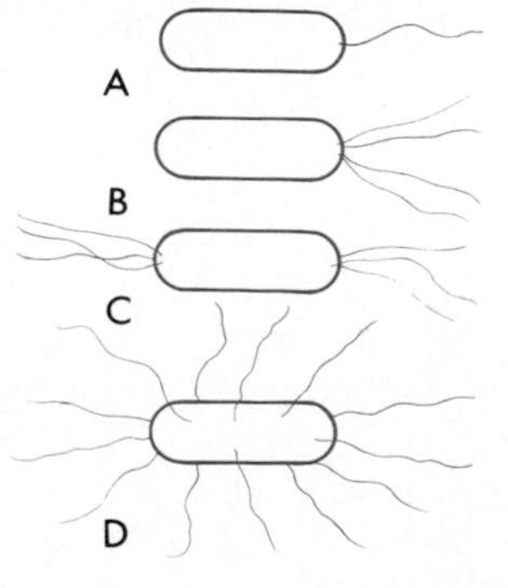

[3]Other, nonpathogenic bacteria do form spores, e.g., *Sporosarcina*.

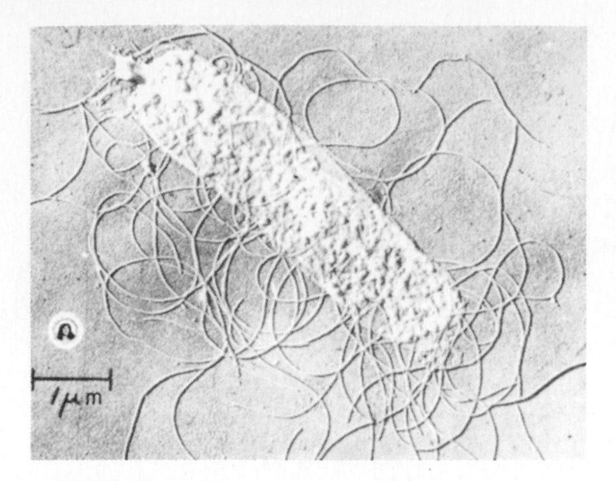

Figure 4–8 Electron micrograph of a common bacterium *(Proteus vulgaris)* showing peritrichous flagellation. (The line represents 1 µm at this huge magnification.) Note the fineness of the flagella in a small granule, suggestive of a blepharoplast. Note also the diffuse, granular character of the cell contents. (Courtesy of Drs. A. L. Houwink, C. F. Robinow, and W. van Iterson. From the collection of the American Society for Microbiology. The small portrait of Antony van Leeuwenhoek is part of the emblem of the American Society for Microbiology.)

more than an empty shell, often called a *sporangium.* The spore may be blown about with dust, which contains spores of many organisms and is always a potential source of contamination. When conditions are favorable for the active, vegetative, and reproductive life of the microorganisms—as in a surgical wound, injured tissue, or unrefrigerated food (moisture, nutriment, and warmth are necessary)—the spore "sprouts" or germinates, changes back to the ordinary vegetative form of the organism, and goes on growing. Only a few species of bacteria of importance in the health professions form spores, but the diseases they cause are often fatal. Among them are anthrax, tetanus (lockjaw), gas gangrene, and botulism (food poisoning)

The Nucleoid. The bacterial nucleus is unlike those of eukaryotic plants and animals because it has no nuclear membrane and no definite form, but it does consist of DNA. In prokaryotic cells, what would be analogous to a nucleus is commonly referred to as a *nucleoid.* Because of the high RNA content in the cytoplasm of bacteria and the lack of a nuclear membrane, it is difficult to demonstrate the nucleoid (also called chromatin body). However, special treatment shows the existence in most bacilli of two or more such bodies per cell (Fig. 4–13). Division of nuclear material occurs first, then cell division. Since the nuclear bodies (nucleoids) are attached to the cytoplasmic membrane by

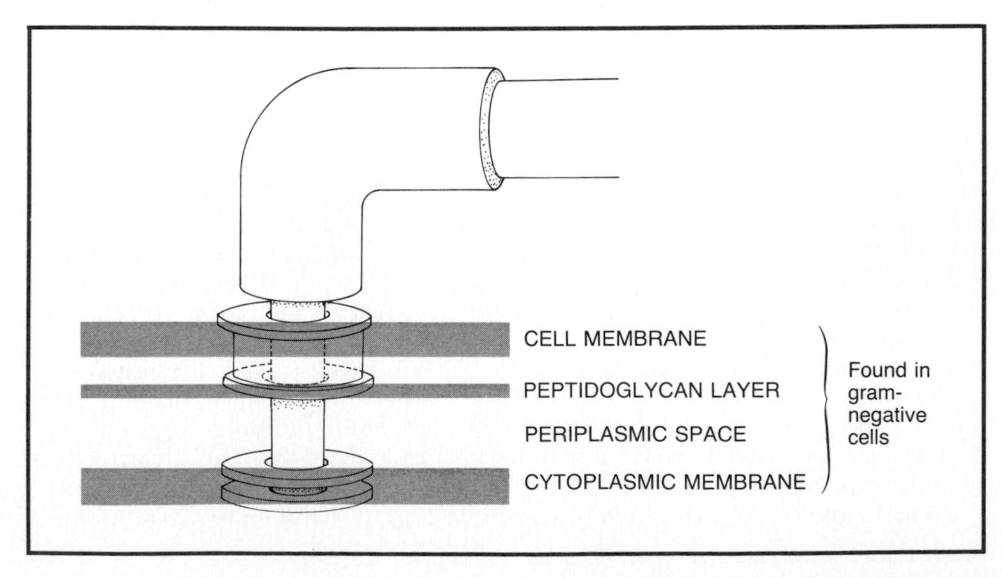

Figure 4–9 Diagrammatic structure of basal body. (After DePamphilis, M. L., and Adler, J.: *J. Bacteriol., 105*:395, 1971).

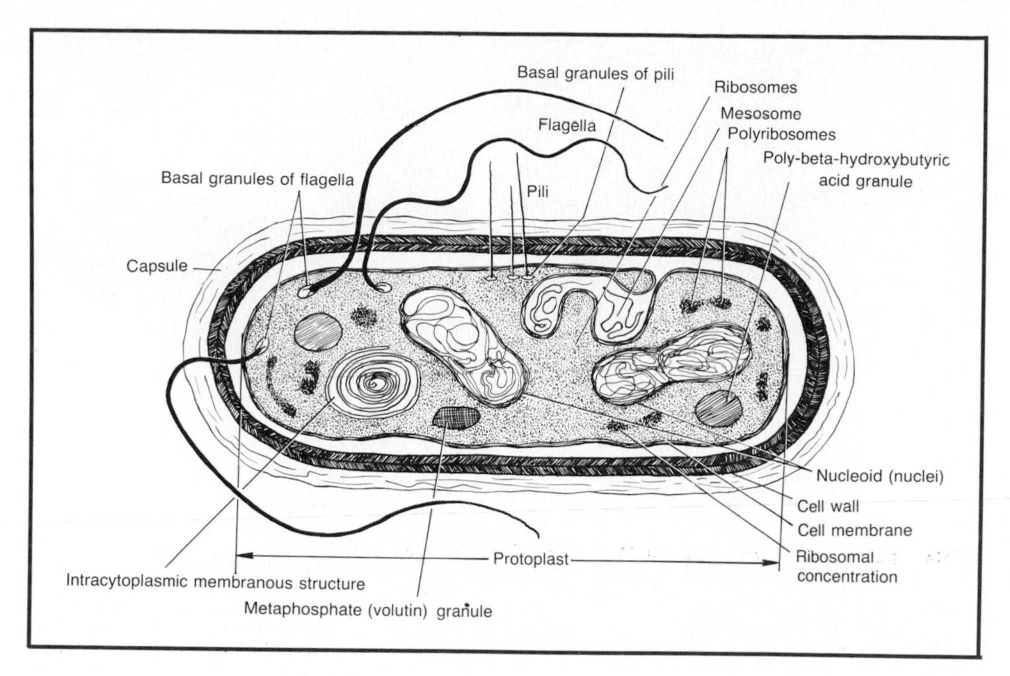

Figure 4–10 Diagram of a typical bacterial (prokaryotic) cell showing all recognized structures. The space (theoretical) between cell membrane and cell wall is exaggerated to show the extent of the protoplast.

mesosomes during cell division, after septum formation the membrane pulls the nucleoids along into each of the daughter cells. No spindle fibers are needed. In experiments *Escherichia coli* was treated with an isotope-labeled nucleic acid constituent, tritiated thymidine. The isotope-containing chemical was taken up into the bacterial chromosome, which made it possible to trace the DNA after cell division with autoradiography (exposure of a photographic film to the isotope). The structure of the labeled chromosome strand was clearly shown (Fig. 4–14). The results indicated that the bacterium contained a circle

Figure 4–11 Various types of bacterial spores. Some of the spores have escaped from the sporangia. Stained with methylene blue, which does not penetrate inside the spore; only the outer surface of the spore is stained.

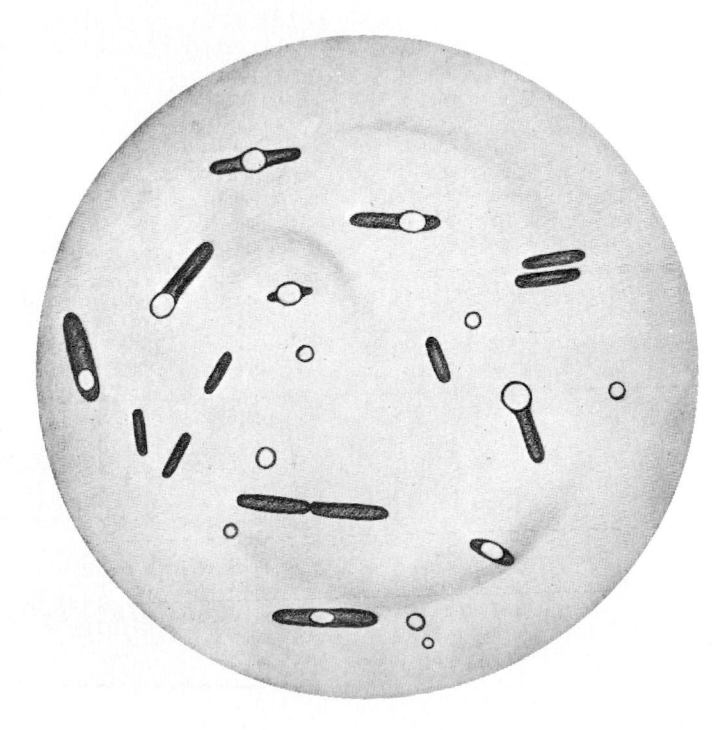

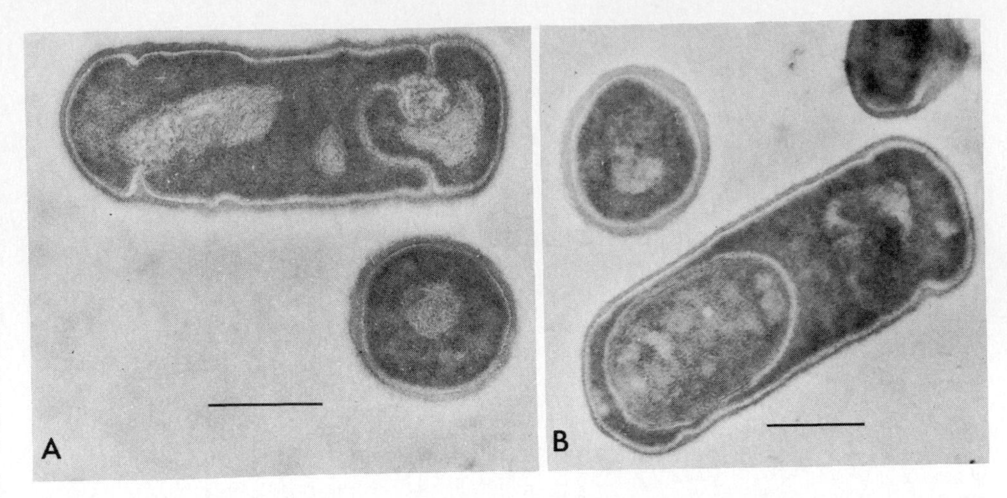

Figure 4–12 Spore formation in *Bacillus subtilis.* Two stages of endospore development within the cell are seen. The three stages of endospore development are: (1) In *A* the invagination of the cell membrane begins to close off the *forespore* from the larger part of the cell. In both parts filaments of DNA are realigned. (2) In *B* a *spore cortex* layer forms over the spore followed by many layers of a *spore coat.* (3) As the mature spore leaves the vegetative cell, the cell is lysed. See the mature spores (round structures) in both *A* and *B.* (From Yousten, A. A., and Hanson, R. S.: *J. Bacteriol., 109*:890, 1972.)

of double-stranded DNA which when spread out would be about 1,000 times as long as the bacterial cell itself.

Bacterial Plasmids. These extrachromosomal elements are small, self-replicating, genetic information–carrying, circular pieces of DNA, found in a wide range of bacterial cells. They may have no function at all, or they may confer resistance to certain antibiotics, resistance to ultraviolet light, phage resistance, toxin production, virulence, or the ability to produce enzymes like coagulase. Plasmid genes may be incorporated into bacterial chromosomes and greatly affect bacterial evolution. They play a major role in the techniques used in *genetic engineering*.

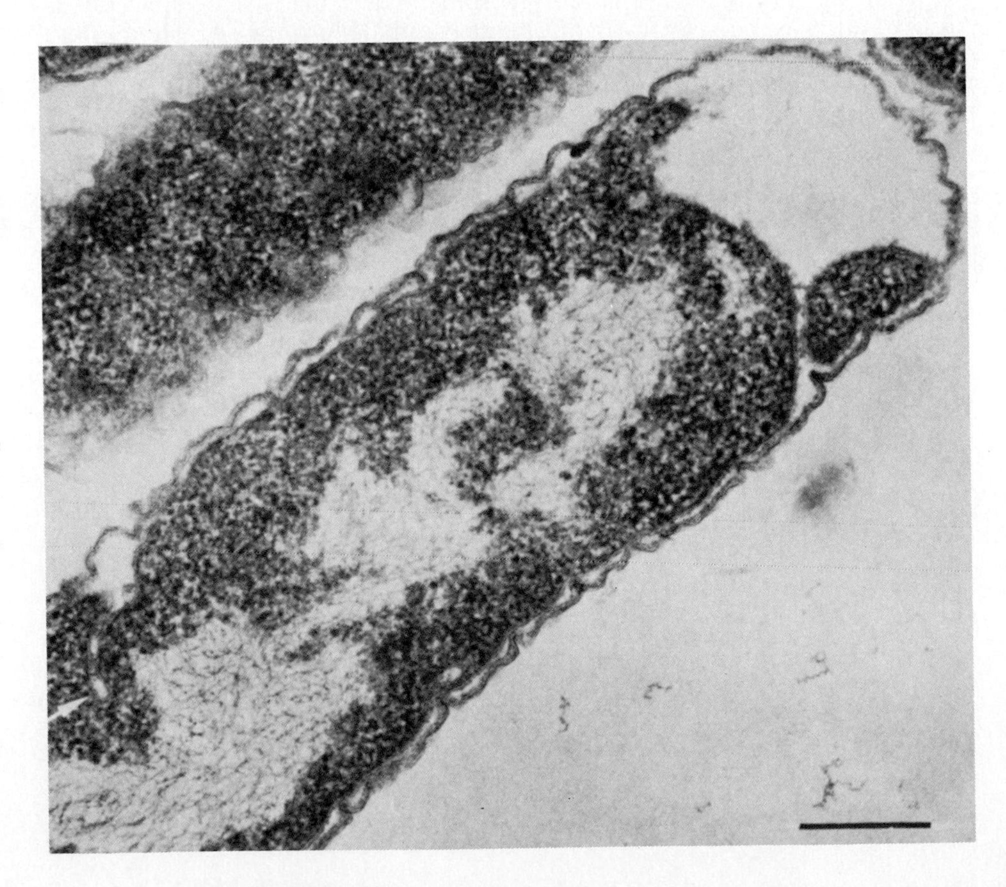

Figure 4–13 Electron micrograph of a cross section of a common nonsporing bacterium (*Escherichia coli),* showing nucleoplasm (light, centrally located amorphous masses) occupying a large portion in the cell. This cell shows the effects of experimentally induced plasmolyses, with the cytoplasm contracted at one pole. This photograph also shows the distribution of ribosomes throughout the cell and the laminated structures of the cytoplasmic membrane and the cell wall. (From Woldringh, C. L., and van Iterson, W.: *J. Bacteriol., 111*:801, 1972.)

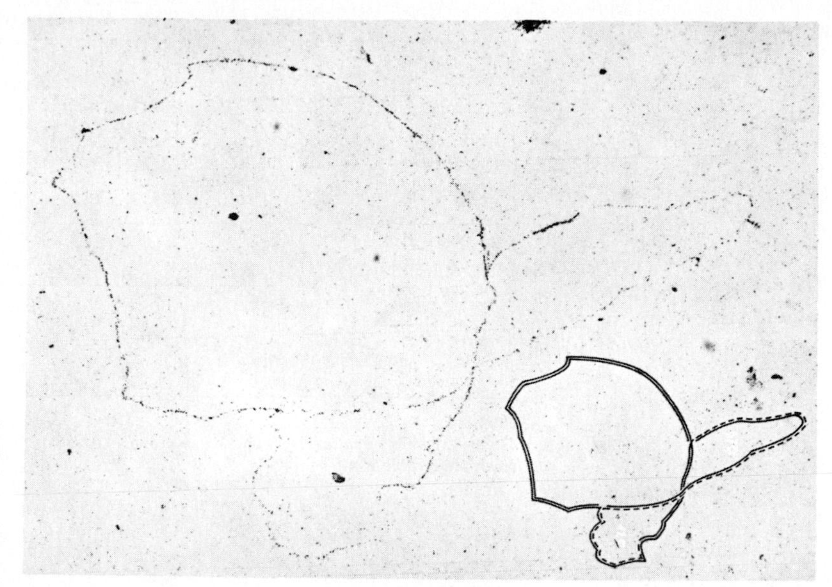

Figure 4–14 Autoradiograph of the chromosome of *Escherichia coli* Hfr K–12, labeled with tritiated thymidine for two generations and extracted with lysozyme in a dialysis chamber. Exposure time was two months. The inset is an interpretive diagram of the replication process. (From Cairns, J.: *J. Mol. Biol.*, *6*:208, 1963.)

Granules of Bacteria. As noted in Chapter 2, cells of microorganisms contain many granular structures; some have a role in reproduction, some in the vital functions of food utilization. In bacteria some granules are obviously stored food material—such as fat, poly-beta-hydroxybutyric acid, and carbohydrates—and they vary greatly in size and number. Granules of complex groups (polymers) of inorganic phosphates are a stored form of phosphorus that is essential in the energy chemistry of the cell. Such granules have long been known as *"volutin"* or sometimes *"metachromatic granules,"* because they take on various colors when stained with certain dyes (Fig. 4–15). In some species, such as the diphtheria bacillus or the bacillus of bubonic plague, they are so consistently present in characteristic numbers and locations in the cell that an experienced bacteriologist can diagnose these diseases merely by microscopic examination of pus and other substances containing the specific bacilli.

Pigments. Many species of bacteria form pigments that range from white through red, pink, yellow, purple, and even black. Some artistically inclined laboratory workers find this quite stimulating. Pigments may be hardly perceptible or very brilliant. Their physiologic significance is usually obscure, but they are of importance to the medical microbiologist because they furnish an additional characteristic for the identification of species and the diagnosis of infections. They appear to have no relation to virulence. Pigments are easily observed in colonies of bacteria grown on solid media in Petri plates or on slants. In some species of alga-like bacteria, the pigments are related to chlorophyll and carry on photosynthesis. They are of no medical importance.

Figure 4–15 A common, harmless bacterium *(Enterobacter aerogenes)* stained by the Albert-Laybourn method for demonstrating volutin granules. Volutin (inorganic phosphate–*polyphosphate* or "polyP") is a reserve food substance. Magnification originally × 3,000. (Courtesy of Drs. J. P. Duguid, I. W. Smith, and J. F. Wilkinson, Dept. of Bacteriology, University of Edinburgh, Scotland. In *J. Path. Bact.*, *67*:289 and *J. Bacteriol.* *68*:450.)

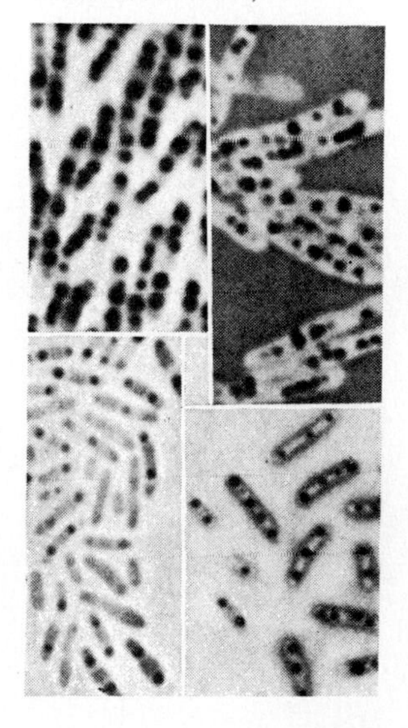

THE MAJOR GROUPS OF BACTERIA

Most of the bacteria that are pathogens are listed in the nine major groups (Parts) in Table 4–2.[4] There are also many nonpathogens in these and in the other ten Parts of bacteria (see Table 3–3). In general, the human bacterial pathogens are morphologically simple and undifferentiated forms: rods, spheres, spheroidal, club-shaped, and fusiform (cigar-shaped). These bacteria exhibit no special anchoring or reproductive structures, as do molds and higher

[4]According to the eighth (1974) edition of *Bergey's Manual of Determinative Bacteriology*.

Table 4–2. Some Bacteria of Medical Significance

PARTS	FAMILY (IF RECOGNIZED)	GENUS	SPECIES	PREVIOUS NAME IF DIFFERENT
Part 5. The Spirochetes	*Spirochaetaceae*	*Treponema*	*pallidum*	
		Borrelia	*recurrentis*	
		Leptospira	*interrogans*	*L. icterohaemorrhagiae*
Part 7. Gram-Negative Aerobic Rods and Cocci	*Pseudomonadaceae*	*Pseudomonas*	*aeruginosa*	
		Alcaligenes	*faecalis*	
		Brucella	*abortus, etc.*	
		Bordetella	*pertussis*	*Haemophilus pertussis*
		Francisella	*tularensis*	*Pasteurella tularensis*
Part 8. Gram-Negative Facultatively Anaerobic Rods	*Enterobacteriaceae*	*Escherichia*	*coli*	
		Edwardsiella	*tarda*	
		Citrobacter	*freundii*	
		Salmonella	*typhi*	*Salmonella typhosa*
		Shigella	*dysenteriae*	
		Klebsiella	*pneumoniae*	
		Enterobacter	*aerogenes*	*Aerobacter aerogenes*
		Enterobacter	*cloacae*	
		Serratia	*marcescens*	
		Proteus	*vulgaris*	
		Yersinia	*pestis*	*Pasteurella pestis*
	Vibrionaceae	*Vibrio*	*cholerae*	*Vibrio comma*
		Haemophilus	*influenzae*	
			vaginalis	
		Pasteurella	*multocida*	
		Cardiobacterium	*hominis*	
Part 10. Gram-Negative Cocci and Coccobacilli	*Neisseriaceae*	*Neisseria*	*gonorrhoeae*	
		Moraxella	*lacunata*	
Part 14. Gram-Positive Cocci	*Micrococcaceae*	*Micrococcus*	*luteus*	*Sarcina lutea*
		Staphylococcus	*aureus*	
			epidermidis	*Staphylococcus albus*
	Streptococcaceae	*Streptococcus*	*pyogenes*	
Part 15. Endospore-Forming Rods and Cocci	*Bacillaceae*	*Bacillus*	*anthracis*	
		Clostridium	*tetani*	
Part 17. Actinomycetes and Related Organisms	Coryneform group of bacteriae	*Corynebacterium*	*diphtheriae*	
	Actinomycetaceae	*Actinomyces*	*bovis*	
	Mycobacteriaceae	*Mycobacterium*	*tuberculosis*	
	Nocardiaceae	*Nocardia*	*farcinica*	
	Streptomycetaceae	*Streptomyces*	*griseus*	
Part 18. The Rickettsias	*Rickettsiaceae*	*Rickettsia*	*prowazekii*	
	Chlamydiaceae	*Chlamydia*	*trachomatis*	
Part 19. The Mycoplasmas (Class Mollicutes)	*Mycoplasmataceae*	*Mycoplasma*	*hominis*	

plants. Except spirochetes, they are not flexible. Only two genera form branched structures, and none grows upon stalks. Usually each cell is independent of the others (except when they may accidentally cling together in various ways after fission). When they are growing in colonies, there is nothing to suggest any functional specialization of certain cells such as occurs in molds and in the higher plants.

The Spirochetes

Some of the characteristics of this order have been mentioned previously. Spirochetes move in an undulating fashion, or by darting and wriggling motions. They progress because of their rotating spiral structure. Unlike most other bacteria, they can bend and twist themselves using contractile elastic protein fibrils; true muscles are believed to be absent. Spirochetes resemble other bacteria in method of nutrition, multiplication, diameter, and prokaryotic cell structure.

The order Spirochaetales contains some very dangerous pathogenic organisms, among them the organism causing syphilis (*Treponema pallidum*); the cause of leptospirosis (including hemorrhagic jaundice), namely, *Leptospira interrogans* (previously called *L. icterohaemorrhagiae*); and the bacteria causing relapsing fever (*Borrelia recurrentis* and several other species of *Borrelia*). Except for *Leptospira,* pathogenic spirochetes are difficult to cultivate and require very special media. They also stain with difficulty and hence are usually observed alive with the darkfield apparatus or with special phase-contrast techniques that will be described later.

Gram-Negative Aerobic Rods and Cocci[5]

Family Pseudomonadaceae. The definition of a species of *Pseudomonas* has by no means been settled. Previously 149 species were recognized; now *Bergey's Manual* lists 29 and the names of more than 200 that are still unresolved. Other workers propose 13 basic species of the pseudomonads.

All but a few soil-living species are simple, unicellular forms, gram-negative, nonspore-forming, and either curved or cigarette-shaped rods. A few are spheroidal. The flagella of motile species of *Pseudomonas* are located only at one or both ends of the cell (*polar* flagella), whereas flagella of motile species of other rod-shaped bacteria may appear at both ends as well as on all sides (*peritrichous* flagella). With only one exception (*Pseudomonas maltophilia*), all species of *Pseudomonas* typically produce cytochrome C oxidase. Tetramethyl-p-phenylenediamine is used to test for this enzyme formation. Whereas *Pseudomonas* gives positive test results, related bacteria in the family *Enterobacteriaceae* give negative test results.

Many species of *Pseudomonas* are of great importance as soil builders, scavengers, and nitrogen fixers. Several species produce fluorescent pigments, e.g., *Pseudomonas fluorescens.* Many are important pathogens of valuable agricultural and horticultural plants and of lower animals. The importance of *Pseudomonas* in human diseases is not fully appreciated. *P. fluorescens* may produce a wide variety of infections, and *Pseudomonas aeruginosa* is a troublesome, nonspe-

---

[5]Aerobic, anaerobic, and modifiers of these terms are used here indicating relative demand for oxygen by the organisms. The reader should consult Chapter 8 for full explanations of these terms.

cific secondary invader of injured tissues, often causing cystitis, otitis media, and infections of ulcers, in which it produces blue-green pus.

Important Species in Other Families. Related to the pseudomonads are the genera of the nitrogen-fixing bacteria *Azotobacter* and *Rhizobium,* both of great economic importance to horticulture. Human pathogens of uncertain affiliation included here are *Brucella abortus, B. suis,* and *B. melitensis,* the causes of undulant fever in humans and brucellosis in animals; *Bordetella pertussis,* the whooping cough organism; and *Francisella tularensis,* the agent of tularemia.

Gram-Negative Facultatively Anaerobic Rods

Family Enterobacteriaceae. These are gram-negative, nonspore-forming short rods, either nonmotile or motile by *peritrichous* flagella. The student will become very familiar with such organisms. They include *Escherichia coli* (the common inhabitant of the lower part of the intestine of warm-blooded animals), the genus *Salmonella* (to which the typhoid fever organism *Salmonella typhi* belongs), *Shigella dysenteriae, Klebsiella pneumoniae, Proteus vulgaris,* and *Yersinia pestis* (the cause of bubonic plague). Many of the relatively harmless organisms used as models in the laboratory belong to the *Enterobacteriaceae.* They provide good examples of pathogens.

The Vibrios. *Vibrio cholerae,* the cause of Asiatic cholera, is common in the Orient but absent from the Western Hemisphere and Europe, unless accidentally introduced by travelers returning from Asia.

Gram-Negative Obligate Anaerobic Short Rods

These short rods have peritrichous flagella and are nonspore-forming. The genera *Fusobacterium, Leptotrichia,* and *Bacteroides* belong to this group of anaerobes. Thanks to the use of the gas chromatograph, some 30 species of *Bacteroides* have been identified. Of these, *Bacteroides melaninogenicus* is found in periodontal pockets of the gum and may even cause them. *Bacteroides fragilis* produces inflammatory diseases and septicemia. It is remarkable that *Bacteroides* species and not *Escherichia coli* are the predominant organisms isolated from human feces, if proper anaerobic methods are applied.

Gram-Negative Cocci

Included in this group of gram-negative flattened diplococci are *Neisseria gonorrhoeae,* the gonococcus, the cause of gonorrhea; and *Neisseria meningitidis,* the meninogococcus, the pathogen of commonly epidemic cerbrospinal meningitis.

Gram-Positive Cocci

The genus *Micrococcus,* quite common in soil, dust, water, our skins, and elsewhere, is not pathogenic to humans, plants, or animals. However, *Staphylococcus aureus* and many species of *Streptococcus* are highly pathogenic. Much of the damage caused to our tissues is due to several different toxins these organisms produce. Pus formation is also a sign of their detrimental activity.

Endospore-Forming Rods

Family Bacillaceae. The gram-positive, spore-forming large rods, belonging to the genus *Bacillus,* are aerobic, and many of them can easily be isolated from the soil. Antibiotics that have been obtained from such organisms include bacitracin, polymyxin, and Coly-Mycin. The only prominent pathogenic species is *Bacillus anthracis,* also called the anthrax bacillus. It was first isolated by Robert Koch from an animal infected with anthrax.

The strictly anaerobic genus in this family is *Clostridium.* Some species cause food poisoning *(Clostridium botulinum),* others tetanus *(Clostridium tetani),* gas gangrene *(Clostridium perfringens),* and various other conditions *(Cl. haemolyticum, Cl. paraputrificum, Cl. putrefaciens, Cl. rectum, Cl. cadaveris).* Clostridia produce many toxins, some deadly even in extremely low concentrations.

Actinomycetes and Related Organisms

The Coryneform Group of Bacteria. The human pathogen *Corynebacterium diphtheriae,* the diphtheria organism, shows a close relationship to the mycobacteria and nocardias by possessing chemically similar cell walls. There are three types of "diphtheroids," as these organisms are also called: (1) the human and animal parasites and pathogens, (2) the plant pathogens, and (3) nonpathogenic *Corynebacteria.*

Actinomycetales. These organisms represent a slightly more advanced stage in evolution than do the other bacteria. Many Actinomycetales branch extensively as molds (Fig. 4–16). In general, Actinomycetales may be cultivated in much the same manner and media as most common bacteria, yeast, and molds, and may be stained and studied in the same way. Harz, in 1878, used the name *Actinomyces (actino,* radially; *myces,* fungus) to describe the radially growing fungus-like organism *(Actinomyces bovis)* that causes "lumpy jaw" in cattle.

The diameter of the branching filaments, or *hyphae,* of the Actinomycetales is unlike that of molds, which are relatively enormous. The diameter of filaments of Actinomycetales, like that of all other bacteria, is only about 2 μm. The length of the hyphae, however, may reach hundreds or thousands of micrometers. The whole structure of these branched, interlacing hyphae is spoken of as a *mycelium* and may form a colony ranging in size from 5 mm to 2.5 cm or more in diameter.

Those biologists who regard the primitive waters of the earth as the place

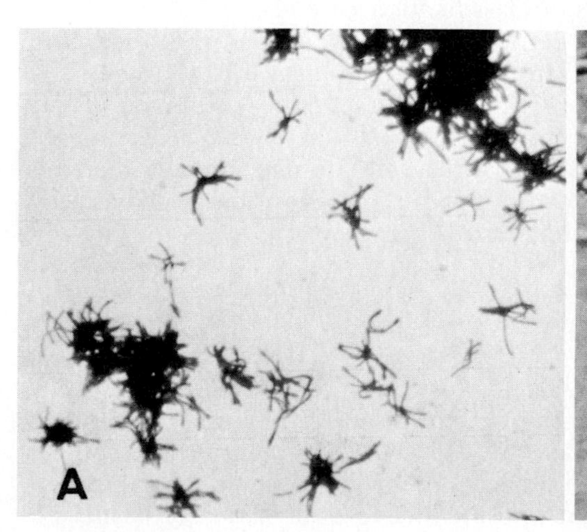

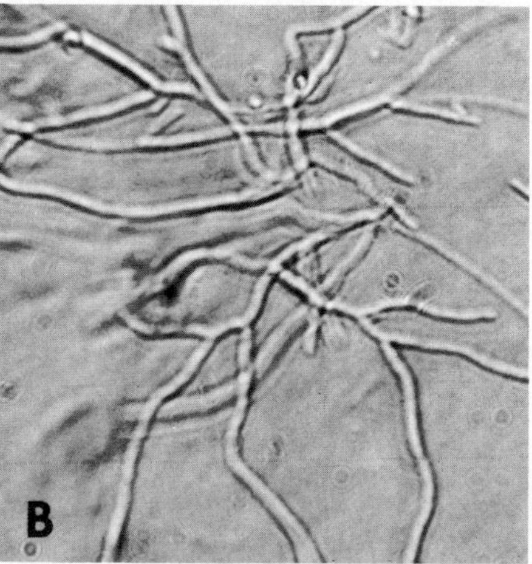

Figure 4–16 A pathogenic actinomycete, *Actinomyces israelii. A,* Smear from anaerobic (thioglycollate broth) culture (× 650). *B,* A very young colony on agar incubated anaerobically with 5 per cent carbon dioxide plus 95 per cent nitrogen at 37 C (× 1,750). Note the true-branching mycelia. (From Howell, A., Jr., Murphy, W. C., III, Paul, F., and Stephan, R. M.: Oral Strains of Actinomyces. *J. Bacteriol. 78*:82, 1959.

in which life originated may find support for this view in the hundreds of species of motile, aquatic, and marine forms of bacteria. Many bacteria are essentially fluid-inhabiting, swimming organisms. The Actinomycetales, however, are mostly terrestrial, largely restricted to the soil, and, with one or two aquatic exceptions, not equipped for swimming. In keeping with a "dry-land" mode of existence, they can grow on a wide variety of solid surfaces exposed to the air, such as old, damp shoes or wallpaper in damp houses, and require relatively little moisture. They grow rather slowly and prefer damp atmospheres and comfortable room temperatures (20 to 30 C) to body temperature (37 C).

The Family Actinomycetaceae. Actinomycetaceae contains three genera, an important one being the genus *Actinomyces,* which branches somewhat more than mycobacteria. The genus *Actinomyces* includes *A. bovis* and *A. israelii,* which are found in the mouths of humans and cattle and sometimes cause ulcerative, invasive, and necrotic suppurative lesions in the jaw ("lumpy jaw" in cattle), tongue ("wooden tongue" in cattle), and adjacent parts by invading the tissues. The disease is called *actinomycosis* and is often fatal.

The Family Mycobacteriaceae. Cells of the Mycobacteriaceae occasionally branch to the extent of forming L, T, Y, V, and similar simple angles, but otherwise the Mycobacteriaceae resemble true bacteria in size and form, being mainly slender, sometimes slightly curved, and often spindle-shaped rods. This family contains only one genus of medical importance, called *Mycobacterium (myco,* moldlike). This genus contains *M. tuberculosis* (the cause of tuberculosis) and *M. leprae* (the cause of leprosy, better called Hansen's disease).

The Family Nocardiaceae. Most species of *Nocardia* are harmless and live in fertile soil on rotting organic matter. They are numerous and widespread and help cause decay. A few species occasionally cause diseases of the lungs or other tissues, sometimes resembling actinomycosis or tuberculosis. Such infections are called *nocardioses.* Among the common pathogenic *Nocardia* are *N. madurae* (causing forms of Madura foot, an actinomycosis-like disease of the feet and legs); *N. asteroides* (causing an infection of the lungs, which may at times be mistaken for tuberculosis); and *N. farcinica* (causing lumpy lesions, or farcy, in tissues of cattle and horses). Actinomycosis and nocardiosis in humans are relatively uncommon in North America and are not highly infectious to attendants, though often fatal to the patient.

The Family Streptomycetaceae. This family of the order Actinomycetales is distinguished by the formation of extensive moldlike mycelia, the long hyphae branching freely. Coiled aerial hyphae are a striking feature. As in true molds, sporelike bodies called *conidia* appear on the ends of these, in long, brilliantly colored chains. These conidia (sometimes called spores) resist drying but do not have the high resistance to temperature characteristic of true bacterial spores and are readily killed by boiling or vigorous disinfection. The forms of the coils of conidia serve as one basis of classification (Fig. 4-17).

To the health worker, only the genus *Streptomyces* is of importance, and only a few species need be mentioned here. One is *S. griseus,* the source of the antibiotic drug streptomycin: another is *S. aureofaciens,* the producer of the antibiotic chlortetracycline (Aureomycin); a third is *S. rimosus,* the source of oxytetracycline (Terramycin). These drugs will be mentioned later. Numerous other species of *Streptomyces* valuable for the production of antibiotics have been discovered. None is an important pathogen of humans.

APPLICATION TO HEALTH

The recognition of scientific names of bacteria and other microorganisms in laboratory reports will enable health workers to bring important reports immediately to the attention of the physician. Knowledge of the variability of bacteria should certainly be helpful in understanding their activities in laboratory studies. Awareness of genetics and the adaptability of microorganisms to

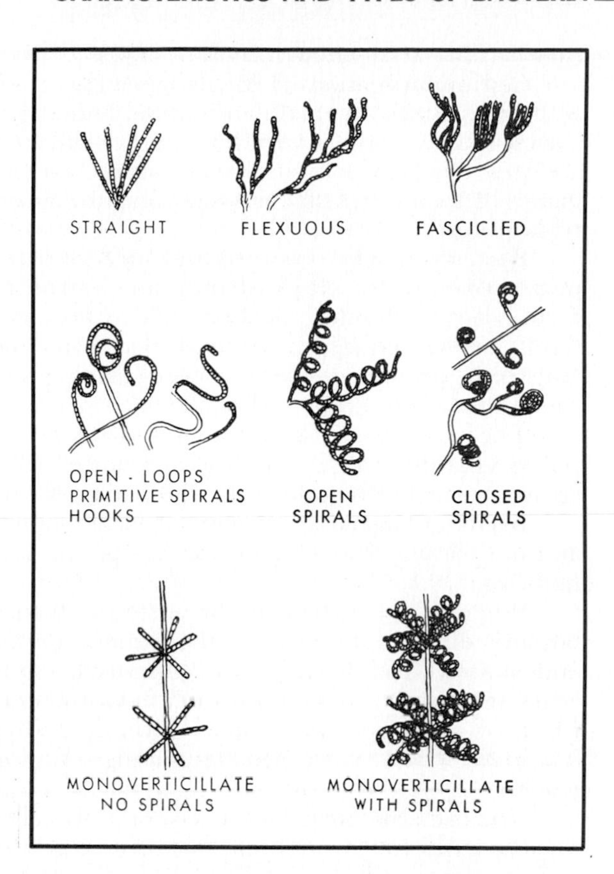

Figure 4–17 Distinctive arrangements of conidial filaments of various types of *Streptomyces*. (Adapted from Pridham, T. G., Hesseltine, C. W., and Benedict, R. G.: *Appl. Microbiol.*, Vol. 6.)

chemotherapeutic agents is essential to intelligent administration of correct doses of sulfonamides and antibiotics at properly spaced intervals. This application of physiology and morphology to the structure of bacteria will be referred to later under the discussion of pathogenic bacteria. The student will soon begin to associate names of organisms with morphology and with specific diseases.

As has been pointed out, the scientific names of microorganisms are clues to the distinctive properties and characteristics of each organism. It would be tedious (but not impossible) to learn by rote the scientific names of all microorganisms important to the care of patients, and it would be pointless without also learning the meaning and significance of these names. For example, the student will quickly learn that recovery of *Bacillus subtilis* from supposedly sterile packages in the operating room is a sign of danger, because *B. subtilis,* though usually regarded as harmless, is a spore-former, and its survival indicates that sterility has not been attained. Recovery of the same organism from the dust in the hospital wards, however, would not be a source of alarm, since the organism is generally nonpathogenic and ubiquitous.

Because of the widespread publicity about staphylococcal infections, the beginning student may view with deep concern reports of skin cultures that show the presence of *Staphyloccus epidermidis.* The more knowledgeable investigator, however, is able to discriminate and recognize that only those strains of *S. aureus* that produce a golden-yellow *(aureus)* pigment and the enzyme coagulase are the primary causes of dangerous cases of staphylococcal infections. One of the best ways of learning microbial nomenclature is to look up the characteristics of each genus and species of organism that you encounter in the classroom, in the laboratory, or in practice. If you do this whenever you are not sure of the distinguishing features of a particular microorganism, a body of information will quickly become a part of your working knowledge, on which you can build the sound practice of your profession.

SUMMARY

Cell multiplication of bacteria involves growth and usually binary fission, with occasionally unequal division and budding. Of more than 1,700 species of bacteria known only a very few are human pathogens of consequence. These are well designated by a nomenclature involving the genus and the species names of each organism, as exemplified by *Salmonella typhi,* a pathogen causing typhoid fever.

Bacteria may be observed and measured through an ordinary compound microscope (or, for fine structure, an electron microscope) and classified as to form: coccus, bacillus, or helicoidal. Cocci may be diplococci, streptococci, staphylococci, tetrads, or sarcinas. Bacilli may be gram-positive spore-forming long rods, gram-negative nonspore-forming short rods, or coccobacilli. Helicoidal bacteria may be rigid or flexible.

The capsules of bacteria are made of gummy, highly species-specific polysaccharide "soluble specific substance" (SSS) materials. They confer increased virulence by protecting the pathogens that have them.

Bacterial flagella are organs of locomotion that extend outside of the cell and are arranged in various species-specific ways. Some bacteria have pili or fimbriae.

When food is exhausted or waste products accumulate, conditions for life become difficult. Species of the genera *Bacillus* and *Clostridium* may form endospores, which permit the organisms to survive in a dormant state for many years, if necessary, until a favorable environment prevails again and the bacteria can grow and reproduce. Spore-forming pathogens—for example *Clostridium botulinum,* which causes food poisoning—may be especially dangerous to humans.

The bacterial nucleus is more correctly called a nucleoid, because it has no nuclear membrane or definite form. Some genetic extrachromosomal elements, circular pieces of DNA called plasmids, exist in the cytoplasm; they play a major role in genetic engineering.

Major groups of bacteria of medical importance are the spirochetes, causing syphilis and leptospirosis; the *Pseudomonas* group, involved in hospital infections; the cholera organism *Vibrio cholerae*; the pathogen of gonorrhea, *Neisseria gonorrhoeae*; streptococci and staphylococci, common agents of numerous mild to very serious infections; the anthrax bacillus; and *Clostridium,* which induces toxic conditions such as botulism, tetanus, and gas gangrene. Other major diseases caused by bacteria include diphtheria, Hansen's disease, and tuberculosis.

REVIEW QUESTIONS

1. Relate the size of bacteria to that of molds, protozoa, and humans.
2. What is a genus? What is a species? What are the rules of the binomial system of naming living organisms?
3. What unit of length is commonly used in microbiology? How is it related to a meter?
4. How do bacteria multiply?
5. How do bacteria receive their food, water, and oxygen and get rid of their waste products?
6. What is the magnification of a compound microscope and of an electron microscope?
7. What are three main morphologic groups of bacteria? Draw an example of each. Draw five subgroups of one of these shapes.
8. What are flagella? Which groups of bacteria may have flagella? What is their function?
9. What are bacterial spores? What organisms form spores? Why?
10. How is a nucleoid different from a nucleus?
11. What are capsules? What is their importance to bacteria and to us?
12. Define plasmids, pigments, aerobic, and anaerobic.

Supplementary Reading

Arai, T.: Actinomycetes. Baltimore, University Park Press,1977.

Buchanan, R. E., and Gibbons, N. E. (eds.): Bergey's Manual of Determinative Bacteriology. 8th ed. Baltimore, The Williams & Wilkins Co., 1974.

Cairns, J.: The chromosome of *Escherichia coli. Cold Spring Harbor Symp. Quant. Biol., 28*:43, 1963.

Frobisher, M., Hinsdill, R. E., Crabtree, K. F., and Goodheart, C. R.: Fundamentals of Microbiology. 9th ed. Philadelphia, W. B. Saunders Company, 1974.

Gerhardt, P., Sadoff, H. L., and Costilow, R. N. (eds.): Spores VI. Washington, D.C., American Society of Microbiology, 1975.

Hirota, Y., Jacob, F., Buttin, G., and Nakai, T.: On the process of cellular division in *Escherichia coli.* I. Asymmetrical cell division and production of deoxyribonucleic acid-less bacteria. *Molec. Biol., 35*:175–192, 1968.

Jawetz, E., Melnick, J. L., and Adelberg, E. A.: Review of Medical Microbiology. 14th ed. Los Altos, Calif., Lange Medical Publications, 1980.

Leifson, E.: Bacterial taxonomy: A critique. *Bact. Rev., 30*:257, 1966.

Leive, L. (ed.): Membranes and Walls of Bacteria. New York, Dekker, Marcel, 1973.

Pease, P.: Evidence that *Streptobacillus moniliformis* is an intermediate stage between a *Corynebacterium* and its L-form or derived PPLO. *J. Gen. Microbiol., 29*:91, 1962.

Pelczar, M. J., Jr., and Chan, E. C. S.: Elements of Microbiology. New York, McGraw-Hill, 1981.

Slack, J. M., and Gerencser, M. A.: Actinomyces, Filamentous Bacteria. Minneapolis, Burgess Publishing Co., 1975.

Stanier, R. Y., and Doudoroff, M.: The Microbial World. 4th ed. Englewood Cliffs, N.J., Prentice-Hall Inc., 1976.

Wittler, R. G., et al.: Isolation of a *Corynebacterium* and its transitional forms from a case of subacute bacterial endocarditis treated with antibiotics. *J. Gen. Microbiol., 23*:315, 1960.

(See also Supplementary Reading list for Chapter 5.)

CHARACTERISTICS OF RICKETTSIAS, CHLAMYDIAS, AND MYCOPLASMAS

The rickettsias and chlamydias are both dealt with in this chapter because, although they possess structural and physiological properties that show their close relationship to other bacteria, they also have several distinctive properties in common, notably, very minute size and, with few exceptions (Bartonellaceae), an inability to grow anywhere except *inside living cells.* In the eighth (1974) edition of *Bergey's Manual of Determinative Bacteriology,* they are classified together in Part 18 as "The Rickettsias": order I, Rickettsiales, families Rickettsiaceae, Bartonellaceae, and Anaplasmataceae; order II, Chlamydiales. These two groups of organisms are now considered to be modified bacteria. They have prokaryotic structure and numerous bacteria-like enzyme systems. Because in most species these enzyme systems are inadequate for independent extracellular life, they can multiply only intracellularly; they are *obligate, intracellular parasites.* In this they resemble viruses. Vertebrate tissues affected by Rickettsiales are chiefly the vascular and reticuloendothelial cells and erythrocytes, with locations depending on species. Most of the Rickettsiales can also invade organs of arthropod hosts, which then become vectors of disease. In the family Rickettsiaceae, only three genera—*Coxiella, Rochalimaea,* and *Rickettsia*—contain human pathogens.

The genus *Coxiella* contains only one species, *C. burnetii,* the cause of Q fever, a moderate to severe raw-milk and cattle barn dust-borne respiratory disease. The genus *Rochalimaea* also consists of only one species, *R. quintana,* causing the louse-borne disease trench fever, which was common during World Wars I and II and which resembles classic typhus fever but is usually less severe. With some exceptions all species of Rickettsiaceae are much alike and are exemplified by the important genus *Rickettsia,* as noted.

THE GENUS RICKETTSIA

These organisms are named for their discoverer, Howard Taylor Ricketts (1871–1910), who lost his life in the study of rickettsial diseases. Three distinct biotypes make up the genus *Rickettsia:* those that cause a form of typhus fever, those causing spotted fever–like diseases, and those that cause scrub typhus, also called tsutsugamushi. Structurally, rickettsias resemble the familiar forms of bacteria but are even smaller, averaging about one tenth to one half their size. They range from around 0.2 to 0.5 μm in diameter up to 0.8 to 2.0 μm in length. Rickettsias are variously shaped, like minute rods, spheres, or diplococci, and may be ellipsoidal or even filamentous. They can just be seen with ordinary microscopes, and thus differ from all but a few "large" viruses (poxviruses) (Figs. 5–1 and 5–2). Typical viruses are far smaller and can be seen only with electron microscopes. It is difficult to stain rickettsias with Gram's stain (most are gram-negative; *Coxiella burnetii* are gram-positive under certain conditions of staining) or with methylene blue, but they are readily stained with

CHAPTER 5

complex mixtures of dyes, e.g., Macchiavello's, Giemsa's, or Wright's stains. The rickettsias are nonmotile and do not form spores. Only one strain, that causing scrub typhus in humans, has been shown to produce a toxin.

No species of *Rickettsia* or *Coxiella* has been cultivated in lifeless media. (*Rochalimaea quintana,* constituting another genus with only one species, is cultivable on artificial media.) In this they resemble chlamydias and viruses. So far as is known, these organisms do not multiply outside the living cells of vertebrates or arthropods such as lice, fleas, and ticks, which can transmit rickettsia and thus cause disease. Most rickettsias also resemble most bacteria in being nonfilterable. An exception is the species of filterable rickettsias that causes Q fever, *Coxiella burnetii,* named for its discoverers H. L. Cox and F. M. Burnet. *Like all other cells,* and *unlike all viruses,* rickettsias contain *both* ribonucleic acid (RNA) and deoxyribonucleic acid (DNA), which occurs in strands. Viruses contain *either* RNA *or* DNA and are classified primarily on this basis. Rickettsias resemble eubacteria in possessing muramic acid in their cell walls. Muramic acid is thought to occur only in eubacteria, rickettsias, and chlamydias. In some species of rickettsias a capsule-like layer has been observed around the cell.

Propagation. Although most of the organisms in the order Rickettsiales will not grow on lifeless media such as are used for yeasts, molds, and bacteria, they will multiply very well in the yolk sac of live embryonic chicks inside the egg. To cultivate *Rickettsia* species for vaccines and other purposes, the yolk sacs are infected with a needle through a minute hole in the disinfected shell

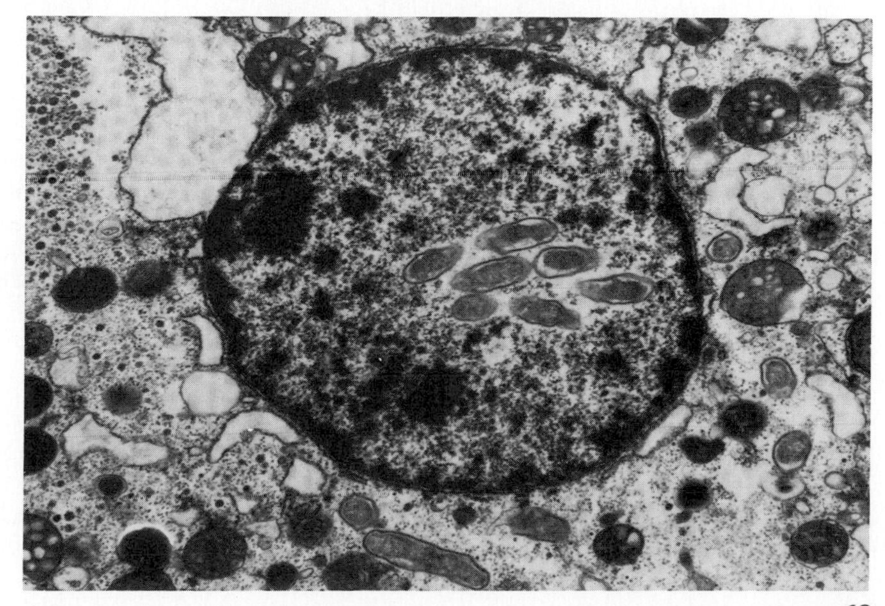

Figure 5–1 Electron micrograph of an ultrathin section of a tick gut epithelium cell showing intranuclear growth of *Rickettsia canada* (× 14,500). Note the bacterium-like size and form. (From Burgdorfer, W., and Brinton, L. P.: Infect. Immunol., *2*:112, 1970. Copyright American Society for Microbiology.)

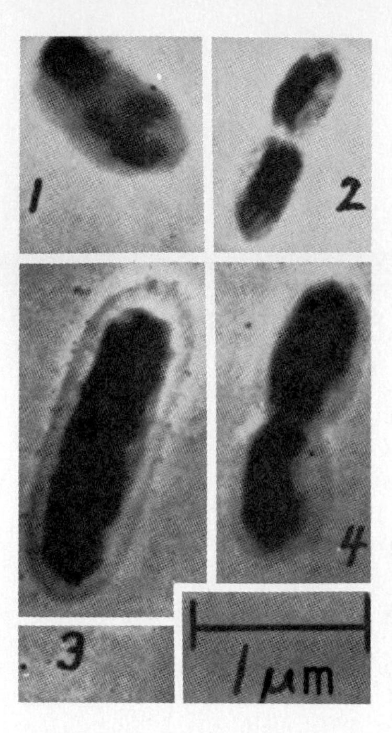

Figure 5–2 Electron micrographs of various representative rickettsias from yolk sac cultures. Note the bacterium-like form and structure, with cell wall, cytoplasm with intracellular granules, and what appears to be binary fission. Note also the small size as compared with bacteria. *1, Rickettsia typhi; 2, Coxiella burnetii; 3, R. rickettsii; 4, R. prowazekii.* (From the collection of the American Society for Microbiology, courtesy of Drs. H. Poltz, J. E. Smadel, T. F. Anderson, and L. A. Chambers. In *J. Exp. Med.*, 77:355, 1943.)

(Fig. 5–3). After inoculation of eggs and several days of incubation, the yolk sacs, with their rickettsias, are removed from the eggs aseptically.[1] After various purifications and treatment with formaldehyde or some other means of killing the rickettsias, the material constitutes a vaccine that is used in immunizing persons likely to be exposed to the diseases caused by specific rickettsias. For

[1]Completely avoiding contamination by any microorganisms from dust, eggshells, and so forth. Surgeons and microbiologists work as aseptically as possible.

Figure 5–3 Methods of inoculating chick embryos. *A*, Injection into the yolk sac for cultivation of rickettsias. *B*, Many microorganisms can infect the chorioallantoic membrane. An artificial air space is created by allowing air from the normal air space to escape through the small opening seen at the right end of the egg as the opening for the needle is made. The collapsed air sac is then sealed up. (Courtesy of E. R. Squibb & Sons. From Kelley and Hite.)

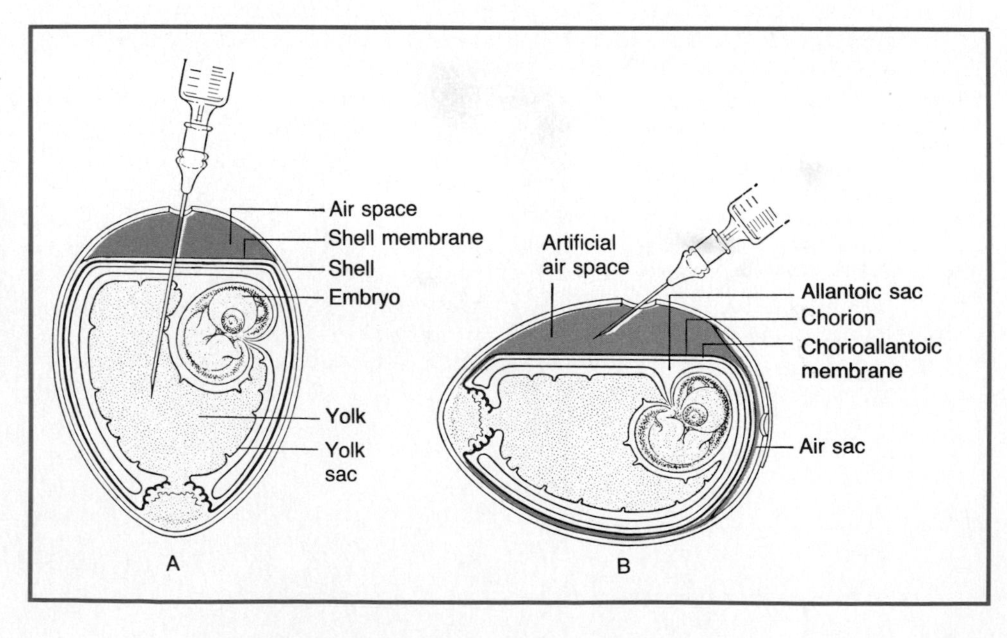

example, persons traveling from the United States to certain areas where typhus fever (louse-borne) is prevalent are generally required to take "shots" (immunizing injections) of such a vaccine. Such antigens are also used in diagnostic serologic tests.

Habitat. Species of *Rickettsia* appear characteristically to inhabit the living cells lining the intestines of arthropods, both bloodsucking and nonbloodsucking. Several pathogenic species inhabit arthropods that bite humans or animals, or both. Nonpathogenic (supposedly!) rickettsias, also, have been found in ticks, lice, bedbugs, spiders, and mosquitoes. The rickettsias, both pathogenic and nonpathogenic species, are found not only in the cells of the arthropod's intestines, but also frequently in the salivary glands, from which they may be transmitted to humans bitten by the arthropod. Since they also occur in the intestinal contents, they appear in the feces. Thus, transmission to animal hosts is often effected not only by bites, but also by rubbing fecal material of arthropods into the skin. This is also true of certain arthropod-borne viral, bacterial, and protozoal diseases. The principal human diseases caused by *Rickettsia* are discussed in Chapter 40.

THE ORDER CHLAMYDIALES

Only one genus is recognized in this order, in the family Chlamydiaceae, the genus *Chlamydia*. It has only two legitimate species: *C. trachomatis* (the TRIC agent) and *C. psittaci*.

Chlamydias, formerly regarded as "large viruses," intermediate in size between "true" viruses and rickettsias, are now, like the rickettsias, considered bacteria that have adapted to *obligate* intracellular existence. Their cellular structure appears to be prokaryotic, though complete details are still lacking. Like all other cellular organisms and totally unlike viruses, they contain both RNA and DNA and have synthetic enzyme systems; thus they are sensitive to the tetracycline antibiotics.

Chlamydias are larger than even the poxviruses (Fig. 6–4) but much smaller than any bacteria. Chlamydias characteristically occur in the form of very minute coccoid or roughly cuboidal bodies. These organisms are visible under ordinary microscopes, and they may be stained like bacteria. Like rickettsias (and many bacteria and viruses), they can be cultured in the chick embryo, especially in the yolk sac. They will not grow in nonliving media, and they are filterable only with difficulty.

Chlamydias appear to parasitize their host cell for most of their energy and to lack most energy-yielding enzyme systems. Their method of multiplication is not binary fission like other bacterial cells, but it is also totally unlike that of viruses. At times, very minute bodies called "elementary bodies" are formed, suggestive of the minute "minimal reproductive units" of mycoplasmas. Like them, chlamydias have a thin, limp, grape-skin–like coating.

Members of this group include the agents of psittacosis and ornithosis; lymphogranuloma venereum;[2] trachoma, a severe infection of the eye with corneal involvement often leading to blindness; and severe conjunctivitis, usually in the newborn (inclusion conjunctivitis), which is commonly contracted from the genitourinary tract of the mother during birth. The inclusion conjunctivitis organism may also cause urethritis in the adult.

---

[2]Lymphogranuloma venereum is a veneral disease causing destructive ulcerations, buboes, and obstruction of the lymph channels in the genitalia and adjacent parts. It is fairly widely distributed among sexually promiscuous persons. This is discussed further in a later section.

THE MYCOPLASMAS

The Class Mollicutes

The first of these curious organisms to be described was isolated in 1898 by Nocard and Roux, French scientists, from the pleural fluids (fluids collecting around the lungs) of cattle with an infectious disease called bovine pleuropneumonia. The organisms were called pleuropneumonia organisms.

PPLO. After the discovery of pleuropneumonia organisms, numerous other organisms resembling them were reported. These were not related to the disease pleuropneumonia but were found in other pathologic conditions and even in sewage, and were spoken of as *pleuropneumonia-like organisms,* or PPLO for short. These organisms are now usually referred to as mycoplasmas.

The mycoplasmas belong to the class *Mollicutes,* which are defined as prokaryotic organisms bounded by a single triple-layered membrane. They lack a true cell wall since they are unable to synthesize muramic acid and diaminopimelic acid, both essential sources of cell wall material.

Mycoplasmas are now considered to be modified bacteria. They are cultivable in the dark on artificial, cell-free media and, like other bacteria, are visible with ordinary microscopes. (See Table 5–1.) In solid media the colonies are minute and have the tendency to grow into the medium; the characteristic growth pattern on the surface makes them look like fried eggs. Unlike other bacteria, however, they have no cell wall and are enclosed only in a thin, limp, triple-layered membrane. Consequently their shape and size are *extremely* variable—that is, they are highly *pleomorphic* (Greek *pleon,* many; *morphe,* form; appearing in many forms). Sometimes they produce elements that, like viruses, are small enough to pass through bacteria-retaining filters. The exact relationship of mycoplasmas to other bacteria is still under investigation by microbiologists. Although many strains of *Mycoplasma,* all of which require cholesterol for growth, have been shown to cause diseases in cattle, goats, rats, chicks, and cats, only two have been proven to be pathogenic in humans, namely *M. pneumoniae* (induces a type of pneumonia) and *M. hominis* (genitourinary tract inflammations). Nonpathogenic species (genus *Acholeplasma*) do not require cholesterol.

L-Forms of Bacteria. Certain cell wall–defective bacteria, known as L-forms, L-bodies, or L-phase variants, are named after the Lister Institute in London where these pleomorphic forms were first isolated. Under special conditions of cultivation (e.g., in the presence of penicillin and some other substances that destroy cell walls or interfere with their synthesis) or as a result, apparently, of genetic mutations, many bacteria grow without a cell wall in a form called *protoplasts* (see Fig. 4–10) or *L-bodies.* Continued studies have shown several supposed mycoplasmas to be merely forms of known species of bacteria

Table 5–1. Characteristics of Infectious Agents from Bacteria to Viruses*

MICRO-ORGANISMS	GROWTH ON ARTIFICIAL MEDIA	TWO NUCLEIC ACIDS	RESPONSE TO ANTIBIOTICS
Bacteria	+	+	+
Mycoplasma	+	+	+
Rickettsia	−	+	+
Chlamydia	−	+	+
Viruses	−	−	±

*From Debré, R., and Celers, J.: Clinical Virology. Philadelphia, W. B. Saunders Company, 1970.

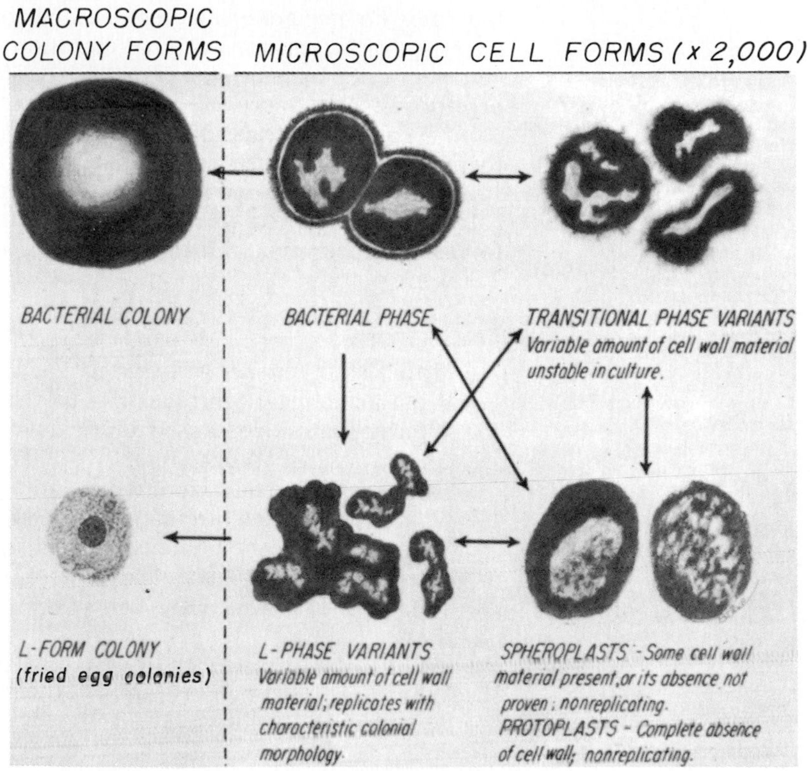

Figure 5–4 Representation of transitions between vegetative bacteria and various cell wall–defective variants. Note the pleomorphic shapes due to missing cell walls. Drawings of the individual organisms are magnified approximately 2,000-fold relative to drawings of the colony forms. (From Cate, T. R.: *In* Lennette, Spaulding, and Truant [eds.]: Manual of Clinical Microbiology. 2nd ed. Washington, D.C., American Society for Microbiology, 1974.)

lacking their cell walls (Fig. 5–4). Many bacteria may readily be induced to assume either cellular or L-body form. These cell wall–defective variants retain the genetic material of the bacterial parent, and they may even secrete the same toxins as the parent. L-forms also produce the characteristic "fried egg" colonies on solid media that are created by mycoplasmas. The reversion of L-forms to their parent bacterial form is not difficult to demonstrate under specific conditions, whereas mycoplasmas retain their stable forms. There are also differences in nutritional requirements (detected only with an electron microscope) and in pathogenicity between L-forms and mycoplasmas. Although many diseases in animals and at least two in humans are caused by mycoplasmas, L-forms may be found in infections that were caused by their parent bacteria.

Whether mycoplasmas are really a separate group of microorganisms, e.g., bacteria that have permanently lost (or never acquired?) the ability to synthesize a cell wall, or are merely L-forms of familiar bacteria that have not been observed to revert to bacterial form, is still a matter of debate.

SUMMARY

Rickettsias and chlamydias are both obligate intracellular parasites. In this they are like viruses, but they are classified with the bacteria, since they resemble them in their structural and physiologic properties, although they are much smaller than the eubacteria.

Rickettsial diseases include typhus fever, scrub typhus or tsutsugamushi, Rocky Mountain spotted fever, and related diseases, all transmitted by either lice, fleas, ticks, or other arthropods. To propagate rickettsias, and produce vaccines (after subsequent chemical treatments), yolk sacs of embryonic chicks inside eggs are injected with the microorganisms.

Chlamydias, like bacteria, may be stained and cultured in chick embryos, but they do not generally require arthropods to transmit diseases. *Chlamydia trachomatis* may cause trachoma, inclusion conjunctivitis, and lymphogranuloma venereum. *Chlamydia psittaci* is the etiologic agent of psittacosis and other ornithoses.

Mycoplasmas (class *Mollicutes*), also known as PPLOs, and the so-called L-forms of bacteria both have defective or nonexistent cell walls. They are both small, highly pleomorphic forms, but seem to be distinctly different organisms. L-forms can be produced without too much difficulty from regular bacterial forms and vice versa, but this is not true of mycoplasmas.

REVIEW QUESTIONS

1. How are rickettsias and chlamydias alike? How do they differ?
2. What are obligate intracellular parasites?
3. What diseases are caused by rickettsias? How are these diseases related to certain insects? What are arthropods?
4. How are rickettsias, chlamydias, and viruses cultivated? Why?
5. Name two species of chlamydia and some diseases they cause.
6. What are two other names for pleuropneumonia-like organisms?
7. Why do some organisms have pleomorphic forms?
8. What are L-forms? How can you explain their existence? How stable are they?
9. What is the importance of mycoplasmas?
10. What are "fried egg" colonies?

Supplementary Reading

Burgdorfer, W., and Brinton, L. P.: Intranuclear growth of *Rickettsia canada,* a member of the typhus group. *Infect. Immunol., 2*:112, 1970.

Carpenter, P. L.: Microbiology. 4th ed. Philadelphia. W. B. Saunders Company, 1977.

Clasener, H.: Pathogenicity of the L-phase of bacteria. *Amer. Rev. Microbiol., 26*:55–84, 1972.

Davis, B. D.: Microbiology. 3rd ed. New York, Harper & Row, 1980.

Fiset, P. F., Myers, W. F., and Wisseman, C. L., Jr.: The Rickettsiales. Handbook of Microbiology. Vol. 1, Organismic Microbiology. Cleveland, CRC Press, 1973.

Hayflick, L.: The Mycoplasmatales and the L-Phase of Bacteria. New York, Appleton-Century-Crofts, 1969.

Holmgren, N. B., and Campbell, W. E., Jr.: Tissue cell culture contamination in relation to bacterial pleuropneumonia-like organism–L form conversion. *J. Bacteriol., 79*:869, 1960.

Klieneberger-Nobel, E.: Pleuropneumonia-like Organisms (PPLO): Mycoplasmataceae. New York, Academic Press, Inc., 1962.

Kramer, M. J., and Gordon, F. B.: Ultrastructural analysis of the effects of penicillin and chlortetracycline on the development of a genital tract *Chlamydia. Infect. Immunol., 3*:333, 1971.

Lennette, E. H., and Schmidt, N. J. (eds.): Diagnostic Procedures for Viral, Rickettsial, and Chlamydial Infections. New York, American Public Health Association, 1979.

Maramorosch, K.: Mycoplasma and Mycoplasma-like agents of human, animal, and plant diseases. *Ann. N.Y. Acad. Sci., 225*:1, 1973.

Moulder, J. W.: The relation of the psittacosis group (chlamydiae) to bacteria and viruses. *Ann. Rev. Microbiol., 20*:107, 1966.

Purcell, R. H., Valdesuso, J. R., Cline, W. L., James, W. D., and Chanock, R. M.: Cultivation of Mycoplasmas on glass. *Appl. Microbiol., 21*:228, 1971.

Sharp, J. T. (ed.): The Role of Mycoplasma and L Forms of Bacteria in Disease. Springfield, Ill., Charles C Thomas, 1970.

Swierczewski, J. A., and Reyes, P.: Isolation of L-forms in a clinical microbiology laboratory. *Appl. Microbiol. 20*:323–327, 1970.

Tamura, A., and Manire, G. P.: Effect of penicillin on the multiplication of meningopneumonitis organisms *(Chlamydia psittaci)*. *J. Bacteriol., 96*:875, 1968.

Theodore, T. S., Tully, J. G., and Cole, R. M.: Polyacrylamide gel identification of bacterial L-forms and *Mycoplasma* species of human origin. *Appl. Microbiol., 21*:272, 1971.

CHARACTERISTICS OF VIRUSES

For many years after the discovery of the tobacco mosaic virus (TMV) by Iwanowski in 1892, viruses were known mainly by the diseases they caused. Common examples of human viral diseases are poliomyelitis, influenza measles, mumps, warts, chickenpox, and the common cold. It is estimated that more than 1,000 viruses are known at present, and new ones are constantly being discovered. In this book only a few selected viruses will be discussed, and some general properties common to all of them will be described.

The study of viruses is now a very large and complex field called *virology*. One of the most studied and best known types of virus is called *bacteriophage* or simply *phage* (*bacterio*, bacteria; Greek *phagein*, to eat). The name is derived from an early and erroneous notion that the virus ate bacteria from within. Some viruses infect plants; others infect animals, both vertebrate and invertebrate; still others, like the bacteriophages, infect bacteria. Viruses pathogenic to algae exist, as do others that infect mycoplasmas.

Viruses resemble rickettsias and chlamydias in not being cultivable outside of living cells: they too are obligate intracellular parasites. Viral proliferation within cells results in injury ranging from cell death and dissolution to abnormal cellular multiplication, often accompanied by distinctive intracellular appearances (see later discussion of cytopathic effect). A single mature, complete individual virus particle of any kind is called a *virion* (Fig. 6–1); it is not a true cell, since it has no autonomous metabolism or life.

Because of their small size, viruses cannot be seen with an ordinary microscope. However, with the electron microscope and special techniques, very high magnifications of viruses may be obtained. Figure 6–2 shows the shapes and relative sizes of different viruses. Clever use has been made of the fact that viruses can pass through clay, porcelain, or plastic (nitrocellulose) membrane filters that can hold back yeasts, molds, most bacteria and rickettsias, and all larger organisms. Today more sophisticated devices (filters) for separating microorganisms by their cell sizes have been made available for the laboratory worker in microbiology. Figure 6–3 shows such a device; others will be discussed in Chapters 9 and 12. High-speed differential centrifugation is also commonly used to separate particles (or viruses) according to size and weight.

VIRUSES IN GENERAL

Identification. A type of shorthand has been developed by virologists called a *cryptogram*. It helps to designate a certain virus by means of a simple formula using a few abbreviations. For example, the vaccinia (cowpox) virus may be written:

[D/2 : 160/5 : X/* : V/O]

The (D) indicates that this is a DNA virus that is double stranded (2), with a nucleic acid weight of 160 million (160), of which 5 per cent (5) is in the virion. The outline of the virion is complex (X), and the outline of the

CHAPTER 6

nucleocapsid is unknown (*). Its hosts are vertebrates (V), but it spreads without a vector (O).

Some of the poxviruses (e.g., fowlpox, goatpox) may be written as:

[D/2 : 160/5–7 : X/* : V/O, Di, Ac, Si]

Vectors may be Diptera (Di), ticks or mites (Ac), or fleas (Si).

One of the flaviviruses causes yellow fever. The cryptogram may be written:

[R/1 : 4/7–8 : S/S : V, I/O, Di, Ac]

(R) is RNA, (S) is spherical, (I) is an invertebrate host. What else can you tell about the yellow fever virus?

As more is learned about viruses, the cryptogram information changes and new rules are added to comply with the new knowledge.

Classification. A binomial classification of viruses has been suggested. A more commonly used system is based primarily on the type of nucleic acid in the virus, the morphology, and other chemical and physical properties (Table 6–1). The nucleic acids of viruses have been intensively studied. Some have single-stranded DNA or RNA, some double-stranded. Apparently all tailed phages possess double-stranded DNA. The genetic strands (genomes) of some

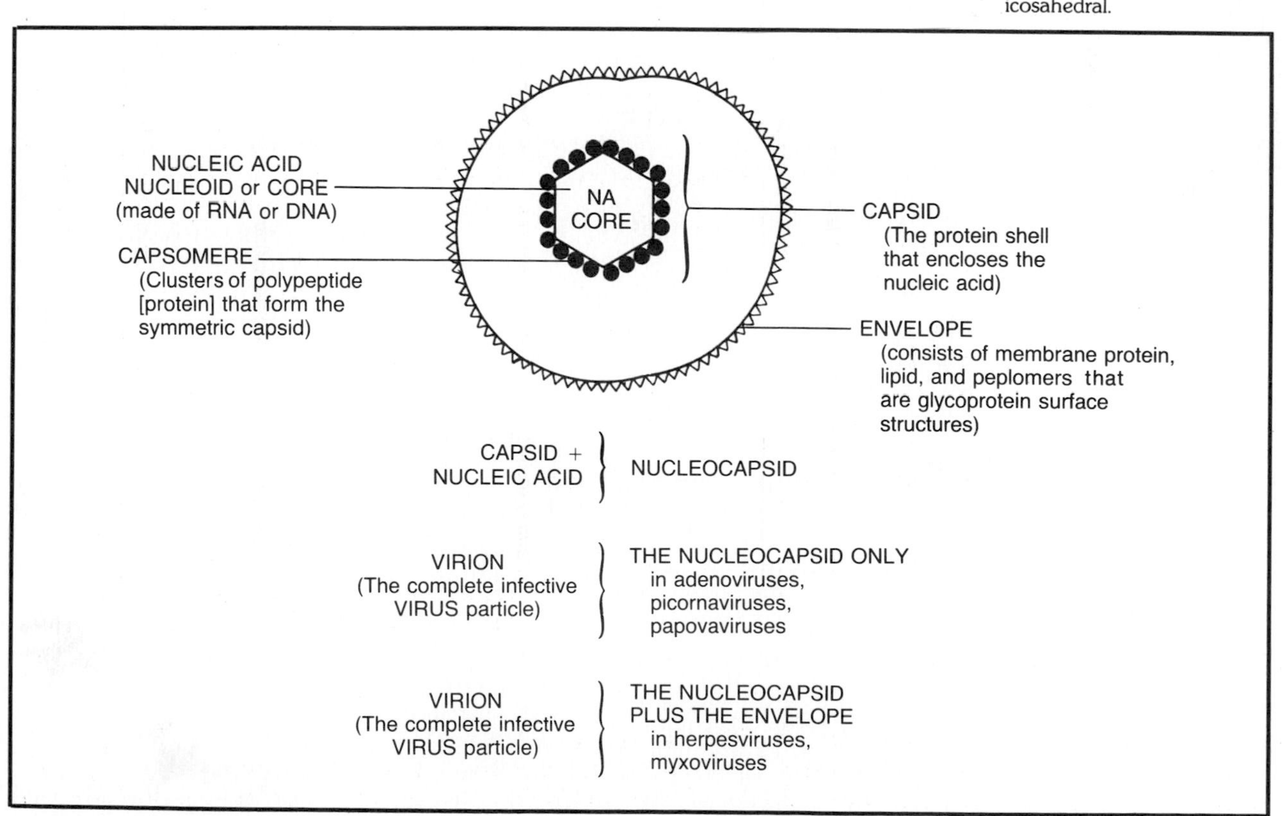

Figure 6–1 Components of the infective virus particle, the virion. Shown is an enveloped icosahedral.

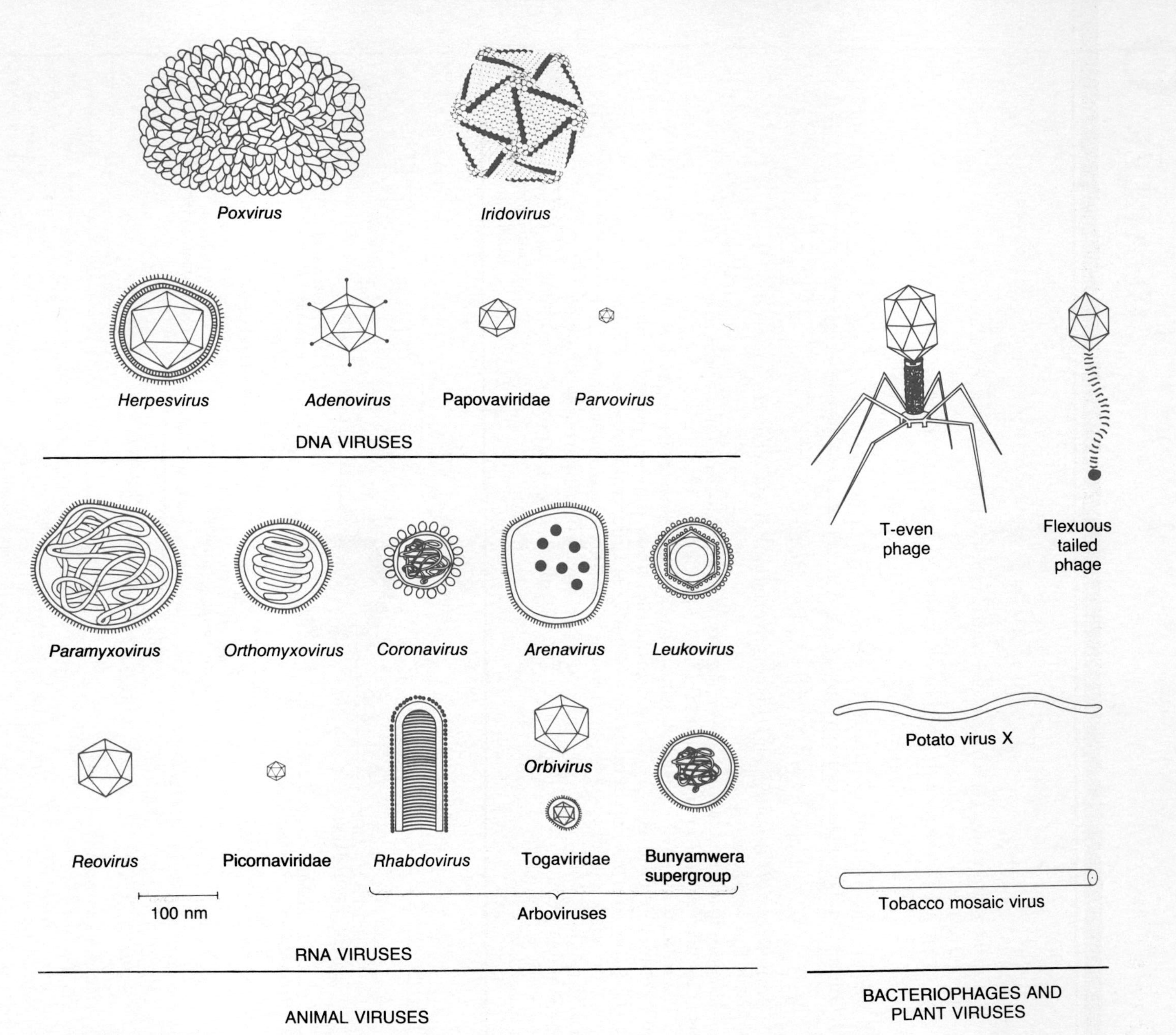

Figure 6–2 Diagram illustrating the shapes and relative sizes of animal viruses, bacteriophages, and plant viruses of the major taxonomic groups (bar = 100 nm = 0.1 μm). (From Fenner, F., et al.: The Biology of Animal Viruses. 2nd ed. New York, Academic Press, 1974.)

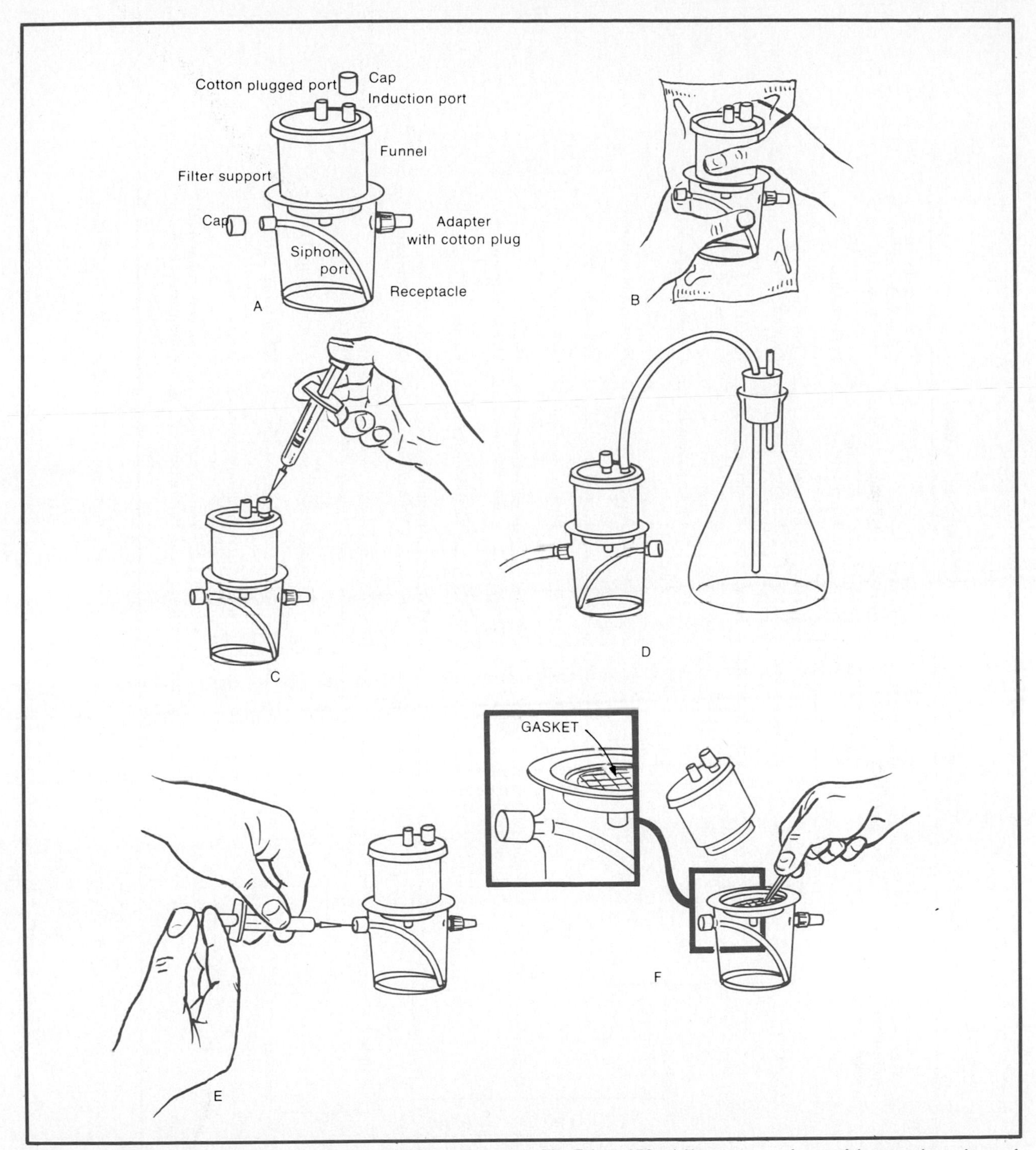

Figure 6–3 Filtration of viral suspensions under sterile conditions. *A,* The Falcon 150 ml filter units can be used free-standing, clamped on ring stand, or connected in series to any vacuum source. I.D. tubing (3⁄8″ or 1⁄4″) may be used. *B,* Before removing filter unit from bag, grasp funnel portion with one hand while holding receptacle with other hand and apply firm pressure in a clockwise direction. Liquids to be filtered may be introduced by pouring directly into funnel, by syringe and needle through cap *(C),* or by tube from a source container *(D).* To remove filtrate, swab cap on siphon port with alcohol, insert needle, and withdraw required volume *(E).* To remove membrane filter *(F),* grasp receptacle and unscrew funnel, then remove the rubber gasket by placing round nose forceps under the inner edge of the ring and lifting. (The gasket may be discarded or plated along with the filter.) Remove membrane filter by grasping the outer edge with flamed forceps. (Courtesy of Falcon Plastics, Division of BioQuest, Los Angles, California.)

Table 6–1. Partial Classification Scheme of Vertebrate Viruses

NUCLEIC ACID CORE	CAPSIDAL SYMMETRY	ENVELOPED	PARTICLE SIZE, nm	STRANDEDNESS OF NA*	ACID-LABILE (pH 3)	VIRUS GENERA	EXAMPLES OF DISEASES CAUSED BY THESE VIRUSES
			20–30	S	No	Enteroviruses	Poliomyelitis
					Yes	Rhinoviruses	Common Cold
			40	S	Yes	Caliciviruses	Swine Diseases
		No	70–75	D	No	Reoviruses	Respiratory and Intestinal
			40–80	D	Yes	Other Diplornaviruses	Colorado Tick Fever
	Icosahedral						
			25–70	S	Yes	Alphaviruses	Equine Encephalitis
		Yes	40	S	Yes	Flaviviruses	Yellow Fever
			80–120	S	Yes	Orthomyxoviruses	Influenza
RNA	Helical	Yes	150–300	S	Yes	Paramyxoviruses	Mumps
			80–130	S	No?	Coronaviruses	Mouse Hepatitis
			60 × 180	S	Yes	Rhabdoviruses	Rabies
	Helical(?)	Yes	80–110	S	Yes	Leukoviruses	Rous Sarcoma
			110–130	S	Yes	Arenaviruses	Choriomeningitis
	Complex	No	250–300	D	Yes	Poxviruses	Smallpox
			100–200	D	Yes	Herpesviruses	Fever Blisters
DNA		Yes	130	D	Yes (Some)	Iridoviruses	African Swine Fever
	Icosahedral		70–90	D	No	Adenoviruses	Respiratory Infections
			45–55	D	No	Papillomaviruses	Warts
		No	45–55	D	No	Polyomaviruses	Leukoencephalopathy
			18–26	S	No	Parvoviruses	Fulminating Rodent Diseases

*NA, nucleic acid; S, single-stranded; D, double-stranded.
(Modified from Wilner, B. I.: Viruses of invertebrates. *In* Handbook of Microbiology. Vol. 1. Cleveland, CRC Press, 1973.)

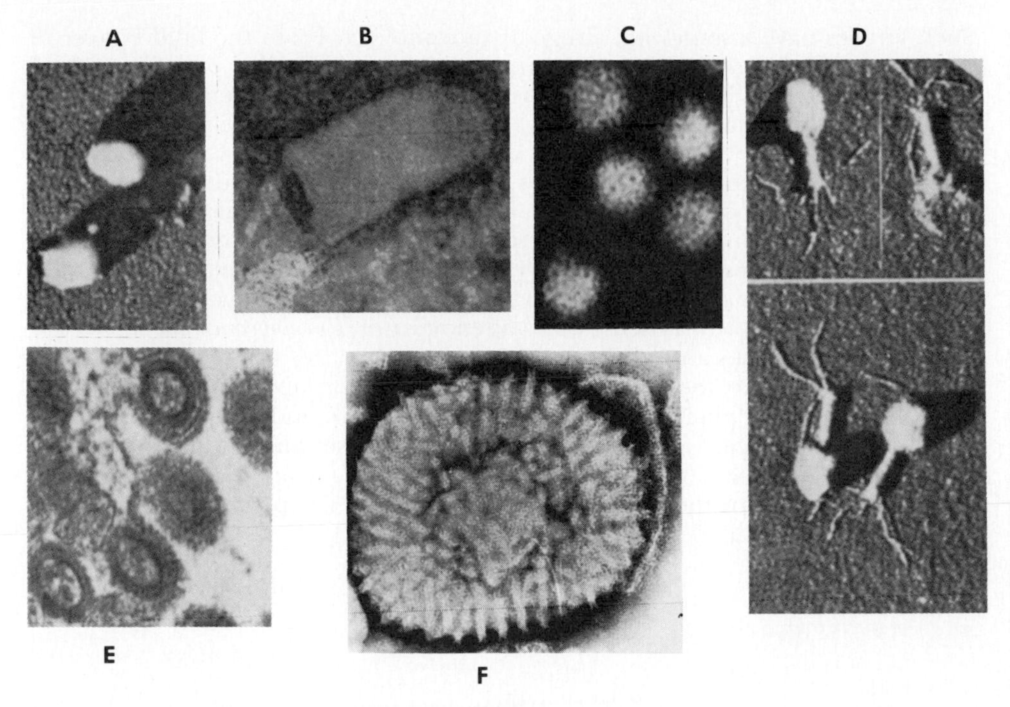

Figure 6–4 Various forms of virion. *A,* "Bun-" or brick-shaped poxvirus (×24,000). (Courtesy of Drs. F. P. O. Nagler and G. Rake: *J. Bacteriol.,* 55:47, 1948.) *B,* Bullet-shaped rabies virus (×30,000). (From Murphy, F. A., et al.: *J. Gen. Virol.,* 3: 289, 1968.) *C,* Icosahedral symmetry of reovirus (about ×10,000). (From Mayor, H. D., and Jordan, L. E.: *J. Gen. Virol.,* 3:233, 1968.) *D,* Bacteriophage showing binal symmetry (×68,000). (Courtesy of Drs. R. C. Williams and D. Frazer: *Virology,* 2:289, 1956.) *E,* Bittner (mouse mammary tumor) virus, showing cores and capsids and the manner in which viruses of this type acquire an envelope from the cell membrane on emerging from the infected cell (×60,000). (From Gay, F. W., Clarke, J. K., and Dermott, E.: *J. Virol.,* 5: 801, 1970.) *F,* A Poxvirus virion related to the virus in *A* but at ×150,000 magnification, showing details of irregular surface structure. (From Dales, S.: *Progr. Med. Virol.,* 7:1–43, 1965.)

viruses have approximately 400 genes (vaccinia virus), whereas others have very few: e.g., a virus of mice has only 8 genes.

Microscopy. Extensive use of electron microscopes has enabled us to visualize many kinds of viruses, including phages (Fig. 6–4). The magnifications achieved in these electronic images (electron micrographs) are in the range of 10,000 to over 1,000,000 diameters, whereas ordinary optical microscopes give clear images up to only about 1,200 diameters.

In the electron micrographs of viruses shown in this book, the appearance of height or thickness is given by "shadowing." This is a technique of casting metallic vapors at an angle across the object to be pictured so that some parts are made opaque (shadowed), thus giving the illusion of perspective in the pictures (Figs. 6–4 and 6–5). The appearance of viruses in electron micrographs is quite striking, and many viruses may be distinguished by their form.

Size. The diameters of some "large" viruses (e.g., smallpox) approach that of a small bacterium, *Escherichia coli* (about 1 μm = 1/25,400 inch or 1,000 nm). The diameter of some of the smallest viruses (e.g., yellow fever) is not much more than that of some protein molecules (e.g., egg albumin: 10 nm or 0.01 μm = 0.00000039 inch). It is easy to understand why viruses were formerly called ultramicroscopic (Fig. 6–6).

The Capsid. All viruses consist of an inner portion (*core* or *nucleoid*) consisting entirely of RNA or DNA, never both. In some viruses the nucleic acid is wound into a ball. The core is enclosed within an outer protective coating or protein shell called a *capsid* (Fig. 6–1). The polypeptide protein units of the capsid (all identical) are known as capsomers,[1] their number form, and arrangement around the core being symmetrical and distinctive of each group of viruses. Figure 6–7 shows capsomers of the tobacco mosaic virus (TMV). The core plus the capsid is called a *nucleocapsid.* The nucleocapsid of many animal viruses (e.g., arbovirus, myxovirus, herpesvirus) is enclosed in a distinctively ether- or chloroform-soluble, sodium deoxycholate-sensitive, lipoprotein-containing *envelope* (also known as *limiting membrane* or *mantle*) that is derived from host-cell membranes during emergence of the virus from the infected cell.

[1]Greek *meros,* part.

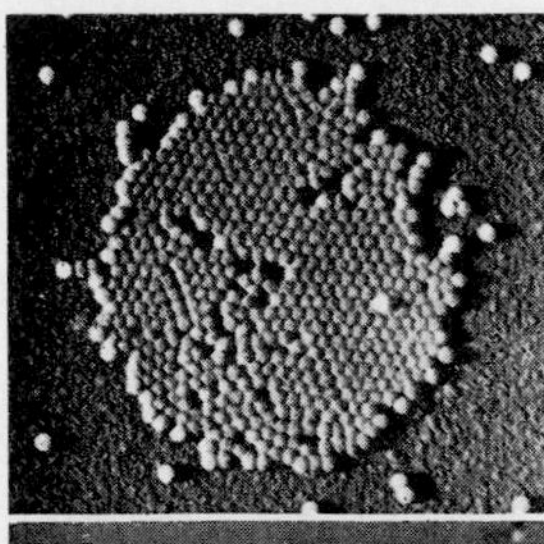

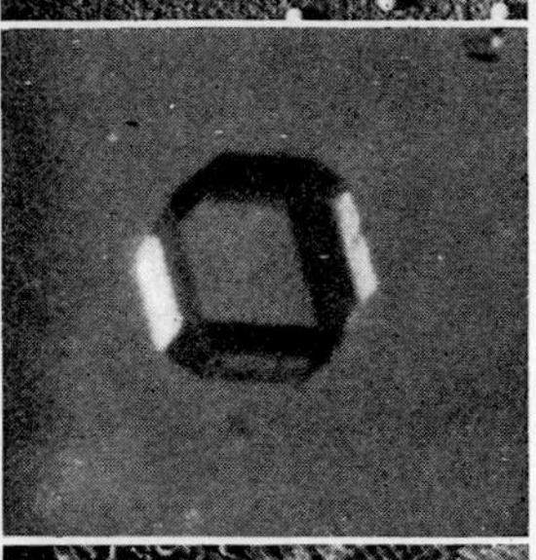

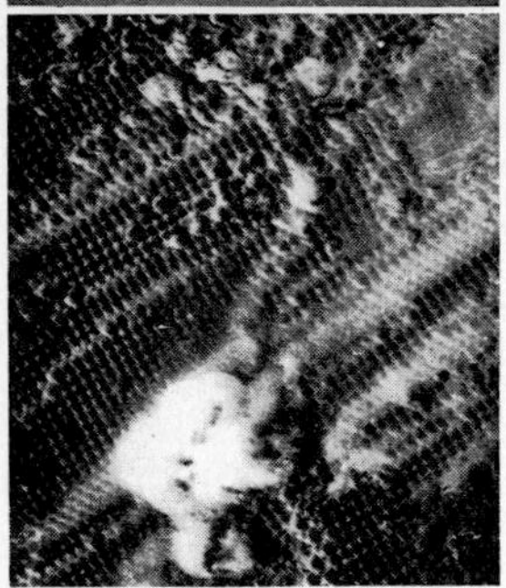

Figure 6–5 Poliovirus particles in a flat array (top) and in a three-dimensional crystal (center). A cut surface of a crystal (bottom) shows it to consist of virus particles in orderly arrays. (Burrows, after Schwerdt.)

Such viruses have a somewhat irregular morphology. From the lipid bilayer of the envelope extend glycoprotein spikes to the outside called *peplomers,* which are glycoproteins. Naked viruses are those that lack an envelope. These structures help to classify viruses that cause human diseases, as shown in Table 6–1.

Shape. Some pathogenic viruses have distinctive polyhedral shapes: icosahedral, cuboidal, spherical, ovoid, bullet-shaped, or helical. These shapes have been divided into three major groups and several subgroups.

Icosahedral viruses. (See Fig. 6–1.) *Naked icosahedral viruses,* once throught to be spherical, have an icosahedral nucleocapsid without an envelope. Examples are papovaviruses, picornaviruses, and adenoviruses. *Enveloped icosahedral viruses* include herpesviruses and togaviruses.

Helical Viruses. (See Fig. 6–7.) *Naked helical viruses* appear to be long rods. Here belong many plant viruses, bacteriophage M13, and the tobacco mosaic virus. *Enveloped helical viruses* include orthomyxoviruses and paramyxoviruses.

Complex Viruses. (See Figs. 6–4*A* and *D*.) These are more complex in structure than any of the other viruses. Here belong the poxviruses, which are also called brick-shaped, and some bacteriophages.

BACTERIOPHAGES

Different in structure (Fig. 6–2) from the human and animal (zoopathogenic) viruses are bacteriophages, commonly called phages. These viruses are pathogenic only for certain microorganisms, chiefly bacteria. Of importance in certain medical, diagnostic, and public health situations, phages are commonly

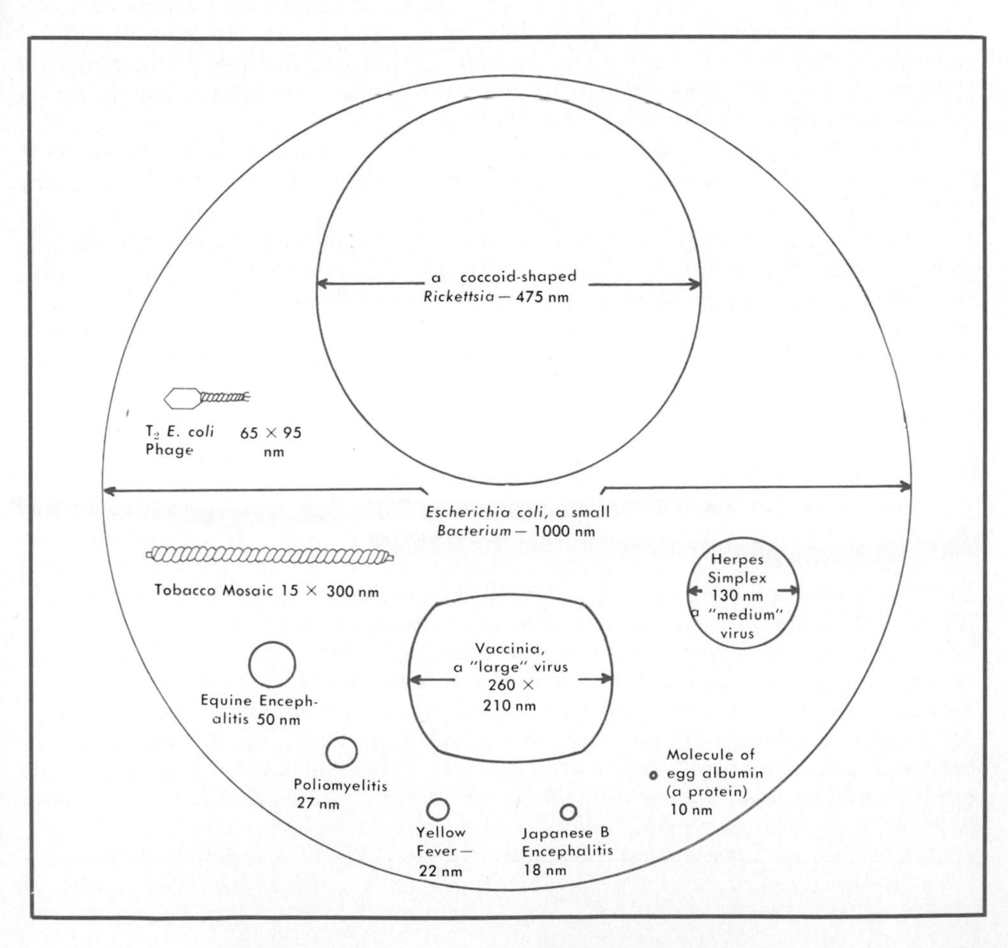

Figure 6–6 Diagrammatic comparison of sizes of viruses and related structures. The largest circle, enclosing the whole, represents the diameter of *Escherichia coli,* a small cylindrical bacterium about 1 μm (1,000 nm) in diameter. The other objects are drawn to approximately the same scale. (Sketch and sizes based on Rivers, T. M., and Horsfall, F. L., Jr.: Viral and Rickettsial Infections of Man. Philadelphia, J. B. Lippincott Co., 1948.)

Figure 6–7 Diagram of the tobacco mosaic virus virion, a helical nucleocapsid, showing the outer protein subunits (capsomers) and the inner helical coil of RNA. (From Klug, A., et al.: *Adv. Virus Res.*, 7:225, 1960.)

found in the intestinal tract. Because they are safe to handle, easily cultivated in ordinary bacterial cultures, and relatively "hardy," phages have for years served as models for thousands of studies of viruses in general. It is essential to know what phages are, their properties, and their possible uses. They do not cause disease in man or lower animals. Phages that attack a wide variety of microorganisms have been discovered: coliphages, brucellaphages, actinophages, phages of yeasts (zymophages), lactic acid bacteria, and blue-green algae. Algal viruses arecalled *cyanophages.* Recently a rod-shaped phage of mycoplasma has been discovered. A not so recent listing of bacteriophages alone recognized 672 types.

Shapes of Bacteriophages. Bacteriophages commonly have tadpole-like shapes with polyhedral "heads" attached to slender tails (Figs. 6–2, 6–4*D*, and 6–8). Since they have two parts with different forms they are said to have *binal symmetry.*

Demonstration of Bacteriophages. If feces or sewage are emulsified in water and the fluid is then passed through a bacteria-retaining filter, the fluid, although entirely free of bacteria, nearly always contains an active agent (phage), invisible with ordinary microscopes and not cultivable in nonliving media, that will destroy young, actively growing bacteria (dysentery bacilli, for example) in broth cultures. This is often called the *Twort-d'Herelle phenomenon* after the British and French scientists who first described it in 1915 and 1917, respectively.

By transferring even a tiny quantity (e.g., one millionth of 1 ml) of the filtered bacteriophage-containing fluid to a new culture of the same species of bacilli, the same process can be made to occur anew, and it will recur through an indefinite number of culture transfers and filtrations. Similar multiplications have been demonstrated with nearly all other known viruses, using cultures of animal cells (complicated and expensive) instead of bacteria (simple and inexpensive).

Infection by Bacteriophage. The tail of a phage virion is protein and has a rather complex structure. At its tip is a *base plate* with *fibrils* that facilitate

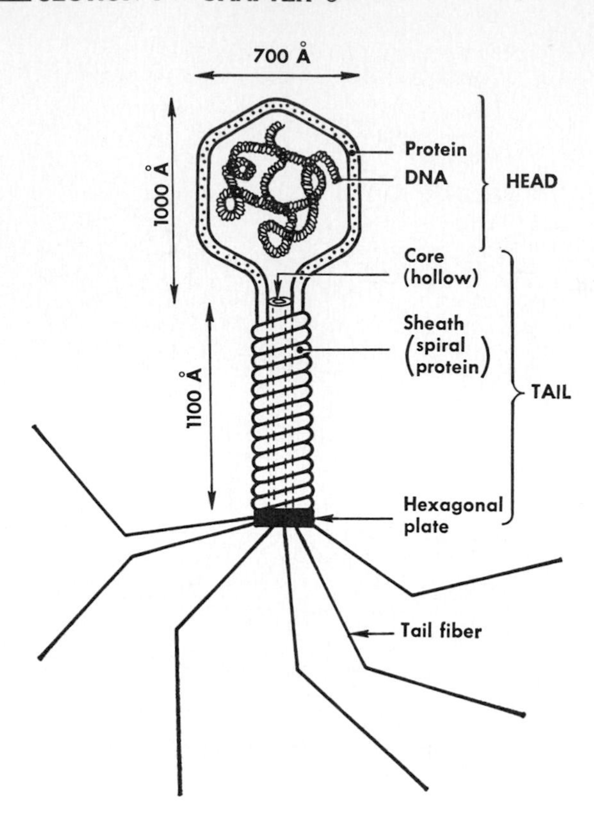

Figure 6–8 Diagrammatic representation of the structures observed in intact and triggered T-even phages of *E. coli.* (From Herskowitz, I. H.: Genetics. 2nd ed. Boston, Little, Brown and Company Inc., 1965.)

attachment of the virion to the bacterial cell that it parasitizes. Once it is attached (Figs. 6–9*A*, 6–10), enzymes at the tip of the tail create an opening in the bacterial cell wall and membrane. Through this opening the nucleic acid of the core passes into the infected bacterial cell. The protein head and tail remain outside, now an inert "ghost." Most animal viruses, unlike phage, appear to enter the cell by passage, intact, through the cell membrane; in many species this occurs by an ingestion or phagocytic process called *viropexis*. The tadpole shape of phages is especially adapted for injection of its nucleic acid core through the bacterial cell walls; it is not commonly found in viruses that infect animal cells, since animal cells do not have cell walls.

By a complicated series of effects the nucleic acid core of any bacteriophage or other virus, once inside a susceptible cell and free of its capsid, dominates the synthetic mechanisms of the cell. New virions are quickly replicated. Infecting viruses or bacteriophages eventually occupy the entire interior of the cell and cause cell rupture or "*lysis* from within." Not all viruses cause immediate death or lysis. Some infected animal cells continue to live for long periods, giving off new virions through the cell membrane. However, when liberated, the new virions attack new cells and continue the process unless inhibited by some extraneous agent, such as antibodies, drugs, or interferon (see page 307).

When multiplying intracellularly, the virions of many viruses tend to aggregate into regularly shaped translucent masses that resemble crystals (Fig. 6–6).

Lysogeny. The nucleic acid core of viruses is like a miniature chromosome. It contains all functional genes (the genome) of the virus and is infective by itself. It has been shown that the genome of the bacteriophage can remain intact in an infected bacterium, apparently as a part of the bacterial chromosome. It is said to have been "reduced" to the *prophage* state. In this state it replicates with the bacterial chromosome through an indefinite series of cell

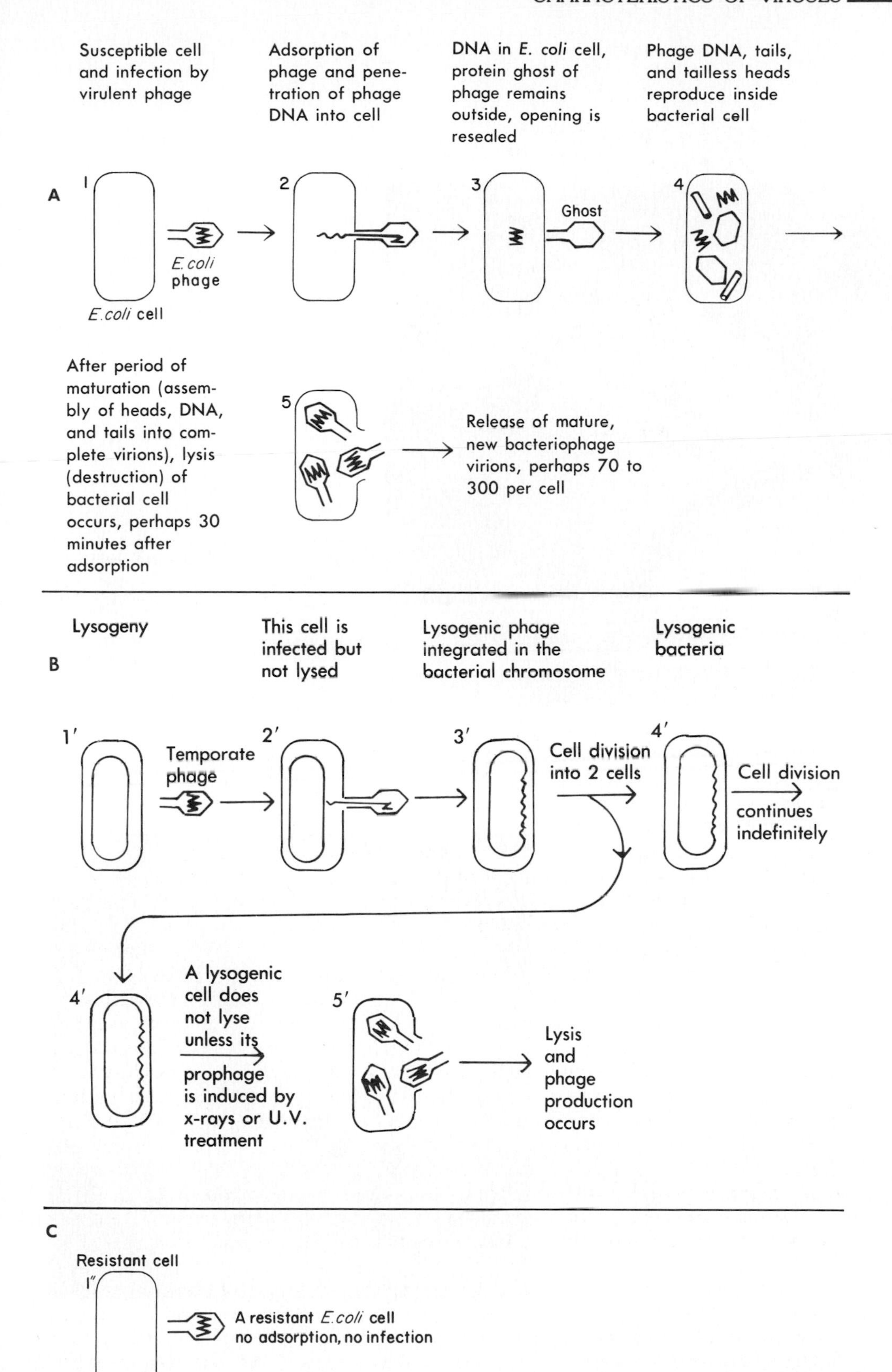
Susceptible cell and infection by virulent phage
Adsorption of phage and penetration of phage DNA into cell
DNA in E. coli cell, protein ghost of phage remains outside, opening is resealed
Phage DNA, tails, and tailless heads reproduce inside bacterial cell
A
1
2
3
4
Ghost
E. coli phage
E. coli cell
After period of maturation (assembly of heads, DNA, and tails into complete virions), lysis (destruction) of bacterial cell occurs, perhaps 30 minutes after adsorption
5
Release of mature, new bacteriophage virions, perhaps 70 to 300 per cell
Lysogeny
This cell is infected but not lysed
Lysogenic phage integrated in the bacterial chromosome
Lysogenic bacteria
B
1′
2′
3′
4′
Temporate phage
Cell division into 2 cells
Cell division
continues indefinitely
4′
A lysogenic cell does not lyse unless its
prophage is induced by x-rays or U.V. treatment
5′
Lysis and phage production occurs
C
Resistant cell
1″
A resistant E. coli cell no adsorption, no infection

Figure 6–9 *A*. Some bacterial cells are susceptible to phage and are destroyed (lysed). *B*, Lysogenic cells are infected, but their progeny are lysed only when exposed to an inducing agent, e.g., radiation. *C*, Resistant cells are not infected.

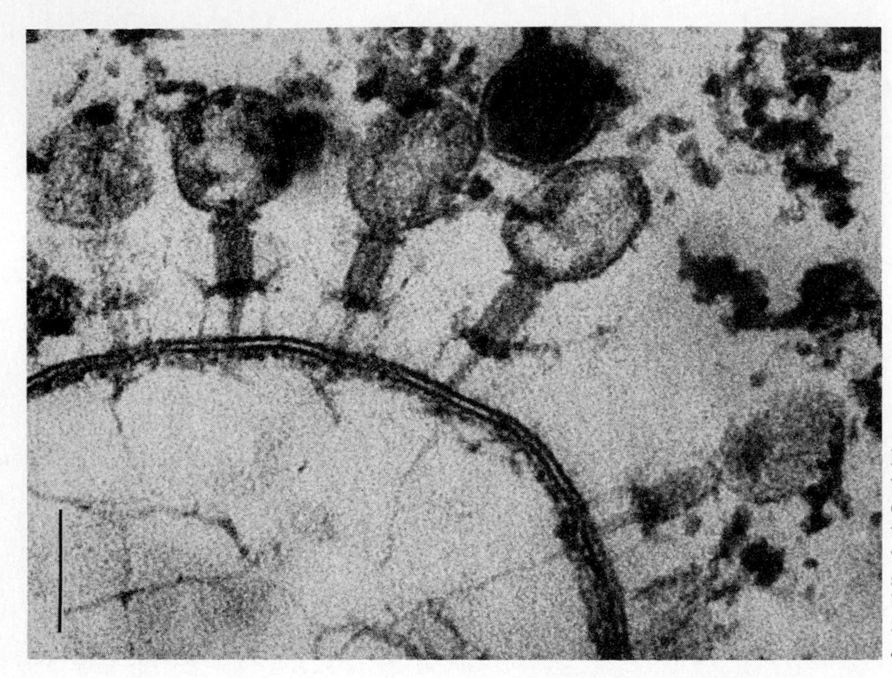

Figure 6–10 Several T4 phages attached to an *E. coli* cell wall. The cell wall is the circular double-membrane. The base plate, tail core, contracted sheath, and head of the virions are clearly visible. Inside the cell are strands that appear to be the DNA molecules being injected. (Bar is 100 mμm) (From Simon, L. D., and Anderson, T. F.: *Virology, 32*:279–297, 1967.)

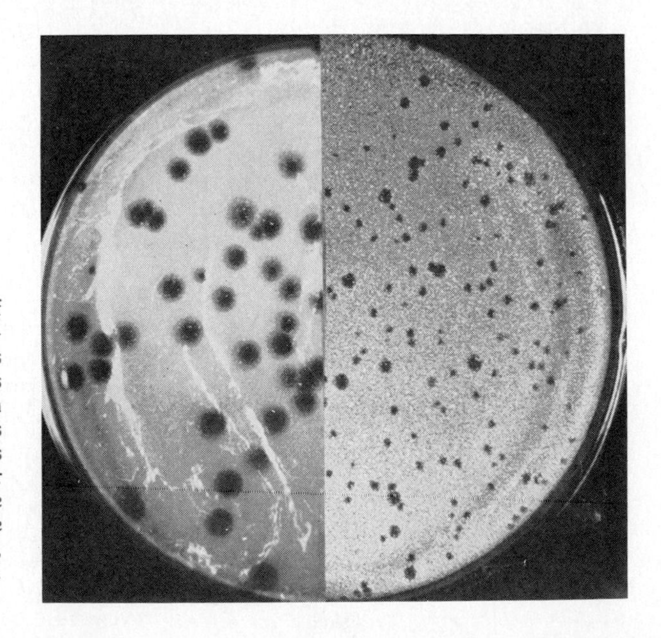

Figure 6–11 Bacteriophage plaques. On the surface of agar of appropriate composition was spread a culture of tubercle bacilli (*Mycobacterium tuberculosis*) mixed with bacteriophage specific for this organism. After incubation, the generalized growth of tubercle bacilli is seen as a whitish film. (Such growth of any bacteria on agar is often called a "lawn." The plaques of phage are seen as dark, circular holes in the "lawn.") Each plaque is, in effect, a colony of phage that has grown at the expense of the tubercle bacilli. Two different races or types of phage are shown here: a large-plaque and a small-plaque race. Different races of phages specific for other organisms show these and other colonial peculiarities. (Courtesy of S. Froman, D. W. Will, and E. Bogen, Olive View Sanatorium, Olive View, Calif. In *Amer. J. Pub. Health, 44*:1326, 1954.)

fissions, sometimes contributing striking new genetic traits (e.g., virulence [toxigenicity] of diphtheria bacilli and of *Clostridium botulinum* or food-poisoning bacilli)—a phenomenon called *viral conversion*—as though it were actually a part of the bacterium. Such a phage is said to be *temperate*. However, under certain circumstances (e.g., exposure to ultraviolet light, x-rays, or other *inducing agents*) it parts company with the bacterial genome, becomes actively *vegetative* and virulent, and causes lysis. A cell containing such a phage is said to be *lysogenic* (Fig. 6–9*B*). The same (or a similar) process appears to occur in some animal tumor virus infections (e.g., murine C-type viruses) and possibly also in herpes, though it has not been shown as clearly as for phage.

PLAQUE FORMATION BY BACTERIOPHAGE. If phage virions are mixed with a growing culture of susceptible bacteria in broth and then spread thinly on an agar surface and incubated, "colonies" of phage are seen as tiny holes (called *plaques*) in the ensuing growth (Fig. 6–11). Each plaque is in reality a colony of phage that has developed from a single virion that initially landed at that spot. The plaque contains lysed bacteria and billions of phage virions. If a sheet of mammalian cells is inoculated with an animal virus infective for those cells, readily visible and often very distinctive plaques are formed. Viral plaques are widely used in diagnostic and other virological investigations (see pages 147 to 148).

Virus Propagation Techniques In order to propagate viruses it is necessary to place them in contact with certain actively growing cells that they can *enter (infect)* and parasitize. They cannot grow independently outside living cells. Plant viruses require specific plant cells for growth; bacterial viruses are specific for certain bacteria; animal viruses are specific for certain animal cells. Three major techniques and types of living material are employed for the cultivation and study of animal viruses.

The Chick Embryo Method. The developing chick embryo (Fig. 5–3), the yolk sac, chorioallantoic membrane, amniotic cavity, or other structures, depending upon the particular virus to be propagated, may be inoculated directly with virus-containing material. This material may be saliva or an infected chick embryo that has been "homogenized" by being forced with the plunger through the opening in a syringe.

Growth of Viruses in Susceptible Hosts. In a second method, susceptible laboratory animals (mice, monkeys, rabbits) are employed. The route of inoculation, age, and genetic background of the animal all affect the growth of the virus.

Mammalian Cells in Culture. The third method is often referred to as *tissue culture*. Animal tissue cells grow readily as a *monolayer* (a "sheet" of cells one cell in thickness) in glass vessels of any desired size or shape. The cultured cells are then inoculated with virus (Fig. 6–12).

Extensive use has been made of living cells of various kinds—e.g., the HeLa strain (from human cervical cancer cells), L (mouse fibroblasts), and monkey kidney cell lines—for the propagation of animal viruses. Human tissues obtained as surgical by-products also provide a *medium* for the propagation of viruses. Although details are different, the essential components of all tissue cultures include a living cell system, a nutritional medium (nutrient solution) adequate for the maintenance and growth of cells, and antibiotics to control bacterial and fungal contaminants. Incubation is in a moist atmosphere rich in CO_2. Certain dyes are sometimes added to the media—e.g., neutral red, which stains living but not dead cells, or phenol red, which detects significant changes in pH.

Special methods of tissue culture (the work of Enders, Weller, Robbins, and others) have led to the preparation of polio (Sabin, Salk) and measles (Enders) vaccines. Numerous others (such as rabies and mumps) are now available. Animal and plant inoculations were once widely used in experimental and diagnostic virology. At present, except for special purposes, cell cultures have replaced live animals to a considerable extent.

Because viruses do not have any metabolic activities of their own that can

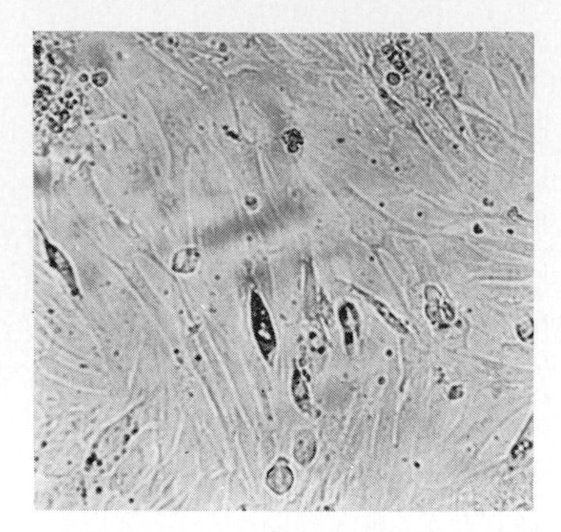

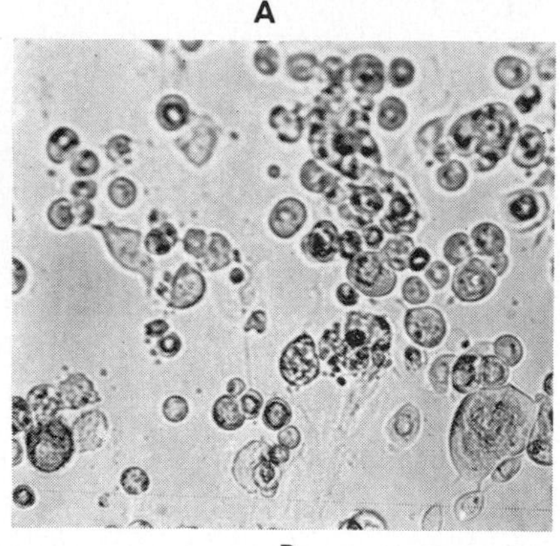

Figure 6–12 The cytopathic effect of a virus on human chorion cells in culture. (The chorion is one of the membranes protecting the embryo in the uterus.) *A,* Normal cells: tissue culture five days old showing a fairly regular, organized network of uniformly elongated, spindle-shaped tissue cells of rather distinctive appearance. *B,* The same sort of cells 72 hours after being inoculated with the virus of herpes simplex ("fever blister"). The cells are largely destroyed. Those remaining have lost their distinctive form and arrangement, have undergone granular rounding, and will eventually become mere shapeless, disconnected, disorganized blobs (×1,500). (Courtesy of Drs. A. M. Lerner, K. E. Takemoto, and A. Shelokov. In *Proc. Soc. Exp. Biol. Med., 95*:76, 1957. Copied by permission of The Society for Experimental Biology and Medicine.)

be demonstrated in the laboratory and multiply only in living cells, authorities do not consider them living organisms. Inside the cell they use cell materials and cause the cell's synthetic mechanisms to synthesize more virus instead of cell constituents. *Viruses are spoken of as active or inactive instead of as alive or dead.*

CYTOPATHIC EFFECT. When viruses grow, they cause damage in the cells: i.e., they are *pathogenic* for the cells. The effect of viruses on cells is said to be *cytopathic* (contraction of *cyto*, cell; *pathogenic*, producing disease in, or damage to). By the use of proper staining methods, the cytopathic changes produced by the viruses in individual cells may be seen and photographed (Fig. 6–12). The viruses themselves are visible only with the electron microscope.

A common response of the host cell to certain viral infections (such as smallpox, rabies, and measles) is the formation of rather prominent and distinct intracytoplasmic or intranuclear masses called *inclusion bodies*. These may be observed microscopically in infected tissues following fixation and staining. Some inclusions are aggregates or colonies of virus particles; others are probably sites of altered cellular metabolism. They may be intranuclear or cytoplasmic. Their morphology, location, and staining properties are often of great diagnostic value.

Heat Resistance. Nearly all viruses are readily inactivated by heat, most of them within 20 minutes by boiling (100 C). There are two exceptions that should always be remembered: the virus of infectious (epidemic) hepatitis ("catarrhal jaundice") and that of homologous serum jaundice (see page 414). Because of the heat resistance of these two closely similar viruses, autoclaving[2] (20 minutes at 121 C) or oven sterilization (90 minutes at 165 C) is the only safe procedure for materials likely to be contaminated with them. These materials include, especially, infectious blood or bloody fluids, blood-contami-

[2]A method of heating in steam under pressure.

nated instruments, and hypodermic syringes and needles (most of which are now disposable). In addition, for epidemic hepatitis, which is transmitted in feces, precautions must be used against sewage-contaminated water, contaminated foods, dishes, clinical thermometers used by patients who have the disease, and other substances discussed in greater detail in Chapters 12, 14, and 15. Numerous viruses that cause diseases in animals or humans are very fragile and do not remain active in body fluids, such as blood and sputum, for more than a few days at room temperature. Viruses that occur in the intestinal tract (e.g., poliomyelitis and epidemic viral hepatitis viruses) may survive for days in sewage. Curiously enough, if infected body fluids or tissues are rapidly dried, preserved with glycerin, or quickly frozen at extremely low temperatures, such as that of solid carbon dioxide ("dry ice," −76 C), even the most fragile viruses remain fully infective but in a static condition for many months, or even years.

Disinfectant Resistance. In general, viruses appear to be somewhat more resistant to some chemical disinfectants, such as phenol, glycerol, and bichloride of mercury, than are most bacteria. In this, however, there are as many exceptions and variations as there are among bacteria. It is especially noteworthy that viruses in general are not affected by therapeutic treatment with sulfonamide drugs or antibiotics. This is because most antibiotics act by inhibiting cellular synthetic or energy-yielding enzymes; viruses do not have any such enzymes to be affected.

DISEASES CAUSED BY VIRUSES

Viruses may be studied by direct observation of infected cells with the electron microscope, or by observing viral growth and its effects in cell cultures, as well as by chemical analyses and physical measurements. One of the most important methods of studying viruses is to observe their effects on cells and tissues of living hosts—bacterial, higher plant, invertebrate, or vertebrate. A virus is recognized primarily by its production of disease in a host or by its cytopathic effect on living cells in cultures (Fig. 6–12) as well as by the virus-specific *antibodies* that it evokes in the host animal.

Viral infections often fail to produce any *recognizable* effect on cells or tissues. Such infections are often called "silent," "subclinical," or "inapparent." They may be detected by tests for certain specific proteins (gamma globulins, *antibodies*) that appear in the blood in response to infections. Such tests are called *serological* or *immunological* tests (see Chapters 9 and 20).

Tissues Affected by Pathogenic Viruses. Five general and not very sharply defined groups of *zoopathogenic* viruses may be recognized on the basis of the body regions they affect pathologically. Table 6–2 shows this for the most common viral diseases in humans. Note that many of these are so-called "childhood diseases."

Dermotropic viruses chiefly cause visible lesions of skin and mucous membranes (epithelial tissues) and include the viruses of:

Smallpox
Measles
German measles (rubella)
Chickenpox
Herpes zoster (shingles)

Pneumotropic viruses cause respiratory diseases such as:

Influenza
Pneumonia or pneumonitis

Neurotropic viruses cause diseases of nervous tissues:

Poliomyelitis (infantile paralysis)
Rabies (hydrophobia)
Encephalitis

Viscerotropic viruses cause diseases that chiefly damage internal organs:

Yellow fever
Hepatitis

Text continued on page 94.

Table 6–2. Representative Viral Diseases in Humans

COMMON NAME OF DISEASE	TECHNICAL NAME OF DISEASE	VIRAL AGENT(S)	BODY REGIONS INVOLVED	MODE OF TRANSMISSION	INCUBATION PERIOD	CHIEF SYMPTOMS OR CHARACTERISTICS
Smallpox	Variola	3 Poxviruses, causing: variola major variola intermedia variola minor	Chiefly skin	Usually dust, droplet, or contact	7 to 17 days	High temperature, chills, headache, backache, and muscular pain; typical smallpox lesions on the body
Measles	Rubeola, epidemic roseola	Only one antigenic type of the paramyxoviruses	Chiefly skin	Dust, droplet, or contact	10 to 14 days	Similar to an ordinary cold; high temperature; spots in the throat; typical skin rash
German measles	Rubella	Togavirus	Chiefly skin	Dust, droplet, or contact	14 to 21 days	Mild symptoms of a cold; rash; enlargement of lymph nodes back of ear; complications; congenital rubella in fetus
Roseola infantum	Exanthema subitum	Togavirus	Chiefly skin	Dust, droplet, or contact	10 to 14 days	3 day rash in infants after high fever, convulsions; may be subclinical
Chickenpox	Varicella	Herpesviruses V-Z (varicella-zoster) virus (causes chickenpox in non-immune children and shingles in partially immune adults;	Chiefly skin	Dust, droplet, or contact	14 to 16 days	Mild symptoms of a cold; the rash may appear as successive crops of vesicles
Shingles	Herpes zoster		Skin and sensory nerves	Dust, droplet	7 to 14 days	Small vesicles surrounded by redness; fever and aches; rash may follow the skin area supplied by the sensory nerve
Fever blisters	Herpes simplex	HSV-1	Skin (nerve cells)	Dust, droplet, or contact		Sores on face or lips; often follow colds, fevers, or severe diseases
Genital herpes	Herpes virus genitalis	HSV-2	Genitalia	Sexual contact		Primary genital infection of vagina; cervical cancer
Mononucleosis ("kissing disease")	Infectious mononucleosis (IM)	EB herpesvirus, EBV_1 (Epstein-Barr virus)	Glands and lymphatic system	Droplet, oral contact	33 to 49 days	High temperature, headache, fatigue, chills, anorexia, sore throat, mild hepatitis, lymphocytosis, CNS involvement
Mumps	Contagious parotitis (parotitis epidemica)	Only one antigenic type A paramyxovirus	Salivary glands and often - reproductive organs; occasionally central nervous system	Droplet	16 to 21 days	Mild symptoms in children, more severe in adults; swelling of the salivary gland; may also involve the testes or ovaries, or the CNS

Warts	Verruca	Papillomavirus	Skin	Uncertain, probably contact	4 weeks to 6 months	Appearance of the characteristic wart-shaped bodies on the skin surface
Flu	Influenza	Types A, B, and C of orthomyxoviruses with subtypes of A	Respiratory tract, including the lungs	Droplet and direct personal contact	1 to 3 days	Fever; aching in muscles of back, arms, and legs; headache; chest pains, which may be complicated by pneumonic type of disease; an epidemic disease
Parainfluenza group (infantile)	Parainfluenza; α respiratory syncytial virus infection	Paramyxoviruses: 4 types of PIV, 1 type of RSV	Entire respiratory tract	Droplet and direct personal contact		From a common cold to croup to laryngotracheobronchitis to pneumonia; most important pathogen for infants and children
Infantile paralysis (polio)	Poliomyelitis	3 Types of polioviruses (belong to the enteroviruses)	Spinal cord and brain	Dust, droplet, food, feces	7 to 21 days	Mild sore throat with respiratory coldlike symptoms followed by all degrees of paralysis—from none to fatal
Hydrophobia	Rabies	Rhabdoviruses	Brain and nerves	Usually bite of rabid animal	10 days to 8 months	Headache, difficulty in swallowing, convulsions, paralysis, violent spells; survival rare
Sleeping sickness (North American)	Equine encephalomyelitis	Group A arbovirus, A alphavirus, several types of togaviruses	Nerves, brain, meninges, and blood	Bites of several types of mosquitoes	4 to 21 days	High fever, convulsions, vomiting, drowsiness, coma, and muscle twitchings
Yellow fever	Yellow fever	Group B arbovirus, A flavovirus, togaviruses	Various internal organs	*Aedes aegypti* mosquito	3 to 6 days	Fever, abdominal pain, vomiting of blood; delirium and prostration; yellow coloration of the skin (jaundice)
Infectious hepatitis or catarrhal jaundice	Epidemic viral hepatitis (IH)	Hepatitis A virus (HAV)	Liver, spleen, and lymph nodes	Food, milk, and water (feces-polluted)	15 to 40 days	Fever, weakness, nausea, vomiting, abdominal cramps; jaundice often is present; enlarged and tender liver
Serum hepatitis	Homologous serum hepatitis (SH)	Hepatitis B virus (HBV) (3 types exist)	Liver, spleen, lymph nodes, and blood	Blood; perhaps the oral route (?)	40 to 180 days	As in infectious hepatitis, but longer incubation and slow onset

The common cold may be caused by any one of 90 types of rhinoviruses; one type appears to be more significant. However, many other viruses may induce clinically similar illnesses.

Table 6–3. Some Properties of RNA-Containing Tumor Viruses (Oncornaviruses) in Animals

VIRUS	ABBREVIATIONS USED	HOST OF ORIGIN	NATURAL TUMORS (HOST OF ORIGIN)	EXPERIMENTAL HOST RANGE		SIZE (nm)	MORPHOLOGY (C OR B TYPE PARTICLES)‡	SITE OF VIRUS MATURATION	PERSISTENCE OF INFECTIOUS VIRUS IN TUMOR
				In Vivo Tumor	*In Vitro Cell Transformation*				
Avian complex									
Leukemia	ALV*	Chicken	Yes	Chicken, turkey	Chicken†		C		
Sarcoma (Rous)	RSV			Avian, rodent, monkey	Avian, rodent, bovine, monkey, human				
Murine complex									
Leukemia	MuLV	Mouse	Yes	Mouse, rat, hamster			C		
Sarcoma	MSV		No	Mouse, rat, hamster	Mouse, rat, hamster				
Murine mammary tumor (Bittner)	MTV	Mouse	Yes	Mouse			B		
Feline complex									
Leukemia	FeLV	Cat	Yes	Cat		70–100	C	Budding at cell membrane	Yes
Sarcoma	FeSV			Cat, dog, rabbit, monkey	Cat, dog, monkey, human				
Other									
Viper		Viper	Yes				C		
Hamster, leukemia?	HaLV	Hamster					C		
Rat, leukemia?	RaLV	Rat					C		
Guinea pig, leukemia		Guinea pig	Yes				C		
Bovine, lymphoma		Cow	Yes				C		
Primate									
Sarcoma, woolly monkey		Monkey	Yes	Monkey	Monkey		C		
Lymphosarcoma, gibbon		Ape	Yes				C		
Mammary carcinoma, monkey (Mason-Pfizer)	M-MPV	Monkey			Monkey		B?		

*The term RAV has been used for ALV strains associated with the defective Bryan strain of RSV (BH-RSV).
†Attained with avian myeloblastosis virus only.
‡"C type" and "B type" viruses differ from each other in certain morphologic, antigenic, and enzymatic properties.

Some Properties of DNA-Containing Tumor Viruses

VIRUS	HOST OF ORIGIN	NATURAL TUMORS (HOST OF ORIGIN)	EXPERIMENTAL HOST RANGE		SIZE (nm)	STRUCTURE	SITE OF VIRUS MATURATION	PERSISTENCE OF INFECTIOUS VIRUS IN TUMOR
			In Vivo Tumors	*In Vitro Cell Transformation*				
Papovaviruses								
Papilloma:								
Human	Human	Yes	Human					Yes
Rabbit	Rabbit	Yes	Rabbit		40–55			
Bovine	Cow	Yes	Cow	Bovine				
Canine	Dog	Yes	Dog					
Polyoma	Mouse	No	Mouse, hamster, other rodents	Mouse, hamster, rat				
SV40	Monkey	No	Hamster	Hamster, mouse, monkey, human		Icosahedral symmetry	Cell nucleus	No
Adenoviruses								
Human types 3, 7, 11, 12, 14, 16, 18, 21, 31	Human	No			70–80	Icosahedral symmetry	Cell nucleus	No
Simian (some)	Monkey	No	Hamster, rat, mouse	Hamster, rat, human				
Bovine type 3	Cow	No						
Avian (CELO)	Chicken	No						
Herpesviruses					100	Icosahedral symmetry	Cell nucleus	No
Human:								
Type 2	Human		Hamster	Hamster				
EB virus	Human		Monkey	Human, monkey				
Monkey (Melendez, lymphoma)	Monkey	No	Monkey					
Avian (Marek, neurolymphoma)	Chicken	Yes	Chicken					
Frog (Lucké, carcinoma)	Frog	Yes	Frog					
Rabbit (Hinze, lymphoma)	Rabbit	No	Rabbit					
Poxviruses					230 × 300	Complex symmetry	Cell cytoplasma	Yes
Molluscum contagiosum	Human	Yes	Human					
Yaba	Monkey	Yes	Monkey					
Fibroma-myxoma	Rabbit, squirrel, deer	Yes	Rabbit, squirrel, deer					

(Reproduced, with permission, from Jawetz, E., Melnick, J. L., and Adelberg, E. A.: Review of Medical Microbiology. 11th ed. Los Altos, Calif., Lange Medical Publishers, 1974.)

Enterotropic viruses, in the fifth group, are those that multiply primarily in the intestinal tract. Among these would be included poliomyelitis viruses, echovirus (Enteric Cytopathogenic Human Orphan), ECBO (Enteric Cytopathogenic Bovine Orphan), NITA (Nuclear Inclusion Type A), reovirus (Respiratory Enteric Orphan), and coxsackieviruses. These are all discussed in Chapter 25.

This classification, although convenient for purposes of reference, is very imperfect and can be misleading, since several viruses cause more than one type of disease and may be localized in several kinds of tissue, besides changing quite unexpectedly. For example, the virus causing "shingles" or herpes zoster, which involves dorsal nerve roots, is a neurotropic modification of one of the "dermotropic" viruses, that of chickenpox.

Oncogenic Viruses

Of about 600 known viruses of vertebrates, about 150 have *oncogenic potential*—i.e., the capacity to initiate cellular changes in animals that lead to the development of tumors. (Cancers are a particular type of tumor.) All tumor viruses have at least one property in common: they can induce tumors when inoculated into susceptible animals. Otherwise they may differ greatly.

RNA tumor viruses, causing neoplastic (abnormal new growth of tissue) disease or solid tumors, have been found in reptiles, birds, and mammals (Table 6–3). These viruses can be transmitted as part of the genome of the vertebrate animal to its progeny. Many *natural* leukemias and sarcomas (special types of tumor) in animal species are caused by oncogenic RNA viruses. The feline leukemia virus (FeLV) was isolated from cats with lymphosarcoma, the most common form of neoplasm in these animals. Kittens inoculated with FeLV develop leukemia. Although FeLV can replicate in human cells in cultures, this alone does not prove that FeLV causes human leukemias.

One variety of mammary cancer in mice has been shown to result from interaction between viral and host genetic factors. The milk factor (the virus) is transmitted *only* in the milk and *only* to genetically susceptible offspring. Human milk contains virus particles that are morphologically similar to those found in mammary tumors of mice. Yet there is no direct evidence that human breast cancer is a virus-induced disease, although there are strong suggestions that this may be so.

DNA tumor viruses are found in the group of papovaviruses, adenoviruses, poxviruses, or, in their natural hosts, malignant tumor–causing herpesviruses (Table 6–3). Herpesviruses are involved in several human diseases. One type of herpesvirus, called the Epstein-Barr virus (EBV), is the cause of infectious mononucleosis. Fortunately this disease is self-limiting: the patient recovers completely. Nevertheless, it strongly resembles an early tumor (lymphoma). Burkitt's lymphoma which is thought to be caused by the EBV, occurs in about half of all children with cancer in Africa, but the EBV etiology of this disease has not been accepted by all workers in the field. Nasopharyngeal cancer of humans may also be caused by the EBV: DNA of this virus has been found in biopsies from patients with nasopharyngeal carcinoma.

Herpes simplex virus type 2 (HSV-2) has been implicated as a possible etiological agent of cervical carcinoma. Occurrence of the disease is associated with sexual promiscuity, but evidence for the role of the virus in this malignancy is circumstantial.

Despite all these studies and others, the viral etiology of human neoplasms (except warts) still has not been demonstrated. Many clinical, pathologic, and epidemiologic similarities exist between human tumors and virus-induced tumors in lower animals. Some human viruses (e.g., adenoviruses) have been used to produce experimental tumors in animals, and some animal tumor viruses have been found to be antigenically related to some human viruses. The problem is very simple but very frustrating. Viruses found in tumors are not necessarily causing them, and tumors free of viruses could have been induced by a virus that is no longer present in the neoplasm. It is difficult to believe that viruses are the cause of malignant tumors in animals but not of similar diseases in humans. The search is on for the *possible* etiology of many neoplasms in humans, and will continue until it is found for all.

Slow Viruses. Certain *"chronic,"* degenerative disorders have been traced to the action of *slow viruses,* whose incubation may involve a long latency between infection and the appearance of clinical symptoms, sometimes as long as 20 years. Among these are kuru, a CNS degenerative disease leading to total paralysis and death, found in cannibals of New Guinea; a related disorder, Creutzfeldt-Jakob disease, a subacute spongiform encephalopathy; subacute sclerosing panencephalitis[3]; progressive multifocal leukoencephalography; and subacute encephalitides caused by herpes viruses. It has been suggested that multiple sclerosis[3] (MS) may be a slow virus disease, without any evidence as to its etiology, and Parkinson's disease also seems a possible candidate. A viral origin has also been proposed for Guillain-Barré syndrome with the suggestion that it is caused by a latent influenza virus. No proof for this exists. Is it possible that slow viruses could cause or contribute to the "normal" aging process? Could we prolong life by preventing viral diseases?

Modifications of Viruses

Zoologic Modification. An important property of viruses is their ability to adapt themselves to animal hosts other than those that they infect under natural conditions. This is sometimes called *"zoologic modification."* The virus of rabies may, as shown by Pasteur, be passed artificially from rabbit brain to rabbit brain so that it becomes extremely pathogenic for these animals; yet, in becoming so adapted, it loses some of its virulence for human beings. The virus of fox distemper may be so modified by repeated transmission in ferrets that, although highly fatal for ferrets, it produces only a very mild disease in foxes and is used successfully in large silver-fox ranches in immunizing the animals against the disease. These "adaptations" probably represent the *selective propagation* of mutants in the viral population best fitted to grow under the altered conditions. As will be described, smallpox vaccination is probably based on zoologic modification.

Histologic Modification. This occurs when a virus is continually induced to infect a tissue that it does not naturally invade. For example, the yellow fever virus, normally infecting the blood, liver, and other viscera (viscerotropic), can be completely altered by injecting it into the brains of mice or other rodents. It becomes neurotropic and may not cause any visible infection if injected into the blood, and may be used for immunization purposes. An almost total loss of both viscerotropic and neurotropic properties will occur if it is cultivated in cell cultures as described previously or passed through chick embryos from egg to egg. The highly successful 17D yellow fever vaccine, now widely used to immunize persons traveling in yellow fever areas, is based on this principle. The Sabin oral vaccine against poliomyelitis is an infective, active-virus vaccine of low virulence for humans, developed in cell cultures of the poliovirus. Salk vaccine is inactivated (noninfective) poliovirus.

---

[3]Other investigators proposed that subacute sclerosing panencephalitis and multiple sclerosis may be due to "latent" measle viruses, differentiating these from "slow" viruses.

Viral Variation and Vaccines. Like the antigens (e.g., proteins) of different animals and plants (even antigens of closely related species of individuals), the antigens of viruses (even of those that cause the same types of disease) are often very different antigenically. If one vaccinates a group of people with influenza vaccine, one may be surprised and disappointed when many of the vaccinated people succumb to influenza during an epidemic. If it is shown, however, that the vaccine virus was type A and that the epidemic virus was type B, then one realizes the important fact that viral vaccines must include all the antigenic possibilities of the particular virus group involved. Influenza vaccine must contain influenza viruses of types A, B, C, and subgroups, and so on; polio vaccines must immunize against poliovirus I, II, III, and so on.

VIROIDS

Viroids are even smaller than the smallest nucleic acid isolated from viruses. They are a group of pathogenic agents causing disease in plants, perhaps also in animals. Viroids appear to be circular single-stranded RNA—*only* the RNA located in the nucleus of their host cells. Viroids cause disease in citruses, chrysanthemums, and spindle tubers, and possibly animal diseases that have not yet been diagnosed as being of viroid etiology.

UNUSUAL VIRUSES

These unusual pathogens are difficult to inactivate, and they are without detectable nucleic acid. Scrapie in sheep fits this classification. Perhaps some of the slow viral diseases are caused by unusual viruses or by viroids. Further investigations may clear up the confusion among these three groups of viral agents.

SUMMARY

Viruses resemble rickettsias and chlamydias in not being cultivable outside of living cells. They are obligate intracellular parasites that can only be seen by means of an electron microscope. The infective virus particle is called the virion. It may consist of the nucleocapsid (nucleic acid and protein shell) only or the nucleocapsid and envelope (protein membrane, lipid, and glycoprotein surface structures).

Some viruses infect plants; others infect animals; others are pathogens of algae or of mycoplasmas; and bacteriophages destroy bacteria.

Cryptograms provide a simple way to identify and code viruses by means of virion shape, size, DNA or RNA content, weight of nucleic acid, and vectors of transmission. Viruses have either single- or double-stranded RNA or DNA; none or only trace amounts of the other nucleic acid may be found.

One of the largest viruses is the smallpox virus; one of the smallest is the yellow fever virus. Zoopathogenic viruses may have various types of polyhedral shapes: icosahedral, cubic, spherical, ovoid, bullet-shaped, or helical symmetry. They may be naked or enveloped.

Bacteriophages, also called bacterial viruses or simply phages, are commonly tadpole-shaped and of binal symmetry. Over 672 types of phages have been listed, presumably only a fraction of those that must exist. They may easily be isolated from bacteria in sewage. They either cause lysis of a susceptible bacterial culture or become prophage in lysogenic bacteria, but they may not affect resistant cells. Lysogeny clearly shows us how viruses may change the genetic makeup of host cells.

Viruses may be grown in intact animals, in mammalian cells in culture (tissue cultures), and in chick embryos. Animal viruses grow in animal cells in monolayers in flasks or dishes, and bacteriophages form plaques on bacterial lawns or agar surfaces in plates.

Many childhood diseases are caused by viruses, such as chickenpox, measles, and smallpox. Other human viral diseases are influenza, hydrophobia (rabies), poliomyelitis, herpes, and the common cold. The reader is urged to study Table 6–2 listing common viral diseases.

Oncogenic viruses have the potential to cause the development of tumors. Many such viruses have been found in animals, but so far no "cancer-causing" virus has been proven in humans. Viruses found in tumors have not necessarily caused them, and tumors free of viruses could have been induced by a virus that is no longer present in the neoplasm.

The ability of viruses to adapt themselves to animal hosts other than those they infect under natural conditions is referred to as zoologic modification. Histologic modification occurs when a virus is continually induced to infect a tissue that it does not naturally invade.

Some pathogenic viruses have been designated as *slow viruses* causing chronic degenerative disorders with prolonged latency, often years, between infection and clinical symptoms; others, called *viroids*, are very small RNA viruses that are plant pathogens; and others called *unusual viruses* are difficult to inactivate and are without detectable nucleic acid.

REVIEW QUESTIONS

1. Why are viruses called ultramicroscopic?
2. Explain the terms icosohedral and cuboidal.
3. Can you draw a virion and list its substructures?
4. Can you draw a bacteriophage and list its substructures?
5. How may viruses be classified? What can you tell about the nucleic acid content of viruses?
6. What organisms may be infected by viruses?
7. Explain how a cryptogram works.
8. What are the relative sizes of rickettsias, chlamydias, bacteria, and different types of viruses?
9. Explain the term filterable viruses.
10. Explain lysogeny, prophage, phage sensitivity, phage resistance, and the reproduction of some viruses.
11. What are plaques? What is a monolayer? What are mammalian cells in culture? What can you tell about HeLa cells?
12. Discuss heat resistance of some viruses and resistance to disinfectants.
13. What tissues may be affected by pathogenic viruses? Why?
14. Name some diseases caused by viruses. Associate them with certain tissues.
15. What is oncogenic potential? What is known about viruses in relation to cancer?
16. Define: slow viruses, zoologic modification, viral variation, viroids, unusual viruses.
17. Briefly compare viruses with bacteria. How are they simular? How do they differ?

Supplementary Reading

Allen, D. W., and Cole, P.: Viruses and human cancer. *N. Eng. J. Med., 286*:70, 1972.

Andrews, C., et al.: Viruses of Vertebrates. 4th ed. New York, Macmillan, 1978. The Williams Wilkins Co., 1972.

Braun, W.: Bacterial Genetics. 2nd ed. Philadelphia, W. B. Saunders Company, 1965.

Debré, R., and Celers, J. (eds.): Clinical Virology: The Evaluation and Management of Human Viral Infections. Philadelphia, W. B. Saunders Company, 1970.

Douglas, J.: Bacteriophages. London, Chapman and Hall, 1975.

Drew, W. L.: Viral Infections: A Clinical Approach. Philadelphia, F. A. Davis Co., 1976.

Dulbecco, R.: Cell transformation by viruses and the role of viruses in cancer. *J. Gen. Microbiol., 79*:7, 1973.

Eklund, M. W., Poysky, F. T., Reed, S. M., and Smith, C. A.: Bacteriophage and the toxigenicity of *Clostridium botulinum* type C. *Science, 172*:480, 1971.

Fenner, F.: The genetics of animal viruses. *Ann. Rev. Microbiol., 24*:297, 1970.

Fenner, F., McAuslan, B. R., Mims, C. A., Sambrook, J., and White, D. O.: The Biology of Animal Viruses. 2nd ed. New York, Academic Press, 1974.

Fraenkel-Conrat, H. L.: Structure and Assembly: Virions, Pseudovirions, and Intraviral Nucleic Acids. New York, Plenum Press, 1975.

Goodheart, C. R.: An Introduction to Virology. Philadelphia, W. B. Saunders Company, 1969.

Hanna, M. G., and Rapp, F.: Immunobiology of Oncogenic Viruses. New York, Plenum Press, 1977.

Haynes, R. E.: Clinically Distinguishable Syndromes Caused by Viruses. Chicago, Year Book Medical Publishers, 1975.

Jawetz, E., et al.: Review of Medical Microbiology. 14th rev. ed. Los Altos, California, Lange Medical Publications, 1980.

Joklik, W., et al.: Zinsser Microbiology. 17th ed. New York, Appleton-Century-Crofts, 1980.

Kaper, J. M. (ed.): The Chemical Basis of Virus Structure, Dissociation and Reassembly. New York, American Elsevier Publishing Co., 1975.

Knight, C. A.: Chemistry of Viruses. 2nd ed. New York, Springer-Verlag, 1975.

Kurstak, E., and Kurstak, C.: Comparative Diagnosis of Viral Diseases. New York, Academic Press, 1977–1978.

Laskin, A., and Lechevalier, H. (eds.): Handbook of Microbiology. Vol. 1, Bacteria. 2nd ed. Cleveland, CRC Press, 1977.

Lwoff, A., and Tournier, P.: *In* Maramorosch, K., and Kurstak, E., (eds.): Comparative Virology. New York, Academic Press, 1971, pp. 1–42.

Maramorosch, L. (ed.): Viruses, Vectors, and Vegetation. New York, Interscience Publications, John Wiley & Sons, 1969.

Matthews, R. E. (ed.): Classification and Nomenclature of Viruses: Third Report of the International Committee on Taxonomy of Viruses. New York, S. Karger, 1980.

McAllister, R. M.: Viruses in human carcinogenesis. *Progr. Med. Virol., 16*:48, 1973.

Melnick, J. L. (ed.): Progress in Medical Virology. (Annual Summaries on Viral Classification and Nomenclature.) New York, S. Karger, 1970–80.

Pearson, G. R.: Epstein-Barr virus immunology. In Klein, G. (ed.): Viral Oncology. New York, Raven Press, 1980.

Rapp, F.: Defective DNA animal viruses. *Ann. Rev. Microbiol., 23*:293, 1969.

Robert, L., et al. (eds.): Burkitt Lymphoma, Hemostasis and Intercellular Matrix. New York, S. Karger, 1976.

Smith, K. M., and Ritchie, D. A.: Introduction to Virology. London and New York, Chapman and Hall, 1980.

Sprunt, K., Redman, W. M., and Alexander, H. E.: Infectious ribonucleic acid derived from enteroviruses. *Proc. Soc. Exp. Biol. Med., 101*:604, 1959.

Stent, G. S.: Molecular Biology of Bacterial Viruses. San Francisco, W. H. Freeman & Co., 1963.

Symposium on Tumor Viruses. Cold Spring Harbor, N.Y., Cold Spring Harbor Laboratory, 1975.
Tooze, J. (ed.): The Molecular Biology of Tumor Viruses. 2nd ed. Cold Spring Harbor, N.Y., Cold Spring Harbor Laboratory, 1980.
Zinder, N. D. (ed.): RNA Phages. Cold Spring Harbor, N.Y., Cold Spring Harbor Laboratory, 1975.

SECTION TWO

MICROORGANISMS AND THE WORLD AT LARGE

Florence Nightingale (1820–1910), frequently credited with the beginning of professional nursing, quickly recognized the importance of sanitation in the prevention and control of disease. When she first arrived at the overcrowded and filthy Barrack Hospital during the Crimean War, the mortality rate was 60 per cent; when she left, it was just over 1 per cent. (From Dolan, J. G.: Nursing in Society. 15th ed. Philadelphia, W. B. Saunders Company, 1983.)

Ferdinand Cohn (1828–1898). At the time of Pasteur's discoveries, Cohn was one of the foremost microbiologists in Germany. He extended Pasteur's findings in his own laboratory, where he worked with algae and fungi and wrote some of the earliest books about bacteria. Following Müller, he classified bacteria into genera and species. Cohn also worked with Robert Koch, who contributed to the evidence needed to prove the "germ theory of disease." (From Wedberg, S. E.: Introduction to Microbiology. New York, Van Nostrand Reinhold Company, 1966.)

GENERAL BACTERIAL PHYSIOLOGY

THE BACTERIA (KINGDOM PROCARYOTAE)

Seventeen years of progressive research following the publication of the seventh (1957) edition of *Bergey's Manual of Determinative Bacteriology* compelled experts in the field of bacterial classification to publish an entirely revised eighth edition in 1974. According to those experts, "All previous classifications seem to have suffered infinite rearrangement due to insufficient information," and we can probably look forward to further reclassification of bacteria in the future (see Chapter 2).

Bacteria are prokaryotic (sometimes spelled procaryotic) microorganisms; some are *autotrophic* and some are *heterotrophic*. Autotrophic organisms are capable of synthesizing their own organic constituents if water, carbon dioxide, inorganic salts, and a source of energy are available. Heterotrophic organisms require complex organic substances called growth factors (e.g., vitamins) for nutrition. Some bacteria are photosynthetic, but bacterial photosynthesis does not produce oxygen (as does that of the blue-green algae[1] and, of course, the green plants). The photosynthetic pigments in some bacteria differ from the chlorophyll of eukaryotic plants and are called bacteriochlorophylls *a, b, c,* or *d.* Most species of bacteria can grow well in the dark, and (like the fungi, including yeasts and molds) they thrive on lifeless, chemically defined, or other inert food materials contained in culture media. Viruses, chlamydias, and rickettsias require living cells for multiplication.

It is not easy for the beginner to realize the importance and complexity of bacteria: many typical species are only 1/25,400 (0.000039) inch (1 μm) in diameter (See Fig. 6–6). It is also hard to understand how bacteria can live and multiply, often without sunlight or air, and survive for years at temperatures hundreds of degrees below zero without water or food! Bacteria, moreover, through mutation and natural selection in different environments, have become adapted to life almost everywhere on the earth. Some can thrive in and at the bottoms of the seas or in and on plants and animals, and can grow and multiply prodigiously, as they have done for millions of years. Not all species live everywhere, however: some are rigidly restricted to certain environments, such as marine depths or mucous membranes of the human body.

Physiology of Bacteria

"An unexpected knowledge of the secret life energies of bacteria has been revealed, through which they rule with demoniacal power over weal and woe, and even over the life and death of man."

Ferdinand Cohn
1828–1898

---

[1]Blue-green algae are now considered to be bacteria. However, since they do not cause diseases, they will not be discussed further in this book.

CHAPTER 7

Metabolism

How Bacteria "Eat." Since even the most complex unicellular organisms (protozoa, eukaryotic algae) have only a cell membrane (and perhaps flagella, slime coating, and a cell wall), nuclear structures, and cytoplasm containing chloroplasts, mitochondria, enzyme granules, granules of stored food, fat, and so forth, it may seem strange that they are able to eat and breathe. In the case of bacteria, which are among the simplest of unicellular organisms, eating involves no taste, chewing, swallowing, or other muscular or nerve activity. The food of bacteria (like that of higher plant cells) must be in solution, that is, dissolved in water. When bacteria are floating in solutions such as blood, beef broth, sewage, sea water, or water in the soil, the substances that they use as food pass into them through the cell wall and membrane. This passage of water through a thin membrane is called *diffusion* or *osmosis.* The membrane is permeable to some but not all of the dissolved solutes. Dissolved waste products in turn pass outward through the cell membrane and cell wall. Cell membranes seem to be permeable only to the right substances. Such a membrane is said to be *semipermeable* or *selectively permeable.* It allows only food substances to pass inward and waste products to pass outward. All substances passing through the cell wall and membrane are presumably in a soluble, not solid, form. Let us not forget the possible exchange of gases; but this is another story, to be discussed later.

If the volume of fluid in which the cells are suspended is small, as in a test tube or a tiny pool, the cells are soon surrounded by a mixture of food substances and waste products, all in solution. The food may then give out, or the accumulation of waste products may halt the growth of the cells, or both factors may affect the cell population.

Nutrition solely by passage of foods dissolved in water through the cell membrane (and cell wall if present) is often called an *osmotrophic* mode of nutrition.

Enzymes. Although all plant and many animal cells are nourished only by soluble substances that are capable of passing through the cell wall or cell membrane (or both) from the surrounding fluids, many kinds of cells, including numerous species of bacteria, can make nutritive use of various solid substances that are neither in solution nor able to pass through the cell membrane. Thus, not only fats, starches, and proteins, but the shells of crabs, the trunks of trees, old rubber tires, petroleum, even (alas!) the paper on which this book is printed may ultimately serve as food for certain bacteria and other microorganisms of the soil or water that can "digest" or, more correctly, *hydrolyze*[2] (and thus solubilize) them. This hydrolysis or "extraneous digestion" is accomplished by means of *enzymes,* or digestive chemicals, that the bacteria (and other cells) secrete (see Fig. 7–1).

---

[2]From Greek, *hydor,* water, and *lysis,* dissolution; a phase of enzymatic digestion in which complex molecules such as starch, protein, or fat are split into simpler molecules such as glucose, amino acids, or fatty acids. Molecules of water are enzymatically introduced into the molecules of starch, protein, or fat and then split, producing simpler molecules that are small enough to pass through the cell wall and cell membrane and can be used as food.

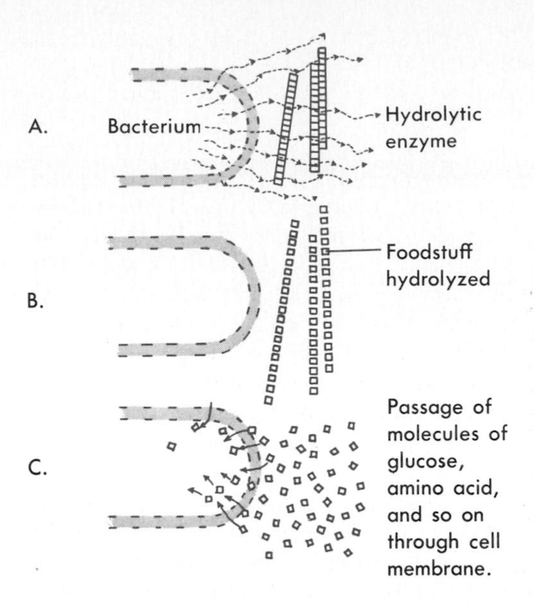

Figure 7–1 Utilization of grossly large and insoluble food particles by extracellular digestion. Sketch *A* shows the bacterium secreting hydrolytic enzyme, which attacks the food particle (cellulose, chitin, starch, protein, fat). In *B* the foodstuff is hydrolyzed and its constituent molecules (glucose, amino acids, fatty acids, etc.) are separated from one another. In *C* the molecules derived by hydrolysis of the food pass (or are actively passed by permease enzymes) readily through the bacterial cell wall and cell membrane into the cytoplasm to be further metabolized.

Enzymes are protein complexes that bring about chemical changes in a great variety of substances. Enzymes have the remarkable power of *catalysis,* that is, of bringing other substances together and greatly speeding up chemical reactions between them, much as a hostess brings a group of people together, introduces them, and persuades them to play bridge—she catalyzes a party.

In general, enzymes are not destroyed in the catalytic process. Thus, very small quantities of enzymes will bring about extensive and continuous reactions. There are enzymes that can so change gelatin that it becomes permanently liquid and will no longer "set." Others cause sugar, water, and oxygen to react so that alcohol and carbon dioxide gas result, yielding energy to the cell. Common bakers' yeast gives off enzymes that do this, and the process is used in making beer, wine, and bread. Glandular cells in the stomach, the intestines, and the pancreas produce hydrolytic enzymes such as pepsin, trypsin, amylase, and lipase, which break down our foods to simple, soluble substances by catalyzing their combination with water (hydrolysis). The products of digestive hydrolysis in our stomachs are absorbed through the cells lining our intestines, eventually enter the circulating blood and lymph, and are used for energy or body substance.

Exo-enzymes. Enzymes that pass outward from the cell are called *exo-enzymes* (*exo,* outside of). These form physicochemical combinations with a great variety of substances, bringing about changes so that even the hard, chemically complex structures referred to previously are reduced to simple substances that dissolve rapidly and pass inward through the walls and membranes of living cells, there to be used as sources of energy and protoplasm (Fig. 7–1). Almost identical digestive processes go on in our own stomachs and enable us to assimilate tough old steaks, popcorn, and raw carrots. Actually, since the cells lining our mouth and digestive system are of the same lineage (ectoderm) as those that become our skin, we also digest on our "outside" (gastrointestinal tract). It is relatively easy to clean out the contents of the G.I. tract from either end, except when we have absorbed toxins (poisons) or bacterial pathogens from the G.I. tract into our bloodstream.

Endo-enzymes. Some enzymes are part of the cells themselves and are not released into the surrounding fluids. They are called *endo-enzymes,* since they remain inside of or on the cell. They are mainly involved in cell synthesis and in liberating energy from foodstuffs and making it available for the cell's use in moving and living. Probably most soluble proteins in a cell are active enzymes. The permeation of food substances through cell wall and membrane by "active transport" due to inducible transport systems was previously attributed to *permease enzymes*. However, since several proteins are involved in this process and since the substrate is not changed enzymatically, the term permease is not considered proper usage anymore.

Specificity of Enzymes. There are probably thousands of different enzymes, each of which catalyzes a single, specific kind of chemical reaction with certain specific substances (called *substrates*) and not with others. Enzymes are therefore said to be *specific* in their action. Specificity may conveniently be thought of as resembling the relationship between a key and a lock. No other key will open that lock (Fig. 7–2). For example, certain enzymes called peptidases hydrolyze proteins but not fats, carbohydrates, or any substance other than a protein. The enzyme β-galactosidase can hydrolyze lactose and related sugars, but not sucrose or other carbohydrates. Oxidases cause only oxidations and not hydrolysis. Enzymes can act only if *coenzymes* (nonprotein molecules that often are vitamin-derived substances) are also present. Coenzymes provide the needed specific physical "fit" between substrate and enzyme (Fig. 7–2).

All cells have certain complex systems of enzymes that catalyze chemical reactions providing the energy and substances required for life. In many cells these systems are virtually identical, just as the engines in automobiles are basically alike, differing only in minor details of arrangement. Not all cells produce *all* kinds of enzymes, however. Some of the enzymes or enzyme-like substances that certain bacteria produce are very poisonous (or *toxic*, from a Greek word meaning "poison"). These cause disease, not as the direct result of growth of the bacteria, but by poisoning the host cells with a toxin. Diphtheria toxin is a good example.

Later we will also learn more about the effect of bacterial enzymes on a variety of substances, such as lactose (milk sugar), saccharose or sucrose (cane sugar), glucose (dextrose), gelatin, milk, coagulated serum, blood, and body tissues. Each species of microorganism has certain enzymatic activities that are

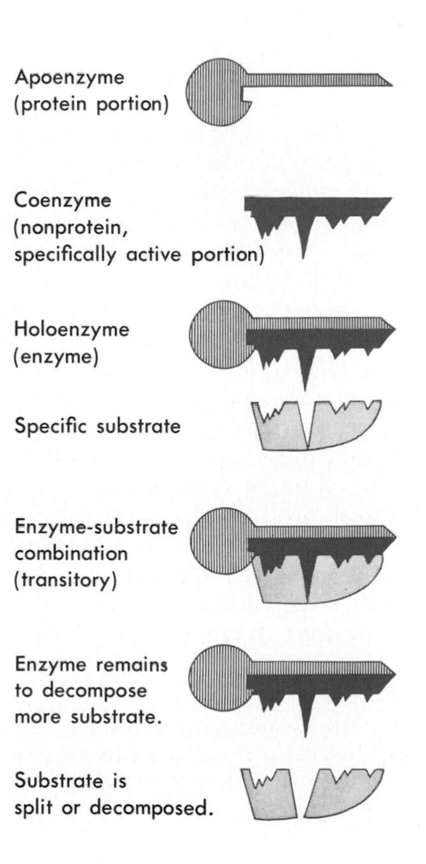

Figure 7–2 A typical enzyme consists of two complex molecular groups: a protein part called the *apoenzyme* (Greek, *apo*, part of) and a nonprotein part, the *coenzyme*. Most coenzymes are chemical derivatives of the so-called vitamins, which in microbiology are often referred to as growth factors. The functional complex formed by an apoenzyme and its appropriate coenzyme is called a *holoenzyme* (Greek, *holos*, whole). Different types of holoenzymes have different molecular structures that depend on the combined structures of the component apo- and coenzymes. These structures vary according to the chemical nature of the substrate with which each holoenzyme uniquely reacts: oxygen (oxidases), water (hydrolases), etc. (see text). Each coenzyme has a distinctive molecular structure that confers a high degree of specificity on the activity of the holoenzyme. As seen in the drawing, the holoenzyme, by virtue of the peculiar structure of both apo- and coenzymes, is restricted to one kind of substrate having a corresponding molecular structure. The relation of enzyme structure and substrate structure suggests a complex key capable of opening only correspondingly formed locks. In the drawing the enzyme has "split," or hydrolyzed, its specific substrate.

more or less characteristic of the species and can therefore be utilized for diagnosis in the laboratory. Thus, it is a relatively easy matter for the microbiologist to distinguish between the bacterium that causes typhoid fever *(Salmonella typhi)* and the common, usually harmless, intestinal bacterium *(Escherichia coli)*. The latter ferments lactose and from it produces lactic acid and other substances, including a mixture of hydrogen and carbon dioxide gas, which is easily seen in special so-called "fermentation tubes" arranged to catch gas (Fig. 7–3). *S. typhi* does not attack lactose at all. The importance of bacterial enzymes in identifying certain specific microorganisms in medical diagnosis is obvious.

Classification of Enzymes. Enzymes are classified according to the kind of chemical reaction they catalyze or the kind of specific substance (substrate) upon which they exert their influence. Enzyme nomenclature also provides a system of classifications involving a number scheme to specifically identify a given enzyme.[3,4]

Each enzyme number contains four elements, separated by periods, with the first number indicating one of the six main divisions of enzymes, the second figure denoting the subclass, the third number indicating the sub-subclass, and the fourth number giving the serial number of the enzyme in its sub-subclass. The six main groups are: (1) Oxidoreductases, (2) Transferases, (3) Hydrolases, (4) Lyases, (5) Isomerases, and (6) Ligases (synthetases).

Selected examples of the numbering system for transferases are:

2.1 Transferring one-carbon groups
- 2.1.1 Methyltransferases (an example of a specific enzyme is E.C.2.1.1.8, *histamine methyltransferase*)
- 2.1.2 Hydroxymethyl-, formyl-, and related transferases
- 2.1.3 Carboxyl- and carbamoyltransferases
- 2.1.4 Amidinotransferases

2.8 Transferring sulfur-containing groups
- 2.8.1 Sulfurtransferases (an example is E.C.2.8.1.1, *thiosulfate sulfurtransferase*)
- 2.8.3 CoA-transferases

The names of most enzymes are derived from the chemical activity or the substrate plus the suffix "-ase." For example, *oxygenases* (or oxygen transferases) and *oxidases* catalyze oxidations that are important sources of energy in all living cells. *Hydrolases* bring about hydrolysis; *dehydrogenases* take hydrogen and electrons away from substrates. Dehydrogenation is an indirect and very common form of biological oxidation. Like direct oxidation, it yields energy to living cells. Oxidoreductases catalyze the transfer of electrons in the energy metabolism of the cell. Transferases catalyze the transfer of molecular groups from one molecule to another. Polymerases catalyze the syntheses of long-chain molecules, such as the combination of the nucleotide chains of DNA. Several enzymes discovered years ago bear unsystematic names, but because these names have become so widely known and have been generally used for such a long time, it would be very inconvenient to change them. *Ptyalin* (starch-digesting enzyme [amylase] of the saliva), *pepsin,* and *trypsin* (protein-digesting enzymes of the stomach and intestines, respectively) are examples of these.

Biological Oxidation. According to Cheves Walling, "Oxygen which bathes all substances exposed to the atmosphere is the most ubiquitous of chemical reagents; through combustion and respiration it provides us with most

[3]The rules for systematic and trivial nomenclature are given in CRC Handbook of Biochemistry. Selected Data for Molecular Biology. Cleveland, CRC Press, 1969, pp. A-33–A-38.

[4]Barman, T. E.: Enzyme Handbook. Vols. 1 and 2. New York, Springer-Verlag, 1969.

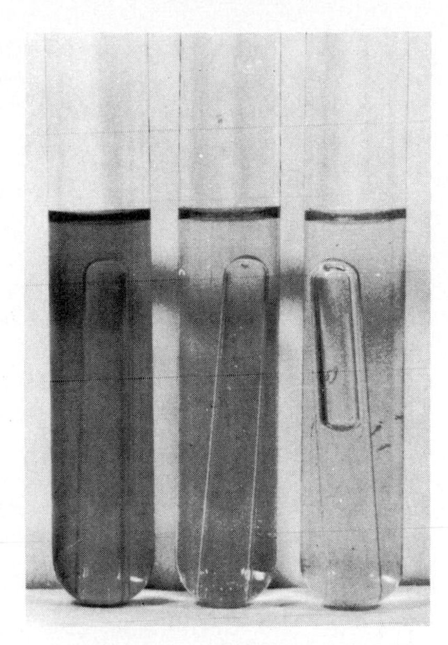

Figure 7–3 Carbohydrate fermentation reactions in glucose broth. The bacteria did not ferment the sugar in the left tube. The liquid is still red, and no displacement of the medium in the tube (inverted to catch gas produced) has occurred. Acid produced in the tube in the center changed the indicator color of the broth from red to yellow. The tube on the right shows acid (the broth also turned yellow) and gas production.

of our available energy, and its ready availability makes it one of our cheapest raw materials."[5] With respect to animals with lungs or gills, the term "breathing" or respiration is quite applicable. Unicellular microorganisms, however, have no lungs or gills. The term *biological oxidation* (biooxidation) is therefore used here, instead of respiration or breathing, to include all of the different energy-yielding (oxidation) mechanisms of all living cells. In some of these oxidative mechanisms the cells use the free oxygen of the air that is dissolved in the water around them.[6] In others the cells use an enzymatic process called *dehydrogenation* by which hydrogen (with its electrons) is taken from the substrate, liberating the energy of the electrons. The hydrogen is then combined with free oxygen, or with oxygen from such substances as $NaNO_3$ or Na_2SO_4, forming H_2O as a waste product. Still others combine the hydrogen with *hydrogen acceptors* other than oxygen: with sulfur, for example, forming H_2S; with nitrogen, forming NH_3; or with carbon, forming CH_4. All the chemical reactions that yield energy to living things are forms of biological oxidation, though oxygen may not be obviously involved. This may be demonstrated as:

$$\underset{\text{Hydrogen donor}}{AH_2} + \underset{\text{Hydrogen acceptor}}{B} \underset{}{\overset{\text{Enzyme}}{\rightleftharpoons}} \underset{\text{Oxidized product}}{A} + \underset{\text{Reduced product}}{BH_2}$$

The substrate AH_2 is oxidized to A, while B, as the hydrogen acceptor, is changed to the reduced product BH_2. Energy is yielded to the cell.

The reader should remember that these reactions involve a transfer of hydrogen (and its electrons) that results not only in oxidation, whether or not oxygen itself is involved, but also in *reduction*. Oxidation is fundamentally a loss of electrons; reduction is the acquisition of electrons. Bacteria perform their electron-energy–transferring reactions by means of a series of carrier systems

[5] *Science, 154*:640, 1966.
[6] By means of oxidases and oxygenases.

that involve numerous enzymes and coenzymes (cofactors) that act as carriers of hydrogen and/or its associated electrons. Important cofactors involved in numerous biochemical reactions that are part of the bacterial metabolism, which often confuse the student because of their abbreviated nomenclature, are listed in Table 7–1.

Sources of Energy

Green Plants and Photosynthesis. All green plants, and a few species of bacteria and protozoa, must obtain their energy from the rays of the sun. They utilize light energy by means of their photosynthetic pigments, the most familiar of which is the green pigment of leaves and grasses called *chlorophyll.* In green plants (*not* bacteria), chlorophyll causes water and carbon dioxide of the atmosphere to combine so that starch, cellulose, and other complex organic substances are formed in the plant. Free oxygen is released to the atmosphere.

Table 7–1. Some Coenzymes Used by Microorganisms*

COENZYME, OR GROWTH FACTOR	COENZYME COMMON SYMBOL	CHEMICAL STRUCTURE OF COENZYME	GROUP TRANSFERRED
HYDROGEN CARRIER:			
Nicotinamide adenine dinucleotide (diphosphopyridine nucleotide)	NAD^+ (DPN^+)	Nicotinamide-ribose-phos-phos-phos-ribose-adenine	Hydrogen
Nicotinamide adenine dinucleotide phosphate (triphosphopyridine nucleotide)	$NADP^+$ (TPN^+)	Nicotinamide-ribose-phos-phos-ribose-adenine	Hydrogen
Nicotinamide adenine dinucleotide phosphate reduced form	$NADPH_2$		Hydrogen
Flavin mononucleotide	FMN	Riboflavin-phosphate	Hydrogen
Flavin adenine dinucleotide	FAD	Riboflavin-phos-phos-ribose-adenine	Hydrogen
GROUP CARRIER:			
Adenosine triphosphate	ATP	Adenine-ribose-phos-phos-phos	Phosphate
Adenosine diphosphate	ADP	Adenine-ribose-phos-phos	Phosphate
Adenosine monophosphate	AMP	Adenine-phosphate	Phosphate
Coenzyme A	CoA	Adenine-phosphoribose-phos-phos-pantetheine (derived from pantothenic acid)	Acyl
Tetrahydrofolate	THF	Pteridine-*p*-aminobenzoate-glutamate	One carbon fragment
Thiamine pyrophosphate	TPP	Thiamine-phos-phos	C_2-aldehyde
Uridine diphosphate	UDP	Uracil-ribose-phos-phos	Isomer of carbohydrate
Pyridoxal phosphate	PALP	Pyridoxal phosphate	Amino and carboxyl

*This list of coenzymes is not complete, but it does contain most important coenzymes used by microorganisms.

†See Figure 7–5.

The chemical change induced by the chlorophyll of *green plants* may be symbolized as follows:

$$(1)\quad nCO_2 + 2nH_2O \xrightarrow{\text{Photosynthesis}} (CH_2O)n + nH_2O + nO_2 + \text{Energy absorbed}$$

In the preceding reaction, solar energy is absorbed and carbon dioxide is fixed from the air. $(CH_2O)n$ represents a carbohydrate like glucose, where n = 6. The reaction in *green plants* may thus also be written as

$$(2)\quad 6CO_2 + 12H_2O \xrightarrow{\text{Photosynthesis}} (CH_2O)_6 + 6H_2O + 6O_2 + \text{Energy absorbed}$$

It is now known that 12 molecules of water are required in this reaction for each molecule of glucose produced. $(CH_2O)_6$ = 1 molecule $C_6H_{12}O_6$.

From the soil, the roots of terrestrial green plants draw up water containing dissolved mineral compounds of nitrogen, sulfur, phosphorus, potassium, and other elements. These, along with the energy furnished by the sun and chlorophyll, are synthesized into the protoplasm and the woody and other structures of the plant. This synthesis, since it proceeds with the aid of light, is called *photosynthesis.* That energy from the sun is actually stored up in the plant is made evident when heat and light and power are released, as in burning wood or coal in a steam engine or when potatoes or sugar is transformed into human energy. The overall chemical change may be represented as the reverse of the photosynthetic reaction, and this can be written in its simplest form, with O_2 being the hydrogen acceptor and H_2O the reduced product.

$$(3)\quad (CH_2O) + O_2 \xrightarrow[\text{biological oxidation}]{\text{Combustion or}} CO_2 + H_2O \text{ (Energy is released)}$$

This reaction as written is *greatly oversimplified* and actually consists of many individual chemical steps, yielding CO_2 and water as the carbohydrate is consumed. We shall show later how this process occurs in muscle tissue as well as in the fermentation process.

Bacteria That Are Photosynthetic. Only a relatively few species of bacteria are photosynthetic, and these are of no known medical interest. The photosynthetic process in bacteria is analogous to that in green plants but may be written as:

$$(4)\quad nCO_2 + 2nH_2A \xrightarrow{\text{Bacterio-chlorophyll}} (CH_2O)n + nH_2O + 2nA$$

Purple photosynthetic bacteria contain *bacteriochlorophylls* that permit a photochemical reaction to proceed from CO_2 to a carbohydrate, but, unlike the reaction based on *chlorophyll* (as in the green plant), oxygen is never produced in this process. H_2A may be any organic or inorganic useful chemical substrate instead of H_2O. Bacteriochlorophyll is also green, but may be masked by red or brown pigments present in the cell.

Some purple and some green bacteria can produce free sulfur by the following photosynthetic reaction:

$$(5)\quad nCO_2 + 2nH_2S \xrightarrow[\text{bacteria}]{\text{Purple sulfur}} (CH_2O)n + nH_2O + 2nS$$

Even plain H_2 can serve as a specific hydrogen donor: thus, H_2A can be any hydrogen donor.

$$(6)\quad nCO_2 + 2nH_2 \longrightarrow (CH_2O)n + nH_2O$$

For example, the species *Rhodopseudomonas sphaeroides,* which contains bacteriochlorophyll a and carotenoids, produces acetone from isopropyl alcohol, as shown in the following reaction:

(7) $$CO_2 + 2 \text{ isopropyl alcohol} \xrightarrow{\text{light}} (CH_2O) + H_2O + 2 \text{ acetone}$$

(This could be any of many organic substances) — (The product depends on the precursor)

Bacteria and Chemosynthesis. Unlike green plants or the photosynthetic bacteria, most bacteria, yeasts, molds, and protozoa have no chlorophyll, and many species are injured by prolonged exposure to sunlight. This is true especially of pathogenic species. The energy for the synthesis of bacterial protoplasm and that of yeasts, molds, and most animal cells comes not from the sun but from *chemical reactions,* such as biooxidation of glucose [see reaction (3), page 000] or, in certain bacteria, some other substances, such as hydrogen sulfide:

(8) $$H_2S + 2O_2 \longrightarrow H_2SO_4 + \text{energy}$$

Such chemically energized self-synthesis is therefore called *chemosynthesis.* Virtually all microorganisms of medical importance are chemosynthetic. Sources of nitrogen, sulfur, phosphorus, and other elements for chemosynthetic organisms are similar to those of green plants and may be organic, inorganic, or both, depending on species. Because bacteria, microscopic plants (like yeasts and molds), and most animal cells (like protozoa) have no roots similar to large plants, they must actually be immersed in solutions of food (osmotrophic nutrition).

Fermentation.[7] Although only a brief outline of fermentation can be given in this chapter, the topic is mentioned later in the book in connection with rising of bread, alcohol production, and spoilage of food. The chemistry of fermentation processes is partially outlined in Figure 7–4. The reader is referred to several references cited at the end of this chapter for a more informative treatment of the reactions involved.

Several well-known fermentation pathways are identified in the literature by a variety of names:

1. Hexose metabolism via the Embden-Meyerhof-Parnas pathway is commonly called *glycolysis,* yeast fermentation, anaerobic glycolysis, anaerobic carbohydrate metabolism of muscle, dissimilation of glucose, or EMP pathway.
2. The hexose monophosphate or *pentose phosphate pathway,* pentose cycle, pentose phosphate shunt, or the Warburg-Dickens-Lipmann pathway is also referred to as the monophosphate shunt or HMP pathway.
3. The Entner-Doudoroff pathway is also known as the ED pathway.
4. The glucuronic acid pathway is also known as the glucuronic acid cycle.
5. The glyoxylate cycle.

Actually only the EMP pathway, the HMP pathway, and the ED pathway, in this order, are well-established routes of carbohydrate metabolism in microorganisms. Much research still needs to be done to ascertain the other cycles and pathways, since they may be much more important in intermediary metabolism than current knowledge indicates.

In fermentation reactions microorganisms use energy-rich compounds like

[7] Fermentation refers to chemical energy-yielding reactions that require organic compounds as electron acceptors.

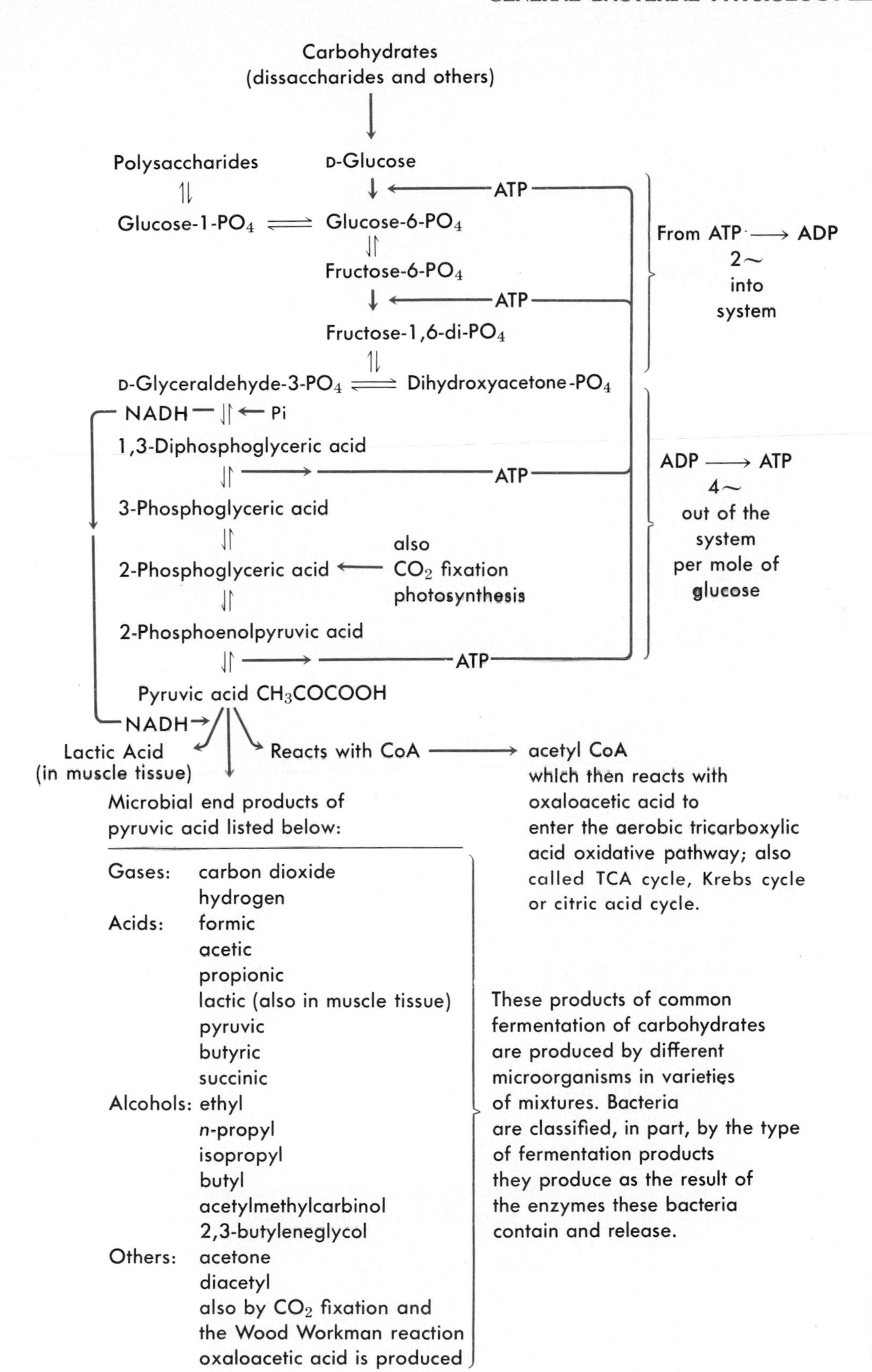

Figure 7–4 Metabolism of carbohydrates. Anaerobic glycolysis fermentation scheme, also called the EMP pathway. For other pathways, also leading to pyruvate, see page 110.

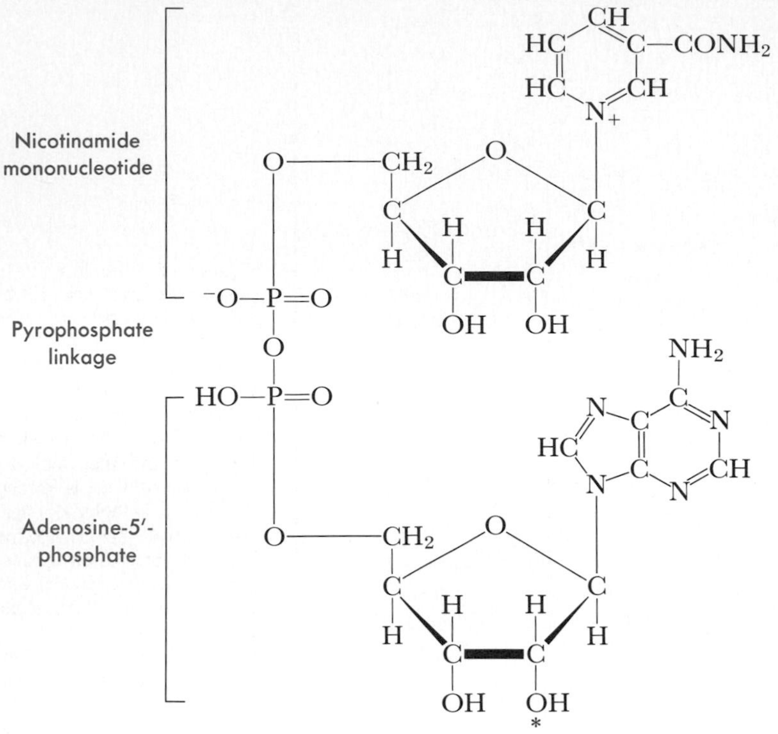

Nicotinamide mononucleotide

Pyrophosphate linkage

Adenosine-5'-phosphate

Figure 7–5 Nicotinamide adenine dinucleotide, also called NAD or, previously, DPN. *This indicates the attachment for H_3PO_4 in NADP, previously called TPN.

ATP (adenosine triphosphate) (Fig. 7–5) for chemical "loans" of energy. This "loan" is then repaid with interest to the system (Fig. 7–6).

Without presenting the numerous enzymes and formulas involved, a fermentation scheme (glycolysis) from glucose to pyruvic acid is shown in Figure 7–4. In this process the hexose (a 6-carbon compound) glucose is broken down into two D-glyceraldehyde-3-phosphate molecules (triose, or 3-carbon compounds), which are then transformed into 2 molecules of pyruvic acid per original mole of glucose. In this process 2 ATP energy-rich phosphate bonds, indicated by the symbol ~, are invested into the pathways and 4 ADP molecules are converted into 4 ATP molecules with a total gain of 2 ~ bonds.

Glycolysis as an anaerobic system releases only a fraction of the potential energy of glucose. Without air, many organisms use pyruvic acid to form some of the end products of the fermentation process, as listed in Figure 7–4. Others, such as *Streptococcus* and *Lactobacillus,* can reduce pyruvic acid to lactic acid, and some enter aerobic pathways such as the citric acid cycle (Fig. 7–7) and proceed to gain further energy through the electron transport system involving cytochrome pigments (Fig. 7–8).

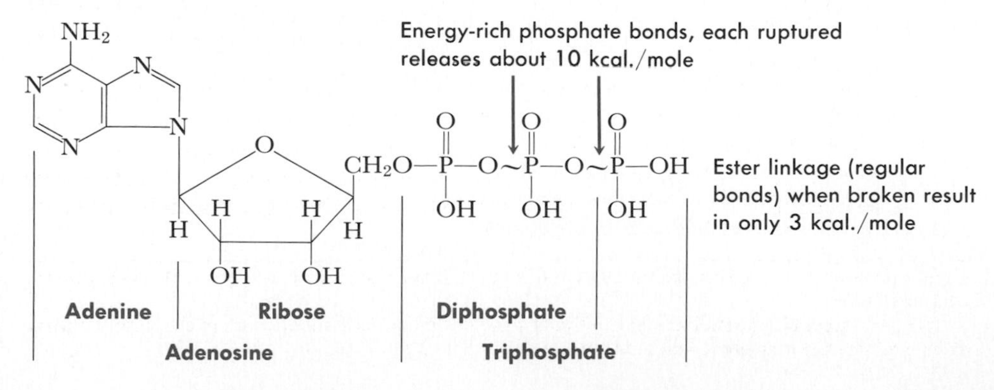

Figure 7–6 Adenosine triphosphate. Other nucleoside triphosphates may replace ATP in certain metabolic reactions. These are: GTP, guanine triphosphate; CTP, cytosine triphosphate; and UTP, uracil triphosphate. One kilocalorie (kcal) = 1,000 calories. The calorie is the unit of energy required to raise the temperature of 1 gram of water at atmospheric pressure 1 C.

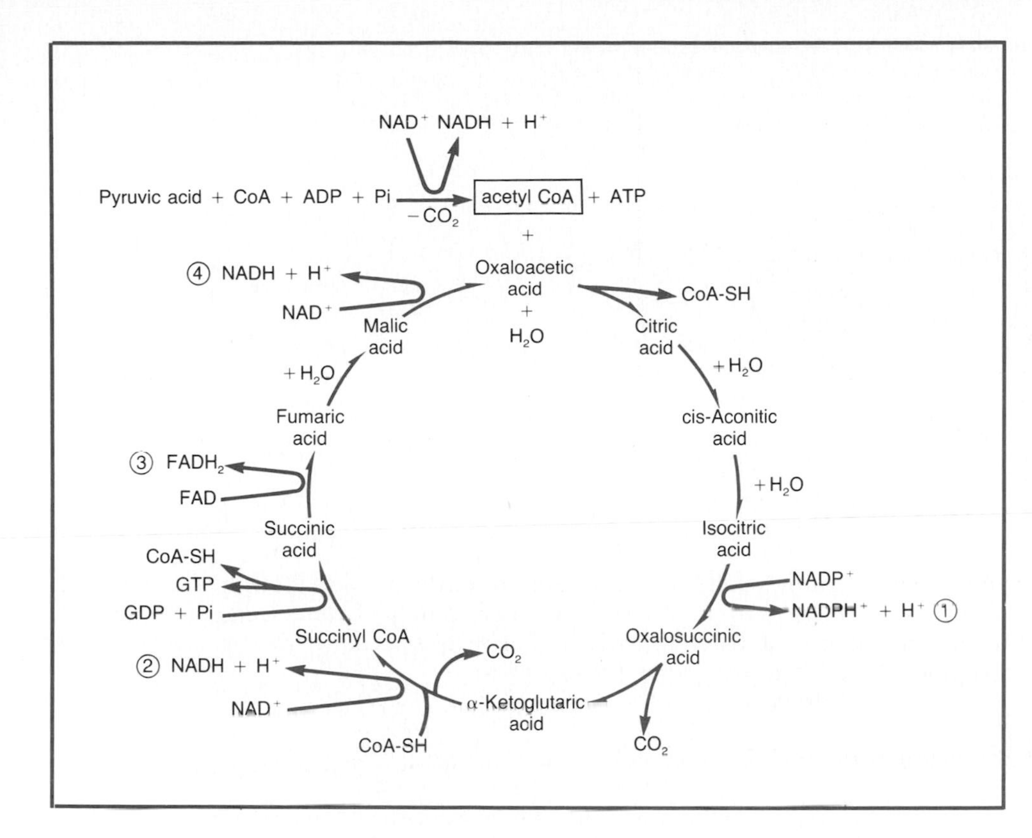

Figure 7–7 The TCA cycle, tricarboxylic acid cycle, Krebs cycle, or citric acid cycle. The $2H^+$ produced in the TCA cycle is carried by the respiratory chain through the cytochrome system to oxygen (see Fig. 7–8). The asterisk in the figure indicates that actually succinyl CoA is formed between α-ketoglutaric acid and succinic acid involving CoA, guanosine triphosphate, guanosine diphosphate, and an energy-rich phosphate bond. Also involved in the TCA cycle is succinyl CoA, between α-ketoglutaric acid and succinic acid. For simplicity it has not been included in this figure.

Summaries of all these pathways are available in sources listed at the end of this chapter. The reader should realize that the estimated gain in energy depends on the enzymes a specific organism possesses, the fermentation pathways used, and the various products of its metabolic reactions.

An approximation of energy released during glycolysis may be close to 35.0 kcal/mole, and the sum of energy released during the tricarboxylic acid cycle is about 57.6 kcal. The energy released from the electron transport is about 594.4 kcal. Thus, 687.0 kcal/mole may be obtained from all the systems discussed in this chapter for the fermentation of glucose.

Some reference sources state that the conversion of glucose to carbon dioxide and water liberates energy as follows:[8]

$$C_6H_{12}O_6 + 6O_2 \longrightarrow 6CO_2 + 6H_2O; \ -\Delta F = 691 \text{ kcal/mole}$$

A total of 38 molecules of ATP are produced in the fermentation exchange of energy.[9] By complete oxidation of glucose to CO_2 and H_2O by way of the EMP scheme and the tricarboxylic acid (citric acid) cycle, a total of 38 $\sim PO_4$ (38 energy-rich phosphates) are released.

The reader should never forget that some microorganisms may never employ the EMP scheme, but instead use the HMP pathway; and fermentation products from pyruvic acid differ, depending on the enzymes produced. Furthermore, the "electron-transport" system may be entered at different sites of phosphorylation, if at all. Microorganisms differ in their genetic determinants

[8]See Mazur, A., and Harrow, B.: Biochemistry: A Brief Course. Philadelphia, W. B. Saunders Company, 1968.

[9]Some sources cite 36 molecules of ATP, but the difference is evidently due to the mitochondria system and does not apply to microorganisms.

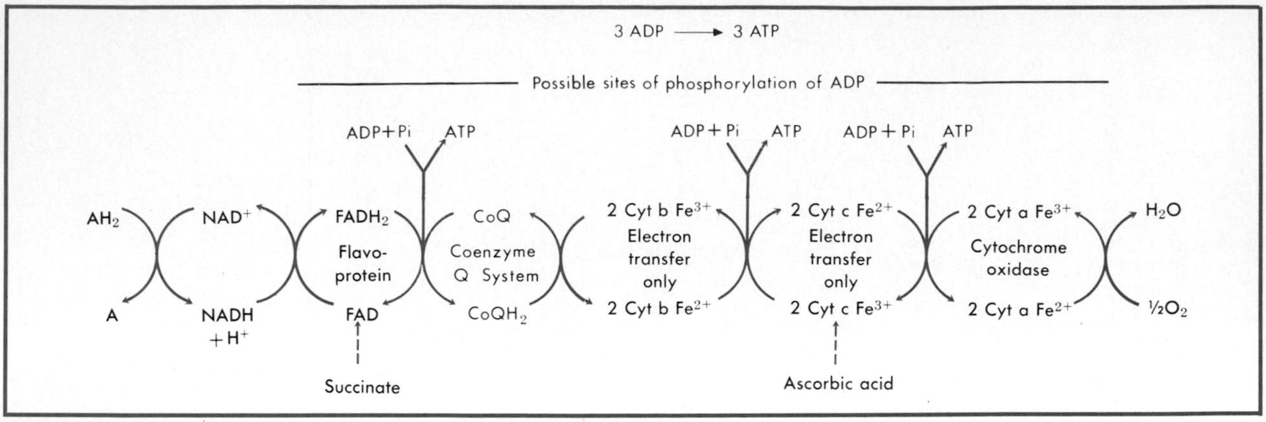

3 ADP → 3 ATP

Possible sites of phosphorylation of ADP

ADP+Pi ATP ADP+Pi ATP ADP+Pi ATP

AH_2 / A — NAD^+ / NADH $+H^+$ — $FADH_2$ Flavo-protein FAD — CoQ Coenzyme Q System $CoQH_2$ — 2 Cyt b Fe^{3+} Electron transfer only 2 Cyt b Fe^{2+} — 2 Cyt c Fe^{2+} Electron transfer only 2 Cyt c Fe^{3+} — 2 Cyt a Fe^{3+} Cytochrome oxidase 2 Cyt a Fe^{2+} — H_2O / ½O_2

Succinate

Ascorbic acid

Figure 7–8 Possible sites of coupling of phosphorylation with electron transport in cytochrome system. Energy is gained in this "electron-transporting" mechanism. Some bacteria have ubiquinone (coenzyme Q), which takes up 2H and becomes ubihydroquinone; others have menaquinone (vitamin K). The respiratory chains of mitochondria and different bacteria do not have identical components; some are yet to be worked out and others are not agreed on. *Escherichia coli,* grown under real aerobic conditions, is known to have a branched pathway after cytochrome b, one through cytochrome 0, the other through cytochrome d. (Adapted from Fruton, J. S., and Simmonds, S.: General Biochemistry. 2nd ed. New York, John Wiley and Sons, Inc., 1963.)

and their enzymes produced and, consequently, in metabolic pathways and products of fermentation. All need energy, but how this energy is obtained determines many of their diagnostic characteristics and their success in the struggle for survival with other forms of life.

Microbial Multiplication

We have seen that yeasts undergo a process of *unequal* fission called budding. In some bacteria also *(Rhodopseudomonas palustris)* budding is a means of multiplication. When two *equal* cells are formed by the division of one cell, the process is called *binary fission.* In most elongated bacteria, fission is *transverse,* i.e., the division occurs at right angles to the long axis. Flagellate protozoa divide lengthwise by *longitudinal* fission.

Multiplication by fission is not confined to independently living microorganisms. Fission occurs in cells that are part of large, complex, multicellular structures: for example, trees, dogs, or humans.

Reproduction and Fission. In trees, dogs, or humans, fission of body cells (not of sex cells) results only in increased size of the plant or animal, or in the replacement of some structural cells that have died (as in injury and the like). Reproduction (i.e., increase in numbers) of large complex organisms is brought about only by the conjunction of sex cells. Among the unicellular microorganisms, only the eukaryotes produce true gametocytes. For example, the microgametocytes and macrogametocytes of malaria protozoa unite in sexual union in the stomach of the mosquito.

In bacterial fission, two cells, and therefore two individuals, result. A "pseudosexual" type of mating that produces genetic recombination, but not multiplication (which will be discussed later in this chapter), occurs in certain species of bacteria.

Mechanisms of Gene Transfer in Microorganisms

In eukaryotes definite sexual phenomena are generally evident. There are contacts of more or less distinctive sex cells (gametes) and dissolution of separating cell walls, followed by intermingling of cell contents, especially genetic material (DNA), that carry hereditary traits and characters (genetic recombination). In bacteria, which have only primitive nuclei (nucleoids), similar but more primitive conjugations have been described in several species (see Figure 7–9).

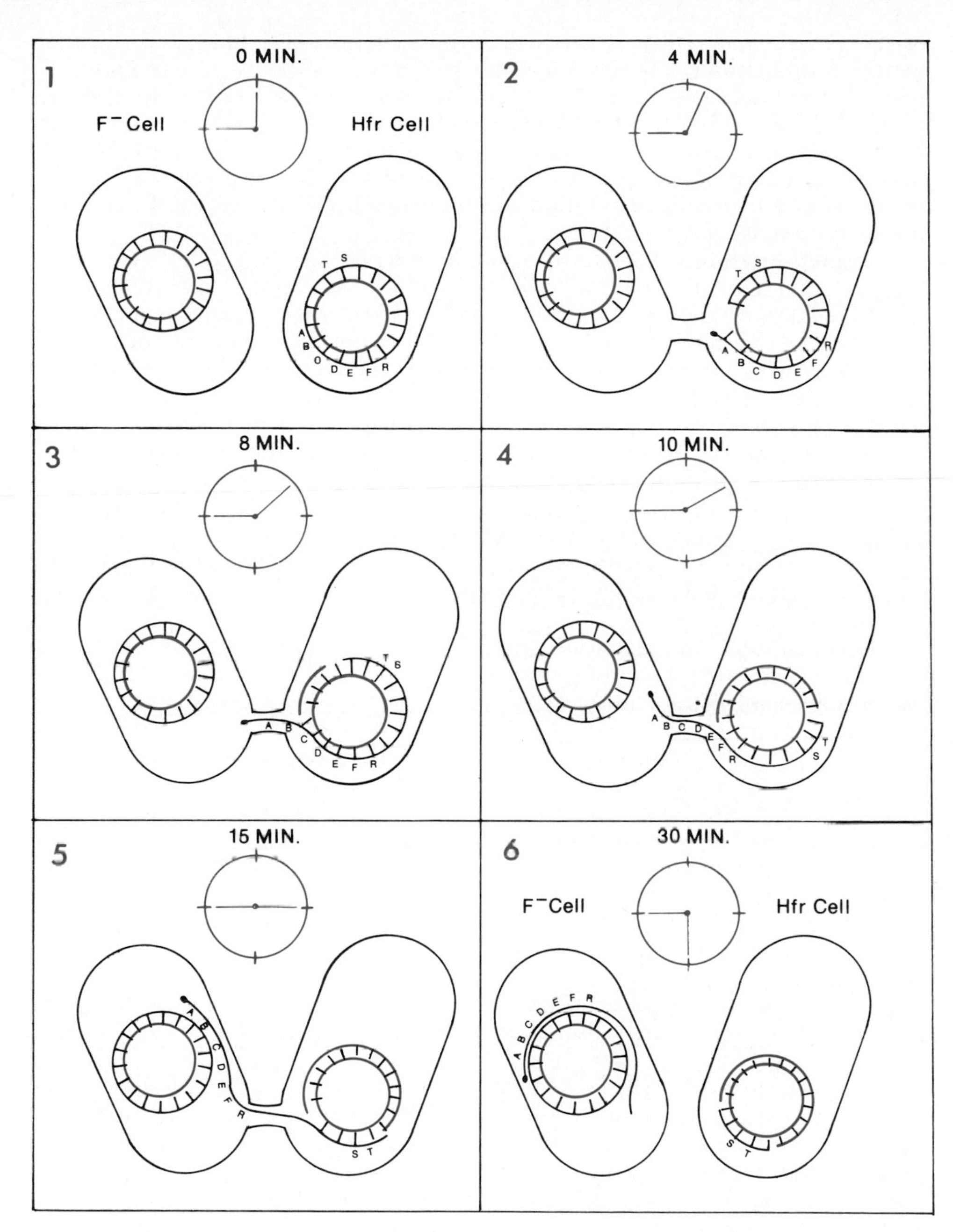

Figure 7–9 Bacterial genes are transferred from one bacterium to another in linear order. Upon surface contact between a donor cell and a recipient cell, a conjugation tube is formed. The donor cell transmits a portion of its chromosome carrying genes A, B, C, D, E, F, R to the recipient cell. A recombinant cell is thus formed, containing genes from both parent types. The donor cell survives the loss of chromatin material because at this time more chromosome substance is being formed in the Hlr cell to replace the lost genetic material. (Modified after Wollman, E. L., and Jacob F.: Sexuality in Bacteria. Copyright © 1956 by *Scientific American,* Inc. All rights reserved.)

In higher (eukaryotic) organisms such as zinnias or cats, sexual processes are necessary for genetic recombinations and for reproduction.[10] Among unicellular eukaryotes, sex, although present in most, is unnecessary since most can multiply continuously by asexual fission. The prokaryotic bacteria not only can reproduce asexually indefinitely, but also can achieve genetic recombination by *transduction* or by *transformation* in the entire absence of sex. Gene recombination also occurs in viruses, although sexual phenomena are entirely absent.

Transduction. In this process certain viruses (bacteriophages) that infect bacteria can invade one bacterial cell after another, carrying with them parts

[10]*Propagation* by cuttings and grafts does not involve genetic changes or sexual processes and is merely continued growth of part of the same individual in a different place.

of the genetic material (genes, DNA) of the bacterial cells. This results in new genetic combinations. The DNA *of the phage itself* may also integrate temporarily with the bacterial DNA, giving the bacterium new traits, but only as long as the phage remains. This is often called viral conversion or "infective heredity." DNA from one *bacterial* cell, transferred by phage to another bacterial cell, imparts new hereditary traits to the second bacterium. The phenomenon is generally called *transduction.* Often the transduced DNA is rejected, resulting in *abortive transduction.*

Transformation. Various microorganisms may be made to "inherit" physiological characteristics artificially by using DNA extracted in test tubes from live or dead microorganisms. Other, closely related microorganisms immersed in such extract acquire the DNA (with heritable properties) of the microorganisms from which the DNA was extracted. Thus, the immersed microorganisms acquire new heritable qualities by in vitro[11] inheritance without benefit of sex. This is called *transformation.* Studies with P^{32}-labeled DNA have shown just how the genetic material from the donor replaces DNA of the recipient bacterial strain in transformation. More than one genetic character can be transferred at one time in the transformation process. The DNA must be double-stranded, either native or renatured; it is not active in transformation if it is denatured or its two strands are separated after being heated for 10 minutes at 100 C. Experimental evidence shows that transformation also occurs under natural conditions.

Conjugation. Conjugation among bacteria is a primitive type of mating process discovered by Lederberg and Tatum in 1946 in *Escherichia coli.* In these organisms conjugation consists of a one-way transfer of genetic material from one cell, the donor, into another, the recipient. The donor cell is analogous to the male in higher biological systems, and the corresponding recipient is analogous to the female. Donor cells contain a cytoplasmic agent (episome or plasmid), which is an extrachromosomal, self-replicating genetic element called a *fertility* (F) or "male" factor; these cells are referred to as F^+ cells. Recipient cells are designated as F^- cells. In mating, the F^+ cells readily transfer the F plasmid (but rarely genetic DNA) to F^- cells, which then become F^+ cells. DNA-containing plasmids other than the F factor seem to be numerous and many have been demonstrated. Plasmids of this type are the Col plasmid (sometimes called "colicinogenic" factor) (see *Bacteriocins* in Chapter 14) and the R plasmids that determine resistance or sensitivity to certain antibiotics. Thus, R donor cells and R recipient cells also exist. (See also *Drug Resistance,* later in this chapter.)

Sometimes the F plasmid (or other conjugative plasmids) becomes integrated into the chromosome of the F^+ cell. This greatly increases frequency of genetic recombination (transfer of genetic DNA from donor cells to F^- cells). Donor cells in which the F plasmid has become integrated into the donor chromosome are therefore called Hfr (high frequency of genetic recombination) cells. The mating process of these cells is often incomplete. The F plasmid in Hfr cells is the last part of the integrated chromosome to be transferred to the recipient cells. The chromosome usually breaks somewhere during the process. Therefore, in Hfr $\times$ F^- matings, although F^- cells frequently become genetic recombinants, they rarely become F^+ or Hfr. Note that Hfr denotes high frequenty of *transfer of genes* but *not* high frequency of mating.

As can be seen in Figure 7–9, it is quite possible to interrupt the mating process by vigorous shaking at measured intervals and then to determine which "genes" (genetic determinants, or markers) have meanwhile been transferred from donor to recipient. It has been found that markers are always transferred in the same order and that the bacterial chromosome is a circular thread of DNA. The thread breaks at mating time and is transferred lengthwise from

---

[11]From the Latin words for "in glass" (in the test tube).

donor to recipient; the F factor is transferred last, if at all. In Figure 7–9 it is seen that the A gene passes into the recipient cell in 8 minutes, the C gene in 10 minutes, and then the D, E, and F genes in that order. After 30 minutes the R gene (e.g., fermentation of galactose) is transferred. When, as is common, the chromosome is broken mechanically at the transfer bridge before completion of the mating, only a portion of the donor genes passes to the recipient. The time intervals may vary in different species.

We have learned much of the arrangement and sequence of markers ("genes") in the bacterial chromosome in part from interrupted mating experiments, and the knowledge thus gained has enabled geneticists to draw circular chromosome or genetic maps of bacteria. A typical genetic map of *Escherichia coli* is shown in Figure 7–10.

Recombination in Bacteria. Lederberg and numerous other microbial geneticists have crossed different strains of bacteria. Some genetic markers that have been used to test for recombination are shown in Table 7–2.

For example, a cross between two *Escherichia coli* parental strains,

$$B^-M^-P^+T^+V_1^s \times B^+M^+P^-T^-V_1^r$$

may result in offspring of 79 per cent B^+ M^+ P^+ T^+ V_1^s and 21 per cent B^+ M^+ P^+ T^+ V_1^r.

From this cross it is concluded that the gene for phage resistance V_1^r is about 21 chromosomal units away from the P and T genes on the *E. coli* chromosome, since V_1^s stays with P^+ and T^+ 79 per cent of the time, which means that the chromosome breaks occur only 21 per cent of the time, resulting

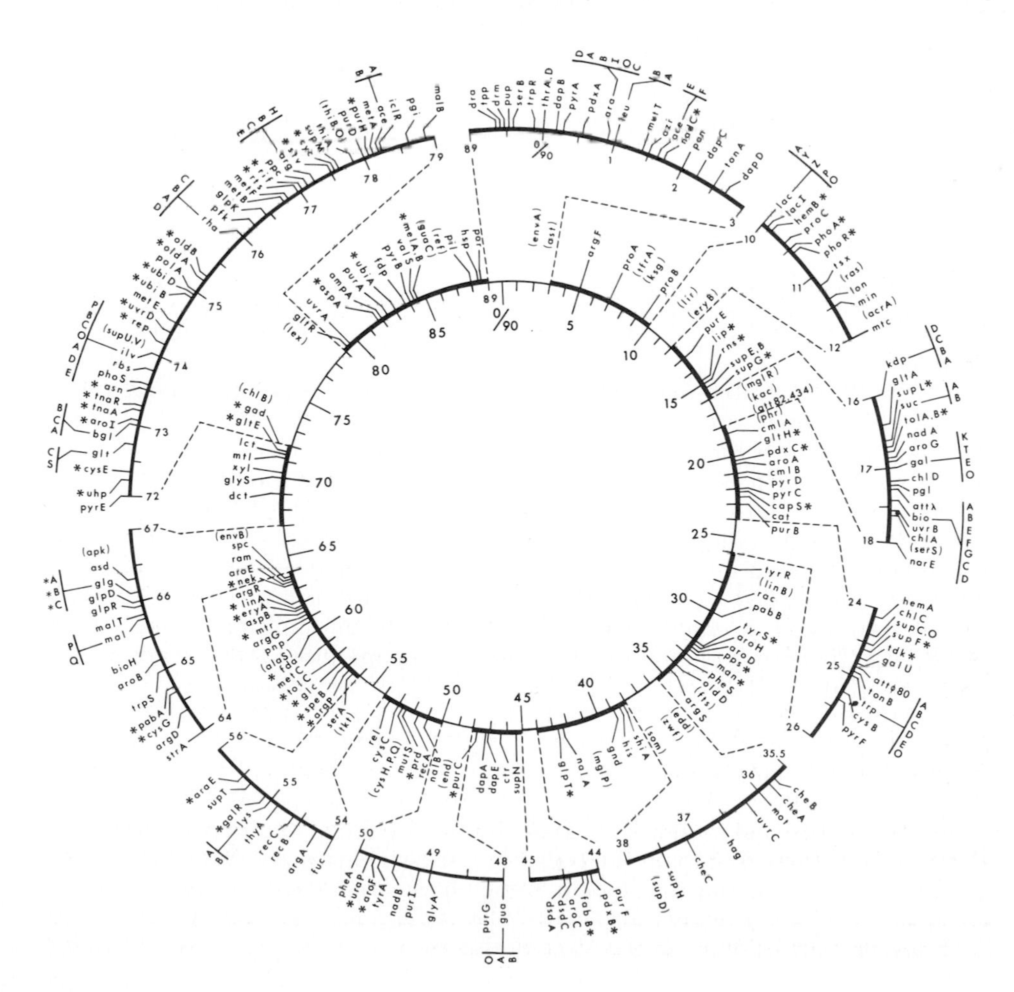

Figure 7–10 Scale drawing of the linkage map of *Escherichia coli*. The inner circle, which bears the time scale from 0 through 90 minutes, depicts the intact circular linkage map. The map is graduated in 1-minute intervals, beginning arbitrarily with zero at the *thrA* locus. Selected portions of the map (e.g., the 10- to 12-minute segment) are displayed on arcs of the outer circle with a 4.5 times expanded time scale to accommodate all of the markers in crowded regions. Markers in parentheses are only approximately mapped at the positions shown. A gene identified by an asterisk has been mapped more precisely than the markers in parentheses, but its orientation relative to adjacent markers is not yet known. (From Taylor, A. L.: Current linkage map of *Escherichia coli*. *Bacteriol. Rev.*, *34*:155–175, 1970. Reproduced by permission of the American Society for Microbiology, Washington, D.C.)

Table 7–2. Symbols Used for Various Loci of *E. coli* K12*

STRAIN	SUBSTANCE	STRAIN
NUTRITIONAL REQUIREMENTS		
(Requirement of a substance is indicated by a superscript$^-$; independence by a superscript$^+$.)		
B^-	Biotin	B^+
B_1^-	Thiamine	B_1^+
C^-	Cystine	C^+
L^-	Leucine	L^+
M^-	Methionine	M^+
P^-	Proline	P^+
T^-	Threonine	T^+
SUGAR FERMENTATIONS		
(Ability to ferment is indicated by a superscript$^+$; inability by a superscript$^-$.)		
STRAIN	SUBSTANCE	STRAIN
Lac^+	Lactose	Lac^-
Gly^+	Glycerol	Gly^-
BACTERIOPHAGE RESISTANCE		
(Resistance to phage is indicated by a superscript^r; sensitivity by a superscript^s.)		
STRAIN	PHAGE	STRAIN
V_1^r	T_1 and T_5	V_1^s
V_6^r	T_6	V_6^s
RESISTANCE TO CHEMICAL AGENTS		
(Resistance to chemical agents is indicated by a superscript^r; sensitivity by a superscript^s.)		
STRAIN	SUBSTANCE	STRAIN
Cla^r	Sodium chloroacetate	Cla^s
A^r	Sodium azide	A^s

*Adapted from Lederberg, J.: *Genetics,* Vol. 32.

in $P^+T^+V_1^r$ new combinations. By this system genes like B, M, P, and others may also be mapped on the chromosome.

Mutation. A mutation is a sudden variation in structure of DNA making up the genes, which is inherited by the progeny of the mutated cell through indefinite generations. A *mutant* is an organism, be it human or bacterium, that has undergone mutation. At this point, it is assumed that the student may remember the definitions of the mutational unit, the *muton;* the recombination unit, the *recon;* and the complementation analysis unit, the *cistron,* all of which roughly refer to the *gene* (see Chapter 2). The term replicon has also been proposed as a common name for the bacterial chromosome and the conjugons (see page 116).

Mutagenic Agents and Cancer

Some of the agents that can alter DNA or change the genetic structure and thus produce mutations in cells are x-rays, ultraviolet light, and other radiation; certain chemicals; possibly some viruses (see page 94); and probably sex hormones. Such agents are said to be *mutagenic,* i.e., mutant-producing. LSD (lysergic acid diethylamide) is said also to be a mutagen. It is interesting to note that many mutagenic agents are also carcinogenic, that is, cancer-producing, or *oncogenic* (producing tumors, especially malignant tumors). A major concern of the human race is how much mutagenic effect is to be feared from radioactive fallout from thermonuclear bombs. Sometimes mutations occur for unknown reasons; then we hide our ignorance of the cause by calling the change a "spontaneous mutation." It seems that no living organism can escape *neutron* bombardment from outer space. Thus, even without any willful exposure to mutagenic agents, genetic material is always exposed to chance irradiation.

Genetic Engineering

Genetic material (DNA) has been synthesized artificially in vitro. The man-made DNA, when applied to certain bacteria, altered their genetic characters just as effectively as though genetic mutation, transformation, or transduction had taken place. May we anticipate that by the use of various forms of predesigned, manufactured DNA, we may direct and control heredity in human beings, with purposeful production of supermen (or frightful monsters)? Stranger things have happened. Some concern exists about possible biohazards that could result from genetic manipulations. It is now possible to insert small bits of DNA from animal tumors, human cancerous tissues, or certain other detrimental genes into other organisms, such as *Escherichia coli*. These re-engineered bacteria could perhaps get into the intestines of laboratory workers, a habitat not too strange to such microorganisms, and cause great harm and spread to other humans. However, many of the fears once expressed by workers in the field have proven to be unfounded. It seems that nature continually carries out "genetic engineering" experiments and perhaps has its own safe-guards built in. "Genetic engineering" may help us overcome inborn errors of metabolism, structural and metabolic defects, diabetes, arthritic diseases, and conditions that arise with aging; also, nitrogen-fixing genes may be built into plants to relieve the world's food problems. *Escherichia coli* x1776, a "safe" strain for genetic manipulations, has been so engineered. Restrictions have been set by certain geneticists on specific types of genetic manipulations with living cells and viruses. These have been made widely available by the NIH in its "Guidelines,"[12] which are brought up to date periodically. Scientists must realize the benefits that free research in genetic engineering will bring. Policies of containment and responsibility will assure progress with the least amount of danger.

Techniques Used in Genetic Engineering. *To isolate specific genes*, restriction enzymes, isolated from various bacteria, are added to extracted DNA (bacterial, human, or plant). They cleave the polynucleotide chain of the DNA after a specific sequence of bases. For example, enzyme Hind III breaks the chain in front of adenine in the adenine-guanine-cytosine-thymine (AGCT) sequence, whereas Eco R1 breaks it before AATT and Hae III breaks GGCC between guanine and cytosine.[13] With the right enzyme, small pieces (hopefully gene size) result. Hydrogen bonds hold the broken pieces of DNA together as if they had sticky ends, provided the temperature is kept low.

To make new gene combinations, DNA from another source, perhaps bacterial plasmids, is divided by the same restriction enzymes and mixed with the other

---

[12]National Institutes of Health: Guidelines for Research Involving Recombinant DNA Molecules. Washington, D.C., U.S. Department of Health, Education, and Welfare, Public Health Service. Released June 23, 1976. New revisions are continually added to the Federal Register by the National Institutes of Health. Last revision on hand July 29, 1980.

[13]Hind III is isolated from *Haemophilus influenzae* Rd and cleaves DNA at ↓AGCT; Eco R1 is isolated from *Escherichia coli* RY13 and cleaves DNA at G↓AATTC; Hae III is isolated from *Haemophilus aegypticus* and cleaves DNA at GG↓CC.

DNA pieces. The temperature is raised, and the sticky ends open up to receive chains of the other genetic material. The enzyme DNA ligase is used to join the DNA ends covalently and permanently, like glue.

Gene transfer from one organism to another may thus be accomplished by having bacteria take up plasmids with foreign DNA inserted in their polynucleotide structures. Cells that take up a plasmid may later be recognized by the original properties inherent in the plasmid, such as resistance to a specific antibiotic. In addition, the plasmid confers on the host bacteria the properties of the foreign genes that were inserted in the plasmid. These genes can be human or of any other origin, perhaps even synthetic DNA material yet to be produced in the laboratory.

This process requires large amounts of specific genes and gene products. These are isolated through careful selection and testing of the bacteria having the desired genes. Cloning techniques are used to manufacture large quantities of bacteria with the specific genetic information selected for mass production (Fig. 7–11).

Vectors for genetic engineering are plasmids and bacteriophages. In the future, mitochondria or plastids may also be used, but this will depend on the development of new techniques. We still do not know how to get foreign DNA into eukaryotic cells of animals and plants. At present, bacteria are more receptive hosts.

VARIABLE BACTERIAL CHARACTERS

Genotype and Phenotype. The final apparent or visible organism, be it fly, mouse, bacterium, or human being, resulting from any given genetic makeup *(genotype)* is called the *phenotype* (*pheno-* is from a Greek word meaning "that which shows") or the *phenotypic expression* of the genotype. An organism is generally modified to some extent by its environment, but its response to the environment is controlled by its genotype. A phenotype, therefore, is generally the result of combined hereditary and environmental factors. For example, some hydrangeas have pink flowers if growing in normal soil, but blue ones if growing in soil containing iron or alum. Such noninheritable, environmentally produced, temporary changes are sometimes called *fluctuations, modifications,* or *phenocopies.*

Among the phenotypic characters of bacteria that commonly vary because of mutation or fluctuation are size, shape, virulence (if pathogenic), color (if pigmented), capsule formation, motility, spore formation (if spore-forming), production of certain enzymes, and colony characteristics. Many other, less readily observable changes occur, such as in chemical (antigenic) composition and nutritional requirements, i.e., the active enzymes. These and other still obscure alterations underlie variations in the virulence of pathogens.

Drug Resistance. A variation of particular importance to the members of the health group is the appearance of bacterial mutants resistant to sulfonamide drugs and antibiotics. This is a problem of major importance in hospitals, especially in communicable disease control, surgical and maternity wards, and nurseries. This aspect of microbial variation directly affects everyone working or living in any hospital. Drug resistance in microorganisms has already been mentioned as resulting from transmission of the R plasmid (page 116) and will be discussed in greater detail later.

Variation and Diagnosis. Although the offspring of a single bacterial cell may vary widely from its parent, there are certain limits beyond which variation does not go under ordinary conditions of laboratory study. Once the common fluctuants or mutants of a species are known, the diagnostic microbiologist should have no difficulty in recognizing that species. Unless a given species of microorganism is subjected to some unusual environmental condition producing

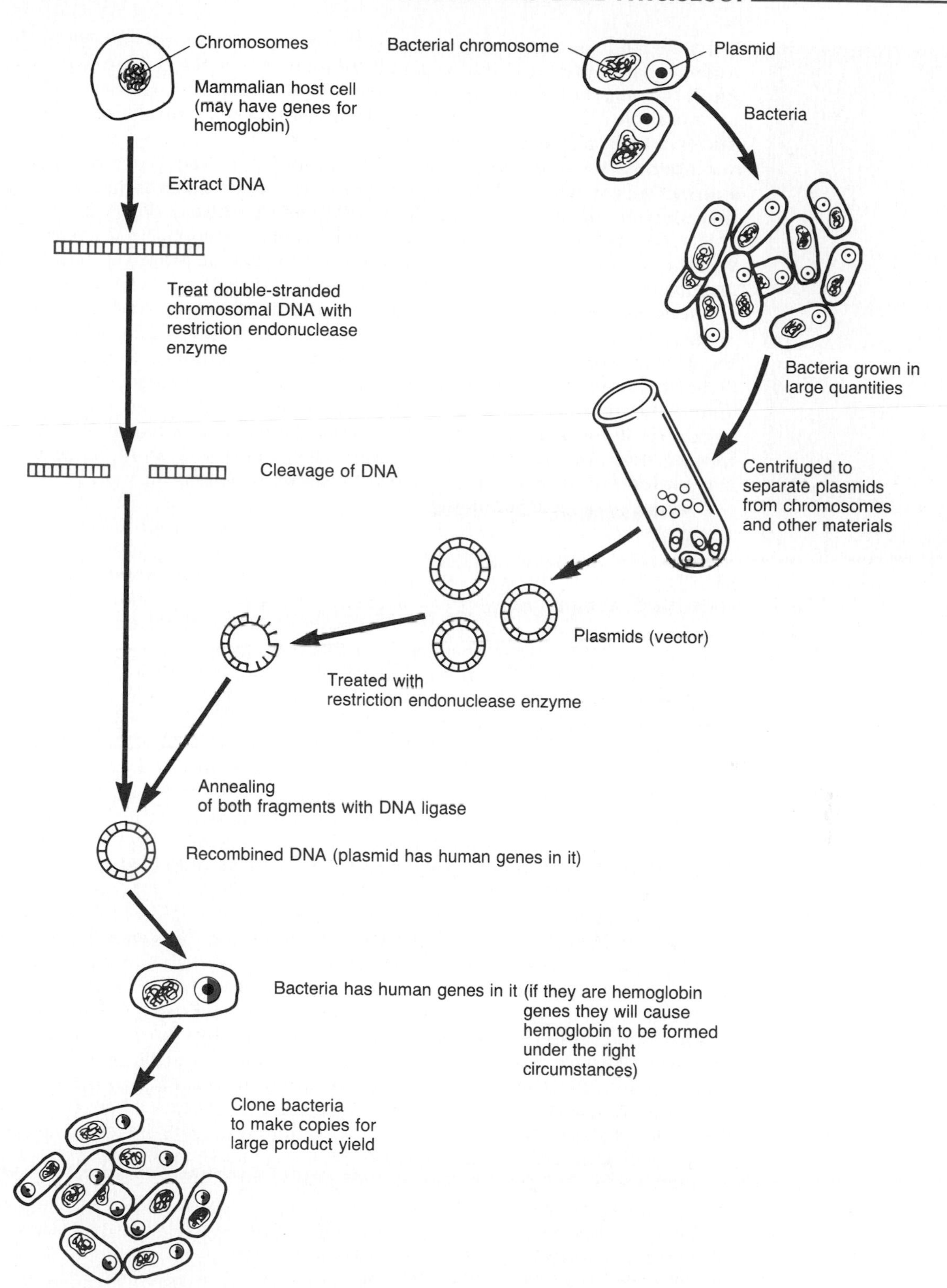

Figure 7–11 Insertion of human hemoglobin genes into restriction enzyme treated plasmids (or virus vectors). The vectors are used to infect bacteria, and the bacteria are grown in large quantities. The vectors are purified, and the genes are removed. Large amounts of plasmid genes for hemoglobin production can thus be produced.

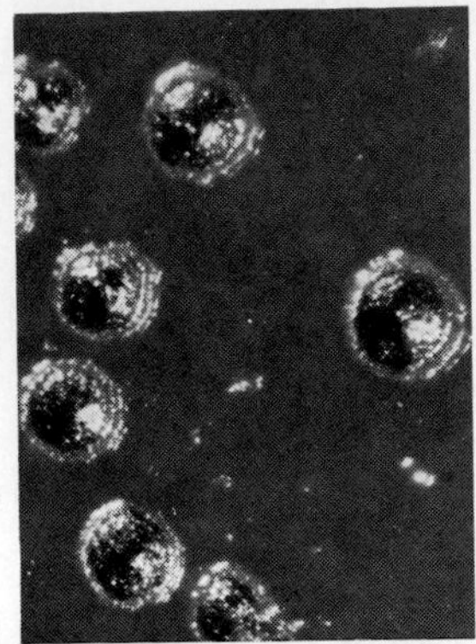

A

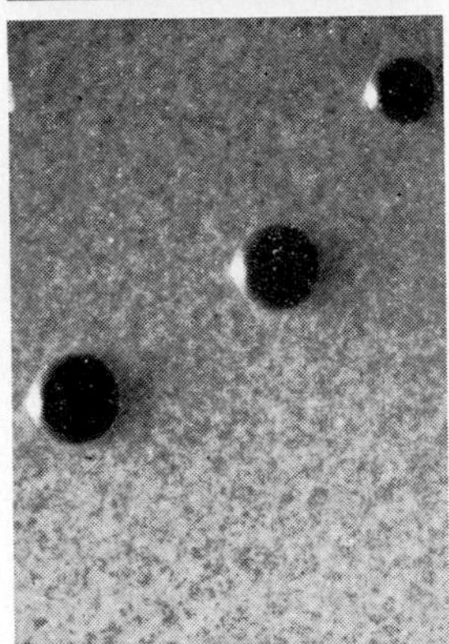

B

Figure 7–12 *A*, Rough (R) colonies of the diphtheria bacillus. *B*, Smooth (S) colonies of the diphtheria bacillus. (About three times actual size.)

a marked fluctuation or mutation of its properties, it tends to retain its well-known characters. Any change in culture medium or pH[14] may cause properties of all the cells in the culture to fluctuate. Certain kinds of variation, due either to mutation or fluctuation, are especially common or important in diagnosis and will be briefly described here.

Colony Variant Types. Four commonly observed types of colonies resulting from variation of the constituent cells when cultivated upon solid media (like nutrient agar) are the rough or "R" type, the smooth "S" type, the mucoid "M" type, and the minute or dwarf colony type. The surface of rough colonies is rather dry and dull, often wrinkled and crumbly or granular, and the edges are irregular (Fig. 7–12 *A*). They are often brittle in consistency. Smooth colonies are pasty or butyrous (buttery) in consistency and have perfectly regular margins, with smooth, glistening surfaces (Fig. 7–12 *B*). Mucoid colonies tend to be smooth and glistening and rather voluminous and watery; they may "run" if the surface on which they grow is tilted. If touched with a wire they spin out long, mucoid threads like saliva or thin taffy. A fourth type of colony variant is the dwarf form. These colonies are extremely minute. S, R, M, and minute colony forms have been observed in many species of bacteria. There are several other, less commonly seen varieties of colony form.

Variations in Virulence.[15] Extensive studies of bacterial variation have revealed that virulence (when present) is usually associated with the S type of colony. The substances and conditions that promote the S (and, consequently, virulent) type of bacteria are present in human and other animal bodies. Therefore, microorganisms in—or freshly isolated from—blood, serum, feces, and other materials from the diseased body are especially dangerous to persons who come into contact with this infectious material. S type variants are frequently encapsulated, and this is usually related to virulence. (Capsules are protective mucoid coatings that help bacteria evade the defensive mechanisms of the body.)

This point is emphasized to the student because health workers usually come into direct contact with bacteria from the infected patient when the microorganisms (pathogens) are most virulent and most capable of causing infection. The pathologist doing an autopsy, the laboratory technician handling infectious tissues, blood, or other body fluid specimens, and the nurse, the surgeon, and the physician dealing with an infectious patient must constantly bear this fact in mind.

After bacteria have been cultivated outside the body on lifeless media such as meat broth for a few days or weeks, they tend to "degenerate" and to lose their invasive character (virulence). They also lose their capsule and form R type colonies. These bacteria usually become the pampered and softened pets of the microbiologist. However, this is not always the case, and no one should assume that a living pathogenic microorganism has lost its virulence at any time except under certain carefully controlled conditions of attenuation of virulence (see Chapter 20). Fatal accidents have resulted from such assumptions!

SUMMARY

Although most species of bacteria can grow in the dark and (like fungi) thrive on lifeless, chemically defined or inert food material, viruses, chlamydias, and rickettsias can only multiply inside living cells.

Bacteria have adapted to life almost everywhere on earth—at the bottom of the sea, in plants and animals, in airless and frigid environments, and on mucous membranes of the human body.

---

[14]The symbol pH, with a number (e.g., pH 6.5), is a measure of acidity. (See Chapter 8.)

[15]Virulence is the degree of ability to produce disease. It is discussed in more detail later in this book.

The food for bacteria must be dissolved in water. A semipermeable membrane selectively permits the *right substances* to enter the bacteria and waste products to be excreted. Enzymes catalyze reactions that *hydrolyze* and solubilize needed substances (food). *Exo-enzymes* pass outward from the cell to do their work, whereas *endo-enzymes* convert chemicals within the cell that are needed for building cell materials or producing energy.

This chapter provides a brief explanation of how enzymes may be classified for better recognition. Most enzymes derive their names from the chemical activities in which they are involved, usually with the addition of the suffix "-ase" (e.g., oxygenase, hydrolase).

Oxidation is fundamentally a loss of electrons; reduction is the acquisition of electrons. Bacteria perform their electron-energy–transferring reactions by means of a series of carrier systems that involve numerous enzymes and coenzymes (cofactors, derivatives of so-called "vitamins"), which act as carriers of hydrogen and/or its associated electrons. Only a few less important bacteria are photosynthetic: they contain bacteriochlorophylls. Most bacteria, yeasts, and molds derive their energy from biooxidation of glucose by chemosynthesis involving fermentation.

Several fermentation pathways are known. Perhaps the most common one is the Embden-Meyerhof-Parnas scheme, also called glycolysis or the EMP pathway. Other routes of carbohydrate metabolism in microorganisms are the HMP pathway and the ED pathway. In these fermentation reactions energy-rich compounds like ATP (adenosine triphosphate) are used for chemical "loans" of energy. This "loan" is then repaid with interest, a gain of energy, to the system. Complete oxidation of glucose via pyruvic acid to CO_2 and H_2O by the EMP scheme and the tricarboxylic acid (citric acid) cycle results in a total gain of 38 energy-rich phosphates, available for use by the organism.

Bacteria can multiply by methods other than binary fission. In a primitive type of mating process called conjugation, the episome or plasmid-containing donor cell, having F^+ (the male fertility factor) or Hfr (high-frequency capacity), injects its genetic DNA into the F^- cell in linear order of the genes on the single circular chromosome. This process enables geneticists to draw genetic maps of bacteria involved in conjugation.

In transduction, bacteriophages carrying DNA of previously infected bacterial cells invade other bacteria. In transformation, living bacteria are treated with DNA extracted from other cells and acquire inheritable properties of the cells from which the DNA was obtained.

In recombination, two parental strains of *Escherichia coli* have been shown to exchange genetic material, so that all possible combinations of both chromosomes (one from each parent) may be obtained in the progeny. This again may be used to determine the relative positions of genes to each other on the bacterial chromosome.

Genetic engineering, in its most restrictive sense, refers to recombinant DNA between species, such as the transfer of selected genes from human cells to *Escherichia coli* cells.

Bacterial colony types, such as S (smooth) or R (rough), are accurate indicators of the virulence of the organism. Virulent (infectious) organisms form an S type colony, and a loss of virulence and capsule is indicated by the type of S colony produced.

REVIEW QUESTIONS

1. What is the Kingdom Procaryotae? What are some of the criteria in classifying bacteria?
2. Define the following terms: autotrophic, heterotrophic, bacterial photosynthesis, metabolism, osmosis, semipermeable membrane, enzymes.
3. Compare enzymes with coenzymes and exo-enzymes with endo-enzymes.
4. What can you tell about the classification of enzymes?

5. What is the function of biological oxidation?
6. Compare hydrogen donors with hydrogen acceptors and oxidation with reduction.
7. Compare photosynthesis in plants with bacterial photosynthesis and biological oxidation. What happens to the energy produced? Why? What does chemosynthesis do?
8. List different schemes of fermentation and explain their purpose. What happens to the ATP? to the glucose? to the energy?
9. Discuss processes in bacteria in which DNA is transferred to other cells. Of what advantage is this to the organisms? What are conjugation, plasmids, the F factor, Hfr, bacterial recombination, and genetic markers?
10. How many chromosomes are there in *Escherichia coli*? What else can you tell about it?
11. What is genetic engineering? What great advances of benefit to humans can be expected from these techniques? Why are some people still concerned about possible dangers associated with recombinant DNA technology? What is *Escherichia coli* x1776?
12. How are colony variations of bacteria related to virulence? What causes colonies to change from S to R? Of what concern is this to workers in the health professions?

Supplementary Reading

Baker, J. J. W., and Allen, G. E.: Matter, Energy and Life. 4th ed. Reading, Mass., Addison-Wesley Publishing Co., 1981.

Baum, S. J.: Introduction to Organic and Biological Chemistry. 3rd ed. New York, Macmillan Company, 1982.

Braun, W.: Bacterial Genetics. 2nd ed. Philadelphia, W. B. Saunders Company, 1965.

Carlberg, D. M.: Essentials of Bacterial and Viral Genetics. Springfield, Ill., Charles C Thomas, 1976.

Davis, B. D.: Microbiology. 3rd ed. New York, Harper & Row, 1980.

Doelle, H. W.: Bacterial Metabolism. 2nd ed. New York, Academic Press, 1975.

Haddock, B. A., and Hamilton, W. A. (eds.): Microbial Energetics. Twenty-seventh Symposium of the Society for General Microbiology, London. New York, Cambridge University Press, 1977.

Hassid, W. Z.: Biosynthesis of oligosaccharides and polysaccharides in plants. *Science, 165*:137, 1969.

Herskowitz, I. H.: Genetics. 2nd ed. Boston, Little, Brown & Co., 1965.

Kornberg, A.: The synthesis of DNA. *Sci. Amer., 219*:64, 1968.

Latner, A. L.: Cantarow and Trumper, Clinical Biochemistry. 7th ed. Philadelphia, W. B. Saunders Company, 1975.

Lengyel, P., and Soll, D.: Mechanism of protein biosynthesis. *Bacteriol. Rev., 33*:264, 1969.

Mahler, H. R., and Cordes, E. H.: Biological Chemistry. New York, Harper & Row, 1966.

Mazur, A., and Harrow, B.: Biochemistry—A Brief Course. Philadelphia, W. B. Saunders Company, 1968.

McGilvery, R. W.: Biochemical Concepts. New York, Holt, Rinehart & Winston, 1975.

Miller, J. H.: Experiments in Molecular Genetics. Cold Spring Harbor, N.Y., Cold Spring Harbor Laboratory, 1972.

Novick, R. P., Clowes, R. C., Cohen, S. N., Curtiss, R., III, Datta, N., and Falkow, S.: Uniform nomenclature for bacterial plasmids. *Bacteriol. Rev., 40*:168, 1976.

Pardee, A. B.: Membrane transport systems. *Science, 162*:632, 1968.
Rose, A. H.: Chemical Microbiology: An Introduction to Microbial Physiology. 3rd ed. New York, Plenum Press, 1976.
Sokatch, J. R.: Basic Bacteriology and Genetics. Chicago, Year Book Medical Publishers, 1976.
Spencer, J. H.: The Physics and Chemistry of DNA and RNA. New York, Holt, Rinehart & Winston, 1972.
Stent, G. S.: Molecular Biology of Bacterial Viruses. San Francisco, W. H. Freeman, 1963.
Umbreit, W. W.: Essentials of Bacterial Physiology. Minneapolis, Burgess, 1976.

Genetic Engineering

Beers, R. F., Jr., and Bassett, E. G. (eds.): Recombinant Molecules: Impact on Science and Society. Tenth Miles International Symposium. New York, Raven Press, 1977.
Biotechnology: Research that could remake industries. *Chemical Week,* October 8, 1980.
Davis, R., Roth, J., and Botstein, D.: Advanced Bacterial Genetics: Laboratory Manual. Cold Spring Harbor, N.Y., Cold Spring Harbor Laboratory Press, 1980.
Grossman, L., and Moldave, K. (eds.): Nucleic Acids Part I. *In* Methods in Enzymology. Vol. 65. New York, Academic Press, 1980.
Old, R. W., and Primrose, S. B.: Principles in Gene Manipulation. *In* Studies in Microbiology. Vol. 2. Berkeley, University of California Press, 1980.
Ptashne, M.: The defense doesn't rest. *Science, 16*:11, 1976.
Rodriguez, R. L., West, R. W., Jr., and Tait, R. C.: Principles of Recombinant DNA Technology. Reading, Mass., Addison-Wesley Publishing Co., in press.
The DNA furor: Tinkering with life. *Time,* April 18, 1977, pp. 32–45.
Wald, G.: The case against genetic engineering. *Science, 16*:6, 1976.
Wu, R. (ed.): Recombinant DNA. *In* Methods of Enzymology. Vol. 68. New York, Academic Press, 1979.

MICROORGANISMS IN THEIR ENVIRONMENT

Several physical and chemical factors are of absolutely critical importance to the growth and reproduction of living cells, be they bacteria or cells of our own bodies. Among the most crucial factors are moisture, nutrients, specific temperature and pH ranges, the presence or absence of oxygen, absence of toxic (poisonous) substances, and the proper osmotic pressure. These factors not only control microorganisms but also form the basis of several extremely important aspects of health care procedures.

MOISTURE

Abundant moisture is essential for life and multiplication of all *vegetative*[1] cells, including those of our own bodies, but is not necessary for spores, cysts, and other *dormant* cells, i.e., cells that exist but do not reproduce.

In the absence of water, microorganisms may survive but cannot grow; hence, many industrial commodities and foodstuffs, such as hay, meats, fruits, vegetables, and soups, are preserved by drying. These substances putrefy and spoil through growth of saprophytic[2] microorganisms only if they are kept in

[1]Actively growing and multiplying; not dormant.

[2]From Greek *sapros*, rotten, and *phyton*, plant; hence a plant living on inanimate, decaying, organic matter. Animals similarly nourished, especially protozoa, are said to be *saprozoic*.

Figure 8–1 Lyophilization (quick freeze-drying) of bacterial cultures at the Microbiology Research Laboratory at the Texas Woman's University. An ampule containing an aqueous suspension of microorganisms previously "quick-frozen" by immersion in alcohol with CO_2 ice (−76 C) can be seen attached to the suction manifold that withdraws the water (sublimates it) at high vacuum. Shown is a Virtis Co. lyophilizer.

moist environments. Although drying prevents growth of microorganisms, it does not necessarily kill them and, in fact, preserves many saprophytic and saprozoic microorganisms in foods. One method used by microbiologists to keep certain microorganisms alive for long periods of time (for many years if desired) is called *lyophilization.* This is the creation of a stable preparation of microorganisms (even the most fragile and sensitive) that are first rapidly frozen and then subjected to a high vacuum to draw off the moisture (Fig. 8–1). Lyophilization is also called quick freeze-drying. The American Type Culture Collection (ATCC) uses this method to store many bacterial strains for years, supplying them to laboratories when requested.[3]

Certain fragile pathogenic microorganisms, such as the gonococcus (*Neisseria gonorrhoeae*, cause of gonorrhea), the spirochete that causes syphilis (*Treponema pallidum*), and vegetative trophozoite forms of the organism causing amebic dysentery (*Entamoeba histolytica*), which can be dried while frozen, are extremely sensitive to drying at temperatures near those of the body. Certain other pathogenic bacteria (tubercle and diphtheria bacilli) and cyst forms of protozoa may survive at room temperatures for long periods, whether frozen or not, in dried mucus, pus, sputum, or feces.

NUTRIENTS

Metabolism. Metabolism is the digestion and utilization of food to synthesize the proteins, fats, carbohydrates, and other substances of which living cells are made and to furnish the energy necessary for life and reproduction. Among bacteria, the biochemical processes constituting metabolism are remarkably like those of all other cells: from your own brain cells down to the most primitive *cellular* microorganisms (viruses are not cellular) imaginable. This is probably because all the higher forms of life are derived by evolution from the same, or similar, primordial ancestor. The cell substance therefore is almost identical chemically in all living things; all cells utilize the same general sorts of foods, metabolizing these foods by means of the same general chemical activities. As in designs of automobile carburetors, fuel pumps, and transmissions, differences in form exist, but the basic functions are the same.

Foods of Microorganisms

Since all cells consist mainly of carbon (C), hydrogen (H), oxygen (O), nitrogen (N), phosphorus (P), and sulfur (S), with lesser amounts of magnesium (Mg), iron (Fe), copper (Cu), sodium (Na), potassium (K), and a few other elements, nutrients of all cells must contain these chemicals. The form in which

[3]The American Type Culture Collection Catalogue of strains (1970) states: "Cultures in the ATCC generally are maintained either in the freeze-dried condition or frozen under liquid-nitrogen refrigeration to preserve them as nearly as possible in the original state, i.e., as deposited. New accessions are routinely freeze-dried or frozen from the first subculture after their receipt." The 1982 catalogue gives direction for reviving freeze-dried cultures.

these nutritional constituents can be used, however, varies widely with different kinds of cells. The elements may be utilized in one or both of two general forms, organic or inorganic.

Organic:[4] Included here are proteins, carbohydrates, fats, and related or derived substances, all of which are fairly complex compounds of carbon, oxygen, and hydrogen, sometimes with phosphorus, nitrogen, sulfur, or other elements. In addition, many species require organic complexes such as vitamins —e.g., nicotinic acid ("niacin"), thiamin, and riboflavin—and use them in much the same manner that human beings do. Indeed, in most fundamental respects, as mentioned before, the composition and metabolism of human cells are much like those of many microorganisms.

Inorganic:[5] Examples of these are sodium chloride, water, carbon dioxide, sulfur, iron, and hydrogen. Some other inorganic nutrients are required in such minute quantities that they are referred to as "trace elements." Trace elements will be mentioned later when microbiologic media for different bacteria are discussed. Some of the trace elements are Zn, Mn, Co, Ca, and sometimes even substances that would be highly toxic in higher concentrations. Organisms require trace elements in the form of inorganic ions.

Nutritional Types

All living organisms may be subdivided on the basis of their sources of life energy. Thus, all green plants, including the eukaryotic green algae (such as seaweeds), the eukaryotic *Cyanobacteria* (blue-green algae), and phototrophic bacteria of the order Rhodospirillales (not medically important), typically obtain all their energy for self-synthesis from sunlight or from an artificial source of radiant energy. Such organisms are said to be *photosynthetic* or *phototrophs.*

All typical animal cells and all fungi, including bacteria and related forms (except the few species of harmless photosynthetic bacteria just mentioned), can live entirely without radiant energy and may even be injured by it. They obtain all their vital energy for self-synthesis from chemical reactions called *biooxidations* (respirations or fermentations) that are independent of light. They are said to be *chemosynthetic* or *chemotrophs.* Each of these groups, phototrophs and chemotrophs, may be subdivided into two groups on the basis of the kinds of materials (not sources of energy) on which their growth depends.

Organisms were formerly classed as *autotrophic*[6] if they could use CO_2 as a sole source of carbon, and *heterotrophic*[7] if they required carbon in organic combination. Because of numerous inconsistencies and overlappings these groupings have been replaced by the terms *photolithotrophic, photoorganotrophic, chemolithotrophic,* and *chemoorganotrophic.*

Phototrophs. Some phototrophs may use only inorganic donors of hydrogen (e.g., H_2O for green plants, H_2S for certain bacteria) for the reduction of CO_2 in photosynthesis; these are called *photolithotrophs.* If they require organic donors of hydrogen for reduction of CO_2 in photosynthesis they are called *photoorganotrophs.* Phototrophic microorganisms are of no known medical significance.

Chemotrophs. Chemotrophs whose energy-yielding chemical reactions (biooxidations) are restricted to oxidation of inorganic substances such as iron or sulfur are said to be *chemolithotrophic.*

Some chemolithotrophic bacteria, common in the soil, rivers, and sea, live

---

[4]Since 1828 chemists have defined organic compounds as those that contain the element carbon.

[5]Inorganic compounds are those that do not contain carbon. However, CO, CO_2 and H_2CO_3 are on the borderline; they are ordinarily dealt with in inorganic chemistry.

[6]From Greek *autos,* self, and *trophe,* feed; self-feeding.

[7]*Hetero* is a Greek word meaning "other."

entirely on inorganic matter. They are restricted to carbon dioxide from the air as a source of carbon, i.e., they are examples of *autotrophs.* For the other requisite elements they may (and some must) use the salts in soil or sea water. They obtain their nitrogen in the form of atmospheric nitrogen or ammonia or other simple nitrogenous compounds. From these simple substances they synthesize their own complex vitamins, carbohydrates, proteins, and fats. These bacteria gain their energy by oxidizing sulfur, iron, nitrogen, and a few other elements. Such bacteria are independent of all other forms of life and may be descendants of some of the earliest living things on this planet. Chemolithotrophic microorganisms are of no known medical significance.

Chemoorganotrophs. Chemotrophs whose oxidation substrates must be organic substances are called *chemoorganotrophs.* Some species of chemoorganotrophic microorganisms in the soil and sewage can digest solid organic substances like wood, crab shells, and old rubber tires. Certain pathogenic species can digest the living tissues of your body. The cells of these various species secrete the appropriate enzymes to liquefy (digest) these solid organic substances and then absorb them. Could you dine on old tires or sawdust? Theoretically you could, easily, if the cells of your alimentary tract secreted the same enzymes as these microorganisms. It is evident that all of the apparently enormous and spectacular metabolic differences between humans and diverse species of microorganisms represent mainly variations in the enzyme makeup of the different species. "One bacterium's meat is another organism's poison." Cats do not enjoy raw carrots, whereas some human beings (and rabbits) relish them. It should be noted that all microorganisms of medical importance are chemoorganotrophs.

Viruses. Viruses receive both their energy and their substance from their hosts (i.e., from infected cells), being neither chemotrophic nor phototrophic; they are said to be *paratrophic.*

Saphrophytes and Saprozoa. From the most primitive chemoorganotrophic (or photoorganotrophic?) species of microorganisms there (probably) evolved species that could utilize more and more complex organic foods, i.e., "higher" forms. Presumably these "more advanced" species of plants and animals depended on the "lower" or more primitive forms for complex, organic foods, derived from the waste products and dead cells of the lower species. The higher forms metabolized the dead organic matter and, thus, began the process of decay so necessary for the disposal of dead bodies, plants, excreta, and other refuse. Such microorganisms are the modern chemoorganotrophs, and they are of many varieties. They occur in every drop of ocean or pond water, ditch mud, soil, and feces—in short, wherever dead organic matter may be found.

Parasites and Pathogens. Presumably, through the ages, the chemoorganotrophic microorganisms "progressed": i.e., they finally became adapted to prey upon other living organisms. They became *parasites,* able to live in and on the bodies of other plants and animals. Because many of them are poisonous or actually invade healthy tissue and utilize valuable food substances, they seriously injure the host in which they live. The result is disease in the infected host: these parasites are *pathogenic.*

It must not be thought that the lines of division between saprophytic, parasitic, and pathogenic species are entirely clear and distinct. Some saprophytes, such as tetanus and botulism bacilli, are highly pathogenic. They do not invade healthy tissues but give off deadly toxins into dead tissues or into foods in which they grow. These toxins produce disease when (and if) they are absorbed by the body. Some parasites are not usually pathogenic. Examples are the numerous species of usually harmless bacteria and fungi found in the respiratory and alimentary tracts and on external mucosa and skin. There are many intergradations, inconsistencies, and overlappings. The major groups named, however, are sharply enough defined for the purposes of the present discussion.

PRESENCE OF TOXIC SUBSTANCES

When living cells metabolize foods, either for cell synthesis or energy or both, by means of their enzyme systems, they give off as waste products various substances such as hydrogen, toxic hydrogen sulfide, carbon dioxide, methane, and many other compounds. These may range from fairly strong sulfuric acid, as in certain sulfur-oxidizing harmless chemolithotrophic bacteria of the soil, to complex organic compounds like toxic alcohols and organic acids, indole, and acetylmethylcarbinol, as in the reactions discussed in Chapter 7. Some of the waste products are very harmful to the bacteria themselves and to other organisms.

Distinctive Wastes. Each species of microorganism is recognized by its food peculiarities, its particular waste products, its motility, form, pigment, and other properties.

In addition to their own deleterious waste products, substances from many other sources may be toxic to microorganisms: for example, dyes, acids, and metallic residues of industrial chemical plants dumped into rivers. Many common substances are so toxic to microorganisms that they serve well as disinfectants: phenol and its derivatives, such as cresols and hexachlorophene, and chlorine compounds like laundry bleach (5 per cent NaOCl, or Clorox).

TEMPERATURE

Mesophiles.[8] Just like roses or cabbages, microorganisms need not only food and water to grow but also a favorable temperature. For example, microorganisms that live in the soil and shallow streams grow best at summer temperature, or about 20 to 30 C (Table 8–1). Bacteria that have become highly adapted to growth in the human body will flourish only at or near body temperature (37 C). These temperatures (from 20 to 37 C) represent the medium or middle range, and microorganisms growing best in such a temperature range are said to be *mesophilic* (Fig. 8–2).

Psychrophiles.[9] Some microorganisms, such as those that live at the bottom of the sea, grow best at temperatures near freezing (4 to 10 C) and are said to be *psychrophilic*. Since relatively few species of microorganisms in the environment of man are extremely psychrophilic, low temperatures are widely used to prevent or greatly retard the growth of many kinds of bacteria and other organisms that cause "souring," "spoiling," and rancidity. Furs, confectionery, milk, vegetables, wood pulp, serums, vaccines, antibiotics, drugs, and many other products may be preserved by efficient refrigeration.

Low temperatures, however, are seldom fatal to most microorganisms. Many species will survive in a dormant state, frozen for years, and can grow vigorously when thawed out. Some can survive the temperature of liquid air at −191 C (−312 F). A method of preserving microorganisms for future study, and bovine and other spermatozoa for artificial insemination, is rapid freezing with "dry ice" (solid CO_2) at −76 C (−105 F). The method is also used to preserve certain diagnostic specimens, sera, and viruses during long-distance transportation. Human tissue cells cultivated in artificial media are routinely kept with liquid nitrogen at −195 C (−319 F) for long periods. Some microorganisms will remain alive at temperatures as low as −269 C (more than 450 below zero F).

Thermophiles.[10] Still other species of microorganisms (certain algae and bacteria) will grow only at high temperatures. Any organisms that grow only at

---

[8]From Greek *mesos*, moderate, and *philos*, loving.
[9]From Greek *psychros*, cold.
[10]From Greek *therme*, heat.

Table 8–1. Table of Equivalents of Celsius (Centigrade, C) and Fahrenheit (F) Temperature Scales*-

C	F	C	F	C	F
−40	−40.0	9	48.2	57	134.6
−39	−38.2	10	50.0	58	136.4
−38	−36.4	11	51.8	59	138.2
−37	−34.6	12	53.6	60	140.0
−36	−32.8	13	55.4	61	141.8
−35	−31.0	14	57.2	62	143.6
−34	−29.2	15	59.0	63	145.4
−33	−27.4	16	60.8	64	147.2
−32	−25.6	17	62.6	65	149.0
−31	−23.8	18	64.4	66	150.8
−30	−22.0	19	66.2	67	152.6
−29	−20.2	20	68.0	68	154.4
−28	−18.4	21	69.8	69	156.2
−27	−16.6	22	71.6	70	158.0
−26	−14.8	23	73.4	71	159.8
−25	−13.0	24	75.2	72	161.6
−24	−11.2	25	77.0	73	163.4
−23	−9.4	26	78.8	74	165.2
−22	−7.6	27	80.6	75	167.0
−21	−5.8	28	82.4	76	168.8
−20	−4.0	29	84.2	77	170.6
−19	−2.2	30	86.0	78	172.4
−18	−0.4	31	87.8	79	174.2
−17	+1.4	32	89.6	80	176.0
−16	3.2	33	91.4	81	177.8
−15	5.0	34	93.2	82	179.6
−14	6.8	35	95.0	83	181.4
−13	8.6	36	96.8	84	183.2
−12	10.4	37	98.6	85	185.0
−11	12.2	38	100.4	86	186.8
−10	14.0	39	102.2	87	188.6
−9	15.8	40	104.0	88	190.4
−8	17.6	41	105.8	89	192.2
−7	19.4	42	107.6	90	194.0
−6	21.2	43	109.4	91	195.8
−5	23.0	44	111.2	92	197.6
−4	24.8	45	113.0	93	199.4
−3	26.6	46	114.8	94	201.2
−2	28.4	47	116.6	95	203.0
−1	30.2	48	118.4	96	204.8
0	32.0	49	120.2	97	206.6
+1	33.8	50	122.0	98	208.4
2	35.6	51	123.8	99	210.2
3	37.4	52	125.6	100	212.0
4	39.2	53	127.4	101	213.8
5	41.0	54	129.2	102	215.6
6	42.8	55	131.0	103	217.4
7	44.6	56	132.8	104	219.2
8	46.4				

*Degrees Celsius (Centigrade, C) may be converted to degrees Fahrenheit (F) and vice versa by the following formulas:

C to F: 1.8 C + 32

To change +54 C to F: (1.8 × 54) + 32 = 129.2 F
−38 C to F: (1.8 × −38) + 32 = −36.4 F

F to C: 0.556(F−32)

To change +54 F to C: 0.556(54−32) = 12.2 C
−38 F to C: 0.556(−38 + −32) = −38.9 C

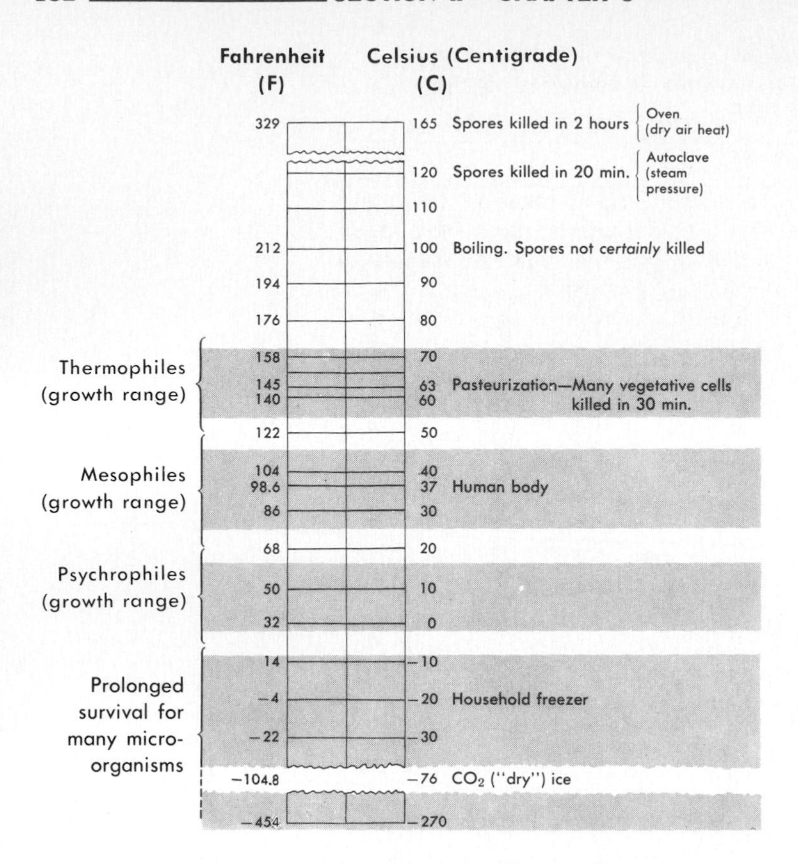

Figure 8–2 Some temperatures of significance in microbiology. Note the slow killing action of dry air heat and the relatively rapid killing action of steam (moist heat) at lower temperature. Boiling kills some but not all spores. Note overlapping growth ranges of thermophilic, mesophilic, and other forms. Strictly speaking, *psychrophiles* cannot grow at temperatures above 20 C (maximal). They grow best at about 15 C (optimal) and can also grow rapidly at 0 C or below. *Mesophiles* that can grow very slowly to a limited extent at temperatures as low as 0 C may be called *psychrotolerant* or *psychrotrophic mesophiles*. Similarly, *thermophiles* that can grow to some extent below their assigned range might be called *mesoduric* or *mesotolerant thermophiles*.

temperatures above about 50 C are called *thermophiles*. For many species there is an inverse relationship between temperature and pH. In neutral or alkaline (pH 8.0) natural springs, some bacteria can grow at boiling temperatures (100 C at sea level). Increases in acidity to pH below about 3.0 are increasingly inhibitory as temperatures rise above about 70 C. Some thermophiles are found in hot sulfurous springs such as those in Yellowstone National Park and in Iceland. Thermophiles are not infectious to man, since they cannot grow at 37 C.

Exposure to temperatures of 75 C or above for more than about 20 minutes is fatal to the actively growing (vegetative) forms of all pathogenic bacteria, higher fungi, and protozoa. This fact is used in various methods of sterilization and disinfection, to be described later. Bacterial *endospores*, however, as previously pointed out, are much more resistant and may survive boiling and even higher temperatures (350 F) for several hours. Certain viruses also are exceptions.

OXYGEN

Obligate Aerobes. Many microorganisms require, for the purpose of biooxidation, free access to the oxygen supply of the atmosphere; they are limited to using solely uncombined oxygen as a hydrogen (electron) acceptor because of the nature of their enzyme systems. They cannot grow without free oxygen. Such microorganisms are said to be *obligate* (or *strict*) *aerobes*. (However, the reader will see shortly that the two terms, obligate and strict, do not have exactly the same meaning.) Examples among bacteria are *Pseudomonas fluorescens*, a common saprophyte of river waters, and *Bacillus anthracis*, the causative agent of the disease anthrax.

Facultative Aerobes. Some microorganisms that can use free oxygen as an electron (hydrogen) acceptor can also use oxygen obtained under anaerobic conditions from easily reducible compounds that contain it, e.g., $NaNO_3$:

$$NaNO_3 + H_2 \longrightarrow NaNO_2 + H_2O$$

Because they can use, as a hydrogen acceptor, either free or combined oxygen, such organisms are said to be *facultative aerobes* (or facultative anaerobes, which are actually aerotolerantly anaerobic). Examples among bacteria are *Salmonella typhi* (cause of typhoid fever) and *Staphylococcus aureus* (cause of boils).

Obligate Anaerobes. However, there are also numerous species of bacteria that are extremely sensitive to the presence of free oxygen and are soon killed by it. These grow in the depths of the soil, in swamp muck, in the animal intestine, and in other places where air cannot penetrate or where a very low oxidation-reduction (O-R) potential is maintained. Such bacteria are called *obligate anaerobes*. Examples are *Clostridium tetani* (cause of lockjaw or tetanus) and *Bacteroides fragilis;* both are inhabitants of the human and animal intestinal tract.

Obligate anaerobes typically do not obtain energy from biooxidation by the use of oxygen, free or combined, but by enzyme-controlled dehydrogenation (Fig. 8–3). The release of hydrogen from an organic substrate also releases electrons (energy). These organisms use, as hydrogen acceptors, organic substances such as pyruvic acid:

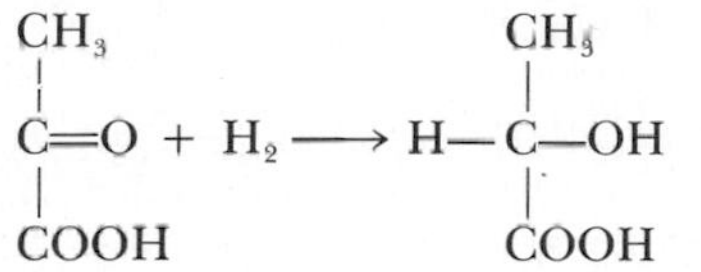

$$\underset{\text{pyruvic acid}}{\begin{array}{c} CH_3 \\ | \\ C{=}O \\ | \\ COOH \end{array}} + H_2 \longrightarrow \underset{\text{lactic acid}}{\begin{array}{c} CH_3 \\ | \\ H{-}C{-}OH \\ | \\ COOH \end{array}}$$

Although some anaerobes, called *aerotolerant*, grow better in the absence of oxygen than in its presence, the obligate anaerobes will not grow where exposed to molecular oxygen in greater than minute amounts.

Another major group of bacteria are the *strict anaerobes*. These are extremely oxygen-sensitive, and they will not survive if the growth environment is not fully chemically reduced and entirely anaerobic.

Cultivation of Anaerobes. When studying anaerobes special means must

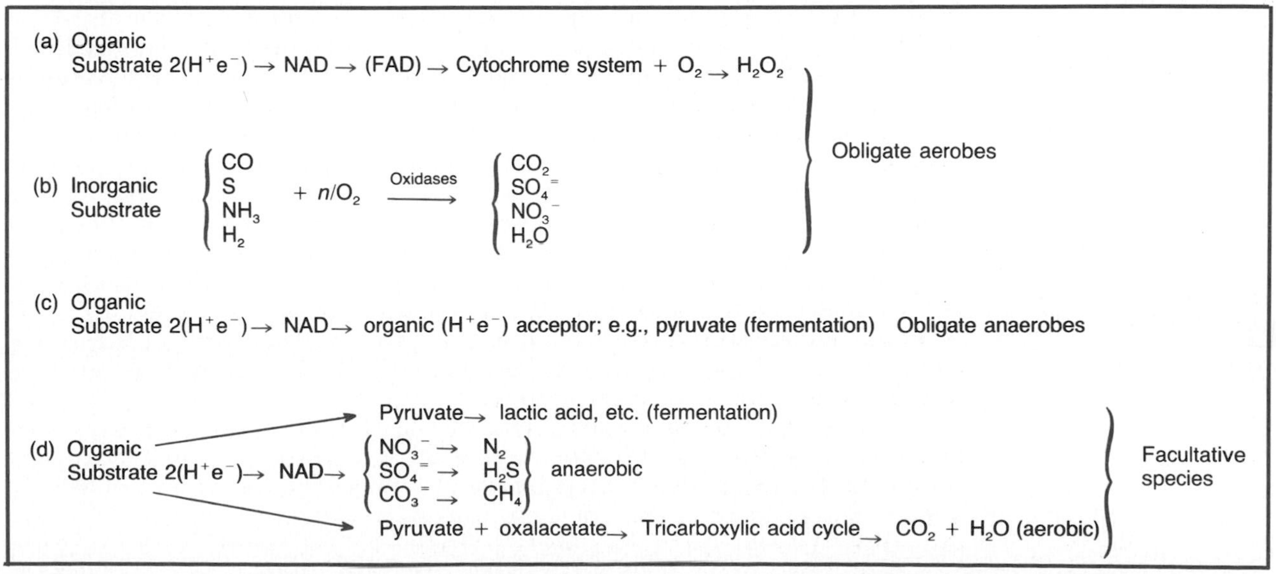

(a) Organic Substrate $2(H^+e^-) \rightarrow NAD \rightarrow (FAD) \rightarrow$ Cytochrome system $+ O_2 \rightarrow H_2O_2$ — Obligate aerobes

(b) Inorganic Substrate $\left\{\begin{array}{l} CO \\ S \\ NH_3 \\ H_2 \end{array}\right\} + n/O_2 \xrightarrow{\text{Oxidases}} \left\{\begin{array}{l} CO_2 \\ SO_4^= \\ NO_3^- \\ H_2O \end{array}\right\}$ — Obligate aerobes

(c) Organic Substrate $2(H^+e^-) \rightarrow NAD \rightarrow$ organic (H^+e^-) acceptor; e.g., pyruvate (fermentation) — Obligate anaerobes

(d) Organic Substrate $2(H^+e^-) \rightarrow NAD \rightarrow$
- Pyruvate $\rightarrow$ lactic acid, etc. (fermentation)
- $\left\{\begin{array}{l} NO_3^- \rightarrow N_2 \\ SO_4^= \rightarrow H_2S \\ CO_3^= \rightarrow CH_4 \end{array}\right\}$ anaerobic
- Pyruvate + oxalacetate $\rightarrow$ Tricarboxylic acid cycle $\rightarrow CO_2 + H_2O$ (aerobic)

— Facultative species

Figure 8–3 Some representative types of mechanisms of biooxidation. Shown are mechanisms of oxidation of (a) organic and (b) inorganic substrates by obligate aerobes; (c) oxidation of organic substrates by obligate anaerobes using organic $H(e^-)$ acceptors (fermentation); (d) the anaerobic use of either organic $H^+(e^-)$ acceptors, e.g., lactic acid (fermentation), or inorganic $H^+(e^-)$ acceptors (depending on species), or terminal oxidation via the tricarboxylic acid cycle, by facultative species. (From Frobisher, M., et al.: Fundamentals of Microbiology. 9th ed. Philadelphia, W.B. Saunders Company, 1974.)

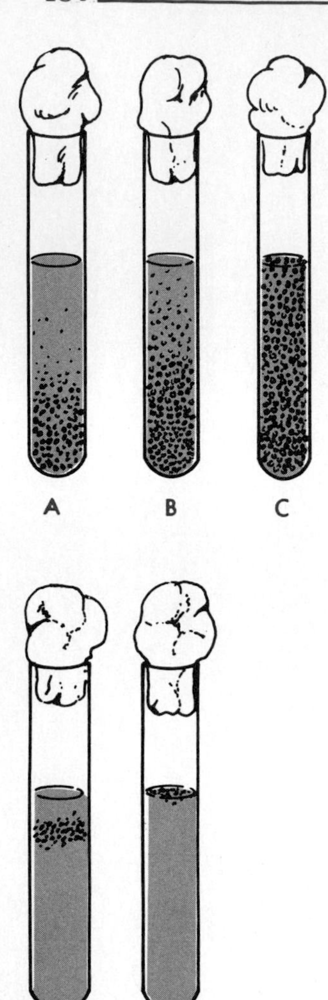

Figure 8–4 Culture tubes containing agar inoculated while still fluid (40 C), gently rotated to distribute the inoculum, then incubated at 36 C after solidifying. The amounts and distributions of growth show different types of oxygen requirements. *A*, Fairly strict anaerobe, like *Clostridium botulinum*. *B*, Less strict anaerobe, like *Cl. perfringens*. *C*., Facultative aerobe-anaerobe, like *Escherichia coli*. *D*, Microaerophilic organism, like *Brucella abortus*. *E*, Strict aerobe, like *Pseudomonas fluorescens*.

be used to exclude the atmospheric oxygen or to maintain a low O-R potential. This may be accomplished in several ways. A simple method is to melt the contents of a tube of beef-extract agar[11] and introduce the desired bacteria into the depths, allowing the agar to harden. Oxygen is thus excluded, and the bacteria will grow in the depths of the tube (Fig. 8–4). Other methods require that the cultures be placed in jars or flasks from which all oxygen is removed by means of vacuum pumps, by chemical combination, or by replacing the oxygen with some inert gas such as hydrogen or nitrogen (Fig. 8–5). Doubtlessly some of these methods will be demonstrated in the laboratory.

One of the simplest and most effective techniques is to add to ordinary culture media a harmless chemical that absorbs oxygen as fast as it diffuses into the medium from the air, such as cysteine or sodium thioglycollate. Thioglycollate broth and agar media are commonly used in medical microbiology (Fig. 8–6).

Microaerophiles. Some microorganisms appear to grow best in a situation where the amount of oxygen is somewhat reduced. These are called *microaerophilic* organisms. (See Figure 8–4*D*.) An example of this type of microorganism is *Brucella abortus*, which usually requires 5 per cent CO_2 for growth.

Other Oxygen Use Terminology. *Capneic* organisms are those that require an increased CO_2 tension rather than a decreased O_2 tension. *Oxybiontic* organisms use molecular oxygen, but *anoxybiontic* bacteria are incapable of using O_2 from the atmosphere. Organisms with oxybiontic metabolisms include aerobes like *Pseudomonas aeruginosa*, facultative anaerobes like *Escherichia coli*, and capneic species like *Haemophilus ducreyi*. As a microaerophilic organism, *Campylobacter fecalis* requires less than 10 per cent O_2. Organisms with anoxybiontic metabolisms include aerotolerant anaerobes like *Clostridium histolyticum*, obligate anaerobes like *Clostridium tetani*, and strictly anaerobic organisms like *Treponema pallidum*.

OSMOTIC PRESSURE

Many substances attract water. They are said to be *hygroscopic* or *deliquescent*. For example, observe grains of table salt or certain candies exposed on a plate during very humid weather. Dissolve a deliquescent substance in water. Let the solution then be enclosed in a tightly sealed sac of material like cellophane or the outer membrane of an egg or a living cell. If the sac or cell is then immersed in distilled water, the deliquescent *solute* (substance in solution) attracts water, which passes inward through the membrane by a process called *osmosis*. The passage of water inward builds up pressure, depending on the nature and concentration of the solute. The cell may burst, a phenomenon called *plasmoptysis*, as in the *hemolysis* (laking) of red blood cells by water. Conversely, if a living cell is immersed in a concentrated solution of a deliquescent substance (e.g., strong brine or sugar syrup), water is drawn out of the cell, which then collapses (*plasmolysis*) and may die. Cells vary greatly in their resistance to unfavorable osmotic conditions, but it is clear that osmotic pressure is of fundamental importance in all cell life. Some (but not all) pathogenic bacteria and most actively growing (nondormant) animal cells are relatively sensitive to continuous exposure to osmotic extremes.

Many saprophytic microorganisms that cause spoilage of food and other products are dehydrated by immersion in strong brines or syrups. Although not necessarily killed, they cannot grow and produce spoilage: they are held in

[11]Agar (a clear, jelly-like substance derived from certain seaweeds) is added to many nutrient solutions (culture media), used for growing microorganisms, to make solidified medium with the consistency of stiff jelly. It liquefies on being heated to boiling and "sets" at about 38 C. In some respects agar resembles pectins such as "Certo" that are used to make jelly.

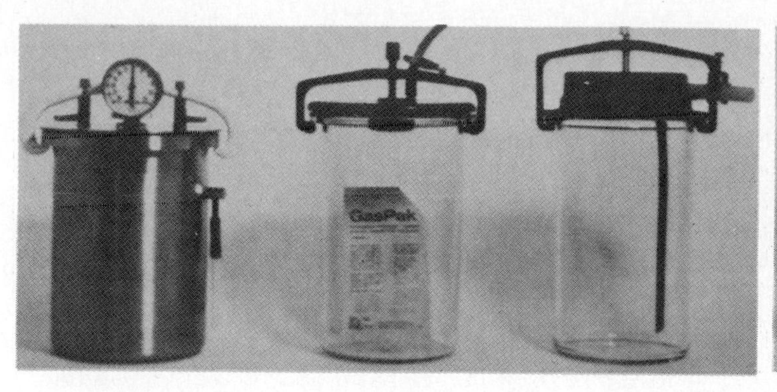

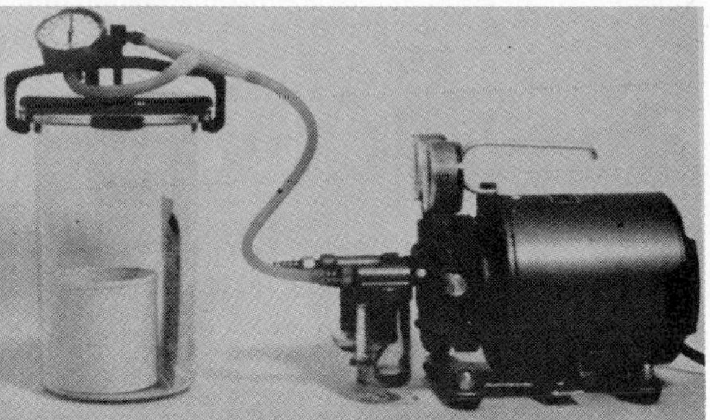

Figure 8–5 Anaerobic jars. (From Anaerobic infections. Courtesy of the Upjohn Company, Kalamazoo, Michigan.)

a dormant state. This principle is used in the preparation of many sorts of pickles, corned beef, salt fish, jams, and jellies. Soaking in brine (dehydration) was a first step used by the ancient Egyptians in the preservation of mummies. The extremely dry climate helped maintain the mummified state afterward.

Halophilic[12] organisms are those that can thrive in strong brines such as those used in "corning" beef and making certain pickles (20 to 30 per cent salt). Others have become adapted to the high salinity of the Dead Sea and Great Salt Lake, which have salinities of about 25 per cent (that of sea water is 5 per cent). These bacteria cannot thrive in ordinary fresh water or even in sea water. None is pathogenic.

ACIDITY AND ALKALINITY

Many microorganisms, especially those that cause disease, are adversely affected by even moderate degrees of acidity or alkalinity. They are so sensitive that it is necessary to adjust the reaction of the media used for their cultivation by adding acid or alkali, as may be required, so that the fluid becomes neutral or faintly alkaline like blood.

Hydrogen Ion Concentration. Many compounds of hydrogen (e.g., water, H_2O), alcohols (e.g., methyl alcohol, CH_3OH), and various acids (e.g., hydrochloric, HCl, and sulfuric, H_2SO_4), when dissolved in water, *dissociate* or *ionize* into one or more hydrogen ions (cations, H^+) and an associated anion (e.g., OH^-). Many acids such as HCl or H_2SO_4 dissociate extensively (up to 90 per cent) into H^+ and an anion (for example, Cl^-, SO_4^{--}) in aqueous solution. Certain other substances, such as sodium hydroxide, dissociate in water into free sodium ions (Na^+) and free hydroxide ions (OH^-). Free hydroxide ions

[12]Salt-loving.

Figure 8–6 The Brewer anaerobic culture dish. The cover fits over a Petri dish so that the flat rim of the cover rests on the agar surface, trapping a very thin layer of air over the surface of most of the agar. (Petri was one of the early students of bacteriology under Robert Koch, circa 1876.) Sodium thioglycollate in the agar then absorbs the oxygen from this thin layer of air so that growth on the agar surface is entirely anaerobic. (Courtesy of Baltimore Biological Laboratories, Baltimore, Md.)

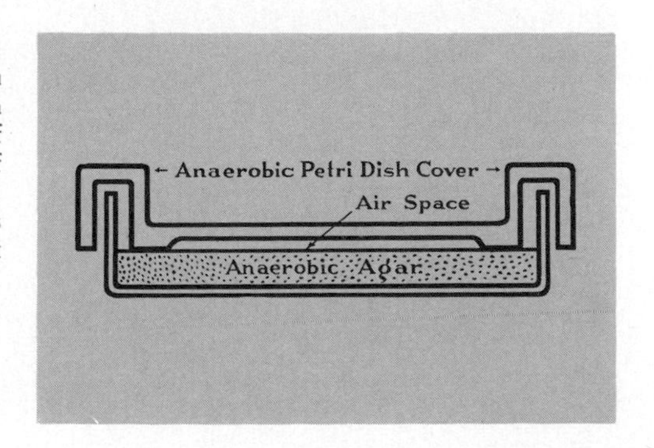

confer the property of alkalinity on aqueous solutions; free hydrogen ions confer the property of acidity.

If an aqueous solution contains a large number of hydrogen ions, it is very acid and is said to have a *high hydrogen ion concentration.* If it is only slightly acid, or if it is alkaline, i.e., contains free hydroxide (OH^-) ions, it is said to have a *low hydrogen ion concentration* (or a correspondingly high OH^- concentration).

pH. Hydrogen ion concentration is usually expressed by numbers used with the symbol pH, which stands for the negative logarithm of the hydrogen ion concentration (expressed as a fraction of a gram of H^+ per liter of water). The hydrogen ion concentration (pH) of most solutions used in microbiology and medical work is easily measured by means of dyes that assume a distinctive color at a certain pH. Dyes are also used in pH paper, with color depending on pH, which is employed for simple but relatively rough measurements. More commonly, exact pH determinations are made by using pH meters, electrical instruments of many different types, some of which are accurate to two decimal places (Fig. 8–7). In the most commonly used scale of hydrogen ion concentrations (Sorenson's) a pH of 14 is the lowest acidity[13] and pH 7 is neutral. pH 0 is the highest acidity in this system and represents 1 g of hydrogen *ions* per liter of solution. Any pH between 0 and 7 is acidic; any pH between 7 and 14 is alkaline. The pH of human blood is about 7.4; that of freshly drawn milk is around 6.8 (Fig. 8–8). Most pathogenic bacteria are inhibited by acidities of near pH 5 and alkalinities around pH 8.5 and prefer a pH close to 7.4.[14] Some free-living chemolithotrophs of the soil thrive at an acidity of pH 1.5.

The preservation of fish, meat, vegetables, and other foods by pickling in vinegar is based, in part, on the sensitivity of spoilage-causing microorganisms

[13] Actually, it is equal to strong alkali.

[14] The numbers used with pH are logarithms of fractions, i.e., negative logarithms. They are therefore inversely related to actual degrees of acidity (concentrations of hydrogen ions). Note carefully in reading, therefore, that a low pH means a high concentration of hydrogen ions or acidity, and vice versa.

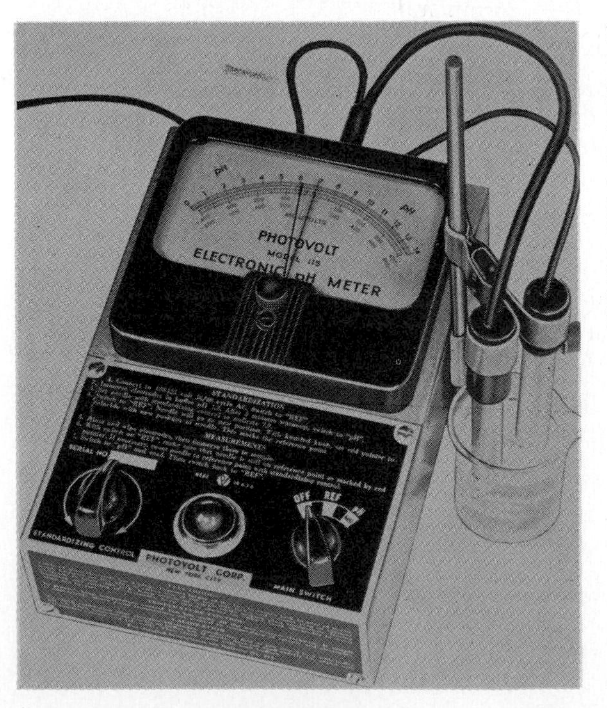

Figure 8–7 A modern pH meter for determining the concentration of hydrogen ions in fluids. The fluid to be tested is placed in the beaker at the right. The two electrodes are then inserted in the fluid. Between the two electrodes an electrical potential develops that is dependent on the pH of the fluid. The instrument measures this potential and expresses it on the dial as pH. The instrument is therefore a sort of calculating voltmeter. (Courtesy of Photovolt Corp., New York City.)

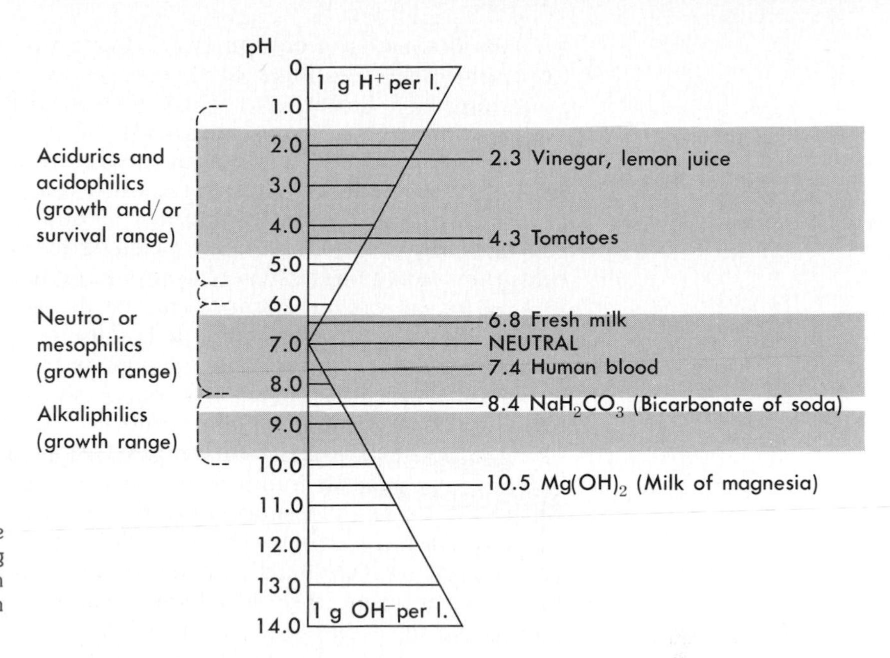

Figure 8–8 Some pH units of significance in microbiology. The slanting lines starting at pH 7.0 indicate increasing concentration of H^+ or OH^- to the maximum of 1 gram of ions per liter.

to the acidity of the vinegar. Salt, sugar, and spices are added for flavor and for osmotic effects. Some spices are said to have a preservative action, but the main reason vinegar-pickled foods "keep" is that the acid prevents bacterial growth. The un-ionized acetic acid of vinegar also is toxic per se.

RADIATIONS

Sunlight and Ultraviolet Rays (UV). Chlorophyll-deficient organisms cannot utilize sunlight as a source of energy; in fact, many species, especially unpigmented varieties of bacteria (which includes most, *but not all*, pathogens), are killed within a few minutes or hours by direct sunlight. The ultraviolet rays of sunlight in the range of about 3,000 Å (300 nm) wavelength are its most bactericidal components. Their greatest lethal effect is exerted on the DNA and enzyme proteins of the cell.

Ultraviolet rays in sunlight are absorbed to a great extent by carotenoid and other pigments. Thus, many pigmented species of bacteria, including many harmless environmental species that are adapted to life in sunlight and air, are protected from UV by their own pigments. Some very dangerous pigmented pathogens, e.g, *Staphylococcus aureus*, are also protected and can resist sunlight. In general, UV rays have very little power to penetrate anything except certain substances like quartz or some plastic or special-glass window materials. Sunlight passing through ordinary window glass loses most of its UV rays and therefore its disinfecting power.

Excessive exposure to UV radiation, solar or artificial, is to be avoided because UV rays can severely damage the eyes and produce detrimental mutagenic, carcinogenic, or oncogenic effects. Sublethal exposures to UV radiation are commonly used in research to induce mutations in many species of microorganisms.

Part of the "purifying" action of sunlight on decaying organic matter is due to the fact that it warms and dries. In addition to its UV contents, certain of the visible rays of sunlight, in the presence of oxygen, induce microbicidally destructive photooxidation of enzymes and other vital cell components, thus helping to suppress odorous decomposition of organic matter.

Various kinds of equipment that produce ultraviolet rays are available and are used in the treatment of skin infections and disinfection of blood plasma, swimming pool water, and so forth. Special lamps producing ultraviolet rays are now used to prevent the growth of surface molds and bacteria in meat-storage warehouses, in research laboratories, and even in some commercial kitchens, as well as to disinfect the air in some operating rooms.

X-Rays and Other Radiations. These are called ionizing radiations because, being of much shorter wavelength than UV (less than 1,000 Å) (100 nm), they have much greater powers of penetration. They can cause disruptive ionizations of various components of the cell, especially DNA and proteins. These effects are commonly lethal; some cells have enzyme-controlled mechanisms for the repair of the damage induced, if it is not too extensive. If exposures are not too intense or prolonged they may, like UV, produce mutations or carcinogenic alterations.

Among the ionizing radiations fatal to microorganisms are x-rays and radium emanations. Radium gives off three kinds of particles or radiations: *alpha rays* (actually helium ions traveling at very high velocities), *beta rays* (negatively charged particles—i.e., electrons—with high velocities), and *gamma rays* (somewhat like "hard" x-rays). Hard x-rays differ from soft x-rays in being of shorter wavelengths and having greater powers of penetration. *Neutrons* are one of the most damaging forms of radiations. Humans as well as microorganisms are continuously being bombarded with neutrons from outer space. Chemicals that are struck may become radioactive and then indirectly cause damage. This may be the principal cause of so called spontaneous mutations. Because of the formation of radioactive substances, neutrons are not used commercially to sterilize foods.

Unicellular organisms vary greatly in resistance to radiation. A protozoan may require 300,000 R (roentgens) to be killed, but 0.01 R is sufficient to affect the growth of *Phycomyces blackesleeanus*, a fungus. The median lethal dose within 30 days ($LD_{50}/30$) of x-ray irradiation for man is about 450 rem (roentgen equivalents, man), for yeast 30,000 rem, for some bacilli 150,000 rem, but for *Paramecium* (a protozoan) as high as 350,000 rem. Although theories exist to explain this range of responses on the basis of nuclear volume and oxygen tension, it is still not certain why some organisms are highly radiosensitive and others distinctly more resistant.

Various radiations, especially ^{60}Co gamma rays and cathode rays (streams of moving electrons) may be used for sterilizing a variety of things, such as canned foods, ready-cooked frozen foods, and surgical materials. The ionizing rays penetrate the closed packages and kill the microorganisms in them without any unwholesome change in the product. The flavors and textures of some foods suffer to varying degrees.

OTHER PHYSICAL INFLUENCES

Electricity and Magnetism. These appear to have little or no direct effect on bacteria, although detrimental effects of extremely high magnetic fields on bacteria have been reported. A powerful electric current passed for a long time through a broth culture generates heat and certain chemical compounds that may kill the bacteria.

Sound Waves. Most sound waves audible to the human ear appear to be without effect on bacteria. Waves of certain frequencies, however, can disintegrate microorganisms of some species under proper experimental conditions.

Microwaves. Microwaves are a type of electromagnetic radiation between far infrared and radio waves. Although most research with microwaves concerns effects on food, and several studies with tissue cultures have been reported, a

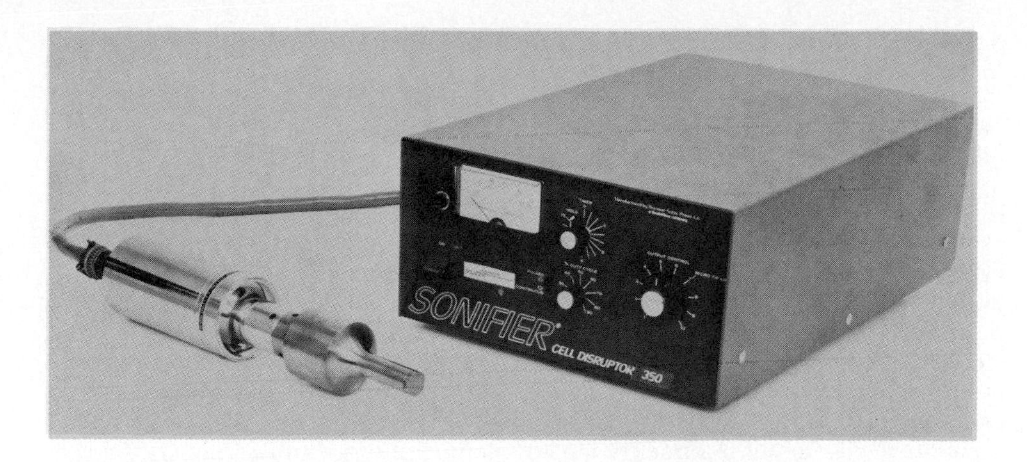

Figure 8–9 Sonic disintegrator. (Branson Sonifier, courtesy of Branson Sonic Power Company, a Smith Kline Company.)

few papers concerned with microwave sterilization appear in the literature. There are also claims that microwaves, applied correctly, can cause mutations in bacteria.

Supersonic Vibrations. If cultures are subjected to certain very rapid vibrations, higher in pitch than the highest capable of being heard by the human ear (*supersonic* or *ultrasonic vibrations*), many microorganisms are entirely disrupted and literally shaken to pieces. This method is utilized in research to investigate intracellular substances without using chemicals to rupture the cells. Figure 8–9 shows a sonic disintegrator, which is frequently used to rupture cells and free enzymes for biochemical studies.

APPLICATION TO HEALTH

To control microorganisms, changes in certain environmental factors may be all that is needed to slow down their growth or to kill them. For example, high temperatures alter the characteristics of the cells so that life is no longer possible. The presence or absence of oxygen may determine whether a pathogenic organism is capable of producing an infection or a toxin detrimental to the host. It must be remembered, however, that what seem to us to be extremely minute quantities of nutrients, moisture, and oxygen may be sufficient to support the life of millions of microorganisms. It is seldom reliable to trust in the removal of food, water, and oxygen for inhibition or destruction of microorganisms. The primary means of reducing or destroying microbial growth are altering temperatures (refrigeration) or subjecting the microorganisms to poisonous substances, i.e., disinfectants (see Section Three). Ultraviolet rays are sometimes used to destroy microorganisms in hospitals (operating rooms, nurseries, communicable disease units) or modern laboratories but are truly effective only when all other methods of control have been adequately carried out.

SUMMARY

The growth and reproduction of all living organisms, including bacteria, depend on environmental factors such as moisture, nutrients, temperature and pH ranges, osmotic pressure, and the presence or absence of atmospheric oxygen. Without abundant moisture, vegetative cells will not reproduce, although dormant cells such as spores or cysts may form. Foods (hay, meat, fruit, soup) may be preserved and protected against certain microorganisms (not molds, however) through dehydration. Freeze-drying, also called lyophilization, is an excellent way to keep bacteria alive but dormant for years. It is also used to protect sera and vaccines against spoilage.

Metabolism is the digestion and utilization of food to synthesize the proteins, fats, carbohydrates, and other substances of which living cells are made and to furnish the energy necessary for life and reproduction. The ways in which nutritional constituents can be used vary with different kinds of cells, but all require inorganic and organic substances as building materials for life.

Organisms that can use CO_2 as a sole source of carbon are called autotrophs; heterotrophs require carbon in organic chemical combinations. These organisms may be photolithotrophic, photoorganotrophic, chemolithotrophic, or chemoorganotrophic, depending on whether they derive their energy through phototrophic reactions via photosynthesis or chemotrophic biooxidations. All microorganisms of medical importance—the parasites and pathogens—are chemoorganotrophs. Viruses receive both their energy and substance from their host cells. Being neither chemotrophic nor phototrophic, they are said to be paratrophic.

The terms psychrophilic (cold-loving), mesophilic, and thermophilic (heat-loving) are used to describe the temperature ranges at which certain microorganisms live best in nature. All human pathogens are mesophiles.

Depending on their responses to free atmospheric oxygen, bacteria are either aerobes or anaerobes (some are microaerophilic). Obligate aerobes must have free oxygen; facultative aerobes can grow in the absence of atmospheric oxygen. A strict anaerobe is destroyed when exposed to even the smallest amount of free oxygen, but an obligate anaerobe is somewhat more aerotolerant.

The acidity or alkalinity of an environment determines enzyme expressions in microbial metabolism and thus may easily inhibit growth. Radiation may result in mutations or even kill the cells exposed to it. Most harmful to microorganisms are x-rays, radium emanations, and gamma rays. Beta rays are less effective, and alpha rays are least harmful. Neutrons from outer space continuously bombard the earth and cause "spontaneous" mutations. Radiation is used effectively to sterilize a variety of prepackaged foods and canned foods. Sunlight and ultraviolet light are also destructive to microorganisms. Special lamps producing ultraviolet rays are used to disinfect the air in meat storage warehouses, research laboratories, kitchens, and operating rooms.

REVIEW QUESTIONS

1. What is the relation of moisture to the growth of bacteria? What foods are preserved by drying? What principle is involved in drying foods? How is drying used in the laboratory? How is lyophilization being used by the ATCC?
2. What is chlorophyll? What is meant by photosynthesis? What is the effect of sunlight on bacteria?
3. What kinds of food are needed by bacteria?
4. Define autotrophic, saprophytic, heterotrophic, photolithotrophic, chemoorganotrophic, parasite, and pathogen.
5. What are the temperature requirements for bacteria? What is meant by mesophilic, psychrophilic, and thermophilic, with respect to bacteria?
6. Describe the requirements of bacteria for oxygen. What is an obligate aerobe? What is the difference between an obligate anaerobe and a strict anaerobe? Give examples of different oxygen tolerances or requirements among bacteria.
7. How is a strict anaerobe cultivated?
8. What is the importance of pH in the growth of bacteria?
9. Why should a health worker be familiar with environmental factors of growth that are important to bacteria?

Supplementary Reading

Bernstein, I. A.: Biochemical Responses to Environmental Stress. New York, Plenum Press, 1971.

Brock, T. D., and Darland, G. K.: Limits of microbial existence: Temperature and pH. *Science, 169*:1316, 1970.

Farrell, J., and Rose, A. H.: Temperature effects on microorganisms. *Ann. Rev. Microbiol., 21*:101, 1967.

Finegold, S. M.: Anaerobic Bacteria in Human Disease. New York, Academic Press, 1977.

Frazier, W. C., and Westhoff, D.: Food Microbiology. 3rd ed. New York, McGraw-Hill, 1978.

Frobisher, M., Hinsdill, R. D., Crabtree, K. T., and Goodheart, C. R.: Fundamentals of Microbiology. 9th ed. Philadelphia, W. B. Saunders Company, 1974.

Harrison, A. P., Jr.: Survival of bacteria. *Ann. Rev. Microbiol., 21*:143, 1967.

Haynes, R. H., Wolff, S., and Till, J. (eds.): Structural Defects in DNA and their Repair in Microorganisms. Radiation Research, Supplement 6. New York, Academic Press, 1966.

Josephson, E. S., Brynjolfsson, A., and Wierlicki, E.: Engineering and economics of food irradiation. *Trans. N.Y. Acad. Sci.*, Ser. II, *30*:600, 1968.

Ladanyi, P. A., and Morrison, S. M.: Ultraviolet bactericidal irradiation of ice. *Appl. Microbiol., 16*:463, 1968.

Mazur, P.: Cryobiology: The freezing of biological systems. *Science, 168*:939, 1970.

Morgan, K. Z., and Turner, J. E.: Principles of Radiation Protection. Reprint ed., Huntington, N. Y., Krieger, 1973.

Nagington, J., and Lawrence, M. F.: Notes on the storage of tissue culture cells with liquid nitrogen. *Monthly Bull. Minist. Health, 21*:162, 1962.

Rose, A. H. (ed.): Thermobiology. New York, Academic Press, Inc., 1967.

Shilling, C. W. (ed.): Atomic Energy Encyclopedia in the Life Sciences. Philadelphia, W. B. Saunders Company, 1964.

LABORATORY STUDY OF MICROORGANISMS

PURE CULTURES

Pure, Mixed, and Contaminated Cultures. A pure culture is one containing only one kind of microorganism. A *mixed culture* contains two or more kinds of microorganisms. A pure culture may be likened to a bed of roses. In such a bed there is nothing but roses and we might say that the gardener has made a *pure culture* of roses. If some geraniums or other kinds of plants were also put in the bed, we might say that the gardener had prepared a *mixed culture.* If weeds got in accidentally, we would say that the culture of roses was *contaminated.* Likewise, when some undesirable microorganisms get into our pure culture by accident, we say that it is contaminated. Sometimes a culture of more than one type of microorganism is called a mixed culture.

Isolation of Pure Cultures of Microorganisms. In order to determine accurately the identity of a specific microorganism obtained from a patient, the microbiologist must first isolate the organism from all others with which it is mixed. Mixtures of microorganisms are common in clinical material such as feces, sputum, and the like. The activities of extraneous organisms (contaminants) in the tests to be made with the organism that is being studied inevitably mask and confuse the results. The contaminants may greatly change the pH; they may produce antibiotics; they may influence, injure, modify, or kill the desired microorganisms in a hundred subtle ways. What is needed is a *pure culture.*

Pure-Culture Methods. In preparing a pure culture of bacteria, or of yeasts or molds, it is very convenient to have a solid surface on which to spread out the mixture of microorganisms, like spreading coins on a table to sort the different kinds. Then we can pick out the kinds we want and put each separately into its own test tube medium to multiply and grow as a pure culture.

Agar. A very useful substance for this purpose is the pectin-like, polysaccharide vegetable gum called agar. If dried, agar is added to a nutrient fluid such as beef broth and boiled at 100 C. This melts the agar. It remains a liquid until the temperature falls below 45 C. From 1.5 g to about 2.0 g of agar per 100 ml of broth is used. When this beef-broth agar mixture is sterilized and then poured into a sterile, flat, covered Petri dish and allowed to cool, it solidifies. We can then spread, on its moist, jelly-like surface, a droplet of the microorganism-containing sample to be analyzed (saliva or sewage, for instance). If we are working with yeasts, molds, or bacteria, and we keep the dish at body temperature (37 C) for 24 hours (a process called *incubation),* it is very easy to obtain a pure culture of microorganisms we wish to isolate from among many others that might be present.

Colonies. The question at once arises: if microorganisms are so small that they are invisible, how can we select one from a mixture of perhaps hundreds on the agar surface? It is simplicity itself. If the agar contains beef broth (or another appropriate nutrient substance), it will nourish the microorganisms. When the dish is put in a warm place, each viable[1] cell multiplies at the site on the agar surface where it initially landed. In 24 to 48 hours or longer, depending

[1]Capable of living and multiplying.

CHAPTER 9

on the species of microorganisms, small, separate masses or *colonies* of the microorganisms are visible to the unaided eye on the surface of the agar. Each colony consists of all the descendants of the single microorganism that landed previously on that particular spot (Fig. 9–1). All of the cells in a single colony, being descendants of a single cell, are of one kind. All that is now necessary to obtain a pure culture from any colony is to touch it with the end of a sterile needle (Fig. 9–2) and transfer the microorganisms adhering to the needle to whatever kind of medium is desired for best growth of the organism. The medium is *incubated* (Fig. 9–3) for as long as is necessary to produce visible growth of a pure culture.

Methods to Obtain Isolated Colonies. One way of spreading the bacteria evenly over the plate is by pipetting a small amount of bacteria in liquid medium, or in water dilution, into a Petri dish. Liquid agar medium at 45 C is poured over the bacteria, and the plate is gently tilted and rotated in a horizontal position to distribute the cells evenly. Another method commonly used is called "streaking" a plate. Let your laboratory instructor show you how to streak! If your method of streaking results in *isolated colonies* on the medium surface (Fig. 9–1), you are doing it right. If not, try again—and remember to streak very closely together. Recently a new machine has become available that rotates a Petri plate under a laterally moving "pen" so that a drop containing bacteria will be spread evenly over the agar surface and distribute the cells to produce perfectly isolated colonies.

Methods of Counting Bacterial Colonies. Frequently it is necessary to determine the number of bacterial colonies on a plate, in order to calculate the total number of original cells in the sample. By knowing exactly how much of a dilution was pipetted to a plate one can easily calculate the number of cells in the sample. Bacterial colony counters are used in all bacterial laboratories to

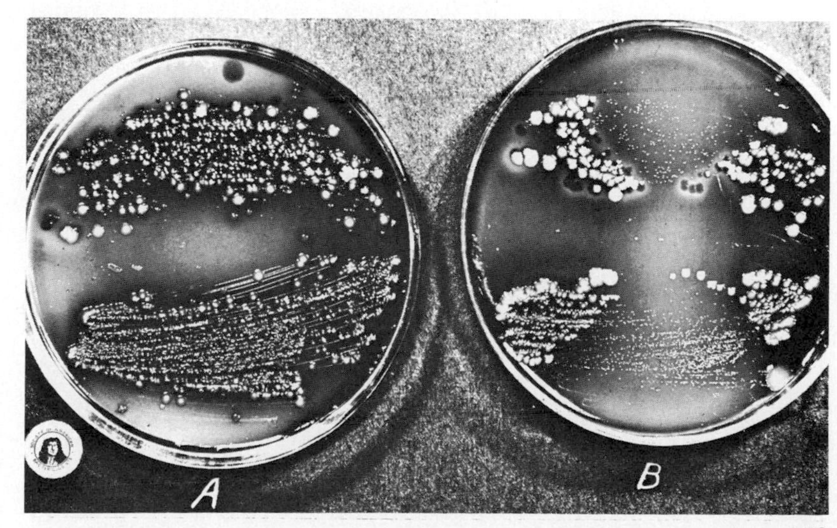

Figure 9–1 Bacterial colonies on blood-agar in Petri plates (about one-half actual size). This picture also illustrates a method of *selective cultivation* of a particular species of bacteria, in this case the organisms of whooping cough, or pertussis. A nasopharyngeal swab with mucus from the patient was passed over both plates, and the mucus was spread over the agar with a sterile loop. At the same time a drop of penicillin solution was placed on two sides of plate *B*. The plates were then incubated at 37 C for three days. Plate *A* shows the customary heavy growth of common bacteria usually found in the nasopharynx. The pertussis organisms *(Bordetella pertussis)* are completely overgrown and obscured. Plate *B* shows two areas where the common organisms have been completely inhibited by the penicillin, permitting *B. pertussis* (tiny white colonies) to grow unhampered by competition of the other organisms. (Courtesy of Dr. William L. Bradford. From the collection of the American Society for Microbiology.)

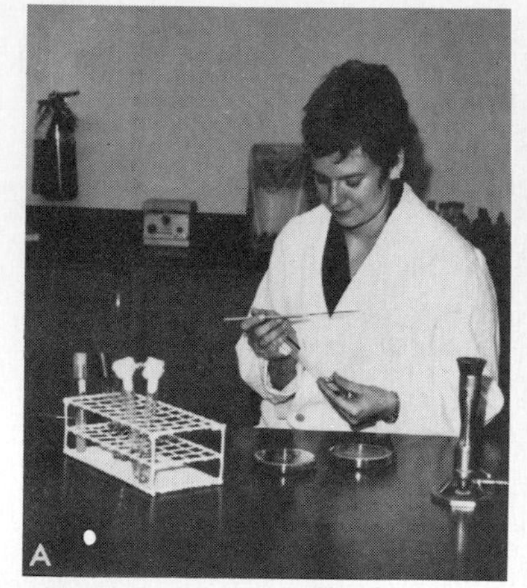

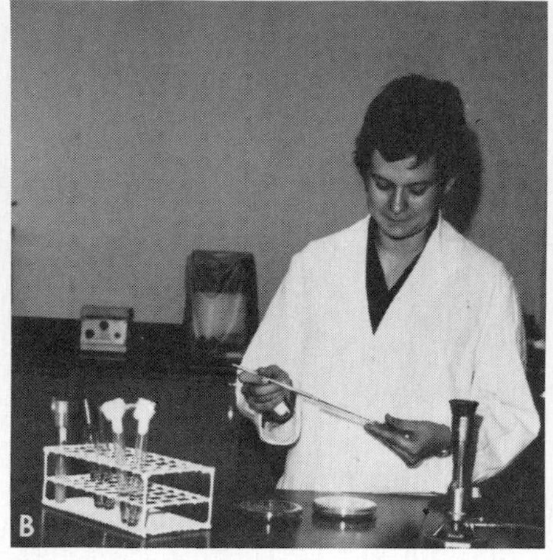

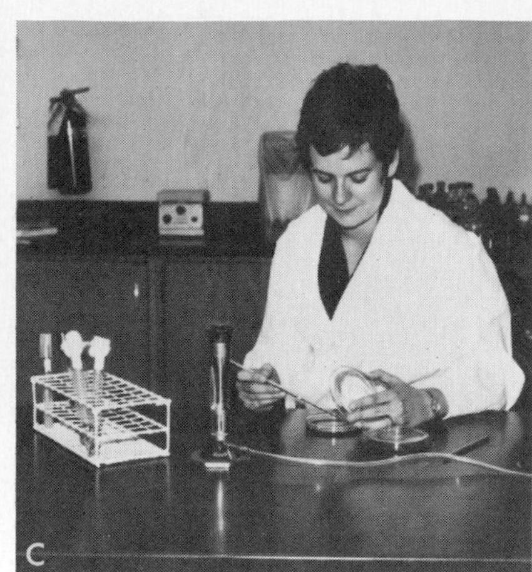

Figure 9–2 Subculturing bacteria. Transfer of bacterial colonies from Petri dish to tubes containing sterile culture media. By this method the desired strains of bacteria are isolated in pure culture. Note the holding of the cotton plug and the needle used, also the Bunsen burner for flaming (sterilizing) the entire wire of the needle. For details of the technique shown here (which may vary in different laboratories), the reader is referred to Fuerst: Laboratory Manual and Workbook for Microbiology in Health and Disease. 7th ed. Philadelphia, W. B. Saunders Company, 1983. (Photographs through the courtesy of Mr. Stephen E. Fisch.)

"count plates," as it is commonly termed. Methods of counting colonies are continually becoming more sophisticated. Completely automatic colony counters are now available. Such a model is shown in Figure 9–4. The technician places the Petri dish on the stage of the counter, presses a button, and takes the reading of the number of colonies per plate from the visual display.

Appearance of Colonies. Microbial colonies vary greatly in appearance. Some look like dewdrops; others are white and glistening like the head of a white pin; still others have various colors (Fig. 9–1). These peculiarities are used in identification. Some of the various forms of colonies arising from a pure culture have been described in the discussion of microbial variation.

When bacteria grow in broth, they may cause cloudiness, or *turbidity;* sometimes they form a surface film called a *pellicle;* at other times they produce a *sediment.*

Prevention of Contamination. To prevent contamination of the pure culture by extraneous bacteria, yeasts, or molds, which are always present in the air, on dust, on the hands of the bacteriologist, and, in fact, everywhere in the general surroundings, all glassware and media used in microbiology are sterilized before use. Precautions employed routinely to prevent contamination of pure cultures consist of cotton plugs, plugs of foam rubber–like substances, or plastic or metal caps on all tubes; glass, plastic, or metal covers on Petri plates; sterilization of needles or pipettes used to inoculate; minimal exposure to air and dust; the use of special so-called "bacteriological hoods," with

Figure 9–3 Pure culture preparation on a commercial scale. The production of pneumonia vaccine begins with carefully controlled growth of the pneumococci in an 800-liter tank, as shown here. (Courtesy of Merck Sharp & Dohme, West Point, Pa.)

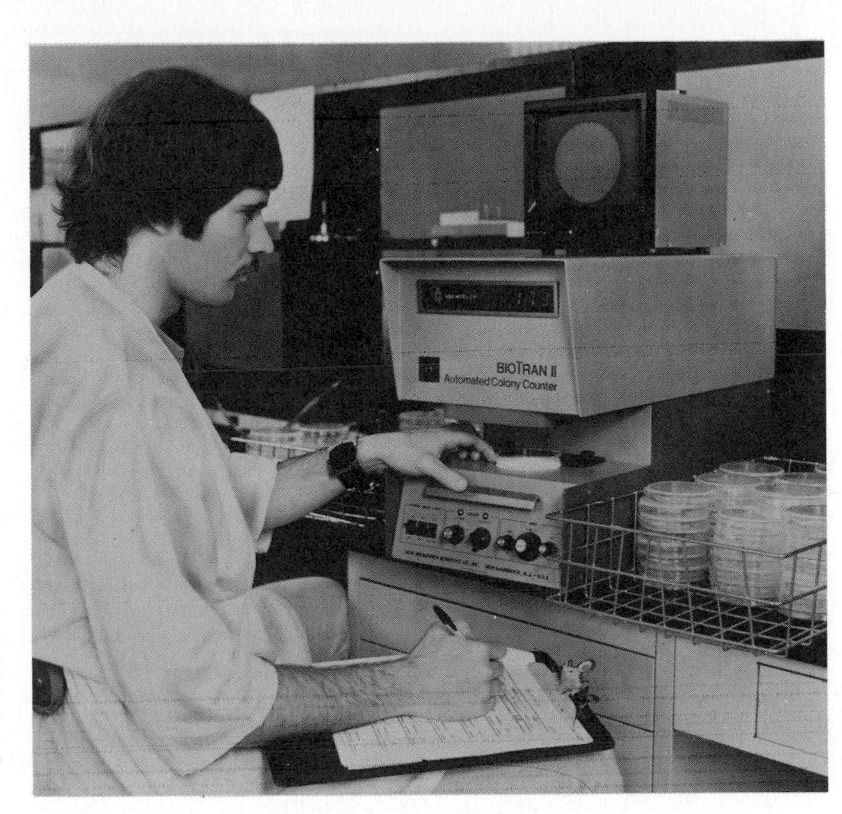

Figure 9–4 With automatic bacterial colony counters such as this, the technician can determine at a glance the number of colonies growing in a Petri dish. The colonies are scanned, the count is registered digitally, and the colonies are marked and shown on a vidicon screen. (Courtesy New Brunswick Scientific Co., Inc., New Brunswick, N.J.)

controlled air flow away from the technician; and rapid work by a skilled microbiologist.

Growth of Bacteria. Under the most favorable circumstances a single bacterium of certain species may divide in 10 to 30 minutes. Calculation will show that in 12 hours a single organism could give rise to over 68 billion cells. If conditions were always suitable, the earth would be overrun with bacteria. Like all other living things, however, they have to struggle for existence, and when conditions are unfavorable they multiply slowly or not at all. Food gives out, noxious waste products accumulate, the acidity increases, and other changes occur so that rapid growth soon ceases. The same principles apply equally to yeasts, molds, and most other *cellular* microorganisms (viruses are not cellular). If the numbers of cells in a pure culture are counted at regular intervals, starting with the inoculum (the cells that are transferred), and the logarithms of the numbers are plotted against time, a curve is obtained showing the changes in numbers with time and altered growth conditions. This is called a "growth curve" (Fig. 9–5), but it is actually a population curve since it is generally equally applicable to most populations, be they protozoa, humans, or pure cultures of living bacteria such as *Escherichia coli.*

Culture Media. The most commonly used culture medium for the growth of heterotrophic species of bacteria is *nutrient broth,* easily converted into *nutrient agar* by the addition of 1.5 to 2.0 per cent agar agar.[2] *Nutrient broth* is essentially meat extract, or weak "beef tea." Several modifications of this medium exist, all based on the following formula:

Nutrient Broth

Meat extract	3 g
Peptone	5 g

[2]Agar agar is often used to designate the agar material added to liquid medium in order to make it solid. By the term agar (above) the microbiologist often means "nutrient agar," a special medium made from nutrient broth with agar agar, the solidifying agent, added.

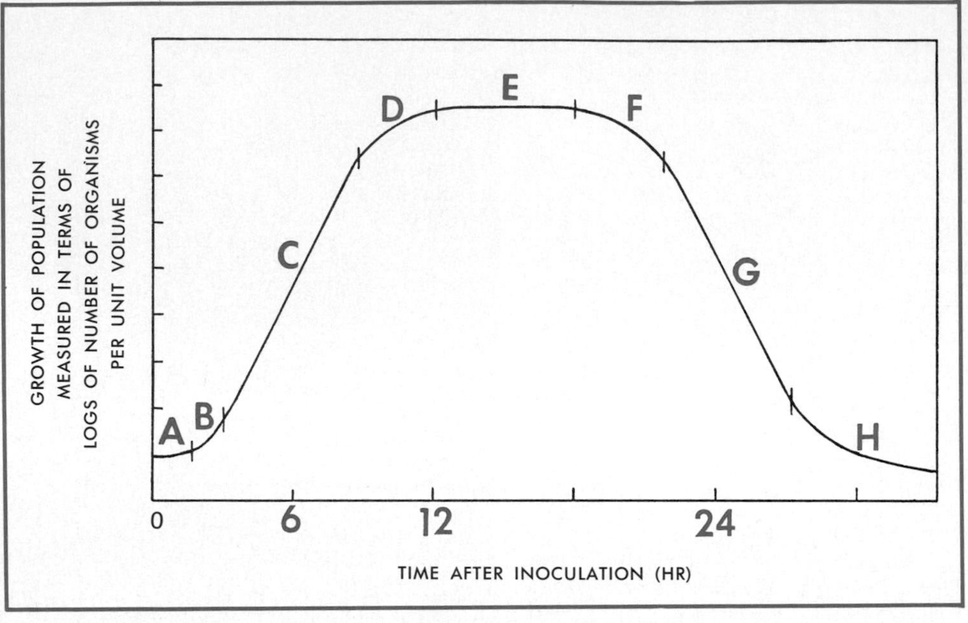

Figure 9–5 Growth curve of cellular organisms under initially optimal conditions of growth. This is a typical population curve whether it applies to bacterial, yeast cell, animal, or human populations. The curve shows changes in numbers of cells or organisms expressed as logarithms (logs) of the actual numbers. For rapidly growing organisms such as *Escherichia coli* the units of time shown could be five hours each. For slow-growing organisms such as tubercle bacilli (*Mycobacterium tuberculosis*) they could be five days each. For humans in the United States, though not unicellular, the time units could be 30 or 40 years each. (Who knows in what part of the growth curve the population of the United States is today?)

At inoculation into a new culture tube at time 0 a certain number of organisms constitute the start of the new population. During phase *A* these organisms become adjusted to the new medium and environment; there may even be a slight decrease in total cell number. This part or phase of the curve (*A*) is referred to as the lag phase. *B* is the phase of accelerated growth. *C* is the logarithmic growth phase, during which numbers increase at a steady rate, such as doubling in each unit of time. The curve in this phase is a straight line. As living conditions become unfavorable, because of exhaustion of food or oxygen and the accumulation of poisonous waste products, growth slows down (*D*, phase of negative growth acceleration), remains briefly at a *steady maximum (E)*, begins to decline (*F*, accelerated death phase; *G*, logarithmic death phase), and finally levels off at a slowly declining minimum, until the culture finally dies out altogether (*H*, phases of readjustment and death). If both living and dead organisms in the culture are counted, the curve for total organisms does not decline. Will a nuclear war make phase *F* of the growth curve of the United States assume a perpendicular position and eliminate phases *G* and *H*?

If a few cells in the culture are transferred to a new tube with medium, the individuals taken from phase *C* will adapt to the new environment quickly, those from phase *E* less quickly, and those from phase *H* perhaps not at all. How will the human population adapt itself to new planets when this test tube called "earth" has become contaminated beyond redemption with radioactive waste and industrial pollution?

These ingredients, in dry form, are dissolved in 1 liter of distilled water. The final pH as adjusted for dehydrated medium is 6.8, although a slightly alkaline pH is desirable in a freshly prepared medium (not commercial).

It is not necessary anymore to prepare beef extract by boiling fat-free beef, weighing out *peptone* (partially digested meat), filtering both, and thus making an *infusion broth.* In the modern laboratory, one simply checks the label on a bottle of dehydrated commercially prepared culture medium, adds water to a weighed amount of dried powder, as directed, and fills test tubes not more than half full (to allow for expansion) with the medium. The tubes are plugged with cotton or closed with plastic or metal caps and sterilized in the autoclave.

Various substances are added to broth or agar media to help the desired microorganisms to grow. This is also helpful in determining what changes are produced in the food substances by the microorganisms. In medical microbiology blood (which is free of microorganisms when carefully obtained from a healthy animal) is frequently added. The broth is then spoken of as "blood broth." With agar added, it becomes "blood agar," a very useful medium for the growth and identification of many human pathogenic bacteria. Various kinds of carbohydrates and similar substances, such as sucrose, melibiose, *d*-mannitol, glucose, starch, or lactose, may be added along with the blood or instead of it, and the broth then takes its name from the substance added: lactose broth, maltose broth, and so forth. Other varieties of media are prepared from eggs, various vegetables, broth of pork or veal, milk, and other substances.

Synthetic Media. For many purposes, such as research and the making of vaccines, it is desirable to use media made up of chemically pure compounds in absolutely exact, reproducible proportions. Such media are called *synthetic* or

Table 9–1. An Example of a Bacterial Assay Medium as Available Commercially in Dehydrated Form for the Quantitative Assay of the Amino Acid *l*-Leucine*†

Ingredient	Amount	Ingredient	Amount
Bacto-Dextrose (Glucose)	50 g	Adenine Sulfate	0.02 g
Sodium Acetate	40 g	Guanine Hydrochloride	0.02 g
Ammonium Chloride	6 g	Uracil	0.02 g
dl-Alanine	0.4 g	Xanthine	0.02 g
l-Arginine Hydrochloride	0.484 g	Thiamine Hydrochloride	0.001 g
Bacto-Asparagine	0.8 g	Pyridoxine Hydrochloride	0.002 g
l-Aspartic Acid	0.2 g	Pyridoxamine Hydrochloride	0.0006 g
l-Cystine, Difco	0.1 g	Pyridoxal Hydrochloride	0.0006 g
l-Glutamic Acid	0.6 g	Calcium Pantothenate	0.001 g
Glycine	0.2 g	Riboflavin	0.001 g
l-Histidine Hydrochloride	0.124 g	Nicotinic Acid	0.002 g
dl-Isoleucine	0.5 g	*p*-Aminobenzoic Acid, Difco	0.0002 g
l-Lysine hydrochloride	0.5 g	Biotin	0.000002 g
dl-Methionine	0.2 g	Folic Acid	0.00002 g
dl-Phenylalanine	0.2 g	Monopotassium Phosphate	1.2 g
l-Proline	0.2 g	Dipotassium Phosphate	1.2 g
dl-Serine	0.1 g	Magnesium Sulfate	0.4 g
dl-Threonine	0.4 g	Ferrous Sulfate	0.02 g
dl-Tryptophane	0.08 g	Manganese Sulfate	0.04 g
l-Tyrosine	0.2 g	Sodium Chloride	0.02 g
dl-Valine	0.5 g	in H_2O to	1 liter

*The growth of *Leuconostoc mesenteroides* is measured acidimetrically or turbidimetrically in this medium with small graduated amounts of l-leucine added. From the results in these "control tubes" a standard curve is constructed for comparison with the "unknown." The unknown amount of l-leucine in a measured sample of material is determined by comparison of growth in the tube of medium containing the sample with the known amounts of l-leucine in the control tubes as indicated on the standard curve.

†From Difco Manual of Dehydrated Culture Media and Reagents for Microbiological and Clinical Laboratory Procedures. 9th Ed. 1965, Detroit, Difco Laboratories.

defined media. They range in complexity from simple, weak, aqueous solutions—such as 1.0 g KH_2PO_4, 1.5 g $NaNH_4HPO_4$, 0.2 g $MgSO_4$, 3.0 g Na citrate (or glucose), or 1 liter water—to very complex solutions containing many amino acids, various vitamins, one or more carbohydrates, purines, pyrimidines, salts, and so on. The formula used depends on the nutritional requirements of the species of organism being cultivated (Table 9–1).

CULTIVATION AND STUDY OF VIRUSES

Cell Cultures. As mentioned in Chapter 6, viruses can be cultivated only in susceptible, actively metabolizing cells. These may be in an intact animal or plant, in cell cultures, or in embryonated eggs. Consequently, any culture of viruses really involves a twofold system: first, the virus itself, and second, the living host cell that will support the reproduction of the virus. The host cell may be a susceptible plant, animal, or bacterium, depending on the particular virus under consideration, or it may be a developing chick embryo, which has already been described in detail as a laboratory host for microorganisms. Cell culture techniques provide an indispensable laboratory or "test tube" (in vitro) device for the investigation of viruses (see pages 87 and 88).

Cell cultures may be grown as a single (mono-) layer or sheet of cells on glass surfaces (e.g., flasks, Petri dishes, test tubes) and maintained in a complex fluid culture medium. Following the inoculation of such cultures, viral multiplication in the growing cells may be detected by microscopic observation of the cytopathic effects (CPE) due to the virus in the cells.

Following inoculation of the sheet of cells with a suspension of virions,[3] the cell culture may be overlaid with a thin film of agar gel to immobilize both cells and virions. After an appropriate incubation period, multiplication of virions may be detected by the presence of a readily observable, clear, cell-free zone around each original virion. This zone is caused by destruction of the cells by the virus growth. The clear area is called a *plaque*. The rest of the plate, without plaques, is covered (seeded) with unaffected cells.

Observation and Study of Viruses. It seems that one of the simplest methods of studying viruses would be to observe them directly. Unfortunately, because of their small size as well as other properties, only some of the larger viruses may be investigated in this way. Most viruses are studied by the use of the *electron microscope,* which will be explained later in this chapter.

Ultracentrifugation. Viruses and many types of molecules, such as DNA or proteins, may be separated, purified, and studied by a number of physical and chemical procedures. One such procedure is differential or density gradient *ultracentrifugation,* or centrifugation at extremely high speeds in layered fluids stratified according to specific gravities. This results in particulate matter forming sediments at different levels, the strata depending on the weights, shapes, and sizes of the particles. Thus, a virus suspension may be purified by this physical process of differential sedimentation.

Ultrafiltration. Another method of isolating viruses and studying their particle size and form is by *ultrafiltration.* Here, suspensions of virus-containing materials (following coarse filtration to remove bacteria and other large particles) are passed through membrane filters. These filters are made of material with pores of a size that will retain particles of a given dimension—for example, a "large" virus—at the same time permitting others, such as "small" viruses, to pass through, thus providing a means of separating viruses according to their particle size.

Immunology. The use of various antigen-antibody reactions (e.g., agglutination and complement fixation reactions, fluorescent antibody staining, and immunoelectrophoresis) as means of detecting the presence of viruses will be considered under the general topic of immunity (Chapter 19).

Other Methods. Other methods and techniques useful in the study and identification of viruses are: determination of type of nucleic acid present (DNA or RNA; see Chapter 6); plaque formation on cell monolayers and other indicator systems; timing of the replication cycle (adsorption, penetration, eclipse, and maturation times of the virus[4]); effects of heat; effects of lipid solvents (e.g., ether or chloroform); and sensitivity to acid (pH 3).

MICROSCOPY

Observation of Morphology.[5] In order to work with any kind of object it is obviously important, although not absolutely essential, to see it. In identifying an unknown microorganism, usual first steps are determination of its form, motion, staining reactions, and other visible characteristics, all of which require a clear view of the individual cells. Although von Leeuwenhoek could see bacteria at magnifications of ×200 to ×300, modern analytical and differential microbiology necessitates the use of the "ordinary" or *optical* compound microscope, which uses ordinary, visible light (Fig. 9–6), and also the electron microscope, which uses no light at all but streams of invisible electrons (Fig. 9–7).

---

[3]The virion is the complete virus particle.

[4]The eclipse phase (when no virus can be recovered) and the maturation time of the virus are known collectively as the *latent period.*

[5]From Greek *morphe,* form; morphology is the study of form and structure.

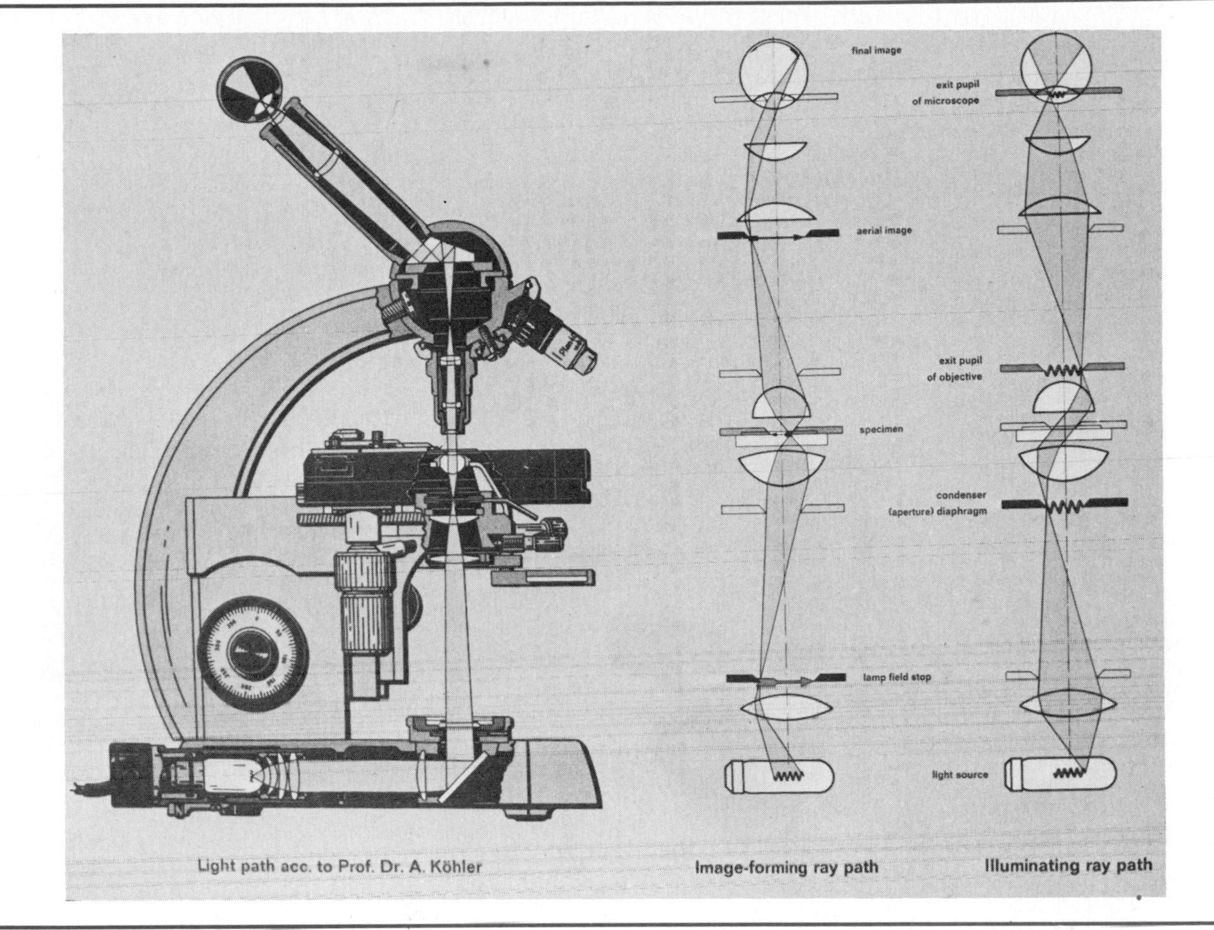

Figure 9–6 Vertical section of a widely used type of compound optical microscope. At the foot is an adjustable built-in light source; above it a *lamp field stop* and an *aperture diaphragm* for light control. Immediately above the diaphragm is a multilens condenser on a vertical-motion rack-and-pinion mount. Above the condenser is the stage with a specimen on a slide. Immediately above the stage, one of several objectives (available on the revolving nosepiece) is in place be neath the barrel of the microscope, the barrel being mounted on a vertical rack-and-pinion arrangement for focusing. At the top is the eyepiece of the ocular lens combination, which magnifies the image produced by the objective lens combination. The final image is formed on the retina by the image-forming rays. (Courtesy of Carl Zeiss, Oberkochen, Württemberg.)

The Optical Microscope. Simple magnifying glasses were in use in ancient times, but the use of combinations of two or more lenses arranged in a tube so that one lens magnifies the image produced by another (the compound microscope) originated, probably, with a Jesuit scholar, Athanasius Kircher (1602–1680). Because the magnified image is seen directly with the eye, microscopes using visible light are called *optical microscopes* (Fig. 9–6), in contrast to microscopes that use invisible x-rays, ultraviolet rays, or electrons (Fig. 9–7).

The lens nearest the object being observed is called the *objective*. In optical microscopes used in microbiology the objective lens is small, has a very short focal distance, like a nearsighted eye, and must be very close to the object. The second lens, which magnifies the image made by the objective, is some inches above the latter. Since the eye is nearer to the second lens, this lens is called an *eyepiece* or *ocular*. To make the image clear and free from distortion, modern microscope manufacturers have added various accessory lenses to both objective and ocular, but the basic principle still holds. The highest magnification given by the ordinary objective is about ×90. This image is magnified usually by a ×10 ocular, giving a total enlargement of the image ×900. Several combinations of other objectives and oculars, giving total magnifications of from around ×50 to somewhat above ×1,000 are usually available with a good microscope. Direct magnifications of ×2,000 yield "fuzzy" outlines. Clear pictures at ×2,000 are obtained only by photographic enlargements of pictures taken at lower magnifications under ideal light conditions. They actually do not show much more detailed structure than the original direct image.

Because of the small lenses used in high-power objectives a strong source

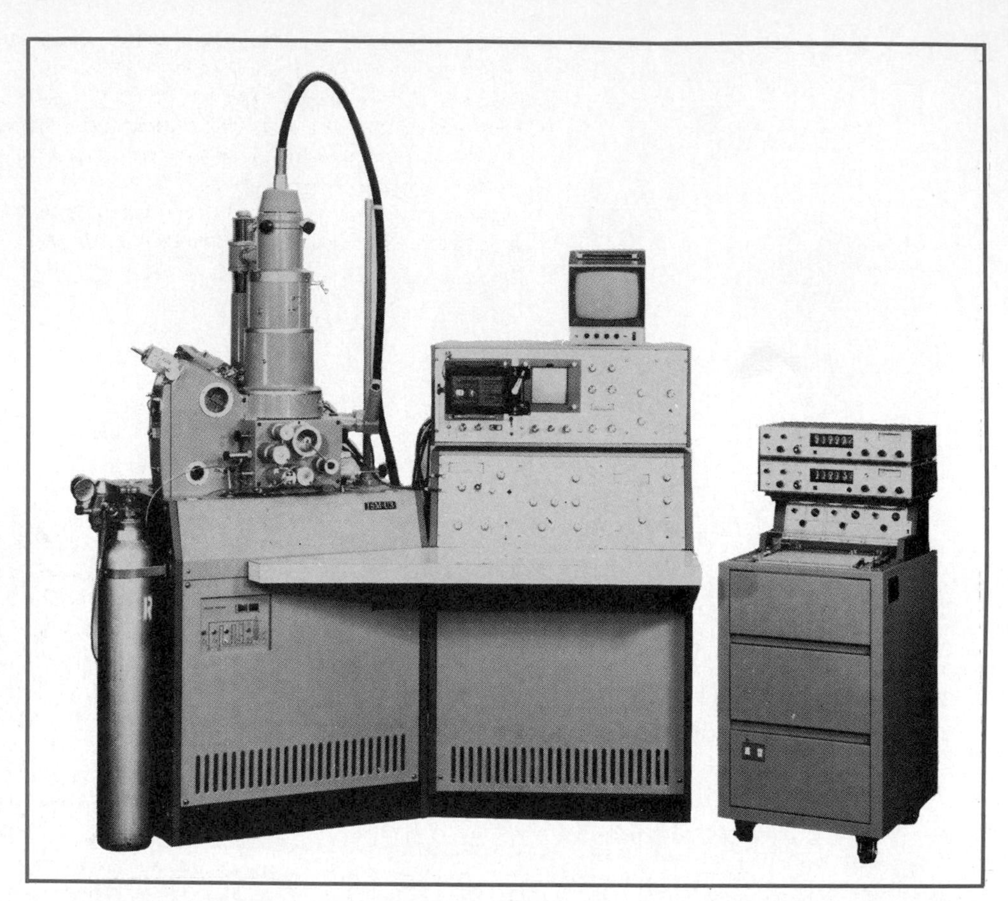

Figure 9–7 A scanning electron microscope, with the electron column, sample port, and vacuum equipment in the left console; controls and cathode ray tubes for viewing in the center console; and a digital counter for various specialized applications to the right. (Courtesy of JEOL USA. Medford, Mass.)

of light is focused upward through the object. All high-power compound microscopes also have a system of lenses between the light source and the object, which collects the light rays and brings them to a focus on the object so that it is well illuminated. This is called a *substage condenser.* An *iris diaphragm* is included with the condenser so that, for the more open, lower-powered objectives, the light may be somewhat diminished in order not to dazzle the observer.

In using a compound microscope the object to be examined, mounted on a glass slide, is placed on the flat metal plate, or *stage,* immediately over the condenser and beneath the objective. The latter is then focused by means of a knob, which raises or lowers it to the correct focal distance. There is usually a knob for coarse adjustment and a small knob for fine, final adjustments. A *mechanical stage* is often provided to hold the slide and move it in various directions smoothly and accurately. It is a convenience but not a necessity. The modern laboratory microscope is quite complicated, and a few directions for its use and care may not be out of place.

Oil Immersion Lens. The objective lens commonly used in medical microscopy is called an *oil immersion* objective (from ×90 to ×100). The procedure for focusing and adjusting it must be learned. First, put a small drop of special optical immersion oil on the object or material to be examined. The oil eliminates reflections and loss of light from surfaces of the lens and from the upper and lower surface of the slide. The object (e.g., bacteria) is usually contained in a "smear" (page 154) of fluid (say blood, saliva, or water) that has been spread, dried, and stained on the glass slide.

Instructions for the use and care of the compound optical microscope are generally given in the classroom or laboratory.

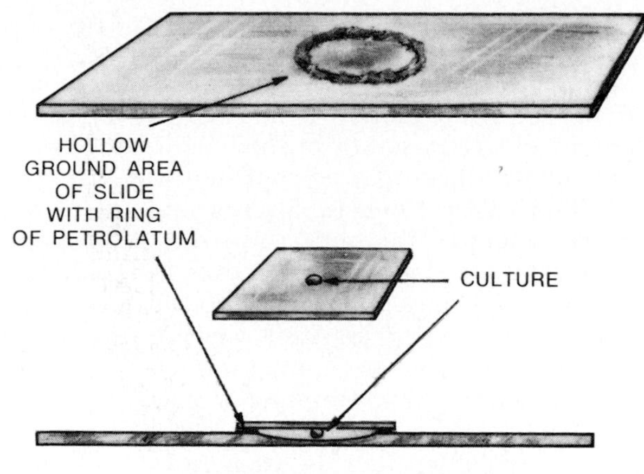

Figure 9–8 Hanging drop preparation.

Hanging Drop. An excellent method for observing the cells of microorganisms in their natural, living state is by means of the *hanging drop* (Fig. 9–8), which will probably be demonstrated in the laboratory. Briefly, it consists in focusing the highest non-oil immersion lens (about ×45) of the microscope (the so-called high-dry lens) on a droplet of fluid containing the microorganisms and observing them there. To prevent drying and accidental spilling, the drop of fluid is placed on a thin square of glass (a coverslip), and this is inverted over a concave depression in a special slide called a hollow ground or depression slide. The diaphragm of the microscope must be partly closed. In a simpler procedure called a "wet mount," the drop may be placed on a plain slide and allowed to spread under a coverslip. Preferable for many purposes are hollow glass slides having capillary inside diameters ranging from 0.40 to 0.05 mm.

Much can be learned about microorganisms by the use of the hanging drop. Some bacteria, for example, have the power of motion (when in the early part of their growth curve), and in a hanging drop they may be seen darting, rolling, squirming about, and bumping into one another (Table 9–2). The power of progressive movement is usually provided by flagella, which propel the bacteria through the liquid. Highly motile bacteria can move at rates approaching 100 μm per second; that is, they can travel a distance equal to their length hundreds of times in a second. An automobile would have to travel nearly a thousand miles an hour to accomplish the same thing. Bacteria and other microorganisms, however, can move only in fluid—not through the air or on dry solid surfaces. In order to travel over any considerable distance—from several feet to thousands of miles—microorganisms must be transported by human beings, birds, dust, airplanes, winds, rivers, and so on.

Table 9–2. Motility or Nonmotility of Some Common Bacteria as Ordinarily Observed*

MOTILE	NONMOTILE
Genus *Clostridium* except	*Clostridium perfringens.*
Genus *Bacillus* except	*Bacillus anthracis.*
Genus *Salmonella.*	Genus *Shigella.*
Genus *Escherichia.*	*Mycobacterium tuberculosis* and related species.
Genus *Vibrio.*	*Corynebacterium diphtheriae* and related diphtheroids.
Genus *Spirillum.*	All cocci, including staphylococci, streptococci, gonococci, meningococci, and pneumococci.
Genus *Treponema,* genus *Borrelia,* and genus *Leptospira.*	

*Only young cultures show motility even in the organisms listed here as motile.

Brownian Movement. Minute particles of any sort, suspended in water, whether motile or not, acquire a rapid, irregular, *nonprogressive,* vibratory motion because they are constantly being pushed about by moving molecules of water. Their motion is like that of a person in a dense crowd of milling people; one is constantly pushed from side to side. Bacteria often show brownian movement, which may be confused with true progressive motility.

The Phase Contrast Microscope. The phase contrast microscope is an improvement of the optical microscope. It does not provide greater magnification, but it does give a more "three-dimensional" view of the material examined. It may be used to look at living, unstained material. Some phase contrast condensers permit a viewing of a certain part of a cell. This is done by simply turning the condenser and increasing or decreasing contrast with respect to other cell components.

Fluorescence Microscopy. Fluorescence is a property of a substance, organic or inorganic, by virtue of which it reflects light rays of a color (wavelength) different from that of the incident rays. Many fluorescent substances give off a bright reddish (wavelength about 800 nm), yellowish (wavelength around 680 nm), or greenish (wavelength around 600 nm) radiance when illuminated with invisible ultraviolet ("black") light (wavelength around 250 nm). Nonfluorescent objects are invisible in ultraviolet light.

If microorganisms are illuminated with ultraviolet light and then observed with the optical microscope through filters that permit only the fluorescent (visible) colors to reach the eye and withhold the eye-damaging ultraviolet rays, fluorescent (e.g., green, yellow) parts of the organisms are readily seen. If nonfluorescent objects such as bacteria are stained on a microscope slide with a fluorescent dye such as auramine and all excess "background" dye is washed away, each bacterial cell, when illuminated with ultraviolet light, is visible as a glowing object in the otherwise dark field. This method is used especially in the diagnosis of tuberculosis.

The Electron Microscope. Scientists who are curious about extremely minute objects, such as molecules, soot particles, tiny crystals, and the minute, inner structural details of bacteria, protozoa, viruses, flagella, or cilia, have been aided greatly by the *electron microscope* (Figs. 9–7, 9–9, and 9–10), which makes possible direct magnifications up to 100,000 diameters. The direct image may be photographed and the photograph enlarged up to ×8, giving final magnifications of up to 800,000. This magnification is so enormous that a human hair, magnified to this degree, would have a diameter of about 1,000 feet,

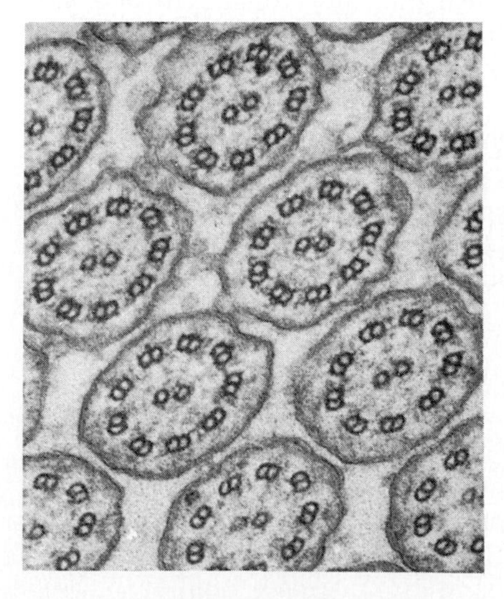

Figure 9–9 Electron micrograph of eukaryotic cilia in cross section. The section passes through the cilium proper, showing the typical structure. Prokaryotic cilia or flagella entirely lack the elaborate inner fibrillar structure of eukaryotic cilia or flagella. (× 72,000.) (Courtesy of J. André and E. Fauret-Fremiet. *In* De Robertis, E. D. P., Saez, F. A., and DeRobertis, E. M. F., Jr.: Cell Biology. 8th ed. Philadelphia, W. B. Saunders Company, 1980.)

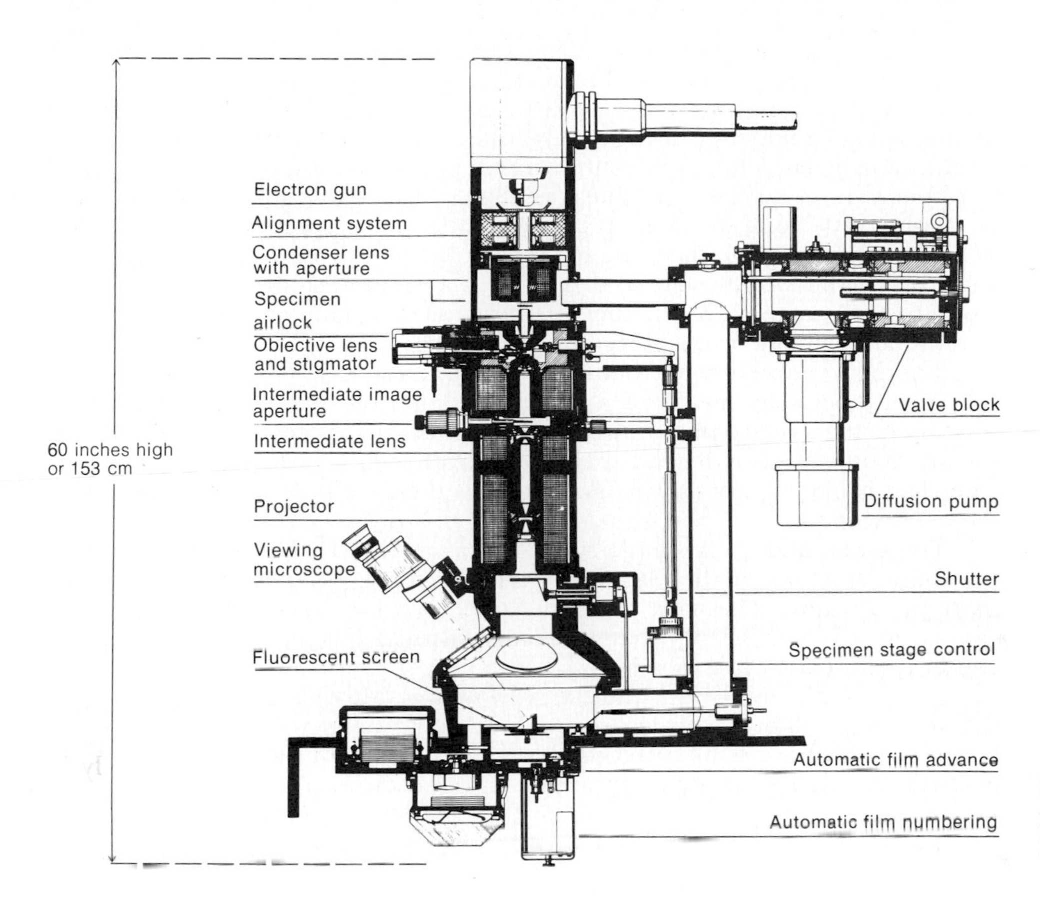

Figure 9–10 Electron microscope. (Courtesy of Carl Zeiss, Inc., Oberkochen, Württemberg.)

several times the diameter of a vehicular tunnel such as those under the Hudson River.

Visible light consists of a spectrum with wavelengths that range from about 400 nm (4,000 Ångström units,[6] violet) to about 770 nm (7,700 Å, red). Objects smaller than bacteria do not reflect such long waves to the eye in a clear pattern (i.e., are not *resolved*) and are therefore not clearly visible with the ordinary microscope.

Electrons create electromagnetic waves of very much shorter length (about 0.005 nm or 0.05 Å), and when focused magnetically, as is done in electron microscopes, are capable of *resolving* objects much smaller than even the smallest bacteria or rickettsias, notably viruses and their internal structures, enzyme granules, and even some large molecules. Electrons are not visible to the human eye. The magnified image produced by electron beams in an electron microscope must therefore be viewed on a fluorescent screen much like that used in clinical x-ray fluoroscopy. Photographs of these images, as mentioned previously, are usually greatly enlarged. These enlargements are called electron micrographs. Several are shown in this book. There are numerous other types of microscopes, such as the scanning electron microscope (Fig. 9–7) and the x-ray, phase contrast, interference, and dissecting microscopes. These are used mainly for research purposes.

---

[6]Å = 1×10^{-8} cm or 0.1 nm. Anders Jonas Ångström was a famous Swedish physicist (1814–1874).

STAINING

Figure 9–11 Hans Christian Joachim Gram (1853–1938). (From Schlesenger, D. [ed.]: Microbiology—1977. Washington, D.C., American Society for Microbiology, 1977.)

In their natural state unicellular microorganisms such as bacteria (not viruses; why?) appear under the microscope as tiny, colorless, translucent spheres, rods, or spirals, which are difficult to observe clearly. In order to see them distinctly and study them closely, they are stained with aniline dyes. To do this, a droplet of the fluid containing them (pus, broth, or blood) is spread in a thin film on a slide. The smear is allowed to air dry and is then warmed by passing it through the Bunsen burner flame. *Do not scorch the film.* It should never be hot enough to burn the hand. The heating dries the microorganisms and fixes the film to the slide so that it will not easily wash off. The film is now ready to be stained. (*Note:* Neither the heat nor the staining can be depended upon to kill pathogenic microorganisms!)

There are numerous methods of staining microorganisms. A widely used, simple method is by means of a dye called *methylene blue.* A few drops of a solution of the dye are put on the film (prepared as just described), allowed to remain about 30 seconds, and then washed off with a *gentle* stream of water. The slide is blotted dry (not rubbed), and is then ready to be examined with the microscope.

The Gram Stain. A simple stain such as methylene blue shows very well the shape and size of the organisms, but there are other methods that give more information. The most generally used stain for identifying bacteria was devised by a Danish scientist named Hans Christian Joachim Gram (Fig. 9–11) and bears his name. It is called a *differential stain* because it divides bacteria into two groups, the *gram-negative* and the *gram-positive* (Table 9–3). (Note, however, that it does not stain all bacteria well!) Many modifications of this method exist, but in general they are all based on the same principle of identifying two types of bacteria, although many morphologic differences are also observed with the same stain.

The *Gram stain* is applied as follows:

> Stain the fixed film for about 3 minutes with a gentian violet (also called crystal violet) solution made alkaline by adding a drop or two of bicarbonate solution. Rinse with "Gram's iodine" or "Lugol's iodine" solution and allow to stand for about 2 minutes. Wash gently with water from a washbottle and de-stain with acetone-alcohol by allowing the mixture to flow over the film, drop by drop, until the drippings show no tint of color (usually less than 10 seconds). Wash again with water.
>
> At this point it is necessary to explain the difference between gram-positive and gram-negative bacteria. Gram-positive bacteria retain the gentian violet and iodine in spite of the alcohol. When viewed under the microscope they appear dark purple. Gram-negative bacteria do not hold the stain and iodine when the acetone-alcohol is applied, and under the microscope appear almost colorless and nearly as invisible as when first put on the slide.
>
> Therefore, in order to make these bacteria visible the organisms are *counter stained* for about 30 seconds with a red dye called safranine or some other dye having a color that contrasts well with the bacteria already stained purple. Alternatively, one may use Bismarck brown (especially recommended for color-blind persons who may not see the contrasts with the other dyes), brilliant green, basic fuchsin (carbol fuchsin), or eosin (red). Wash, blot, or drain dry and examine under the oil immersion lens of the microscope.

The purple bacteria are gram-positive; the red, brown, or green are gram-negative. Sometimes, in material such as feces, both kinds can be seen and differentiated in the same smear. The differentiation is not absolute, and it is sometimes very difficult to be sure whether a bacterium is gram-positive or gram-negative. The Gram stain may be considered reliable only when the

Table 9–3. Some Common Bacteria Listed According to their Gram-Staining Reaction

GRAM-POSITIVE	GRAM-NEGATIVE
All common members of the genus *Bacillus,* including *Bacillus subtilis, Bacillus anthracis, Bacillus cereus,* and *Bacillus polymyxa.*	All intestinal bacilli of the typhoid-dysentery-paratyphoid group, i.e., the Enterobacteriaceae, including *Salmonella typhi, Shigella dysenteriae, Escherichia coli,* and others.
All members of the genus *Clostridium.*	*Vibrio cholerae*
All streptococci (especially in blood, serum, or media containing these).	All in the genus *Neisseria.*
All micrococci and staphylococci.	All brucellas, including *Brucella abortus.*
Pneumococcus.	All in the genera *Haemophilus* and *Bordetella*
Corynebacterium diphtheriae and related diphtheroids.	All in the genera *Pasteurella* and *Yersinia.*
Mycobacterium tuberculosis and related species.	The genus *Bacteroides*
Yeasts.	

bacteria stained were obtained from a "young" culture, perhaps best 18 hours after inoculation and proper incubation, and not made acid by fermentation.

Negative Staining. Not all microorganisms readily take up ordinary stains. Among those that do not are some protozoa and spirochetes such as *Treponema pallidum,* the cause of syphilis. A very convenient method of demonstrating the outward form and size of such microorganisms is by means of the so-called *negative staining* process.

The suspension of microorganisms to be examined is mixed with a small amount of black dye called nigrosin or with India ink. The mixture is then smeared on a slide and allowed to dry. On examination with the microscope the microorganisms are seen to be unstained and appear colorless on a black background. The dye does not actually stain them at all (Fig. 9–12).

Different methods of staining, like the "acid-fast" stain and others, will be discussed later in connection with certain bacteria for which they are especially appropriate.

Spirochetes and similar microorganisms are also beautifully demonstrated in their living, motile state by a special means of illumination called the darkfield. This is described in the section dealing with spirochetes.

Figure 9–12 Negative staining or relief demonstration. The background is darkened by nigrosin. The bacilli (agriculturally important *Azotobacter chroococcum*) are unstained and transparent (×978). (Courtesy of Dr. Robert L. Starkey, College of Agriculture, Rutgers University, New Brunswick, N.J.)

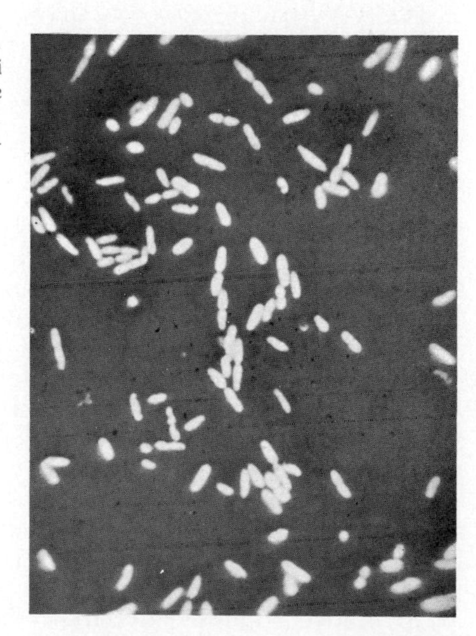

DIFFERENTIAL DIAGNOSTIC METHODS

In addition to microscopic examination, preparation of pure cultures, staining and examination of smears, and determination of motility, arrangement (chains, clumps, pairs), and other features, the diagnostic microbiologist also observes the way in which an organism grows on different food substances, the appearance of colonies, and their size, shape, color, and other characteristics. Usually the microbiologist also studies antibiotic resistance and the enzymes each organism produces that are peculiar to its species and that differentiate it from other organisms. These data, combined with its appearance under the microscope and its staining reaction, usually give the diagnostician definite clues as to the species of microorganism with which he is dealing and may help in the diagnosis of a patient's illness.

Study of Enzymes. In order to observe the activity of enzymes, the microbiologist induces the organisms under investigation to grow in various kinds of specially prepared media in which they can multiply rapidly and exhibit their enzymatic activity to the best advantage. Many distinctive enzyme activities are readily studied by testing for the by-products of enzyme action in pure cultures. As has been mentioned, bacteria, yeasts, and molds, by means of their enzymes, can cause a great variety of chemical changes very quickly under favorable conditions. Species differ greatly and may be identified by their chemical reactions and the end products that are formed. For example, in suitable culture media certain species such as paratyphoid bacilli can form acids and gases by fermenting glucose. Others (which cause gas gangrene) clot milk, liquefy gelatin, and bring about the changes that we call putrefaction, fermentation, and disease. Not all microorganisms can do all these things, but the ability of each kind of organism to produce certain chemical changes is known.

Rapid Methods for Cultural Identification. At least two types of rapid procedures for such studies are often used. In the first, instead of awaiting the relatively slow growth of large numbers of microorganisms in the test cultures, one prepares a very heavy suspension of the desired cells by collecting the billions that grow overnight on an agar plate or slant. Two drops of such a suspension contain an enormous number of enzymatically active, young bacterial cells. These are added to tubes containing a test substance, such as lactose, in a suitable medium such as agar or broth. The billions of young, active cells bring about the desired changes in the test media within two to six hours. The tubes with the carbohydrate fermentation broth medium may be arranged to catch gas produced by the microorganism as was shown in Figure 7–3. If an agar medium is used in test tubes, gas becomes evident as bubbles and cracks in the agar.

A second type of rapid procedure consists of placing previously prepared dried tablets or disks of filter paper, saturated with a test substance such as lactose, on the surface of agar medium previously inoculated with a heavy suspension of bacteria. On incubation, lactose in the paper disk is attacked by the growing organisms. Acid is formed and produces a color change in an indicator, such as phenol red, in the paper disk or in the agar around the disk. Reagent-impregnated paper strips are available that give a color reaction when a single pure colony producing a certain enzyme is smeared on them.

Various ingenious improvements of both of these procedures are constantly being devised, especially in clinical and industrial microbiology where rapid and massive microbial action is essential. One of these methods permits eleven tests in one simple procedure. It is called the "Enterotube" (Chapter 24). A similar system is the Oxi/Ferm tube for rapid identification of oxidative-fermentative gram-negative rods. Various sophisticated automated devices are available for special purposes, such as detection of bacteria and isolation and study of anaerobes.

Membrane Filter Methods. When microbiologists want to detect and identify bacteria in fluids such as milk, drinking water, or a patient's blood, they ordinarily take a sample of 1 to 50 ml of the fluid. This is put into tubes and flasks of culture media and incubated. Growth and identification of the bacteria that may be in the specimen may take one to ten days or more.

It often happens that no bacteria are found in small samples, but if larger samples had been examined, bacteria certainly would have been found. "The wider the net, the more the fish."

Very large volumes (for example, quarts of milk or gallons of water) in test tubes and flasks might yield better results but are awkward and expensive to handle. It is easier to pass large fluid samples through small, fine sieves or filter disks called *membrane filters* (Fig. 9–13) about 5 cm in diameter, thus collecting all the bacteria in the sample in one small area and discarding the bulky fluid. When the filter disk is removed from its holder, with the microorganisms on it from the fluid that passed through it, it is laid on a pad already saturated with an appropriate medium. The pad is placed in a Petri dish in an incubator, and each live bacterium on the filter disk will grow into a colony where it can be counted, isolated, and studied. If we include test substances in the medium in the pad, we can make some tentative inferences as to the identity and numbers of the bacteria on the disk according to the visible changes they produce (such as changes in color) in the test substances (Fig. 9–14). The use of membrane and other filters is generally demonstrated in the laboratory.

Membrane filters are also available that may be inserted into a syringe; fluid is then pushed through the filter under pressure. The advantages of this rapid method of filtration are obvious. It should also be noted that this method permits the injection of filter-sterilized but previously nonsterile solutions in animal experiments.

Animal Inoculation. In studying microorganisms suspected of being pathogenic, it is frequently necessary to inject or inoculate them into animals (Fig. 9–15). This is the only way in which their action on the living body can be studied directly and exactly, since we cannot ordinarily experiment on

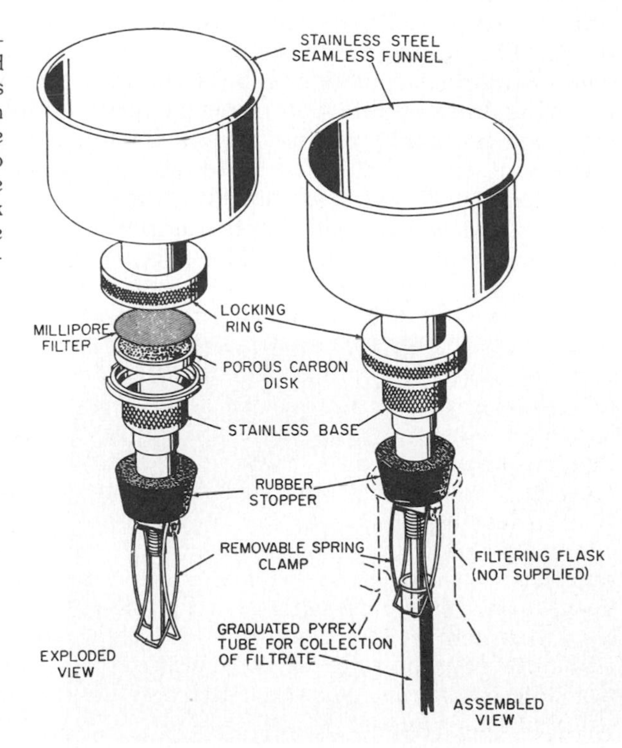

Figure 9–13 Use of a membrane filter to sterilize a fluid. The "exploded view" on the left shows the various parts separately. The porous carbon disk serves merely to support the fragile membrane (Millipore) filter. It plays no role in the filtering process. The whole assembly, mounted in a receiving flask (dashed lines, right), is sterilized before use. (Courtesy of Millipore Filter Corporation, Bedford, Mass.)

Figure 9–14 Disposable plastic dish containing a membrane filter disk through which a sample of river water was passed. After filtration of the water sample, the filter disk with the bacteria from the sample on it was placed in the dish on a pad saturated with culture medium. The whole was then incubated overnight at 35 C. Each bacterium on the filter disk, supplied with nutrient from the pad wet with culture medium, has multiplied into a visible colony. The colonies thus formed are seen here as dark spots. The gridmarks on the filter are to locate and identify specific colonies. (Actual size). (Courtesy of Millipore Filter Corporation, Bedford, Mass.)

human beings. Animals react to many kinds of microorganisms in much the same way as humans react, and the organisms often cause the same kinds of changes in their bodies. Diagnosis of disease may often be made only by this means. (It should be pointed out, however, that some microorganisms that cause fatal infections in humans will not infect animals and vice versa, and that different animals may react differently to a given bacterium.) Microbiology and medicine could never have developed without the study of the action of microorganisms on animals and the experiments that have been performed on them. Animals commonly used for this purpose are guinea pigs, white mice, and rabbits.

Furthermore, the inoculation of living animals may be useful in making a more rapid diagnosis because some microorganisms grow better and more quickly in the bodies of animals than in test tubes. For instance, the bacilli of "rabbit fever" (tularemia) and of tuberculosis grow slowly and not always with certainty on culture media in the laboratory, but by inoculating a guinea pig with the suspected material a positive diagnosis may often be obtained more surely. The animal will develop tularemia or tuberculosis, and an autopsy will demonstrate the condition. It is of the greatest advantage to a patient who is suffering from an infection to determine as soon as possible what microorganisms are causing his trouble, and the inoculation of animals often helps toward this end.

Serologic Tests. Occasionally an infection in human beings or in experimental animals is so mild that it is imperceptible by ordinary means. It is said to be "subclinical" or "silent." The only way of demonstrating that an infection

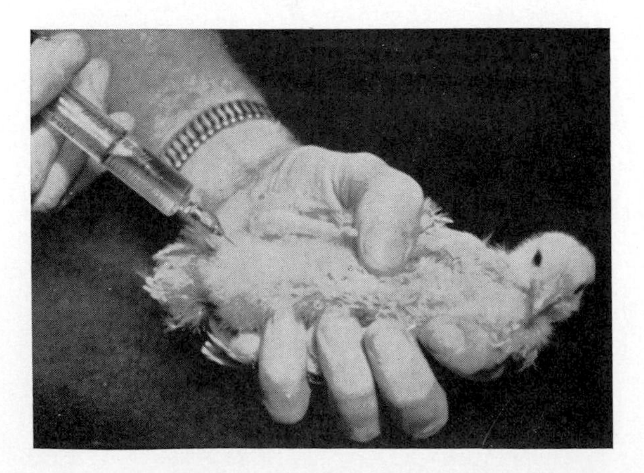

Figure 9–15 Inoculation of a chick to determine the pathogenicity or nonpathogenicity of bacteria isolated from a child with a sore throat. The material (broth culture of diphtheria bacilli) is being injected into the peritoneal cavity of the chick. If the bacilli are virulent, the chick will be dead in 24 hours. If they are not virulent the chick will live, perhaps to become a broiler! (Courtesy of U.S. Public Health Service, Communicable Disease Center, Atlanta, Ga.)

has occurred is to test the blood (or the *serum*, the fluid part of blood after clotting) for distinctive protein substances (types of globulins, especially gamma globulins or immunoglobulins) called *antibodies*. Antibodies (immunoglobulins) are produced by the body tissues in response to the infection. Each type of infection usually evokes antibodies that react with the *specific* infectious agents (antigens) that stimulated their production by the tissues. Thus, the detection of specific antibodies can have great diagnostic value. Many of these serologic tests for antibodies are used daily in every diagnostic laboratory. Examples include the *agglutination* test, the *precipitin* test, and the *complement fixation* test. They are used in the diagnosis of bacterial, viral, and rickettsial diseases, syphilis, enteric diseases, and fungal and protozoal infections. Many of these tests are named for the individuals who devised them (e.g., the Widal, Kahn, Eagle, Hinton, Wassermann, and Bordet-Gengou tests), although lately it has become more fashionable and efficient to use abbreviations for the names of

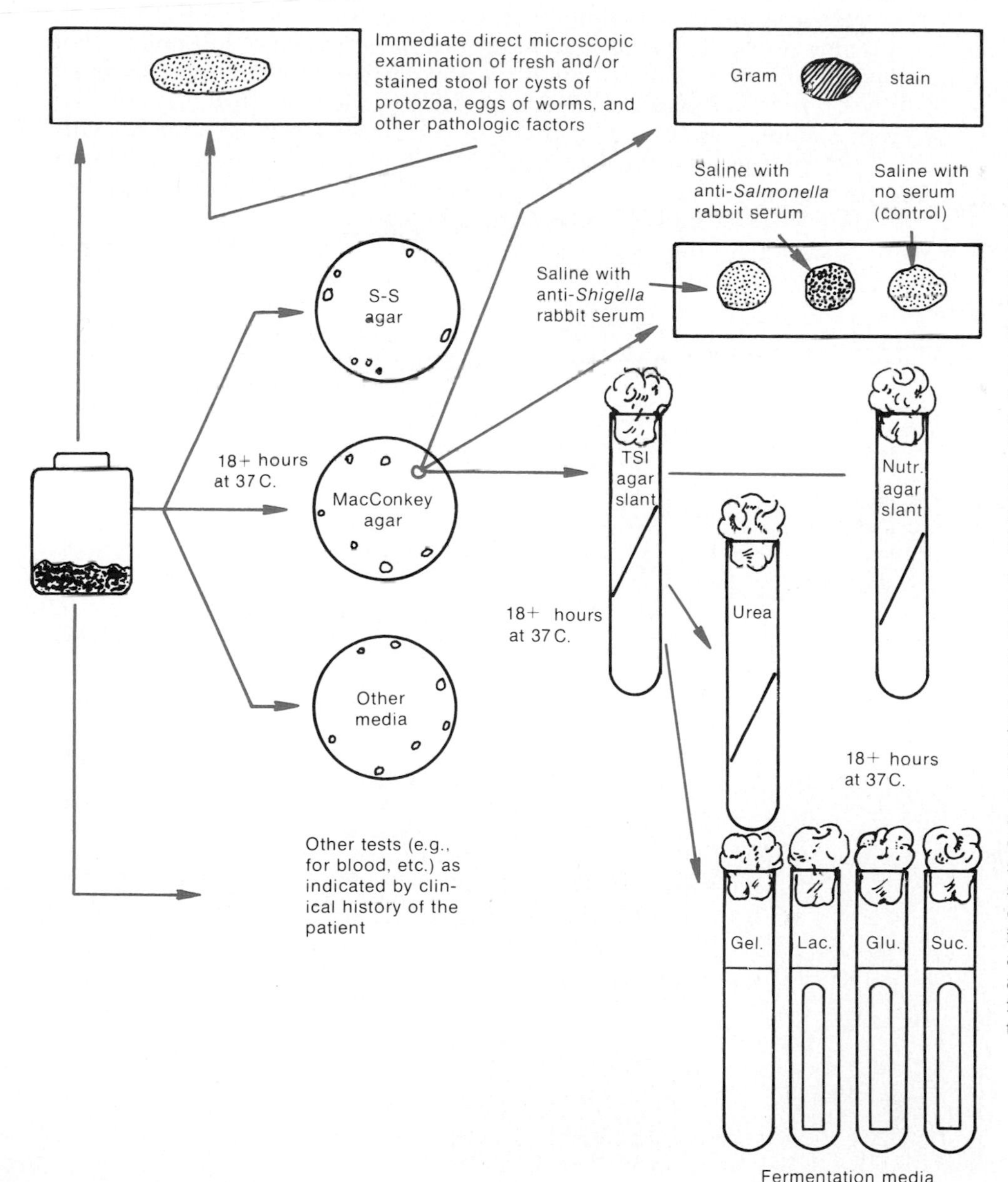

Figure 9–16 Steps in a typical examination of stool for pathogenic microorganisms (not including viruses). The specimen is examined immediately for protozoa and helminths by direct microscopic observation. It may also be subjected to tests for blood. Inoculated onto one or more plates of selective media, such as S-S agar (*Salmonella-Shigella* agar), it is incubated for 18 hours or more at 37 C. Colonies thought to be those of pathogenic enterobacteria are examined microscopically in a Gram-stained smear. If gram-negative rods, they are tested for agglutination with group- or genus-specific (polyvalent) agglutinating sera. They are also transferred to a triple-sugar-iron (TSI) agar slant and incubated for 18 hours at 37 C. This medium gives distinctive reactions suggestive of *Shigella, Salmonella*, and other groups of *Enterobacteriaceae* (Chapter 24). Growth from this slant is transferred to tubes containing urea media, nutrient gelatin, and broth with lactose, glucose, sucrose, and many other substances to determine the species of bacterium being dealt with. These cultures are incubated for about 18 hours at 37 C. A plain agar slant is also inoculated to be used for further serologic and cultural tests if desired.

some tests, e.g., VDRL (Venereal Disease Research Laboratories of the U.S. National Communicable Disease Center).

A Routine Laboratory Procedure. To illustrate the practical application of some of the procedures outlined in this chapter, let us suppose the microbiologist in your hospital laboratory receives a specimen of feces from an adult patient with gastroenteritis and fever of four days' duration. One immediately suspects infection by one of the organisms that (in the geographical location of your hospital) commonly cause enteric disease in adults: possibly the ameba *(Entamoeba histolytica)* that causes dysentery, or perhaps some dysentery bacilli (one species of *Shigella*[7]), or a species of the genus of typhoid-like *Salmonella*[8] bacilli. There are many other possibilities that can be considered: a viral infection, a fungal infection, no infection at all but accidental poisoning, or a condition requiring surgery (appendicitis). We will observe the microbiologist at work on the three most likely of the various possibilities. (See Fig. 9–16.)

After the patient's name, the room or ward, physician, date, and other identifying data are recorded, a minute portion of the stool specimen is streaked on a Petri plate containing a medium that is selective for *Shigella* and *Salmonella,* i.e., it contains nutrients for *Shigella* and *Salmonella* and also substances that inhibit virtually all the millions of other intestinal bacteria. These are discussed more fully later. Often plates of several media designed for this purpose, but of different formulas, are used to increase the chances of isolating a pathogenic bacterium. After overnight incubation of the plates at 35 C the microbiologist searches for certain colorless, grey, or black colonies, which are recognized (from training and experience) to be those of *Shigella* or *Salmonella.*

With a sterile needle portions of some of the suspected colonies are transferred, in pure culture, to slants of agar medium, each colony to a different slant in a separate tube. A bit of one colony may also be mixed in two drops of water (or saline solution) on two separate microscopic slides. From one drop a smear is made, dried, and Gram stained. If one finds on microscopic observation

[7]Genus of bacteria named for their discoverer, the Japanese microbiologist Kiyoshi Shiga.
[8]Genus of bacteria named for the American bacteriologist Daniel Salmon.

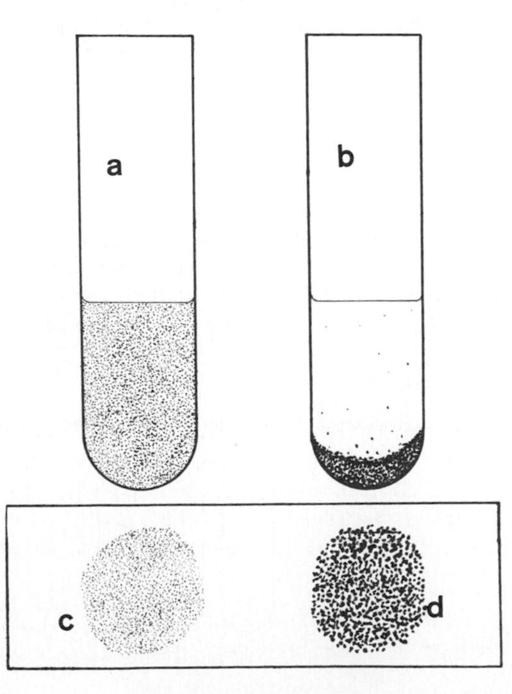

Figure 9–17 Agglutination in tubes and on a slide. In tube *a* bacteria were smoothly suspended in sterile saline solution containing nonspecific serum (in this example, *Shigella* serum). The bacilli remained evenly suspended throughout the fluid even after incubation for two hours at 35 C and 18 hours at 4 C. In *b* the saline solution was mixed with serum containing agglutinins specific for the bacteria (in this example, anti-*Salmonella* rabbit serum). After incubation as noted above, the bacteria in tube *b* had settled to the bottom of the tube in flocs. On the slide at *c* the bacteria were smoothly mixed in a few drops of serum-saline as in *a*. The saline at *d* contained *Salmonella*-agglutinating serum. The *microscopic* appearance of agglutinated bacteria is seen in Figure 9–18.

that gram-negative, nonspore-forming rods, and no other types, are present, presumably one has a pure culture. These could be *Shigella* or *Salmonella,* since both are gram-negative, nonspore-forming bacilli and look exactly alike. The hypothesis is not conclusive, however, because dozens of harmless species look exactly like the pathogens. Further data are necessary.

Of the other drop of bacterial suspension, one portion (A) is mixed with a small amount of diluted serum from a rabbit that has received many injections of dead *Shigella* bacilli. A second portion (B) of the same drop is mixed with a small amount of diluted serum from a *Salmonella*-injected rabbit. This may be done on a glass slide or in test tubes (Fig. 9–17). Now, let us say that the serum of the *Salmonella*-infected rabbit causes the bacilli in portion B to clump together (or agglutinate) like minute curds of sour milk. This clumping is due to antibodies called *agglutinins* in the animal serum. These tiny masses of agglutinated bacilli may be visible to the naked eye (Fig. 9–17) but are seen better with the microscope (Fig. 9–18). The bacilli in portion A remain unaffected. The microbiologist may also observe the still wet saline suspension of the bacilli with the microscope and see that the bacilli are actively swimming about.

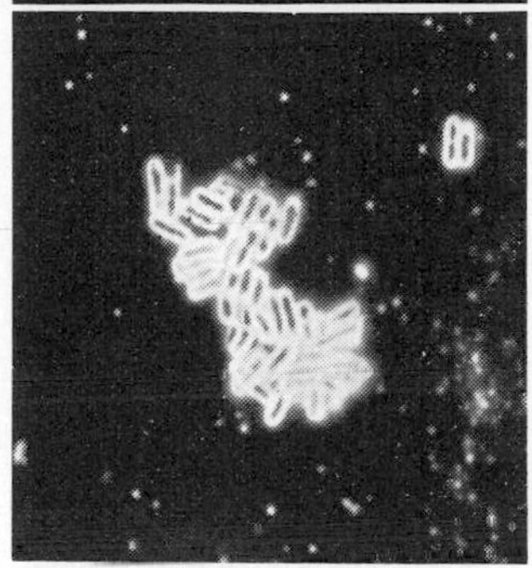

Figure 9–18 Agglutinated bacteria as seen with the microscope by darkfield illumination. Two different types of agglutination are shown here: above, *flagellar, flocculent,* or *H*; below, *somatic, granular,* or *O*. (After Adrianus Pijper.)

It is now known that one can practically eliminate the possibility of *Shigella* because the bacilli were not agglutinated by *Shigella* serum; they were agglutinated by *Salmonella* serum, and *Shigella* are *very rarely* motile, whereas *Salmonella* are, as a rule, actively motile. A tentative diagnosis may now be made—*Salmonella* infection or salmonellosis. One cannot say, however, what species of *Salmonella* is present. Hence, it is best to determine this and, since a single test may be misleading, to confirm the preliminary diagnosis by additional examinations.

The species of *Salmonella* can be determined by further agglutination tests using successive lots of rabbit serum, each especially prepared to agglutinate a single, different species of *Salmonella.*

After some hours of incubation of the pure culture, one can confirm and extend the tentative diagnosis of salmonellosis by transferring a bit of the growth from one of the previously inoculated agar-slant pure cultures to tubes of sterile broth containing test substances such as lactose, gelatin, glucose, or other carbohydrate fermentation media. After incubating these test cultures for 24 hours or longer, the reactions obtained are noted and either compared with those on a table of known test reactions in a book on diagnostic bacteriology[9] or, in the case of an adept and expert diagnostician (such as we hope our reader will become!), these reactions are known by heart, and we can arrive at a final diagnosis immediately by inspection of the test cultures (see Table 24–2, page 363). Here again, other simplified test systems like the Enterotube may be useful. This description of a diagnostic procedure is a simplified one, but it gives an idea of a routine, relatively easy laboratory diagnosis.

We might add that the stool specimen of this patient, on direct microscopic examination, also revealed the cysts of the ameba of dysentery (Fig. 3–3); the patient thus suffered from a double infection and had a very bad time. Expert medical treatment and skillful care, however, cured the patient of both infections, restored him to health, and prevented the spread of the infection to personnel in the hospital and to visitors.

SUMMARY

A pure culture of microorganisms contains only one kind of organism; each cell is genetically like every other cell. Clinical materials such as sputum, feces, and pus usually provide mixed cultures, as may be expected. Both pure cultures and mixed cultures may be accidentally contaminated with outside

[9] Such as *Bergey's Manual of Determinative Bacteriology.* (See Supplementary Reading.)

organisms. By streaking solid media it is possible to isolate pure colonies which are picked with a sterile needle and inoculated into liquid media. Microorganisms may also be spread on agar surfaces by other means, perhaps in order to determine the number of bacterial colonies on a plate, either by direct count or by means of automatic electronic bacterial colony counters.

A plot of the logarithm of the number of cells in a bacterial population against time results in a bell-shaped curve. The slopes of the curve depend on the type of medium used and other environmental factors affecting growth, as well as the bacterial strain being tested and its nutritional requirements.

Wherever possible synthetic (chemically defined) media are used, since they permit greater reproducibility of test results.

Viruses may be grown in cell cultures or on thin films of cells on plates. Clear zones resulting from the destruction of cells by the virus are called *plaques*. Each plaque is thought to be the progeny of a single virus particle. Viruses can also be studied with the electron microscope, by ultracentrifugation, by ultrafiltration, by the use of antigen-antibody reactions, and by physical tests.

The optical microscope permits clear magnifications to about $\times 1{,}000$. It is used for the observation of stained preparations, of cell motility in hanging drops, of brownian movements, and of fluorescence in color reflections of light by means of fluorescent microscope attachments. With phase contrast improvements, "three-dimensional" images can be produced by a phase contrast microscope. The electron microscope permits magnifications from $\times 100{,}000$ up to possibly $\times 800{,}000$.

The student will learn how to stain bacterial cells in the laboratory. Simple stains include the negative stain, the methylene blue stain, and, for differential purposes, the Gram stain. Gram-positive bacteria stain purple and gram-negative ones stain red, and other contrasts are possible with other stains. The acid-fast stain is used to identify the organism causing tuberculosis.

Other methods are also employed to differentiate bacteria (possibly pathogens) for diagnostic purposes. Much can be learned about their enzymes from certain tests involving growth and fermentation on specific chemical-containing media (e.g., lactose broth, liquid gelatin, glucose broth), as well as membrane filter methods, animal inoculations, and serologic tests.

REVIEW QUESTIONS

1. Define pure culture, mixed culture, contamination, bacterial colonies, streaking, and plate count.
2. How would you obtain an isolated colony? A pure culture?
3. Explain the bacterial growth curve. How does it differ from other population curves?
4. What is a synthetic medium? A defined medium?
5. Define viral plaque, ultracentrifugation, ultrafiltration, and cell monolayer.
6. What can the optical microscope do and what can it not do? Why?
7. Explain the differences between bacterial locomotion and brownian movement, fluorescent microscopy and electron microscopy.
8. Discuss staining and its many practical uses. Why is a stain necessary? What are the advantages of the phase contrast microscope?
9. What are some of the methods used for differential diagnosis of infectious diseases?

Supplementary Reading

Alexander, M.: Microbial Ecology. New York, John Wiley & Sons, Inc., 1971.

B. B. L. Manual of Products and Laboratory Procedures. 5th ed. Baltimore, Baltimore Biological Laboratory, Inc., 1968.

Balows, A., and Hausler, W. J., Jr. (eds.): Diagnostic Procedures for Bacterial, Mycotic and Parasitic Infections. Washington, D.C., American Public Health Association, 1979.

Bondi, A., Bartola, J., and Prier, J. E. (eds.): The Clinical Laboratory as an Aid in Chemotherapy of Infectious Disease. Baltimore, University Park Press, 1977.

Buchanan, R. E., and Gibbons, N. E. (eds.): Bergey's Manual of Determinative Bacteriology. 8th ed. Baltimore, The Williams & Wilkins Co., 1974.

Difco Manual of Dehydrated Culture Media and Reagents for Microbiological and Clinical Laboratory Procedures. 9th ed. Detroit, Difco Laboratories, 1965.

Finegold, S. M., Martin, W. J., and Scott, E. G.: Diagnostic Microbiology. 5th ed. St. Louis, C. V. Mosby Co., 1978.

Frobisher, M., Hinsdill, R. D., Crabtree, K. T., and Goodheart, C. R.: Fundamentals of Microbiology. 9th ed. Philadelphia, W. B. Saunders Company, 1974.

Gerhardt, P., et al. (eds.): Manual of Methods for General Bacterology. Washington, D.C., American Society for Microbiology, 1981.

Holt, S. C., and Beveridge, T. J.: Electron microscopy, its development and application to microbiology. *Can. J. Microbiol., 28*:1, 1982.

Jawetz, E., et al.: Review of Medical Microbiology. 14th rev., 12th ed. Los Altos, Calif., Lange Medical Publications, 1980.

Lennette, E. H., and Schmidt, N. J. (eds.): Diagnostic Procedures for Viral, Rickettsial and Chlamydial Infections. Washington, D.C., American Public Health Association, 1979.

Lennette, E. H., et al. (eds.): Manual of Clinical Microbiology. 3rd ed. Washington, D.C., American Society for Microbiology, 1980.

Lillie, R. D.: Conn's Biological Stains. 9th ed. Baltimore, The Williams & Wilkins Co., 1977.

Lynch, M. J., Raphael, S. S., Mellor, L. D., Spare, P. D., and Inwood, M. J. H.: Medical Laboratory Technology and Clinical Pathology. 2nd ed. Philadelphia, W. B. Saunders Company, 1969.

Pirt, S. J.: Principles of Microbe and Cell Cultivation. New York, Halsted Press, 1975.

Prier, J.: Modern Methods in Medical Microbiology. Baltimore, University Park Press, 1976.

Skinner, O. C.: Introduction to Diagnostic Microbiology. New York, Bobbs-Merrill Co., 1975.

MICROORGANISMS IN OUR ECOLOGICAL SYSTEM

BACTERIAL ECOLOGY

The study of the relationship between living organisms and their environment is called *ecology*, a term introduced by Haeckel in 1869. A "biotic community" includes *all* the organisms living in any given environment; an *ecosystem* includes the biotic community and the inanimate (or physically limiting) parts and factors that make up the total environment. As an example of a microbiological ecosystem we may cite the human colon, with its enclosed biotic community of billions of microorganisms of hundreds of different species: bacteria, protozoa, yeasts, and so on. The colon is not an inanimate environment: it contains and supports life within it and is the physically limiting boundary of a very complex universe or ecosystem. Numerous other types of microbial ecosystems are found in various kinds of soil, lakes, and waterways and in the sea.

Microorganisms are almost ubiquitous on the surface of the earth and are therefore important factors in the human ecosystem. Their ecological relationships with humans and all other forms of life are extremely complex and important. Some microorganisms may rapidly kill people by thousands or millions (e.g., the 1971 cholera epidemics in East Pakistan, now Bangladesh); others are of tremendous value to humanity in industries and of vital importance in agriculture and related human activities.

MICROORGANISMS AS BENEFACTORS

Health personnel—doctors, nurses, medical technologists, dietitians, attendants, and so forth—are perhaps inclined to think of all microorganisms as causes of disease and as enemies of human life, but comparatively few species are known to be pathogenic to humans. Microorganisms may be divided into two groups according to their activities: the harmful ones and the useful ones. The first kind, the *pathogenic* microorganisms (viruses, bacteria, protozoa, fungi, and, rarely, algae), cause disease merely because they are able to live on or in the bodies of human beings, animals, or plants and, unfortunately, damage the *host* (infected plant or animal) in greatly varying degrees.

The second kind live in the outside world and are harmless to humans. These are *saprophytic* or *saprozoic,* unless they are pathogenic for plants or some animals. Many of them, especially yeasts, molds, and bacteria, are used to manufacture alcohol, lactic acid, butter, cheese, solvents for paints and oils, antibiotics such as penicillin, and other products, and to increase soil fertility. Of more than 1,700 known kinds of bacteria, only about 70 cause disease in human beings, and of these only a dozen or so are, as a rule, really dangerous.

Many of the pathogenic microorganisms can thrive only in the body, finding the outside world cold, dry, and unfriendly, and they soon die if they are cast forth into it. Conversely, microorganisms adapted only to life in the outside world find conditions in the human body unsuitable to their growth and cannot multiply there. There are, however, numerous exceptions in each group. Beware also of microorganisms that are usually harmless but that may

CHAPTER 10

suddenly become adapted to growth in the body and cause disease and even death. More and more common are infections caused by "harmless" species of *Pseudomonas,* when antibiotics have destroyed the harmful pathogens.

Before studying the disease-producing microorganisms, with which medicine and public health are chiefly concerned, we should get a wider view of the activities of useful microorganisms and their place and importance in the world.

Decay, Putrefaction, and Fermentation. *Decay* (or rot) is a general term denoting gradual decomposition of organic matter such as dead animals and plants and their wastes, on and in the soil. Strictly speaking, *putrefaction* and *fermentation* are decomposition of proteins and carbohydrates, respectively, under *anaerobic* conditions. The terms decay and putrefaction are often loosely used interchangeably. Under natural conditions a great variety of saprophytic microorganisms, including yeasts, molds, bacteria, and algae, may be involved in all three processes as they transform organic refuse into useful plant foods; if they are disturbed our ecological system suffers. Industrial waste may kill animals and plants not only directly but also indirectly by preventing production of their (and our!) food by microorganisms in the soil.

When tissues of dead animals or plants are buried in the ground, microorganisms from the soil and those already in the animal's intestine (or on the plant) enter the tissues. There, by means of their various enzymes, microorganisms cause the fats, proteins, and carbohydrates of the tissues to disintegrate. The gases (carbon dioxide, ammonia, hydrogen sulfide, and so on) and water that are formed pass off into the earth or air. Other substances that the microorganisms produce by decomposition of the animal's or plant's body contain nitrogen, phosphorus, sulfur, and other necessary elements combined in water-soluble molecular forms. In this way the dead matter disappears; the complex organic molecules of which it was made up during life are broken apart by enzyme action and are used over again to nourish plants. Animals and humans then use the plants for food. After their death they, in turn, are changed into food for plants. It is through this decomposing work of microorganisms that undue accumulation of dead animal and vegetable matter is prevented and the earth is kept fit for living beings. Without microbial activity, higher life would be impossible. The alternating cycle of the elements between animate matter and inanimate matter is carried on largely by the saprophytic yeasts, molds, bacteria, and related microorganisms.

The conditions most favorable for putrefaction, fermentation, and decay in the outside world are those that are most suitable for the growth of microorganisms living in soil, rivers, lakes, and oceans. Summertime usually provides the best temperature for rapid turnover of organic wastes.

PURIFICATION OF SEWAGE AND FRESH WATER SUPPLY

Until recently every well-designed waste water plant (sewage disposal plant) was merely a man-made device to control and exploit the decomposing activities of certain groups of saprophytic microorganisms. Only now are industrial processes of a different nature beginning to replace this type of installation. People can live in large communities under sanitary and healthful conditions only when there is some way of taking care of their wastes. Great epidemics have been caused in the past by the accumulation of human excreta in cities, with resultant contamination of available water supplies by pathogens of the intestinal tract, organisms causing typhoid, cholera, and so on. In ancient times, and even more recently in some developing countries, it has been the custom to throw the contents of household "slop pails" out the window to lie in the

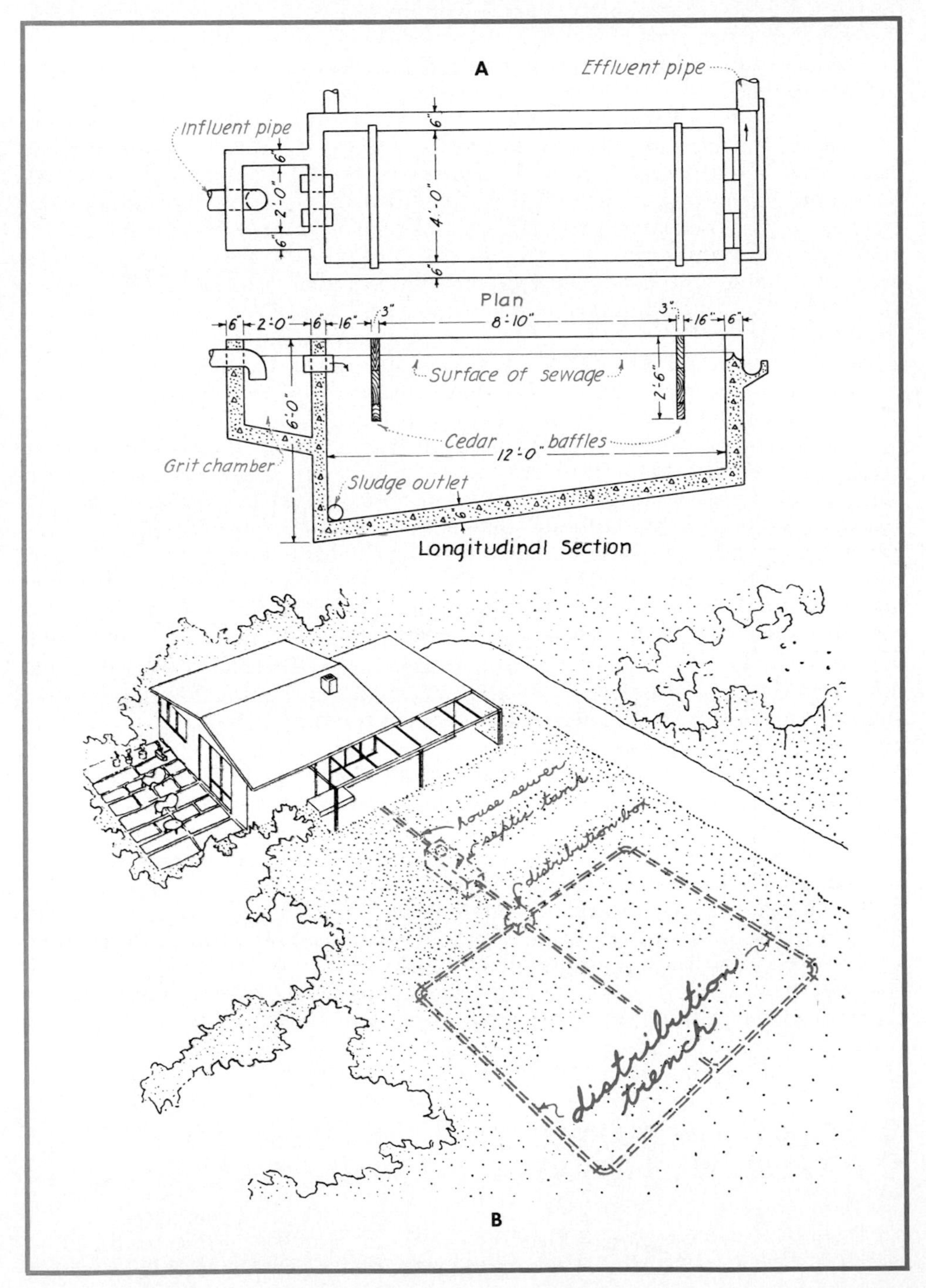

Figure 10–1 *A,* Septic tank arrangement for home or other small construction. The sewage is brought first to the *grit chamber.* Insoluble grit and heavy extraneous objects are caught here. The bulk of the sewage flows into the larger chamber. Here solid organic matter settles; grease forms a scum on top and may eventually have to be removed. The settled solids undergo anaerobic microbic digestion in the sump at the bottom of the tank. They are finally reduced to largely inorganic *sludge,* which is pumped out from the sludge outlet. Sometimes several such tanks are connected in tandem. In these tanks sewage fluid is cleared of odor and solid material before it flows onto the filter beds or, if available, into a stream. (Bulletin No. 16, Engineering Experiment Station, University of Washington.)

B, A common form of sewerage layout for a rural or suburban home, showing location of septic tank and drain-tile system for disposal of the fluid effluent from the septic tank. (U.S. Housing and Home Financing Agency, Division of Housing Research, Construction Aid # 5, Superintendent of Documents, Washington, D.C. 20402.)

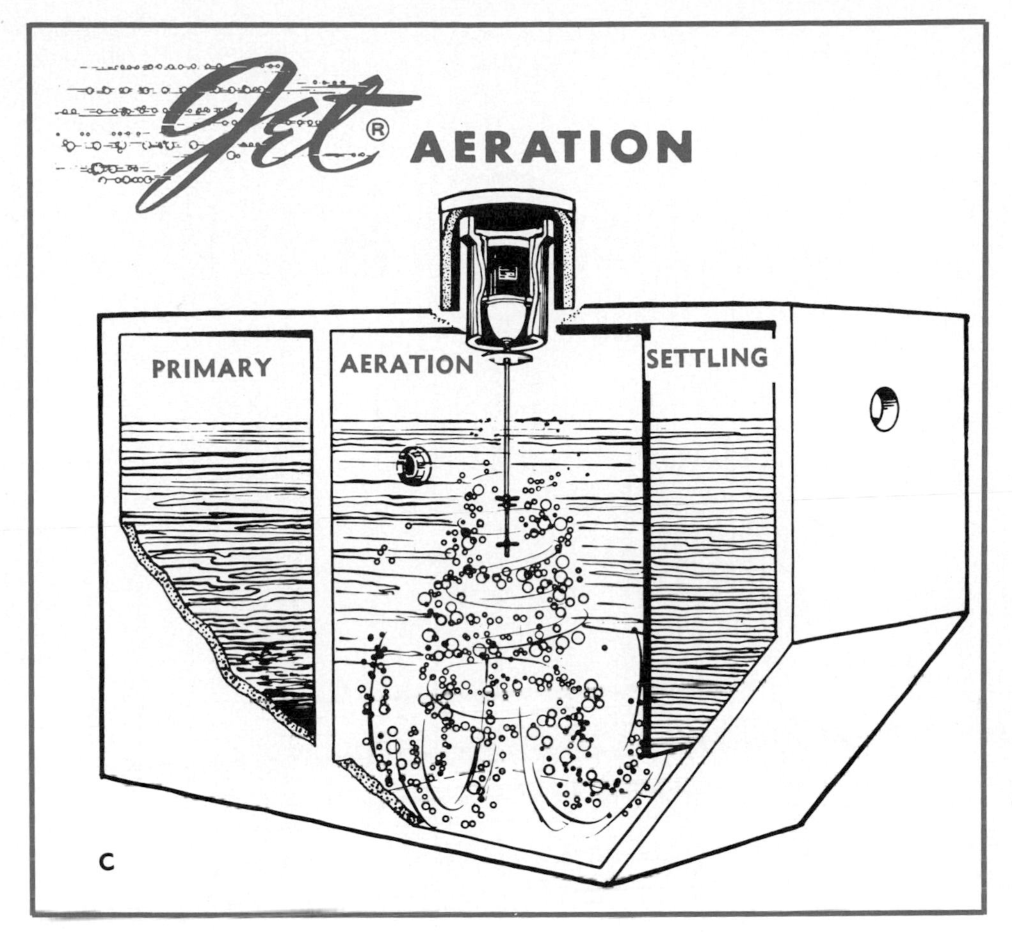

Figure 10–1 *(Continued) C,* A compact aerating sewage disposal unit. This is designed to provide a clear, colorless, and odorless effluent within 24 hours under normal conditions of use. (Courtesy of Jet Aeration Company, Cleveland, Ohio.)

streets. The community water supplies, if they exist as such, may then become contaminated with feces. Some of the plagues and pestilences that we read of in historic writings were undoubtedly spread in this way. Today the activities of soil and water microorganisms are used to "purify" sewage. The microorganisms utilize the organic substances of the sewage as food and convert them into harmless, inoffensive materials that are food for plants. A common type of septic tank for suburban or rural use is shown in Figure 10–1*A* and *B*. A more modern, actively aerating home disposal system is shown in Figure 10–1*C*.

Municipal Sewage. Sewage is simply defined as water after use; most of it is just polluted water. A much used method for purifying municipal sewage is first straining out extraneous objects by passing the "raw" (untreated) sewage through metal screens or racks, then allowing it to flow very slowly through large tanks. In Figure 10–2*A* these are of a type known as *Imhoff tanks.* In such tanks the solid matter in suspension settles to the bottom. This solid material is slowly decomposed through the hydrolytic action of microbial enzymes (produced mostly by anaerobic and facultative bacteria). It eventually forms a sort of mud (*humus* or *sludge*), rich in plant food, which is pumped out, dried, and frequently used for garden fertilizer. *Milorganite* is a familiar example. It is sterilized before packaging.

The fluid part of the sewage is sprayed on the surface of large beds (*trickling filters* in Figure 10–2*A, B*) of coarse gravel, during which process it becomes fully aerated. On the surfaces of the pieces of gravel or sand a slimy film develops. This film consists of the growth of aerobic microorganisms,

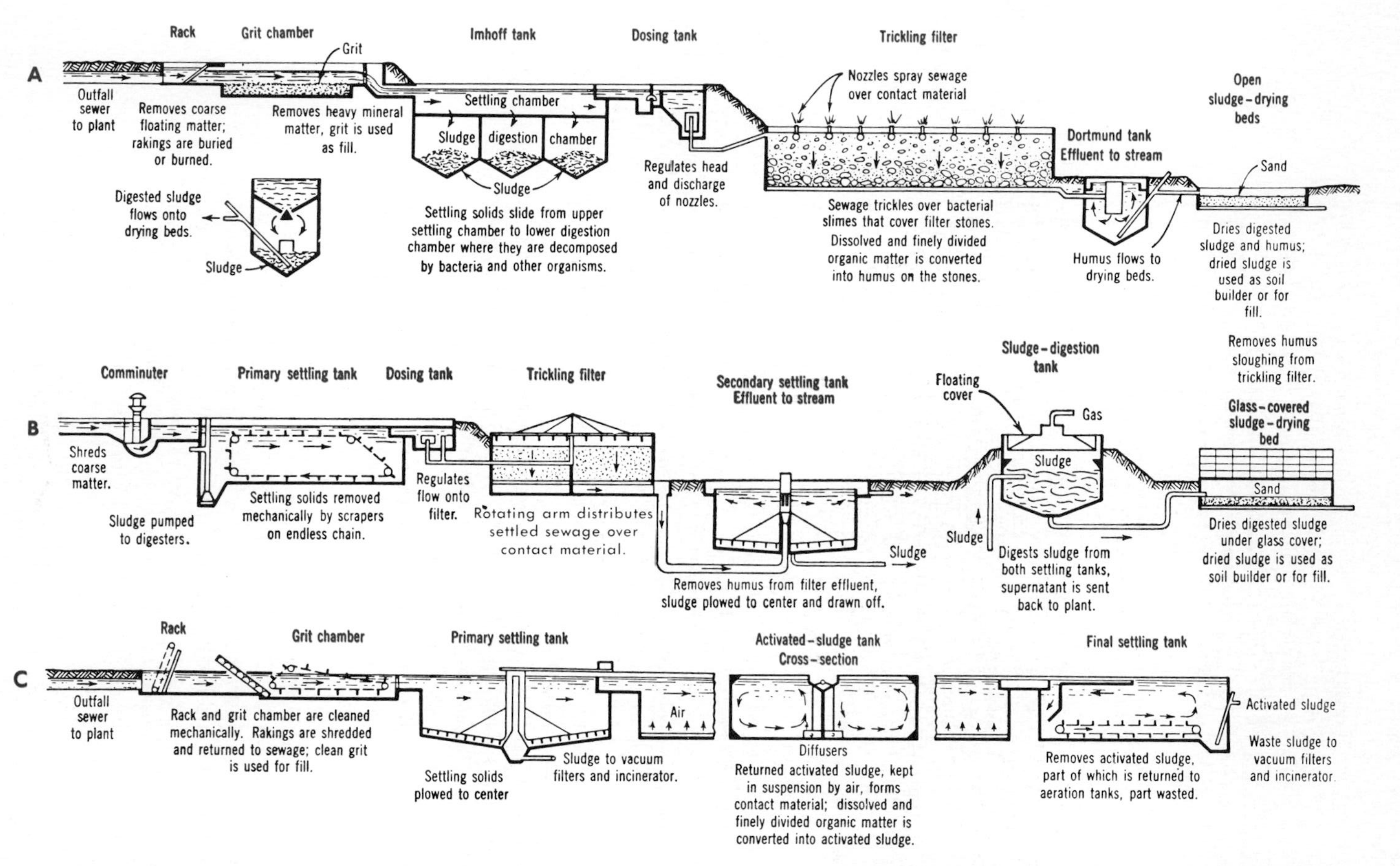

Figure 10–2 Three types of sewage disposal systems accomplishing the same thing: they differ mainly in the methods used to aerate the sewage. In A are seen a coarse screen (rack) and settling basin for grit, followed by a series of Imhoff tanks. The fluid then is admitted under control (dosing tank) to a trickling aerating filter. The aerated fluid then passes through a secondary tank (Dortmund tank) from which residual solids (humus) are removed to sludge-driving beds. In *B* the solids are comminuted mechanically and collected mechanically from a primary settling tank. The aerating filter is of the sparger type. A secondary settling tank removes residual sludge or humus to a digestion tank that has a "floating" cover to trap sewer gas for use as fuel. The sludge is finally removed to a covered sludge drier. In *C* the partly clarified fluid is aerated and treated by the activate sludge process instead of by trickling filters. (Fair and Gever, *in* Water Supplv and Wastewater Disposal, New York, John Wiley & Sons, Inc.)

which get their nourishment by decomposing and oxidizing offensive materials in the sewage as it trickles slowly through the gravel. Any solid matter (*humus* in Fig. 10–2*A*, *B*) is collected in a final sludge-digestion tank (e.g., Dortmund tank in Fig. 10–2*A*) and pumped onto sludge-drying beds. The fluid part, now largely deodorized and much cleansed, is collected in drains and led away to a convenient water course or run into fields for irrigation with fertilization.

Many waste water treatment plants combine these operations in efficiently designed "package plants," consisting of a single compact unit. A large number of rapidly growing communities have discovered that their present methods of sewage disposal are hazardous. These communities spend millions of dollars to protect their inhabitants from potentially infectious microorganisms that enter and exit from the human gastrointestinal tract. Yet the construction process of newer and better disposal plants is never finished. As one plant is completed it is being redesigned to take care of a greater load *ad infinitum.*

Aerobic Treatment. In *activated sludge* processes, aeration of the sludge and the sewage is accomplished by violently agitating the sewage with large volumes of air (Fig. 10–2*C*). Solid matter is torn into small granules or particles. The particles of sludge contain millions of active *aerobic* microorganisms that use the air to oxidize and decompose rapidly the offensive matter in the sludge. Aeration is the key objective in the form of sewage disposal shown in Figure 10–2*A*, *B*, *C*.

Anaerobic Treatment. After aerobic treatment of sewage, *anaerobes,* such as methane-forming bacteria, may act in a digester on the remaining solids and produce methane (CH_4) and carbon dioxide (CO_2). This process is less efficient than the aerobic method, but it does produce energy (methane) rather than use it; thus, it will be employed more often in future waste water treatment plant designs.

Pressure Cooking of Sewage with the Wet Air Oxidation Unit. Some other facilities make use of a process that is entirely different, yet also familiar to the microbiologist. It consists of a giant "pressure cooker" that works like an autoclave. Once exposed to the unit's great heat, all organisms are killed in the sewage. Usually this step follows the aerobic digestion process.

The purpose of the Wet Air Oxidation Unit (Fig. 10–3) is to partially decompose the organic material in sewage sludge and to render the remaining sludge sterile and readily drainable on the sludge-drying beds, producing a sludge cake that is nonputrescible and unrecognizable as being of sewage origin. The filtrate is nontoxic and highly biodegradable, and it is recycled to the treatment plant. When sewage sludge and air are mixed and retained in a reaction vessel at proper temperature and pressure conditions for a sufficient period of time, oxidation and degradation of organic compounds in the sludge take place, and the residue is sterilized. This process operates continuously, provided that air, water, and combustible material in proper quantities are furnished and products of combustion are removed. Steam is injected into the mixture after entering the reactor to bring the sludge and air up to required oxidation temperature. This process seemed very promising when it was first introduced, but it uses a great deal of energy and thus is uneconomical for modern sewage treatment. It is being replaced with anaerobic treatment installations.

Drinking Water. Many saprophytic microorganisms are indigenous to the waters of rivers, lakes, springs, and oceans. They are often present in drinking water and are harmless to the human body. We take considerable numbers of them into the body with food, water, and milk every day. Water polluted with "raw" (untreated) sewage, however, usually contains pathogenic microorganisms, among which may be the typhoid or dysentery bacilli, cholera vibrios, polio and hepatitis viruses, the amebas that cause dysentery, and others. The water of streams, rivers, and lakes is so likely to be contaminated with sewage or fresh feces that it is always unsafe to drink it without disinfection. Dug or open wells are also dangerous, as they frequently receive drainage from

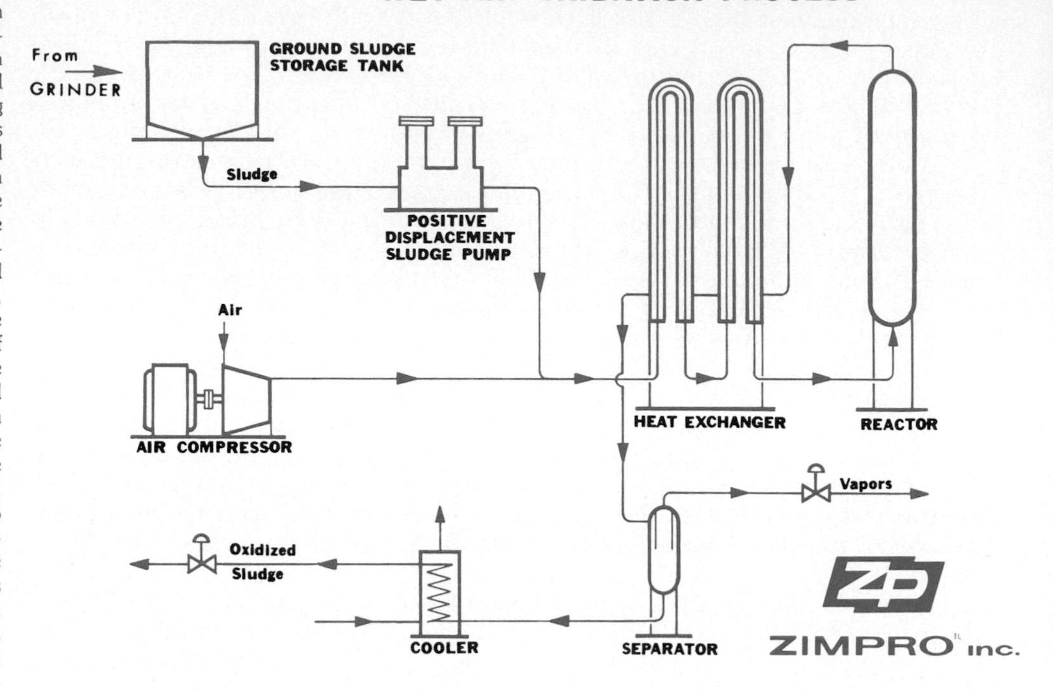

Figure 10–3 A simple schematic of the sludge oxidation and dewatering system. Sewage sludge is pumped through the grinder to the ground sludge storage tank (existing digester), and from there it is pumped by the centrifugal sludge feed pump to the high pressure pump where the sludge pressure is raised to the system pressure, approximately 350 psi. Compressed air is introduced at this point, and the slude and air mixture is passed through a series of heat exchangers where the mixture temperature is raised to 280 to 310 F. Steam from the boiler is injected near the bottom of the reactor, into the heated sludge and air mixture, increasing its temperature to the required reaction temperature of about 350 F. The resultant mixture passes through the reactor, which provides sufficient retention time to allow the oxidation and *sterilization* to be accomplished. The reactor products (gas, steam, and oxidized sludge) then pass back through the heat exchangers to preheat the incoming sludge and air. The reactor products leave the heat exchangers at 110 to 150 F. In the separator, the gas and steam portions of the reactor products are separated from the liquid and solid portions. The gas and steam are depressurized through the pressure control valve and pass through the vapor diffuser located in the bottom of the primary clarifier feed manhole. The oxidized sludge (liquid and solid reactor products) passes through the oxidized sludge cooler, where it is cooled by effluent water. It is then depressurized through the level control valve and discharged into the drying beds. The liquid or effluent, which is drained from the oxidized sludge, is recycled through the treatment plant. (From Operating Manual, Zimpro Wet Air Oxidation Unit, Denton, Texas.)

cesspools, barnyards, and sink drains. Water from drilled wells is usually, but not always, safe; polluted water may reach the wells through crevices in the rocks or be drawn into them by continuous pumping. There are no simple tests that tell whether or not water is bacteriologically safe to drink. Positive tests for synthetic detergents in drinking water reveal pollution of the water with household or other wastes, provided detergent is used and the pollution is recent.

Except for water from approved sources, the only safe water is that subjected to boiling or treatment with chlorine a few hours before use. Tablets of hypochlorite or other chlorine compounds for this purpose are available from campers' outfitters or pharmacists. Ordinary laundry bleach (5 per cent sodium hypochlorite; *Clorox,* for example), purchasable in any grocery store, makes an excellent disinfectant for water as well as for many other things such as dishes, laundry, urine, and feces. The manufacturers usually provide ample instructions for use as disinfectant on the label of the bottle. The individual who is faced with the problem of disinfection in the home will do well to remember this. There are also excellent iodine disinfectants, *Wescodyne* (an iodophor; see page 221), for example.

Municipal Water Reclamation. In many large cities, water is usually first allowed to clarify by the process of *sedimentation,* or settling out, while the water is stored for weeks or months in reservoirs or lakes. Certain organisms in natural waters like protozoa or *Bdellovibrio,* a small bacterial lytic parasite, destroy enteric pathogens and thus help in the purification process. The water is then usually subjected to some sort of screening process to remove dead fish, leaves, and other solid matter. The water is pumped (from different intakes) into the "water works," "filtration plant," or "water purification plant," now often called a *"water reclamation plant"* (Fig. 10–4). Previous chemical experimental testing determines which chemicals are required to be mixed with the raw water to bring about a *"floc."* This means that the organic substances in the water flocculate with the mixture of added chemicals, e.g., alum, ferrous sulfate, lime, aluminum hydroxide, and others in the mixing chamber. Different waters require different treatments. This floc settles in the "floc settling chamber,"

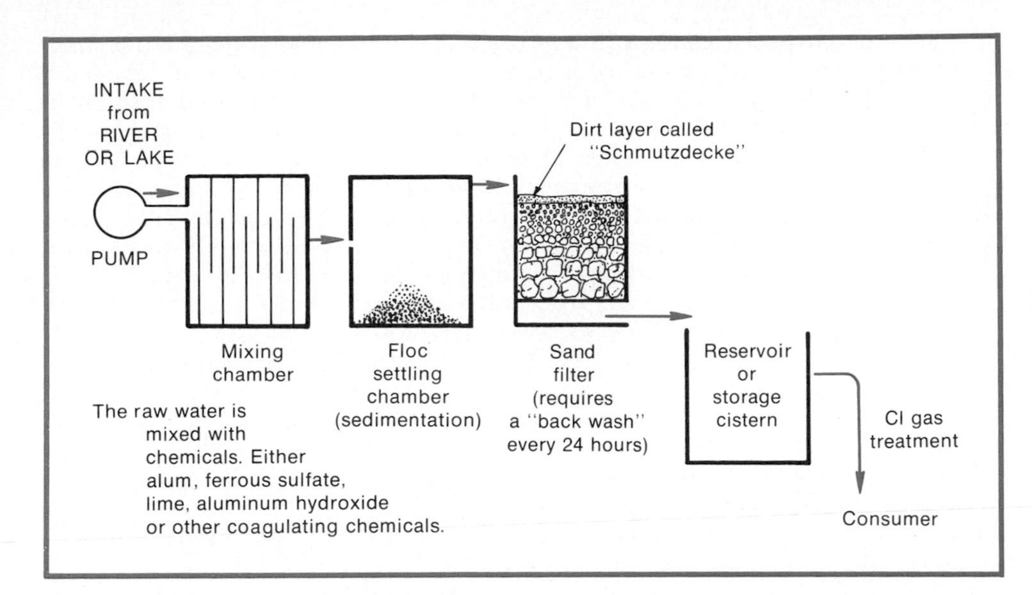

Figure 10–4 Shown here is a schematic diagram of one of the simplest and oldest established water purification systems, as used in many towns in the United States. It was first introduced in St. Louis to purify water from the Mississippi River. All other "water reclamation" systems are variations and/or improvements of the one shown here.

also called the *sedimentation basin,* which may have a capacity of a million gallons or more to permit the mud, silt, and floc to collect on the bottom. The water then flows onto a *filter bed* consisting of layers of sand built up over layers of gravel. This may be many feet thick and cover an area of many thousands of square feet. Any remaining microorganisms, floc, and other solid particles are removed by the sand, and the filtered water then resembles spring water purified by filtration through the earth. Under the gravel are drains that carry the filtered water to storage cisterns. Before release for consumption, the water is chlorinated to retain 0.1 to 0.2 ppm of chlorine at the last outlet in the supply system. Fluorine is often added.

From time to time reports appear in the literature expressing concern about pathogenic viruses found in chlorinated effluent from treated sewage and even in the water supply. Evidently some viruses can survive the final chlorination process, but they are usually few in number, highly diluted, and thus noninfectious.

All water treatment processes are developments in engineering based on principles of microbiology. In our modern civilization about 125 gallons of fresh water are used per person per day. Obviously this is not all for drinking or even bathing; most of it serves industrial needs associated with our way of life.

Bacteriological Examination of Drinking Water. Certain bacteriological tests, prescribed for legal purposes by the American Public Health Association, are carried out daily in every public health department laboratory. They are used to keep watch on the probable safety of public water supplies and public swimming pools. These tests are designed to reveal the presence of certain well-known and generally harmless intestinal bacteria known as "coliforms," of which *Escherichia coli* is particularly significant in determining pollution with feces or sewage. When cultivated in lactose broth these bacteria produce gas (CO_2 and H_2). In so-called fermentation tubes, the gas bubble is easily seen. Some bacteria unrelated to feces or sewage also produce gas under the same conditions. It is necessary, therefore, when gas occurs in the cultures, to determine whether it is due to coliform or nonintestinal microorganisms. Ingenious selective media[1] are used in cultivating desired organisms.

[1]Media containing substances such as antibiotics, certain dyes, and salts that inhibit unwanted microorganisms but permit the desired species to grow freely; also, media containing all nutrients required by the desired species, but lacking in one or more essential growth factors (e.g., vitamins) necessary for the growth of unwanted species.

For this purpose samples of water (or other fluid to be tested for intestinal [coliform] bacteria) in amounts of from 0.01 to 50 ml, depending on suspected intensity of pollution, are placed in fermentation tubes containing lactose broth or lactose broth with sodium lauryl sulfate, which inhibits virtually all noncoliform organisms (Fig. 10–5). After 24 hours (or 48 hours, if necessary for gas to appear) at 35 C, a drop is transferred from one or more cultures showing any amount of gas to tubes of brilliant green–lactose bile broth (BGLB), which inhibits nearly all noncoliform organisms. If gas appears within 48 hours at 35 C, the test for coliform organisms is said to be "confirmed." To complete the test, however, transfers are made from tubes of BGLB showing gas to plates of Endo or EMB agar, which are selective for coliform organisms. If, after 24 hours at 35 C, colonies that have the distinctive coloration of coliform colonies appear on these media, such colonies are transferred to fresh fermentation tubes of lactose broth and to slants of plain nutrient agar. If gas appears in any fermentation tube within 48 hours at 35 C and *only* gram-negative, nonspore-forming rods appear on the corresponding agar slant, then coliform organisms have been isolated in pure culture from the sample. The test is "completed." Water tested is reported as either potable (safe to drink) or polluted (potentially dangerous). By arrangement of the tests in multiple series, and by calculating from the numbers of highest dilutions yielding coliforms, the results can be made roughly quantitative ("most probable numbers").

In the preceding procedure the transfer from the initial lactose or lactose-lauryl sulfate broth tubes to confirmatory BGLB tubes may be bypassed, as shown in Figure 10–5, and the initial broth culture spread directly on Endo or EMB agar plates. This short cut is now the usual method for the confirmed test.

MEMBRANE FILTER METHOD. In still another approved procedure the samples of fluid are passed through a membrane filter, which is then incubated in contact with special media (e.g., M-Endo) selective for the coliform group. Typical coliform organisms produce colonies of characteristic pink or red color with a green metallic sheen (Fig. 10–6). This method is more rapid than the fermentation methods previously outlined and has the advantage of simplicity,

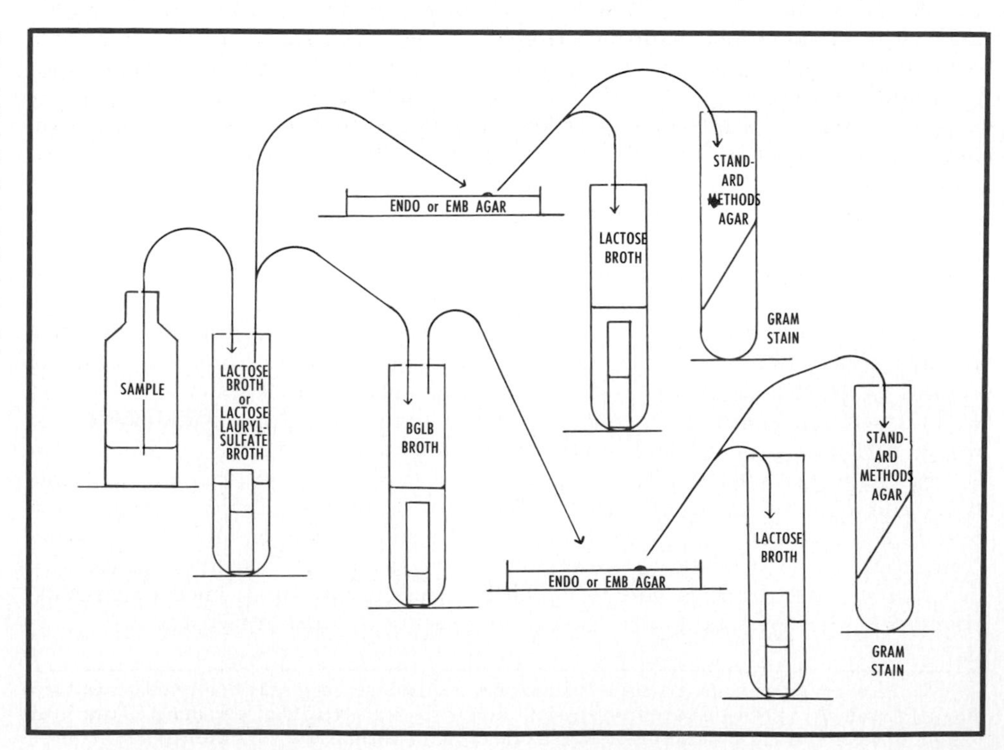

Figure 10–5 Representative steps in determination of coliform organisms in water, milk, and other liquids. The media and procedures are standardized according to criteria set by the American Public Health Association and affiliated organizations. For explanation see text. (Adapted from Standard Methods for the Examination of Water and Wastewater. 13th ed. New York, American Public Health Association, 1971.)

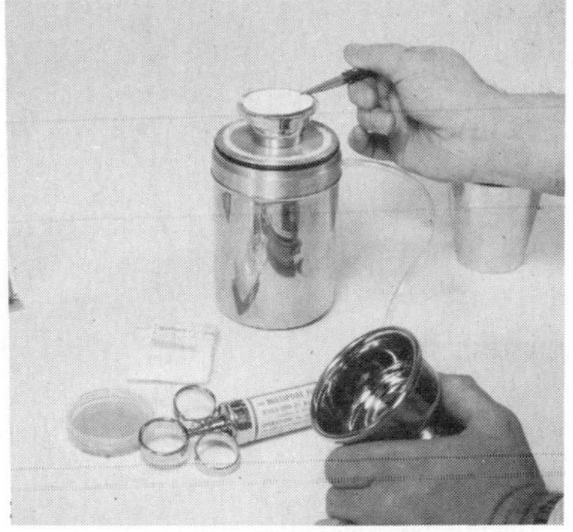

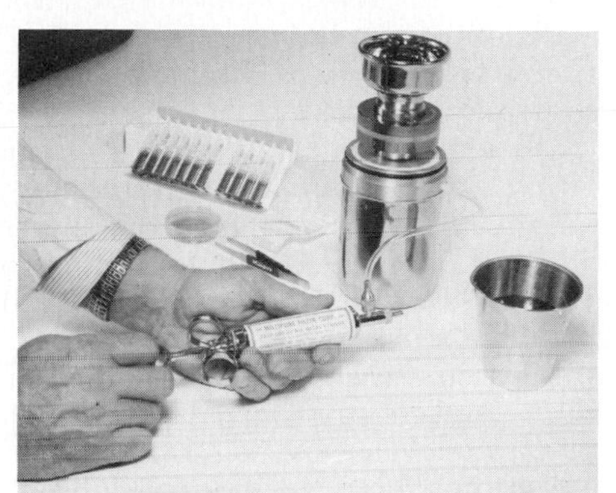

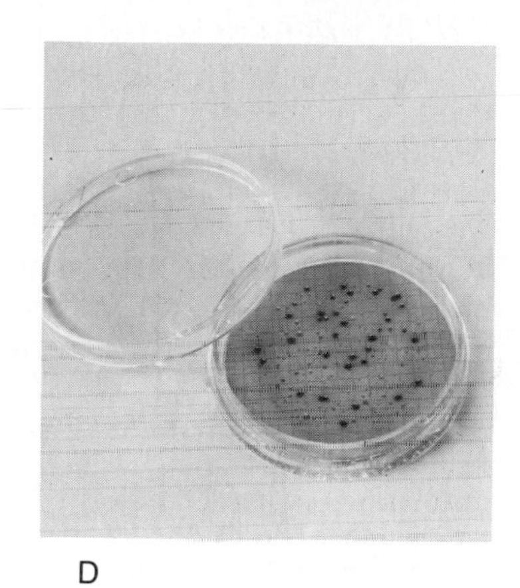

Figure 10–6 Equipment and materials used in rapid bacteriologic examinations of drinking water using membrane. *A,* Vacuum pump to draw water through filter, mounted in the funnel assembly on the large flask; box of sterile filters, medium, and tweezers. *B,* Smaller water-sample filtering equipment for "on the spot" sampling. After sterilizing the apparatus, a sterile, type HA Millipore filter is placed aseptically on the filter holder base. *C,* Several short strokes of the syringe are used to pull the water sample through the filter. *D,* Growth of distinctive coliform on medium-saturated filter after incubation, obtained with either form of apparatus. (From Millipore Corporation: Microbiological Analysis of Water, 1969.)

which is valuable in field work. The membrane filter method, using media specially adapted to the needs of the organisms sought for, is widely applicable in microbiology. If *Escherichia coli* is found in water, it is probable that the water contains feces. It therefore may contain typhoid, dysentery, or cholera organisms, as well as poliomyelitis and hepatitis viruses and other intestinal pathogens, since these are discharged with the feces from patients, convalescents, and *carriers.*[2]

MICROORGANISMS IN THE AIR

Under ordinary conditions numerous microorganisms may be found in the air all about us. Many of these are the spores of molds and yeasts, conidia of *Streptomyces,* and spores of bacteria of the genera *Bacillus* and *Clostridium.* Spores and conidia are admirably adapted to survive floating about on dust in the air for weeks or years.

In considering microorganisms in the air one must remember that they cannot fly and have no power to leave any surface of their own volition. Microorganisms are, however, great "hitchhikers." Indeed, the number of microorganisms in the air usually depends on the amount of dust, since most of them are riding on dust particles. They are usually of the harmless kinds found in soil and soon die in the dry air and sunlight. However, the air of

[2]Carriers are persons, lower animals, or plants that harbor pathogenic microorganisms without showing any perceptible evidences of disease but are able to transmit the infection to others; the condition may be transitory, intermittent, or continuous.

dark, badly ventilated rooms, if they are not kept clean, may contain many pathogenic bacteria, especially when occupied by persons who have such organisms in their noses and mouths.

With the recent increase in air pollution, smog problems, industrial growth of cities, car exhausts, and even smoking exhalations by people who should know better, bacteria find it easier to be carried over great distances from host to host. This is one price we have to pay for civilization.

In recent years much attention has been given to air as a means of disease transmission, especially in hospitals. It is known that every drop of saliva and nasal exudate, even from healthy persons (as well as health personnel and friendly hospital visitors!) contains microorganisms capable of causing disease. Among these are staphylococci, pneumococci, streptococci of scarlet fever, puerperal sepsis, and septic sore throat, and sometimes diphtheria bacilli, tubercle bacilli, and numerous viruses (polio, influenza, adenoviruses, and so on). By means of bacteriological examination of air, the presence of all of the bacteria just listed has been demonstrated in sick rooms, hospital wards, and even such carefully guarded places as operating rooms.

Droplet Infection. The air of classrooms, theaters, schools, and public buses must swarm with microorganisms, especially in winter when sneezing and coughing add to the general pollution of the atmosphere by bacteria- and virus-laden sprays of saliva and mucus. Transmission of disease by droplets of saliva or mucus is often called "droplet infection." These sprays contaminate dust, and, when dry, the dust carries the bacteria about (Fig. 10–7).

Droplet Nuclei. The droplets of saliva and the bacteria contained in the dried particles of mucus float about in the air and are inhaled like dust. Such dried particles are spoken of as *droplet nuclei.* These land on floors, furniture, lips, hands, surgical wounds, foods, and so on. The possibilities are obvious.

MICROORGANISMS IN FOOD

Many saprophytic microorganisms, and some dangerous pathogenic bacteria, grow well in moist and not too acid foods at warm room temperature. Such foods include milk, cooked cereals, macaroni dishes, custards, and soups. The bacteria come from dust, dishes, hands, ingredients in the foods, water, and utensils. We eat large numbers of saprophytic microorganisms in foods every day with no bad results. Many microorganisms flourish so well in some

Figure 10–7 The atomization of droplets into the air during sneezing. A violent, unstifled sneeze, not quite completed. Photo taken in a strong light. (Courtesy Dr. M. W. Jennison, Syracuse University, Department of Plant Sciences, Syracuse, N.Y.)

foods, such as soups, stews, broth, gelatin, potatoes, and milk, that these substances are often used for cultivating such organisms in the laboratory. The growth of certain microorganisms in some kinds of food is advantageous. The good flavors of butter and cottage cheese are produced by the action of certain harmless streptococci purposefully added to them. Varieties of cheese are largely determined by the kinds of microorganisms (bacteria and molds) present in them. For this reason it is customary to add "starters" (materials containing the desired bacteria) to cream before it is churned into butter, and to milk that is to be made into cheese. Pickles and sauerkraut are fermented by certain flavor-producing bacteria *(Lactobacillus, Leuconostoc).*

Protein Decomposition. Excessive growth of saprophytic microorganisms in food, or in any other product for that matter, results in spoilage. The anaerobic decomposition of proteins (mussels, egg white, fish, and similar foods) by microorganisms *(putrefaction)* is usually accompanied by very bad odors due to the formation of ammonia, hydrogen sulfide, mercaptans, and other volatile substances. Ordinarily, putrefied materials are disagreeable, but we sometimes eat "putrefied" milk in the form of Limburger and Liederkranz cheeses! Putrefied materials are not necessarily dangerous unless they contain pathogenic microorganisms.

Carbohydrate Decomposition. The series of changes that microorganisms bring about in carbohydrates under anaerobic conditions is called *fermentation.* The chief substances formed as a result of fermentation are acids of different kinds (such as lactic acid), alcohol, and gases such as carbon dioxide. Many bacteria, as well as yeasts and molds, cause fermentation.

There are different kinds of fermentation. One of the most familiar is the production of alcohol by yeast from the sugar of fruit juices, as in the making of wine and hard cider. A familiar microbial action (which, being strictly aerobic, is not true fermentation [an anaerobic process]) is the change of cider or wine into *vinegar* when bacteria of the genus *Acetobacter* form acetic acid from alcohol of cider or wine. These aerobic bacteria collect on the surface as scum called "mother of vinegar."

The souring of milk is usually due to the formation of lactic acid by certain streptococci *(Streptococcus lactis),* which ferment the sugar of milk (lactose). These harmless streptococci, along with many other saprophytic bacteria, may gain entrance to the milk from dust in the barns, the cow's skin, milker's hands, and unsterilized buckets. Fresh market milk commonly contains thousands of harmless saprophytic bacteria per cubic centimeter, which cause it to spoil if it is not refrigerated. In modern dairies much contamination is avoided by drawing milk from the udder directly into sterilized milking machines and, thence, via a closed system of sterile pipes, directly into refrigerated tanks, "untouched by human hands." Also, some milk that reaches the market today is actually made from dried powder; all nonspore-forming microorganisms in this milk have been killed by heat.

The rising of bread, also, is due to fermentation. The yeast cells multiply in the dough and decompose sugar derived from the flour starch, forming alcohol and carbon dioxide gas. The bubbles of gas, imprisoned in the dough, raise ("leaven") the bread. Baking dries and firms the bread and drives off the alcohol.

A few microorganisms are sometimes present in eggs even before they are laid. Microorganisms can also pass through the shell after the egg is laid. Many people have, in ignorance, washed dirt from the outside of an egg through the shell into the egg.

The spoiling or decay of foods is caused by the growth of various saprophytic microorganisms in them. Tainted meat, rancid butter, rotten eggs, and decaying fruit and vegetables are all the result of the growth of microorganisms. Spoiled foods, although unpleasant, are not necessarily harmful or infectious.

Fat Decomposition. Chemically, fats are organic salts or esters, being commonly composed of glycerin (also known as the tri-alcohol glycerol, one of the components of all fats) and one or more fatty acids, such as butyric, oleic, or stearic acid. One type of decomposition of fats by microorganisms is hydrolysis, which separates glycerol and butyric acid and similar volatile fatty acids. These acids are the principal factors in the odor and taste of rancidity.

Contamination of Foods. Sometimes pathogenic bacteria get into various foods such as milk, meat, sandwich fillings, salads, and puddings. The bacteria come from the unclean hands, respiratory droplets, and secretions of cooks and other food handlers, or from flies, roaches, rats, or mice that come to exact their tribute from (and to pollute) the kitchen. If food, having become polluted, is held at room temperature and is not promptly and sufficiently cooked after being contaminated, the pathogenic bacteria may multiply in the food and cause disease in the people who eat it. The food, far from appearing spoiled, may seem to the eye, nose, and taste to be perfectly wholesome. The bacteria may, under certain circumstances (especially lack of refrigeration and/or insufficient cooking), give off potent poisons into the food, which, when swallowed, cause distress and sometimes death. These matters will be discussed more fully later.

Moist food should never be allowed to stand unrefrigerated unless it has been thoroughly cooked and remains in a closed vessel. No moist food should be allowed to remain uncovered and unrefrigerated for more than a short time before cooking or eating. It is important to remember this, for protection not only of others but also of yourself and your family.

INDUSTRIAL MICROBIOLOGY

There is hardly a major industry that does not employ highly specialized microbiologists. In fact there are probably more "industrial microbiologists" working today than medical microbiologists. Industrial microbiologists are needed in the vast food industry, especially for canning and dairy processing, for drug and vaccine production, for manufacture of fabrics and fibers, in the oil industry, in paper manufacturing, in water reclamation, in the cosmetic field, in paint, wood, stone, plastic, and metal preservation, in space microbiology, for antibiotic and other assays in laboratories, and in the giant fermentation and chemical industries.

Industrial fermentations are often carried on in great vats or tanks holding thousands of gallons. Here the skill and knowledge of the microbiologist, engineer, and chemist are pooled for the common good. From the fermentation of carbohydrates are formed not only ethyl alcohol and lactic, citric, and acetic acids, but also isopropyl, methyl, and butyl alcohol, glycerin, antibiotics (over 84 different ones at a recent count), vitamins, enzymes (over 15 of commercial value), nucleic acids and derivatives, amino acids, insecticides, steroids and hormones (remember "the pill"), and many other substances of great value and importance, depending on the species of microorganisms present and the culture medium. The fermentative, putrefactive, synthetic, and other enzyme-controlled powers of microorganisms are also utilized in the manufacture of products such as rubber, coffe, cocoa, tobacco, linen, spices, leather, stock feed, pickles, and drugs.

Nitrogen Fixation. Our atmosphere consists roughly of 80 per cent nitrogen and 20 per cent oxygen, both essential to life. Humans are able to use oxygen directly for subsistence, but we have always been totally unable to utilize atmospheric nitrogen (until relatively recent advances were made in chemical engineering). We have been wholly dependent on "lower" forms of life to prepare nitrogen for us by combining it with other elements, mainly oxygen, hydrogen, and carbon. The process of combining nitrogen of the atmosphere with other elements is called *nitrogen fixation.*

Nonsymbiotic Nitrogen Fixation. It is worth noting that the evolution of our entire grand and glorious nuclear and space age has depended in good part on certain humble bacteria of the soil that "fix" nitrogen from the air by at least two distinct methods: nonsymbiotic and symbiotic. Nonsymbiotic nitrogen fixation is the direct combination of atmospheric nitrogen as part of the substance of a living cell without the cooperation of any other organism. For example, atmospheric nitrogen can be built up directly into complex enzymes or nucleic acid by bacteria of the genus *Azotobacter* (*azo,* nitrogen) and by several other microorganisms, including some algae, certain eukaryotic fungi, and bacteria of the genus *Clostridium.* These useful species abound in all fertile soils. A farmer, allowing a field to lie fallow or unplanted, permits *Azotobacter, Clostridium,* and some other microorganisms to accumulate nitrogen from the air as a gift of nature. Nitrogen in the form of commercial fertilizers is very expensive.

Symbiotic Nitrogen Fixation. The little nodules on the roots of *leguminous* plants such as clover, beans, peas, and alfalfa contain multitudes of bacteria belonging to the genus *Rhizobium.* These, growing together with the plant, have the power of taking nitrogen out of the air and combining it into substances essential for the growth of both the bacteria and the plants. This fixed nitrogen is released from the soil on the death of the plants. The process is *symbiotic* (living together for mutual benefit), since neither the *Rhizobium* (root-living) nor the plants could effectively accomplish the nitrogen fixation alone, whereas together they act for mutual advantage (Fig. 10–8). Crops of beans, peas, alfalfa, and clover do not grow without the indispensable *Rhizobium.*

Figure 10–8 Nodules on the roots of Alsike clover. These contain symbiotic nitrogen-fixing bacteria of the genus *Rhizobium.* (Swingle, D. B.: Plant Life. New York, Van Nostrand Company, Inc., 1942.)

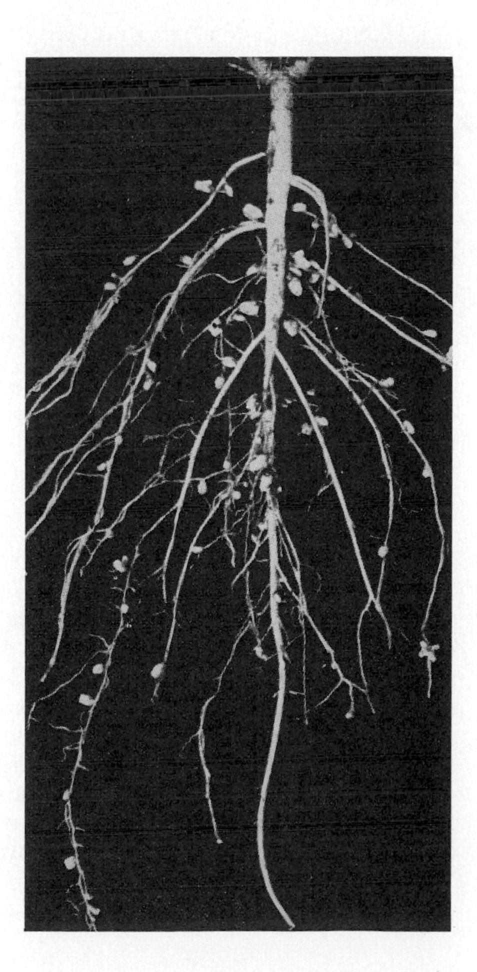

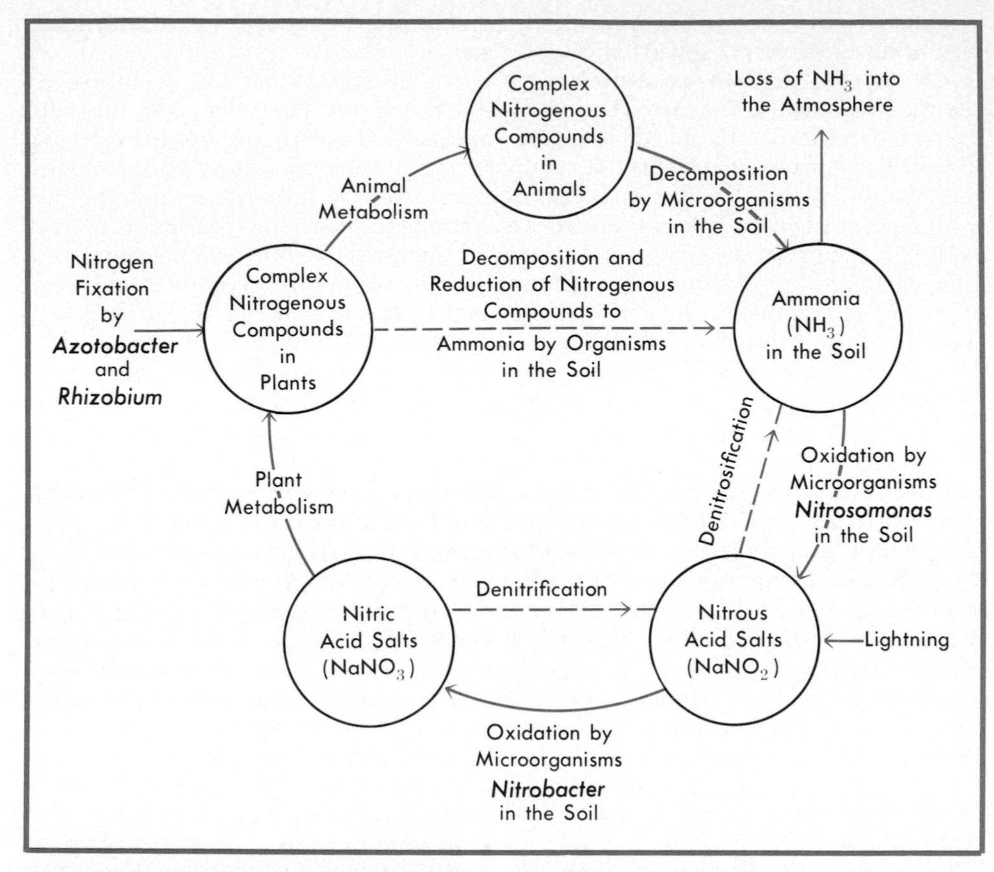

Figure 10–9. The nitrogen cycle. At the right, ammonia (NH_2) and oxides of nitrogen are brought into the cycle. Ammonia is nitrogen in its most reduced form. Oxides of nitrogen are carried to the soil in rain. Ammonia is derived principally from decomposing organic matter (center of diagram). Some escapes to the atmosphere.

Proceeding in a clockwise manner, the process of nitrosification, carried on by soil bacteria (e.g., *Nitrosomonas)*, oxidizes the NH_3 to nitrites. Other soil bacteria (e.g., *Nitrobacter)* oxidize the nitrites to nitrates, in which form the nitrogen is available to plants (left of diagram). Facultative and anaerobic bacteria of the soil are constantly acting to reverse these processes, as indicated by the lines marked "denitrification" and "denitrosification." After nitrogen is at last incorporated in plants as proteins and other compounds, it is converted into animal tissues (top of diagram). When plants and animals die, and their wastes decay, the saprophytic microorganisms in the soil convert the nitrogen back into the form of ammonia, and the cycle recommences.

At the left of the cycle are shown the means by which atmospheric nitrogen is converted directly into living matter by the processes of nonsymbiotic and symbiotic nitrogen fixation carried on by soil bacteria, such as *Clostridium Azotobacter,* and *Rhizobium.* Once it is in this form, it follows the same course in the cycle as do other vegetable proteins.

Nitrification. Nitrogenous compounds released into the soil by leguminous plants may be in part taken up by other plants and partly decomposed by putrefaction, yielding ammonia (NH_3). In the form of ammonia, nitrogen is almost useless to green plants. To be readily available to plants the nitrogen of ammonia is first oxidized by the bacteria to nitrous acid (HNO_2); this in turn is oxidized by other species to nitric acid (HNO_3). These important functions are carried on by soil bacteria of the family *Nitrobacteraceae.* The nitric acid they form immediately combines with various substances to make nitrates, which can be used by plants. Many commercial fertilizers contain some nitrates.

Oxides of nitrogen are also produced in small amounts by lightning flashes and are washed into the soil by rain. The process of changing nitrous acid into nitric acid is called *nitrification.* The nitrifying bacteria, as well as the nitrogen-fixing microorganisms, are clearly of immense importance to agriculture. The great store of nitrogen in the air would be useless for the needs of many living things if it were not for the nitrifying and nitrogen-fixing microorganisms (Fig. 10–9).

Below a soil depth of four feet, microorganisms become less numerous, and at a depth of 10 to 12 feet there are usually none.

MICROORGANISMS IN AND ON HUMAN BEINGS

Numerous microorganisms find their optimum (and for several pathogenic species their only) habitat in or on the bodies of humans or of animals. The healthy human body harbors millions of microorganisms on the skin and in the mouth, eyes, ears, genitourinary tract, and intestines—in short, on every surface that comes in contact with the outside world, with respired air, or with food (Table 10–1, Fig. 10–10). The human body is a complete ecosystem. Most of these microorganisms rarely or never cause disease under normal conditions, but some may, under certain circumstances (e.g., in wounds or after surgery), gain entrance to the deeper parts of the body, where bacteria are not usually present, and produce what we call an infection. Each region of the body in contact with the exterior normally has its characteristic microorganisms.

The skin carries large numbers of bacteria, picked up from the various things with which it comes in contact. In addition, *Staphylococcus aureus,* the common cause of boils, carbuncles, breast abscess, infantile impetigo or pemphigus, and other conditions, is found at times in the hair follicles and sweat ducts. The skin cannot be made absolutely free of bacteria even with the most thorough scrubbing and the application of antiseptics: although the bacteria in the outer layers of the skin can be removed, it is impossible to get rid of those in the deeper layers, the sweat and sebaceous glands, and the hair follicles. Therefore gloves, which can be made sterile, are worn in the operating room and in other situations where sterile conditions are essential.

Inhaled air contains particles of dust, some of which have microorganisms attached. The air around persons who have been coughing or sneezing may be filled with a a bacteria-carrying mist of nasal secretion and saliva droplets. To demonstrate this for yourself, watch someone sneeze who is sitting near the window of a darkened room where the sun is shining brightly against one windowpane (Fig. 10–7). Hold a Petri plate with nutrient agar in front of the mouth of someone coughing, sneezing, or even talking, and then incubate it. It will prove that we inhale and exhale considerable numbers of microorganisms all the time.

The inside of the nose is especially adapted for dealing with these microorganisms. It contains a complicated, scroll-like arrangement of bones (the turbinates), covered with mucous membrane. As the air passes over these surfaces it is not only warmed and moistened (nature's air-conditioning apparatus), but the microorganism-laden dust sticks to the moist surface and is finally carried to the outside in the nasal secretion. In a healthy nose few microorganisms can gain a permanent foothold. If the nasopharynx is stopped up by adenoids or by other obstructions, or if the normal defense mechanisms are held in abeyance by diseases such as influenza, measles, or whooping cough, microorganisms may find suitable conditions for growth and cause such diseases as sinusitis, rhinitis, and pneumonia.

Table 10–1. Microorganisms that May be Found More or Less Regularly in or on Apparently Normal Persons

Scalp:
- *Staphylococcus epidermidis*†
- *Propionibacterium acnes (Corynebacterium acnes)*†

Conjunctivae:
- *Corynebacterium xerosis*
- *Staphylococcus epidermidis*
- *Haemophilus* sp.*-

Ears:
- *Mycobacterium phlei*
- *Corynebacterium* sp.
- *Staphylococcus epidermidis*†

Nose, mouth, and pharynx:
- *Streptococcus pyogenes**
- *Streptococcus salivarius*
- *Streptococcus mitis*
- *Streptococcus faecalis*†
- *Streptococcus pneumoniae (Diplococcus pneumoniae)**
- *Corynebacterium pseudodiphtheriticum*
- *Corynebacterium xerosis*
- *Corynebacterium diphtheriae**
- *Haemophilus influenzae**
- *Treponema vincentii (Borrelia vincentii)*†?
- *Treponema buccale (Borrelia buccalis)*†
- *Leptotrichia buccalis (Fusobacterium fusiforme)*†?
- *Trichomonas* sp. *(tenax?)*
- *Staphylococcus aureus**
- *Staphylococcus epidermidis*
- *Branhamella catarrhalis (Neisseria catarrhalis)*
- *Neisseria meningitidis**
- *Mycobacterium phlei*
- *Lactobacillus casei*
- *Lactobacillus fermenti*‡
- *Candida albicans**
- Spores of *Bacillus,* yeasts, and molds
- Any microorganisms of food or air
- Filamentous actinomyces-like organisms around teeth (*Leptotrichia buccalis?*)
- Influenza and adenoviruses*

Axillae:
- *Mycobacterium smegmatis*
- *Mycobacterium phlei*
- *Staphylococcus epidermidis*†
- *Corynebacterium* sp.

Smooth (hairless) skin:
- *Staphylococcus epidermidis*†
- *Staphylococcus aureus**
- *Propionibacterium acnes (Corynebacterium acnes)*† etc.
- Any organisms of surrounding air, clothing, etc.
- Spores of *Bacillus,* molds, etc.

Colon:
- *Escherichia coli** and coliform group
- *Shigella* sp.*
- *Salmonella* sp.*
- *Bacteroides fragilis**
- *Bacteroides serpens**
- *Clostridium perfringens*†
- *Clostridium tetani*†
- *Streptococcus faecalis* and enterococci†
- *Alcaligenes faecalis*
- *Bifidobacterium bifidum (Lactobacillus bifidus)* (esp. infants)
- *Lactobacillus acidophilus*
- *Pseudomonas aeruginosa*
- *Proteus vulgaris*†
- *Candida albicans**
- *Giardia lamblia*
- *Trichomonas hominis*
- *Entamoeba coli*
- *Entamoeba histolytica**
- Enteroviruses*
- Poliovirus*
- Virus of epidemic hepatitis*

Perianal folds and area:
- *Mycobacterium smegmatis*
- *Escherichia coli** and coliforms
- Enterococci
- *Clostridium* sp.†
- *Lactobacillus* sp.
- *Corynebacterium* sp.
- *Staphylococcus epidermidis*†
- Spores of fungi and yeasts

Genitalia:
- *Mycobacterium smegmatis*
- *Treponema refringens* (esp. male)
- *Corynebacterium* sp.
- *Lactobacillus acidophilus* (vagina)
- *Trichomonas vaginalis**

Hands:
- Variable, depending on materials being handled; commonly microorganisms of skin, respiratory tract, feces, and perianal region

*Primary invaders.

†Secondary invaders, opportunists, or lesser pathogens. *Note:* Fungi that cause dermatomycoses (skin infections) are not included in the table, although, in an unsuspectedly high percentage of our population, apparently normal persons have these organisms on their scalp, skin, ears, and feet.

‡This species has been renamed, but it is not clear if it is *Lactobacillus curvatus* or *Lactobacillus brevis.*

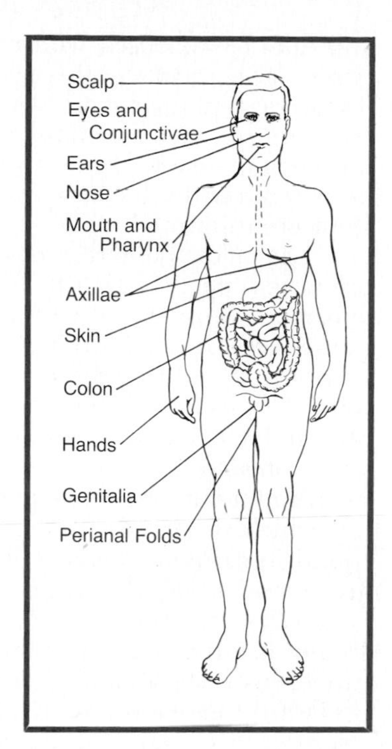

Figure 10–10 Areas of the body where microorganisms are normally found. See Table 10–1 for the types of microorganisms expected to be present.

The mouth and throat constantly contain numerous kinds of microorganisms. These live and multiply in the secretions of the nose and oral cavity and among the teeth; hence, good oral and dental hygiene is desirable. Commonly found are streptococci *(Streptococcus pyogenes)* of several pathogenic varieties, staphylococci, pneumococci *(Streptococcus pneumoniae)*, and certain spirochetes *(Treponema vincentii*, formerly called *Borrelia vincentii)*. Also occasionally present are diphtheria bacilli *(Corynebacterium diphtheriae)*, *Haemophilus influenzae* (*not* the cause of influenza), the influenza virus, adenoviruses, and tubercle bacilli, as well as saprophytes from food, dust, and fingers. Microorganisms of many types grow readily in a neglected mouth because conditions of warmth and moisture are favorable, and bits of food and desquamated epithelium around the teeth provide nourishment.

In patients with febrile and various other diseases, the mechanisms ordinarily holding oral microorganisms in check are weakened and decompositions occur, giving rise to the oral fetor or halitosis common in such patients. Bacteria may aid in decay of the teeth. The tartar that collects around the gums is composed partly of microorganisms, bits of foods, and cast-off cells from the tongue. The mouth would contain more microorganisms than it does if it were not for the saliva, which acts as a continuous mouthwash. The microorganisms are swallowed with the saliva, and many are killed by the acidic gastric juice. Persons who have been in contact with meningitis, scarlet fever, pneumonia, or diphtheria patients may carry the bacteria causing these diseases in their mouths or throats, although they may not develop any symptoms.

The normal conjunctivae usually contain bacteria of a harmless kind, which are related to the diphtheria bacillus and are called "diphtheroids." One common species is known as *Corynebacterium xerosis*. Similar bacteria are found in the nose and throat *(Corynebacterium pseudodiphtheriticum)*, in the wax of the ears, on the skin, and around the genitalia.

Harmless bacteria called *Mycobacterium smegmatis*, closely related to the tubercle bacillus *(Mycobacterium tuberculosis)*, are common on and around the genitalia. They are important because they sometimes get into urine specimens and may be confused with tubercle bacilli in attempting bacteriological diagnoses. This explains the necessity of taking urine specimens intended for bacteriological examination by means of a sterile catheter into a sterile tube.

The stomach normally contains few live microorganisms because of the disinfectant action of the strongly acid gastric juice. After meals the acidity has been reduced by the food and many bacteria may grow and pass through the stomach unharmed.

The neonatal intestinal contents are free of microorganisms, but these organisms enter with the first feeding. The intestine of the adult contains enormous numbers of microorganisms, and billions are thrown off every day in the feces. In connection with these immense numbers of organisms it should be remembered that the intestine is about 26 feet long, and conditions of food, warmth, and moisture in it are ideal for microorganisms, especially the anaerobic and facultative types. They are more abundant in the large than in the small intestine.

Prominent among the intestinal bacteria is the relatively harmless *Escherichia coli,* which, as already mentioned, is viewed as an index of fecal pollution when found in drinking water. It has been estimated that for each *E. coli* cell there are about 1,000 *Bacteroides* species cells in the human intestine, but these anaerobes are much more difficult to isolate and identify. Other important species are the organisms that cause gas gangrene (*Clostridium perfringens* and numerous related species) and *Clostridium tetani,* the cause of lockjaw. In some backward places where ignorance and filth are widespread, fecal material (containing *Cl. tetani*) gets into the stump of the umbilical cord of infants at birth and causes tetanus neonatorum (tetanus of the newborn). In some sections of aboriginal Africa, it is said that over 50 per cent of the infants die of this disease. Many harmless bacteria are also present in the intestine, including *Streptococcus faecalis,* one species of the group called enterococci. Always present in feces and fresh sewage, it is sometimes used by microbiologists instead of *E. coli* as an indicator of fecal pollution.

The vagina normally contains certain distinctive nonpathogenic bacteria, including *Streptococcus faecalis* and *Lactobacillus* species. The uterus is normally free of bacteria. Immediately after childbirth, however, it is an excellent place for growth of pathogenic bacteria, especially *Streptococcus pyogenes,* which may be carried in during parturition, or afterward on dirty hands, instruments, or dressings, and cause puerperal fever. This is one of the reasons why sterile precautions are so necessary at childbirth.

Microorganisms in the Blood. In certain diseases the causative microorganisms circulate in the blood for periods ranging from hours to days, sometimes intermittently. This is true in typhoid fever, brucellosis (undulant fever), syphilis, malaria, yellow fever, typhus fever, and numerous others. When the microorganisms are bacteria, the patient is said to have a *bacteremia* (presence of bacteria in the blood). Similarly, when the microorganisms are viruses, the patient is said to have a *viremia.* Microorganisms may also sometimes be present in the blood of perfectly healthy persons. The vessels that drain the blood from the lower intestines into the liver (portal veins) often contain a few intestinal microorganisms. These are quickly removed from the blood by certain cells in the blood vessels, especially in the liver and spleen.

The Mononuclear Phagocyte System (MPS). These cells of the *reticuloendothelial system* (or RES as it is also called) constitute part of the lining of the blood vessels. They have the power of engulfing minute foreign particles in the blood (bacteria, cell fragments). They are *phagocytic.*[3] The entire group of these phagocytic tissue cells throughout the body should be distinguished from the freely wandering phagocytic cells called leukocytes (Chapter 18). Sometimes

[3]From the Greek *phagein,* to eat, and *cyte,* cell; phagocytes are therefore "cells that eat."

after the extraction of a tooth, a local injury, or some unnoticed trauma, a few dangerous pathogenic bacteria may circulate in the blood for a few minutes before being taken up and killed by the phagocytic cells.

Microorganisms in Blood Banks. Temporary bacteremias would be of little or no significance were it not for the danger that the bacteria might get into blood being drawn for blood banks or transfusion. Some of the organisms normally on the skin can also get into blood bank blood when a *venipuncture* (placing the needle in the vein) is made. With properly sterilized equipment and clean, expert technique, this very rarely happens. If the bacteria are numerous in blood bank blood, however, they can cause reactions that are always serious and sometimes fatal.

Precautions with Blood Bank Blood. The blood in a blood bank container must not be exposed to contamination from any source whatever. The blood must be kept refrigerated. It must not be allowed to stand at room temperature for more than a few minutes before being stored or after removal from the refrigerator for use in a patient. Some of the most troublesome bacteria in blood can grow well even at the low temperature at which blood is stored. Nearly all of them can grow rapidly at room temperature. Bacteria sometimes found in blood bank blood and blood products include species of *Pseudomonas, Salmonella, Escherichia, Staphylococcus, Bacillus,* and *Corynebacterium.* Syphilis organisms do not survive in blood bank blood for more than two days. Malaria organisms die in about five days. A constant danger is the virus of homologous serum jaundice (infectious hepatitis B), against which special safeguards must be constantly maintained. Blood bank blood must be used within the date limit. It must *never* be used if there is evidence of hemolysis (i.e., if the clear, fluid part looks red), discoloration, or other abnormality. If there is any doubt in your mind about the blood, notify your superior at once.

APPLICATION TO HEALTH FIELDS

The beginning student of microbiology frequently views with alarm the fact that microorganisms are around us constantly. As has been pointed out, however, relatively few of these are harmful. A knowledge of where *pathogenic* microorganisms are likely to be found is extremely helpful so that you can be constantly on the alert to protect yourself and others from infection.

SUMMARY

Microorganisms are almost ubiquitous on the surface of the earth and are therefore most important in the human ecosystem, as well as in all other forms of life. Those that can live on or in the bodies of human beings, plants, or animals may be among the relatively few *pathogens* (viruses, bacteria, protozoa, fungi) causing harm, whereas the great majority are *saprophytic* or *saprozoic,* harmless and often greatly beneficial. Many of them, especially yeasts, molds, and bacteria, are used to manufacture alcohol, lactic acid, butter, cheese, solvents of all sorts, antibiotics, and products to increase soil fertility.

Three processes carried out by microorganisms are *decay,* the decomposition of organic waste materials (animal or plant); *putrefaction,* the decomposition of proteins; and *fermentation,* the breakdown of carbohydrates into alcohols, acids, and gases (CO_2). The conditions most favorable for putrefaction, fermentation, and decay in the outside world are those that are most suitable for the growth of microorganisms living in the soil, rivers, lakes, and oceans.

Every waste water plant (sewage disposal plant) receives "water after use" of housing and industrial origin. Besides human excreta, the water may contain microorganisms that could cause typhoid fever, cholera, or amebic dysentery. Sewage may be treated by a combination of systems, aerobically and anaerobically or by wet air oxidation, in Imhoff tanks with trickling filters, activated sludge, and final chlorination.

Purification (reclamation) of drinking water may go through the following steps: intake to the mixing chamber, where the water is mixed with coagulating chemicals, the floc settling chamber, the sand filter, a storage cistern, and final chlorination before it is released to the consumer.

Tests designed to check the potability of drinking water and the safety of swimming pools make use of the coliform organism *(Escherichia coli)* as an indicator of fecal pollution. Water that contains coliform organisms is considered polluted: it is potentially dangerous because it contains feces and thus may harbor intestinal pathogens as well.

The number of microorganisms in the air depends on the amount of dust particles they may be attached to. Those in hospital wards and sick rooms and public places are naturally more numerous and more dangerous than in clean, quiet surroundings, especially where sneezing and coughing adds to the general pollution of the atmosphere as "droplet nuclei."

Our food is also excellent food for microorganisms. Some microorganisms are highly desirable in making cheeses, pickles, sauerkraut, buttermilk, bread, and even tenderized meat. Also, let us not forget fermented beverages such as wine. However, some microorganisms do spoil foods, and others, as pathogens, may give off potent poisons into food, which when ingested cause distress and sometimes even death.

Useful microorganisms have been employed in practically every major industry for the benefit of all humans. Industrial microbiology is an active field for highly specialized microbiologists. One type of bacteria *(Rhizobium)* is used to fix nitrogen from the air to enhance soil fertility.

Each region of our body in contact with the exterior environment normally has its characteristic microorganisms, which are usually harmless. They are on and in the skin, the mouth, and the nose; we inhale them into the lungs and swallow them; and our intestines swarm with special inhabitants, some beneficial to us. Those that reach the blood usually do not survive thanks to the reticuloendothelial system, and our organs are free of microorganisms as long as no infection has occurred.

REVIEW QUESTIONS

1. What is ecology? What does an ecosystem include? In what environments does one find microorganisms?
2. What is putrefaction? fermentation? decay? What sorts of bacteria may cause each? Of what value are these bacteria to mankind?
3. What is the basis of sewage purification? What is a trickling filter? How does it operate?
4. What goes on in a septic tank? How do natural processes of putrefaction and fermentation in soil compare with the action of sewage disposal plants?
5. Give one other name for a sewage disposal plant. By what other name may a water purification plant be called?
6. Outline the nitrogen cycle. What is symbiotic nitrogen fixation? nitrification? Name genera of bacteria involved in each. Why are these processes essential to mankind?
7. When may drinking water transmit human disease? Name two important waterborne diseases. What is the significance of the presence of *Escherichia coli* in water? How is its presence determined? What is the value of chlorine in water purification? What simple means of water disinfection are generally

available? Compare aerobic with anaerobic treatment of waste water. What is the purpose of the filter in water purification?
8. What types of diseases are transmitted by droplets or particles in the air? Name three such diseases.
9. Describe some methods used to control airborne infection. What properties must be possessed by bacteria in order to be transmitted in an infective state by dust?
10. Are bacteria in foods harmful? advantageous? Give examples. How may foods transmit disease? How may this be prevented?
11. Mention some industrial uses of bacteria. What jobs may a microbiologist find in industry?
12. Mention several locations in or on the healthy human body where bacteria constantly occur. What sorts of bacteria are found in each? Why is it that they do not ordinarily cause disease? When may they do so?
13. What is bacteremia? When may it occur? How is it related to blood transfusions? How may it be detected?
14. What precautions must be taken in handling blood bank blood?
15. What is the reticuloendothelial system?

Supplementary Readings

Aaronson, S.: Experimental Microbial Ecology. New York, AIBS Books, Academic Press, 1970.
Alexander, M.: Introduction to Soil Microbiology. New York, John Wiley & Sons, Inc., 1961.
Allen, M. J.: Microbiology of ground water. *J. Water Pollut. Control Fed., 53*:1107, 1981.
Allen, M. J.: Microbiology of waste water. *J. Water Pollut. Control Fed., 53*:1109, 1981.
Barth, E. F. (ed.): Advanced Waste Treatment and Water Reuse Symposium, Dallas, Texas, Jan. 12–14, 1971, Session Two. Sponsored by Environmental Protection Agency.
Berg, G., et al. (eds.): Viruses in Water. Washington, D.C., American Public Health Association, 1976.
Bernstein, I. A.: Biochemical Responses to Environmental Stresss. New York, Plenum Press, 1971.
Brody, A. L.: Flexible Packaging of Foods. Cat. No. 0103/106. Cleveland, The Chemical Rubber Co.
Carr, J. G., Cutting, C. V., and Whiting, G. C.: Lactic acid bacteria in beverages and foods. Proceedings of Long Ashton Research Station Symposium, University of Bristol, September 19–21, 1973. New York, Academic Press, 1975.
Curds, C. R., and Hawkes, H. A. (eds.): Ecological Aspects of Used-Water Treatment. New York, Academic Press, 1975.
Current List of Water Publications, 1965–1970. Cincinnati, Robert A. Taft Sanitary Engineering Center, Office of Information, Ohio Basin Region, Federal Water Quality Administration, U.S. Department of the Interior.
Doetsch, R. N.: Introduction to Bacteria and Their Ecobiology. Baltimore, University Park Press, 1973.
Fair, G. M., Geyer, J. C., and Okun, D. A.: Water and Wastewater Engineering. Vols. 1-2. New York, John Wiley & Sons, Inc., 1966–1968.
Foster, E. M., Nelson, F. E., Speck, M. L., Doetsch, R. N., and Olson, J. C., Jr.: Dairy Microbiology. Englewood Cliffs, N.J., Prentice-Hall, Inc., 1957.
Frazier, W. C., and Westhoff, D.: Food Microbiology. 3rd ed. New York, McGraw-Hill Book Co., Inc., 1978.
Frobisher, M., Hinsdill, R. D., Crabtree, K. T., and Goodheart, C. R.: Funda-

mentals of Microbiology. 9th ed. Philadelphia, W. B. Saunders Company, 1974.

Furia, T. E. (ed.): CRC Handbook of Food Additives. 2nd ed. Cleveland, CRC Press, 1980.

Furia, T. E.: Fenaroli's Handbook of Flavor Ingredients. 2nd ed. Cleveland, CRC Press, 1975.

Geldrerch, E. E.: Microbiology of Water. *J. Water Pollut. Control. Fed., 53*:1083, 1981.

Gray, W. D.: The Relation of Fungi to Human Affairs. New York, Henry Holt & Co., 1960.

Gray, W. D.: The Use of Fungi As Food and in Food Processing. Cat. No. 0104/106. Cleveland, The Chemical Rubber Co.

Greenberg, A., et al. (eds.): Standards Methods for the Examination of Water and Wastewater. 14th ed. Washington, D.C., American Public Health Association, 1975.

Gregory, P. H.: Microbiology of the Atmosphere. New York, Interscience Publishers, Inc., 1962.

Hassall, K. A.: World Crop Protection. Vol. 2, Pesticides. Cleveland, CRC Press, 1969.

Hentges, D. J.: Resistance of the indigenous intestinal flora to the establishment of invading microbial populations. *In* Microbiology—1975. Washington, D.C., American Society for Microbiology, 1975, pp. 116–119.

James, G. V.: Water Treatment. 4th ed. Philadelphia, International Ideas, 1971.

King, C. J.: Freeze-Drying of Foods. Cat. 0105/106. Cleveland, The Chemical Rubber Co.

Loesche, W. J.: Bacterial succession in dental plaque: Role in dental disease. *In* Microbiology—1975. Washington, D.C., American Society for Microbiology, 1975, pp. 132–136.

Miller, B. M., and Litsky, W. L.: Industrial Microbiology. New York, McGraw-Hill Book Company, 1976.

New Technology for Treatment of Wastewater by Reverse Osmosis. Environmental Protection Agency, Water Quality Office, Cincinnati, Ohio. Washington, D.C., U.S. Government Printing Office.

Odum, E. P.: Fundamentals of Ecology. 3rd ed. New York, Holt, Rinehart & Winston, 1971.

Perlman, D.: The fermentation industries. *American Society of Microbiology News, 39,* 1973.

Pipes, W. O.: Microbiology of waste water treatment. *J. Water Pollut. Control Fed., 53*:1142, 1981.

Rhodes, A., and Fletcher, D. L.: Principles of Industrial Microbiology. London, Pergamon Press, 1966.

Rose, R. E.: Effective Use of Millipore Membrane Filters for Water Analysis. Water and Sewage Works, 1966.

Rosebury, T.: Microorganisms Indigenous to Man. New York, McGraw-Hill, 1962.

Rossmoore, H. W.: Microbes, Our Unseen Friends. Detroit, Wayne State University Press, 1976.

Roy, D., Tittlebaum, M., and Meyer, J.: Microbiology detection occurrence and removal of viruses. *J. Water Pollut. Control Fed., 53*:1138, 1981.

Rushing, N. B., et al.: Growth rates of *Lactobacillus* and *Leuconostoc* species in orange juice as affected by pH and juice concentration. *Appl. Microbiol., 4*:97, 1956.

Slade, F. H.: Food Processing Plant. 2 vols. Cleveland, CRC Press, 1967.

Stapley, J. H., and Geyner, F. C. H.: World Crop Protection. Vol. 1, Pests, Diseases and Controls. 1969.

Stiles, H. M., Loesche, W. J., and O'Brien, T. C. (eds.): Microbial Aspects of Dental Caries. Vols. 1-3. New York, Information Retrieval, Inc., 1976.

Teitell, L., et al.: The effect of fungi on the direct current surface conductance of electrical insulating materials. *Appl. Microbiol., 3*:75, 1955.

Underkofler, L. A., Barton, R. R., and Rennert, S. S.: Production of microbial enzymes and their applications. *Appl. Microbiol.*, *6*:212, 1958.
Wallis, C., and Melnick, J. L.: Concentration of viruses from sewage by adsorption on Millipore membranes. *Bull. WHO*, *36*:219–225, 1967.

References on Filtration

Manual for Swimming Pool Operators. Texas Beach and Pool Association, 1100 West 49th Street, Austin, Texas.
Manual for Waterworks Operators. Texas Water and Sewage Works Association, 2202 Indian Trail, Austin, Texas.
Minimum Standards for Public Bathing Places. American Public Health Association, 1790 Broadway, New York, N.Y.
Municipal and Rural Sanitation. New York, McGraw-Hill Book Company.
Swimming Pools *(PHS publication No. 665)*. Washington, D.C., U.S. Government Printing Office.
Swimming Pool Operation. Circular No. 125. Springfield, Ill., State of Illinois, Department of Health.

SECTION THREE

INHIBITION, REMOVAL AND DESTRUCTION OF MICROORGANISMS

Selman A. Waksman (1888–1973), who won the Nobel Prize for his work, isolated streptomycin from *Streptomyces griseus* and reported on its antibiotic activity in 1944. Of particular significance was its effectiveness against human strains of tubercle bacilli, and it became the primary drug for the treatment of tuberculosis. Streptomycin is also effective against several species of bacteria that are unaffected by penicillin. (From *Scientific Contributions of Selman A. Waksman* edited by H. Boyd Woodruff. Copyright © 1968 by Rutgers, the State University. Reprinted by permission of Rutgers University Press.)

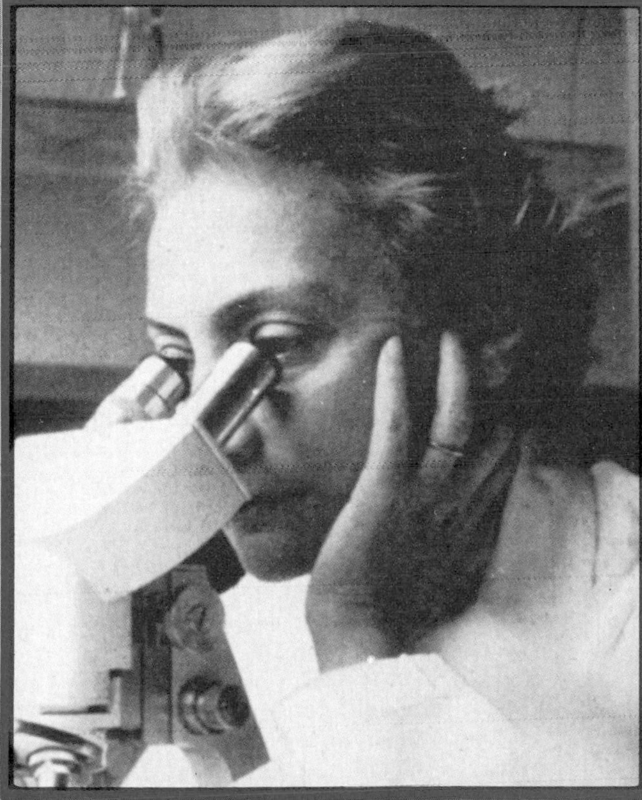

Amalia Voureka Fleming (1909–). A physician and bacteriologist in her own right, Amalia Voureka married Alexander Fleming, her mentor, in 1953. In particular, her work at St. Mary's Laboratory on the mutations of microbes earned recognition in the *Lancet*. (Courtesy of Keystone.)

DEFINITIONS AND BASIC PRINCIPLES OF STERILIZATION AND DISINFECTION

HISTORICAL ASPECTS

Until it was discovered that microorganisms cause disease, the need for sterilization, disinfection, and sanitation was not recognized. Instruments had to be developed to see microorganisms; they had to be isolated, cultivated, and studied in pure culture in the laboratory; and the effects of various substances and physical conditions on them had to be tested. Someone has said, "If we know more about science today, it is because we stand on the shoulders of geniuses who preceded us." Thousands of years ago the Egyptians and Hebrews used certain laws of sanitation that we can understand today because of our comparatively recent discoveries in the field of microbiology. Alexander the Great was advised by Aristotle to have his armies boil all drinking water and bury their excreta and the dung from their beasts of burden. The ancient Romans built aqueducts to bring pure water to their cities. This was amazingly practical wisdom in a world that was completely unaware of microorganisms.

In the seventeeth century Molière said, "Presque tous les hommes meurent de leurs remèdes et non de leurs maladies."* But even as recently as the early nineteenth century, hospitals were dingy, dirty buildings, not the gleaming structures of the twentieth century. Patients were often placed on mats on the floor or at best on low beds close together; the dead and dying were often crowded together in the same beds. The poor went to these "pest houses," in which nearly as many patients died as recovered (Fig. 11–1). These same conditions still exist in some parts of the world today. One of your author's former students, who as a nurse returned to her home country, confirmed this in a letter detailing conditions in her hospital.

Picture a surgeon making "rounds" in a first-rate hospital as recently as 1870. Typically, he wears a black frock coat, so he may wipe his hands on it and the blood may not be seen. Linen thread is wound around one of his buttons to provide suture material. His scalpel is sticking out of his breast pocket to be used periodically for lancing infections. If the scalpel is not sharp enough, it can be sharpened conveniently on the heel of his boot. The doctor washes his hands only when they are obviously soiled with pus and blood. In those days, hospitals were truly pest houses.

During this period, Semmelweis (c. 1848) insisted that doctors wash their hands before delivering babies; and Oliver Wendell Holmes (c. 1843) also insisted on handwashing in maternity practice. Pasteur proved (c. 1863) that infections were caused by microorganisms; Koch discovered (c. 1865) many microbial causes of disease and improved the criteria for determining the cause of disease; and Lister first used dilute carbolic acid (phenol) (c. 1867) to prevent and treat infections in surgery. With these men, the modern practice of asepsis, antisepsis, and disinfection began. Earlier attempts at disinfection relied largely on aromatic substances and on chemical deodorants. Even today, the odor of antiseptics is associated (sometimes erroneously) with cleanliness. You may buy Lysol spray with two different odors, one for the kitchen and one for the bathroom!

*"Nearly all men die as a result of their remedies and not of their maladies."

CHAPTER 11

One of the earliest systematic and purposeful uses of heat to destroy microorganisms in foods (although the microorganisms themselves were still unknown) was in response to a demand by Napoleon Bonaparte, who needed better food for his armies. A process of preserving canned food by heat was developed by a French scientist, Nicholas Appert, about 1805. The process, then called "Appert-izing," involved steam, heat, and pressure: essentially, it was autoclaving, a method of sterilization used today in the canning industry, public health clinics, and every hospital.

Another early use of heat for destruction of microorganisms was devised by Pasteur. In 1863 he found that spoilage of beer and wine could be prevented by subjecting them to temperatures of 50 to 60 C for a few minutes. In 1898 this procedure was adopted in Denmark for disinfection of milk. The process later became known as *pasteurization*.

Koch used a method of intermittent heat at 100 C to make culture media sterile. Tyndall, a British scientist, perfected the method of fractional sterilization in order to destroy spore-forming bacteria, and it has since become known as tyndallization. About 1880 Charles Chamberland,[1] a collaborator of Pasteur, constructed an apparatus, similar to the modern pressure cooker, that was an early autoclave.

[1]He also invented the porous porcelain *candle filter* to "separate pure water" from microorganisms. It was very useful for the isolation of viruses (Ivanowski [1892], Loeffler and Frosch [1898], Beijerinck [1898], and Walter Reed [1900]).

Figure 11–1 A ward in the Hôtel de Dieu of Paris. Facsimile of a wood engraving at the head of a sixteenth century manuscript. Note the attendants sewing a corpse into its shroud (lower left), patient and corpse in the same bed (upper left), more than one patient in a bed (upper right), and the generally unsanitary surroundings. (Reproduced from Hoffbauer: Paris, À Travers les Âges, I. Courtesy of Modern Hospital Publishing Co.)

In the health profession one must constantly bear in mind that microorganisms are always present unless special precautions have been taken to kill or eliminate them. It is also necessary to realize that, although most microorganisms in our environment are harmless and we take millions of them into the body every day, pathogenic species are often mixed with them. In many circumstances it is essential to destroy or eliminate these pathogenic microorganisms, and to this end certain surgical, public health, hospital, and laboratory techniques have been devised.

As in many technical procedures, the results obtained depend largely on the intelligence and skill of the person doing the work and only secondarily on the apparatus. Elaborate and expensive equipment is available in hospitals, but the ingenious person realizes that simple equipment found in the home, such as a bottle of grocery store hypochlorite solution (Clorox or Purex), a household refrigerator, a pan of boiling water, a "double boiler," an antiquated oven, or a pressure cooker, if intelligently used, can kill or inhibit microorganisms as effectively as the gleaming, expensive equipment available in the most streamlined hospital.

DEFINITIONS

The following terms are often used erroneously and sometimes interchangeably. The student must learn to use them *correctly*.

Disinfection. This is any process that destroys, neutralizes, or inhibits *pathogenic* organisms. The term is easily understood by simply separating the word into "dis" and "infect": disinfection simply means doing away with infection. It is sometimes used incorrectly as a synonym for sterilization. An example of disinfection (but not sterilization) by heat is the pasteurization of milk. This destroys the pathogenic organisms present in milk. The milk is disinfected; but it is not sterile since it still contains many living (though harmless) microorganisms, as is evidenced by the putrefaction of pasteurized milk. Another example of disinfection without sterilization is the use of iodine or some other disinfectant on the skin in preparation for surgery. The application of a suitable chemical on the skin destroys most harmful bacteria present on the surface, but, because of the layered and pitted structure of the skin and the presence of bacteria and spores under the layers and in the hair follicles, such disinfectants never kill all microorganisms. In careful surgical procedures the surgeon discards the scalpel he has used to cut through the outer skin and then uses a second, sterile scalpel for cutting underlying tissues.

Disinfectant. A disinfectant is an agent that destroys, neutralizes, or inhibits the growth of pathogenic microorganisms. Common disinfectants include heat (nonsporicidal temperatures of 60 C to 80 C or short periods [3 to 10 min.] of boiling) and chemicals like iodine or phenol (carbolic acid). The term generally applies to preparations, usually liquids, intended for use on inanimate objects, as distinguished from living tissues (see *antisepsis*). Liquid[2] chemical disinfectants are not expected to kill bacterial spores and rarely do so. This is an important fact and should never be forgotten or overlooked. The suffix "cide" indicates "killer." Thus, a *germicide* is a germ killer; a *bactericide* is a bacteria killer and is similar to a germicide but is restricted to bacteria and does not affect their spores. A *fungicide* kills fungi and a *virucide* inactivates viruses; a *homicide* is the killing of a human being.

Of course, a disinfectant is not selective in "doing away with" pathogens: it destroys nonpathogens as well. However, after the infectious organisms have been effectively removed or destroyed, some nonpathogenic microorganisms

[2]Some gaseous disinfectants can kill spores and are, in fact, sterilizing agents (see page 193).

may still remain. This condition cannot be called sterile: it is only disinfected, or aseptic if on the body.

Sterilization. This is achieved by any process that *completely removes or destroys all living organisms* in or on an object. Any process designed to sterilize must be adequate to destroy bacterial spores. Few chemical methods of inhibiting and killing microorganisms are true sterilization procedures, since spores resist ordinary chemical disinfectants. Various methods of heating are therefore commonly employed for sterilization. The use of *ethylene oxide* and *beta-propiolactone,* which are germicidal gases or vapors capable of killing bacterial spores, is replacing heat in some selected applications, however. *One should never talk of sterilizing one's hands, because as long as any tissue or body part is living it cannot, by definition, be called sterile. Sterile means free of all life.*

Bacteriostasis. Bacteriostasis is the process of inhibiting (stopping or greatly slowing down) the growth of bacteria. Freezing and drying are two common methods. Some of the sulfonamides, and antibiotics like chlortetracycline, are bacteriostatic (inhibitory) rather than bactericidal (killing) in their action. Usually a substance is said to be bacteriostatic if the bacteria are alive but dormant after an hour or so of contact with it.

Selective Bacteriostasis. In diagnostic work substances are often used that inhibit one type of microorganism but permit another type to grow. A selective bacteriostatic agent would act against some types of microorganisms but not affect others. For example, in cultivating viruses (Chapters 6, 9), antibiotics are used to prevent the growth of fungi and bacteria. The antibiotics selected for this purpose have no effect on viruses or on the tissue cells in the culture.

A good example of bacteriostasis is the use of the dyes *eosin, methylene blue,* and *basic fuchsin. Sodium desoxycholate,* a salt derived from bile, is used for the same purpose. These bacteriostatic agents are mixed with nutrient agar designed for the isolation of certain species of intestinal bacteria, such as *Salmonella typhi* (cause of typhoid fever) and *Escherichia coli.* These organisms grow readily in the presence of the bacteriostatic substances used, although nearly all the other organisms in feces are more or less inhibited.

Antisepsis. This is any process that prevents or combats infection or sepsis (the growth of pathogenic bacteria in living tissues) by killing or inhibiting the bacteria. This term is not as widely used as the related word "antiseptic." Because the word *antiseptic* has been much exploited for commercial purposes, the term has been defined in the Federal Food, Drug and Cosmetic Act, which states that "the representation of a drug in its labeling as an antiseptic shall be considered to be a representation that it is a germicide (i.e., kills bacteria), except in the case of a drug purporting to be, or represented as, an antiseptic for inhibitory use as a wet dressing, ointment, dusting powder, or such other use as involves prolonged contact with the body."[3] This definition states that an antiseptic must kill or inhibit bacteria, and thus prevents unscrupulous manufacturers from flooding the drug stores of the country with worthless "antiseptic" lotions, toothpastes, salves, mouthwashes, and other products. In practice *antiseptics are diluted disinfectants* that come in contact with the body, but the reverse is *not necessarily true.*

Sanitization. Santization is a useful term for the process of making objects free from pathogenic organisms and esthetically clean as far as organic material (saliva, mucus, and feces) is concerned. This term has been applied to the process used in restaurants and other eating establishments for cleaning dishes. In hospitals, nurses and others have for years talked about "sterilizing bedpans," although most of the procedures in use only *sanitize* bedpans. It would be the rare medical or surgical patient, indeed, who needed a sterile bedpan, but

[3]U.S. Department of Health, Education, and Welfare, Food and Drug Administration: Federal Food, Drug and Cosmetic Act and General Regulations for Its Enforcement. Revision of June 1958. Washington, D.C., Government Printing Office.

certainly all patients should be given sanitized bedpans. Bedpans from patients with enteric infections, such as typhoid fever, poliomyelitis, or amebic dysentery, must be thoroughly disinfected.

GENERAL INFORMATION

Before inhibition and destruction of microorganisms are discussed in specific detail, one should recall the following relevant facts: (1) all microorganisms are composed chiefly of complex organic materials that are largely protein in nature; (2) some vegetative cells are capable of producing resistant forms, such as the extremely thermostable spores of all species of *Bacillus* and *Clostridium* and certain protozoa that form cysts; (3) cells of the tubercle bacillus are shielded by a protective wax coating; and (4) microorganisms differ in susceptibility to adverse changes in their environment.

In addition, students must remember that, in practice, they are applying disinfection and sterilization techniques in settings quite unlike the laboratory, where the effectiveness of a particular process is studied with pure cultures of microorganisms, to which protein material may or may not be added. In practice, disinfection and sterilization are required in situations in which mixtures of various types of microorganisms are almost inevitable, where resistant forms of pathogens frequently occur, where living human tissue as well as feces, mucus, pus, and blood may be involved, and where delicate materials or instruments may be required. The health worker must evaluate each situation and decide on the best possible procedure, drawing upon a thorough knowledge of the principles of microbial inhibition, removal, or destruction (Fig. 11–2).

The following are several means of accomplishing these objectives:

I. *Inhibition* of microorganisms by
    A. Low temperatures (refrigeration, Dry Ice [−78 C], liquid nitrogen [−180 C], etc.)
    B. Desiccation (drying processes)
    C. Combinations of desiccation and freezing ("freeze-drying")
    D. High osmotic pressure (syrups, brines, etc.)
    E. Bacteriostatic chemicals and drugs
        1. Certain dyes such as eosin, methylene blue, crystal violet, and desoxycholate
        2. Chemotherapeutic drugs such as sulfonamides and antibiotics

II. *Removal* of microorganisms (especially bacteria) by
    A. Passing fluids containing them through very fine filters
    B. High-speed centrifugation ("slinging" the bacteria to the bottom of tubes held in the rotor of a centrifuge)

III. *Destruction* by
    A. Fire (burning of used dressings, etc.)
    B. Heat (autoclaves, pressure cookers, ovens, etc.)
    C. Chemicals (liquid disinfectants or gaseous agents)
    D. Radiations (such as x-rays, gamma radiation, ultraviolet light)
    E. Mechanical methods (crushing, shattering, ultrasonic vibrations, etc.)

In time other means will also be used routinely to destroy specific microorganisms selectively, including the use of bacteriophages, bacteriocins (such as colicins; see Chapter 14), or immunologic methods.

Destruction by Heat and Chemical Agents

The rate and effectiveness of heat and chemicals in destroying microorganisms are determined by the characteristics of the microorganisms, properties

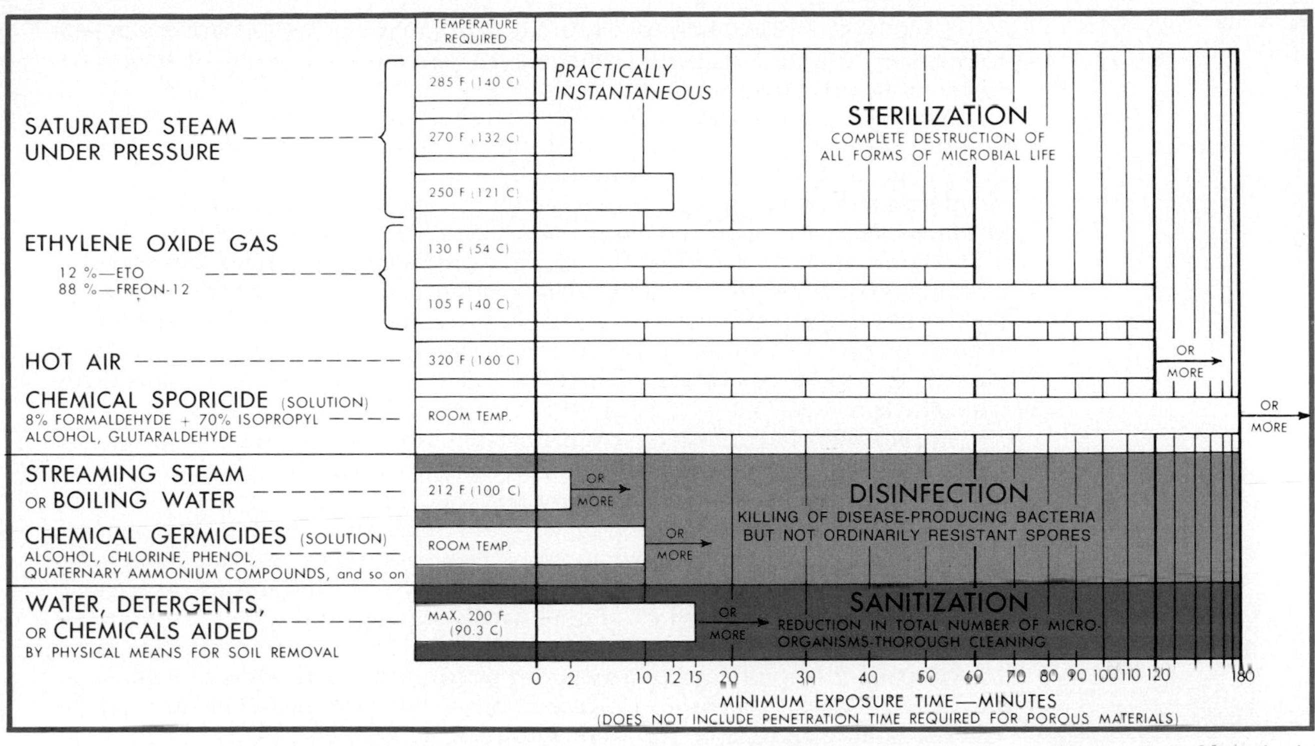

Figure 11–2 Methods for controlling microbial life. *Note:* Chemical sporicide is recommended in practice for high-level disinfection, not routine sterilization. (Courtesy of Research and Development Section, American Sterilizer Co.)

of the agent used, and factors influencing the interaction between the agent and the microorganisms.

Characteristics of the Microorganisms. Important in this connection are the degrees of resistance of various vegetative organisms to agents used for their destruction, conditions for spore formation, and numbers of organisms present.

Resistance of Nonspore-Forming Bacteria. The tubercle bacillus is an example of a nonspore-forming organism relatively resistant to certain common forms of destruction, especially chemical disinfectants. This is partly the result of a waxy substance that is either a part of the cell wall or closely associated with the cell in some other way. The gonococcus and the meningococcus, conversely, are examples of organisms that are very fragile and very susceptible to all common methods of destruction. *All of the nonspore-forming pathogenic bacteria are readily destroyed by boiling at 100 C for 5 to 10 minutes.*

Resistance of Bacterial Spores. Spore-forming bacteria, conversely, are highly resistant to destruction by heat and chemicals because of their spores. Many bacterial endospores can resist boiling and baking temperatures for an hour or more, and only two or three chemicals suitable for general use as disinfectants can kill bacterial spores within a practical period of time, at reasonable cost, and with certainty. Such spores can be killed efficiently (for most purposes) only by adequate exposure to steam under pressure (autoclaving: 121 C for 20 to 30 minutes or 132 C for 3 to 5 minutes) or by two hours of exposure to relatively high temperatures (165 C) in an oven.

In many hospital and industrial applications and for certain materials that can be ruined by heat, two effective chemical sterilizing agents have been adopted: *ethylene oxide* gas and *beta-propiolactone* vapor. Certain microbicidal radiations, such as ultraviolet or cathode rays, have been employed experimentally for sterilization but are not yet in general use.

The ascospores of yeasts and the conidia of molds and of Streptomycetaceae, Actinoplanaceae, and other spore-bearing Actinomycetales are much less resistant to heat and as a rule are killed after 15 to 30 minutes of boiling at 100 C, although possibly not in 10 minutes.

Numbers of Microorganisms. When large numbers of microorganisms are present in material such as feces, sputum, and some other body discharges, more than minimal amounts of disinfectant or a longer application of heat or disinfectant must be allowed. The clumping together of organisms in masses in mucus, feces, or other discharges may interfere with the penetration of the disinfecting agent to all cells of the organisms present. Also, in any large group of organisms, some of the individuals are more resistant than the majority. Furthermore, the inert organic material in feces, mucus, and similar substances affords protection by coating the microorganisms with a tough layer of coagulum and also by combining with the disinfectants and thus diverting them from the bacteria.

Properties of the Agent Used for Destruction. If heat is used for sterilization it may be either "dry heat," as applied in an oven, or "moist heat," as applied by boiling in water or using steam.

With chemical disinfectants, one should take into account the chemical nature of the substance, the concentration necessary for best effect, its solubility, the ability of the agent to affect the vital parts of microorganisms, its surface tension, its destructive, toxic (poisonous), irritating, and other undesirable properties, and its cost and availability.

"Chemical Nature." This refers to the constituents in a disinfectant that may be inhibitory or lethal to microorganisms. For example, chlorine is of no value as a disinfectant in the state in which it occurs in table salt (sodium chloride), but it is one of the most effective disinfectant substances in a free form as moist gas, in alkaline aqueous solution, or in bleaching solution (NaOCl). Certain organic acids are often used to preserve foods—e.g., lactic acid (used as a preservative of fermentation products), propionic acid (retards mold growth in bread), and acetic and benzoic acids or their salts (used in fats, fruit juices, and soft drinks). Phenol, or "carbolic acid," has acidic and other properties (e.g., surfactant) that render it especially efficacious for some uses. In concentrations above 1 per cent it exerts a coagulative action and can be very destructive to human tissues.

Substances strongly acidic or basic are likely to be highly bactericidal but may be too corrosive for general use as disinfectants. Salts of some heavy metals, such as silver, copper, and mercury, are effective disinfectants because they coagulate protoplasm (protein), or are poisonous to certain enzymes in living cells, or both. Some agents oxidize vital portions of organisms (e.g., potassium permanganate, hydrogen peroxide). Some active disinfectants react especially with certain molecular groups in the cell structure: e.g., amino, hydroxyl and sulfhydryl groups, DNA, and RNA. The mode of action depends on the chemical nature of the disinfectant. In many instances the mode of action is multiple, in some it may still be obscure.

Necessary Concentration. In selecting or dealing with chemical disinfectants, it is desirable to know the most effective concentration of each for a specific microorganism. The use of an unnecessarily high concentration is wasteful and often irritating or destructive to human tissues. The use of very low concentrations fails to disinfect or requires prohibitively long exposures for adequate results. For example, a 2 per cent solution of saponated cresol is effective against most vegetative nonspore-forming forms of bacteria within 30 minutes. A 20 per cent solution of this disinfectant is no more effective but is very irritating, poisonous to tissues, expensive, and wasteful. Solutions under 0.1 per cent are of some value, but ample time must be allowed for their action.

Intensity of Heat. The susceptibility of different organisms to moist heat varies greatly. Some thermophilic ("heat-loving") bacteria inhabit hot springs with temperatures approaching boiling. So far as is known, with the exception of the viruses of hepatitis (infectious or epidemic hepatitis and serum hepatitis), no organisms of significance in medicine resist more than 5 to 10 minutes of boiling at 100 C unless they form spores. Bacterial spores may resist boiling for hours. Only temperatures like that of saturated steam under 15 pounds of pressure (autoclaving: 121 C for 15 to 20 minutes) are effective against the spore-forming organisms. If oven baking (dry heat) is used instead of steam, a much higher temperature must be applied for a longer time (165 C for two hours). The viruses of hepatitis may survive boiling (100 C) for 30 minutes and possibly longer (see Chapters 26 and 39) but not autoclaving or baking as described.

Conversely, some nonspore-forming (vegetative) pathogenic organisms, e.g., diphtheria bacilli, are susceptible to temperatures as low as 55 C applied for 10 minutes. The pasteurization procedure (62.8 C [145 F] for 30 minutes or 71.7 C [161 F] for 15 seconds) is based on the fact that most organisms that cause epidemic diseases and that occur in milk are nonspore-forming and are destroyed by such exposure.

Moisture and Coagulation. It is clear from the earlier discussion that dry heat sterilizes less efficiently than moist heat. When moisture is present, the proteins, protein complexes, and other unstable constituents of living cells are more readily coagulated and hydrolyzed than when dehydrated. Egg albumin (egg "white"), if nearly dehydrated, coagulates only at 170 C (340 F). In a fully hydrated condition albumin is readily coagulated at much lower temperatures (about 75 C). The same is true of the proteins in microorganisms.

Water Solubility. Since the action of some disinfectants depends on ionization, which occurs only on solution in water, their water solubility is of obvious importance. The resistance of bacterial spores to heat and chemicals may be partially explained by the dehydrated form of the proteins and other organic complexes of the spores. Thus, neither heat nor chemicals can readily affect them.

Effect on Vital Parts of Microorganisms. Certain noncoagulative agents are of obvious importance if they either inhibit vital microbial cellular constituents or kill the microorganisms. This is, in part, the basis of the action of antibiotics and sulfonamide drugs. Among the most vital parts of microorganisms are the enzymes and related structures in the cell, by means of which they obtain energy and synthesize cell substances. In rapidly growing cells, penicillin and cephalosporin inhibit the transpeptidation reaction, which interferes with the synthesis of peptidoglycan (the center layer in the cytoplasmic membrane) and consequently with cell transport. Enzymes are readily affected by heat and by most chemical and physical agents that affect proteins. If these vital enzyme systems are interfered with, the organisms die or are held inert by bacteriostasis. Suboptimal but nonlethal temperatures, osmotic pressures, or pH may prevent the functioning of certain enzymes, but these do not necessarily kill the organisms. Thus, unfavorable physical factors in cell environments may inhibit without destroying; such factors are microbistatic.

For example, a temperature of 37 C (body temperature) is advantageous to the growth of pathogenic microorganisms because their oxidative enzymes and other biochemical functions proceed well at this temperature. Temperatures above 50 C are usually harmful to them because many enzymes do not function at this higher temperature. Freezing temperature decreases or stops enzyme activity, slows down chemical reactions, and destroys some susceptible organisms. Many pathogenic organisms, however, particularly the *Salmonella* and *Shigella* groups (causes of typhoid fever, food infection, and dysentery), may survive for months frozen in ice.

Surface Tension. An exceedingly important property of disinfectants is the surface tension of their solutions. So vital is this property that millions of dollars are spent annually by large manufacturing companies in producing disinfectants with solutions of low surface tension.

Without going into a detailed explanation of the physics of surface tension, we may gain some understanding of it from illustrations of its effect. Surface tension is responsible for the property of "wetness" of a fluid. Mercury is a fluid (at ordinary temperatures) with very high surface tension (466 dynes per cm). Because of its high surface tension it refuses to come into intimate contact with (or wet) most objects and draws itself away into tiny spheres; therefore, it would refuse to spread and contact bacterial or other surfaces if one tried to use it as a disinfectant. Since alcohol has very low surface tension (about 28 dynes per cm), an alcoholic solution of iodine (tincture of iodine) spreads very well, comes into intimate contact with rough, dry surfaces, and gets into narrow spaces where bacteria may escape aqueous solutions with surface tensions of 77 dynes per cm. An aqueous solution of iodine (e.g., Lugol's solution) wets much less efficiently than an alcoholic tincture of iodine unless a surface-tension reducer is added to the aqueous solution. This is done in aqueous iodine disinfectants of low surface tension, such as *Wescodyne* and *Ioclide*.[4]

Many household detergents[5] and soaps (although not disinfectants per se), have the power to lower the surface tension of aqueous solutions and thus bring them into immediate contact with microorganisms. Some surface-tension reducers are also disinfectants, e.g., alcohols and substances like cresol. Saponated cresol solution is aqueous cresol, with soap added as a surface-tension reducer. The whole group of quaternary ammonium compounds, of which *Zephiran* is a well-known representative, are surface-tension reducers and disinfectants at the same time. Surface-tension reducers are often called "surface-active" substances or, more briefly, "surfactants." They are characteristically effective in producing foaming or "suds," the housewife's delight and the sanitarian's headache. Foaming suds were, and perhaps still are, one of the most annoying and difficult chemical contaminants to control in our public waters. The use of "biodegradable" detergents readily decomposed by microorganisms in sewage has eliminated much of this difficulty.

ADSORPTION.[6] An important property of water-soluble surface-tension reducers in general, including disinfectants that reduce surface tension, is called *adsorption*. Surface-tension reducers accumulate at surfaces and tend to adhere there. Water-soluble surface-tension–reducing disinfectants such as cresol and its many derivatives (among them hexachlorophene) and quaternary ammonium compounds (among them Zephiran and Ceepryn) tend to be adsorbed and remain in very thin films on the surfaces of hands and objects and thus maintain disinfectant action there. Special precautions are necessary in the use of disinfectants based on phenolic compounds (see page 223). Thanks to adsorption, floors, surgeons' and nurses' hands, and other surfaces, if washed with appropriate and properly prepared solutions of surface-tension–reducing disinfectants, tend to remain disinfected for some time, perhaps hours.

SURFACE-TENSION REDUCERS IN MEDIA. A derivative of a long-chain fatty acid, sold under the name Tween 80, permits the growth of many microorganisms in liquid media. This is due to the action of Tween 80 as a nondisinfectant surface-tension reducer. For example, the tubercle bacillus grows and multiplies quite profusely when Tween 80 is present in Dubos' medium, an aqueous solution that would not otherwise come into effective contact with (i.e., wet) the waxy tubercle bacilli.

Toxicity. The living substance of all cells, in prokarytes or eukaryotes, is chemically very similar. Chemical agents that are toxic for the cells of micro-

---

[4]These and similar disinfectants are often called *iodophors* because the iodine is in loose chemical combination with an organic substance, which thus "carries" the iodine (phor is from the Greek word *phoros*, carrying).

[5]From the Latin *detergere*, to wash away. The basis of the cleansing action of most commercial detergents is lowered surface tension. These act mainly by *emulsifying* lipids (fats, grease, and so on). Lipid *solvents* such as alcohol and gasoline act as detergents because they *dissolve* lipids.

[6]Do not confuse with *absorption*, which means the taking up, by capillary action, of fluids by finely divided, powdery, or fibrous materials such as absorbent cotton or dry plaster.

organisms may therefore also be toxic for human cells. Violently poisonous or corrosive chemicals may be used for the disinfection ("sanitization") of inanimate objects like bedpans, pails, or other containers. Disinfection (antiseptic) processes to be carried out on or in living human tissues, however, must be selected with due and careful consideration for toxicity. If disinfectants (for example, phenol at high concentration, strong lye, or sulfuric acid) are used, not only is damage done to the patient but the dead tissues become an excellent medium for growth of pathogenic microorganisms. An example of the misuse of poisonous disinfectants is the once common practice of pouring strong (7 per cent) tincture of iodine into open cuts. This is highly irritating and toxic to the exposed tissues and does more harm than good. The *ideal disinfectant* for use in contact with living animal tissue is *nontoxic to animal cells and tissues, nonirritating, not destructive to materials, odorless, inexpensive, highly water-soluble, chemically stable, easily available, and lethal to all pathogenic microorganisms.* Such an ideal substance is not known at present, but several available disinfectants and drugs approach this ideal in some respects. In selecting one of these disinfectants one has to analyze the job it will have to do and the conditions under which it will be expected to work.

Factors Influencing the Interaction Between Agent and Microorganism. Important in this connection are temperature, absence of organic material, contact, time, surface tension, and pH.

Temperature. Temperature is an important factor in improving the efficiency of some chemical disinfectants. For example, a warm solution is likely to be more effective than an ice-cold one, because warming lowers surface tension and heat increases the speed of the chemical reactions involved in disinfection.

Organic Material. Organic materials such as blood, serum, mucus, pus, feces, and foods are important obstacles to the inhibition and killing of microorganisms for two main reasons:

(1) Agents such as bichloride of mercury, which coagulate protein, will coagulate the organic material of pus, mucus, or feces to form a coating surrounding the microorganisms. This leaves live microorganisms inside the coagulated mass, protected from the disinfectant. It is thus clear that whenever large masses of organic material are to be disinfected, the disinfection process will be more efficient if it is possible to break up the masses mechanically. For example, when feces from a patient who has typhoid fever or another intestinal communicable disease are disinfected, large masses should be broken into the smallest possible particles before the disinfectant is added unless they are to be deposited directly into a sewer. Chlorinated lime, phenol, or saponated solution of cresol would be chosen rather than bichloride of mercury because each is less coagulative.

(2) Many chemical disinfectants combine as readily with extraneous organic materials such as blood and mucus as with vital organic complexes in the cell. If large amounts of organic material are present, these may take up or inactivate the disinfectant, leaving little or nothing to kill the microorganisms. For this reason, in addition to breaking apart solid masses of mucus and feces, it is necessary to add extra large amounts of disinfectant, enough to combine with both the organic matter and the microorganisms.

Contact. It is obvious that for any sterilizing or disinfecting agent to be effective it must come into contact with the microorganisms to be killed. This is true of both chemicals and heat. In the practical application of any method of killing microorganisms, all parts of the article or substance to be sterilized or disinfected must be in contact with the agent. This means complete immersion of the object in the chemical with no air bubbles that might prevent contact of the agent with the object. This also means correct packing of materials in an autoclave or oven so that all parts can be contacted by the steam or hot air. As has been indicated previously, contact of chemical disinfectants with microorganisms is directly related to the presence of organic material and suface tension.

Some substances that dissolve lipids, e.g., alcohol, dimethyl sulfoxide (DMSO),[7] can penetrate through lipid-rich cell walls and membranes of microorganisms and thus make good "carriers" for disinfectants that are soluble in them.

Clean, smooth instruments can be sterilized or disinfected in a shorter period of time than soiled, contaminated objects that have crevices that may be difficult to penetrate. Instruments that have been oiled recently cannot be disinfected by an aqueous solution of disinfectant. This applies also to rectal thermometers lubricated with petrolatum or petroleum jelly because aqueous solutions do not penetrate films or droplets of oil or grease. Cleaned clinical thermometers can be sterilized by sporicidal gases that do not require too high a temperature (e.g., beta-propiolactone), but this is not used for individual thermometers. Electronic thermometers with disposable covers are now used more frequently.

Time. Time of contact or application is extremely important in disinfection and sterilization. The time necessary to disinfect various substances is dependent on contact, concentration, penetration, and the other factors just listed. For this reason, no chemical can be expected to act instantaneously, and only moist heat at 100 C or above acts very rapidly. Some organisms (not spores) are killed in a few seconds by temperatures around 80 C; others may resist 100 C for many minutes. Chemicals require time to make contact with the organisms and to react with them.

As a general rule, at least 10 minutes of boiling (at sea level) or one hour of contact with disinfectant is necessary. Many circumstances will change these recommendations. Neither boiling nor ordinary chemical disinfectants may be depended on to sterilize or to kill spores.

Figure 11–2 compares methods that are commonly used to destroy microbial life. The reader should note that sterilization by steam under pressure takes less time than sterilization by ethylene oxide or hot air. Also in Figure 11–2, disinfection and sanitization are graphically related to sterilizing processes. Not shown are methods such as filtration processes that may be used to sterilize heat-sensitive media. Filtration is discussed in detail in Chapter 12.

pH. The acidity or alkalinity of a solution will affect the efficiency of a chemical agent. Cationic bactericides are more active at high pH (alkaline) and anionic compounds at low pH (acidic), apparently owing to greater penetration of undissociated forms of disinfectants and to increases in opposite charges in cell constituents. In general most bactericides work best at a neutral or slightly alkaline pH. Acid solutions are more likely to be bactericidal if heated, since heat tends to increase dissociation of acids.

SUMMARY

It has been only somewhat more than 100 years since humanity has found ways to protect itself against many infectious diseases that once killed millions, caused untold suffering, and spread constant fear. Thanks to Pasteur, Koch, Semmelweis, and other great scientists, we can now guard ourselves against most pathogenic microorganisms and the diseases they cause.

Disinfectants were developed to remove or destroy infectious organisms and others that are also sensitive to these chemicals. Milk that is pasteurized is disinfected, but not made sterile, which means free of all life. A bactericidal agent kills bacteria, a fungicide kills fungi, and a virucide destroys viruses.

A bacteriostatic agent inhibits the growth and reproduction of microorganisms, but it does not kill. Low temperatures and many antibiotics are bacteriostatic.

Antiseptics are disinfectants that come in contact with the body, like

[7]Caution: DMSO penetrates human tissues very easily and carries within various dissolved, possibly undesirable and toxic chemicals.

mouthwashes, eyedrops, shampoos, skin antiseptics, and douches. In practice, antiseptics are diluted disinfectants, but the reverse is not necessarily true.

Sanitization simply means cleaning; it does not mean sterilization.

One may inhibit the growth of microorganisms by refrigeration or freezing, desiccation, osmotic pressure, or the bacteriostatic action of chemicals; one may remove them by filtration or high-speed centrifugation; or one may destroy them through heat, fire, gaseous sterilization, radiation, or mechanical crushing. Bacterial spores are resistant to many common forms of destruction, especially chemical disinfectants and the temperature of boiling water. Of course, the type of organism, the relative concentrations, the temperatures, and the time of exposure greatly influence the effectiveness of the treatments. The pasteurization process is equally effective at 62.8 C (145 F) for 30 minutes and at 71.7 C (161 F) for 15 seconds.

Factors that determine the effectiveness of chemical agents on bacteria include water solubility, certain enzymes that may be affected, the composition of the cytoplasmic membrane, surface tension of the solution, pH, and adsorption. The ideal disinfectant would be nontoxic to animal cells and tissues, nonirritating, not destructive to materials, odorless, inexpensive, highly water-soluble, chemically stable, easily available, lethal to all pathogenic microorganisms, and quick in action. Since no such chemical exists, one simply has to select the best agent for the job to be done from among the disinfectants available.

REVIEW QUESTIONS

1. Describe some of the practices used by surgeons in hospitals up to about 100 years ago. Why and how did they change?
2. Define sterilization, disinfection, sepsis, asepsis, antiseptic, and sanitization. What is the difference between disinfection, sanitization, pasteurization, and sterilization?
3. When is an article or a substance sterile?
4. How do the characteristics of microorganisms affect the methods used in their inhibition and destruction? What are the characteristics involved?
5. Define bactericidal and bacteriostatic.
6. What organisms are killed only by autoclaving or long oven baking? Why are they heat-resistant?
7. Of what importance is the chemical nature of a disinfectant?
8. What are the properties of an "ideal" chemical disinfectant?
9. What is the role of surface tension in chemical disinfection?
10. What factors are of prime importance in chemical disinfection?
11. Why does organic matter interfere with the action of many chemical disinfectants?
12. What is the role of hydration in sterilization? What is accomplished by coagulation? Name two types of agents that cause coagulation. Why are spores resistant to heat?
13. What precautions are essential in the application of heat as a sterilizing agent?
14. What is ethylene oxide? What is it used for? What other agent may be used in its place?
15. Relate time and temperature to pasteurization.
16. Why is a "skin knife" used separately for the first incision in surgery?

Supplementary Reading

(See the list of readings following Chapter 15.)

STERILIZATION BY HEAT, RADIATION, CHEMICALS, AND FILTRATION

DRY HEAT

Incineration. This is an excellent procedure for disposing of materials such as soiled dressings, used paper mouth wipes, sputum cups, and garbage. One must remember that if such articles are infectious, they should be thoroughly wrapped in newspaper with additional paper or sawdust to absorb the excess moisture. Disposable plastic liners for waste containers are inexpensive and may be easily closed on top to prevent scattering of refuse. The wrapping protects persons who must empty the trash cans, and it assures that the objects do not escape the fire, but it may also protect the microorganisms if incineration is not complete!

Adequate instructions should be given to workers responsible for burning disposable materials to insure complete burning. For example, a sputum cup containing secretions from a patient who has active tuberculosis is filled with paper or sawdust to absorb excess moisture. The cup is then placed in a plastic bag with shredded absorbent paper to prevent spilling. If it is burned only on the outside, a soggy mass of dangerous infective material is left on the inside. Other possibilities will occur to the imaginative student.

Ovens. Ovens (Fig. 12–1) are often used for sterilizing dry materials such as glassware, syringes and needles (Fig. 12–2), powders, and gauze dressings. Petrolatum and other oily substances must also be sterilized with dry heat in

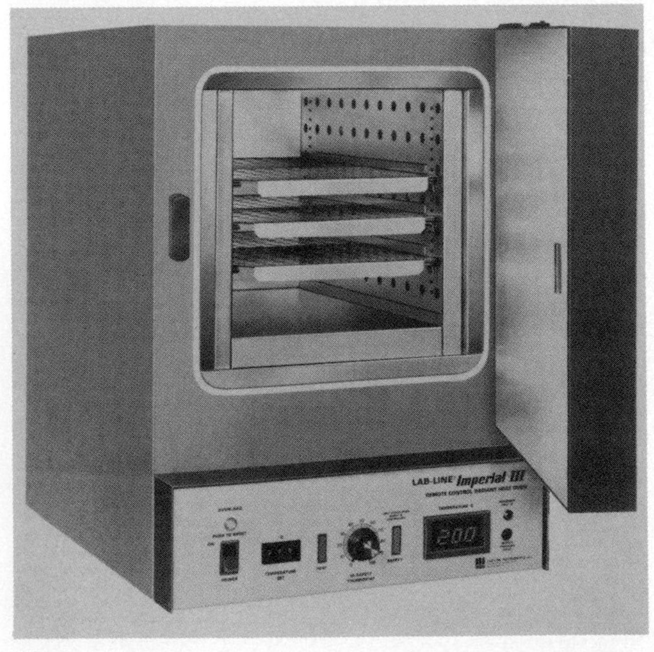

Figure 12–1 A type of sterilizing oven for microbiology. A motor in the bottom of the oven (not visible) forces streams of hot air through the chamber. (Courtesy of Lab-Line Instruments, Inc., Melrose Park, Ill.)

an oven because moist heat (steam) will not penetrate materials insoluble in water.

In order to insure sterility the materials in the oven must reach a temperature of 165 to 170 C (329 to 338 F), and this temperature must be maintained for 120 or 90 minutes, respectively. This destroys all microorganisms, including spores. However, the oven must be maintained at that temperature for the entire time. This means that the oven door must remain closed during the sterilizing time: opening the door will cool the articles below effective temperatures so that sterilization cannot be assured. Also, *hot* glassware will shatter immediately in contact with cool air. It is usual practice in a microbiology laboratory to let an oven cool completely before it is opened.

It is practical to load the oven with glassware, pipettes, and so forth in the afternoon or evening and turn it on. In the morning, the oven is turned off and by lunchtime it is unloaded. This routine assures sterile glassware, once the setting of the temperature is regulated. A home oven, set at 330 F (moderate temperature), can be used as well as an oven built for laboratory or hospital equipment. It is wise to check the temperature in the oven with an oven thermometer (available at household supply stores).

Items may be secured in brown wrapping paper with a string, but never with a rubber band. Some types of plastics, like the one used in connecting hoses, are heat-stable in an oven, but most plastics cannot be sterilized in this way.

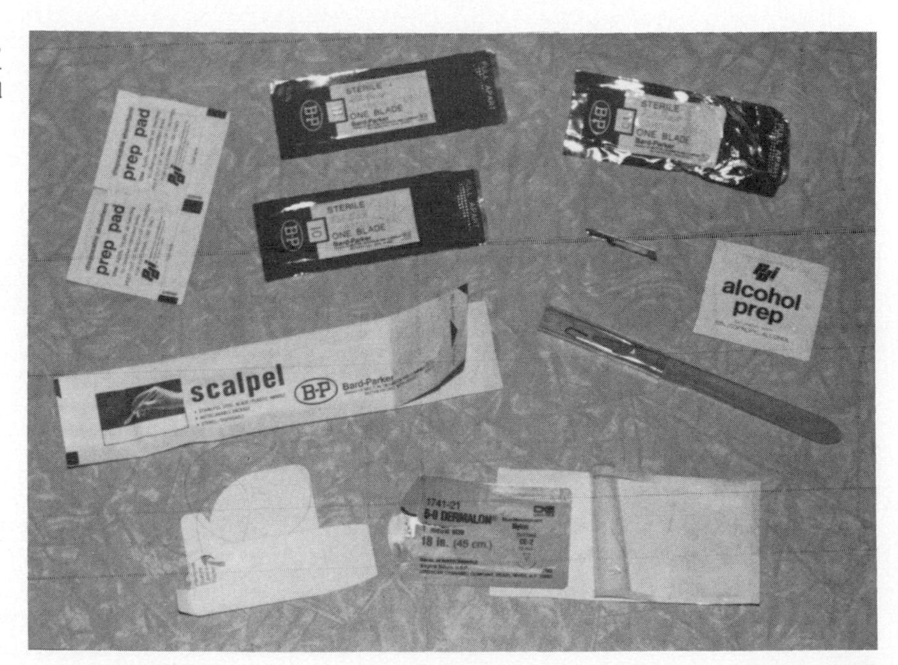

Figure 12–2 An assortment of sterile, packaged, disposable instruments—scalpels, knife blades, sutures, and alcohol wipes—are now used in surgery.

MOIST HEAT

Boiling Water. Boiling water can never be trusted for absolute sterilization procedures because its maximum temperature is 100 C (at sea level). As indicated previously, spores can resist this temperature. Boiling water can generally be used for contaminated dishes, bedding, and bedpans: for these articles neither sterility nor the destruction of spores is necessary except under very unusual circumstances. All that is desired is disinfection or sanitization. Exposure to boiling water kills all pathogenic microorganisms in 10 minutes or less, but not bacterial spores or hepatitis viruses. At altitudes over 5,000 feet the boiling time should be increased by 50 per cent or more because water there boils at temperatures of only about 95 C or below (Fig. 12–3).

Live Steam. Live steam (free flowing) is used in the laboratory in the preparation of culture media or in the home for processing canned foods. It must be remembered that steam does not exceed the temperature of 100 C unless it is under pressure.

To use free flowing steam effectively for sterilization, the fractional method must be used. Fractional sterilization, or tyndallization (mentioned in Chapter 11), is a process of exposure of substances (usually liquids) to live steam for 30 minutes on each of three successive days, with incubation during the intervals. During the incubations, spores germinate into vulnerable vegetative forms that are killed during the heating periods. This is a time-consuming process and is not used in modern laboratories. The use of membrane (Millipore) filters or similar rapid methods makes the preparation of heat-sensitive sterile solutions much easier.

Compressed Steam. In order to sterilize with steam certainly and quickly, steam under pressure in the autoclave is used (Figs. 12–4 to 12–7). An autoclave is essentially a metal chamber with a door that can be closed very tightly. The inner chamber allows all air to be *replaced* by steam until the contents reach a temperature far above that of boiling water or live steam. The temperature depends on the pressure, commonly expressed in pounds per square inch, often written as psi. Steam under pressure hydrates rapidly and therefore coagulates very efficiently. Also, it brings about chemical changes somewhat

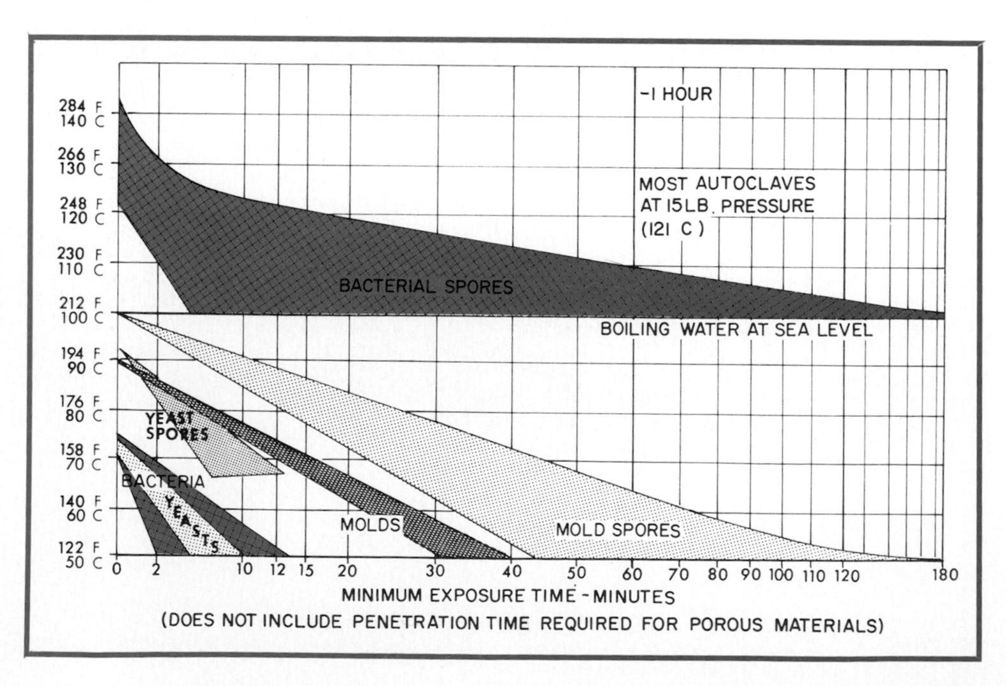

Figure 12–3 Temperatures required to kill various classes of microorganisms. Note the higher temperatures needed to destroy bacterial spores, whereas vegetative cells cannot survive even temperatures below boiling for 15 minutes or less. The autoclave should kill all spores at 121 C in 15 minutes or at higher temperatures in less time. (Courtesy of American Sterilizer Company, Erie, Pa.)

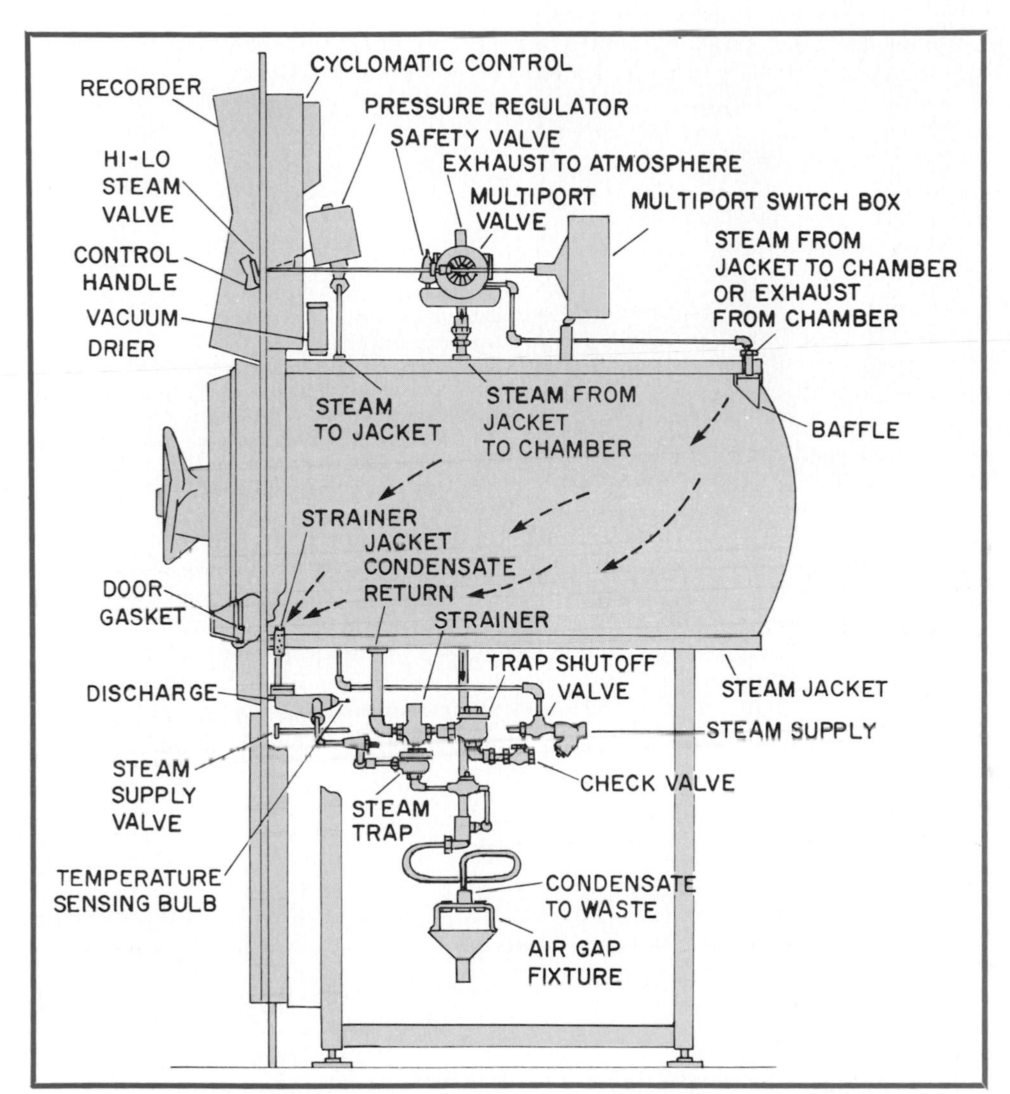

Figure 12–4 Diagrammatic illustration of steam-jacketed sterilizer. Steam is admitted to the jacket through the steam supply valve and pressure regulator (or Hi-Lo valve). The condensate formed during heating up of the jacket leaves through the jacket condensate return, strainer, trap, and checkvalve. When the jacket pressure reaches the selected pressure, indicated on the jacket gauge, steam is admitted to the chamber through the multiport valve and enters behind the baffle. Condensate formed during heat-up of the contents of the chamber leaves through the strainer, then goes to the discharge, past the temperature sensing bulb, through the steam trap, condensate-to-waste, and air gap fixture. When the chamber pressure equals the jacket pressure and registers on the gauges in the Cyclomatic Control, and the temperature on the recording thermometer reaches the operating temperature, the automatic timer is activated for the desired exposure time. At the end of the exposure, the Cyclomatic Control automatically moves to the selected exhaust (fast and dry, fast exhaust, or slow exhaust), and steam leaves the chamber through the multiport valve to the atmosphere. (Courtesy of American Sterilizer Company, Erie, Pa.)

Figure 12–5 Rectangular autoclaves for steam sterilization in a large pharmaceutical manufacturing establishment. Note the heavy steel doors swinging easily on trunions, the steel door-fasteners simultaneously manipulated by the central wheel, and the gauges recording time, temperature, and pressure. Operation is almost wholly automatic and electrically controlled. These enormous autoclaves are no different in principle from the physician's tiny office model seen in Figure 12–6. (Courtesy of Wilmot Castle Co., Rochester, N.Y.)

like digestion, called *hydrolysis.* These characteristics give it special advantages in sterilization.

By first allowing all the air in the chamber of the autoclave to escape and be replaced by the incoming steam, the spaces in the interior of masses of material may be brought quickly into contact with the steam. The escape of air is absolutely essential since sterilization depends on the water vapor. Whenever air is trapped in the autoclave, sterilization is inefficient. One must be sure that:

1. All the air is allowed to escape and is replaced by steam.
2. The pressure of the steam reaches at least 15 pounds to the square inch (psi) and remains there. (In most automatic autoclaves it is now 18 pounds, permitting sterilization to be accomplished in less time.)
3. The thermometer reaches at least 121 C without downward fluctuation for 15 minutes. (Less time is required when 18 pounds of pressure is used.)

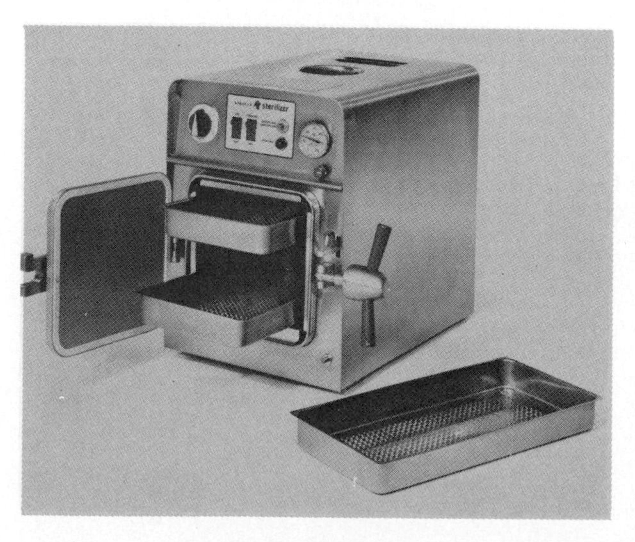

Figure 12–6 Small-size, large-capacity electrically powered steam sterilizer for medical offices, clinics, laboratories, and similar facilities. The timer and thermostat automatically control the selected temperature and exposure period. The thermometer provides visual assurance that the desired chamber temperature is being maintained. A bell signals the completion of the sterilization cycle. Air and entrained vapor are automatically and continuously evacuated from the chamber through the air elimination trap. An automatically timed quick-dry cycle may be selected following a sterilizing cycle. Operating instructions are affixed to the top of the unit. (Courtesy of American Sterilizer Company, Erie, Pa.)

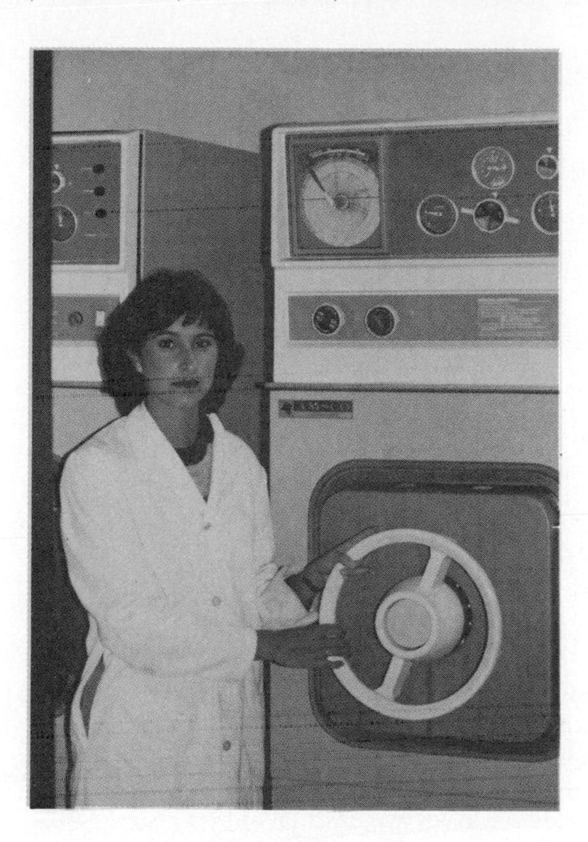

Figure 12–7 AMSCO Medallion Small Rectangular Laboratory Sterilizers have Cyclomatic Controls, temperature-indicating recording, and controlling instruments calibrated in degrees Celsius, a chamber trap bypass valve for nonpressure processing, and a pointer for selecting desired sterilizing temperatures. After selecting the proper processing cycle, sterilizing temperature, and exposure time, no further attention is necessary; an end-of-cycle alarm signifies completion.

If these conditions are met and if the masses or bundles are well separated and not too large, the autoclaved material will be sterile.

The actual amount of water present as steam in the pressure chamber is usually small; consequently, the articles sterilized are not wet with much condensed steam when they are removed from the autoclave. All modern autoclaves are arranged so that all the steam is removed by vacuum after the sterilization period, to prevent dampening the articles inside.

The automatic autoclave, as shown in Figures 12–5 and 12–7, or one with a round chamber, as used in many laboratories, has the following settings:

1. Manual—used when the electrical power is off. The operator must then set and time all cycles.
2. Slow exhaust—used for a *wet load,* for media or water (for dilutions).
3. Fast exhaust—used for killing microorganisms quickly on and in glassware that is to be washed.
4. Fast exhaust and dry—used for pipettes, Petri plates, or dressings, a so-called *dry load.*

After closing the door tightly, the operator sets the autoclave control to the desired setting, to the time interval that is necessary for the maximum preset temperature and pressure, and to ON. Lights go on as the autoclaving moves from pretimed cycle to cycle. Finally a bell rings, and the operator turns the setting to OFF and opens the door carefully. Asbestos gloves protect the hands, when hot sterile materials are unloaded, but watch the right elbow—the inside of the open door is very hot.

The exhaust trap inside the autoclave must always be cleaned before starting a load, since dirt in the trap may delay the time needed for the various cycles. It is best to do this when the autoclave is still cold. Why?

Since the effectiveness of an autoclave is dependent upon the penetration of steam into all articles and substances, the preparation of packs of dressings

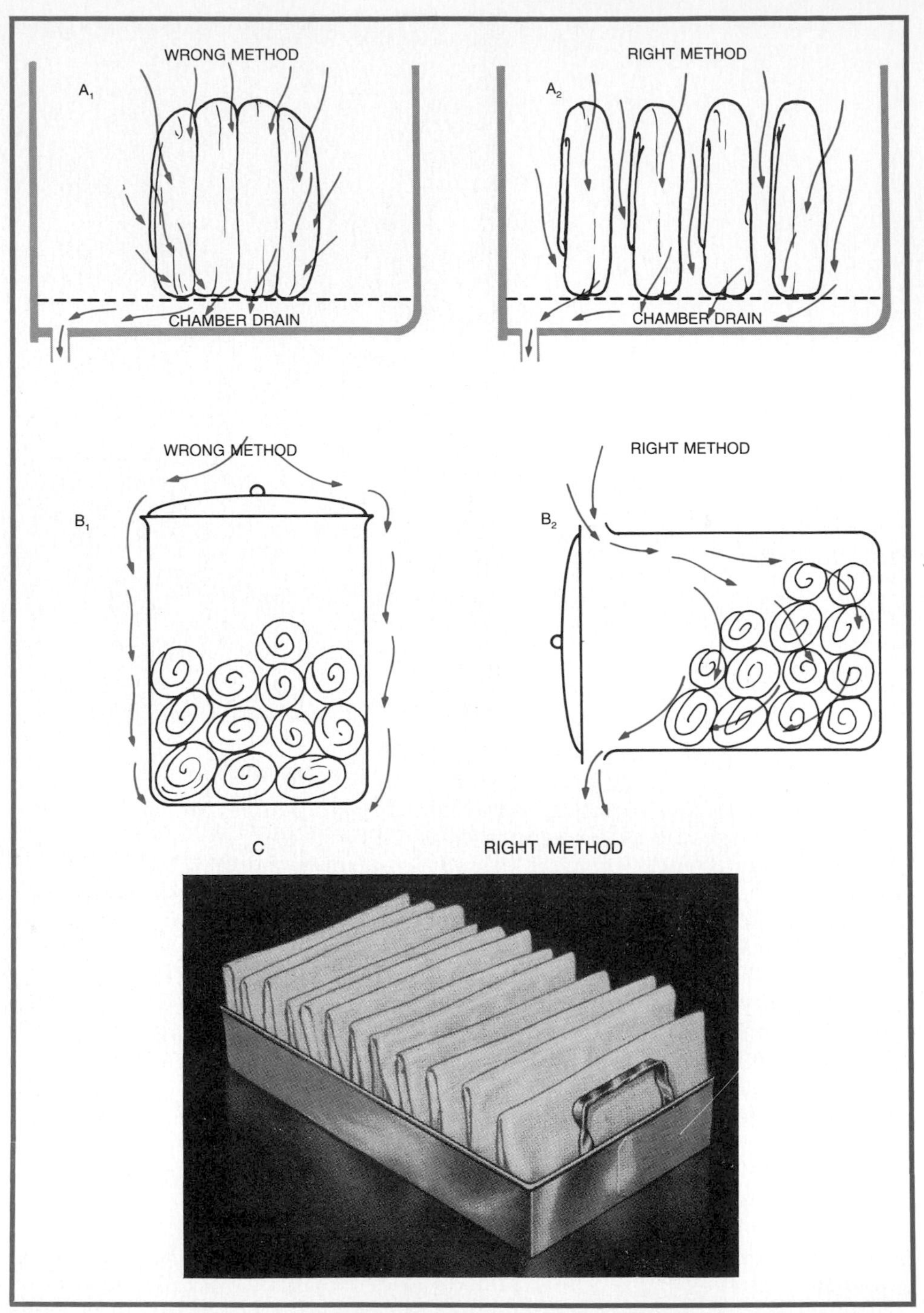

Figure 12–8 Figures A_1, A_2, and C show arrangements in the autoclave. B_1 and B_2 may be in the autoclave or another type of pack sterilizer. A_1 shows improper arrangement of four packs, tightly held together, while A_2 shows the same packs slightly separated from each other. Steam will now permeate the entire mass quickly, and in the much shorter period of exposure needed there will be no oversterilized outer portions. B_1 and B_2 indicate the correct and the incorrect ways of placing jars of dressings in the sterilizer. Right side up, even with the cover removed, all air is trapped within the jar. Resting on its side, with the cover held loosely in place, air will drain out and steam will promptly take its place, as indicated by the arrows. *(Surgical Supervisor.* Courtesy of American Sterilizer Co.) *C,* The instrument sterilizer tray with wire-mesh bottom makes an excellent container for gloves. The glove packs should rest on edge in the sterilizer, stacked loosely, never more than one tier deep.

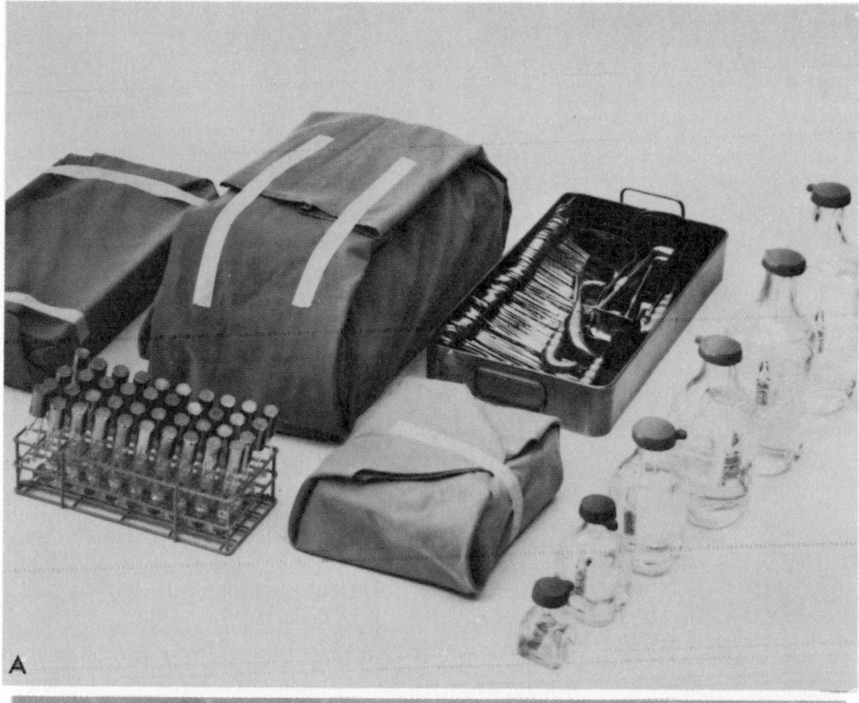

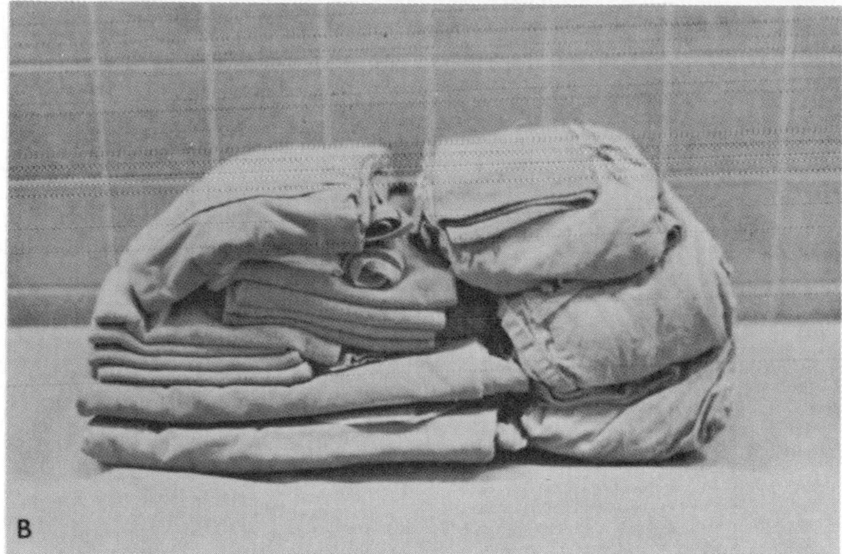

Figure 12–9 *A*, Preparation of items for steam sterilization. (Courtesy of AMSCO/American Sterilizer Co.) *B*, Standard gown and sheet pack. Note alternation of layers for better ventilation and steam penetration. (From LeMaitre, G. D., and Finnegan, J. A.: The Patient in Surgery. 4th ed. Philadelphia, W. B. Saunders Company, 1980.)

is very important, and the correct placement of articles in the autoclave is essential to adequate sterilization (Figs. 12–8 and 12–9).

Substitution of an autoclave for an oven by admitting steam only to the jacket and keeping the chamber dry is not advisable when sterilization is necessary because the temperature thus achieved (100 C) does not kill spores. The dryness of such an atmosphere may actually preserve some pathogens that would be quickly killed in a moist atmosphere.

Cleaning Instruments. When sterilizing solutions, the pressure must be allowed to fall gradually so that the solutions will not boil. If the pressure falls rapidly, violent boiling occurs. Advantage is taken of this fact in autoclaving used surgical instruments. They are immersed in water in a perforated tray.

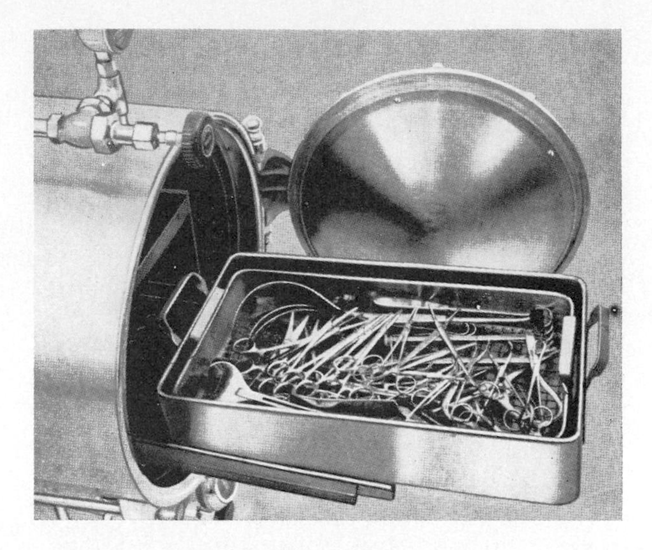

Figure 12–10 Method of sterilizing and washing used instruments. The lower tray is filled with water. The instruments in the perforated tray thus soak in hot water while being autoclaved. At expiration of the autoclaving time the pressure is suddenly released, permitting the water to boil violently, thus subjecting the instruments to a strong flushing action. *(Surgical Supervisor.* Courtesy of American Sterilizer Co.)

After autoclaving the pressure is reduced suddenly. The water boils violently and washes the instruments clean (Fig. 12–10).

CLEANING BY ULTRASONIC ENERGY. Machines are now available for cleaning surgical instruments, syringes, and so on by extremely rapid (ultrasonic) vibrations. These can clean and dry hundreds of instruments (perfectly) every five to ten minutes. They do not sterilize.

Indicators. Many institutions always include some sort of indicator inside bundles being sterilized, such as dyes that change color when the necessary temperature has been maintained for the required time. On glassware and bundles, labels are placed that read NOT STERILE before autoclaving or after insufficient autoclaving but read STERILE if sterilization has been fully effective. Another device, similar in principle, is cellulose tape having on it a chemical indicator that changes color when properly heated in the autoclave. One can use wax pellets that melt only at the necessary temperature but may not indicate lapse of time. Strips of paper containing bacterial spores can be dropped into broth in culture tubes after the sterilizing procedure. If the sterilizer has been properly operated, these broth cultures should remain sterile, even after seven days of incubation, since all spores have been killed. This method does not give immediate indication of faulty operation, but it does constitute an absolute and permanent record.

Most modern autoclaves have a self-recording thermometer that plots the temperature the instrument has reached and the time of sterilization required for each "load." A permanent record provided in this way often proves to be very valuable.

STERILIZATION WITHOUT HEAT

For many years heat was the only dependable and practicable means of destroying bacterial endospores. Now, at least three other means of killing microorganisms are available. These are use of the gas *ethylene oxide,* the vapors of *beta-propiolactone* (BPL), and certain *electromagnetic radiations* (especially electron beams or cathode rays). The method of ultrasonic vibrations, although quite effective in destroying certain microorganisms, is not a practical means of large-scale sterilization. Besides, it produces a heating effect. At present, we can only dream of an ultrasonic "dishwasher" that sanitizes dirty dishes, preferably without any water!

Sterilization with Radiations

Some of the electromagnetic radiations mentioned in Chapter 8 are in use or are being developed for general sterilization purposes.

Ultraviolet Light. This is satisfactory for the sterilization of smooth surfaces and of air in operating rooms; unfortunately, UV radiation has virtually no power of penetration. Mercury-vapor lamps emitting 90 per cent UV radiation at 254 nm are used to decrease airborne infection. Ultraviolet lamps are also used to suppress surface-growing molds and other organisms in meat packing houses, bakeries, storage warehouses, and laboratories. Sunlight is a good, inexpensive source of ultraviolet rays, which can induce genetic mutations in microorganisms. In excess, it can cause burns and even cancer.

X-Rays. X-rays penetrate well but require very high energy and are costly and inefficient for sterilizing. Their use is therefore mostly for medical and experimental work and the production of mutants of micoorganisms for genetic studies.

Neutrons. Neutrons are very effective in killing microorganisms but are expensive and hard to control, and they involve dangerous radioactivity.

Alpha Rays (Particles). Alpha rays are effective bactericides but have almost no power of penetration.

Beta Rays (Particles). Beta rays have a slightly greater power of penetration than alpha rays but are still not practical for use in sterilization.

Gamma Rays. These rays are high-energy radiations now mostly emitted from radioactive isotopes such as cobalt-60 or cesium-137, which are readily available by-products of atomic fission. Gamma rays resemble x-rays in many respects. The U.S. Army Quartermaster Corps has used gamma rays and other radiations to sterilize food for military use. X-rays or gamma rays must be applied in 2 mrad[1] to 4 mrad doses to become a reliable sterilizing treatment of food. Foods exposed to effective radiation sterilization, however, undergo changes in color, chemical composition, taste, and sometimes even odor. These problems are only gradually being overcome by temperature control and oxygen removal.

Cathode Rays (Electrons). These are used mainly to kill microorganisms on surfaces of foods, fomites, and industrial articles. Since electrons have limited powers of penetration, they are at present not very useful for surgical sterilization. However, as a result of research on proper dosage and packaging, cathode rays are being developed for general purposes such as food processing. This may completely revolutionize the food canning and frozen food industries as well as surgical sterilization techniques.

Pharmaceutical and medical products are adequately sterilized by treatment with a radiation dose of 2.5 mrad. The Association of the British Pharmaceutical Industry has reported that benzylpenicillin, streptomycin sulfate, and other antibiotics are satisfactorily sterilized by this method. In addition, package radiation at dose levels of 2.5 mrad has become common procedure for the sterilization of disposable Petri plates, pipettes, syringes, needles, rubber gloves, tubing, and so on.

Sterilization with Chemicals

Ethylene Oxide. This is a gas with the formula CH_2CH_2O. It is applied in special autoclaves under carefully controlled conditions of temperature and

[1]One mrad is 1/1,000 of a rad. A rad is 100 ergs of absorbed energy per gram of absorbing material (retained by matter).

humidity. Since pure ethylene oxide is explosive and irritating, it is generally mixed with carbon dioxide or another diluent in various proportions: 10 per cent ethylene oxide to 90 per cent carbon dioxide (sold as Carboxide), 20 per cent ethylene oxide to 80 per cent carbon dioxide (sold as Oxyfume), or 11 per cent ethylene oxide to 89 per cent halogenated hydrocarbons (sold as Cryoxcide and Benvicide). Each preparation is effective when properly used. Oxyfume is very rapid in action but is more inflammable and more toxic than Carboxide; however, Carboxide requires high pressure. Cryoxcide is more toxic and more expensive, but it is more convenient and requires less pressure. Other mixtures of ethylene oxide (e.g., with Freon) are also commercially available. All are more costly and time-consuming than autoclaving with steam.

Ethylene oxide is generally measured in terms of milligrams of the pure gas per liter of space. For sterilization, concentrations of 450 to 1,000 mg of gas/liter are necessary. Concentrations of 500 mg of gas/liter are generally effective in about four hours at approximately 136 F (58 C) and a relative humidity of about 40 per cent. Variations in any one of these factors require adjustments of the others. For example, if the concentration of gas is increased to 1,000 mg/liter, the time may be reduced to two hours. Increases in temperature, up to a limit, also decrease the time required. At a relative humidity of 30 per cent, the action of ethylene oxide is about 10 times as rapid as at 95 per cent. The use of ethylene oxide, although as simple as autoclaving, generally requires special instructions (provided by the manufacturers) for each particular

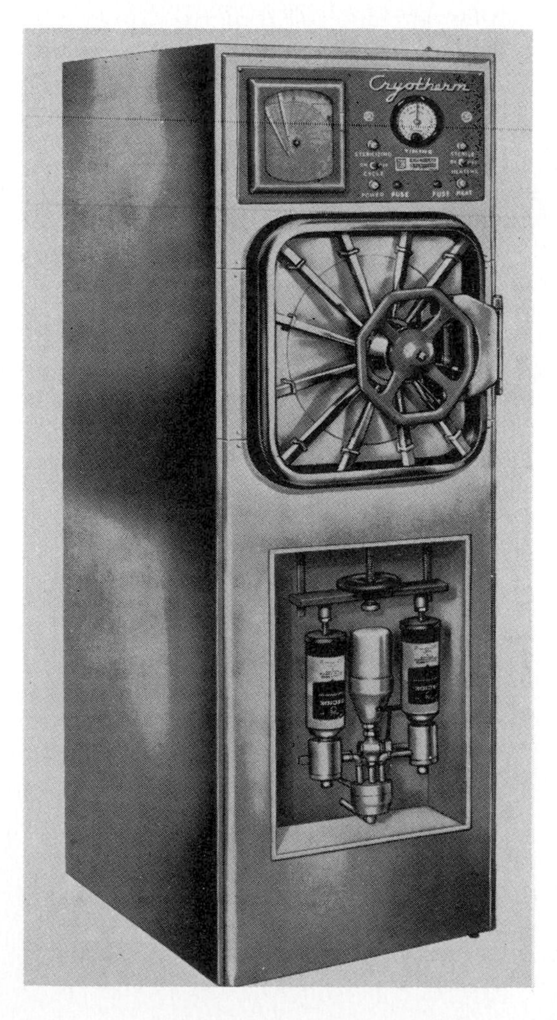

Figure 12–11 Autoclave arranged for sterilization by sporicidal ethylene oxide gas (Cryoxcide). Below the steel door are seen the containers of gas and a water inlet so arranged that the vapors can be mixed and humidified (much as gasoline, air, and water vapor are mixed in a modern automobile carburetor) before admission of the sterilizing vapor mixture to the chamber of the autoclave. Above the door are the dials controlling sterilizing temperature and time, vacuum pump, gas and humidity inlet, and a maximum and minimum recording thermometer. The operation is almost completely automatic. (Courtesy of American Sterilizer Co.)

situation. At present ethylene oxide is used largely by commercial companies that dispense sterile packages of a variety of products.

In general, seven steps are involved after loading and closing the sterilizing chamber:

1. Draw out nearly all air with a vacuum pump.
2. Admit a measured amount of water vapor.
3. Admit the required amount of ethylene oxide gas mixture.
4. Raise the temperature to the required degree.
5. Hold for the required time; turn off the heat.
6. Draw out the gas with the vacuum pump.
7. Admit filtered and sterilized air to the chamber.

A fully automatic ethylene oxide autoclave requiring only proper supervision is available (Fig. 12–11).

Beta-propiolactone (BPL). At about 20 C this substance is a colorless liquid. It has a sweet but very irritating odor. It is unstable at room temperatures but may be refrigerated at 4 C for months without deterioration. Aqueous solutions effectively inactivate some viruses, including those of poliomyelitis and rabies, and also kill bacteria and bacterial spores. The vapors, in concentrations of about 1.5 mg of lactone per liter of air with a high relative humidity (75 to 80 per cent), at about 25 C, kill spores in a few minutes. A decrease in temperature, humidity, or concentration of the lactone vapors increases the time required to kill spores.

Beta-propiolactone is not inflammable under ordinary conditions of use. It is, however, very irritating and may cause blisters if allowed in contact with skin for more than a few minutes. It is not injurious to most materials. BPL appears to act by forming chemical compounds with cell proteins. The necessity for high humidities during its use and also its cost are disadvantages. Its activity at room temperatures is a distinct advantage. BPL does not penetrate as well as ethylene oxide and is therefore more suitable for disinfecting surfaces (e.g., rooms, buildings, and furniture) by fumigation.

Aqueous solutions of BPL can be used to sterilize biological materials such as virus vaccines, tissues for grafting, and plasma.

Sterilization by Filtration

Many fluids may be sterilized without the use of heat, chemicals, or radiations. This is accomplished mechanically by passing the fluids to be sterilized through very fine filters. Only fluids of low viscosity that do not contain numerous fine particles in suspension (e.g., silt, erythrocytes), which would clog the filter pores, can be satisfactorily sterilized in this way. The method is applicable to fluids that are destroyed by heat and cannot be sterilized in any other way, such as fluids and medications for hypodermic or intravenous use, as well as culture media, especially tissue culture media and their liquid components, e.g., serum.

Several types of filters are in common use. The Seitz filter, consisting of a mounted asbestos pad, is one of the older filters used. Others consist of diatomaceous earth (the Berkefeld filter), unglazed porcelain (the Chamberland-Pasteur filters), or sintered glass of several varieties. A widely used sintered glass filter is shown in Figure 12–12. The Sterifil aseptic filtration system shown in Figure 12–13 (*A* to *D*) consists of a tubelike arrangement that sucks up fluid around all sides of the tube into a Teflon hose-connected receiving flask. The advantage of this is that the filter is very inexpensive and can be thrown away when it clogs up.

A widely used and practical filter is the membrane, molecular, or Millipore filter. It is available in a great variety of pore sizes, ranging from 0.45 μm for

Figure 12–12 Bacteriologic filter of the sintered glass type. Plastic tubing is attached to (A) and at the other end to a pump that pulls a vacuum that forces the liquid down the filter into the flask. One may position a trap between the flask and the pump to prevent water from accidentally entering the pump.

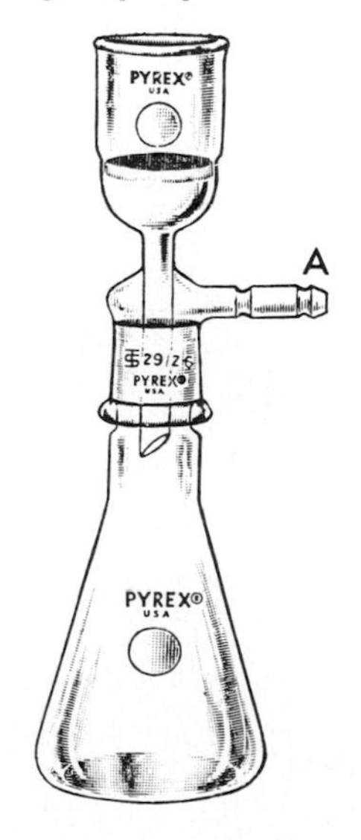

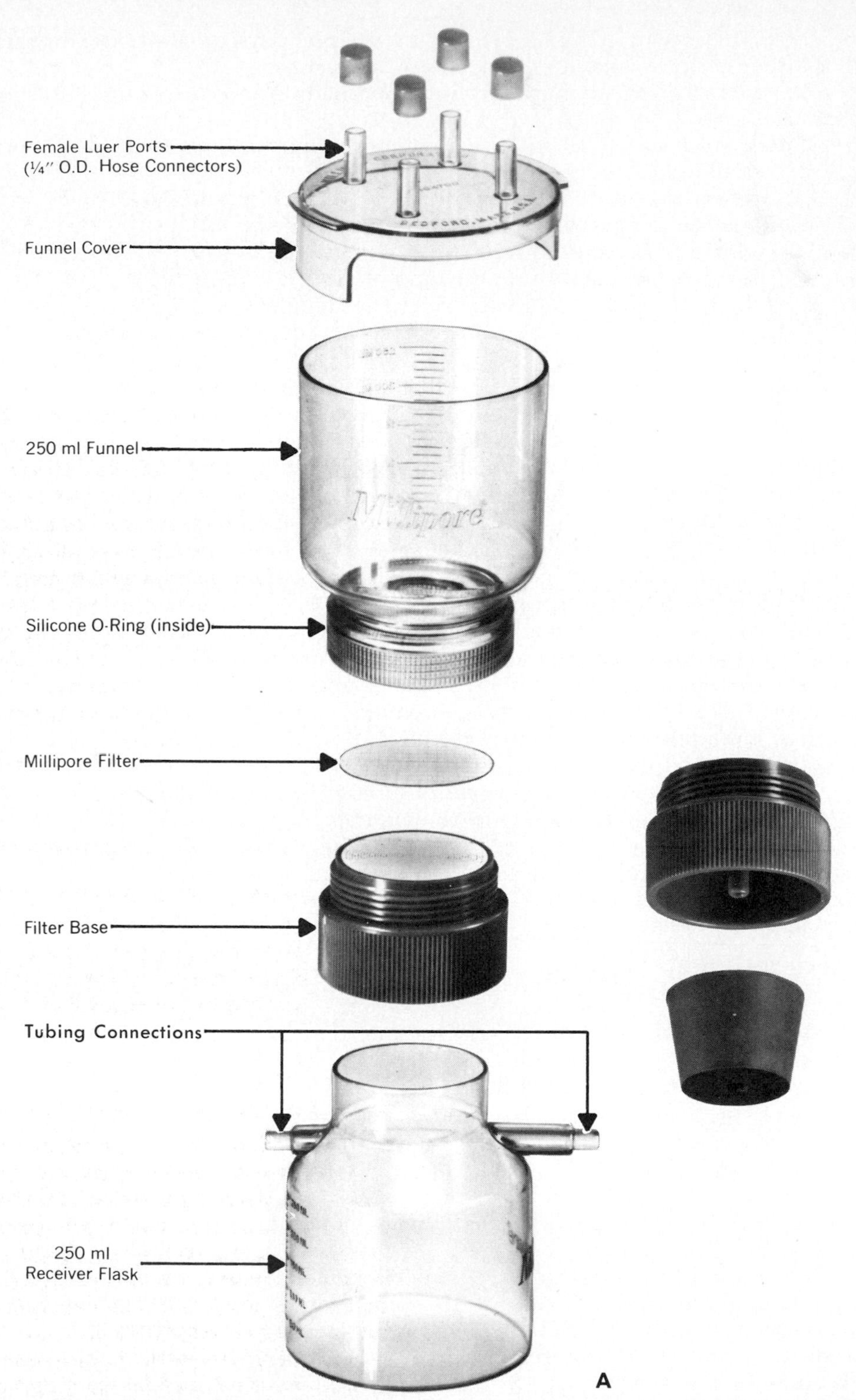

Figure 12–13 Sterifil asceptic filtration system. *A,* Apparatus disassembled, showing all parts of the unit. *B,* Solutions may be introduced aseptically through a sterilized two-way valve inserted in a port and connected by an adapter to sterilized ¼-inch tubing. Solutions may also be introduced from a syringe by inserting a needle through one of the rubber cap coverings of the ports (not shown). A vacuum pump (*C*) or a plastic syringe (*D*) may be used to pull solutions through the filter. (Courtesy of Millipore Corporation).

Illustration continued on opposite page

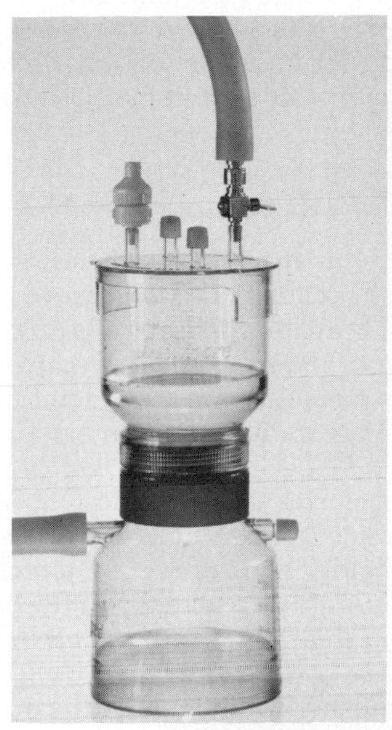

B

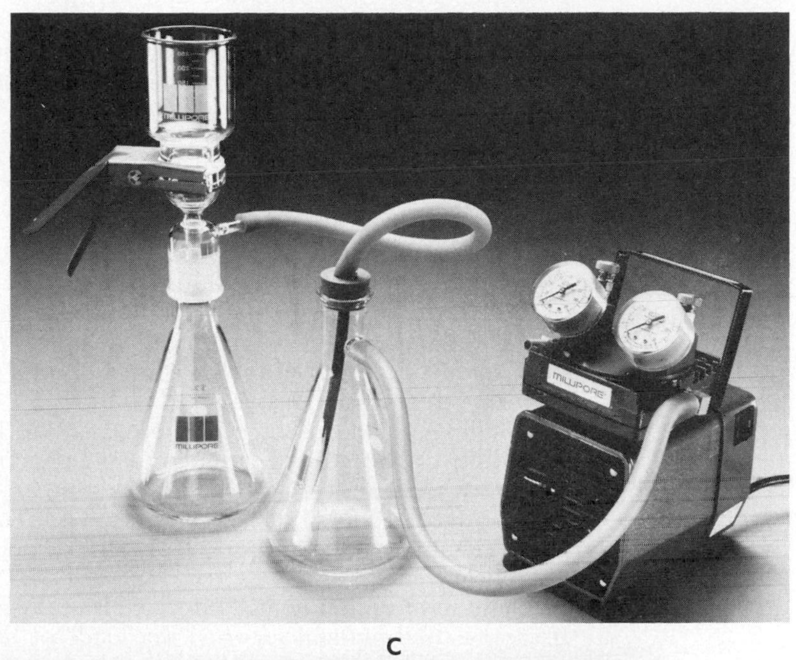

C

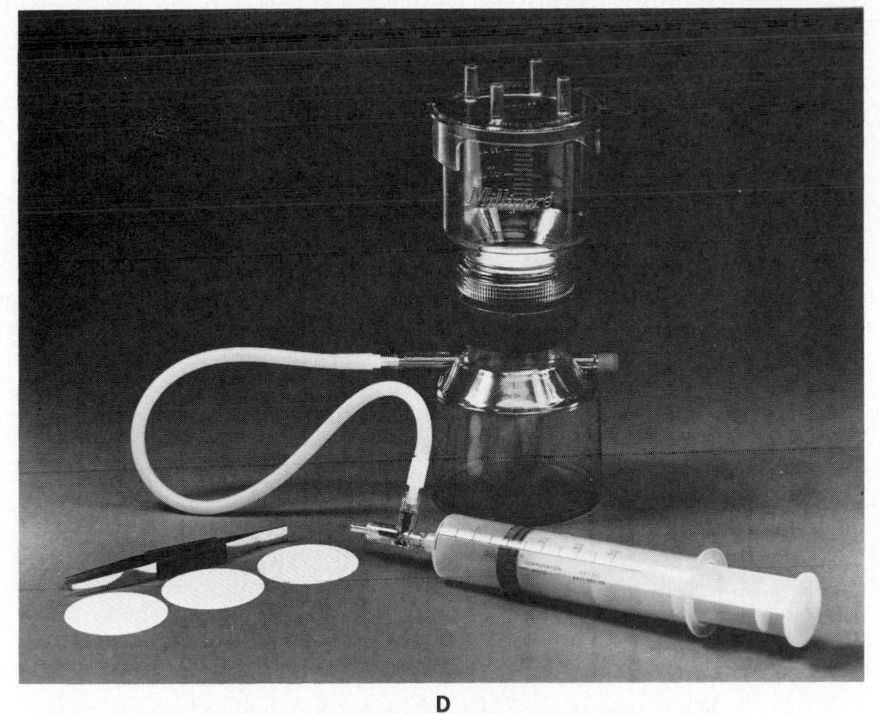

D

Figure 12–13 *Continued*

virus studies to 0.01 μm. These filters consist of paper-thin, porous membranes of material resembling cellulose acetate (plastic). One common form of these special filters is shown in Figure 9–11. In general, the porcelain, clay, paper, or plastic filtering element is held in some supporting structure, and the fluid to be filtered is forced through the filter into a receptacle by a vacuum or by pressure. The filter, support, and receptacle are assembled and autoclaved before use. Further details concerning these procedures need not be given here, since sterilization by filtration is rarely used without adequate information pertaining to the specific filtration problem, which would describe the advantages of one type of filter over another.

SUMMARY

The most complete way to dispose of infectious materials is incineration, although precautions must be taken to prevent spilling and to assure that everything is fully burned. Things that cannot be incinerated are sterilized to free them not only of pathogens but of all living organisms. For dry materials, glassware, syringes, dressings, filters, and pipettes, this may be done in a sterilizer oven at 165 C (329 F) for 2 hours. This treatment destroys fungi, bacteria, spores, and viruses.

Boiling water cannot be expected to kill bacterial spores, unless applied according to the tyndallization method (fractional sterilization). In order to sterilize with steam certainly and quickly, steam under pressure in an autoclave may be used at 15 psi for 15 minutes. The temperature reached should be at least 121 C, sufficient to destroy all bacteria, their spores, and all other microorganisms.

The modern automatic autoclave is really only a glorified pressure cooker. It can be used manually, for a "wet load" of liquid materials, for a "dry load" containing glassware, or with fast exhaust—to kill microorganisms quickly so that contaminated dishes, plates, flasks, and pipettes can be safely washed.

Ethylene oxide (in a mixture with carbon dioxide called Carboxide) and beta-propiolactone (BPL) are used routinely in hospitals for gas sterilization of all types of surgical and other materials.

Ultraviolet light, employed routinely for sterilization of the air in operating rooms, cannot penetrate like x-rays into materials, but it is readily available in sunlight, which therefore has great bactericidal powers. Commercially, the use of UV in restaurants and so forth is impressive, but, like x-rays, they are inefficient for effective sterilization.

Although they are highly bactericidal, neutrons are expensive to produce and difficult to control. Alpha and beta rays are not practical for use, but gamma rays are used widely to sterilize foods and pharmaceuticals. Cathode rays (electrons) are applied to food canning, frozen food, and surgical sterilization, and are also used to sterilize disposable Petri plates, pipettes, tubing, and most packaged materials.

Many fluids may be sterilized without the use of heat, chemicals, or radiation by means of filters. Several types of filters are in common use, such as the Seitz, the Berkefeld, the Chamberland-Pasteur, sintered glass of many varieties, and membrane filters such as the Millipore. The great advantage of modern bacterial filters is that they are disposable.

REVIEW QUESTIONS

1. What precautions should be taken when infectious material is incinerated?
2. What temperature should a sterilizing oven reach and for how long? What may be sterilized in this way?
3. Describe the adaptation of a kitchen oven to sterilization in the home. (Give time and temperature.)

4. How effective in sterilization is live steam when not under pressure?
5. Describe three methods of sterilization with moist heat. Why is moist heat more effective than dry heat? Why is boiling at high altitudes less effective than at sea level?
6. What precautions are essential to effective use of an autoclave? (Give time, temperatures, and pressures.) What is the role of air in autoclaving? Of steam? What does "psi" stand for?
7. How may fluids be sterilized without the use of heat or chemicals?
8. What indicators may be used as proof of adequate sterilization?
9. Explain the uses of ethylene oxide and of beta-propiolactone.
10. Discuss sterilization with radiation.
11. What filters can you name? What are some of the uses of filtration in microbiology?
12. Imagine that you are an operating room nurse. You are responsible for sterilizing all supplies used in the operating room. Describe how you would sterilize each of the following, giving your reasons for the choice of methods: *(a)* linens, *(b)* dressings, *(c)* instruments, *(d)* scalpels and scissors, *(e)* talcum powder, *(f)* Vaseline gauze, and *(g)* rubber gloves.

Supplementary Reading

Lawrence, C. A., and Block, S. S. (eds.): Disinfection, Sterilization, and Preservation. 2nd ed. Philadelphia, Lea & Febiger, 1977.

LaMaitre, G. D., and Finnegan, J. A.: The Patient in Surgery. 4th ed. Philadelphia, W. B. Saunders Company, 1980.

Perkins, J. J.: Principles and Methods of Sterilization. 2nd ed. Springfield, Ill., Charles C Thomas, 1980.

Sykes, G.: Disinfection and Sterilization. 2nd rev. ed. Philadelphia, J. B. Lippincott Co., 1965.

Brochures

Beta-propiolactone (BPL). 1959. Wilmot Castle Co., Rochester, N.Y. 14618.

Data on Membrane Filters. The Millipore Filter Corporation, Bedford, Mass.; Karl Schleicher and Schull, Keene, N. H.; AG Chemical Co., P.O. Box 65C, Pasadena, Calif. (bibliography included); Arthur H. Thomas Co., P.O. Box 779, Philadelphia, Pa., 19105.

Stierling, H., Reed, L. L., and Billick, I. H.: Evaluation of Sterilization by Gaseous Ethylene Oxide. Washington, D.C., U.S. Dept. of Health, Education, and Welfare, Public Health Monograph No. 68, 1962.

DISINFECTION

Disinfection is a process that destroys, removes, or inactivates infectious (pathogenic) microorganisms. It may also eliminate noninfectious microorganisms but not necessarily thermoduric (capable of withstanding high temperatures) vegetative organisms, bacterial endospores, or hepatitis viruses. Disinfection may be accomplished by exposure in aqueous fluid at moderate temperatures as in pasteurization (63 C for 30 minutes) but is generally understood to mean application of a microbicidal chemical to an inanimate object. Cysts of protozoans and conidia (spores) of fungi usually are somewhat more resistant to unfavorable temperatures and chemical disinfectants than vegetative cells of the same species; tubercle bacilli are slightly more resistant to several chemical disinfectants (but not to heat) than other pathogenic bacteria.

It would be impossible to include here a discussion of all the chemicals used for disinfection. There are too many, and new ones are put on the market every day. An attempt will be made to group some of these disinfectants and to discuss those most commonly used.

Table 13–1 evaluates common types of bactericidal agents. The chart may be useful to the reader, who should remember that, in general, pure compounds have been used in the studies on which Table 13–1 is based, whereas in commercial usage disinfectants are often compounded in mixtures for greater effectiveness.

THE HALOGENS

The halogens derive their name from a Greek word *(halos)* meaning salt; hence, *halogen* means salt-former. They readily form salts, e.g., sodium chloride. Three of the halogens (chlorine, iodine, and bromine) are among the best bactericidal agents. They act mainly by forming protein-halogen (saltlike) compounds in living cells, and they kill quickly. Bromine is rarely used because of its cost and toxicity. In commonly used concentrations of organic and inorganic halogen-releasing compounds, vegetative bacteria, fungi, and viruses are destroyed, but not the tubercle bacillus or bacterial spores.

In a special category is the most reactive of the halogens, namely, fluorine. As sodium fluoride it is frequently used as an additive in community water supplies to prevent dental caries in children, building resistance to tooth decay that carries over into adult life.

Chlorine

Chlorine, like other halogens, is effective as an oxidant.

$$Cl_2 + H_2O \rightleftharpoons HCl + HOCl$$

Hypochlorous acid (HOCl) is a strong oxidizing agent, and hydrochloric acid (HCl) lowers the pH. Both are detrimental to bacterial growth.

CHAPTER 13

Table 13–1. Disinfectants and Antiseptics*

Class	Disinfectant	Antiseptic	Other Properties
GAS			
Ethylene oxide (450–800 mg/liter at 60 C during autoclaving)	+2 to +4†	0	Toxic; good penetration; requires relative humidity of 30% or more; dry spores highly resistant.
LIQUID			
Glutaraldehyde, aqueous (2%)	+3	0	Sporicidal; active solution unstable; toxic.
Formaldehyde (8%) + alcohol (70%)	+2	0	Sporicidal; noxious fumes; toxic; volatile.
Formaldehyde (3%), aqueous (8%)	+1 to +2	0	Sporicidal; noxious fumes; toxic
Phenolic compounds (1–3%)	+3	±	Stable; corrosive; little inactivation by organic matter; irritates skin.
Hexachlorophene (1–3%)	0	+2	Bland; insoluble in water, soluble in alcohol; not inactivated by soap; weakly bactericidal.
Chlorine compounds (500–5,000 ppm available Cl)	+1	±	Flash action; much inactivation by organic matter; corrosive; irritates skin.
Alcohol (best concentration, 70%)	+2	+3	Rapidly bactericidal; volatile; flammable; dries and irritates skin.
Iodine (0.5%) + alcohol (70%)	0	+4	Corrosive; very rapidly bactericidal; causes staining; irritates skin, flammable.
Iodophors (75–150 ppm available iodine)	+1	+3	Somewhat unstable; relatively bland; staining temporary; corrosive.
Iodine, aqueous (1%)	0	+2	Rapidly bactericidal; corrosive; stains fabrics; stains and irritates skin.
Quaternary ammonium compounds (1:750–1:500)	+1	+2	Bland; inactivated by soap and anionics: absorbed by fabrics.
Mercurial compounds (1:1,000–1:500)	0	+1	Bland; much inactivated by organic matter; weakly bactericidal.

*Instruments, apparatus, and other objects should be cleansed to remove gross organic soil prior to the use of chemical disinfectants that coagulate protein in order to get good penetration of crevices and porous material. Instruments, as well as rubber and plastic tubing, must be rinsed or flushed with sterile water before coming into contact with skin and especially mucous membrane to avoid irritation. For the same reason, aeration is necessary after exposure to ethylene oxide.

Some of the above material was suggested, modified, or contributed by G. F. Mallison, U.S. Public Health Service, Center for Disease Control.

†Maximal usefulness is denoted by +4, less useful by +3, etc.

(From Lennette, E. H., Spaulding, E. H., and Truant, J. P. (eds.): Manual of Clinical Microbiology. 2nd ed. Washington, D.C., American Society for Microbiology, 1974.)

Chlorine Gas. Chlorine is widely used to disinfect municipal water supplies and swimming pools. In the form of chlorine gas it is effective but highly toxic and requires special equipment for its use.

Chloride of Lime. Chloride of lime or calcium hypochlorite, $Ca(OCl)_2$, is used in a 1 to 5 per cent aqueous solution and is an excellent general disinfectant. It releases chlorine, which when free in appreciable concentrations is toxic for all living things. Although the presence of extraneous organic material somewhat decreases the efficiency of chloride of lime (why?), it is still effective for use on feces from a patient who has an intestinal infection. Usually a 5 per cent solution is used. It is mixed with an estimated equal amount of

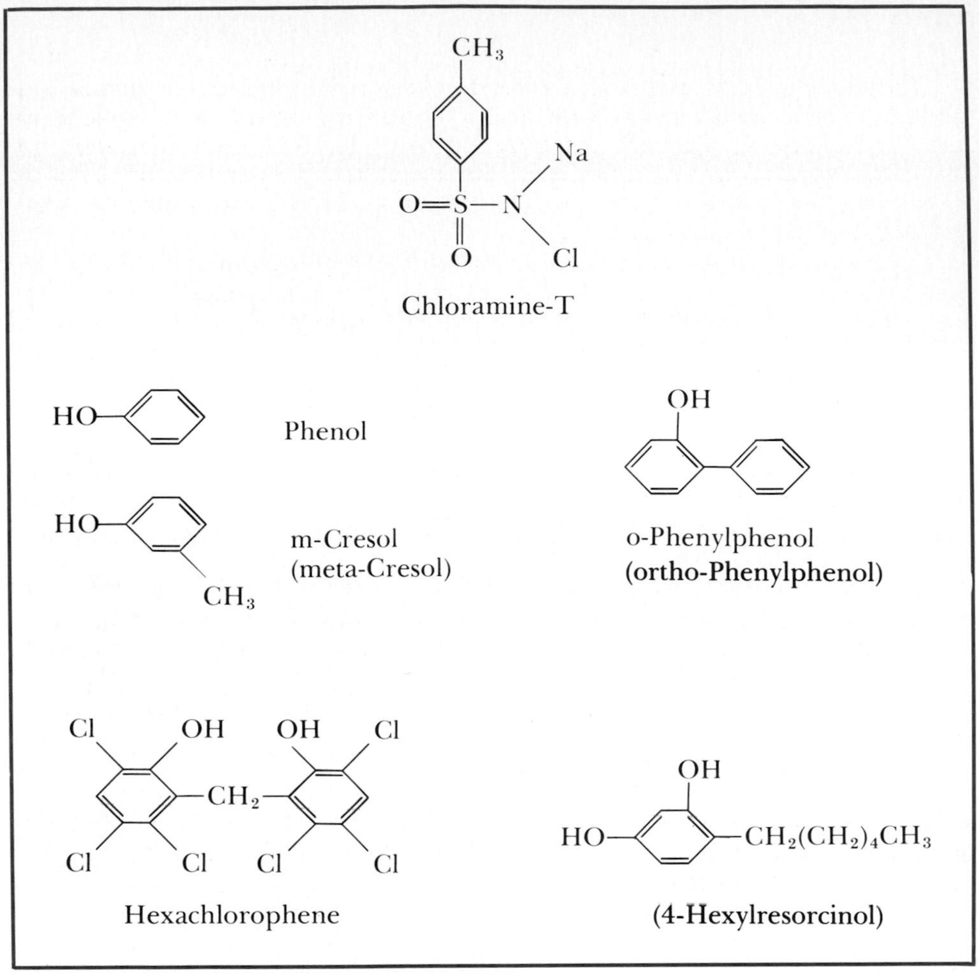
CH3
Na
O=S—N
O
Cl
Chloramine-T
HO
Phenol
OH
HO
m-Cresol
(meta-Cresol)
CH3
o-Phenylphenol
(ortho-Phenylphenol)
Cl
OH
OH
Cl
CH2
Cl
Cl
Cl
Cl
Hexachlorophene
OH
HO
CH2(CH2)4CH3
(4-Hexylresorcinol)

Figure 13–1

feces or urine. The mixture is allowed to stand in a covered container (such as a pail) for one hour before being discarded into the sewer.

Although some local and state health authorities permit the disposal of excreta from patients with enteric infections directly into the sewerage system, it seems a better practice to disinfect the excreta before disposal because of the danger of spattering and of possible leaks or other defects in plumbing systems.

Sodium Hypochlorite. Sodium hypochlorite (NaClO) solution (5.25 per cent) is available in every grocery store as a laundry bleach. (Clorox, Purex, and crystals of Comet are representative products.) This is one of the most generally useful convenient forms of chlorine, and it is a highly efficient disinfectant and deodorant. Unless diluted, it is irritating to skin and mucous membranes and is used mainly for laundry, floors, and inanimate objects of all sorts. It can be used to disinfect drinking water, to deodorize, and for many other purposes. Manufacturers generally give full details for use on the label. Sodium hypochlorite may leave an odor on the hands.

Organic Compounds of Chlorine. Azochloramide and chloramine-T (Fig. 13–1), unstable organic compounds of chlorine, and monochloramide (NH_2Cl) are for many purposes more convenient than chloride of lime or sodium hypochlorite. Their action is slower. They and several other similar compounds that give off free chlorine are useful for general sanitization in dairies and restaurants. The manufacturers give reliable directions for use.

Iodine

Iodine is a very useful and effective bactericide. Besides acting as an oxidant, it irreversibly reacts with and iodinates the amino acid tyrosine in proteins. Iodine in alcoholic solution (tincture) is often used for cuts and abrasions, disinfection of clinical thermometers, and preparation of the skin for surgery. For the last of these purposes a surgical scrub with Betadine[1] soap followed by Betadine solution is used.

Iodine is very poisonous and can cause serious burns of the skin unless used properly. The 2 per cent alcoholic solution should be applied in one coat, allowed to air dry, and then covered with sterile gauze When used for surgery, iodine solution should not be allowed to run down the sides of the body so that it can concentrate under the patient. If the patient is lying on a rubber mat on an operating room table, this concentration of iodine tincture on the back can cause serious burns.

Although aqueous solutions of iodine have a high surface tension, they appear to be very effective for skin abrasions, and the irritating effect of alcohol is avoided. A commonly available preparation for such uses is 2 per cent iodine in 50 per cent alcohol containing KI (mild solution of iodine).

Organic Compounds of Iodine. Like chlorine, iodine is available in organic combination. It is sold in solutions containing a surface-tension reducer. These serve most of the purposes for which solutions of chlorine are used. Being less volatile, the iodine remains longer and has less odor, and one can tell by the color of the solution approximately how much iodine remains in it. In so-called *iodophors,* organic compounds, iodine is loosely combined with some surface-active agent. *Wescodyne* and *Ioclide,* representative iodophors, are not irritating to the skin (except in cases of iodine hypersensitivity) but may cause a slight, temporary tan discoloration if used in strong solutions. These products are probably sporicidal under certain conditions of use. Other iodophors are *Betadine, Hi-Sine,* and *Iosan.*

Fluorine

In the form of sodium fluoride (NaF), fluorine is used in the prevention of dental caries. Applied topically to teeth it may be antibacterial in effect. It is often added to otherwise deficient water supplies (fluoridation) to reduce tooth decay in children by about half. It is thought to produce fluoroapatite, which is resistant to acids, during tooth formation. We find fluorine also in the Freons (as coolants in refrigeration) or in plastics such as Fluon or Teflon. There are many ways in which fluorine reacts with and disintegrates organic compounds and thus destroys microorganisms, especially fungi. Yet, except for the incidental use in water, fluorine is not practical for disinfection purposes. Thus, only two of the halogen elements, chlorine and iodine, are commonly used as antiseptics and disinfectants.

COMPOUNDS OF HEAVY METALS

Heavy metals are commonly applied in two forms of compound: inorganic and organic. Examples of each are described on page 222.

---

[1]*Betadine* (povidone-iodine, also named PVP-I for 1-vinyl-2-pyrrolidinone polymers, iodine complex) is an organic compound of iodine (an iodophor). Povidone (polyvinylpyrrolidone) is a plasma expander owing to its surface action, and with iodine it is an excellent antiseptic.

Mercury and silver salts must be used at very low concentrations because of their high toxicity to human tissues. In these concentrations they combine with sulfhydryl groups of proteins.

Bichloride of Mercury. This is a common inorganic form of mercury compound. It was formerly widely employed in dilutions of between 1:1,000 and 1:5,000 but has been gradually replaced by other, more efficient disinfectants.

Organic Compounds of Mercury. Several organic compounds of mercury are used for disinfection of skin and for superficial applications. Among these are *Mercurochrome, Merthiolate, Metaphen,* and phenylmercuric nitrate. These organic forms of mercury are less irritating and less coagulative than bichloride of mercury. They are mainly bacteriostatic, especially when the amount of mercury is small. Their action can be reversed by agents that precipitate mercury, especially sulfur compounds like sodium thioglycolate and hydrogen sulfide:

$$H_2S + Hg \rightleftharpoons H_2 + HgS$$

No doubt the artificial red color of some organic mercurials makes a child feel proud of his "battle scars," but many bacteriologists question the actual benefit derived from the application of these disinfectants to a wound.

Silver Nitrate. Silver nitrate is most frequently used as a 1 per cent aqueous solution in the eyes of newborn babies to prevent gonorrheal (and other) infections (ophthalmia neonatorum). The excess silver nitrate solution must be carefully removed from the eyes by washing with physiological saline solution after instillation. If the silver nitrate is allowed to remain, it may cause serious irritation to the delicate membranes of the eye. The use of silver nitrate has markedly decreased the number of cases of blindness caused by ophthalmia neonatorum. In some places the use of penicillin has been substituted for or combined with silver nitrate, with acceptable results. The use of one or the other is generally required by law for newborn infants.

Fused silver nitrate is also used in the form of a pencil as a styptic, to stop bleeding. This is a fine way to keep your minor cuts aseptic.

Organic Compounds of Silver. *Argyrol* is an example of an organic (protein) compound of silver. It is used occasionally in 5 to 20 per cent aqueous solutions for the treatment of infections of the mucous membranes of the eye, nose, and urethra. In the concentrations mentioned it is nonirritating to mucous membranes. Other organic silver compounds sometimes used are *Silvol, Neo-Silvol,* and *Protargol.*

ALCOHOLS

Ethyl Alcohol. This is used very widely as a skin disinfectant, for "cold sterilization" of instruments, and as a disinfectant for clinical thermometers. Normally, alcohol is considered to be a protein coagulant, but it does not sterilize.[2] When used mixed with water for disinfection, the most effective concentration of alcohol is near 70 per cent. The presence of water hydrates and thus aids coagulation by the alcohol. Alcohol also affects the physical structure of lipid membranes. It should be noted that near 100 per cent (200 proof) ethyl alcohol is not a good disinfectant, either externally or internally. Like bichloride of mercury, it forms protective coatings of coagulum around microorganisms.

---

[2]Under certain conditions, cells "killed" with alcohol have been revived.

The rapid application of ethyl alcohol to the skin prior to a hypodermic injection probably accomplishes nothing more than cleaning the area and removal of some of the surface bacteria. The time is undoubtedly too short for disinfection. An alcohol rinse or soak, following scrubbing of the hands before surgery, probably reduces the numbers of bacterial inhabitants to some degree. The use of ethyl alcohol as a solution for "cold sterilization" of instruments, syringes, or needles should not be practiced, except for special purposes, because spore-forming organisms or resistant viruses may be present. If thermometers are adequately wiped, or washed with soap and water to remove all organic material, and if these thermometers are then completely immersed for a minimum of ten minutes in 70 per cent ethyl alcohol, they will probably not transmit the bacterial pathogens commonly occurring in the mouth or rectum, especially if the alcohol contains 0.5 to 1.0 per cent iodine or one of the quaternary ammonium compounds. These thermometers are not necessarily sterile.

Isopropyl Alcohol. Isopropyl alcohol ("rubbing alcohol") is commonly used in a 70 per cent solution and is as effective as ethyl alcohol for ordinary purposes, especially if fortified with iodine or another suitable disinfectant. It is cheaper and more easily obtainable than ethyl alcohol.

Methyl Alcohol. Because of the very toxic nature of methyl alcohol its use as a disinfectant is discouraged.

PHENOLIC DISINFECTANTS

Phenol ("carbolic acid") is an efficient bactericide in a 1 to 2 per cent solution. However, it is very corrosive to animal tissue and, if used in pure form, it is too expensive for common use. Many derivatives of phenol are now available on the market. They are mainly coagulative in action. Pure phenol is used mostly as a standard for testing disinfectants (see page 227). *"Crude carbolic,"* an impure phenol product, is useful for general purposes. It is apt to cause brown stains.

Phenol Derivatives. There are scores of these compounds, some of which have wide usefulness as disinfectants; others are too corrosive, too expensive, or otherwise not so useful. Most of these act in somewhat the same manner as phenol, i.e., probably by combined coagulative, toxic, and dissolving action. The exact mechanisms are not clear and differ with diverse compounds and probably with different species of microorganisms. Many of these compounds are surface-tension reducers and tend to remain adsorbed as thin films on surfaces to which they are applied. Their action is therefore prolonged compared with that of alcohols or halogens, which tend to volatilize quickly. Two general classes of phenol derivative may be noted: the *cresols,* which are like phenol with methyl ($-CH_3$) groups attached, and *bis-phenols,* which are composed of two phenol groups or phenol derivatives joined directly or through some other radical.

Cresols. These are more effective than pure phenol for most general purposes.[3] Saponated solution of cresol (used in a 1 to 5 per cent solution) is a good example of an effective cresol disinfectant. The soapy solution of cresol has a far lower surface tension than does the aqueous solution. This disinfectant is used for feces, soaking contaminated instruments, and general cleaning and disinfecting purposes. Cresols, like phenol (Fig. 13–1), are corrosive to living tissue and are not used in prolonged contact with tissues. Cresols exist in ortho-, meta-, and para-cresol forms, or may be purchased as commercial mixtures of all three.

[3]The bactericidal power of cresol is about four times that of phenol against *Salmonella typhi.*

Hexylresorcinol. Hexylresorcinol (4-hexyl-1,3-dihydroxybenzene), another derivative of phenol, may be used as a urinary antiseptic and against worm infestations in animals. Preparations of hexylresorcinol, mixed with glycerol, are good general antiseptics, in part because they are strong surface-tension reducers. A familiar oral preparation ("ST 37") has a surface tension of 37 dynes per cm.

Bis-phenols. Like the cresols, these retain much of their disinfectant action in the presence of soap and other surface-tension reducers and are commonly used in mixed preparations containing surface-active compounds. Prominent among the bis-phenolic disinfectants are Lysol[4] and several made with *hexachlorophene* (see Fig. 13–1). One great advantage of Lysol over some cresols commonly used is that it does not foam when used in 2 per cent solution for scrubbing and washing surfaces. Various combinations of hexachlorophene with soap and other ingredients are sold commercially as Gamophen, Hex-O-San, pHisoHex, Surgi-Cen, and Surofene. Fostex, pHisoHex, pHisoderm, Hexagerm, and others have been used as skin washes and shampoos. In some hospitals hexachlorophene took the place of topically applied alcohol. For nurses or dentists who must wash their hands many times every day, or for patients' body care, hexachlorophene-containing preparations were less unpleasant to use than many more irritating, but perhaps more effective, antiseptics. The high toxicity of hexachlorophene stems from the fact that it is *absorbed through the intact skin* and may reach a level of 0.38 ppm in the blood when used extensively on the body, as in bathing babies. It has been shown to cause brain damage in rats. Do not use hexachlorophene in vaginal deodorants or bathe infants in hexachlorophene washes. Your physician may prescribe pHisoHex to be used as directed.

ORTHOPHENYLPHENOL. Orthophenylphenol is used in proprietary mixtures such as *Amphyl* and *O-Syl. Chlorothymol* is another phenol derivative that has disinfectant properties similar to those just mentioned. Some bis-phenol preparations are said to be effective after surgical washup because they tend to sustain the disinfectant action on the skin. Opinions are divided on this. The use of any disinfectant must always be regarded as a supplement to, and not as a substitute for, cleanliness and aseptic technique.

Pine Oil. Derived by steam distillation from weathered pine boles and roots, this product is emulsified in water with soap or resin. It is effective against many gram-negative bacteria such as *Salmonella typhi* (phenol coefficient of 5; see page 227), but has very limited bactericidal action against certain gram-positive pathogens. It is sometimes used for janitorial or household purposes on floors, walls, and bathrooms and has a pleasant piny odor. Several proprietary products containing it are available at household supply stores.

Disinfectant Soaps. As stated earlier, phenol derivatives, especially cresols and bis-phenols, are often mixed with soaps. Such combinations appear to be fairly effective, and there are several on the market. Soap, if present in excess, may coat the bacteria, displacing the disinfectant and thus protecting the bacteria. Some residual surface action of the phenol derivative is maintained by daily use of the preparation. However, it is quickly removed by ordinary soap and water and alcohol washes. Phenolic compounds must not remain long in contact with the skin or other tissues, especially over large areas.

OXIDIZING AGENTS

Other useful antiseptics are H_2O_2 (hydrogen peroxide) used in 3 per cent solution for a mouthwash or gargle, and another oxidizing agent, potassium

[4]Lysol is a mixture of orthophenylphenol, cresylic acid, and soap.

permanganate ($KMnO_4$). Since peroxides are strongly mutagenic, their use in the mouth should be discouraged; instead, plain salt water is highly recommended as a gargle. Sodium perborate and zinc peroxide are other oxidizing agents, used as antiseptics in the mouth and on the skin, respectively.

A cleansing mouth antiseptic is Gly-Oxide (carbamide peroxide). It is very effective against minor oral irritations and inflammations. Ten drops on the tongue mixed with saliva are expectorated after three minutes—a simple, quick method of oral hygiene.

ALKYLATING AGENTS

Compounds like formaldehyde (an aqueous solution of 37 to 40 per cent gaseous formaldehyde is marketed as Formalin), ethylene oxide, and glutaraldehyde react with cellular constituents by substituting alkyl groups for hydrogen atoms. These strong disinfectants are sporicidal and therefore also sterilizing agents. Use of formaldehyde and glutaraldehyde as disinfectants is limited to excreta and utensils, owing to their odor and toxic properties. Have you ever dissected a frog or fetal pig that was kept in Formalin? This preservative of dead animals is very effective but at times quite corrosive to the skin tissues of the hands of the students. The use of ethylene oxide was discussed in Chapter 12.

DETERGENTS AND QUATERNARY AMMONIUM DISINFECTANTS

Most household detergents are surface-tension reducers. For their activity, these depend on their ability to emulsify liquids (i.e., grease). Exceptions are certain "dry cleaners" that dissolve, adsorb, or decompose lipids. On the basis of their chemical structure the surface-tension–reducing (surfactant) detergents are divided into three main groups. *Anionic* detergents include soaps, sodium and potassium salts of long-chain fatty acids, and some organic (alkyl) sulfates such as sodium lauryl sulfate, in which the surface-tension–reducing property resides in the negatively charged part of the molecule, the anion. *Nonionic* detergents are mainly complex ethers and polyglycerol soaplike compounds represented by Tween 80. *Cationic* detergents, the *quaternary ammonium chloride derivatives,* commonly called "quats," include such compounds as Zephiran, Ceepryn, Phemerol Chloride, Diaparene Chloride, C.T.A.B., and Roccal. Some of the anionic detergents have limited disinfectant properties, but the most important detergents, when disinfection is the consideration, are the cationic quats. These combine both disinfectant and surface-tension–reducing properties in the cation.

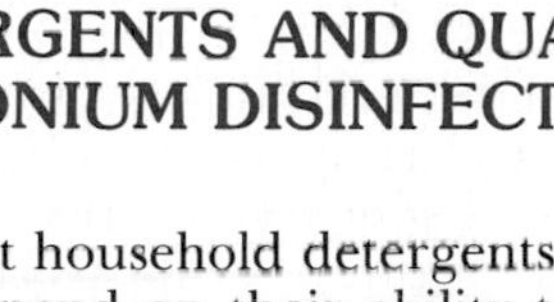

$$\left[\begin{array}{c} R_1 \\ | \\ R—N—R_2 \\ | \\ R_3 \end{array}\right]^+ \cdot Cl^- \qquad \left[\begin{array}{c} \quad\quad\quad\quad\quad O \\ \quad\quad\quad\quad\quad \| \\ C_{12}H_{25}—O—S—O— \\ \quad\quad\quad\quad\quad \| \\ \quad\quad\quad\quad\quad O \end{array}\right]^- Na^+$$

I 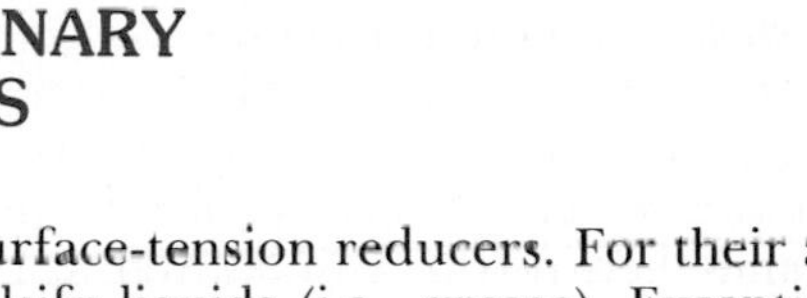 II

I, General formula for cationic ammonium chloride derivatives (quats). In ammonium chloride, the central nitrogen (N) of ammonia has four hydrogen atoms attached, but in the quats these are replaced by various organic radicals (R) such as $-CH_3$, and a chlorine atom is added. II, Formula of an anionic detergent, sodium lauryl sulfate.

The quaternaries are used principally for general external purposes as disinfectants and sanitizing agents. They have many other important properties: they are only slightly injurious to animal tissues, seem to be effective against many microorganisms in high dilutions, are stable, have no odor, do not stain, are not corrosive, dissolve easily in water, are easily obtainable, and are inexpensive.

In practice, the cationic quaternaries should never be mixed with anionics such as soaps because of the incompatible charges of the ions. For this reason quats are much less effective in "hard" water and iron-rich water: however, they tend to retain their activity in the presence of organic matter like pus, blood, or feces. There is good evidence that alcoholic solutions (tinctures) of quats are very effective against tubercle bacilli. Quats are of little known value against most fungi; some viruses rich in lipids may be susceptible to them.

SOAP

Household soap is a good detergent or cleansing agent, largely because it is a powerful surface-tension reducer. It is a highly effective emulsifier of fats and oils. It thus aids in the *mechanical removal of bacteria,* especially from oily surfaces like the skin. Many widely advertised detergents are not soaps but surface-tension reducers. Tide, Cheer, and Electra Sol (used in automatic dishwashers) are representative. They have cleaning and emulsifying properties like those of soap but are not disinfectants. An important ingredient in most of them was the potent surface-tension reducer alkyl benzene sulfonate ("ABS"). This could not be decomposed by the microorganisms in sewage disposal plants. The result was pollution of rivers and lakes with detergents, as evidenced by excessive foaming, a major problem in sanitary engineering. First in Great Britain and then in the United States, the manufacturers of detergents were quietly urged by their governments to stop using "microorganism-resistant" chemicals in their products. Today, the "foaming cleanser" actually foams less and is also inferior to the earlier product. It can, however, easily be decomposed by sewage microorganisms (i.e., it is "biodegradable") and can be safely disposed of by running the washing machine water into the sewer—our eventual fresh water supply. Polyphosphates such as sodium tripolyphosphate were subsequently used in many preparations sold on the market. However, phosphates too were found to be undesirable pollutants and were replaced. The most recent casualties to be eliminated from cleaning agents are *enzymes.* We don't know enough about their activities in our environment to dispose of them in sewage in large quantities.

Soap, in addition to being a surface-active detergent, is bactericidal to some degree. It is especially effective against *Treponema pallidum,* the cause of syphilis. This organism is rich in lipids. The disinfectant value of a soap depends to a great extent on the chemical nature of its fatty acid radical, i.e., the kind of fat it was made from.

DYES

Dyes have been used as antibacterial agents for many years. Their effectiveness depends greatly on their concentration. In general, crystal violet or malachite green inhibits most gram-positive organisms but not gram-negative bacteria. Crystal violet is also used as an antifungal agent. Acridine dyes (acriflavine) have a wide spectrum of antibacterial action; they are useful against infections caused by gram-positive organisms. Antibacterial dyes are commonly employed in selective culture media for preferential growth of the tubercle bacillus, *Escherichia coli, Brucella* species, and so on.

Dyes as disinfectants are not as practical as was once expected; however, their use led to the formulation of chemotherapy by Ehrlich (page 231) and the therapeutic applications of the sulfonamides, which originated with the dye Prontosil, employed first by Domagk (1895–1964; Nobel prize winner) in 1935.

AEROSOL SPRAYS

Not to be forgotten are "instant first-aid sprays" like Medi-Quik, Un-Burn, and Bactine. Read the label of contents of these medications: the chemicals mentioned will be familiar if you have read this chapter.

STRENGTH OF DISINFECTANTS

Phenol Coefficient. Certain kinds of disinfectants, especially aqueous solutions of substances chemically related to phenol, are often rated according to their bactericidal activity compared with that of phenol itself. According to this system they are said to have a certain *phenol coefficient.* In the determination of a phenol coefficient the conditions of time, temperature, and concentration during the comparison are very carefully standardized according to specifications of the United States Food and Drug Administration. The standard procedure is based upon the rate at which a certain dilution (1 in 90) of pure phenol (carbolic acid) kills the cells of a certain strain of *Salmonella typhi* at 20 C in 10 minutes. If, under the standard conditions, a given disinfectant kills *S. typhi* in a dilution of 1 in 180 as compared with a phenol dilution of 1 in 90, it is said to have a phenol coefficient of 2 (180 = 2 × 90) or, as a formula:

$$\text{phenol coefficient} = \frac{\text{dilution of disinfectant}}{\text{dilution of phenol}}$$

If a phenol coefficient is greater than 1, the disinfectant tested is better than phenol under the conditions of the test; if it is less than 1, the disinfectant is not as good as phenol.

This procedure is useful but gives a very limited and often inexact idea of the value or activity of a disinfectant. It does not test under "field conditions." For example, a given disinfectant may be 50 times as active as phenol against typhoid bacilli, yet almost worthless against staphylococci.

"Practical tests" are more valuable procedures since they simulate practical conditions. For example, the disinfectant may be tested in feces or blood instead of in distilled water, it may be used at very low temperatures or at body temperatures instead of at exactly 20 C, and it may be given little time or many hours to act instead of 5, 10, or 15 minutes. What would be its "phenol coefficient," or effectiveness, then?

EVALUATION OF DISINFECTANTS

In the United States the "official" tests for the evaluation of disinfectants in the laboratory are known as the "A.O.A.C. tests."[5] These tests evaluate

[5]Association of Official Analytical Chemists: Official Methods of Analysis. 10th ed. Washington, D.C., Association of Official Analytical Chemists, 1970.

proprietary germicides and determine if these agents meet certain minimal acceptable standards, as claimed by the manufacturers, with respect to *sporicidal, fungicidal,* and *tuberculocidal* activities.

Surface-active disinfectants or antiseptics may best be tested by "in-use tests" under actual field conditions. These may be tests on the skin after topical applications or washing the hands with the antiseptics. The determinations are usually made for residual organisms by touching a plate of culture medium or streaking the test area with a swab and counting the bacteria before and after the use of the antiseptics. It is much more difficult to standardize "in-use tests" than to perform A.O.A.C. tests.

SUMMARY

Disinfection is a process that destroys, removes, or inactivates infectious microorganisms. It also affects nonresistant microorganisms that are not pathogenic. The student must know that disinfection and sterilization are most certainly *not* the same process.

Among the halogens fluorine is used only to prevent dental caries, whereas chlorine, often as chlorine gas, acts as a strong oxidant in municipal water supply and swimming pools. Chlorine is also used as chloride of lime, calcium hypochloride, and sodium hypochloride (laundry bleach), and in organic compounds such as chloramine-T. Also very useful and effective as bactericides are iodine and its numerous derivatives, such as the iodophors Betadine, Wescodyne, and Hi-Sine.

Compounds of heavy metals used as disinfectants are bichloride of mercury, silver nitrate (1 per cent aqueous solution in the eyes of newborn babies to prevent infections causing blindness), Argyrol, Protargol, and other organic compounds of silver.

Ethyl alcohol is a good disinfectant if used in 70 per cent concentration, but it does not affect bacterial spores or viruses. Also readily available is isopropyl alcohol, used externally, and methyl alcohol, which is very toxic.

An early, universally applied general disinfectant is phenol. It is mainly coagulative in action. Derivatives are numerous and widely employed: they include Lysol, the cresols, hexylresorcinol, hexachlorophene, Amphyl, and pine oil.

Soap is a good cleansing agent because it reduces surface tension, but much better are the quaternary ammonium disinfectants called "detergents." In combination with other chemicals or soap they are very conspicuous in the supermarket, and every homemaker depends on them for most cleaning tasks where disinfection is desirable—washing dishes, clothing, floors, hands, bathtubs, and hair.

For some disinfection purposes, oxidizing agents like hydrogen peroxide or potassium permanganate serve best. For others, alkylating agents such as formaldehyde, dyes, or aerosol sprays may be most useful.

The effective strength of a disinfectant may be determined by comparing it with the effectiveness of phenol under specific conditions. If the phenol coefficient is greater than 1, the unknown is better than phenol; if less than 1, the opposite is true. In the United States the "official" tests for the evaluation of disinfectants with respect to sporicidal, fungicidal, and tuberculocidal activities are known as the "A.O.A.C. tests."

REVIEW QUESTIONS

1. In what situations are strongly coagulative disinfectants not desirable? Give two examples.
2. In application to (a) wounds, (b) surgical instruments, (c) hands, and (d)

infectious stools, compare the usefulness of the following: (1) alcohol, (2) tincture of iodine, (3) 2 per cent saponated cresol solution, (4) strong chlorine water or bichloride of lime, (5) bichloride of mercury, (6) 0.5 per cent carbolic acid, and (7) hydrogen peroxide.

3. Name three bacteriostatic substances and a common use for each. Why are fluids of high osmotic pressure bacteriostatic? Why is drying bacteriostatic?
4. Describe a good technique for disinfecting rectal thermometers.
5. What are quaternary ammonium compounds, and what are their uses?
6. Discuss the sources, advantages, disadvantages, and uses of several disinfectants, such as halogens, alcohols, phenolic compounds, detergents, and soap.
7. When should nurses wash their hands? Why?
8. Describe an appropriate handwashing technique for (a) communicable disease nursing and (b) preparation for surgery.
9. Describe an adequate technique for handling, cleaning, and sterilizing clinical thermometers, scalpels, surgical instruments, gowns, and catheters.
10. What is a transfer forceps? What is its principal property? When should it be sterilized?
11. Describe a simple procedure for terminal disinfection.
12. How does chlorine gas disinfect water? Where would chlorine be used, and why?
13. What is the best concentration of alcohol to use in disinfection?
14. Why is the nurse the most important person in the hospital with respect to guarding the personnel and the public from infection?
15. In home nursing what implements may be used in place of the hospital (a) hot-air sterilizer, (b) autoclave, (c) bedpan flusher, (d) pasteurizer, and (e) instrument boiler?

Supplementary Reading

Bartlett, P. G., and Schmidt, W.: Surfactant-iodine complexes as germicides. *Appl. Microbiol.*, *5*:355, 1957.

Lawrence, C. A., and Block, S. S. (eds.): Disinfection, Sterilization, and Preservation. 2nd ed. Philadelphia, Lea & Febiger, 1977.

Rockwell, V. T.: Surgical hand scrubbing. *Amer. J. Nurs.*, *63*:75, 1963.

Sommermeyer, L., and Frobisher, M.: Laboratory studies on disinfection of oral thermometers. *Nurs. Res.*, *1*:32, 1952.

Sommermeyer, L., and Frobisher, M.: Laboratory studies on disinfection of rectal thermometers. *Nurs. Res.*, *2*:85, 1953.

Spaulding, E. H., and Gröschel, D. H. M.: Hospital disinfectants and antiseptics. *In* Lennette, E. H., *et al.* (eds.): Manual of Clinical Microbiology. Washington, D.C., American Society for Microbiology, 1974.

Sykes, G.: Disinfection and Sterilization, 2nd rev. ed. Philadelphia, J. B. Lippincott Co., 1965.

Top, F. H. (ed.): Control of Infectious Diseases in General Hospitals. New York, American Public Health Association, 1969.

U.S. Department of Health, Education, and Welfare (Center for Disease Control): Investigation of hospital use of hexachlorophene and nursery staphylococcal infections. *Morbid. Mortal.* Vol. 21, Nos. 5 and 30, 1972.

Wade, N.: Hexachlorophene; FDA temporizes on brain-damaging chemical. *Science*, *174*:805, 1971.

ANTIMICROBIAL AGENTS AND CHEMOTHERAPY

. . . antibiotic substances are, so to speak, charmed bullets which strike only those objects for whose destruction they have been produced.

Paul Ehrlich

INTRODUCTION

One of the most important means of controlling microorganisms is by inhibition or complete suppression of their growth, as opposed to killing them through sterilization or vigorous disinfection. Prolonged inhibition of bacterial growth is called *bacteriostasis*; when applied to fungi, *fungistasis*; and so on. Inhibition of microorganisms is the basis of million-dollar industries and is most valuable in clinical medicine. We may consider inhibition of microorganisms in three categories:

1. Inhibition of spoilage-causing microorganisms may be achieved by nonspecialized but reliable means, such as drying foods, making hay, preservation in brines and syrups, refrigeration, and addition of acetic acid (vinegar) and other preservatives to certain foods.
2. Inhibition of contaminating microorganisms in culture media is essential so that the organisms causing infections in the patient may be isolated and grown in the diagnostic laboratory. This has previously been discussed as *selective bacteriostasis.* Dyes and certain chemicals are commonly used as microbiostatic agents in such situations. In cultivating viruses in cultures of living tissue cells, antibiotics such as penicillin, cycloheximide, and gentamicin are often added to the culture media. These have no effect on viruses but inhibit contaminating bacteria and fungi in the cultures.
3. Inhibition of microorganisms in humans or lower animals (or plants) is desirable so that the normal defensive mechanisms, especially the phagocytic cells of animals, can more readily clear the body of the invading microorganisms. This process is part of the general field of *chemotherapy* and is employed in human and veterinary medicine and in horticulture.

Antibiotics are substances originally isolated from filtrates of media in which bacteria, molds, or species of *Streptomyces* had grown. Over the last few years chemical modifications of these substances have produced numerous new antimicrobial agents synthesized in laboratories and more properly referred to as antibacterial, antiviral, antifungal, or antineoplastic chemotherapeutic agents.

CHEMOTHERAPY

As generally used today, this term means the treatment of infections[1] by means of substances ("chemicals") or drugs that (ideally) in relatively minute

[1]This implies that malignancies are also infections (by cancer cells) that are treated with antineoplastic chemotherapeutic agents.

CHAPTER 14

concentrations kill or inhibit the growth of the infecting microorganisms but do not adversely affect the host. The term chemotherapy was coined by Paul Ehrlich, discoverer of the first antibacterial chemotherapeutic agent, active specifically against the spirochete *Treponema pallidum,* the cause of syphilis. The drug was an organic arsenic compound sold as Salvarsan (606, chemically named dioxydiaminoarsenobenzene dihydrochloride).

Mechanism of Chemotherapy. Whereas older drugs such as quinine (against malaria) and arsenic (against syphilis) were used on an entirely empirical basis, most modern chemotherapeutic drugs depend on the fact that the drug combines specifically with substances directly involved in protein (enzyme) synthesis essential to the pathogen. In many instances the enzyme combines with the drug instead of with its normal substrate. To be effective in this way, the drug must so closely resemble the molecular form of the normal substrate (or *metabolite*) of the enzyme that the cell is "fooled" into accepting the drug instead of the normal metabolite. The drug is said to be an *antimetabolite* or an *antagonist* of the normal substrate.

As an example of a *metabolite antagonist* consider the chemical structure of the drug sulfanilamide in comparison with para-aminobenzoic acid, or PABA (Fig. 14–1). PABA is a normal portion of the molecule known as the vitamin *folic acid.* Folic acid is in turn a vital part of an important coenzyme in many microbial cells. The structural resemblance between PABA and sulfanilamide is perfectly clear even to one who is not an organic chemist. When sulfanilamide is given to a patient, the drug antagonizes (replaces) PABA in the folic acid in the cells of the infecting microorganism. The enzyme of which folic acid is a part "stops in its tracks"! This is because the sulfonamide drug is an impostor: it takes the place of PABA but cannot carry on its functions. The cell stops growing, but it remains alive. It is in a state of *microbistasis.* The result of this for the patient will be explained later.

The blocking action of the drug can often be reversed by providing a generous dose of PABA if the reversing agent is not too long delayed. Clearly the organism is not dead, since it can be reactivated.

Although the action of some chemotherapeutic agents, including antibiotics, is not yet as fully clarified as the action of sulfonamide drugs, many are known, on experimental bases, to act in a similar manner, i.e., through specific molecular combinations that produce microbistatic effects. Each drug, however, probably acts on a different chemical entity in the cell, and each drug is highly discriminatory in its action on different species of cells.

Chemotherapy and Phagocytosis. In spite of the fact that many antimicrobial chemotherapeutic agents do not immediately *kill* the infecting organisms, they are effective because *inhibition* of the microorganisms permits phagocytes to overpower them. Since the action of phagocytes is greatly facilitated and enhanced in the presence of specific antibodies, the value of chemotherapeutic drugs is heightened if the patient has some specific antibodies in his blood. He may have them as a result of previous infection, vaccination, or the purposeful injection of specific antiserum along with the chemotherapeutic drug. Phagocytosis is discussed in more detail on pages 269 and 291 to 292.

This "pinpointed" attack on microorganisms is in contrast to the indiscrim-

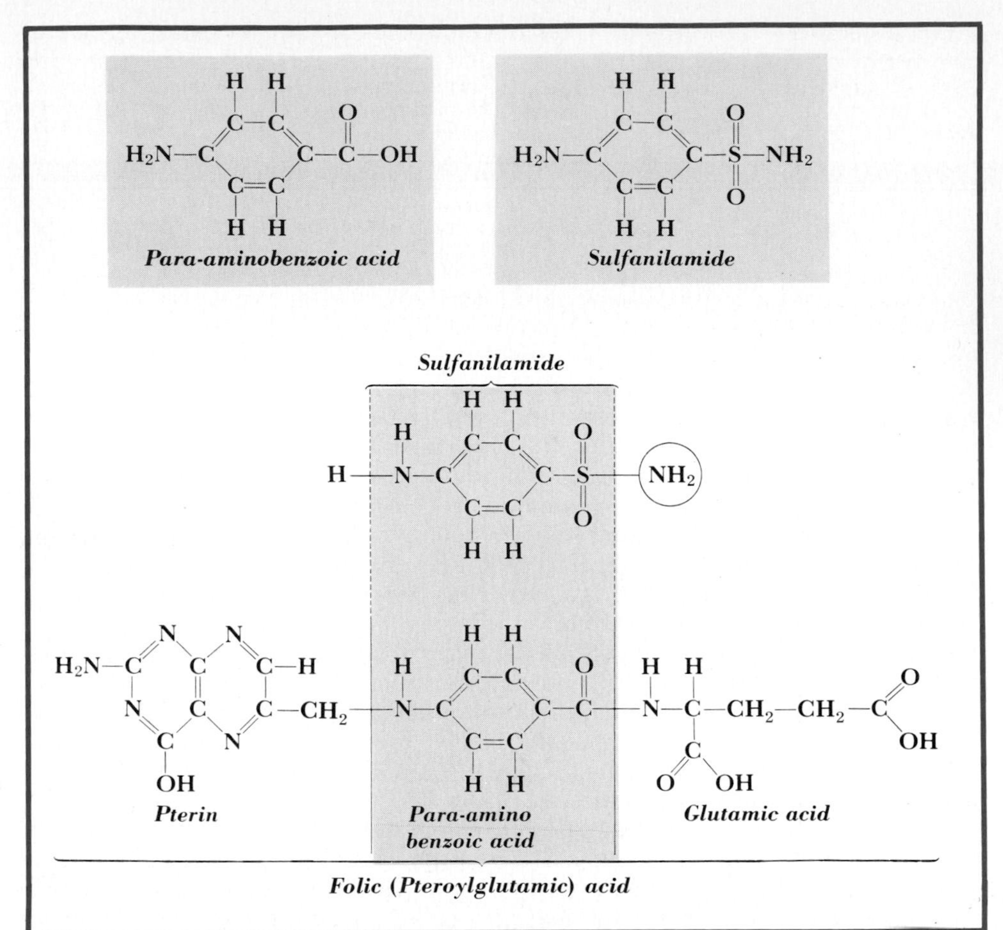

Figure 14–1 Structural resemblance between molecules of para-aminobenzoic acid (PABA) and sulfanilamide (above), and the potential antagonism of PABA by sulfanilamide in the folic acid molecule (below).

When sulfonamide replaces PABA in the folic acid molecule (pteroylglutamic acid). the enzymatic functions involving folic acid cease.

inate, broad combining power of nonspecific, general disinfectants like iodine, chlorine, or the cresols. These tend to form physicochemical combinations with almost anything and everything, from feces and stable floors to human blood and tissues, and will kill the host as readily as the parasite.

Selective Toxicity. *Selective toxicity* refers to a desired property of antibiotic substances, namely selective destruction of pathogenic organisms without damage to the cells of the infected host. Only the penicillins and cephalosporins have been found to possess this ideal property; however, other antimicrobial chemotherapeutic agents may still be used selectively with great benefit to the host and only minor detriment to the tissues of the patient. Table 14–1 lists commonly used antimicrobial agents in relation to their modes of action as they are presently understood.

Toxic Effects. Chemotherapeutic agents that have been released for general use by the Food and Drug Administration are generally much more toxic for the parasite than for the host. Minute doses have insignificant, or at least not fatal, effects on the host, but they help destroy the parasites. It is to be borne in mind, however, that excessive or improperly timed doses of antibiotics or other chemotherapeutic agents (including "miracle drugs") or administration by improper routes or without medical supervision can in some instances result in severe illness or death.

Some antibiotics are allergens (page 346), and patients may have severe allergic responses to them. Penicillin is notable but not by any means unique in

Table 14–1. The Modes of Action of Commonly Used Antimicrobial Agents

Mode of Action	Antibacterial Agents	Antifungal Agents
Inhibition of cell wall synthesis	Penicillins[1] Cephalosporins[1] Cycloserine[3] Vancomycin Bacitracin Ristocetin	
Inhibition of active transport or changes of cell membrane permeability	Polymyxins[2] Colistin	Amphotericin B Nystatin Natamycin Ketoconazole[5]
Inhibition of protein synthesis	Aminoglycosides: Streptomycin Gentamicin Neomycin Tobramycin Kanamycin Chloramphenicol Clindamycin Lincomycin Fucidin[1] Macrolides[4]: Erythromycin Oleandomycin Spectinomycin Tetracyclines	
Inhibition of nucleic acid synthesis	Rifamycins: Rifamide Rifampin Novobiocin Nalidixic acid Pyrimethamine Trimethoprim Sulfonamides	Griseofulvin[1]
Inhibition of cell wall synthesis and nucleic acid synthesis	Novobiocin	

[1]These antibacterial agents are produced by molds.
[2]These antibacterial agents are produced by bacteria. All others not designated [1] or [2] are products of *Streptomyces* species.
[3]Cycloserine acts differently on cell-wall synthesis than penicillins or cephalosporins. It blocks alanine racemase and thus D-alanine incorporation.
[4]Macrolides interfere with bacterial protein synthesis at ribosomes.
[5]This is a very recent effective antifungal agent.

this respect. Physicians generally inquire of patients whether they have had any previous allergic reactions before giving antibiotics.

CHEMOTHERAPEUTIC AGENTS

There are many antimicrobial chemotherapeutic agents on the market. Today the most widely used are the sulfonamides (synthetic) and the antibiotics (obtained first from living organisms before chemists learned how to synthesize them). The reader should distinguish between antibacterial, antifungal, antiprotozoal, antiviral, and antineoplastic chemotherapeutic agents, all more or less specific for treatments of diseases caused by specific pathogenic agents.

Sulfonamides

These were discovered around 1935 by Gerhard Domagk, a German chemist and Nobel Prize recipient who investigated the poisonous action of a certain aniline (coal tar) dye, Prontosil. This substance, in staining bacteria, eventually killed them. Sulfanilamide (from which virtually all sulfonamide drugs are derived) was found to be the active part of Prontosil. Although not a dye, sulfanilamide (see Fig. 14–1) is derived from benzene (coal tar) like many dyes and acts in much the same way as Prontosil—i.e., by metabolite antagonism. Fundamentally, sulfonamides are a type of wholly synthetic antibiotic.

The group of "sulfa drugs," as they are often called, includes sulfathiazole, sulfadiazine, sulfamerazine, and many others. One of their chief drawbacks at first was their toxic side effects, but the drugs now available avoid these difficulties to a large extent (e.g., sulfisoxazole [Gantrisin] and sulfaethylthiadiazole).

The discoveries leading to clinical use of sulfonamide drugs gave an entirely new impetus to the chemotherapeutic treatment of infections. Many infections that previously had always been fatal or that always required a prolonged nonspecific treatment followed by a long convalescence could be cured rapidly with these drugs. In general, sulfonamides are more effective against gram-positive than against gram-negative organisms, but there are important exceptions, notably the gram-negative *Neisseria* (gonococcus, meningococcus), Usually sulfonamides are ineffective against viruses, rickettsias, fungi, and protozoa. Since the end of World War II sulfonamide drugs have been superseded by antibiotics (such as penicillin) in many clinical situations, but they still have important uses, especially in urologic infections (e.g., cystitis and some cases of meningococcal meningitis) and for illustrating basic principles of chemotherapy.

Drug Resistance or Drug Fastness. As previously explained, microorganisms often undergo genetic mutations. These mutations generally result in alterations in enzyme production in the mutant cells. Let us suppose that mutation occurs in a few of the cells of a species of microorganism (say, *Staphylococcus aureus*) that is multiplying in a patient being treated with penicillin. The mutation may (and, unfortunately, frequently does) so alter a few cells that they are no longer susceptible to the action of the penicillin. They are said to be *antibiotic-resistant* or *drug-fast.* Resistance may similarly develop in many species of microorganisms to any chemotherapeutic drug. The resistant cells proceed to grow and finally predominate in spite of the drug. This is very dangerous for the patient and for everyone around who may acquire the organisms, as tragically illustrated by the development of sulfonamide-resistant gonococci and meningococci of group B and streptomycin-resistant tubercle bacilli.

The emergence of such drug-resistant microorganisms can be avoided to some degree by using adequate (often high) doses of antibiotics for treatment. This keeps a sufficient quantity of the chemotherapeutic drug in the bloodstream to affect even the most resistant members of the infecting organisms before they can start growing. Drug fastness of infecting bacteria can be detected and measured by laboratory tests.

Maintenance of Blood Levels. Since many chemotherapeutic drugs tend to be eliminated from the blood and tissues rather rapidly, the amounts in the blood *(blood levels)* tend to fall to levels at which the resistant individuals among the infecting bacteria can grow, unless the dose is repeated at proper intervals. If delayed, the patient soon has to combat a drug-fast infection. He may not survive. It is of the greatest importance, therefore, that any person responsible for giving repeated doses of such drugs sees that the patient receives the *drug on time,* in order that the blood levels do not fall dangerously low.

Antibiotic Substances[2]

The name antibiotic means "against life," an unfortunate choice of words! "Antibacterial chemotherapeutic agent" is much more specific and clear. Since antibiotics are most widely used to combat microorganisms, it might be more accurate to call them "anti*micro*biotics." The principal action of some antibiotics (e.g., penicillin) is *bactericidal.* Others (e.g., chloramphenicol) are *bacteriostatic,* as are the sulfonamides; rather than kill pathogenic microorganisms, they stop

[2]A number of important bacteria are mentioned in this discussion, but it is not necessary that the student become familiar with them at this time. For quick reference look them up in the index.

Table 14–2. Some Commonly Used Antibiotics*

Antibiotics†			
Generic Name	***Trade Name***	**Description**	**Source**
Penicillin	Penicillin G Oxacillin Ampicillin Methicillin Carbenicillin	Gram-positive bacteria; *Treponema; Neisseria*	*Penicillium notatum* and *Penicillium chrysogenum*
Fumagillin		*Entamoeba histolytica*	*Aspergillus fumigatus*
Paromomycin	Humatin	*E. histolytica*	*Streptomyces rimosus*
Streptomycin		*Mycobacterium tuberculosis;* gram-negative bacteria	*Streptomyces griseus*
Dihydrostreptomycin		Like streptomycin	Streptomycin; also some species of *Streptomyces*
Tetracycline	Achromycin	Broad-spectrum	Chlortetracycline
Oxytetracycline	Terramycin	Broad-spectrum	*Streptomyces rimosus*
Chlortetracycline	Aureomycin	Broad-spectrum	*Streptomyces aureofaciens*
Chloramphenicol	Chloromycetin	Broad-spectrum	*Streptomyces venezuelae*
Erythromycin	Ilotycin, Erythrocin	Broad-spectrum (not Enterobacteriaceae)	*Streptomyces erythreus*
Lincomycin	Lincocin		
Clindamycin	Cleocin	*Bacteroides* infections	
Carbomycin	Magnamycin	Like erythromycin	*Streptomyces halstedii*
Oleandomycin	Matromycin	Broad-spectrum	*Streptomyces antibioticus*
Neomycin B	Flavomycin	Mycobacteria	*Streptomyces fradiae*
Viomycin	Viocin	Like penicillin	*Streptomyces floridae; Streptomyces funiceus*
Oligomycin		Fungi of plants	*Streptomyces diastatochromogenes*
Amphotericin B	Fungizone	*Candida* sp. & other fungi	*Streptomyces nodosus*
Kanamycin	Kantrex	Broad-spectrum	*Streptomyces kanamyceticus*
Amikacin	Amikin	Against enteric infections	
Nystatin	Mycostatin	Pathogenic fungi	*Streptomyces noursei*
Cycloheximide	Actidione	Saprophytic fungi	*Streptomyces griseus*
Griseofulvin	Grisactin	Pathogenic fungi	*Penicillium griseofulvum*
Miconazole	Monistat	Antifungal drug	
Bacitracin		Like penicillin	*Bacillus subtilis*
Polymyxin B	Aerosporin	Gram-positive bacteria	*Bacillus polymyxa*
Pyocyanin§		Miscellaneous	*Pseudomonas aeruginosa*
Gentamicin	Garamycin	Enteric infections	
Tobramycin	Nebcin	Enteric infections	
Novobiocin		Staphylococci, *Proteus*	*Streptomyces niveus*
Colistin (Polymyxin E)	Coly-Mycin	Gram-negative bacteria	*Bacillus colistinus*
Ristocetin	Spontin	Resistant staphylocci	*Nocardia lurida*
Rifampin		Meningitis carrier stage	
Vancomycin	Vancocin	Resistant staphylococci	*Streptomyces orientalis*
Demethylchlortetracycline	Declomycin	Broad-spectrum	Synthetic; also from *Streptomyces aureofaciens*
Cephalosporin, cephalothin	Keflin	Against gram-pos. and gram-neg. bacteria	*Cephalosporium* mold
Isoniazid‡	Nydrazid	Specific against the tubercle bacillus	Synthetic

*Several not listed here are valuable commercially, agriculturally, and horticulturally.
†Several of these antibiotics are in reality mixtures consisting of related compounds such as the penicillins, polymyxin A, B, C, D, carbomycin A and B, the rifamycins, cephalosporins, and so on.
‡This is not an antibiotic (see page 241).
§Not used medicinally. One of the first known antibiotic substances.

their growth so that phagocytes and other defensive mechanisms, with the help of specific antibodies when present, can dispose of the pathogens.

Source of Antibiotics. Unlike the artificially synthesized sulfonamide drugs, most antibiotics are produced in cultures of certain microorganisms, principally species of *Streptomyces* and *Bacillus,* and molds of the genera *Penicillium* and *Aspergillus.* Some well-known antibiotics, and the names of the organisms from which they are obtained, are listed in Table 14–2. New antibiotics are constantly being sought and found in many plants and animals, terrestrial and marine. In the manufacture of antibiotics the desired antibiotic-producing organism is cultivated in large tanks of suitable liquid medium at specified temperature, pH, and aeration for a predetermined length of time. The culture is then centrifuged and filtered to remove the organisms. The filtered fluid, containing the antibiotic, is then subjected to purification and concentration processes. At length, these yield crystals of the pure antibiotic, which is tested for potency and sterility and then packaged and distributed. Usually, after further studies, organic chemists learn how to make these antibiotics synthetically, and this often becomes the preferred method of production.

Use of Antibiotics. The action of antibiotics varies greatly. Some may be given by mouth only; some are used exclusively in ointments; others, which can be destroyed by gastric acidity, are suitable specifically for intravenous or intramuscular (i.e., parenteral) use; and some may be administered by any route. Some are effective chiefly against gram-positive organisms; whereas others also work against certain gram-negative ones. Several have a wide range of activity and hence are called *broad-spectrum antibiotics.*

The specific uses, dosages, and contraindications of antibiotics are given in courses in pharmacology and materia medica, and are not detailed here. However, the student should be aware of several factors that affect the ability of any particular antibiotic or "antibiotic drug combination" to help the patient to overcome infection. These are: (1) insufficient, irregular, or wrong dosage; (2) resistance or sensitivity of the microorganism to the antibiotic; (3) inactivation of the antibiotic by the host's proteins or flora; (4) poor host defense; (5) poor penetration of the antibiotic into tissues or cells; (6) toxicity to the patient; (7) allergy of the patient to the antibiotic; and (8) likelihood of the pathogen becoming resistant to the antibiotic (Table 14–2).

Classes of Antibiotics. Some antibiotics are more bactericidal than bacteriostatic (Table 14–3) in the concentrations generally used. These factors may affect the rate at which the drugs act. Bactericidal antibiotics that may be used together effectively are:

Antibiotics	
Penicillin Streptomycin Bacitracin Polymyxin B	When used together, these are often synergistic,[3] never antagonistic. For example, penicillin and streptomycin combined are valuable in bacterial endocarditis.

Other antibiotics are definitely bacteriostatic, especially in the low concentrations generally used in therapy:

Antibiotics	
Tetracyclines Cloramphenicol Erythromycin Carbomycin Neomycin Oleandomycin Novobiocin	These are neither antagonistic nor synergistic. They sometimes work well in combination with antibiotics of the first group.

[3]The combination is more effective than the sum of both.

Table 14–3. Bactericidal and Bacteriostatic Chemotherapeutic Agents

Bactericidal	Bacteriostatic
Penicillins	Tetracyclines
Streptomycin	Erythromycin
Kanamycin	Trimethoprim
Gentamicin	Chloramphenicol
Amikacin	Lincomycin
Tobramycin	Clindomycin
Neomycin	Sulfa drugs
Vancomycin	
Polymyxin B	
Colistin	
Bacitracin	
Cephalosporins	

Antibiotics of the second group are generally of the broad-spectrum type. There are many other broad-spectrum antibiotics, and new ones appear on the market frequently, e.g., gentamicin and cephalosporin derivatives.

Two important properties of microorganisms in relation to antimicrobial substances are their usual response to antibiotic therapy and the probability that the pathogens will become drug-fast or antibiotic-resistant. Note that drug fastness is a property of the microorganism and not of the drug! (See Table 14–4.)

Some Representative Antibiotics

The Penicillins. All penicillins and cephalosporins[4] are inhibitors of cell wall synthesis of gram-positive bacteria; they are bactericidal rather than bacteriostatic.

[4]Cephalosporin C is related to penicillin G in biosynthesis. The cephalosporins are not as useful as the penicillins except when allergic reactions to penicillin occur. *Cephalothin* is a broad-spectrum antibiotic.

Table 14–4. Response of Some Microorganisms to Antibiotics

MICROORGANISM	USUAL RESPONSE TO PROPER ANTIBIOTIC TREATMENT	FREQUENCY OF APPEARANCE OF RESISTANCE IN STRAINS IN PATIENTS
Pneumococcus Meningococcus Gonococcus Beta streptococci *Shigella* *Haemophilus* *Treponema*	Rapid	Rare
Staphylococcus	Rapid	Frequent
Alpha and gamma streptococci Coliforms *Proteus* *Pseudomonas* Tubercle bacilli	Slow or incomplete	Frequent
Brucella *Salmonella typhi* Rickettsias	Rapid. Frequent relapses require repeat treatments.	Rare

Figure 14–2 Sir Alexander Fleming (right) with Dr. Waksman (left) in 1949. (From Woodruff, H. B. [ed.]: Scientific Contributions of Selman A. Waksman. New Brunswick, N.J., Rutgers University Press, 1968.)

Penicillin, the first clinically effective antibiotic, was discovered by Sir Alexander Fleming in 1929 (Nobel Prize recipient, 1945, with coworkers Florey and Chain) (Fig. 14–2). As produced in cultures of *Penicillium notatum* or *Penicillium chrysogenum,* penicillin is generally a mixture of related substances called penicillin X, G, F, dihydro F, and K. Penicillin G sodium or a modified form is mostly used. Various semisynthetic penicillins are shown in Figure 14–3*A*. Penicillins are most effective against infections with gram-positive bacteria, but work equally well against the gram-negative gonococcus and certain spirochetes, especially those causing syphilis *(Treponema pallidum)* and related tropical diseases called yaws *(T. pertenue)* and bejel *(T. pallidum).*

In addition to the five natural penicillins just mentioned, several valuable derivatives have been produced by chemical and biosynthetic processes. Among these are compounds of penicillin G with benzathine (benzathine penicillin G), benzathine penicillin V, and procaine penicillin G. These, being relatively insoluble suspensions, are slowly absorbed but give prolonged action at lower blood levels than the more rapidly absorbed penicillins; they are therefore not suitable for the treatment of acute infections. Also, since penicillin G is destroyed by gastric acidity, these compounds are not administered orally but must be injected.

Penicillinase. One of the dangerous enzymatic properties often found in staphylococci (and in some other pathogens, notably *Shigella*) is production of the enzyme *penicillinase,* which *destroys penicillin.* Obviously, any organism that can excrete this enzyme can protect itself from penicillin: i.e., it is at least partially penicillin-resistant. The power to produce penicillinase often appears as a result of genetic mutation. Penicillinase is a β-lactamase and can act on molecules like penicillin that contain the β-lactam ring (Fig. 14–3*A*). Penicillinase in staphylococci is due to a gene carried on plasmids.

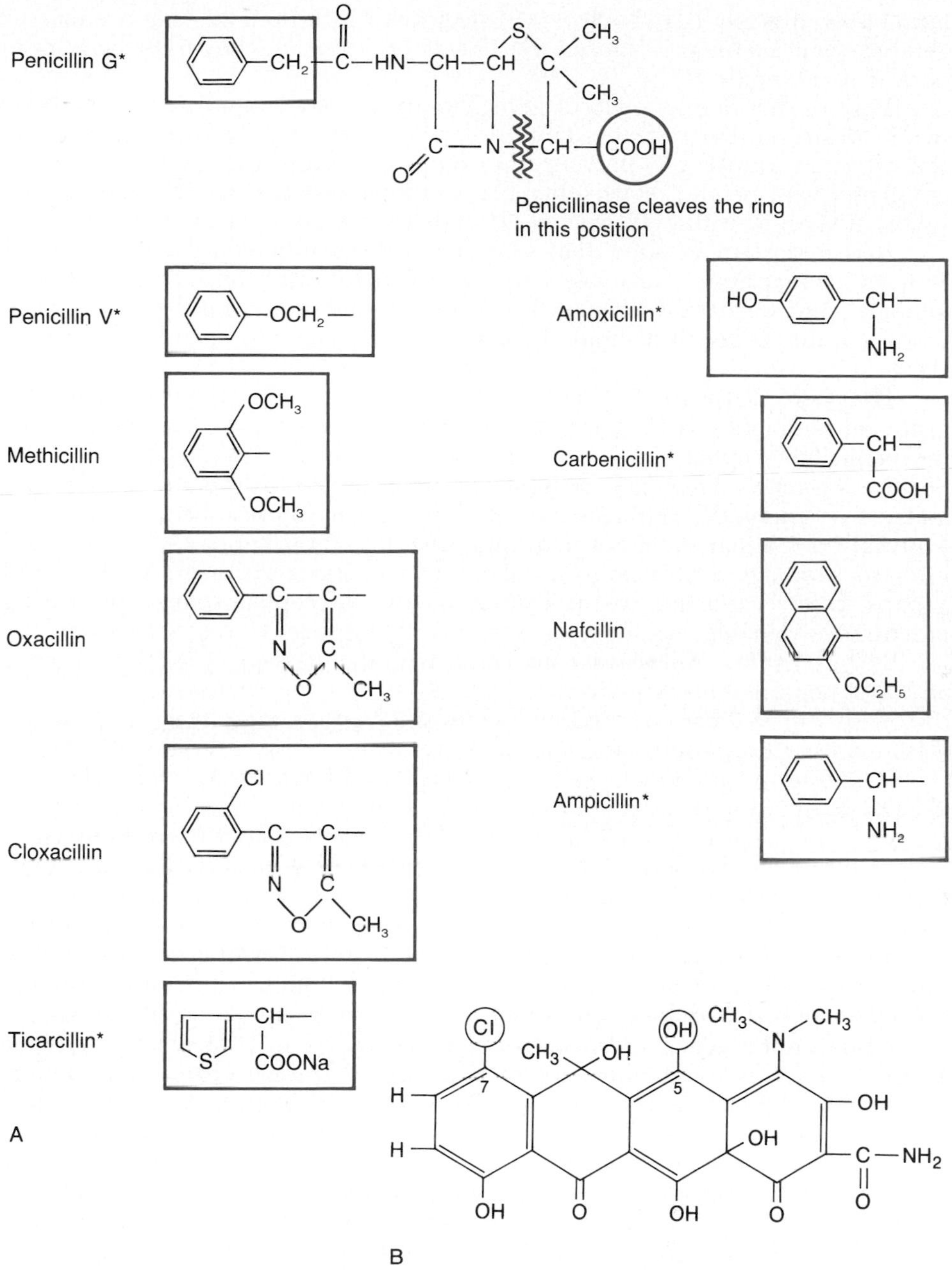

Penicillin G*

Penicillinase cleaves the ring in this position

Penicillin V*

Amoxicillin*

Methicillin

Carbenicillin*

Oxacillin

Nafcillin

Cloxacillin

Ampicillin*

Ticarcillin*

A

B

Figure 14–3 *A,* Different penicillins result when the side chains shown in the red squares are different. Other changes occur when the hydrogen in the carboxyl group of penicillic acid, as marked with the red circle, substitutes with potassium or sodium to make a salt of penicillin. Penicillinase (β-lactamase), produced by bacteria, can only break the β-lactam ring in those penicillins marked above with an asterisk; the other derivatives are resistant to this enzyme. *B,* Composite formula of the tetracycline group of antibiotics. In tetracycline each of the numbered carbon atoms (5 and 7) has an attached hydrogen atom. In chlortetracycline, chlorine (circled) replaces the hydrogen of the number 7 carbon atom; in oxytetracycline an OH group (circled) replaces the hydrogen of the number 5 carbon atom.

Adaptive or Induced Enzymes. Sometimes bacteria, notably staphylococci, appear to be stimulated to produce penicillinase by mere contact with the drug, not by genetic mutation. Thus, treatment with penicillin may, by itself, induce production of the enzyme. Enzymes ordinarily not produced by an organism but produced *in response to contact* of the organism with a specific substrate are said to be *adaptive* or *induced enzymes.* They represent activation of a latent genetic potentiality. There are many such enzymes in various organisms.

Several semisynthetic penicillins (e.g., methicillin and cloxacillin, also called oxacillin) have been developed that are effective against penicillinase-producing bacteria, notably certain strains of *Staphylococcus aureus* and certain gram-negative intestinal pathogens resistant to ordinary penicillin. Ampicillin is used orally in bacterial meningitis, salmonellosis, and enterococcal infections. Some of the semisynthetic penicillins are resistant to gastric acidity and may be administered orally (see Table 14–5). For deep infections like septicemias or cerebrospinal meningitis, caused by *Neisseria meningitidis,* they must be administered parenterally.

Most of the soluble forms of penicillin are excreted rapidly from the body in the urine, so that repeated large doses are necessary. In order to prolong the effect of a dose and prevent its prompt excretion, the antibiotic may be combined with peanut oil or other preparations designed to delay absorption by the tissues, or with oral probenecid to delay excretion by the kidneys.

It is important to know that some penicillin solutions, as well as the dry powder, are relatively unstable and rapidly deteriorate on exposure to air, sunlight, and warmth. Such penicillin preparations should not be dissolved or opened until needed and should be kept in the refrigerator (not frozen) in the dark.

The Cephalosporins. *Cephalosporin C,* the first type discovered in this group of antibiotics, is produced by the mold *Cephalosporium acremonium.* Like the penicillins, cephalosporins prevent the synthesis of the cell wall of gram-positive bacteria. They are resistant to penicillinase. Although a relatively ineffective antibiotic, cephalosporin C is of great interest because it has a formula very similar to that of penicillin. Like penicillin, cephalosporin C lends itself to production of a large number of synthetic derivatives. Various acyl groups may be attached to the central residue of cephalosporin, producing much more valuable cephalosporins than cephalosporin C, e.g., *cephalothin.*

Streptomycin. *Streptomycin* interferes with the message of the genetic code and with peptide synthesis. Streptomycin, obtained from *Streptomyces griseus* and discovered by Selman A. Waksman (Nobel Prize winner) (Fig. 14–2), and *dihydrostreptomycin,* a derivative, are effective against human strains of tubercle bacilli. These drugs, now largely superseded by isoniazid (see page 241), are used with rifampin or ethambutol or both as an adjunct to bed rest and other medical and surgical therapy in the treatment of tuberculosis. Streptomycin is also effective against several species of gram-negative bacteria that are unaffected by penicillin: *Brucella* (cause of undulant fever), *Shigella* (dysentery bacilli), and others. It also controls some gram-positive species. Streptomycin should be used in combination with other antibiotics because of rapid development of resistance in the pathogens within 48 hours. Thus, other chemotherapy is now the preferred treatment, except for plague and tularemia.

Chloramphenicol. *Chloramphenicol* (Chloromycetin), discovered by Paul R. Burkholder, is an inhibitor of protein synthesis. It is effective against a large

Table 14–5. Some Properties of Improved Penicillins and Related Antibiotics

	Resistant to		
Generic Name	***Penicillinase***	***Gastric Acid***	**Trade Name***
Methicillin	+	−	Staphcillin
Cloxacillin	+	+	Prostaphlin
Nafcillin	+	+	Nafcil
Ampicillin	−	+	Polycillin
Amoxicillin	−	+	Amoxil
Benzathine-penicillin G	−	†	Bicillin C-R
Penicillin G-K	−	†	Several
Procaine-penicillin G	−	†	Several
Cephalothin Na‡	+	−	Keflin
Cephaloglycin ‡	+	+	Keflex

*Though only one product is named here there are, on the market, several similar products with different names.

†Injected only, never given orally.

‡These are cephalosporins, antibiotics related to the penicillins.

number of bacterial infections, both gram-positive and gram-negative. It has been found to be of particular value in typhoid and paratyphoid infections, as well as in the treatment of some rickettsial and a few chlamydial diseases. There is evidence that it adversely affects the blood-forming organs. Its use should therefore be carefully followed by appropriate daily examinations of the blood.

The Tetracycline Group (Broad-Spectrum Antibiotics). The antibiotics called *Achromycin, Terramycin,* and *Aureomycin* are chemically related. All belong to the tetracycline group of compounds. Their chemical relationship and structures are explained in Figure 14–3*B*. Tetracyclines inhibit protein synthesis by blocking tRNA activity.

Although each is derived from a different species of *Streptomyces,*[5] all tetracylines have similar, but not identical, antibiotic properties. They are effective against many gram-negative and gram-positive species of bacteria, some chlamydias, and some rickettsias. They are typical of the *broad-spectrum antibiotics* and have similar ranges of therapeutic activity. In spite of their similarities, however, in any given patient any one of these drugs may at times show surprising irregularities, giving unexpectedly brilliant results or failing. This is a very important fact to remember about all antibiotics.

The Polymyxins. *Polymyxins,* peptide antibiotics, are active against gram-negative bacteria by inhibition of cell membrane function. They are also used against gram-positive bacterial infections, but their use is limited owing to their toxicity.

Isoniazid. *Isoniazid* is a synthetic bactericidal compound specifically effective against *Mycobacterium tuberculosis.* The reader should note that isoniazid is not an antibiotic (see page 235). In structure it is related to nicotinamide (DPN or NAD), and it affects pyridoxamine (vitamin B_6) metabolism.

Antifungal Antibiotics. *Nystatin* or *fungicidin* (Mycostatin) is of particular interest because it is one of the very few therapeutically useful antibiotics that are effective against pathogenic fungi. It is often combined with broad-spectrum antibiotics.

Amphotericin B is used internally against deep or systemic mycoses. Being very toxic to both fungal and mammalian cells, it must be used only under skilled medical direction. This drug is reasonably safe for *topical* application in superficial mycoses (e.g., candidiasis).

Cycloheximide (Actidione), another antifungal antibiotic, is effective against *Cryptococcus neoformans,* a pathogenic yeast, and many destructive fungi of plants. Cycloheximide is also used in the laboratory to suppress saprophytic fungi that contaminate cultures of bacteria and viruses.

Griseofulvin (Fulvicin) is a fungistatic agent that has proved to be quite effective against superficial mycoses (diseases of the hair, the skin, and the nails caused by molds). Its action is due to nucleic acid interference by the drug.

Ketoconazole is an imidazole derivative that acts on the fungal membrane It is a new, effective broad-spectrum antifungal agent that seems to act where griseofulvin does not meet expectations. It is also orally administered.

Antineoplastic Antibiotics. *Actinomycin A, B, C,* and *D* and *mitomycin A, B,* and *C* are products of *Streptomyces* species. These antibiotics inhibit DNA synthesis not only in bacteria, fungi, and even viruses, but also in mammalian cells. For this reason their use is principally in the selective treatment of certain cancers. Some are dangerously toxic, like amphotericin B.

Resistance R-factors. Resistance to different antimicrobial agents may be due to genes located on plasmids. In gram-negative bacteria these genes may affect exposure to penicillins, streptomycin, chloramphenicol, tetracyclines, and sulfonamides; in gram-positive bacteria they can affect the potency of penicillins, erythromycin, and heavy metals. Genes on these plasmids produce enzymes that inactivate the antibiotics.

---

[5]Tetracycline is now derived commercially mainly by the artificial alteration of the chlortetracycline molecule.

Effects of Excessive or Improper Use of Antibiotics. Antimicrobial substances, although often referred to as "miracle drugs," are not without their disadvantages. Sensitivities or allergies that cause serious reactions to the drugs may develop in human beings. This is especially true of penicillins: several produce serious toxic side effects if their administration is not carefully controlled.

One unfortunate side effect of the administration of antibiotics in large, prolonged dosages is seen in a disturbance of normal host-parasite relationships. Microorganisms that are ordinarily restricted harmlessly to the skin or mucous membranes find an opportunity to set up an infection, possibly because competing microorganisms that usually hold them in check are suppressed by the antibiotic. Fungi, especially *Candida albicans,* often present in the normal intestine or vagina, sometimes cause distressing gastrointestinal or vaginal *superinfections* in such circumstances. For this reason drugs like *amphotericin B* and *5-fluorocytosine* are of great importance. Staphylococci, ordinarily not pathogenic in the intestine, sometimes grow there excessively, producing severe and even fatal enteritis when competing microorganisms of the gut are suppressed by antibiotics in preparation for surgery of the gastrointestinal tract. Antibiotics are not harmless and should not be used without medical supervision.

Several new drugs are being developed that are aimed at organisms that have become resistant to conventional antibiotics. An example of these is *noxalactam* (Moxam), effective in treating meningitis and pneumonia. It appears to be safe, but only time will tell.

Antibiotics should be used only when definitely indicated and not for any and every infection. In some critical cases, e.g., acute meningitis, the doctor may have to make a diagnosis on the basis of the clinical picture before laboratory reports are available. In these cases, the doctor may have to choose the antimicrobial agent most likely to be effective before the causative agent is known. Even then, it is advisable to attempt to find the causative organism rather than to continue blind treatment with antibiotics. It is also essential to know whether the particular strain of organism causing the infection is or is not wholly resistant to the drug chosen for use. This can be determined in the bacteriology laboratory by procedures called "sensitivity testing."

Sensitivity Testing

Tube Dilution Method. In this method the drug to be tested against an infectious organism is added, in a series of dilutions (1:5, 1:10, 1:20, and so on), to test tubes containing an appropriate culture medium for the bacterium under examination. Each tube is then inoculated with the organism whose resistance to the drug is being tested. After incubation of all the tubes, a comparison of the growths is made and the resistance of the organism to each drug is evaluated (Fig. 14–4).

Disk Plate Method. In this procedure, an agar plate with suitable medium is heavily inoculated with the infecting microorganism whose sensitivity is in question. Previously prepared commercially available paper disks, saturated with various concentrations of several antibiotics, are released from an automatic disk dispenser onto the agar surface. The plate is then incubated. The drugs diffuse from the disks into the agar. If the organism is not resistant to the drugs, zones of complete inhibition of the growing organism are found around the disks containing the effective drugs (Fig. 14–5).

Agar Dilution Method. This is similar to the tube dilution method in principle. The antibiotics are diluted, not in tubes of broth but in tubes of melted, cooled (45 C) agar. The agar dilutions are poured into plates. When the agar is solid, the organisms to be tested, perhaps several on a single plate, are streaked on segments of the agar surfaces and the plates are incubated.

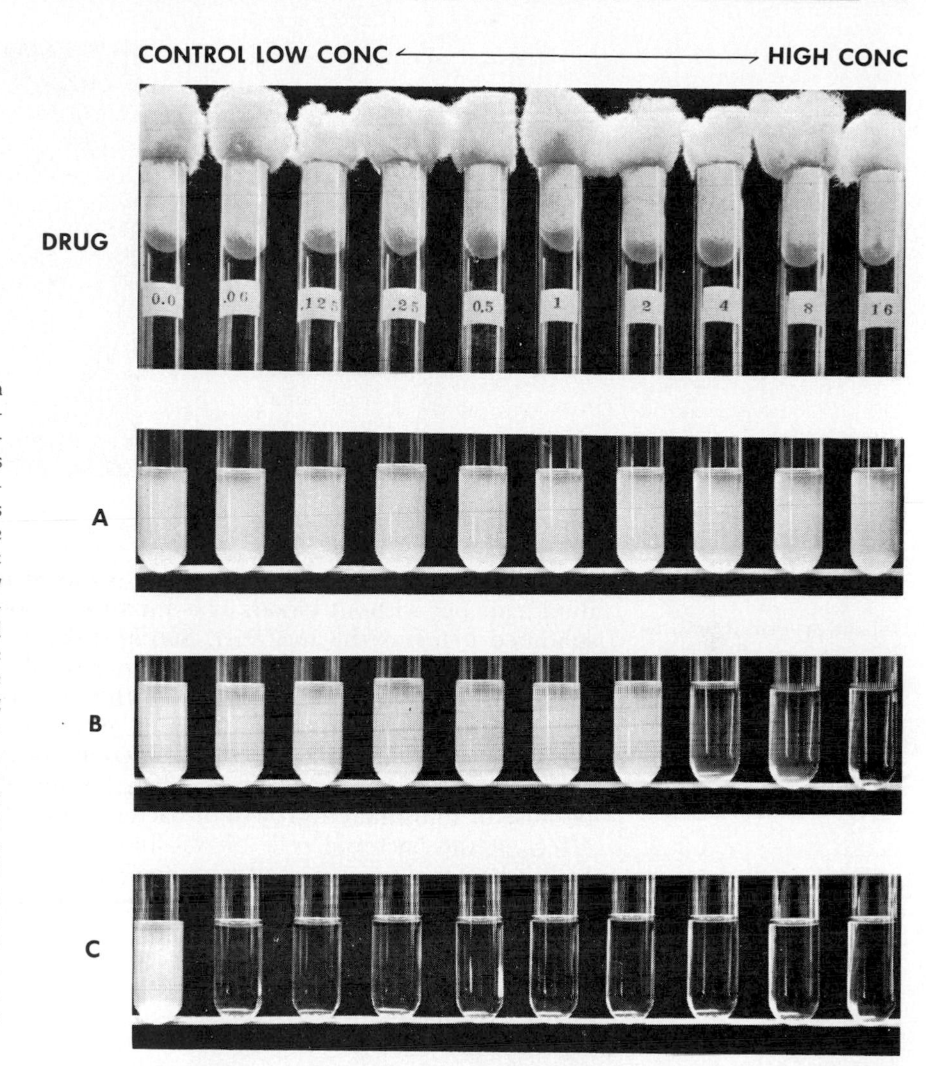

Figure 14–4 Testing sensitivity of a bacterium to antibiotics or other chemotherapeutic agents by the "tube dilution" method. In the top row, tubes of broth culture contain an antimicrobial "Drug" in increasing amounts, as shown by the figures on the tubes. The second row contains "Agent A" in the same amounts, while the third row contains "Agent B" in the same quantities. The bottom row contains graded amounts of "Agent C." All tubes were inoculated with "Organism X" from the same patient and the same culture at the same time, and incubated together. Growth, as shown by white turbidity, has occurred even in the highest concentration of Agent A, showing that Organism X (perhaps from a patient very ill with this infection) is not in the least affected by this drug. The drug of choice will be Agent C, which prevents growth of Organism X even in the lowest concentration. Agent B is slightly effective. The left-hand tube in each row is a "control," containing medium but no antibiotic. (Courtesy of Abbott Laboratories, North Chicago, Ill.)

Effectiveness of an antibiotic is indicated by failure of a bacterial culture to grow (Fig. 14–6).

Simplified Disk Procedure. Sterile filter paper disks are saturated with a suspension of live bacterial spores and a harmless dye, like methylene blue, which is decolorized when reduced by bacterial growth in a closed chamber. The disks are dried. For use, one or more disks are laid on a glass or plastic strip and wetted with a few drops of sterile water. Air is excluded by enclosing the disks in a small, tight glass chamber or covering them with transparent plastic tape. When incubated, the spores germinate in one to three hours, and their growth reduces the methylene blue, which becomes colorless. If the disk is wetted with test fluid that contains enough of an antibiotic to which the organism is sensitive, no growth occurs and the disks remain bright blue. By various experimental adjustments the method may be made roughly quantitative.

Blood-Agar Plate Method. This rapid and effective procedure is representative of several basically similar methods that use sensitive indicators of bacterial growth. In this example, blood is used as the indicator. These procedures depend on the drug's suppressing the discoloring effect of growing bacteria on fresh blood. Blood is mixed with melted agar of suitable nutrient composition at 45 C. This mixture is poured into a Petri dish to a depth of about 2 mm and allowed to solidify. This is the blood or *base layer*. A thick

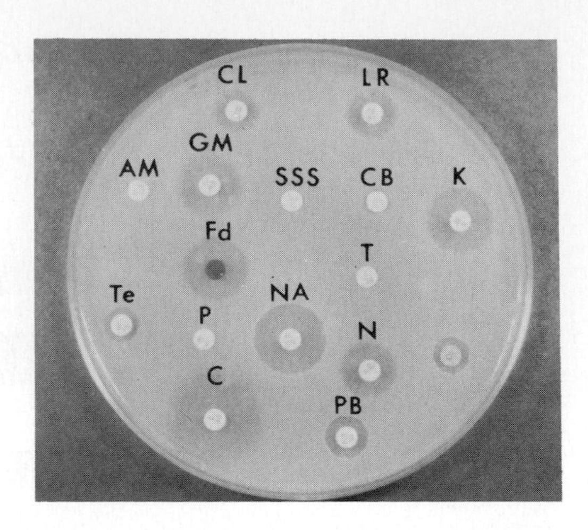

CL – Coly-Mycin
AM – Ampicillin
TE – Terramycin
P – Penicillin
C – Chloramphenicol
PB – Polymyxin B
N – Neomycin
Fd – Nitrofurantoin
T – Tetracycline
K – Kanamycin
LR – Cephaloridine
GM – Garamycin
SSS – Triple Sulfa
CB – Carbenicillin
NA – Nalidixic Acid

Figure 14–5 Testing sensitivity of a bacterium to antibiotics or other chemotherapeutic agents by the "disk method." The entire surface of agar medium in a Petri plate is inoculated with the organism to be tested. Paper disks of uniform thickness containing graded amounts of the agent to be tested (or the same amount of different agents if a comparison is desired) are then placed on the surface of the agar. The agent diffuses into the agar and prevents growth of the bacterium in a zone around the disk. The width of the zone indicates, roughly, the sensitivity of the organism to the agent or agents being tested, though the *presence* or *absence* of a zone is of greater significance. (About one-half actual size.) (Courtesy of Linda Kaye Hickey; work done at the Texas Woman's University Microbiology Research Laboratory.)

suspension of the bacteria to be tested is mixed in a separate tube with similar fluid agar but without blood. This mixture is poured on the base layer. When solidified it forms the *seed layer.* Sterile disks of paper saturated with desired antibiotics are spaced at suitable intervals on the surface of the seed layer, as in the disk plate method. The plate is then incubated.

Examined after two to six hours, zones of bright red, unchanged blood are seen under and around the disk containing the effective antibiotic. Elsewhere the blood is discolored and may be hemolyzed (erythrocytes destroyed) because of uninhibited growth of bacteria. This is a very rapid test because the effect of the bacterial cells on the blood color is readily visible many hours before bacterial growth itself may be seen.

In a simplified adaptation of this principle filter paper pads containing

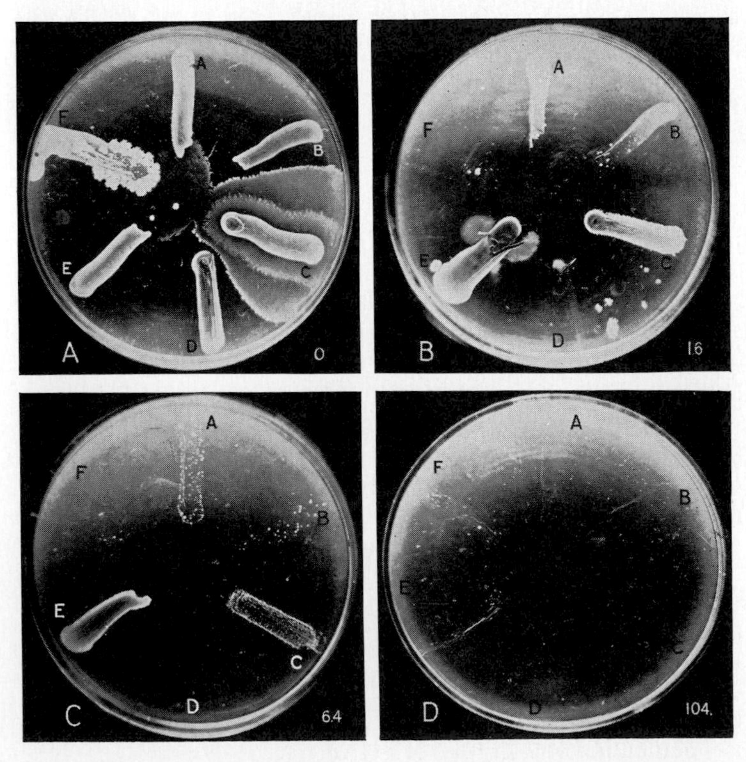

Figure 14–6 Testing sensitivity of several species of bacteria to three different concentrations of streptomycin by the agar dilution method. Plate A contains no antibiotic and is inoculated with *(A) Escherichia coli, (B) Salmonella typhi, (C) Proteus* sp., *(D) Klebsiella pneumoniae, (E) Pseudomonas aeruginosa,* and *(F) Mycobacterium tuberculosis.* In Plate B, containing 1.6 mg of streptomycin per ml of agar, organisms *D* and *F* have been virtually eliminated. Larger amounts of the antibiotic, as shown in Plates C and D, successively eliminate all but a few resistant colonies of the other organisms. (What is the significance of these resistant colonies?) Note the great sensitivity of this strain of *M. tuberculosis* to streptomycin. (Courtesy of Merck & Co., Inc.)

dried medium, an antimicrobial agent, and an indicator dye are inoculated with about 0.5 ml of an aqueous suspension of the microorganisms under investigation. Growth (or inhibition of growth) is indicated after a few hours of incubation by change (or no change) in the color of the indicator dye.

Bacteriocins

Bacteriocins are bactericidal glycolipoproteins, lipopolysaccharide-protein complexes, or related proteins, produced by certain species of bacteria. They exert their lethal action against other bacteria of the same family, but not of the same species. *Colicins* produced by *Escherichia coli* are one type of bacteriocin. Many other organisms produce these apparently specific antibiotic substances—for example, *megacins* from *Bacillus megaterium* and *pyocins* from *Pseudomonas aeruginosa.* Bacteriocins resemble bacteriophages in many respects. They may be incomplete bacteriophages. They constitute a type of antibiotic agent that may become important in the treatment of certain diseases when we understand more about their structure as opposed to the genetic code of the pathogen and host cells. Bacteriocin production is controlled by an episome, which is a plasmid containing a Col factor, analogous to the F or fertility factor.

SUMMARY

Microorganisms may be destroyed or removed before infection through the use of disinfectants, or they may be killed or inactivated after infection by antimicrobial chemotherapeutic agents, also referred to as antibiotics. It is clearly better to prevent disease than to have to treat it. However, we are fortunate to have now an impressive number of drugs that will stop the growth of specific microorganisms without causing major injury to the host. Even if some injury occurs, it represents a fair trade-off for the benefit derived. We are always seeking selective toxicity: destruction of the pathogenic organism without damage to the cells of the infected host. So far, only penicillins and cephalosporins accomplish this.

Two frequent problems are (1) the development of drug-resistant microorganisms that diminish the value of effective antibiotics with time, and (2) severe allergic responses of patients to certain antibiotics.

Antimicrobial chemotherapeutic agents may be antibacterial, antifungal, antiviral, or (most desirably) broad-spectrum antibiotics, destroying not only bacteria, but also rickettsias, chlamydias, and even some pathogenic viruses.

The first effective antibiotics developed were wholly synthetic. They were the sulfonamides, often called "sulfa drugs." Their antibacterial activity is easily explained by their structural resemblance to para-aminobenzoic acid, a constituent of the folic acid molecule, which sulfa drugs replace, thus inhibiting bacteria that need folic acid (a B vitamin) for their metabolism.

Most antibiotics are produced in culture media by the growth of certain microorganisms, principally species of *Streptomyces, Bacillus,* and molds of the genera *Penicillium* and *Aspergillus.* After chemists establish the structure of these antimicrobial agents, the antibiotics are often synthetically prepared in the laboratory, and more effective derivatives can be made and tested against specific disease-causing agents.

The action of antibiotics varies greatly and depends on methods of administration, dosage, toxicity to the patient, resistance of the pathogen, and often synergism or antagonism with other drugs. Of course, the severity of the disease is a major factor in the effectiveness of treatment.

All penicillins and cephalosporins are bactericidal inhibitors of cell wall synthesis of gram-positive organisms. Because penicillins are highly effective and the least toxic antimicrobial agents known, they are used in preference to other antibiotics whenever possible. Penicillinase (β-lactamase) and other enzymes are often produced in antibiotic-resistant bacteria under the control of genes located on plasmids, circular pieces of DNA inside the bacteria.

Other highly useful antimicrobial agents are streptomycin and other aminoglycosides, chloramphenicol, the tetracyclines (broad-spectrum antibiotics such as Aureomycin, Achromycin, Terramycin, and doxycycline), and numerous other life-saving medications listed in Tables 14–1, 14–2, and 14–3. Improved penicillins are shown in Table 14–5.

Although they are not antibiotics, bacteriocins (colicins, etc.) are bactericidal glycolipoproteins produced by some bacteria and destructive to others.

REVIEW QUESTIONS

1. What is an antimicrobial agent? How would you differentiate between an antiviral agent and an antibacterial agent?
2. Why should we prefer the term "antimicrobial chemotherapeutic agent" to "antibiotic agent"? Should we use the term "antimicrobiotics"? Explain!
3. What is selective toxicity? How is it of benefit to the host?
4. Differentiate between disinfectants, chemotherapeutic drugs, and antibiotics.
5. What groups of chemicals were the first effective antibiotics? What metabolites are they structurally related to?
6. What other reasons can you give for the modes of actions of different antimicrobial agents? Cite specific examples.
7. Why are many bacteriostatic chemotherapeutic agents as effective in fighting disease as bactericidal agents?
8. What are some of the factors to be aware of when an antibiotic is administered to a patient?
9. List several "improved" penicillins. How are they related to penicillin G? Why can some of them be given by mouth whereas others have to be injected?
10. Who was Sir Alexander Fleming? Selman A. Waksman?
11. What is drug fastness? Which organism develops this property? When? How can a nurse help prevent drug resistance?
12. How are most new antibiotics obtained?
13. What are the tetracyclines?
14. Define the followings: penicillinase, R-factors, plasmids, cephalosporins, broad-spectrum antibiotics, isoniazid, synergism, chloramphenicol.
15. List some antifungal antibiotics.
16. What are antineoplastic antibiotics?
17. Explain some simple methods used for testing sensitivity of microorganisms to antibiotic drugs. Which one would you select to be used first if you had an infection?
18. What are bacteriocins?

Supplementary Reading

Basch, H., Erickson, R., and Gadebusch, H.: Epicillin: In vitro laboratory studies. *Inf. Immun., 4*:44, 1971.

Bauer, D. J.: The Specific Treatment of Virus Diseases. Baltimore, University Park Press, 1977.

Becker, I.: Antiviral Drugs: Mode of Action and Chemotherapy of Viral Infections of Man. New York, S. Karger, 1976.

Bodey, G. P., and Stewart, D.: In vitro studies of semisynthetic α-(substituted-ureido) penicillins. *Appl. Microbiol., 21*:710, 1971.

Braude, A. I.: Antimicrobial Drug Therapy. Vol. VIII in the series Major Problems in Internal Medicine. Philadelphia, W. B. Saunders Company, 1976.

Braude, A. I. Associated Editors: Davis, C. E., and Fierer, J.: Microbiology. Philadelphia, W. B. Saunders Company, 1982.

Brown, M. R. W. (ed.): Resistance of *Pseudomonas aeruginosa.* New York, John Wiley & Sons, 1975.

Dulaney, E. L., and Laskin, A. I. (eds.): The problems of drug-resistant pathogenic bacteria. *Ann. N.Y. Acad. Sci., 182*:1, 1971.

Falconer, M. W., Schram Ezell, A., Patterson, H. R., and Gustafson, E. A.: The Drug, the Nurse, the Patient. 6th ed. Philadelphia. W. B. Saunders Company, 1978.

Freitag, J. J., and Miller, L. W. (eds.): Manual of Medical Therapeutics. 23rd ed. Boston, Little, Brown, and Co., 1980.

Hash, J. H. (ed.): Methods in Enzymology. Vol. 43, Antibiotics. New York, Academic Press, 1975.

Hook, E. W., Mandell, G. L., Gwaltney, J. M., and Sande, M. A. (eds.): Current Concepts of Infectious Disease. New York, John Wiley & Sons, 1977.

Jawetz, E., Melnick, J. L., and Adelberg, E. A.: Review of Medical Microbiology. 14th ed. Los Altos, Calif., Lange Medical Publications, 1980.

McHenry, M. C., and Cerat, G. A.: Bacteremia (Septicemia). *In* Conn, H. F. (ed.): Current Therapy 1977. Philadelphia, W. B. Saunders Company, 1977.

Mitsuhashi, S. (ed.): Drug Action and Drug Resistance in Bacteria. Baltimore, University Park Press, 1971–1975.

Patterson, H. R., Gustafson, E. A., and Sheridan, E.: Falconer's Current Drug Handbook 1980–1982. Philadelphia, W. B. Saunders Company, 1980.

Riva, S., and Silvestri, L. G.: Rifamycins: A general review. *Am. Rev. Microbiol., 26*:199, 1972.

Simon, H. J., and Yin, E. J.: Microbioassay of antimicrobial agents. *Appl. Microbiol., 19*:573, 1970.

Skinner, F. A., and Hugo, W. B. (eds.): Inhibition and Inactivation of Vegetative Microbes. New York, Academic Press, 1976.

Zähner, H., and Mass, W. K.: Biology of Antibiotics. New York, Springer-Verlag, 1972.

STERILIZATION AND DISINFECTION IN HEALTH CARE

In this chapter we shall describe some procedures that apply the principles underlying disinfection and sterilization to the practice of nursing and related medical activities.

IN THE MEDICAL AND SURGICAL WARDS OF A GENERAL HOSPITAL

Handwashing. All health personnel should wash their hands carefully after the care of each patient, before serving food, before preparing and dispensing medicines, before doing each surgical dressing, after dressing an infected wound, after handling bedpans and urinals, and before going to the dining room for their meals. Do not forget that all persons should wash their hands *every time* after they use a restroom.

The manner in which handwashing is done is far more important than the time taken for the procedure. Adequate soaping, care in washing all areas of hands, mechanical friction, and frequent rinsing and resoaping all help to prevent the transfer of microorganisms by means of hands. Nurses sometimes say that they "do not have time to wash their hands." These same nurses would not question the time required to scrub their hands before assisting with a surgical operation.

Thermometer Technique. There seems to be little justification for attempting to take temperatures on a 30-bed ward with fewer thermometers than there are patients, but unfortunately this is sometimes necessary. Whether thermometers are left at the bedside or kept in a central location on each ward depends on availability and the individual organization of nursing care. Let us not forget that what may seem simple and common at one hospital may not really be so at others. Would you put two patients into one bed? A former nursing student wrote back to us that this is exactly what she encountered when she returned to her own country.

Oral Thermometers. A recommended procedure for disinfecting oral thermometers is: wipe the thermometer clean with gauze or cotton saturated with a mixture of equal parts of tincture of green soap and 95 per cent ethyl alcohol, rinse well with clear water, and immerse the entire thermometer in a solution of 0.5 to 1 per cent iodine in 70 per cent ethyl or isopropyl ("rubbing") alcohol for 10 minutes. Plain 70 per cent alcohol may be used with good results, but it is not as effective without the iodine. Possibly the organic iodine disinfectants referred to previously would be as effective, though further experience with them is needed. Formaldehyde, bichloride, and phenolic disinfectants should not be used because they have been shown to be less effective against tubercle bacilli. Tinctures of quaternaries appear to be as good as the alcoholic iodine solutions and are effective against tubercle bacilli. After disinfection, thermometers should be thoroughly rinsed. They may then be kept in a clean, dry container until used again. They are not infectious but are not necessarily sterile (why?). Improved, electronically registering thermometers have disposable plastic covers over the bulb. These types of thermometers (Fig.

CHAPTER 15

15–1), now in common use in modern hospitals, come with two different probes, one oral, the other rectal. The disposable covers over these probes protect the patients.

Rectal Thermometers. These should be lubricated before use with a water-soluble lubricant. Petrolatum or other oily lubricants are difficult or impossible to remove with soap and cool water. After use, rectal thermometers can be cleaned and disinfected in the same way that oral thermometers are treated. Petrolatum and oils entirely prevent proper cleaning and disinfection. Where facilities are available, well wiped thermometers may be sterilized with germicidal vapors.

Syringes and Needles. These articles are used for intramuscular or intravenous injection or for withdrawal of blood. Most of the syringes now in use are presterilized, nontoxic, nonpyrogenic, *disposable plastic syringes.* They are quite inexpensive, come in all sizes that are in common use, and are supplied either with attached needles or with separate needles individually packaged for easy attachment. These presterilized syringes and needles are safe and are never used more than once. This also assures that the needles are sharp, quite a relief to anyone who ever was injected with a poorly "resharpened" needle having a hook at the tip. The older type glass syringes were frequently boiled rather than autoclaved or sterilized in a hot air oven. Boiling is not the method of choice, since spores and some viruses, especially those of epidemic hepatitis and of homologous serum jaundice, may resist boiling for indefinite periods.

All syringes should be handled carefully so that the plunger and the inside of the barrel remain sterile. This seems obvious, yet persons have been seen to touch with their fingers the part of the syringe that is pushed back in and thus comes in direct contact with the solution to be injected. In mounting a separately sterilized needle on a syringe, the sterile needle is carefully held by the shank with sterile forceps or, better, placed mechanically on the syringe without handling. The shaft and point of the needle are preferably in a sterile tube. It is not good technique to moisten sterile gauze with a disinfectant, handle it with fingers (depositing numerous organisms from the hand), and then place this moist gauze over the sterile needle.

A small bottle of skin disinfectant (such as 1 per cent iodine in 70 per cent alcohol) should be carried, with the loaded syringe and the properly mounted and protected needle, to the patient. The sterile syringe and needle need no disinfectant if properly protected with a sterile glass tube or gauze.

An excellent method of handling sharp instruments that cannot be sterilized by heat is to mount them on convenient racks in jars containing a disinfecting solution such as the Bard-Parker germicide (Parker, White and Heyl, Inc., Danbury, Conn.): isopropyl alcohol, 65.26 per cent; methyl alcohol, 2.75 per cent; formaldehyde, 8 per cent; hexachlorophene, 0.5 per cent; water and various other components, 23.49 per cent (patented). This solution is said to kill spores in about three to five hours.

Surgical Dressings. Dressings from wounds, whether aseptic or infected, should never be removed by hand because organisms from the lesion may easily have penetrated the outer layer of the gauze. These dressings should be

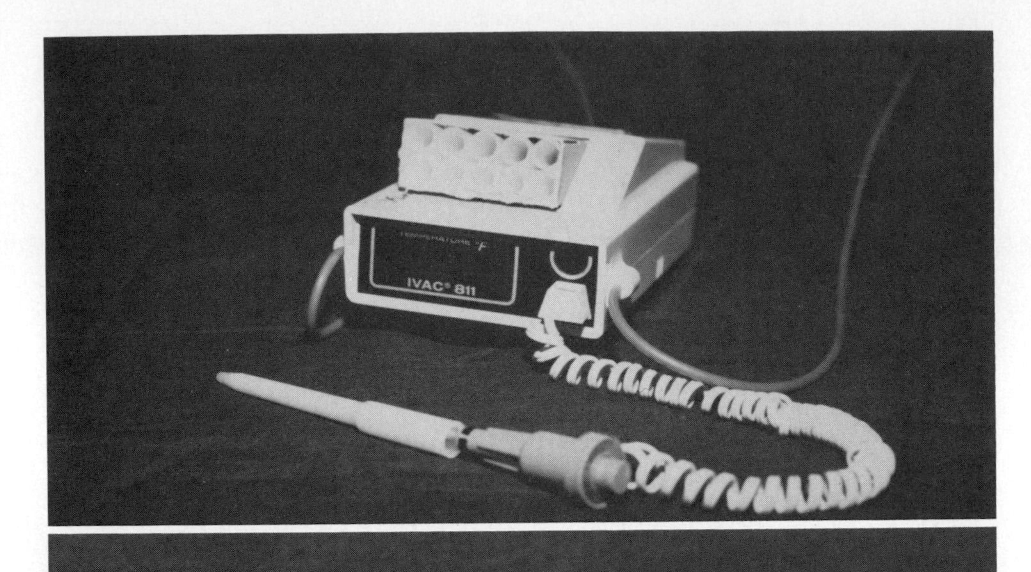

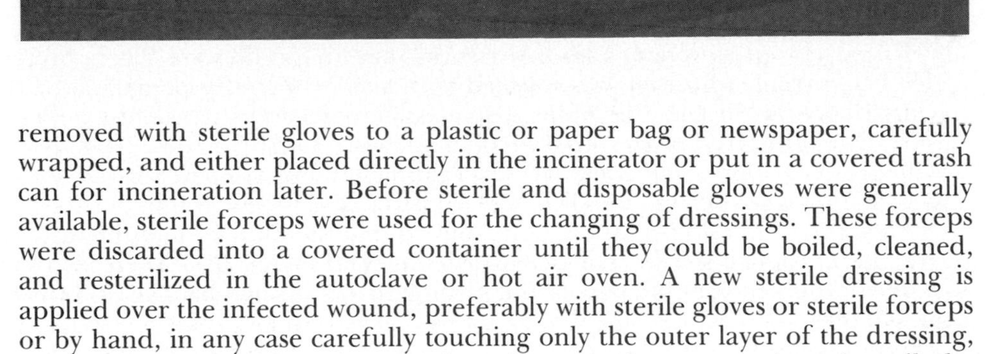

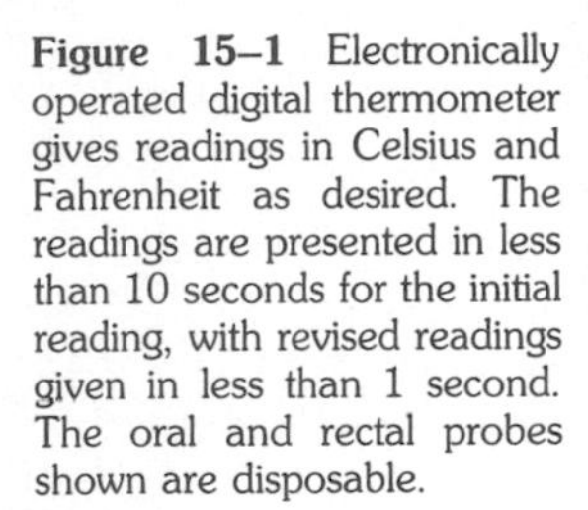

Figure 15–1 Electronically operated digital thermometer gives readings in Celsius and Fahrenheit as desired. The readings are presented in less than 10 seconds for the initial reading, with revised readings given in less than 1 second. The oral and rectal probes shown are disposable.

removed with sterile gloves to a plastic or paper bag or newspaper, carefully wrapped, and either placed directly in the incinerator or put in a covered trash can for incineration later. Before sterile and disposable gloves were generally available, sterile forceps were used for the changing of dressings. These forceps were discarded into a covered container until they could be boiled, cleaned, and resterilized in the autoclave or hot air oven. A new sterile dressing is applied over the infected wound, preferably with sterile gloves or sterile forceps or by hand, in any case carefully touching only the outer layer of the dressing, thus keeping the layer that touches the wound uncontaminated until the moment of contact.

Dressings removed from clean (noninfected) wounds may be removed by touching only the outer layers of the gauze. The health worker's hands should never touch any cut, incision, or wound without sterile gloves because of the possibility of infecting the wound or the health worker's hands.

The Dressing Cart. An instrument that was used extensively before the general use of sterile gloves, always found on a surgical dressing cart (Fig. 15–2) or tray, was the transfer forceps. Used correctly, this instrument greatly increased the efficiency of the surgical dressing procedure. If used incorrectly, it became a source of gross contamination. The transfer forceps is a sterile instrument and was used only to transfer sterile instruments or dressings from one sterile field to another. If at any time this instrument touched unsterile surfaces or objects, it had to be discarded until resterilized. The transfer forceps was kept in a tall jar with enough chemical disinfectant present in the jar to cover adequately any part of the instrument that came in contact with sterile instruments or dressings. Sterile gloves have replaced transfer forceps, but even more widespread is the use of wrapped sterile instruments, gauze pads,

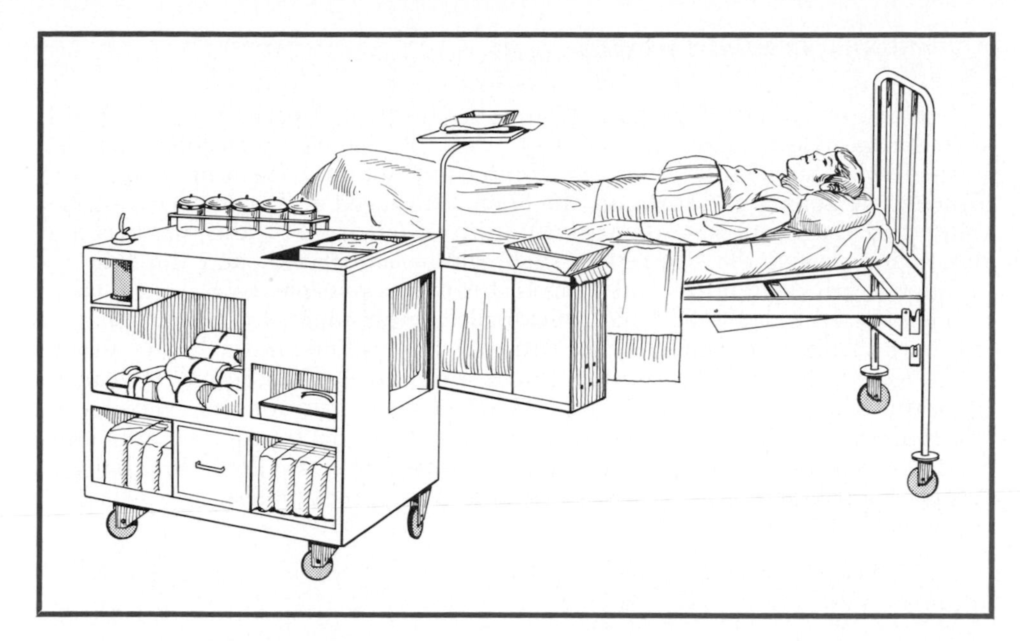

Figure 15–2 A surgical dressing cart for use at patient's bedside. Note the wide-mouthed opener fitted over the paper bag to receive soiled dressings. (Why?) Note the tray that can be held above the patient to provide a sterile field for instruments and dressings. Soiled instruments are discarded into the instrument pan. They are later decontaminated, cleaned, and resterilized. There should be no confusion between sterile and unsterile equipment. (Redrawn from Walter, C. W.: The Aseptic Treatment of Wounds. New York, Macmillan Co., 1948.)

bandages, and so on, which greatly facilitate the work of the health worker. Disposable materials have simplified many procedures and certainly are safer to use. However, they have also contributed to the higher cost of medical care.

Catheterization. The equipment for catheterization (removal of urine from the urinary bladder) is sterilized in the autoclave. The hands are washed thoroughly immediately before carrying out the procedure of catheterization, and sterile gloves are always worn. The external meatus is washed gently with a mild disinfectant solution. This mechanically removes a large percentage of the organisms that may be present; it does not sterilize. The rubber or plastic catheter (glass and metal catheters are no longer used) is adequately lubricated with a sterile, water-soluble lubricant and then gently inserted into the urethra. Anything that may traumatize these delicate tissues will increase the possibility of infection.

Reverse Precautions. In certain medical conditions the doctor will request to have his patient put on "reverse precautions." This order indicates that the doctor wants special precautions taken to protect the *patient* from any kind of infection. He will order reverse precautions for a premature baby, a patient with anuria who has a transplanted kidney, a patient with severe and extensive burns, or a debilitated patient. Patients on steroids (e.g., patients who are receiving tissue or organ transplants) are more susceptible to infections. In brief, reverse precautions will be ordered for any patient whose resistance is low and who is very susceptible to infection.

To someone who understands what needs to be accomplished, the procedures for establishing this routine become obvious. All personnel wash their hands thoroughly before entering the room. Gowns, masks, and caps are worn by everyone who comes in contact with the patient. Visitors are restricted; no one who has the slightest infection is permitted to be in contact with the patient. Everything that enters the room is either sterile or scrupulously clean. Cleaning and dusting are done with a vacuum cleaner or with moist, disinfectant-soaked mops and cloths, to avoid stirring up and disseminating the dust.

Modern techniques sometimes enclose the patient and his bed and equipment entirely within a plastic cubicle or "bubble." Access to the patient is via built-in plastic arm covers, air tubes, telephones, and so on. Such procedures have been used in heart and other organ or tissue transplant work.

TERMINAL DISINFECTION

After any patient is discharged from the hospital, especially a patient with a communicable disease, the room or ward unit must be cleaned, the bed washed with soap and water or disinfectant solution, bedside equipment sanitized by boiling and cleaning, the linen autoclaved and/or washed with hot water and detergent and then ironed with steam ironers, the mattress and pillow autoclaved, the blankets washed and/or autoclaved, and the unit re-equipped with clean linens and utensils. Insofar as possible, each private room and each ward unit should have individualized equipment. In hospitals that do not have facilities for autoclaving mattresses and pillows, these articles should be protected with special covers that can be removed and laundered (or cleaned) after the discharge of each patient. There is no excuse for the exchange of undisinfected blankets between patients. Much of the transfer of antibiotic-resistant staphylococci in hospitals, now recognized as a constant serious danger, has been attributed to carelessness in taking these precautions.

IN THE OPERATING ROOM

All objects or substances that are to come in contact with a surgical wound must be sterile. Instruments, drapes, gowns, and gloves are all sterilized in the autoclave or the hot air oven. A few instruments used in surgery would be ruined by heat sterilization: chemical disinfection must therefore be used (see sections on ethylene oxide and beta-propiolactone, pages 211 to 213). Notable in this group are delicate instruments for surgery of the eye. Some sutures in nonboilable packages may not be subjected to heat. These are disinfected with gas, radiation, or a chemical solution (e.g., phenol, 2 per cent; ethyl alcohol, 70 to 75 per cent). The package is opened in a sterile towel, and the sutures are transferred to the sterile field with sterile forceps.

In preparing for a surgical operation or for a delivery, the nurse must thoroughly clean all instruments; open instruments that are hinged so that all surfaces will have contact with the gas, steam, or dry heat; and sterilize by using gas, the autoclave, or the hot air oven. The nurse is responsible for the sterilization of linens and dressings and for her own technique. She carefully covers her hair with a cap so that no hair or dandruff escapes, she covers her nose and mouth with a gauze mask to prevent infecting the patient, and she scrubs her hands and forearms with soap for 10 minutes, being sure to wash each area thoroughly. Soap containing hexachlorophene (3 per cent) may still be used routinely in many hospitals. Its use is now restricted to special situations (see warning on page 224).

Some hospitals have omitted the use of the scrub brush because of excessive irritation and depend entirely on friction produced by gauze or by the hands. Following the soap and water scrub, all soap should be thoroughly rinsed from the hands and forearms with clear running water. If hexachlorophene was not used, the hands and arms may be immersed in a mild disinfectant solution (e.g., 70 per cent ethyl alcohol, or aqueous Zephiran, 1:1,000). The nurse puts on a sterile gown and sterile rubber gloves and is ready to assist in setting up instruments and linens. She dresses the surgeons and helps with the operation as needed. She must be alert at all times to maintain sterility on the sterile field; she must guard against contamination and report to the surgeon any breaks in technique not observed by other members of the operating team (Figs. 15–3 to 15–7).

The patient is prepared by thoroughly washing the skin of the operative area with soap and water, shaving this area to remove all hairs, and applying ether, acetone, or other solvent to remove fatty skin secretions. In the operating

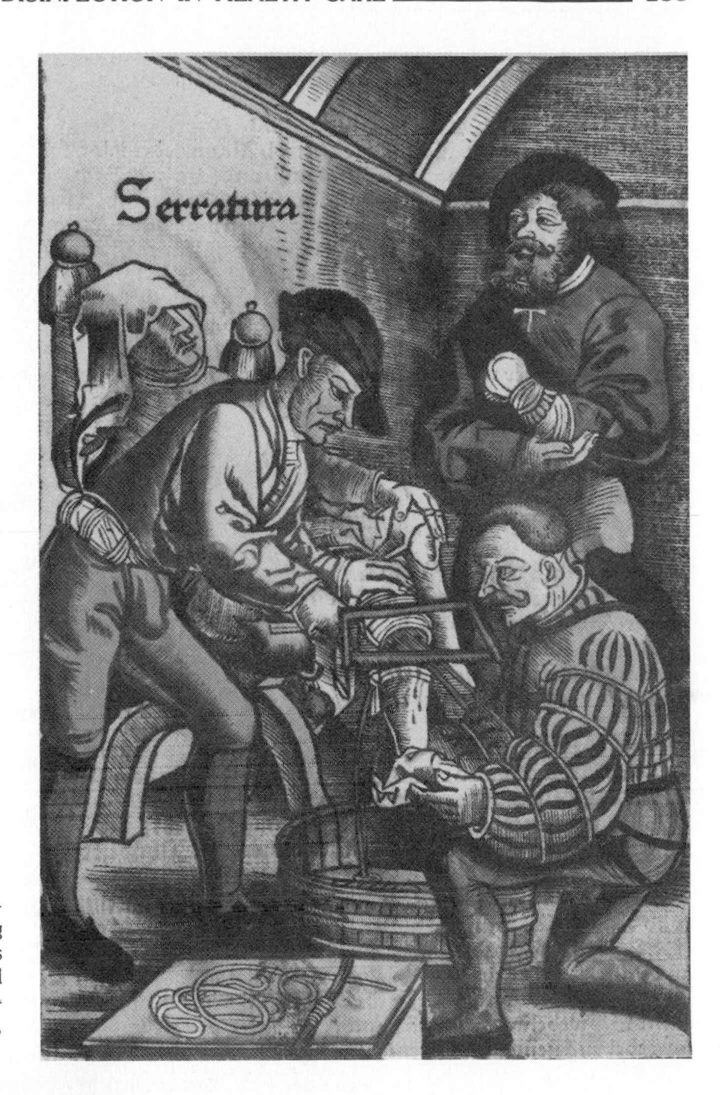

Figure 15–3 First illustration of amputation. Bandages constrict, at most, superficial veins. Two arteries are spurting uncontrolled. Man in background is wearing Gersdorf's pig's bladder dressing over his forearm stump. Bacteria had a field day under such conditions! (From Hans von Gersdorf's Feldtbuch der Wundt Artzney, Frankfurt, 1529. Trent Collection, Duke University Medical Center Library.)

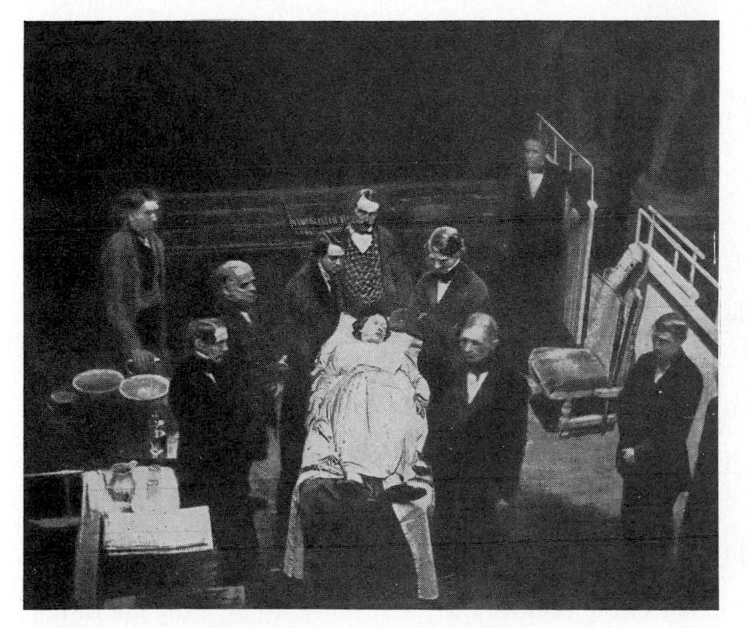

Figure 15–4 How an operation was performed in preantiseptic days. The frock coats were often hung on pegs in the operating room and used by successive visitors and surgeons at different (often infectious) operations! This patient had a tumor of the jaw. The picture is of historical interest: it shows one of the first public demonstrations of ether anesthesia in the Massachusetts General Hospital, October 16, 1846. (From Warren, J. C.: The Influence of Anesthesia on the Surgery of the Nineteenth Century. Boston, Merrymount Press, 1906.)

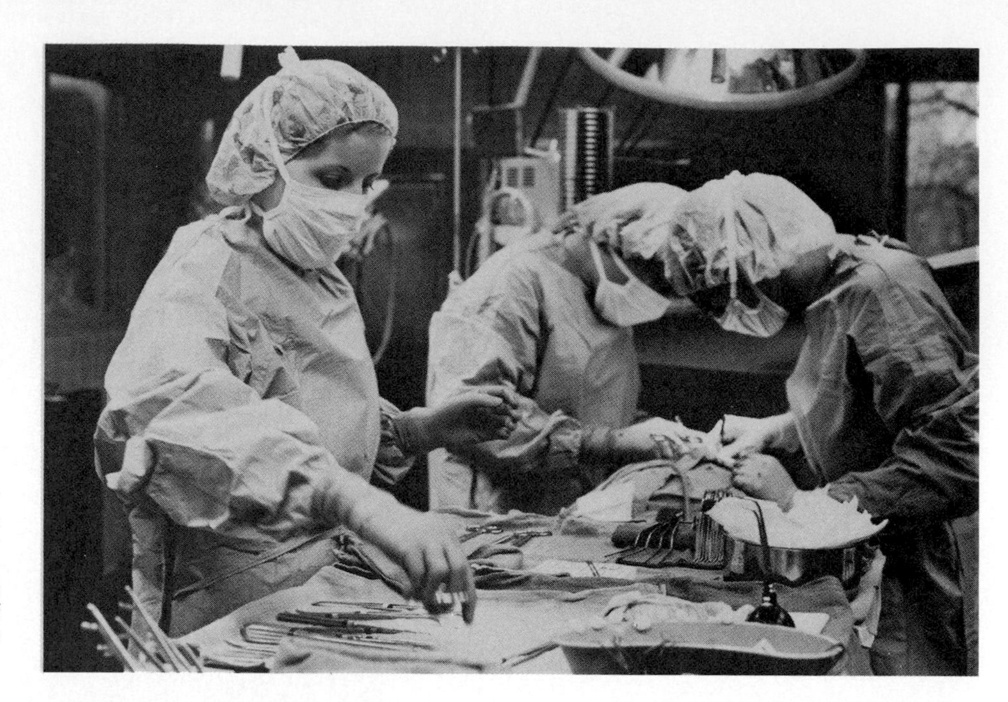

Figure 15–5 Modern surgery. The masks, caps, gowns, rubber gloves, sheets, and instruments have all been sterilized. (Courtesy of Loyola University Medical Center of Chicago, Foster G. McGaw Hospital.)

room, the area is again washed well with a detergent solution for 10 minutes, or with soap and water. Then an antiseptic agent is applied (e.g., Betadine solution, organic mercurial disinfectant, or organic iodine disinfectant). It is important to remember that the cleaning of an operative area of the skin is probably more effective in the removal of microorganisms than the chemical disinfectant is in killing them. The actual cleaning process removes not only microorganisms but also superficial organic material, improving the penetration of the disinfectant.

As is evident, no attempt has been made to cover in detail the field of operating room and delivery room technique. Everyone who works in any capacity in the operating room must have a thorough understanding of the principles of microbiology relevant to the transfer and control of microorganisms.

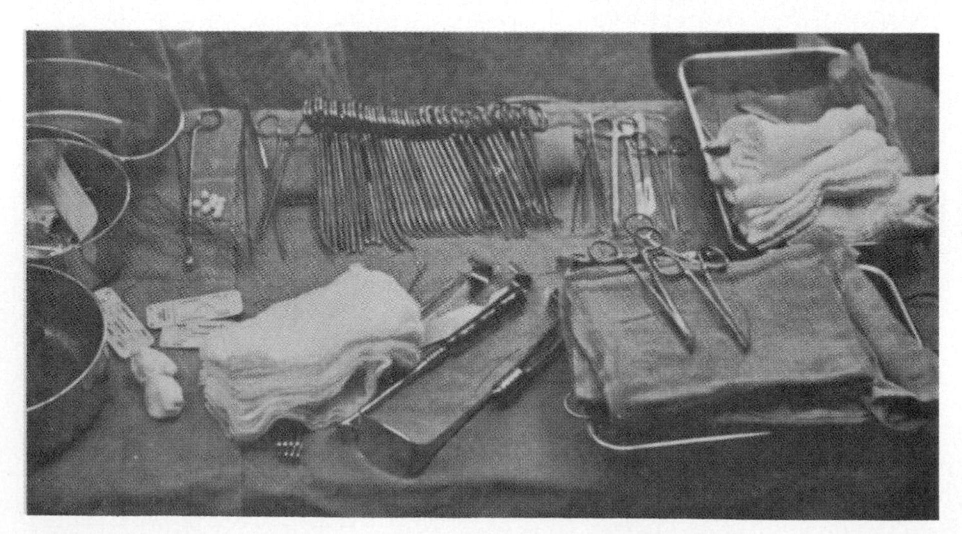

Figure 15–6 The instrument reserve table. It contains most of the instruments and equipment that will be used as the operation proceeds. The drapes and instruments are all sterilized. (From LeMaitre, G. D., and Finnegan, J. A.: The Patient in Surgery. 4th ed. Philadelphia, W. B. Saunders Co., 1980.)

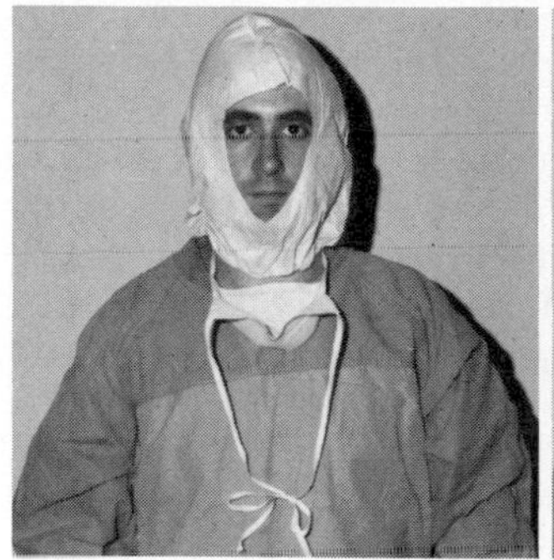
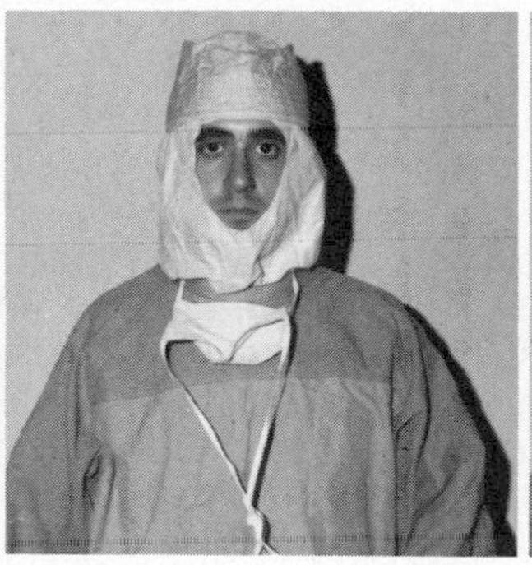
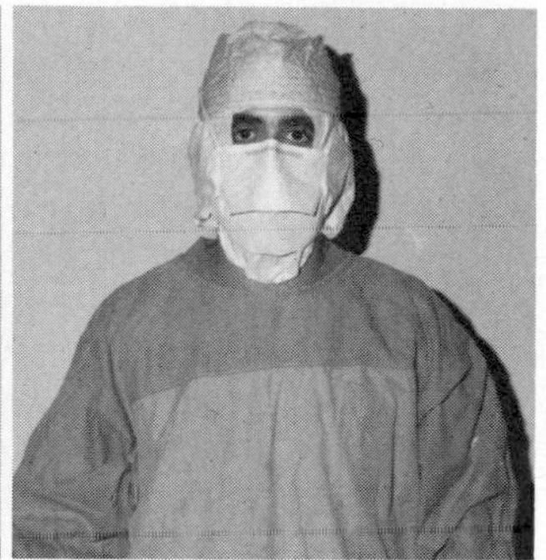

Figure 15–7 Modern surgical face masks. These masks are designed to accommodate present-day hair styles and beards. (Courtesy of Jan F. Fuerst, M.D., Harbor General Hospital, Torrance, California.)

IN THE COMMUNICABLE DISEASE UNIT OR HOSPITAL

When a patient has a communicable disease, the microbiological problems are mainly those of keeping the infecting organisms confined to that patient and to his unit or cubicle, and of preventing transfer of pathogenic organisms from the patient to other patients, health personnel, visitors, or the outside world.

Bedside Equipment. It is essential that the same bedside equipment (such as bedpans and thermometers) be used for the entire infectious phase of the disease. Equipment used only occasionally must be adequately disinfected or sterilized between patients. There is rarely an excuse for using stethoscopes, otoscopes, and other instruments for examining successive patients without disinfecting these objects after use. Adequate immersion in a disinfectant, or boiling, is far more effective than merely wiping with 70 per cent isopropyl or ethyl alcohol; however, not all instruments can be boiled or immersed. Thorough wiping with saponated cresol solution (0.5 to 1 per cent) followed by a wipe with clean water and drying is a useful method for handling such instruments, though the cresol may impart an odor to hands and instruments and leave an irritating residue. Gas sterilization is recommended when applicable.

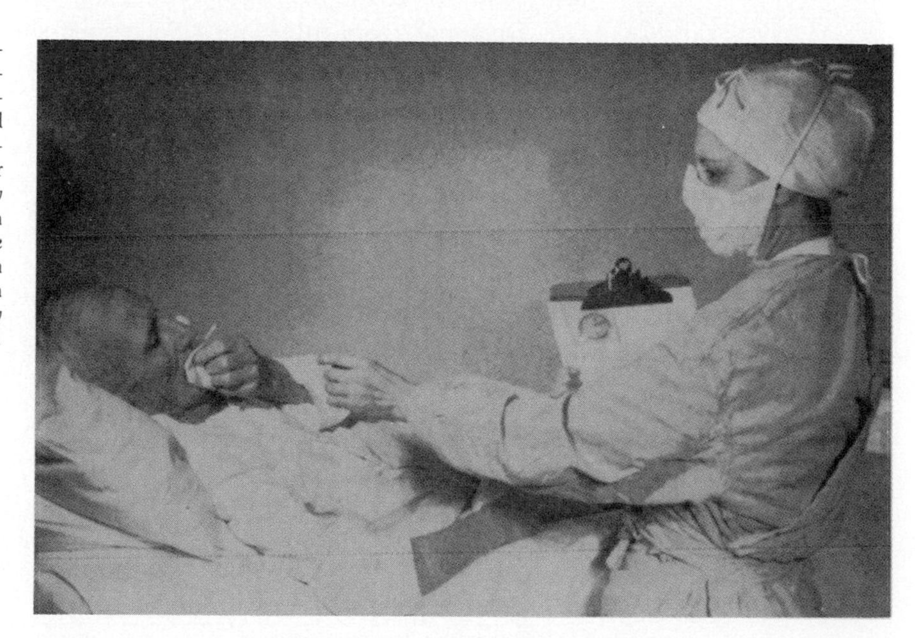

Figure 15–8 Nurse receiving a thermometer from a patient who has a transmissible respiratory infection (e.g., tuberculosis). Note the nurse's cap, gown, and mask. Patient has wiped saliva and sputum from the thermometer with paper wipe so that thermometer can be easily read and nurse does not need to touch contaminated end of thermometer. The patient is ready to protect the nurse from a possible cough by covering his mouth and nose with a paper wipe. (Courtesy of Esta H. McNett: *Am. J. Nursing,* Vol. 49, No. 1.)

Dishes. The dishes of *all* patients, and certainly those of any patient who has a communicable disease, should at least be scraped to removed uneaten food, boiled 5 to 15 minutes to destroy pathogenic organisms, and washed for the next use. This is of course best done with the electric dishwasher. The food may be contaminated and should also be burned or disinfected before disposal as garbage. Other waste from patients' trays should also be burned. Some communicable disease units and hospitals have adopted attractive paper or plastic plates, cups, and other dishes and trays to eliminate the time-consuming problem of disinfecting these items. The paper dishes are burned and only the metal-ware, if not also disposable, needs disinfection and washing. Patients with diseases caused by spore-forming bacteria or hepatitis viruses require special care.

Secretions and Excretions. If a disease-causing microorganism is known to be transferred by respiratory secretions, paper tissue is used by the patient to receive the secretions (Fig. 15–8), and then the wipes are burned. If the respiratory secretions are copious, as in tuberculosis, they are received into a covered watertight container and the entire container is carefully wrapped and burned. In some instances, sawdust, fine newspaper strips, or another absorbent material may be used in the container to absorb the excess moisture. If a container other than a plastic liner is used (glass or metal), it and its contents are immersed in 5 per cent saponated solution of cresol or strong chlorine bleaching solution for at least one hour. The containers may then be washed and reused. Receptacles to receive specimens for microscopic examinations should be *new* (why?).

Figure 15–9 A bedpan washer for use in a patient floor utility room or a central cleaning area. The door is opened by depressing the pedal, and the bedpan is placed in a rack mounted on the inside of the door. The cleaning cycle is automatic. (Courtesy of Sybron Corporation, Rochester, N.Y.)

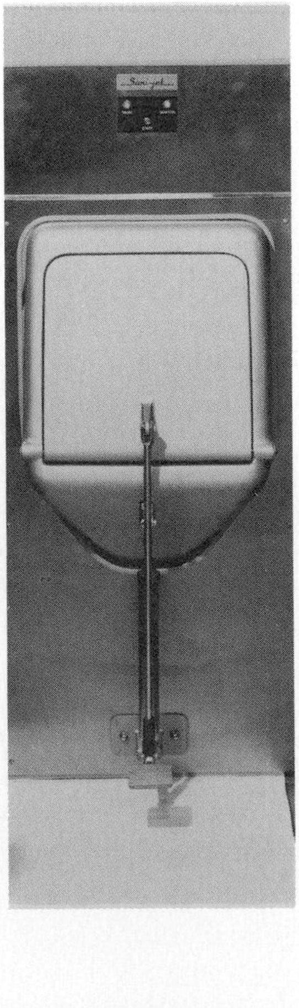

Most modern hospitals use disposable plastic bags inside bedside trash cans and also for large garbage cans.

Urinary or intestinal excretions from a patient with a communicable disease may be disinfected in a tightly covered pail (so that insects cannot touch the infectious matter) by contact with chlorinated lime (5 per cent), strong chlorine bleaching solution, or 5 per cent saponated solution of cresol for at least one hour. The contents should be disposed of in a sewerage system if available. The bedpan or urinal should then be autoclaved or at least boiled for 5 to 15 minutes so that the pathogenic organisms will be destroyed. A very convenient device, available in many hospitals, is the bedpan flusher. This washes the pan and then throws jets of steam upon it, which disinfect if allowed sufficient time for action *but do not sterilize* (Figs. 15–9 and 15–10).

Dressings from infectious skin lesions should be burned. Bed and body linens used by patients who have communicable diseases are probably adequately disinfected by processing through routine laundry procedures if the water is hot enough and sufficient contact is allowed. If there is any question about the adequacy of linen disinfection, it should be soaked in saponated cresol (2 per cent) or laundry bleach solution overnight, then autoclaved or boiled.

Other applications of the principles of disinfection and sterilization will be apparent as the health worker becomes more familiar with portals of entry and exit, methods of transfer of organisms, and other factors involved in transmission of communicable disease. These are discussed later. Details of sterilization of instruments and other hospital supplies are tabulated in Appendix B.

NOTE ON BOILING

Boiling (100 C at sea level) for 10 minutes kills the vegetative or actively multiplying cells of all pathogenic bacteria, including rickettsias, chlamydias, and mycoplasmas, and appears to inactivate most viruses[1] if they are fully

[1]It does not destroy the viruses of epidemic hepatitis or serum hepatitis.

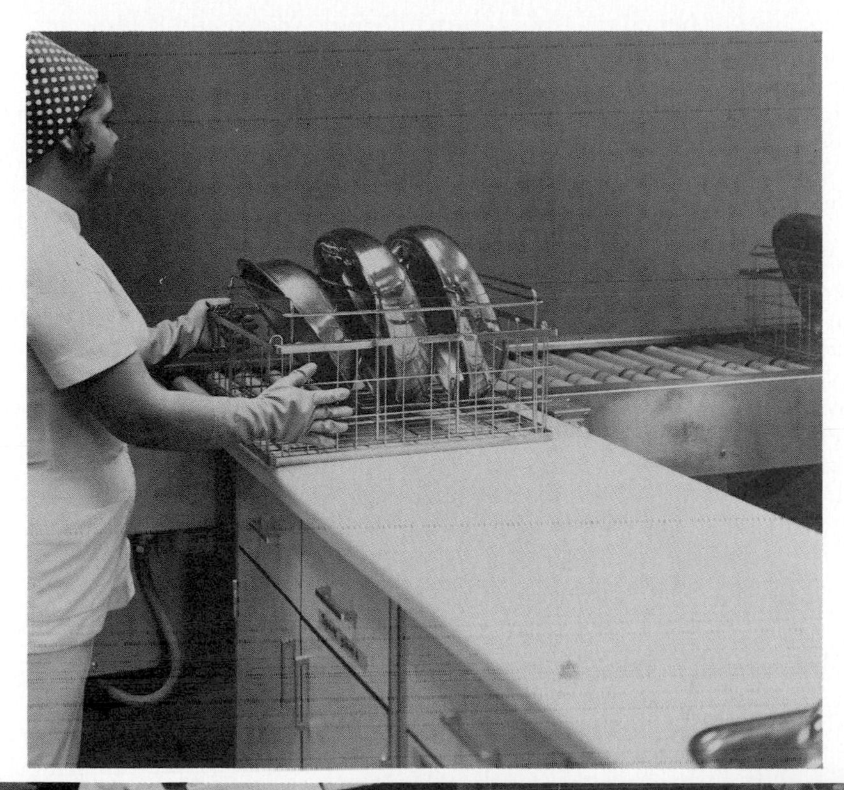

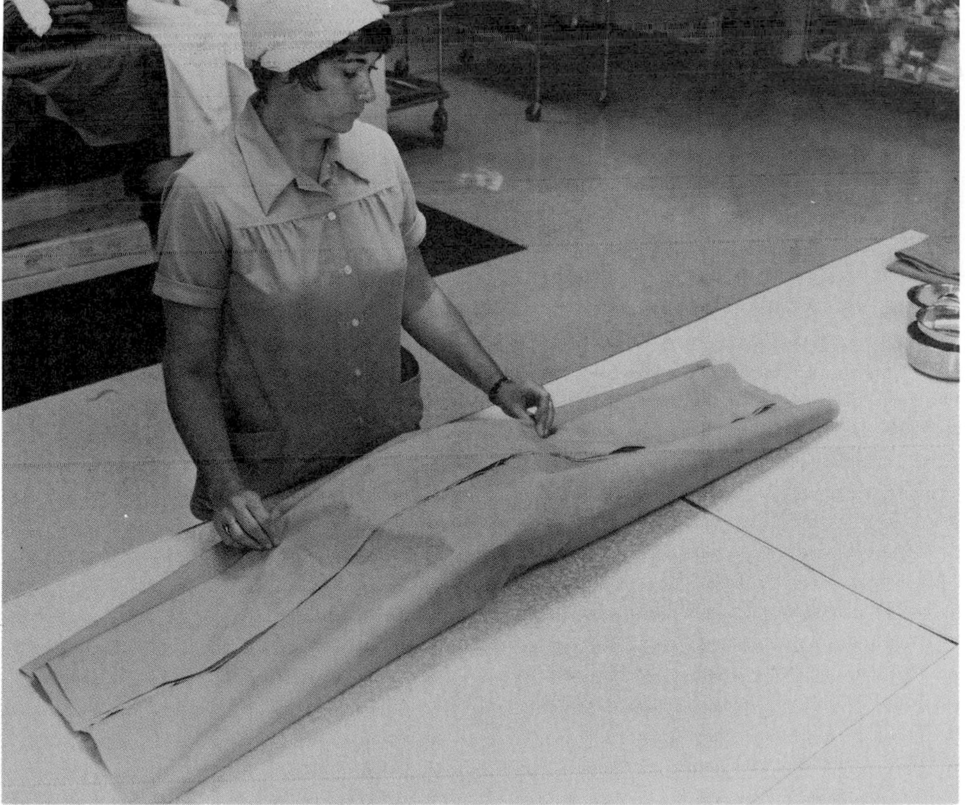

Figure 15–10 *Above,* Loading bedpans into an automated washer/sterilizer. *Below,* Wrapping a bedpan in preparation for autoclaving. (Courtesy of Sybron Corporation, Rochester, N.Y.)

exposed and not protected by clots of blood, masses of food, feces, mucus, and so on. This period of boiling also appears to kill or severely injure cysts of protozoa, ova of helminths, spores of pathogenic yeasts, and conidia of molds.

However, boiling for 10 minutes or even 100 minutes does not guarantee safety where bacterial spores (especially of tetanus, gas gangrene, and anthrax bacilli) and the highly thermoduric hepatitis viruses are concerned. Spores of tetanus and gas gangrene organisms can be assumed to be present in all soil, feces, and dust and on instruments or bandages contaminated with dust, blood, exudates, or exposed materials of any kind. Anthrax spores are rare except in areas where animals infected with these organisms live. The virus of epidemic (so-called "infectious") hepatitis can be assumed to be present in feces and sewage and anything (including hands!) contaminated therewith. The virus of serum hepatitis can be assumed to be present in any undisinfected blood, exudates, serum, or plasma, on instruments and objects contaminated therewith, and possibly also in feces and sewage.

Where these spores and these viruses are *known* not to be present, boiling for ten minutes at 100 C is a reasonably safe means of disinfection (not sterilization) of glassware and instruments if they are not grossly soiled with blood, feces, food, and the like. Boiling for 30 minutes greatly increases the margin of safety. These facts may be summarized as follows:

Boiling at 100 C for 10 minutes inactivates or kills:

1. Vegetative forms of all bacteria and fungi pathogenic for humans
2. Rickettsias and chlamydias pathogenic for humans
3. Helminths and their ova
4. Many viruses (except hepatitis viruses)
5. Bacterial toxins (except staphylococcal enterotoxin)

Boiling at 100 C for 10 minutes does not inactivate or kill:

1. Bacterial spores
2. Hepatitis viruses (possibly some others)
3. Ascospores and conidia of some fungi pathogenic for humans[2]
4. Staphylococcal enterotoxin

Care and well-informed judgment must be exercised in deciding when boiling may be used with safety and how long it should be continued. Factors such as the source, amount, and nature of soiling or contamination, uses to which the boiled articles are to be put, altitude above sea level, and pH of suspending fluid (if any) must influence the decision.

When in doubt, autoclave! (or use sporicidal gases or chemicals if necessary and feasible).

DISINFECTION IN COMMUNITY HEALTH PRACTICE

The principles of microbiology applied to practice in the hospital apply equally to the care of the patient in the home. The community health worker is usually more aware of the importance of handwashing after visiting each patient than is the worker in the hospital, because the former must go from home to home between patients. The community health worker, like the hospital

[2]Accurate data are not available for all species. Many fungal spores and conidia may be killed by less than 100 C.

worker, must be sure to wash all areas of the hands thoroughly so as not to be a carrier of infection from one patient to another.

For the visiting worker starting from headquarters, syringes and needles—unless in preassembled, packaged, and commercially presterilized plastic units—should be sterilized in an autoclave or a hot air oven and carried in sterile containers in the worker's bag. In an emergency, away from hospital or headquarters facilities, syringes and needles may be disinfected for ordinary purposes by boiling in a covered saucepan for 30 minutes. A home oven sterilizes as well as a hospital oven, and a pressure cooker holding as little as one or two quarts of water can serve perfectly in lieu of an autoclave. There is never justification for simply rinsing syringes and needles with a disinfectant solution before their use for injection. Presterilized, packaged, disposable needles and plastic syringes are now in general use in many areas of the world, although not yet in some developing nations.

Obviously the community health worker cannot carry a thermometer for each patient visited. A satisfactory procedure is to carry a screw cap tube of glass or metal cushioned with cotton at the bottom and around the inside. This tube is then filled with 70 per cent alcohol (or other suitable disinfectant) into which the thoroughly washed and wiped thermometer may be placed until used again. The worker may carry two such outfits in order to provide longer disinfection for each thermometer. Electronic thermometers with supplies of sterile tip covers are convenient and time-saving though bulky and still expensive. Disposable thermometers have become less expensive and save time.

Pasteurization may be carried out at home or in camp very easily. A double boiler or similar arrangement of buckets is adapted so that water in the outer container is heated to about 70 C and the temperature of the milk in the inner container held at between 63 and 65 C for 30 minutes. The milk is then cooled rapidly and refrigerated. Babies' bottles may be heated in an inexpensive apparatus purchasable in most department stores.

Isolation technique can be set up admirably in the home if the principles of microbiology and disease transmission are understood by the health worker. The patient with a communicable disease is isolated in a unit (preferably in one room) and equipment is kept inside that room for that person only. Any objects, such as dishes, are removed from the unit in a covered container and are disinfected by boiling for 30 minutes, by soaking in 5 per cent saponated solution of cresol for one hour, or by other effective methods. Linens and bedding are rolled tightly, with the soiled surface inward, and placed in plastic or tightly closable bags. Members of the family can usually be taught the principles of isolation technique, and if they clearly understand those principles they can safely care for patients.

Improvisation Techniques. An ingenious public health worker can use the simple domestic equipment at hand to accomplish the desired results. (One must have a clear idea of what needs to be done!) As already mentioned, one can substitute a pressure cooker for an autoclave, the oven of a cooking stove for a laboratory oven, a covered pan of boiling water for disinfection of dishes, and laundry bleach for disinfectant. One may also adapt a wash boiler containing disinfectant or boiling water for disinfection of linen, use a covered pail for disinfection of body secretions or excretions, and make numerous other adaptations without sacrificing safety.

SUMMARY

Principles of sterilization and disinfection in health care must be thoroughly understood by all health personnel to protect the patient and everyone in contact with the patient. This chapter delineated in some detail procedures in the medical and surgical wards of a general hospital, after a patient discharge, in the operating room, in the communicable disease unit or hospital, and in community health practice. The importance of proper handwashing, thermometer technique, use of syringes and needles, surgical dressings, catheterization, and "reverse precautions" can hardly be overestimated. The reader cannot expect to learn this from a brief summary but should *study this chapter thoroughly.*

All bedside equipment and dishes in hospitals can spread pathogenic organisms. Secretions and excretions must be carefully wrapped and burned. The bedpan and urinal require special precautions and techniques in cleaning (which do not sterilize).

The lesson to remember is that boiling at 100 C for 10 minutes *cannot* be trusted to destroy (1) bacterial spores, (2) hepatitis viruses and possibly others, (3) some spores of human pathogenic fungi, and (4) staphylococcal enterotoxins. Boiling will destroy vegetative forms of bacteria and fungi, rickettsias and chlamydias, helminths and their ova, many viruses, and most bacterial toxins.

The community health worker visits each patient and thus must carry disinfection and proper health practices into the homes of people who often do not have the necessary facilities and do not understand how diseases may be transmitted. The tools of the health worker must of necessity be simple and portable yet just as effective in protecting the patient's health as hospital equipment. The community health worker must be able to improvise should existing facilities be insufficient.

REVIEW QUESTIONS

1. Describe an appropriate handwashing technique for (a) communicable disease nursing and (b) preparation for surgery.
2. When is it permissible to rely upon five to fifteen minutes of boiling to eliminate pathogenic microorganisms? When should boiling not be relied on to eliminate pathogens?
3. In home health care what implements may be used in place of the hospital (a) hot air sterilizer, (b) autoclave, (c) bedpan flusher, (d) pasteurizer, and (e) instrument boiler?
4. Imagine that you are a public health worker in a summer camp. Three cases of infectious hepatitis occur. What procedures would you use to isolate the patients until they could be transferred to a hospital? What procedures would you recommend for protecting the remainder of the camp?
5. Imagine that you are the head nurse on a hospital ward. A patient is admitted with excessive secretions of the respiratory tract. What procedures would you carry out before you have a diagnosis? If the patient has pulmonary tuberculosis, what difference would this make in your precautionary techniques? Why?
6. You are a head nurse on a surgical ward. A patient develops symptoms of typhoid fever three days after the removal of his gallbladder. What procedures would you carry out to protect the other patients, the medical and nursing staff, the dietary workers, and the housekeeping staff?
7. Imagine that you are a public health worker who has to use improvised techniques. Give some specific examples.

Supplementary Reading

Benenson, A. S. (ed.): Control of Communicable Diseases in Man. 12th ed. New York, American Public Health Association, 1975.

Falconer, M. W., Schram Ezell, A., Patterson, H. R., and Gustafson, E. A.: The Drug, the Nurse, the Patient. 6th ed. Philadelphia, W. B. Saunders Company, 1978.

Finland, M., Jones, W. F., Jr., and Barnes, M. W.: Occurrence of serious bacterial infections since introduction of antibacterial agents. *J.A.M.A.*, *170*:2188, 1959.

Harding, W. le R., Balcom, C. E., and Van Belle, G.: Control of Infections in Hospitals. Toronto, University of Toronto Press, 1968.

Lawrence, C. A., and Block, S. S. (eds.): Disinfection, Sterilization, and Preservation. 2nd ed. Philadelphia, Lea & Febiger, 1977.

Patterson, H. R., Gustafson, E. A., and Sheridan, E.: Falconer's Current Drug Handbook 1980–1982. Philadelphia, W. B. Saunders Company, 1980.

Perkins, J. J.: Principles and Methods of Sterilization. 2nd ed. Springfield, Ill., Charles C Thomas, 1980.

Rockwell, V. T.: Surgical hand scrubbing. *Amer. J. Nurs.*, *63*:75, 1963.

Sommermeyer, L., and Frobisher, M.: Laboratory studies on disinfection of oral thermometers. *Nurs. Res.*, *1*:32, 1952.

Sommermeyer, L., and Frobisher, M.: Laboratory studies on disinfection of rectal thermometers. *Nurs. Res.*, *2*:85, 1953.

Top, F. H. (ed.): Control of Infectious Diseases in General Hospitals. New York, American Public Health Association, 1969.

Brochures

Beta-propiolactone (BPL). 1959. Wilmot Castle Co., Rochester, N.Y. 14618.

Stierling, H., Reed, L. L., and Billick, I. H.: Evaluation of Sterilization by Gaseous Ethylene Oxide. Washington, D.C., U.S. Dept. of Health, Education, and Welfare, Public Health Monograph No. 68, 1962.

SECTION FOUR

INFECTION, IMMUNITY AND ALLERGY

Lady Mary Wortley Montagu (1689–1762) is known for her early recognition of the value of immunization against smallpox. As the wife of the English Ambassador to Turkey, she observed that the local inhabitants inoculated children against smallpox by "ingrafting," which usually resulted in mild infections of not more than 30 pustules that did not leave marks and did not kill. Lady Montagu had her own children inoculated with variola and popularized such vaccination in England. (Courtesy of the National Library of Medicine.)

Paul Ehrlich (1854–1915) proposed and established the concepts of chemotherapy (antibacterial chemotherapeutic agents) and developed arsphenamine (Salvarsan, 606), the first effective cure for syphilis. Ehrlich also explained the action of toxins and proposed the side-chain theory explaining receptors on antibodies, which is of great significance in immunology. (From Marguardt, M.: Paul Ehrlich. New York, Henry Schuman, Inc., 1951.)

THE FUNCTIONS AND COMPONENTS OF NORMAL BLOOD

The blood is of fundamental importance in the production, transmission, diagnosis, prevention, and cure of infections and diseases caused by microorganisms. In order to have an adequate understanding of these conditions, it is necessary to review the composition and some of the essential functions of blood. This fluid either directly or indirectly serves as a means of intercommunication among the various cells and tissues in the body, transporting food, hormones, oxygen, carbon dioxide, and other important substances to and from cells and removing waste products. It may even carry destruction to tissues if viruses, bacteria, or their toxins invade the body. Blood also contains substances and living cells that help the body combat disease caused by microorganisms.

FUNCTIONS OF THE BLOOD

The arterial and venous networks of the vascular system and the blood itself constitute the great transportation system and supply lines of the body. Blood is pushed continuously through the arterial system by the left ventricle of the heart under a normal arterial blood pressure of about 130 to 150 mm of Hg. In the smallest, most remote branches of the system, the microscopic *arterioles* and *venules* or *capillary vessels,* the walls of the blood vessels are only one cell in thickness. Through these thin, membrane-like walls, nutrients, hormones, and other substances pass by diffusion and filtration with and to the fluid *(lymph)*[1] that surrounds the capillaries and the tissue cells. The lymph (the interstitial or tissue fluid) constitutes a sort of culture medium for the tissue cells. Lymph is actually blood plasma minus most blood proteins: it consists of the fluid from about 2.8 liters of liquid plasma. The nutrients are brought from the gastrointestinal tract and the liver by the arterial blood and passed into the lymph.

Waste metabolic products (acids, carbon dioxide, and other substances) from the tissue cells diffuse out into the lymph and from the lymph into the bloodstream and are carried back to the right auricle of the heart via the capillary venules and venous system, which are a direct continuation of the arterioles.

The blood is then forced through the lungs by the right ventricle. Oxygen from the air is taken up by the red blood cells, carbon dioxide is given off by both red blood cells and plasma, and other important exchanges occur. The blood is then returned to the left ventricle and circulated to the tissues again. Lymph seeps into a sort of drainage system—the lymphatic channels—passes through a system of filters *(lymph nodes)* where particles such as cell detritus, bacteria, and the like are phagocytized by phagocytes, and is returned to the venous blood via a large lymph channel, the thoracic duct.

The physiology of the blood is very complex and need not be further

[1] Latin *lympha,* water.

CHAPTER 16

detailed here. The relation of the capillary blood vessels to tissue cells and to the tissue fluids surrounding them is shown diagrammatically in Figure 16–1. One leukocyte is shown passing between the cells of the capillary wall by *diapedesis* (Greek, *dia,* through; and *pedan,* to leap; hence, "leaping through").

THE COMPOSITION OF BLOOD

The body of an average adult male, weighing about 160 pounds, contains about 5.8 liters of blood; the body of a female contains about one liter less. Of this, the blood cells constitute roughly 2.7 liters and the liquid (plasma) about 3.1 liters. The cells and plasma may be measured and separated by adding an anticoagulant (e.g., sodium oxalate or citrate) and centrifuging in a calibrated, flat-bottomed, tubular vessel called a *hematocrit.* The cells are heavier and settle to the bottom: their volume may be read from the calibrations on the tube. The clear, oxalated or citrated, supernatant plasma may be drawn off into a separate vessel.

Plasma. The fluid portion of blood (the plasma) is a yellowish, transparent fluid, a complex mixture of proteins and other substances, including food for

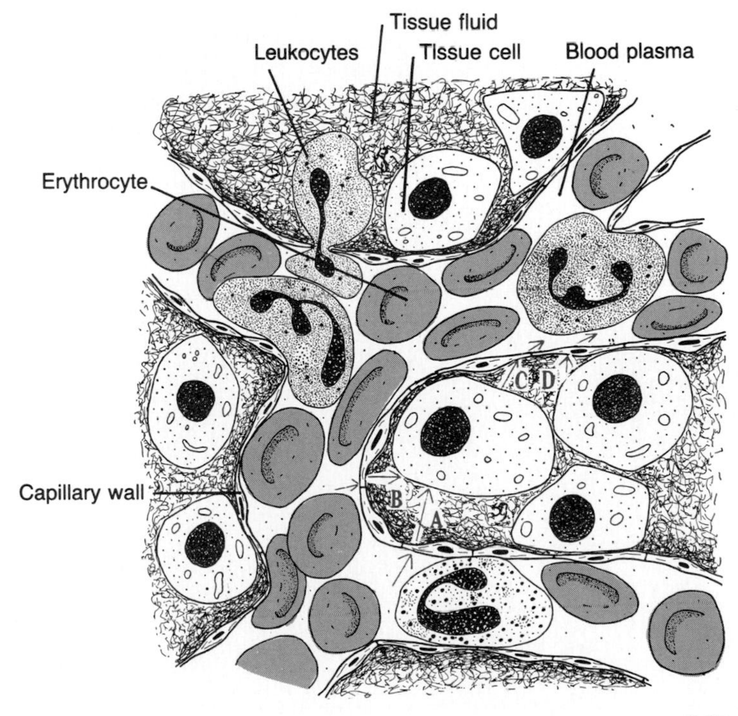

Figure 16–1 Relationships between intercellular lymph or tissue fluid, the tissue cells, and the blood plasma. At *A,* soluble food passes from the plasma, through the microscopically thin wall of the capillary blood vessel or arteriole, into the tissue fluid, and thence into the tissue cells. (Blood flow in this drawing is from left to right.) At *C,* wastes from the tissue cells are leaving in the reverse manner. At *B,* oxygen brought by the erythrocytes is diffusing through the arteriole wall and lymph to the tissue cells, and, at *D,* carbon dioxide from the tissue cells is leaving in the same manner via a venule. A leukocyte is moving out of the capillary into the tissue space by passing between two cells of the capillary wall, illustrating the phenomenon of diapedesis. (Modified from Hegner and Stiles: College Zoology. New York, Macmillan Co., 1959.)

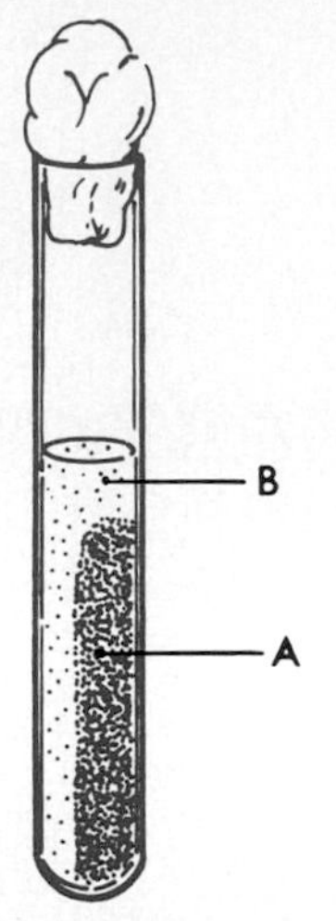

Figure 16–2 Tube with clot (*A*), which has shrunk, enmeshing virtually all the blood cells and leaving the pale yellow serum (*B*) quite clear.

the body tissues and waste products excreted from the tissues. It consists of about 91 per cent water, 1 per cent minerals (CO_2, salts of Na, K, Ca, Mg, and P, and minute amounts of several others), 7 per cent proteins, and about 1 per cent of various organic substances, such as uric acid, urea, amino acids, glucose, lipids (fats and fatlike substances), and hormones. *Plasma* is obtained when blood is prevented from clotting by using heparin, sodium oxalate, or sodium citrate in the syringe, test tube, and vessel. When blood does clot, it separates into *serum* (the liquid) and *clot* (containing the cells) (Fig. 16–2).

Important among the plasma proteins are albumin and various globulins, which, together with electrolytes and other components, help maintain the water balance of the body (Table 16–1). Of the protein components, *gamma globulins* are exceedingly important since all specific antibodies (Chapter 19) are gamma globulins. *Fibrinogen,* another plasma protein, plays an important role in the blood-clotting mechanism, being converted from its normally soluble state to insoluble *fibrin,* which will be discussed later in this chapter. Plasma, which makes up approximately 55 per cent of whole blood, serves as the fluid medium for the transportation of the formed elements: red cells, white cells, and platelets, which together make up the remaining 45 per cent. The normal pH of plasma is 7.4; that of whole blood is between 7.35 and 7.45.

The Origin of Blood Cells. All blood cells originate from undifferentiated embryonic mesenchymal cells. By the time of birth, the bone marrow has developed into the only source of pluripotent (hematopoietic, blood-forming) "stem cells" and a major source of lymphocyte precursors. It is not clear whether lymphocytes have the same origin as the other blood cells. From the bone marrow stem cells clones of cells differentiate and ultimately appear in the circulating blood as red cells, platelets, and various types of white cells. The earliest cells of each cell line have similar morphologic characteristics and cannot be differentiated one from the other by appearance alone.They are given specific names such as "myeloblast," "lymphoblast," or "rubriblast," depending on the tissue in which they are found, the cells with which they are

Table 16–1. Major Protein Fractions of Human Blood as Determined by Electrophoresis. (See accompanying diagram.)

PLASMA PROTEIN FRACTION	PERCENTAGE OF TOTAL PROTEIN
Albumin	52.0–68.0
α_1 globulin	2.4– 4.4
α_2 globulin	6.1–10.1
β globulin	8.5–14.5
γ globulin*	18.7–21.0

*Contains the antibodies.

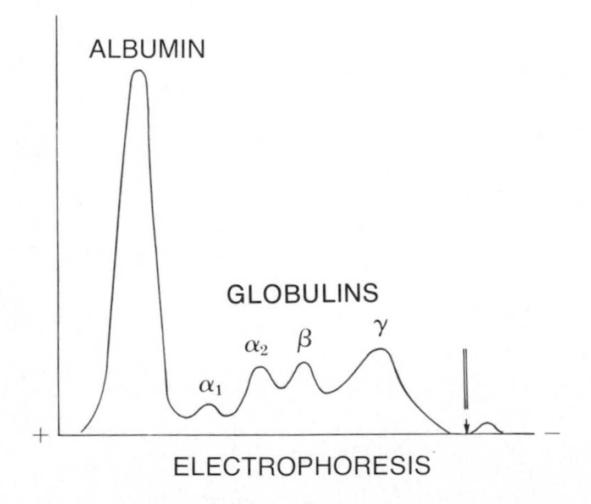

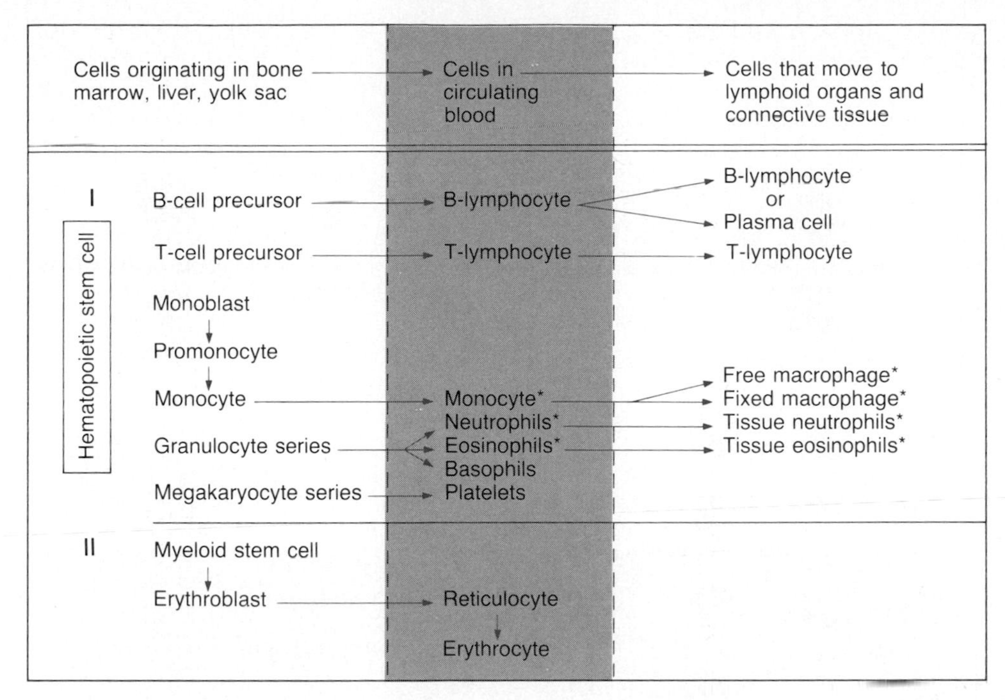

Figure 16–3 Differentiation and localization of cells involved in immune processes. Cells originating as pluripotent bone marrow stem cells differentiate (I) via the lymphoid stem cells to the T-cell and B-cell lines and (II) via the myeloid stem cells to other cells. (Adapted from van Furth, R.: Mononuclear Phagocytes in Immunity, Infection and Pathology. Oxford, Blackwell Scientific Publications, 1975; and Bellanti, J. A.: Immunology: Basic Processes. Philadelphia, W. B. Saunders Company, 1979, p. 28.)
*Phagocytic cells in the blood and tissues.

associated, and the definitive cells that they are destined to produce (Fig. 16–3).

The Red Blood Cells (RBC's). These are also called *erythrocytes.*[2] They are formed mainly in the bone marrow and float in the plasma (Fig. 16–4). The red color of the blood comes from hemoglobin, which the erythrocytes contain. Hemoglobin is an iron-containing pigment of the class known as *respiratory pigments.* They have the property of forming a loose chemical combination with oxygen when exposed to air (as in the lungs) and of releasing the oxygen to body tissues that need it. Hemoglobin functions as an oxygen carrier. Bacteria do not contain hemoglobin, but many species have a yellowish, iron-containing respiratory pigment of similar function called *cytochrome* (from Greek *kytos,* cell; *chroma,* color).

The red blood cells of mammals are formed mainly in bone marrow. They are initially nucleated cells called *proerythroblasts.* These develop into *normoblasts* and finally into erythrocytes. When they become old and look deformed they are called *poikilocytes.*

Human erythrocytes, circulating in the blood, are non-nucleated biconcave disks about 7.5 μm in diameter with a life span of approximately 130 days. They are constantly being replaced from hemopoietic tissues in bone marrow. In the normal adult male, erythrocytes number approximately 5.3 million per cubic millimeter of blood; in normal adult females, there are about 4.3 million per cubic millimeter.

When red cells break up, the hemoglobin escapes into the surrounding fluid. This process is called *hemolysis* (from Greek *haima,* blood; *lysis,* disruption or dissolution). Certain bacterial toxins (poisons) have the power of hemolyzing erythrocytes both in the body (producing anemia) and in the test tube. As we shall see later, this capacity to produce hemolysis is used in the laboratory as one method of identifying certain organisms, especially those *(Streptococcus pyogenes)* that cause scarlet fever and "strep throat" (tonsillitis).

The White Blood Cells (WBC's). These cells, of which there are several types, contain no hemoglobin and are collectively called *leukocytes.*[3] They are

[2]From Greek *erythros,* red; and *kytos,* cell.
[3]From Greek *leukos,* white.

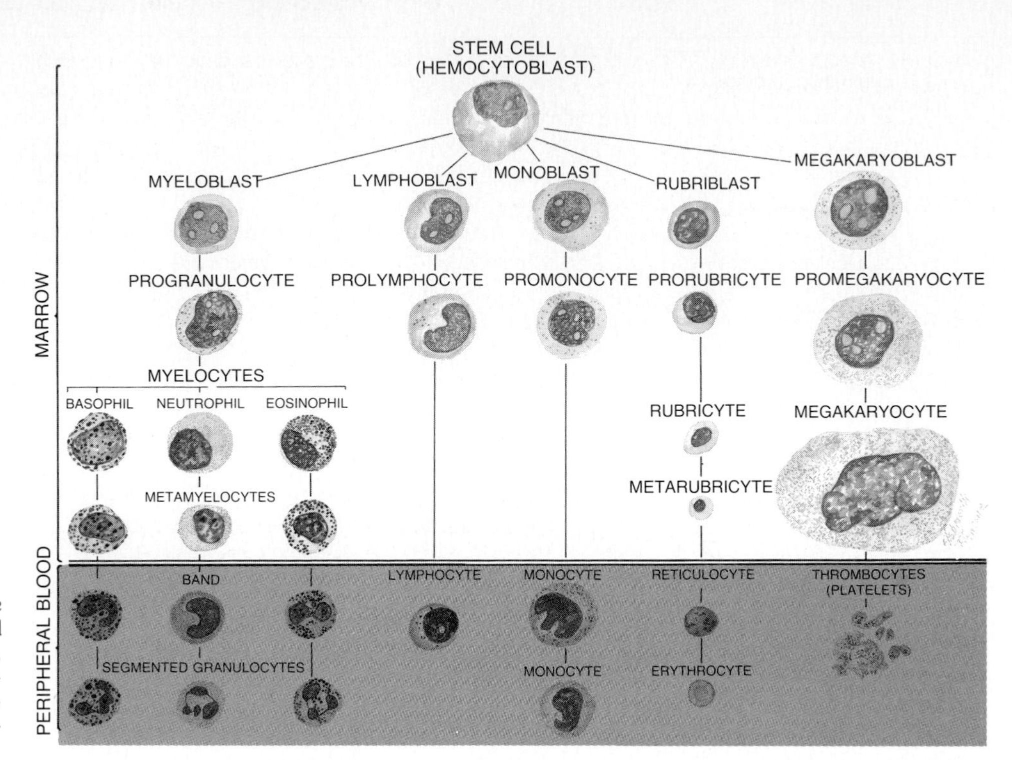

Figure 16–4 Stages in the formation of the peripheral blood cells. (From Jacob, S. W., et al.: Structure and Function in Man. 5th ed. Philadelphia, W. B. Saunders Company, 1982.)

larger and less numerous than RBC's (Fig. 16–4). There are between 5,000 and 9,000 per cubic millimeter of normal blood. Some types (the *granulocytes:* polymorphonuclear neutrophilic, eosinophilic, and basophilic) are derived from bone marrow cells *(myelocytes)*. Other leukocytes mature in the thymus *(thymocytes),* and still others *(lymphocytes)* come from lymphoid tissues such as the spleen, adenoid tissues, and lymph nodes ("lymph glands").

Granulocytes. The granulocytic myelocytes, especially polymorphonuclear neutrophils or polymorphonuclear leukocytes ("polys" for short), have an important function as scavengers and "policemen" in the blood. The first part of the name, "polymorphonuclear," means "with a many-lobed nucleus." The nucleus, when stained, is seen to consist of two to four large, connected lobes and is very distinctive. The cell designation "neutrophil" indicates that certain stainable granules in the cells do not exhibit special affinity for either acidic or basic stains. Other granulocytes have large, round nuclei and contain granules that are either strongly basophilic (basophilic leukocytes) or acidophilic (eosinophilic[4] leukocytes). These *basophils* and *eosinophils* are relatively rare in the blood, except in certain diseases. Any foreign particles, such as bacteria or dead tissue cells that enter the blood or tissues, are promptly engulfed by the "polys" and certain other cells of the vessel linings that also have the power of ingesting solid particles directly (Fig. 16–5 and 16–6).

The polymorphonuclear leukocytes resemble remarkably the primitive protozoa called amebas, which can be found in soil, sewage, and stagnant water. They have no constant shape and can move about by sending out temporary swellings or thin, finger-like cytoplasmic extensions called pseudopodia ("false feet") and then flowing into them (see Chapter 3). They can thus intrude themselves into minute crevices (such as exist between tissue cells of the body) and are often called *wandering cells.* Most importantly, from the standpoint of natural defense against disease, they can extend pseudopodia around minute

[4]Eosin is one of the acidic dyes, i.e., it tends to dye the more basic (generally, the cytoplasmic) parts of the cell.

particles, draw them in, and engulf them. The particle passes through the cell membrane to the leukocyte's interior. If the particle is a bacterium, it may be killed there and liquefied by digestive enzymes inside the white cell. However, leukocytes are often destroyed by bacteria. The whole process of ingestion by these leukocytes and also by certain *reticuloendothelial cells* that line the blood vessels *(fixed phagocytes)* is called *phagocytosis.*[5]

By means of ameboid movements, leukocytes are able to move out of the blood vessels, passing between the cells lining the vessels. They can thus travel about through the tissues. When some undesirable particles (such as bacteria in a boil or a splinter of wood) set up inflammation, the leukocytes are attracted to the spot by certain substances in the affected tissue. They congregate there, sometimes in enormous numbers. The white, creamy material in a boil or other infected spot consists largely of dead leukocytes ("pus cells"), dead bacteria, some of the tissue cells, and lymph and serum from blood and tissues, all mixed together. This material is called pus (Fig. 16–7).

The *reticuloendothelial system* (RES), also known as the *mononuclear phagocyte system* (MPS), consists of fixed phagocytic cells, the garbage collectors of the body, that are the Kupffer cells of the liver, dendrite and sinusoidal cells of the spleen, glial cells of the central nervous system, alveolar cells of the lungs, and "activated" endothelial cells. Fixed phagocytes may also be called *histiocytes, macrophages,* or *mononuclear phagocytes.* (See Fig. 16–3).

Monocytes and *lymphocytes* (large and small) are white blood cells with round nuclei, commonly seen in inflammatory exudates. The former are active in phagocytosis; the small lymphocytes and the plasma cells are important in the production of specific antibodies (gamma globulins) (Fig. 16–3).

In normal adult blood the proportions of the different types of WBC's are as follows:

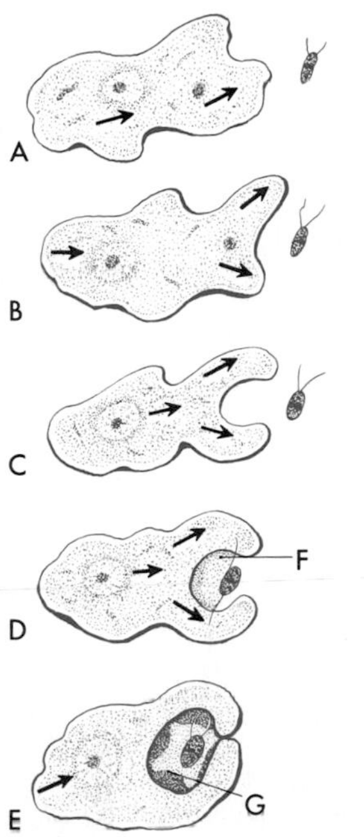

	Normal Blood Values (%)
Granulocytes, also called *Polymorphonuclear leukocytes* or *PMN leukocytes*	
1. Eosinophils (phagocytic cells)	0–6
2. Basophils	0–1
3. Neutrophils, also called polymorphonuclear neutrophils (phagocytic cells)	50–75
Agranular leukocytes	
1. Lymphocytes	20–45
2. Monocytes (phagocytic cells) (macrophages[6])	2–10

Leukocyte Counts. Leukocytes may vary in kind and numbers. An increase in the percentage of polymorphonuclear leukocytes indicates infection, inflammation, or toxin absorption. Noninfectious conditions—some of a normal nature, such as labor or exercise, and some not so normal, such as convulsions or vomiting—also tend to increase the neutrophil percentage.

The percentage of lymphocytes increases in mumps, brucellosis, whooping cough, influenza, tuberculosis, goiter, and certain mycoses, and often after exposure to excess sunlight. A relative increase in monocytes is most distinctive of typhoid fever, but also arises in certain parasitic infestations, Rocky Mountain spotted fever, Hodgkin's disease, and some types of microbial intoxication.

Figure 16–5 Diagrammatic representation of an ameba ingesting a flagellate (unicellular animal). In *A,* the ameba is moving toward the flagellate by means of pseudopod (indicated by arrow). In *B* the pseudopod has divided into two smaller pseudopodia preparatory to surrounding the flagellate. The process is completed in *C, D,* and *E. F* and *G* show the formation of a vacuole around the flagellate. Any other portion of the ameba could have extruded itself around the flagellate as well as that indicated in the picture. Leukocytes behave in the same manner. (From Schaeffer, A. A.: Ameboid Movement. Princeton, Princeton University Press, 1920.)

[5]*Phago* is from the Greek *phagein,* to eat. Phagocytosis, therefore, is the process of being eaten by a cell. A *phagocyte* is a cell that eats cells and other particulate matter.

[6]Macrophages are large phagocytic cells. They may be leukocytes in the blood or tissue cells.

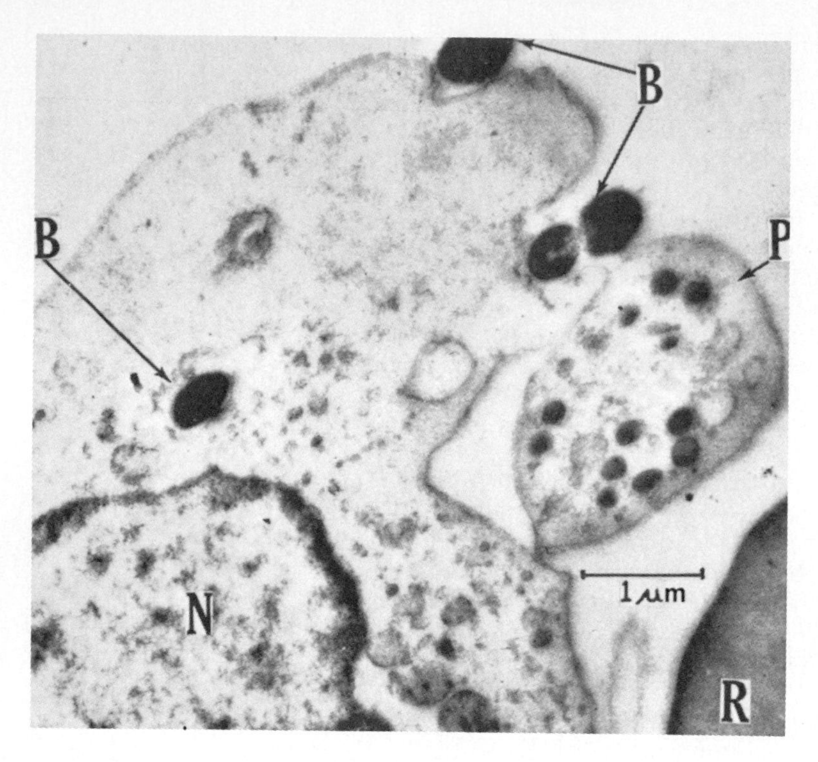

Figure 16–6 Electron micrograph of an ultra-thin (around 350Å) section of a part of a human polymorphonuclear leukocyte, showing phagocytosis of virulent *Staphylococcus aureus*. *N* is part of one lobe of the nucleus of the leukocyte; note the well-defined nuclear membrane. *B* are cells of *S. aureus;* note that one bacterium is undergoing fission. The bacteria at the upper right are being grasped by the finger-like pseudopodia of the leukocyte and will be "ingested" or drawn completely inside the leukocyte, surrounded by a vacuole wall such as is seen to be forming around the bacterium at the left of center. Note the granular structure of the cytoplasm of the leukocyte and its differentiation from the nucleoplasm. *P* is a platelet; *R* is part of a red cell. Note the line (µm) indicating 1 micrometer. (×27,500.) (Courtesy of Drs. J. R. Goodman and R. E. Moore, V. A. Hospital, Long Beach, Calif.)

Eosinophil percentages increase in many helminthic and protozoal infections, allergic conditions such as bronchial asthma, myelogenous leukemia, scarlet fever, and other diseases. Little is known about basophils, except that they play an important role in allergy. Their percentage may increase with myelogenous leukemia.

In some infectious processes—acute appendicitis, for example—the number of white blood cells increases from the normal 5,000 to 9,000 per cubic millimeter to 15,000 or even 20,000 per cubic millimeter. An increased number of leukocytes in the blood is called a *leukocytosis*.

The *differential count* (percentages of the various types of WBC's) also changes during infectious processes. The proportion of polymorphonuclear leukocytes especially increases, sometimes to 80 or 90 per cent, with the others diminishing in proportion.

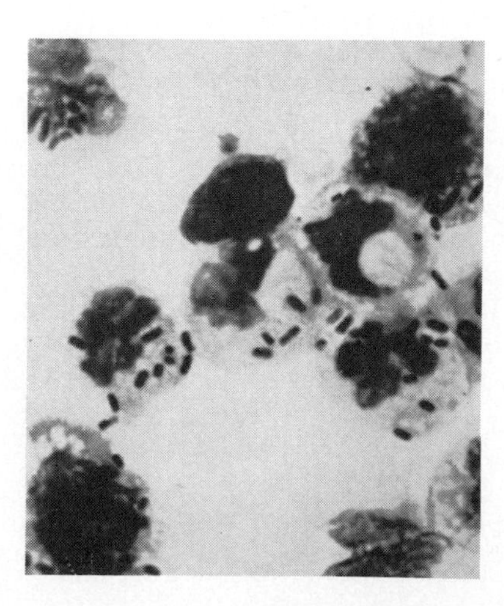

Figure 16–7 Stained smear of pus from the lung of a mouse inoculated with a species of pathogenic bacilli. The bacilli are seen to have been engulfed by the leukocytes (phagocytized) in large numbers. This is an excellent illustration of one of the most important defensive mechanisms. (×1,000.) (Courtesy of M. R. Smith and W. B. Wood, Jr., Washington University School of Medicine, St. Louis, Mo. In *J. Exp. Med.*, Vol. 86.)

Certain infections always cause leukocytosis, which also may be due to pregnancy or intoxication. Some *pyogenic* (Greek, *pyon,* pus; *gennan,* to produce) pathogens may produce leukocytosis, whereas others may not affect leukocyte concentrations. Among the pyogenic conditions are infections caused by staphylococci, gonococci, streptococci, and pneumococci; among the nonpyogenic conditions are typhoid fever, measles, influenza, tuberculosis, and syphilis. In some diseases in the second group—typhoid fever and influenza, for example—there is a diminution of leukocytes below normal numbers. Such a condition is not uncommon in disease and is called *leukopenia* (Greek, *leukos,* white; *penia,* scarcity).

Blood Platelets. These are non-nuclear, irregularly shaped bodies (Fig. 16–4), varying in size but much smaller than erythrocytes and occurring in low numbers, about 300,000 per cubic millimeter of blood. They are derived by fragmentation from the largest cells of the bone marrow, which are called megakaryocytes. Platelets play a role in the process of coagulation of blood by producing thromboplastin, one of the components of fibrin. They are sometimes called *thrombocytes.* Because platelets easily stick together (agglutinate), they may plug up small breaks in the circulatory system and thus prevent seepage of blood.

Blood Clotting. *Hemostasis* is the mechanism by which bleeding from an injured blood vessel is controlled or stopped. It involves not only blood clotting but also constriction of the damaged blood vessels and the action of extrinsic factors of skin, muscle, and elastic tissues.

Blood clotting is a very complex phenomenon involving at least 15 clotting factors (now designated as I, II, IIa, and so forth). The clotting process is closely related to fibrinolysis (a system that prevents clotting in blood vessels and breaks down clots) in that the *contact factor* (also known as *Hageman factor* or clotting factor XII) is involved in the initiation of both processes.

Blood plasma contains several substances, such as fibrinogen (factor I) and other "blood proteins," which are intrinsic factors formed chiefly in the liver (see Tables 16–1 and 16–2): these cause blood to clot when it leaves the body. Inside the body, certain substances in the plasma normally keep the blood in a fluid condition. Once the blood escapes, however, the coagulation mechanism is initiated and clotting takes place.

When tissues are injured and cells are punctured or sliced through, thromboplastin is released, platelets are disintegrated in the wound, and the following process occurs:

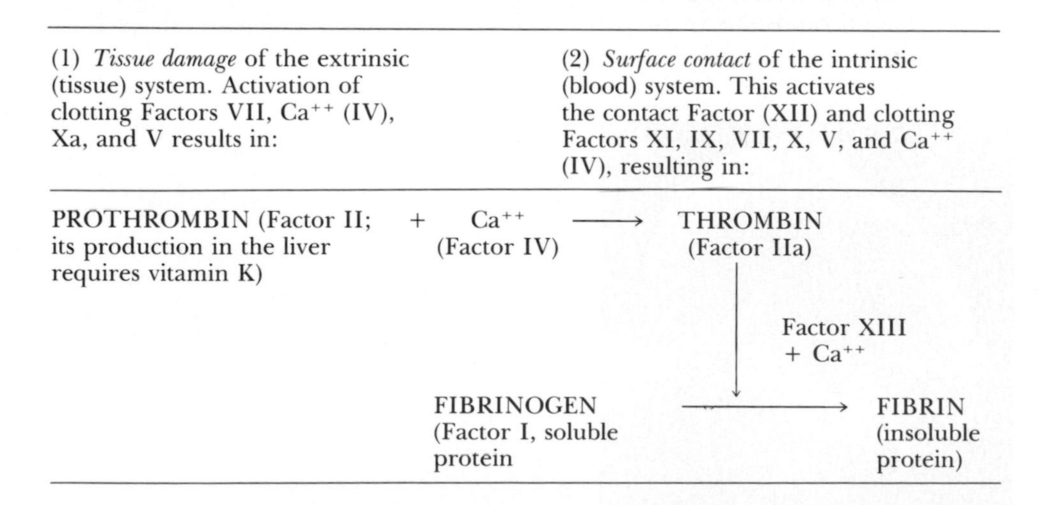
(1) *Tissue damage* of the extrinsic (tissue) system. Activation of clotting Factors VII, Ca^{++} (IV), Xa, and V results in:
(2) *Surface contact* of the intrinsic (blood) system. This activates the contact Factor (XII) and clotting Factors XI, IX, VII, X, V, and Ca^{++} (IV), resulting in:
PROTHROMBIN (Factor II; its production in the liver requires vitamin K)
+
Ca^{++} (Factor IV)
THROMBIN (Factor IIa)
Factor XIII + Ca^{++}
FIBRINOGEN (Factor I, soluble protein
FIBRIN (insoluble protein)

The fibrin, in the form of an elastic, spongy, interlacing network of protein fibers, traps the red and white blood cells (Fig. 16–8) to form a blood clot, normally in 2 to 6 minutes. As the clot shrinks (retraction) it pulls the red and white cells with it, exuding a clear, straw-colored fluid called serum, which is plasma minus the components of the clot (Fig. 16–2).

Table 16–2. Factors Involved in Blood Coagulation

Factor Designation	Name of Factor	Factor Originating Tissue	Factor Source
Factor I	Fibrinogen	Liver	A plasma protein
Factor II	Prothrombin	Liver	A plasma protein
Factor III	Thromboplastin	Found in blood and in cells	Released by injured cells
Factor IV	Calcium	In bones, food, and milk	As Ca^{++} in plasma
Factor V, VI	Proaccelerin	Liver	A plasma protein
Factor VII	Proconvertin	Liver	In plasma
Factor VIII	Antihemophilic factor (AHF)	Liver ? Endothelial cells?	In plasma
Factor IX	Plasma thromboplastin component (PTC)	Liver	In plasma
Factor X	Stuart-Prower factor	Liver	In plasma
Factor XI	Plasma thromboplastin antecedent (PTA)	Liver	In plasma
Factor XII	Hageman factor	Unknown	In plasma
Factor XIII	Fibrin stabilizing factor (FSF)	Unknown	In plasma
Platelet factor	Cephalin	Bone marrow	In platelets
Vitamin K	Menadione, etc.	Intestinal bacterial synthesis and normal diet	In plasma

If, however, heparin, sodium oxalate, or sodium citrate is used (in the syringe or test tube), the calcium of the blood reacts with the substance and is therefore not available to react with the prothrombin: no thrombin is formed. In consequence, the fibrinogen does not become fibrin and the blood cannot clot.

Hemophilia (bleeding disease) occurs in sons who have received the defective X chromosome from their mothers that results in deficiency of clotting Factor VIII.

Figure 16–8 Scanning electron micrograph of an erythrocyte enmeshed in fibrin. Part of a thrombus (blood clot) found on the inner surface of an intravenous catheter implanted proximal to the heart (about ×20,500.) (Emil Bernstein and Eila Kairinen, Gillette Company Research Institute, Rockville, Maryland.)

Table 16–3 Inheritance of Human Blood Types

BLOOD GROUP OF INDIVIDUAL AS DETERMINED BY BLOOD TYPING	PRESENCE IN ERYTHROCYTE OF AGGLUTINOGENS (ANTIGENS)
PHENOTYPE	GENOTYPES*
A	I^AI^A or I^AI^O
B	I^BI^B or I^BI^O
O	I^OI^O
AB	I^AI^B

*It should be noted that one I gene comes from the mother and one I gene from the father. It is impossible to transmit a gene to one's natural offspring if one does not have that gene.

THE HUMAN A, B, AB, O BLOOD TYPES

Every human individual is born with a blood type of the A, B, AB, or O blood group series that is inherited from his parents. There are numerous other blood group factors or substances: Le^a, Le^b, MN, P, S, Rh, and so on. Since 23 chromosomes of each set of 46 are normally transmitted from *each* parent, the genes on two homologous chromosomes determine the genotype of an individual for any trait, including blood groups. For simplicity, in the A, B, and O groups three alleles (different expressions of a gene) have been postulated to explain this inheritance: they are I^A, I^B, and I^O. In reality, subtypes of these alleles are known and play an important role in genetics. Some cause difficulties to the offspring somewhat like those resulting from the Rh factor, which will be discussed later in this chapter. Table 16–3 shows a sample diagram explaining the inheritance of these blood types. We may assume that a person's own *serum* does not contain agglutinins against his own erythrocytes, but only against the absent agglutinogen(s). Isohemagglutinins do not agglutinate (clump) one's own RBC's. As can be see from Table 16–4, serum from a person with blood type A produces isohemagglutinins[7] (antibodies) against type B erythrocytes (anti-B); O serum contains anti-A and anti-B antibodies; but AB serum contains no antibodies against any RBC's at all. The term *iso-* implies

[7]These agglutinate erythrocytes of another individual of the *same species.*

Table 16–4. Isohemagglutination. International System of Blood Groups*

RECIPIENT'S *Blood Group and Sera (Antibodies)*	BLOOD DONOR'S *Erythrocyte (Antigens) of Blood Group* *AB*	*B*	*A*	*O*
AB (contains no antibodies)	–	–	–	–
B (contains anti-A antibodies)	aggl**	–	aggl	–
A (contains anti-B antibodies)	aggl	aggl	–	–
O (contains anti-A and anti-B antibodies)	aggl	aggl	aggl	–

*This table is an oversimplification, since blood subtypes like A_1 or A_2 in A_1B or A_2B, and the "Bombay type," exist in this system. For example, A_2B may occasionally form anti-A_1 antibodies.

**aggl means agglutination; – means no agglutination. For example, a person with blood type AB cannot donate blood to a person with blood type B, who has natural anti-A antibodies.

that these hemagglutinins (antibodies) are inherited, i.e., are from the same species and not acquired through previous exposure or immunization.

A person's blood type is a physiologic constant determined by the basic Mendelian laws of inheritance. By properly selecting donors and recipients of compatible groups, it is possible to avoid hemagglutination following transfusions. Most hospitals have blood and a list of donors of various groups available. Blood banks store citrated blood donations already grouped and held in the refrigerator until needed. A match between patient and donor RBC's, serum, and plasma is always highly advisable; this is referred to as "cross matching."

Blood Grouping. *Cells* of the recipient, usually prepared by adding a drop of the patient's blood to a small amount of physiologic salt solution (0.9 per cent NaC1), are mixed on a microscope slide (or white porcelain plate) with specially prepared rabbit sera containing pure anti-A or anti-B antibodies. Agglutination, if any, is usually prompt and readily visible microscopically or macroscopically (or both) (Fig. 16–9).

If the cells of the recipient are agglutinated by both anti-A and anti-B rabbit sera, the cells are of group AB; if the cells are agglutinated by neither serum, they are of group O. If anti-A serum agglutinates them and anti-B serum does not, the cells are of group A; if anti-B serum agglutinates them and anti-A does not, the cells are of group B. The *serum* of the recipient is then tested against the red cells of the donor blood to be sure that no mistake has been made and to eliminate certain irregularities in reaction that sometimes occur.

The microscopic reactions are shown schematically in Figure 16–9. Note that in this diagram agglutination is by specially prepared rabbit sera, containing antibodies against group A or group B red blood cell antigens of persons with any one of the four possible blood types.

"Universal" (?) Donors. A person of group O is sometimes inaccurately (and unfortunately) called a "universal" donor because his cells are not usually agglutinated by serum of recipients of any other group. However, he must receive only O blood because his serum agglutinates cells of all other groups in the A-B-O system.

Actually, there is no such thing as a safe "universal" donor (without direct cross matching between recipient and donor), because although the erythrocytes of a person of group O do not contain A or B antigens (i.e., are not agglutinated by A or B group sera), there are several other antigen-antibody systems in human blood such as the Rh factor, M-N-S types, P types, and so on, which

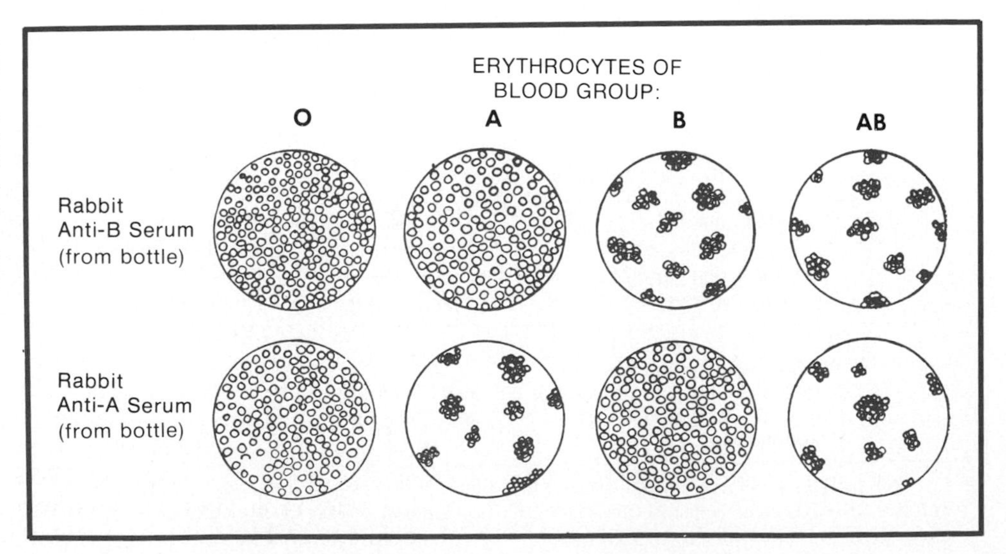

Figure 16–9 Hemagglutination in blood grouping (blood typing in the laboratory). Anti-B serum and anti-A serum have been mixed on glass slides with erythrocytes of groups O, A, B, and AB. Anti-B serum agglutinates erythrocytes of groups B and AB; anti-A serum agglutinates erythrocytes of groups A and AB.

may cause serious transfusion reactions regardless of the A, B, O, or AB group (see Table 16–4).

H Antigen. Although typical O cells have neither A nor B isoagglutinogens, the serum of O type persons contains antibodies against *both* A and B blood types. The I^A gene determines the A antigen, and I^B gene the B antigen (Table 16–3), and the I^O gene determines neither A nor B. O cells nevertheless have a specific heterogenetic or *H* agglutinogen that reacts (like the heterogenetic Forssman or heterophil antigen) with sera from several species of lower animals as well as with sera from humans. When the H antigen (also A and B antigens) is absent from certain unusual O type persons first observed in Bombay ("Bombay types"), the serum contains not only A and B antibodies but also H antibodies that agglutinate O cells.

In one study of a population in the United States the following relative composition of blood subtypes of the A, B, AB, O group was recorded.

	(%)		(%)
A_1	34	A_1B	3
A_2	10	A_2B	1
B	9	O	43

Rh Factor

In 1940 it was discovered that if erythrocytes of *Rh*esus (Rh) monkeys were injected into rabbits, the serum of the rabbits soon acquired anti-Rh agglutinins. These antibodies could also agglutinate the erythrocytes of 85 per cent or more of human beings. The antigen in the red cells of humans with which the anti-Rh rabbit serum reacted was designated the Rh agglutinogen, and individuals possessing it are labeled Rh positive (+). The remaining 15 per cent, i.e., person with red cells that do not react with anti-Rh serum because of the absence of the Rh agglutinogen, are designated Rh negative (−). Later studies showed that the Rh agglutinogen is a complex one, corresponding to several antigenic determinations or *blood factors,* the most important of which is called the **Rh_o (Rh)** factor. Some of the other, related factors are: **Hr_o, hr′, rh′, hr″, rh″,** and so on. (The *same* factors are also designated as Rho, rh, Rho′, rh′, Rho″ and rh″, as shown in Table 16–5.) Here we need consider only the **Rh** factor as being representative.

Nomenclature. Some confusion in names has arisen because of the use of various symbols and ideas to express these antigen-antibody relationships. The Wiener Rh system and the Fisher C-D-E system, although based on different genetic theories and not properly interchangeable are nevertheless sometimes used as approximate equivalents (see Table 16–5).

In the Fisher system a person with any group of agglutinogens containing D (Rh_o) or a common variant of D called "weak D" or D^u is Rh positive. All others would be Rh negative. Thus, an Rh positive individual could be genotype CcDDee or ccDdEe or any other possible combination, as long as he or she has six determining genes, one of which is D (or Rh_o) for the Rh factor. The A-B-O and Rh blood group systems, complicated as they seem here, are actually very much more complex. For example, besides genes C, D, E, c, e, and perhaps d, genes F, f, G, D^u, E^u, E^w, V antigen, and LW antigen have also been demonstrated in the Rh system. Some of these contribute to serious reactions between bloods of recipients and donors, making it desirable always to do direct cross matching (cells of donor + serum of recipient and cells of recipient +

Table 16–5. Symbols and Frequency of Rh Types and Reactions of Cells with Antisera

	RH TYPE OF CELLS*			REACTION WITH ANTISERA				
	Fisher	*Wiener*	APPROXIMATE %	*Anti-D*	*Anti-C*	*Anti-E*	*Anti-c*	*Anti-e*
Rh negative 15%	cde	(rh)	13.5	–	–	–	+	+
	Cde	(rh′)	1.0	–	+	–	+	+
	cdE	(rh″)	0.5	–	–	+	+	+
	CdE	(rh′rh″)	rare	–	+	+	+	+
Rh positive 85%	cDe	(Rh_0)	2.0	+	–	–	+	+
	CDe	(Rh_0')	53.0	+	+	–	+	+
	cDE	(Rh_0'')	15.0	+	–	+	+	+
	CDE	($Rh_0'Rh_0''$)	15.0	+	+	+	+	+

*The Fisher types are also known as the Race-Sanger types. (After Fisk, R. T.: A Manual of Blood Grouping and Rh Typing Serums. Los Angeles, Hylands Laboratories, 1956.)

serum of donor) before any transfusions. The Rh status of recipient and donor should also be known.

Rh Factor and Transfusion. If an Rh negative person receives a blood transfusion from a person who is Rh positive, the recipient develops antibodies (in about two weeks) that agglutinate the red cells of all Rh positive persons. Now, if a subsequent transfusion is given from an Rh positive donor and the Rh antibody concentration (titer) of the recipient is sufficiently high, all the red cells of the donor are agglutinated and later hemolyzed, and a severe transfusion reaction occurs, often with a fatal outcome. Thus, it is of vital importance to determine the Rh status as well as the blood group (A, B, and so on) of donors and recipients before each transfusion.

Erythroblastosis Fetalis. Blood group antigens (factors) that may lead to erythroblastosis fetalis are, in order of occurrence: Rh $>$ A and B $>$ other antigens of blood groups, such as K and k, S and s, Fy^a and Jk^a. Only the Rh system will be discussed here. If an Rh negative woman becomes pregnant by an Rh positive man, the chances are good that the child will inherit Rh antigen in its blood cells. During pregnancy, if the fetus is Rh positive, fetal red cells often cross the placental barrier and pass into the blood of the mother. Acting as an antigen, the Rh^+ cells from the fetus may then stimulate the maternal tissues to produce Rh antibodies. These antibodies, as soluble components of blood plasma, freely diffuse back into the fetal circulation. In an *initial* pregnancy, the mother's Rh antibody titer may be too low to produce any harmful effects in the fetus. With subsequent pregnancies, however, the increase in concentration of Rh antibodies may be sufficient to destroy the red cells of the new fetus. Thus the baby, possessing both the Rh antigen (i.e., Rh^+) and Rh antibodies may be born dead (uncommon) or with hemolytic disease of the newborn (erythroblastosis fetalis), which is not uncommon. A similar situation may occur if the mother, being Rh negative herself, has previously received a transfusion from an Rh positive donor.

If the Rh agglutinogen should represent a complicating factor in pregnancy, a precautionary measure frequently employed is the periodic examination of maternal blood for a *rising titer* of Rh agglutinins.

A newborn child with erythroblastosis fetalis receives what is called a "blood exchange." This is a slow dilution procedure in which a small amount of blood is withdrawn from the baby and replaced by the same type of blood without the Rh antibodies in it. This process may be repeated for hours, as long as is necessary. Some uninformed students visualize blood exchange as draining the baby of all blood and replacing it with new blood. If this were done, the baby would be dead.

Rh immune globulin (RhIG), now available, is derived from human donors whose blood is rich in Rh antibodies. Injected intramuscularly into Rh negative women lacking Rh antibodies within 72 hours after delivery of an Rh positive infant or abortion, the RhIG suppresses the formation of Rh antibodies, giving protection during a *subsequent* Rh positive pregnancy or abortion. The protection may be needed for each subsequent pregnancy or abortion.

SUMMARY

The blood serves as a means of intercommunication among the cells and tissues of the body, transporting food, hormones, oxygen, and carbon dioxide to and from cells and removing waste products. Blood also contains substances and living cells that help the body combat diseases caused by microorganisms. The lymph nodes act as a system of filters removing cell detritus, bacteria, dirt, and other particles from the circulatory system.

When blood is permitted to clot, the liquid is called serum; where heparin, oxalate, or citrate are used to prevent clotting, the liquid portion of the blood is the plasma, obtained after centrifugation. Plasma contains the plasma proteins, (α_1, α_2, β, and γ globulins, fibrinogen, and albumin) minerals, uric acid, urea, amino acid, glucose, lipids, hormones, CO_2, and water.

All blood cells originate from undifferentiated "stem cells" in the bone marrow. Normoblasts originate in the bone marrow, give up their nuclei, and become erythrocytes (and blood cells) in the blood, where they transfer oxygen from the lungs to the tissue cells. When they become old they are called poikilocytes. These non-nucleated biconcave disks are 7.5 μm in diameter and live approximately 130 days. Several types of leukocytes (white blood cells) exist. They are larger than the erythrocytes and less numerous. The granulocytes, also called polymorphonuclear leukocytes, are phagocytic cells—the eosinophils, basophils and neutrophils—and are formed in the bone marrow. The agranular leukocytes, the lymphocytes (T-lymphocytes and B-lymphocytes) are derived either from the thymus (T-lymphocytes) or from lymphoid tissue, spleen adenoids or lymph nodes (B-lymphocytes). The phagocytic monocytes (macrophages) originate from promonocytes in the bone marrow. They attack certain microorganisms and tumor cells and present antigens to the lymphocytes.

Macrophages are large, phagocytic cells, either leukocytes in the blood or stationary fixed phagocytic cells in the "mononuclear phagocyte system" (MPS), also called the reticuloendothelial system (RES), or (as cells) the histiocytes, the garbage collectors of the body. Small lymphocytes and plasma cells are important in the production of specific antibodies (gamma globulins). Infections alter the differential count of leukocytes and often may be diagnosed in this way.

Blood platelets are formed in the bone marrow from megakaryocyte cell fragments. They play a role in blood clotting. At least 15 clotting factors are involved in stopping blood flow from a wound. If any of these are missing blood will not clot in the normal time of 2 to 6 minutes. In cases of hemophilia, the bleeder's disease, this may even lead to death.

Every human individual is born with either A, B, AB, or O type blood, as well with Rh+ or Rh− type. A person that has blood type B has the antingen for B on the surface of the erythrocytes and antibodies against A in the serum. Type AB has both A and B antigens on the erythrocytes and no antibodies in the serum, whereas type O has anti-A and anti-B serum. Erythroblastosis fetalis is a condition in the newborn child (other than the first born) of an Rh+ father and Rh− mother that requires a "blood exchange." This dilutes out the antibodies against the Rh+ factor (antigen) that a baby receives from a mother previously sensitized against an Rh+ child.

REVIEW QUESTIONS

1. What are the functions of blood? of lymph?
2. Compare the composition of plasma with that of whole blood, serum, and lymph.
3. What plasma proteins can you name? What are they? How can they be separated?
4. What major blood cells are there? Where do they originate? What is their function?
5. What are phagocytes, B-lymphocytes, T-lymphocytes, and plasma cells?
6. Identify MPS, RES, and macrophages.
7. What are the relationships between antigens, antibodies, and blood groups?
8. Explain why blood clots and why it may not clot properly.
9. Who is a universal donor?
10. Why is blood typing so important?
11. Explain the immunological basis of erythroblastosis fetalis and of transfusion reactions involving the A-B-O system.

Supplementary Reading

Altman, P. L., and Dittmer, D. S. (eds.): Biology Data Book. 2nd ed. Bethesda, Md., Federation of American Societies for Experimental Biology, 1972–1974.

American Hospital Supply Corporation: Blood Group Immunology. Miami, Fla., Dade Division, American Hospital Supply Corp., 1976.

Blood Group Antigens and Antibodies as Applied to Compatibility Testing. Raritan, N.J., Ortho Diagnostics, 1967.

Blood Group Antigens and Antibodies as Applied to the Hemolytic Disease of the Newborn. Raritan, N.J., Ortho Diagnostics, 1968.

Blood Group Antigens and Antibodies as Applied to the ABO and Rh Systems. Raritan, N.J., Ortho Diagnostics, 1969.

Carpenter, P. L.: Immunology and Serology. 4th ed. Philadelphia, W. B. Saunders Company, 1977.

Clarke, C. A.: The prevention of "Rhesus" babies. *Sci. Amer., 219*:46, 1968.

Davie, E. W., and Fujikawa, K.: Basic mechanisms in blood coagulation. *Ann. Rev. Biochem., 44*:799, 1975.

Delaat, A. N. C.: Primer of Serology. New York, Harper & Row, 1976.

Gilmore, C. P.: The Scanning Electron Microscope: World of the Infinitely Small. New York, New York Graphic Society, 1972.

Gershowitz, H., and Neel, J. V.: The blood group polymorphism: Why are they there? *In* Aminoff, D. (ed.): Blood and Tissue Antigens. New York, Academic Press, 1970.

Hermann, M.: Rh-prophylaxis with Immunoglobulin Anti-D Administered During Pregnancy and After Delivery. Stockholm, Scandinavian Association of Obstetricians and Gynaecologists, Almqvist and Wiskell, 1976.

Kieth, L., Cuva, A., Houser, K., et al.: Suppression of primary Rh-immunization by anti-Rh. *Transfusion, 10*:142, 1970.

Levine, P.: Prevention and tretment of erythroblastosis fetalis. *Ann. N.Y. Acad. Sci., 169*:234, 1970.

Mohn, J. F., et al. (eds.): Human Blood Groups. New York, S. Karger, 1977.

Morgan, W. T. J.: Molecular aspects of human blood-group specificity. *Ann. N.Y. Acad. Sci., 169*:118, 1970.

Oatley, C. W.: The Scanning Electron Microscope: Part I, The Instrument. Cambridge, Cambridge University Press, 1972.

Queenan, J. T. (ed.): The Rh problem. *Clin. Obstet. Gynecol., 14*:491, 1971.

Race, R. R., and Sanger, R.: Blood Groups in Man. 6th ed. Philadelphia, J. B. Lippincott, 1975.
Raphael, S. S. (ed.): Lynch's Medical Laboratory Technology. 3rd ed. Philadelphia, W. B. Saunders Company, 1976.
Wiener, A. S.: Modern blood group nomenclature. *J. Amer. Med. Technol.*, *30*:174, 1968.

INFECTIONS AND ANTIGENS

Blood and Infection. Many species of pathogenic microorganisms can enter the tissues and/or the blood and grow,[1] often with fatal effects to the victim. Sometimes microorganisms present in infected tissue may enter leukocytes and "hitchhike" to other tissues, carried by the blood. If blood circulates through tissues in which pathogenic microorganisms are producing poisonous substances (toxins)—as in diphtheria, tetanus, and some diseases caused by streptococci—the toxins enter the blood and are distributed, with detrimental effects, to all the tissues in the body.

In diseases in which the microorganisms causing the condition are transported by blood or cells, the pathogen may establish itself in other organs as well, causing such complications as brain abscesses, osteomyelitis, and middle-ear infections (otitis media). Numerous blood-borne diseases (malaria, yellow fever, plague, and others) are regularly transmitted by various extraneous means, especially mosquitoes, from person to person over great distances.

INFECTION

Host-Parasite Relationships

In the course of evolution many species of living organisms, forced to coexist, have established a variety of interrelationships. The field of science that investigates these relationships in the context of their environment is called *ecology*. Some living organisms are mutually helpful (*symbiosis* or *mutualism*); some are mutually antagonistic (*antibiosis*). Others coexist with various degrees of tolerance or intolerance. Some organisms thrive at the expense of others (*predation* and *parasitism*). Most infectious diseases represent this last type of ecologic relationship.

When one creature (a *parasite*) attempts to thrive at another's (a *host's*) expense, the host nearly always reacts in some manner to evade, remove, or kill the parasite. If the host fails to eliminate the parasite, the two often become mutually adapted (sometimes only after generations of contact and dwindling antagonism) and establish a sort of armed truce in which they live, more or less peaceably, side by side (*commensalism*). Many curious host-parasite relationships and adaptations are exhibited in the processes we call *infection, the carrier state, immunity*, and *allergy*.

There is nothing vindictive or purposeful in microorganisms, any more than the growth of weeds or poison ivy in a flower garden is vindictive or purposeful. For convenience in discussion, however, an infectious disease may be regarded as a struggle between an attacking force (the parasite) and a defending body (the host). Both combatants have their means of "offense" and "defense." We shall first consider the struggle from the side of the parasitic microorganisms.

Microorganisms may gain entrance to the tissues of the body and grow in or on them, injuring them and producing a reaction in the host. This is an

---

[1]This condition is called *septicemia* (Greek, *septikos*, putrefactive, and *haima*, blood; hence, "blood poisoning").

CHAPTER 17

infection. (Some authors distinguish between *infection* and *disease*, the former term being used to describe the mere presence of organisms, the latter term designating the detrimental effect upon the host. The need for such a distinction arises from the widespread use of the phrases "inapparent [or *subclinical*] infection" and "latent infection," in which no detrimental effect is detectable.) The sensed or demonstrable (clinical) aspects of this reaction to infection are commonly fever, chills, nausea, and headache. Some infections are so mild that they cause hardly any discomfort; others are fatal. A pimple, for example, is a trivial reaction to an infection of the skin; typhoid fever is a serious and—without antibiotic treatment—often fatal infection of the blood, intestine, and other organs.

Factors in Infection

Dose of Infecting Organisms. An important factor in determining whether infection shall or shall not occur is the number of infecting organisms. A single coccus or a single bacillus, even though highly virulent, may not be able to establish itself in the healthy body. If a hundred or a million such organisms were to be implanted, they might well overcome even strong resistance. Mildly virulent or even supposedly harmless organisms may set up an infection if their number is sufficiently large, especially if tissue resistance is lowered by other factors, such as disease, fatigue, drugs, or alcoholism.

Pathogenicity. *Pathogenicity* is the ability to produce pathological changes that are referred to as disease. It is more a qualitative term, whereas *virulence* is a quantitative concept.

Virulence of Organisms. Virulence is the ability of a microorganism to establish, maintain, and extend an infection and to damage the body (thereby producing a disease). Virulence depends upon at least two other properties, *aggressiveness* and *toxigenicity*, each of which is contingent on a number of other properties. Actually, exact knowledge of all the factors, and their interrelationships, that contribute to virulence is still incomplete.

Aggressiveness. An aggressive organism is one that is not unfavorably affected by the normal resistance of the body or too rapidly ingested by histiocytes and circulating phagocytes, and can therefore enter the blood or tissues and grow there vigorously at the expense of the victim. Degrees of aggressiveness vary greatly. For example, the mucous membranes of the upper respiratory tract are inhabited by large numbers of bacteria that are not equipped to penetrate the unbroken membrane or overcome normal resistance. Many are potentially dangerous pathogens. They remain on the surface without causing an infection because they are readily held in check by the normal protective mechanisms of the body. Once these organisms gain entrance to the deeper tissues, because of injury or weakening of the defensive mechanisms by age, disease, and so on, they may exhibit great "aggressiveness" simply because they find conditions favorable to good growth, like weeds in a fertile field, and because there is nothing to oppose them. Other more aggressive bacteria require only a very slight opportunity to establish themselves even in normal and healthy tissues, which they can invade in spite of the protective system opposing them.

Organisms initially of relatively low aggressiveness may rapidly become highly aggressive in reaction to the defensive mechanisms of the host. They appear to undergo various physiologic and biochemical changes, some known (such as genetic mutations and enzyme induction), others unknown, that enable them to evade the host's defensive mechanisms. One of the most readily demonstrable of these changes is the formation of a sort of protective armor in the form of a thick, usually viscous, polysaccharide capsule, often referred to as SSS (soluble specific substance). Thus, microorganisms in the blood or body fluids of patients are particularly likely to be highly dangerous. This is of vital importance to visitors in hospitals and to nurses, laboratory technologists, bacteriologists, physicians, and other health personnel.

Toxigenicity. This is the capacity to produce a toxin or poison. For example, if a toxin is produced by a bacterium, the organism does not need to be very aggressive (i.e., invasive) in order to be virulent. It can damage the body by means of its toxin and can paralyze or destroy the powers of resistance of the cells in which it localizes. It may be unable to invade beyond very superficial tissues, but if only a few very toxigenic bacteria lodge in the throat on the tonsils or in small foci, such as abscesses at the roots of the teeth or in wounds made by the puncture of a nail, they may kill their victim by producing "poison" that is absorbed and carried all through the host's system.

This is exactly what happens in diphtheria and in tetanus ("lockjaw"). The bacteria that cause these diseases lack aggressiveness almost entirely. Yet, by maintaining a slight foothold in the crushed, dead tissues of a mangled limb (tetanus) or on the surfaces of the tonsils (diphtheria), they can fatally poison the patient. They are extremely virulent because of their toxigenicity, not because of their aggressiveness. Highly invasive or aggressive organisms, such as the streptococci that cause septicemia ("blood poisoning") and the staphylococci that cause such diseases as meningitis and osteomyelitis, are particularly virulent because they produce several kinds of toxins, some of which destroy leukocytes. Many other destructive substances, e.g., lipid-digesting enzymes *(lecithinases)* and enzymes that digest intercellular "cement" substance *(hyaluronidases)*, are produced by a variety of virulent microorganisms.

Toxins

Toxigenic microorganisms produce their toxic effect in one or both of two ways.

Endotoxins. These extremely poisonous substances (although somewhat less toxic than the exotoxins) are components of the cell structure, chiefly of gram-negative bacteria. *Endotoxins* (from Greek *endon*, inside) are the lipopolysaccharide complexes of the cell walls; therefore, to study these toxins it is necessary to break up the cells. Also, unlike exotoxins, the endotoxins do not form toxoids when treated with detoxifying chemicals such as formaldehyde. Examples of endotoxin-producing bacteria are the typhoid bacillus (*Salmonella typhi*), the dysentery organism *(Shigella dysenteriae)*, the gonococcus *(Neisseria gonorrhoeae)*, and even relatively avirulent strains of *Escherichia coli*. Endotoxins often produce fever in the host, whereas exotoxins do not.

Exotoxins. Toxigenic bacteria may also form toxic proteins that diffuse from the cell through the cell membrane and wall. Such toxins are called *exotoxins* (from Greek *exo*, outside). Examples of bacteria that produce exotoxins are *Corynebacterium diphtheriae, Clostridium tetani, Clostridium perfringens*, and *Staphylococcus aureus*. Some bacteria form both exotoxins and endotoxins.

Most exotoxins are relatively heat-labile proteins. Just how they injure the tissues of the body is not yet fully understood. Some combine with certain vital tissues—nerves, heart muscle, liver, and so on—in some noxious manner. The "neurotoxin" of *Shigella dysenteriae* destroys the small blood vessels in the brain and spinal cord. Others injure the phagocytes. Some are types of digestive enzymes and digest living cells or parts of cells. Some of these destroy erythrocytes *(hemolysis)*. It should be noted that the production of toxic substances is not limited to microorganisms, but is frequently observed in plants (such as some "toadstools," e.g., *Amanita phalloides*) and animals (e.g., poisonous fish and spider and snake venoms).

Specificity of Toxins. The toxin of each microorganism differs chemically from all others. Thus, diphtheria toxin, an exotoxin, injures the kidneys, nervous tissues, and heart particularly. Tetanus toxin (also an exotoxin) injures certain nerves and thus produces the spasms of lockjaw, and botulinum toxin produces paralysis. As noted previously, staphylococci produce several toxins. One of these, called *enterotoxin*, which happens to be an exotoxin, produces violent vomiting and diarrhea when swallowed (staphylococcal food poisoning). The term *enterotoxin* implies that the poison acts in the intestine.

Tissue-Damaging Enzymes. Many tissue-damaging enzymes are secreted by bacteria. They are not real toxins but are detrimental to the host. Streptococci produce fibrinolysins (streptokinase); hyaluronidase affecting connective tissue results from staphylococci, streptococci, clostridia, and others; and collagenase comes from clostridial species and *Bacteroides*. Hemolysins and leukocidins destroy blood cells, and certain proteases act on antibodies and prevent phagocytosis.

Variations in Virulence

Virulence may vary. A highly pathogenic organism may become attenuated or lessened in virulence (or pathogenicity) by growing in certain unfavorable situations, by contact with certain substances, or because of genetic mutations. Bacteria maintained in laboratory culture media may lose some or all of their virulence for animals or human beings. Like hothouse plants, they diminish their native power to fend for themselves when treated to a protected existence in an artificially suitable environment. Bacteria produce less capsular material in cultures propagated on laboratory media; and the morphology of their colonies changes from smooth to rough. However, one can never safely say that an organism has entirely lost its virulence unless very careful and extensive tests have been made to establish this fact. By passing organisms of relatively low virulence from one animal or person to another, their virulence may be greatly increased. These apparent changes in microbial virulence probably result from faster growth of, and eventual predominance by, certain individual mutants among the millions of bacterial cells, such mutants being the most capable of multiplying under the condition of cultivation (or infection or chemotherapy) to which the entire microbial population is subjected.

In the next two chapters two types of resistance to infection will be discussed: *nonspecific* resistance (Chapter 18) and *specific* resistance or "immunity" (Chapter 19).

ANTIGENS

Any substance that, upon entering the blood or tissues of the body, elicits the formation of a specific antibody and/or a sensitized T-lymphocyte with which it will react is called an *antigen* or *immunogen*. Specifically, an antigen is an immunogen when its function is to induce an immune response in the host. As already noted, the body responds to antigens with the production of antibodies, which act as specific "antidotes" to the antigens and react with them to destroy or remove them. It should be explained that, unless otherwise noted, the discussions of antigens, the reaction of body cells to antigens, and the formation and role of antibodies in Chapters 18 through 20 refer to *initial contact* of the body cells with an antigen. This contact, in addition to inducing specific immunity or tolerance, may also produce a state of *hypersensitivity* or *allergy* on repeated later contact with the same antigen (see Chapter 21).

For some time it was thought that, with a few important exceptions, the only substances that could act as antigens were water-soluble *proteins* or non-protein substances combined (conjugated) with proteins, and that if a substance was a protein or a protein conjugate, it could probably act as an antigen. Indeed, proteins and protein conjugates are the most active of all antigens, which include macromolecules such as nucleoproteins and lipoproteins. It is now recognized, however, that *carbohydrates*, such as pneumococcal polysaccharides, carbohydrate–amino acid compounds, the yeast polysaccharide zymosan, synthetic *polypeptides*, or even *lipoidal* substances (e.g., sterols or lecithin) may be antigenic in certain animals or in humans. Even *nucleic acids* such as DNA and RNA have been shown to be antigenic in rabbits, most likely because of specific functions of the purine and pyrimidine bases.

Antigens need not be poisonous or of microbial origin. Such substances as egg white (ovalbumin), serum, milk, snake venom, dead or living bacteria, bacterial exo- and endotoxins, plant or animal tissues, or their derivatives (chiefly proteins or protein conjugates) may act as antigens when introduced subcutaneously or into deeper body tissues. Most substances taken into the normal gastrointestinal tract as food do not act as antigens because their molecular structure is quickly destroyed by the digestive juices. To act antigenically the specific substance must gain entrance to the blood and other tissues in a chemically unaltered state. Entrance is commonly by infection, by injection, or parenterally (from Greek *para*, beyond or outside of; *enteron*, intestine), i.e., by routes other than the gastrointestinal tract. There are certain exceptions, such as food allergies (Chapter 21) and contact skin reactions (for example, poison ivy). A specialized class of immunogens called allergens (such as pollens) may be inhaled and thus reach the blood by way of the lungs.

Antigen Classification. Substances that are normally native to the body components do not generally bring about an antibody response. We refer to these chemicals as *self*, whereas foreign materials, such as antigens, may be spoken of as *nonself* or *foreign*.

Antigens may be classified in several different ways (Table 17–1). Exogenous types of antigens are those coming from the exterior of the organism—from viruses, bacteria, pollen, and drugs—whereas endogenous antigens tend to be present in organs or tissues. Among the antigens in organs, many are shared by other tissues, but some are organ-specific, so that one can obtain anti-kidney serum from a rabbit properly sensitized with nonself (foreign) kidney extract. All members of a species share certain tissue antigens that are unique for that species. However, antigens in certain tissue components may be very similar (if not alike) in very diverse species, for example the collagens found in bone, cartilage, and connective tissue.

One heterologous antigen—present in many diverse species of animals, plants, and bacteria, and closely related to a substance found in type A blood —is called the *Forssman antigen*, an example of a heterophil antigen. In many species it is found either in the tissues or in the erythrocytes, but not in both. The heterophile antibody response is used in a clinical test to diagnose infectious mononucleosis. The virus causing the disease contains the Forssman antigen, which is detected in the test.

Blood group antigens (see Chapter 16) and histocompatibility antigens, involved in tissue or organ transplantation rejection mechanisms, are called *isoantigens*. Isoantigens are inherited and exist in the tissues of some members of a species, but not all. These genetically determined polymorphic antigens are present on erythrocytes, on leukocytes, on platelets, on the surface of cells, and in serum proteins, while histocompatibility antigens are found in fixed tissues of the body.

Haptens. As shown by Landsteiner, Pauling, Boyd, and others, numerous substances not ordinarily thought of as antigens (i.e., not immunogenic)—

Table 17–1. Classification of Antigens

Source	Type	Example	Clinical Significance
Exogenous	Several	Microorganisms, pollen, drugs, pollutants	Susceptibility to infection, immunologically mediated disease (bronchial asthma)
Endogenous			
Xenogeneic (Heterologous)	Xenoantigen (Heteroantigen)	Forssman antigen, certain tissue antigens that crossreact with exogenous antigens (e.g., renal and beta-hemolytic streptococcus)	Pathogenesis of certain diseases (e.g., glomerulo-nephritis, rheumatic fever)
Autologous (Self component)	Autoantigen	Organ-specific antigens (e.g., thyroid antigen)	Autoimmune disease (e.g., Hashimoto's thyroiditis)
Allogeneic (Homologous)	Alloantigen (Isoantigen)	Blood group, histocompatibility antigens (HL-A)	Hemolytic disease of the newborn, transfusion reactions, transplantation immunity

From Bellanti, J. A.: Immunology: Basic Processes. Philadelphia, W. B. Saunders Company, 1979.

including drugs such as aspirin and penicillin, synthetics such as perfumes, and rubber, all of which lack the chemical structure and molecular weight to make them antigenic *per se*—can, nevertheless, be conjugated with proteins and thus become fully and specifically antigenic. As a class, such substances are called *haptens* or *partial antigens* or, if derived from larger molecules, *residue antigens*. Many of them, even if entirely separated from the proteins that made them antigenic, can combine with the specific antibodies. In such combinations they do not typically cause a precipitin or other demonstrable antigen-antibody reaction; however, their combination with an antibody inhibits combination of the antibody with the true, complete, specific antigen. This is called *hapten inhibition*. Many haptenic substances can be of great importance as inducers of the *allergic state*, i.e., as *allergens* (Chapter 22).

SUMMARY

Infections result from an interrelationship of living species in which one is the host and the other the pathogenic parasite. Adaptations by the host to parasites and to foreign chemical substances in general may express themselves as infection, the carrier state, immunity, or allergy. Infection occurs when microorganisms gain entrance to the tissues of the body and grow in or on them, injuring them and producing a reaction in the host. Specifically, the mere presence of the organism is the infection, and the resulting injury is the disease. Pathogenicity is the ability to produce disease in a host. It is more a qualitative term, whereas virulence is a quantitative concept.

Toxigenicity is the capacity to produce a toxin or poison. Some bacteria are harmless by themselves, but their toxins may be deadly. Endotoxins are lipopolysaccharides, constituents of cells that often produce fever in a host when the bacteria carrying them are broken up. Exotoxins diffuse from the cell and injure the host but do not induce fever. Enterotoxins are endotoxins or exotoxins that are toxic in the intestine of the host.

Any substance that, upon entering the blood or tissues of the body, elicits the formation of a specific antibody and/or a sensitized T-lymphocyte with which it will react is called an antigen or immunogen. An antigen is an immunogen when it functions to induce an immune response in the host. Although proteins and protein conjugates are the most active antigens, carbohydrates such as pneumococcal polysaccharides, polypeptides, lipoidal substances, and nucleic acids—from viral, microbial, plant, animal, or even synthetic sources—may be antigenic. Antigens may enter the body through infective routes, as foods, from touch, or through breathing.

Substances that are normally native to the body components do not generally bring about an antibody response. We refer to these chemicals as *self*, whereas foreign materials, such as antigens, are *nonself* or *foreign*.

An example of a heterophile antigen, present in many diverse species of animals, plants, and bacteria, is the *Forssman antigen*. The inherited blood group antigens and histocompatibility antigens, involved in tissue and organ transplantation rejection mechanisms, are called *isoantigens*.

Haptens are antigenic determinants attached to carrier molecules (proteins) that stimulate the formation of specific antibodies.

REVIEW QUESTIONS

1. In the course of evolution different relationships between species have developed. Explain and give examples of these interrelationships.
2. Differentiate between parasite and pathogen, infection and disease, pathogenicity and virulence, aggressiveness and toxigenicity, and exotoxins, endotoxins and enterotoxins.
3. Define toxins. Give examples of specific toxins and organisms producing them. What are tissue damaging enzymes?
4. Define antigens. What are their relationships to antibodies and to immunogens? List examples of different groups of antigens. What are heterophile antigens? What is the Forssman antigen? What are histocompatibility antigens and what role do they play in transplantations of tissues and organs?
5. Name groups of antigens that are inherited.
6. What are haptens?

Supplementary Reading

Ajl, S. J., Montie, T. C., et al. (eds.): Microbial Toxins. Vol. 2A, Bacterial Protein Toxins. New York, Academic Press, 1971.

Bellanti, J. A.: Immunology II. Philadelphia, W. B. Saunders Company, 1978.

Bernheimer, A. W.: Cytolytic toxins of bacteria. *Science, 159*:847, 1968.

Burnet, F. M. (ed.): Immunology, Readings from Scientific American. San Francisco, Freeman and Co., 1976.

Burrows, W.: Textbook of Microbiology. 20th ed. Philadelphia, W. B. Saunders Company, 1973.

Davis, B. D., Dulbecco, R., Eisen, H. N., and Ginsberg, H. S.,: Microbiology. 3rd ed. New York, Hoeber Medical Division, Harper & Row, 1980.

Eisen, H. N.: Immunology. New York, Harper & Row, 1974.

Fudenberg, H., et al.: Basic Immunogenetics. 2nd ed. New York, Oxford University Press, 1978.

Fudenberg, H., et al. (eds.): Basic and Clinical Immunology. 3rd ed. Los Altos, Calif., Lange Medical Publications, 1980.

Gell, P. G. H., Coombs, R. R. A., and Lachman, P. J. (eds.): Clinical Aspects of Immunology. 3rd ed. Oxford, Blackwell Scientific Publications, 1975.

Good, R. A., and Fisher, W. (eds.): Immunobiology. Sunderland, Mass., Sinauer Associates, 1974.

Joklik, W. K., et al. (eds.): Zinsser Microbiology. 17th ed. New York, Appleton-Century-Crofts, 1980.

Kabat, E. A.: Structural Concepts in Immunology and Immunochemistry. 2nd ed. New York, Holt, Rinehart & Winston, 1976.
Origins of Lymphocyte Diversity. Cold Spring Harbor Symposium on Quantitative Biology, Vol. 41. Cold Spring Harbor, N.Y., Cold Spring Harbor Laboratory, 1977.
Parker, C. W. (ed.): Clinical Immunology. Philadelphia, W. B. Saunders Company, 1980.
Roitt, I.: Essential Immunology. 3rd ed. Oxford, Blackwell Scientific Publications, 1977.
Rubenstein, E., and Federman, D. D.: Medicine. New York, Scientific American, Inc., 1980.
Samter, M. (ed.): Immunological Diseases. 3rd ed. Boston, Little, Brown & Co., 1978.
Schwartzman, S., and Boring, J. R., III: Antiphagocytic effect of slime from a mucoid strain of *Pseudomonas aeruginosa*. *Inf. Immun.*, *3*:766, 1971.
Smith, H.: Biochemical challenge of microbial pathogeneicity. *Bacteriol. Rev.*, *32*:164, 1968.
Smith, H., and Pearce, J. H. (eds.): Microbial Pathogenicity in Man and Animals. 22nd Symposium, Society of General Microbiology. London, Cambridge University Press, 1972.
Smith, I. M.: Death from staphylococci. *Sci. Amer.*, *218*:84, 1968.
Streilein, J. W., and Hughes, J. D.: Immunology, a Programmed Text. Boston, Little, Brown & Co., 1977.
Urbaschek, B., Neter, E., and Urbaschek, R.: Infections with Gram-Negative Bacteria and Mode of Endotoxin Actions. New York, Springer-Verlag, 1975.
Villee, C. A.: Biology. 7th ed. New York, Holt, Rinehart & Winston, 1977.

Books on Immunologic Methods

Bloom, B. R., and David, J. R. (eds.): In Vitro Methods in Cell-Mediated and Tumor Immunity. New York, Academic Press, 1976.
Bloom, B. R., and Glade, P. R. (eds.): In Vitro Methods in Cell-Mediated Immunity. New York, Academic Press, 1971.
Hudson, L., and Hay, F. C.: Practical Immunology. Oxford, Blackwell Scientific Publications, 1976.
Lynch, M. J., Raphael, S. R., Mellor, L. D., Spare, P. D., and Inwood, M. J. H.: Medical Laboratory Technology and Clinical Pathology. 2nd ed. Philadelphia, W. B. Saunders Company, 1969.
Rose, N. R., and Friedman, H. (eds.): Manual of Clinical Immunology. 2nd ed. Washington, D.C., American Society for Microbiology, 1980.
Weir, D. (ed.): Handbook of Experimental Immunology. 3rd ed. Oxford, Blackwell Scientific Publications, 1977.
Williams, C. A., and Chase, M. W. (eds.): Methods in Immunology and Immunochemistry. New York, Academic Press, 1968–1976.

Immunology Series

Advances in Immunology. New York, Academic Press.
Contemporary Topics in Immunobiology. New York, Plenum Press.
Contemporary Topics in Molecular Biology. New York, Plenum Press.
Perspectives in Immunology. Proceedings of Brooklodge Symposia. New York, Academic Press.
Proceedings of the International Congress of Immunology. Progress in Immunology I (1971), II (1974), III (1977).
Proceedings of the Leucocyte Culture Conferences. New York, Academic Press.
Progress in Allergy. New York, Academic Press.
Transplantation Reviews. Copenhagen, Munksgaard.

NONSPECIFIC RESISTANCE TO INFECTION

GENERAL MECHANISMS OF DEFENSE

The human race and other species of animals have survived these many centuries because they are equipped with various defensive physiological mechanisms. These help to prevent damage caused by biological agents such as microorganisms or their toxins that gain entrance to the body. Some of these defensive mechanisms are of a general nature and serve to protect against many types of harmful agents. They are the basis of *nonspecific immunity (resistance)* and are discussed in this chapter. Other defensive mechanisms are specific in that each is effective against a certain noxious agent and no other. The latter mechanisms involve *specific immunity (resistance)* and will be discussed in Chapter 20.

The various defensive mechanisms of the mammalian body (especially the human body) are outlined in Figure 18–1. Unless we are born with immunity (nonspecific, congenital) to a certain disease, we may *acquire* resistance to it either naturally, by being infected or receiving maternal antibodies *in utero,* or artificially, from being vaccinated against it or injected with antitoxin. Acquired resistance may be *active* artificial (resulting from vaccination), when the individual produces his or her own immune response, or *passive* artificial (on injection of antitoxin), when another individual or another species has produced the antitoxin.

Species Resistance

Some species of animals have certain diseases that are peculiar to them and are resistant to infective conditions that affect animals of some other species. For example, the lower animals never have measles or typhoid fever under natural conditions. Birds are resistant to the kind of tubercle bacilli that infect cattle and human beings. Again, humans rarely become infected with the bird (avian) type of tubercle bacilli or with fowlpox. Under natural conditions none of the lower animals contracts syphilis or gonorrhea: this is reserved for humans.

Species resistance to many mammalian diseases, especially in birds and fish, is probably due largely to differences in body temperature, although chemical and physiologic variations in diverse animals undoubtedly also play a part. A microorganism adapted to living and multiplying in one species often finds it difficult or impossible to adapt itself to life in another, finding there a most unfavorable environment. It is virulent in one host and harmless in another.

An individual of any race or species, plant or animal, because of general good health and robust condition, may be resistant to certain diseases to which members of that race or species are usually susceptible. This is largely a relative resistance, entirely nonspecific; it may vary from day to day and constitutes a resistance not to any particular disease but to disease in general. For example, a man in robust health may resist infection by pneumococci or influenza viruses discharged onto his face by a coughing patient. If, however, he becomes exhausted and weakened by starvation, cold, overwork, or some chronic disease,

CHAPTER 18

he may readily succumb to these and a variety of other infections that he could easily resist if he were healthy. Tuberculosis is an excellent example of this type of infection. Because of the interplay of so many factors that contribute to resistance and immunity, it is almost impossible to determine an infective dose of a specific pathogenic microorganism.

The First Line of Defense

The first line of defense against infective microorganisms consists of the *mechanical barriers* of the body.

The Skin. The *unbroken skin* is a good mechanical protection against microorganisms, which, with few exceptions, cannot penetrate it. The skin is a thick, scaly covering with bactericidal secretion, but if it is injured or torn, even microscopically, microorganisms may enter the underlying tissues. Here they may find favorable conditions for growth and then multiply, causing an inflammation, usually with the formation of pus. The danger of neglected accidental wounds and cuts is that they may be the starting point of a serious infection, especially in medical personnel who come into contact with infectious material from patients.

Burns of the Skin. Extensive burns impair the ability of the skin to retain water in the tissues and to oppose entrance of bacteria, which, finding warm, moist, and nutritious conditions, grow rapidly. Twenty-four hours after injury,

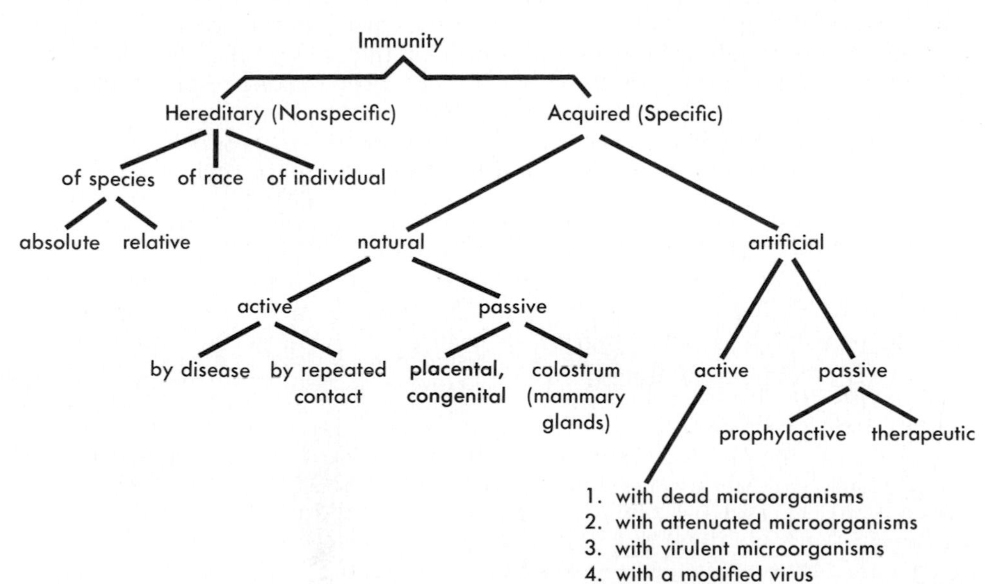

Figure 18–1 Classification of immunity. (After Professor Andrés Soriano. Adapted from Smith, D. T., Conant, N. F., and Overman, J. R.: Zinsser Microbiology. 13th ed. New York, Appleton-Century-Crofts, 1964.)

the bacterial count per gram of eschar (scab, crust) may be as high as 100 million.

Silver nitrate, in a concentration of 0.5 per cent, has been used in the past on wounds infected with *Staphylococcus aureus* and *Pseudomonas aeruginosa,* two pathogens frequently associated with severe burns. Penicillin and oxacillin are administered to all patients as a precaution against lymphangitis from *Streptococcus pyogenes* and *Staphylococcus aureus,* organisms that multiply in the lymphatic vessels.

In another method of treating burns, Sulfamylon,[1] a drug effective against most of the bacteria commonly found in burn wounds, is utilized as a 10 per cent emulsion in a cream base. Daily the wound is cleansed and loosened eschar is removed. Deep dermal wounds can now be kept relatively free of infection without resorting to skin grafting.

A substitute for skin destroyed by burns is a double layer of velour, a synthetic "skin," which is flexible and easily applied. Prosthetic skin has remained in place as long as eight months without evidence of sepsis or rejection.

Mucous Membranes. The moist surfaces of mucous membranes provide an excellent defense against potential invaders, which stick to its surface, are opposed by "normal" microorganisms present, and are destroyed or eliminated in many different ways. These membranes are in the eye, the lungs, the gastrointestinal tract, and the vagina. Tobacco smoke, alcohol, narcotics, and other unhealthy influences are known to diminish the effectiveness of mucous membranes.

The Eye. The *conjunctivae* are protected by the motion of the eyelids and the constant washing of the tears, which carry microorganisms down into the nose. Tears contain a bactericidal substance or enzyme called lysozyme, which presumably helps protect the eyes. Certain bacteria, however, may find the conjunctivae a portal of entry. Gonococci sometimes infect the eye, as do certain organisms of the genus *Haemophilus* that cause "pinkeye." Occasionally pneumococci and chlamydias infect the conjunctivae.

The Lungs. The *lungs* are safeguarded by the complicated arrangement of the nasal passages, with tiny hairs near the external openings and a moist mucous membrane. The mucous membranes are covered with a slimy secretion in which dust, bacteria, and other foreign particles are caught like flies on flypaper. The secretion is then removed by sneezing or other means. Leukocytes that have engulfed bacteria may usually be found in the normal nasal secretion. The bacteria that get by and pass into the lungs are generally carried out again by mucus and coughing and by the action of the cilia of the cells lining the trachea and bronchi (Fig. 18–2). If the bacteria are virulent respiratory pathogens, they may evade these defensive mechanisms.

The Gastrointestinal Tract. The *gastrointestinal tract* is protected to some extent by the acidity of gastric juice and by other secretions, such as mucus and

[1]Mafenide or α-amino-p-toluenesulfonamide, also known as 4-homosulfanilamide.

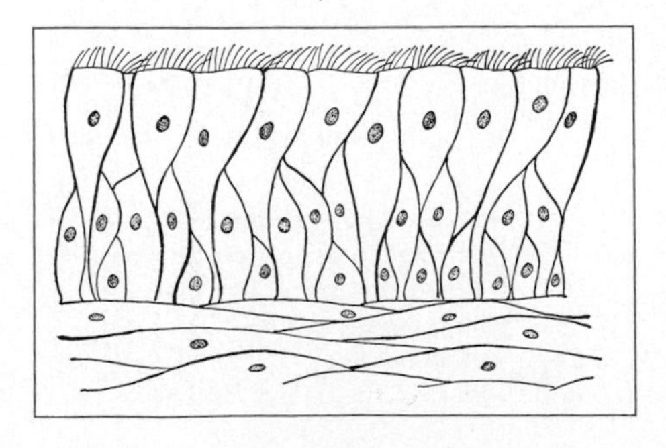

Figure 18–2 Section through the wall of a bronchus (leading to the lung), showing the ciliated epithelial cells. These cilia are in constant waving motion, carrying dirt and bacteria upward toward the mouth. The walls of the trachea and nasal sinuses are similarly lined with ciliated epithelium. (About ×1,000.)

bile. The acidity of the stomach, however, is often much diminished temporarily by foods such as milk and eggs, so that its protective action may be circumvented. In the intestine, phagocytosis and mucus are important defensive mechanisms. Bile, a secretion of the liver and the juice for digesting fats, is lethal to many bacteria.

The Genitalia. The *mucous membranes of the genitalia* are protected from most bacteria by a rather thick layer of epithelium (skinlike cells) and by acid and mucous secretions. Several organisms can, however, invade the body by means of the genital surfaces. Among these are, of course, the organisms of gonorrhea *(Neisseria gonorrhoeae)* and syphilis *(Treponema pallidum),* as well as *Streptococcus pyogenes, Streptococcus pneumoniae,* and even *Corynebacterium diphtheriae.*

Phagocytosis and the Second Line of Defense

One of the most important defensive mechanisms comprises leukocytes (wandering cells) and certain phagocytic tissue cells (the mononuclear phagocyte system). Phagocytosis was discovered in 1884 by Elie Metchnikoff, who noted that infective organisms that find their way into the body are engulfed and digested by certain cells that he named phagocytes (Figs. 18–3 and 18–4).

The Phagocytic Process. The foreign particle is detected by the surface receptors of the phagocyte and adheres to the surface of the phagocyte's plasma membrane. This somehow induces *ingestion* by invagination of the plasma membrane, resulting in a *phagocytic vacuole* that surrounds the particle. The phagocyte degranulates partially, losing granules in direct proportion to the particle ingested (Fig. 18–4). *Digestion* in the intracellular *phagolysosome* (phagocytic vacuole) depends on many factors: hydrolytic enzymes, pH 4 due to lactic acid production, reactive oxygen intermediates, hydrogen peroxide, and lysozyme to hydrolyze peptidoglycan in the cell wall of gram-positive bacteria. Consequently, some pathogens and nonpathogens may be fully destroyed within

Figure 18–3 Elie Metchnikoff (1845–1916), discoverer of phagocytosis. (From Bellanti, J. A.: Immunology: Basic Processes. Philadelphia, W. B. Saunders Company, 1979.)

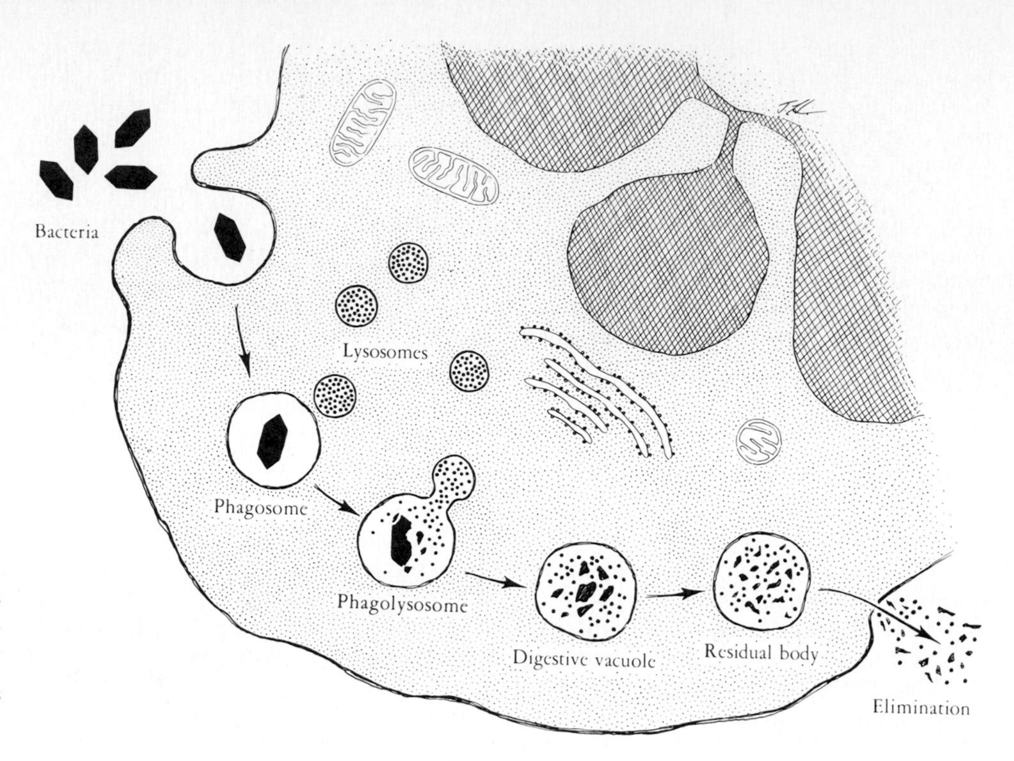

Figure 18–4 Schematic representation of phagocytosis showing ingestion process and intracellular digestion. (From Bellanti, J. A.: Immunology II. Philadelphia, W. B. Saunders Company, 1978.)

10 to 30 minutes, yet others, like the tubercle bacillus (*Mycobacterium tuberculosis*), are not affected and even multiply and finally kill the phagocyte.

Leukocytes as Phagocytes. Leukocytes may dispose of a comparatively large mass of "nonself" material by removing it piecemeal. The "core" of a boil is gradually carried off by leukocytes, as are also silk and catgut ligatures and, in part, the exudate in various inflammatory conditions. The leukocytes are attracted by anything abnormal or unusual in a tissue or in the blood, such as bacteria, a slight injury, a hemorrhage, a poison, any dead or useless tissue, or a foreign body. This attraction to the site of injury is called *chemotaxis*, and is due to the release of chemotactic factors, which are biological substances. Some phagocytes secrete enzymes that enable them to reduce digestible foreign matter to harmless liquids or in other ways dispose of it. Thus, phagocytes of all kinds constitute a *second line of defensive mechanisms* of the body, after the mechanical barriers.

Macrophages

Macrophages are large phagocytic cells, easily demonstrated histologically by the uptake of carbon. They either wander throughout tissue spaces as phagocytic monocytes (lymphocytes) or are fixed cells (histiocytes). Macrophages do not synthesize or secrete antibodies, but they are important in both cellular and humoral immunity because they present antigen to T-cells and B-cells (see Fig. 19–2) to start the immune response. Most particulate antigens reach the immune system in the phagocytized state inside phagocytes. Macrophages have a very sticky surface that traps antigens before they are ingested by the cells. If antigen is present in macrophages, its immunologic potential in stimulating differentiation of B-cells to antibody-producing plasma cells is greatly enhanced. Similarly, macrophages adherent to glass, plastic surfaces, or cells in culture stimulate other lymphocyte responses.

Fixed macrophages may be found in vascular or lymphatic epithelium. (Kupffer cells of the liver, the spleen, lymph nodes, and so forth). Although the life span of macrophages may be several weeks, neutrophil microphages live in tissues for only a few days.

Macrophages of the Mononuclear Phagocyte System. Phagocytic tissue cells or *histiocytes* (Greek, *histion,* tissue) differ from leukocytes in being relatively fixed in the tissues; they do not travel about freely. Histiocytes belong to a large group of cells lining the blood vessels and are present in the spleen, the liver, and the bone marrow. These fixed macrophages are part of the mononuclear phagocyte system (MPS) also known as the reticuloendothelial system (RES). The reader should remember that phagocytic cells ingest many things besides bacteria—bits of harmful or useless material of various kinds, such as inhaled coal dust or the remains of red blood cells. They are the garbage collectors of the body.

Microphages

Microphages are small phagocytic cells, the most numerous sort being called neutrophilic granulocytes. These neutrophils of the blood remove undesirable particles from the circulation. When these microphages move into pulmonary capillaries or into the spleen, the liver, or other tissues where macrophages are active, the work of cleaning up performed by the macrophages is much more efficient than phagocytosis done by microphages.

Antibodies

Antibodies are globulins (globular proteins) that have the properties of immunoglobulins and are released by certain cells of the blood and tissues into the blood in response to antigenic stimulation. All specific antibodies are in the *gamma globulin* fraction of blood plasma (see Chapter 16). Antibodies may act in one or more of several different ways as described in the next chapter.

Opsonic Action. Some antibodies have no directly injurious effect on microorganisms but render them more susceptible to phagocytosis by coating their surface. These specific "*opsonins*"[2] are directed against surface antigens of bacteria, such as the capsular SSS of pneumococci, and thus alter the surfaces of the organisms so that the phagocytes more readily engulf them. It is now thought that several of the antibodies to be discussed involve such an *opsonizing effect* (or *opsonic action*).

INFLAMMATION

Inflammation is a defensive response of living body tissues to any irritating or injurious agent. The inflammatory reaction is a complex process involving several physiologic processes that tend to remove the cause of the irritation and to repair the damage done, i.e., to heal the wound or lesion. Injury or irritation resulting in inflammation may be due to mechanical factors (such as blows, cuts, or breaks), chemical agents (acids, excessive applications of mustard, bee stings, pepper, gases), physical agents (such as excessive heat or cold or ultraviolet light), or living agents (pathogenic microorganisms, worms, and so forth).

---

[2]The name is based on the Greek word *opsōnein,* to buy food.

The four obvious features of any acute inflammation, frequently referred to as its "cardinal signs," are *redness* or hyperemia (rubor), *swelling* (tumor), *heat* (calor), and *pain* (dolor). These may be observed readily in a boil or in an inflamed joint. The redness results from the increased amount of blood in the dilated local blood vessels. The pain is caused by compression and injury of the sensory nerve branches in the tissue. The swelling (often called *edema*) is caused by the dilatation and increased permeability of the local blood vessels, resulting in the collection of fluid under some slight pressure in the spaces between the tissue cells (inflammatory exudation of fluid).

Hyperemia. Because of the increased blood, an inflamed area is said to be *hyperemic*. The dilatation of the vessels locally causes the blood flow to slow down in that particular area, and in severe injury blood flow may actually be stopped. Then it forms clots, or *thrombi,* in the tiny vessels all around the point of injury. These often act to prevent microorganisms and their poisons in the inflamed site from entering the general blood circulation.

Pus Formation. As the blood flow slows or stops, the leukocytes leave the vessels and begin to make their way into the area of damage, passing between the cells of the vessel walls by the process of ameboid motion called diapedesis. As previously pointed out, these leukocytes become *pus cells.* Inflammation, especially if acute and intense, is usually accompanied by the formation of pus. A mild inflammation may heal without the production of visible pus. In an intense inflammation, such as the venereal disease gonorrhea, very large amounts of pus appear.

Temperature Changes. The increased heat of inflammation is more apparent than real, being essentially the warmth of the deep body blood brought rapidly to the surface by local dilatation of the capillary blood vessels. Thus, the inflamed area actually feels hot compared with the usual cool temperature of the body surface. Complicated chemical changes go on in the infected sites, and these may also contribute to the warmth of the inflamed area.

Defensive Fibrin. The inflammatory exudate referred to previously is an important defensive factor. Like plasma, which it closely resembles, it contains fibrinogen and other important elements needed in the formation of the blood clot. The inflammatory area is thus soon filled and surrounded by a continuous network of fibrin fibers. At the periphery of the infected area, connective tissue cells *(fibroblasts)* begin to grow among these fibers. The entire inflammatory area is therefore soon surrounded by a wall or sac of fibrin and fibrous connective tissue that tends to confine the inflammatory process and prevent its spread. The area is said to be "walled off." A closed sac of this type, filled with pus and microorganisms, is called an *abscess,* pimple, boil, or furuncle. Several coalesced or intercommunicating abscesses are called a *carbuncle.*

Enzymes. Various microorganisms produce or activate enzymes that tend to digest the fibrin obstruction. The fluid portion of the exudate in the inflammatory area or sac also contains enzymes and antibodies from the blood that help combat the microorganisms. These enzymes may eventually digest the wall of the abscess nearest the exterior, finally making an opening through which the abscess drains. A boil "coming to a head" is a good example of this process.

Complement. At least 20 different plasma proteins are involved in the complement system and function in immunologically induced inflammation. This complex process is described in some detail in Chapter 20.

Types of Inflammatory Process

An *acute inflammation* is of short duration, lasting perhaps for a number of days. It is usually, but not always, fairly intense.

A *chronic inflammation* usually lasts for weeks, months, or years. It may be either mild or severe, extensive or restricted.

Catarrhal inflammation, which may be either acute or chronic, is a form that affects principally a mucous surface. It is marked by discharge of mucopus and epithelial debris as in colds or influenza.

In *fibrinous inflammation* the inflammatory exudate contains fibrin in large amounts. This is often seen in very intense and acute inflammations and is characteristically found on certain inner surfaces, such as the lining of the abdominal cavity (peritoneum), the covering of the heart (pericardium), or the coverings of the lungs and lining of the thoracic or chest wall (pleurae). The fibrin strands may cause roughness on the surfaces, with resultant pain on motion, as in pleurisy. Fibrin may also cause adherence (fibrous *adhesions*) between structural surfaces or following surgery, with resulting disturbances of function.

In *serous inflammation* there is more of the serum-like fluid and less of the fibrin. Sanguinoserous (*sanguino-* means bloody) and other combined types of exudate may occur. The blood seen in such exudates may have escaped from local capillary vessels injured by the pathogenic agent.

If much pus is formed, we may describe the process as *purulent* or *suppurative.* A fluid may be *sanguinopurulent* or *fibrinopurulent,* and so on.

Healing and Scar Formation

As soon as fibrin forms other defensive mechanisms begin to bring the infection or injury under control. Tissue cells (fibroblasts) and other tissue elements (blood vessels and so on) begin to grow into the area of healing from the periphery, and form new healthy tissue. This early healing process is called *organization,* and the tissue it forms is solid, tough, connective tissue when mature. This tissue, in repair situations, constitutes *scar tissue.* The process of scar formation is called *cicatrization.*

SUMMARY

Human beings are protected from diseases caused by microorganisms and other infectious agents by nonspecific and specific defensive mechanisms. The first line of defense is the skin, which may be breached by burns, cuts, and other injuries. Protection is afforded by blinking the eyelids, secretions such as tears and lysozyme in the eye, the mucous membranes and nasal passages with their hairs and slimy secretions, the cilia of the cells in the trachea and bronchi, and sneezing and coughing from the lungs. The genitalia are similarly protected by acid and mucous secretions.

Phagocytosis is the second line of defense. In 1884, Elie Metchnikoff noted that infective organisms that find their way into the body are engulfed and digested by phagocytic cells. These phagocytes are "wandering" leukocytes (polymorphonuclear neutrophils [microphages], monocytes [macrophages], and "fixed" macrophages of the mononuclear phagocyte system [MPS], or reticuloendothelial system [RES]). The fixed macrophages are the histiocytes; they are the garbage collectors of the body. Pathogenic microorganisms with slimy capsular surfaces are not easily phagocytosed, but when covered with opsonins (antibodies directed against the surface antigens of the bacteria) the antibody/antigen complex is more easily ingested by the phagocytic cells.

Infection can produce inflammation, a defensive response of living tissue to irritating or injurious agents. The four obvious features of acute inflammation are swelling, heat, redness, and pain. Hyperemia (redness) stops movement of bacteria away from the injury, pus is produced from leukocytes, defensive fibrin fibers form, and soon the inflammatory process is confined. Finally, enzymes force an opening and bring the boil "to a head."

Acute inflammation lasts only a few days, whereas chronic inflammation may persist for weeks or even years. Inflammation may be catarrhal, fibrinous, serous, purulent, or suppurative.

Healing and scar formation are part of the defensive system, keeping microorganisms away from tissues they may infect. The early healing process is called organization; scar formation is cicatrization.

REVIEW QESTIONS

1. Define nonspecific, specific, active, passive, natural, and acquired immunity.
2. What is the first line of defense against infectious microorganisms? Give examples of anatomical structures that protect us against infection. How are they effective against microorganisms?
3. What is the phagocytic process? What cells are phagocytes? How do they help in the destruction of pathogens? What is a phagolysosome? Who was Elie Metchnikoff?
4. What is the MPS? The RES? What are histiocytes, microphages, and macrophages?
5. What are opsonins? How does opsonic action work?
6. How does inflammation help in the defensive process? Name the four obvious features of any acute inflammation. List types of inflammatory processes. How long does each type of inflammation last?
7. Of what value is healing and scar formation in defending the body against infection?

Supplementary Reading

Bellanti, J. A.: Immunology: Basic Processes. Philadelphia, W. B. Saunders Company, 1979.

Benacerraf, B., and Unananue, E. R.: Textbook of Immunology. Baltimore, Williams & Wilkins Co., 1979.

Berry, L. J., Smythe, D. S., Colwell, L. S., Schoengold, R. J., and Actor, P.: Comparison of the effects of a synthetic polyribonucleotide with the effects of endotoxin on selected host responses. *Inf. Immunol., 3*:444, 1971.

Bowen, W. H., Genco, R. J., and O'Brien, T. C.: Immunologic Aspects of Dental Caries. Washington, D.C., Information Retrieval, Inc., 1976.

Carpenter, P. L.: Immunology and Serology. 4th ed. Philadelphia, W. B. Saunders Company, 1977.

Cooper, M. D., and Dayton, D. H. (eds.): Development of Host Defenses. New York, Raven Press, 1977.

Cooper, M. D., and Warner, N. L. (eds.): Contemporary Topics in Immunobiology. Vol. 3. New York, Plenum Press, 1974.

Golub, E. S.: The Cellular Basis of the Immune Response: An Approach to Immunology. Sunderland, Mass., Sinauer Associates, 1977.

Herrmann, E. C., Jr., and Stinebring, W. R. (eds.): Second Conference on Antiviral Substances. *Ann. N.Y. Acad. Sci. 173* (Art I): 1–844, 1970.

Hirsch, J. G.: Phagocytosis. *Ann. Rev. Microbiol., 19*:339, 1965.

Kabat, E. A.: Structural Concepts in Immunology and Immunochemistry. 2nd ed. New York, Holt, Rinehart & Winston, Inc., 1976.

Metchnikoff, E.: Lectures on the Comparative Pathology of Inflammation. London, Kegan Paul, Trench, Trüber and Co. 1893; reprinted New York, Dover Publications, 1968.

Pollard, M. (ed.): Antiviral Mechanisms. New York, Academic Press, 1975.

Spector, W. G. (ed.): The acute inflammatory response. *Ann. N.Y. Acad. Sci., 116*(Art 3):747, 1964.

Streilein, J. W., and Hughes, J. D.: Immunology—A Programmed Text. Boston, Little, Brown & Company, 1977.

LYMPHOCYTES AND ANTIBODIES

HUMORAL AND CELLULAR IMMUNITY

It is not surprising that in the millennia-long battle for survival of complex living organisms, such as humans, a warning and quick-action immune system evolved, analogous to a fire department, whose function it is to detect fires, to fight them, and to help prevent future ones by installing devices that will quench them.

The immune response in the human body to antigenic stimulation is analogous to a new fire. The defensive response occurs in cells (lymphocytes) of the lymphoid system (page 299). Initially undifferentiated (unspecialized), several types of lymphocytes develop, each with distinct functions, as in a fire department where each fireman begins as a novice and acquires specialized skills (Fig. 19–1).

Even though many details are not yet fully understood, one may simplify a rather complex concept in the following way:

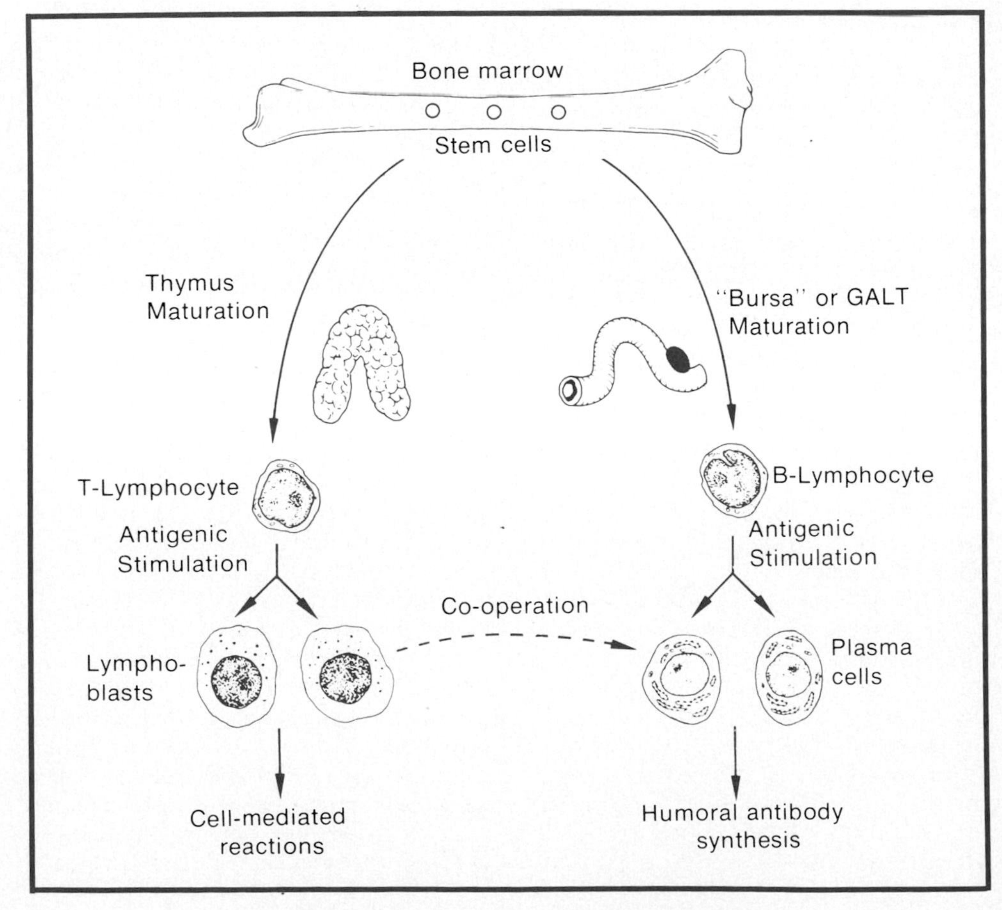

Figure 19–1 Bone marrow stem cells undergoing maturation in the thymus or the bursal tissue (gut-associated lymphoid tissues [GALT] in mammals) to become T- or B-lymphocytes, respectively. (From Frobisher, M., et al.: Fundamentals of Microbiology. 9th ed. Philadelphia, W. B. Saunders Company, 1974.)

CHAPTER 19

1. In *humoral immunity* (see section on B-cells that follows) globulin molecules called antibodies are produced by these cells to oppose foreign complex molecules, referred to as antigens.

2. In *cellular immunity* lymphocytes (see section on T-cells that follows) destroy abnormal (strange or foreign) *cells,* which are made up of various antigens (surface, somatic, and so on).

B-Lymphocyte and T-Lymphocyte Mediated Immunity

Although derived from the same type of stem cells in the bone marrow during embryonic development, lymphocytes differentiate into *B-lymphocytes* (after the *bursa* of Fabricius in birds) and *T-lymphocytes,* also called thymocytes after the *thymus* gland (Fig. 19–2). Lymph nodes and the spleen contain areas that are either thymic-dependent, containing T-cells, or bursal-dependent (thymic independent), containing B-cells (Fig. 19–3). Morphologically, B-cells and T-cells are indistinguishable: both are nonphagocytic agranular leukocytes from 5 to 8 μm in size.

However, when sheep erythrocytes are mixed with human lymphocytes *in vivo* the T-lymphocytes (E^+ cells) clump with the erythrocytes to form "*E rosettes*" easily seen with immunofluorescence microscopy. B-lymphocytes are E^- cells and do not bind sheep erythrocytes, but they are surface membrane immunoglobulin positive ($SmIg^+$), whereas T-lymphocytes are $SmIg^-$. Variations and subpopulations of these lymphocyte types exist and are of great importance in disease conditions, such as systemic lupus erythematosus and autoimmune diseases.

B-Cells. The B-lymphocytes, the basis of *humoral* immunity, make up about 20 per cent of circulating lymphocytes. Most B-cells have about 100,000 *immunoglobulin* molecules (IgM and IgD mostly, with IgG or IgA in some cells) as surface receptors, which react with the hapten determinant of the antigen. B-cells survive for only a few days or weeks. This system of cells determines the development of antibodies in mammals. It consists of "*g*ut-*a*ssociated *l*ymphoid *t*issue" cells (abbreviated as GALT) found in the Peyer's patches (intestine), the tonsils, and the vermiform appendix (see Fig. 19–4). B-cells can differentiate and become large lymphocytes called plasma cells, which then synthesize *specific antibodies.* Alternatively, B-cells as well as T-cells may develop into "memory cells" and store specific immunologic information.

The immature lymphocyte may detect the presence of antigen and binds this antigen to the immunoglobulins on its surface. The B-cell now divides, differentiates, and via a plasmablast, becomes either a *plasma cell* in 4 to 5 days or a *memory cell.* The plasma cell gives off specific *circulating antibodies* to fight the specific foreign chemical antigen, but the memory cell waits until a later time when the same specific antigen reenters the system. In the mammalian host, B-lymphocyte populations are mainly responsible for the production of specific immunoglobulins and for general antibody formation. Studies with an

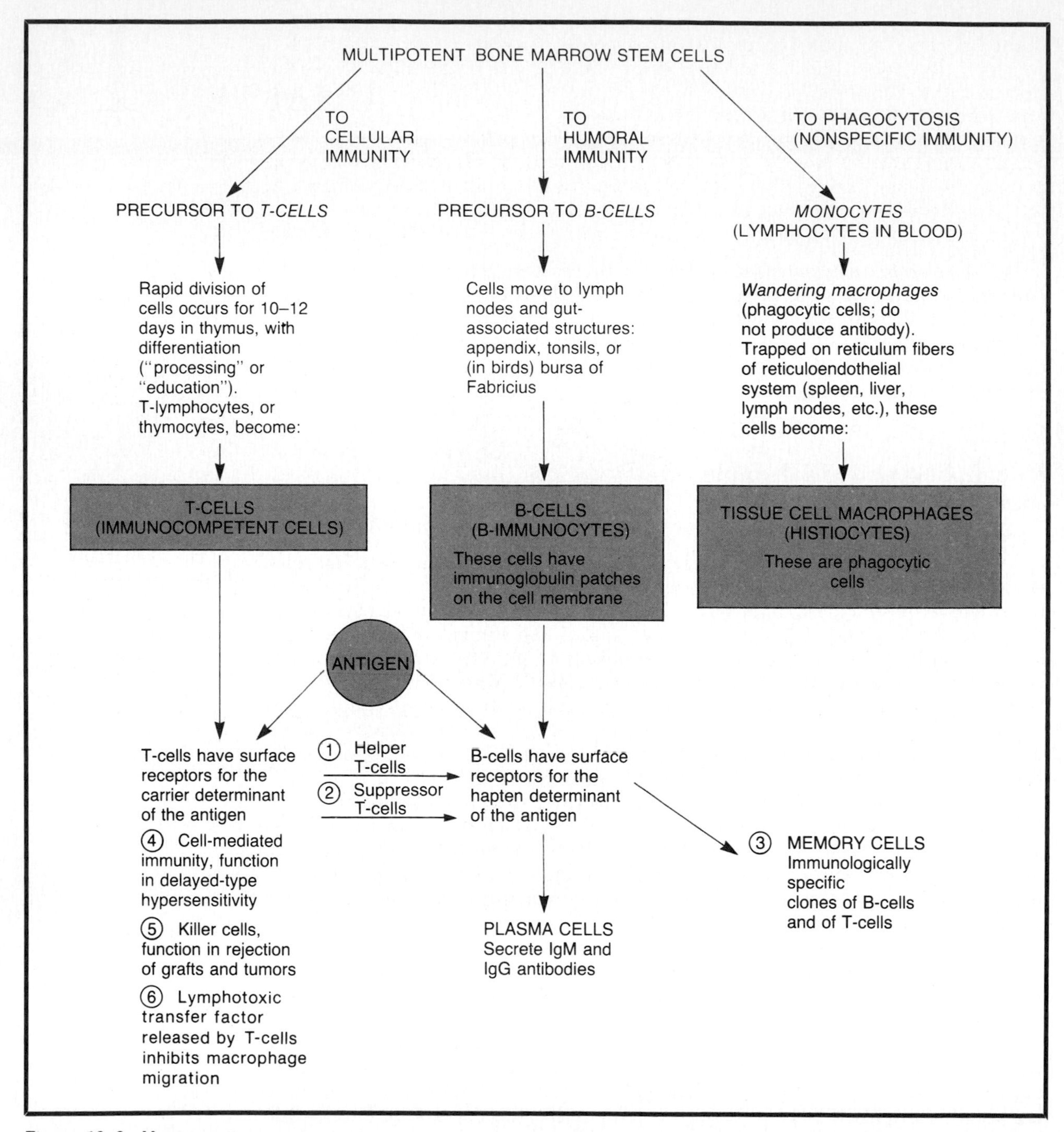
MULTIPOTENT BONE MARROW STEM CELLS
TO CELLULAR IMMUNITY
TO HUMORAL IMMUNITY
TO PHAGOCYTOSIS (NONSPECIFIC IMMUNITY)
PRECURSOR TO T-CELLS
PRECURSOR TO B-CELLS
MONOCYTES (LYMPHOCYTES IN BLOOD)
Rapid division of cells occurs for 10–12 days in thymus, with differentiation ("processing" or "education"). T-lymphocytes, or thymocytes, become:
Cells move to lymph nodes and gut-associated structures: appendix, tonsils, or (in birds) bursa of Fabricius
Wandering macrophages (phagocytic cells; do not produce antibody). Trapped on reticulum fibers of reticuloendothelial system (spleen, liver, lymph nodes, etc.), these cells become:
T-CELLS (IMMUNOCOMPETENT CELLS)
B-CELLS (B-IMMUNOCYTES)
These cells have immunoglobulin patches on the cell membrane
TISSUE CELL MACROPHAGES (HISTIOCYTES)
These are phagocytic cells
ANTIGEN
T-cells have surface receptors for the carrier determinant of the antigen
① Helper T-cells
② Suppressor T-cells
B-cells have surface receptors for the hapten determinant of the antigen
③ MEMORY CELLS Immunologically specific clones of B-cells and of T-cells
④ Cell-mediated immunity, function in delayed-type hypersensitivity
⑤ Killer cells, function in rejection of grafts and tumors
⑥ Lymphotoxic transfer factor released by T-cells inhibits macrophage migration
PLASMA CELLS Secrete IgM and IgG antibodies

Figure 19–2 Heterogeneity of lymphocytes.

antigen consisting of a carrier protein and a hapten show that the B-lymphocytes react with the hapten, resulting in the release of hapten-specific antibodies by the B-cell derived plasma cells, whereas T-lymphocytes respond to the carrier determinant of the antigen (Fig. 19–5).

T-cells. The T-lymphocytes are the basis of *cellular immunity.* They produce lymphokines, interferon, transfer factor, and other cell products, but not antibodies (Fig. 19–2). T-cells constitute nearly 80 per cent of circulating small lymphocytes, with a life span of months to years. Although they originate in the bone marrow, these cells are under the control of the thymus gland during early development. Precursors of T-cells that enter the thymus, or are under the influence of thymus hormone in other tissues, reproduce rapidly for 10 to 12 days; they may then be called thymocytes. After leaving the thymus, the now immunocompetent T-cells circulate in the blood and thoracic duct and enter the lymph nodes, the spleen, and so on.

In response to antigenic stimulation by *abnormal cells,* the defensive, "educated," immunocompetent T-cells become sensitized, proliferate, and develop progeny with specific reactive sites, but these are not antibodies and do not at any time produce or develop into circulating immunoglobulins.

T-lymphocytes that function in *cellular immunity* have a special surveillance

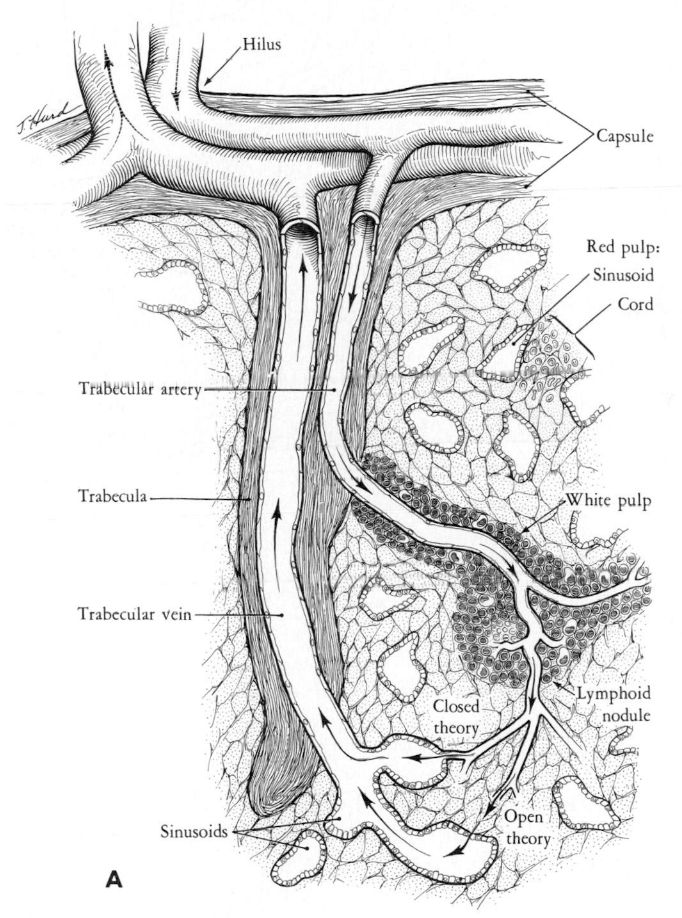

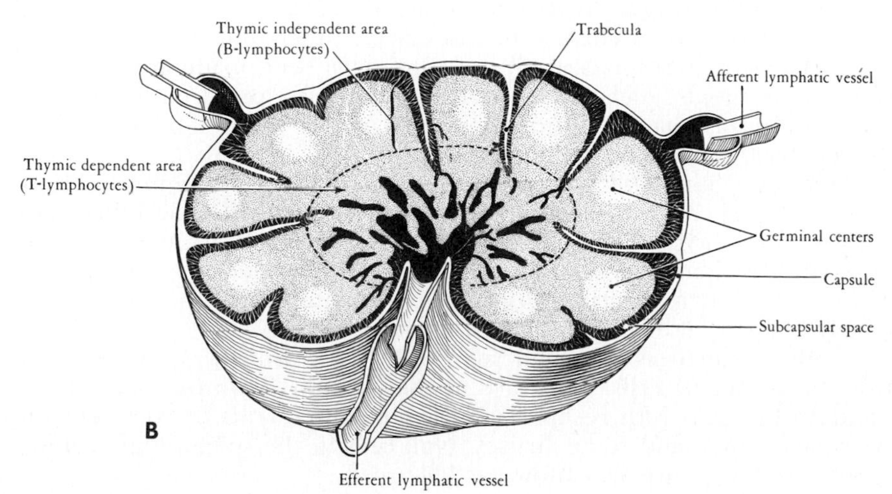

Figure 19–3 In *A*, a schematic representation of the spleen, note the red pulp, which contains large numbers of erythrocytes, and the white pulp, where most of the lymphocyte production in the spleen occurs. The lymphoid nodules in the white pulp contain both B-lymphocytes and T-lymphocytes. In the lymph node depicted in *B* are shown the thymic dependent area (red) where T-lymphocytes are mostly found and the thymic independent area containing B-lymphocytes in the germinal centers. The inner core (medulla) contains plasma cells. (From Bellanti, J. A.: Immunology II. Philadelphia, W. B. Saunders Company, 1978.)

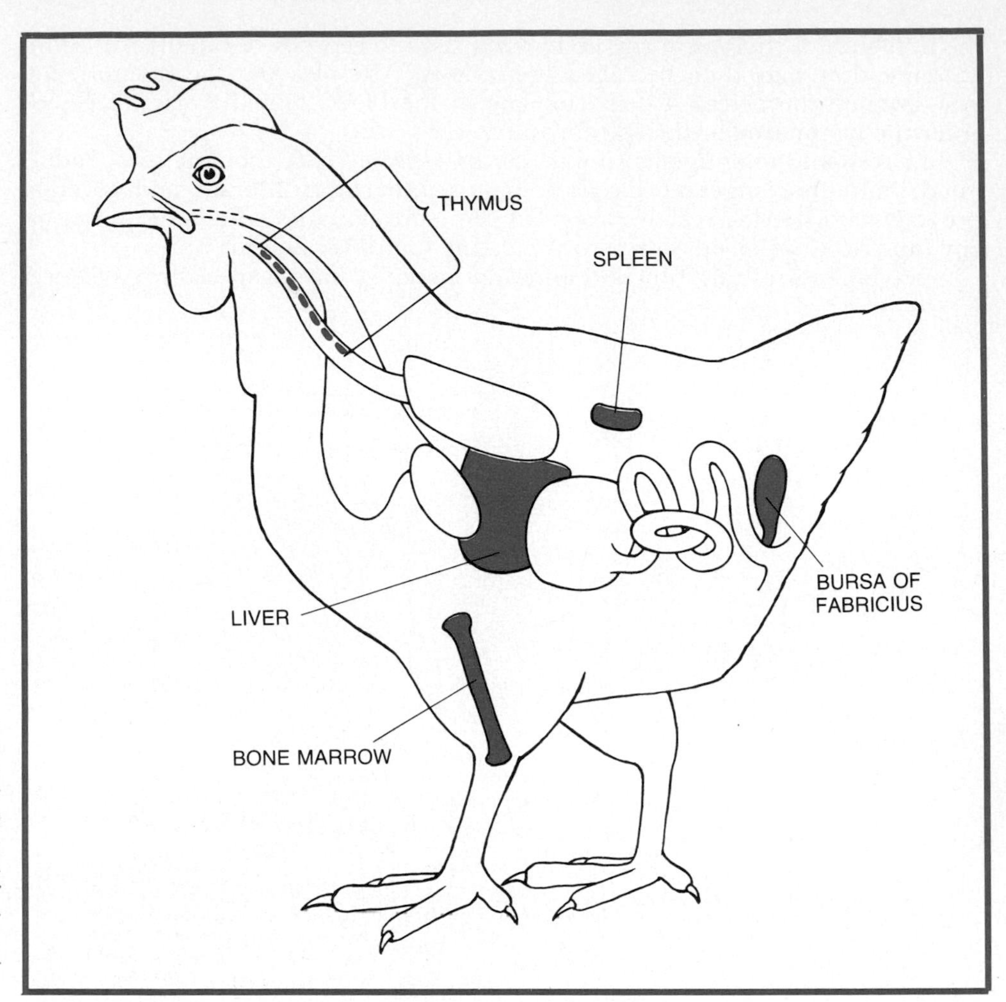

Figure 19–4. In birds, B-cells are derived from the bursa of Fabricius, a pouch attached to the intestine near the cloaca. It is thought to influence the immunoglobulin-secreting cells of humoral immunity in a fashion similar to gut-associated lymphoid tissue (GALT) in mammals. (Modified courtesy of *Scientific American.*)

function in detecting and destroying *abnormal* (nonself) *cells.* Abnormal cells are opposed by the defensive lymphocytes because abnormal cells contain antigenic components (foreign chemicals) that differentiate them from normal (self) cells. Abnormal cells may be (1) foreign cells, as found in tissue transplants from other persons or animals; (2) cells that have been exposed to radiation damage or to foreign chemicals (e.g., contact dermatitis); (3) bacteria (as well as viruses) present in the body and in chronic diseases (e.g., tuberculosis) that produce delayed hypersensitivity responses; (4) cells that have been made abnormal by certain infectious agents, such as bacteria, fungi, viruses, protozoa, or worms; and (5) neoplastic cells (malignant tumors), which may be rejected by the body—but unfortunately this does not always occur.

Plasma Cells. Stimulated by antigen the B-lymphocyte becomes a "blast cell," enlarges from 7 or 8 μm up to 20 μm in diameter, and becomes metabolically more active. The developed *plasma cell* is RNA-rich, has an eccentric round nucleus and a well-developed endoplasmic reticulum, and is distended with granular material consisting of the antibodies it produces. The incompletely formed antibodies are called *Russell bodies.*

Null Cells. Null cells (German for "zero") are probably a mixture of immature hematopoietic cells. They are neither T- nor B-lymphocytes. They are E^- and $SmIg^-$ cells with receptors for some components of complement, the Fc portion of IgG, and some viruses. Null cells in the spleen may destroy tumors. Some null cells are also known as killer cells.

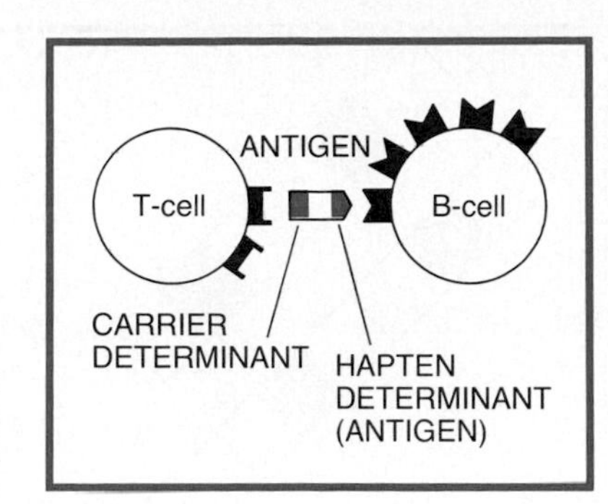

Figure 19–5 The hapten portion of the antigen reacts with the hapten-specific determinant of the B-cell, which has recognized it, while the carrier determinant of the conjugated carrier protein reacts with the T-cell.

Mast Cells. Mast cells are tissue cells similar to basophil leukocytes of the blood. They contain heparin, serotonin, and histamine, and they play an important role in immediate hypersensitivity reactions.

ANTIBODIES (IMMUNOGLOBULINS)

Antibodies are very specialized proteins called immunoglobulins that are produced in response to antigenic stimulation. However, some antibodies (hemagglutinins), like those against specific blood types, are present in the serum at the time of birth (Chapter 16). Antibodies are only produced by vertebrates.

Classes of Immunoglobulins

Different classes of immunoglobulins can be identified and separated through electrophoresis (Table 16–1) and other separation techniques from the gamma-globulin fraction of human serum.

Certain cells—especially plasma cells derived from B-lymphocytes that differentiated in the GALT, the fetal liver, or the bone marrow, respond to infection by producing *antibodies* which are soluble protein molecules also called *gamma globulins.* There are several forms of gamma globulins, designated alphabetically as IgG, IgA, IgM, IgD, and IgM. In most normal and hyperimmune individuals, over 85 per cent of the immunoglobulins are IgG type proteins. In molecular structure the simplest of the immunoglobulins is IgG. Each molecule is Y-shaped and consists of two long, intertwined, heavy (H) polypeptide chains that diverge near their centers to form the branches of the Y (Fig. 19–6). To each branch of the Y is attached a shorter, lighter (L) polypeptide chain. Immunologic specificity depends largely on the amino acid sequences in the branched parts of the molecule. IgG consists of two light and two heavy polypeptide chains, held together by disulfide bonds. Four structures of IgG are known: IgG1, IgG2, IgG3, and IgG4. IgA exists as a dimer or trimer, and IgM is a pentamer (Fig. 19–7). The different classes of immunoglobulins have been studied from analyses of the protein chains making up the molecules (Table 19–1). Interested readers should consult texts of immunology for further information. They will encounter a rapidly expanding field of knowledge.

Differences in the molecular structure of the various antibodies are of great practical as well as theoretical interest, as illustrated by the fact that a perfectly healthy, newborn baby of a syphilitic mother may have syphilitic IgG in its serum, the antibodies having passed through the placenta from its mother. In contrast, if the baby's serum contains syphilitic IgM, it shows that the neonate

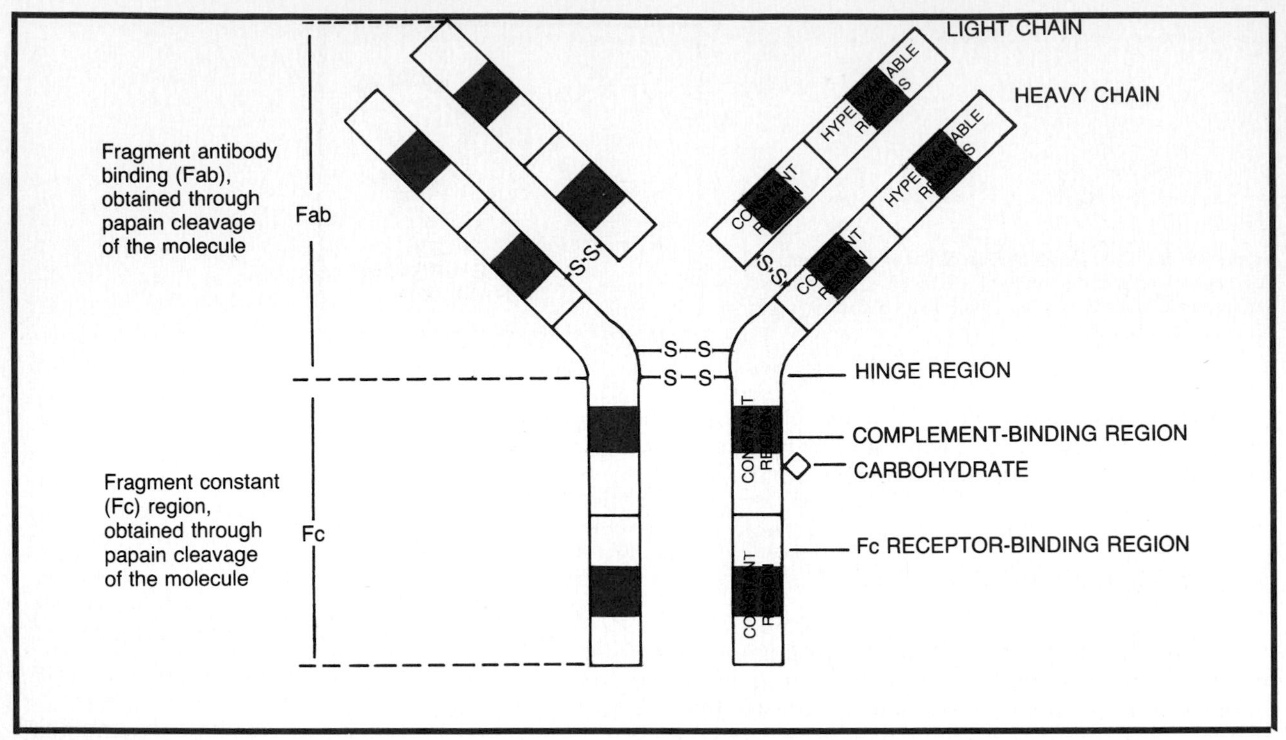

Figure 19–6 The immunoglobulin molecule shown above consists of two heavy chains and two light chains, each made up of many amino acid constituents. The chains are joined through —S—S— bonds and contain intrachain disulfide bonds in the regions (called domains) marked in color. As indicated, part of the molecule consists of constant regions (with no variations in amino acid sequences). The variable region contains several hypervariable regions, which are probably the antigen-binding sites of the immunoglobulin.

has an active spirochetal infection (congenital syphilis). The IgM of the baby is not derived from the mother because IgM cannot pass through the placental membrane. The two forms of Ig may be differentiated by a special technique of the FTA-ABS fluorescent antibody test for syphilis.

The gamma globulins act as a defensive barrier against microorganisms and their toxins. (This action was explained in Chapter 18.) The blood transports antibodies and defensive phagocytic blood cells throughout the tissues, like a supply train that distributes soldiers and ammunition to an embattled army. Other, stationary phagocytic cells called *histiocytes* line the inside of the blood vessels and, like snipers in war, destroy invaders (microorganisms) or other foreign substances that pass by. These cells constitute the *mononuclear phagocyte system (MPS),* also called the *reticuloendothelial system.* Tests made with antibodies and with blood cells are of immense importance in the diagnosis—and consequently the treatment—of disease. An important role in the activities of both antibodies and phagocytes is played by a group of blood proteins that together are called *complement.* Complement and various other blood components will be discussed later.

Clonal Selection

Even though two immunoglobulins may have different constant regions in their heavy chains, their variable regions may provide identical specificity for an antigen (Fig. 19–6). This explains why different clones of immunoglobulins are formed in response to antigenic stimulation by one antigen.

Each person has a large number of specific B-lymphocytes, and each lymphocyte has one type of specific receptor on its surface for one type of antigen. After the antigen binds to the B-lymphocyte, the plasma cells develops and antigen-specific antibodies are released.

It is thought that during fetal development all possible lymphocyte recep-

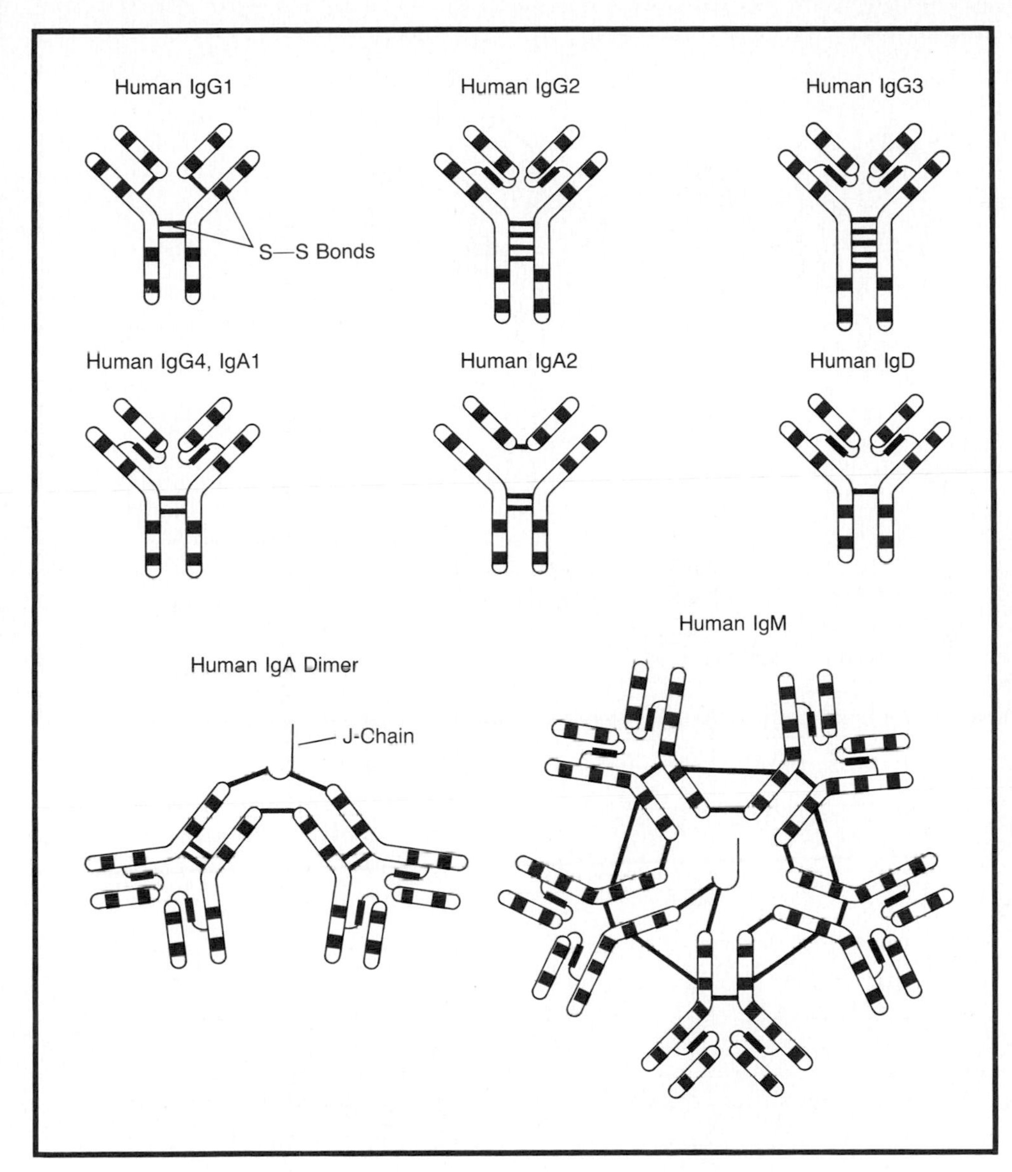

Figure 19–7 Human immunoglobulins may appear as monomers (forms of IgG, IgA1, IgA2, IgD), dimers (IgA), or pentamers (IgM). The structure depends on the amino acid sequence and the disulfide bonding patterns of the molecules. Squares represent intrachain bonds; bars indicate interchain bonds. (Data adapted from *Scientific American Medicine;* Gally, J. A.: *In* Sela, M. (ed.): The Antigens. New York, Academic Press, 1973; Frangione, B.: Immunogenetics and Immunodeficiency, 2nd ed. London, Medical and Technical Publ. Co., 1975; and Kalab, E. A.: Structural Concepts in Immunology and Immunochemistry, 2nd ed. New York, Holt, Rinehart and Winston, 1976.)

tors for antigens are produced, for self and nonself, but that some time before birth all self antigen receptors are eliminated, leaving only nonself receptors to recognize foreign nonself substances. Failure to eliminate self-antigen receptors may result in autoimmune conditions, to be discussed in Chapter 21.

The Immunoglobulins in Serum

Since each plasma cell produces only immunoglobulins of the same chemical structure, only one type of light chain and one type of heavy chain is produced (Table 19–1). The first antibody activated in infection is IgM, but on further stimulation by the same antigen IgG is formed by the same plasma cell. This is called the *IgM-IgG switch,* and it involves the genes in control of the variable region of the heavy chain of IgM and the genes of the constant region of IgM. Consequently, IgM and later IgG have identical antigen specificity.

Table 19–1. Properties of Human Immunoglobulins

Major Class of Immunoglobulin	Subclasses Known	Subunits, Type of Chains			Structure	Sedimentation Constant	Half-Life in Serum (days)	Description
		Light	*Heavy*	*Others*				
IgG 1.25 g/100 ml serum	IgG1 IgG2 IgG3 IgG4	κ (kappa) or λ (lambda)	γ (gamma 1) (gamma 2) (gamma 3) (gamma 4)		monomers	7 S	23	Formed after IgM. Only IgG1 and IgG3 bind to monocytes and one complement component. Found in serum and amniotic fluid.
IgA 0.28 g/100 ml serum	IgA1 IgA2	κ (kappa) or λ (lambda)	α (alpha 1) (alpha 2)	J in dimer SP*	monomers or dimers	7 S monomer 11 S dimer	6	Synthesized by plasma cells. In secretions, tears, saliva, colostrum, gastrointestinal tract, vagina, and mucous membranes.
IgM 0.12 g/100 ml serum	IgM	κ (kappa) or λ (lambda)	μ (mu)	J	pentamer	19 S	5	The first Ig produced against infection. It is 20 to 1,000 times as effected as IgG. Useful for diagnosis.
IgD 0.3–30 mg/100 ml serum	IgD	κ (kappa) or λ (lambda)	δ (delta)		monomer	7 S	3	First found as a myeloma protein.
IgE up to 0.2 mg/100 ml serum (minute trace)	IgE	κ (kappa) or λ (lambda)	ε (epsilon)		monomer	8 S	2.5	Plays important role in hay fever, asthma, anaphylaxis.

*SP is the secretory piece. Produced by epithelial cells, it combines with the dimer. The SP polypeptide chain makes the IgA resistant to proteolytic enzymes.

INTERFERON

Interferon is the name given to any or all of a group of closely related glycoproteins (first described in 1957 by Isaacs and Liebermann) of relatively low molecular weight (human, 26,000; chick interferon, 38,000) that are produced by animal cells in response to infection by *double-stranded* viral DNA or RNA, intact viruses, rickettsiae, bacterial endotoxins, protozoa, or even synthetic polynucleotides. Viral infection stimulates new synthesis of interferon; nonviral agents may release preformed interferon of a slightly different nature. It should be noted that *interferons are not virus-specific but host cell–specific*—e.g., they are ineffective in mice if produced in chicks, although effective in humans if produced in human or monkey kidney cells. Interferon is most effective in inhibiting virus reproduction in the cells of the species in which it originated.

Although the nomenclature of *human interferons* is currently under review, at least four types are recognized: (1) *leukocyte* interferon, stimulated by viruses or synthetic polynucleotides; (2) *fibroblast* interferon from human connective tissue, similarly stimulated; (3) *lymphoblastoid* interferon, resulting from long-term stimulation by virus and leukocyte-fibroblast mixture; and (4) *immune* interferon from T-lymphocytes. The *classical* type I interferon is leukocyte, fibroblast, or lymphoblastoid interferon and is acid pH stable and heat labile, whereas the *immune* type II interferon is acid pH unstable but heat stable. Type I is released into the blood within 2 hours after injection of endotoxins, and thus appears to be preformed; type II takes 18 hours after stimulation.

Interferon was at first thought to interfere (hence its name) with replication of the virus in the cell, probably by blocking an early eclipse activity of the viral mRNA synthetase system. It is now also thought that interferon evokes synthesis of a second protein that may or may not be antiviral per se. Interferon has no effect whatever on extracellular virus or on intracellular replication of the virion once it has developed beyond the early eclipse phase.

Interferon is neither an enzyme nor a coenzyme. Unlike specific antibodies, it is not an immune globulin (Ig) and is not specific in its action against any particular virus but acts against viruses in general. It has no immunologic relationship to the inducing virion or other inducing agent. The only specific aspect of any interferon is the apparent limitation of its protective activity to animals of the same species as the animal whose cells produced it. Interferons from closely related species (e.g., some primates and humans) will cross protect to some extent. Interferon released by an infected cell is taken in by other cells of the same species and thus tends to limit the spread of infection in the body, in a tissue culture, or in a passive recipient of the interferon-containing fluid.

Interferon removed from animals or from tissue cultures (exogenous interferon) has been used both prophylactically and therapeutically in animals, with encouraging results (Fig. 19–8). The injection of interferon inducers into

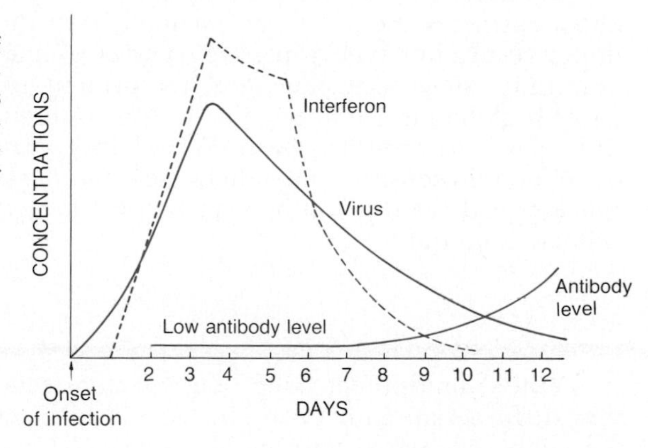

Figure 19–8 Note that appearance of interferon in the blood is almost immediate after onset of a virus infection. Its concentration increases rapidly. Interferon quickly prevents further virus multiplication. Antibodies form much later and more slowly.

animals at the time of, or just prior to, infection with a virus, with a view to stimulating the cells to produce their own protective or curative interferon (endogenous interferon), has also given very promising results, especially against influenza infections during epidemics. After ways had been found in Finland to produce interferon in small quantities using human leukocytes and Sendai viruses, experiments began in Sweden to treat bone cancer with interferon. Other treatments of many types of malignancies were attempted: some were successful, some promising, and many did not work. These experiments are still in progress and they must be evaluated with caution. Most of the interferon used now comes from cell cultures; soon it will be produced using genetic engineering techniques, growing *Escherichia coli* equipped with genes for interferon production.

Despite the apparent success in interferon research, production, and application, many problems remain. We still need to learn what diseases interferon is effective against, how long its effects will last, what dosages should be administered for what lengths of time, what side effects or toxicities interferon produces, and what treatments it can be combined with.

COMPLEMENT

Complement is a complex of about 11 interrelated and interacting serum proteins designated C1q, C1r, C1s, C4, C2, C3, C5, C6, C7, C8, C9, and so on. They are very important in complementing the action of specific antibodies and are discussed more fully in Chapter 20.

THE PROPERDIN SYSTEM

Properdin (Factor P) belongs to a group of serum constituents that together apparently play an important role in primary, nonspecific resistance to infection. Besides properdin (from Latin *perdere,* to destroy), there is a glycine-rich beta-glycoprotein (C3 proactivator of complement) and other serum protein components in the system, called A, B, and O°. It seems that the properdin system provides an alternative pathway for activating the C3 factor of complement.

Properdin itself is an enzyme-like agent with a molecular weight of 223,000 daltons. The amount of properdin in the blood seems to be directly related to the degree of nonspecific resistance to numerous agents. (The normal serum concentration is 25 μg/ml.) It is thought to participate in the destruction of certain gram-negative bacteria, the lysis of certain erythrocytes, and possibly the inactivation of viruses. Injection of properdin-rich serum from animals having naturally high levels of properdin in their serum appears to increase resistance to infection. (See Chapter 20.)

Transfer Factor. A local injection of synthesized lymphoid cells (T-cells) or an extract of *transfer factor* obtained from the cells of a recently immunized donor results in a type of passive transfer of immunity. However, this is cellular immunity, since antibodies are not present in the serum as they are when gamma globulins are injected as serum components to produce passive immunity. A *transfer factor* has been isolated and partially purified from lymphocytes of tuberculin-sensitive individuals (see *The Tuberculin Test,* page 482), information derived from it will be very helpful in getting a better understanding of cellular immunity.

SUMMARY

When multipotent bone marrow stem cells migrate to the thymus gland they differentiate into T-lymphocyte (T-cells) active in cellular immunity; when the stem cells move into lymph nodes and gut-associated structures (bursal-dependent, GALT) they become B-lymphocytes (B-cells) active in humoral immunity (the antibody-forming system). Studies with an antigen consisting of a carrier protein and a hapten show that the B-lymphocytes react with the hapten, resulting in plasma cells that release specific antibodies (hapten-specific), whereas the T-lymphocytes respond to the carrier determinant of the antigen (the carrier protein). Both T-cells and B-cells have surface receptors for their respective determinants of the antigen and both, after stimulation with antigen, form immunologically specific clones of cells called T-memory cells and B-memory cells. These are lymphocytes that will recall earlier encounters with a specific antigen.

T-cells may be helper cells or suppressor cells to the B-cells. T-lymphocytes also function in the delayed type of hypersensitivity, as killer cells in the rejection of grafts or tumors, and in releasing lymphotoxic transfer factor that inhibits macrophage migration. The T-lymphocytes are the basis of cellular immunity. They produce lymphokines, interferon, transfer factor, and other cell products, but *not* antibodies. T-cells constitute nearly 80 per cent of circulating small lymphocytes, with a life span of months to years. The special function of T-lymphocytes is to detect and destroy abnormal cells.

A B-lymphocyte, after attracting a specific antigen to its surface, divides, differentiates, and, via a plasmablast, becomes either a plasma cell in 4 to 5 days or a memory cell. The plasma cell, larger than the B-cell, is very active metabolically and forms Russell bodies, which develop into antibodies. The plasma cell releases specific circulating antibodies (immunoglobulins) from the cell when they are fully formed. All antibodies produced from one plasma cell are identical and react specifically with the antigen type that originally stimulated the B-lymphocyte.

Antibodies are specialized proteins called immunoglobulins (Ig's) that are produced in response to antigenic stimulation. These soluble protein molecules, also called gamma globulins, have been designated IgG, IgA, IgM, IgD, and IgE in order of quantities found in the serum. IgG consists of two light and two heavy polypeptide chains held together by disulfide bonds, and it takes the form of IgG1, IgG2, IgG3, and IgG4. IgA exists as a dimer or trimer, and IgM is a pentamer. The different regions on the chains have been studied well and much is known about the amino acid compositions of the constant regions, the variable regions, and the hypervariable regions as they relate to the functions of the immunoglobulins.

Interferon refers to a group of closely related glycoproteins that are produced by animal cells in response to infection by double-stranded viral DNA or RNA, intact viruses, rickettsiae, bacterial endotoxins, protozoa, or synthetic polynucleotides. Viral infection stimulates new synthesis of interferon, which is not virus-specific but host-specific. Type I interferon is leukocyte, fibroblast, or lymphoblastoid interferon; it is acid pH stable, heat labile, and released into the blood within 2 hours after injection of endotoxin. Type II interferon is acid pH unstable but heat stable, and it appears 18 hours after stimulation. Only very minute amounts of interferon are produced naturally. However, through genetic engineering the supply is increasing. Interferon has shown great promise in treating certain types of malignancies, especially bone cancer. Only time will tell what its use against viral diseases and neoplastic conditions will be.

Transfer factor refers to cellular immunity when a local injection of sensitized T-lymphocytes or an extract of transfer factor obtained from the cells of a recently immunized donor results in a type of passive transfer of immunity.

REVIEW QUESTIONS

1. Compare humoral immunity with cellular immunity. What are bone marrow stem cells, T-lymphocytes, B-lymphocytes, the bursa of Fabricius, GALT, plasma cells, and memory cells?

2. Which cells have surface receptors for antigen? Explain.
3. What method is available to distinguish B-cells from T-cells? What do the designations $SmIg^+$, $SmIg^-$, E^+, and E^- mean?
4. Describe the different functions of T-cells and B-cells. What is an immunocompetent T-cell?
5. How are antibodies formed? What cells are responsible?
6. Define immunoglobulin. What classes of Ig's can you name? Discuss the structural details of an immunoglobulin.
7. Of what significance are the constant regions as opposed to the hypervariable regions on Ig chains?
8. Which Ig is first formed after an infection? Which is the second to be formed? How?
9. What is interferon? What different types are known? What are possible uses for interferon? What problems still exist?
10. What is the properdin system? What is the transfer factor?

Supplementary Reading

Bellanti, J. A.: Immunology: Basic Processes. Philadelphia, W. B. Saunders Company, 1979.

Berry, L. J., Smythe, D. S., Colwell, L. S., Schoengold, R. J., and Actor, P.: Comparison of the effects of a synthetic polyribonucleotide with the effects of endotoxin on selected host responses. *Inf. Immunol., 3*:444, 1971.

Boehmer, H. von: Specificity of Effector T Lymphocytes. Copenhagen, Munksgaard, 1976.

Carpenter, P. L.: Immunology and Serology. 4th ed. Philadelphia, W. B. Saunders Company, 1977.

Colby, C., and Morgan, M. J.: Interferon induction and action. *Ann. Rev. Microbiol., 25*:333, 1971.

Cooper, M. D., and Warner, N. L. (eds.): Contemporary Topics in Immunobiology. Vol. 3. New York, Plenum Press, 1974.

Davis, B. D.: Microbiology. 3rd ed. New York, Harper & Row, 1980.

Fitzpatrick, F. W., and Di Carlo, F. J.: Zymosan (reproperdin). *Ann. N.Y. Acad. Sci., 118*(Art. 4):233, 1964.

Golub, E. S.: The Cellular Basis of the Immune Response: An Approach to Immunology. Sunderland, Mass., Sinauer Associates, 1977.

Herrmann, E. C., Jr., and Stinebring, W. R. (eds.): Second Conference on Antiviral Substances. *Ann. N.Y. Acad. Sci., 173*(Art. I):1–844, 1970.

Kabat, E. A.: Structural Concepts in Immunology and Immunochemistry. 2nd ed. New York, Holt, Rinehart & Winston, 1976.

Kazar, J., Gillmore, J. D., and Gordon, F. B.: Effect of interferon and interferon producers on infections with a nonviral intracellular microorganism, *Chlamydia trachomatis. Inf. Immunol., 3*:825, 1972.

Loor, F., and Roelants, G. E. (eds.): B and T Cells in Immune Recognition. New York, John Wiley & Sons, 1977.

Naff, G. B.: Role of the properdin system. *In* Microbiology—1975. Washington, D.C., American Society for Microbiology, 1975.

Porter, R. R.: Chemical Aspects of Immunology. Carolina Biology Reader 85. Burlington, N.C., Carolina Biological Supply Company, 1976.

Schwick, H. G., and Bräuer, H.: Exempla Immunologica, Bildatlas zu Antikörperrektionen. Frankfurt am Main, Behring Institute, Behringwerke, 1980.

Streilein, W. J., and Hughes, J. D.: Immunology—A Programmed Text. Boston, Little, Brown & Company, 1977.

Wright, R. K., and Cooper, E. L. (eds.): Phylogeny of Thymus and Bone Marrow: Bursa Cells. Proceedings of an International Symposium on the Phylogeny of T and B Cells. New York, North Holland Publishing Co., 1976.

SPECIFIC RESISTANCE TO INFECTION

ACTIVE IMMUNITY

An individual may become immune to certain infectious diseases by actually having the disease and recovering from it. The resulting immunity is entirely *specific*; that is, a person who recovers from diphtheria is usually immune to subsequent attacks of diphtheria, but can still contract typhoid or smallpox.

Students frequently ask which diseases induce permanent immunity in patients who recover from them. A complete reply would involve a considerable list of illnesses. Among the well-known diseases of humans, lasting immunity usually results from measles, chickenpox, whooping cough, bubonic plague, poliomyelitis, smallpox, Rocky Mountain spotted fever, diphtheria, scarlet fever, tularemia, typhus fever, typhoid fever, and yellow fever, among others. Confirmed second cases of yellow fever, smallpox, and measles have rarely been reported, though some reinfections have been observed following diphtheria, scarlet fever, cholera, typhoid fever, mumps, and several other diseases. In some of these the diagnosis of either the first or the second case has been questioned; in other instances, there is no doubt of the repetition. Little if any lasting immunity results from the group of infections called "common colds" or from erysipelas, furunculosis, gonorrhea, septic sore throat, herpes simplex, or pneumococcal pneumonia (if caused by a strain of *Streptococcus pneumoniae* other than that which induced the initial infection). Immunity to infectious diseases can vary and is not necessarily either absolute or unchanging. It is, in part, a result of the production by plasma cells of soluble protein molecules called *antibodies* or *immunoglobulins* and of T-lymphocyte–mediated immunity in bacterial diseases in which the parasite is intracellular.

In some diseases, the stimulus by the infecting organism is long-lasting (pages 303 and 304), so that the antibody-producing cells continue to make antibodies long after the infection has disappeared, sometimes during the entire life of the patient. In some instances this may be due to continuous subclinical infection. Antibodies can often be found in the blood many years after recovery from clinical and even subclinical cases of some diseases by appropriate test methods, some of which will be described later. In other infections (for example, typhoid fever) detectable antibody production subsides completely, but "memory cells," the progeny of T-cells and of B-cells (or possibly B-cells themselves), retain the information in a sort of "dormant" state. Memory cells remain "alert" and conditioned to react with the specific antigen whenever it is encountered again. On renewed contact with the specific antigen, they quickly become reactivated and multiply rapidly as plasma cells. Within a few hours large amounts of the required specific antibody are produced. Immunologic memory may last many years and, after secondary stimulation with antigen, produce a short latent period and a high antibody titer.

Subclinical or Inapparent Infections. In order to acquire specific active immunity against a disease one need not contract that disease in a severe (or *clinical*) form. On the contrary, many persons develop immunity to such diseases as diphtheria, scarlet fever, and yellow fever without ever being aware of any definite attack of illness. Their youth, health, genetic determinants, or racial

CHAPTER 20

stock may have enabled them to withstand the invaders to the extent that no severe illness occurred, yet the body cells were sufficiently stimulated to produce antibodies. Possibly the organisms were somewhat less virulent than usual, or the person's physical condition contributed to his or her resistance. Either way, this antibody response in the absence of recognizable clinical symptoms is usually ascribed to a *subclinical, latent,* or *inapparent* infection.

Specific Antibody Formation

The formation of specific antibodies (immunoglobulins) is a response of lymphocytes and possibly other cells of the body to certain substances that are of particular chemical composition and that, except for certain rare cases, come from sources outside the body proper. The foreign substances that stimulate antibody formation are given the general name of *antigens.* (See Chapters 17 and 19.)

Antitoxins. Plant, animal, or bacterial *exo*toxins that are *proteins* and *soluble in water* stimulate the formation of antibodies called *antitoxins.* These, by means of a physicochemical interaction, inactivate or neutralize the exotoxins that stimulated their formation. For example, when diphtheria organisms establish themselves in the throat and grow there, they excrete a powerful exotoxin into the blood. The antibody-producing cells of the body respond by producing *diphtheria antitoxin.* An analogous reaction occurs with all other soluble protein toxins, no matter how they gain entrance (undigested) to the blood—by infection, by hypodermic injection, or by absorption. Each toxin, however, stimulates the production of antibodies only against itself and not against some other toxin. For example, the poison of the cobra will engender antitoxin (antivenin) only against the cobra venom and not against diphtheria toxin or tetanus toxin. Each antibody is strictly specific against its own antigen.

Immunologic Specificity. In an attempt to explain this specificity, Paul Ehrlich, a famous German physician and medical immunologist, tried to describe the specific relations of antigens and antibodies graphically. He thought of these substances as particles or complex protein molecules having chemical groups called side arms or "side chains," which could fit together only by complex chemical interactions. Today Ehrlich's basic idea is still applicable, but we now know that the chemical side chains as conceived by Ehrlich include in addition all chemical, physical, atomic, and electrical configurations of the molecular surfaces and structures. Some of these chemicophysical relationships between antigen and antibody are very complex—much like a complicated Yale lock and the elaborately shaped and highly specific key needed to unlock it. Specificity is characteristic of all immunologic reactions and enzyme functions.

Sensitizing Antibodies and Complement

There are thousands of different antibodies. Only some, however, are antitoxins. The antigen-antibody combinations often do become manifest in a variety of other ways. Instead of reacting with soluble molecules of protein (exotoxins), some may react with antigens that are parts of whole cells.

Cytolysins. A good illustration of this type of antibody is cytolysin (from Greek *kytos,* hollow vessel, cell; *lysis,* dissolution). The production of sensitizing antibodies is stimulated by organized structures such as living bacterial cells (as contrasted with unorganized matter such as soluble toxins, cobra venom, egg white, and serum). The cytolysins, as their name indicates, help dissolve the cells that stimulated their production.

In causing the destruction of cells such as erythrocytes of another species, the cytolysin must first combine with the specific *antigenic determinant* of the cell. The antibody is said to be *adsorbed* by the cells: this is a specific physicochemical surface combination. This combination is not by itself sufficient to dissolve or lyse the cells. It sensitizes them, however, to the action of substances called *complement.*

Complement. Complement, a distinct set of proteins with different physical, chemical, and biological properties that work in sequence reactions, has also been called alexin. It is enzyme-like in activity and always present in fresh, normal blood. The activity of a component of complement may be destroyed by heating it to 56 C for a short period of time. Complement is not produced as a result of the introduction of an antigen. It complements the work of lysis—hence its name. It is nonspecific because it helps all cytolysins and some other antibodies (e.g., opsonins) to complete their work.

The complement system involves at least 20 different plasma proteins (Fig. 20–1). It is chiefly involved in immunologically induced inflammation and functions against tissue injury and in nonspecific resistance to infection. Active in proper sequence in the complement pathway are three components of C1 protein (C1q, C1r, C1s), then C4, C2, C3, C5, C6, C7, C8, and C9, with molecular weights between 95,000 to 390,000 daltons in serum concentrations of around 60 μg/ml to 1,200 μg/ml for C3. In the alternative pathway of activation are properdin (factor P), factor D (an activated serine esterase), and factor B (Fig. 20–2). Other control proteins are activated C1 inhibitor, β1H, and C3b inactivator.

The activation of complement involves the following sequence:

1. An antibody binds to the antigen (foreign blood cell or bacteria).
2. Subunits of C1 protein bind to two IgG or one IgM (the antibody).
3. C1s splits C2 into C2a and C2b and C4 into C4a and C4b. (A kinin-like factor is produced from C4 that contracts smooth muscles, increases vascular permeability, and resists the effects of antihistaminic drugs.)
4. C2a binds to C4b, which is attached to the cell surface. C4b2a is C3 convertase, which splits C3, and C3b attaches to the cell membrane. This makes the cell susceptible to phagocytosis by leukocytes. It also destroys undesirable bacteria, but if the surface of erythrocytes, platelets, or leukocytes is affected by C3b, this may constitute a highly detrimental autoimmune disorder. (C3a acts like an anaphylatoxin. It induces the contraction of smooth muscle, increased vascular permeability, and the release of histamine from mast cells [connective tissue cells with yet unknown physiological function])
5. C5 splits into C5a and C5b, and C5b binds to C6 and C7.This complex attaches to the cell surface.
6. C8 and C9 attach to the complex, which damages the cell membrane and leads to lysis of the cell (hemolysis for RBCs or bacteriolysis for bacteria).

The alternative pathway involving properdin is not given here. It does not require antibodies but involves direct reactions of C3b with the cell surface. IgA is sometimes involved.

The complement components that activate the release of mediators (histamine, heparin, and serotonin in some animals) from *mast cells* bring about increased vascular permeability; contraction of smooth muscles; chemotaxis of neutrophils, eosinophils, and mononuclear cells; increase of susceptibility to phagocytosis; lysis of cell membranes; solubilization of immune complexes; the destruction of bacteria; and neutralization of viruses.

Hemolysins. If an animal is injected several times with the erythrocytes of another species of animal, the serum of the recipient animal acquires the power of destroying the RBC's of the donor species. This is due to the formation of cytolysins that combine with the donor RBCs and sensitize them to *complement* in the blood of the recipient. Further reactions are very complex as described previously. They involve stepwise incorporation of complement fractions to produce a build-up of cell membrane antigen with antibody and the fractions of complement, with Ca^{++} and Mg^{++} as cofactors. The red blood cell or bacterial cell develops small holes in the membrane, each 100 Å in diameter, through which the hemoglobin (or bacterial cell contents) escapes into the surrounding fluid. This process is called lysis or *cytolysis* or, in the case of RBC's, *hemolysis*. The hemolytic cytolysins are called *hemolysins*. Similarly, the specific sensitizing antibodies that mediate the dissolution of bacterial cells are designated *bacteriolysins*. Gram-positive bacteria, including mycobacteria, are not susceptible to the action of complement.

Fixation of Complement. When complement is involved in a reaction between antigen and antibody, it combines with them, presumably by adsorption on the surface of the combined antigen-antibody molecule. It is apparently used up or "fixed." *Complement fixation* (CF) is said to have occurred.

The *Wassermann test* (or Kolmer-Wassermann test) is an outstanding application of CF for diagnostic purposes.[1] The German bacteriologist and physician August von Wassermann (1866–1925) showed that the organisms of syphilis, like other bacteria, stimulate antibody production in the host. These antibodies "fix" complement when combined with their antigens, the syphilis organisms, or, as we shall see later, a nonspecific cardiolipin antigen. The fixation of complement can then easily be demonstrated by a simple test (hemolysis) for free (unfixed) complement. The CF technique has many other diagnostic and research applications. For example, it may be used to diagnose infections with the TRIC agents, the *Chlamydia* of trachoma, and inclusion conjunctivitis, or to determine antibodies present in individuals infected with adenoviruses, arboviruses, influenza, or rickettsial diseases.

Immobilizing Antibodies. Once infection with the spirochete *Treponema pallidum* occurs, syphilis develops. As the disease progresses specific antibodies appear in the blood. These can be demonstrated by CF tests, as just mentioned. They can also be demonstrated by mixing a small amount of serum from the patient with a suspension of living *T. pallidum*. The spirochetes, ordinarily characterized by an active twisting, undulating, corkscrew-like motion, are seen to lose this mobility within a short time in the presence of syphilitic serum. They die soon afterward. The test is entirely specific and is highly reliable when properly performed. Similar specific immobilizing antiodies can be shown to occur in numerous other infections.

It is of particular interest to note here that the *Treponema pallidum* immobilization (TPI) test was the first practical and specific blood test for syphilis, and that complement is a necessary component of the immobilization test. This suggests that IgG is the immobilization antibody.

The Immune-Adherence Phenomenon. Bacteria of several widely different species, in the presence of specific antibody and complement, adhere to the erythrocytes of the host as though the microorganisms had become sticky. This immune-adherence phenomenon is important because sticky bacteria held thus by erythrocytes fall an easy prey to leukocytes. The immune-adherence phe-

[1]The TPCF test is now used in its place in diagnostic laboratories. The same basic principle applies to it as to the Wassermann test (see Chapter 35).

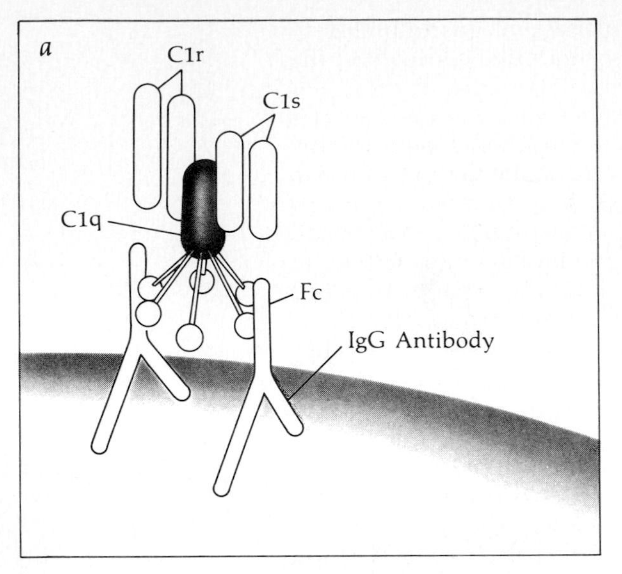

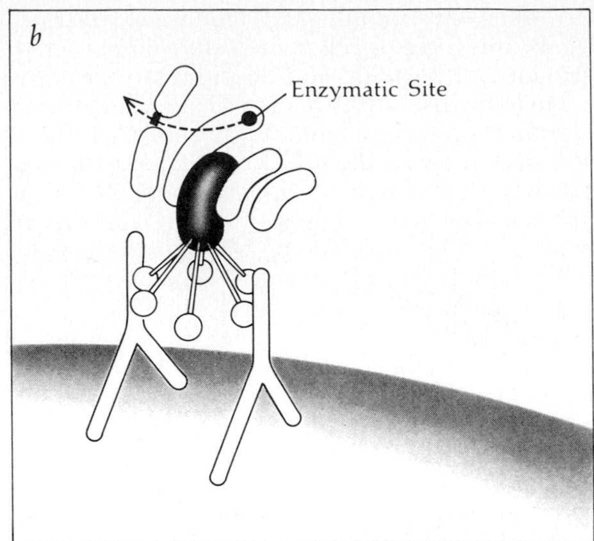

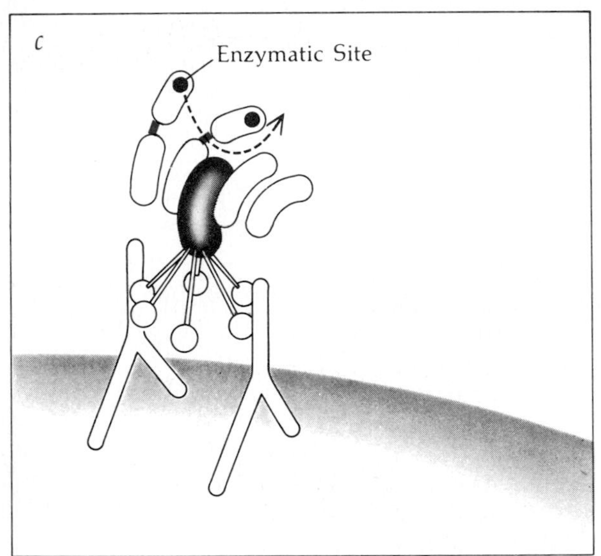

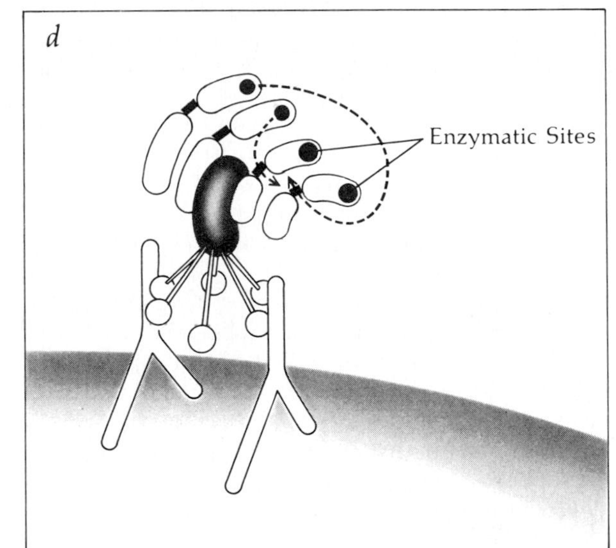

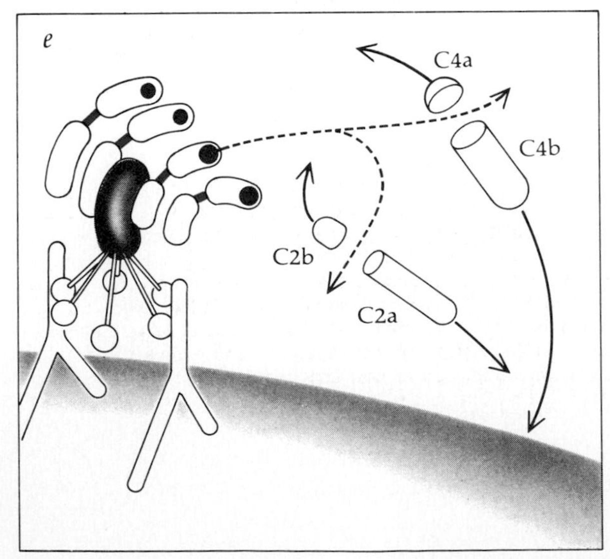

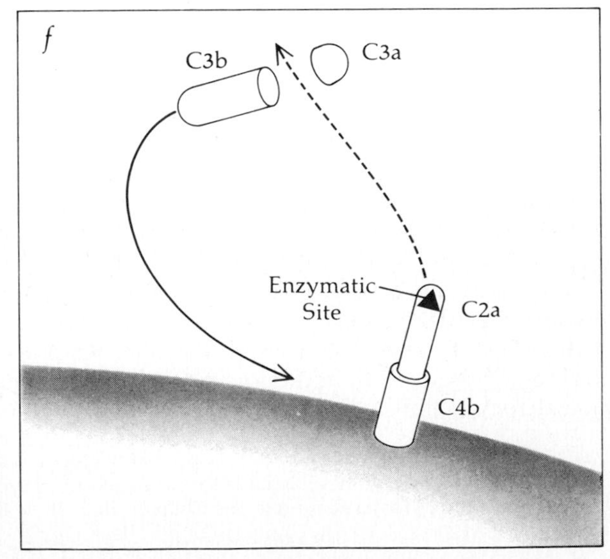

Figure 20–1 *See legend on opposite page*

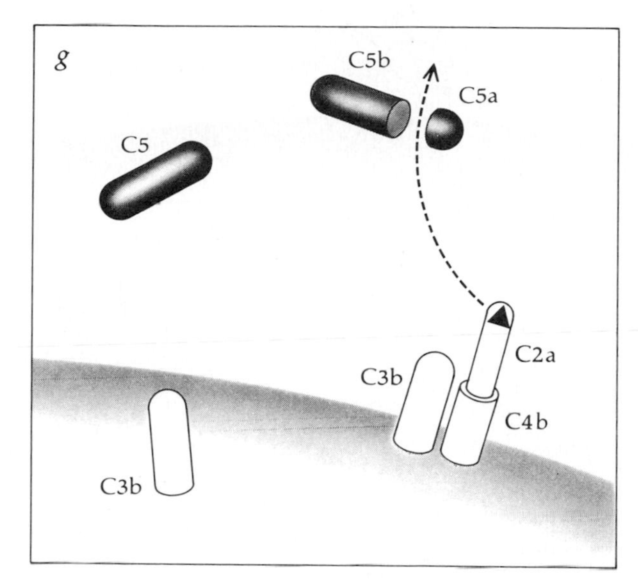

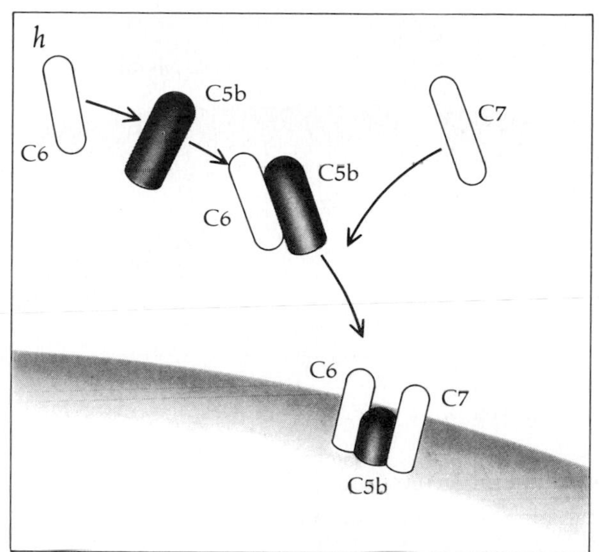

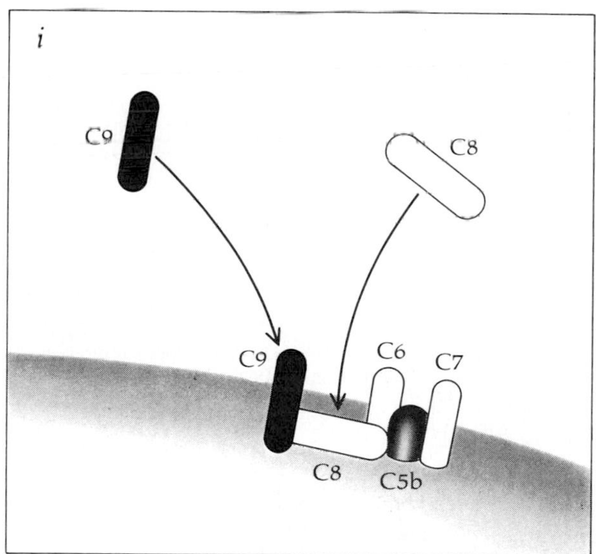

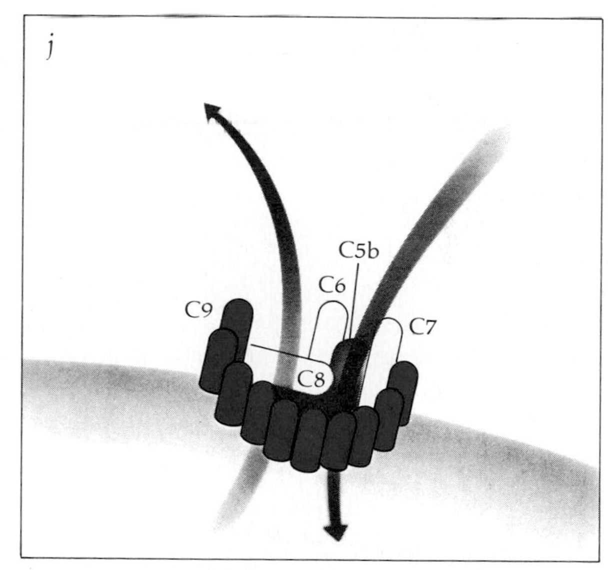

Figure 20–1 At the start of the classical complement activation pathway *(a)* antibodies recognize and bind to the antigen (e.g., foreign cells or bacteria). The complement component C1, consisting of subunits C1q, C1r, and C1s, binds to two adjacent IgG antibody molecules. Binding by the C1q subunit to the antibody alters its configuration, which results in conformational changes in C1r, exposing a proteolytic site on one C1r molecule *(b)*. This proteolytic site then acts on the other C1r molecule, exposing a similar enzymatic site *(c)*. These enzymatic sites act on C1s, exposing two additional enzymatic sites *(d)*. The $\overline{C1s}$ now cleaves two other complement components, C4 and C2 *(e)*, allowing the C4b fragment to attach to the cell membrane. C2a then binds to C4b, producing $\overline{C4b2a}$ (C3 convertase), which then cleaves C3 *(f)*. C3b can attach to the cell membrane. When C3b attaches itself close to the C3 convertase, it alters the specificity of the enzyme site on the C2a, which can now split C5, and is thus called a C5 convertase *(g)*. C5b binds to C6 and C7 *(h)* and attaches to the cell surface. C8 and C9 *(i)* now attach to the C5b67. The insertion of this late complex damages the cell membrane and permits a rapid flux of ions that ultimately leads to cell lysis *(j)*. (© 1983 *Scientific American Medicine*. All rights reserved.)

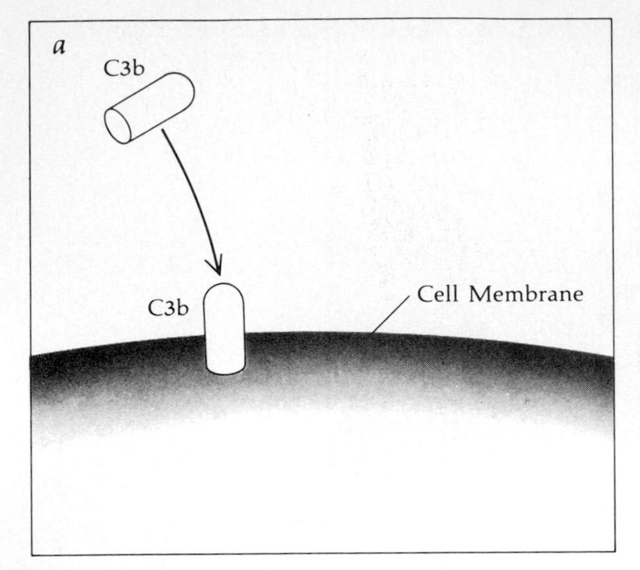

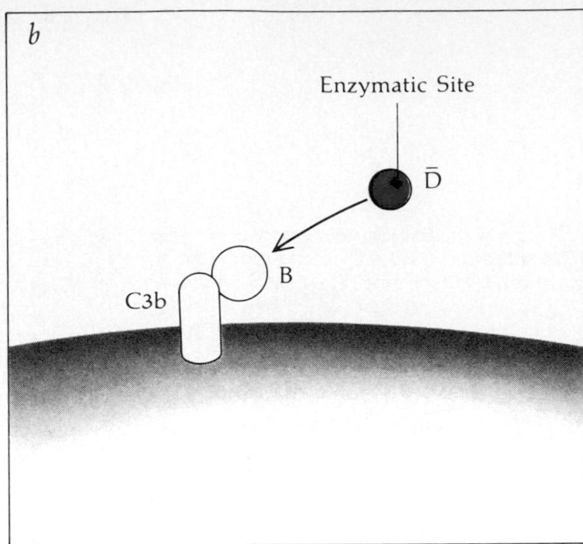

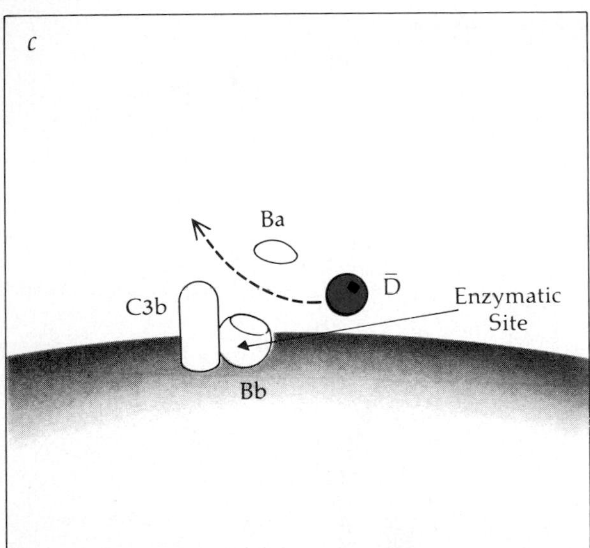

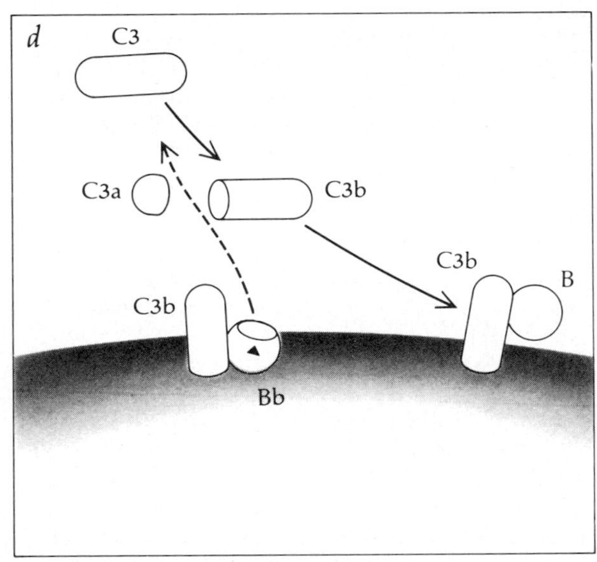

Figure 20–2 The alternative, or properdin, pathway is triggered when complement component C3b becomes attached to certain surfaces *(a)*. Component B then attaches to C3b and is activated by a serine esterase, $\overline{D}$ *(b, c)*. The enzymatic site on Bb now cleaves more C3, resulting in more C3b, which attaches to the membrane *(d)*. More B attaches to the C3b and a chain reaction follows, leading to the amplification process. It is thought that there is a constant amount of C3b interacting with B. C3b is generally broken down by C3b inactivator, and $\overline{C3bBb}$ by β1H. Only when $\overline{C3bBb}$ is stabilized on certain cell surfaces and becomes resistant to these control proteins does the amplification process proceed efficiently. Why certain cell membranes stabilize $\overline{C3bBb}$ whereas others do not is a question presently being investigated. (Courtesy of *Scientific American*.)

nomenon is thus a specific opsonic function, aiding an important defensive mechanism as well as offering a good diagnostic test.

Precipitins. Constant amounts of antibody mixed with increasing quantities of soluble antigen yield precipitates that can be measured quantitatively. The homologous antibody combines in suitable proportions with the molecules of soluble proteins and cause the protein molecules to gather in minute but visible flakes or granules, clouding the fluid in which they are suspended. This flocculent turbidity is called a precipitate (Fig. 20–3). Antibodies that act in this way against protein molecules are called *precipitins,* and the antigen-antibody reaction itself is termed *precipitation.* These reactions, like all antigen-antibody reactions, are specific. Although not essential to precipitate formation, complement is fixed if it is present, because it is readily absorbed by any minutely particulate matter, such as soot and clay.

Precipitin Reactions. These are used for many purposes when the identity of a protein antigen or antibody is in question For example, we may use a group of known antigens to detect unknown antibodies in a patient's serum and thus diagnose disease. We may also have available a collection of sera, each

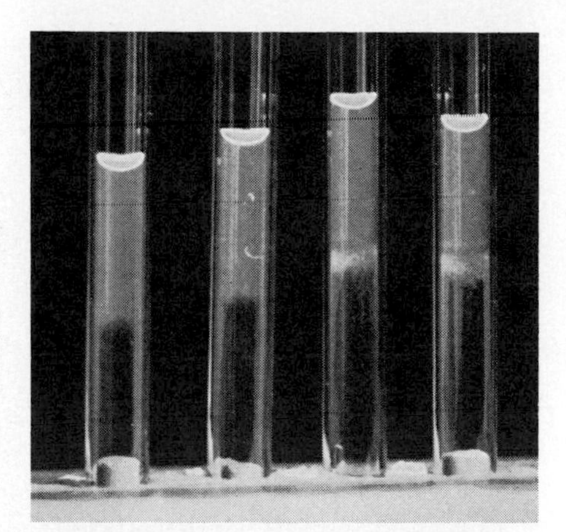

Figure 20–3 Precipitin test in narrow-bore glass tubes. The antigen solution is the cloudy zone above; the antiserum is the clear zone below. At the interface between them, in the two right-hand tubes, a definite white floc or precipitate has formed, representing a reaction between specifically related antigen and antibody. The same antigen fails to react with a different (nonspecific) serum in the two left-hand tubes. (×5.) (Preparation by Dr. Elaine L. Updyke. Photo courtesy of the Communicable Disease Center, U.S. Public Health Service, Atlanta, Ga.)

containing a known type of precipitin for a different protein, and use these antibodies to identify organisms and various protein substances. In forensic medicine, it is possible to determine whether a blood stain is of human or animal origin. The test may even be done centuries after the stain has dried, as in archaeologic remains. If the blood spot has come from other than human sources, it will not yield a precipitate with the serum of an animal immunized against human protein. Table 20–1 shows a typical scheme for a complete precipitin test with controls.

The Precipitin Test in Gel. The test may be done in a Petri dish by placing serum containing precipitins in a depression on the surface of a layer of agar, near another depression containing fluid antigen. The antigen and antibodies diffuse outward from their respective depressions. Along the line where the ratios of antibody to antigen are optimal a white precipitate appears in the agar if antigens and antibodies are mutually specific (Fig. 20–4). This is the basis of the "Ouchterlony" method. There are various modifications of this procedure, using gels on slides (the most common method) or test tubes, single diffusion, double diffusion, and so on. Widely used are the Oudin procedure, the Preer method, and the Oakley-Fulthorpe diffusion in one dimension. Often the precipitin reaction in agar is combined with mobility studies of the proteins

Table 20–1. Scheme for a Complete Precipitin Test*

	Tube Number									
	Tests					*Controls*				
	1	2	3	4	5	6	7	8	9	10
Upper Layer (Antigen)	Extract 1:HT	Extract 1:10T	Extract 1:T	Extract 1:H	Extract 1:10	Extract 1:H	0.9% NaCl	Sub-strate extract	Heter-ologous blood 1:T	Known human blood 1:T
Lower Layer (Antibody)	Anti-serum	Anti-serum	Anti-serum	Anti-serum	Anti-serum	Normal rabbit serum	Anti-serum	Anti-serum	Anti-serum	Anti-serum
Expected Results						—	—	—	—	+

*H for hundred and T for thousand are common abbreviations used in bacteriology for designating certain dilutions. Thus, 1:HT represents a 1:100,000 dilution, and 1:T stands for 1:1,000. The controls as shown must give the expected results or the tests will not constitute valid data. For example, if the antigen is human blood on a man's shirt in medicolegal testing, tubes 6 and 7 would test for the antibody specificity, tube 8 would check the possibility that a piece of the man's shirt could be responsible for a false-positive test (this has to be ruled out), tube 9 would be blood from other than human sources, and known human blood would be the positive control in tube 10.

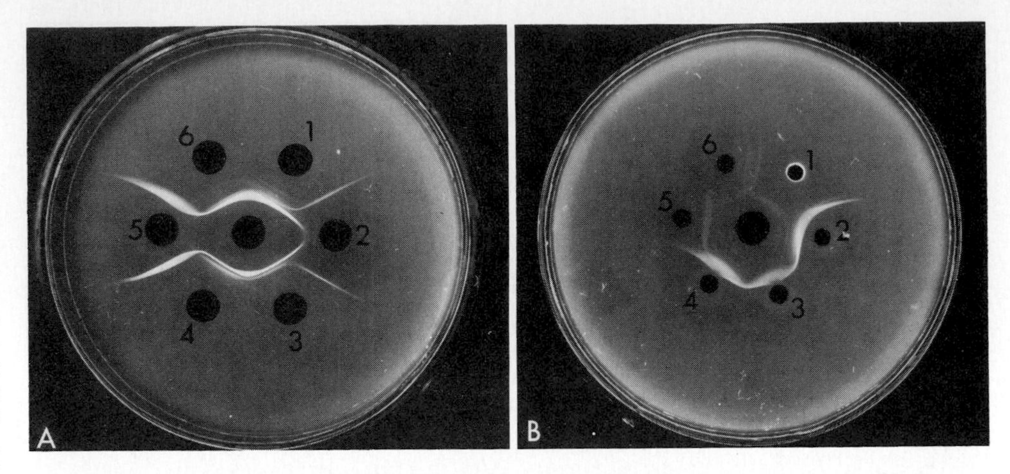

Figure 20–4 *A*, Immunodiffusion pattern using Auto-Gel 7-mm size cutters. Goat anti-human serum is in the central cup; (1) and (4) contain normal human serum, (2) goat anti-human gamma G, (3) and (6) human urine, and (5) goat anti-human albumin. *B*, Immunodiffusion pattern using Auto-Gel 4-mm size cutters. Normal human serum is in the central cup; (1) contains goat anti-human alpha² macroglobulin, (2) and (4) goat anti-human serum, (3) goat anti-human albumin, (5) goat anti-human gamma globulin, and (6) goat anti-human glycoprotein. (Courtesy of Grafar Corp., Michigan.)

(antibody, antigen) in one electric field. This technique is called "immunoelectrophoresis" (Fig. 20–5). It permits the identification of each constituent with great resolving power and has unlimited applications in immunochemistry research.

Agglutinins. Agglutinins are similar to precipitins in their action, but they bring about the clumping of entire organized cells rather than molecules of protein. Under the influence of agglutinins the bacterial or other cells act as though their surfaces had become sticky, and they seem to be glued together in clumps—hence the traditional term *agglutination.* Actually the cells are held together by multivalent antibodies. This reaction, due to the much larger antibody-antigen complex, is macroscopic, whereas the precipitin reaction particles are almost at the microscopic level of visibility. The connecting of cells by "bridges" of coupling antibody molecules has an opsonic effect, since agglutinated cells are more rapidly and readily engulfed by phagocytes than single cells. The production of agglutinins can be stimulated by injections of various kinds of cells: among those that have the greatest practical importance to the diagnostician are bacteria and erythrocytes.

A number of useful diagnostic tests are based on bacterial agglutinins. For example, if a person has typhoid fever, his blood will be found to contain agglutinins against typhoid bacilli. By mixing a little of his serum with a culture or suspenson of the bacteria, which are kept growing in the laboratory for such tests, the action of the agglutinins can be observed. After a short interval all the bacilli, evenly distributed throughout the culture at first, gather together in fairly large clumps and flocs and settle to the bottom of the test tube, leaving the suspending fluid clear (Fig. 20–6).

Diagnoses of various undetermined infections can be made in this way. If a patient's serum agglutinates a certain kind of organism, then he probably is or has been infected with that organism. Sometimes a person's serum yields an

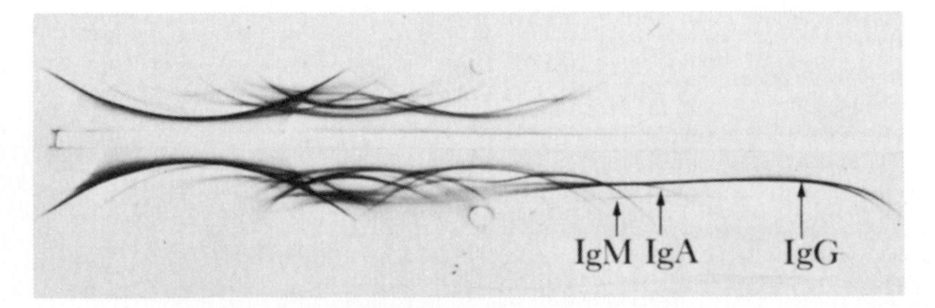

Figure 20–5 Immunoelectrophoresis of serum. *Top,* Congenital agammaglobulinemia. Note complete absence of IgM, IgA, and IgG bands. *Bottom,* Normal serum showing normal amounts of IgM, IgA, and IgG. (From Bellanti, J. A.: Immunology II. Philadelphia, W. B. Saunders Company, 1978, p. 651.)

agglutination reaction not because he has an infection at present but because he has previously received injections (vaccine) of the organism that his serum agglutinates.

Diagnostic Importance of Rising Titers. In order to distinguish between antibodies due to present infection and those due to past infection or to vaccines given long since, it is necessary to take two specimens of the patient's serum, one near the onset of the infection and one 10 to 12 days later. If the patient's cells are producing antibodies in response to the present infection, the *titer* (concentration of antibodies in the serum) will be low in the first specimen and definitely higher in the second. If few or no antibodies are present in either specimen, the patient is failing to produce any. This may be because of a congenital defect (agammaglobulinemia), excessive radiation or other immunosuppressive measures, or because the infection is due to some organism other than that being used as antigen. If antibodies are present but fail to show any change in titer between first and second specimens, then they were probably present before the infection began and are unrelated to it.

Hemagglutination. In addition to the specific agglutination of bacteria described previously, there are some curious and important agglutination phenomena called *hemagglutination* because they are related to red blood cells. These may be considered under four headings: *cold hemagglutinins, passive* or *indirect hemagglutination, viral hemagglutination,* and *isohemagglutinins.*

COLD HEMAGGLUTININS. These are substances that appear in the blood of person with certain respiratory diseases, such as certain "atypical" pneumonias of mycoplasmal *(M. pneumoniae)* origin, and also in the blood of those infected by certain protozoan parasites (trypanosomiasis). These hemagglutinins are of unknown significance beyond the fact that, since they rarely appear in association with other diseases, they are of value in aiding diagnosis. The sera of the patients, in dilutions as high as 1:10,000, agglutinate their own erythrocytes (autohemagglutinins) when cooled to 2 C but do not do so at 37 C. This agglutination at low temperature gives rise to the term *cold hemmaglutination.*

INDIRECT (PASSIVE) HEMAGGLUTINATION REACTION. This very useful procedure illustrates the fact that antigens at the surface of a cell determine its immunologic specificity. Erythrocytes (sheep, in this example) are washed free from their serum. The cells are treated with tannic acid or formaldehyde to solidify them and toughen the surfaces. The erythrocytes are again washed and then suspended is a saline solution containing any desired *soluble* antigen—for example, an antigen extracted from tuberculosis bacilli. This antigen is adsorbed on the surfaces of the erythrocytes and covers then as a coating. The cells are again washed to remove excess antigen and are resuspended in saline solution.

A series of dilutions are now made with serum containing antibodies specific for the antigens with which the erythrocytes are coated (in this example tuberculosis antigen). Into each serum dilution is introduced a drop of saline suspension of antigen-coated erythrocytes, which behave as though they were tubercle bacilli. Within a short time the coated sheep erythrocytes are specifically agglutinated. Tuberculosis antibodies have no visible effect whatever on the uncoated, tannic acid–treated erythrocytes of a sheep. It is interesting that totally inert particles of plastic, gum arabic, latex, and so on can be similarly coated with antigen and agglutinated by specific antibodies. The method is very accurate and is being widely adapted for diagnostic purposes, e.g., in syphilis.

VIRAL HEMAGGLUTINATION. Agglutination of chicken erythrocytes is brought about by several respiratory viruses, such as the influenza virus. Note that this is not antibody agglutination. The virus itself appears to adhere to the RBCs of the chick at about 5 C and agglutinate them. The virus can be separated (eluted) from the cells by mixing with saline solution, followed by incubation for two hours at 37 C. Still other viruses, such as those of yellow fever and encephalitis (mosquito-borne viruses), cause agglutination of avian erythrocytes at temperatures around 30 C. These hemagglutination tests, which

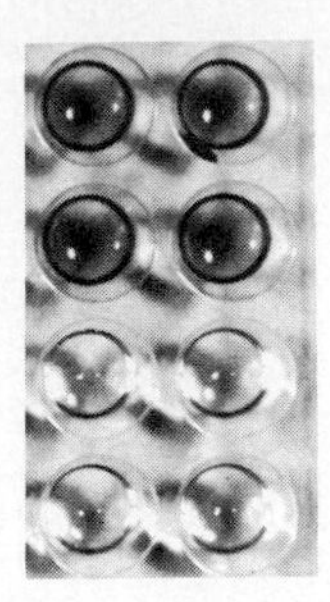

Figure 20–6 Tests for hemagglutination-inhibiting (HI) antibodies against influenza virus. HI antibodies were present in the sera tested in the four lower tubes (viral hemagglutination completely inhibited); viral hemagglutination occurred in all other tubes (no influenza HI antibodies present). The bottoms of the tubes are viewed from directly above.

may involve the use of RBCs derived from many different species, are important in the study of viruses and in the diagnosis of viral diseases.

In some infections, antibodies may occur that prevent viral agglutination of erythrocytes. They are called hemagglutination-inhibition (HI) antibodies. They are of great practical importance in diagnostic virology, being one of the principal means of identifying infecting viruses (Fig. 20–6).

ISOHEMAGGLUTININS. These important antibodies are genetically determined in the serum of most normal persons. Isohemagglutinins are responsible for the A, B, AB, O, and other blood groupings discussed in Chapter 16.

Synopsis of Antibody-Antigen Reactions. As a convenient summary, antibodies may be listed by the name of the kind of reaction they produce, with the type of immunoglobulin involved given in parentheses:

1. *Antitoxins:* form flocculent precipitates with soluble ions and detoxify ("neutralize") the toxin; toxoids precipitate similarly. Precipitates may or may not be visible. (IgG, IgM, IgA.)
2. *Precipitins:* reaction with soluble antigens of any sort to form visible precipitates; similar to antitoxins. (IgG, IgM, IgA.)
3. *Agglutinins:* aggregate cells to form clumps. The reaction is visible macroscopically because of the size of the clumps. (IgG, IgM, IgA.)
4. *Opsonins:* react with bacterial or other antigenic cell surfaces to facilitate phagocytosis; (various antibodies).
5. *Lysins or cytolysins:* react with antigens of the cell and (in the role formerly called "amboceptor" or sensitizer") with complement to cause lysis of the cell. (IgG, IgM.)
6. *Complement fixation:* occurs when antigen-antibody complexes (e.g., precipitates, cell aggregates, cytolysis by antibodies) absorb all the complement present so that it is not available for any second reaction—e.g., lysis of specifically sensitized cells in the Wassermann reaction. (IgG, IgM.)
7. *Blocking antibodies:* have different immnologic characteristics or functions; they react with antigens and block their reactive sites for competent antibodies, thus inhibiting secondary reactions—e.g., agglutination and precipitation.
8. *Neutralizing antibodies:* react with infectious agents, usually viruses to destroy ("neutralize") their infectivity.

Fluorescent[2] Antibody Staining. This method, sometimes called FA staining, is one of the most interesting and valuable advances in the field of microbiology. It involves the fluorescent labeling of antibodies so that their combination with specific antigens may be detected visually.

Although the FA staining procedure is technically complex, the principle involved is relatively simple. The first step is the separation and concentration of specific antibody globulins from the serum in which they occur. These globulins are then combined with a fluorescent dye, commonly fluorescein isothiocyanate or rhodamine B. The antibodies are then said to be *labeled* or *conjugated.* When illuminated with ultraviolet light they emit a brilliant glow (Fig. 20–7). Various labeled antibody preparations may be obtained commercially.

Let us suppose that a bacterial smear, section of tissue, or tissue culture containing a particular antigen is flooded with a soluton of fluorescent-labeled antibody specific for the antigen. The antibody combines as usual with the corresponding antigen. The preparation is then washed to remove excess (unattached) antibody. When the smear or section is viewed with a microscope using ultraviolet light as a source of illumination (Fig. 20–8), the antigen is revealed as a brilliantly fluorescent particle glowing on a relatively dim or dark background (Fig. 35–8).

[2]Fluorescence is the property of reflecting light rays having a color (wavelength) differing from that of the incident rays. Fluorescent objects are particularly brilliant in ultraviolet light.

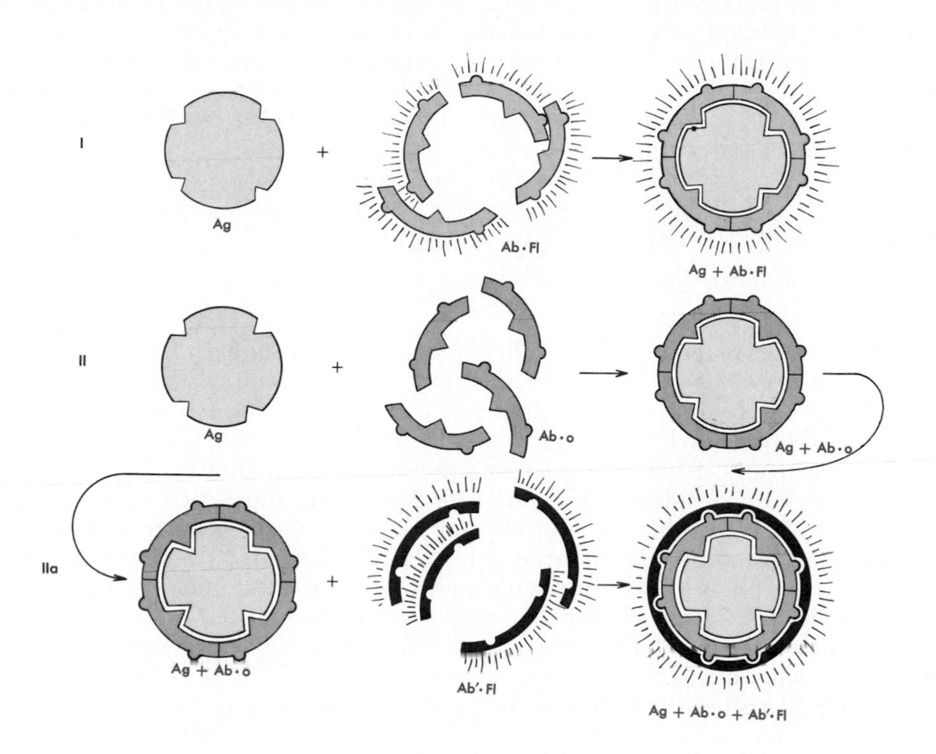

Figure 20–7 Direct and indirect fluorescent-antibody staining. In the direct method (I) antigen (Ag) is allowed to react with its specific antibody which has previously been conjugated with a fluorescent dye (Ab·Fl). When viewed in the microscope with ultraviolet illumination the antigen-antibody combination (probably on the surface of bacteria or virus particles) glows brilliantly.

In the indirect method (II) the antigen is combined with ordinary (nonfluorescent) specific antibody (Ab·o). Viewed with ultraviolet light no fluorescence is seen. It is impossible to say whether specific combination of Ag and Ab has occurred. The invisible Ag + Ab·o combination is now treated (IIa) with fluorescent antibody specific for *any* gamma globulin (from a different species of animal) (Ab'·Fl). All antibodies, including Ab·o, are gamma globulins. Ab'·Fl therefore combines with Ab·o on the surface of Ag, causing the particles of Ag + Ab·o + Ab'·Fl to glow in ultraviolet light.

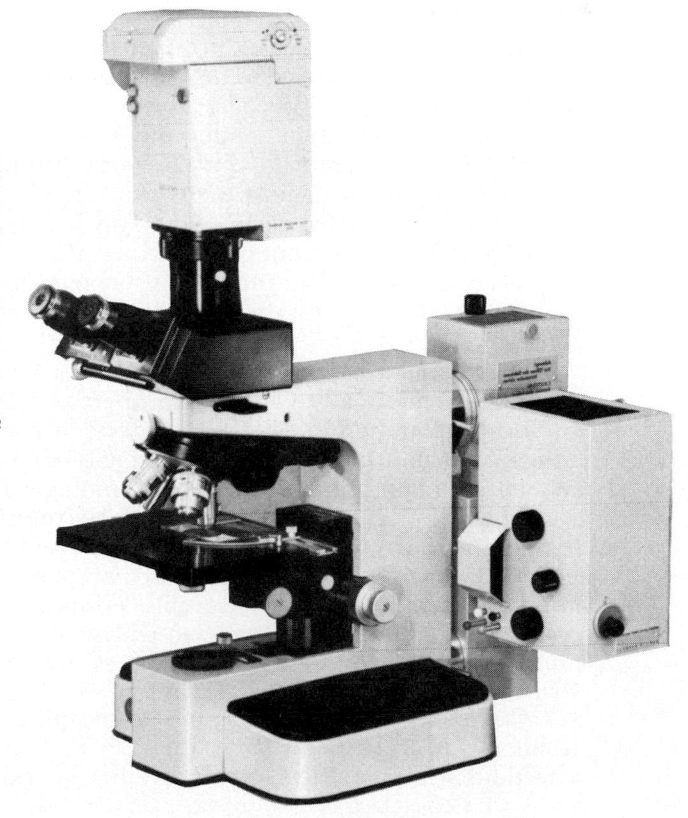

Figure 20–8 Leitz ORTHOPLAN microscope set up for transmitted and reflected fluorescence microscopy, as well as bright field observations. Attached to the back of the microscope are two lamp housings. The larger one contains a mercury burner HBO 200 (for transmitted light) and the smaller one the HBO 100 (for reflected light). On top of the fluorescence microscope is the automatic 35 mm ORTHOMAT W camera. (This illustration was provided through the courtesy of Mr. Albert Heydemann and the E. Leitz Co., Rockleigh, N.J.)

This staining procedure has many variations, including the direct method, an indirect method (Fig. 20–7), and the detection of complement fixed in situ, where the antigen-antibody reaction has taken place. It has greatly simplified problems of medical diagnosis and research by making it possible to detect, and localize quickly, extremely minute amounts of antigen, even virus particles inside infected cells.

LYMPHOCYTE HYBRIDOMAS AND MONOCLONAL ANTIBODIES

The fusion of two selected cell populations to produce hybrid cells with one nucleus makes it possible to "build" homogeneous antibody-producing cells in culture. An example of such "antibody factories" is a mouse hybridoma.

A lymphocyte hybridoma results from the fusion of a malignant mouse myeloma cell and a lymphocyte from the spleen of a mouse that has been immunized with a specific antigen. The hybrid of these two cells has the enzymes of the lymphocyte and the immortality of the cancer cell. It can produce specific antibodies in large quantities if it is cloned; thus, for practical purposes, it is an antibody-producing machine. The antibodies have a titer in the millions and are specific for the antigen originally selected to produce these monoclonal antibodies that are homogeneous and available in quantity. These pure antibodies have practical applications in the study of immune mechanisms, in serological testing, in hormone and endocrine investigations and in specific transfers of immunity to disease. It is expected that, in the years to come, new uses of monoclonal antibodies will be developed that the author could not even imagine when this book was written.

SUMMARY

Many infectious diseases produce specific immunity that does not protect against other infections. Some diseases confer no immunity at all, others produce some immunity, while another group usually induces lifelong immunity, although "permanent" immunity is not entirely certain.

Many plants, animals, and bacteria produce proteins called exotoxins; antibodies produced against them are called antitoxins. Toxins and antitoxins are very specific toward each other. This was explained by Paul Ehrlich on the basis of the surface configurations of these two reacting substances. The cytolysins, are antibodies (hemolysins against RBCs, bacteriolysins against bacteria) that absorb into the cells and lyse then, but only with the aid of complement.

Complement consists of a set of at least 20 plasma proteins with different physical, chemical, and biological properties. They work in sequence reactions and bring about lysis of a cell (erythrocyte or bacteria) after the initial antibody-antigen reaction begins the complement "chain reactions." The complement system has a determining role in immunologically induced inflammation.

The complement components activate the release of mediators (histamine, heparin, and serotonin in some animals) from connective tissue *mast cells* with the following results: increased vascular permeability; contraction of smooth muscles; chemotaxis of neutrophils, eosinophils, and mononuclear cells; increased susceptibility to phagocytosis; lysis of cell membranes; solubilization of immune complexes, the destruction of bacteria; and the neutralization of viruses.

Complement has been used in complement-fixation tests, such as the Kolmer-Wassermann test and the TPCF, for the diagnosis of syphilis and infections with the TRIC agent. Numerous other tests have been developed involving antibody-antigen reactions; most common are precipitation reactions (with many variations) and agglutination tests such as hemagglutination and fluorescent antibody staining. These methods have many very practical uses, not the least of which is diagnosis of infectious disease.

A lymphocyte hybridoma is the result of fusion of a malignant cell and a lymphocyte. It produces specific antibodies in large quantities.

REVIEW QUESTIONS

1. Explain why some diseases appear to confer permanent immunity, some produce temporary immunity, and others produce no immunity at all.
2. What are antitoxins? What did Paul Ehrlich say about them? What are cytolysins?
3. Describe the functions of complement. What is it? What role does C3 play? What is hemolysis? What is properdin?
4. What is an anaphylatoxin? What do mast cells do?
5. Of what use is complement fixation in the diagnosis of diseases? Which diseases?
6. What are immobilizing antibodies?
7. Explain the main differences between precipitation and agglutination tests. State some uses for each.
8. What is a CF reaction?
9. Name different types of hemagglutination.
10. Discuss different antibody-antigen reactions.
11. What are lymphocyte hybridomas?

Supplementary Reading

Cunningham, A. J.: Understanding Immunology. New York, Academic Press, 1977.

Fearon, D. T.: Activation of the alternative complement pathway. *In* Critical Reviews in Immunology, vol. 1. Boca Raton, Fla., CRC Press, 1980.

Gergely, J.: Antibody Structure and Molecular Immunology. New York, American Elsevier Publishing Co., 1975.

Haber, E.: Symposium on the Future of Antibodies in Human Diagnosis and Therapy. New York, Raven Press, 1977.

Hobart, M. J., and McConnell, I.: The Immune System. A Course on the Molecular and Cellular Basis of Immunity. Philadelphia, J. B. Lippincott Co., 1976.

Mayer, M. M.: Complement, Past and Present. New York, Academic Press, 1978.

Müller-Eberhard, H. J.: Complement. *Ann. Rev. Biochem., 44*:697, 1975.

Nysather, J. O., Katz, A. E., and Lenth, J. L.: The immune system; its development and functions. *Am. J. Nursing, 76*:1614, 1976.

Porter, R. R., and Reid, K. B. M.: The biochemistry of complement. *Nature, 275*:699, 1978.

(See also the list of readings following Chapter 21.)

ACTIVE AND PASSIVE IMMUNITY

ACTIVE ARTIFICIAL IMMUNITY

As described in Chapter 20, immunity that is actively acquired in the ordinary course of life as a result of clinical or subclinical infectious disease is generally effective and often durable. However, one can never be sure whether such natural processes will occur, when they will occur, or whether the infection necessarily involved will be severe or fatal. In order to develop immunity safely, with certainty, at desired times and in specific people, scientists have developed *artificial* processes, such as vaccination, which simulate nature but are under human control.

Injection of Living, Attenuated, or Harmless Organisms

Active specific immunity may be acquired "artificially" by imitating nature's method of mild, subclinical infection. Living organisms are actually injected or otherwise put into the body, but either they are previously weakened (*attenuated*) by various processes so that they cannot cause a severe infection, or else a less virulent species, so closely related to the actual disease agent that both share the same immunogenic properties, may be used. Either method induces a very mild disease that produces the desired immunity.

Vaccination Against Smallpox. The *vaccinia* virus, closely related to the *variola* virus that causes smallpox in human beings, produces in cows a mild disease called cowpox, which brings about a pustular eruption on the udder. Although the virus of cowpox is distinct from that of smallpox, the two are immunologically closely related. Vaccination with the *vaccinia* virus is sufficient to give protection from smallpox for one to ten years or longer, depending on the individual and the environmental conditions. Means of producing active artificial immunity to smallpox have been studied for centuries.

Edward Jenner, an English country doctor, noted that persons working around cows that had cowpox developed sores on their fingers like those on the cow's udders, and that these persons never caught smallpox during an epidemic. He published his observations in 1798, after he had been studying and experimenting on the subject for 20 years. He introduced on a large scale the practice of inoculating people with cowpox matter. This, in an improved form, is what we now call vaccination (from Latin *vacca*, cow).

Smallpox was widely prevalent in Jenner's time. Today vaccination is no longer recommended, but it would be in order for persons proceeding to or returning from areas with new outbreaks of smallpox and for persons at special risk (e.g., certain health personnel). The United States Public Health Service no longer advocates *routine vaccination* of American youngsters. This decision is based on the greatly reduced probability of importation of cases (in the U.S., only one since 1949) from foreign countries, the effectiveness of measures to detect and prevent spread of imported cases, and the complications and risk involved in vaccination.

CHAPTER 21

By surveillance and control measures, with vaccination only in special areas, the World Health Organization (WHO) eradicated smallpox from its last strongholds in Afghanistan and Bangladesh. However, there were setbacks in 1977 in the war-torn areas of Ethiopia and Somalia. In theory it should be possible to eradicate the smallpox virus, since humans are its only natural host and no arthropod vectors are involved. Constant surveillance is now necessary: an enormous nonimmune population could develop and may demand a crash defensive program if even one case of smallpox should occur in any population center of the world.

Vaccine virus was once commonly prepared by inoculating healthy calves (or sheep in England) with living cowpox organisms either from an animal previously inoculated with vaccine virus or from a human vaccination sore. In a few days the calf developed vesicles or sores similar to those following vaccination in human beings. The material from these was scraped out and mixed with glycerin (see Fig. 21–1). Freshly prepared vaccine of this sort always contained bacteria. It was therefore disinfected with phenol, which, curiously, does not inactivate this particular virus. The vaccine was tested for bacterial contamination and for its immunizing power (*antigenicity*) by animal experimentation. A single dose of virus suspension was put in a small, hermetically sealed tube. This was cleaned with alcohol gauze before use, broken while being grasped with sterile gauze, and emptied by means of a small rubber bulb. To prevent rapid deterioration, the virus was stored at temperatures of about 1 C until used. Any alcohol or disinfectant gaining entrance to the tube from gauze or from the patient's skin inactivated the virus.

To overcome problems of storage, bacterial contamination, and possible instant need, liquid vaccine preparations have been replaced all over the world with freeze-dried (*lyophilized*) vaccine. It is stable, easily stored in large doses, and resistant to hot climates.

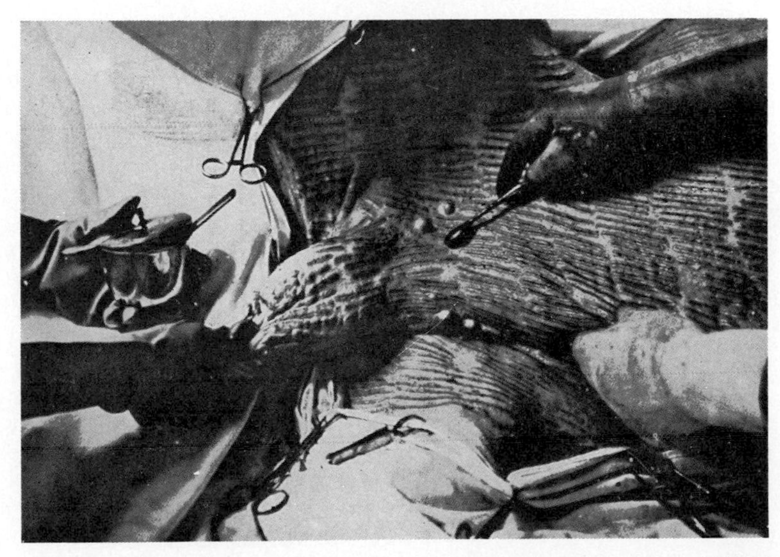

Figure 21–1 The collection of cowpox lymph from the skin of a calf. The skin has been shaved, cleaned, disinfected, and inoculated in long parallel scratches (clearly seen in the picture) with cowpox lymph. Typical pustules have developed along the scratches, and the lymph from these is being collected with surgical cleanliness. (From Monteiro and Godinho.)

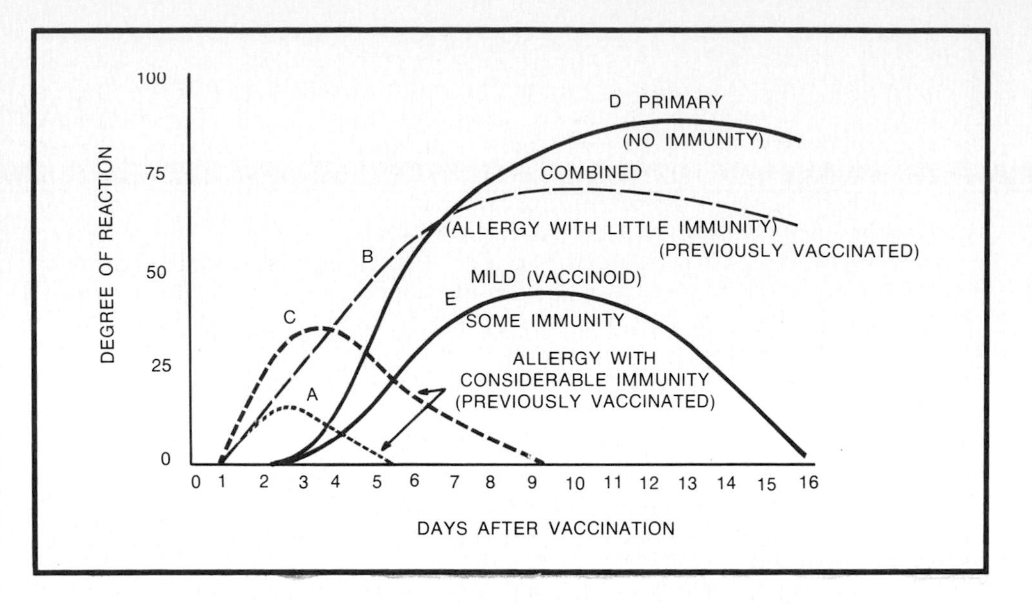

Figure 21–2 Types of reaction to smallpox vaccination. Curves *A* and *C* represent rapid, superficial reactions beginning during the first 24 hours, sometimes showing small vesicle formation but terminating rather quickly after a mild course, without scar formation. These occur only in previously vaccinated persons (or in those convalescent from smallpox), and the quick reaction of short duration and mild degree is characteristic of immunity with allergy to the smallpox virus, active or inert. Allergy may exist when immunity has all but lapsed, as shown in curve *B*. This begins as an allergic reaction but goes on to a real "take" with papule, vesicle, pustule, and eventual scar formation, much like the primary "take" in a wholly susceptible person without allergy, shown in curve *D*. Curve *E* is a milder reaction in a person without allergy but with a considerable degree of immunity. This is a "vaccinoid" reaction, usually without scar formation. Note that the reactions in the absence of allergy (curves *D* and *E*) do not begin until the second or third day. It is occasionally difficult to differentiate the types of reaction. (Adapted from M. Mitman, River Hospitals, Joyce Green, Dartford, Kent, England. In *Monthly Bull. Minist. Health,* 2:100, 1952.)

Within 24 hours after a known exposure to smallpox, human vaccinia-immune globulins (VIGs) from previously immunized individuals are recommended for passive immunization, should it be needed. VIGs are available from the U.S. Center for Disease Control, Atlanta, Georgia, as a prophylactic measure for those for whom vaccination is contraindicated but who still are thought to need protection.

Immune Reaction in Vaccination. In an immune person there is usually a more or less mild, rapidly developing, and transitory (about 24 to 96 hours) reaction recognized as probably allergic in nature (Fig. 21–2).

Pasteur's Preventive Immunization Against Rabies. As developed by Pasteur around 1885, immunization against rabies consisted of a series of 7 to 14 daily antigenic injections of live, *attenuated* rabies virus in the form of dried spinal cord of rabies-infected rabbits, the virus being attenuated by propagation in rabbits and by the drying process. A later method of immunization consisted of injections of rabbit spinal cord containing phenol-inactivated virus (Semple vaccine). These rabbit-tissue vaccines sometimes provoke an allergic response to the rabbit tissues carrying the virus, with distressing, sometimes fatal demyelinating encephalomyelitis. Also, there is little viral antigen in them in proportion to the bulk of rabbit tissues. To avoid this allergic response, an attenuated virus (the Flury strain) is cultivated repeatedly in successive, live, embryonated[1] duck eggs.

Vaccines Against Rabies Now in Use. The two types of postexposure prophylaxis against rabies should be used concurrently. (1) *Vaccines* induce active immunity, requiring 7 to 10 days to develop antibodies that persist for one year or longer. (2) *Globulins* provide rapid passive immunity of only a few weeks' duration.

The antirabies vaccines being used now are: *human diploid cell rabies vaccine* (HDCV), consisting of an inactivated vaccine from rabies virus grown in human diploid cell tissue culture and treated with tri-n-butyl phosphate or beta-propiolactone; and *duck embryo vaccine* (DEV), also inactivated and then lyophilized. *Rabies immune globulin, human* (RIG) is antirabies concentrated gamma globulin obtained from hyperimmunized human donors; whereas *antirabies serum, equine* (ARS) is a refined concentrated serum from hyperimmunized horses. Based on relative risk of adverse reactions, the preferred choices for usage are HDCV and RIG. (For further information on rabies, see Chapter 33.)

---

[1]Incubated long enough (eight to 14 days) to contain a partly developed duckling.

Other Viral Vaccines. Among the most effective immunizing agents made with active, attenuated viruses are Sabin's oral poliomyelitis vaccine and the 17D yellow fever vaccine, based on the first cultivation of yellow fever virus in chick-embryo tissue cultures by Smith and co-workers in 1937. These vaccines are good illustrations of the principles just outlined. It should be noted that, like the yellow fever vaccine, vaccines prepared with avian eggs, including those against rabies, rubeola, rubella, mumps, and influenza, can cause severe allergic reactions in persons hypersensitive to avian proteins.

One of the best active-virus vaccines is that against rubella (German measles), which, when it occurs in women during the first three or four months of pregnancy, frequently kills or deforms the fetus. The United States Public Health Service Advisory Committee on Immunization Practices (1972) recommends the active vaccine for all children between the ages of one year and puberty, and for all older females of child-bearing age who are found to be susceptible by serologic (agglutination-inhibition) test. These women are cautioned not to become pregnant during the three months following vaccination, since this presents a 40 per cent risk of developing a temporary arthritic condition within a month after the vaccination. Males of any age over one year may be vaccinated to prevent spread of the virus to nonimmune pregnant women during epidemic periods. Vaccination of *pregnant* women is absolutely contraindicated because the virus can infect and damage the fetus.

Since there seems to exist only one antigenic type of measles virus (*rubeola*) a live attenuated vaccine is very effective in preventing the disease (Fig. 21–3). This vaccine has been given to more than 70 per cent of all children in the United States since 1979, resulting in the disappearance of major epidemics that used to occur every two to three years and afflicted 98 per cent of all children by the age of ten.

Vaccines against influenza infections are used extensively when outbreaks

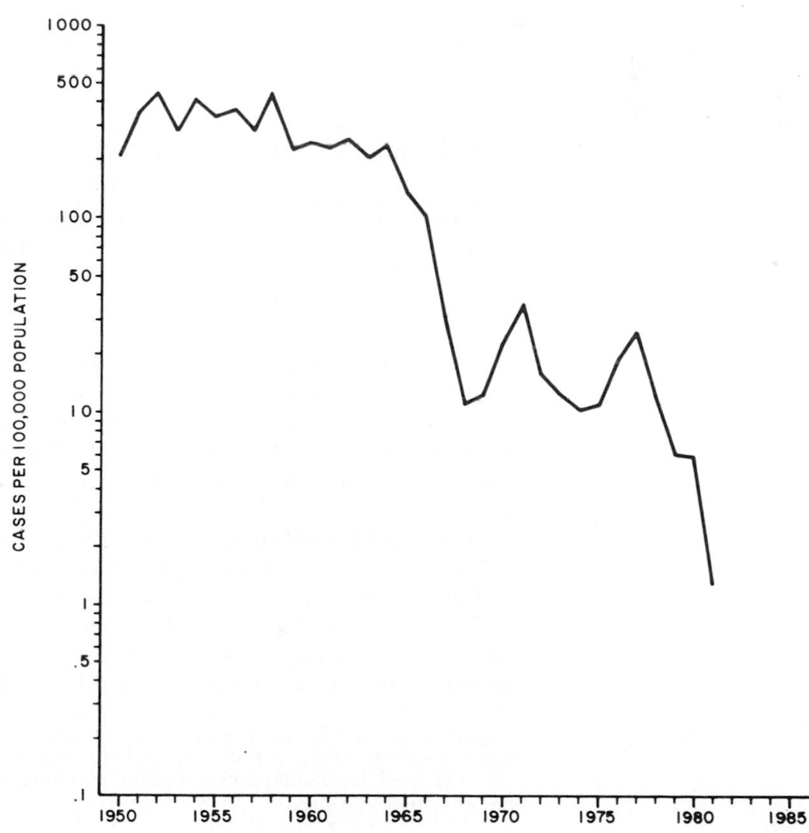

Figure 21–3 Incidence of reported measles, United States, 1951 to 1981. (Courtesy of the U.S. Department of Health and Human Services, Public Health Service.)

of the disease are expected, but they are not as effective in preventing the disease as we would wish. The problem is the complex variability of the antigens involved. Influenza virus types A, B, and C are immunologically unrelated. The avian, equine, and swine influenza viruses are all related to the human A type, but antigenic shifts produce new variants that result in new outbreaks, as in Hong Kong and Singapore. We are still seeking long-lasting immunity against influenza, so that yearly immunizations will no longer be necessary.

BCG Vaccination. The effectiveness, safety, and value of a tuberculosis vaccine consisting of live tubercle bacilli having low virulence for humans were first demonstrated by Calmette and Guérin, two French scientists, who introduced this method in 1921. The material injected consists of cultures of tubercle bacilli grown in a special medium containing bile, which lowers the virulence of the bacilli. It is called BCG (bacille Calmette-Guérin) vaccine. It should be noted that the use of BCG is still controversial. Its use is definitely not indicated in tuberculin-positive individuals (see Chapter 32). Large-scale studies conducted in England, in Puerto Rico, among the North American Indians, and in other places proved the BCG vaccine to be between 13.4 and 83.0 per cent effective as an immunizing agent. It is used especially for groups frequently exposed to the disease—i.e., nurses, doctors, and inhabitants of areas where tuberculosis is prevalent.

BCG is one of the few examples of living bacterial vaccines used for human beings. Veterinary use of attenuated bacterial vaccine includes *Brucella abortus* to prevent contagious abortion or Bang's disease in cattle (still quite prevalent in Texas) and cultures of attenuated *Bacillus anthracis,* the efficacy of which was first demonstrated by Pasteur in 1881.

BCG is available as fluid or dehydrated (lyophilized) cultures. The latter are reconstituted with appropriate sterile fluid. Either type of culture is usually injected very superficially into the skin of the deltoid region by a multipointed needle (*multiple puncture method*). The WHO statement on tuberculosis indicates that only the intradermal route assures that an effective dose of vaccine is given. In addition, in hot climates, freshly reconstituted, freeze-dried, heat-stable BCG vaccine must be used.

Injections of Dead Organisms

In a second method of producing active artificial immunity, the actual organisms that cause the disease are injected into the person needing the protection, whose defensive mechanisms react as usual by producing antibodies. In this procedure, however, the microorganisms, instead of being attenuated, weakened, or of low virulence, have been killed or inactivated by heat or some chemical disinfectant. The best known viral agents of this nature are the Salk polio vaccine, based on the first cultivation of polio virus in primate cells in 1949 by Enders, Robbins and Weller (Nobel Prize recipients), and the egg-derived vaccines against human influenza and swine influenza. (For data concerning the U.S. 1977 immunization program against influenza, see Chapter 24.)

Bacterial Vaccines. Immunizing preparations consisting of suspensions of killed bacteria are properly spoken of as *bacterins* but are more frequently called *bacterial vaccines* or simply vaccines.[2]

Many types of bacterial vaccines are used. The commonest are typhoid vaccine and whooping cough (pertussis) vaccine. Typhoid vaccine is prepared by cultivating fully virulent typhoid bacilli on broad agar surfaces, The fresh

[2]The term "vaccine" can, with its Latin derivation, be properly applied only to the cowpox material used for immunization against smallpox. However, it is widely used for any artificial immunizing procedure.

young growth is washed from the agar with physiological salt solution, and the suspension of bacilli thus obtained is heated at a temperature of approximately 60 C for about one hour. In some laboratories, the bacilli are killed with formaldehyde instead of heat. The suspension is then diluted to a suitable concentration (to contain about one billion organisms per milliliter), a small amount of disinfectant (tricresol or phenol, 0.25 per cent) is sometimes added as an additional precaution, and the vaccine is put up in ampules. It is ready for injection after suitable tests to prove that the bacteria are dead, that no extraneous bacteria are present, and that the vaccine produces immunity in animals. It is customary to give three successive injections of this vaccine at intervals of about a week and an annual "booster dose" (see pages 333 and 334).

Mixed Vaccines. Sometimes several kinds of bacteria are mixed together in a single suspension. Typhoid vaccine preparations are available containing also dead paratyphoid bacilli of two types. In some countries dysentery bacilli and the vibrios of cholera are included. Vaccines containing more than one kind of organism are spoken of as *mixed vaccines.* In preparing whooping cough vaccines, it is common practice to mix in antigens for diphtheria and for tetanus and sometimes other antigens, e.g., inactivated (Salk) polio virus. The "quad" vaccine of the United Nations World Health Organization (see Table 21–1) is very effective against whooping cough, diphtheria, polio, and tetanus and saves many extra "shots" or injections. Also, each antigen in the mixture helps the others to be more effective, although the exact reason for this is not clear. It should be noted that vaccines consisting of active viruses, e.g., Sabin polio, measles (rubella and rubeola), mumps, yellow fever, and smallpox vaccines, require different time sequences of immunization and cannot be administered mixed with other vaccines.

Sensitized Vaccines. Sometimes organisms intended for use in a vaccine are treated before injection with the serum of a person or animal who is immune to those bacteria. The bacteria combine with the antibodies in the serum so that they are acted upon quickly by the blood, tissues, or phagocytes of the person into whom they are injected and presumably immunize more rapidly and advantageously. Such a vaccine is called a *sensitized vaccine.*

Autogenous Vaccines. It sometimes seems advisable to prepare a vaccine from organisms isolated from the patient himself. This is often done with staphylococci isolated from boils. Such a vaccine is called an *autogenous vaccine.* In infections with staphylococci that are resistant to all available antibiotics, this is the only really effective way to clear up the condition, which may have spread all over the body.

Injection of Bacterial Exotoxins

A third method of producing active immunity artificially differs from the two preceding methods in that no organisms, living or dead, come into contact with the body. In certain diseases—for example, diphtheria and tetanus—the principal damage is done by the exotoxins that the bacteria give off into the blood. These toxins engender antibodies, called *antitoxins,* very readily. Therefore, if a person receives carefully controlled injections of small amounts of toxin, he will soon be able to withstand large doses of the same toxin. His body develops antitoxin just as though he had had the disease naturally. There is great danger from the toxin, however, even when mixed with antitoxin, as was the practice at the turn of the century.

Toxoids. Rather than inject these potent and dangerous toxins, it is now the custom to inject substances derived from the toxins by heating them or combining them with formaldehyde. These modified toxins are called *toxoids.* They are not poisonous but they act as specific antigens. After proper purification of the active immunogen and packaging under strictly sterile conditions,

Table 21–1. Preventive Vaccines Against Infectious Diseases of humans*

Practical Schedule for Active Immunization in Children

Age	*Vaccine (Immunogen)*	*Diseases*
2–3 Months	DTP (toxoids of diphtheria and tetanus, and pertussis [whooping cough] bacterial antigen)	Diphtheria Tetanus Pertussis
	TOPV (Trivalent Oral Poliovirus Vaccine, active [Sabin])	Poliomyelitis
4–5 Months	DTP	
6–7 Months	DTP TOPV	
12 Months‡	Measles, attenuated vaccine (Edmonston strain given with gamma globulins, Schwarz strain without)	Rubeola
15–19 Months	Smallpox vaccine (active vaccinia virus) — elective DTP TOPV	Smallpox
1–2 Years*	Mumps vaccine (active, attenuated)	Mumps
4–6 Years*	DTP Smallpox vaccine — elective TOPV	
1–10 Years*	Rubella vaccine (active, attenuated)	Rubella
12–16 Years*	TD (tetanus toxoid and diphtheria toxoid, adult type) Smallpox vaccine — elective	

Prevention of Diseases Not Included in the Above Schedule

These vaccines may be given electively wherever indicated	Active virus (attentuated)	Yellow fever
	Inactivated virus	Rabies, influenza
	Killed *Rickettsia prowazekii*	Typhus fever
	Killed bacteria (*Salmonella typhi, S. paratyphi, S. schottmuelleri*)	Typhoid fever and paratyphoid fever
	Purified polysaccharide of *Streptococcus pneumoniae*†	Pneumonia
	Purified polysaccharide of *Neisseria meningitidis* (groups A and C)	Cerebrospinal meningitis
	Cell extract of *Vibrio cholerae*	Cholera
	Cell extract of *Yersinia pestis*	Plague
	Infectious attenuated mycobacteria (*BCG, bacille Calmette-Guérin*)	Tuberculosis

*A tuberculin test may be administered in high risk populations. It can help in preventing the disease, but, of course, it is not a vaccine.

†A new pneumonia vaccine, developed in 1977, is effective against 14 different types of pneumococcus that cause 90 to 95 per cent of infections. It should be given to children with defective or missing spleens, those suffering from sickle cell anemia, and the elderly, all of whom are highly susceptible to pneumonia.

‡Alternatively, 15 months rubeola-mumps-rubella is often recommended.

It is hoped that vaccines against infections, serum hepatitis, and herpes virus diseases will soo be added to the above list.

they are among the safest and most widely used immunizing agents. (See Chapter 30.)

Primary and Secondary Antigenic Stimuli

In most of the processes of active artificial immunization described, several injections of the immunizing agents (antigens or immunogens) are used. The response of the body to initial contact with most antigens (a so-called *primary stimulus*) is relatively slow, requiring 2 to 10 weeks to reach fully effective antibody titers. The antibodies and the immunity of a person who has been immunized may decline to a very low level over several years. It is then desirable to provide reimmunization through a *secondary stimulus*. It has been found that the body cells, on reimmunization, respond very much more rapidly than at first, often within a few hours (Fig. 21–4). For this reason only a small single dose of the antigen (a so-called "booster dose") is needed to re-establish high-grade immunity.

Booster Doses. This rapid reaction to secondary stimuli is seen in methods of preventing polio, tetanus, diphtheria, and salmonellosis (typhoid or paratyphoid fever). Persons having once received an initial course of three 0.50 ml

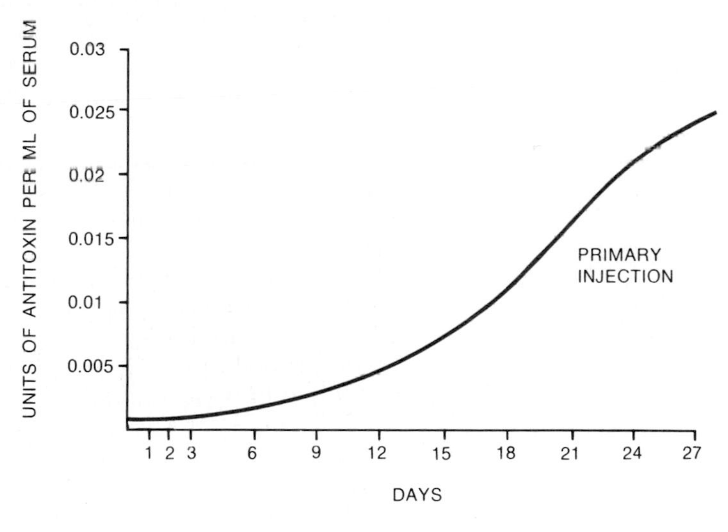

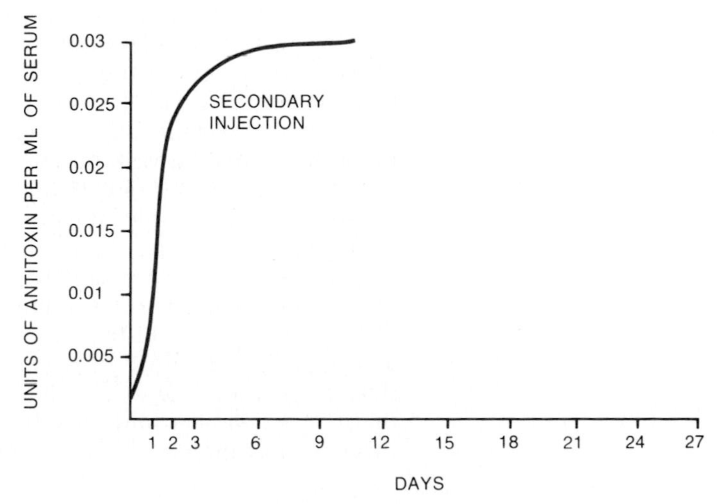

Figure 21–4 *A*, The production of antitoxin following injection of antigen (primary stimulus). It is a slow response, requiring 2 to 10 weeks to reach fully effective antibody titers. *B*, After a secondary stimulus, months or years later, there is a much more rapid and pronounced response of antitoxin production. The rate and extent of the secondary reaction vary with different antigens, individuals, and species.

subcutaneous injections of typhoid (*Salmonella*) vaccine need only take an annual booster dose of about 0.1 ml *intracutaneously,* or a 0.25 to 0.50 ml subcutaneous injection every three years. The tiny intracutaneous dose very rarely produces sore arms or severe general reactions. The fourth and later trivalent oral polio vaccine (TOPV) doses are based on the same principle. Similarly, inductees into military service and others, including children, receive an initial immunization against tetanus consisting of injections of tetanus toxoid. If these persons are later wounded, they immediately receive a small booster dose of tetanus toxoid, which helps prevent tetanus. Together with prompt medical treatment of wounds and the use of antibiotics, the booster dose procedure has greatly reduced the incidence of tetanus.

For international travel to yellow fever–infected areas a yellow fever vaccination is recommended. A yellow fever vaccine booster will protect for ten years; a cholera vaccine booster lasts only six months.

Slow Response to Primary Stimulus. Because the production of antibodies after the primary antigenic stimulation is relatively slow, the first injections must be started weeks or months before a person is likely to be exposed to disease, e.g., on starting school. Active immunization is too slow to be of value as a *therapeutic* measure once disease is established or as a preventive measure in the face of *immediate* exposure. There are, however, methods of developing immediate immunity to some diseases, which are called *passive immunization.*

PASSIVE IMMUNITY

Artificial Passive Immunity

Need for Passive Immunity. In cases of exposure to such diseases as diphtheria and tetanus, it is necessary that a large supply of antibodies appear in the blood immediately. The chief damage in these diseases is caused by the very rapid poisonous action of bacterial exotoxins. When the patient is already ill, there is no time to lose in waiting for him to develop an active immunity, natural or artificial. The patient must immediately receive an ample supply of ready-made antibodies. Immunity resulting from injections of these ready-made antibodies is called *passive immunity* because the patient passively receives them and his tissues play no part in producing them.

One may purchase syringes or ampules already filled with antitoxic serum, prepared for just such emergencies, at any well-stocked pharmacy or health department. Such antibody-containing serum is obtained from animals, usually horses, that weeks or months previously had received repeated injections of the special antigen against which antibodies are desired.

Preparation of Therapeutic Sera. Among the most important therapeutic sera are those against the diphtheria and tetanus bacilli. Innumerable lives have been saved through their application. The use of immune antiviral serum to help combat rabies has already been mentioned. Passive immunity is also involved in dealing with measles, viral hepatitis, and several other diseases.

A common method of making therapeutic antitoxic sera is as follows: A horse (or rabbit) is injected subcutaneously or intravenously, at intervals of five to seven days, with gradually increasing amounts of broth in which the desired organisms have grown and which contains their dead cells (now more often cell extracts) and/or their toxins. The horse is the animal usually chosen for antitoxins since, because of its size, it yields a large amount of desired serum. The first injected dose of toxin is very small, not enough to make the animal sick. By the end of the treatment the animal may take, without symptoms, many times what would have been a fatal dose at the beginning. At the end of three or four months, when the horse has developed a large amount of antibody in its blood, it is bled, with aseptic precautions, from the jugular vein; six to eight quarts of blood are removed at one time. The blood serum separates from the clot on standing and contains the antitoxic gamma globulins needed in medical practice. The horses used for the production of curative sera are kept under the best hygienic conditions and remain well and comfortable throughout the treatment, which is not painful. As noted previously, toxoids are now often used as antigens instead of dangerous toxins.

Sera must be kept in a dark, cool place, as they gradually lose their strength when exposed to light and warmth. Under suitable conditions their potency is retained for a year or more. A difficulty with the use of such sera is that the equine proteins act as antigens and induce a marked allergic state toward any proteins from horses.

In order to reduce the number and severity of *allergic protein reactions,* processes have been devised to remove as much of the nonspecific and unneeded proteins—i.e., *horse* (equine) proteins—from therapeutic sera and leave as much of the antitoxin as possible. Such purified material is called *concentrated* or *purified antitoxin* or *immune serum globulin(s)* (ISG). It is much superior to the unconcentrated serum and is the preparation generally sold as antitoxin.

Injections of antibody-containing sera into an infected patient, before the toxin or virus or other microorganisms in the patient have done too much damage, practically always result in an improvement of condition and usually a cure. It is extremely important to remember that therapeutic sera will stop the damaging process but will not repair damage already done. Sera given very late in any disease are practically without effect and may then do harm.

In the United States, the production and sale of therapeutic sera bacterial vaccines (bacterins), smallpox vaccine, and other immunizing agents for human use (also for use in animals) are regulated by law. The establishments that manufacture these agents are licensed, and samples of products for human use are tested at the National Institutes of Health at Bethesda, Maryland. Products that do not come up to its standards may not be sold. The U.S. Food and Drug Administration, in laboratories located in many states, also checks these medications upon requests of consumers. Also, the departments of health of the various states make sure that state laws are not being violated. All of these laboratories, as well as the U.S. Center for Disease Control at Atlanta, Georgia, collaborate when necessary. Veterinary products are similarly controlled in cooperation with the U.S. Department of Agriculture.

Passive Immunity in the Prevention of Disease. Passive immunity is used in the prevention, as well as in the cure, of disease. For example, if it is known or suspected that a person is soon likely to be exposed or has very recently been exposed to certain diseases, under some circumstances (to be determined by the physician) it is advisable to inject a small quantity of serum, or some derivative of serum containing the appropriate antibodies, as a preventive or prophylactic[3] measure. Diseases against which this form of prophylaxis is commonly used are measles, diphtheria, and tetanus. Effective sera are also available for some other infections (e.g., rabies and hepatitis due to hepatitis virus A). Still others are yet under investigation.

Before the immunizing value of tetanus toxoid had been widely demonstrated, tetanus antitoxin was routinely administered to persons injured in street and farm accidents. In such accidents, dirt containing spores of tetanus bacilli gets into the wound, where tetanus bacilli can grow and produce their toxin. Today toxoid immunization and booster doses are used more than antitoxin in preventing tetanus.

Allergy in Passive Immunity. If a susceptible individual is not allergic to horse serum, the injection of antitoxin made with horse serum will produce such allergy. This is a serious deterrent to the indiscriminate use of horse

[3]From Greek *pro,* before; *phylassein,* to guard or protect. Prophylaxis is a measure used in advance to prevent infection or disease.

serum, either therapeutically or prophylactically. Chapter 22 will deal with this type of response and similarly induced allergic conditions.

Human Gamma Globulins. To avoid the use of serum from lower animals, with resulting allergic complications, and also because certain antibodies are more readily available from human sources, human blood is often used to provide therapeutic and prophylactic antibodies. Immune globulins are separated from the blood of immune humans and are injected into other persons to provide *passive immunity* against several diseases, especially measles and infectious hepatitis and, less often, herpes zoster, pertussis, and others. For example, very young or "delicate" children and pregnant women, in whom rubella is a dangerous disease, often receive gamma globulin of some child or adult who is convalescing or has recently recovered from measles. This prevents or modifies the disease in the expectant mother and in the young or fragile child until he is older and can better withstand the infection. Usually this sort of passive protection is used during epidemics, times of special danger, or just before long journeys. Artificial active immunization is certainly preferable if time permits.

In high-risk areas it is recommended that prophylactic immunoglobulin against hepatitis (viral type A) be given every 4 to 6 months. Pregnancy is not a contraindication to using immunoglobulin. Only very limited amounts of vaccinia immune globulin (VIG) are still being held at the Center for Disease Control for treatment of life-threatening vaccination complications.

Human immunoglobulins are of special value for patients whose immune response is being suppressed, as in heart and other organ transplantation. If the immunoglobulins administered to a person are especially selected for specific antibodies, we may refer to them as HIGG (hyperimmune gamma globulins).

As previously mentioned (page 305), the chemistry of the gamma globulins has been extensively investigated; the designations IgG, IgM, IgA, and IgE are used, as well as their subunits the L (light) and H (heavy) chains. Indeed, the amino acid sequences of these chains in humans and many animals have been carefully mapped in normal sera and in sera taken during abnormal conditions (myeloma protein,[4] rabbits immunized against pneumococci, and similar instances).

Agammaglobulinemia.[5] Some persons suffer from a genetic difficulty (or total inability) to form gamma globulins. a condition called *agammaglobulinemia.* As a result, they lack antibodies, are often extremely susceptible to infectious disease, and may not respond well, or at all, to vaccinations. A similar condition results from *hypoglobulinemia,* or a reduced amount of globulin in the blood.

Natural Passive Immunity

An expectant mother who has antibodies against infectious diseases confers a share of these antibodies on her unborn child through the placenta. This is *natural passive immunity.* Of the various types of gamma globulins, the IgG molecules can pass through the placenta and are consequently found in the blood of the infant at birth. In the sera of most normal people and in hyperimmune individuals, IgG makes up over 85 per cent of all immunoglobulins.

The immunoglobulins have two important advantages over commercial antibodies, such as diphtheria or tetanus antitoxins prepared from the serum

---

[4]Found in patients with multiple myeloma; secreted by neoplastic plasma cells.

[5]This formidable word is really a graceful Latin phrase: *a,* without or lacking; *gamma globulin* has already been explained; *emia,* in the blood. Pronunciation is just as the word is divided here, and spoken thus comes trippingly from the tongue! In hypoglobulinemia, the prefix *hypo,* meaning little or "below par," is substituted for *a* and *gamma* is left out.

of immunized horses. First, as already noted, antibodies from animal sources make the recipient allergic to the proteins of the animal. Second, such antibodies are completely rejected and destroyed in the body within two to four weeks. Human maternal antibodies do not cause allergy and protect the child for six months to one year after birth.

After that, the child becomes susceptible to many infectious diseases. It is therefore advisable to begin active immunization of the child early (second month to sixth month) against pertussis and diphtheria. Tetanus toxoid may be given either simultaneously or later; and still later (or simultaneously) immunization may be begun against other diseases, including poliomyelitis. The use of combinations of several antigens (but not active viruses) in one or two "shots" is now common and effective (Table 21–1).

It is important to note that immune gamma globulins (IgG) in the young infant imply immunity in the mother. The woman who anticipates becoming pregnant but who does not have good antibody titers to diphtheria, tetanus, pertussis, poliomyelitis, rubella, and possibly salmonellosis would do well to become actively immunized before pregnancy in order to confer passive immunity on her child.

SUMMARY

Active specific immunity may be acquired artificially from mild, subclinical infections through vaccination with attenuated (weakened) or less virulent organisms. Vaccination against smallpox, a disease caused by the variola virus, is achieved by injecting the immunologically related virus of cowpox, the vaccinia virus, into the nonimmune host. Smallpox, a killer of millions in past epidemics, is now thought to have been eliminated worldwide. However, should the disease ever appear again, vaccinia-immune globulins (VIGs) from previously immunized individuals are recommended for passive immunization.

Postexposure prophylaxes against rabies include *vaccines* that induce active immunity (HDCV or DEV) and *globulins* that provide rapid passive immunity (RIG or ARS). Preferred choices for usage are HDCV and RIG.

Preventive vaccines against infectious diseases of humans are given according to recommended time and dosage schedules to children from two months to 16 years of age and to international travelers. Available vaccines include DTP (diphtheria tetanus pertussis), TOPV (trivalent oral polio vaccine), rubeola-mumps-rubella, smallpox vaccine (less readily available), and vaccines against yellow fever, influenza, rabies, typhus fever, typhoid and paratyphoid fever, pneumonia, cerebrospinal meningitis, cholera, and plague. Influenza vaccines still present a problem owing to antigenic shifts among the different virus types that cause this disease. The tuberculosis vaccine, BCG (bacille Calmette-Guérin), a live bacillus is not given routinely in the United States, but only to high-risk individuals. For veterinary uses attenuated bacterial vaccines include *Brucella abortus* against contagious abortion in cattle and *Bacillus anthracis* against anthrax. For human uses, mixed bacterial vaccines (e.g., DTP) are preferred for obvious reasons.

Toxins mixed with antitoxins used to be administered to stimulate antitoxin production; now toxoids are used as active immunogens instead. The antibody response to a primary stimulus from an antigen or toxoid is always slow, whereas the secondary injection brings about a much more rapid formation of antibodies.

For immediate protection human gamma globulins are essential; for more certain specificity HIGG (hyperimmune gamma globulins) may be used. Pregnancy is not a contraindication to using immunoglobulins.

See the figure on page 339 for a general model of immunologic reactions.

REVIEW QUESTIONS

1. How may active specific artificial immunity be acquired? How is passive immunity conferred? Cite examples of each.
2. Define vaccinia, variola, attenuated, vaccinia-immune globulins, BCG, bacterial vaccines, mixed vaccines, toxins, toxoids, TOPV, DTP, human gamma globulins, HIGG, Sabin's vaccine, 17D yellow fever vaccine, influenza, and booster.
3. List the two types of postexposure prophylaxis against rabies and two examples of each type.
4. Recommend a time schedule and a series of vaccines that may be given to children to protect them against disease. List some diseases a traveler should be vaccinated against.
5. Explain the differences between primary and secondary antigenic stimuli and antibody responses.
6. How may a therapeutic horse antitoxin be prepared? Which diseases can this procedure be used to combat?
7. What is agammaglobulinemia?
8. Is it dangerous to give immunoglobulins to a pregnant woman?

Supplementary Reading

Beutner, E. H. (ed.): Defined immunofluorescent staining. *Ann. N.Y. Acad. Sci., 177*:1, 1971.

Brady, J. A., et al. (eds.): Chronic Infections, Neuropathic Agents, and Other Slow Virus Infections, Current Topics in Microbiology and Immunology, vol. 40. New York, Springer-Verlag, 1967.

Eichhorn, M. M.: Rubella: Will vaccination prevent birth defects? *Science, 173*:710, 1971.

Friedman, H. (ed.): Immunological tolerance to microbial antigens. *Ann. N.Y. Acad. Sci., 181*:1–315, 1971.

Grubb, R.: The genetic markers of human immunoglobulins. *Mol. Biol. Biochem. Biophys.*, vol. 9. New York, Springer-Verlag, 1970.

Humphrey, J. H., and White, R. G.: Immunology for Students of Medicine. 3rd ed. Oxford, Blackwell Publications, 1970.

Kawamura, A., Jr. (ed.): Fluorescent Antibody Techniques. 2nd ed. Baltimore, University Park Press, 1977.

Kochwa, S., and Kunkel, H. G. (eds.): Immunoglobulins. *Ann. N.Y. Acad. Sci., 190*:1–584, 1971.

Public Health Service Advisory Committee on Immunization Practices. *Morbid. Mortal., 21*:25, 1972.

Roitt, I.: Essential Immunology. 3rd ed. Philadelphia, J. B. Lippincott Co., 1977.

Streilein, W. J., and Hughes, J. D.: Immunology—A Programmed Text. Boston, Little, Brown & Company, 1977.

Tunevall, G.: Periodicals Relevant to Microbiology and Immunology. New York, John Wiley & Sons, 1969.

Weir, D. M.: Handbook of Experimental Immunology. 3rd ed. St. Louis, C. V. Mosby Co., 1978.

WHO Expert Committee on Rabies, 5th Report. Geneva, World Health Organization, Technical Report Series No. 321, 1966.

WHO Expert Committee on Tuberculosis. 8th Report. Geneva, World Health Organization, Technical Report Series No. 290, 1964.

Wiener, A. S.: Elements of blood group nomenclature with special reference to the Rh-Hr blood types. *J.A.M.A., 199*:985, 1967.

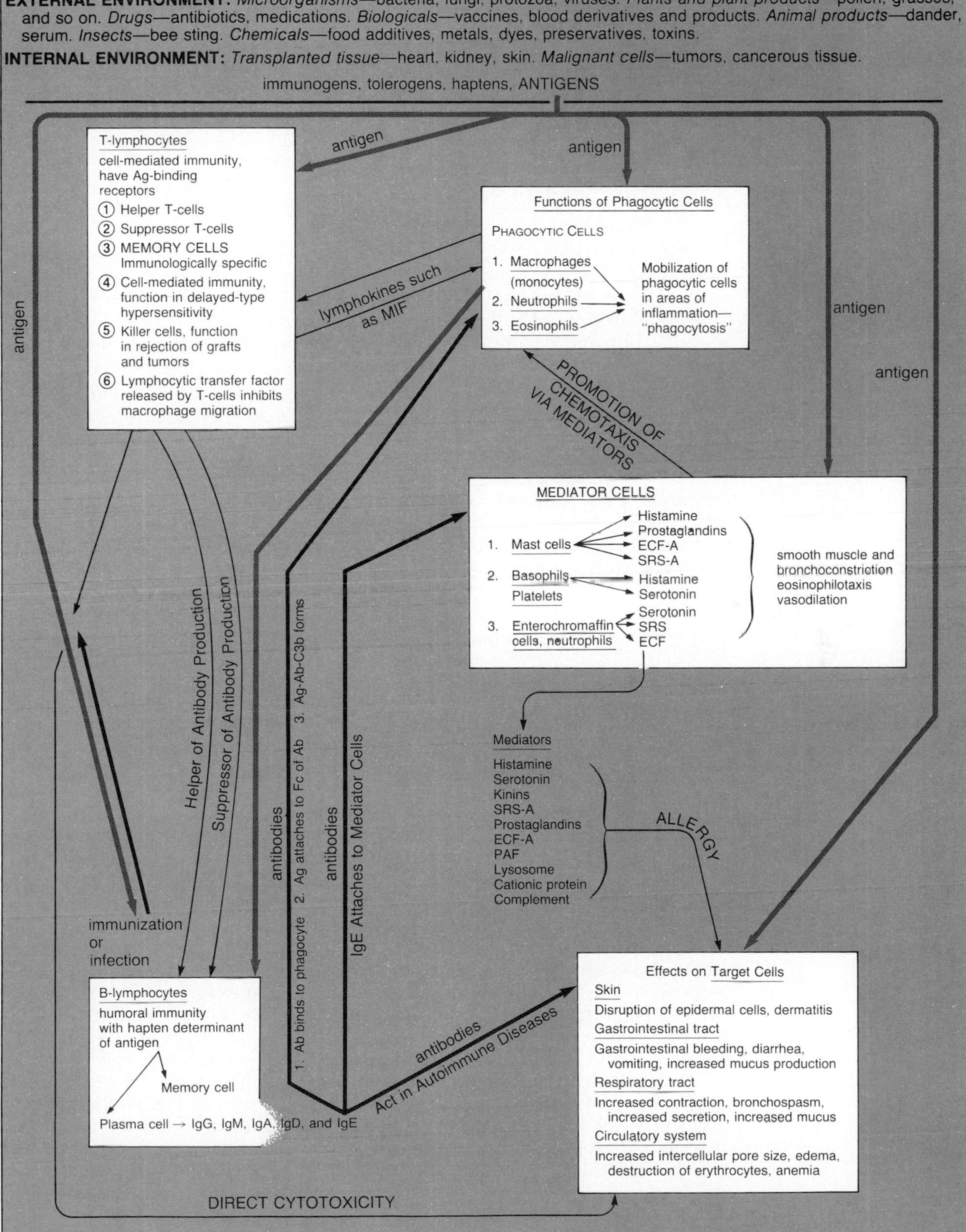

GENERAL MODEL FOR IMMUNOLOGIC REACTIONS
EXTERNAL ENVIRONMENT: Microorganisms—bacteria, fungi, protozoa, viruses. Plants and plant products—pollen, grasses, and so on. Drugs—antibiotics, medications. Biologicals—vaccines, blood derivatives and products. Animal products—dander, serum. Insects—bee sting. Chemicals—food additives, metals, dyes, preservatives, toxins.
INTERNAL ENVIRONMENT: Transplanted tissue—heart, kidney, skin. Malignant cells—tumors, cancerous tissue.
immunogens, tolerogens, haptens, ANTIGENS
antigen
antigen
antigen
antigen
antigen
T-lymphocytes
cell-mediated immunity, have Ag-binding receptors
1 Helper T-cells
2 Suppressor T-cells
3 MEMORY CELLS Immunologically specific
4 Cell-mediated immunity, function in delayed-type hypersensitivity
5 Killer cells, function in rejection of grafts and tumors
6 Lymphocytic transfer factor released by T-cells inhibits macrophage migration
lymphokines such as MIF
Functions of Phagocytic Cells
PHAGOCYTIC CELLS
1. Macrophages (monocytes)
2. Neutrophils
3. Eosinophils
Mobilization of phagocytic cells in areas of inflammation— "phagocytosis"
PROMOTION OF CHEMOTAXIS VIA MEDIATORS
MEDIATOR CELLS
1. Mast cells
Histamine
Prostaglandins
ECF-A
SRS-A
2. Basophils
Platelets
Histamine
Serotonin
3. Enterochromaffin cells, neutrophils
Serotonin
SRS
ECF
smooth muscle and bronchoconstriction
eosinophilotaxis
vasodilation
Mediators
Histamine
Serotonin
Kinins
SRS-A
Prostaglandins
ECF-A
PAF
Lysosome
Cationic protein
Complement
ALLERGY
Helper of Antibody Production
Suppressor of Antibody Production
antibodies
1. Ab binds to phagocyte 2. Ag attaches to Fc of Ab 3. Ag-Ab-C3b forms
antibodies
IgE Attaches to Mediator Cells
immunization or infection
B-lymphocytes
humoral immunity with hapten determinant of antigen
Memory cell
Plasma cell → IgG, IgM, IgA, IgD, and IgE
antibodies
Act in Autoimmune Diseases
Effects on Target Cells
Skin
Disruption of epidermal cells, dermatitis
Gastrointestinal tract
Gastrointestinal bleeding, diarrhea, vomiting, increased mucus production
Respiratory tract
Increased contraction, bronchospasm, increased secretion, increased mucus
Circulatory system
Increased intercellular pore size, edema, destruction of erythrocytes, anemia
DIRECT CYTOTOXICITY

ALLERGY AND AUTOIMMUNE DISEASES

INTRODUCTION

The word allergy is derived from two Greek words: *allos,* altered, and *ergon,* action. It refers to an altered state of reactivity that develops toward an antigen 10 to 15 days after the first contact with (or injection or dose of) that antigen. This dose of antigen is commonly called a *sensitizing dose.* There is no *perceptible* change or reaction following sensitization. A second dose, given intravenously 10 to 15 days later, if large enough, typically evokes a violent allergic reaction and is accordingly called a *toxic* or *shocking dose.* The reaction following the toxic dose is due to the formation of certain antibodies and/or the proliferation of certain reactive cells.

The increased reactivity induced by a sensitizing dose of antigen is often referred to as *hypersensitivity.* Substances inducing allergic hypersensitivity are called *allergens.* Although allergy is a valuable defensive mechanism against infectious disease in some instances and in some individuals, it does more harm than good when it is overactive, like an overzealous servant who, attempting to warm the living room with a good fire, burns the house down.

TYPES OF ALLERGY

All allergic reactions are due to highly specific antigen-antibody or antigen-cell reactions. With a few exceptions, persons having allergic reactions must have had sensitizing (theoretically immunizing) contact with at least one antigen or allergen at least 10 days previously for antibodies to have formed or cells to be activated.

Allergic reactions may be classified under two general headings: *immediate* hypersensitivity, due to IgE antibodies, and *delayed* hypersensitivity, due to the action of T-cells. Both are related to immunity and are the responses of certain tissue cells to antigens or haptens. Because the "immediate" reactions are sometimes deceptively slow and the "delayed" reactions are sometimes not delayed very much, it is customary to speak of the *immediate-type* or *delayed-type* of reaction to avoid implications of the rates at which they occur. Antigens (allergens) commonly (but not invariably) responsible for immediate-type allergic reactions are not all cellular and include mainly proteins, simple or conjugated: serum, egg white, fish proteins, meats, pollens, animal dander, feathers, and so on. Antigens involved in delayed-type allergy are typically, but not necessarily, cells of microorganisms or of large plants or animals.

A fundamental difference between immediate-type and delayed-type allergy is that immediate-type allergy in humans is due to IgE antibodies that circulate in the blood and therefore can be *passively transferred* in serum from a sensitized person to a normal person. Antibodies in the serum of the donor quickly attach themselves to reactive cells in the recipient, making him or her also hypersensitive to whatever specific allergen is involved. In marked contrast, delayed-type hypersensitivity, which is due to the action of T-cells and not to

CHAPTER 22

antibodies in the plasma, is not transferable in serum from a sensitized donor to a normal person, but is transferable with T-cells (activated lymphocytes). It is helpful to think of the two types of hypersensitivity as (1) antibody-mediated, the immediate allergic response, and (2) cell-mediated, the delayed allergic reaction.

Antibody-Mediated Immediate-Type Allergic Reaction

Noncytotoxic Allergic Reactions. In humans the reacting antibodies are all or mostly IgE (sometimes called the "anaphylactic antibody" because it is involved in anaphylaxis, the most massive, dramatic, and dangerous of the immediate-type reactions). Most (about 95 per cent) of the Fc portion of IgE is bound to mast cells and basophilic lymphocytes, and most of it does not circulate freely in the plasma. In immediate-type allergic reactions there is little accumulation of phagocytic or other lymphocytes as occurs in delayed type allergy. The antigen-antibody complexes that are formed with IgE *do not fix complement* and are therefore *not cytotoxic:* i.e., tissue cells are not destroyed. Tissues that contain IgE accumulate eosinophils (nasal, bronchial, or intestinal mucosa). Enzymes released from eosinophils help to destroy SRS-A (slow-reacting substance of anaphylaxis) and histamine.

Contact of specific antigen with IgE-coated mast cells[1] or basophilic granulocytes releases several physiologically active substances in 1 to 10 minutes (see Table 22–1). Their release causes strong contractions of smooth muscle, which is abundant in the blood vessels, gastrointestinal tract uterus, bronchioles, and elsewhere, depending on the species. Local capillary vessels (no muscle fibers) become dilated, followed by exudation of fluids, thus producing edema and swelling. Heparin is also released, reducing the coagulability of the blood. Besides heparin, mast cells release mediators that may cause anaphylaxis, hives, hay fever, or allergic asthma: these include histamine, serotonin, eosinic chemotactic factor (ECF-A), neutrophil chemotactic factor (NCF-A), SRS-A, and platelet-activating factor (PAF), as well as a number of others.

If the affected smooth muscle fibers and capillary vessels are in the gastrointestinal tract, diarrhea and cramps may result; if they are in the gravid uterus, abortion is known to occur; in the bronchioles they can lead to asthma; and if they are present in the walls of the cutaneous blood vessels, painful or itching bumps or wheals known as "hives" may appear, especially after taking in certain foods or allergenic serum or drugs. Skin reactions are called *cutaneous allergies.*

Anaphylaxis. The term anaphylaxis was invented by Richet around 1902 to describe what he viewed as a sort of immunologic betrayal—i.e., increased

[1]Mast cells are nonmotile connective tissue cells found next to capillaries throughout the body, especially in the lungs.

Table 22–1. Some Substances Involved in Anaphylactic Reactions

PHARMACOLOGICALLY ACTIVE AGENT					
Name	*Structure*	SOURCE	CHEMICAL PRECURSOR	ACTIVITY	SPECIES
Histamine*†	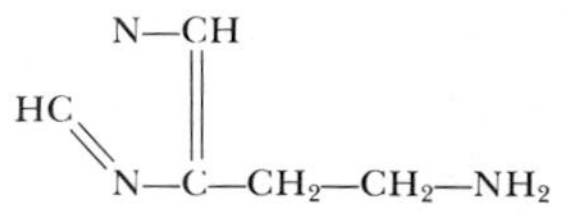 N—CH ‖ HC=N—C—CH_2—CH_2—NH_2	Mast cells,‡ platelets, basophilic leucocytes, etc.	L-histidine		Guinea pig Rabbit Dog Human
Serotonin (5-hydroxytryptamine)	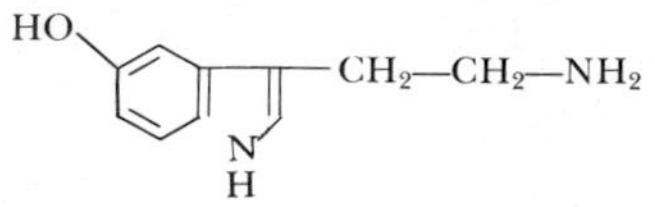 HO–(indole ring, N–H)–CH_2—CH_2—NH_2	Mainly in platelets, enterochromaffin cells	L-tryptophan	Increases vascular permeability, lowers blood pressure, contracts smooth muscles	Rabbit Rat Mouse Dog?
Bradykinin*† Lysyl-bradykinin*† other kinins	arg-pro-pro-gly-phe-ser-phe-arg lys-arg-pro-pro-gly-phe-ser-pro-phe-arg (peptides)	Plasma, salivary glands, some tumors	α-Globulin in plasma		Guinea pig Rabbit Rat Mouse Dog Human?
Heparin	A polysaccharide polymer consisting of hexosamine, hexuronic acid, and sulfuric acid ester groups	Mast cells†	Heparin	Anticoagulant	Dogs Other species?
SRS-A* (slow-reacting substance of anaphylaxis)	An acidic lipoprotein	May come from mast cells‡ in human lung tissue during anaphylaxis	Two lipids and platelet-activating factor (PAF)	Slow, prolonged effect on only certain types of smooth muscles; may be cause of bronchospasm in allergic asthma in humans	Guinea pig Rabbit Human
ECF-A (eosinic chemotactic factor of anaphylaxis)	val-gly-ser-glu-ala-gly-ser-glu	Mediator released from mast cells		Promotes eosinophilia of immediate allergic reactions (deactivates eosinophils)	
Prostaglandins and other mediators, also N-acetyl-β-glucosaminase, arylsulfatase A, PAF, chymase, heparin, and others					

*Found in humans in shock (the kinins?)
†Action blocked by antihistamines.
‡Mast cells are connective-tissue cells very rich in heparin and in histamine (an amine formed by decarboxylation of the amino acid histidine), which causes strong vasoconstriction.

susceptibility as a result of immunizing injections, instead of increased resistance (Greek *ana,* against; *phylaxis,* protection; opposite of prophylaxis). Fortunately, anaphylaxis is not common in humans, but it is well illustrated by the guinea pig. A protein (for example, egg white) is usually entirely harmless when injected into an animal the first time, except that it produces a state of allergy or hypersensitivity: i.e., it is a *sensitizing dose.* It may produce severe and sometimes fatal reactions called *anaphylactic shock* in guinea pigs or rabbits if injected a second time, that is, as a *toxic* or *shocking dose,* about two weeks later. Anaphylactic shock in the guinea pig is characterized by lowered blood pressure and body temperature, difficult breathing, weakness, and finally convulsions and death. Guinea pigs show this phenomenon with particular violence. Other animals may present a different picture, depending on numerous anatomic and physiologic differences. If the second (shocking) dose of the protein is given before the end of the two-week period—during the so-called *"refractory period"*—no anaphylactic reaction is produced. This is the period during which antibodies (IgE in humans) are being formed.

In humans, edema and inflammation are characteristically present in both immediate and delayed type allergy, though occurring with different degrees of intensity and developing at different rates. Edema and inflammation may affect any of the mucosal surfaces. When the edema, as sometimes happens in immediate-type reactions, occurs quickly in tissues around vital passages, such as the larynx and glottis, rapid death from asphyxiation can ensue. Stings of bees and hornets in persons allergic to their venoms are notorious in this respect. Injections of 0.5 to 1 ml of 1:1,000 epinephrine (Adrenalin)—a potent *antihistamine* (inhibitor of histamine), vasodilator (capillaries and arterioles), and decongestant—quickly reduce these swellings. The drug should always be immediately available when administering biologicals hypodermically, especially serum and vaccines, some antibiotics, and similar preparations containing proteins and certain haptenic substances (e.g., some drugs) to which allergy is common. Numerous antihistamines are sold in tablet form, some without a physician's prescription. Since they give only minimal or inadequate relief, it is best to follow the advice of your physician if you suffer from allergy.

Serum Sickness. Although typically this occurs 6 to 18 days after a therapeutic dose of serum, it is not a true delayed-type reaction. Antibodies accumulate in the plasma during this period. When they reach a certain concentration, they react with the proteins of whatever residual portion of the therapeutic serum may still remain in the body. The result is an immediate-type allergic reaction, though modified in severity. Minor cytotoxic effects sometimes occur. Serum sickness is characterized by itching hives (*urticaria*), fever, joint pains (arthralgia), smooth muscle contraction, and some lesser discomforts. If serum is to be given parenterally, a preliminary skin test for hypersensitivity to horse serum may be made. If the test is positive, the patient should be *desensitized* as described later. Persons who have had previous injections of serum usually require desensitization.

Atopy. Some individuals inherit not the true allergic state itself, but an increased tendency to become allergic on contacting certain allergens. They therefore tend to develop immediate-type allergic reactions such as asthma, hay fever, and food intolerances, and perhaps liability to certain forms of arthritis and rheumatic heart disease. These persons are sometimes called atopic.

Desensitization. In immediate-type allergy, especially in systemic anaphylaxis, if the hypersensitive person or animal survives the toxic dose, temporary insensitivity results because the antibodies available for an immediate reaction have been used up. The allergic state may reappear if more antibodies are produced. Desensitization may be accomplished by the physician by giving a series of small cutaneous doses of the allergen several hours apart. These combine with antibodies in several small steps, and "let the patient down" gradually. In delayed-type allergy, desensitization is difficult and often impossible because antibodies are not involved.

Clinical treatment or, better yet, prevention of allergic reactions in humans may be brought about by repeated injections of small but gradually increasing doses of allergen spaced at regular intervals, often weekly. This desensitization process prevents anaphylactic reactions by stimulating the formation of "blocking antibodies" that degrade mediators and consequently block their accumulation. Persons allergic to horse serum or penicillin may receive small doses of these substances in closely spaced injections for desensitization.

Cytotoxic Immediate-Type Allergic Reactions. These are complement-dependent. Reacting antibodies include IgG and IgM, which can form large, insoluble, complement-fixing complexes (precipitates) with specific antigen.

These complexes locally attract great numbers of macrophages and other lymphocytes, which release lysosomal enzymes, lymphotoxins and kinins (proteolytic), blood-coagulating factors, and some other agents that, together, cause prompt swelling and edema within 2 to 6 hours, followed by cytolysis and hemorrhagic necrosis. The principal manifestations of cytotoxic allergic reactions are tissue-destructive lesions such as the Arthus phenomenon (an artificial lesion). The Arthus reaction is an antibody- and complement-dependent allergic reaction brought on by repeated injections of horse serum several weeks apart, resulting in localized inflammations with tissue damage. In severe form, it was first observed in rabbits.

Cell-Mediated Delayed-Type Allergic Reaction

Typically, in this type of allergy complement and immunoglobulins are not involved, and the reactions are not cytotoxic, although necrosis may occur if the amount of antigen is excessive.

Table 22–2. Lymphokines Released by T-Cells (Activated Lymphocytes) and Their Effect on Cells*

LYMPHOKINES	AFFECTED CELLS
Macrophage-activating factor (MAF) Migration-inhibiting factor (MIF) Chemotactic factor	Macrophages (convert to "killer" cells when MAF lymphokines are stimulated by T-cells)
Mitogenic factor T-cell–replacing factor Chemotactic factor Suppressor factor	Lymphocytes
Leukocyte-inhibiting factor (LIF) Chemotactic factor	Neutrophilic leukocytes
Chemotactic factor Migration-stimulating factor	Eosinophils
Chemotactic factor	Basophils
Lymphotoxin (cytotoxic and cytostatic factors)	Other cells
Osteoclast-activating factor Interferons Colony-stimulating factor (CSF)	

*After Bloom, B. R., et al.: Genetic approaches to the mechanism of microphage functions. *Fed. Proc.*, *37*:2765, 1978.

Delayed Hypersensitivity Reactions. Delayed allergy reactions are dependent on nonantibody-forming thymus-activated immunocompetent T-cells. The *sensitized* lymphocyte (T-cell) detects newly present antigens upon new cellular stimulation (possibly due to infection), and substances called *lymphokines* are released from the competent lymphocyte to destroy the abnormal cells (Table 22–2) and also some that are not abnormal. The T-cell releases a cytotoxic substance that destroys not only the target cell but also other cells that happen to be in the area. Delayed hypersensitivity reactions are apparent only after about 18 hours, persist for days, and are not suppressible by antihistamines, since histamine is not present. It is quite likely that many subgroups of T-cells exist, possibly with different varieties of lymphokines. Several types of these substances are now known.

The combined function of all lymphokines seems to be to summon help for the defender of the normal cells so as to destroy the abnormal cells more effectively. How is this done? Besides their nonspecific cytotoxic effect on living cells, the T-cell–produced lymphokines prevent macrophages (phagocytes) from migrating, whereas other leukocytes are attracted to the area where they are needed. Other nonsensitized lymphocytes are induced to undergo repeated cell divisions to produce cells that are needed, which stimulates B-cells directly. Also, in some way the T-cell functions in the production of interferon (nonspecific virus inhibitor: see page 307).

The Tuberculin Reaction. Delayed-type allergy is well illustrated by the tuberculin reaction and is often called the tuberculin type of allergy. Persons infected by tubercle bacilli, whether perceptibly ill or not, become allergic to proteins in tubercle bacilli. Their hypersensitive state can be demonstrated by injecting a minute amount of sterile protein derived from tubercle bacilli (*tuberculin*) intracutaneously on the forearm. This constitutes the *tuberculin test* and is widely used in the study of tuberculosis. Tuberculin preparations are of two main types: "O.T." (old tuberculin) and "P.P.D." (purified protein derivative). The local swelling, inflammation, and (rarely) necrosis resulting from the tuberculin test appear only after a delay of 36 to 48 hours; the area is firmly indurated, in great part due to large numbers of mononuclear lymphocytes. These do not appear in immediate-type reactions.

Many other slowly progressive bacterial infections besides tuberculosis (notably syphilis, yaws, leprosy, brucellosis, tularemia) induce delayed-type allergy. For this reason delayed-type allergy is often called infection-type allergy; whereas pollens (e.g., ragweed), poison ivy, and other nonbacterial antigens are commonly associated with immediate-type allergy.

Homograft Reactions. Another manifestation of delayed-type hypersensitivity is the cytotoxic, tissue-rejection reaction. Unless tissues for transplantation (skin grafts, hearts, kidneys, and so on) are "self" or *autoantigens*—i.e., derived from one's own body (difficult in the case of heart transfer!) or from an identical (monozygotic) twin—the immune mechanisms of the body tend to reject them sooner or later. Even *isoantigens*—i.e., those derived from a donor of the *same species* as the recipient—are eventually rejected as "foreign" or "nonself." In general, tissues or antigens from *any* source other than one's own body are immunologically "resented" by the body as "nonself." As noted later, autoantigens may also be resented under certain conditions. There is no question that the knowledge gained in immunology from organ transplants will reduce rejection risks as more of these operations are performed and more is learned about the immune system.

Cortisone Medication. Many medications that are used in delayed-type allergy and some similar conditions contain cortisone. Cortisone suppresses inflammation and edema, which are the underlying causes of itching in hives and in many other pruritic conditions, some not related to allergy. Cortisone is useful in rheumatic fever, which presumably is an infection-type allergy due to antigens of beta-type, Group A streptococci (*S. pyogenes;* see Chapter 29). Similarly, cortisone relieves (but does not cure) certain conditions thought by some to be delayed-type allergy to one's own tissues, the so-called *auto*(self)*immune diseases:* rheumatoid arthritis, multiple sclerosis, thrombocytopenic purpura, acquired hemolytic anemia, lupus erythematosus, and so forth. The inciting tissues in such conditions are called *autoantigens.*

Immunosuppression. Any immune response may be slowed or abolished by various immunosuppressive measures that prevent antibody and lymphocyte production, such as x-radiation, treatment with cobalt-60, certain DNA base analogues, and antibiotics. Most of these severely depress bone marrow, the source of immunologically important lymphocytic cells. Such agents may be used to suppress immune reactions against tissues in organ- and tissue-transplant patients, as well as in cancer therapy. The patients may indeed be helped to resist cancer or to retain transplanted tissues or organs for a longer time, but usually, being deprived of their immune responses, they are highly vulnerable to bacterial and other infections and require special isolation methods.

Drug Idiosyncrasies. Certain plant perfumes, cosmetics, drugs, rubber goods, dyes, and cigarette smoke may cause immediate allergy-like responses, such as hives, dermatitis, and asthma. These immediate reactions, although closely simulating allergy, may often be basically *different from true allergy* and not primarily dependent on specific antibodies at all. Such reactions are usually limited to certain individuals and are often referred to as *drug idiosyncrasies.*

One should always differentiate between *drug hypersensitivity* and *drug intolerance.* True hypersensitivity is often induced by antibiotics such as penicillin, the sulfonamides, aspirin, and barbiturates: reactions are the result of previous sensitization, with the drug presumably acting as a protein conjugate. Sensitivity to penicillin increases at a steady rate in our population; however, the drug is still given too freely, too often when the patient could do better without it. Drug intolerance is a physiologic, metabolic reaction of the body, involving cell-mediated delayed responses, and is of greater intensity than normal but without the expected pharmacologic activity of the drug.

Immunity Against Viral Diseases. Current evidence suggests that antibodies limit or prevent viral infection, whereas cell-mediated immunity functions by eliminating virus-infected cells. Patients with lower cellular immunity due to disease or treatment with immunosuppressive drugs would be highly sensitive to cytomegalovirus, herpes, varicella zoster, and measles infections. These viruses take their envelopes from the membranes of infected cells.

SKIN TESTS FOR HYPERSENSITIVITY

When a small quantity of an antigen is scratched or injected intradermally into a person or an animal hypersensitive to that allergen, a large, red, swollen area will develop at the injection site after an interval; the time, type of reaction, and duration depend on whether one is dealing with immediate-type or delayed-type allergy. These reactions persist for an hour or two in immediate-type allergy and for several days in delayed-type allergy. Such local allergic reactions are extremely useful in determining which antigen is causing a patient's troubles. Tests for hypersensitivity to horse serum, ragweed, and other pollens and antigens are made in this way.

The most widely used test for bacterial allergy is the tuberculin test already discussed (see also Chapter 32). Analogous tests for allergy to certain fungi (*Histoplasma, Coccidioides*) are also widely used. In almost all deep mycoses also, delayed skin reactions occur (see Chapter 38). These are of diagnostic value, and all are due to cell-mediated hypersensitivity.

Table 22–3. Some Autoimmune Disorders in Man

ORGAN OR TISSUE	DISEASE	ANTIGEN	DETECTION OF ANTIBODY*
Thyroid	Hashimoto's thyroiditis (hypothyroidism)	Thyroglobulin	Precipitin; passive hemagglutination; IF on thyroid tissue
		Thyroid cell surface and cytoplasm	IF on thyroid tissue
	Thyrotoxicosis (hyperthyroidism)	Thyroid cell surface	Stimulates mouse thyroid (bioassay)
Gastric mucosa	Pernicious anemia (vitamin B_{12} deficiency)	Intrinsic factor (I)	Blocks I binding of B_{12} or binds to $I{:}B_{12}$ complex
		Parietal cells	IF on unfixed gastric mucosa; CF with mucosal homogenate
Adrenals	Addison's disease (adrenal insufficiency)	Adrenal cell	IF on unfixed adrenals; CF
Skin	Pemphigus vulgaris	Epidermal cells	IF on skin sections
	Pemphigoid	Basement membrane between epidermis and dermis	If on skin sections
Eye	Sympathetic ophthalmia	Uvea	Delayed-type hypersensitive skin reaction to uveal extract
Kidney glomeruli plus lung	Goodpasture's syndrome	Basement membrane	IF on kidney tissue; linear staining of glomeruli
Red cells	Autoimmune hemolytic anemia	Red cell surface	Coombs' antiglobulin test
Platelets	Idiopathic thrombocytopenic purpura	Platelet surface	Platelet survival
Skeletal and heart muscle	Myasthenia gravis	Muscle cells and thymus "myoid" cells	IF on muscle biopsies
Brain	Allergic encephalitis	Brain tissue	Cytotoxicity on cultured cerebellar cells
Spermatozoa	Male infertility (rarely)	Sperm	Agglutination of sperm
Liver (biliary tract)	Primary biliary cirrhosis	Mitochondria (mainly)	IF on diverse cells with abundant mitochondria (e.g., distal tubules of kidney)
Salivary and lacrimal glands	Sjögren's disease	Many: secretory ducts, mitochondria, nuclei, IgG	IF on tissue
Synovial membranes, etc.	Rheumatoid arthritis	Fc domain of IgG	Antiglobulin tests: agglutination of latex particles coated with IgGs, etc.
	Systemic lupus erythematosus (SLE)	Many: DNA, DNA-protein, cardiolipin, IgG, microsomes, etc.	Precipitins, IF, CF, LE cells

*IF = immunofluorescence staining, usually with fluorescent anti-human Igs; CF = complement fixation.

(From Davis, B. D., Dulbecco, R., Eigen, H. N., and Ginsberg, H. S.: Microbiology. 3rd ed. New York, Harper & Row, 1980, based on Roitt, I.: Essential Immunology. Oxford, Basil Blackwell, 1971.)

APPLICATION TO HEALTH

Knowledge about allergy helps one to understand untoward reactions when certain individuals come into contact with antigenic (allergenic) substances, such as sera, antibiotics, some drugs, vaccines, or certain foods. With this knowledge a person can sometimes prevent a severe allergic reaction following injection of an antigenic substance, and may save a life. For example, if a patient is allergic to eggs, and the physician, not knowing this, has ordered a viral vaccine prepared in chick embryos (e.g., yellow fever or swine-influenza vaccine), the person giving the injection should immediately inform the doctor about the allergy to avoid a severe, possibly fatal allergic reaction. A patient who receives a dose of the antitoxic serum derived from horses may also have a serious and possibly lethal reaction, especially if he has had previous injections of horse serum. The serum, although given to provide him with immediate protection against a specific disease, is potentially deadly, since he is allergic to horses. It is important to note that the patient's reaction to the antitoxic serum is not due to the antibodies in it *per se* but to the equine proteins in the serum. He would react to any horse serum, whether antibodies were present in it or not. Other allergies may actually be against a hapten and not necessarily against the protein; thus, the same drug may elicit a different response if obtained from other manufacturers.

The person who is knowledgeable about the potentialities of allergic responses will always have a supply of epinephrine or effective antihistamines readily available whenever antigens or antibodies are being given by injection *(parenterally)* since such drugs control immediate-type hypersensitivity. Patients must be observed constantly for immediate- and delayed-type reactions to antigenic or allergenic substances. If a patient responds allergically to any substance, he should be informed so that he can avoid future injections of this particular substance.

AUTOIMMUNE DISEASES

There are numerous autoimmune disorders, all of which are only partially understood. Some are associated with the histocompatibility complex (HLA), such as myasthenia gravis and Graves' disease. Some are immunodeficiencies involving antibodies such as X-linked agammaglobulinemia, secretory piece deficiency, IgM production without gamma light chain, and kappa chain deficiency. Hormonal defects may be related to hypoparathyroidism or to diseases such as DiGeorge's syndrome. True autoimmune disorders of man (Table 22–3) are due to immune responses to "self" antigens, resulting in autoimmune diseases with serious pathological conditions. The affected individual makes antibodies or produces T-cells to antigens that are normal constituents of his own body.

Autoimmune diseases are clearly a failure in homeostatic function to differentiate "self" from "nonself." This may be due to a change in self–component-making, new contacts with lymphocytes, new crossreacting antigens, development of new clones of B- or T-cells, or chronic viral infections that produce chronic allergic disorders. If the immune system reacts to self-cells resulting from neoplastic transformation (tumors), this is highly desirable, but if the immune system attacks normal self-tissue, autoimmune disease may result.

A "new" disease, *acquired immune deficiency syndrome (AIDS),* has received much publicity. It appears to be transmitted sexually in homosexual males and in blood transfusions, doubles in incidence every 6 months, and has a 40% mortality rate.

SUMMARY

After a sensitizing dose and a later shocking dose of an antigen, a violent allergic reaction called hypersensitivity can be observed in animals and humans. Allergic reactions may be classified either as immediate hypersensitivity, due to IgE antibodies, or as delayed hypersensitivity, due to the action of T-cells. Immediate hypersensitivity is antibody-mediated and delayed allergic reaction is cell-mediated.

Contact of specific antigen with IgE-coated mast cells (in nonmotile connective tissue) or basophilic granulocytes releases several physiologically active substances called mediators, which may cause anaphylaxis, hives, hay fever, or asthma. Mediators include histamine, heparin, ECF-A, NCF-A, SRS-A, and PAF. Epinephrine (Adrenalin) or antihistamine compounds are used as vasodilators, as decongestants, or to reduce swelling, but only temporary relief may be expected.

Desensitization of persons hypersensitive to horse serum is effected by a series of small cutaneous doses of allergen several hours apart. This results in the formation of blocking antibodies, which degrade mediators and "let the patient down" gradually.

In delayed hypersensitivity reactions T-cells are involved, but not antibodies. T-cells release cytotoxic substances called lymphokines that destroy not only the undesirable target cells but also other cells in the area. Some lymphokines are MAF, MIF, LIF, and CSE.

A good example of a delayed-type allergy is the reaction produced by the tuberculin test. Another manifestation involves the rejection of skin grafts and organ transplants. Immunosuppressive measures prevent rejection of tissue grafts and are useful in cancer therapy, but they do present the danger of infection.

Skin tests for hypersensitivity to horse serum, antibiotics, ragweed, other pollens and fungi are of great diagnostic value and can save lives.

Autoimmune diseases are a failure of homeostatic functions to differentiate between "self" and "nonself" antigens. They include rheumatoid arthritis, Hashimoto's thyroiditis, male infertility due to antibodies against sperm, and allergic encephalitis.

REVIEW QUESTIONS

1. Define allergy and hypersensitivity.
2. What is the relationship between allergy and immunity?
3. What is the role of histamine in allergy? What do antihistaminic drugs do?
4. Give examples of two types of hypersensitivities. What types of allergens may produce each?
5. Explain the role of each of the following in allergy: T-cells, IgE, activated lymphocytes, antibodies, mediators, mast cells, anaphylaxis, blocking antibodies, and lymphokines.
6. Define and give examples of serum sickness. How and why does desensitization work?
7. How is the tuberculin reaction related to delayed hypersensitivity?
8. Explain the problems associated with homografts, immunosuppressive measures, autoimmune diseases, cortisone medications, and drug idiosyncrasies.

9. What is the importance of skin tests for hypersensitivity?
10. Why might immunosuppressive drugs render some individuals highly sensitive to certain virus infections?
11. Name some autoimmune diseases. What causes them?

Supplementary Reading

Allergy Foundation of America: Allergy: Its Mysterious Causes and Modern Treatment. New York, Grosset & Dunlap, Inc., 1967.

Becker, E. L., and Austen, K. F.: Anaphylaxis. *In* Mueller-Eberhard, H., and Mischer, P.: Immunopathology. Boston, Little, Brown & Co., 1967.

Bloch, K. J., Lee, L., Mills, J. A., et al.: Gamma heavy chain disease—An expanding clinical and laboratory spectrum. *Am. J. Med., 55*:61, 1973.

Borel, Y.: Autoimmunity and Self-Nonself Discrimination. Copenhagen, Munksgaard, 1976.

Burnet, F. M.: Immunology, Aging, and Cancer. San Francisco, W. H. Freeman and Co., 1976.

Castro, J. E.: Immunology for Surgeons. Lancaster, England, MTP Press, 1976.

Chase, M. W.: The allergic state. *In* Dubos, R. J., and Hirsch, J. G.: Bacterial and Mycotic Infections of Man. 4th ed. Philadelphia, J. B. Lippincott Co., 1965.

Craven, R. F.: Anaphylactic shock. *Am. J. Nurs., 72*:718, 1972.

Criep, L. H.: Allergy and Clinical Immunology. New York, Grune & Stratton, 1976.

Edelman, G. M.: Antibody structure and molecular immunology. *Science, 180*:830, 1973.

Epstein, W. L.: To prevent poison ivy and oak dermatitis. *Am. J. Nurs., 63*:113. 1963.

European Congress of Allergology and Clinical Immunology: Allergy. Frankland, A. W., and Ganderton, M. A. (eds.). Brooklyn Heights, N.Y., Beekman Pubs., 1975.

Henny, C.: T-cell mediated cytolytsis. *Contemp. Top. Immunobiol., 7*:245, 1978.

Hung, T. C., and Rauch, H. C.: Antibody response to synthetic encephalitogenic peptide. *Mol. Immunol., 17*:527, 1980.

Krahlenbuhl, J. L., Blaykovec, A. A., and Lysenko, M. G.: In vivo and in vitro studies of delayed-type hypersensitivity to *Toxoplasma gondii* in guinea pigs. *Inf. Immunol., 3*:260, 1971.

Leskowitz, S.: Immunologic tolerance. *Bioscience, 18*:1030, 1968.

Levine, B. B.: Immunochemical mechanisms of drug allergy. *Ann. Rev. Med., 17*:23, 1966.

McGovern, J. P., Smolensky, M. H., and Reenberg, A.: Chronobiology in Allergy and Immunology. Springfield, Ill., Charles C Thomas, 1977.

Milgrom, F., Centeno, E., Shulman, S., and Witebsky, E.: Autoantibodies resulting from immunization with kidneys. *Proc. Soc. Exp. Biol. Med., 116*:1009, 1964.

Milstein, C., and Munro, A. J.: The genetic basis of antibody specificity. *Ann. Rev. Microbiol., 24*:335, 1970.

Nakahara, W.: Recent Advances in Human Tumor Virology and Immunology. Baltimore, University Park Press, 1972.

Nathenson, S.: Histocompatibility. New York, Academic Press, 1976.

Notkins, A. L., Mergenhagen, S. E., and Howard, R. J.: Effect of virus infections on the function of the immune system. *Ann. Rev. Microbiol., 24*:525, 1970.

Ptak, W., Rozyck, D., Askenase, P. W., et al.: Role of antigen-presenting cells in the development and persistence of contact hypersensitivity. *J. Exp. Med., 151*:362, 1980.

Ruddle, N., and Waksman, B. H.: Cytotoxic effect of lymphocyte-antigen interaction in delayed hypersensitivity. *Science, 157*:1060, 1967.

Shaffer, J. H., and Sweet, L. C.: Allergic reactions to drugs. *Am. J. Nurs., 65*:100, 1965.

Uhr, J. W.: Delayed hypersensitivity. *Physiol. Rev., 46*:359, 1966.

Waksman, B. H.: Cellular hypersensitivity and immunity: Inflammation and cytotoxicity. *In* Parker, C. W. (ed.): Clinical Immunology. Philadelphia, W. B. Saunders Company, 1980, vol. 1.

SECTION FIVE

PATHOGENIC MICROORGANISMS

Waldemar Mordecai Wolff Haffkine (1860–1930) is known for his development of a vaccine effective against *Yersinia pestis*, the causative organism of plague. Of all the antiplague vaccines, Haffkine's is by far the best known and was the most widely used. (From *The Brilliant and Tragic Life of W. M. W. Haffkine* by Selman A. Waksman. Copyright © 1964 by Rutgers, the State University. Reprinted by permission of Rutgers University Press.)

Sara Elizabeth Branham (1888–1962) became the world's foremost authority on the genus *Neisseria*, especially the meningococcus. *Branhamella catarrhalis*, formerly known as *Neisseria catarrhalis*, is named in her honor. (Courtesy of the American Society for Microbiology and Dr. Margaret Pittman.)

FACTORS IN TRANSMISSION OF COMMUNICABLE DISEASES

The occurrence of communicable disease implies: (1) a source of infection, i.e., the presence of active pathogenic agents in or on some sort of transmissible matter, animate or inanimate; (2) a vector, animate or inanimate, capable of transmitting those agents to a host capable of harboring them; (3) some means of egress or *portal of exit* of the pathogenic agent from the infected body; and (4) the arrival of the pathogenic agent at a *portal of entry* from where it can gain a foothold in a new host.

PORTALS OF ENTRY

These are certain routes or pathways by which some microorganisms normally enter the body and cause infections. The particular route in a given instance depends on the kind of microorganism and, to some extent, the kind of vector involved. The most important portals of entry are cuts or abrasions in the skin (including those due to bites of arthropods and other animals), the mucous membranes of the respiratory tract (nose, throat, tonsils, and lungs), the eyes, the mouth and gastrointestinal tract, the anus, the ears, and the genitourinary tract. Direct introduction of microorganisms into the bloodstream is also a common route of infection when unsterile needles or cutting instruments are used. (The modes of transmission of disease-causing microorganisms are shown in Figure 23–1.)

Abnormal Portals of Entry. In this discussion only normal and natural portals of entry are considered. Many microorganisms, even so-called harmless saprophytes, may cause a rapidly fatal infection if injected with a hypodermic needle into the brain or peritoneal cavity, or if sprayed into the nose and lungs. Such treacherous doings, although valuable experimentally, are like admitting a hostile army through a secret postern gate into a walled city, bypassing the normal portal of entry.

Many disease-causing organisms have their *obligate* portals of entry and can

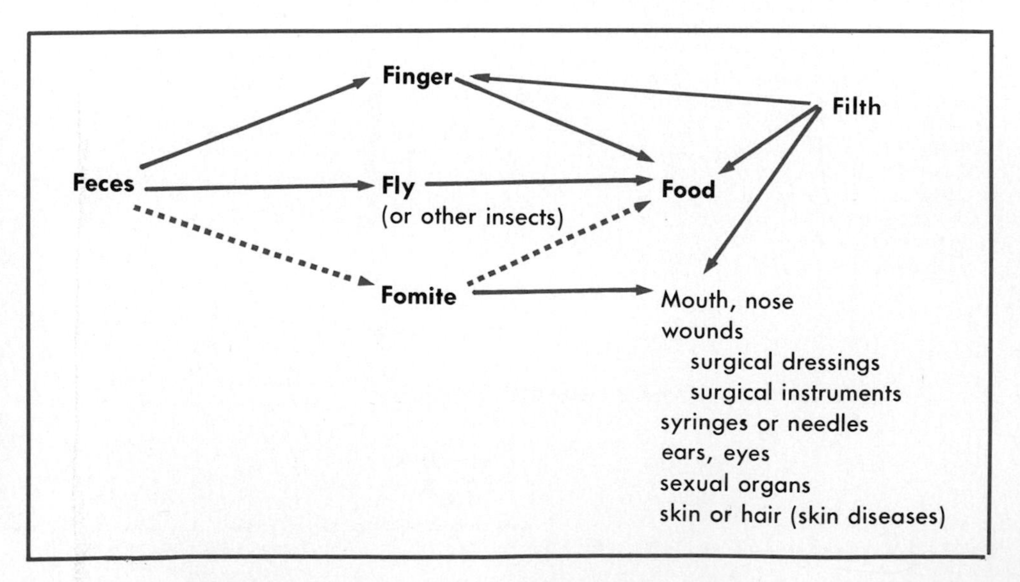

Figure 23–1 Transmission of disease-causing microorganisms from feces to food with hands, insects, and fomites as vectors in our daily life. Fomites are nonliving objects—inanimate vectors that transmit pathogens. Examples of fomites are money (coins and paper currency), pencils, pens, eating utensils, plates, glasses or bottles, toys (especially dolls), postage stamps, stationery, door handles, bed linens, pillows, towels, thermometers, books, magazines, newspapers, dentists' and physicians' instruments, toilet items, clothing, and many other objects used in daily life. We are also "stuck" in our own *filth*—dirt in food establishments, in the kitchens, below counters, behind refrigerators, and animal droppings and dead insects in dust, and perhaps on the chair you sit on now. In general, filth carries microorganisms, but it is not a true fomite.

cause an infection only if they enter via such a portal. Thus, dysentery bacilli rubbed into a skin wound would probably not be responsible for any infection, whereas the same organisms, if swallowed, might produce fatal dysentery. Conversely, the streptococci or staphylococci that cause boils do not produce infections when swallowed, but if rubbed into the skin or a wound, they can produce a severe, even fatal infection. Some microorganisms can enter the body tissues through almost any portal.

Path of Organisms in the Body. From the various portals of entry or sites of initial infections, organisms may pass into the circulating blood and start a secondary or *metastatic* infection in some of the internal organs or in the membranes of the brain and spinal cord (spinal meningitis). Organisms entering through the nose or mouth may take one of several paths. They may be swallowed and thus reach the stomach and intestines. In most instances, they will be killed by bile and other digestive juices. If they are pathogens of the intestinal tract *(enteric pathogens)*, such as those that cause typhoid fever, dysentery, cholera, or poliomyelitis, they will survive contact with the gastrointestinal juices and gain entrance to the tissues via the gastrointestinal tract, especially in the presence of foods that temporarily reduce the acidity of the gastric juice.

If they are respiratory pathogens, such as pneumococci, diphtheria bacilli, tubercle bacilli, scarlet fever or septic sore throat streptococci, or the bacilli of whooping cough, they may locate in and on the tonsils, or they may pass on to the lungs, as they do in pneumonia and tuberculosis (Fig. 23–2).

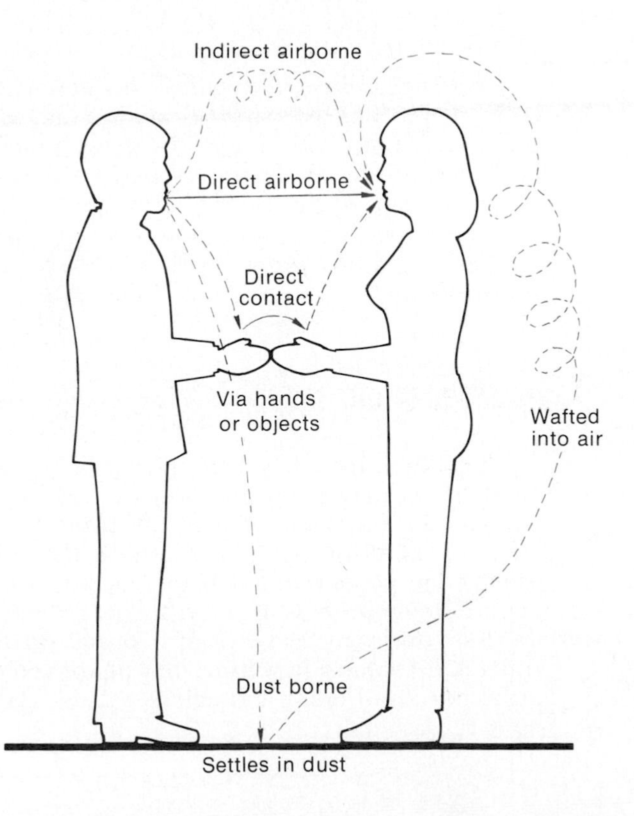

Figure 23–2 Transmission of infectious agents from the respiratory and oral tract of one person to another by direct contact or indirectly via airborne particles. Among the agents of disease thus transmitted are viruses of influenza, colds, measles, and mumps; streptococci; pneumococci; meningococci; tubercle and diphtheria bacilli; some fungi; and occasionally other pathogens such as plague and anthrax bacilli. Every educated person should visualize these pathways of transmission and create obstructions (such as gowns and masks, correct hand washing, correct food handling procedures, correct sputum disposal, and proper sweeping methods) to protect others and oneself against diseases transmitted via these and other routes. (After F. Schwentker. Courtesy of American Sterilizer Co.)

Table 23–1. Transmission of Communicable Diseases: Portals of Entry and Portals of Exit

Type of Infection	Infectious Agents and Diseases They Cause	Portals of Entry	Portals of Exit
Gastrointestinal Tract (Food- and Water-borne Diseases)	A. BACTERIA 1. Enterobacteriaceae, *Salmonella, Shigella* (typhoid, dysentery, etc.) (gram-negative rods) 2. *Brucella* (undulant fever) 3. *Leptospira* (leptospirosis) 4. *Clostridium* (botulism) (gram-positive rods) B. VIRUSES 1. Poliomyelitis 2. Coxsackie 3. Echo 4. Hepatitis A (epidemic hepatitis) C. PROTOZOA 1. *Entamoeba histolytica* (dysentery, etc.) 2. *Trichomonas hominis* (enteritis, etc.) 3. *Giardia lamblia* (enteritis, etc.) D. HELMINTHS 1. Hookworm 2. *Ascaris* 3. Pinworms	Mouth	Fecal route, some urinary
Oral and Respiratory Tracts (Droplet Infections)	A. BACTERIA 1. Gram-positive cocci: *Streptococcus pneumoniae* (pneumonia), *Streptococcus pyogenes* (scarlet fever) 2. Gram-negative cocci 3. Gram-positive rods: *Neisseria meningitidis* (epidemic meningitis), *Corynebacterium diphtheriae* (diphtheria), *Mycobacterium tuberculosis* (tuberculosis) 4. Gram-negative rods: *Haemophilus influenzae* (laryngitis, etc.), *Bordetella pertussis* (whooping cough) 5. Spirochetes: *Treponema pallidum* (syphilis, Vincent's angina) 6. Psittacosis organisms B. VIRUSES 1. Smallpox 2. Mumps 3. Measles 4. Chickenpox 5. Rabies 6. Myxoviruses 7. Adenoviruses, rhinoviruses, etc. 8. Poliovirus C. FUNGI: Some deep mycoses D. HELMINTHS: Whipworm, hookworm, *Ascaris* flukes, pinworm, tapeworm	Oral route	Oral route

Table continued on opposite page

The organisms of syphilis and gonorrhea find their principal portal of entry in the mucosal surfaces and glands of the genitourinary tract, though they not infrequently infect the eyes (gonorrheal ophthalmia), lips, and oral cavity (syphilis) and are transmitted via these portals.

It is tragic when several young people use a single nonsterile hypodermic needle to inject drugs into each other via a blood vessel, after the same needle has been used by a person with hepatitis or malaria. This "drug scene" is truly bad medicine!

PORTALS OF EXIT

Equally important is the portal by which microorganisms leave the body on their journey from one host or patient to another. A person attending a patient with infectious disease can protect himself and safeguard the environment by knowing when and where the microorganisms of the disease are leaving the patient and how to deal with them. In the course of an infectious disease, microorganisms often follow a devious and complicated path through the host and may leave from a portal different from the one they entered. Indeed, infectious organisms may be conveniently grouped on the basis of their usual portals of entry and exit (see Table 23–1).

Table 23–1. Transmission of Communicable Diseases: Portals of Entry and Portals of Exit *Continued*

Type of Infection	Infectious Agents and Diseases They Cause	Portals of Entry	Portals of Exit
Venereal and Related Infections	A. BACTERIA 1. *Treponema pallidum* (syphilis) 2. *Neisseria gonorrhoeae* (gonorrheal infection) (gram-negative rods) 3. *Haemophilus ducreyi* (chancroid) 4. *Calymmatobacterium granulomatis* (granuloma inguinale) 5. *Chlamydia trachomatis* VR 121 (lymphogranuloma venereum) B. PROTOZOA 1. *Trichomonas vaginalis* (vulvovaginitis, etc.) C. VIRUS: HSV-2 (venereal herpes)	Genital tract Genital tract, eyes	Genital tract
From The Soil Into Cuts in the skin	A. BACTERIA (gram-positive rods) 1. Genus *Clostridium* (anaerobes) a. Gas gangrene group b. *Cl. tetani* (tetanus) c. *Cl. botulinum* (food poisoning) 2. Genus *Bacillus* (aerobes) a. *B. anthracis* (anthrax) B. FUNGI 1. *Coccidioides immitis* (coccidioidomycosis) 2. *Histoplasma capsulatum* (histoplasmosis) 3. *Sporotrichum* (sporotrichosis) 4. *Blastomyces*, etc.	Wounds, cuts, and scratches in the skin	Wounds, (not a significant source of transmission)
Pathogens of Humans Usually Transmitted in Blood	A. BACTERIA (gram-negative rods) 1. *Yersinia pestis* (bubonic plague) 2. *Francisella tularensis* (tularemia) 3. *Borrelia* (relapsing fever) B. RICKETTSIAS 1. Rocky Mountain spotted fever, typhus, etc. C. VIRUSES 1. Yellow and dengue fevers 2. Other arboviruses D. PROTOZOA 1. *Plasmodium* (malaria) 2. *Trypanosoma* (trypanosomiasis) 3. *Leishmania* (leishmaniasis) E. HELMINTHS 1. Filarias (filariasis)	Mainly by sanguivorous arthropods, bites	Blood sucked by arthropods
	A. VIRUSES Notably those of epidemic hepatitis and of homologous serum hepatitis, i.e., hepatisis viruses A and B, which may be circulating in the blood at the time the blood is drawn or the instruments used B. BACTERIA THAT FREQUENTLY CAUSE BACTERIEMIA: *Brucella, Staphylococcus, Neisseria, Pasteurella, Leptospira, Treponema, Salmonella, Streptococcus*	Mainly by artificial vectors (e.g., hypodermic needles, syringes, autopsy instruments, surgical instruments, etc., and by some blood derivatives—plasma, serum, whole blood, etc.)	From wounds (blood) to fomites

VECTORS

Vectors of infectious organisms may be animate or inanimate.

Inanimate Vectors. These may be food, body discharges (feces, saliva, pus, and the like), water, bandages, dressings, instruments, hypodermic needles, bedding, eating utensils, and other inanimate objects called *fomites* contaminated with infectious discharges. However, fomites are not always so obvious as vectors of disease transmission. Think of fomites when you lick a stamp, put a pencil in your mouth, borrow a book, eat at a restaurant, or use the rest room. When one knows the portal of exit of any pathogen, the possible vectors are almost self-evident.

Animate Vectors. These vectors are generally arthropods or mammals. Anopheline mosquitoes are notorious vectors of malaria; dogs and other mammals are vectors of rabies and also of worms (helminths), toxoplasmosis, and numerous other infectious diseases. Domestic animals often transmit ringworm (the tineas) infections, which are not worms at all, but disease conditions caused by fungi (see Chapter 38). The *hands* of patients, doctors, nurses, and other persons in contact with an infectious patient or his fomites may also be classed as animate vectors. The control of each type is a specialized study in itself. Of course, humans themselves may be regarded as one of the most dangerous vectors of disease because they cannot be eliminated with DDT or muzzled and put on a leash. Humans are the only significant living vector of such diseases as poliomyelitis, infectious mononucleosis, measles, venereal herpes, and syphilis, and they are the most frequent live vector of human tuberculosis.

The whole philosophy of communicable disease management, and indeed of the entire profession of preventive medicine as it relates to communicable disease, is based largely on knowledge of these three factors (portals of entry, portals of exit, and vectors) and their ramifications. The nurse or other health worker who grasps these simple basic principles will be well prepared in the field of communicable disease control.

PREVENTION OF DISEASE TRANSMISSION

It is evident that the transfer of microorganisms from one person to another may be blocked at one or more of three points just mentioned.

Portal of Entry. The entrance of pathogens into the body is (theoretically) always preventable if their portal of entry is known. For example, the respiratory portals may be blocked by masks; cuts and scratches on the hand may be covered by bandages or surgical gloves. The alimentary tract is best protected by appropriate selection and disinfection (as by pasteurization of milk, chlorination of water, and cooking) and sanitary preparation of food. The skin may be protected from biting arthropods by clothing, repellents, and screens. The tissues and blood may be made resistant by vaccination and by the prophylactic use of antibodies and chemotherapeutic drugs.

Portal of Exit. Unless the vector is a blood-sucking arthropod, the transmission of most infectious agents may be stopped at the portal of exit by immediate collection and proper disposal of whatever bodily excretions, secretions, discharges, and tissues may contain the pathogens: feces, urine, saliva, mucus, pus, tissue drainages, bandages, and the like.

All these matters will be discussed in more detail in following chapters. The portal of exit of pathogens is the point at which one can most effectively block routes of disease transmission.

Vector. Transmission of infection may theoretically be entirely blocked by effective vector control. In some cases this has been dramatically effective, as when yellow fever and malaria were abolished from certain areas such as Cuba, Italy, and the Panama Canal Zone by elimination of the vector mosquitoes. The transmission of such enteric diseases as cholera and typhoid fever by municipal water supplies has been completely blocked in the Western Hemisphere and in Europe by expert sanitation of water supplies and efficient sewage disposal. Cholera, once a worldwide scourge, no longer exists in endemic form in these areas. Water-borne typhoid fever is virtually extinct in the United States. Other examples of large-scale prevention of disease transmission will occur to the thoughtful student. Means other than vector control have also been successful. What do you remember about global control of smallpox?

SUMMARY

Communicable diseases are caused by pathogenic agents transmitted either directly or by means of a vector, animate or inanimate. The routes by which some microorganisms enter a susceptible host are portals of entry, which include the nose, throat, tonsils, lungs, anus, genitourinary tract, and cuts or abrasions in the skin, as well as stabs into the bloodstream itself. Some portals of entry are obligate for certain pathogens, but other microorganisms can enter the body through almost any portal.

Disease-causing pathogens (many from feces) may be transmitted by fomites (inanimate vectors, such as forks, dishes, glasses, door handles, towels, or surgical instruments), direct contact (such as shaking hands, touching a wound, kissing or sexual intercourse), and contacts with animals, including insects. Some microorganisms are carried in dust, and others may be transmitted by indirect contact through sneezing, coughing, or spitting.

Portals of exit are just as important in disease transmission. They too are specific for certain microorganisms and must be understood to prevent disease from spreading.

REVIEW QUESTIONS

1. Describe different portals of entry and portals of exit for disease transmission.
2. What are communicable diseases? List some inanimate and animate vectors of disease.
3. Define direct contact, indirect contact, and airborne infection.
4. What are fomites? Make a list of fomites and show how they affect our daily life.
5. Associate several types of diseases with the organ systems they affect.
6. List some arthropods that are vectors of disease.
7. How can an understanding of portals of exit help to protect against infectious diseases?
8. How may a harmless saprophyte cause a rapidly fatal infection?

Supplementary Reading

Braude, A. I.: Microbiology. Philadelphia, W. B. Saunders Company, 1982.

Finegold, S. M., Martin, W. J., and Scott, E. G.: Bailey and Scott's Diagnostic Microbiology. 5th ed. Saint Louis, C. V. Mosby Company, 1978.

Freeman, B. A.: Burrows' Textbook of Microbiology. 21st ed. Philadelphia, W. B. Saunders Company, 1979.

Frobisher, M., Hinsdill, R. D., Crabtree, K. T., and Goodheart, C. R.: Fundamentals of Microbiology. 9th ed. Philadelphia, W. B. Saunders Company, 1974.

Lennette, E. H., Spaulding, E. H., and Truant, J. P.: Manual of Clinical Microbiology. 2nd ed. Washington, D. C., American Society for Microbiology, 1974.

Rose, N. R., and Friedman, H.: Mannual of Clinical Immunology. 2nd ed. Washington, D. C., American Society for Microbiology, 1980.

Schuhardt, V. T.: Pathogenic Microbiology. Philadelphia, J. B. Lippincott Company, 1978.

SALMONELLOSIS, SHIGELLOSIS, AND CHOLERA

THE FAMILY ENTEROBACTERIACEAE

General Description. The name of this family of bacteria is derived from the fact that nearly all species in it more or less constantly inhabit the intestines of humans or animals or both. Important pathogenic members of the family include the bacilli of typhoid and paratyphoid fevers (salmonellosis) and bacillary dysentery (shigellosis), as well as some related bacilli, notably *Escherichia coli.* The Enterobacteriaceae are gram-negative, nonspore-forming straight rods, aerobic and facultative anaerobes ranging from 1 to 2 μm in diameter and from 3 to 10 μm in length. All species ferment glucose and all, microscopically, are indistinguishable. Most species are motile with peritrichous (not polar) flagella; however, the dysentery bacilli (*Shigella*) are characteristically nonmotile. Several strains of *Salmonella, Shigella, Escherichia, Klebsiella, Enterobacter,* and *Proteus* possess fimbriae or pili.[1] All of these bacilli grow readily on simple, blood-free culture media, including many common foods such as milk, salad dressings, and sandwich fillings, over a range of temperatures from 15 to 40 C. Most of them can resist cold well and survive in soil, sewage, ice, water, milk, and foods for periods varying from hours to weeks, depending on the species and the environment. Since these bacteria do not form spores, however, they are readily killed by 5 minutes of boiling, by the commercial processes of pasteurization, and by standard disinfectants. Drying and direct sunlight destroy them fairly quickly.

Subdivisions. These gram-negative enteric bacilli are classified by a system proposed by Edwards and Ewing, differing from the classification given in the eighth edition of *Bergey's Manual* in some minor ways, as may be seen in Table 24–1. These systems and the British usage (after Topley and Wilson, 1975) agree in part with recommendations made by the International Subcommittee on Enterobacteriaceae. It should be noted that although *Bergey's Manual* provides "officially" accepted nomenclature of classification of bacteria, the system of Edwards and Ewing is now commonly used in most clinical laboratories in the United States.

As shown in Table 24–1, the family Enterobacteriaceae is subdivided into five tribes and these into 12 genera. Special emphasis is given to the pathogenic species in each genus. Most important among these groups are the tribe Escherichieae (named for the German scientist Escherich), containing some harmless saprophytes but also some troublesome pathogens; the genus *Shigella* (named after the Japanese microbiologist Shiga); the tribe Salmonelleae (named after the American bacteriologist Salmon), containing the typhoid, paratyphoid, and dysentery bacilli; and the tribe Proteeae (named for the pleomorphic Greek god Proteus), containing some troublesome pathogens and also some species that are occasionally confused with *Salmonella.*

Because of the importance of the Enterobacteriaceae in sanitation and disease, the identification of genera and species is essential. Table 24–2 shows

[1]Pili, as readily seen with the electron microscope, are smaller than flagella. Pili are not antigenically related to flagella. See page 54.

PART A PATHOGENS TRANSMITTED FROM THE INTESTINAL AND/OR URINARY TRACT

CHAPTER 24

one of several schemes used to differentiate these organisms. Many suppliers of microbiologic media have recently manufactured their own systems for performing the necessary biochemical tests. Three of these commercially available test systems (the Enterotube, API-20, and the r/b system) are listed for comparison with the standard common tests in Table 24–3 (see also Figs. 24–1 and 24–2). Other systems are Inolex, PathoTec, Auro Tab and Minitek; still others will doubtless be devised.

We may classify these organisms by relating them to three general types of clinical conditions as follows:

Salmonella enteritis infections (salmonelloses):
- Typhoid fever and similar generalized infections by *Salmonella* species
- Gastroenteritis, infections usually confined to the bowel (food infections)

Shigella infections (shigelloses):
- Bacillary dysentery

Escherichia and *Proteus* infections:
- Gastroenteritis and infant diarrhea
- Urinary tract infection

The tribe *Yersiniae,* with the species *Yersinia pestis* (previously *Pasteurella pestis*), the cause of bubonic plague, is now classified with the enteric organisms in the family *Enterobacteriaceae* on the basis of "numerical methods" of nomenclature. However, for practical reasons *Yersinia* will be discussed later, in Chapter 39.

The Salmonelloses

Salmonellosis is the medical term for infection with any species of *Salmonella.* Certain species of *Salmonella* are more likely than others to invade the blood and to cause the more severe salmonelloses. *Salmonella typhi,* causing typhoid fever, is one of these. Others are *S. paratyphi* and numerous related species. The name *S. paratyphi A* indicates "typhoid-like" and was given to certain salmonellas causing typhoid-like conditions. The term "paratyphoid bacillus" is still often used for nearly all salmonellas except *S. typhi.* Today, except for gradations in clinical severity, the same basic pathogenesis is recognized in all *Salmonella* infections.

The term salmonellosis, therefore, includes typhoid fever, paratyphoid fever, and infection by any of over 1,900 very similar subspecies or serotypes of *Salmonella.*

Typhoid Fever. Typhoid fever is a good representative of enteric infections in general: bacterial, viral, and others. It is the type of infection transmitted by feces or urine. With regard to *means of transmission,* what is true of typhoid is, with a few minor modifications, also true of *Salmonella* food-borne infections (sometimes incorrectly called "food poisoning"), bacterial dysentery, epidemic viral hepatitis, and amebic dysentery.

Table 24–1. Pathogenic and Potentially Pathogenic Species in the Family *Enterobacteriaceae*

EDWARDS AND EWING CLASSIFICATION (1972)		BERGEY'S CLASSIFICATION (1974)		
(1) Tribe	I Escherichieae	(1) Tribe	Escherichieae	Important as indicators of fecal pollution of water and foods. Certain types of *E. coli* cause enteritis, urinary tract infections, and secondary infections.
Genus	I *Escherichia*	Genus	*Escherichia*	
Species	1 *E. coli*	Species	1 *E. coli*	
Genus	II *Shigella*	Genus	*Shigella*	Cause of bacillary dysentery or shigellosis.
Species	1 *S. dysenteriae*	Species	1 *S. dysenteriae*	
	2 *S. flexneri*		2 *S. flexneri*	
	3 *S. boydii*		3 *S. boydii*	
	4 *S. sonnei*		4 *S. sonnei*	
(2) Tribe	II Edwardsiellae			
Genus	I *Edwardsiella*	Genus	*Edwardsiella*	May cause diarrhea.
Species	1 *E. tarda*	Species	1 *E. tarda*	
(3) Tribe	III Salmonelleae	Genus	*Salmonella**	*S. typhi:* Cause of typhoid fever (a form of salmonellosis). Other species of *Salmonella:* Cause various forms of salmonellosis other than typhoid fever, especially food infection.
Genus	I *Salmonella*	Species	1 *S. cholerae-suis*	
Species	1 *S. typhi*		2 *S. hirschfeldii*	
	2 *S. paratyphi A*		3 *S. typhi*	
	3 *S. paratyphi B* or (*S. schottmuelleri*)		4 *S. paratyphi-A*	
			5 *S. schottmuelleri*	
	4 *S. paratyphi C* or (*S. hirschfeldii*)		6 *S. typhimurium*	
	5 *S. typhimurium*		7 *S. enteritidis*	
	6 *S. cholerae-suis*		8 *S. gallinarum*	
	7 *S. enteritidis*		9 *S. salamae*	
			10 *S. houtenae*	
Genus	II *Arizona*			Cause salmonellosis-like conditions.
Species	1 *A. hinshawii*		11 *S. arizonae*	
Genus	III *Citrobacter*	Genus	*Citrobacter*	Confused with *Salmonella* species and *S. arizonae*; may not be truly pathogenic.
Species	1 *C. freundii*	Species	1 *C. freundii*	
	2 *C. intermedius*		2 *C. intermedius*	
(4) Tribe	IV Klebsielleae	(2) Tribe	Klebsielleae	Pneumonia, enteritis, septicemia, peritonitis, urinary infections, etc.
Genus	I *Klebsiella*	Genus	*Klebsiella*	
Species	1 *K. pneumoniae*	Species	1 *K. pneumoniae*	
Genus	II *Enterobacter*	Genus	*Enterobacter*	Urinary tract infections, septicemias.
Species	1 *E. cloacae*	Species	1 *E. cloacae*	
	2 *E. aerogenes*		2 *E. aerogenes*	
	3 *E. hafniae*	Genus	*Hafnia*	Found in sewage, feces, water, soil.
		Species	1 *H. alvei*	
Genus	III *Serratia*	Genus	*Serratia*	Found in water, soil, food, occasionally pathogenic in humans.
Species	1 *S. marcescens*	Species	1 *S. marcescens*	
		(3) Tribe	Erwinieae	Many are plant pathogens; some may have become opportunistic pathogens in humans.
Genus	IV *Pectobacterium*	Genus	*Erwinia*	
Species	1 *P. carotovorum*		3 groups with 14 species classified	
(5) Tribe	V *Proteeae*	(4) Tribe	Proteeae	*P. vulgaris* and *P. mirabilis* may cause infections of urinary tract. *P. mirabilis* and *P. morganii* cause diarrhea, especially in infants.
Genus	I *Proteus*	Genus	*Proteus*	
Species	1 *P. vulgaris*	Species	1 *P. vulgaris*	
	2 *P. mirabilis*		2 *P. mirabilis*	
	3 *P. morganii*		3 *P. morganii*	
	4 *P. rettgeri*		4 *P. rettgeri*	
Genus	II *Providencia*		5 *P. inconstans*	Sometimes found in human urine and feces.
Species	1 *P. stuartii*			
		(5) Tribe	Yersiniae	*Y. pestis* (previously *Pasteurella pestis*) is the cause of bubonic plague.
		Genus	*Yersinia*	
		Species	1 *Y. pestis*	

**Salmonella* is now divided into four subgenera. Subgenus I includes *S. cholerae-suis, S. typhi,* and all other species that are not in the other subgenera. Subgenus II has the species *S. salamae*; subgenus III the species *S. arizonae*; and subgenus IV the species *S. houtenae*. The 11 species are subdivided by the Kauffmann-White scheme into more than 1,900 subspecies.

Table 24–2. Differential Characters of some Enterobacteriaceae and *V. cholerae**

GROUPS	GENERA	SPECIES	REACTIONS										
			Glucose	*H_2S*[a]	*Indole*	*Phenylalanine deaminase*	*Urease*	*Dulcitol*[b]	*Lactose*[c]	*Lysine*[d, e] *decarboxylase*	*Simmons Citrate*[f]	*Ornithine decarboxylase*	*Motility*
Escherichia	*Escherichia*	*coli*	⊕	–	+	–	–	v	⊕	–	–	v	±
Shigella	*Shigella*		+	–	∓	–	–	v	–	–	–	∓	–
	Edwardsiella	*tarda*	⊕	+	+	–	–	–	–	+	–	+	+
Salmonella			⊕	+	–	–	–	+	–	+	+	+	+
Arizona			⊕	+	–	–	–	–	v	v	+	+	+
Citrobacter			⊕	+	–	–	v	v	v	–	+	∓v	+
Klebsiella	*Klebsiella*	*pneumonia*	⊕	–	∓	–	+	∓	⊕	+	+	–	–
Enterobacter	*Enterobacter*	*cloacae*	⊕	–	–	–	±	∓	⊕	–	+	+	+
		aerogenes	⊕	–	–	–	–	–	⊕	–	+	+	+
		hafniae	⊕	–	–	–	–	–	∓	+	v	+	+
		liquefaciens	⊕	–	–	–	v	–	v	v	+	+	+
Serratia	*Serratia*	*marcescens*	⊕	–	–	–	v	–	–	+	+	+	+
Proteus	*Proteus*	*vulgaris*	⊕	+	+	+		–	–	–		–	+
		mirabilis	⊕	+	–	+		–	–	–		+	+
		morgani	±	–	+	+		–	–	–		+	±
		rettgeri	∓	–	+	+		–	–	–		–	+
Providencia	*Providencia*		±	–	+	+		–	–	–		–	+
	Vibrio	*cholerae*	+	–	+		–	–	–	+	±	+	+

*Modified from *Enterotube®*, Roche Diagnostics, Nutley, N.J.

+ Positive
– Negative
± Most are positive
∓ Most are negative
v Various biochemical types
⊕ Acid and gas produced

[a]*S. enteritidis* (bioserotype *paratyphi-A*) and some others may be H_2S negative.
[b]Most salmonellae ferment dulcitol in 24 hours; *S. typhi, S. enteritidis* (bioserotypes *paratyphi-A* and *pullorum*), and *S. cholerae-suis* often ferment slowly or not at all.
[c]*Shigella sonnei* and *V. cholerae* ferment lactose only after 24 hours.
[d]After 24 hours reactions may become positive.
[e]*S. enteritidis* (bioserotype *paratyphi-A*) does not decarboxylate lysine.
[f]*S. cholerae-suis* utilizes citrate slowly; *S. typhi* and *S. enteritidis* (bioserotype *paratyphi-A*) not at all.

Table 24–3. A Comparison of Four Different Systems of Biochemical Tests for the Enterobacteriaceae and Closely Related Species

FERMENTATION IN DURHAM TUBES AND OTHER COMMON TESTS*	API-20** TESTS	ENTEROTUBE TESTS***	R/B ENTERIC**** DIFFERENTIAL SYSTEM
1. Lactose	1. Lactose	1. Lactose	1. Lactose
2. Arginine*	2. Arginine		
3. Lysine*	3. Lysine	3. Lysine	3. Lysine
4. Ornithine*	4. Ornithine	4. Ornithine	4. Ornithine
5. Phenylalanine*	5. Phenylalanine	5. Phenylalanine	5. Phenylalanine
6. Citrate*	6. Citrate	6. Citrate	
7. H_2S formation*	7. H_2S formation	7. H_2S formation	7. H_2S formation
8. Urease*	8. Urease	8. Urease	8. (Urease-additional)
9. Methyl Red*			
10. Indole*	10. Indole	10. Indole	10. Indole
11. V.P.*****	11. V.P.		
12. Gelatin*	12. Gelatin		
13. Glucose	13. Glucose	13. Glucose	13. Glucose
14. Mannitol	14. Mannitol		14. (Manitol-additional)
15. Inositol	15. Inositol		
16. Sorbitol	16. Sorbitol		16. (Sorbitol-additional)
17. Rhamnose	17. Rhamnose		17. (Rhamnose-additional)
18. Sucrose	18. Sucrose		
19. Melibiose	19. Melibiose		
20. Arabinose	20. Arabinose		
21. Dulcitol		21. Dulcitol	21. (Dulcitol-additional)
	22. Amygdaline		
23. Motility*			23. Motility
24. Raffinose			24. (Raffinose-additional)
25. Maltose			
			26. (Acetate-additional)
			27. (DNase-additional)

*Tests other than fermentation tests.

**API-20 is a system of 20 test capsules on one card. The test medium is dehydrated in these capsules. Distilled water is used to rehydrate. API-20 was supplied by Analytab Products, Inc., New York, N.Y.

***Enterotube is a set of eight media contained in one tube. It was supplied by Roche Diagnostics, Nutley, N.J.

****Diagnostic Research, Inc.

*****V.P. is the abbreviation for the Voges-Proskauer test.

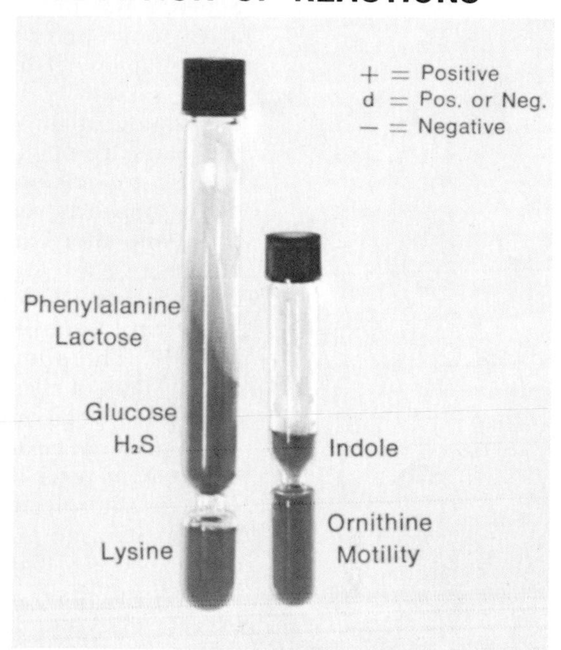

Figure 24–1 Types of tubes used in the r/b system and some of the reactions they can demonstrate. (Courtesy of Diagnostic Research, Inc., Roslyn, N.Y.)

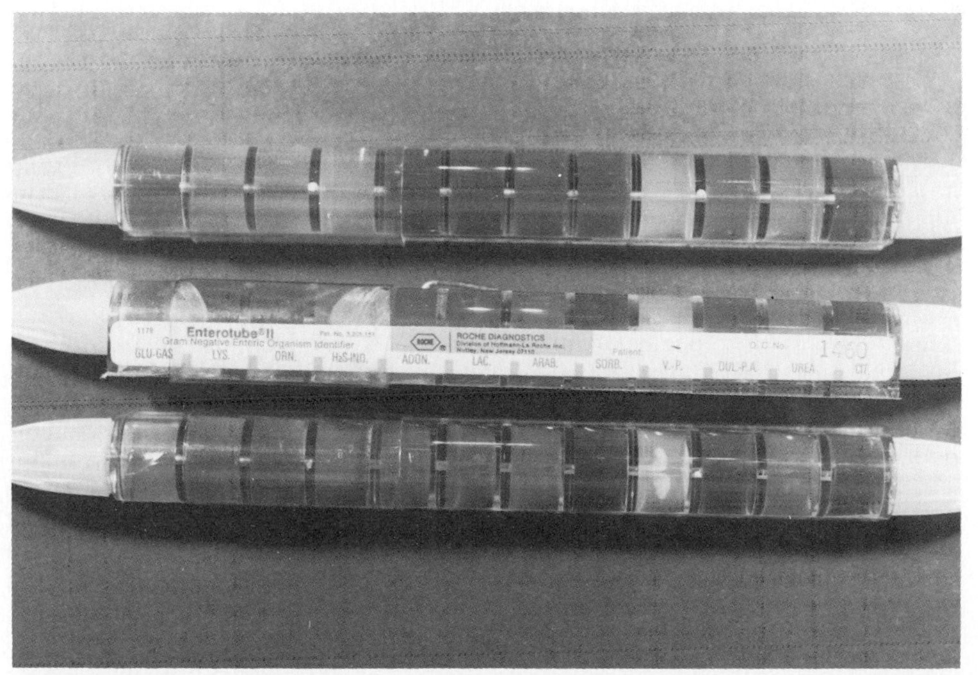

Figure 24–2 The Enterotube II is a prepared, sterile, multimedia tube for the rapid differential identification of gram-negative bacteria of the family Enterobacteriaceae. It is a product of Roche Diagnostics, Division of Hoffmann-La Roche, Inc., Nutley, N. J. The top Enterotube is the control and was not inoculated. The middle and bottom Enterotubes have been inoculated by touching a colony of *Serratia marcescens* with the needle and reinserting the needle through the glucose, lysine, and ornithine layers. The tip of the needle is left in the H_2S/indole compartment.

Although typhoid fever is no longer a common disease in the United States, it was one of the major causes of death five or six decades ago and still is, in some lands. The fact that a disease is now rare is no indication that it cannot again become the scourge that it once was. We are too apt to forget that if the sanitary precautions (sewerage systems, water purification plants, pasteurization of milk, food and restaurant inspections) arduously built up over the years are neglected, even for an hour, enteric infections of many sorts are waiting to strike—dysentery, cholera, amebiasis, hemorrhagic jaundice, epidemic hepatitis, and other once terrifying specters of disease and death. It is necessary, therefore, to know something not only of the causative organisms but also of the structure of environmental sanitation and the nature of the menace that it holds at bay.

There were only 510 cases of typhoid fever reported in the United States in 1980. The number of typhoid cases has remained relatively constant since 1970. Many of these infections developed during foreign travel and some were contracted in laboratories.[2]

Salmonella typhi has also been called *Salmonella typhosa, Eberthella typhosa,* or the typhoid fever bacillus. It was discovered by a German physician (Eberth) in 1880, in the spleens of typhoid victims at autopsy. It appears to be a peculiarly human parasite and is rarely or never found in lower animals, unlike many other species of *Salmonella.* Although the bacillus develops chiefly in the human body, it can easily survive in the outside world in water or food polluted by the excreta of typhoid patients or carriers. Experiments have shown that it may survive for several weeks in river water, 12 days in sewage, 4 months in butter, and 39 days in ice cream. It is not killed immediately by freezing, and if polluted water is used for making ice, or if these organisms are present in milk used for other dairy products, some of the typhoid bacilli may survive in the food for days. Impure ice and ice cream are therefore real sources of danger. Fresh cheese made of contaminated milk can transmit typhoid bacilli. The typhoid (and other *Salmonella*) bacilli appear to pass through the stomach uninjured and may multiply in the intestine, shortly afterward infecting patches of lymphatic tissue (Peyer's patches) in the intestinal wall.

Usually *S. typhi* (and occasionally other *Salmonella* species) appears in the circulating blood in the first week of the disease. Salmonellosis, especially that form of it caused by *S. typhi* (typhoid fever), therefore soon becomes a *septicemia.* In typhoid fever, ulcers of the small intestine are practically always present. The bacilli, especially *S. typhi* but also other salmonellas if they invade the blood, may localize in the periosteum (membranes covering bones), liver, gallbladder, bone marrow, spleen, or kidney, and may even cause meningitis or pneumonia. These complications, rather than the original intestinal involvement of the disease, are usually the cause of death.

Salmonella typhi begins to appear in the stools from the broken-down intestinal ulcers after the first week of the disease. By the second week it is passed in enormous numbers. Other *Salmonella* organisms are usually present in the stools from the first signs of enteritis; much less frequently, and later, they are found in the blood. Billions may be passed in a single bowel movement. The bacilli usually decrease during convalescence and finally die out. In about 10 per cent of patients they continue for 8 to 10 weeks, and in a small percentage, perhaps 2 to 4 per cent, they persist indefinitely. The first group of persons are *temporary convalescent carriers,* and the second group, *permanent convalescent carriers.* The carrier state may also occur temporarily, and possibly permanently, in persons who have no knowledge of having had an attack of salmonellosis, either typhoid or paratyphoid fever, since many infections are so

[2]Between 1977 and 1980 25 cases of laboratory-acquired typhoid fever were reported by the U.S. Public Health Service. Nine of these cases were contracted by students conducting laboratory exercises in medical technology or microbiology courses. These infections were not associated with outbreaks or carriers.

mild that they pass without notice. Typhoid Mary is a classic case of a permanent carrier. She was a cook whose trail of employment across the United States was followed by outbreaks of the typhoid fever that she left behind. She was finally forbidden by court order to obtain any employment in which she could transmit her infection to others.

Because of lesions in the urinary tract, *S. typhi* may appear in the urine about the fifteenth day of the disease, often in great numbers. It may continue for weeks, months, or, in rare cases, for years. There are thus urinary as well as fecal typhoid carriers. Species of *Salmonella* other than *S. typhi* rarely appear in the urine.

Transmission of Enteric Infection

All of the Enterobacteriaceae discussed in this chapter are discharged from the body in the feces; *S. typhi* also occurs in the urine. These organisms gain entrance to the body through the mouth by anything (food, water, milk, hands) polluted with sewage, fecal material, or urine. It is distressing to think how readily and frequently careless, unsanitary people ingest food contaminated by feces and/or urine and pollute the environment of other people every day of their lives. Each case of enteric infection, including those due to enteric viruses, enteric protozoa, and some helminths, comes from a previous case or from a carrier, in that the excretions of a diseased person or a carrier are in some way taken into the mouth of someone else who in turn becomes infected.

Enteric infections of any kind (viral, bacterial, and others) are frequently spread by close association with patients. Individuals who care (or really *don't* care) for these patients may soil their hands with contaminated feces and neglect to wash them properly before going to meals or preparing foods. Thus, these individuals may infect themselves or transfer enteric disease to others.

Carriers, which may be convalescent, temporary, or permanent, are extremely important in the spread of all enteric (and many other) diseases. These people seem to be perfectly healthy. Typhoid carriers have been reported who had recovered from the disease as long as 64 years before they were found to have *S. typhi* in their stools. In typhoid carriers the source of the bacilli in the feces is probably the gallbladder, in which the organisms may sometimes continue to live and multiply for years as comfortably as in a culture tube in the incubator. From the gallbladder they are carried to the intestine with the bile. There is no certain way known at present to prevent a patient from developing into a carrier. Methods of curing carriers include prolonged treatments with ampicillin, which is about 60 to 80 per cent effective. This and removal of the gallbladder often cure the infection.

Carriers of any intestinal infection are a continual source of danger if they have anything to do with caring for patients (especially infants) or with the preparation of food or milk. Many epidemics of salmonellosis (food infection) have been traced to them. The history of such an epidemic, which occurred in a California town some years ago, illustrates what may happen. A large number of typhoid cases broke out in a certain town within a few days of each other. It was found that all the patients had attended the same supper about three weeks previously, and the one dish they had all eaten was creamed spaghetti, insufficiently baked. The woman who had contributed this to the supper had had typhoid fever many years before, and examination of her feces by the State Board of Health proved that she was still a carrier. In this particular case she had contaminated the spaghetti with typhoid bacilli from her fingers while preparing the food. There were 93 cases of typhoid altogether, including the persons infected by the carrier and those cases that developed from contact with the first cases. About 10 persons died.

A less severe outbreak of *Salmonella* gastroenteritis was traced to a New Year's Eve party at which a buffet dinner was served. Among approximately

540 guests, 116 persons were interviewed for clinical and food histories. Of these, 51 (44 per cent) were ill with watery diarrhea, cramps, vomiting, headache, and fever; three persons were hospitalized. The mean incubation period was 22½ hours, and the illness lasted from two to four days. *Salmonella st. paul* was recovered from stool specimens from five of the patients. Food histories clearly incriminated a turkey salad made from commercially prepared, precooked frozen turkey, rolls, and salad ingredients, but unfortunately none of the served food was available for culture. The kitchen employees denied symptoms of gastroenteritis, but six of 14 stool cultures obtained from them yielded *Salmonella st. paul.* All six employees with positive cultures had helped prepare the turkey salad and admitted they had eaten a portion during preparation.

The vast majority of carriers of *Salmonella* and other enteric pathogens are never discovered. As a rule, unless a special search is made by laboratory methods, only those proven to have caused other cases of the disease are known.

Flies can act as distributors of salmonellas and other enteric pathogens. In areas without sewerage systems, flies contaminate their feet with infected feces in unscreened privies or other places of convenience to which they have access, and then crawl over food, depositing the bacilli on it, or drop into milk and inoculate it with *Salmonella* or *Shigella* (dysentery bacilli), which can readily multiply there. Rats, mice, and many other vertebrates, domestic and wild, frequently harbor and distribute various species of *Salmonella* (except *S. typhi*). Poultry are often vectors.

Milk and other foods contaminated by flies, by carriers of enteric bacterial pathogens, by people who have the disease, or by feces or urine of rats account for scattered outbreaks of food infections and occasionally for epidemics. *Salmonella* and *Shigella* grow rapidly in milk without changing its appearance or taste. Under improper conditions of milk production (which is the usual state of affairs in many parts of the world), the organisms may get into milk from the hands of a dairyworker, who is possibly a carrier, or by washing cans or bottles in polluted water. All of these seem to be remote dangers, yet the most extensive North American epidemic of typhoid fever of this century occurred in a large Northern city and was probably caused by its milk supply. In this epidemic there were 4,755 cases of typhoid fever and 453 deaths. It seems that for some reason certain sanitary precautions surrounding the production of the milk failed to function properly for a few hours, and the result was this appalling loss of life. Proper and constantly maintained control of public milk supplies and environmental sanitation is an absolute necessity.

Numerous epidemics have also been traced to ice cream, cheese, and butter made from milk containing *Salmonella* bacilli. Such an outbreak of typhoid fever among tourists in a fashionable European resort received newspaper publicity in 1963.

Water polluted with sewage was formerly the most frequent source of typhoid fever. Other enteric infections (viral, bacterial, protozoal, and helminthic) may also be transmitted by water. In the United States waterborne epidemics due to *S. typhi* are now virtually nonexistent. Most large North and South American cities have provided themselves with good water supplies. In the United States, typhoid fever or any other enteric infection that may occur in a city is usually due to a carrier or to sources other than water. In country districts and in villages, however, polluted water still accounts for some cases of this disease as well as other enteric infections. In such places wells and privies are often situated near each other and feces may be washed by rains or carried by animals over the ground and into an open well (Fig. 24–3). Enteric pathogens of various sorts may occasionally be carried into the well through the soil by underground seepage, especially through crevices in rocks. In the United States, streams, lakes, and springs are nearly always polluted.

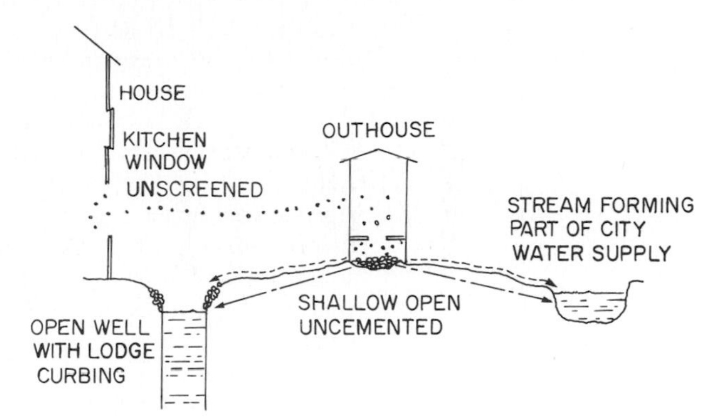

Figure 24–3 An unsanitary outhouse may pollute a city water supply (over-the-surface washings and underground seepage), a household well (surface washings and underground seepage), and a kitchen (flies).

Diagnosis

In most cases of salmonellosis and shigellosis with early diarrhea, the organisms can be found by the bacteriologist in the feces during the first days of the infection.

In typhoid fever, early enteritis is not as prominent, and the bacilli usually appear first not in the stools but in the blood. The nurse or health worker may be asked to assist in obtaining specimens for diagnosis and should have some knowledge of what is involved.

Blood Culture. The diagnosis of typhoid fever is made in the first week of the disease by means of a *blood culture.* In typhoid fever a blood culture may be positive before stool cultures become positive. Blood taken with a 5-ml syringe is transferred to 100 ml of bile broth or another favorable medium. After incubation, plates of nutrient or selective agar are streaked and incubated. Appropriate types of plating media used for blood cultures are Wilson-Blair bismuth sulfite agar, SS (*Salmonella-Shigella*) agar, EMB (eosin–methylene blue) agar, or MacConkey agar. Blood agar may be included. Characteristic colonies are sought and transferred in pure culture for further study. Usually the diagnosis may be made in this way only during the first week, since after seven or eight days the bacilli generally disappear from the blood. The typhoid bacilli may then be cultivated from the stools. In other forms of salmonellosis, blood cultures are less commonly positive, and then usually only in later stages.

Stool Culture. In diagnosing enteric bacterial infections in general, a drop of fresh feces is spread over the surface of agar plates. Except for blood agar, these contain certain substances (e.g., sodium desoxycholate and citrate, or dyes such as eosin and methylene blue) that inhibit nearly all organisms except Enterobacteriaceae. These media also contain lactose. The inhibitory substances serve two purposes: they inhibit bacteria other than Enterobacteriaceae and indicate which colonies have fermented the lactose and produced acid. When acid is formed from the lactose, the dyes in the medium change color. The acid-forming (lactose-fermenting) colonies can thus easily be recognized. Table 24–2 shows that none of the species of *Salmonella* or *Shigella* forms acid promptly from lactose. It is therefore easy to differentiate their colorless colonies at a glance from those of *Escherichia coli* and related nonpathogens, which ferment the lactose promptly and therefore have deeply colored (red or violet) colonies.

The desired colonies (easily recognized) are subcultured on triple sugar iron (TSI) agar slants for more study in pure culture (Table 24–2). As seen in Figure 24–4, TSI agar and other media and tests described there are used for further identification of genus and species of infective organisms.

A highly selective medium like Kauffmann brilliant green agar inhibits most other organisms, including *Escherichia* and *Proteus* and even some *Shigella,* but not most *Salmonella.* Other moderately selective useful media are Hektoen

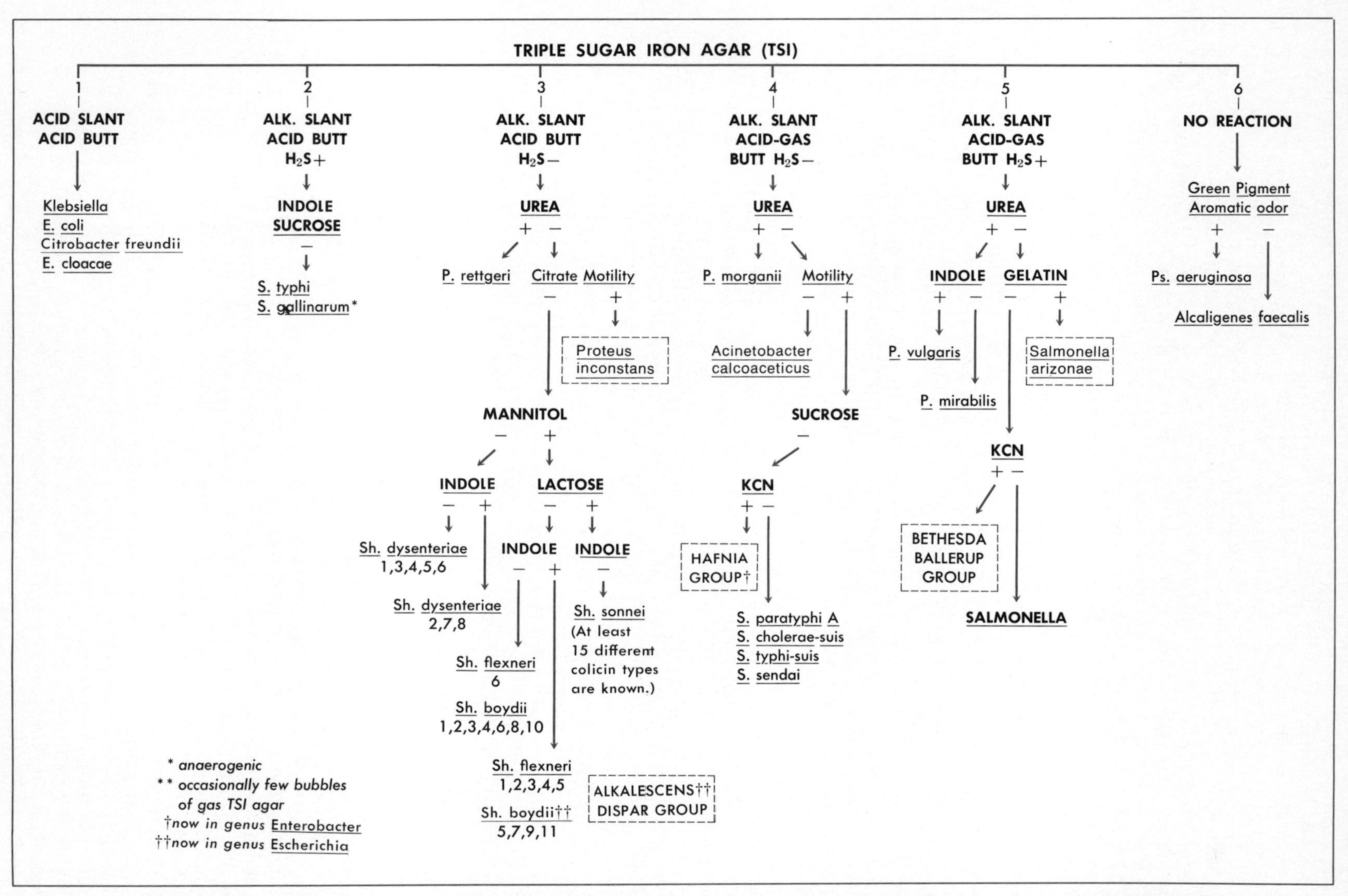

Figure 24–4 Key to partial differentiation of enteric bacilli. TSI agar contains three sugars—glucose, lactose, and sucrose—and phenyl red indicator and ferrous sulfate to demonstrate hydrogen sulfide production. (After Joklik, W. K., and Smith, D. T.: Zinsser Microbiology. 15th ed. New York, Appleton-Century-Crofts, 1972.)

enteric (HE) agar, desoxycholate-citrate medium, and xylose lysine desoxycholate (XLD) agar.

Urine Culture Isolation. About 25 per cent of typhoid fever cases yield *Salmonella typhi* from urine cultures. Other species of the Enterobacteriaceae—such as *Escherichia, Proteus, Klebsiella,* and other *Salmonella* species—may also be isolated from this source. It is recommended to centrifuge urine suspected of containing *Salmonella* and then to transfer the sediment to enrichment media. Selective and inhibitory plating media are also employed with success. Carriers are similarly detected by cultivating the bacilli from their feces, their urine, or both.

Agglutination Reactions. In salmonellosis and especially in typhoid fever, antibodies begin to appear in the blood about five days after onset. The antibodies most easily demonstrated are *agglutinins.* In most cases of salmonellosis an attempt is made to demonstrate the presence of these specific agglutinins in order to confirm the diagnosis This test is often called the Widal test after the French physician Georges Widal (1862–1929), who first published information on the subject. Diagnostic agglutination reactions have already been described.

The Widal reaction may lead to erroneous diagnosis unless the history of the patient is known, because the reaction is often positive in carriers, in persons who have had typhoid fever, and also in persons who have been injected with dead typhoid bacilli (typhoid vaccine) for the prevention of the disease. In those who have had typhoid fever or who have been immunized against it, a positive Widal reaction may sometimes be obtained during any febrile disease, such as malaria or influenza. Such a nonspecific Widal reaction is said to be an *anamnestic*[3] *reaction.* Similar nonspecific increases in specific antibody titers may also occur in other infectious diseases. Supplementary tests, using separately the antigens of the flagella (H antigens) and of the cell body (O or somatic antigens), are often employed to eliminate such errors. The tests are often called "H and O agglutination tests." Increase in O agglutinins is more suggestive of a current *Salmonella* infection; H agglutinins often appear anamnestically.

Agglutinins against the Vi antigens also occur in the blood of typhoid patients, especially during the presence of the bacilli in the body. Vi antigens are somatic antigens of a special type. They are often, but not invariably, associated with *Vi*rulence. The Vi agglutinins that they evoke in the serum of the patient or carrier tend to disappear after the bacilli dissociate from the carrier. The detection of Vi agglutinins is therefore of value in detecting carriers. Unfortunately, although generally furnishing good supporting or indicatory information, none of these tests is infallible. Associated with the Vi antigen is the V to W variation. W cells have lost the Vi antigen. The V type of organism is virulent and does not agglutinate with O antiserum. The W form agglutinates in O antiserum and it is avirulent; it cannot be typed with Vi phages. Pigmentations of solid media are also different for V and W colonies.

The Kauffmann-White Scheme. It is interesting to note that there are now over 1,900 known kinds of *Salmonella,* each differing in its content of various antigens. One aspect that may be confusing is that each type of *Salmonella* has a species name, usually representing the place where it was first isolated (Table 24–4)—*S. manhattan, S. california, S. heidelberg, S. mikawasima,* and *S. homosassa.* Many are grouped and classified according to a numbered and lettered antigen system, called the *Kauffmann-White scheme* (see *Bergey's Manual,* eighth edition, pages 301 to 317 for a listing of *Salmonella* serotypes as based on the Kauffmann-White scheme). Three ecologic divisions of *Salmonella* are (1) *Salmonella* serotypes pathogenic for humans (*S. typhi,* the paratyphoid organisms), (2) serotypes of nonhuman hosts, and (3) *Salmonella* serotypes without specific hosts. Most human infections are caused by the third group,

[3]From Greek word *anamnesis,* to remember or recall; thus, any antigenic stimulus may reactivate or *recall* a previous antigenic stimulus.

Table 24–4. Some Serotypes of the Kauffmann-White Antigenic Scheme

		H ANTIGENS*	
TYPE	O ANTIGENS	*Phase 1*	*Phase 2*
	Group A		
S. paratyphi A	1, 2, 12	a	—
	Group B		
S. tinda	1, 4, 12, 27	a	e, n, z_{15}
S. schottmuelleri	1, 4, [5], 12	b	1, 2
S. typhimurium	1, 4, [5], 12	i	1, 2
S. heidelberg	4, [5], 12	r	1, 2
	Group C_1		
S. hirschfeldii	6, 7, Vi	c	1, 5
S. thompson	6, 7	k	1, 5
	Group C_2		
S. newport	6, 8	e, h	1, 2
	Group D_1		
S. typhi	9, 12, [V_1]	d	—
S. enteritidis	1, 9, 12	g, m	—
S. sendai	1, 9, 12	a	1, 5
	Group E_1		
S. oxford	3, 10	a	1, 7
S. london	3, 10	1, v	1, 6

*After Bailey, W. R., and Scott, E. G.: Diagnostic Microbiology. 3rd ed. St. Louis, C. V. Mosby Co., 1970.

[ ] May be lacking.

which accounts for about 80 per cent of all *Salmonella* diseases. The classification is of great interest to epidemiologists and other professional experts, but all *Salmonella* and *Shigella* are transmitted in the same general ways, and diseases caused by them are prevented by the same general precautions and require the same general type of treatment.

Food Infection: Paratyphoid Fever

In the United States, important causes of food-borne enteric *infection* (not food *poisoning*) are *Salmonella schottmuelleri* (previously called *S. paratyphi B*), *S. choleraesuis, S. enteritidis, S. typhimurium,* and others. Less common in this country, but often found in Europe and elsewhere, is *Salmonella paratyphi A*.

The usual history of a food infection outbreak due to *Salmonella* parallels the experience with spaghetti previously recounted. A group of people are all infected at the same meal by a certain dish of food prepared by an unsuspected *Salmonella* carrier and left in a warm room for some hours, thus incubating the bacilli. Twelve to 24 hours after the food has been eaten, there develop cases of gastroenteritis of various degrees of severity, ranging from mild diarrhea (common) to severe (uncommon) or fatal (rare) attacks.

As previously noted, hogs, cattle, poultry, and eggs are sometimes infected with *Salmonella* bacilli, and if the improperly cooked flesh or eggs are eaten, infection is apt to occur. Dealers and cooks may be infected by handling the raw meats. Animals such as dogs, rats, and mice are often carriers of paratyphoid bacilli, especially *Salmonella typhimurium,* and thus can be vectors of food infections.

Salmonella infections are frequent. Excluding typhoid fever, 33,715 cases of salmonellosis were reported in the United States during 1980, compared with 16,841 in 1966. Why should the number of cases of salmonellosis double in 14 years, when the population of the United States only increased 13.66 per cent? Let us hope that better reporting is part of the reason; perhaps improvements in diagnosis account for the rest. Some epidemiologists have estimated that as many as two million cases of *Salmonella* infections may occur in the United States each year.

Immunization Against Salmonellosis

It is possible to produce a certain degree of *artificial active* immunity to salmonellosis by giving injections of dead typhoid or paratyphoid bacilli (bacterin or vaccine). A common procedure is to give three doses of such vaccine one week apart, the first of one billion bacilli, the second and third of two billion. The vaccine usually contains a mixture of typhoid and paratyphoid bacilli and is often called "TAB" vaccine, in reference to *Salmonella typhi, S. paratyphi A,* and *S. paratyphi B* (now known as *S. schottmuelleri*). The injections are often followed within 24 hours by considerable swelling and soreness of the arm. There are sometimes general symptoms, headache, nausea, and a rise in temperature, but these are not serious and they pass off in a day or so. The most suitable time for giving the inoculations is late Friday or Saturday afternoon; the person should remain quiet for the rest of the day and should not undertake hard work on the following day.

The immunity is only relative and is probably transitory. For this reason a booster dose of 0.1 ml given intradermally every three years, or more often if indicated by special conditions of exposure, is recommended by many to bolster waning immunity.

The 1981 U.S. Public Health Service recommendations advised against TAB vaccine in combined form, since it was found to cause undue reactions. Typhoid fever vaccine should be given alone. The effectiveness of paratyphoid A vaccine also remains a point of contention.

Vaccination is not at any time an absolute protection against salmonellosis, since a recently vaccinated person may develop the disease if given a large dose of virulent bacilli. The treatment will, however, confer some protection against the moderate doses that one is likely to get in ordinary chance infection. A vaccinated person must not relax precautions against polluted water, food, milk, and patients and should be as scrupulously careful about washing hands and observing all other measures for self-protection as though no inoculation had been given.

Clinical Management

Antibiotic treatment of *Salmonella* infection is often contraindicated, since plasmid–mediated antibiotic resistance even to ampicillin and chloramphenicol may exist.

Shigellosis

Dysentery is a pathologic term for any form of diarrhea due to intestinal irritation. It may have any one of several causes, from eating unripe apples to Asiatic cholera.

Bacillary Dysentery. This is an infectious enteric disease caused by dysentery bacilli (*Shigella*). The term *bacillary* is used to distinguish dysentery due to these bacteria from dysentery due to protozoa, commonly called *amebic* dysentery or *amebiasis,* and from viral and other forms of dysentery. A shorter and more specific term for bacillary dysentery is *shigellosis.*

Shigellosis may vary greatly in severity, ranging from a very mild and transitory intestinal disturbance, to a more acute inflammatory gastrointestinal involvement, to severe and fatal dysentery. In severe cases, the dysentery bacilli cause ulcers of the large intestine, the appearance of blood, mucus, and pus in the stool, and the most intense and painful diarrhea, nausea, fever, dehydration, and toxic symptoms. *Shigella* organisms do not, as a rule, invade the blood, like some salmonellas, but are usually limited to the large intestine. Shigellas form a potent *endotoxin* that causes the diarrhea and the consequent prostration and loss of weight associated with the disease. *Shigella* organisms are cast off in large numbers in the stools and may persist for months in convalescents.

Most cases of shigellosis in adults in the United States are relatively mild. Many originate from persons with a transitory diarrhea so mild that they do not suspect the nature of their disease, take no special sanitary precautions, and ascribe their trouble to "something they ate." The same infection may prove fatal to young children. Adults with diarrhea should, therefore, avoid contact with children.

Dysentery Bacilli. There are four main groups or types of dysentery bacilli, with more than 30 known serotypes. The first species to be described, now called *Shigella dysenteriae,* was discovered by a Japanese physician, Shiga, during an epidemic in Japan in 1898. A similar type, discovered in Europe, is sometimes called *Shigella ambigua* or Schmitz bacillus. Another species was discovered in 1906 by an American physician, Flexner, in the Philippine Islands and is often called the "Flexner dysentery bacillus" or *Shigella flexneri.* A number of similar varieties have since been discovered that are sometimes collectively called *Shigella paradysenteriae.* Still another type, common in the United States, first discovered by Duval and described by Sonne, is now called the Sonne dysentery bacillus (*S. sonnei*). A large group, headed by *S. boydii,* comprises species like that first discovered by a British worker, Boyd.

To simplify matters, the four main groups now recognized are: *dysenteriae, flexneri, sonnei,* and *boydii.* All species in these groups are pathogenic to humans. They may be differentiated by cultural (Table 24–5) and immunologic tests. For purposes of treating the patient, differentiation of the organisms generally is not necessary and they may be considered as a single species.

Like other members of the Enterobacteriaceae, *Shigella* species grow well in simple culture media and are easily killed by heat, drying, and disinfectants. They remain alive outside the body on food, in clothing, or in the soil or water for only a short time, as compared with *Salmonella*; however, they can grow in milk and other foods without changes in appearance or taste.

Transmission. Shigellosis is spread in much the same manner as salmonellosis—that is, by feces, fingers, flies, milk, foods, and any articles that have been in contact with the feces of a patient or carrier. *Shigella* does not survive for long periods in feces or sewage. Thus, although waterborne epidemics of dysentery occur, they are less frequent than waterborne epidemics of typhoid fever. *Shigella* species are not commonly found in animals, which is equally true for many species of *Salmonella.* In the United States 19,041 cases of shigellosis were reported in 1980, 8.4 cases per 100,000 population.

Diagnosis. Diagnosis of dysentery in the laboratory is made in much the same manner as the diagnosis of salmonellosis except that blood cultures are of little value because dysentery bacilli rarely invade the blood. Repeated stool specimens are often necessary. They must be fresh. (Why?) Agglutination tests with patients' serum are not very satisfactory in shigellosis. Stool leukocyte smears showing many polymorphonuclear leukocytes are often useful in diagnosis, even though they are not specific.

Dysentery has not disappeared from modern communities as has typhoid fever. Like some mild forms of salmonellosis, shigellosis is constantly with us. Both are spread largely by the feces-soiled hands of unknowing and unsuspected ambulatory cases and carriers or by water from contaminated wells.

Toxin Production. It is thought that *Shigella* produces endotoxins that

Table 24–5. Biochemical Differentiation of Genus *Shigella**

Acid only in glucose**				
Mannitol negative		Mannitol positive†		
Indole negative	Indole positive	Lactose negative		Lactose positive‡
		Indole negative	Indole positive	Indole negative
Sh. dysenteriae 1	*Sh. dysenteriae* 2	*Sh. flexneri* 6	*Sh. flexneri* 1	*Sh. sonnei*
3	7	*Sh. boydii* 1	2	
4	8	2	3	
5		3	4	
6		4	5	
9		6	*Sh. boydii* 5	
10		8	7	
		10	9	
		12	11	
		14	13	
			15	

*From Finegold, S. M., Martin, W. J., and Scott, E. G.: Bailey and Scott's Diagnostic Microbiology. 5th ed. St. Louis, C. V. Mosby Co., 1978, p. 164; adapted from Edwards, P. R., and Ewing, W. H.: Manual for Enteric Bacteriology. Atlanta, National Communicable Disease Center, 1951.

***Sh. flexneri* 6 varieties may be aerogenic (Newcastle and Manchester).

†Certain cultures of *Sh. flexneri* 4, *Sh. flexneri* 6, and *Sh. boydii* 6 may not produce acid from mannitol.

‡Lactose fermentation is delayed with *Sh. sonnei* (usually 4 to 7 days). Closure of the fermentation tube with a tighly fitting stopper will hasten the reaction.

destroy epithelial cells. The pathogenic process appears to be somewhat complex. *S. dysenteriae* type 1 also secretes a potent enterotoxin, which may be identical with *"Shiga neurotoxin."* This is heat-labile and causes paralysis, diarrhea, and death when injected into animals.

Prevention of Shigellosis. Methods of preventing the spread of bacillary dysentery in adults are generally the same as for salmonellosis, except that dysentery vaccines, still being developed, will probably be oral, live, and bivalent (against *S. flexneri* and *S. sonnei*). Resistance develops quickly to sulfonamides, tetracyclines, and other antibiotics used to eliminate *Shigella*. Presently, ampicillin appears to be the best treatment.

Prevention of Shigellosis in Infants. This depends on encouraging breast feeding, on teaching mothers and hospital personnel correct methods of caring for babies and preparing their food, and on the general improvement of milk and water supplies and pasteurization of all milk. The breast-fed baby not only gets milk that is especially suited to his or her needs but also gets it in nearly bacteria-free condition as long as the breasts are kept clean and the mother does not contaminate them with her hands.

For artificial feedings, there are at least four forms of commercial formula available besides homogenized milk; two of these have to be mixed with sterilized or disinfected water (boiled for 25 minutes). These preparations are (1) powder, (2) liquid concentrate in a can, (3) canned formula, which requires no additional water, and (4) ready-to-use disposable bottles, to which only sterilized nipples need to be attached. For the first three preparations, bottles and nipples still have to be presterilized. In most hospitals, formula in prepackaged bottles fitted with nipples is used: all one has to do is break the seal and give the bottle to the baby. Warming the milk or formula before giving it to the baby is not considered necessary and has been almost entirely discontinued, but the milk must be brought to room temperature. Such unheated formula may be kept at room temperature for four hours safely. The feedings should be prepared by a person with clean hands, and the bottle, after it is prepared, should be kept away from flies and dirt and refrigerated until used. Hospital nurseries generally have very strict procedures for preparing and handling baby formulas. Aseptic technique during the preparation of food for babies in hospitals is essential.

Babies should be kept away from anyone having an intestinal disturbance. A person who is caring for a child with summer diarrhea or "summer complaint" should not prepare food for other children or for adults. The baby's crib and carriage should be screened during the fly season. The diapers of infants suffering from diarrhea should be disinfected at once by placing them in a bucket with a 2 per cent saponated cresol solution or another good disinfectant. Disposable, absorbing diapers with plastic liners are now available in most drug and department stores.

Children's clinics have done much to combat dysentery in infants. Community health nurses and other health workers connected with these clinics visit homes and teach mothers correct methods of preparing baby food and caring for infants.

One practice should be discontinued: it is the use of the unsanitary pacifier. If this behavior is not established in infancy, it is not necessary later on to break a bad habit. Perhaps fewer people would smoke if oral gratification had not been overly encouraged in early childhood.

The Escherichieae

The organisms of this tribe have the general properties of the Enterobacteriaceae. *Escherichia* may be differentiated from *Salmonella* and *Shigella* by various biochemical tests, especially by rapid (24 to 48 hours) fermentation of lactose, as shown in Table 24–2, and by the characteristic IMViC, reactions (+ + − − for *E. coli*). This means that typical *E. coli* gives positive Indole and *M*ethyl-red tests and a negative reaction in *V*oges-Proskauer[4] and *C*itrate tests.[5]

Various species of *Escherichia* are of special importance to medical and sanitation personnel because certain serotypes cause enteric and urinary tract infections and are transmitted in water and milk. The genus *Escherichia* contains numerous antigenic groups based on O, H, and K antigens (see Chapter 28). *Escherichia coli* occurs in enormous numbers in normal feces and is widely distributed in the intestinal canal of animals and humans. Ordinarily it does no harm.

Pathogenic Escherichia. The majority of *Escherichia coli* in the gastrointestinal tract are not pathogenic there, but they may cause disease when displaced to the urinary tract, the meninges, or the liver and gallbladder routes. A study by Rantz points out that about 45 per cent of urinary infections are acquired in the hospital as the result of catheterization; other studies give even higher percentages. Nonintestinal *E. coli* strains that may cause diarrhea are enterotoxigenic (which may produce "traveler's diarrhea"), enteroinvasive (resembling shigellosis), or enteropathogenic (causing infant diarrhea).

These pathogenic *E. coli* can be distinguished from less harmful varieties by immunologic studies (similar to those used in establishing the Kauffmann-White scheme) of their antigenic structure. Some of these strains of *E. coli* are designated as O119:B14,O111:K58:H2, and so on. The numbers and letters refer to antigens in them. Now 14 enteropathogenic *E. coli* serotypes are generally recognized. Many more will be discovered.

These organisms are particularly troublesome in children's institutions and

[4]Production of red color with NaOH in peptone-glucose broth, due to acetylmethylcarbinol formation.

[5]Utilization of citrate as a sole source of carbon in a mineral medium.

nurseries. They are spread about by hands and fomites, as are other enteric pathogens, and at times they are very difficult to eradicate. Only the most scrupulous attention to clean technique keeps them under control.

One of the difficulties is the handling of diapers in such situations. In one useful study it was found that for the small hospital or the home a three-step process using home laundry equipment yields an adequate supply of clean, dry, noninfectious diapers, as follows:

1. On removing the diaper, drop it carefully into a can containing enough Watkin's solution[6] (4 oz to 1 gal of water) to cover it, close tightly, and soak at least two hours. (Probably 2 per cent saponated cresol would serve as well.)

2. Wash in an automatic machine that rinses, washes with detergent, and rinses three times.

3. Dry immediately in an automatic dryer.

To overcome the problems associated with diaper handling almost all cities in the United States now have diaper services for mothers and institutions. These commercial services pick up the soiled diapers in plastic bags and return sterile, fresh diapers, at quite reasonable fees. However, disposable diapers, now in ever increasing use, will in time eliminate all reusable diaper services.

Sanitary Significance of Escherichieae. Since *Escherichia coli* is always present in feces, and since other species of the Escherichieae tribe frequently accompany it and closely resemble it, this tribe is frequently referred to as "the coliform (i.e., *E. coli*-like) group." The presence of any of them in water, milk, or food in considerable numbers strongly suggests pollution with feces. They are easily cultivated and recognized and usually remain alive in food and water for considerable periods of time. The coliform group is therefore commonly sought after in bacteriologic examinations of water, milk, and food as evidence of fecal pollution (and potential infection), rather than *Salmonella* or *Shigella*. These latter may be present only intermittently and in small numbers, and therefore may be difficult or impossible to find. Also, sewage pollution can introduce enteric viruses, protozoa, and other intestinal pathogens besides *Salmonella* and *Shigella*.

Enterotoxic E. coli. Some strains of *E. coli* produce two kinds of soluble exotoxin. One is heat-labile and antigenic (LT—labile toxin) and acts somewhat like cholera toxin, causing severe diarrhea in humans; the other is heat-stable (ST—stable toxin). An organism may have one or both of these toxins. These enterotoxic strains should be considered a distinct group of enteropathogenic *E. coli*: they differ serologically from the common H and O varieties that do not produce any soluble exotoxin.

Examination of Water. Since the coliforms rapidly ferment lactose with the production of acid and gas, measured portions of the suspected water (or other material to be examined) are put into broth containing lactose. If gas is present after 24 or 48 hours of incubation, it is not necessarily due to coliform species. Several other organisms also produce gas from lactose. Plates or tubes of *selective* medium are therefore inoculated from the lactose-broth tubes showing gas. After incubation, colonies resembling those of the coliform group are selected for pure-culture study and are subjected to simple biochemical tests that easily identify them (Table 24–2).

Such tests, or specific filtration procedures giving similar results, are carried on in every public health laboratory. For official purposes methods are prescribed by the American Public Health Association and affiliated groups. These tests usually constitute a part of the laboratory work in a course for which this textbook is written; thus, the student will receive proper instructions in using *E. coli* as an indicator of fecal pollution.

Besides the coliform group, enteric streptococci (*S. faecalis* and related enterococci) may also be used as test organisms in detecting fecal pollution.

[6]A mixture of cresols with isopropyl alcohol and soap.

Infections of the Genitourinary Tract

E. coli and several related species of bacteria found in feces can be the cause of serious infections of the genitourinary tract (Table 24–6).

Proteus species (Tribe Proteeae) are a genus of Enterobacteriaceae occasionally found in the normal intestine but more commonly present in sewage and also decomposing organic matter of nonfecal origin. *Proteus* is sometimes found in cases of infant diarrhea. It is distinguished from all other Enterobacteriaceae by rapid (two to four hours) decomposition of urea (production of the enzyme urease) and other characteristics shown in Table 24–2 and Figure 24–4.

Pseudomonas aeruginosa in many respects resembles *Proteus*, as just described, but differs in having polar flagella and in some other details. It is not one of the Enterobacteriaceae, but is classed in the family Pseudomonadaceae. It is characterized especially by producing two pigments: a water-soluble, yellow, fluorescent pigment, and a chloroform-soluble, sky-blue pigment called pyocyanin (*cyan* is from a Greek word meaning blue; compare with *cyan*otic). The mixed blue and yellow pigments of *Ps. aeruginosa* give cultures of the organism or the pus in which it is growing a blue-green-color, hence the older name *Bacillus pyocyaneus* (blue-green pus). Like *Proteus*, it is a frequent invader of injured tissue. In fact, *Pseudomonas* infections are becoming more and more of a problem in hospitals, responding to none of the older antibiotics. Some newer antibiotics give promise of controlling *Ps. aeruginosa*, e.g., tobramycin, gentamicin, butyrosine.

Proteus species and *Ps. aeruginosa* are often found on the normal skin in the perianal regions and around the genitourinary openings. As a result, they are frequently introduced into the urethra and bladder and adjoining organs by instruments such as cystoscopes and catheters and during surgical operations in these regions, in spite of the most careful technique. They may also infect in the absence of instrumentation, especially in patients with strictures or other structural abnormalities of the genitourinary tract.

Many strains of these bacteria have, through long contact with most antibiotics during treatment of patients, become highly resistant to these drugs. Consequently, they cause very stubborn and troublesome infections (such as cystitis, pyelitis, urethritis, and bacteremia) with a fairly high fatality rate. Antibacterial chemicals such as the nitrofurans, methenamine, mandelic acid, and nalidixic acid are sometimes effective in such conditions.

A useful system to assist the medical technologist in identifying species of *Pseudomonas* or other oxidative-fermentative gram-negative rods is the OXI/FERM TUBE. It is similar to the Enterotube and just as easy to use.

Table 24–6. Bacteria Common in Infections of the Genitourinary Tract

BACTERIUM	MORPHOLOGY	GRAM STAIN	MOTILITY	SPORES	COMMON HABITAT	LACTOSE	GLUCOSE	UREASE	GELATIN	PIGMENT	OTHER DISTINCTIVE CHARACTERISTICS
Escherichia coli	rod	neg.	+	−	intestine	⊕[1]	⊕	−	−	none	
Enterobacter aerogenes	rod	neg.	+	−	intestine	⊕	⊕	±	−	none	
Proteus sp.	rod	neg.	+	−	intestine	−	⊕	+	+	none	spreading growth on agar plates
Pseudomonas aeruginosa	rod	neg.	+	−	intestine	−	−	−	+	blue & yellow	
Enterococci (*Streptococcus zymogenes*)	coccus	pos.	−	−	intestine	+[2]	+	−	+	none	beta hemolysis (group D)
Streptococcus faecalis	coccus	pos.	−	−	intestine	+	+	−	−	none	alpha hemolysis ("viridans")
Staphylococcus aureus	coccus	pos.	−	−	skin, nares	+	+	−	+	yellow	

[1]Produces acid and gas.
[2]Produces acid or some other distinctive reaction indicated by column heading.

ASIATIC CHOLERA

Vibrio cholerae. In *Bergey's Manual*, Part 8 (gram-negative facultatively anaerobic rods) contains two families: the Enterobacteriaceae and the Vibrionaceae. *Vibrio cholerae* (Fig. 24–5) belongs to the Vibrionaceae and is a strongly aerobic, facultatively anaerobic, short comma-shaped rod with a single polar flagellum. *V. cholerae* resembles *Salmonella* in several respects (Table 24–2) and is included here as an important and dangerous enteric pathogen. *Vibrio cholerae* reduces nitrates to nitrites. Grown in nitrate-peptone broth, it gives a *cholera-red* color when sulfuric acid is added. The reaction is due to production of nitroso-indol, which gives a red color in the presence of H_2SO_4. Once thought to be distinctive of *V. cholerae* and therefore called the "cholera-red test," the color reaction is now not considered to be of diagnostic significance. Unlike most of the Enterobacteriaceae (except *Enterobacter cloacae* and some *Proteus* species), *V. cholerae* is strongly proteolytic.

V. cholerae causes Asiatic cholera, the chief symptoms of which are intense diarrhea, prostration, and emaciation. Death often occurs within two or three days of onset, but may result within hours unless treatment is started promptly. It is not unusual for a patient with severe cholera to lose 20 liters of water per day. The treatment, consisting of infusion of water and electrolytes into the veins, results in quick and remarkable relief of the symptoms of the disease. Oral tetracycline is also helpful. Without treatment the death rate is between 20 and 50 per cent. More recently, it has been found that fluid may be replaced orally following intravenous rehydration.

The vibrio may be cultivated on the same sort of nonselective media as are used for Enterobacteriaceae. An excellent medium, thiosulfate citrate bile salts sucrose (TCBS) agar, is available commercially. The pH should be near 8.6. Different from other fecal organisms, vibrios have a distinct opaque yellow colonial morphology on TCBS. On meat extract agar, the organism forms translucent gray colonies. It is nonspore-forming and, like Enterobacteriaceae, easily killed by boiling, pasteurization, and standard disinfectants. Finkelstein and his colleagues demonstrated that cholera vibrios produce an "exo-enterotoxin," which they called choleragen. This toxin is an antigenic protein com-

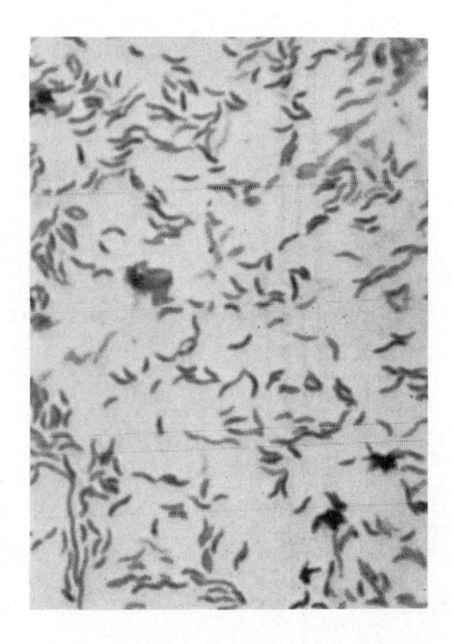

Figure 24–5 A typical *Vibrio*, morphologically like *V. cholerae*, cause of Asiatic cholera. Note the short, comma-shaped rods and the long, wavy forms. The particular species shown here is *V. fetus*, cause of infectious abortion in farm animals. It is closely related to *V. cholerae* (× 1,500). Courtesy of Dr. Wayne Binns, Department of Veterinary Science, Utah State Agricultural College, Logan, Utah.)

posed of noncovalently associated subunits, A and B, which are responsible for the biologic activity and for the binding of the toxin to host-cell membrane receptors, respectively. The toxin is a potent and broad-spectrum activator of the host-cell enzyme adenylate cyclase, and the symptoms of cholera result from the elevated levels of cyclic adenosine monophosphate (cAMP). This toxin is an antigenic protein that appears to be a very potent permeability factor (compare enterotoxin *E. coli*). It has been shown to produce clinical manifestations of cholera in laboratory animals as well as in human subjects in the absence of living vibrios. A toxoid has been prepared from the toxin and is presently under trial in southeast Asia and Pakistan. Older vaccines, using killed cholera vibrios, have not been very effective.

Cholera and Its Transmission. Asiatic cholera is transmitted from person to person in the same manner as *Salmonella typhi* and its spread is prevented by the same measures. Like *S. typhi*, it is a distinctively human pathogen. Large outbreaks of cholera are usually waterborne. Like the dysentery bacilli, cholera vibrios do not ordinarily invade the blood, but generally remain within the intestine. The vomitus of cholera patients is infectious, and soiled clothing and bedding remain infectious while damp. This is important in concurrent disinfection. In endemic[7] and epidemic[8] areas food-borne infection is common, and healthy carriers are said to be frequent sources of infection, though there is disagreement on this point. Very mild cases occur that are unrecognized sources of infection, as are carriers.

Cholera has long been present in India and other Eastern countries. In the past it has many times been carried from the East along caravan and shipping routes over almost the entire earth. During the nineteenth century there were five great epidemics, some of which reached the United States. Millions died. Epidemics of cholera occurred in the Philippines in 1961 and in Taiwan in 1962. The widespread El Tor-type cholera began in Southeast Asia during 1961. Clinically, it seemed to be no different from the classic cholera type. The causative El Tor *Vibrio*[9] can be differentiated from the classic *V. cholerae* only by laboratory methods. It is commonly referred to as the El Tor biotype. Other types of *Vibrio* less virulent than *V. cholerae* or the El Tor *Vibrio* are known, e.g., the Ubon El Tor type from Thailand and some enteritis-causing, antigenically different groups. Numerous saprophytic species have been described; some species are pathogenic for sheep and other animals.

There is no known endemic cholera at present in Europe or the Western Hemisphere. This is true only because of constant watchfulness by public health agencies in this and other countries. Still, in the first eight months of 1970 the number of registered deaths due to cholera listed in the World Health Statistics Report were 3,577 in Ghana, 19,280 in India, 8,066 in Pakistan, 293 in Brazil, and 5 in the United States—the last being cases imported from Vietnam. In 1981 two cases, with one fatality, occurred on the Gulf Coast of Texas, possibly due to shrimp caught in the area. New outbreaks in Jordan and other Middle Eastern countries occurred in later 1977. By September there were 2,217 cases in Syria, 397 in Jordan, 26 in Lebanon, 17 in Saudi Arabia, 11 in Iran, and 3 in Iraq. The Gilbert Islands reported a total of 380 cases and 17 deaths.

Prophylactic vaccination with killed cultures of the cholera vibrio may be of some value. The immunity lasts for six months or less and is not very potent. The newer cholera toxoid may prove more effective. Antibiotics help but are not curative. An orally administered living attenuated vaccine, developed by Finkelstein and his associates, has been shown to elicit immunity in human

[7]Always present in a population at a relatively constant rate of prevalence.

[8]An epidemic is a sudden marked increase in prevalence of a disease, which later declines to the usual rate of prevalence.

[9]El Tor is a town on the west side of the Sinai peninsula.

volunteers. Other enterotoxic enteropathies, e.g., diarrhea caused by enterotoxigenic *E. coli*, have been shown to be caused by toxins related immunologically and by modes of action to the cholera enterotoxin.

Other Vibrios. *Vibrio parahaemolyticus* is closely related to *V. cholerae* and has come into prominence since it was first suspected by the Japanese microbiologists Fujino, Okuno, and associates as a cause of a severe form of gastroenteritis contracted by eating *shirasu* (a form of fish eaten in Japan), later also discovered in the United States and elsewhere. The disease is associated especially with the eating of marine and estuarine seafoods. Most strains differ from *V. cholerae* in these respects: failure to grow in 1 per cent tryptone broth without about 8 per cent NaCl; failure to ferment sucrose; production of soluble hemolysin for goat erythrocytes; and growth at 42 C. In many ways *V. cholerae* and *V. parahaemolyticus* resemble Enterobacteriaceae but do not produce indophenol oxidase. *V. parahaemolyticus* is found in the stools of patients and in crabs and shrimp.

CAMPYLOBACTERIOSIS

Campylobacter jejuni is a gram-negative, comma-shaped, short rod. In this way it somewhat resembles the cholera organism. *Campylobacter* has only recently been awarded the attention it seems to deserve. It is now believed that *C. jejuni* is the most common cause of bacterial diarrhea in humans, especially children; this high incidence exceeds all infections attributed commonly to *Salmonella* or *Shigella*. Clinically, *C. jejuni* enterocolitis can mimic either granulomatous or idiopathic ulcerative colitis. It has been suggested that stool cultures from patients with diagnosed colitis should be routinely examined for *C. jejuni*. Infections with *C. jejuni* have been shown to occur with great frequency in immunosuppressed individuals and in those with other illnesses. Although transmission of campylobacteriosis is still not very well undertood, it is clear that the portal of entry for the organism must be oral. The organism has been found in water and in milk and must be transmissible from sick animals as well. Much attention will be given in future clinical studies to *Campylobacter jejuni*, which may or may not confirm the present evaluation of the high incidence of its pathogenicity.

OTHER INTESTINAL ORGANISMS

The microflora of the human intestine include many other organisms besides Enterobacteriaceae. *Bacteroides* species (gram-negative short rods that are obligate anaerobes, see Chapter 37) are the predominant microorganisms in normal stool. Anaerobic streptococci and staphylococci, together with clostridia, are easily isolated, and others like yeasts and more than 65 species of intestinal viruses have been identified.

PREVENTION OF ENTERIC INFECTIONS

It is clear, from what has been said, that infectious organisms from the intestinal tract—such as cholera vibrios, *Salmonella*, *Shigella*, bacilli that cause undulant fever and intestinal tuberculosis, the viruses of poliomyelitis and viral hepatitis, pathogenic protozoa, and some worms (helminths)—can be controlled most effectively at the bedside as they come from the patient (or healthy carrier) and before they are scattered.

The feces and urine must be disinfected (or otherwise disposed of in a hygienic manner), as well as everything that could possibly be contaminated with them, such as bed linen, clothing, and wash water. Bedpans must be given special attention immediately after use by the patient. If a mechanical disinfecting flusher or other approved means of disposal is not available, the feces or urine should be transferred to a covered vessel and thoroughly disinfected with saponated cresol, chloride of lime, or strong chlorine laundry bleach. The bedpan is flushed and boiled if possible or soaked in disinfectant solution, after which it is thoroughly washed and returned to the patient. Disposable paper (papier-mâché) bedpans are now available: special apparatus is required for their disposal. The patient's dishes should be boiled or disinfected in an electric dishwasher. Attractive disposable dishes are readily available and currently used in many hospitals that have patients with infectious diseases.

The patient must be isolated as in any case of infectious disease: no visitors should be allowed in the sickroom. If the patient is at home and if milk is taken from a dairy using bottles, the milkman should be notified not to collect empty bottles, as they may carry the infectious organism back to the dairy. The person who cares for the patient should not prepare food for anyone else, even for himself. Cholera, salmonellosis, and shigellosis patients should, if possible, be treated in hospitals, in isolation quarters designed to exclude flies. The presence of flies in or around a hospital is evidence of carelessness.

The greatest danger of infection is not, however, from the known *Salmonella* or *Shigella* or cholera patient, but from a person in an infectious stage before a diagnosis has been made and from persons with such mild symptoms that diagnosis is never made. Every patient having any intestinal disturbance with or without fever should be put on enteric precautions until a diagnosis can be made. This measure would help prevent the spread of cholera, shigellosis, and salmonellosis. It is better to isolate unnecessarily than too late.

No convalescent from salmonellosis or shigellosis should be taken out of isolation until consecutive specimens of feces and urine, taken at intervals of six days and not earlier than one month after onset, have been found to be negative for the causative organisms. This is often difficult, however, because convalescents sometimes continue to harbor enteric bacilli for long periods. These persons may be released with suitable injunctions for the protection of others and under health department supervision. Even with repeated examinations under the most favorable conditions, some carriers will be missed.

Control of Carriers. The control of carriers of any infectious disease is difficult but very important. The most important point is their occupation. It is imperative that the carrier of enteric pathogens shall not engage in any occupation requiring the handling of food and milk, that the feces shall be disinfected (or otherwise disposed of in such a manner as to avoid infecting others with them), and that the carrier shall conscientiously and thoroughly wash and, if possible, disinfect his hands after using the toilet. The health authorities keep in touch with known carriers by means of visits and reports.

Individual Precautions. For one's individual protection certain precautions should be taken. If one suspects that the water supply is polluted, drinking water should be boiled. Campers, picnickers, and vacationers should be careful about their drinking water.

A good precaution when boiling is not feasible is to provide oneself with commercially available tablets of chloramine (e.g., Halozone, Globaline) for disinfecting water. These are available in drug stores and sporting goods stores and are very effective when used as directed. A good emergency substitute is to add 20 to 30 drops of one of the numerous household laundry bleaches that contain about 5 per cent NaOCl, such as Clorox, to a quart of water at least 30 minutes before use. Laundry bleaching fluids or strong solutions of hypochlorite are also excellent general disinfectants for feces, floors, and bedpans. The labels on the bottles give detailed directions for various uses, including infectious disease applications.

SUMMARY

The family Enterobacteriaceae consists of gram-negative, short, straight, aerobic and facultatively anaerobic rods that inhabit the intestine of man and animals. All species ferment glucose and grow readily on simple blood-free culture media, mostly common foods like milk, salad dressing, and cream fillings. They do not form spores, and they are easily killed by boiling.

The Edwards and Ewing system of classifying Enterobacteriaceae is used in most clinical laboratories. The most important pathogenic genera in the *Enterobacteriaceae* are *Salmonella* (causes typhoid fever and other salmonelloses, such as food infection), *Shigella* (causes shigellosis, or bacillary dysentery), and *Escherichia* and *Proteus* (which can cause gastroenteritis and urinary tract infections).

Only one species, *Salmonella typhi*, causes typhoid fever, a salmonellosis that may be transmitted by human carriers (convalescent, temporary, or permanent), feces, urine, food, or water. *Salmonella typhi* breaks down intestinal ulcers, passes into the stool, invades the blood, and localizes in many body organs, such as the liver, gallbladder, bone marrow, and spleen. Meningitis, pneumonia, or other complications are the usual cause of death. Blood culture, urine cultures, or stool cultures may help in the diagnosis of typhoid fever as well as other salmonelloses, shigelloses, and related diseases. The Kauffmann-White scheme now lists over 1,900 kinds of *Salmonella*. Some of these cause food infection, or paratyphoid fever. Vaccines against salmonellosis exist and are often recommended.

Shigellosis is spread in much the same manner as salmonellosis—that is, by *f*eces, *f*ingers, *f*lies (insects in general), *f*omites, or *f*ood. Toxins are produced by some *Shigella* species. *Escherichia coli* is an indicator of fecal pollution; it may also cause infections when displaced to the urinary tract, the meninges, or the liver and gallbladder routes. Diarrhea, especially "traveler's diarrhea," may be due to *E. coli* enterotoxigenic strains.

Vibrio cholerae causes Asiatic cholera, which involves intense diarrhea, prostration, and emaciation. It results in severe dehydration and death unless fluids and salts can be restored to the body by massive infusions of water and electrolytes.

REVIEW QUESTIONS

1. What organisms belong to the family Enterobacteriaceae? How do you know it is a family? What suffix indicates a tribe?
2. Differentiate between the Edwards and Ewing classification, Bergey's and others.
3. List each genus in the Enterobacteriaceae and the diseases it may cause. How are these diseases transmitted?
4. What tests and test systems can you name to differentiate between genera or species of the enteric bacteria?
5. What is the Kauffmann-White scheme?
6. What is the EMViC reaction?
7. What disease is caused by *Salmonella typhi*? How does it differ from disease conditions caused by other *Salmonella* or *Shigella*?
8. How are some enteric diseases transmitted?
9. What are the three types of human carriers of enteric diseases?

10. Compare blood cultures, stool cultures, and other cultures.
11. Why is *Escherichia coli* important? What infections can it cause? Why?
12. Discuss the pathology of cholera. How is it transmitted?
13. What other bacteria are found in the human intestine?

Supplementary Reading

Benenson, A. S. (ed.): Control of Communicable Diseases of Man. 12th ed. New York, American Public Health Association, 1975.

Blaser, M. J. and Reller, L. B.: *Campylobacter* enteritis. *N. Engl. J. Med., 305*:1444, 1981.

Bockemuhl, J., Amedone, A., and Triemer, A.: *Vibrio parahaemolyticus* gastroenteritis during the El Tor cholera epidemic in Togo (West Africa). *Am. J. Trop. Med. Hyg., 24*:101, 1975.

Brown, M. R. W.: Resistance of *Pseudomonas aeruginosa*. New York, John Wiley & Sons, 1975.

Center for Disease Control: Diarrheal illness on a cruise ship caused by enterotoxigenic *Escherichia coli. Morbidity and Mortality Weekly Reports. 25*:229, 1976.

Clarke, P. H., and Richmond, M. H. (eds.): Genetics and Biochemistry of *Pseudomonas*. New York, John Wiley & Sons, 1975.

Drinking Water Disinfection. Washington, D.C., Superintendent of Documents, Government Printing Office, Public Health Service Publication No. 387.

Edwards, W. M., et al.: Outbreak of typhoid fever in previously immunized persons traced to a common carrier. *N. Engl. J. Med., 267*:742, 1962.

Evans, D. J., Jr., and Evans, D. G.: Inhibition of immune hemolysis: Serological assay for the heat-labile enterotoxin of *Escherichia coli. J. Clin. Microbiol., 5*:100, 1977.

Ewing, W. H.: Differentiation of *Enterobacteriaceae* by Biochemical Reactions. Atlanta, Center for Disease Control, 1973.

Ewing, W. H.: Enterobacteriaceae infections. *In* Diagnostic Procedures for Bacterial, Mycotic and Parasitic Infections. 5th ed. New York, American Public Health Association, 1970.

Ewing, W. H.: Isolation and Identification of *Salmonella* and *Shigella*. Atlanta, Center for Disease Control, 1972.

Ewing, W. H., Davis, B. R., and Martin, W. J.: Biochemical Characterization of *Escherichia coli*. Atlanta, Center for Disease Control, 1972.

Finegold, S. M., et al.: Diagnostic Microbiology. 5th ed. St. Louis, C. V. Mosby Co., 1978.

Finkelstein, R. A.: Cholera. *CRC Crit. Rev. Microbiol., 2*:553, 1972.

Finkelstein, R. A.: Cholera enterotoxin. *In* Microbiology—1975. Washington, D.C., American Society for Microbiology, 1975.

Gemski, P., Jr., and Formal, S. B.: Shigellosis: An invasive infection of the gastrointestinal tract. *In* Microbiology—1975. Washington, D.C., American Society for Microbiology, 1975.

Hirschhorn, N., and Greenough, W. B., III: Cholera. *Sci. Amer., 225*:15, 1971.

Hugh, R., and Sakazaki, R.: Minimal number of characters for the identification of *Vibrio* species, *Vibrio cholerae*, and *Vibrio parahaemolyticus J. Conf. Publ. Health Lab. Dir., 30*:133, 1972.

Kourany, M., and Vasquez, M. A.: The first reported case from Panama of acute gastroenteritis caused by *Vibrio parahaemolyticus. Am. J. Trop. Med. Hyg., 27*:638, 1975.

Rettig, P. J.: *Campylobacter* infections in human beings. *J. Pediatr., 94*:855, 1979.

Rhoden, D. L., Tomfohrde, K. M., Smith, P. B., and Balows, A.: Evaluation of the improved Auxotab 1 system for identifying *Enterobacteriaceae. Appl. Microbiol., 26*:215, 1973.

Rubin, R. H., and Weinstein, L.: Salmonellosis: Microbiologic, Pathologic, and linical Features. New York, Stratton Intercontinental Medical Book Corp., 1977.

Sazie, E. S. M., and Titus, A. E.: Rapid diagnosis of *Campylobacter* enteritis. *Ann. Intern. Med., 96*:62, 1982.

Smith, P. B., Rhoden, D. L., Tomfohrde, K. M., Dunn, C. R., Balows, A., and Hermann, G. J.: R/b enteric differential system for identification of Enterobacteriaceae. *Appl. Microbiol., 21*:1036, 1971.

Tiehan, W., and Vogt, R. L.: Waterborne *Campylobacter* gastroenteritis—Vermont. *Morbidity and Mortality Weekly Report, 27*:207, 1978.

Tomfohrde, K. M., Rhoden, D. L., Smith, P. B., and Balows, A.: Evaluation of the redesigned Enterotube-A system for the identification of *Enterobacteriaceae. Appl. Microbiol., 25*:301, 1973.

Vibrio parahemolyticus—Louisiana. *Morbid. Mortal., 21*:341, 1972.

Washington, J. A., II, Yu, P. K. W., and Martin, W. J.: Evaluation of accuracy of multitest micromethod system for identification of Enterobacteriaceae. *Appl. Microbiol., 22*:267, 1971.

Winblad, S. (ed.): *Yersinia, Pasteurella* and *Francisella. In* Contributions to Microbiology and Immunology, vol. 2. Basel, S. Karger, 1973.

BRUCELLOSIS AND LEPTOSPIROSIS

The two groups of bacteria discussed in this chapter, while differing greatly in many characteristics and therefore widely separated toxonomically, have in common some means of transmission from host to host in body fluids and excreta. Means of control may therefore be discussed in parallel.

Brucella, cause of brucellosis or undulant fever in humans and Bang's disease in cattle, is sometimes present in the feces of patients, and *Leptospira,* cause of leptospirosis (hemorrhagic jaundice, canicola fever, Weil's disease), occurs in the urine of patients. These facts necessitate the same general precautions for control of infections by *Brucella* and *Leptospira* as are used in all feces-borne and urine-borne diseases. Certain species of *Clostridium* are also invariably present in feces, but they are common in the soil and are more fully described in connection with the soil-borne diseases, tetanus, and gas gangrene (Chapter 37).

GENUS BRUCELLA

Brucella is a coccobacillus or short rod (0.5 to 0.7 by 0.6 to 1.5 μm), nonmotile, gram-negative, nonspore-forming, microaerophilic, and rather fastidious in its growth requirements. It does not require blood derivatives for growth but is much more invasive than the otherwise somewhat similar hemophilic bacteria (*Bordetella bronchiseptica, Haemophilus influenzae*) and generally penetrates to all the tissues of the infected body and circulates in the blood. In humans it is the etiologic agent of undulant fever or brucellosis; in cattle, swine, and goats it causes infectious abortion and tends to localize, especially in the udders of lactating animals.

These bacteria were first discovered in 1887 by Sir David Bruce, a British scientist: hence the name of the genus. Bruce discovered the bacilli originally in goats and in British soldiers who had drunk goat's milk on the island of Malta (the disease was thus called Malta fever). Since then, several varieties of the organism have been found in other parts of the world, causing slightly different diseases according to whether the variety primarily infects cattle, hogs, or goats. Since it is commonly found in the milk of infected animals, unpasteurized (or *uncertified*) milk is an important means of transmission. Infected animals often secrete the organisms in milk only intermittently.

Species of Brucella. The variety infecting cows characteristically causes abortions in these animals when they are pregnant for the first time and is called *Brucella abortus.* The variety infecting hogs is called *Brucella suis,* and that infecting goats, *Brucella melitensis.*[1] The infectivity is not highly specific. Any of the types may infect other animals, including horses, sheep, and dogs, as well as humans. A fourth species, *Brucella neotomae,* was isolated in 1957 from the desert rat. It does not seem to affect humans. However, *Brucella canis,* a dog pathogen, has been known to infect laboratory workers. Sixteen cases of *Brucella*

[1]Malta was known to the ancient Romans as *Melita* because of the fine *mel* (honey) to be had there; hence *melitensis,* meaning "of Melita."

CHAPTER 25

canis infections have been reported in humans between 1965 and 1974 in the United States. This can become a problem in kennels and similar places. *Brucella ovis* causes abortions in ewes and epididymitis in rams.

Characteristics of Brucellas. When first isolated, the various species of *Brucella* often grow very slowly. Cultures of infectious milk, blood, cerebrospinal fluid, urine, or tissues used for diagnosis should therefore be incubated for at least four weeks and subcultured and examined every day or two to see if growth has started. *Brucella abortus* and *Brucella ovis* will grow at first only in a special atmosphere containing about 5 to 10 per cent carbon dioxide. The other varieties of *Brucella* can use carbon dioxide but can also grow well aerobically. Blood cultures are of special value in diagnosing human cases. The three varieties of *Brucella* pathogenic to humans can be differentiated only by very careful laboratory tests. Some of these differentiations are shown in Table 25–1. The best growth may be obtained on tryptose, trypticase, liver infusion, or brucella agar at pH 7.0 to 7.2 or slightly more acid, in 10 per cent carbon dioxide at 37 C.

In the infected body the organisms tend to be removed from the blood by phagocytic cells lining the blood vessels (MPS, RES). Cultures are therefore often made from bone marrow (taken usually from the sternum or "breast bone") or with material drawn with a trochar from a lymph node. The nurse or health worker should be prepared to assist in obtaining such material for diagnostic culture. The procedure requires local anesthesia.

Brucellosis (Undulant Fever)

The disease in humans, most commonly called undulant fever (also Malta fever or Mediterranean fever[2]), is frequently characterized by a long preliminary stage lasting days or weeks, during which there is an increasing weakness and later backache, stiffness of the joints, progressive loss of weight, and a long list of other, less definite, and highly variable symptoms. The incubation period may be as long as six weeks or as short as four or five days. There are usually severe night sweats and remittent daily fever or repeated undulatory attacks of fever, each lasting several days, with remissions between the waves. In its milder forms, which are the more usual, the malady may be regarded by the victim as "grippe" or "intestinal flu." About 2 per cent of the cases are acute and fatal, but the great majority of persons afflicted, after being ill for days, weeks, or months, eventually recover. A chronic stage of the disease may develop with the patient suffering psychoneurotic symptoms, as well as pains, fever, weakness, and other conditions hard to diagnose; open sores or lesions may also be present. IgG may be demonstrated with a high titer, but no *Brucella* organisms are found.

[2]The disease is also known as Bang's disease, after Bernhard Bang, who isolated the organism in Denmark from cattle suffering infectious abortion.

Table 25–1. Physiological Characteristics of Typical Strains of *Brucella**

SPECIES	PREFERRED HOST†	5% CO_2 REQUIREMENT‡	H_2S PRODUCTION	HYDROLYSIS OF UREA	GROWTH ON MEDIUM CONTAINING:		AGGLUTINATION TESTS
					Thionine	*Fuchsin*	
*Brucella melitensis*①	Goats	–	– or ± for 4 days	Slow or negative	+	+	Has specific antigens
*Brucella abortus*②	Cows	+	+ for 2 days only	Slow or negative	–	+	Share antigens, but different from *B. melitensis* (*B. abortus* and *B. suis*)
*Brucella suis*③	Hogs	–	+ for 4 days	Rapid	+	–	
Brucella canis	Dogs	–	–	?	+	–	Aglutination in anti-rough serum, resembling only *B. ovis;* all other strains are negative

*Biotype variations exist.
†Actually any type may infect other animals or man.
‡On first isolation.
①Biotypes I–III are known.
②Biotypes I–IX are known.
③Biotypes I–IV.
Other biotypes are the species *Brucella neotomae* (rats) and *B. ovis* (sheep).

The bovine type of infection *(Br. abortus)* in humans is likely to be mild, and many cases pass unnoticed and undiagnosed. The porcine and caprine varieties (*Br. suis* and *Br. melitensis)* cause much more severe infections as a rule.

Transmission of Undulant Fever. Brucellosis is only rarely transmitted from human to human. The organisms from animal sources find portals of entry via the gastrointestinal tract and through cuts and scratches in the skin. On gaining entrance to the body, the organisms travel from the portal of entry by way of the blood and lymph channels. The bacilli can be found in lymph nodes in all the various organs. In female farm animals, the bacilli tend also to localize in the udder and, in pregnant animals, in the uterus, placenta, and similar locations. The fetus, membranes, and fluids discharged during abortion due to *Brucella* are highly infectious, as is the milk of infected animals and occasionally of women. The organisms also sometimes appear in the feces and urine of both infected animals and humans; hence this discussion in the section on intestinal infections. In man and male farm animals, inflammation of the testis (orchitis) is not uncommon and often results in sterility.

Since the organisms invade the blood, they are present in all the tissues, and for this reason the disease is particularly common among butchers, employees of slaughterhouses, stock raisers, veterinarians, packinghouse employees, government meat inspectors, and rendering plant workers. Persons who take unpasteurized milk or other dairy products from cows or goats that have not been carefully tested for the disease are in danger of contracting the infection, and generally do so sooner or later. It is apparent that human cases of brucellosis have declined in the United States, from the reported (diagnosed) 3,510 cases in 1950 to 183 cases in 1980.

Serologic Diagnosis of Undulant Fever. In addition to a blood culture (which is the most conclusive test, if positive), an agglutination test, performed almost exactly like the Widal test (see page 371), is one method of diagnosis. It does not distinguish between species of *Brucella.* The serum of patients and of infected farm animals usually agglutinates the brucellas in significant dilutions, as does the sweet whey from junket made with milk of infected animals. In many confirmed cases of brucellosis, however, agglutinins appear in the serum only in low concentrations or not at all, and they are found sometimes in supposedly normal persons. There is no real "diagnostic titer" of agglutinins, though in practice a positive reaction with serum diluted above 1:320 or 1:640 is generally regarded as indicating infection, either past or present. It is much more significant if there is a rise in titer between two tests done about 10 days apart. This is true of all serologic tests and should be remembered by the student (Chapter 20). It shows that the tissues are actively responding to the immunologic stimulus of an infection. The student should compare this rise in titer with, and carefully differentiate it from, the anamnestic reaction and the secondary antigenic stimulus. Antibodies in human sera have also been detected by means of the FA (fluorescent antibody) test.

It is worth noting that *Vibrio cholerae* and *Brucella* share a common O antigen, so that cross-agglutinations can occur and may cause confusion. Sometimes blood or the cream from suspected milk is injected into guinea pigs for diagnostic purposes. These animals are very susceptible to infection with *Brucella* and show characteristic lesions. Cultures inoculated with blood or cream are also often used for diagnostic purposes, with good results. Until the machine age of milk processing, the "ring test" was widely used and could be easily performed in any barn by practically untrained personnel. To 5 ml of whole unhomogenized milk, 0.2 ml of a heavy suspension of killed, purple-stained *Brucella* cells is added. The two are mixed well and allowed to stand. In the presence of antibodies against *Brucella,* the stained organisms adhere to the fat globules and rise with them to the surface, forming a purple ring of cream, a positive "ring test." This is diagnostic in cows infected with brucellosis but may also occur in infected cows that have been vaccinated. Milk known to be infectious by any pathogen should not be used. Isolation of specific, causative microorganisms in pure culture is the surest diagnosis of any infection.

Prevention of Undulant Fever. The disease may be prevented by using only pasteurized dairy products and by carefully avoiding any infected animals and their flesh or discharges.

Vaccines have been found to be of some value in preventing loss of calves and other farm animals by abortion, although the infection may occur. There is no good vaccine for human beings.

The patient with undulant fever is not highly infectious unless there are draining lymph nodes or other open lesions. Dressings and clothing soiled with such drainage should be carefully disinfected and not allowed to touch other articles. The organisms are occasionally present in the urine and feces, and these should be given the same sort of treatment as stools from typhoid fever patients.

Treatment of Undulant Fever. Tetracyclines alone are usually adequate; for severe cases chlortetracycline (Aureomycin) and streptomycin are used. Treatment with tetracycline and streptomycin together has been recommended. Treatment must be continued for at least three weeks because of the intracellular location of the organisms; relapses are frequent.

GENUS LEPTOSPIRA

Leptospires are the smallest and most delicate cell forms of spirochetes (6 to 20 μm by 0.1 μm). They are tightly coiled, slender, and curved into a hook at one or both ends (Figs. 25–1 and 25–2). They may be cultivated in simple mineral solutions (e.g., Stuart's) supplemented with about 10 per cent of rabbit serum and incubated at about 20 to 35 C. In cultures, urine, or infected tissues, they may readily be seen by means of the darkfield microscope, wriggling and twirling with fascinating energy. At the present time only one species of this genus is recognized in *Bergey's Manual:* it is named *Leptospira interrogans.* The reason for this is not that different species do not exist but simply that the

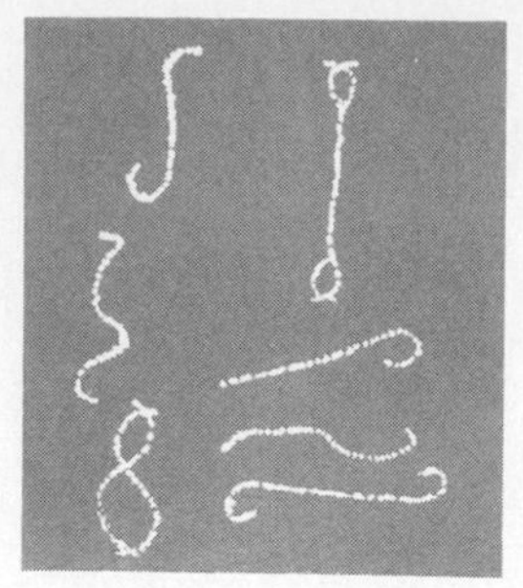

Figure 25–1 *Leptospira interrogans,* serotype *icterohaemorrhagiae.* Appearance of organisms in the darkfield (×1000). (From Joklik, W., and Smith, D. T.: Zinsser Microbiology. 15th ed. New York, Appleton-Century-Crofts, 1972.)

experts do not agree on proper designations for them. Some workers suggest two species, *L. interrogans* for the pathogens and *L. biflexa* for the 16 serogroups and more than 150 serotypes of harmless *Leptospira,* which multiply in sewage, stagnant water, and feces. These harmless types grow well at temperatures of 5 to 10 C, several degrees below the minimum (13 to 15 C) for pathogenic leptospires. The pathogens generally exhibit much greater ability to hydrolyze leucyl naphthylamide than the saprophytes. Other differences have been noted.

All species, both saprophytes and pathogens, are quite fragile and easily killed by heat, drying, and standard disinfectants; however, all can survive for considerable periods in streams, ponds, and other bodies of fresh water. If pathogenic species happen to be present in such waters, they can (and often do) infect animals or uninformed persons drinking the polluted water or swimming or wading in it. Like *Brucella,* the leptospires enter the body via the gastrointestinal tract or cuts or scratches in the skin. Since the pathogenic leptospires usually damage the kidneys especially, they often occur in the urine as well as in the blood and other tissues of infected humans and animals. The presence of the spirochetes in bodies of water polluted with human or animal urine or tissues is therefore easily understood. The carcasses of dead infected animals sometimes pollute streams and ponds. Animals likely to be infected (and infectious) include dogs, cattle, swine, mules, rats, and numerous wild mammals.

Leptospirosis

The pathogenic leptospires, after entry, invade the entire body by way of the blood. Infection with any species of *Leptospira* is properly called *leptospirosis.* Leptospires are discussed here because one of their important portals of entry is the mouth and because their usual portal of exit is the urinary tract (via urine): hence, like the Enterobacteriaceae, they ure transmitted by sewage-polluted or urine-fouled water and food. They do not ordinarily occur in the feces.

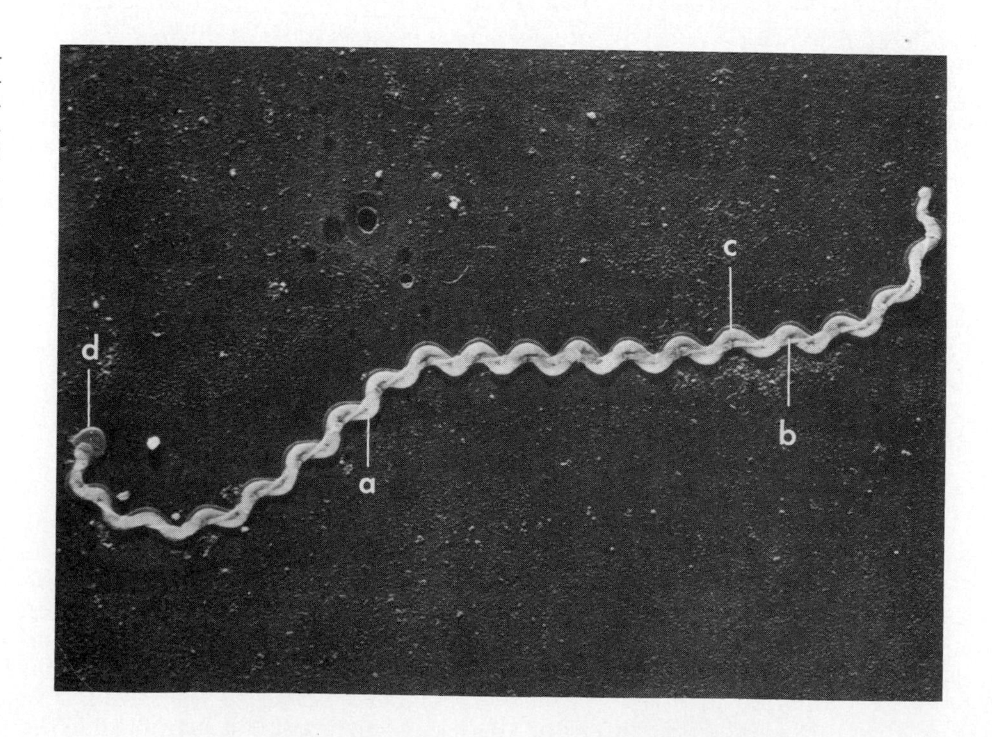

Figure 25–2 *Leptospira interrogans,* serotype *icterohaemorrhagiae.* Spray preparation shadowed with chromium. Note protoplasmic spiral (*a*), axial filament (*b*), covering sheath (*c*), and end bulb (*d*). ×12,500. (From Simpson, C. F., and White, F. H.: Electron microscope studies and staining reactions of leptospires. *J. Infect. Dis., 109*:243, 1961. Copyright 1961, University of Chicago Press.)

The different serotypes of *Leptospira* that infect humans are so much alike that they can be differentiated only by skilled laboratory workers. Once, different species names were given to those infecting dogs, cattle, swine, and humans. The student needs to know only one, now called *Leptospira interrogans* but still appearing in the literature under its old name *Leptospira icterohaemorrhagiae* (now called serotype *icterohaemorrhagiae*).[3] It causes Weil's disease or infectious hemorrhagic jaundice in humans.

Clinical manifestations of leptospirosis are varied, ranging from lesions of the eye to meningitis, hepatitis, jaundice, high fever, "black vomit,"[4] and other symptoms, depending somewhat on the infecting species.

Weil's Disease. One of the most familiar forms of leptospirosis in man was originally called Weil's disease, after A. Weil, a physician of Wiesbaden, Germany. It was once confused with yellow fever (a viral disease), partly because of the intense jaundice due to hepatitis and other clinical features (including "black vomit") found in both diseases. This form of leptospirosis is quite widespread in some regions, including the United States. It often occurs in persons who spend much time (especially with bare feet or leaky shoes) in wet, poorly drained places, such as badly constructed mines, trenches during war, rice fields, sewers, and the bilges of boats, where there is human urine and where rats abound. It may also occur in persons living at home, however, if they come into contact with human cases of leptospirosis or with water or food contaminated with the urine of rats or other animals.

The pathogen of Weil's disease occurs in the blood and urine of patients and may be cultivated from them in serum media. The organisms survive and possibly multiply in polluted water and may easily be cultivated from the water. Leptospirosis is readily transmitted from human to human by polluted water. The disease is often severe and not infrequently fatal, though many mild cases occur. Leptospirosis at the present time is a disease of low frequency in the United States. In 1971, 62 cases were reported and in 1980, 85. During the first 18 weeks of 1982, only 20 cases were reported.

Rats and Leptospirosis. Rats commonly acquire the infection, and in any large number of rats properly examined (kidney examinations by darkfield microscopy), a certain proportion (often 50 per cent) will always be found to harbor these organisms. Studies conducted from 1958 to 1968 and reported in 1971 showed that up to 46 per cent of the brown rat (*Rattus norvegicus*) population is infected with serotype *icterohaemorrhagiae*. It is thus evident that rats transmit the disease, polluting sluggish streams, mines, ships, trenches, and establishments in which poultry, fish, or meats are cleaned, as well as food in kitchens and water in shallow wells. When rats or other infected animals (such as dogs) die in wells, boats, ponds, and streams, their bodies make the water infectious.

Other Types of Leptospirosis. The same general sequence of events holds true for other forms of leptospirosis. Leptospirosis causes serious losses in dog kennels; in swine and cattle leptospirosis costs farmers many thousands of dollars annually, and carries with it the risk of infecting the farmers themselves. In the wild, animals like racoons may also harbor leptospires and induce new infections. Formerly regarded as a rare disease, leptospirosis has been found to be frequent and widespread in both humans and animals in many countries other than the United States.

Diagnosis. Darkfield examination of urine of patients with leptospirosis often reveals the leptospires. They are not generally found in the blood by this means. Guinea pigs are highly susceptible and inoculation of them with

---

[3]Others are serotype *pomona* (swine), serotype *canicola* (dog), and so forth.

[4]Black vomit is a lay term for vomitus containing blood blackened by acid gastric juice. The blood in black vomit comes from hemorrhages in the alimentary tract, which are common in Weil's disease. Damage to the liver causes jaundice, hence the term "hemorrhagic jaundice" for Weil's disease.

infectious blood or urine usually produces a typical case of leptospirosis. Cultures made with blood or urine also reveal the organisms.

Initially, leptospirosis is often mistakenly diagnosed as aseptic meningitis, "viral infection," encephalitis, hepatitis, or just some fever. This underscores the importance of proper and early laboratory tests.

Shortly after the onset of infection with *Leptospira,* agglutinins and cytolytic antibodies begin to appear in the blood. These increase in concentration until the serum of the animal or person is capable of protecting him against large doses of leptospires. Stable, formalized *Leptospira* serotype suspensions are available and are used satisfactorily in rapid slide agglutination tests for serologic diagnosis. Also used is the immune adherence method and the FA (fluorescent antibody) test.

Sanitary Measures. Because of the frequent occurrence of leptospiremia in taking blood specimens for "blood counts," cultures, and serologic studies, precautions must be taken to see that cotton pledgets, needles, syringes, and other items soiled with blood are properly disposed of. The urine of patients with leptospirosis, as well as clothing and bedding soiled with it, should be disinfected with the usual precautions. The patient must avoid contamination of the surroundings with urine. Ordinary cleanliness of hands will usually suffice. Rats should be exterminated. When managing cases of leptospirosis under home or field conditions, water should be boiled or chlorinated before drinking, unless it is from a supply of known purity. Food possibly contaminated with excreta of rats, or from unsanitary sources, should not be eaten.

SUMMARY

Pathogens of *Brucella* and *Leptospira,* although very different taxonomically, are similarly transmitted from host to host in body fluids and excreta. Named brucellosis when it infects cattle, hogs, goats, or sheep, or Malta fever after the island of Malta, the disease is also known as undulant fever in humans. Undulant fever is contracted through cuts and scratches in the skin or, more commonly, via the gastrointestinal tract, from where the *Brucella* organisms travel by way of the blood and lymph to various organs. It localizes in lymph nodes all over the body and is present in all tissues. Butchers, employees in slaughterhouses, veterinarians, and meat industry workers are especially exposed to these infections. In female farm animals, the bacilli tend to localize in the udder and, in pregnant animals, in the uterus, placenta, and genital area. Milk, feces, urine, the flesh of the animals, and cow fetuses aborted because of brucellosis, with their membranes and discharges, are all highly infectious. *Brucella abortus* causes contagious abortion in cattle; *Brucella suis* causes brucellosis in hogs; *Brucella melitensis* is a pathogen of goats; and any *Brucella* species can cause disease in other animals, including humans. At present, there are no vaccines to prevent undulant fever, but several antibiotics are effective treatments, and pasteurization of dairy products and inspection of animals to be slaughtered have greatly reduced the risk for the general population.

Leptospires are the smallest and most delicate coiled spirochetes. The only pathogenic species recognized at present is *Leptospira interrogans,* a pathogen of cattle, swine, dogs, mules, numerous wild mammals (especially the rat), and humans. Leptospirosis can be contracted from streams, ponds, and other bodies of polluted water in which the excreta, urine, or dead bodies of infected animals were deposited. This type of pollution is especially common in mine shafts, ships, trenches, and boat houses where rats reside. Drinking, eating, or cuts in the skin allow *Leptospira interrogans* to invade the body. Leptospirosis in humans is most often referred to as Weil's disease. It was once confused with yellow fever (a viral disease), partly because it produces intense jaundice and "black vomit."

REVIEW QUESTIONS

1. In what ways are *Brucella* and *Leptospira* similar? How do they differ?
2. What disease does *Brucella* produce in animals, and what are the species that are most likely to cause these infections?
3. What are the different names by which brucellosis is known, and which one is the name given to the disease of humans?
4. How is the disease transmitted from animal to animal? How are humans most likely to be infected? What are the symptoms of the disease? What is done to prevent it and to cure it?
5. What diagnostic test may be performed in the laboratory to diagnose brucellosis?
6. Who established the etiology of the disease brucellosis, and where did he do it?
7. What is the morphology of *Leptospira interrogans*? What animals may be infected with it?
8. What are the modes of transmission of leptospirosis among animals and among humans? What name is used to refer to human *Leptospira* infection?
9. List the symptoms that present when humans are afflicted with the disease. Why was it confused with yellow fever?
10. Are either of the two diseases mentioned in this chapter very common? If yes, why? If not, why not?

Supplementary Reading

Alexander, A. D., Wood, G., Yancey, F., Byrne, R. J., and Yager, R. H.: Cross-neutralization of leptospiral hemolysins from different serotypes. *Infect. Immun., 4*:152, 1971.

Alston, J. M., and Broom, J. C.: Leptospirosis in Man and Animals. Baltimore, The Williams & Wilkins Co., 1958.

Babudieri, B.: The agglutination-absorption test of leptospira. *Bull. WHO, 44*:795, 1971.

Burton, G., Blenden, D. C., and Goldberg, H. S.: Naphthylamidase activity of *Leptospira. Appl. Microbiol., 19*:586, 1970.

Center for Disease Control: Brucellosis Surveillance, Annual Summary, 1972. Atlanta, Center for Disease Control, 1974.

Galton, M. M., et al.: Leptospirosis, Epidemiology, Clinical Manifestations in Man and Animals, and Methods in Laboratory Diagnosis. Washington, D.C., Government Printing Office, Public Health Service Pub. No. 951, 1962.

Hendricks, S. L., et al.: Brucellosis outbreak in an Iowa packing house. *Amer. J. Pub. Health, 52*:1166, 1962.

McKiel, J. A., Rappay, D. E., Cousineau, J. G., Hall, R. R., and McKenna, H. E.: Domestic rats as carriers of leptospires and salmonellas in Eastern Canada. *Can. J. Pub. Health, 61*:336, 1971. *Morbidity and Mortality,* Vol. 20, No. 53, 1971.

Myers, D. M., and Varela-Diaz, V M.: Selective isolation of leptospiras from contaminated material by incorporation of neomycin to culture media. *Appl. Microbiol., 25*:781, 1973.

Vedros, N. A., Smith, A. W., Schonewald, J., Migaki, G., and Hubbard, R. C.: Leptospirosis epizootic among California sea lions. *Science, 172*:1250, 1971.

INTESTINAL PROTOZOA AND HELMINTHS

PROTOZOA: THE AMEBAS

Numerous species of the phylum Protozoa inhabit soils and waters of the earth and the intestinal tract. Some are ciliates, some are flagellates, others are amebas. Most of these animals are harmless and may be observed by microscopic examination of normal feces. Common harmless species of enteric (intestinal) amebas are *Entamoeba coli, Entamoeba hartmanni, Endolimax nana, Iodamoeba bütschlii* and *Dientamoeba fragilis*, which is common in the mouth (Fig. 26–1).

Entamebas, like all typical animal cells, have no rigid cell walls and, in the active, multiplying, or trophozoite stage, have no particular or invariable form. Entamebas in the trophozoite stage are continually changing their shapes from round or oval to very irregular forms with protrusions and finger-like processes sticking out from various portions (Fig. 3–2). With their fluid cytoplasm constantly in motion (*cytoplasmic streaming*), they can move by "flowing" into any of these protrusions, which are called *pseudopodia*. This kind of motion is called *ameboid movement*. It is seen also in the blood cells called phagocytes. By means of pseudopodia, both phagocytes and amebas can engulf solid particles of food, a type of nourishment called *phagotrophic nutrition* (Chapters 3 and 18).

AMEBIASIS AND AMEBIC DYSENTERY

Infection by any species of the amebas is properly called *amebiasis*.[1] The most harmful of the species pathogenic to humans bears the name *Entamoeba histolytica (ent*, inside; *histo*, tissue; *lytic*, dissolving). The organisms are usually ingested as dormant *cysts* in feces or sewage-polluted food or water. They soon undergo *excystation* and grow into fragile, actively multiplying *trophozoites*, which primarily attack the lining (mucosa) of the intestine, usually the large bowel. These entamebas, by means of tissue-destroying enzymes, burrow into and, in places, undermine the intestinal lining and cause ulcers (amebic dysentery).[2] There is little inflammatory reaction unless, as is common, secondary bacterial infection develops.

Entamoeba histolytica in the trophozoite form often burrows through the intestinal lining and deep into the intestinal wall. Occasionally, rupture of the intestine occurs as a result. The patient may then die of peritonitis caused by escape of the bacteria of the feces into the abdominal cavity.

The amebas may also invade the intestinal lymph and blood vessels and then are carried to the liver, lungs, brain, and other organs, where they can become localized and produce large amebic abscesses (Fig. 26–2).

---

[1]The term amebiasis includes amebic dysentery and other disease processes due to invasion of the liver and other organs by amebas.

[2]The student must distinguish between amebic dysentery and dysentery due to other causes, such as *Shigella*.

CHAPTER 26

Like many other infectious diseases, amebiasis is often chronic and may be present with little definite symptomatology for a long time. Carriers of *E. histolytica* are common in some areas of low-grade sanitation. Thus, amebic infection is often unknowingly disseminated by persons with mild cases of the disease or by carriers. Amebiasis is common in all warm regions and is frequently found in temperate zones around the world.

Entamoeba histolytica is eliminated only in the feces and may appear in one or both of two forms.

Cysts. Most amebas have the property of forming rounded, dormant, thick-walled, drought-resistant cysts 10 to 20 μm in diameter (Fig. 26–1) that remain alive and dormant for hours or days in the lumen of the colon, in feces or in moist, polluted soil or water. They are slowly killed by drying and exposure to sunlight, and they are susceptible to heat and vigorous disinfection. Unless active diarrhea is in progress, which quickly flushes the trophozoites from the bowel before encystment can occur, it is the cyst form that is commonly found in stools. These cysts are transmitted from person to person by the well-known (and nauseatingly common!) fecal-oral route of transmission of all types of intestinal infection. Cysts of less than 10 μm in diameter (ranging downward to 3.5 μm) may be found in feces; they are most likely nonpathogenic *Entamoeba hartmanni* or a subspecies called *Entamoeba hitolytica* var. *hartmanni*.

Trophozoites. The actively multiplying, fragile, trophozoite form of *Entamoeba histolytica* is excreted only in the watery stools of acute amebic dysentery, is not resistant outside the intestine, and quickly dies when cooled or dried. Trophozoites of *E. histolytica* frequently ingest red blood cells; those of other species rarely or never do. These facts are of diagnostic value.

Trophozoites multiply only by binary fission. They may be easily studied in culture at 37 C under partial anaerobiosis. If they produce disease (about 10 per cent of infections), they attack the lining of the cecum or lower ascending colon and more distal portions.

Transmission

Anything recently contaminated with infected human feces from a "cyst-passer" (chronic case or carrier) may transmit the cysts. Transmission on fruits and vegetables is said to be a common occurrence in the Orient and other places where human sewage and feces ("nightsoil") are used for fertilizer. Fresh vegetables—lettuce and celery, for instance—may then very readily have live ameba cysts upon them when eaten. Although opinions differ concerning the transmission of amebiasis by fruits and vegetables, it is wisest not to eat uncooked foods in the Orient or the tropics (Fig. 26–3). Some serious outbreaks of amebic dysentery in the United States have been caused by sewage-polluted water supplies.

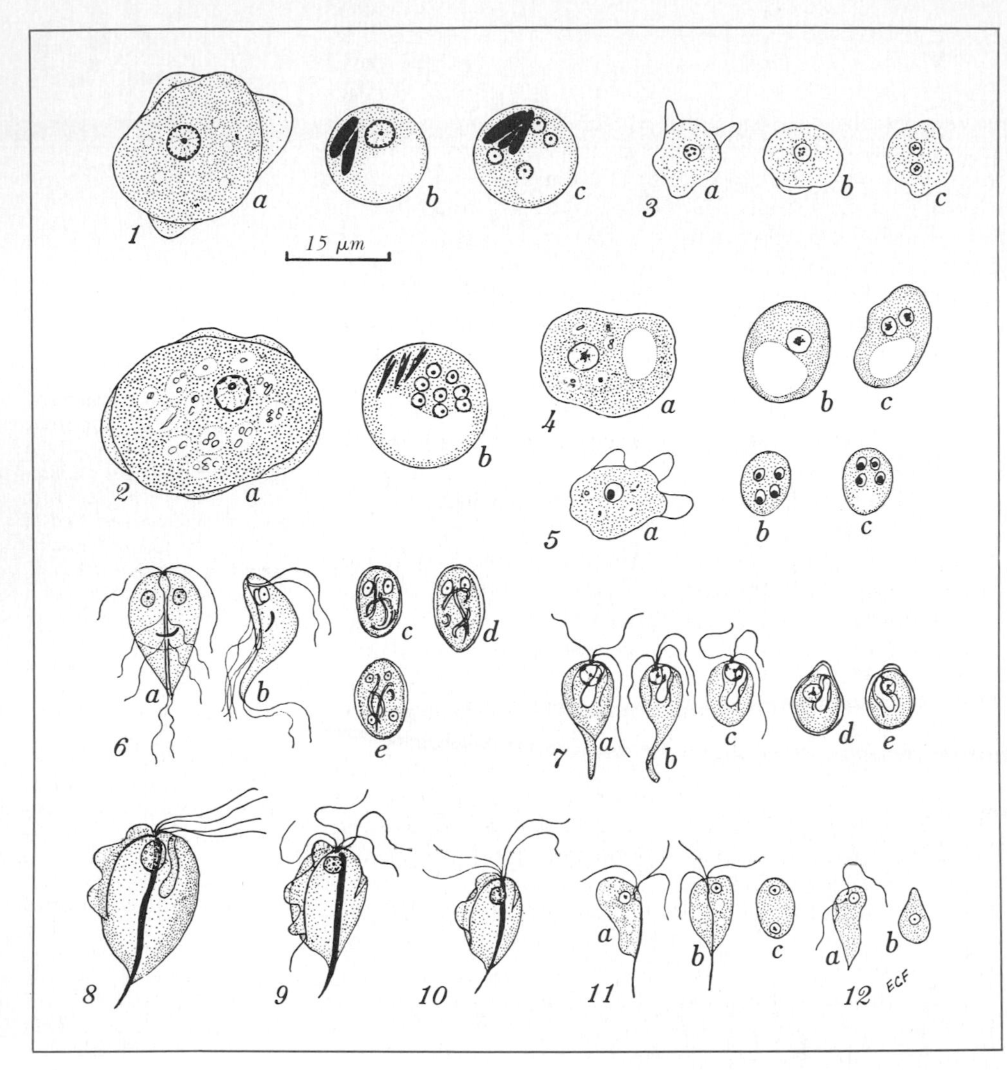

Figure 26–1 Trophozoites and cysts of the more common amebae and flagellate protozoa of the digestive tract (iron-hematoxylin staining). All drawn to scale.

1. *Entamoeba histolytica: a,* trophozoite; *b,* uninucleate cyst; *c,* mature cyst
2. *Entamoeba coli; a,* trophozoite; *b,* mature cyst
3. *Dientamoeba fragilis: a–c,* representative trophozoites
4. *Iodamoeba bütschlii; a,* trophozoite; *b,c,* mature cysts
5. *Endolimax nana: a,* trophozoite; *b,c,* mature cysts
6. *Giardia lamblia: a,* ventral view; *b,* lateral view of trophozoite; *c,d,* immature cysts; *e,* mature cyst
7. *Chilomastix mesnili: a–c,* trophozoites; *d, e,* cysts
8. *Trichomonas vaginalis:* trophozoite, from the vagina
9. *Trichomonas hominis:* trophozoite, from the cecum
10. *Trichomonas tenax:* trophozoite, from the mouth
11. *Enteromonas hominis: a,b,* trophozoites; *c,* cyst
12. *Retortamonas intestinalis: a,* trophozoite; *b,* cyst

(From Faust, E. C., et al: Animal Agents and Vectors of Disease, 3rd ed. Philadelphia, Lea & Febiger, 1968.)

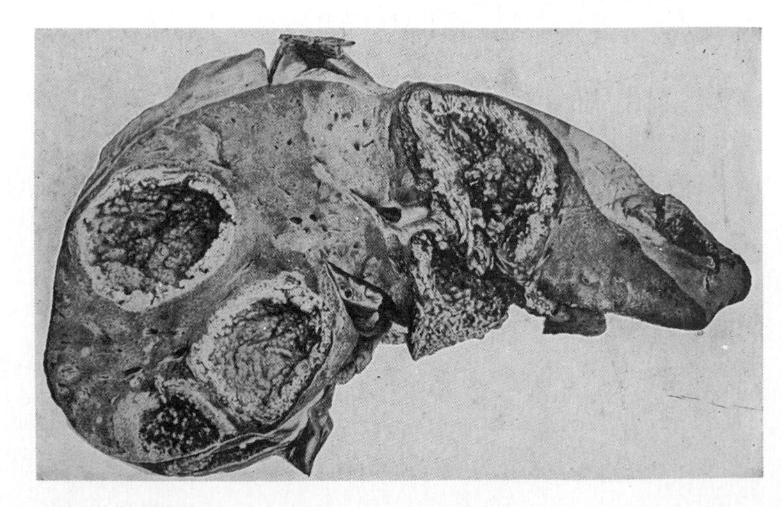

Figure 26–2 Multiple amebic abscesses of the liver. (From Mense's *Handbuch der Tropenkrankheiten.)*

Although amebiasis is more prevalent in areas where unsanitary disposal (or *no* disposal!) of feces is the rule, carriers of *Entamoeba histolytica* exist in all populations and are dangerous sources of infection. In many areas of the world up to 50 per cent of the populations is infected, but even in well-sanitized cities 1 to 5 per cent may carry the organism. In the United States, Craig found 11 per cent of 59,336 persons examined to be carriers of *Entamoeba histolytica*. In some groups, as among certain North American Indians, the number of carriers may rise to as much as 25 per cent. Feces-soiled hands and flies appear to be the major vectors. Carriers who handle food may transmit cysts to the food via their soiled hands.

In regions and in countries where disposal of sewage and feces is effective, little amebic dysentery exists. Unlike *Salmonella* and *Shigella*, the amebas do not

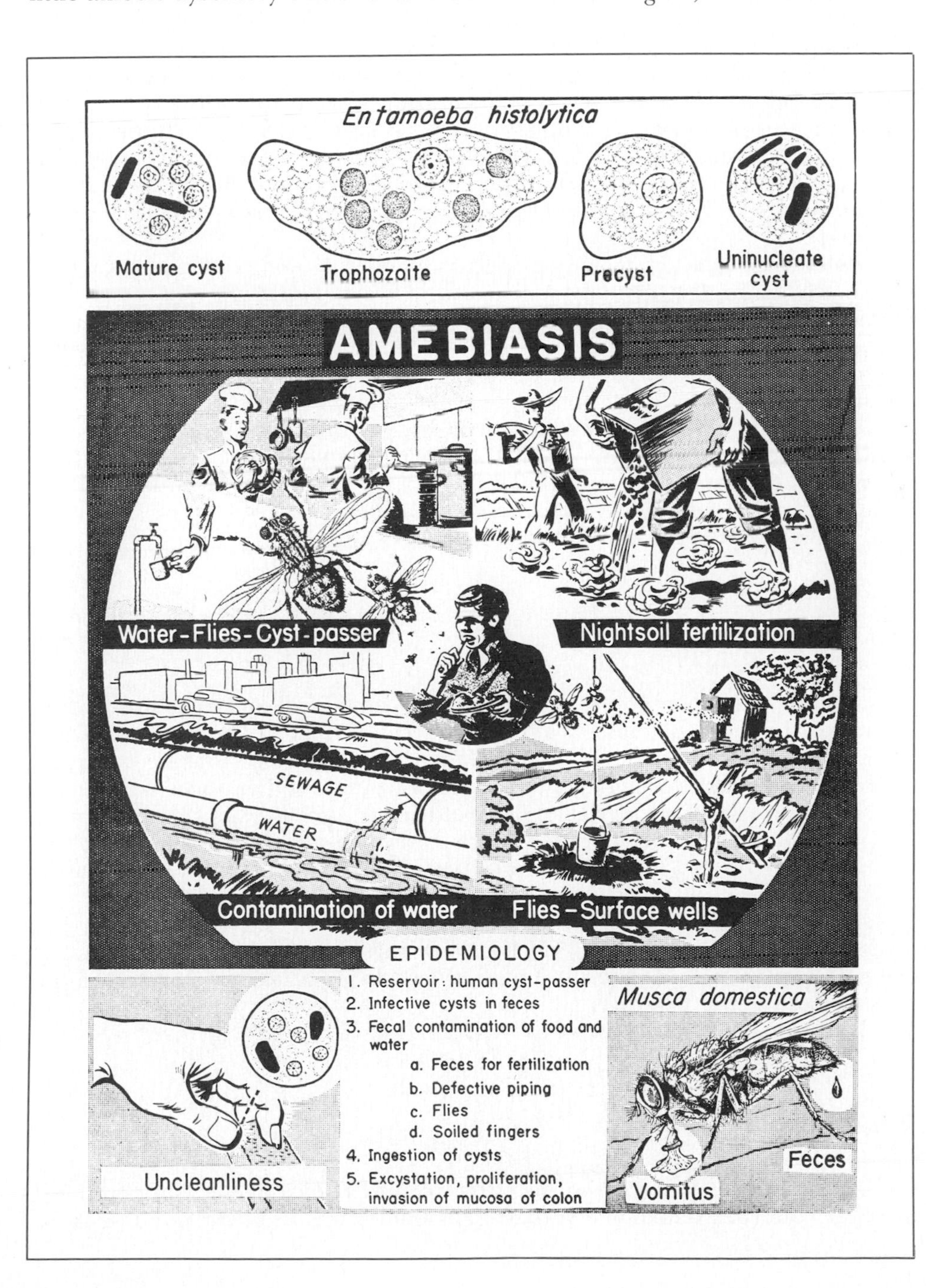

Figure 26–3 Epidemiology of amebiasis and numerous other enteric infections. At the top are shown forms of *Entamoeba histolytica*. At lower right and left are seen two of the main methods of transmission: fingers soiled with excreta, and flies from infected feces. These are avenues of transmission of many intestinal infections. In the central panel are shown means by which various vectors of intestinal diseases (including amebiasis) operate: carriers or cyst-passers who handle foods contaminate foods and utensiles; in the Orient human excreta ("nightsoil") is used to fertilize vegetables eaten raw; water mains are contaminated by leaky sewerage; flies spread feces everywhere. (From Hunter, G. W., III, Swartzwelder, J. C., and Clyde, D. F.: Tropical Medicine. 5th ed. Philadelphia, W. B. Saunders Company, 1976.)

multiply in food. It is difficult to cultivate them except on specially prepared media. They usually require bacteria as food. Yet in 1976 in the United States 2,906 cases of amebiasis were reported, of which about 11 per cent died. In 1980 the number of cases had risen to 5,271.

In March 1981 the Colorado State Department of Health reported an outbreak of amebiasis that occurred from 1977 to 1980 and resulted in seven deaths. Eighty-seven individuals, including thirteen with cases of amebiasis had received colonic irrigation—a series of enemas performed by machine to "wash out" the colon—a practice that has recently gained unfortunate popularity. When tested, the machine was found to be contaminated with *Escherichia coli*, the indicator of fecal pollution.

Diagnosis of Amebic Dysentery. Diagnosis by clinical means is often difficult. The disease may be diagnosed in the laboratory by microscopic examination of the patient's fresh, warm feces. The diagnostically distinctive cysts, and occasionally vegetative forms, may then be found. The cysts are about 12 μm in diameter. When mature, they contain four distinct spherical nuclei in a finely granular cytoplasm. Bars of deeply staining material (chromatoid bars) are likewise diagnostically distinctive in their size and arrangement. The cysts are readily recognized by those trained to observe them. Complement fixation, precipitin, and counterelectrophoresis tests are also valuable in diagnosis.

Entamoeba coli. In making diagnoses by microscopic examination of stool specimens, one must carefully distinguish between *Entamoeba coli*, a harmless species, and the pathogenic *Entamoeba histolytica*. This is usually not difficult but requires experience, as is seen in Figure 26–1. An important improvement in diagnostic procedure is the use of polyvinyl alcohol fixative (or merthiolate-iodine-formaldehyde) and preservative for mounting the protozoa on the microscope slide. In basic principle it is like mounting them in transparent plastic, much as flowers and similar items for ornaments and costume jewelry are mounted. Cysts of *E. coli* differ from those of *E. histolytica* in that they are about 18 μm in diameter and contain eight (sometimes 16 or 32) nuclei, larger than those of *E. histolytica*.

Prevention

Although isolation of the patient who has amebic dysentery is not usually required, the potential infectivity of his feces should be recognized. Feces should be disinfected with chlorinated lime (5 per cent) or saponated solution of cresol (5 per cent) for a minimum of one hour unless other, more satisfactory methods of disposal are available. Flies and other insects should be eliminated from the patient's unit so that they do not carry feces from the patient to food. As in the care of all patients with enteric infections, all attendants should wash their hands carefully so that they do not inadvertently become carriers. The patient should also wash his hands after defecation. Remember that after the acute diarrheal phase, during which it is mainly the fragile trophozoites that are passed, the formed stools are often even more dangerous because of the presence of the durable cysts.

Disinfectant dips for fruits and vegetables have no proven value.

Treatment

Metronidazole (Flagyl) was once the drug of choice in the United States; in other countries diloxanide furoate (Furamide) was used for nondysenteric intestinal amebiasis. Since metromidazole has been found to be carcinogenic and mutagenic, diloxanide furoate is now also used in the United States together with chloroquine or diiodohydroxyquin in very low doses, since it destroys vision.

OTHER INTESTINAL PROTOZOA

The student who is interested in more detailed study of other Protozoa that have adopted the human intestinal tract as a place of residence will find the subject a fascinating and profitable one. Because the modes of transmission (fecal-oral) and controlling problems are alike for all, we here merely name and briefly describe only two of the common species. Like the entamebas, most of the intestinal flagellates and ciliates occur in both trophozoite and cyst form; the forms of the latter are diagnostically distinctive.

Giardia lamblia. In the trophozoite stage, these protozoans (which belong to the group of Mastigophora or flagellates) are fantastic in appearance (Fig. 26–1). They are found in stools of many normal persons in the cyst form. Trophozoites are flushed out only by rapidly moving diarrheic stools. They may at times cause irritations of the gallbladder and upper intestinal tract. Infection with *Giardia* is spoken of as giardiasis or flagellate diarrhea. Although annoying, they are not generally regarded as very dangerous pathogens, although they may cause persistent (10 days plus) diarrhea. They occur in some wild animals (e.g., beavers), which sometimes pollute water reservoirs. In areas where water is not properly filtered and chlorinated, epidemics have occurred.

Trichomonas hominis. The trophozoites of all common species of *Trichomonas* are egg- or pear-shaped, with three to five free flagella at the rounded anterior end and another forming the edge of an attached, undulating membrane that extends the length of the animal. All have a conspicuous nucleus and a pointed, spike-tipped posterior (Fig. 26–1). *T. hominis* is probably nonpathogenic but may at times cause a mild intestinal infection called trichomoniasis. Other trichomonads are the pathogen *T. vaginalis*, found in the genitourinary tract (Chapter 36) and *T. tenax,* a harmless inhabitant of the mouth. *T. hominis* does not form cysts as does *E. histolytica* or *G. lamblia.*

HELMINTHS (WORMS)

The term helminth means worm, but it is now generally restricted to pathogenic worms. Many helminths are parasitic in valuable agricultural plants, others in wild and domestic animals and humans. The most common human intestinal helminths are included in two large groups: the phylum Nematoda or roundworms (for example, hookworms) and the phylum Platyhelminthes, including Trematoda (flukes) and Cestoda or cestodes (tapeworms). Each of these groups contains parasites of human blood and tissue (Chapter 41).

Adult helminths are far from microscopic (see Table 26–1). As compared with unicellular microorganisms, they have very complex organic structures and life cycles. Some spend a part of their developmental period in the soil, in the sea, or in various hosts, such as fish, hogs, rats, snails, and the tissues of humans.

The fact of chief importance at this point is that the microscopic eggs (Fig. 26–4) or other developmental forms of most intestinal helminths are eliminated in the feces (or urine). Diagnosis is commonly made by finding the eggs or parts of worms (e.g., segments of tapeworms) in stools of patients. The portal

Table 26–1. Some Helminths

COMMON NAME	PRINCIPLE ENDEMIC AREAS	SCIENTIFIC NAME	STAGE USUALLY PRESENT IN FECES	REQUIRES FURTHER DEVELOPMENT IN
Nematodes				
Hookworm	moist tropics and warm temperate zones	*Necator americanus; Anyclostoma duodenale*	immature eggs	warm, moist soil
Pinworm (seat worm)	worldwide, especially temperate zones	*Enterobius vermicularis*	adults that lay eggs on skin	none; anus-to-mouth transmission
Whipworm	worldwide, except cold and arid areas	*Trichuris trichiuria*	immature eggs	warm, moist soil
Giant roundworm	moist warm and temperate zones	*Ascaris lumbricoides*	active, immature larvae	warm, moist soil
Filaria worms	moist tropics and subtropics	*Loa loa, Onchocerca, Wuchereria bancrofti*	none; transmitted by insects	—
Trichina worm	pork-eating peoples	*Trichinella spiralis*	none; occurs in tissues	hogs, rats
Cestodes				
Dwarf tapeworm	children, worldwide warm areas	*Hymenolepis nana*	mature eggs	none; anus-to-mouth transmission
Beef tapeworm	beef-eating peoples	*Taenia saginata*	mature eggs in segment of worm	muscles of cattle
Pork tapeworm	peoples eating poorly cooked pork	*Taenia solium*	as above	muscles of hogs
Fish tapeworm	peoples eating raw or poorly cooked fish	*Diphyllobothrium latum*	mature eggs	cool, fresh water; in water flea
Dog tapeworm	areas where man-dog contact is close	*Echinococcus granulosus,* etc.	eggs in dog feces	muscles and tissues of man
Trematodes				
Blood fluke	tropics	*Schitosoma mansoni*	mature eggs (in feces)	fresh water; in snails
Blood fluke	mainly Africa	*Schistosoma haematobium*	mature eggs (in urine)	as above

of entry for most intestinal worms is via the mouth or through the skin. For example, beef-tapeworm larvae are ingested in rare beef and pork tapeworm larvae in underdone pork (Fig. 26–5). Hookworm larvae develop from eggs in feces that have been deposited on moist, warm soil and penetrate the skin of bare feet. For convenience, data pertinent to prevention and control of the more important types of these worms are given in Table 26–1.

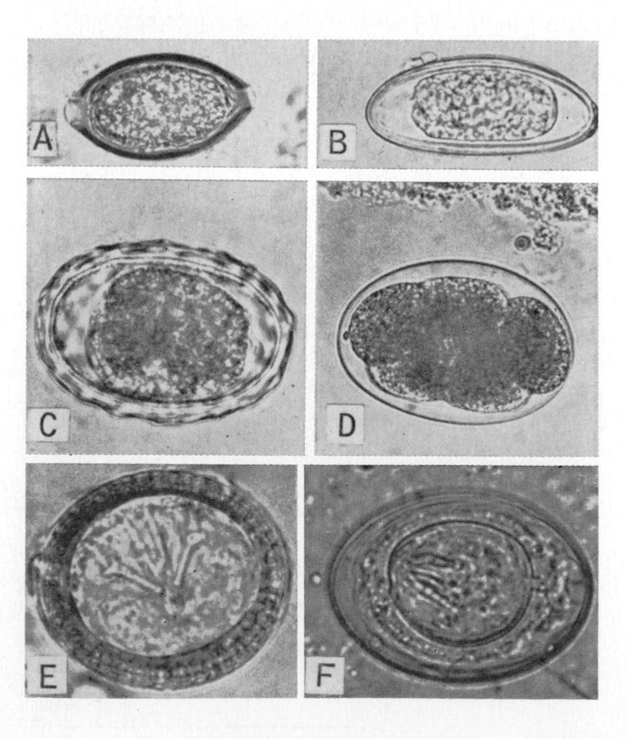

Figure 26–4 Some common nematode eggs: *A,* Whipworm, *Trichuris trichiura; B,* pinworm, *Enterobius vermicularis; C,* large roundworm. *Ascaris lumbricoides,* fertilized egg; *D,* hookworm. Some cestode eggs: *E,* Human tapeworm, *Taenia sp.* (×750); *F,* dwarf tapeworm, *Hymenolepis nana* (×750). (*A, B, C,* and *D,* courtesy of Dr. D. L. Roudabush, Ward's Natural Science Establishment, Rochester, N.Y. Photos by T. Romaniak. *E* and *F,* courtesy of Photographic Laboratory, AMSGS. Photos by Milt Cheskis.)

Pathogenic for Humans

PRINCIPAL SOURCE OF INFECTION OF HUMANS	USUAL PORTAL OF ENTRY TO HUMANS	IMPORTANT MEANS OF CONTROL	USUAL MEANS OF DIAGNOSIS IN LABORATORY
feces-contaminated soil	via skin in contact with soil	avoid skin contact with soil; sanitary disposal of feces; chemotherapy	microscopic examination of stool for eggs
eggs in perianal area. dust, etc.	mouth	frequent bathing; sanitary disposal of feces, chemotherapy	find eggs on perianal skin
infected soil, food, water	mouth	sanitary disposal of feces; avoid feces-contaminated soil	demonstration of eggs in stool
as above	mouth	as above	as above
mosquitoes, etc.	mosquito bite	mosquito control; chemotherapy	demonstration of microfilariae in blood
undercooked pork	mouth	adequate cooking of pork; sanitary garbage disposal	microscopic examination of excised muscle
eggs from feces-soiled clothing, etc.	mouth	cleanliness of perianal region and underwear; chemotherapy	find eggs in feces
undercooked or raw beef	mouth	prevent sewage pollution of pastures; avoid rare beef; chemotherapy	find segments (occasionally eggs) in feces
undercooked pork	mouth	sanitize hog-raising; avoid undercooked pork	as above
undercooked or raw fish	mouth	prevent sewage and fecal pollution of fresh waters; avoid undercooked fish	find eggs in feces
eggs in feces of dogs and other canines	mouth	avoid feces of dogs, especially in areas where they are numerous	immunological tests
sewage-polluted waters where snails abound	skin in contact with polluted water	sanitary sewage disposal; kill snails; do not go into polluted streams, pools, etc.	find eggs in feces
as above	as above	as above	find eggs in urine

Figure 26–5 Epidemiology of the taeniases (tapeworm infections). *Taenia saginata* is the beef tapeworm; *T. solium*, the pork tapeworm. Tropical Medicine, 5th ed. (From Hunter, G. W., III, Swartzwelder, J. C., and Clyde, D. F.: Philadelphia, W. B. Saunders Company, 1976.) The scolex is the head of the mature worm, by means of which it fastens itself to the intestinal mucosa; a proglottid is a sexually mature segment of the long "tape" of the worm as passed in the feces. It contains fertile eggs (Fig. 26–4).

FIVE GENERAL PREVENTIVE RULES

Five general rules may be applied to avoid most helminthic infections (and many other enteric infections as well): (1) avoid direct contact (e.g., with bare feet) with feces-contaminated soil, especially in warm, moist areas: (2) do not eat (in *any* area) "rare" or uncooked meats, fish, or shellfish; (3) avoid contact with feces-soiled clothing or persons, (4) avoid eating, drinking, or contact with raw food, vegetables, or water contaminated with feces, urine, or sewage of human or animal origin; and (5) avoid arthropod bites in all areas, especially in warm climates.

SUMMARY

Amebiasis, a protozoal disease, includes any infection caused by amebas, the most prominent of which is *Entamoeba histolytica,* the pathogen of amebic dysentery. This amebiasis, like others, is common in warm regions but is also present in temperate zones around the world. People become infected with cysts or trophozoites present in feces or sewage-polluted food or water. The disease may be chronic, but ulcers can undermine the intestinal lining and the trophozoite may burrow through the tissues, causing rupture of the intestine, peritonitis, and possibly death. The more resistant form of the protozoa is the cyst, which is flushed out when the host has diarrhea. The trophozoites multiply rapidly, but they are fragile forms of *Entamoeba histolytica.* Microscopic examination of feces is still the best diagnostic test; it helps to rule out *Entamoeba coli,* a harmless inhabitant of the intestine.

The fecal-oral mode of disease transmission also applies to infections with *Giardia lamblia* and *Trichomonas hominis,* both of which are less virulent pathogens than *Entamoeba.*

The term helminth means worm, but it is now generally restricted to pathogenic worms. Although many helminths are parasitic in valuable agricultural plants or in animals, the human intestinal pathogens concern us most. Much data on helminths pathogenic for humans are summarized in Table 26–1. The reader is well advised to examine this table. Five general preventive rules listed in this chapter, when rigorously applied, may prevent infections with helminths as well as other enteric infections. Briefly, these rules cite skin contact, the eating of uncooked or rare food, contact with feces, and arthropod bites as the most likely causes of these infections.

REVIEW QUESTIONS

1. What are the differences and similarities between amebic dysentery and bacillary dysentery? How is *Entamoeba histolytica* differentiated from *Entamoeba coli*? When is this differentiation essential?
2. How is amebic dysentery transmitted, and how may this disease be avoided? What damage is done to the patient by *Entamoeba histolytica*? How important are carriers in the transmission of amebic dysentery?
3. Describe the disease caused by *Giardia lamblia.*
4. Define the term helminth. How has its meaning changed over the years?
5. State five preventive rules that apply to helminthic infections. How do they also protect against enteric infections?
6. What are nematodes, cestodes, and trematodes? What species belong to them, and what diseases do they cause?
7. Explain the risks you take if you eat rare (undercooked) meat. It has been said that pork is more dangerous to eat than other meats. Is this true or false? Explain.

Supplementary Reading

Blecka, L. J.: Concise Medical Parasitology. Menlo Park, Cal. Addison-Wesley Publishing Company, 1980.

Burke, J. L.: The clinical and laboratory diagnosis of giardiasis. CRC. *Crit. Rev. Clin. Lab. Sci., 3*:373, 1977.

Center for Disease Control: Waterborne giardiasis outbreaks. *Morbid, Mortal. Weekly Rep., 26*:169, 1977.

De Léon, E.: Trichomoniasis. *In* Marcial-Rojas, M. D. (ed.): Pathology of Protozoal and Helminthic Diseases. Reprint ed. Melbourne, Fl., Robert E. Krueger, 1975.

Faust, E. C., Beaver, P. C., and Jung, R. C.: Animal Agents and Vectors of Human Disease. 4th ed. Philadelphia, Lea & Febiger, 1975.

Hunter, G. W., III, Swartzwelder, J. C., and Clyde, D. F.: Tropical Medicine. 5th ed. Philadelphia, W. B. Saunders Company, 1976.

Marcial-Rojas, M. D.: Pathology of Protozoal and Helminthic Diseases. Reprint ed. Melbourne, Fl., Robert E. Krieger, 1975.

Most, H. (ed.): Health Hints for the Tropics. 6th ed. Supplement to *Trop. Med. Hyg. News,* 1967.

Muller, R.: Worms and Disease: A Manual of Medical Helminthology. London. Heinemann, 1975.

WHO Expert Committee on Helminthiases: Soil-transmitted Helminths. Geneva, WHO Technical Report Series No. 277, 1964.

Wilmot, A. J.: Clinical Amoebiasis. Philadelphia, F. A. Davis Co., 1962.

ENTERIC VIRAL INFECTIONS

In addition to various kinds of bacteriophages, many viruses capable of infecting mammalian cells are found in the intestinal tracts of men and animals. One special group of viruses isolated from the intestine of humans is called the enteroviruses.

THE ENTEROVIRUSES

These viruses were classified in 1963 as picornaviruses. Picornaviruses are small (*pico,* small) and contain an RNA core *(rna).* The family *Picornaviridae* consists of three genera, the *Enteroviruses* (*entero,* intestine), the *Rhinoviruses* (*rhino,* nose), causing the common cold, and the *Caliciviruses* (*calici,* cap), found in swine and feline infections. The *Enteroviruses* include the *polioviruses,* the *echoviruses,* and the *coxsackieviruses* (see Table 6–1). All the enteroviruses that infect man multiply primarily in cells of the human gastrointestinal tract, and when they produce clinically recognizable disease it usually involves the central nervous system (CNS, brain and spinal cord). Possibly some of the reoviruses may be enteroviruses since they appear to multiply in the intestine and, like poliovirus, appear in feces or oral secretions and, in animals at least, affect the CNS. Suggested interrelationships of some enteroviruses are indicated in Figure 27–1. For a proposed (but still unsettled) classification of viruses, consult Table 6–1.

To assist your memory:

ECHO stands for Enteric, Cytopathogenic, Human, Orphan.[1]
ECBO stands for Enteric, Cytopathogenic, Bovine, Orphan.
NITA stands for Nuclear Inclusion Type A.
REO stands for Respiratory, Enteric, Orphan.
CHINA stands for Chronic Infectious Neuropathic Agents.

These viruses are all members of a large "spectrum" of viruses. They have certain properties in common, but each is different from the others. They can be differentiated from one another by means of immunologic tests and other laboratory procedures, as well as by the clinical conditions and pathologic changes they produce.

Another virus, that of *infectious hepatitis* (IH) or *epidemic viral hepatitis* (hepatitis virus A), must also be considered with the intestinal viruses because of its fecal-oral route of transmission. Strictly speaking, it is not one of the group called enteroviruses because it apparently multiplies in the liver, not in the intestine or CNS.

The viruses discussed in this chapter are among the smallest of viruses and are relatively stable and durable in environments outside the human body. For example, all are resistant to the fermentation, acidity, putrefaction, and other

[1]"Orphan" viruses are those not associated with any particular disease.

CHAPTER 27

conditions occurring in feces and sewage, and they appear to be more or less regularly transmitted in such materials. All are relatively resistant to certain disinfectants, and some (e.g., coxsackievirus and possibly epidemic hepatitis virus) may survive ordinary pasteurization processes. All are wholly resistant to common antibiotics. Enteroviruses are regularly found in sewage even after it has received some treatment.

Any of these viruses, on gaining entrance to a host, may produce little or no obvious disease, or "flulike" or coldlike episodes, or gastrointestinal distress of varying severity and symptomatology (especially coxsackie-, echo-, reo-, and polioviruses). Such conditions are rarely specifically diagnosed, and their cause usually remains unknown unless special laboratory investigations are made.

Enteroviruses, although commonly remaining unrecognized in the gastrointestinal or respiratory tracts, sometimes invade the blood (temporarily) and the CNS. They thus manifest one of their outstanding properties that distinguishes them from the hepatitis viruses: they are strongly *neurotropic*. Neurotropic viruses have marked affinity for the nervous tissues and characteristically cause disease of the CNS. The enteroviruses most commonly and distinctively cause anterior poliomyelitis (especially polioviruses), numerous polio-like conditions (especially the coxsackieviruses), and meningitis or encephalomyelitis (especially the echoviruses). The conditions caused by these viruses are frequently not clearly distinguishable from each other by clinical means.

Poliomyelitis

This disease is sometimes called "infantile paralysis," which is unfortunate since it occurs in persons of all ages and rarely causes permanent paralysis; however, most cases occur in the first 30 years of life and a large proportion before adolescence. More often the term "polio" is used by both layman and professional to designate this disease. Poliomyelitis is an acute febrile disease, and like most viral diseases it is usually sudden in onset, with headache, chills, nausea, and fever. Patients often exhibit extreme irritability and pain when being moved, and characteristically the muscles of the neck are held rigid. If much invasion of the nervous system occurs, especially of the parts called "the bulbar region" (medulla oblongata)—the exception rather than the rule—it results in injury or destruction of the motor cells of the anterior horns of the cord (Fig. 27–2), producing paralysis or "paralytic polio." Depending on where the lesions occur, one may speak of spinal poliomyelitis, bulbar poliomyelitis, or encephalitic poliomyelitis.

Following the fever there is sometimes paralysis of the legs and, in some cases, also of the arms and other muscles. This may or may not result in permanent injury, depending on the extent of damage to the nerves and the rehabilitation treatment. If the disease extends to the nervous mechanisms controlling respiration, death may quickly ensue from respiratory failure unless

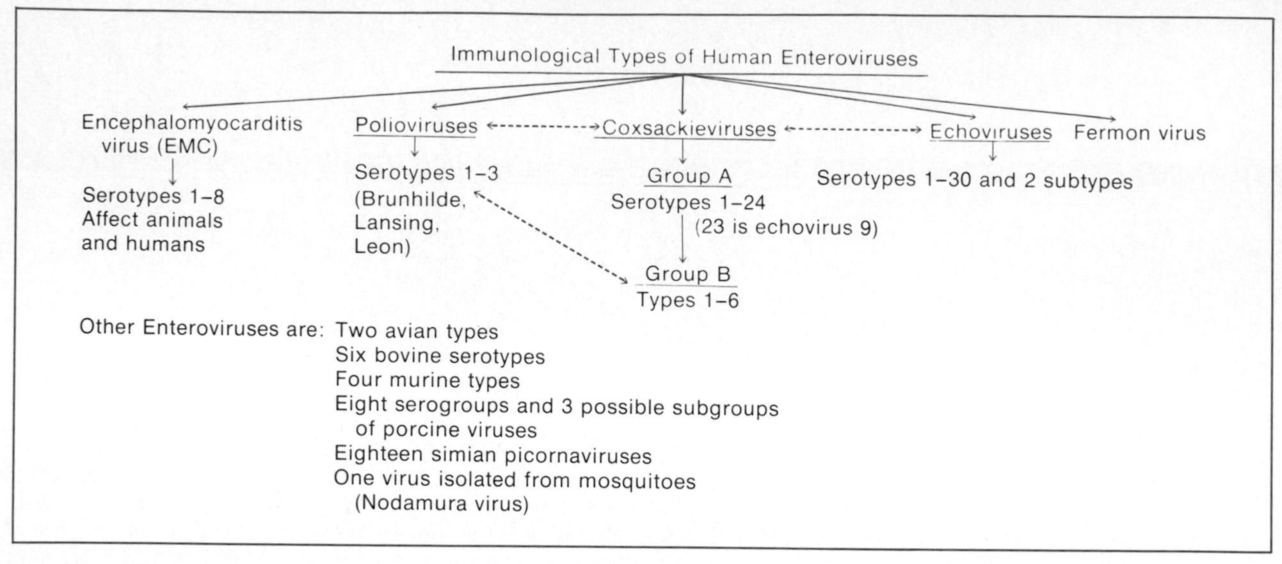

Figure 27–1 Suggested immunologic relationships between some enteroviruses. (Modified after Syverton, J. T.: *Am. J. Trop. Med. Hyg., 8.*)

some apparatus such as a respirator (formerly "iron lung") or a "rocking bed" is used to replace the action of the muscles of respiration. For each case of paralytic poliomyelitis, probably a hundred inapparent or mild, nonparalytic infections occur. These mild cases produce such signs and symptoms as slight fever, muscle soreness (myalgia), malaise, drowsiness, headache, constipation, and sore throat. They are not diagnostic and often pass unnoticed, ranging in severity from fairly severe "flulike" attacks to no apparent symptoms in healthy carriers. There are apparently thousands of such unrecognized infections with poliovirus every year. They confer immunity and do no harm to the individual, but they disseminate the virus widely. Humans are the only natural host for the poliomyelitis virus.

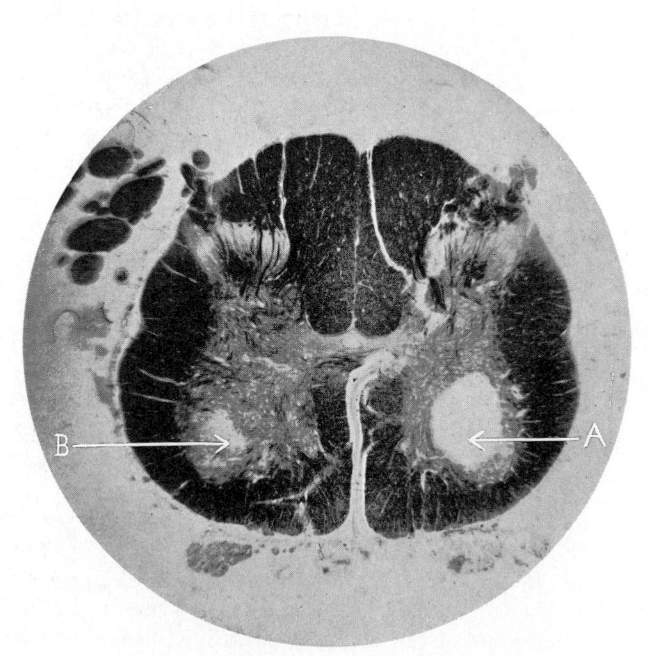

Figure 27–2 The results of acute anterior poliomyelitis. A cross section of the spinal cord in the lumbar region. The gray matter of a large area in the left anterior horn (*A*) has been completely destroyed by the inflammation, leaving a hole. The result of this lesion would be paralysis of the left leg. In the right anterior horn is an area (*B*) of partial degeneration, as evidenced by the light spot. (From Wechsler, I. S.: Clinical Neurology. 9th ed. Philadelphia, W. B. Saunders Company, 1963.)

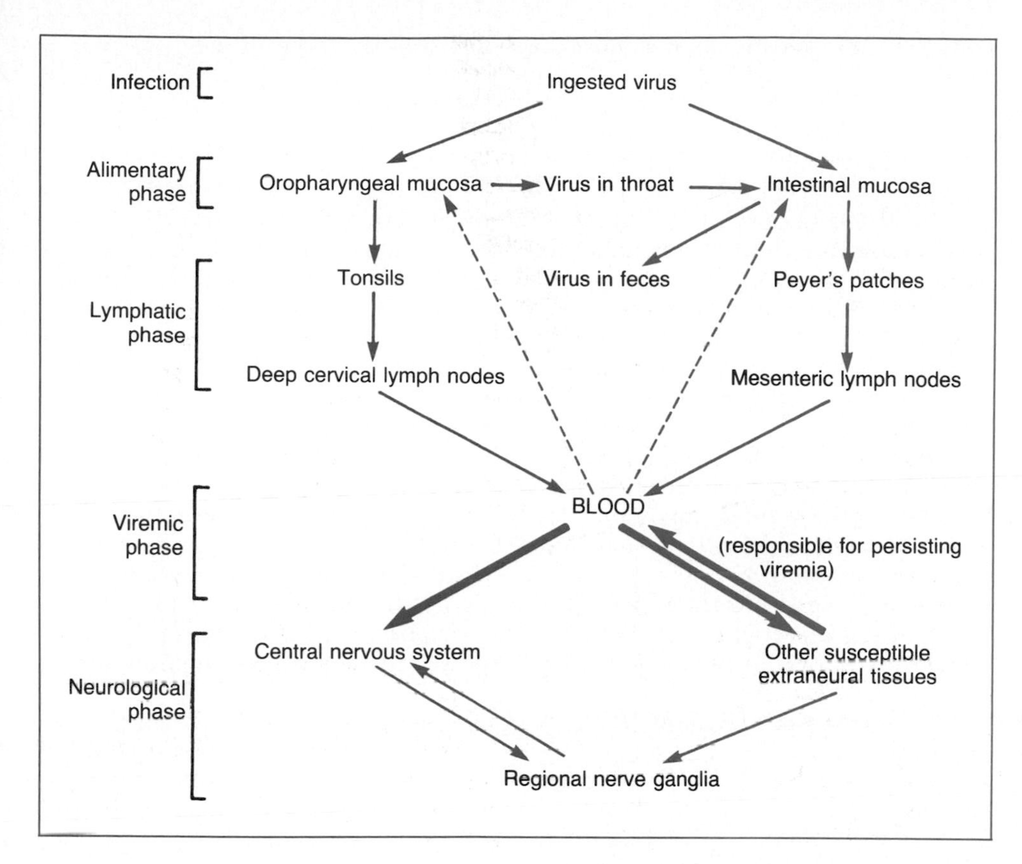

Figure 27–3 Pathogenesis of poliomyelitis. Model is based on a synthesis of data obtained in man and chimpanzees. (From Davis, B. D.: Microbiology. Hagerstown, Md., Harper and Row, 1980. Adapted from Sabin, A. B.: *Science, 123*:1151, 1956; Bodian, D.: *Science, 122*:105, 1955.)

The Poliomyelitis Virus. The poliovirus is thought to multiply in the Peyer's patches, the small intestine, the lymph nodes in the neck, and the tonsils. By way of the circulatory system the CNS may be infected, with the poliovirus spreading along axons to lower motor neurons involving the spinal cord or the brain (Fig. 27–3). Exposed nerve fibers after tonsillectomy may easily be infected by viruses present in the throat. The virus seems to be quite widespread. Its principal portal of entry appears to be the gastrointestinal tract, via the mouth. In the vast majority of infections it remains in the alimentary tract and is cast off in the feces. The virus is nearly always present in municipal sewage, especially during epidemic periods. This proves three things: (1) the virus is not sensitive to life in the outside world, (2) it must be abundant in the feces of patients and unrecognized carriers, and (3) it apparently is readily transmitted from person to person by personal and familial contact. It is probably only after the virus invades beyond the alimentary tract that it enters the walls of the nasopharynx and thus becomes transmissible by oral and nasal secretions during the acute phase of the illness. Prevention of spread is therefore doubly difficult since it involves the problems of both enteric and respiratory vectors.

Types of Poliomyelitis Virus. There are three main immunologic types of poliovirus: I, II, and III. Type I is known as the *Brunhilde* strain, type II as the *Lansing* strain, and type III as the *Leon* strain. Immunity to one type does not confer immunity to another. Formerly, a large proportion of urban adults had antibodies to all three types in their serum, yet had never had any disease recognizable as polio. They undoubtedly had had subclinical, immunizing *infections.* Such gratuitous immunizing infections are now much less common owing to the decline in incidence of poliomyelitis. Protection by vaccine is needed.

Polio Vaccines

In 1949, Enders, Weller, and Robbins (Nobel Prize winners) discovered how to cultivate the virus of poliomyelitis in tissue cultures made with various living cells such as those from human kidneys, monkey kidneys, human intestinal tissues, human embryonic tissues, and, most interestingly, human cancer (HeLa) cells. The virus multiplies in and kills the tissue cells. The two vaccines used are inactivated polio vaccine (IPV) and oral polio vaccine (OPV).

The Salk Vaccine (IPV). The vaccine developed by Dr. Jonas Salk and first demonstrated by public trial in 1954–55 was prepared by cultivating poliovirus types I, II, and III in cultures of live monkey kidney tissue cells. After several days of incubation the dead tissue cells and other detritus are removed and the fluid, containing the poliovirus, is collected. Formaldehyde is added in about 1:4,000 concentration to inactivate the virus. After allowing one week at about 37 C for the formaldehyde to act, the excess formaldehyde is removed. Tests are then made to determine that the virus is really inactive and that the material is not contaminated with bacteria. These and other tests being satisfactory, preservative is added, the vaccine is put into ampules, and it is then ready for use.

Use and Effectiveness of IPV. Various schedules of inoculation with Salk polio vaccine are recommended. One advises four doses: three about four to six weeks apart and a fourth about seven months later. Booster doses are variously recommended at intervals of one to five years. Persons of all ages may be vaccinated. High-risk patients[2] are vaccinated with IPV but not with OPV.

The vaccine does not necessarily prevent *infection* by poliovirus, although incidence of the disease in clinical and severe form rapidly declined to a very low level after Salk vaccine was generally utilized. Gastrointestinal infections by extraneous or "wild" active polio strains frequently occur after vaccination; however, the antibodies in the blood and tissues of vaccinated persons interpose a barrier between the wild virus and the central nervous system. Thus, the most important effect of the vaccine is to prevent severe *paralytic polio,* especially the crippling and deadly "spinal paralysis" or bulbar polio. The vaccine appears to be at least 90 per cent effective in preventing paralytic polio. The mild infections in vaccinated persons appear to have the beneficial, immunizing effect of booster doses of vaccine (Fig. 27–4).

Vaccination in Pregnancy. An important aspect of polio vaccination is the protection of pregnant women as early in pregnancy as possible. Expectant mothers appear to be more susceptible to polio than other adults, and the disease is much more dangerous in pregnancy. Immunization of the mother with vaccine also probably gives some passive protection to the infant for a few months.

The Sabin Oral Polio Vaccine (OPV). Studies by Sabin on attenuation of the virus of poliomyelitis led to production of an active but harmless virus vaccine. This is the trivalent oral polio vaccine (OPV), the vaccine of choice for infants and all individuals when there are no contraindications to the vaccination. The vaccine now may be made with virus cultivated in human cells. OPV has been administered to millions of people of all ages. Routinely, the vaccine is given in a small amount of prepared culture fluid in candy, food, or beverage.

[2]High-risk patients include those with immunodeficiency diseases such as agammaglobulinemia or hypogammaglobulinemia; those with altered immunity due to leukemia or other malignancies; and those receiving corticosteroids, antimetabolites, radiation therapy, or certain types of drug therapy.

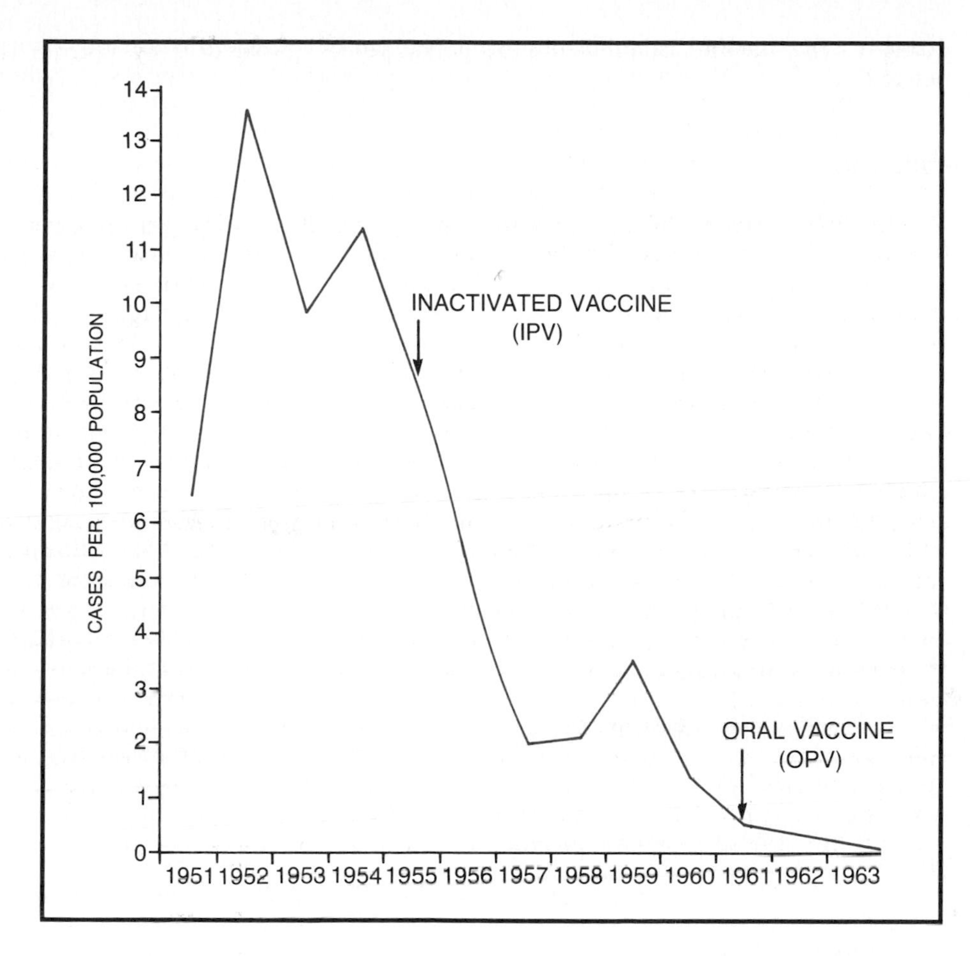

Figure 27–4 Reported annual case rates of paralytic poliomyelitis in the United States, 1951–1963. The case rates since 1963 have been less than 0.06 per 100,000 population. (From the Centers for Disease Control, *Morbidity and Mortality Weekly Report,* 1980 Annual Summary.)

For simplicity of administration, trivalent vaccine (mixture of three types) is commonly used in various schedules—e.g., two to four doses at six- to eight-week intervals beginning at the age of two to three months, with a final dose at one to two years of age. However, one dose of each virus may be given separately (to avoid mutual interference of the viruses), at intervals of three to six weeks, to persons over three months of age (to avoid inhibition of the virus by residual maternal antibodies in infants).

Following the oral administration of the Sabin vaccine, multiplication of the virus occurs in the alimentary tract and virus is discharged in the feces, with no perceptible symptoms. This not only immunizes the recipient but, like ordinary ("wild") virus, for a short time spreads among the recipient's contacts, probably by the fecal-oral route, thus disseminating the benefits. In addition to the excellent antibody response, the intestinal infection apparently results in an immunization of the intestinal tract as well, thus preventing reinfection. With the widespread use of this vaccine, poliomyelitis, previously a dreadful scourge in the world, is rapidly being forgotten, at least in the United States. Still, in 1976 14 cases of poliomyelitis were reported in the United States, 12 of which were paralytic; these probably occurred in unvaccinated individuals. In 1980 there were eight cases of paralytic poliomyelitis, four of which were imported from Mexico.

It is now abundantly clear that the Sabin vaccines are among the safest of the active (live) antigens in use.

Because the incidence of poliomyelitis in adults in the United States and

Canada is at present negligible, authorities have advised that the vaccine should be given only to those under 18 years of age, except in outbreaks or other circumstances in which adults might be exposed to unusually high risk.

Tissue Culture Diagnosis. The tissue culture technique is widely used to isolate the virus from feces, oral secretions, and similar sources. This permits not only epidemiologic investigations but also diagnosis by actual demonstration of the virus in specimens from the patient.

Transmission of Poliomyelitis

The possibility of transmission by sewage-polluted water exists, although the spread of epidemics of poliomyelitis is not suggestive of a waterborne disease like typhoid fever. Epidemiologic studies suggest, rather, transfer by feces-contaminated hands and objects, in much the same manner as bacillary dysentery is spread. Young children playing together constitute an ideal situation for this type of dissemination, especially in areas of low-grade sanitation and cleanliness. Flies have been shown to harbor virus in their gut for as long as 48 hours, but they are probably not important in its transmission.

Polio and Surgery. It has been shown that the virus finds entrance through the tissues around the gums, tonsils, and pharynx very easily, especially if there is injury such as would result from tonsillectomy, adenoidectomy, or tooth extractions. It is therefore recommended that such surgical procedures not be done during seasons of prevalence of poliomyelitis. Such seasons are now rare in the United States. Furthermore, paralysis appears more frequently in arms or legs that have recently been the site of injection of materials such as pertussis vaccine, antibiotics, or diphtheria toxoid. These injections certainly increase the risk of paralysis, possibly also of infection.

Prevention

The control of the poliomyelitis virus in caring for the now rare paralytic patient involves proper disposal of respiratory secretions during the first two to three weeks from onset and adequate disposal and disinfection of stools (as done in enteric infections) for four weeks from onset. These are arbitrary figures. There is little basis for exact rules, and the disinfection of feces from poliomyelitis patients is not performed by all communicable disease units or hospitals. There is every indication, however, that feces and feces-contaminated objects are very important vectors of the infection. Hands, as well as dishes and other objects used by the patient, should therefore be thoroughly disinfected to prevent the spread of the disease. It is known that poliomyelitis virus is inactivated by boiling and by adequate exposure to chlorine (chloride of lime, laundry bleach).

In addition to caring for poliomyelitis patients, the health team has the responsibility of encouraging all people to become vaccinated as recommended by local health officials.

Coxsackieviruses

The first coxsackievirus was isolated in 1948 by Dalldorf and Sickles from patients ill with a polio-like disease in the town of Coxsackie, New York. This virus resembles the poliomyelitis virus in size and in resistance to destruction in sewage. The coxsackieviruses, like the polioviruses, are found in feces as well as in nasopharyngeal washings and nervous tissue, and they produce diseases clinically resembling poliomyelitis. The coxsackieviruses include at least two groups, A and B, differentiated mainly by their effects on infant mice. Each group consists of numerous subgroups and types (Fig. 27–1): group A has 24 subgroups, and group B six serotypes. Like polioviruses, the coxsackieviruses are widespread and cause extensive epidemics, often simultaneously with poliomyelitis outbreaks. Simultaneous infections with both a coxsackievirus and a poliovirus are not uncommon.

Diseases attributed to specific serotypes of coxsackieviruses include herpangina, a children's disease with fever, sore throat, and abdominal pain; aseptic meningitis and mild paresis; a neonatal disease causing feeding difficulties and possible death; colds; minor summer illnesses; hand, foot, and mouth disease; myocardiopathy; and pleurodynia, a disease affecting the hearts of infants and fetuses.

Vaccines can doubtless be prepared against infection by several, if not all, coxsackieviruses. Because of the considerable number of different antigenic types, however, protection against all coxsackievirus diseases is not practicable at present.

Enteric Viruses in Urine. As will be pointed out in Chapter 33, at least one virus (mumps) previously thought to be associated only with the respiratory tract is now known to appear in the urine of patients. Similarly, coxsackievirus B has been shown to occur in urine, probably due to viral damage to the kidney. This requires that nursing precautions in coxsackievirus B infections (possibly in all intestinal viral infections) include handling not only feces and nasal secretions but also urine as infectious materials. In light of these findings, it is conceivable that any virus that injures the kidney may appear in the urine. This phase of the transmission of viruses requires further investigation.

Echoviruses

Studies of intestinal viruses have led to the discovery that there are many viruses in the bowel, as has been mentioned previously. Attention was drawn to some of these viruses not because they caused disease but because they destroyed tissue culture cells. After several such agents had been described by several workers, they were collectively designated echoviruses, i.e., "enteric" viruses producing cytopathogenic (cell-damaging) effects (see Fig. 6–11), of human origin, and "orphan" in relationship. (They are not orphans any longer but well-known viruses that may be associated with conditions ranging from mild respiratory diseases to very severe ones of the CNS [see page 405].) To shorten this description they were named ECHO *(Enteric, Cytopathogenic, Human, Orphan)* viruses.

Thirty-six different prototypes of echoviruses have been described. Undoubtedly more echoviruses will be found, and some will be reclassified into other groups. Several have been definitely associated with diseases, such as aseptic (nonbacterial) meningitis (types 2–6, 9, 14, 16) and summer diarrhea of infants (type 18 and others). The term *reoviruses* was also suggested as a new designation for the ECHO type 10 virus and others antigenically related to it, evidence of the close relationship between the two groups.

These viruses are found chiefly in, and cause diseases of, children and young people. The diseases are often influenza-like, frequently with a blotchy, red rash; they are not usually fatal. They often resemble nonparalytic polio, which in turn often resembles "flu."

Prevention of echovirus diseases centers on control of feces-borne and possibly urine-borne infection by methods already described, and also on control of respiratory seccretion-borne infection, especially if upper respiratory tract symptoms are present. Thus, echoviruses and numerous intestinal-respiratory viruses present a double, sometimes triple problem of control.

DIABETES

Diabetes mellitus is a disorder of carbohydrate metabolism. Although our knowledge of this and closely related diseases is extensive, its cause is still clouded in mystery and confusion. Very recent work seems to indicate that infection with coxsackieviruses may lead to diabetes. Studies in mice have shown that the encephalomyocarditis virus may damage the beta cells of the islets of Langerhans in the pancreas and thus cause diabetes. Autopsies of insulin-dependent early diabetics disclose virus-induced autoimmune responses, which supports the viral theory. This discovery has tremendous implications and is one more indication of how rapidly science progresses and how much there is yet to be learned.

VIRAL HEPATITIS

The term hepatitis is drawn from pathology and means inflammation of the liver due to any cause, mechanical, chemical, or biological. Several viruses are known to cause hepatitis in animals (e.g., in dogs, sheep, cattle, horses, swine, mice, ducks, and canaries), but there is no evidence that any of these is etiologically related to viral hepatitis in man. Viral hepatitis occurs in all parts of the world.

Two kinds of human hepatitis virus are the infectious hepatitis virus or hepatitis virus A (HAV) and the virus of homologous serum jaundice or hepatitis virus B (HBV). Since both are infectious, virus A is more accurately referred to as the cause of epidemic viral hepatitis; infection with virus B does not occur in truly epidemic form. Other viral hepatitis agents now recognized are HCV and HDV, both viruses that cause non-A and non-B hepatitis. Another virus, that of *yellow fever* (see Chapter 40), is not a true hepatitis virus, but the student should remember that it too causes jaundice, involving the liver, as do hepatitis virus A and virus B.

Epidemic Hepatitis (Infectious Hepatitis, IH). The virus of epidemic or *infectious hepatitis* (IH), or *epidemic jaundice,* called *hepatitis virus A,* once gaining entrance to a host, usually does not remain unnoticed in the gut but causes overt disease. It invades the blood and liver and often produces prolonged (weeks or months) inflammation in that organ (hepatitis), with resulting fever, regional pain, gastrointestinal distress, and, less commonly, the jaundice that gives the disease one of its older names—*catarrhal jaundice* (Fig. 27–5). Relapses are frequent. Intestinal carriers may occur, but if so they appear to be much less common than carriers of the enteroviruses and related enteric-respiratory viruses. Persons who carry hepatitis virus in their blood (healthy carriers) appear to be fairly common and are a real menace if selected for blood donations. Immune pooled gamma globulin is of value in prophylaxis of infectious hepatitis (IH) due to virus A: under proper conditions, it protects for a period of four to six months. Like poliovirus, virus A is excreted in feces. In 1980, 29,087 cases of infectious hepatitis were reported in the United States, whereas there were 19,015 cases of disease caused by virus B. Another 11,894 cases of reported hepatitis were unspecified as to type (Fig. 27–6); some of these were due to HCV or HDV.

Homologous Serum Hepatitis, SH (Hepatitis Virus B). We have de-

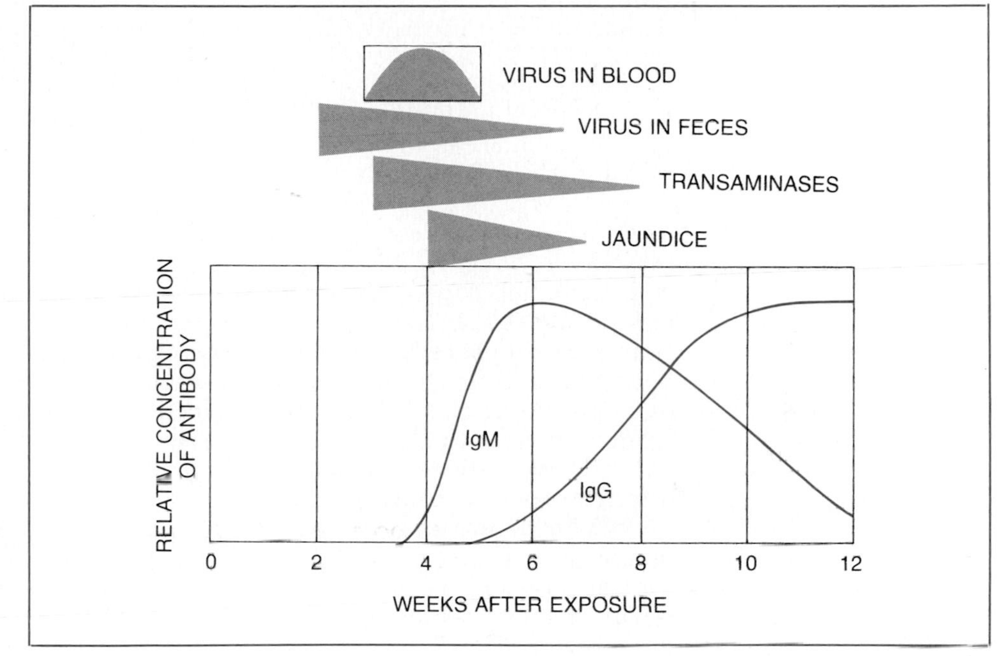

Figure 27–5 Immunologic and biologic events associated with viral hepatitis type A. (After Hollinger, F. B., and Dienstag, J. L.: Manual of Clinical Microbiology, 3rd ed. Washington, D.C., American Society for Microbiology, 1980.)

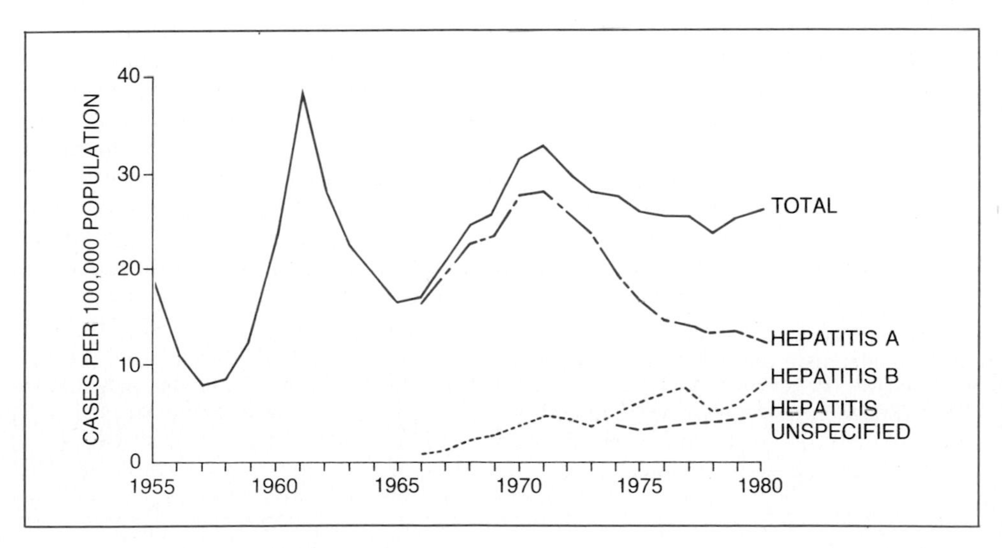

Figure 27–6 Reported annual hepatitis case rates in the United States, 1955–1980. Viral hepatitis continues to be one of the four most frequently reported infectious diseases in this country. The total number of viral hepatitis cases reported for 1980 was higher than the two previous years. Increases were observed for both hepatitis B and unspecified hepatitis, and there was a slight decrease in the number of hepatitis A cases. Persons 15 to 39 years of age were most affected by all types of hepatitis. (From the Centers for Disease Control, *Morbidity and Mortality Weekly Report,* 1980 Annual Summary.)

scribed one cause of hepatitis as *hepatitis virus A*. The second viral agent of hepatitis in humans, though closely related to and possibly a variant of virus A, is nevertheless antigenically distinct from it and, so far as is known, is not one of the intestinal viruses. It is called the virus of *homologous serum hepatitis* (SH), *serum jaundice, transfusion jaundice,* or *long incubation hepatitis,* or simply hepatitis virus B. The exact relationship between viruses A and B is not known, though the two are closely related and similar. Virus B (HBV) is a 42-nm double-shelled particle (the Dane particle) containing protein, circular double-stranded DNA, and DNA polymerase. The virus appears to be as yet inseparably associated—in the blood of serum hepatitis patients and carriers—with a very tiny (20 to 22 nm in diameter) particle consisting of two or more proteins. It may be separate or attached on the surface of the HBV. Having been found in association with serum hepatitis in Australia, it is known as the *Australia antigen,* also called Au/SH antigen, HBsAg, or HAA. This antigen is not identical to the virus itself: it contains neither DNA nor RNA. It is thought that HAA may consist of aggregates of viral protein that appear in the blood toward the end of the incubation period. Since it is diagnostic for SH, many tests for it exist: complement fixation, immunoelectrophoresis, fluorescent antibody tests, counterelectrophoresis, radioimmunoprecipitation, and gel immunodiffusion. Such tests are of enormous importance in the prevention of serum hepatitis due to transfusion of SH virus in the blood of prospective donors. Other virus-related antigens that have been identified are HBcAG, the hepatitis B core antigen, and HBeAG, hepatitis *e* antigen, which has some relation to the infectivity of blood from a carrier.

Virus B usually occurs in human blood or serum: hence it is called homologous serum hepatitis virus. HBsAg has been found in saliva, semen, vaginal washings and menstrual flow. (Could it be sexually transmitted?) The disease it causes differs in several clinical details from virus hepatitis: hepatitis due to virus B has a longer incubation period than that due to virus A, and its mortality rate ranges up to about 20 per cent.

Virus B was once thought to be transmitted only by artificial means: syringes, needles, dressings, blood plasma and serum, and so on. It was a mystery how, with no natural vector, it could have become widespread in the human race.[3] HBsAg (though not the SH virus itself) has been found in the feces and urine of infected persons, which supports epidemiologic data indicating that transmission of SH by sewage-polluted water may occur under some conditions. Virus B may be carried in the blood, apparently for many years.

Non-A, Non-B Hepatitis. Post-transfusion hepatitis B has been reduced 30 to 50 per cent as the result of surface antigen testing of blood. The hepatitis A virus has a short incubation period and thus is rarely transmitted by transfusions. Consequently, it is now thought that 50 to 70 per cent of post-transfusion hepatitis is due to HCV or HDV. Testing for increased blood alanine aminotransferase which is associated with these viruses, should reduce

[3]Some imaginative persons have suggested that the bloody swords, daggers, bayonets, and general quarrelsomeness and bloodthirstiness of our ancient (and not so ancient) ancestors disseminated the virus widely. Bloodletting was probably another means of transmission by way of the instruments used by premodern "physicians."

Table 27–1. Hepatitis Risks from Blood Transfusions

Type of Transfusion	Percentage of Patients Developing Hepatitis
All transfusions	2.8
Transfusions from voluntary donors	1.5
Transfusions from paid donors	5.3

the danger further. The risk of hepatitis from transfusions is noted in Table 27–1.

Prevention

Passive immunization against HAV has been used effectively, as stated earlier. FDA-approved hepatitis B vaccine (Hepatavax-B) has been found to be more than 90 per cent effective in preventing infection, if given in three doses. Non-A and Non-B post-transfusion hepatitis may also be reduced through prophylaxis.

With respect to virus A, precautions applicable to all enteric infections are required. The same dangers exist for viruses A and B concerning transmission by blood and certain blood derivatives, syringes, instruments, and so on.

One additional point must be stressed because it has bearing on all routines of disinfection and sterilization. *Hepatitis viruses are exceptionally resistant to heat.* They can resist boiling for 10 minutes, probably longer. They are also somewhat resistant to chemical disinfectants. Dried blood still remains infective!

All health workers must keep these viruses constantly in mind when dealing with any materials likely to be contaminated with human excreta or blood or blood derivatives. Syringes, needles, instruments, and bloody dressings may transmit the virus via tiny scratches on the hands or because of inadequate sterilization; this can be avoided by using disposable syringes and needles. Thermometers used by patients with viral hepatitis, and usually other communicable diseases, are isolated with the patient and discarded after the patient is discharged. Disposable plastic thermometers are now in use in most modern hospitals. Each patient should have his own bedpan. This applies to any infectious disease borne by feces and urine.

SUMMARY

The enteroviruses multiply in cells of the human gastrointestinal tract and later involve the central nervous system (CNS, brain, and spinal cord). These viruses include the polioviruses, the echoviruses, and the coxsackieviruses. The infectious hepatitis virus (IH) has a fecal-oral route of transmission like the enteroviruses, but unlike them it multiplies in the liver, not in the intestine or CNS. Enteroviruses are regularly found in treated sewage. They are resistant to fermentation, acidity, putrefaction, and pasteurization and thus survive conditions that destroy other viruses.

Poliomyelitis (infantile paralysis) is an acute febrile disease with sudden onset, producing paralysis and possibly death. However, many inapparent, mild, nonparalytic infections occur. These may pass unnoticed, but they confer immunity. There are three main types of poliomyelitis virus—the Brunhilde, Lansing, and Leon strains—and two very successful vaccines—the Salk vaccine (IPV—inactivated polio vaccine) and the Sabin vaccine (OPV—oral polio vaccine). High-risk patients should receive the IPV, although the OPV is considered very safe.

The two groups of coxsackieviruses, like the polioviruses, are found in feces as well as in nasopharyngeal washings and nervous tissue. They produce diseases clinically resembling poliomyelitis.

ECHO (enteric cytopathogenic human orphan) viruses are intestinal pathogens that cause an influenza-like, blotchy red rash condition, mostly in children, that is usually not fatal.

Infectious hepatitis (IH), also called epidemic hepatitis or catarrhal jaundice, gains entrance to a host by way of the gastrointestinal tract, often from an intestinal carrier who should not handle food. Unlike IH, serum hepatitis (SH) is thought to infect blood directly. It is also known as transfusion jaundice, homologous serum hepatitis, or long incubation hepatitis. However, SH is also now attributed to sewage-polluted water as well as to nonsterile syringes, needles, knives, and plasma. The virus of IH is known as HAV (A virus), and the virus of SH is HBV. All types of hepatitis, as well as yellow fever and other liver ailments (such as injury or cirrhosis of the liver), are known to produce jaundice, a yellow discoloration of the skin and eyeballs. All health workers should remember that hepatitis viruses are exceptionally resistant to heat and that dried blood remains infective if it contains any hepatitis viruses.

REVIEW QUESTIONS

1. Name and discuss different enteroviruses, polioviruses, echoviruses, and coxsackieviruses. Where do you expect to find enteroviruses and why?
2. Name three strains of polioviruses and two vaccines used to prevent poliomyelitis. Why does this disease occur so much less frequently now than it once did? Can we stop vaccination against poliomyelitis, as we did with smallpox? If not, why not?
3. How did echoviruses get their name? Is this name still justified? What type of diseases do echoviruses produce?
4. Differentiate between infectious hepatitis, serum hepatitis, yellow fever, and other jaundice-causing conditions. What is jaundice? List the viruses that cause hepatitis.
5. What precautions need to be taken with patients who have hepatitis?
6. What have you learned about sewage in this chapter?

Supplementary Reading

Burnet, F. M., and White, D. O.: Natural History of Infectious Disease. 4th ed. New York, Cambridge University Press, 1972.

Carver, D. H., and Seto, D. S. Y.: Production of hemadsorption-negative areas by serums containing Australia antigen. *Science, 172*:1265, 1971.

Gerin, J. L., Holland, P. V., and Purcell, R. H.: Australia antigen: Large-scale purification from human serum and biochemical studies of its proteins. *J. Virol., 7*:569, 1971.

Goodheart, C. R.: An Introduction to Virology. Philadelphia, W. B. Saunders Company, 1969.

Horsfall, F. L., Jr., and Tamm, I. (eds.): Viral and Rickettsial Infections of Man. 4th ed. Philadelphia, J. B. Lippincott Co., 1965.

Huckstep, R. L: Poliomyelitis: A Guide for Developing Countries. Edinburgh, Churchill Livingstone, 1975.

Luria, S. E., et al.: General Virology. 3rd ed., New York, John Wiley & Sons, Inc., 1978.

Lwoff, A., and Tournier, P.: The classification of viruses. *Ann. Rev. Microbiol., 20*:45, 1966.

Melnick, J. L., and Wenner, H. A.: Enteroviruses. *In* Lennette, E. H. and Schmidt, N. J. (eds.): Diagnostic Procedures for Viral, Rickettsial, and Chlamydial Infections. New York, American Public Health Association, 1979.

Pumper, R. W., and Yamashiroya, H. M.: Essentials of Medical Virology. Philadelphia, W. B. Saunders Company, 1975.

Stanley, W. M., Valens, E. G., et al.: Viruses and the Nature of Life. New York, E. P. Dutton & Co., Inc., 1961.

Viral Hepatitis: Report on a Working Group. Copenhagen, Regional Office for Europe, WHO, 1976.

WHO Expert Committee on Hepatitis: Second Report. Geneva, WHO Technical Report Series No. 285, 1964.

WHO Scientific Group: Human Viral and Rickettsial Vaccines. Geneva, WHO Technical Report Series No. 325, 1966.

Zuckerman, A. J.: Human Viral Hepatitis. New York, American Elsevier Publishing Co., 1975.

FOODS AS VECTORS OF MICROBIAL DISEASE; SANITATION IN FOOD HANDLING

FOOD POISONING AND FOOD INFECTION

Most foods contain living bacteria unless they have just been thoroughly heated or made sterile in some other way. Many foods, especially those that are not very acid, salty, or syrupy, serve as good culture media for bacteria. Usually the microorganisms encountered in appropriately handled and prepared food are harmless. Many such bacteria can grow rapidly in food that is not properly refrigerated or cooked, and they may spoil the food. If pathogenic bacteria are present, they may cause food poisoning or food infection or both (Table 28–1).

The term *food poisoning* is often used as though it were synonymous with the term *food infection*. However, the two are fundamentally different with regard to both cause and prevention, and they should be carefully distinguished by professional personnel.

Food Poisoning

As medically defined, food poisoning is an illness caused by ingestion of food containing: (a) naturally poisonous substances (e.g., certain kinds of scombroid fish,[1] or fungi such as *Amanita phalloides* or ergot growing on rye); (b) food accidentally poisoned, as with rat or roach poison; (c) poisons of plant origin such as tung nuts, rhubarb leaves, and water hemlock; or (d) microbial toxins. The first types of food poisoning (a,b,c) are not discussed here. Microbial food poisoning (d) is an illness caused by ingestion of food containing toxins formed by extensive growth of toxigenic microorganisms in the food *prior* to its being ingested. The high potency of these toxins is indicated in Table 28–2.

Growth of toxigenic microorganisms *in the patient* (i.e., infection) is not essential to food *poisoning* and rarely occurs. Examples of classic food poisoning follow.

The Botulism Organism. *Clostridium botulinum* resembles all other clostridia (e.g., *Cl. tetani* and *Cl. perfringens*) in having complex organic nutritional requirements, in being a gram-positive, motile, saprophytic, spore-bearing anaerobic rod, and in being found in the soil. Formation of its large endospore gives it a snowshoe-like form, the basis of the name of the genus (from Greek *kloster*, spindle). Like *Cl. tetani*, it cannot invade healthy tissues.[2] How then, does it produce disease?

Clostridium botulinum, an obligate anaerobe, can grow well in such anaerobic

---

[1]Scombroid fish poisoning, usually attributed to the family *Scombridae*, occurred in 1980 in Detroit and Chicago when restaurants serve mahimahi *(Coryphaena hippurus)*, also called blue dolphin.

[2]A few cases of botulism resulting from growth of *Cl. botulinum* in soil-contaminated wounds have been reported.

CHAPTER 28

places as the center of large sausages (*botulus* is the Latin for sausage) and in canned foods that are not too acid, such as canned spinach, asparagus, beans, meat, or corn. Spores of the organism are enclosed in the cans with soil on improperly washed vegetables or other food. If the spores are not killed by the "processing" or autoclaving at the canning factory or in the home, they may germinate and multiply vigorously inside the container. Some of these spores are among the most heat-resistant organisms known. A related and very similar species has been called *Cl. parabotulinum.*

When growing (in sealed containers), *Cl. botulinum* gives off extremely potent toxins. Botulinal toxins, when swallowed, frequently prove fatal. The appearance, taste, or odor of the poisoned food may or may not be bad. Food often appears to be normal and gives little or no hint of the presence of toxins that cause the forms of food poisoning called botulism.

Seven types of *Cl. botulinum* are known. Of these, types A, B, E, and F cause botulism in humans. Their corresponding toxins A and B are most commonly found in cases of poisoning. Type E toxin is associated with botulism resulting from eating contaminated fish.

Botulism. This disease is due to the effect of botulinal toxin on nerves that activate muscles. The first symptoms—nausea, dizziness, gastrointestinal distress, and diarrhea—appear a few hours to one week after eating, depending on the severity of the poisoning. Unlike tetanus toxin, which irritates motor nerve cells, producing tetanic convulsions, botulinal toxin blocks nerve terminals (myoneural junctions) so that paralysis occurs. This paralysis progresses downward from the eye to the face, throat, speech, swallowing, arms, and so on. Prognosis depends largely on the amount of the toxin swallowed. When the thoracic muscles become fully involved, respiration is impossible and, in the absence of artificial respiration, death supervenes.

Prevention of Botulism. Protection against botulism is easy because the toxin (not the spores!) is destroyed by 10 minutes of boiling. For absolute safety, all home canned foods should be boiled, *after* opening the can, for at least 10 minutes before eating. Botulism from commercially canned foods is uncommon in the United States. However, every once in a while some isolated case reminds the public that this danger is still with us. Through the media we are told what cans are to be withdrawn from sale as being unsafe. The canners themselves take excellent precautions against botulism by clean preparation and thorough autoclaving of all canned goods. Any cans that show even slightly bulging ends (due to gas from bacterial fermentation within) should be discarded, as well as any cans showing other evidences of fermentation, acid or gas formation, or leaks. Never eat or even taste sour, spoiled, or discolored food from cans or jars. Unless the food is first thoroughly cooked, it is inadvisable to throw such foods to dogs, hogs, or chickens, since they also may die of botulism.

BOTULISM ANTITOXIN. There is an efficient antitoxic serum that may be used to prevent botulism. This trivalent botulinum antitoxin against A, B, and E toxins is made available to physicians by the National Center for Disease

Table 28–1. Most Commonly Found Food-Borne Pathogens

ENTEROVIRUSES and epidemic hepatitis virus (Chapter 27).
FAMILY ENTEROBACTERIACEAE (Chapter 24)—Gram-negative, nonspore-forming, intestinal rods; easily killed by heat and disinfectants.
Genus *Salmonella:* Typhoid, paratyphoid, and "food-poisoning" bacilli,
Genus *Shigella:* Dysentery bacilli; much like typhoid bacilli.
Genus *Escherichia:* Enterotoxigenic *E. coli.*

FAMILY BACILLACEAE (Chapter 37)
Genus *Clostridium:* Gram-positive, spore-forming, soil-derived rods; resistant to heat and disinfectants. *Cl. botulinum* causes botulism, a dangerous form of food poisoning; *Cl. perfringens* causes acute but transitory gastroenteritis.

GRAM-POSITIVE COCCI (Chapter 29)
Genus *Staphylococcus;* Causes one form of food poisoning (enterotoxin).
Genus *Streptococcus:* Beta type sometimes contaminates milk and other foods.

GRAM-NEGATIVE AEROBIC COCCOBACILLI (Chapter 25)
Genus *Brucella:* Gram-negative, nonspore-forming, animal-derived rods easily killed by heat and disinfectants, cause brucellosis (undulant fever) in man.

FAMILY MYCOBACTERIACEAE (Chapter 32)
Genus *Mycobacterium:* Bovine and human tuberculosis via meat and milk.

PROTOZOA, like *Entamoeba histolytica* (Chapter 26).

HELMINTHS, like tapeworms (Chapter 26).

Control in Atlanta, Georgia. Unfortunately, diagnosis of botulism is often delayed. As in tetanus, after definite symptoms have appeared, the patient's chances of recovery are not very good, even with large doses of serum. There are seven serologic types of botulinal neurotoxins. A, B, and E are most common in human outbreaks, F less so. C type causes botulism in fowl, and D most often affects cattle. For therapeutic and preventive purposes polyvalent serum, effective against several types of botulinal toxins, is generally used.

In managing cases of botulism, no isolation precautions are indicated and no special attention to infectious disease technique is needed, since no infection exists.

Four Categories of Botulism. Cases of human botulism in the United States are now classified into four categories: (1) food-borne botulism, due to the ingestion of botulinal toxin in contaminated food; (2) infant botulism, caused by the absorption of botulinal toxin into the blood from the intestinal tract of infants containing multiplying *Cl. botulinum* cells; (3) wound botulism, resulting from *Cl. botulinum*, producing toxin in an infected wound; and (4) cases of botulism in individuals other than infants where the cause of the toxicity is unknown. In 1980, 89 cases of botulism were reported in the United States: 18 of these were food-borne and 68 were infant botulism.

Clostridium perfringens. This gram-positive rod resembles *Cl. botulinum* in being distributed in soil, forming heat-resistant spores, and growing well in many foods under anaerobic or partly anaerobic conditions. It is commonly present in feces. It is not as strict an anaerobe as *Cl. botulinum* and is discussed more fully as a cause of wound infection (gas gangrene, Chapter 37). It is mentioned here because some strains appear to cause food poisoning characterized by acute gastroenteritis, usually beginning 8 to 14 hours after eating and lasting for 6 to 12 hours.

The *Cl. perfringens* toxins A, B, C, D, and E presumably are destroyed by heating, in this resembling *Cl. botulinum* toxin. Poisoning by *Cl. perfringens* type A strains (at least nine antigenic types are known), which produce an acute poisonous enterotoxin, may therefore be avoided by clean preparation and prompt eating or refrigeration of foods. Poisoning by *Cl. perfringens* is often associated with cooked meats. In preparing canned foods the same precautions

Table 28–2. Toxicity of Bacterial Endo- and Exotoxins Compared with Some Other Poisons*

POISON	CHEMICAL NATURE	MOLECULAR WEIGHT	TOXIC DOSE (μg)	TOXICITY (PER UNIT WEIGHT TEST ANIMAL) COMPARED WITH STRYCHNINE
Strychnine	$C_{21}H_{22}O_2N_2$	334	10 (lethal, mouse)	1
Tetrodotoxin (fish poison)	$C_{11}H_{17}O_8N_3$	319	0.2 (lethal, mouse)	50
Crotactin (snake poison)	protein	?	1 (lethal, mouse)	10
Ricin (plant poison)	protein	36,000	0.1 (lethal, mouse)	100
Bacterial endotoxins	protein-poly-saccharide-lipid	1,000,000	100 (lethal, mouse)	0.1
			200 (lethal, sensitive strain rabbit).	5
Botulinum A toxin (crystalline)	protein	1,130,000	0.00002 (lethal, mouse)	500,000
Botulinum B toxin	protein	1,000,000?	0.00002 (lethal, mouse)	500,000
Botulinum D toxin	protein	1,000,000?	0.00001 (lethal, mouse)	1,000,000
Tetanus toxin (crystalline)	protein	67,000	0.00002 (lethal, mouse)	500,000
Cl. perfringens epsilon toxin	protein	40,500	0.1 (lethal, mouse)	100
Diphtheria toxin (crystalline)	protein	72,000	0.05 (lethal, guinea pig)	5,000
Dysentery neurotoxin	protein	82,000	0.002 (lethal, rabbit)	500,000
Staphylococcal leucocidin:				
F Component (crystalline)	protein	32,000	0.004 /ml	
S Component (crystalline)	protein	38,000	0.004 /ml (lethal, macrophages	
Staphylococcal enterotoxin	protein	24,000	2 (emesis, monkey)	
Staphylococcal alpha toxin	protein	44,000	1 (lethal, mouse)	10
Staphylococcal delta toxin	protein	68,000	2/ml (50% hemolysis of 1% red cells)	
Streptococcal erythrogenic toxin	protein	27,000	0.0000005 (skin reaction, man)	
Bacillus megatherium megacin	protein	51,000	0.01/ml (inhibits bacterial growth)	

*The reported values of toxic doses have been rounded off. (From Florey, Lord: General Pathology. 4th ed. Philadelphia, W. B. Saunders Company, 1970.)

should be used as for prevention of botulism. The *Cl. perfringens* enterotoxin is heat-labile and resembles *Escherichia coli* enterotoxin in that it produces excessive self-limiting diarrhea.

Staphylococcal Food Poisoning. Another important and common cause of food poisoning is the toxigenic *Staphylococcus aureus.* Staphylococci grow in many foods, especially precooked hams, milk, custards, cream fillings, and salads. Some strains of these staphylococci produce powerful exotoxins (*enterotoxin,* Chapter 29) when they grow in food. When the food is eaten, food poisoning results, with severe but transitory gastroenteritis, nausea, vomiting, diarrhea, and marked weakness or prostration. Severity depends on the amount of toxin swallowed. No infection occurs. These symptoms usually appear within 2 to 12 hours after eating the toxin. This permits differentiation from salmonellosis, which produces very similar symptoms, but only after a necessary incubation period, usually 12 to 24 hours. Staphylococcal food poisoning is rarely or never fatal per se.

Five serologically different staphylococcal enterotoxins are known, designated as types, A to E. Most cases of food poisoning are caused by types A and D. Enterotoxin B has been implicated in a condition called staphylococcal pseudomembranous enterocolitis.

The enterotoxigenic species of staphylococci are usually, but not always, of the *aureus* type. They ferment lactose and mannitol, digest gelatin, and, in addition to enterotoxin, produce an enzyme called *coagulase,* which causes citrated blood plasma to coagulate. In contrast to the heat-labile botulinal toxin, staphylococcal enterotoxin can resist 100 C for at least 30 minutes and probably longer. Once formed in food, *cooking does not necessarily destroy it,* although the staphylococci that produced it may easily be killed by heat.

Other Enterotoxin-Producing Organisms. Recent reports indicate that *Bacillus cereus* grown on fried rice produces an enterotoxin that causes food poisoning of brief duration.

Salmonella is often thought to produce "food poisoning" since it causes *gastroenteritis,* often with sudden and violent onset. However, since it takes large numbers of the organism in the gastrointestinal tract to cause the *infection,* any toxin that may be produced is of minor consequence in the pathology of the salmonellas.

Vibrio parahaemolyticus, a halophilic (salt-loving) marine bacterium found in shellfish, crustaceans, and other seafood, causes about half of all cases of bacterial food poisoning in Japan. An enterotoxin has not yet been isolated from this pathogen.

Mycotoxins. These are poisonous by-products of the growth of certain eukaryotic fungi in foods. They have caused serious economic losses to stock and poultry growers when their animals were fed moldy products, mainly vegetables such as ground peanuts, rice, and corn. The role of mycotoxins in human disease may be more important than is generally known but is still to be fully evaluated. Among the mycotoxins are the *aflatoxins,* produced in foods by *Aspergillus flavus,* and a group of toxins, the *rubratoxins,* produced by *Penicillium purpurogenum* and *P. rubrum.* Others are continually being discovered.

Food Infection

By definition, infection is the multiplication of any pathogenic microorganism in human, plant, or lower animal tissues. In food infection, unlike food poisoning, there need be no preformed toxins in the food: only a few living microorganisms can cause an infection. Once ingested by a susceptible person, these organisms begin to multiply. Not until they have grown to considerable numbers in the body do they cause symptoms of food infection. The principal sources of pathogens in foods are infected animals, infected persons who handle or inspect foods, or contaminated ingredients in the foods.

Contamination by Animals. Animals may contaminate food in at least two ways. They may pollute it with their excreta (as do flies, roaches, rats, and mice), or flesh that is used as food may come from infected animals directly. For example, the meat and dairy products of tuberculous cattle can be highly infectious; flesh and milk of cattle, swine, or other animals with brucellosis may cause human brucellosis if the meat or milk is not properly handled. Hunters and market workers not infrequently contract "rabbit fever" (tularemia, Chapter 39) from infected wild rabbits they handle. Pork, poultry, and eggs are often infected with *Salmonella* and have caused numerous epidemics of salmonellosis.

Salmonella typhi, cause of typhoid fever, is often found in oysters and other seafoods if they are taken from sewage-polluted fishing grounds, a practice prohibited by health authorities. Epidemic viral hepatitis is now known to be transmitted in the same way. *Salmonella typhimurium,* a cause of one form of food infection, is a common (though not the only) *Salmonella* parasite of mice and rats. Excreta of these animals can sometimes get into or on food. Rats also transmit, in their urine, species of *Leptospira* (Chapter 25) that cause hemorrhagic jaundice or meningitis; both are forms of leptospirosis. Vermin should not be tolerated around food or elsewhere. Trichina worms *(Trichinella spiralis)* are almost always present in pork, and tapeworms are not uncommon in beef and raw fish. These foods may be made safe by *thorough* cooking.

Contamination of Foods by Persons. In addition to the pathogens just mentioned, which are commonly derived from infected animals, directly or indirectly, food handlers may contaminate food.

The organisms of respiratory diseases—e.g., colds, influenza, scarlet fever, diphtheria, and tuberculosis—are often present in saliva droplets and on the hands of food handlers, especially if those persons are carriers of the pathogens. Toxin-producing staphylococci (see page 436) often are found in respiratory secretions and in pimples and boils on the hands and forearms of food handlers, and thus may gain entrance to food. Shoppers in stores or cafeterias where unwrapped foods are displayed in open cases obviously transmit microorganisms to foods by sneezing, talking, and coughing over them and by handling them with soiled hands. A thumb in the plate that is handed over to the customer is the rule and not the exception in some cafeterias.

Intestinal pathogens such as the virus of epidemic hepatitis, *Salmonella, Shigella,* and cysts of *Entamoeba histolytica* may be on the hands of untidy kitchen workers and other food handlers who are carriers.

Other Food Microorganisms. Also on the list of food pathogens are opportunistic pathogens, certain saprophytic bacteria such as *Proteus* species (commonly *Proteus mirabilis,* among others). *Clostridium perfringens* (Chaper 37), and *Pseudomonas aeruginosa,* certain intestinal cocci (e.g., *Streptococcus faecalis*), and some others that can grow in food and decompose it. Some of the products of decomposition of food by these bacteria may be irritating to the intestine. There is little reason to suppose that ordinarily these organisms (except *Cl. perfringens*) or their products cause any but the mildest irritations of the gastrointestinal tract, if any at all. Certain strains of *Escherichia coli,* including enterotoxin producers (Chapter 24), cause enteritis, especially in infants, and can grow in many foods, including unhygienically prepared babies' formulas.

Because spoiled food is esthetically objectionable and unpalatable and may cause gastroenteritis, it should never be eaten, especially not by infants, children, or ill patients. The mere fact that large numbers of bacteria are present in food does not in itself imply that the food is harmful. It is the kind of bacteria that determines the harmfulness of the food. Various foods are prepared by the cultivation of certain bacteria: sauerkraut, buttermilk, alcoholic beverages, and cheeses are examples.

"Ptomaine Poisoning." Ptomaines are products of protein decomposition that is so extensive that the food is partly liquefied and offensive to sight, taste, and olfactory sense. Such putrefied matter is never accepted as food. The condition that used to be called "ptomaine poisoning" is now known to be, in the vast majority of cases, either infection by *Salmonella* species or poisoning due to enterotoxigenic microorganisms. Ptomaine poisoning, as once commonly thought of, *does not exist.*

Food Infections vs. Food Poisoning. All pathogenic microorganisms are poisonous because they consist of poisonous components, such as a cell wall. Therefore, their presence in *any* part of the body proper is a sort of poisoning (e.g., "blood poisoning," streptococcal *infection* of the blood). Possibly because of this, but more probably because of similarity of symptoms and means of transmission, food infections are often loosely called food poisonings.

Several types of "food poisoning" are due to infection by *Salmonella* species, such as *S. arizonae* (see Chapter 24). These may result in enteritis that can range in intensity from mild diarrhea to severe (rarely fatal) "poisoning," the latter commonly due to *S. typhimurium.* Serotype 1 of *Shigella dysenteriae* (Shiga's bacillus) produces, in addition to its endotoxin (cell wall), a potent exotoxin. Shiga dysentery is thus primarily a food *infection* complicated by poisoning due to formation of toxin *in the patient, not* in his *food* (see page 375).

Similarly, as previously mentioned (see Chapter 24), enterotoxins active in the upper small intestine and resulting in acute diarrhea, especially in infants, are produced in the patient following infection by certain strains of *Escherichia coli.* At least 14 different serotypes of this enterotoxigenic type of *E. coli* are known.

Another type of "food poisoning," including about half of all cases reported in Japan, is due to infection by *Vibrio parahaemolyticus,* a halophilic marine bacterium common in seafoods (see page 381).

In time several viruses may also be implicated as etiologic agents of food-borne diseases that resemble food poisoning rather than food infections.

PREVENTION OF FOOD INFECTION AND FOOD POISONING

Cooking. Cooking may or may not prevent food infection and food poisoning. This is because of the peculiar heat relationships of the factors involved. *Vegetative cells* of *Clostridium botulinum, Cl. perfringens, Staphylococcus aureus, Salmonella, Shigella, Vibrio cholerae* (in the Orient), cysts of *Entamoeba histolytica,* enteroviruses, and toxins of *Cl. botulinum* and (presumably) of *Cl. perfringens* are all inactivated by 10 minutes of boiling (100 C). But *spores* of several types of *Cl. botulinum* are among the most *heat-resistant* organisms known to exist and will survive oven baking and steam pressure cooking (autoclaving), unless prolonged and at high temperature. *Cl. perfringens* spores are also very thermostable. *Staphylococcus aureus* enterotoxin will resist boiling for an hour or more (Table 28–3). The hepatitis virus is also known to resist boiling longer than one would normally boil food.

It must be remembered that in solid food such as roasts, heat penetrates slowly. For example, the outside of a roast or the top of a pudding may be browned, yet the center may never be more than lukewarm. Steaming or boiling is more effective in disinfecting or sterilizing food than is baking or frying. The thrifty housewife thoroughly heats leftover food that cannot be refrigerated and that she is afraid will not "keep" overnight. If it must be kept unrefrigerated longer than overnight, it should be reheated at 100 C for 10 minutes before being eaten. Food that is to be consumed uncooked should be selected with care.

Never serve or eat opened or handled (i.e., *potentially contaminated*) food that has remained in a warm place (i.e., incubated) for more than four hours. These precautions apply, of course, to moist, nonacid foods suitable for the growth of staphylococci, *Cl. perfringens, Salmonella,* and *Shigella.* These organisms

Table 28–3. Relative Heat Resistance of Common Food-Borne Microbial Pathogenic Agents

Pathogenic Agent	Resistance to a Temperature of 100 C or More for 30 Minutes*
Salmonella, Shigella, Vibrio cholerae, enteroviruses, *Coxiella, Mycobacterium,* cysts of *Entamoeba histolytica*	–
Staphylococcus:	
enterotoxin	+
cells	–
Clostridium botulinum and *Cl. perfringens:*	
toxin	–
cells	–
spores	+
Hepatitis A Virus (HAV)	– (killed after 1 min.)
Hepatitis B Virus (HBV)	– (killed after 10 min.)

*+ = resistant; – = inactivated.

will not multiply in acid foods such as stewed tomatoes, rhubarb, or pickles, or in dry foods like salted nuts and cookies, but they will have a field day in gravies, cheese, spaghetti, soups, whipped cream, creamed potatoes, bread or rice pudding, and the like.

Microwave ovens do not permit long exposure of microorganisms in food to radiation or heat. Consequently, the microorganisms are practically not affected and are certainly not killed under the conditions of microwave application.

Refrigeration. Efficient and prompt refrigeration of food soon after preparation or possible contamination not only prevents unncessary and wasteful bacterial spoilage but also goes far to retard excessive growth of any harmful bacteria that may have gained access to the food after it was handled.

Babies' bottles can be prepared for feedings for a 24 hour period and then refrigerated with complete safety, provided that they are sterilized by heat to begin with. Special facilities are usually provided in hospitals and clinics, either for aseptic preparation, using sterilized materials and equipment, or for sterilizing the bottled formulas after preparation. Proper handling and use of commercially prepared products eliminates this problem entirely.

Freezing. This is another method of inhibiting growth of microorganisms. Frozen foods are not necessarily sterile, but most will "keep" for months, and many for years, because the metabolism of organisms present has been halted. Not only foods but also many medicinal supplies and solutions liable to deterioration or microbial spoilage at room temperatures are preserved by freezing as well as by refrigeration. (Caution: aqueous fluids frequently break glass containers on freezing!)

Supervision of Food Handlers. Essential in the prevention of food infection is employment only of noninfectious food handlers. This is not always as easy as it might appear. Many cities require that all food handlers in hotels, restaurants, and institutions pass an examination for general cleanliness, carrying of enteric infections, and the presence of boils or sores due to staphylococci, streptococci, or diphtheria, tuberculosis, or syphilis. No person harboring infectious disease organisms transmissible by food or fomites should be allowed to work in a kitchen or dining place. The general enforcement of such regulations is difficult.

Clean hands and sanitary habits are of prime importance in handling food since bacteria are commonly transferred from hands to food. Eating food as it is prepared should be discouraged, especially if the cook sticks his fingers in his mouth and then touches the food others will eat or uses his spoon to taste your food. Hands should always be washed thoroughly before preparing or serving a meal and after using the toilet and blowing or picking the nose. The reason is obvious. The problem of clean hands is a pressing and difficult one in institutions such as state mental hospitals, where patients help in kitchens and dining rooms—not to mention university cafeterias that employ students who have not taken a microbiology course.

Proper Dishwashing. The method of dishwashing has great bacteriologic importance and is regarded seriously by health departments and administrators of institutions. Practices in some "soda-lunch" counters and small restaurants are revolting to any educated person. One often notices a stale cigar taste or sees lipstick on his glass and not infrequently finds "eggy" saliva adhering to his eating utensils. The mere swishing of tableware through a basin of lukewarm, filthy water laced with saliva is obviously useless from the standpoint of preventive medicine or public health, but it is very common practice. Also, washing dishes in a pan of lukewarm water with a dirty dishcloth, even with a dash of detergent or soap powder, merely distributes bacteria over them.

An abundance of tolerably hot water with detergent should be used for initial washing. The proper washing of spoons, forks, cups, and glasses is especially important. In the absence of a disinfectant, the cleaned dishes should be immersed in almost boiling water (180 F [82 C]) for several minutes to disinfect them. If a chlorine disinfectant is used, the dishes should be rinsed in a basin of cool, clean water containing at least 50 parts per million of free chlorine. Hot water drives off the chlorine. Chlorine content is easily determined by health department inspectors (Fig. 28–1). Poorly washed spoons or forks

SANITIZING WITH HOT WATER
REQUIRES 2 STEPS

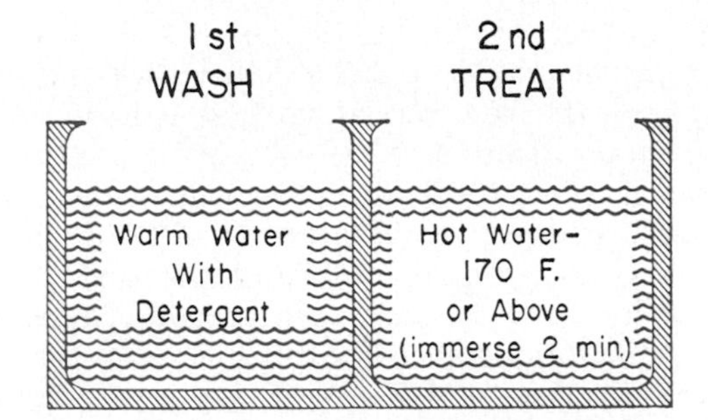

SANITIZING WITH CHEMICALS
REQUIRES 3 STEPS

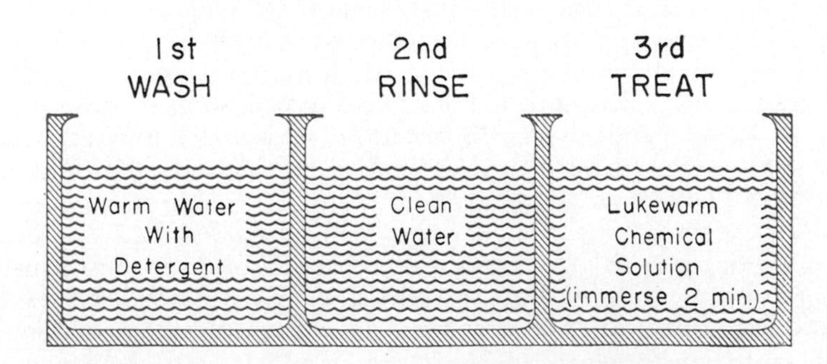

Figure 28–1 Methods of sanitizing dishes washed by hand. The water in the third sink in the chemical method (lower diagram) should contain a commercial disinfectant or at least 50 parts per million of free chlorine, preferably 100–200 ppm. If the water in the sink is too hot, the heat drives off the chlorine. (From U.S. Public Health Service, Publication No. 83.)

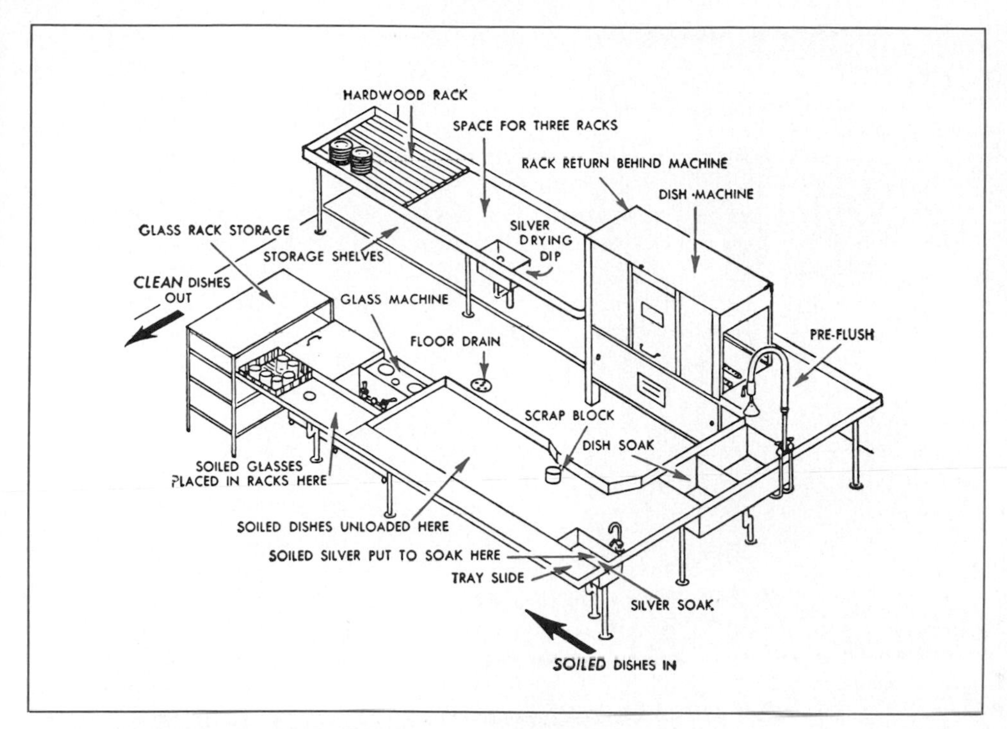

Figure 28–2 One form of modern, sanitary dishwashing equipment. The working bench is of stainless steel. Soiled dishes are piled on the bench in the foreground. They are sorted and scraped, the larger scraps of food dropping into a barrel beneath the counter. Glasses are rinsed over rotating brushes, dipped into disinfectant, and placed in trays in a rack (left). Silverware soaks in a pan of special detergent solution (right foreground). The dishes, arranged in baskets, are soaked and then given a preliminary rinse with *hard* streams of *hot* water (right). They then pass through a machine dishwasher (center, background). The silver, after soaking, passes through the same process as the dishes and is self-dried after a dip into a drying agent. Afterward all utensils are stacked and stored in dust-proof cabinets. Eating utensils handled in this way should be free of all pathogens. (Courtesy of John L. Wilson and William M. Podas, Economics Lab., Inc. In *Modern Sanitation.*)

can transmit pneumonia, diphtheria, tuberculosis, the organisms associated with Vincent's angina, scarlet fever, septic sore throat, and influenza from one person's mouth to another's.

Machine dishwashing is far more effective than hand methods, and much hotter water can be used (Fig. 28–2). In managing cases of disease transmitted by saliva and sputum, the dishes and eating utensils of the patient should be boiled.

Dish towels, if used, should be clean! Since dish towels cannot long stay clean if used, it is better to let glasses or dishes air-dry. We really cannot digest cotton-lint from dish towels; they should be avoided.

Some institutions subject tableware to bactericidal ultraviolet light. One must remember, however, that ultraviolet light does not pass through glass and that it kills bacteria only on those surfaces directly exposed to it for some time.

Bacteriologic Examination of Tableware. It is relatively easy to determine approximate numbers of bacteria on tableware before and after washing, storage, and handling. In one simple procedure a swab, moistened with sterile broth or water, is rubbed over a measured area of the tableware and then shaken violently in a tube containing 5 ml of broth. A colony count on agar medium is made of the bacteria that were released from the swab into the broth. There are many modifications of this procedure: in several, a water-soluble, filamentous material (calcium alginate) is substituted for the cotton.

MILK

Bacteria in Milk. Bacteria of many kinds grow well in milk. As sold on the market, milk may be processed to be sterile, or it may contain thousands of harmless bacteria per milliliter. Mere numbers of bacteria, however, are not alarming: it is the kind that is important.

Milk as it is drawn from a healthy cow contains a few bacteria, but unless modern, "closed" methods of milking are used (Fig. 28–3), it acquires many more from the cow's body, the dust and dirt in the barn, the hands, the milking machine, and the milk cans. Some of this contamination is unavoidable, but it

Figure 28–3 An automatic milking stall. Milk is conveyed directly from the cows' udders in a completely enclosed glass, plastic, or stainless steel pipe (above) through a filter (glass containers, below) and into a refrigerated bulk cooler. The milk is never open to contamination from the air or environment. After milking, the assembled units are connected with an automatic washer. By pressing a button the entire system is prerinsed, washed, and rinsed again automatically. (Courtesy of Alfa-Laval, Inc.)

can be greatly reduced by care in the cleanliness of the cows, milkers, barns, cans, and other equipment, which is now routine in all good dairies.

When it has just been drawn, very clean milk may contain as few as 100 bacteria per ml. By the time milk reaches the city consumer 24 to 48 hours later, the number of bacteria ranges from about 3,000 per ml in the very best milk to millions per ml in unacceptable milk. Good, pasteurized, Grade A milk contains not over 20,000 bacteria per ml and should contain as few as 10,000. The presence of large numbers of bacteria in milk shows either that it has been handled under unclean conditions or that it is stale and has not been kept cool. It may also have been mixed in a tank-truck with high-count milk or come from infected cows even though it does not contain large numbers of bacteria. Although drawn under good conditions from healthy cows, it may have been inoculated with pathogens by milk handlers who were carriers of enteric and/or respiratory pathogens (see Table 28–4). Powdered milk is now often used by the consumer and also in institutions: the water added must be of acceptable sanitary quality. Evaporated and condensed milks are sterile if properly processed.

Public Health Supervision of Milk Supplies. Good milk is so important that boards of health and medical milk commissions in all states and progressive communities regulate the conditions under which it shall be produced and sold. The farms where milk is produced are inspected, and sanitary codes, established in cooperation with state and federal experts, are enforced in the care of the cows and the handling of the milk.

In general, milk from sick animals may not be used, nor may milk be marketed if there is a case of infectious disease transmissible by milk among the family or employees on the farm. All dairy cattle should be checked for tuberculosis by the tuberculin test and for brucellosis by an agglutinin test. Milk must be refrigerated in shops and sold only by the bottle or carton. In most states only pasteurized Grade A milk,[3] powdered milk, or presterilized bottle milk is permitted to be sold for household use. *Certified Raw Milk* ("baby milk") is produced by specially licensed and very carefully inspected and controlled dairies.

Bacteriologic Examination of Milk. The bacteria in milk are readily made visible and may be counted if a smear of 0.01 ml of the milk is made over an area exactly 1 cm square, stained with methylene blue, and examined under the microscope. The smear shows whether the predominating species are streptococci (potential pathogens) or other varieties. Pus cells, revealing that the udders are infected, are also easily detected. Since the volume and area of the milk smear are known, it is easy to calculate the numbers per unit volume. This procedure is widely used by creameries and health departments in daily estimations of the quality of the milk. It is known as the "Breed count" (after the famous American bacteriologist Robert Breed) or the *direct microscopic method.* One drawback of the Breed count is that it does not enable the examiner to distinguish between living and dead bacteria. New or specially cleaned slides must be used. The stained slides may be stored for permanent records.

A widely used method of estimating the number of living bacteria in milk is to dilute the milk appropriately (1:10, 1:100, 1:1,000) and mix 1 ml of the diluted sample with about 12 ml of melted nutrient agar of a prescribed composition in a sterile Petri dish. After proper incubation (usually 48 hours at 32 C), readily visible colonies will have developed in the plate, each colony presumably having originated from a live bacterium in the milk. To count the colonies and make the appropriate arithmetical calculations is then a simple matter. This method is equally applicable to estimating the number of bacteria in blood, water, or any other fluid if suitable medium is used. The procedure is often called a "plate count."

[3]Grade A milk is produced under legally specified conditions. The plate count is required to be lower than 30,000 per ml after pasteurization.

Table 28–4. Sources of Pathogens in Cow's Milk

I. Udder of cow is infected *prior* to milking:
  a. *Mycobacterium bovis*
  b. *Brucella* sp.
  c. *Streptococcus pyogenes* (Group A)*
  d. *Coxiella burnetii*
  e. *Staphylococcus aureus**

II. Worker contaminates milk *after* milking:
  a. *Corynebacterium diphtheria*
  b. *Salmonella* sp.
  c. *Shigella* sp.
  d. *Staphylococcus aureus*
  e. *Streptococcus pyogenes* (Group A)*

**Streptococcus* (Group A) appears in both parts of the table because infected workers can both infect the cow's udder (resulting in mastitis) and contaminate the milk *after* it is drawn. This may be true also of *Staphylococcus*.

A peculiarity of this method is that, although each colony theoretically represents a single bacterium, actually many bacteria tend to cling together in clumps of from two to 100 or more, and a single colony may represent one of these clumps rather than a single cell. Furthermore, in order to avoid complicated methods, a relatively simple medium is generally used for counting bacteria in milk or water, and many bacteria, especially pathogens and anaerobic species, will not grow in it. For these reasons the Breed count is always higher than the plate count. A combination of the two methods is probably best. It is evident that the enumeration of bacteria in milk or water is merely approximate, but the method has given us some of our most useful information about water and milk supplies.

The *reductase test* is an indicator of milk quality. Methylene blue will decolorize in milk rapidly if the milk contains excessive numbers of bacteria, but will not decolorize in sooner than 6 hours in Grade A Raw Milk for pasteurization.

Pasteurization. All microorganisms causing transmissible disease that ordinarily occur in dairy products are killed by the usual commercial processes of pasteurization, i.e., 63 C (145 F) for 30 minutes by the vat or holding process, or 72 C (161 F) for 15 seconds by the "flash" or high-temperature, short-time (H.T.S.T.) process. Many harmless bacteria in milk survive so that pasteurization *disinfects* but does not sterilize. After pasteurization, the milk must be refrigerated promptly and bottled under hygienic conditions.

The Phosphatase Test. This is a simple means of checking for improper pasteurization of milk. The enzyme phosphatase, normally present in fresh milk, is slightly more resistant to denaturation at the time and temperature needed for proper pasteurization than are the pathogens that pasteurization must destroy. Presence of phosphatase in supposedly pasteurized milk indicates improper pasteurization, illegal admixture of unpasteurized milk, or other irregularity. A negative phosphatase test and a negative test for excessive numbers of coliform organisms in milk (as in water) are additional precautions to safeguard your health.

Care of Milk in the Household. Milk should be kept covered and cold. The mouth of the bottle or carton should be wiped with a clean cloth before the milk is poured out. One never knows what contaminating influence (e.g., inquisitive dogs or hungry cats) has been at work on the tops of milk bottles standing on porches and doorsteps.

The tops of milk bottles should not be touched with fingers. Milk bottles should never be taken into a sickroom. The use of waxed paper cartons is doing much to eliminate the evils of the glass milk bottle. Their advantages are obvious. A disadvantage is that they sometimes leak and may thus become contaminated.

Diseases Transmitted by Milk. Pathogens transmitted by milk may be listed under two headings, as shown in Table 28–4.

The bacteria causing bovine tuberculosis (*Mycobacterium bovis,* formerly *M. tuberculosis* var. *bovis*) gain entrance to the milk from infected cows; as do those causing brucellosis (*Brucella* species) and *Coxiella burnetii,* the cause of Q fever. *C. burnetii* may also be transmitted by ticks and dust.

The organisms causing septic sore throat and scarlet fever *(Streptococcus pyogenes)* may get into milk directly from some infected person who handles the milk, or indirectly by means of an infection in the cow's udder, which in turn comes from an infected milker. The bacilli of shigellosis, salmonellosis, and diphtheria gain entrance to milk only after it is drawn and only from infected human beings, since the udders of cattle are not normally infected by these organisms (Table 28–4).

All these microorganisms, excepting *C. burnetii* and possibly *M. bovis* and *Brucella* species grow very well indeed in milk and, if the milk is kept in a warm place, may soon make it exceedingly dangerous. Milk from an udder heavily infected by *Staphylococcus aureus* (a common cause of bovine mastitis) may contain enough preformed enterotoxin to cause severe poisoning in spite of pasteurization.

The pasteurized and carefully guarded milk supplies of large cities are rarely the means of spreading infection. Milk consumed in rural districts and in backward areas of the world is sometimes produced under very unsanitary conditions, handled carelessly, and not pasteurized or (worse) ineffectually pasteurized, creating a false sense of security.

SUMMARY

Most microorganisms encountered in properly handled and prepared food are harmless, although improper storage may lead to food spoilage. Spoiled food is usually not consumed if it smells or looks bad, but often it seems entirely fresh. The pretty yellow "sugar" crystals on a cream pie may be pure colonies of *Staphylococcus aureus,* causing food poisoning. If pathogenic bacteria are present on food, they may cause *food poisoning* or *food infection* or both. Food poisoning is due to the ingestion of toxins produced by microorganisms or of poisonous fish, plants, or chemicals.

Clostridium botulinum, an obligate anaerobe that produces heat-resistant spores, also gives off extremely potent exotoxins that, when swallowed, frequently prove fatal. The four categories of botulism are food-borne botulism, infant botulism, wound botulism, and cases of unknown origin. *Clostridium perfringens,* which causes gas gangrene, may also cause acute gastroenteritis leading to food poisoning. Five serologically different enterotoxins are known to cause food poisoning due to *Staphylococcus aureus,* which strikes many people but is usually not life-threatening.

Food poisoning is caused by toxins produced by an organism; food infections are caused by pathogens in food coming from infected animals or contaminated by infected persons. Infected animals may transmit tuberculosis, brucellosis, and many other infectious diseases in their meat or their milk. Intestinal pathogens such as the virus of epidemic hepatitis, *Salmonella* and *Shigella* species, and cysts of *Entamoeba histolytica* may be on the hands of food handlers who are carriers or simply untidy.

However, not all bacteria in food are objectionable. Sauerkraut, buttermilk, alcoholic beverages, and cheeses are all products of bacterial metabolism.

The prevention of food infections involves adequate cooking, refrigeration, freezing, dishwashing, supervision of food handlers, and (of course) "common sense," which is expected of anyone who passes a course in microbiology.

Since milk is perhaps the "perfect food," bacteria of many kinds grow well in it, especially the pathogen of tuberculosis, which is killed through pasteurization.

REVIEW QUESTIONS

1. Are bacteria in food harmful or advantageous? Give examples. How many disease-causing microorganisms are transmitted in food? How may this be prevented?
2. Mention some industrial uses of bacteria.
3. What is the difference between food poisoning and food infection? What is botulism?
4. What is the role of staphylococci in food poisoning? How does staphylococcal enterotoxin differ from botulinal toxin?
5. What is the value of cooking in preventing food infection and food poisoning? How may thoroughly cooked food transmit disease-causing microorganisms?
6. Why should food be protected from insects? Why should it be protected

from persons who cough and sneeze, from carriers of enteric diseases, and from rats and mice?

7. What disease-causing organisms are transmitted from cows to humans via milk, from humans to cows, and from humans directly to milk?
8. Does market milk contain bacteria? Describe the bacteriologic examination of milk. By what means is milk graded?
9. What procedure is commonly used to disinfect milk? How is it done?
10. What procedure would you suggest to wash dishes safely? What should not be done?

Supplementary Reading

Annelis, A., Grecz, N., Huber, D. A., Berkowitz, D., Schneider, M. D., and Simon, M.: Radiation sterilization of bacon for military feeding. *Appl. Microbiol., 13*:37, 1965.

Chesbro, W. R., and Auborn, K.: Enzymatic detection of the growth of *Staphylococcus aureus* in foods. *Appl. Microbiol., 15*:1150, 1967.

Cockburn, W. C., Taylor, J., Anderson, E. S., and Hobbs, B. C.: Food Poisoning. London, The Royal Society of Health Journal, 1962.

Dack, G. M.: Food Poisoning. 3rd ed. Chicago, University of Chicago Press, 1956.

Food Service Sanitation Manual. Washington, D.C., Government Printing Office, Public Health Service Publication No. 934, 1962.

Foster, E. M., and Sugiyama, H.: Recent developments in botulism research. *Health Lab. Sci., 4*:193, 1967.

Frazier, W. C., and Westhoff, D.: Food Microbiology. 3rd ed. New York, McGraw-Hill Book Co., 1978.

Frobisher, M., Hinsdill, R. D., Crabtree, K. T., and Goodheart, C. R.: Fundamentals of Microbiology. 9th ed. Philadelphia, W. B. Saunders Company, 1974.

Hall, H. E., and Lewis, K. H.: *Clostridium perfringens* and other bacterial species as possible causes of food-borne disease outbreaks of undertermined etiology. *Health Lab. Sci., 4*:229, 1967.

Ingram, M., and Roberts, T. A. (eds.): Botulism 1966. London, Chapman and Hall, 1967.

Joint Editorial Committee: Standard Methods for the Examination of Dairy Products. International Commission on Microbiological Specifications for Foods. Microorganisms in Foods. Their Significance and Methods of Enumeration. 2nd ed. Toronto: University of Toronto Press, 1978. 14th ed. New York, American Public Health Association, 1978.

Natori, S., Sakaki, S., Udagawa, S. I., Ichinoe, M., Saito, M., Umeda, M., and Ohtsubo, K.: Production of rubratoxin B by *Penicillum purpurogenum Stoll. Appl. Microbiol., 19*:613, 1970.

Schlinder, A. F., Palmer, J. G., and Eisenberg, W. V.: Aflatoxin production by *Aspergillus flavus* as related to various temperatures. *Appl. Microbiol., 15*:1006, 1967.

Smith, L. D. S.: Botulism: The Organism, Its Toxins, the Disease. Springfield, Ill., Charles C Thomas, 1977.

Strong, D. H., Canada, J. C., and Griffiths, B. B.: Incidence of *Clostridium perfringens* in American foods. *Appl. Microbiol., 11*:42, 1963.

Subcommittee on Methods of the Microbiological Examination of Foods: Recommended Methods, 2nd ed. New York, American Public Health Association, 1966.

Varga, S., and Anderson, G. W.: Significance of coliforms and enterococci in fish products. *Appl. Microbiol., 16*:193, 1968.

Vermilyea, B. L., Walker, H. W., and Ayers, J. C.: Detection of botulinal toxins by immunodiffusion. *Appl. Microbiol., 16*:21, 1968.
Welt, M. A.: When food is irradiated. *Science, 160*:483, 1968. See also McKinney, M. E., and Schweigert, B. S., same source.
White, A., and Hobbs, B. C.: Refrigeration as a preventive measure in food poisoning. *Roy. Soc. Health J., 83*:111, 1963.
WHO Expert Committee and F. A. O.: Microbiological Aspects of Food Hygiene. WHO Technical Report Series No. 398, 1968.

INFECTIONS BY PYOGENIC COCCI

TRANSMISSION OF RESPIRATORY TRACT INFECTION

Pathogenic microorganisms occurring in the respiratory tract (including the sinuses, inner ear, conjunctival sacs, and adjacent tissues) are regularly transmitted via oral and nasal secretions and discharges from infected eyes and ears. Oral and nasal secretions inevitably contaminate the atmosphere during talking, laughing, sneezing, coughing, and similar actions that produce droplets and droplet nuclei. In addition, oral, nasal, and conjunctival secretions are spread by hands, eating utensils, drinking glasses, and improperly maintained swimming pools, which may be contaminated by feces and urine as well as by respiratory secretions. It is clear that we constantly face a formidable, unceasing, and virtually ubiquitous onslaught of infection.

Pathogens of the respiratory tract, and airborne pathogens in general, are among the most difficult to control. Indeed, except under the most carefully restricted circumstances, such as the use of special breathing apparatus, masks, germicidal vapors in closed rooms, irradiated isolation cubicles or rooms, and "life islands" or "bubbles" used for patients being treated with drugs that depress the immunologic responses (e.g., in heart and other organ transplants), we have virtually no control over airborne diseases in ordinary daily human contacts.

Infection, Natural Immunization, and Carriers. On the comforting side, daily experience shows that the majority of normal, healthy persons generally do not suffer serious or even perceptible infectious disease except during epidemics. Fortunately, as a result of previous subclinical infections (or vaccinations) and robust health, they generally become actively immunized (or *re*immunized) to many of the organisms with which they daily come into contact. Unfortunately, they also often become carriers of pathogenic microorganisms; they remain healthy but transmit their pathogens to others, perhaps to ill or aged persons or young children, who may be infected with a serious disease as the result.

It is fortunate that none of the common respiratory pathogens forms heat-resistant endospores.[1] All nonspore-formers (except certain viruses) are readily killed by five minutes of boiling and by such standard disinfectants as 2 per cent saponated cresol, household chlorine bleaches (5 per cent NaOC1), and low–surface-tension iodine solutions.

CONTROL OF RESPIRATORY INFECTIONS

There are many preventive measures that may be effectively employed by health workers against respiratory pathogens. These include the thorough washing (using disinfectant soap) of hands known to be soiled with respiratory

[1]Pulmonary anthrax ("wool-sorter's disease") is an exception. This disease is rare, except in the wool and hides industry.

tract discharges; adequate sanitization of dishes and eating utensils, most readily accomplished in institutions and homes by machine dishwashers with hot water (185 F or 85 C) as a final rinse; disinfectant laundering of bedding of patients with any infectious disease (i.e., washing in hot water, soaking in disinfectant, and hot ironing); disinfection (not merely "airing") of blankets and mattresses by autoclaving, sporicidal vapors, or gas; use (and proper disposal!) of clean handkerchiefs or paper tissues for sneezing and coughing; use of proper sputum containers for tuberculosis patients (see discussion of tuberculosis); disinfection of air and dust; and the use of masks, dust control, and ventilation. Only the last four require discussion here.

Masks. These should be sufficiently thick and should cover the nose and mouth completely. They must never be reversed and never reused after dangling about the neck. Masks should be changed when perceptibly moist. When discarded after use they should be placed in a closed container. Boiling for at least 5 minutes, followed by washing with hot (80 C) water and soap is sufficient routine disinfection for reusable masks. Better yet are sterile, disposable, inexpensive masks, packaged for immediate effective use. Some modern masks are made of molded plastic or metal to which are adapted removable filter units of very fine-pored foam rubber, fibered glass, and like materials (see Fig. 15–8). Units or entire masks that are completely disposable are highly effective, but each type should be adopted only after actual experimental trial.

The traveler in Japan will notice many people who walk through Tokyo to their daily tasks wearing a face mask. They are concerned about either catching a cold or, more likely, giving their respiratory infection to other people. This is indeed being very polite, civilized, and alert to the ever-present danger of the airborne respiratory pathogens around us.

Dust Control. Since dust (including droplet nuclei) is obviously an important factor in disease transmission by air, methods of suppressing dust have been developed. One consists of using "sweeping compounds" that apply an imperceptible film of oil to all floors, window sills, and furniture in barracks, hospital wards, and similar places where infectious dust accumulates. The dust sticks to the oil and is easily removed without being stirred up. Modern vacuum cleaning systems are devised to draw dust not into a "dust bag," which often scatters much of the dust, but into a built-in vacuum-duct systems with intakes in each room that conduct the dust through airtight tubes to an incinerator or dust-tight container. Devices that combine wet-scrubbing with vacuum cleaning are also excellent means of floor dust control. Probably the perfect dust control system remains to be invented.

Ventilation and Climate Control. When ventilation is thorough, it is a very effective means of removing airborne pathogens and reducing dosage of infectious organisms to a virtual vanishing point. In cold climates ventilation with outdoor air raises problems of heating costs. Ventilation by recirculation of indoor air requires expensive blowers, dust filters, and incineration; however, these are engineering details with which we need not be concerned in this book.

Disinfection. Air and dust may be disinfected by several means, among which are irradiation with ultraviolet light, surface disinfection, and microbicidal vapors. These methods have been discussed in Chapters 10, 12, and 13. Passage of air through filters and ducts at high temperature (virtual incineration) is also used.

PATHOGENIC COCCI OF THE RESPIRATORY TRACT

Classification and Description of Cocci. Cocci arrange themselves in different ways as they multiply. Those forming predominantly chains are called streptococci; those usually arranging themselves in irregular clusters are called staphylococci if *facultative* aerobes and micrococci if *strict* aerobes; those that form pairs are called diplococci (however, this characteristic is not strictly a determining classification of the organism) (see Fig. 4–3). All cocci are gram-positive except the important genus *Neisseria,* containing the gonococci and meningococci and some *Neisseria*-like but anaerobic species of some medical importance called *Veillonella,* which may be gram-negative or variable, possibly depending on growth conditions. The genus *Neisseria* is discussed further on page 449. The pathogenic gram-positive cocci described in this chapter are:

Staphylococcus aureus

Streptococcus pyogenes (also known as *Streptococcus hemolyticus)* and closely related species (often called the hemolytic streptococci or "hemolytic strep")

Streptococcus salivarius, S. faecalis, S. mitis, and related species (previously called *Streptococcus viridans)*

Streptococcus pneumoniae (often called the pneumococcus, formerly called *Diplococcus pneumoniae).*

(See Chapter 37 for anaerobic coccal infections, *Peptococcus, Peptostreptococcus,* and *Veillonella.*)

The Staphylococci

Staphylococcus aureus is named for its golden yellow pigment.[2] A similar species called *Staphylococcus epidermidis* (epidermis-inhabiting) produces a chalky white pigment. These pigments are most distinctive in colonies on solid media. Neither produces pigment in broth or when grown anaerobically. The only other species now recognized is *Staphylococcus saprophyticus* (from Greek *sapros,* putrid), which grows on dead tissues. *S. aureus* and *S. epidermidis* grow well on infusion agar or in peptone broth or milk at temperatures of 25 to 40 C. They are facultative anaerobes that grow more rapidly and abundantly under aerobic conditions. Although they form no spores, they are somewhat resistant to drying and can therefore remain alive for several months under favorable conditions in dust or elsewhere outside the body. One or both species are often present in the normal upper respiratory tract.

Infections by staphylococci commonly cause much pus formation and hence are said to be *pyogenic (pyo,* pus).

Staphylococcal Toxins. Typical strains of *Staphylococcus aureus* produce a metabolic product called *leukocidin,* which kills white blood cells. Many strains also produce a poison that is called *enterotoxin* because it causes gastroenteritis. In addition to this, *S. aureus* usually secretes substances that cause necrosis of skin tissues (dermonecrotic toxins), hemolysis of red blood cells, and coagulation of oxalated or citrated[3] plasma *(coagulase).* Coagulase-producing staphylococci, regardless of color, are especially dangerous, and the *coagulase test* is frequently done in the diagnostic laboratory.

---

[2]From Latin *aurum,* gold. The Greek word *staphyle* means a bunch of grapes.

[3]Sodium oxalate or citrate is usually added to blood specimens for chemical examination to prevent clotting.

A positive coagulase test for either *Staphylococcus aureus* or any similar strain is taken as an indication that the organism being tested is pathogenic. Nonpathogenic strains do not produce coagulase. It should be noted that staphylococci that cause food poisoning by means of an enterotoxin also usually are coagulase-positive. There are several *S. aureus* hemolysins, separable by electrophoresis. Strains isolated from exfoliative (scale-forming) dermatitis, impetigo, or pemphigus (watery blisters on the skin) of the newborn generally produce a toxin called *exfoliatin.*

Phosphatase production has also been associated with pathogenicity of staphylococci; however, it does not always agree with coagulase activity. Nontoxic metabolites produced are deoxyribonuclease, hyaluronidase, staphylokinase, lipase, gelatinase, and protease.

Isolation of Staphylococci

Pigmented and nonpigmented strains of *Staphylococcus aureus* produce the enzyme coagulase[4] and ferment mannitol, whereas *S. epidermidis* and *S. saprophyticus* do not. Also, *S. aureus* can tolerate a high salt concentration of close to 10 per cent. Consequently, mannitol salt agar is often used as a selective medium for the isolation of *S. aureus* from feces and other sources. In 48 hours, salt-tolerant colonies of staphylococci appear on mannitol salt agar surrounded by a yellow halo. Vogel and Johnson agar permits early detection of coagulase-positive colonies of *S. aureus.* On this red medium, a yellow zone surrounds the black colonies of *Staphylococcus.*

Staphylococcal Infections

Staphylococcus epidermidis, the white, coagulase-negative species, is widely distributed on the bodies of human beings and animals. These cocci are usually harmless residents. If the skin is injured by a cut or a scratch, however, the cocci can get into the underlying tissues and may cause an infection.

Pimples and acne lesions are common forms of infection in which these organisms are found. Infection of stitches following surgical operation (commonly called "stitch abscess") is most frequently caused by these staphylococci. As a rule, staphylococcal infections are not highly dangerous, but cocci designated as *S. epidermidis* have been reported in some cases of fatal septicemia. Microorganisms are devious and treacherous! *Staphylococcus saprophyticus* has been implicated in some infections of the urinary tract, although this appears to be an exception to its saprophytic nature.

Staphylococcus aureus, the dangerous coagulase-producing species, typically produces boils and carbuncles. These cocci can also invade the whole system, causing fatal septicemia, meningitis, endocarditis, puerperal sepsis, toxic shock syndrome, pneumonia, destructive abscesses in the internal organs called *multiple abscesses,* infections of the bones called *osteomyelitis,* and septic arthritis. *Staphylococcus aureus* is of special significance in hospitals, notably in surgical and maternity wards and nurseries, as the cause of epidemics of antibiotic-resistant wound infections, severe abscesses of the lactating breast (mastitis), and distressing ulcers on infants (pemphigus or impetigo neonatorum[5] associated with exfoliatin-producing strains), as well as other serious infections (nosocomial or "in-the-hospital" infections). As previously mentioned, staphylococci often become antibiotic-resistant and cause hospital epidemics. These outbreaks have received much serious consideration in medical as well as lay literature.

---

[4]Coagulase causes blood plasma to clot.

[5]From Greek *neos,* new; *natus,* born.

Toxic Shock Syndrome. Toxic shock syndrome has been associated with focal staphylococcal infections in children, males, and nonmenstruating females; but 98 per cent of all cases occur among women using tampons during menstruation (Fig. 29–1). Although vaginal cultures from menstruating women with toxic shock syndrome contain *Staphylococcus aureus*, a suspected staphylococcal exotoxin has not yet been isolated. The acute illness is associated with hypotension, shock, sore throat, vomiting, diarrhea, and skin eruption succeeded by desquamation of the soles and palms. Evidently the microorganism multiplies in vaginal discharges, with ulceration and vaginitis conditions resulting from superabsorbent tampons. In light of a 5.6 per cent case-fatality ratio, the Food and Drug Administration recommended certain changes in the use of tampons and advocated napkins wherever risks exist. Since August 1980 the incidence of toxic shock syndrome has steadily declined.

Phage Types of Staphylococci. Specific lytic group types have been established by staphylococcal phage typing. More than 22 of these bacterial strains have been classified into six groups. For example, Group II, which causes many skin lesions, reacts only with phage 71. The international staphylococcal phage typing system is not yet completely accepted; some laboratories do some additional typing. This is useful in identifying epidemic types.

Plasmids in Staphylococci. The function of plasmids—small extrachromosomal DNA fragments in bacterial cells—in controlling antibiotic sensitivity and resistance has been discussed previously (Chapter 7). *Staphylococcus* plasmids have been found that carry a gene for penicillinase production, making the strain resistant to penicillin. These plasmids are transferred to other cells through bacteriophage transduction.

Drug-fast Staphylococci. A drug-fast organism is one that is resistant to a certain drug. Bacteria that are not drug-fast are sensitive to the action of the drug, which is an antibiotic. Every hospital is concerned with drug-fast staphylococci because of the constant contact of the patients with visitors and personnel who are unknowningly carriers of *Staphylococcus aureus*. All hospital personnel should be examined periodically for various pathogenic microorganisms or subclinical infections of any sort, and as much as possible kept from contact with patients. Prevention of the spread of respiratory microorganisms requires rigid precautions in washing hands, disposing of infectious dressings, disinfecting bedding, dishes, and thermometers, maintaining cleanliness of the

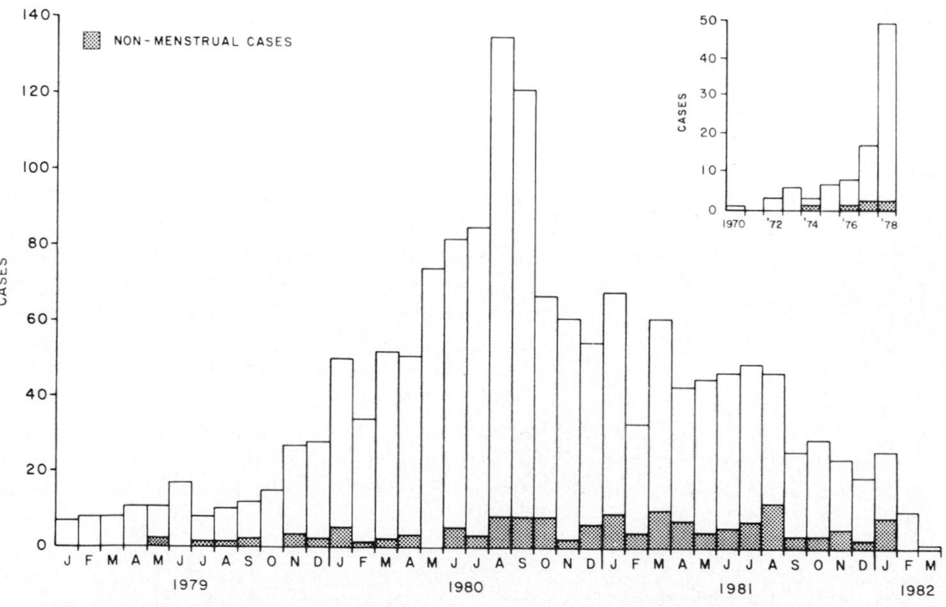

Figure 29–1 Confirmed cases of toxic shock syndrome in the United States, January 1970–March 1982. (From the Centers for Disease Control, *Morbid. Mortal. Weekly Rep.*, *31*:201, 1982.)

rooms and wards, wearing gowns and masks, and other details of expert barrier technique. Complete details are given in the report of the American Hospital Association, *Prevention and Control of Staphylococcal Infections in Hospitals;* see also Top, F. H., et al.: *Control of Infectious Diseases in General Hospitals* (New York, American Public Health Association, 1967).

In handling staphylococcal infections, the greatest care should be exercised not to let the pus come into contact with anything that could distribute it. Staphylococci and, indeed, all organisms in pus are particularly virulent. Dressings of boils and other lesions may be dropped into a pot of disinfectant or into boiling water, or wrapped carefully in several thicknesses of newspaper, fastened securely, and burned completely. Probably the most important single factor in the transmission of staphylococcal infections in hospitals is the dissemination of the organisms on lint from bedding and contaminated dust in the environment of infected patients, especially patients (or personnel) with nasal infections.

Staphylococcal Food Poisoning. It is obvious that *Staphylococcus aureus,* though commonly occurring in the respiratory tract, can cause a great variety of pathologic conditions and infect almost any part of the body via many portals of entry and exit. It can cause boils of the nose, hands, arms, or face, from which it can be transferred to food. The cocci grow readily in many not-too-acid foods such as "pre-cooked" hams, milk, custards, salads, sandwich fillings, and creamed foods. Many strains give off a potent toxin *(enterotoxin*[6]*)* as they grow in the food. Any one of four antigenically different enterotoxins, when swallowed, causes gastroenteritis: nausea, vomiting, cramps, and diarrhea. Thorough cooking destroys the organism *but not the toxin.* The best safeguard against staphylococcal food poisoning is prompt refrigeration, cooking, or eating of all handled foods and discarding any that are stale or "spoiled." Contaminated food, i.e., any food that has been handled by or exposed to people who are coughing, sneezing, or talking, may develop large amounts of the enterotoxin if allowed to stand at kitchen temperatures for more than three to five hours. Staphylococcal food poisoning is rarely fatal, but it has spoiled many a dance date, especially in colleges.

The Streptococci

The streptococci comprise a large group of organisms, among which are many useful and harmless species and also some of the most deadly pathogens. Most species of streptococci are catalase-negative, do not ferment inulin, do not reduce nitrates, and are not soluble in bile salts. If a carbohydrate is fermented by streptococci, large amounts of lactic acid are produced, without gas. The pathogenic streptococci grow readily in the laboratory but for best development require blood or serum media and body temperature (37 C). They are facultative anaerobes and do not form spores. They are killed by pasteurization, by five minutes of boiling, and by disinfectants such as 1 per cent saponated cresol, 5 per cent sodium hypochlorite, and iodine disinfectants. They may remain alive for some time when dried: for several days or weeks in sputum, in exudate from lesions, or in droplet nuclei. Streptococci may be easily differentiated, both from staphylococci and as to group, by the scheme shown in Figure 29–2. Of special value is the growth on blood agar.

Blood-Agar Types of Streptococci. All streptococci may be assigned to one of three types, *alpha, beta,* and *gamma,* differentiated, when grown *aerobically,*

[6]Note that *enterotoxin* refers to an exotoxin that affects the *enteric* tract (from Greek *enteron,* intestine). The word enterotoxin must not be confused with *endotoxin,* which means a toxin that remains inside the bacterial cell (from Greek *endon,* inside of) and is generally associated with the cell wall.

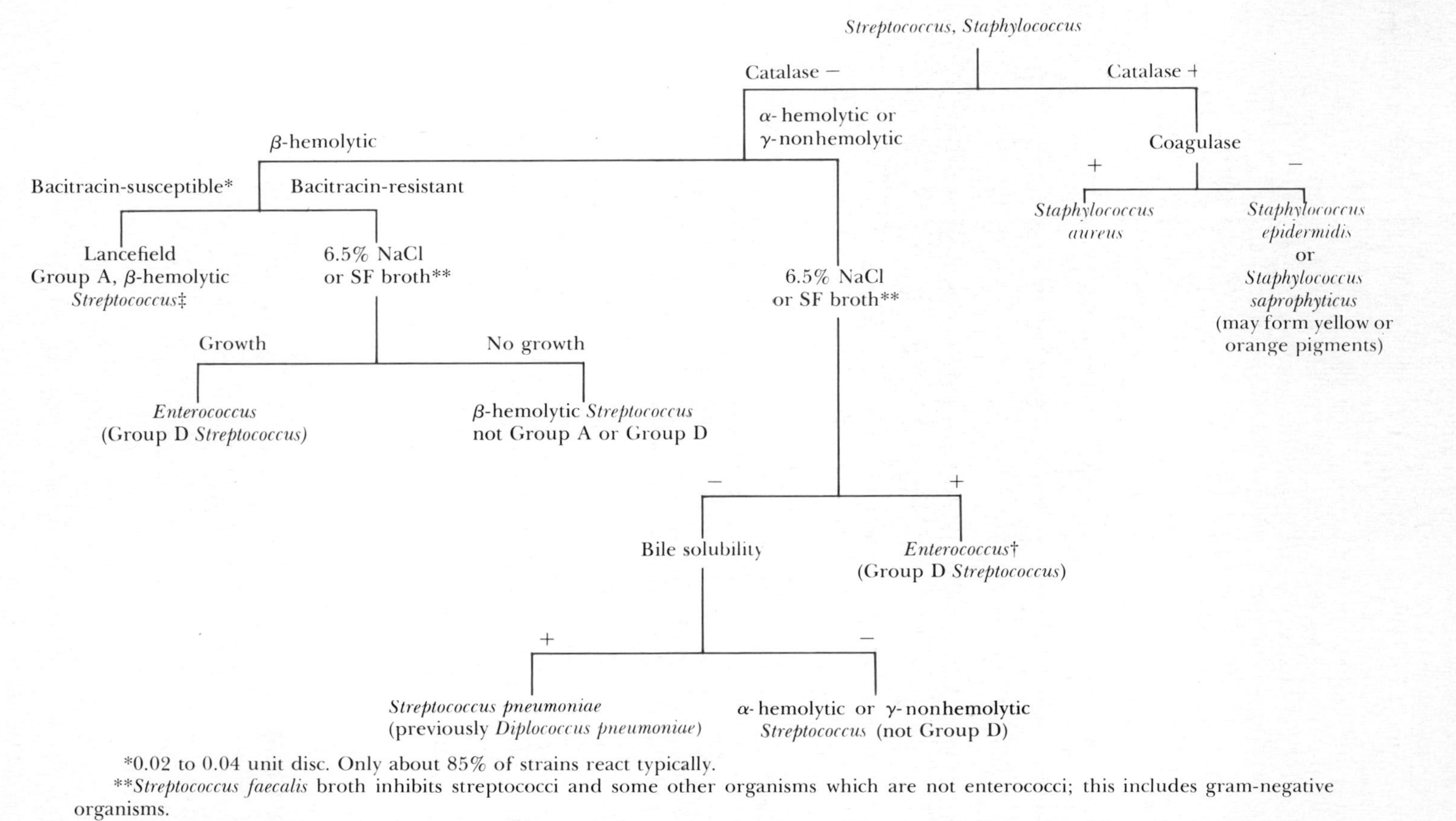

*0.02 to 0.04 unit disc. Only about 85% of strains react typically.

***Streptococcus faecalis* broth inhibits streptococci and some other organisms which are not enterococci; this includes gram-negative organisms.

†This category may contain some non-catalase-producing strains of *Aerococcus*. These can be distinguished from *Enterococcus* by their inability to grow at 45 C.

‡May be confirmed by the Lancefield precipitin test or the direct fluorescent-antibody technique, using group-specific serum.

Figure 29–2 Differentiation of pathogenic gram-positive cocci. (Adapted from CRC Manual of Clinic Laboratory Procedures. 2nd ed. Cleveland. The Chemical Rubber Co., 1970.)

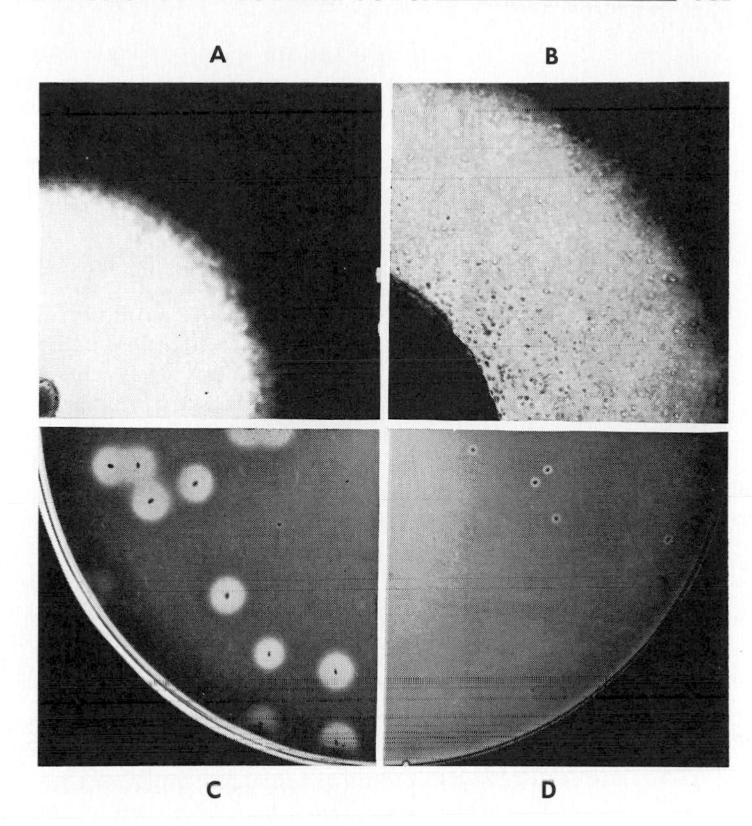

Figure 29–3 Colonies of hemolytic streptococci in blood agar. *A*, One β-type colony enlarged to show edge of colony at lower left and absence of erythrocytes in clear hemolyzed zone. *B*, One α-type colony enlarged to show edge of colony at lower left, with many intact erythrocytes in hemolyzed zone. *C*, Clear zones of complete hemolysis around colonies of β type (*Streptococcus pyogenes*, Lancefield Group A, natural size. *D*, Small hemolytic zones of α-type colony, *S. mitis*, natural size. (Preparations by Dr. Elaine L. Updyke. Photo courtesy of U.S. Public Health Service, Communicable Disease Center, Atlanta, Ga.)

by the action of their subsurface colonies on sheep erythrocytes of blood-agar medium in Petri dishes.

***Alpha-Type* (α-Hemolytic) Streptococci.** These so-called viridans (viridis, green) species of streptococci produce a brown to green zone due to the discolored erythrocytes around their blood-agar colonies, with or without an outer clear, colorless area of hemolysis. The zones of discoloration may vary from less than one to several millimeters wide. The outer clear area around the greenish zone may best be seen after the cultures are stored in the refrigerator. Such colonies are said to be of the *alpha type,* and the streptococci producing them are called alpha-type hemolytic streptococci, viridans streptococci, or *Streptococcus viridans* (not a proper species) (Fig. 29–3).

***Beta-Type* (β-Hemolytic) Streptococci.** These produce perfectly clear, colorless zones around their subsurface colonies in blood agar. These zones are due to *hemolysin* secreted by the streptococci in the colony. The clear, colorless zone of hemolysis is called *beta-type* hemolysis, and the streptococci are said to be beta-type hemolytic streptococci (Fig. 29–3).

In general, the beta-type hemolytic streptococci are *pyogenic* and are pathogenic for humans and/or animals. The principal human pathogen is *Streptococcus pyogenes.* This and several very similar variants cause scarlet fever, septic sore throat, puerperal sepsis, "blood poisoning" (streptococcemia), and other conditions, many of which are similar to those caused by *Staphylococcus aureus.*

Double-Zone Beta-Type Streptococci. These streptococci are of bovine origin and are often found in dairy products. The hemolysis produced is of the beta type, but when the plates are refrigerated a second ring of hemolysis appears, separated from the first one by a ring of red erythrocytes; this is commonly called "hot-cold hemolysis."

Gamma-Type (Nonhemolytic) Streptococci. Some streptococci produce no visible change in the blood agar around their colonies. These are called *gamma-type* streptococci, or indifferent or nonhemolytic streptococci. The colonies produced are usually small, gray, and translucent. These streptococci are not commonly pathogenic—some species are found in milk—but sometimes they are isolated from the same sorts of lesions as the alpha-type streptococci.

Infections by Alpha-Type Hemolytic Streptococci

Several species of alpha-type hemolytic streptococci are pathogenic for humans and/or lower animals, including *S. salivarius* (reputedly harmless), *S. mitis,* and *S. faecalis* (common in feces), all of which are found in saliva. In general, they produce less acute infections than do the beta-type streptococci, but are nonetheless dangerous because the infections tend to become chronic.

Tooth Abscesses. Streptococci of the alpha type are often found in abscesses in the roots of teeth. These abscesses may cause no definite symptoms and tend to be detected only in the dentist's x-ray picture, yet the organisms and their toxins can cause serious damage to various organs. Alpha-type streptococci may also cause infections of joints, resulting in at least one type of arthritis. (See also Chapter 31.)

Bacterial Endocarditis. This disease of the heart, often caused by alpha-type streptococci, may also be caused by beta-type streptococci, pneumococci, staphylococci, and some other invasive organisms. It usually results from bacteremia, sometimes following extraction of abscessed teeth or other injuries to vascular tissues or joints. The causative organisms come via the blood from infected teeth, tonsils, chronic sinus infections, the intestines, or various other foci of infection elsewhere in the body.

The microorganisms cause inflammation of the heart valves, especially damaged valves (Fig. 29–4). This is followed by shrinkage and thickening (scarring) of the valves, causing them to leak and producing "murmurs." The clinical picture is very bleak unless proper treatment is given soon. Care in treating children's throat diseases and the removal of badly diseased tonsils and teeth in both children and adults probably diminish the number of streptococcal infections of heart valves.

Infections by Beta-Type Hemolytic Streptococci

Capsules of Streptococci. Beta-type hemolytic streptococci are frequently encapsulated, especially in infectious material such as pus, exudate, or sputum. These capsules may be stained or they may be demonstrated by mixing the material containing them with a little India ink and observing them under the microscope (see Fig. 4–5). As mentioned previously, these capsules contribute to infectivity.

Serologic Groups.[7] The beta-type hemolytic streptococci can be divided into 18 or more serologic groups (often called Lancefield groups, after their discoverer Rebecca Lancefield) on the basis of *carbohydrate* antigenic substances contained in the cell walls of the cocci. For example, the cell walls of all group A streptococci contain a certain complex, group-distinguishing carbohydrate antigen in common; group B streptococci contain another group-distinctive carbohydrate antigen, and so on up to group O. It is easy to make these distinctions by means of the precipitin test. With few exceptions, Lancefield's groupings apply only to beta-type streptococci. The terms beta-type streptococci of group A and "group A streptococci" or "hemolytic strep" are virtually synonymous with *Streptococcus pyogenes,* the most dangerous of the beta-hemolytic streptococci.

The outstanding value of this means of differentiation lies in the fact that group A streptococci cause serious, acute human diseases such as scarlet fever, erysipelas, septicemia, puerperal sepsis, wound infections, and streptococcal sore throat. Most group B streptococci are entirely harmless to humans, but

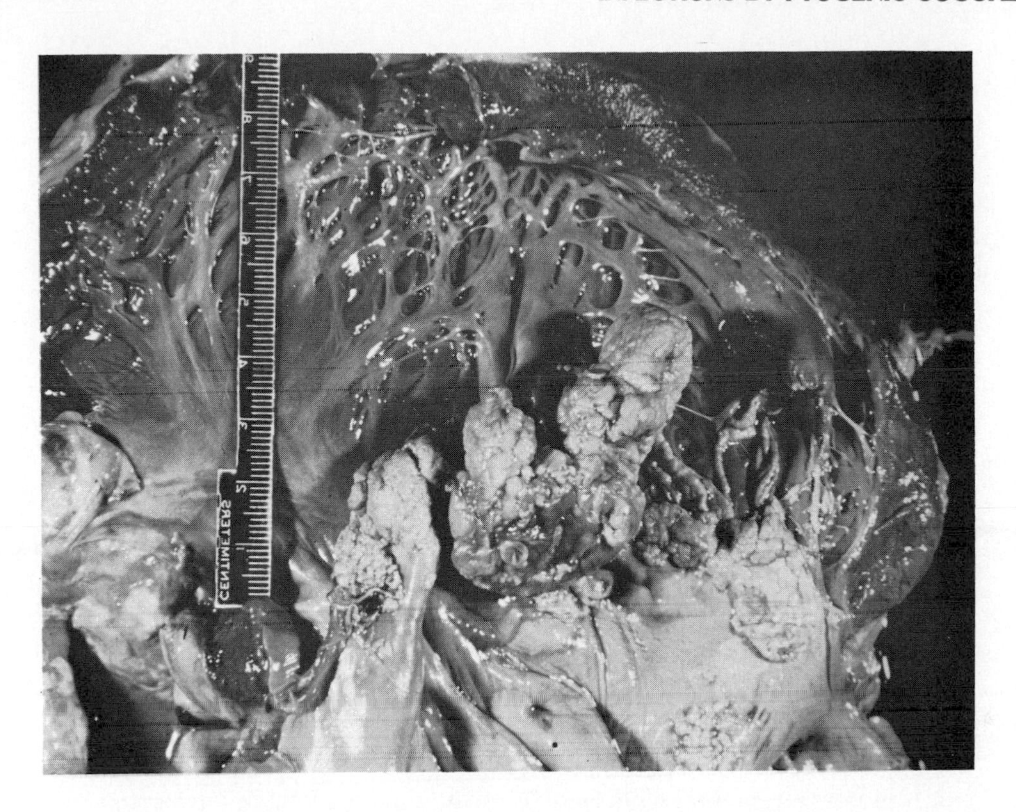

Figure 29–4 Endocarditis due to *Streptococcus faecalis* from the intestine. The heart has been cut open and we look into the left auricle and ventricle. On the leaflets of the mitral valve and the wall of the auricle are irregular, granular clusters of inflammatory tissue called "vegetations." These heal leaving nodules of scar tissue, which causes deformities of the valves that prevent their closure or narrow the opening, with resulting defective circulation. (From Robbins, S. L.: Pathologic Basis of Disease. Philadelphia, W. B. Saunders Company, 1974.)

they cause mastitis in cattle. Groups C and G cocci cause infections of humans such as tonsillitis, sinusitis, and similar conditions, but these are usually not as dangerous as scarlet fever and other group A diseases. Group D streptococci are of minor importance as pathogens, occurring only among the intestinal streptococci or "enterococci." The other groups need not concern us here. (See Table 29–1.)

Table 29–1. Classification of Streptococci Based on Hemolysis of Sheep Red Blood Cells*

Type of Hemolysis	Representative Species	Lancefield's Group	Habitat and/or Diseases Caused
1. Pyogenic Group, beta-hemolytic	*Streptococcus pyogenes* (formerly *S. hemolyticus* or *S. scarlatinae*)	A	Present in various lesions and inflammatory exudates
	Streptococcus equisimilis	C	Erysipelas, puerperal fever
2. Viridans Group, alpha-hemolytic	*Streptococcus mitis* (formerly *S. viridans*)		Human respiratory tract
	Streptococcus pneumoniae		Inflammatory exudates; lobar pneumonia
	Streptococcus bovis		Human endocarditis; found in some human feces
	Streptococcus salivarius (may be gamma type nonhemolytic, or beta-hemolytic, with Lancefield's Group K)		Saliva, human feces
3. Fecal Group, alpha- or beta-hemolytic; some strains are gamma type with Lancefield's Group N	*Streptococcus faecalis*		Urinary tract infections, subacute endocarditis
4. Lactic Acid Group, gamma type (nonhemolytic)	*Streptococcus lactis*		Contaminant in milk and dairy products

*Only selected strains are included in the table. None of these are pathogens of animals.

Serologic Types of Group A. Group A streptococci are further divided into some 60 or more *types,* each containing a different *protein* called an M antigen. These M types are differentiated by precipitin tests and may be still further subdivided within the group A streptococci by T-agglutination.

The terminology seems a little confusing but may be simplified by thinking of a streptococcus as living in a street (blood-agar type), an apartment house (Lancefield group), an apartment (M protein type), and in a certain room (T and R[8] antigen types). For example:

Streptococcus pyogenes:
Beta type (blood agar)
  Lancefield group A (carbohydrate antigenic substance)
    Serologic type 14 (M protein antigenic substance)
      Slide agglutination T and R antigen types (have not been correlated with virulence, as have the M types).

The serologic differentiations of beta-type hemolytic streptococci are matters of professional information, but clinical and nursing procedures are alike for all, differing only according to whether the streptococci are infecting the brain, the respiratory tract, the peritoneum, the parturient uterus, or other body areas.

The capsule of group A *Streptococcus* consists of hyaluronic acid of a type similar to human connective tissue hyaluronic acid. It is not antigenic.

Group A hemolytic streptococci produce several extracellular toxins and enzymes that partly explain the pathogenicity of these streptococci. Some of the products are streptolysin S and O; diphosphopyridine nucleotidase; erythrogenic toxins A, B, and C; deoxyribonucleases A, B, C, and D; proteinase; hyaluronidase; streptokinases A and B; nucleotidase; proteinase; esterase; and amylase. There are probably others.

Most Lancefield group A strains are susceptible to concentrations of the antibiotic bacitracin that do not inhibit most strains of other groups. They are quite susceptible to sulfonamides but may become resistant; tetracycline resistance is about 20 per cent. Groups B and C are moderately susceptible to antibiotics and sulfonamides. Group D is resistant to sulfonamides and tetracyclines but is susceptible to streptomycin and most penicillins.

Cellulitis. This is an infection of the skin and underlying tissues and is one of the most dangerous conditions caused by beta-type streptococci. The group A cocci give rise to a very severe, rapidly spreading infection, often with much swelling, attended by marked general symptoms due to potent exotoxins. The infection spreads in subcutaneous and lymphatic tissues. The "red streaks" or "blood poisoning" often seen developing up the arm or leg from a focus of streptococcal infection represents spreading of the infection via lymphatics (lymphangitis). These conditions require immediate medical attention. *Erysipelas* is another form of acute infection of the skin and underlying tissues due to beta-type streptococci of group A. It can be rapidly progressive and is frequently fatal unless promptly treated. The streptococci may enter the tissues through a very tiny break in the skin—for example, when a pathologist pricks his finger through a rubber glove while performing an autopsy on a corpse that harbors a streptococcal infection.

Streptococcal *impetigo* cannot be distinguished clinically from staphylococcal impetigo. Both are inflammatory skin diseases. It is best to remove the crusts that form over the pustular lesions and to keep the skin clean with soap and water to help clear up superficial infections.

Beta-type hemolytic streptococci of groups A, C, and G may also cause severe *tonsillitis (septic sore throat)* and are frequently present in the healthy nose

---

[7]Distinguish clearly between the Greek α and β (alpha and beta), which refer to *types as determined on blood-agar plates,* and the Roman capitals A, B, C, and so on, used to designate serologic groups of *beta-type* hemolytic streptococci.

[8]R surface antigens are found in M protein types 2, 3, 28, and 48, but not in type 14.

and throat: hence their inclusion in the group of respiratory pathogens. Like staphylococci, they sometimes invade the blood, causing "blood poisoning" (septicemia), and are often found in infections of the lungs *(pneumonia* and *empyema), meningitis, sinusitis,* and *mastoiditis.* Fortunately, infections by beta-type hemolytic streptococci are efficiently controlled by most of the broad-spectrum antibiotics.

Rheumatic Heart Disease. Rheumatic fever is a very serious disease with resulting heart damage that often cripples and frequently kills, occurring as a sequel to infection with group A hemolytic streptococci. Rheumatic heart disease is still one of the principal causes of death and disability from infectious disease in children in some developing countries, as it once was in the United States. It is probably due to antigenic relationships between group A streptococcal cell membrane antigen and human heart antigens. Antibodies produced against *Streptococcus* cause rheumatic fever in up to 3 per cent of all infections. Relapses (reinfections by the streptococci) in rheumatic fever patients may be, in large part, prevented by special treatment schedules with sulfonamide drugs or antibiotics.

Acute rheumatic fever occurs only after streptococcal pharyngitis. This condition should be treated with penicillin, erythromycin, or clindamycin. Another complication resulting from group A *Streptococcus* infections is glomerulonephritis, a sequela affecting kidney functions.

Antistreptolysin in Rheumatic Fever. The hemolytic toxins (streptolysins) of group A streptococci are of two sorts: streptolysin S, unstable in the presence of heat and acids, and streptolysin O, unstable in oxygen. Streptolysin O is strongly antigenic. In persons infected with group A streptococci, antibodies to streptolysin O are generally found and easily measured. The measurement (titration) of antibodies against streptolysin O, especially when the titer rises sharply over a period of two weeks or so during or after an illness such as sore throat, affords not only a valuable diagnostic test but also a measure of the reaction of the patient. Well over 80 per cent of rheumatic fever patients have a considerable titer of antistreptolysin O.

Puerperal Infections. Beta-type hemolytic streptococci may be carried into the parturient uterus from skin or external parts that have been insufficiently cleansed, on contaminated dressings or instruments, or on the improperly disinfected hands of the doctor or nurse—especially if the doctor or nurse is a carrier of hemolytic streptococci, a common situation. Such infections are preventable in practically every case. Postpartum mortality due to "childbed fever" or streptococcal puerperal fever caused by beta-type hemolytic streptococci was formerly as high as 50 per cent but is now very uncommon, thanks to clean obstetric techniques (remember the pioneer work of Ignaz Semmelweis) and antibiotics; however, infections sometimes develop when labor is complicated and operative interference is necessary. A woman should, if possible, go to a hospital to have a baby, because there are better facilities for treating any emergencies that may arise, and it is easier to carry out aseptic technique there than in the home.

Puerperal fever and *erysipelas* patients should be isolated to prevent the spread of the infection to other patients. The discharges, packs, and bandages are highly infectious. The recent postpartum patient has an open wound in the uterus and in some instances in the vagina and perineum following episiotomy. These are all very good portals of entry for streptococci, staphylococci, and other microorganisms. Nurses, doctors, and other personnel with head colds or sore throats should not be allowed to care for obstetric or indeed any patients. Personnel who have infectious lesions anywhere are real hazards to all patients.

Bronchopneumonia. This may be a serious complication of other diseases. Frequently it determines a fatal outcome in patients weakened by other disease. Bronchopneumonia may be caused by one or more of a variety of organisms, common among which are staphylococci and alpha- and beta-type hemolytic streptococci. Several species of microorganisms that can cause bronchopneumonia are usually in the patient's own mouth and often gain entrance to the lungs by accidental inhalation of saliva and mucus, as during surgical anesthesia.

Organisms that can cause bronchopneumonia are found in large numbers in the droplets of sputum or saliva given off by normal persons talking, sneezing, or coughing. Patients who are especially likely to develop bronchopneumonia—that is, those having chronic diseases or passing through acute diseases—must be protected from both the bacteria in their own mouths and those from visitors, other patients, and any persons (including medical and hospital personnel) having respiratory infections.

Scarlet Fever and Septic Sore Throat. Within recent years in the United States, these streptococcal diseases have declined greatly in frequency and severity. This is due, in part, to sanitation of dairy products. These diseases usually occur in epidemics, and formerly were often traceable to unpasteurized milk from cows having mastitis due to group A streptococci introduced into the cows or their milk by infected milkers. Pasteurization kills the undesirable streptococci. Septic sore throat and scarlet fever are really different manifestations of the same infection. Some strains of group A streptococci form a toxin that produces the rash seen in scarlet fever. Once a person becomes immune to this *erythrogenic* (rash-producing) toxin, he is no longer subject to rash but may still contract infection with group A streptococci of some other M-antigen type. The infection then develops without rash and therefore is called by different names, a common one being septic sore throat or "strep throat."

Hence, although scarlet fever is not common in adults, most of whom have some immunity to the erythrogenic toxin, septic sore throat may occur in persons of any age. The desquamation or scaling of scarlet fever is not seen in persons with septic sore throat, since it is due to the erythrogenic toxin. Usually, in an epidemic of septic sore throat, cases of scarlet fever occur; the same streptococci cause both conditions. Not infrequently, in severe cases, the cocci invade the blood and may cause bacterial endocarditis. This is usually a sign of waning resistance.

An antitoxin for the scarlet fever rash, analogous in principle to other antitoxins, neutralizes the erythrogenic toxin but does not kill streptococci. There is also a test (the *Dick* test: intradermal injection of a minute dose of scarlet fever toxin into the forearm) that shows who is susceptible to the erythrogenic toxin, and a toxoid for giving active artificial immunity against the toxin. Neither is now very widely used. A diagnostic skin test, based on the Schultz-Charlton reaction, is sometimes made by injecting intradermally, at the site of a rash, a small dose of scarlet fever antitoxin. If the rash is due to scarlet fever, it promptly fades or is *blanched* (the "blanching reaction").

Transmission. In mode of transmission from person to person, scarlet fever is an excellent model of most diseases of the upper respiratory tract and should therefore be discussed in detail. The scarlet fever (and septic sore throat) patient can spread the streptococci from the time of the very first symptoms and even earlier, and he or she remains infectious for a considerable but indefinite time after recovery, two to four weeks or even months. The vectors are *saliva* and *mucus* from the mouth and nose. These are nearly always present on any patient's hands and face to some degree. The dangers of handshaking and of kissing (alas!) are only too obvious.

Convalescents who are released from isolation too soon may transmit disease-causing organisms, since they may remain carriers for weeks or months. A patient can spread the disease as long as there is any discharge from the nose, the ears, or any other part of the body containing the streptococci. Patients with streptococcal infections can be made noninfectious within 24 hours by treatment with penicillin. This is not necessarily a cure and may have to be repeated or replaced with a different antibiotic if the streptococci are penicillin-resistant.

Like many other communicable diseases, scarlet fever and septic sore throat are spread also by healthy carriers and by persons with mild, unrecognized cases who sometimes exhibit minimal symptoms, e.g., only a slight sore throat or a temperature associated with a faint rash. These infected individuals keep the disease from disappearing in a community and are the starting points of epidemics. Health workers and informed teachers are always on the watch for these overlooked cases. A good method of finding them is follow-up work by the school nurse in investigating absent pupils: this discloses children with mild cases of a variety of childhood diseases who have not been seen by a physician.

Prevention

As has been indicated, the discharges from any lesion infected with streptococci (or, indeed, with any infectious organisms) are highly dangerous, as the organisms can be transferred to clean wounds, scratches, cuts, the postpartum mother, and possibly to normal mucous membranes, such as the nose, throat, and eyes. Moreover, pathogens fresh from the body are often encapsulated and particularly virulent. Dressings from all septic wounds, including those infected with streptococci, should be adequately wrapped, securely fastened, and burned. Any instruments or objects contaminated by such wounds should be sterilized or at least well disinfected by boiling or with chemicals.

In all streptococcal infections, the portal of entry and the portal of exit must be remembered and considered. All exudates, secretions, and dressings should be received into paper "wipes" or gauze and placed in a plastic bag at the bedside. The bag should be discarded after clamping carefully at the top, and incinerated. Bedding, dishes, and other objects that have had contact with the mouth or nose should be adequately disinfected.

Anaerobic Streptococci and Staphylococci

These organisms belong to the genus *Peptostreptococcus* or *Peptococcus* and grow only anaerobically. Many pathogenic conditions affecting the uterus, sinus, and ear are caused by these organisms. They often occur in gangrenous lesions. These organisms are strongly proteolytic, and a foul odor is characteristic of pus resulting from most *Peptostreptococcus* infections.

Streptococcus pneumoniae

Pneumococci (*Streptococcus pneumoniae*) are gram-positive, encapsulated cocci that characteristically occur in pairs but often grow in chains; thus, they are in fact streptococci. The capsule is a particularly distinctive part of pneumococci and confers on them three important properties: smoothness of colony form (S), antigenic type specificity similar to the M-type specificity of streptococci, and virulence. Without the capsule, pneumococci are in the R (rough) phase, have little or no virulence, and have no antigenic type specificity. Pneumococci grow on blood or serum media at 37 C under the same conditions as other streptococci and, like them, are quickly killed by boiling and by common disinfectants. In blood-agar plates they produce the alpha type of reaction, often with considerable hemolysis, and cause many of the same kinds of infection as do the group A hemolytic streptococci. Pneumococci were originally observed in association with pneumonia: hence their name.

Infections by Pneumococci

Lobar Pneumonia. This is an acute infectious disease most frequently caused by *Streptococcus pneumoniae.* The student should distinguish carefully between bronchopneumonia and lobar pneumonia. *Bronchopneumonia* localizes in scattered patches throughout the lungs and may be caused by a wide variety of organisms, including not only the staphylococci and streptococci but also *Klebsiella pneumoniae,* a gram-negative short rod. *Lobar pneumonia* derives its name from the fact that one or more entire lobes often become wholly consolidated if left untreated. It is caused most often by the pneumococcus and sometimes by various streptococci or Friedländer's pneumobacillus *(Klebsiella pneumoniae).* Formerly one of the leading causes of death, lobar pneumonia is no longer so lethal, partly because of the use of antibiotics.

Pneumococci, like group A streptococci, may also invade the pleural cavity, causing empyema; the meninges, resulting in meningitis; and the peritoneal cavity, with developing peritonitis. In addition, they sometimes cause puerperal sepsis, sinusitis, septicemia, and bacterial endocarditis. *Streptococcus pneumoniae* is found in the saliva and sputum of patients with lobar pneumonia and also occurs in normal persons.

Pneumococcal pneumonia is the most common type of bacterial pneumonia, with an estimated 500,000 cases per year in the United States and a 5 to 10 per cent mortality rate. Bacterial meningitis, due to the pneumococcus, has a fatality rate of up to 40 per cent.

Serologic Types of Pneumococci. Extensive studies of pneumococci have shown that, like group A streptococci, they may be divided into over 80 serologic types, called types I, II, III, and so on.

These types, like those of group A hemolytic streptococci, are differentiated by serologic methods, all of which depend on the fact that the capsules of pneumococci contain an antigenic substance *(soluble specific substance,* SSS, or *capsular polysaccharide)* that is chemically different in each type. Each type of capsular polysaccharide calls forth specific antibodies (precipitins and agglutinins) against itself. Formerly an important diagnostic procedure in preparation for type-specific serum therapy, determination of type or "typing" is now rarely done because of reliance on chemotherapy.

Identification of Pneumococci. Pneumococci are bile-soluble (Fig. 29–2). Optochin (Taxo P Sensi-Disks) disks on blood agar produce a zone of inhibition on a pneumococcus lawn after 8 hours of incubation at 37 C, but no inhibition of other streptococci. Most pneumococci typically ferment the carbohydrate inulin with acid (no gas) production. In addition, mice injected with virulent pneumococci die within 24 hours. Finally, the *Neufeld quellung reaction* is a test for the identification of the pneumococcal types: type-specific antiserum causes "specific" swelling (from German *Quellung,* swelling) of the capsules, which may easily be seen under the microscope.

Transmission. The discussion of transmission of scarlet fever is fully applicable also to pneumococcal and most other respiratory infections. Man seems to be the only natural host of the pneumococcus, which has as its natural habitat the nasopharynx. The carrier rate varies between 5 and 40 per cent and is especially high during the winter, when droplet infection increases in confined environments.

Prevention

The care of the patient with lobar pneumonia has changed markedly with the use of sulfonamides and antibiotics, which can render the patient noninfectious in 48 hours.

The use of masks for the protection of medical attendants has been almost entirely abandoned. The organisms causing lobar pneumonia are found so frequently in the respiratory secretions of apparently healthy individuals that emphasis is now placed on maintaining good general hygiene (rest, adequate diet, and so on) as offering better overall protection than the doubtful procedure of wearing a mask.

Only 14 SSS serotypes account for more than 80 per cent of pneumococcal infections. The *vaccine* released in 1978 for clinical use against *Streptococcus pneumoniae* contains capsular polysaccharides from all these 14 types, including type 3 pneumococcus, the most virulent one. It is thought that this vaccine may prevent not only pneumonia and meningitis but also pneumococcal otitis media infections, which affect approximately one million children in the United States each year.

The Respiratory Neisseria

The genus *Neisseria* is named for the German physician Neisser, who in 1879 discovered the gonococcus *(Neisseria gonorrhoeae),* the cause of gonorrhea.[9] The gonococcus is, of course, associated not with the respiratory tract but with the genital tract and the venereal disease gonorrhea. It is unique among *Neisseria* in that it is not characteristically found in the upper respiratory tract; however, it is so much like the respiratory *Neisseria* that the gonococcus and the related respiratory species are distinguishable only by special laboratory tests. We shall therefore describe the gonococcus with the respiratory *Neisseria,* reserving discussion of gonorrhea to Part C of this section (see page 536).

Among the respiratory *Neisseria* there is only one important pathogen: *N. meningitidis,* cause of epidemic spinal meningitis. It is almost exactly like *N. gonorrhoeae* except for habitat.[10] There are several other respiratory tract species of *Neisseria,* all of which are relatively harmless; however, these "harmless" *Neisseria* species sometimes become important because they can cause a gonorrhea-like vulvovaginitis in preadolescent girls and may be confused with the gonococcus.

Distinctive Morphology of Neisseria. All species of *Neisseria* are gram-negative diplococci, 0.8 μm in diameter, the cocci in pairs pressed together like two coffee beans with their flat surfaces apposed (Fig. 29–5). The pairs are often encapsulated, though the capsules are not prominent.

Physiologic Properties. The species of this genus, except *N. gonorrhoeae* and *N. meningitidis,* can grow well on ordinary laboratory media, without blood or serum, at room temperature (20 to 25 C) and exposed to air. The gonococci and meningococci require a 37 C temperature and special culture media containing heated blood and special nutrients adapted for their growth. Their growth is best in a moist atmosphere containing about 10 per cent carbon dioxide. Thus, the two major pathogens are readily distinguished from the lesser ones.

The Oxidase Reaction. Colonies of any species of *Neisseria* on "chocolate agar" (blood agar heated at 90 C for ten minutes) are small, watery, and not very distinctive. They may be made quite conspicuous, however, by moistening the growth on the agar with a 1 per cent solution of an *oxidase* indicator, tetramethyl-para-phenylene-diamine. This turns first red, then black, in contact with oxidase-producing colonies, among which *Neisseria* species are outstanding (Fig. 29–6). (Compare with the tetrazolium test for *Corynebacterium diphtheriae,* page 457.)

Unlike other species of *Neisseria,* gonococci and meningococci have little

---

[9]From Greek, *gone,* seed or semen; *rhein,* to flow. It was formerly thought that the copious pus formed in acute gonorrhea in the male was actually a flow of semen, hence the name.

[10]Some cases of meningitis have been ascribed to the gonococcus and some cases of gonorrhea-like disease to meningococci.

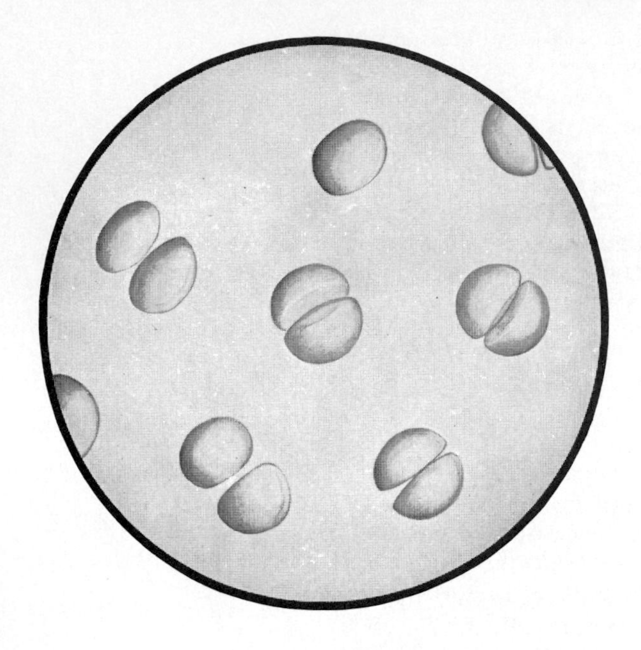

Figure 29–5 Diagrammatic enlargement of *Neisseria gonorrhoeae.* The magnification is in the range of the electron microscope. Note the rounded, hemispheric appearance of each cell of the pairs.

resistance to heat, light, drying, or disinfectants, and die quickly outside the body. When kept moist and protected from light, they *may* live on sheets, on clothing, or in pus for several hours, but they *rarely* do so. They are easily killed by chemicals containing silver, and hence silver-containing drugs like silver nitrate, Argyrol, or Protargol are often used in gonorrheal infections, especially for the destruction of gonococci in the eye (see *Gonorrheal Ophthalmia,* page 538). Other ordinary disinfectants are quite effective in disinfecting hands, clothing, and utensils that might be contaminated with either of these organisms. Both, especially the gonococcus, are sensitive to penicillin, although many strains have become resistant.

The Meningococcus

The meningococcus, *Neisseria meningitidis,* is the most dangerous pathogen among the upper respiratory *Neisseria.* There are several immunologic groups and types of *N. meningitidis:* A, B, C, Y, and W-135 are those found most frequently in clinical infections. Type specificity resides, as in pneumococci, in the chemical nature of the polysaccharide capsular substance. A universally accepted classification of groups and serotypes is still lacking.

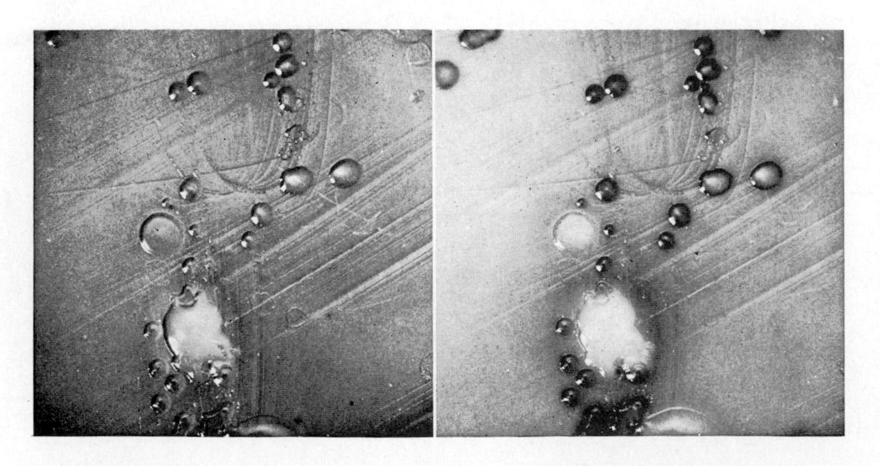

Figure 29–6 The oxidase test for the identification of meningococcus (or any *Neisseria*) colonies. Mixed culture on blood agar. *Left,* colonies of meningococci and contaminants before the application of tetramethyl-*p*-phenylene-diamine solution. *Right,* the same colonies after the application of the reagent. Note that the meningococcus colonies (dark) show the development of color first about the edges, and also a slight discoloration of the medium. (×5). (From Burrows, W.: Textbook of Microbiology. 20th ed. Philadelphia, W. B. Saunders Company, 1973.)

N. meningitidis causes a very dangerous and often fatal inflammation of the coverings (meninges) of the brain and spinal cord. The disease is called epidemic cerebrospinal meningitis, meningococcemia, or cerebrospinal fever. *N. meningitidis* is very pyogenic, and in this resembles the gonococcus. Gram-negative anaerobic cocci of the genus *Veillonella* also occur in pairs, masses, or chains and are found in the respiratory tract or genitourinary structures of humans, but they are probably nonpathogens.

Meningococcus Meningitis. Since this disease, like influenza, is transmitted by respiratory secretions, it usually occurs in epidemics, and is therefore often called *epidemic meningitis.* This form of meningitis should be differentiated from sporadically occurring cases of meningitis, which frequently are caused by blood-borne streptococci, pneumococci, tubercle bacilli, *Salmonella,* and numerous other bacteria, usually secondary to other severe infections. Such cases of meningitis never occur as epidemics.

Epidemic meningitis occurs in persons of all ages and sexes and is spread chiefly by the oral and nasal secretions and droplets of sputum of *carriers.* In many outbreaks, the mode of transmission is obscure, but it obviously requires fairly close contact (as in very crowded rooms), since the meningococci die quickly outside the body. The carrier rate (i.e., the percentage of persons who carry the cocci) in some population groups, such as military organizations, may be as high as 50 per cent, yet morbidity (illness) rates may be only 2 to 5 per cent, indicating a high rate of dissemination but a low degree of susceptibility of persons in general, or a low virulence of the meningococci being spread. Virulence, however, apparently may suddenly increase, causing a serious epidemic. The connection between cases is sometimes impossible to trace, as is true of all infections spread by carriers. Persons directly associated with a patient (e.g., doctors, nurses, and the patient's family) contract the disease only infrequently, although they often become carriers. Recent studies indicate that deficiencies in complement fractions (C6, C7, C8) increase the risk of meningococcal or gonococcal bacteremia.

Meningococcal meningitis is particularly serious in children. Of the 1,000 to 2,000 cases reported in the United States yearly and 30,000 or more cases in epidemics around the world, more than 50 per cent are children under five years of age. In 1980, 2,840 meningococcal infections were reported in the United States. The last big epidemic occurred in 1943, with 18,223 individuals stricken.

Meningococcal Vaccines. Meningococci of groups A, B, and C are most frequently the cause of epidemic meningitis. Cell-wall polysaccharides of cocci of groups A and C (unfortunately not group B, which seems to be unencapsulated) have been found to be antigenic in humans. Vaccines consisting of group A or C or both have been used to immunize hundreds of thousands of children in Egypt, the United States, and Brazil, with only rare untoward reactions and otherwise very satisfactory results. They are recommended at present especially for persons at unusual risk of infection, particularly during epidemics.

The meningococcus enters the body via the nasopharynx. In many instances, the meningococci remain localized in the nasopharynx, producing no symptoms whatever or no more than a rhinitis with purulent discharge. Such cases are regarded as mere "colds" or "catarrh" and probably serve to infect many persons. Once localized in the nasopharynx, the cocci may proceed no further, which appears to be the usual event, or they may invade the blood. The disease is known to be frequently a *septicemia* in its early stages, and blood-broth cultures are often positive for the meningococci. In many instances the organisms invade no farther than the blood. Probably only a small percentage of blood infections result in meningitis. Erythematous areas appear on the skin, giving rise to an older name for the disease, "spotted fever."

Diagnosis. In any case in which there is clinically a suspicion of meningitis (fever, headache, stiff neck, vomiting, rash), it is of extreme importance to

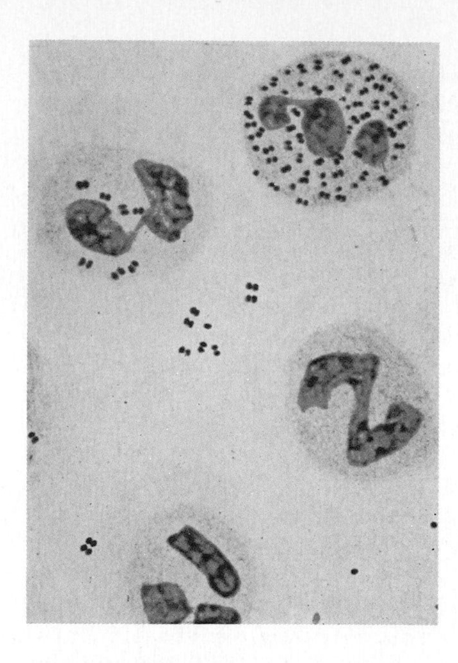

Figure 29–7 Meningococci in spinal fluid stained by Gram's method. The leukocytes have engulfed (phagocytized) large numbers of the diplococci. (From Ford, W. W.: Bacteriology. New York, Hafner Publishing Co., 1939.)

make a lumbar puncture very early. The diagnosis is made or confirmed, usually from a Gram-stained smear of the *centrifuged sediment* of the spinal fluid. The organisms are found in the phagocytes and exactly resemble gonococci (Fig. 29–7). Cultures may also be made, in the same manner as for gonococci, using the spinal fluid sediment. The results of cultures are too slow in preparation to be of value for immediate diagnosis and therapy of such rapidly progressive diseases as bacterial meningitis. Death can occur in only 48 hours.

Therapy. Because sulfadiazine was highly effective for both oral prophylaxis and intravenous therapy in meningococcal meningitis, it was very widely used until most strains of meningococci became resistant to it. This made necessary the use of antibiotics like 24 million units per day of intravenously administered penicillin G or, in penicillin-sensitive patients, broad-spectrum antibiotics like the tetracyclines or (preferably) chloramphenicol. None of these is as effective as sulfadiazine used to be.

Prevention

The precautions in cerebrospinal meningitis are those observed in dealing with pneumonia, scarlet fever, and all infections transmitted by oral and nasal secretions. Immediate chemoprophylaxis is highly indicated in cases of close contact with patients. The carrier state may best be eliminated with rifamycin, minocycline, or sulfadiazine.

SUMMARY

Oral and nasal secretions containing pathogenic microorganisms are spread through talking, laughing, sneezing, and coughing, on hands and eating utensils by fomites in general, and by personal contacts with susceptible hosts. Pathogens of the respiratory tract, and airborne pathogens in general, are among the most difficult to control. Fortunately, owing to previous subclinical infections, normal, healthy persons generally do not suffer seriously from infectious disease, except during epidemics. At those times and others, respiratory infections may be prevented or partially controlled by wearing masks, wet-scrubbing or vacuum cleaning of dusty areas (including droplet nuclei on surfaces), ventilation and climate control effective against airborne pathogens, and use of disinfection. Among the pathogenic cocci of the respiratory tract, *Staphylococcus aureus* is both common and dangerous. It produces a golden yellow pigment on agar media, is pyogenic, and causes gastroenteritis because of its enterotoxin. It also causes hemolysis of erythrocytes and coagulation of plasma (due to coagulase), destroys white blood cells (with leukocidin), causes necrosis of skin tissue, and is generally toxin-producing and pathogenic. Toxic shock syndrome is one manifestation of its wide range of infectivity. It was deadly to many women until its mode of action in superabsorbent tampons was discovered. However, food poisoning caused by *Staphylococcus aureus* because of its enterotoxin fortunately is rarely fatal.

Streptococci comprise a large group of organisms, only some of which are deadly pathogens. These and others can be screened by growing them on sheep erythrocytes on blood-agar medium in Petri dishes, where they either produce alpha- or beta-type hemolysis or grow as nonhemolytic colonies. The alpha-hemolytic streptococci produce mostly chronic conditions such as tooth abscesses, infections in arthritic joints, tonsillitis, and sinusitis and not exclusively (besides beta-type streptococci, pneumococci, staphylococci, and so on) bacterial endocarditis. Encapsulated beta-hemolytic streptococci cause more acute diseases. They are classified by a complex serologic scheme into Lancefield groups, of which group A is most pathogenic; it, too, is subdivided further into 14 types. Diseases caused by beta-hemolytic streptococci are severe tonsillitis, septic sore throat, puerperal sepsis, septicemia, meningitis, mastoiditis, and scarlet fever, often leading to rheumatic fever (rheumatic heart disease).

Lobar pneumonia is caused by *Streptococcus pneumoniae*. This organism has over 80 serologic types, differentiated by the soluble specific substance (SSS) capsular polysaccharide. Pneumococcal pneumonia is the most common type of bacterial pneumonia, with an estimated 500,000 cases per year in the United States and a 5 to 10 per cent mortality rate. Bacterial meningitis, caused by the pneumococcus, has a mortality rate of up to 40 per cent.

Closely related to the gonococcus is a dangerous pathogen of the upper respiratory tract called the meningococcus, or *Neisseria meningitidis*. It causes a very infectious and often fatal inflammation of the coverings (meninges) of the brain and spinal cord referred to as meningococcus meningitis, epidemic meningitis, or cerebrospinal meningitis.

Vaccines now exist against epidemic meningitis and pneumonia.

REVIEW QUESTIONS

1. What is a coccus? How may cocci be classified? Are all cocci pathogenic?
2. Give the morphologic and cultural characteristics of *Staphylococcus epidermidis* and *Staphylococcus aureus*. How do these two organisms differ? What infections can each of these organisms cause? What is the importance of staphylococcus enterotoxin?
3. Describe the appearance of alpha, beta, and gamma types of streptococci in blood-agar plates.
4. What is meant by Lancefield's groups of streptococci? Which groups are of most importance in disease production? How are these related to alpha, beta, and gamma blood-agar types?
5. In what sorts of infection are alpha-type streptococci found? How are they transmitted? What is their usual habitat?
6. What organisms cause (*a*) erysipelas; (*b*) puerperal fever; (*c*) septicemia; (*d*) bronchopneumonia; (*e*) septic sore throat; (*f*) scarlet fever?

7. Is there a relationship between septic sore throat and scarlet fever? Discuss it.
8. What is bovine mastitis? How is it transmitted? How is it related to (*a*) scarlet fever; (*b*) septic sore throat?
9. Discuss the various means by which scarlet fever is transmitted. How may transmission be prevented?
10. What is the Dick test?
11. What measures of concurrent disinfection must be taken in scarlet fever and septic sore throat?
12. Discuss the duties of the nurse and other health workers in dealing with streptococcal infections of (*a*) the respiratory tract; (*b*) the skin (erysipelas); (*c*) the postpartum female.
13. Mention different clinical manifestations of infection with pneumococci. Why are pneumococci named *Streptococcus pneumoniae*?
14. What is the relationship of pneumococci to streptococci? What is their normal habitat? How do they resemble and differ from alpha-type streptococci?
15. What is meant by pneumococcus typing? Describe the method used in the quellung reaction. What is the Neufeld test? Upon what facts is it based? When and why was it performed?
16. What is type-specific pneumococcus serum and what was it used for?
17. What is the role of bacterial capsules in the property of virulence?
18. Who was Neisser and what did he contribute to microbiology? Describe the genus of bacteria bearing his name.
19. What is the relationship between the meningococcus and the gonococcus?
20. What is the etiologic factor in meningitis? What is the etiologic factor in *epidemic* meningitis? How is cerebrospinal meningitis transmitted?
21. What is the relation of carriers of meningococci to cases? What are the details of concurrent disinfection in cases of meningococcus meningitis?
22. Why is it important with respect to respiratory diseases and epidemic meningitis that crowded living conditions be avoided?
23. What is the difference in origin between bronchopneumonia and lobar pneumonia?
24. How is lobar pneumonia transmitted? What groups of persons are especially susceptible? How may transmission be prevented?
25. What therapeutic agents have largely supplanted type-specific serum in the treatment of patients with pneumococcus pneumonia?
26. Discuss rheumatic heart disease and its causes.

Supplementary Reading

Arbuthnott, J. P.: Staphylococcal toxins. *In* Microbiology—1975. Washington, D.C., American Society for Microbiology, 1975.

Catlin, B. W.: Nutritional profiles of *Neisseria gonorrhoeae, Neisseria meningitidis,* and *Neisseria lactamica* in chemically defined media and the use of growth requirements for gonococcal typing. *J. Infect. Dis., 128*:300, 1973.

Davis, B. D.: Microbiology. 2nd ed. New York, Hoeber Medical Division, Harper & Row, 1973.

Facklam, R. R., Padula, J. F., Thacker, L. G., Wortham, E. C., and Sconyers, B. J.: Presumptive identification of Group A, B, and D streptococci. *Appl. Microbiol., 27*: 107, 1974.

Goldschneider, I.: Vaccination against meningococcal meningitis. *Conn. Med., 34*:335, 1970.

Jeljaszewicz, J.: Staphylococci and Staphylococcal Diseases: Proceedings. Stuttgart, Fischer Verlag, 1976.

Kloos, W. E., and Schleifer, K. H.: Simplified scheme for routine identification of human *Staphylococcus* species. *J. Clin. Microbiol., 1*:82, 1975.

Kondo, I., Sakurai, S., Sarai, Y., and Futaki, S.: Two serotypes of exfoliatin in *Staphylococcus* strains isolated from patients with scalded-skin syndrome. *J. Clin. Microbiol., 1*:397, 1975.

Lennette, E. H., et al. (eds.): Manual of Clinical Microbiology. 3rd ed. Washington, D.C., American Society for Microbiology, 1980.

Males, B. M., Rogers, W. A., Jr., and Parisi, J. T.: Virulence factors of biotypes of *Staphylococcus epidermidis* from clinical sources. *J. Clin. Microbiol., 1*:256, 1975.

Mej'are, B.: On the Facultative Anaerobic Streptococci in Infected Dental Root Canals in Man. Lund, Sweden, Gleerup, 1975.

Orth, D. S., and Anderson, A. W.: Polymyxin-coagulase-deoxyribonuclease-agar: A selective isolation medium for *Staphylococcus aureus. Appl. Microbiol., 20*:508, 1970.

Otaka, Y.: Immunopathology of Rheumatic Fever and Rheumatoid Arthritis. Tokyo, Igaku Shoin, 1976.

Sanders, W. E., Jr.: Interactions between streptococci and other bacteria in the throat. *In* Microbiology—1975. Washington, D.C., American Society for Microbiology. 1975.

Sternberg, G. M.: The etiology of croupous pneumonia. *Nat. Med. Rev., 7*:175, 1897.

Stollerman, G. H.: Rheumatic fever and streptococcal infection. New York, Grune and Stratton, 1975.

Top, F. H., et al. (eds.): Control of Infectious Diseases in General Hospitals. New York, American Public Health Association, 1967.

Wilkinson, H. W., Facklam, R. R., and Wortham, E. C.: Distribution by serological type of group B streptococci isolated from a variety of clinical material over a five-year period (with special reference to neonatal sepsis and meningitis). *Infect. Immun., 8*:228, 1973.

DIPHTHERIA

CORYNEBACTERIUM DIPHTHERIAE

Corynebacterium[1] *diphtheriae,* the bacterium that causes diphtheria[2] (also called croup), is characteristically confined to the respiratory tract. Unlike streptococci, staphylococci, and pneumococci, which may infect many different tissues or organs, it is rarely found in any other part of the body.[3] Its mode of transmission is representative of respiratory pathogens in general. Formerly one of the leading causes of death in all parts of the world, diphtheria has now all but disappeared in the United States. Infants three to five months old are immunized with one preparation containing diphtheria toxoid, tetanus toxoid, and pertussis vaccine (DTP) (see page 332).

Diphtheria bacilli were discovered in 1883–84 by two German physicians, Klebs and Löffler, in bacteriologic specimens from the throats of patients. These bacilli are therefore sometimes spoken of as Klebs-Löffler or K-L bacilli. The disease was first called *diphtheritis* by Bretonneau, who recognized it as a distinct malady in 1826.

Physiologic Characteristics. *C. diphtheriae* is nonmotile and does not form spores. As a nonspore-former, it is easily killed by commercial pasteurization (63 C for 30 minutes or 72 C for 15 seconds) and standard disinfectants. However, it survives drying and exposure to light better than many nonspore-bearing pathogenic organisms and can be disseminated in dust. A few hours of exposure to direct sunlight kills most nonspore-forming pathogenic bacteria, including *C. diphtheriae.*

Morphologic Characteristics. These gram-positive bacilli have an especially characteristic appearance when stained with slightly alkaline *methylene blue* (Löffler's stain). They are seen to be very pleomorphic, sometimes slender, pointed, and curved, with deep blue or reddish-purple granules of volutin[4] at the ends and often several transverse bars. In laboratory cultures four hours old or more, the bacilli usually assume curious shapes, often resembling clubs: hence their generic name. Sometimes the volutin granules at the ends make them resemble dumbbells or exclamation points (Fig. 30–1). These morphologic and staining peculiarities are very useful in making bacteriologic diagnoses of diphtheria.

Laboratory Diagnosis. The bacilli grow well aerobically in the laboratory at 35 C on Löffler's medium[5] and on blood or serum agar (or broth) with glucose. In making diagnostic throat cultures on these media, a sterile swab is rubbed gently over tonsils or white patches on the patient's throat and then rubbed on the surface of Löffler's medium in culture tubes or on Pai's medium.[6] These cultures are incubated at about 35 C and examined microscopically after about 4 hours and again after 8, 18, and 24 hours. When smeared on a slide,

---

[1]From Greek, *koryne,* club. Club-shaped rods are one of the numerous distinctive forms of diphtheria bacilli.

[2]From Greek, *diphthera,* membrane. This refers to the distinctive, membrane-like film (pseudomembrane) formed in typical cases of diphtheria.

[3]During World War II it caused serious trouble to soldiers in Burma by infecting cutaneous ulcers (cutaneous diphtheria). Vulvovaginal diphtheria may also occur.

CHAPTER 30

stained with methylene blue, and examined with a microscope, the diphtheria bacilli in the cultures are usually seen to be mixed with other microorganisms, which also grow on these rich media unless a *selectively inhibitory* substance, potassium tellurite (K_2TeO_3), is added. Mixed or pure, the distinctive diphtheria bacilli can usually be recognized by a skilled bacteriologist. Hence, it is possible to tell within 4 to 24 hours whether diphtheria-like bacilli are present in a patient's throat.

Pure Cultures. Pure cultures of *C. diphtheriae,* which are necessary for conclusive diagnosis, may be isolated from primary Löffler cultures or directly from throat swabs by spreading the material on blood or serum agar containing K_2TeO_3. Tellurite inhibits much contaminating growth and is reduced to a dense black color by colonies of *C. diphtheriae.* The colonies on tellurite medium are greyish-black or metal-gray, without a sheen. On Tinsdale's medium there is a brownish-black halo. In the absence of tellurite, the colonies may be recognized by treating them with a solution of a tetrazolium salt, which is reduced first to a distinctive color and then to black. The tetrazolium test is analogous to the oxidase test used in identifying colonies of *Neisseria* in the diagnosis of meningitis or gonorrhea (Chapters 29 and 36). *C. diphtheriae* is catalase-positive, reduces nitrate to nitrite, and is gelatin- and urea-negative. Acid is produced from glucose and maltose in 48 hours, but sucrose is rarely fermented.

Throat Cultures. The procedure in preparing diagnostic throat cultures, whether for *C. diphtheriae* or for other microorganisms, is identical. The culture medium, which should be appropriate for the organisms to be cultured, must be moist and fresh, and the tube should be properly labeled with the patient's name, the date, and the hour. If the patient has been on antibiotic therapy or sulfonamides within five days, this information should be sent to the laboratory with the culture, since these drugs may inhibit growth in the cultures.

Pathogenesis of Diphtheria

Typically, diphtheria commences as an acute inflammation of the pharynx caused by the diphtheria bacillus. The organisms are almost entirely lacking in aggressiveness. They usually remain localized in the tonsils, if present,[7] and in the upper respiratory mucous membranes. There the bacilli secrete an exotoxin comparable in deadliness to the exotoxins of *Clostridium botulinum* and *Cl. tetani,* all of which are much more poisonous than cobra venom.

The exotoxin irritates the tissues, which give forth a fibrinous exudate that coagulates into a tough, leathery, greyish-white *pseudomembrane.* This is not

[4]Volutin is a complex metaphosphate, probably stored food reserve. It is commonly associated with nucleic acids and occurs in many species of bacteria. It stains very intensely with basic dyes.

[5]Heat-coagulated serum containing 20 per cent of 5 per cent glucose infusion broth.

[6]Coagulated egg.

[7]Presence of tonsils greatly predisposes to the harboring and carrying of diphtheria bacilli, hemolytic streptococci, and other pathogens.

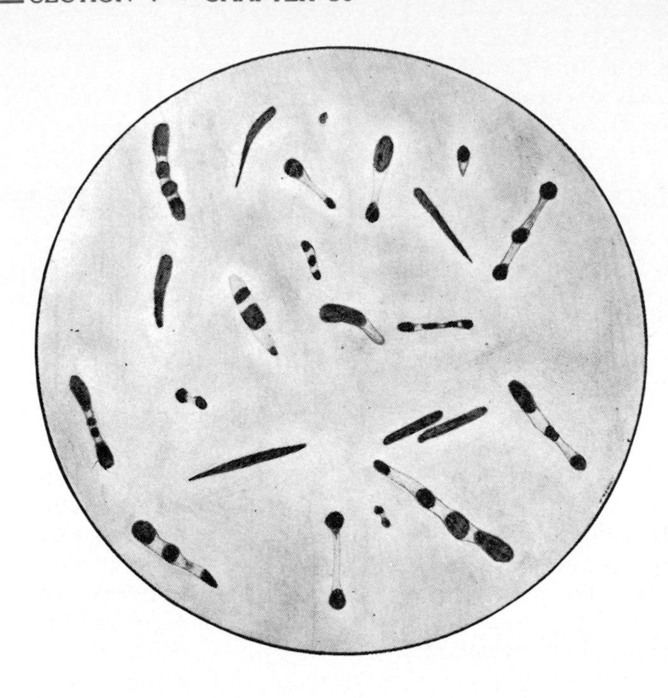

Figure 30–1 *Corynebacterium diphtheriae* (drawing from pure cultures). These have been stained with Löffler's alkaline methylene blue solution. Note the great variation in length, the pleomorphism, and the volutin arranged as bars and granules and sometimes filling the entire cell. (× approx. 2,500.)

always present or typical in color or consistency. This membrane, along with swelling due to inflammation, may occlude the air passages, especially if the larynx is invaded (laryngeal diphtheria). Death may be caused mechanically by asphyxiation in a few hours unless the obstructed larynx is bypassed by inserting an air tube into the trachea from the outside *(tracheotomy).* Other serious symptoms, and often death, arise from the effect of the toxin on the heart muscle (myocarditis and sudden "heart failure"), nerves (paralysis), kidneys (nephritis with albuminuria), and adrenal cortex (circulatory failure). In 10 to 25 per cent of diphtheria patients the exotoxin causes *myocarditis,* a lethal inflammation of the heart muscles.

When bacilli grow on the tonsils and walls of the throat (pharynx) of a susceptible person, they cause the usual (pharyngeal) form of diphtheria. The organisms may grow also in the nose, producing *nasopharyngeal diphtheria,* and in the larynx, causing what used to be known as "membranous croup" but is now generally called *laryngeal diphtheria.*

In children under five years of age, diphtheria is likely to be more severe than in older children or in adults. A considerable percentage of deaths from the disease occurs in children under five years old. The number of cases of diphtheria and other contagious children's diseases usually rises in the fall because bringing children together in school increases opportunities for infection, and also because fall and winter are the seasons of various infections of the nose and throat that may allow diphtheria bacilli to become established. Some cases may occur in any season, although the disease is now rare in the United States, thanks to active artificial immunization with toxoid (see page 462).

Clearly diphtheria is not confined to childhood; cases may also occur in older persons of all ages, including doctors, nurses, and other health workers who have not taken proper precautions.

Transmission of Diphtheria

Diphtheria, typical of diseases of the respiratory tract, is spread by *saliva* and *nasal discharges*; by direct contact, as in kissing; by coughing into another's face; or, as is frequently the case with children, by articles that go directly from the mouth of one child into that of another, such as pencils, soft drink bottles, or drinking cups. Fingers may also carry the bacilli. These methods of transmission convey all sorts of microorganisms of the upper respiratory tract. *C. diphtheriae* may live quite a long time in saliva and mucus deposited on dishes, tableware, and toys. Bits of the fibrinous pseudomembrane that forms in the diphtheria patient's throat may be thrown out in coughing and dry on the furniture or the floor. In such fragments of membrane the bacilli may remain alive for weeks and be distributed as dust. Oral and nasal mucus and exudate from sores in the nose or on the lips of patients also contain the bacilli.

Carriers. Diphtheria carriers were the first kind discovered, and much of our knowledge of carriers in general has come from the study of diphtheria. There are three kinds of diphtheria carriers: (1) convalescents, (2) those who have contact with diphtheria patients, and (3) those who are not known to have had any association with the disease (so-called "casual carriers").

Convalescents. Convalescents usually get rid of the bacilli in a few days or weeks, but in some cases the bacilli may remain in the throat and nose for a long time after the patient has recovered. In about 5 per cent of the cases they persist for two months, and in about 1 per cent they remain indefinitely. Most health departments require two or three negative cultures (that is, cultures that do not show the diphtheria bacillus) from both the throat and nose, taken at intervals of 24 to 48 hours, before a patient may be released from isolation.

ANTIBIOTICS AND RELEASE CULTURES. In addition to antitoxin, most physicians administer antibiotics to patients with diphtheria, often before a diagnosis has been made. Although antibiotics do not neutralize diphtheria toxin, they control growth of numerous other bacteria (such as streptococci and pneumococci) in the throats of diphtheria patients, thus decreasing complications.

These antibiotics may continue to be present in oral and nasal secretions for five days to a week after the last dose. They often suppress growth of diphtheria bacilli in diagnostic or quarantine-release cultures. If the laboratory is informed that the patient has received penicillin, the bacteriologist can eliminate it by the use of penicillinase in the cultures. If the patient has received a sulfonamide drug, PABA is added to the culture media. (Explain these devices.)

Even though a convalescent remains a carrier for a long period, he or she is much less infectious two weeks after the complete healing of all lesions. Experience shows that such carriers may be readmitted to school without any untoward results. Antibiotics cure most carriers.

Contact Carriers. These may be schoolmates of the patient, members of the family, or persons who have taken care of the patient. These people may be more dangerous than the patient or convalescent because they go about freely. Contact carriers must observe the same precautions as are taken in actual cases of diphtheria to avoid scattering the bacilli by means of saliva and nasal secretion.

Casual Carriers. The third class of carriers, those who have had no known contact with diphtheria cases, are discovered accidentally when routine throat cultures are made. Careful studies have shown that it is the person with the clinical case and his or her family-contact carriers who are most likely to cause the disease. Healthy children found carrying diphtheria bacilli in the schools seldom give rise to epidemics.

Subclinical Infections. The chief sources of diphtherial infections are carriers and subclinical and missed cases. Diphtheria is not always severe. Many children have mild sore throats not recognized as diphtheria, and the infection is ignored. Because of their own resistance they do not succumb, but another child whom they infect may die. The carrier state is usually temporary and is often cured by removal of tonsils. In children's wards and schools, health personnel should be on the watch for nasal discharges, especially those that irritate the nostrils or are tinged with blood, as these may be due to subclinical diphtheria infections. Hoarse voice, "croupy" cough, or difficulty in swallowing or breathing should also arouse immediate suspicion of diphtheria.

Milk as a Vector of Diphtheria. *Corynebacterium diphtheriae* does not change the appearance or taste of milk and grows well in it. The milk may be contaminated somewhere on its way from the cow to the consumer by a person with a subclinical infection or by a carrier.

The characteristics of a milk-borne epidemic of any disease (e.g., scarlet fever, enteric infections, and diphtheria) are that the cases break out all at once; they are all on the route of supply to the customers of one dairy or firm; and a majority of the patients are children.

The Structure and Action of Diphtheria Toxin

Diphtheria toxin is a polypeptide with a molecular weight of 62,000 daltons. Active toxin consists of two chains linked by a disulfide bridge. One part of the molecule functions in penetrating susceptible cells, the other blocks incorporation of amino acids into protein. The toxin is produced only by strains that are lysogenic for phage β[8] and only at specific concentrations of available iron. To guinea pigs the toxin is lethal and erythrogenic at 1.5×10^4 MLD/mg (see below). The enzyme amino-acyl transferase is specifically inactivated by the diphtheria toxin, and its action is primarily on cardiac tissue. Adsorption of toxin occurs over a wide pH range and does not take place in the absence of salts. Initial binding of toxin to the cell is electrostatic in nature, involving positively charged surface groups. No specific receptors are required for the attachment of toxin to cells.

Virulence of Diphtheria Bacilli

One way of testing for virulence is by inoculating a pure culture of the suspected bacilli into animals. State health departments are prepared to make this test using guinea pigs, rabbits, and young chicks. Certain characteristic changes are produced that can be recognized by those specially trained in such work.

An alternative method is available that does not utilize animals but depends on the production of toxin by the bacilli while growing on a special antitoxin-containing agar. It is called the *in vitro toxigenicity test.* The presence of diphtheria toxin is made evident by distinct white lines that appear in the agar. The white lines are in reality a precipitation reaction between toxin from the growing bacilli and diphtheria antitoxin previously incorporated into the agar (Fig. 30–2).

Therapeutic Use of Diphtheria Antitoxin

For therapeutic purposes the dosage of any antitoxin is determined by the seriousness of the symptoms, the stage of the disease at which treatment is begun, and the age of the patient. The sicker the patient and the later the stage of the disease, the larger the dose must be. It is best given intramuscularly;

---

[8]Microorganisms are said to be *lysogenic* when they are carriers of a bacteriophage that does not immediately destroy the carrier cells (a so-called *temperate phage*). Temperate phages often confer new genetic properties (e.g., toxigenicity) on the carrier cells (see *Transduction,* page 117).

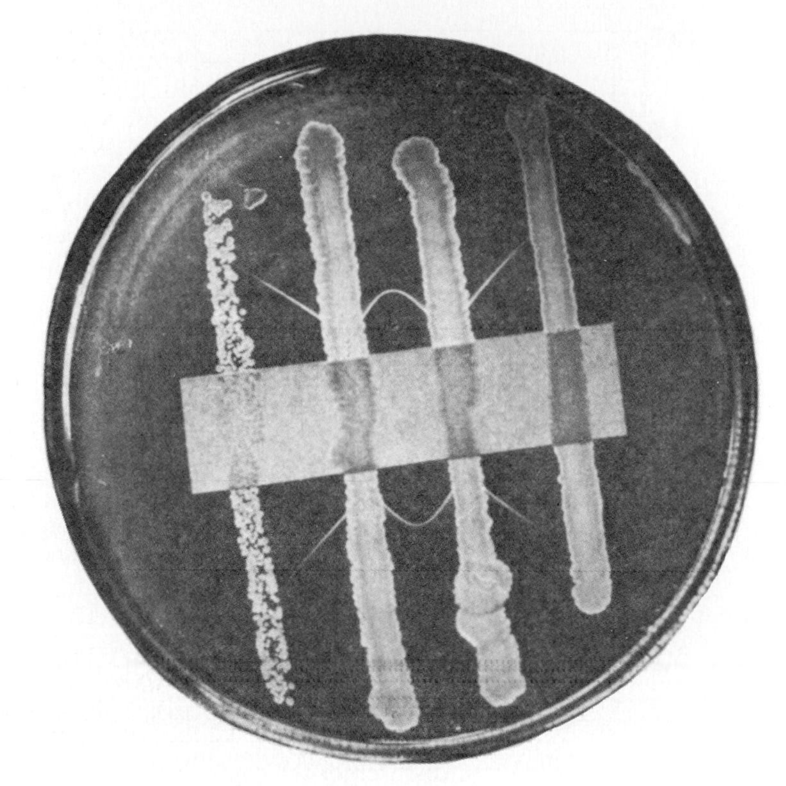

Figure 30–2 The in vitro test for toxigenicity. Serum-agar was poured into the dish at about 45 C. Before it hardened, the strip of filter paper, saturated with diphtheria antitoxin, was pressed to the bottom of the agar in the dish. After the agar hardened, it was inoculated on the surface in long streaks at right angles to the paper. As the growth developed at 37 C, toxin diffused from the culture into the agar. Simultaneously, antitoxin diffused from the paper strip. Where the toxin and antitoxin met in proper concentration for reaction, precipitation occurred. This is seen as thin white lines between the growths of the cultures. This reaction with diphtheria antitoxin is produced only by virulent (i.e., toxigenic) *C. diphtheriae*. (From King, E. O., Frobisher, M., and Parson, E. I.: *Am. J. Public Health 39*:1314, 1949.)

only in desperate cases is antitoxin of any kind given intravenously. The antitoxin combines with the toxin circulating in the blood and renders it harmless. It does not repair damage to heart, nerves, and kidneys once the toxin has combined with these tissues. Severe allergic reactions must be constantly guarded against in administering any serum for any purpose.

The Unit of Antitoxin. The potency of diphtheria antitoxin is stated in units. The commercial "unit" of antitoxin is the amount that counteracts slightly over 100 minimal lethal doses (MLD) of diphtheria toxin. An MLD of toxin is the least amount of toxin necessary to kill four of five 250 gram guinea pigs in four to five days. A unit of antitoxin, therefore, neutralizes a little over 100 MLD of toxin. The unit is used for expressing the dosage of antitoxin just as milligrams or milliliters are used for ordinary medicines. Some other units, used mainly in the preparation and testing of commercial diphtheria antitoxin, are shown in Table 30–1. A strong, artificially concentrated diphtheria antitoxin

Table 30–1. Units of Diphtheria Toxin and Antitoxin

ALD—The average lethal dose; the same as the LD/50. The dose of toxin that kills *half* of a group of 250-gram guinea pigs in 96 hours.

LoD—The largest amount of toxin that, when mixed with one unit of antitoxin, does not cause symptoms of diphtheria toxication in guinea pigs.

L+D—The least amount of toxin that, when mixed with one unit of antitoxin, kills 250-gram guinea pigs in 96 hours.

MRD—Minimum reaction dose; the highest dilution (i.e., least amount) of toxin that gives a typical skin reaction on intracutaneous injection.

LrD—The highest dilution (i.e., smallest amount) of toxin that, when mixed with one unit of antitoxin, gives a typical skin reaction.

TCID—Tissue culture infective dose.

LfD—The flocculation unit is the amount of toxin that will combine most rapidly with one unit of antitoxin in the in vitro Ramon flocculation test.

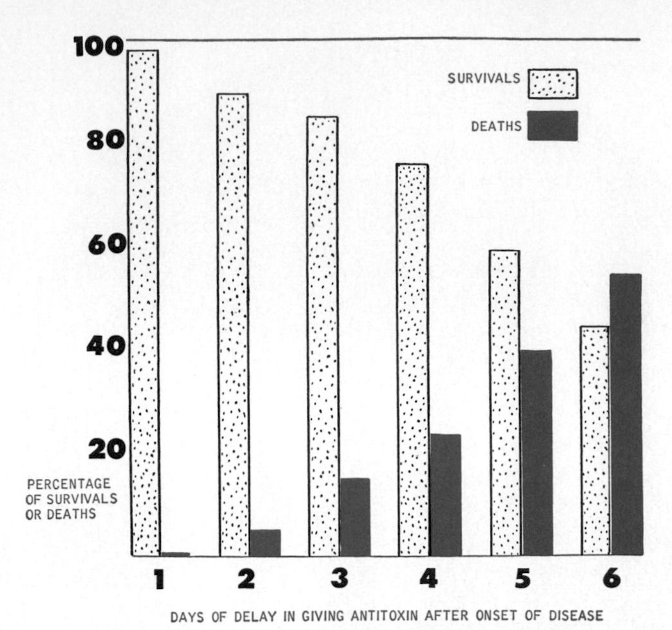

Figure 30–3 Extent to which delay in giving antitoxin decreases chances for survival in diseases due to toxigenic bacteria such as diphtheria bacilli, the food poisoning bacillus (*Clostridium botulinum*), and the tetanus organism (*Clostridium tetani*). Prompt diagnosis often depends on proper handling of diagnostic specimens. The percentages given are merely illustrative and may differ greatly in individual instances.

contains from 2,000 to 2,500 units per ml. The dosage may range from 20,000 to 100,000 units, depending on age, weight, severity of the attack, and time after onset. Similar basic considerations apply in connection with the use of other antitoxins, although units and dosages may be expressed somewhat differently for each.

Necessity for Early Diagnosis and Therapy. If sufficient quantities of specific antitoxin are given early in any disease (first day after onset or sooner, if possible), a large percentage of patients with uncomplicated intoxications or infections, including diphtheria, recover (Fig. 30–3). In any infectious or toxic process the effectiveness of serum therapy depends on counteracting the toxin and/or the infectious agent before the pathogenic agent (toxin or microorganism) has had the opportunity to injure the body cells. A few hours lost in beginning treatment may make the difference between life and death. In certain cases of malignant or "bull neck" diphtheria (so called because of the enormously swollen cervical lymph nodes), antitoxin appears to be of no value whatever. The reason for this is not known.

Immunity to Diphtheria

Immunity to diphtheria depends in part on the presence of antitoxin in the blood and in part on the ability of a person to manufacture such antitoxin quickly. This latter ability usually results from previous active immunization by either natural infection or artificial injection of toxoid (primary stimulus). The tissues act as though they have become sensitized (allergic) to diphtheria toxin and bacilli and react very strongly and quickly to any new infection by diphtheria bacilli or to a secondary or "booster" injection of toxoid, thus warding off disease.

Natural Immunity. In some urban areas many adults and young children have at least 0.02 unit of antitoxin in their blood as a result of mild or unrecognized (subclinical) infections with *C. diphtheriae* or the action of still obscure natural mechanisms. Such persons are therefore Schick-negative (page 464), i.e., immune to the disease, even though exposed to it. In rural areas natural factors of immunization are often much less effective. Infants born of immune mothers usually have maternal antitoxin in their blood for only about three months.

Passive Artificial Immunity. A dose of antitoxin (10,000 units) will usually prevent the development of diphtheria in children under 10 years old exposed to the disease and not previously immunized with toxoid. This is called a prophylactic or preventive dose, but the protection given by it lasts only two to three weeks. This is a good example of the temporary nature of *passive artificial immunity.* Active immunization with toxoid should be started at the same time.

The use of antitoxin is sometimes followed by serum sickness. This is not due to the antitoxin but is a manifestation of allergy to the proteins in the horse serum containing the antitoxin. The use of horse serum sensitizes the patient, usually for life, to all other antisera derived from horses (for example, to tetanus antitoxin), so that it should not be used prophylactically unless the need is clearly evident.

Active Artificial Immunity. As previously pointed out, it is possible to stimulate the human body artificially to produce any specific antitoxin by the injection of toxoid made from that toxin. The procedure is commonly used in prophylaxis of diphtheria and also of tetanus (lockjaw).

Fluid Toxoids. Whether making tetanus toxoid or diphtheria toxoid, these antigens consist primarily of broth culture of the appropriate species of bacteria, filtered to remove the bacteria. The filtrate (filtered broth culture) contains the exotoxin of the bacilli. This is treated with formaldehyde to remove toxicity. Specific antigenicity remains. The product is called "fluid toxoid."

Alum-Precipitated Toxoid. If alum is added to fluid toxoid, it forms a precipitate to which the toxoid is adsorbed and is carried to the bottom of the flask as a white sediment. The supernatant broth is then poured off, leaving the concentrated toxoid in the sediment relatively free from the impurities of the culture broth. The precipitate is resuspended in physiologic saline solution, and the milky suspension thus formed is called *alum-precipitated toxoid (A.P.T.)* or *toxoid, A.P.,* which is injected after suitable tests for sterility, potency, antigenicity, and nontoxicity.

ACTION OF ALUM-PRECIPITATED TOXOID. A.P.T. is more effective than fluid toxoid because the antigenic stimulus from a deposit of alum toxoid is prolonged. This is because the alum, being insoluble, is not absorbed for some time but remains for days or weeks in the tissues. It slowly and steadily releases the attached toxoid into the blood in a long-continued, very effective antigenic stimulus simulating actual infection.

Triple Toxoids. It is common practice to combine diphtheria and tetanus toxoids (A.P.) with pertussis (whooping cough) vaccine (bacterin) in a "triple vaccine" or "triple toxoid" called DTP; three immunizations are thus accomplished with one series of injections. The first injection is given three to five months after birth, with one or two others after intervals of four to five weeks, another at about one year, and another upon entering school. Various other injection schedules are also in use; the choice must rest with the physician. The three mixed antigens—tetanus, diphtheria, and pertussis—act better together than any of the three alone. One product that has been recommended also is OPV (Sabin vaccine) for the prevention of poliomyelitis. All are very effective.

Primary and Secondary Stimulus. The two or three initial doses of such immunizing agents in infancy constitute a primary stimulus. Even though the antibodies called forth into the blood by such a stimulus may disappear with time, the person who has had such a primary stimulus responds very quickly to a *secondary stimulus,* which may consist of actual infection (tetanus, diphtheria, pertussis, or polio) or a "booster" dose of toxoid or other specific antigen. The previously immunized person is usually able to respond quickly with the production of specific antibodies (see Fig. 21–4).

Immunized persons, especially those who have not had a booster dose at all or not for many years (frequently adults), sometimes contract diphtheria, pertussis, tetanus, or polio, but the infection is generally mild or subclinical. Fatality from diphtheria, pertussis, or tetanus in immunized persons is very rare.[9] It is important, therefore, that every child receive at least a primary stimulus with a multiple antigen early in life, a "booster" stimulus a year later, and another usually at the time of entering school.

Infants are protected from a number of diseases by antibodies derived from their mothers' blood. These persist for three to four months after birth provided the mothers are immune. Diphtheria is one of these diseases. For this and other reasons it is customary not to immunize children against diphtheria until they are two months of age or older. After that period their susceptibility increases and artificial immunization is necessary.

The Schick Test. The "Schick test" is named for the Viennese physician who first made practical use of it. It is a test for the presence or absence of significant amounts of diphtheria antitoxin in the blood. The test is performed by injecting into (not under) the skin an extremely small amount of diphtheria toxin (1/50 MLD). If the person has no antitoxin in his blood (i.e., is not immune), within 24 hours a small red area appears at the place of injection, which gradually spreads for a centimeter or more. This is caused by the irritating effect of the toxin on the cells of the skin. In five days the area becomes slightly pigmented, and later a few flakes of epidermis may scale off. The test should not be read earlier than the fifth day after the injection because there is often a temporary irritation (pseudoreaction) caused by foreign substances always present in the toxin. The red area due to such impurities is nearly always gone by the end of five days. An area that persists after five days and undergoes the changes described is a positive reaction and means that the person has less than 0.02 unit of diphtheria antitoxin in the blood and may be susceptible to diphtheria, especially if the individual is under 15 years of age.

If one has at least 0.02 unit of antitoxin in the blood, it will counteract the toxin injected in the Schick test, and the skin will remain normal. This is a negative Schick reaction and shows that the person will not contract diphtheria under ordinary conditions of exposure. The test is harmless and painless and leaves no scar.

Immunization. It is scarcely necessary to say that every member of the health team should be protected from the beginning of training, whether their work includes the care of contagious diseases or not. From one half to one fourth of urban adults give a positive Schick test. Most of them were probably once Schick-negative, but antitoxin tends to disappear from the blood with passage of time and the Schick reaction reverts to positive. However, diphtheria is uncommon among adults unless they are heavily and constantly exposed. Their resistance depends on the fact that at some time nearly all received a primary antigenic stimulus: doses of toxoid or mild or unrecognized infections. Reinfection acts as a secondary stimulus. This same principle appears to apply to many infectious diseases and should be clearly understood and remembered.

It is truly remarkable how this killing disease of small children, once feared by every mother, has been brought under control by workers in the field of immunology. In the early years of this century many thousands of cases and deaths from diphtheria occurred annually in the United States. During the first 38 weeks of 1977 only 69 cases of diphtheria were reported. In 1965, 18 deaths were reported in the entire nation; in 1974, only 5; in 1980, only 3.

DIPHTHEROIDS

Closely related to *Corynebacterium diphtheriae* are a number of organisms having somewhat similar morphology. Two common species, apparently incap-

[9]Persons with congenital agammaglobulinemia are exceptions. They usually fail to respond to vaccines in general, and are likely to have low resistance to infectious diseases.

able of producing serious disease, are *Corynebacterium pseudodiphtheriticum* (Hofmann's bacillus) and *Corynebacterium xerosis.* These are frequent in the normal throat and nose. They grow in much the same manner as *C. diphtheriae* and are often mistaken for it in throat cultures by inexperienced diagnosticians: hence the name diphtheroid, which means "diphtheria-like." They may usually be differentiated by the fact that they are less granular and more uniform in size and shape. In pure culture they exhibit entirely different biochemical and physiologic properties. They give negative reactions in toxigenicity tests. Other related species are *C. ulcerans* (now thought to come between *C. diphtheriae* and *C. pseudotuberculosis,* not fully recognized as a species), *C. haemolyticus* (this may be identical to *C. pyogenes,* but neither is fully recognized as species), and *C. equi.* Some of these are serious pathogens of domestic animals; others, not listed here, are plant pathogens.

The most common infection caused by the diphtheroids is bacterial endocarditis: it is an expected complication after heart surgery, with a 35 per cent death rate. Diphtheroids are also implicated in osteomyelitis, brain abscesses, meningitis, hepatitis, and pneumonia. Anaerobic diphtheroids often cause infection in reconstructive surgery, such as hip joint replacement, where foreign materials are implanted in tissues of the body. A common diphtheroid found in soft tissue infections is the aerotolerant anaerobe *Propionibacterium acnes* (formerly *Corynebacterium acnes*). Based on cell wall antigens, 62 strains of this genus have been identified.

THE GENUS LISTERIA

Listeria monocytogenes, a gram-positive bacillus that may also exist in coccoid forms, was formerly classified with the Corynebacteriaceae but has only a morphologic resemblance to them. The species name is derived from the appearance of large numbers of monocytes in the blood of experimentally infected animals. It bears no relation to infectious mononucleosis in man, a disease caused by a virus.

Listeria causes more infections (called *listeriosis* [singular] or listerioses [plural]) and deaths than formerly suspected. It may be isolated from cerebrospinal fluid, blood, vaginal smears, and other sources, apparently as a contaminant; however, the organism has been definitely established as the etiologic agent in certain forms of meningoencephalitis, endocarditis, abortion, abscesses, and also septicemia. Although *L. monocytogenes* is found in soil where plants are growing, it occurs also in the intestines of many animals and birds and, in rare cases, in human feces. Infants infected with *Listeria* have a 50 per cent mortality rate. Meningitis occurs in up to 70 per cent of all listeriosis patients. Listeriosis sometimes results as a complication in debilitated patients with alcoholism, diabetes, or neoplastic diseases; it responds well to the common antibiotics.

SUMMARY

Although diphtheria, a once dreaded disease, is now fully under control in the United States, vaccination of infants three to five months old with DTP must be continued. The diphtheria toxoid used in this vaccine is obtained from the diphtheria bacillus *Corynebacterium diphtheriae.* Pai's and Löffler's media, which contain potassium tellurite, are used to grow the organism selectively, inhibiting other bacteria. *C. diphtheriae* is obtained from throat cultures, consisting of bacilli from the tonsils and upper respiratory mucous membranes. The exotoxin produced by the organism irritates the tissues, which give forth a fibrinous exudate that coagulates into a tough, leathery, grayish-white *pseudomembrane.* With swelling and inflammation, this membrane occludes air passages and asphyxiates the patient (usually a child) unless a tracheotomy is performed. Diphtheria is typically transmitted by carriers, often in schools. Antitoxins given intravenously have been used to render the diphtheria toxin harmless, but they do not repair damage to the heart, nerves, and kidneys after the toxin has attacked these tissues.

Active artificial immunity may be obtained through fluid toxoids or alum-precipitated toxoid. The same general methods involving toxoids are employed in the preparation of effective vaccines whether toxins come from *Corynebacterium*, the tetanus organism, staphylococci, or any other toxin-producing microorganism.

The Schick test is used to test for susceptibility to diphtheria. It is similar to the Dick test, which tests for antibodies to scarlet fever.

Diphtheroids are organisms related to *C. diphtheriae*, such as *C. pseudodiphtheriticum*. They cause osteomyelitis, brain abscess, meningitis, hepatitis, pneumonia, and various infections.

Listeria monocytogenes, a gram-positive bacillus, causes listerosis, which most often attacks infants and debilitated patients with alcoholism, diabetes, or neoplastic diseases. It responds well to antibiotics.

REVIEW QUESTIONS

1. Who discovered the diphtheria bacillus and when?
2. What are the morphologic and cultural characteristics of *Corynebacterium diphtheriae?* How may one obtain a satisfactory throat culture in suspected cases of diphtheria?
3. How are diphtheria bacilli destroyed? How are they transmitted?
4. On what media may the diphtheria organisms be grown?
5. What are the characteristics of diphtheria?
6. What is the importance of carriers and subclinical cases in the transmission of diphtheria?
7. What is the importance of antibiotic therapy in cultures for release of patients and carriers?
8. What is the value of the toxigenicity test for diphtheria?
9. What is the value of diphtheria antitoxin in the treatment of diphtheria?
10. What kinds of immunity can be produced against diphtheria? How is each utilized to prevent the disease?
11. What is "fluid toxoid"? When is it used? What is A.P.T.?
12. Explain the occurrence of diphtheria in supposedly immunized individuals.
13. What is the Schick test? How can it be used?
14. What are diphtheroids?
15. Why is diphtheria no longer the danger that it once was? Can we stop giving DTP vaccinations to children? Explain.
16. Explain ways to measure effective concentrations of toxins.
17. List diseases caused by *Listeria*.

Supplementary Reading

Burrows, W.: Textbook of Microbiology. 20th ed. Philadelphia, W. B. Saunders Company, 1973.

Center for Disease Control: Diphtheria outbreak—Canada. *Morbidity and Mortality Weekly Reports, 25*:355, 1976.

Craig, J. P.: Diphtheria: Prevalence of inapparent infection in a nonepidemic period. *Amer. J. Public Health, 52*:1444, 1962.

Drazin, R., Kandel, J., and Collier, R. J.: Structure and activity of diphtheria toxin. *J. Biol. Chem., 246*:1504, 1971.

Duncan, J. and Groman, N. B.: Activity of diphtheria toxin (intoxication in HeLa cells). *J. Bacterol., 98*:963, 1969.
Gill, D., Pappenheimer, A. M., Jr., Brown, R., and Kurnick, J.: Mode of action of diphtheria toxin. *J. Exper. Med., 129*:1, 1959.
Gray, M. L., and Killinger, A. H.: *Listeria monocytogenes* and *Listeria* infections. *Bacteriol. Rev., 30*:309, 1966.
Herman, G. J.: Diphtheria. Diagnostic Procedures. 5th ed. New York, American Public Health Association, 1970.
Lennette, E. H., et al. (eds.): Manual of Clinical Microbiology. 3rd ed. Washington, D.C., American Society for Microbiology, 1980.
Monis, B., and Reback, J. R.: Tetrazolium salts and identification of *Corynebacterium diphtheriae. Proc. Soc. Exp. Biol. Med., 111*:81, 1962.
Steigman, A. J., and Epting, M. H.: Diphtheria in children. *Amer. J. Nurs., 57*:467, 1957.

LARYNGOTRACHEITIS, CONJUNCTIVITIS, WHOOPING COUGH, AND "TRENCH MOUTH"

Many species of saprophytic gram-negative rods or coccobacilli may occasionally be found in the upper respiratory tract, accidentally carried there by inhaled dust, hands, foods, and other such means. However, there are three genera (*Haemophilus, Moraxella, Bordetella*) that are typical respiratory tract pathogens and are characteristically transmitted only in respiratory secretions. They include the so-called influenza bacillus, the bacillus of "pink-eye" or infectious conjunctivitis,[1] and the bacillus of whooping cough or pertussis.[2] These three species have many properties in common and formerly belonged to the same family.

All are very small short rods or coccobacilli, ranging from 0.3 to 0.6 μm in diameter and from 3 to 8 μm in length. All are gram-negative, nonmotile, and nonspore-forming. They are fragile and are easily killed by common disinfectants, pasteurizing, or boiling. They grow best at 37 C in atmospheres containing about 10 per cent carbon dioxide and on media containing blood or serum and extracts of meat. They are highly specialized parasites of the mammalian body and have special nutritive requirements.

Genus Haemophilus

Haemophilus influenzae is an aerobic facultative anaerobe. Bipolar staining and pleomorphism are often encountered. It is particularly adapted to live and grow in the body. Outside the body it soon dies unless provided with media containing whole blood or blood derivatives: *heme* (known as the "X" factor) and nicotinamide adenine dinucleotide (NAD) (called the "V" factor). The X and V factors may be purchased as such and added to media for *Haemophilus.* The genus name is derived from its requirement for blood (from Greek *haima,* blood; *philus,* loving); its species name is based on the fact that it was once thought to cause influenza. *H. influenzae does not cause influenza;* a *virus* is the etiologic agent of this disease.

H. influenzae is found in the nose and throat of many normal persons and is transmitted in droplets by sneezing and coughing; hence, it is a typical respiratory tract organism. So are *Haemophilus parainfluenzae* and *Haemophilus parahaemolyticus,* which may rarely cause pharyngitis or subacute bacterial endocarditis. *H. influenzae* is present in and appears to cause many inflammatory conditions of the respiratory tract, such as bronchopneumonia and sinusitis, and it has been found in pure culture in the blood, the conjunctivae, cerebrospinal fluid, osteomyelitis, otitis, and pus in joints. It may sometimes cause meningitis and endocarditis and also infections of the sinuses and eyes, especially following colds and influenza. Chronic illness and old age are predisposing

[1]The eyes, being connected with the upper respiratory tract via the tear ducts, can contribute infectious microorganisms to the oronasal secretions.

[2]*Per* is from the Latin for much; *tussis* from the Latin for cough. Pertussis is therefore "much cough."

CHAPTER 31

conditions in infections due to *H. influenzae,* which are usually secondary rather than primary.

There are six encapsulated types of *H. influenzae,* differentiated by soluble specific capsular substances, much as are pneumococci and meningococci. These types are a, b, c, d, e, and f.[3] The type is determined by methods like those used in typing pneumococci, e.g., the quellung reaction and the precipitin test. These organisms also resemble pneumococci in being bile-soluble. Type b bacilli appear to be especially virulent, causing many severe throat infections, especially *laryngotracheitis,* a serious and often fatal disease in infants, causing obstruction of the air passages by severe edema and mucous secretions. A type-specific antiserum as well as sulfonamide drugs and appropriate antibiotics, especially tetracycline, have been used effectively in treating such infections.

H. influenzae infections of the respiratory tract are treated, as far as nursing precautions and isolation techniques are concerned, like other upper respiratory infections (pneumonia, scarlet fever, diphtheria, and so on).

"Pink-Eye". One of the etiologic agents of this disease, more accurately called acute or angular conjunctivitis or inflammation of the lining of the eyelids, is named *Haemophilus aegyptius*[4] (the Koch-Weeks bacillus). This organism has the general characteristics of *H. influenzae,* which it closely resembles. It requires heme and NAD for growth.

Genus Moraxella

Organisms of this genus occur in the same kinds of infections in which species of *Haemophilus* are found. The genus *Moraxella,* named for Victor Morax, a famous French ophthalmologist, includes about eight species of very small, gram-negative, nonspore-forming, aflagellate, aerobic coccobacilli that generally require media containing serum but not heme or NAD. As a group they are not active in attacking carbohydrates, though species differ slightly in this respect (e.g., *M. kingii*[5] ferments glucose). All produce oxidase and several form catalase. Some are actively proteolytic. Among the best known species are *M. lacunata* (the Morax-Axenfeld bacillus) and *M. kingii.*

Acute conjunctivitis due to species of *Haemophilus* and *Moraxella* is very contagious and spreads rapidly in families, institutions, and schools. "Pink-eye" is also frequently caused by streptococci and staphylococci. Transmission appears to be by means of direct contact with the infected eyes or by fingers, towels, washcloths, and handkerchiefs soiled with secretions from the eyes. Dust may also transmit the organisms, but as they are not highly resistant to drying, this is thought not to be important. The disease is easily treated and is usually not serious, although it is extremely annoying, unsightly, and temporarily disabling. It may last for three or four days to two weeks.

[3]Other strains are unencapsulated.
[4]Species incertae sedis. *Bergey's Manual.*
[5]Named after Elizabeth O. King, an American bacteriologist.

Prevention and Treatment

The eyes of the patient are usually bathed with mild disinfectant or antibiotic solutions to reduce the infection and the inflammation. A child with pink-eye must be kept from contact with other children. The school or institution nurse who has had experience with an outbreak of pink-eye will realize the importance of segregating the child, his bed linen, his handkerchiefs, his washcloths, his towels, and anything that may be soiled with discharge from the eyes, directly or by way of the hands. Conjunctivitis of this origin tends to clear up spontaneously, but while it lasts the same care should be given as for gonorrheal ophthalmia.

Other causes of acute conjunctivitis are at least one virus, gonococci, and the so-called TRIC agent: *Chlamydia trachomatis* (trachoma and inclusion conjunctivitis). All these infections are more severe than pink-eye and are discussed more fully in connection with their respective etiologic agents.

Genus Bordetella

The Bacillus of Whooping Cough. This organism, for years called *Haemophilus pertussis,* has many of the distinctive properties of *H. influenzae.* It was discovered in 1906 by two Belgian scientists, Bordet and Gengou. It is now called *Bordetella pertussis* or the Bordet-Gengou bacillus. The bacillus grows slowly and with difficulty outside the body, appearing on the surface of glycerin-potato-blood agar (Bordet-Gengou or B-G medium) only after five to ten days' incubation at 37 C. The addition of penicillin to this medium is recommended to inhibit gram-positive organisms that may cause contamination. The tiny colonies of *B. pertussis* are described as being "like droplets of mercury" (Fig. 31–1). They are hemolytic. *B. pertussis* is as delicate and, when first isolated, as fastidious about its growth requirements as *H. influenzae.* After growth in artificial media it no longer requires heme or NAD. Both antigenic analyses (encapsulated types I to IV) and direct fluorescent antibody staining of smears promise more rapid diagnosis.

A related organism, *Bordetella parapertussis,* produces a similar though milder disease, parapertussis. The following discussion of pertussis applies equally to parapertussis.

Another pertussis-like disease is caused by *Bordetella bronchiseptica* (formerly

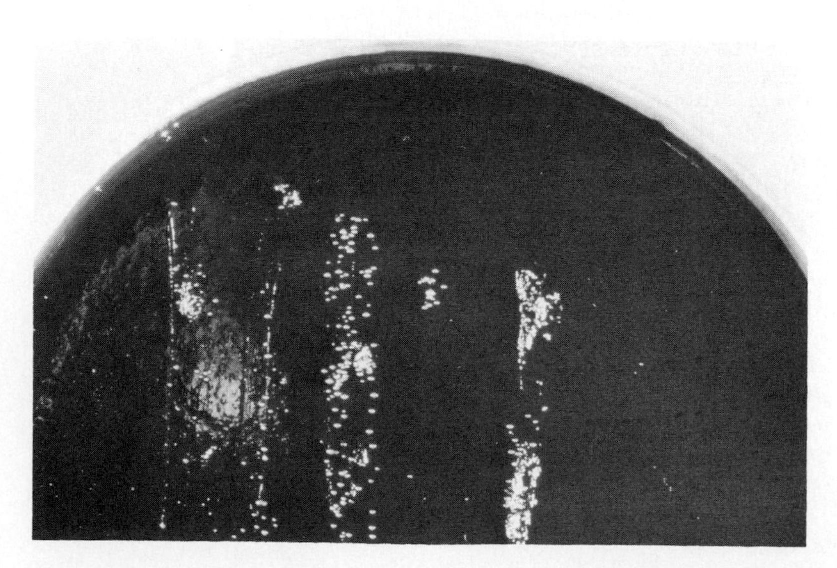

Figure 31–1 Typical colonies of a 4-day growth of *Bordetella pertussis* on Bordet-Gengou medium. Note the colonial growth that resembles "droplets of mercury." Penicillin in the medium (0.5 μ/ml) suppresses other organisms but allows *B. pertussis* to grow. Natural size. (From Schneierson, S. S.: Atlas of Diagnostic Microbiology. North Chicago, Ill. Abbott Laboratories.)

classified as *Haemophilus bronchiseptica* or *Alcaligenes bronchisepticus*). It differs markedly from other *Bordetella* in being actively motile and able to grow luxuriantly and rapidly at 25 C on plain peptone media. It is most commonly found in respiratory diseases of dogs and rabbits.

Pertussis

Nature and Transmission. Whooping cough, also called pertussis, comes with or without whooping, and whooping comes with or without pertussis. Although mass immunization and improvements in care and treatment have reduced the incidence and mortality, this disease still occurs and is still extremely dangerous, especially during infancy. *B. pertussis* produces an inflammation of the trachea and bronchi and consequently is found in the sputum, saliva, and nasal discharges. It is transferred directly from person to person by droplet infection during coughing and by articles freshly soiled with nose and mouth secretions. A spray of sputum and saliva may be thrown out four or five feet during the violent attacks of coughing.

It is possible to obtain a culture of pertussis bacilli if a Petri plate containing a suitable medium (Bordet-Gengou or substitute) is held before the mouth of a coughing patient. Such plates are called "cough plates" and are useful in bacteriologic confirmation of diagnosis, but they are subject to excessive contamination and difficulties in obtaining a productive cough from infants. The nasopharyngeal swab is simpler.

It is particularly advantageous to add penicillin to the B-G medium. This prevents the growth of most of the unwanted staphylococci and other organisms of the nasal and oral secretions, but permits good growth of colonies of *B. pertussis*. The same principle (selective cultivation) is used in diagnosing other bacterial infections.

Whooping cough begins with the symptoms of an ordinary cold. Like measles, diphtheria, and scarlet fever, it is highly infectious during the first few days of the actual disease when the causative organisms are present in the sputum in enormous numbers and before anything more serious than a "cold" is suspected. Health workers should therefore be extremely suspicious of "colds" in children and insist that the patients be kept at home for several days. Six weeks after the onset of whooping cough it is rarely possible to demonstrate *Bordetella pertussis*, even though the child may still cough violently. Carriers of whooping cough are therefore not a problem in prevention of the disease. New cases of whooping cough doubtless arise from mild cases, sometimes in adults, that are not thought to be whooping cough, or from cases in the early, "snuffly" stage before they are diagnosed.

Complications. Whooping cough, like measles, is a serious disease because of the other infections that complicate or follow it. The most important of these are bronchitis, pneumonia, and tuberculosis. Whooping cough is responsible for many fatal cases of bronchitis and pneumonia in young children. About 97 per cent of the fatal cases occur in children under five years of age, and about 70 per cent in those under one year. Care is directed toward preventing exposure of susceptible children, especially those under five years of age, to known cases. The usual general precautions for respiratory diseases are applicable. The organisms are easily killed by standard disinfectants and by drying and sunlight. Before the advent of active artificial immunization it was a common sight to see small children with "nurses" in public parks sitting in the sun. Often the child would start to cough until blood would come up, not a pretty sight to see, and one that is not easily forgotten.

Antibiotic Therapy. It is doubtful that antibiotic therapy changes the course of pertussis. Ampicillin and tetracyclines have been used with some claim of shortening the period of communicability of the organism.

Whooping Cough Vaccine. A vaccine made of killed whooping cough bacilli is widely used and effectively prevents or mitigates the disease. It is not effective as a curative measure or if it is used too close to the time of exposure. Four to six weeks must be allowed for immunity to develop after injection. This is true of most methods of active immunization.

Alum-Precipitated and Mixed Vaccines. Pertussis vaccine is prepared by precipitating the killed bacilli, suspended in saline solution, with alum or aluminum hydroxide. The precipitated bacilli, resuspended, constitute "pertussis vaccine, alum-precipitated." It may be used alone, but it is common and approved practice to mix it with diphtheria toxoid, also alum-precipitated, and often also with alum-precipitated tetanus toxoid (triple antigen). The advantages of using alum-precipitated antigens have been pointed out previously. Some preparations may also contain polio vaccine. In children, the pertussis vaccine should be started one month after birth or even earlier. Immunization against pertussis is usually not required past six years of age.

The incidence of pertussis in the United States has declined from over 120,000 cases in 1950 to 1,010 in 1976, but went up to 1,730 in 1980. Deaths from this disease have been greatly decreased from over 1,500 in the 1950s to 8 in 1975, thanks to DTP (diphtheria-tetanus-pertussis) vaccinations.

Passive Immunity. Transitory protection may be given to highly susceptible children under five years of age by the use of serum or, better, gamma globulin from immune persons. Hyperimmune gamma globulin (HIGG) is used to hasten recovery, prevent complications, and reduce mortality. The children who need this type of protection are those who have not received pertussis vaccine, those who are debilitated by other disease or malnutrition, and those who are likely to be exposed to the infection.

Prevention. For as long as four weeks from onset, the child with whooping cough expels large numbers of the organisms from the nose and mouth when coughing and sneezing. A susceptible child is almost certain to contract the disease if he or she spends any time in a closed room with a person who is discharging the organisms by coughing. Even outdoors, close contact should be avoided for at least four weeks after the onset of the disease and for six weeks if possible. Adults are not commonly victims of pertussis, but unless they are known to be immune they should avoid contact with known cases of the disease. Sometimes people become infected (and infectious) following contact with a patient, yet manifest only the symptoms of a cold. All fomites of the patient contaminated with sputum or vomitus, as well as the vomitus itself, should be disinfected. *Bordetella pertussis* is a fragile organism and does not live very long in the outer world; nevertheless, it survives long enough to be effectively transmitted by droplets, saliva, and so on.

It was reported in 1970 that a pertussis-like syndrome, indistinguishable from pertussis, was caused by adenoviruses types 1, 2, 3, and 5. This viral etiology could be of considerable importance and needs further investigation.

ORAL SPIROCHETES

"Trench Mouth." *Treponema (Borrelia) vincentii,* a spirochete, is anaerobic and difficult to cultivate in the laboratory. Related species, obtained from the oral cavity of humans, are *Treponema denticola* and *Treponema orale.* Although commonly present in the normal mouth in relatively small numbers, *Treponema vincentii* is implicated in a painful, ulcerative disease of the gums, cheeks, and throat often called "trench mouth" (because it was common among troops in the trenches during World War I), Vincent's angina, or ulcerative gingivostomatitis. Vincent's angina is characterized by a superficial, gangrenous ulceration or stomatitis, often with an offensive odor (*oral fetor*). It might be an infectious

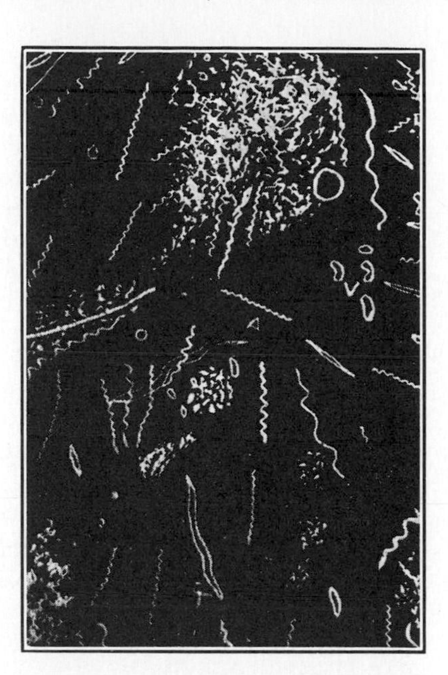

Figure 31–2 Darkfield demonstration of a mixture of organisms commmonly found in ulcerous infections of the respiratory tract. Note especially the spirochetes *(Treponema vincentii)* and fusiform bacilli *(Leptotrichia buccalis)* characteristic of Vincent's angina (trench mouth) (From Joklik, W., and Smith, D. T.: Zinsser Microbiology, 15th ed. Englewood Cliffs, N. J., Prentice-Hall, 1972.)

disease, or it may be that certain vitamin deficiencies are the basic cause of the lesions seen in Vincent's angina and that the spirochetes then multiply in the lesions as secondary invaders or opportunists. It has also been suggested that herpes simplex virus may be involved, or that injury to mucous membranes may be a prerequisite to initiate conditions for the disease. The actual relation of the spirochetes to the cause of the disease is not clear. In any event, the spirochetes are always found in enormous numbers in the ulcers, always associated with long, thin, spindle-shaped or cigar-shaped bacilli called *fusiform bacilli.* (Previously called *Fusobacterium fusiforme,* these may now be classified as *Leptothrix buccalis.*) The organisms may be demonstrated by staining material from the ulcers and examining it microscopically. Both spirochetes and fusiform bacilli stain well and are gram-negative (Fig. 31–2).

The organisms can be transmitted by improperly disinfected drinking glasses and eating utensils, kissing, and articles that pass, undisinfected, from one infected mouth to another. Vigorous disinfection of the mouth usually cures the infection. The antispirochetal drugs used for the treatment of the treponematoses also give prompt relief in some cases, and penicillin is most effective. The disease is not very dangerous as a rule, although it can cause great discomfort. The ulcerous lesions are sometimes mistaken for the membrane found in diphtheria.

Other species, such as *Borrelia recurrentis* (related to *Treponema vincentii),* causing relapsing fever, will be discussed later in Chapter 39.

Syphilis. *Treponema pallidum,* the spirochete that causes syphilis, occurs in the mouth in some cases of secondary syphilis and also in later stages of untreated syphilis. It may be transmitted by kissing and by objects freshly contaminated with oral secretions. Further discussion of syphilis appears in Chapter 35.

SUMMARY

Respiratory pathogens of three genera, *Haemophilus, Moraxella,* and *Bordetella,* are characteristically transmitted only in respiratory secretions. They grow best at 37 C in microaerophilic conditions on enriched media and are easily killed by common disinfectants, pasteurization, or boiling.

Haemophilus influenzae requires blood, blood derivatives, or heme for growth. It is not the cause of influenza, which is a viral disease, but it causes complications in existing respiratory infections and has been found in the blood of patients with terminal disease. It also may be present in the cerebrospinal fluid, in osteomyelitis, in otitis, in pus in joints, and in conjunctivitis. *Haemophilus influenzae* is also known to cause often fatal laryngotracheitis in infants and meningitis or endocarditis following colds and influenza. *Haemophilus aegyptius* and *Moraxella* are pathogens of acute conjunctivitis, also known as pink eye, a very contagious disease. Fortunately, it is easy to cure, differing from eye infections with the TRIC agent, viruses, the gonococcus, or fungi.

A disease well under control because of successful vaccinations (usually with DTP [diphtheria-tetanus-pertussis]) is whooping cough, also named pertussis. *Bordetella pertussis* has very specific growth requirements; it grows slowly on Bordet-Gengou medium or on glycerin-potato-blood agar. Whooping cough is a serious disease, inflaming the trachea and bronchi and causing great difficulties to the child suffering from it. The patient may suffocate, hemorrhage, or develop complications such as bronchitis, pneumonia, and tuberculosis.

Treponema vincentii, an anaerobe, is implicated in a painful, ulcerative disease of the gums, cheeks, and throat often called "trench mouth," or Vincent's angina. It is an ulcerative gingivostomatitis. Another organism, *Leptothrix buccalis,* may also be involved in the infection.

REVIEW QUESTIONS

1. Name three species of *Haemophilus.* What does the word *Haemophilus* mean?
2. What is the importance of *Haemophilus influenzae*?
3. What is the cause of "pink eye" and how can it be prevented?
4. What is the cause of whooping cough? How is this organism isolated in suspected cases of whooping cough? What is the importance of this disease? What is the value of vaccine in preventing whooping cough?
5. If you were responsible for the care of a child with whooping cough, what precautions would you use to protect yourself and others?
6. Name different causes of acute conjunctivitis.
7. What can you tell about the nutrition of *Bordetella*? On what media does it grow?
8. What are the symptoms of whooping cough? What are complications associated with this disease?
9. What is the cause of trench mouth? What are its symptoms? How is trench mouth related to syphilis?

Supplementary Reading

Abbott, J. D., Preston, N. W., and Mackay, R. I.: Agglutinin response to pertussis vaccination in the child. *Brit. Med. J., 1*:86, 1971.

Brooks, G. F., and Buchanan, T. M.: Pertussis in U.S.A. *J. Infect. Dis., 122*:123, 1970.

Brooksaler, F. S.: The pertussis syndrome. *Texas Med., 67*:56, 1971.

Burnett, G. W., and Shuster, G.: Oral Microbiology and Infectious Diseases. 4th ed. Baltimore, The Williams & Wilkins Co., 1978.

Burrows, W.: Textbook of Microbiology, 20th ed. Philadelphia, W. B. Saunders Company, 1973.

Connor, J. D.: Evidence for an etiologic role of adenoviral infection in pertussis syndrome. *N. Engl. J. Med., 283*:390, 1970.

Holwerda, J., Brown, G. C., and Pickett, G.: Symposium on pertussis immunization in honor of Dr. Pearl L. Kendrick. *Health Lab. Science, 8*:206, 1971.

McGuire, C. D., and Durant, R. C.: The role of flies in the transmission of eye disease in Egypt. *Amer. J. Trop. Med. Hyg., 6*:569, 1957.

National Communicable Disease Center: Pertussis in the United States. *J. Infect. Dis., 122*:123, 1970.

Nelson, J. D.: Whooping cough—Viral or bacterial disease? *N. Engl. J. Med., 283*:428, 1970.

Stanfield, J., Bracken, P. M., Waddell, K. M., and Gall, D.: Diphtheria-tetanus-pertussis immunization by intradermal jet injection. *Brit. Med. J., 2*:197, 1972.

Top, F. H., and Wehrle, P.: Communicable and Infectious Diseases. 8th ed. St. Louis, C. V. Mosby Co., 1976.

Van Bijsterveld, O. P.: New *Moraxella* strain isolated from angular conjunctivitis. *Appl. Microbiol., 20*:405, 1970.

TUBERCULOSIS AND HANSEN'S DISEASE

GENUS MYCOBACTERIUM

The order *Actinomycetales* consists of eight families of branching, moldlike bacteria, including the two families to which *Nocardia* and *Streptomyces* belong and the family *Mycobacteriaceae,* which contains the tubercle and leprosy bacilli—genus *Mycobacterium.*[1]

All mycobacteria are distinguished from virtually all other microorganisms[2] by the property of acid-fastness, which will be explained later in this chapter. Hence, they are often called "acid-fast bacilli." They are all rods, usually slender and curved, often pointed and beaded, but occasionally showing forms more or less branching. This branching character and their bacterium-like size and structure certify their membership in the order of *Actinomycetales.* Mycobacteria are obligate aerobes and are generally considered to be gram-positive, but take the stain with difficulty, if at all, and their Gram reaction is meaningless.

Several species of mycobacteria, whose pathogenic status is still not fully clarified, are often found in tuberculosis-like pulmonary conditions or associated with true tuberculosis. These mycobacteria are sometimes called "atypical acid-fast bacilli" or *unclassified* or *anonymous* mycobacteria, terms used for mycobacteria other than *M. tuberculosis* and *M. bovis.* Most are nonpathogenic, saprophytic bacteria that live in the soil and on plants (e.g., *Mycobacterium phlei*) and on the human skin *(M. smegmatis)* (Table 32–1). Runyon groups were established by E. H. Runyon, an expert on mycobacteria.

By far the most important members of the genus *Mycobacterium* to humans are *M. tuberculosis,* the cause of tuberculosis in humans, and *M. leprae,* the cause of Hansen's disease (leprosy). *M. tuberculosis* is commonly transmitted via the respiratory tract.

THE TUBERCLE BACILLUS

One of the most important discoveries in medicine was that of *Mycobacterium tuberculosis,* first seen and cultivated in the laboratory by Robert Koch in 1882. For hundreds of years previously, however, it had been known that tuberculosis was infectious. In 1865 Jean Villemin, a French pathologist, had produced tuberculosis in animals by inoculating them with material from patients who had died of tuberculosis.

Physiologic Properties. Tubercle bacilli and, indeed, all mycobacteria are obligate aerobes,[3] nonspore-forming, and nonmotile. Special methods of staining, the *Ziehl-Neelsen acid-fast stain* or *Kinyoun's acid-fast stain,* are generally used.

The Acid-Fast Stain (Ziehl-Neelsen Method)

1. Prepare the smear (e.g., sputum) as usual and dry.
2. Flood with a solution of carbolfuchsin (a red dye).

---

[1] *Myco* is from the Greek *mykes,* fungus or mold.

[2] Except a few species of the closely related *Nocardia,* named for the French microbiologist Edmund Nocard (see Fig. 1–10).

[3] *Mycobacterium tuberculosis* grows best on tissue that is highly aerobic, which explains its affinity for the human lung.

CHAPTER 32

3. Heat the slide gently so that the solution steams. Do not allow to boil. Do not allow to dry.

4. After three to five minutes, wash off the stain with a gentle stream of water.

5. Apply alcohol containing 5 per cent hydrochloric acid to the slide for one or two minutes. Wash.

6. Counterstain with methylene blue.

7. Wash and blot.

The acid alcohol removes the red fuchsin from everything except the acid-fast bacilli. These retain the red stain in spite of the acid alcohol; everything else appears blue. The Ziehl-Neelsen stain is a *differential* stain since it differentiates acid-fast bacilli from other kinds. It is one of the most widely used methods for detecting tubercle bacilli in sputum. The reason for heating the slide is that many workers believe that the acid-fast bacteria have a waxy cell wall. This must be softened so that the stain can soak in rapidly. Once in and cooled, it stays there in spite of the application of acid and alcohol. Tubercle bacilli are more acid-fast than are most other mycobacteria. *Nocardia* species are also somewhat acid-fast, but it is not too difficult to distinguish them from the tubercle bacilli.

Fluorescence Microscopy. Acid-fast bacilli are easily detected with fluorescence microscopy because the fluorescent dye is also acid-fast.

Fluorescence is a property of numerous substances, as a result of which they reflect light waves having a wavelength (color) different from that of the incident rays. A fluorescent object may be invisible in ordinary light yet glow (or fluoresce) with a bright yellowish luminosity when viewed under ultraviolet ("black" or "invisible") light.

If tubercle bacilli in sputum on a microscope slide are coated with a fluorescent stain containing auramine O and rhodamine B, the excess stain is washed from the rest of the slide, and the slide is then illuminated with ultraviolet light and examined under the microscope, the tubercle bacilli are easily seen as brightly glowing golden rods on a dark field. An ordinary microscope is readily equipped for fluorescence microscopy by means of a special mirror, appropriate light filters, and a good source of ultraviolet light in place of the usual illumination. Do not confuse this application of fluorescence in microscopy with fluorescent-antibody staining, described in Chapters 20 and 35 (see Figs. 20–7 and 35–7). Fluorescence microscopy with a blue light source is fully effective and *greatly superior* to the Ziehl-Neelsen technique. The Truant fluorescence technique permits the observation of mycobacteria and acid-fast *Nocardia* as bright, yellow-orange fluorescent bacilli on a dark background.

Cultivation and Differentiation. *Mycobacterium tuberculosis* grows slowly. Its generation time is between 12 and 18 hours. Consequently, infections caused by the organism tend to be subacute and chronic. In the laboratory it takes three to eight weeks to successfully isolate the tubercle bacillus, a very slow process.

All mycobacteria (with the exceptions of *M. leprae* and *M. lepraemurium*) grow on appropriate culture media exposed to air. Rate of growth, pigment production, and optimal growth temperatures are important primary differential characters (Table 32–1). Among many solid media for mycobacteria are Löwenstein-Jensen culture, an egg-based medium, and Middlebrook 7H10, a synthetic agar medium. The latter requires an atmosphere containing 7 per cent carbon dioxide. Glycerin infusion agar or any of several media made with mixtures of eggs, milk, and potato are commonly used for cultivating tubercle bacilli for diagnostic purposes. Most *nonpathogenic* species, especially if exposed to light at about 25 C, generally produce luxuriant, brilliant orange, red, or yellow growth in three to five days.

Human and bovine species of tubercle bacilli grow much more slowly on such media, requiring about two to three weeks or longer to develop perceptible growth. They grow only at about 37 C. The growth is usually rough, granular, and very pale yellow or cream-colored; experienced microbiologists can recognize it with some assurance. Certain improved methods and commercially available media—especially fluid media (e.g., Dubos' broth) containing surface-tension reducers, such as polysorbate 80 (Tween 80), that enable the nutrient fluid to wet the waxy bacilli—speed up the growth of *M. tuberculosis.* Oleic acid-albumin agar containing polysorbate 80 is especially favorable for tubercle bacilli.

Species of Tubercle Bacilli

There are several species of tubercle bacilli. Some of these grow well at low temperatures and infect cold-blooded (poikilothermic) animals like frogs, snakes, and turtles. There are at least three species pathogenic to mammals: *M. tuberculosis,* infect humans; *M. bovis,* infectious to cattle and humans; and *M. microti,* the so-called vole bacillus, causing tuberculosis in voles (a kind of field mouse) and other rodents; it is closely related to *M. tuberculosis.* There is also an avian variety *(Mycobacterium avium),* growing best at about 40 C, which infects birds and can be a real problem to poultry workers. It rarely causes disease in humans, but it sometimes infects pigs. The several species are morphologically indistinguishable.

The two kinds of bacilli, human and bovine, can be differentiated by laboratory tests but are very similar in most respects. If a cow has tuberculosis of the udder, the milk will contain the bacilli. Children may be infected by drinking the milk of tuberculous cows. This method of infection accounts for much tuberculosis of the lymph nodes of the neck and abdomen and also of the bones in children, especially in some countries where pasteurization of milk and precautions against tuberculosis in cattle are not as rigorous as in the United States. Tuberculosis of organs other than the lungs (extrapulmonary tuberculosis), however, can also result from infection with human tubercle bacilli.

Differentiation of Mycobacteria. Characteristics that differentiate various mycobacteria are listed in Table 32–1. No single test provides a wholly reliable diagnosis, but determinations of virulence in guinea pigs and rabbits are among the most dependable diagnostic procedures for *M. tuberculosis. M. bovis* infects both guinea pigs and rabbits, *M. tuberculosis* only guinea pigs. Neither infects fowls, which are susceptible to *M. avium. M. tuberculosis* and *M. microti* are the only species of *Mycobacterium* known to produce niacin.

It is important for the diagnostician to know the characteristics that differentiate the various kinds of mycobacteria, and especially those differentiating tubercle bacilli from harmless acid-fast saprophytes, which are widely distributed in soil, dust, the dung of domestic animals, dairy products, and hay as well as on the surface of the human body. Since they occur in dairy products and dust, their occasional presence in saliva is readily understood. This can cause confusion when sputum is examined microscopically for tubercle bacilli, since all mycobacteria closely resemble one another in appearance. Saprophytic mycobacteria may be found also in urine and gastric contents, which are often examined for tubercle bacilli.

Tubercle Bacilli and Disinfection. In spite of the fact that tubercle bacilli do not form spores, they are more resistant to some disinfectants and drying than are the previously discussed nonspore-forming pathogens. In dried sputum kept in the dark, they may live for several months or withstand temperatures of about 70 C for one hour. In particles of sputum-infected dust, they may remain alive for eight to ten days if not exposed to sunlight. Dried organisms are often somewhat more resistant to heat and chemical disinfectants than are the same organisms in a moist or fully hydrated condition. Exposure to direct sunlight, kills tubercle bacilli in a few hours (mainly because of the ultraviolet rays). Pasteurization (63 C for 30 minutes or 72 C for 15 to 30 seconds) also kills them in milk.

It is difficult to kill tubercle bacilli in sputum by means of ordinary disinfectants because they are protected by mucus and their waxy membrane. Disinfection of sputum with solutions like 5 per cent saponated cresol, full-strength chlorine laundry bleach, or organic aqueous iodine disinfectant may require from two to five hours. Bichloride of mercury and alcohol are unsatisfactory because they coagulate the sputum around the bacilli and are thereby excluded from contact. The use of formaldehyde is also unreliable for this purpose.

TUBERCULOSIS IN HUMANS

Clinical Forms

The symptoms produced by tuberculous lesions are varied and may simulate other diseases. In the diagnosis of any chronic infection, as well as some acute ones, the possibility of tuberculosis (and also mycotic infections) must be considered. Although about 90 per cent of all tuberculosis infections involve the lung, the CNS, the skin, bones, joints, kidneys, bladder, genital organs, intestines, liver, middle ear, eye, abdomen, and peritoneum can also be infected.

Primary Tuberculosis and Tubercles. *Primary tuberculosis* is the disease process resulting directly from the first entry of tubercle bacilli, whether by inhalation, ingestion, or otherwise. It may occur at any age but is most common in infants, children, and young adults. Humans, especially infants, readily become infected with tubercle bacilli, but usually are highly resistant to progression of the disease under good living conditions, especially between ages 5 and 15. When the local *lymph nodes* become involved, the process is called a *primary complex* (Fig. 32–2).

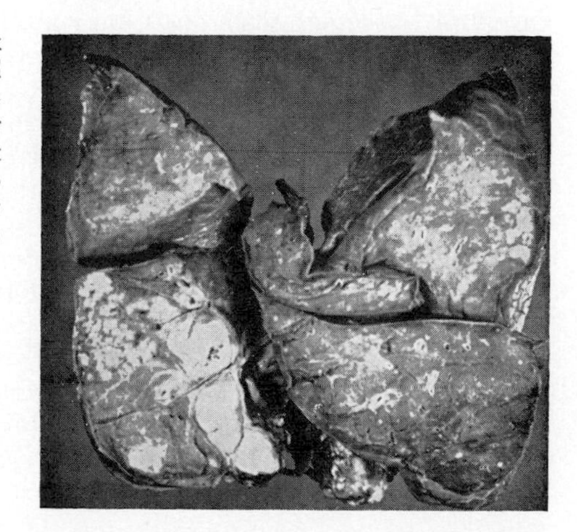

Figure 32–1 Tuberculous lesion (primary) in apex of left upper lobe (right upper corner) with associated massive involvement of regional lymph nodes (primary complex). Large wedge-shaped lesion in lower half (lateral portion) of right lower lobe. This last lesion (tuberculous pneumonitis) is secondary to bronchial erosion from a tuberculous node. (Courtesy of Drs. Charles Dunlap and James B. Gray: *In* Nelson, W. E., et al. (eds.): Textbook of Pediatrics. 9th ed. Phiadelphia, W. B. Saunders Company, 1969).

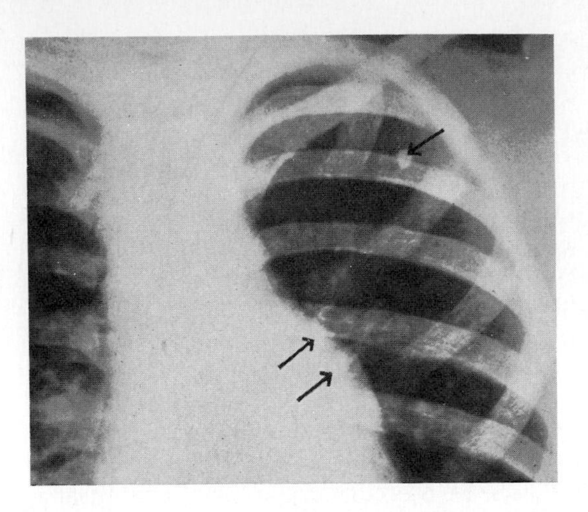

Figure 32–2 Primary infection with the tubercle bacillus. The child, when two years of age, was in contact with a tuberculous mother, who died of the disease. Six years later, at the age of eight years, an x-ray of the child shows a prominent calcified lesion in the left upper lobe and several calcified hilar lymph nodes. This is a healed primary complex. (From Rubin, E. H.: Diseases of the Chest. Philadelphia, W. B. Saunders Company, 1947.)

Within a few days after tubercle bacilli first locate in the body of a susceptible person (or animal) they cause a tissue reaction that results in the formation of a distinctive kind of lesion, usually in the lungs, called a *tubercle.* A tubercle consists of one or more tubercle bacilli surrounded by a small mass of pus and phagocytes. They are later enclosed within distinctive multinucleate tissue cells called *giant cells,* all surrounded by connective tissue cells (fibroblasts), which form a tough retaining wall or sac with relatively little fluid. This is the *proliferative* type of lesion, a type likely to heal without further progress and without remarkable symptoms of any kind. The disease is called "childhood tuberculosis." The whole lesion is surrounded by a zone of inflammation. The cells and structure of tubercles are so characteristic that a diagnosis of tuberculosis can be made from them alone, even though the bacilli may not be found with the microscope. A single, very early tubercle is a grey mass, the size of a pin head or smaller, which feels hard to the touch (Fig. 32–1). Masses of tubercles usually occur together in an organ, and the resistance of a healthy child is generally sufficient to stop their process. Tubercle formation is commonly associated with the development of allergy to the tubercle bacilli.

When virulence is high, or dosage large and continuous, or resistance low (as in undernourished children), the bacilli continue to multiply, producing further inflammation and the exudation of fibrinoserous fluid. This is the *exudative* type of lesion, one likely to progress. The bacilli continue to grow, enlarging the tubercles and killing the tissue at the center, which becomes coagulated into a cheesy mass. When the process has extended to this point, *caseation* is said to have occurred. If the process continues, numbers of such caseated abscesses may encounter each other as they expand, finally fusing together to form one large, caseous mass.

If the infection is in a lung, the necrotic process often invades and erodes through the wall of a bronchiole, and then the caseous contents, along with numerous tubercle bacilli, are coughed up with the sputum. This is a very dangerous stage of the disease for other people, since the sputum is then highly infectious. Tuberculous sputum and the pus from tuberculous abscesses contain dead tissue, pus, and large numbers of tubercle bacilli. Sputum is often swallowed, especially by young children, and the gastric contents may often reveal tubercle bacilli when sputum examinations are persistently negative or cannot be made. Intestinal or generalized infection may then occur.

If a tuberculous abscess erodes through the wall of a blood vessel, a *hemorrhage* occurs, which may prove fatal.

Tubercles frequently heal, especially in the nonprogressive, primary form of the disease, in which case they become surrounded by thick envelopes of scarlike tissue. Calcium salts may be deposited in them; i.e., they undergo *calcification,* which is readily seen in x-ray pictures (Fig. 32–2). As we shall see later, many of us are carrying healed, calcified tubercles somewhere in our bodies.

Classification of Tuberculosis Infections. There are various classifications of tuberculosis, some based on the extent of the disease, others on bacteriologic findings. Tuberculosis may be diagnosed as *active, quiescent,* or *inactive.* In active form the patient casts off bacilli in the sputum or other excretion. In quiescent and inactive cases bacilli are not given off and lesions are not progressive or are healing. Such infections are not an immediate danger to others. An active case may become inactive if healing takes place; an inactive case may become reactivated and again dangerous to others. The progress of the disease represents a constantly changing balance between the resistance of the patient and the virulence of the organism.

Reinfection (or Adult) Type Tuberculosis. This type occurs when tubercle bacilli reenter the tissues after a primary infection has healed or partially healed. The bacilli may be reintroduced from an outside source (reinfection or superinfection) or from a partially healed tubercle (postprimary progression). Most persons are constantly being reinfected from outside sources with no obvious ill effects whatever. If resistance is low, however, reinfection or postprimary progression may result in serious developments associated with allergy to tubercle bacilli.

Allergy in Tuberculosis

In many chronic or prolonged microbial infections such as syphilis, undulant fever, tularemia, relapsing fever, fungal infections, and tuberculosis, the patient develops *an allergic condition* specific for the particular antigens involved. Generally benign and protective in character, in some individuals this allergy becomes excessive and does great harm. In tuberculosis, syphilis, and some other chronic infections, the ulcerative, softening character of lesions developing late in the disease, or after the lapse of many years in reinfection, is believed to be the result of excessive allergy of the tissues. Such lesions tend to break down rather than heal.

Allergic tissues tend to arrest pathogenic organisms and to prevent their spread throughout the body, although the organisms may severely damage the arresting tissues. Allergy as an arresting agent must, therefore, be regarded as an important defensive mechanism. Whether or not it stops further progress of the disease depends on the nature of the allergic response and the resistance of the patient. Allergy is probably the chief defense of normal adults against repeated reinfection with tubercle bacilli as well as against many other infections.

Allergy and Koch's Phenomenon. The arresting action of allergy in tuberculosis is well illustrated by the *Koch phenomenon.* (Do not confuse it with Koch's postulates.) If a normal guinea pig is inoculated (usually in the right groin) with virulent tubercle bacilli, the bacteria form a local abscess, and then proceed almost unopposed from the abscess to the lymph nodes of the abdominal cavity, to the spleen, the liver, the lymph nodes of the thorax, the lungs, and the kidneys, and the pig finally dies of disseminated tuberculosis in about six to eight weeks. Now, if in the second or third week of this progressive disease a second injection of tubercle bacilli is made into the left groin, there is a strong local allergic tissue reaction. The bacilli are held in the site where they are injected and do not progress further, although they may cause a local abscess. The tissues are highly defensive because of the allergy to the first infection. We do not understand why the bacilli of the first infection are not also held in check.

Similarly, human beings who have had a mild, unrecognized infection with tubercle bacilli, one perhaps long since healed, are generally much more resistant to tuberculosis than persons who have never had any contact with tubercle bacilli. The first group are moderately allergic (hence resistant) to the bacilli, as shown by the fact that they react not excessively to the tuberculin test (see the following discussion). The second group may readily be made allergic (resistant) to tubercle bacilli by giving them a very mild infection, as is done in BCG[4] vaccination.

The Tuberculin Test. Intracutaneous injection of diluted antigenic extracts of tubercle bacilli (called *tuberculin*) into the skin of urban adults evokes a red, indurated, itching wheal in most of them after about 48 hours. It disappears after a few days, and the test is entirely harmless if small doses of tuberculin are used. The reaction is called a *tuberculin reaction.*

Tuberculins of various types were first developed by Koch (notably "old tuberculin" or O.T.), who thought he had devised a vaccine against the disease and had great hopes of using tuberculin for the cure of tuberculosis. Modern tuberculins are made from *purified protein derivatives* (P.P.D.) of tubercle bacilli. It has since been found, however, that tuberculin alone will not cure or prevent the disease. In fact, if too much of it is introduced into the body of a person allergic to the protein, a very severe, generalized, and even fatal allergic reaction may occur, instead of merely a local one on the skin. If the patient has live tubercle bacilli in his tissues, the reaction may give the organisms a fresh start and thus greatly injure him. It is therefore recommended by some authorities that *tuberculin-positive individuals should not be vaccinated with BCG.*

Healed tuberculous infection in an adult, even though the individual may never have been aware of it, still leaves him in an allergic condition toward tubercle bacillus protein. Thus, a positive tuberculin reaction in an older person does not necessarily indicate active or even inactive tuberculosis. Resulting from a healed childhood infection, the tuberculin reaction may remain positive for life or it may revert to negative, although the individual may still have resistance. *The tuberculin test is used mostly to detect early primary tuberculosis in children.*

Many young children, as well as some adults in nontuberculous environments, give a negative tuberculin reaction. Allergic sensitivity appears in primary infection at any age within three to ten weeks after infection. Therefore, in children who have only recently developed tuberculin reactivity for the first time, the infection is probably present in a more or less progressive form, since they are not old enough to have recovered completely. The reaction remains positive whenever viable tubercle bacilli are present in the body, whether or not active disease is present.

The presence or sudden appearance of a positive tuberculin reaction does not necessarily mean that the person is going to develop clinical tuberculosis. That depends on continued reinfection, malnutrition, poor living conditions, and other factors. The tuberculin test merely serves as a warning that proper measures should be taken before it is too late.

Methods of Performing the Tuberculin Test. Tuberculin may be applied in a number of ways, all of which are designed to bring the tubercle bacillus protein into intimate contact with the body cells. Von Pirquet described a method of scratching the tuberculin into the skin—the *von Pirquet test* or *scarification.* It is a simple and widely used procedure.

In a popular modification called the *Heaf test,* the tuberculin (P.P.D.) is first spread over a small area of skin. An instrument sometimes called a Ster-needle gun is then pressed against the skin, and a triggered, spring-drive plunger with six short, solid needle points drives the P.P.D. painlessly into the skin. The needle-tipped heads are replaceable for sterilization by heat. In the *Tine test* the points are precoated with O.T. and dried.

The injection of minute, accurately measured amounts of tuberculin (P.P.D.) *intradermally* was devised by *Mantoux,* and his test, quantitatively the most accurate procedure, is designated by his name. *Vollmer* demonstrated the

---

[4]Bacille Calmette-Guérin. See page 487.

method of applying tuberculin to the surface of the skin on patches of gauze or tape. This is the *patch test*. The patch is left on for 48 hours, and the test is read 48 hours after its removal. The test is convenient but considered not very accurate.

Standardization of both O.T. and P.P.D. has been set by international agreement, based on a biologic assay with "Seibert's P.P.D. Lot # 49608" (called P.P.D.-S) under specified conditions.

Diagnosis of Tuberculosis

Microscopic. A tentative diagnosis of pulmonary tuberculosis is generally made on the basis of x-ray and clinical findings and is confirmed whenever possible by microscopic examination of the sputum. Smears of the material are made on slides and stained by the Ziehl-Neelsen method, Kinyoun's acid-fast stain, or fluorochrome to see if acid-fast bacilli are present. It is frequently necessary to make many examinations before the bacilli are finally found; therefore, in a suspicious case, one negative report is insufficient.

Cultural. Since the bacilli in sputum or other material may be present in such small numbers that they cannot be found by routine microscopic examination and since (especially in urine, feces, and gastric contents) it is impossible to differentiate tubercle bacilli from acid-fast saprophytes microscopically, it is absolutely necessary (for first diagnosis, at least) to make cultures from the pathologic material, identify the bacilli by their growth characteristics, and, if in doubt, prove their pathogenicity by injection of the specimen into animals, usually guinea pigs. Some of the media commonly used in medical laboratories for the diagnosis of tuberculosis have already been mentioned. Many contain a dye such as crystal violet or malachite green. The dye inhibits growth of many microorganisms that often contaminate pathologic material (such as sputum) in which tubercle bacilli are found but does not affect the tubercle bacilli. Most of the contaminating bacteria in sputum and other exudates are often killed first by mixing the material with an equal volume of Zephiran-trisodium phosphate and 4 per cent sodium hydroxide and neutralizing the pH after 20 minutes.

To avoid the destruction of a large percentage of tubercle bacilli from sputum and bronchial secretions by strong alkali, a mild decontamination and digestion procedure has been developed. This method uses the mucolytic agent N-acetyl-L-cysteine (NALC) and 2 per cent sodium hydroxide. Another mild method employs dithiothreitol at pH 7.0, followed by 1 per cent NaOH or Zephiran.

It should be noted that tubercle bacilli may be isolated from gastric specimens, urine, cerebrospinal and other body fluids, and tissues removed surgically or at necropsy, as well as from sputum. Tubercle bacilli can frequently be cultivated directly from *untreated* sputum if a small amount of penicillin and cycloheximide are placed on the surface of the medium to inhibit growth of extraneous microorganisms.

Animal Inoculation. After isolation from pathologic materials pure cultures may be used for confirmatory tests (Table 32–1) and for injection into animals. When only a few tubercle bacilli are present in sputum, urine, pus, or other material, it may be impossible to find them with the microscope or even by means of cultures. Another method of discovering them is to inject a guinea pig with the material. Lesions produced by the bacilli are easily recognized.

Urine Specimens. In examining urine for tubercle bacilli, the *sediment* is usually collected for staining or injection. In urine, acid-fast bacilli other than tubercle bacilli are sometimes found in stained smears *(Mycobacterium smegmatis, M. phlei)*. These bacilli are usually present beneath the prepuce or on the labia as harmless saprophytes. They look exactly like tubercle bacilli but are readily differentiated by tests shown in Table 32–1.

Table 32–1. Distinctive Properties of Various Mycobacteria*†

SPECIES OR SUBGROUP	NIACIN PRODUCED	NITRATE REDUCTION		CATALASE PRODUCED *mm.*			TWEEN HYDROLYSIS (DAYS)		TELLURITE REDUCTION IN 3 DAYS	PIGMENT FORMATION		GROWTH ON 5 % NaCl	GROWTH IN LESS THAN 7 DAYS AT 37 C	PRODUCE ARYLSULFATASE ENZYME +3 DAYS	GROWTH ON MACCONKEY AGAR
		>1+	*>3*	*>40*	*>50*	*68°+*	*+5*	*+10*		*Dark*	*Light*				
M. tuberculosis●	++	++	++	++	–	–	∓	∓	–	–	–	–	–	–	–
M. bovis	–	–	–	++	–	–	–	–	–	–	–	–	–	–	–
Runyon Group I (Photochromogenic) (lemon yellow pigment)															
M. kansasii●	–	++	++	++	++	++	++	++	–	–	++	–	–	–	–
M. marinum●	–	–	–	±	∓	–	++	++	–	–	++	–	∓**	–	–
Runyon Group II (Scotochromogenic) (yellow orange to dark red)															
M. sp. *scrofulaceum*	–	∓	–	++	++	++	–	–	–	++	++	–	–	–	–
M. sp. *gordonae*	–	–	–	++	++	++	+	++	–	++	++	–	–	–	–
M. flavescens	–	++	+	++	++	++	++	++	–	++	++	±	–	–	–
Runyon Group III (no pigments in light)															
M. avium‡●	–	–	–	++	–	++	–	–	±	–	∓	–	–**	–	–
M. intracellulare‡●	–	–	–	++	–	++	–	–	+	–	–	–	–	–	∓
M. xenopei	–	–	–	++	–	++	–	–	–	∓§	∓§	–	–	±	–
M. gastri	–	–	–	++	–	–	++	++	–	–	–	–	–	–	–
M. terrae complex	–	++	+	++	++	++	++	++	–	–	–	–	–	–	–
M. triviale	–	++	+	++	++	++	+	++	–	–	–	++	–	∓	–
Runyon Group IV (rapid growers) (some scotochromogens)															
M. fortuitum	–	±	±	++	++	++	–	∓	++	–	–	++	++	++	++
M. smegmatis	–	++	∓	++	++	++	++	++	++	–	–	++	++	–	–
M. phlei‡	–	++	∓	++	++	++	++	++	++	++	++	++	++	–	–
M. vaccae‡	–	++	+	++	++	++	±	++	+	++	++	++	++	–	–
M. chelonei	V	–	–	++	++	++	–	–	–	–	–	–	++	++	±

*Modified after Current Item No. 165, Laboratory Program, 1968. National Communicable Disease Center. Courtesy George Kubica, former Chief, Mycobacteriology Unit, NCDC, Atlanta.

†Key to percentage of strains reacting as indicated: ++ = 85 per cent or more; + = 75–84 per cent; ± = 50–74 per cent; ∓ = 15–49 per cent; – = <15 per cent; V = variable.

‡With tests listed the pairs of organisms so indicated cannot be separated; colonial morphology on 7H-10 may be helpful in the case of *M. phlei–M. vaccae.*

§Pigment increases with age.

***M. ulcerans* and *M. marinum* grow best at about 32 C; *M. avium* and *M. intracellulare* at 41 C.

●*Mycobacteria* that are of clinical significance. The others are rarely or never significant.

In order to avoid the presence of confusing contaminants in urine specimens to be examined for tuberculosis, it is necessary to draw the urine through sterile catheters into sterile flasks, avoiding contamination with material from the external genitalia, hands, or instruments. The specimen is placed in a sterile container clearly indicating the patient's name, the date, and whether the specimen is from the *right* or *left* kidney. The patient's life may depend on this!

X-rays. As noted earlier, diagnosis of tuberculosis of the lungs is commonly made by means of x-rays. Caseous masses, cavities, and calcified tubercles give more or less distinctive appearances, which can be recognized by those trained in x-ray diagnosis. However, so many other diseases cause shadows on x-ray plates of the chest, which may be confused with those due to tuberculosis, that microscopic, cultural, and animal inoculation studies should always be made if x-ray plates are suspicious and clinical data inconclusive. Among the diseases causing confusion in this way are histoplasmosis, coccidioidomycosis, nocardiosis, blastomycosis, cancer, pneumonias due to viral agents, and Q fever. Skin tests with tuberculin, coccidioidin, histoplasmin, and similar antigens from other fungi are of great value in differentiating some of these conditions.

Transmission of Tuberculosis

It is clear from the foregoing that in tuberculosis of the lungs the *sputum* is the chief source of infection, although in tuberculosis of the kidneys and bladder (relatively infrequent) the bacilli may also be present in the *urine* in large numbers. The bacilli are found in the *feces* in tuberculosis of the intestine (uncommon) and in the pus from tuberculous abscesses. It has been estimated that a single patient who is raising a considerable amount of sputum may discharge, in 24 hours, 500 million to three billion bacilli. A coughing or sneezing tuberculous patient throws out a spray of sputum containing tubercle bacilli, and this may be inhaled directly or in the form of droplet nuclei by other people. Thus, pulmonary tuberculosis is primarily an airborne disease. It means almost certain infection for infants in contact with the patient. In the infant, tuberculosis is very likely to take the form of a generalized, acute, fatal infection, with meningitis.

Most of the following discussion applies in situations in which effective chemoprophylaxis with drugs like isoniazid (INH or Nidrazid) is not available. Proper chemoprophylaxis, as with INH, quickly renders most patients noninfective, although cure usually depends on prolonged use of combinations of drugs—e.g., isoniazid with ethambutol or rifampin, or all three. The drugs are toxic, and their use must be supervised by a physician.

Unless immediate chemotherapy with antituberculosis drugs is available, uninfected children should whenever possible, be taken from a home in which there is an active case of tuberculosis, or the patient should be removed as soon as the diagnosis of tuberculosis is made. The community health worker can explain to families the need for such separations. The difficulties are often financial or stem from lack of proper institutions for the care of the tuberculous patient.

Tuberculous infection can occur through the mouth, especially in children. For example, a careless or ignorant person with active pulmonary tuberculosis kisses children, contaminates the floor or furniture with sputum, or expectorates into the street or other public places. Children playing around the room or in the street get the bacilli on their hands and thus into their mouths. Oral infection probably accounts for many of the nonpulmonary tuberculous infections of childhood. As previously noted, children may also be infected orally by drinking unpasteurized milk from tuberculous cows.

Although probably a vector of only secondary importance, food may be contaminated by handling it with fingers soiled with sputum or by flies that have crawled over tuberculous sputum. The tuberculous food handler may cough or sneeze over it. The common drinking cup or imperfectly washed spoon or fork may also carry the bacilli.

Among the great difficulties in preventing the spread of a disease like tuberculosis are the ignorance and indifference of the public. In 1966 there were about 60 times as many deaths from tuberculosis as from poliomyelitis in spite of the fact that the tuberculosis death rate per 100,000 population had dropped from about 225 in 1911 to about 4 in 1966. Figure 32–3 shows the decrease in deaths due to tuberculosis since 1950. Note, however, that the number of active cases is still too high, considering what is known about this disease. In 1980, 27,749 new active cases occurred in the United States, and in 1982, 25,728. An estimated 3,000 to 5,000 deaths per year still result from tuberculosis, decidely too many with the excellent chemotherapeutic agents available for effective treatment of the disease.The geographical distribution of tuberculosis in the United States is shown in Figure 32–4.

Tuberculosis Infections

Tubercle bacilli are so widespread, especially in city streets, public transportation, theaters, and other places where people congregate, that no one in an urban community escapes eventual contact with the organisms. It is well known that 50 to 80 per cent of city-dwelling adults have at some time been slightly infected with the tubercle bacillus (probably in childhood). The healed

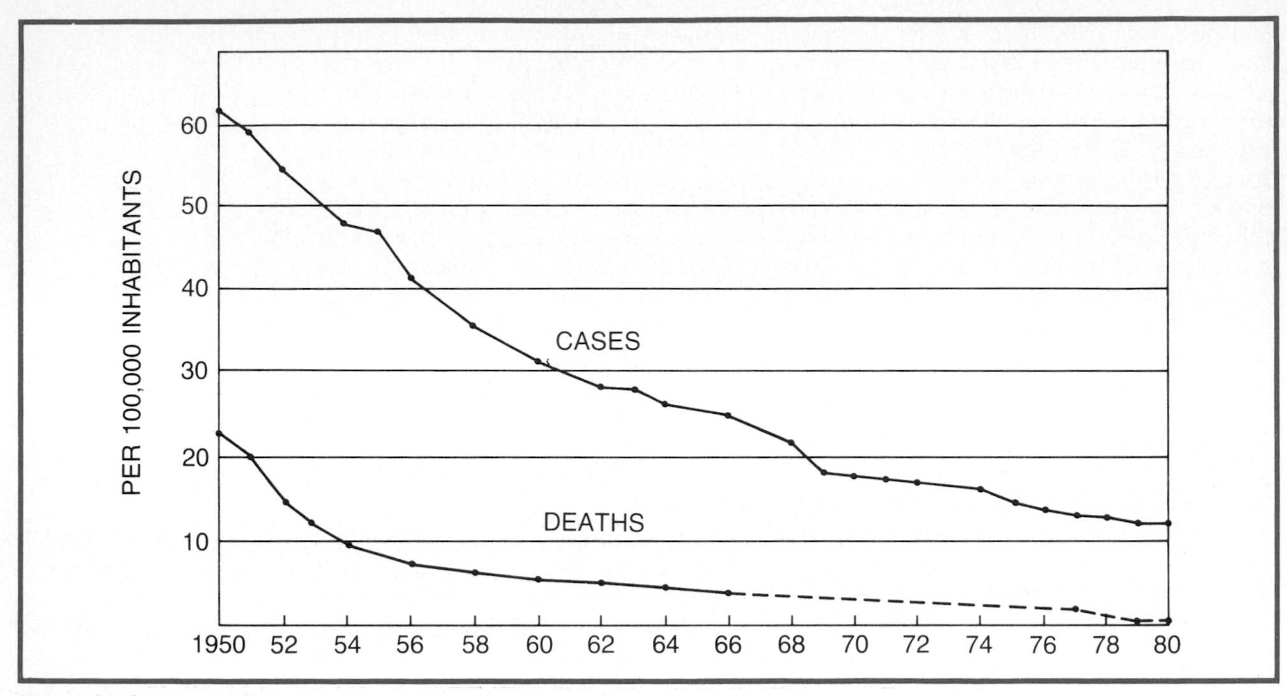

Figure 32–3 Approximates the data for 1966 to 1980 for tuberculosis. Based on data from *Morbidity and Mortality Weekly Reports.*

lesions are found in the lungs in a large proportion of adults upon whom autopsies are performed following death due to wholly unrelated causes. Probably most of these persons never showed any recognized evidence of tuberculous infection. In 1958 it was estimated that over 8,000 persons in New York City had active, open tuberculosis, unknown to anyone, and the situation has not improved much since then.

There appears to be a very delicate balance between health and disease in tuberculosis. In most people the balance is heavily weighted in their favor by robust health (both physical and mental), a well-balanced diet, rest, exercise, and recreation. In others, insufficient rest and inadequate food may tip the

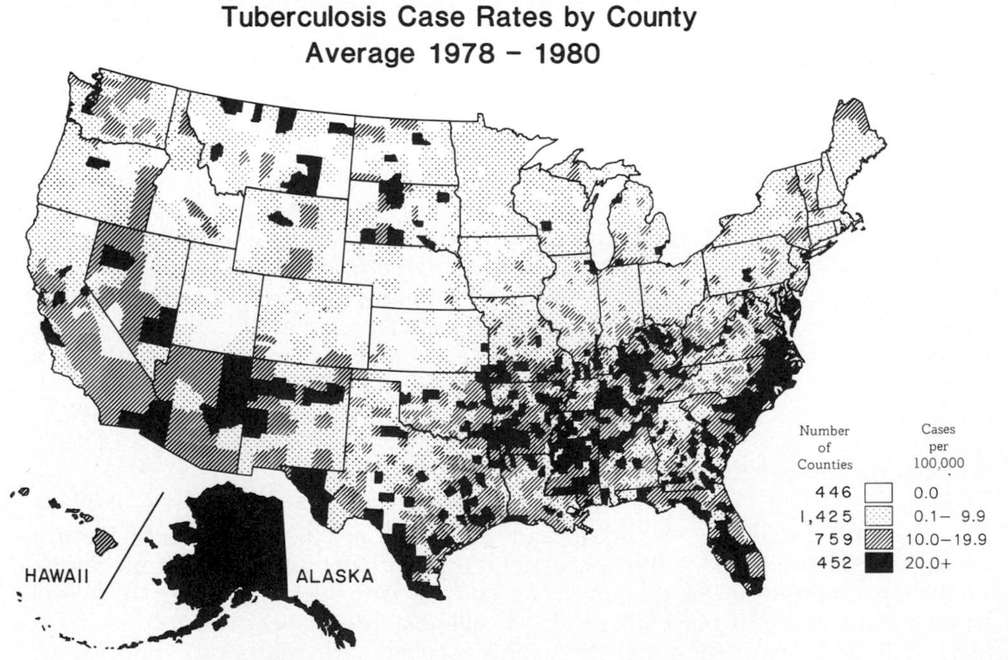

Figure 32–4 New active cases of tuberculosis per 100,000 population, 1978–1980. (From Tuberculosis in the United States, 1980. Atlanta, Georgia, Centers For Disease Control, Tuberculosis Control Division, 1983.)

scales in favor of the tubercle bacillus. Treatment, aside from the use of chemotherapy and surgery, is aimed largely at keeping the balance of health in favor of the patient. Resistance is kept at a high level by good food, fresh air, sufficient rest, and recreation—everything, in short, that keeps one in good condition. It is diminished by anthing that decreases bodily vigor—other infections (such as measles, whooping cough, and influenza); frequent child-bearing; continuous overstrain and fatigue, poor living conditions of all kinds, the abuse of drugs, tobacco, alcoholism, and malnutrition.

Socioeconomic Conditions

Tuberculosis is closely connected with social and economic conditions in a community. It is much more frequent among the poor than among the well-to-do. Low standards of living mean lack of isolation and insufficient rest, sunlight, fresh air, cleanliness, medical care, and food. The death rate from tuberculosis in any community clearly indicates what the social conditions are.

During and after World War II conditions in war-torn countries in Europe created an enormous increase in tuberculosis. This has also been true of almost every other war. Wars commonly bring an influx of refugees into more prosperous countries and with them frequently comes tuberculosis.

Roughly half of the world's population has been infected with *Mycobacterium tuberculosis,* but good living conditions and education hold tuberculosis in check in the "developed" countries. However, the danger of active infection always lurks for the undernourished, alcoholics, drug addicts, former tuberculosis patients, and people with debilitating diseases of all types.

Prevention and BCG

A vaccine against tuberculosis was prepared in 1923, in France, by Calmette and Guérin, from an attenuated strain of *M. bovis.* It is called *BCG* (bacille Calmette-Guérin). Its value has been and still is disputed by some health workers. Outstanding authorities have confirmed its effectiveness as an antigen. UNICEF has given it to over 226 million children, both tuberculin-positive and tuberculin-negative, in 95 countries, with gratifying results.

The physical examination of candidates for a school of medicine or nursing should include an intradermal tuberculin test and chest x-rays if the tuberculin test is positive. This identifies persons who would be unusually susceptible to the disease. It also detects previously unsuspected tuberculous infection and allows early treatment of the disease. It must be remembered that a positive tuberculin reaction has no diagnostic significance in persons vaccinated with BCG, since the vaccine induces tuberculin hypersensitivity.

Tuberculous patients are sources of disease for persons who care for them unless the patients are under a regimen of treatment with INH,[5] ethambutol, rifampin, and/or streptomycin. PAS (para-aminosalicylic acid), for many years the drug of choice, is only used as a standby drug now; it is only weakly tuberculostatic. It is especially important to examine, by x-ray, sputum smear, tuberculin test, and physical examination, patients who will be in an institution for long periods, such as psychiatric patients and those with chronic disease. Hospital employees, especially orderlies, maids, and cooks, should be similarly examined

The majority of hospitalized patients with tuberculosis have the pulmonary form. If they are in active stages of the disease, the sputum coughed up from the lungs must be destroyed. Unless the patient has copious pulmonary

[5]INH is isoniazid (isonicotinylhydrazine), a drug of choice against the tubercle bacillus. It is best used simultaneously with other drugs.

secretion, he or she is instructed to cough into disposable paper wipes and place these wipes in a disposable plastic or paper bag pinned to the bed or fastened to the bedside stand. When the bag is approximately three-quarters full of wipes, it is removed, carefully closed at the top to prevent spilling of the wipes, wrapped in plastic bags, and burned. If an incinerator is not readily available, the packages should be placed in marked trash cans. The personnel who are responsible for emptying these cans must be instructed in the importance of complete incineration of this material. Sputum cups are used only when absolutely necessary because they are difficult to disinfect completely, and special care must be taken to guarantee complete destruction of all the material in a closed incinerator system. All patients with excessive pulmonary secretions should be considered tuberculous until proved otherwise. Disregarding this simple precaution is probably the single most important cause of pulmonary tuberculosis among nurses and others caring for patients with chronic respiratory conditions.

With the advent of chemotherapy and chemoprophylaxis, hospitals are not used as much as previously to *keep* patients. The principle role of the hospital now is to care for the *acutely* ill and release the ambulatory patient who cooperates in regular chemotherapy.

Unless continually receiving INH, rifampin, ethambutol, streptomycin, or other effective chemotherapy, each patient with an active case of tuberculosis must be taught to live so as to avoid being a danger to others, and should be isolated whenever possible. If he is not isolated, proper precautions must be taken to protect others. The patient must not cough or expectorate openly. A separate set of dishes, which should be boiled after use, is a necessity, and the patient should sleep alone.

Every tuberculous patient must be trained to avoid exposing others, to avoid all contact with small children and elderly or sick persons, to adequately cover the nose and mouth while coughing or sneezing, and to wash his hands. This instruction is an important function of the health team, and it can also be carried out successfully at home.

HANSEN'S DISEASE

The disease "leprosy" has a historical stigma attached to it that has been completely repudiated by modern research. Therefore, it is now preferable to call the disorder by its more modern name, "Hansen's disease."[6] The microbiologist and other knowledgeable professional people should lead the way to enlightenment and progress in this as well as other areas.

Although Hansen's disease is not primarily a respiratory disease, it is included here because it is apparently transmitted from extensive open lesions that develop in the upper respiratory tract, especially in the nasal septum, and because it is caused by the acid-fast *Mycobacterium leprae* bacillus. The lesions of leprosy involve the "cooler" tissues of the body, such as the skin, peripheral nerves, nose, eyes, pharynx, larynx, and testicles.

Hansen's disease is one of the "historical diseases," ranking in importance with bubonic plague, cholera, typhus, and yellow fever; it has been known for centuries and has played a major role in the history of mankind. In ancient and medieval times Hansen's disease had an impact on religious activities, architecture, politics, science, literature, and many other aspects of human life. Although not now common in the United States, it is nevertheless present, almost unknown to people whose medieval ancestors were haunted by fear, loathing, and hatred of the leper.

[6]Gerhard Armauer Hansen (1841–1912), a Norwegian physician, was the first to recognize leprosy as a clinical entity (in 1873) caused by the bacillus he described. This was nine years before the discovery of the tubercle bacillus by Robert Koch.

M. leprae can be seen in histologic sections of infected tissues from lepers. It is morphologically indistinguishable from other mycobacteria. Many workers have cultivated species of mycobacteria from leprous tissues and called them "*M. leprae*," but these have all the characteristics of the saprophytic mycobacteria. Experimental transmission of Hansen's disease, even with fresh leprous tissues, is very difficult. Shepard, in 1960, was able to grow *M. leprae* in the foot pads of mice. It produced infections and could be transferred for many passages without loss of virulence. In World War II, two American marines developed Hansen's disease in areas of the skin that had previously been tattooed. Thus, without having proved Koch's postulates for *M. leprae* even the skeptic now accepts it as the causative agent of Hansen's disease.[7]

Contrary to centuries-old notions. Hansen's disease is *not* highly contagious. The natural method of transmission of the disease is most obscure. Although lesions and discharges, including nasal discharges when the nasal tissues are involved, contain the bacilli in large numbers, prolonged, close contact appears to be necessary. Persons under 25 are more likely to contract the disease than are older people. As in tuberculosis, children are very susceptible to infection from leprous parents; however, the incubation period appears to range from a few months to as long as 20 years.

In certain regions of the world—tropical and subtropical Asia and Africa, Polynesia, and South America—Hansen's disease is endemic, that is, always present. It is estimated that there are now at least 11 million cases of Hansen's disease in the world, with about 30,000 in Central and South America and virtually none in Europe. In 1959 there were about 500 known cases in the United States, most residing in the U.S. Public Health Service Hospital at Carville, Louisiana, and possibly 1,000 unknown cases. In this country the disease spreads only in certain restricted areas along the Gulf Coast. In 1976, 145 new cases were reported in the United States; in 1980, 223; and in 1982, 231. An authority has estimated that there are 4,200 cases in the United States today, owing to large-scale immigration from Vietnam, Cuba, Mexico, and the Philippines. Screenings of refugees cannot disclose a disease that may take years to show symptoms.

Two Clinical Types of the Disease. On the basis of laboratory tests. Hansen's disease is divided into two distinct types: the "lepromatous" and the "tuberculoid." The tuberculoid type is more benign and progresses more slowly than the more severe lepromatous involvement.

Hansen's disease is most distinctive in its lepromatous phase. Nodules, gross deformities, and ulcerations occur in the skin, with thickening, discolorations, and wrinkling. Nerves are often affected so that wounds go unnoticed; the individual has no pain sensations, and disfigurations that are revolting and terrifying (to the ignorant and superstitious) occur (Fig. 32–5). It was largely because of this disfigurement that the ancient world so feared lepers. Many disfiguring nonleprous diseases (various fungal and protozoal infections, yaws, and bejel) were doubtless confused with Hansen's disease.

Modern surgery and chemotherapy with sulfones, such as DDS (dapsone) and 4,4′-diacetyldiaminodiphenyl-sulfone (DADDS), a repository or long-acting drug, are working wonders to combat this scourge of the Dark Ages. Rifampin is about four times as effective as DDS; however, it is much more expensive ($600.00/year vs. $2.00/year) and causes flulike side effects. In a country like India, where many "lepers" live, the cost of a drug like rifampin is an obstacle that may well prohibit its use for many if not most patients.

Armadillos are susceptible to Hansen's disease; and a vaccine has been prepared from *Mycobacterium leprae* obtained from these animals. The initial results are encouraging, but it has not yet been tested on humans, and it will be many years before its effectiveness can be evaluated.

---

[7]Recent studies have shown that the nine-banded armadillo can be infected with *M. leprae,* and subsequently infected animals were found in the wild in southern Louisiana.

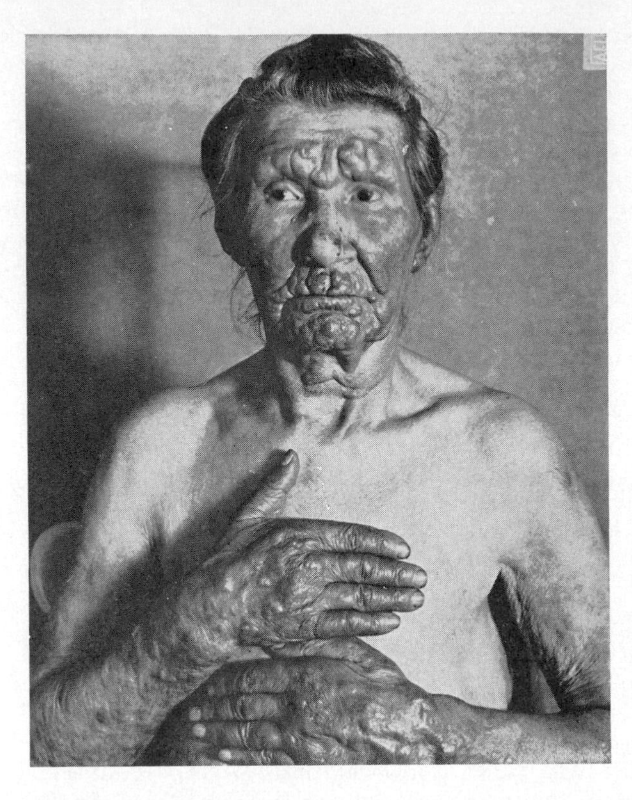

Figure 32–5 An advanced lepromatous case with leonine face, both furrowed and nodulate, and with marked involvement of the forearm and hands, less of the upper arms, and still less of the body. (From Hunter, G. W., III, Swartzwelder, J. C., and Clyde, D. F.: Manual of Tropical Medicine. 5th ed. Philadelphia, W. B. Saunders Company, 1976.)

Figure 32–6 A mass (sulfur granule) of *Actinomyces bovis* from pus in an abscess. This organism causes a disease called lumpy jaw in cows. It grows around the jaw bone in pus and sulfur-like granules. The granule has been crushed and shows the central mass of mycelia and the radially arranged filaments. (Courtesy of Weed, L. A., and Boggenstass, A. D.: *Am. J. Clin. Pathol.*)

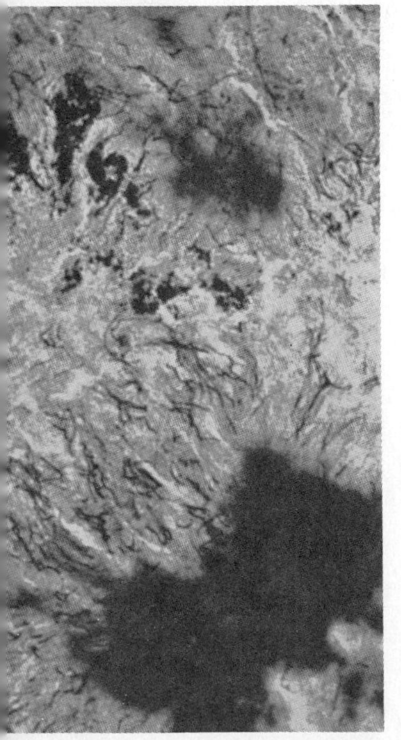

Hansen's Disease and Tuberculosis. There are several striking similarities between Hansen's disease and tuberculosis. The causative agent of each is a species of *Mycobacterium;* both tend to be slowly progressive and chronic in many patients and to be most readily acquired in early childhood; progress of both diseases in patients appears to be related to personal resistance and to factors involving allergy. With respect to allergy, the *lepromin test* is analogous to the tuberculin test; the results of the tests have similar significance. Most significantly, BCG can induce lepromin sensitivity, and there is encouraging evidence that BCG vaccination can be of great value in immunization against this historic specter.

From the standpoint of community health today in the United States, Hansen's disease is not a serious problem. To those concerned with the treatment and control of leprosy, however, it is a dedication for life.

ACTINOMYCOSIS

Actinomyces. This is a genus of bacteria closely related to the order *Mycobacteriaceae* (to which the tubercle bacillus belongs) but classified in the order Actinomycetales, which resemble molds in forming branched, filamentous structures. The organisms are much smaller than true molds, having diameters of 0.5 to 3 μm like other bacteria, and are prokaryotic in structure. No other bacteria normally exhibit branching to the extent that it occurs among the Actinomycetales. The degree of branching varies greatly in different species, being almost negligible in the previously discussed genus *Mycobacterium,* especially in *M. tuberculosis.*

Actinomyces israelii. This organism is part of the normal flora of the human mouth. It may, however, cause disease in man. It is rigidly anaerobic and gram-positive but not acid-fast. Under unusual conditions, such as trauma following the extraction of a carious tooth, it becomes a pathogen when imbedded deep in the tissues. Swellings and abscesses form and are filled with pus. Sinus tracts may drain to the surface of the face. The surrounding tissue becomes hard. Colonies of the organism appear in the pus as yellowish granules (called "sulfur granules" because of their color) ranging up to a millimeter or more in diameter. Each granule consists of a central mass of very fine, branching, tightly interlaced and tangled, moldlike filaments (Fig. 32–6). The tips of the branches project radially at the periphery of the mass, like spines on a sea urchin or a chestnut burr. This radial arrangement of the filaments in the granule gave rise to the name *Actino* (radial) *myces* (moldlike). These radial structures are not formed when the organism is grown in culture. *A. israelii* may also cause sulfur granules in lung abscesses. The reader should note that the condition called actinomycosis, associated with "lumpy jaw" in cattle, is due to *A. bovis* and is not identical to the human disease caused by *A. israelii;* nevertheless, the two conditions and the causative organisms are similar.

Actinomycosis is not contagious, though infections following *human bites* are known. This is, therefore, one of several reasons for keeping saliva out of a deep, penetrating wound. Actinomycosis is considered here not because it is a respiratory disease but because it is transmitted through saliva, as are the true respiratory diseases.

RESPIRATORY FUNGOUS DISEASES

Several important and widespread infections of the respiratory tract are caused by fungi, notably histoplasmosis and coccidioidomycosis. These and some other fungus infections are rarely transmitted from person to person by oronasal secretions, but arise mainly from soil. One important and widely distributed fungous disease that is not transmitted from soil but presumably often from the mouth or intestinal or vaginal tract is *candidiasis* due to *Candida albicans.* All these organisms are discussed with other fungi in Chapter 38.

SUMMARY

The order *Actinomycetales* consists of eight families of branching, moldlike bacteria, including the genera *Nocardia, Streptomyces,* and *Mycobacterium,* to which the species *Mycobacterium tuberculosis* and *Mycobacterium leprae* belong. *M. tuberculosis,* the tubercle bacillus, is an aerobic, nonspore-forming, nonmotile rod that requires the "acid-fast" stain (Ziehl-Neelsen or Kinjoun's acid-fast stain or others) for identification. Only *Mycobacterium* species and *Nocardia* take on this differential stain.

Robert Koch established *M. tuberculosis* as the etiologic agent of tuberculosis. Human and animal species of the pathogen, although slow growing, can be identified in the laboratory on Löwenstein-Jensen, Middlebrook 7H10, or Dubos' medium, and virulence can be determined in guinea pigs. Mycobacterial isolates come from meat, milk, or milk products, from possibly tuberculous cows, seldom from feces or urine but chiefly from patient sputum discharges—often in airborne droplets—dried sputum containing dust, or from harmless acid-fast saprophytes widely distributed in soil, dung, and dust and on surfaces of the human body. The tubercle bacillus is destroyed by pasteurization (72 C for 15 to 30 seconds), but it is quite resistant to disinfection and to drying owing to a protective waxy layer around the organism.

The three stages of human tuberculosis are active, quiescent, and inactive. These depend on the resistance of the patient and the infectivity of the organism. Humans, especially infants, readily become infected with tubercle bacilli but usually are highly resistant to progression of the disease under good living conditions. Tubercles form in the lungs, and these masses of mycobacteria surrounded by connective tissue and calcium become, like vaccinations, protective to the host, unless the defensive system breaks down, causing the infection to develop into an active case. Although about 90 per cent of all tuberculosis infections involve the lungs (as seen in x-rays), the CNS, skin, bones, joints, kidneys, bladder, genital organs, intestines, liver, middle ear, eye, abdomen, and peritoneum can also be infected. Tuberculosis tends to be a chronic disease. Allergy, as demonstrated by Koch's phenomenon, is probably the chief defense of normal adults against repeated reinfection with tubercle bacillus.

The tuberculin test consists of intracutaneous injections of antigenic extracts of tubercle bacilli into the skin. The test is used mostly to detect early primary tuberculosis in children. Methods used are the Heaf test, the Mantoux test, the Vollmer test, also known as the *patch test,* and others.

Tuberculosis is closely connected with social and economic conditions of populations. Most people have been slightly infected with the tubercle bacillus, and the lung lesions (tubercles) may be found at autopsy. Generally, better living conditions and "education" hold tuberculosis in check in "developed" countries. The constant danger of active infection lurks for the undernourished, alcoholics, drug addicts, and people with debilitating diseases of all types. BCG, a vaccine against tuberculosis, has been given to over 226 million children by UNICEF and to high-risk populations where its use seems indicated.

The disease "leprosy" has a historical stigma attached to it that has been completely repudiated by modern research. This relatively *noninfectious* disease should therefore be called Hansen's disease, after the man who discovered *Mycobacterium leprae* in lesions of the disease in 1873. However, it was not until recently that this organism was established as the etiologic agent of lepromatous lesions in footpads of mice and in the armadillo.

The disease consists of two clinical types—the more benign tuberculoid form and the more severe lepromatous involvement. The lesions involve the "cooler" tissues of the body and develop after a long incubation period (from a few months to over 20 years). Over 11 million people are currently afflicted with Hansen's disease, which is endemic in tropical and subtropical regions of the world, including a possible 1,500 cases in the United States. Modern surgery and chemotherapy are working wonders to combat this scourge of the Dark Ages.

REVIEW QUESTIONS

1. What important families belong to the order *Actinomycetales*?
2. What are the special characteristics of the mycobacteria? How is an acid-fast stain done? How are mycobacteria cultivated? Why is it important to differentiate between saprophytic and pathogenic members of the genus *Mycobacterium?* What kinds of specimens are used when looking for the tubercle bacillus in humans?
3. What are the facts known about the resistance of tubercle bacilli? How many varieties of tubercle bacilli are there? Are they all of equal importance to man? Discuss.
4. What are tubercles? Describe them.
5. How may tuberculosis be transmitted?
6. How are cases of tuberculosis classified? What methods are used in the diagnosis of tuberculosis? What is the relationship between tuberculous infection and clinical tuberculosis?
7. What is the importance of tuberculous infections in childhood?
8. Discuss allergy in tuberculosis. What is the tuberculin test? What methods are used to perform the tuberculin test?
9. How may tuberculosis be prevented? What is the relationship between tuberculosis and social conditions?
10. Outline the program that you should follow in order to safeguard yourself against tuberculosis.
11. What precautions would you set up for a patient who has pulmonary tuberculosis in order to protect yourself and others? What would you teach the patient about protection of others?

12. What is BCG? How effective is it?
13. Name three "historic diseases."
14. Why should leprosy be called Hansen's disease?
15. How may Hansen's disease be transmitted? How contagious is it? What are its symptoms?
16. What are the two distinct types of Hansen's disease?
17. Why is it that rifampin cannot be given to everyone who needs it?
18. When and in what tissues does *Actinomyces israelii* cause infections?

Supplementary Reading

Almeida, J. O., and Bechelli, L. M.: Immunological problems in leprosy research. *Bull.* WHO, *43*:870, 1970.

Alvarez, W. C.: Tuberculosis often a geriatric disease. *Geriatrics 26*:82, 1971.

Becker, B.: Leprosy vaccines. *J.A.M.A., 216*:1038, 1971.

Binford, C. H., Meyers, W. M., and Walsh, G. P.: Leprospy. *J.A.M.A., 247*:2283, 1982.

Bullock, W. E.: Studies of immune mechanisms in leprosy. *N. Engl. J. Med., 278*:298, 1968.

Chapman, J. S.: The Atypical Mycobacteria and Human Mycobacteriosis. New York, Plenum Medical Book Co., 1977.

Collins, F. M.: The immunology of tuberculosis. Am. Rev. Respir. Dis., *125*:42, 1982.

Gruft, H., Gaafar, H. A., and Kaufmann, W.: Identification of mycobacteria—What constitutes adequate examination? *Am. J. Public Health, 60*:2055, 1970.

Lowell, A. M.: Tuberculosis in the World. Atlanta, U.S. Dept. of Health, Education, and Welfare, Public Health Service. Center for Disease Control, 1976.

Murohashi, T., and Yoshida, K.: Cultivation of *Mycobacterium leprae* in cell-free semi-synthetic soft agar media. *Jpn. J. Bacteriol., 24*:202, 1969.

Newman, R., Doster, B., Murray, F. J., and Ferebee, S.: Rifampin in initial treatment of pulmonary tuberculosis. A U.S. Public Health Service tuberculosis therapy trial. *Amer. Rev. Resp. Dis. 103*:461, 1971.

Noussitou, F. M. et al.: Leprosy in Children. Geneva, World Health Organization, 1976.

Runyon, E. H., Kubica, G. P., Morse, W. C., Smith, C. R., and Wayne, L. G.: Mycobacterium. *In* Lennette, E. H., et al. (eds.): Manual of Clinical Microbiology. Washington, D.C., American Society for Microbiology, 1980.

Russell, D. A., Shepard, C. C., McRae, D. H., Scott, G. C., and Vincin, D. R.: Treatment with 4,4′-diacetyldiaminodiphenylsulfone (DADDS) of leprosy patients in the Karimui, New Guinea. *Amer. J. Trop. Med. Hyg., 20*:495, 1971.

Silcox, V. A., and David, H. L.: Differential identification of *Mycobacterium kansasii* and *Mycobacterium marinum. Appl. Microb., 21*:327, 1971.

Smith, D. T.: Which children in the United States should receive BCG vaccination? *Clin. Pediat., 9*:632, 1970.

UNICEF News: The fight against leprosy. *52*:12, 1968.

Wayne, L. G.: Microbiology of tubercle bacilli. *Am. Rev. Respir. Dis., 125*:31, 1982.

Weg, J. G.: Tuberculosis and the generation gap. *Amer. J. Nurs., 71*:495, 1971.

WHO Chronicle: Therapy on leprosy. *24*:374, 1970.

WHO Chronicle: Leprosy: Progress and problems. *25*:178, 1971.

Young, W. D., Jr., Maslansky, A., Lefar, M. S., and Kronish, D. P.: Development of a paper strip test for detection of niacin produced by mycobacteria. *Appl. Microb., 20*:939, 1970.

RESPIRATORY VIRAL, MYCOPLASMAL, AND CHLAMYDIAL INFECTIONS

VIRAL INFECTIONS

In considering viruses the reader will no doubt wonder at the wide assortment of clinical entities that are associated with a given group of viruses, such as the "respiratory viruses," and the occurrence of any given virus in more than one category of clinical entities. Poliomyelitis is but one example of this problem of viral classification. This agent has been described as a *neurotropic* virus because of its frequent and striking affinity for nervous tissue; however, the presence of the virus in secretions of the respiratory tract warrants its mention in this chapter. Yet, the initial site of poliovirus multiplication is the alimentary tract, and therefore the poliovirus is now classified as one of the intestinal viruses or *enteroviruses.* It is most widely and frequently disseminated in feces (Chapter 27).

The classification of viruses and viral diseases changes frequently, primarily because of our rapidly increasing knowledge of the chemical, physical, and biological properties of viruses.

SMALLPOX (VARIOLA)

It is hard to realize today that smallpox was once one of the most prevalent and dreaded diseases in the world. Before the days of vaccination, 95 persons of every 100 contracted it, and about one fourth of those died. Many who recovered were blinded or disfigured. Smallpox is readily transmitted by contact with patients or their fomites, as is measles, and both rank among the most contagious of human diseases. In both diseases the causative virus is present in oral and nasal secretions. In smallpox it is also found in the pox fluid and in the scales and crusts that form later. It is quite resistant to drying and may persist for some days in scales, bedding, clothing, and dust. The portal of entry of the variola virus is the mucous membrane of the upper respiratory tract.

Smallpox has been widespread in China and other Eastern countries from antiquity, and has repeatedly swept over Europe in great epidemics. About 1,500 deaths due to smallpox were reported in Bangladesh in early 1975, and in the first five months of 1977 more than 800 cases were detected in Mogadishu the capital of Somalia.

On May 8, 1980 the World Health Organization of the United Nations announced that smallpox had been effectively eliminated all over the earth. Many laboratories have destroyed their stocks of vaccine virus, on the assumption that they will no longer be needed.

Two clinically and epidemiologically distinct forms of smallpox have been recognized: (1) a mild form with a less than 1 per cent mortality rate termed *variola minor* or *alastrim,* and (2) the classical, severe form, *variola major,* which resulted in an overall mortality of 15 to 30 per cent. A third form, *variola intermediate,* has also been recognized. Some 40 related viruses are known: goatpox, sheep-pox, monkeypox (which can also infect humans and may be prevented by cowpox vaccination), pigeonpox, camelpox, and others.

Figure 33–1 shows the number of smallpox cases that occurred in Bombay,

CHAPTER 33

India, between 1966 and 1967. There were 840 cases and 346 deaths. One person who left Bombay via air on February 21, 1967 imported smallpox into Regensburg, West Germany; another similarly brought the disease to Prague, Czechoslovakia, on March 4. Others transmitted smallpox still later from Bombay to Hanover, West Germany, and elsewhere. The imported cases did not cause epidemics because the local populations were largely protected by vaccination.

Until 1971 everyone entering the United States from foreign countries had to show proof of recent vaccination or be vaccinated. This protective regulation has been relaxed by governmental agencies. Let us hope we shall not regret it. Measures now applicable to international travelers are specified in "Health Information for International Travel," available from the National Center for Disease Control, Atlanta, Georgia, 30333.

Immunity to Smallpox. Smallpox was the first disease to which immunity was intentionally obtained by artificial means, and also, if the experts are right, the first disease to be eliminated from this planet.

Modern Method of Vaccination. The principles and techniques of smallpox vaccination apply equally to vaccinations in general. The skin is washed with soap and water and then with alcohol, which is allowed to evaporate. Disinfectants such as iodine should not be used on the skin as they will inactivate the virus in the vaccine. The fluid or lymph containing the virus is expelled from the tube onto the skin, and, with a sterile needle, two shallow punctures or slight scratches are made through the drop of virus and into the vascular layer of the skin. Vaccination is a small surgical operation and should be done only by a person who understands surgical cleanliness. The wound may become infected like any other wound. Occasional complications of vaccination are often due to infection.

VACCINIA. After vaccination of a previously unvaccinated person there is an incubation period of three or four days, during which there are no noticeable changes. Then, if the vaccination "takes," the mild, immunizing infection called *vaccinia* occurs. An eruption appears at the place of inoculation and goes through a series of characteristic stages during a period of about two weeks (papule, vesicle, pustule, scab, and finally healing), leaving a distinctive scar. About the seventh day there is often a reaction on the part of the whole body, lasting several days and often shown by fever, loss of appetite, general discomfort, and headache. During this time specific antibodies are formed in the body.

REACTIONS IN IMMUNE PERSONS. (See Fig. 21–2.) If the vaccinated person is completely immune, a small red papule, sometimes with a tiny vesicle or blister, appears within 10 to 72 hours and heals within five to seven days. This is spoken of as an *immune* (or *immediate) reaction.* If the person is only partly immune, what looks like a small and rapidly evolving "take" with pustule and scab develops, beginning within one to five days and lasting a week to 10 days. This is an *accelerated* or *vaccinoid reaction* or *reaction of partial immunity.* If no reaction of any kind occurs, it is likely that the vaccine virus was inactive, the person vaccinated was not allergic to it, disinfectants on the skin inactivated the virus, or some other error occurred, and the vaccination should be repeated.

Protection Given by Vaccination. Full protection lasts for up to about 7 years but is almost completely gone in 10 years. In areas where smallpox was

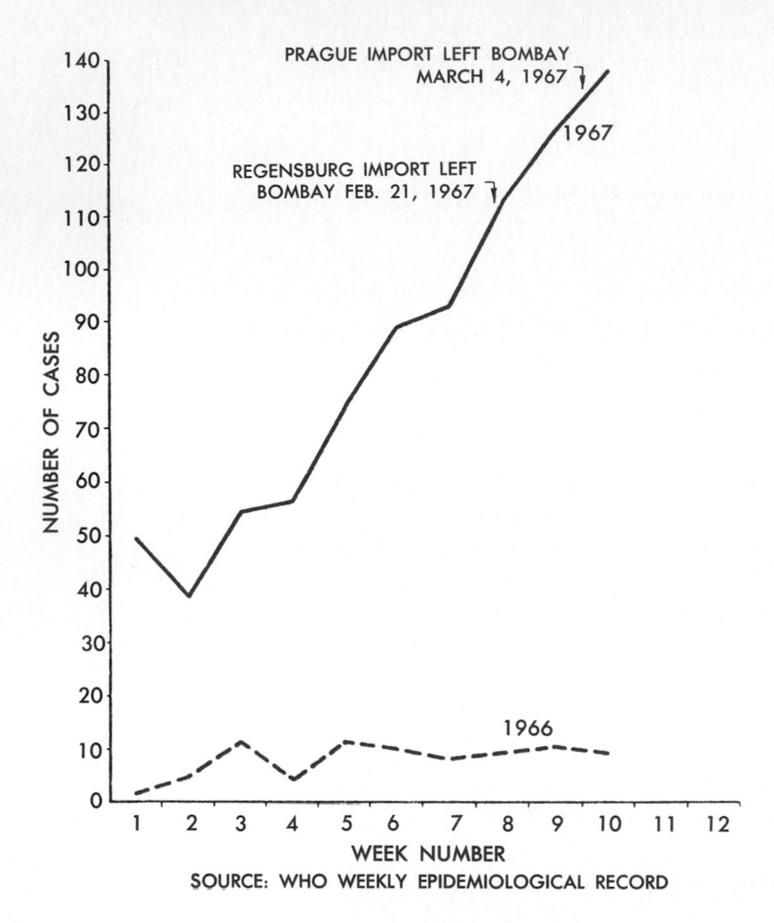

Figure 33–1 Smallpox cases by week of report—Bombay, India, 1966 and 1967. One import was transported to Prague, another to Regensburg (see text). (From *Morbidity and Mortality,* Vol. *16*, 1967.)

endemic doctors used to recommend that every child be vaccinated between the ages of six months and two years and again when about 12 years old (Fig. 33–2). Applicants to most schools of medicine, nursing, or other health professions were required to have a recent vaccination against smallpox before being accepted. Compulsory vaccination in the United States was waived in the case of certain skin disorders, pregnancy, altered immune states due to immunosuppressive drugs (steroids, antimetabolites, and so on), radiation therapy and any other alterations of normal physical conditions.

Compulsory Vaccination. During a sudden epidemic of smallpox in Tabriz, Iran, in 1957, 123 cases of smallpox and 17 deaths had occurred by the time the outbreak was recognized, and the disease was spreading rapidly when the epidemic was reported. By prompt action of the authorities, 92 per cent of the population were vaccinated in three weeks, and the epidemic was promptly stopped. Thus, smallpox can be absolutely prevented by the organized, compulsory vaccination of a whole population. There are many other illustrations of this.

Obviously, it was far less dangerous to have a case of smallpox accidentally introduced into an immune community than into a susceptible one. In addition to long-range preventive measures, all contacts of known cases were vaccinated as soon as possible after the diagnosis was made. A great deal of detective work often had to be done to identify these contacts, particularly if the patient traveled extensively.

MOLLUSCUM CONTAGIOSUM

The *Molluscum contagiosum* poxvirus resembles the smallpox virus. It causes small nodular tumors on the body but not on the palms of the hands or the soles of the feet. Boys contract this infectious disease more often than girls. It may last from two weeks to eighteen months. The virus has not been propagated in tissue culture and a vaccine has not been prepared, nor has antiviral therapy been developed.

MEASLES (RUBEOLA)

Measles is one of the most common and perhaps the most contagious of human diseases caused by viral agents. True measles, also known as rubeola, may be confused by the layman with rubella (German measles, "three-day" measles), or with roseola infantum,[1] which is another related disease. To resolve this confusion (or to compound it even more), a suggestion has been made to use the Latin term morbilli instead of rubeola for true measles, also called the "red measles."

A person who has not had measles, exposed at any age, is almost certain to contract the disease. The reason that rubeola is chiefly confined to children is that most persons have had it before reaching adult life. When rubeola was a common disease, its clinical features, especially the rash, were well recognized by physicians and parents. Since the incidence of measles has been remarkably decreased by vaccination (Fig. 33–3), it is sometimes important to confirm the diagnosis with laboratory tests.

In the United States about 300,000 cases of measles were reported in 1964, 75,290 in 1971, and 41,126 in 1976. In 1977 the number of reported cases increased to 57,345, partly owing to negligence about vaccination. In 1980 it was down to 13,506 or 5.96 per 100,000 population. A national effort was undertaken to eliminate indigenous measles from the United States by October 1, 1982, and only 1,697 cases were reported in 1982.

Transmission of Measles. The measles virus does not appear to survive long outside the body. It is presumably transmitted by the same means as smallpox. A very important fact is that the organism is present in the nasal secretion and saliva before the rash breaks out and for about five days after the rash appears. The first symptoms of the disease resemble those of an ordinary cold, and the rash does not appear until three or four days later. A number of other diseases may begin in the same way—for example, whooping cough and scarlet fever. Rubeola may be easily diagnosed during the pre-skin eruption period by a fever of 104 to 105 F and *Koplik's spots,* which appear on

[1]Rubeola is also known as epidemic roseola, not to be confused with roseola infantum.

Figure 33–2 Children of one family who were brought to the Municipal Hospital of Philadelphia with the mother and father, who had smallpox. The child in the center had been considered too young to be vaccinated. The other children had been vaccinated a year before; they remained free from the disease, although for several weeks they lived in smallpox wards. (From Schamberg, J. F., and Kolmer, J. A.: Acute Infectious Disease. 2nd ed. Philadelphia, Lea & Febiger, 1928.)

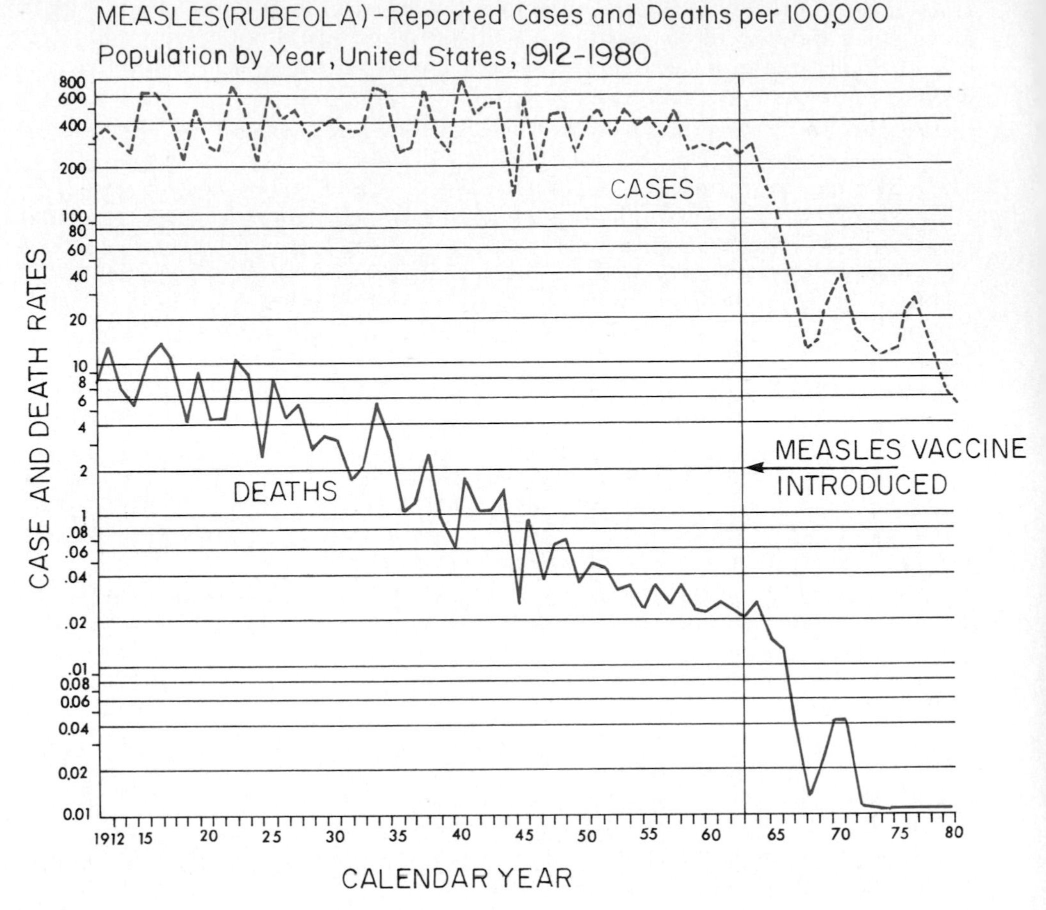

Figure 33–3 Measles incidence in United States, 1912–1980. (Adapted from *Morbidity and Mortality*; Annual Supplement, 1970.)

the mucous membranes in the mouth. These spots are bright red but have bluish white center specks. As in mumps, urine is infectious.

Measles is most easily transmitted during this early stage and during the rash; there is little danger to the patient's spreading the disease after the temperature has returned to normal.

Complications Attending Measles. Measles itself, although it may make a patient very ill, seldom causes death. Complications, however, may be very serious or even fatal unless properly treated. The most common of these are bronchopneumonia and abscess in the ear, caused by pneumococci or streptococci. Measles is sometimes regarded lightly, but it is potentially a serious disease because of the infections that often accompany or follow it. As with many other infections, the younger the child, the more serious the disease. Less than half of the cases but about 75 per cent of the deaths occur among children five years and under. The death rate in the tropics is 1.5 to 7.0 per cent, much higher than in temperate areas. Since the introduction of antibiotics, the death rate from bacterial complications in measles has been reduced to a very low level.

One disease in which the measles virus evidently survives in the brain for many years is subacute sclerosing panencephalitis, called SSPE. This serious disease is fortunately very rare. It may be classed with the "slow" viral diseases due to CHINA (*CH*ronic, *I*nfectious, *N*europathic *A*gents) viruses.

Prevention of Measles. The best means of prevention is by vaccination with measles vaccine. However, when a case of measles appears among unvaccinated children, the greatest hope of limiting the spread of the disease lies in isolating all persons having symptoms of a cold or known to have been in contact with the patient while he had the early or later symptoms. A great deal can be done to decrease the number of deaths due to the disease by preventing complications through good nursing, antibiotics, and protecting very young children from infection. Since children are now being immunized by vaccination against measles, just as they were against smallpox, measles, too, may very soon be one of the "conquered" diseases.

In the hospital, complete isolation of measles cases is desirable. In an open ward it is practically impossible to prevent the spread of measles, even with the most painstaking technique. Rigid terminal disinfection is not necessary for the short-term measles virus, but streptococci that often accompany them are vigorous and long-lived. It is best, therefore, to disinfect the bedding and clothing and to give the room a thorough cleaning with soap and water and disinfectant.

Gamma Globulin. Until recently, gamma globulin (obtained from the serum of measles-immune persons) was commonly used to protect children under four years of age when exposed to the disease. This protection, like all passive immunity, is temporary but is valuable for young children, in whom complications are apt to be severe. A very specific preparation of human gamma globulins, known as "hyperimmune gamma globulins" (HIGG), is administered in preferred cases. Human gamma globulins are also available under the name of "immune serum globulins" (ISG) for protection (not cures!) against hepatitis virus A, poliomyelitis, and chickenpox (varicella) and as an aid to chemotherapy in some other infectious diseases. To avert measles, the gamma globulin must be given within three days after exposure. If the material is injected in small doses between the fourth and sixth days after exposure, it does not prevent the disease but modifies its course, making it very mild. This mild attack, however, probably gives permanent protection. Children under nine months of age contain maternal antibodies if the mother has had measles. In spite of the maternal antibodies, gamma globulin may be administered as a precautionary measure. Vaccination against measles makes this practice unnecessary except in special situations (see below).

Active Immunization. The adaptation of the Edmonston strain of measles virus (named for the patient from whom this strain was isolated) to grow in chick-embryo tissue cultures decreases its pathogenicity for humans while retaining its capacity to stimulate the production of specific antibodies. This virus, developed by Dr. John F. Enders and his coworkers, provides a vaccine analogous to the Sabin oral active-virus polio vaccine. The active measles vaccine differs from the Sabin active polio vaccine in that the measles vaccine must be injected instead of being given orally, and the infection produced by it is not transmissible. One dose of measles vaccine (0.2 to 0.25 ml in a subcutaneous injection) may be given to susceptible young persons and children from 15 months up to 12 years of age, but to those most likely to be exposed and adversely affected by measles it may be given as early as 6 months after birth. Since use of the Edmonston vaccine sometimes produces fever and rash characteristic of measles and may require the simultaneous injection of measles-immune gamma globulin, it is usually preferable to use a "further attenuated measles vaccine," such as one prepared with the Schwartz strain of virus, which does not necessitate the administration of gamma globulin. The immunity induced by the vaccines appears to be durable, prompt, and very effective.

Vaccines prepared with inactivated (killed) measles virus are not used because they are less effective and are associated with particularly severe cases in those later contracting the disease. Normal infants and children are now frequently vaccinated at 15 months of age with a combined measles-mumps-rubella vaccine.

Contraindications to the Use of Active Virus Vaccines. Pregnant women, persons with debilitating diseases, and patients on drugs that reduce immune responses or adversely affect bone marrow should *never* be vaccinated with *any* active virus vaccine.

GERMAN MEASLES (RUBELLA)

This viral disease is fairly common, but in itself it is rarely serious. It may be called the "three-day measles" because, although the duration of the disease is from one to five days, the rash usually evolves over a three-day period. Rubella is probably transmitted in much the same way as measles. The incubation period is about ten days to three weeks. It not infrequently affects adults. When it occurs in an expectant mother during the first three months of pregnancy, it often causes severe damage to the fetus since it appears to have a special affinity for embryonic tissues. Deformities or defects are particularly likely to occur in the heart, ears, brain, and eyes.

Because an attack of German measles[2] in childhood is rarely harmful and probably confers lifelong immunity, it has been recommended by some that girls be purposely exposed to the disease between the ages of six and eight years in order to avoid contracting it during pregnancy later in life. Vaccination of all children is now the preferred procedure.

Passive immunization of nonimmune women during the first three months of pregnancy to protect the unborn child has been recommended. Gamma globulin (antibodies) from the blood of immune persons is used. The results are variable.

Rubella Vaccine. An extremely important advance toward the control and elimination of rubella was made in 1962 when the virus of rubella was cultivated in tissue culture. A single dose of active attenuated rubella virus vaccine (RA 27/3, HPV 77, or Cendehill) was found to protect between 90 and 95 per cent of susceptible children when exposed to the German measles. In the United States vaccination of all children between 15 months of age and puberty will assure protection against rubella in pregnant women. This will eliminate the need for therapeutic abortions to avoid *congenital rubella* in some 20 to 25 per cent of infants born to women who had rubella during the first trimester of pregnancy. *Under no circumstances* should a pregnant female be vaccinated with *any* vaccine containing *any* active, though attenuated, virus. Women of childbearing age should not be vaccinated against rubella unless shown to be susceptible and unless they promise to prevent pregnancy for at least two months following vaccination.

EXANTHEMA SUBITUM (ROSEOLA INFANTUM)

This acute viral disease is often confused with rubella because it, too, sometimes causes a three-day rash. Roseola typically occurs in infants about 12 months old. The high temperature and the sudden onset are frequently accompanied by convulsions. The fever remains high for three or four days. After the fever, the rash appears and lasts for one to two days. Allergies may confuse the diagnosis. Many infections remain subclinical.

CHICKENPOX (VARICELLA)

The herpesvirus causing this annoying malady is contained in the early-stage "pock" or eruption and is probably transmissible by this; however, the respiratory secretions seem to be the most important vector. Chickenpox is extremely contagious. The disease is usually mild, although severe and (rarely) fatal cases occur, more often in adults than in children. Pustules develop in successive lots. Complications are usually not serious.

---

[2]This disease was first described in Germany as "*rötheln,*" which means "like roses."

In otherwise healthy children, chickenpox is almost always benign and self-limiting, although complications with bacterial infections do occur. An experimental live-virus vaccine has been used safely and effectively in Japan but will require extensive evaluation before licensure for use in nonselected populations. In 1980, 190,894 cases of varicella were reported in the United States.

HERPES ZOSTER ("SHINGLES")

This painful disease is due to a herpesvirus identical to the chickenpox virus and therefore called the varicella zoster (VZ) virus. The dermal lesions are often very similar. Patients with herpes zoster are commonly over 40 and usually have a history of clinical childhood varicella.

Herpes zoster affects the dorsal nerve roots and areas of skin to which the corresponding sensory nerves extend. There seems to be a tendency for the skin lesions to occur where there is continuous irritation, such as caused by restricting clothing. It is usually, but not always, confined to one side. The pain and lesions may occur in successive patches and crops, as in chickenpox, anywhere from the neck to the lumbar region, or even the legs and arms. Itching and pain are often intense. Brownish spots may remain after healing. The disease is rarely fatal but can cause much intense discomfort.

The mode of transmission appears to be like that of measles and chickenpox. An adult with herpes zoster can infect children, who usually develop chickenpox. Children with chickenpox can infect adults, who may develop herpes zoster. The two occasionally occur together in children. Zoster-immune gamma globulins have been reported to prevent the disease.

HERPES SIMPLEX (FEVER BLISTERS)

The herpes simplex virus, *Herpesvirus hominis* or HSV-1, causes local eruption of "fever blisters" or "cold sores" around the mouth, face, and mucous membranes. Because of the similarity of the name, the disease may be confused with the previously described herpes zoster. Clinical conditions more severe than fever blisters have been attributed to the virus of herpes simplex, such as conjunctivitis, invasions of the central nervous system, asymptomatic conditions in the neonatal infant, and, as will be discussed in Chapter 36, acute vulvovaginitis (venereal herpes caused by HSV-2). It is then transmissible by coitus and other forms of sexual contact. For a discussion of infectious mononucleosis caused by the Epstein-Barr herpesvirus (EBV_1) and malignancies due to herpesviruses, the reader is referred to pages 90 to 94.

The herpes simplex virus is continually present as a latent infection in the sensory nerves of skin tissues of more than 1 per cent of the adult population. Eruptions result from extreme changes in temperature (either hot or cold), trauma, or fever. Exposure to ultraviolet irradiation, menstruation, or emotional stress also may cause fever blisters to form. An ointment called Zovirax gives relief, but it is not a cure.

CYTOMEGALOVIRUS INFECTIONS

Cytomegaloviruses (CMV) are also known as salivary gland viruses. They are herpesviruses and are infectious to humans and other mammals. Usually CMV causes congenital infections of the newborn, often involving the lungs, liver, blood-forming organs, and brain, resulting in mental retardation and disability of motor functions. These congenital anomalies occur in about 10,000 infants in the United States every year. In older patients CMV infection may manifest itself as hemolytic anemia or resemble mononucleosis.

CMV may be transmitted in whole blood transfusions and organ transplants; it may also be acquired through sexual intercourse. The virus has been isolated from semen and from cervical secretions.

MUMPS (CONTAGIOUS PAROTITIS[3])

This disease (also called parotitis epidemica), due to one of the paramyxoviruses, is characterized by sudden onset, with swelling and pain in the salivary glands, usually only the parotids; however, the sublingual and submaxillary glands and numerous other glandular tissues may also be involved. In many cases of mumps these glandular involvements do not occur or are so slight as to pass unnoticed. Thus mild, unrecognized cases of mumps (i.e., inapparent infections) occur, which actively immunize the patient but also serve to spread the disease. In 1982, 5,196 cases were reported.

The virus may also seriously affect the ovaries or, more commonly, the testes in sexually mature persons, resulting in sterilization. It not infrequently invades the central nervous system, causing meningoencephalitis, which may result in permanent damage to the brain.

Skin Sensitivity. As in many other viral infections, a degree of skin sensitivity follows infection with the mumps virus. This sensitivity may be detected by inoculating persons previously infected by mumps virus with an extremely minute quantity of the heat-inactivated virus and then observing the wide zone of erythema resulting at the site of inoculation within 20 to 24 hours. Unfortunately, the dermal test cannot be depended on as an indicator of immune status.

Transmission. The virus is transmitted in saliva and by objects recently contaminated with saliva. The virus may be present in the saliva from seven days before the onset of symptoms until nine days after glandular swelling subsides.

Mumps virus has also been repeatedly demonstrated in the urine of mumps patients within two weeks from onset and often longer. This necessarily alters the older concept of mumps as a disease requiring only those preventive precautions applicable to respiratory secretion-borne pathogens. Urine of mumps patients, as well as their respiratory secretions, must be dealt with as infectious. It may be that other viruses infecting the respiratory tract also appear in the urine.

Prevention

Persons of any age with mumps should be isolated, as should any patient with communicable respiratory disease. There is no effective specific treatment. Patients virtually always recover promptly and without serious complications, except when the ovaries, testes, or central nervous system is involved. Hyperimmune mumps gamma globulin has been tested, but it may even increase the incidence of the disease. Vaccines consisting of attenuated or inactivated viruses are available and may be included in routine immunization programs, with a revaccination schedule required. The attenuated "live" virus vaccine is preferred when not contraindicated. It may be given simultaneously with measles, rubella, and trivalent oral poliomyelitis vaccines.

[3]Avoid the error of the red-faced student who confused parotitis with parrot fever!

RABIES (HYDROPHOBIA)

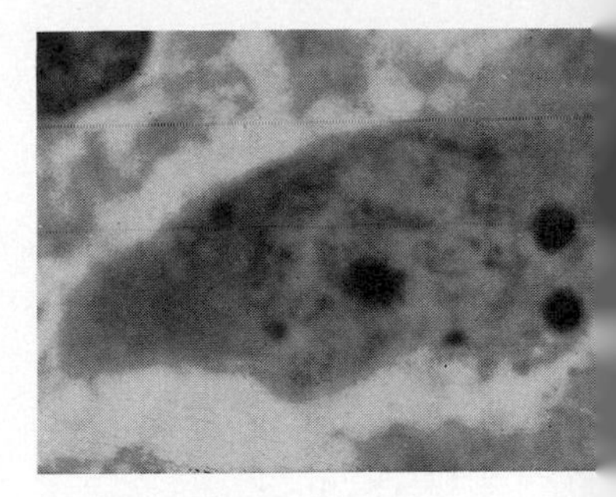

Figure 33–4 Negri bodies in a nerve cell from a section of the brain of a rabid cat. The three dark, spherical bodies within the cytoplasm of the cell are Negri bodies. A mixture of dyes like Seller's stain has been applied to improve visibility. (From Schleifstein.)

Although caused by a distinctively neurotropic rhabdovirus and apparently not a disease of the respiratory tract, rabies is transmitted, like mumps, with respiratory secretions in saliva. Humans can also be infected by way of the respiratory tract by dusty guano in bat caves. Hence, for convenience, it is included in this section.

Rabies is primarily a disease of mammals (an encephalitis) and is transmitted to humans usually as a consequence of the saliva of rabid animals gaining entrance to wounds caused by bites or scratches. The disease may be transmitted to and by a great many species of biting mammals: dogs, cats, skunks (most common), rodents of all sorts, foxes, horses, cattle, swine, and so on. Vampire and other species of bats are important vectors in certain parts of Europe, South America, and the United States.

Rabiesvirus particularly affects the central nervous system, to which it is conveyed from the wound chiefly by the path of the nerve trunks. The disease in humans is terrible: the patient, usually with clear consciousness, passes through periods of depression and excitement, attended by painful spasms, to a stage of paralysis and death. The name hydrophobia means fear of water. Human beings, as well as other mammals with the disease, have this symptom. The muscles involved in swallowing develop such painful spasms upon attempts to swallow food or drink that the patient comes to fear even the sight or suggestion of them. Once the symptoms appear, death is almost inevitable. No specific treatment for this disease is known, but strict supportive measures, such as control of intracranial pressures, have prolonged survival of several patients and may even have cured some reported cases.

Diagnosis of Rabies. The brain cells of mammals killed by rabies contain certain cytoplasmic inclusions called *Negri bodies* after the Italian investigator who discovered them in 1903 (Fig. 33–4). These inclusions contain viral particles and rabies antigens. The examination of the brains of rabies-suspected animals for the presence of Negri bodies is an extremely important method of diagnosis, and it takes only an hour or so to complete the examination. A small bit of the portion of brain called the hippocampus major, or Ammon's horn, is smeared thinly on a slide, stained (preferably with Seller's stain[4]), and examined for the characteristic intracytoplasmic inclusion bodies.

Failure to find Negri bodies does not necessarily exclude the possibility of rabies. If there is any doubt, injection of the brain into other animals is highly advisable because mice, rabbits, and guinea pigs are very susceptible to the disease. Inoculation of the brains of suspect animals into the brains of such rodents is a much more reliable means of diagnosis than microscopic examination, but it requires several days or weeks for observation of the test animals. The fluorescent rabies antibody (FRA) test makes a prompt diagnosis possible. It is preferable to the more lengthy animal inoculation tests, which are not always successful.

Immunity to Rabies

Avianized Rabies Vaccine. To avoid the allergic reactions that too often follow repeated injections of Semple-type rabies vaccines (rabbit nervous tissues), an attenuated virus (the Flury strain) is cultivated repeatedly in successive live embryonated avian eggs, commonly duck eggs. After the contents of the eggs (containing relatively large amounts of virus) have been collected, the virus is inactivated (for human use) by various means—e.g., beta-propiolactone, phenol, ultraviolet radiation—and finally is dehydrated.

For persons frequently exposed to rabies (veterinarians, doghandlers, and so on), an initial series of several immunizing doses of vaccine, followed by

[4]Seller's stain: *a*, 5 ml of basic fuchsin (1 per cent solution in methyl alcohol). *b*, 10 ml of methylene blue (1 per cent solution in methyl alcohol). Mix *a* and *b*; keeps well in tightly stoppered bottle. Stain smear five to ten seconds; wash with water; blot dry. Examine with oil-immersion lens.

booster doses at suitable intervals and immediately after any known exposure, is recommended. For dogs and other animals, one injection of the active but attenuated virus annually or triannually is widely used to protect against rabies. It is very effective. For cats a high-egg-passage (HEP) strain of active Flury virus is used.

The principles and techniques employed in vaccinating against rabies are all based on Pasteur's early work (1881–1886). They are not a therapeutic procedure but an immunizing measure, used after receiving bites from suspect animals, to develop immunity before rabies has time to develop.

Methods in Current Use to Prevent Hydrophobia. To assist the reader in identifying preferred current methods used to produce immunity to hydrophobia, the following outline is presented (see also Table 33–1).

1. *Duck embryo vaccine* (DEV) is an inactivated virus treated with beta-propiolactone. Twenty-three injections are required for its administration. DEV, reconstituted with fluid, is safer and more effective than the Semple-type vaccine.

2. *Human diploid cell rabies vaccine* (HDCV) is an inactivated virus from human diploid cell tissue cultures. Its use requires five injections. Since 1980 Merieux human diploid cell rabies vaccine (M-HDCV) has been distributed in many states for postexposure rabies prophylaxis, replacing DEV. It produces fewer adverse reactions, and a smaller dose yields higher antibody levels, but some systemic allergic reactions have been reported.

3. *Antirabies serum, equine* (ARS) is a concentrated serum from hyperimmunized horses.

4. *Rabies immune globulin, human* (RIG, HRIG) is a hyperimmunized human concentrated serum fraction of gamma globulins. It is now commercially available. In cases of severe bites by known rabid animals, especially bites near the head and on the trunk and hands, the use of HRIG is of great value and can prevent a horrible death when the action of vaccines would be too slow. commercially. When the injection is followed after an interval of several days by a course of vaccine treatment, the protection obtained is excellent. Injection of HRIG simultaneously with vaccine tends to lower the effectiveness of the vaccine but is recommended nevertheless. Note that HRIG must be given immediately after the bite if possible—at the latest, within 72 hours of the bite. Half the dose should be injected below and around the bite(s) area.

The use of immunologic preparations does not preclude the necessity for local treatment of the bites. Prompt, thorough washing with soap or detergent and the cauterizing of deep, narrow puncture wounds are always important in preventing the development of the disease. Tetanus prophylaxis with toxoid is also recommended.

Control of Rabies. Until 1971 it was generally believed that death was inevitable after symptoms of rabies had developed. However, in that year a boy, bitten by a bat believed to be rabid, developed symptoms like those of rabies. The boy was treated with many specific drugs to deal with each symptom as it developed. The boy survived—the first such case in the known history of rabies.

The best way to control rabies is to vaccinate and control all pet dogs and cats, destroy all stray dogs and cats, and require that others be kept from wandering at large. This sounds like a simple program, but at present it is impossible in the United States. In Great Britain it is an accomplished fact. Wild mammals of most species, including several species of bats, can also contract and spread rabies.

Only four cases of hydrophobia in humans were reported in the United States in 1979, and none in 1980 and 1982. During 1980, 6,421 notifiable rabies infections occurred in animals, and in 1982, 6,066.

Practical Procedure. The bite of a rabid animal does not necessarily mean that rabies will develop, since the virus may not be in the saliva at the

Table 33–1. Rabies Immunization Regimens, March 1980**

Pre-Exposure: Pre-exposure rabies prophylaxis for persons with special risks of exposure to rabies, such as animal care and control personnel and selected laboratory workers, consists of immunization with either human diploid cell rabies vaccine (HDCV) or duck embryo vaccine (DEV) according to the following schedule.

Rabies vaccine	*No. of 1-ml doses*	*Route of administration*	*Intervals between doses*	*If no antibody response to primary series, give:**
HDCV	3	intramuscular	1 week between 1st and 2nd; 2-3 weeks between 2nd and 3rd†	1 booster dose†
DEV	3 or 4	subcutaneous	1 month between 1st and 2nd; 6-7 months between 2nd and 3rd† **or** 1 week between 1st, 2nd, and 3rd; 3 months between 3rd and 4th†	2 booster doses,† 1 week apart

Postexposure: Postexposure rabies prophylaxis for persons exposed to rabies consists of the immediate, thorough cleansing of all wounds with soap and water, administration of rabies immune globulin (RIG) or, if RIG is not available, antirabies serum, equine (ARS), and the initiation of either HDCV or DEV, according to the following schedule.‡

Rabies vaccine	*No. of 1-ml doses*	*Route of administration*	*Intervals between doses*	*If no antibody response to primary series, give:**
HDCV	5§	intramuscular	Doses to be given on days 0, 3, 7, 14, and 28†	an additional booster dose†
DEV	23	subcutaneous	21 daily doses followed by a booster on day 31 and another on day 41† **or** 2 daily doses in the first 7 days, followed by 7 daily doses. Then 1 booster on day 24 and another on day 34†	3 doses of HDCV at weekly intervals†

*If no antibody response is documented after the recommended additional booster dose(s), consult the state health department or CDC.

†Serum for rabies antibody testing should be collected 2-3 weeks after the last dose.

‡The postexposure regimen is greatly modified for someone with previously demonstrated rabies antibody.

§The World Health Organization recommends a 6th dose 90 days after the 1st dose.

**From *Morbidity and Mortality Weekly Report, 29*:280, 1980.

time of the bite, or at least not present in effective concentrations. Also, thick clothing worn at the site of the bite may prevent the inoculation of virus into the tissues. It is important, however, that bites of animals suspected of being rabid and bites of unknown "bite and run" animals be immediately and thoroughly cleansed with soap and water. A soft syringe may be used to force soapy water to the depths of the wound. When the disease is still in incipient stages in dogs or cats, the saliva is infectious and may infect if transferred by any means to cuts or scratches on the hands. Bites or infections on hands, arms, or face are more dangerous than those on the legs and feet because the virus is then inoculated closer to the brain. The incubation period of the disease is variable: from two to eight weeks or longer in human beings, possibly as long as 12 months, and from one to eight weeks in dogs. If it seems reasonably probable that the biting animal was rabid, the treatment should be started without delay. However, the very painful Pasteur anti-rabies injections, which were the method used until only a few years ago, should not have been administered out of hysteria and in the absence of real danger of rabies. Even now, injections required are numerous and painful.

Do not kill the animal unless it is diagnosed rabid by a competent veterinarian. If it is caught, hold it (well caged!) for observation for 10 days. Rabid dogs rarely survive 10 days after development of symptoms. If the dog has rabies, it will show unmistakable symptoms before this time and may then be killed. Rabies can then be definitely established by examination of the dog's brain for Negri bodies and by mouse inoculation, or by means of fluorescent antibody staining of brain tissues. If the dog shows absolutely no symptoms at the end of 7 days, it may be released, and the victim may then discontinue the treatment. His mind will be much more at ease to know that the animal is free from rabies than if it had been killed with the suspicion still upon it. If the animal has already been killed, send the head, packed in ice, immediately to the laboratory of the State or City Department of Health or to a competent veterinarian for examination. Wild animals suspected of having rabies should be killed immediately and examined for the disease. Any *unprovoked* bite by a wild or domestic animal should be regarded as *highly* suspicious.

Rabies in Dogs and Cats. It is important to be able to recognize rabies in a dog or cat. The first sign of rabies in a dog is usually an alteration of its disposition: it develops a nervous, furtive manner, becomes restless, and seems afraid. Because of its vague apprehensions it may seek human company and appear unusually affectionate. Later it acts in a definitely abnormal manner, develops a peculiar wolflike bark or howl, and becomes snappish and even more restless, sometimes running away for days at a time, during which it may bite animals or persons. This is "furious rabies." In "dumb rabies" the furious phase is much less evident. In any case, difficulty in eating develops, and drooling ("foaming at the mouth") occurs because swallowing is difficult. The dog may bite any animal or person and often chews sticks and stones and may swallow them. Paralysis and eventually death supervene.

In cats, rabies is usually less difficult to recognize than in dogs because cats do not attempt to retain the affections of their owners in spite of the disease, as do dogs. The early symptoms are much like those in dogs, manifested as irritability and restlessness, but the disease soon develops into the furious stage as a rule, and the cat becomes entirely wild and ferocious, with dilated pupils, a salivating, open mouth, and extended claws. Cats are especially dangerous because of their claws, which become contaminated with saliva, and because cats give allegiance to no one.

Transmission of rabies from one human being to another is very rare. Since the virus is present in the saliva of human patients, precautions must be taken to prevent its entrance into cuts and scratches on the nurse, the physician, or other persons. The use of face masks in caring for humans who have rabies is probably indicated. The disease may also be transmitted by affectionate licking by rabid animals.

ANTERIOR POLIOMYELITIS (INFANTILE PARALYSIS)

Mention has already been made of the dramatic and frequently fatal paralytic effects resulting from invasion of the CNS (central nervous system) by the poliomyelitis virus. (Fortunately paralysis is now very rare thanks to the use of oral trivalent polio vaccines.) Because of this, the disease was formerly regarded as intimately related to nervous tissues. Although this is true, we now know that poliomyelitis is primarily an infection of the gastrointestinal tract (caused by one of the *enteroviruses*) and that, like other enteric infections, it is transmitted partly in and by feces. It is discussed more fully in Chapter 26 with other diseases of the intestinal tract.

In this section on respiratory viruses it suffices to say that poliovirus is also present in the respiratory (pharyngeal) secretions of many patients, probably for several days before onset of symptoms and for some days (up to two weeks?) afterward. The respiratory (pharyngeal) secretions of patients in early stages, and objects contaminated with them, must therefore be treated as highly infectious. Whether feces or respiratory tract secretions are the more important vectors of the virus (a debated question), both are acknowledged to be effective vectors.

Precautions must be based in part on the modes of transmission. With respect to proper handling of respiratory secretions, in poliomyelitis the precautions already described for diseases of the respiratory tract apply fully. With regard to fecal transmission, consult the section on enteric diseases. The same applies to the reoviruses and some echoviruses.

INFLUENZA

In 1918–1919 a terrible epidemic of influenza swept over the entire world, causing more than 10 million deaths; in the United States alone at least 450,000 died. There is relatively little direct evidence of the true nature of this disease. It is often referred to as pandemic influenza. We do not know with certainty whether this epidemic was the same as the infection now called influenza. Serologic studies indicate that it was due, at least partly, to the virus of swine influenza.

Early in 1976 several cases of an influenza-like illness occurred among soldiers at Fort Dix, New Jersey, and one man died. After serologic studies indicated that the disease might be "swine flu," a massive immunization program (later discontinued) was initiated in order to forestall a repetition of the 1918–1919 epidemic (Fig. 33–5).

Transmission. The mode of transmission of influenza is like that of other respiratory infections. Probably crowded indoor (low humidity) conditions, with droplet infection, favor epidemics.

Influenza Virus Types. There are at least three main types of human influenza viruses, all causing epidemics of the same (clinical) disease. The first types to be differentiated were called types A (1933) and B (1940). Later type C was discovered. There are numerous subtypes, which are distinguishable mainly on the basis of antigenic and immunologic properties. A whole set or family of such variants of type A was found in 1957 in the worldwide epidemic commonly called Asian influenza ("Asiatic flu"). These viruses are termed variants of Far East (FE) or Asian influenza (A). Three subtypes of A are now designated as A0, A1, and A2. During the winter of 1968–1969, the Asian influenza virus, "Hong Kong flu" (A2) variant, was widespread in the United States, though fatality rates were relatively low. In 1972 the "London flu epidemic," due to influenza virus A/England/42/72, spread widely and rapidly

Figure 33–5 Swine flue virus being injected into eggs. Three days after the injection the culture is removed and filtered through a bed of salt. The virus is then inactivated, and the culture is tested for sterility and potency and converted into doses. (Courtesy of Wyeth Laboratories, Radnor, Pa.)

all over the world by air travel. Another commonly recognized strain is B/Vic/98926/70, the B Victoria virus. In 1976 and early 1977 isolates of A/Victoria-like virus and Hong Kong virus were reported all over the United States.

The viral envelope of influenza viruses consists of two major proteins, hemagglutinin and neuraminidase. Changes in the structure of these two proteins, especially hemagglutinin, which has four separate antigenic sites, are thought to produce the "new types" of virus encountered with every new major epidemic of influenza. When two different viruses infect the same cell, recombinants are formed, with major changes in hemagglutinin and/or neuraminidase sufficient to create new forms of infectious influenza viruses and thus, possibly, new epidemics. It has been estimated that it takes about 11 years for a new influenza virus to develop with a new protein envelope of changed hemagglutinin and neuraminidase.

Complications. Uncomplicated influenza is usually not fatal; however, it can lead to bronchopneumonia, which is responsible for most of the deaths associated with the disease. The bronchopneumonia may be caused by pneumococci, streptococci, the influenza bacillus, or other organisms, alone or in combinations. Thousands of deaths in army camps in 1918 were due to the bronchopneumonia accompanying influenza. Apparently, the virus of the pandemic of 1918 so lowered the local resistance of the lungs that almost any bacterium could invade the lung tissues and cause pneumonia. Chemotherapy is very effective in controlling bacterial complications of influenza as well as of measles, but it cannot cure or prevent the viral infection in either disease.

Active Immunization. It is possible to prevent influenza by means of vaccines. They are made of virus propagated in chick embryos, purified, and inactivated with formalin (see Figs. 5–3 and 33–5). In preparing influenza vaccine it is important to use a mixture of viruses representing all the antigens of the various groups, since vaccine for only one type offers little or no protection against the others. In 1975 and early 1976 representative commercial vaccines contained A/Port Chalmers, A/Scotland, and B/Hong Kong virus components. The vaccines are purified and concentrated antigen, some from virions disrupted with ether or a surfactant (polysorbate 80). The immunity is short-lived, lasting from two to 12 months, and vaccinated persons may contract influenza unless vaccination is repeated frequently. The infecting virus strains tend to vary continually, so that a vaccine effective today may be ineffective next month.

Treatment

If influenza develops, it is possible in the great majority of cases to avoid the complicating pneumonia by proper care: first, by keeping the patient in bed from onset and during convalescence; and second, by giving protection from the bacteria that are known to cause pneumonia. This is done by means of chemotherapy and by carrying out the same precautions as are prescribed for bronchopneumonia. Care must be taken to keep susceptible persons from contact with the patient. Fomites should be disinfected as in any respiratory disease. Careful washing of the hands and proper wearing of a gauze mask may protect health workers, although opinions differ on the value of masks.

THE "COMMON COLD"

The term "common cold" has been used for many years to designate infectious diseases of the upper respiratory tract characterized by various all-too-familiar signs and symptoms in multiple combinations, including adenopathy, inflammation of the mucous membranes of the nose, pharynx, and bronchial tree, fever, headache, copious watery nasal exudate, muscular pains, sinusitis, lacrimation, conjunctivitis, general malaise, sneezing, and coughing. As previously mentioned, many of these appear also in the early stages of measles and several other diseases: there is nothing very distinctive about them. Probably most common respiratory infections, unless unusually severe, are assumed to be "flu" or a "common cold." If the patient recovers in a few days, these conditions are given no laboratory study and an exact diagnosis is never arrived at. Clinically most of these infections are indistinguishable. We now know that we have to deal with a considerable number of viruses that may be divided among several distinct but related groups.

The "Cold" Viruses. As a result of our increasing knowledge of viruses, great progress has been made in the isolation and identification of agents implicated in the syndrome designated as the common cold or "afebrile respiratory illnesses." Among these agents are 89 or more serotypes and one subtype of *rhinoviruses*[5] (other rhinoviruses are still being found), several strains of coronaviruses, parainfluenza viruses, respiratory syncytial viruses, echoviruses, and even polioviruses. Coldlike illness may also be caused by numerous members of the myxovirus group, including classic influenza and mumps viruses. With all these many causes of the common cold, it is indeed naive to expect to find "a" cure for the variety of infections it represents.

Parainfluenza of Infants and Children. Among the true myxoviruses, the group of paramyxoviruses includes the mumps virus, four serotypes of the parainfluenza viruses (PIVs), and the closely related respiratory syncytial viruses (RSVs), the last belonging to the group of pseudomyxoviruses.

The PIVs and the RSVs produce a disease syndrome in infants and children that is often severe and not infrequently fatal, especially in the absence of maternal antibodies. Respiratory distress such as occurs with the common cold, pharyngitis, laryngitis, bronchitis, and bronchopneumonia, and probably also serious involvements of the liver and central nervous system, may all be due to these viruses. Epidemics of infection due to PIVs occur in late fall, whereas RSVs usually produce epidemics in late winter and early spring. Note that the

[5]*Rhino* is from the Greek for "nose."

parainfluenza viruses, although related to the influenza viruses, are nevertheless quite distinct.

Other causes of acute respiratory infections, some discussed elsewhere in this book, include *Mycoplasma pneumoniae,* the cause of primary atypical pneumonia, and some 33 serotypes and one subtype of adenoviruses.[6]

Although often neglected, infections with respiratory viruses can lead to serious difficulties. The inflammation of the nose and throat that accompanies such infections prepares a hospitable place for pneumococci, streptococci, influenza bacilli, and other organisms to lodge and create more serious disease. Furthermore, if a person who is carrying streptococci, pneumococci, diphtheria bacilli, or other pathogenic organisms in his throat has a respiratory virus infection, his sneezing and coughing will distribute not only the virus but the other organisms as well.

Transmission of Respiratory Viruses. Nasal discharges and saliva carry the organisms, which are spread by coughing, sneezing, kissing, carelessness about handkerchiefs (disposable or otherwise), hands soiled with nasal secretion, imperfectly washed dishes and eating utensils, and the use of common towels and drinking cups. These are methods of transmission common to all infectious diseases of the respiratory tract. The health worker may therefore disregard the virologic complexities of serologic grouping and nomenclature, since all of the respiratory viruses and bacteria are transmitted and controlled in the same way (see Chapter 29).

PNEUMONIA CAUSED BY MYCOPLASMA (PLEUROPNEUMONIA-LIKE ORGANISMS)

Mycoplasma, or PPLO, was discussed in general in Chapter 4. Some *Mycoplasma* species occur normally in the mouth and other parts of the body. (*Mycoplasma salivarius* and *Mycoplasma orale* are examples.) Although many of these organisms infect animals and mammalian cells in tissue cultures, only those organisms that are known to produce respiratory infections in man are discussed here.

Atypical Pneumonia in Humans. Many cases of pneumonia that defy clinical diagnosis are too easily dismissed as "virus pneumonia." In fact, *Mycoplasma* infections are often the cause of these truly pneumonia-like diseases. It has been estimated that 20 per cent of all pneumonias are due to *M. pneumoniae,* an organism easily spread by respiratory secretions. With fluorescent antibody staining and improved cultural methods, mycoplasmal pneumonia is now recognized more frequently. The disease caused by *Mycoplasma pneumoniae* responds to treatment with erythromycin and tetracycline, although the true viral pneumonias do not. Penicillin and the cephalothins are of no value in mycoplasmal infections. Why?

Transmission is as in other infectious diseases of the respiratory tract. The disease is self-limiting and fatality is rare.

LEGIONNAIRES' DISEASE

A pneumonia-like disease outbreak (182 cases) at a convention of the American Legion in Philadelphia in 1976 resulted in death for 16 per cent of the affected legionnaires. The gram-negative bacillus causing legionnaire's disease has been named *Legionella pneumophila.* It is unlike any other rickettsia-

[6]From *Aden,* Greek for "lymph node."

like organism, which cannot be cultured on media, in that it can be grown on supplemented Mueller-Hinton medium as well as in embryonated eggs. Four antigenic types of the organism are known. Dust from air conditioning systems or ground excavations is an environmental source of *Legionella*, as may be cooling towers. It does not seem to be spread through direct contact.

After an incubation period of two to ten days, the patient suffers from high fever with abrupt onset, chills, dry cough, diarrhea, prostration, and delirium. Multilobar lung involvement and bacteremia may spread to affect liver function and cause brain abscess metastatic infections, resulting in a 10 to 20 per cent death rate, especially during epidemics. Diagnosis is then usually made from lung autopsy.

Legionnaires' disease is thought to cause 5,000 to 10,000 cases of pneumonia per year in the United States. For the year 1980, 441 "sporadic cases" of *legionellosis* were reported by the Center for Disease Control and in 1982, 556.

L. pneumophila is susceptible to antibiotics like erythromycin, tetracycline, and rifampin.

Other Legionelloses

Legionella pneumophila research led to the rediscovery of microorganisms with strange names like HEBA and TATLOCK. These pneumonia-causing agents were reclassified, and TATLOCK became *Legionella micdadei*; others were renamed *L. bozemanii, L. dumoffii, L. gormanii,* and *L. longbeachae. Legionella* species appear to infect patients who have been exposed to water, are possibly immunosuppressed, or have chronic lung disease.

PSITTACOSIS (PARROT FEVER)

The causative agent of psittacosis[7] is *Chlamydia psittaci,*[8] previously known as *Miyagawanella psittaci.* Pneumonitis of cats and pneumonitis of birds are other respiratory infections attributed to the same species. Psittacosis is discussed more fully in Chapter 34.

SUMMARY

Before the days of vaccination, 95 of every 100 persons contracted smallpox, one fourth of those died of the disease, and many who recovered were blinded or disfigured. The use of the cowpox vaccine prepared by Jenner against smallpox showed clearly that protection against specific diseases was possible and practical. It was so effective that it is now thought that smallpox has been eliminated from this planet and vaccination is not considered necessary any more. Let us hope that this is true.

Vaccinations against rubeola and rubella are now recommended. Although measles (rubeola) seldom causes death, complications may be very serious, such as bronchopneumonia or an abscess in the ear, as well as subacute sclerosing

[7]Psittacidae is the scientific name of the zoologic group containing parrots, lovebirds, and similar species. The name of the virus is derived from the fact that it causes a disease of psittacine birds ("parrot fever") that is transmissible to humans.

[8]The only other species of this genus is *Chlamydia trachomatis,* the cause of trachoma and other chlamydial diseases of the eye.

panencephalitis. Exposed children under 4 years of age may be protected against the measles with gamma globulin or HIGG. Under no circumstances should a pregnant female be vaccinated with any vaccine containing any active, although attenuated, virus, especially the rubella virus. A fetus infected with rubella in the first three months of pregnancy may develop deformities or defects in the heart, ears, brain, or eyes. To prevent this, all children should now receive early rubella vaccinations.

Varicella (chickenpox) is not a serious disease, but the virus that causes it (varicella zoster, VZ virus) is also the infectious agent of shingles in adults, who usually had a history of chickenpox in childhood.

Herpes simplex (fever blisters) is caused by HSV-1, whereas genital herpes is due to HSV-2. The herpes virus erupts from the sensory nerves of skin tissues whenever extreme changes in temperature, trauma, emotional stress, sun exposure, or menstruation causes the fever blisters to form. An ointment called Zovirax gives relief, but it is not a cure.

About 10,000 infants are affected every year in the United States with congenital anomalies caused by cytomegaloviruses. These viruses are related as salivary gland viruses to the virus of contagious parotitis (mumps). There is no effective treatment for mumps, although vaccines are available.

Rabies (hydrophobia in man) is an encephalitis and is usually the consequence of a bite or scratch by a rabid animal. The central nervous system is affected, and after "hydrophobia," death is almost inevitable. The disease is caused by a relatively slow virus, so that during the incubation period (2 to 8 weeks, possibly as long as 12 months), preventive counter measures can be taken. Recommended are vaccines such as DEV, HDCV, or M-HDCV and immune globulins such as RIG, HRIG, or ARS.

The viruses of poliomyelitis (infantile paralysis) caused frequently fatal paralytic effects from invasion of the central nervous system. However, the virus invades the body as an enterovirus. Effective vaccinations control this disease now.

The influenza virus is distinguishable mainly on the basis of antigenic and immunologic properties. The viral envelope undergoes continuous changes, resulting in new subtypes of the virus and, consequently, new outbreaks of the "flu." This limits the effectiveness of vaccination greatly. Uncomplicated influenza is usually not fatal; however, it may lead to pneumonia and thousands of deaths, as occurred in 1918 because of the bronchopneumonia accompanying influenza.

At present, there is no cure for the common cold, but then this malaise may be a variety of early stages of diseases to which immunity exists. If the patient recovers quickly he or she had "a cold" or the "flu"; if not, a specific disease develops. However, there are also some 89 agents called *rhinoviruses* that do cause colds.

Legionnaires' disease has received special notice as the result of its outbreak in Philadelphia in 1976. It is a dangerous, potentially lethal respiratory disease with abrupt onset caused by *Legionella pneumophila* and is now known to be more common than once believed.

REVIEW QUESTIONS

1. What common mode of transmission exists in many viral, mycoplasmal, and chlamydial infections?
2. How is the *Molluscum contagiosum* poxvirus related to the smallpox virus?
3. How are cowpox and smallpox related? What other pox diseases can you name?
4. What is the current view of vaccination against variola? Why?
5. When is measles most likely to be widely transmitted by a patient?

6. Why may measles result in death? What is gamma globulin and what is it used for? Need it be given to infants under 6 months of age? Explain.
7. How should a case of measles be managed to prevent the spread of disease? Explain the use of vaccines and of gamma globulin in relation to rubeola.
8. What is the relationship between rubeola, rubella, and roseola?
9. Why has it been suggested that 6- to 8-year-old females be purposely exposed to rubella? How do you feel about this? Explain the proper use of rubella vaccine.
10. What is the relationship between herpes zoster and chickenpox?
11. What do cytomegaloviruses do? How are they related to parotitis?
12. What animals may transmit rabies? How may the disease be contracted from them?
13. What are Negri bodies? What is their status as a diagnostic sign compared with the inoculation of mice?
14. What is the basic principle of the Pasteur "treatment"? How may dogs be made resistant to rabies? How do we now prevent hydrophobia?
15. What should be done in case of a bite by an animal suspected of being rabid: To the bite? To the animal? To the patient? In what parts of the body is the bite most dangerous?
16. What changes in behavior in pet dogs should be regarded with suspicion with respect to rabies?
17. Define DEV, HDCV, ARS, RIG, and HRIG.
18. What is the pathologic meaning of anterior poliomyelitis? For what disease and condition was an "iron lung" used? What has taken its place in modern hospitals? Why is the term "infantile paralysis" a misnomer?
19. What is the etiologic factor in anterior poliomyelitis?
20. How may the disease be transmitted? What is the role of carriers?
21. What materials are currently available for producing immunity to poliomyelitis? What kinds of immunity does each produce? What problems do these substances raise? Discuss.
22. Why were tonsillectomies not performed during periods of unusual prevalence of poliomyelitis?
23. What is the role of subclinical infections in preventing poliomyelitis?
24. What is the cause of the common cold? What is the role of bacteria in sequelae to colds? What diseases in children, in the early stages, resemble colds? Of what importance is this fact to health care workers?
25. What is the cause of epidemic influenza? Distinguish between epidemic, endemic, and pandemic influenza.
26. How does the cause of serious sequelae in influenza resemble that in colds, measles, and whooping cough?
27. How is influenza vaccine prepared? How many viruses should it contain? How may the immunity conferred by it be maintained?
28. What is peculiar about *Legionella pneumophila* and the disease it causes?
29. What are the chief sources of infection in psittacosis? How may the disease be avoided?

Supplementary Reading

Adams, D. H.: Slow Viruses. Reading, Mass., Addison-Wesley Publishing Co., 1976.

Baer, G. M.: The Natural History of Rabies. New York, Academic Press, 1975.

Benenson, A. S.: Control of Communicable Diseases of Man. 11th ed. New York, The American Public Health Association, 1970.

Beveridge, W. I. B.: Influenza: The Last Great Plague: An Unfinished Story of Discovery. New York, Prodist, 1977.

British Society for Antimicrobial Chemotherapy: Chemotherapy for Herpes Simplex Virus Infections. London, Academic Press, 1977.

Center for Disease Control: Changes in Rabies Control—New York City, Philadelphia. *Morbidity and Mortality Weekly Reports, 24*:82, 1975.

Debre, R., and Celers, J.: Clinical Virology: The Evaluation and Management of Human Viral Infections. Philadelphia, W. B. Saunders Company, 1970.

Dixon, C. W.: Smallpox. Boston, Little, Brown & Co., 1962.

Enright, J. B.: Geographical distribution of bat rabies in the United States, 1953–1960. *Amer. J. Public Health. 52*:482, 1962.

Gibbs, R. C.: Possible mechanisms for maintaining immunity to varicella-zoster virus. *Amer. J. Dis. Child., 120*:456, 1970.

Haire, M., and Hadden, D. S. M.: Rapid diagnosis of rubella by demonstrating rubella-specific IgM antibodies in the serum by direct immunofluorescence. *J. Med. Microbiol., 5*:237, 1971.

International Association of Microbiological Societies, Symposium on Actinomyces, Kyoto, 1974: Actinomyces: The Boundary Microorganisms. Baltimore, University Park Press, 1977.

Jackson, G. G.: Viruses Causing Common Respiratory Infections in Man. Chicago, University of Chicago Press, 1975.

Jacobs, J. W., Peacock, D. B., Corner, B. D., Caul, E. O., and Clarke, S. K. R.: Respiratory syncytial and other viruses associated with respiratory diseases in infants. *Lancet, 1*:871, 1971.

Kenny, M. T., Albright, K. L., and Sanders, R. P.: Microneutralization test for determination of mumps antibody in vero cells. *Appl. Microbiol., 20*:371, 1970.

Kilborne, E. D.: The Influenza Viruses and Influenza. New York, Academic Press, 1975.

King, D. W.: Herpes Virus: Epidemiology, Molecular Events, Oncogenicity, and Therapy: Symposium on Herpes Virus, National Institute of Allergy and Infectious Diseases: Rabies. Bethesda, Md., National Institute of Allergy and Infectious Diseases, 1975.

Koechli, B., Martin du Pan, R., and Douath, A.: Protection of nonimmune volunteers against rubella by intravenous administration of normal human gamma globulin. *J. Infect. Dis., 126*:341, 1972.

Krugman, S., Muriel, G., and Fontana, V. J.: Combined live measles, mumps, and rubella vaccine. *Amer. J. Dis. Child., 121*:380, 1971.

Kurstack, E., and Kurstak, C.: Comparative Diagnosis of Viral Diseases. New York, Academic Press, 1977–1978.

Lennette, E. H., and Schmidt, N. J. (eds.): Diagnostic Procedures for Viral, Rickettsial, and Chlamydial Infections. New York, American Public Health Association, 1979.

National Communicable Disease Center: Smallpox imported into Germany from India: *Morbidity and Mortality, 16*: No. 12, 1967.

Oxford, J. S., Potter, C. W., McLaren, C., and Hardy, W.: Inactivation of influenza and other viruses by a mixture of virucidal compounds. *Appl. Microbiol., 21*:606, 1971.

Parkman, P. D., Meyer, H. M., and Hopps, H. E.: Production and laboratory testing of experimental live rubella virus vaccine. *Canad. J. Pubic Health, 62*:30, 1971.

Prineas, J.: Paramyxovirus-like particles associated with acute demyelization in chronic relapsing multiple sclerosis. *Science, 178*:760, 1972.

Pumper, R. W., and Yamasheroya, H. M.: Essentials of Medical Virology, Philadelphia. W. B. Saunders Company, 1975.

Quick rabies test. *Medical World News,* November 12, 1971:15.

Recommendations of the Public Health Service Advisory Committee on Immunization Practices. *Morbidity and Mortality, 21*:1, 1972.

Schlessinger, D. (ed.): Microbiology–1977. Washington, D.C., American Society for Microbiology, 1977.

Schmidt, N. J., Lennette, E. H., and Magoffin, R. L.: Immunological relationship between herpes simplex and varicella-zoster viruses demonstrated by complement fixation, neutralization and fluorescent antibody test. *J. Gen. Virol.,* *4*:321, 1969.

U.S. Congress: House Committee on Interstate and Foreign Commerce, Subcommittee on Health and the Environment: Swine Flu Immunization Program. Washington, D.C., U.S. Government Printing Office, 1976.

CHLAMYDIAL DISEASES

The family Chlamydiaceae—previously called Rickettsiaformis, Prowazekia, Miyagawanella, and Bedsonia—has been described in Chapter 5. Of the more than 30 different species or varieties that have been distinguished from time to time on the basis of differing antigenic composition, animal hosts, and diseases caused, only one genus, consisting of two species, is recognized at the present time. (However, other genera and *species incertae sedis* do exist, and the taxonomy of *Chlamydia* is likely to change again.) The two recognized species are *Chlamydia trachomatis* and *Chlamydia psittaci.* Both are gram-negative obligate intracellular pathogens of humans and other animals, cultivable in avian embryos and mammalian tissue-cell cultures *but not extracellularly. C. trachomatis* is not inhibited in eggs by sulfadiazine.

Like *Rickettsia,* the genus *Chlamydia* is known for its complex, obligate intracellular growth cycle, which makes it a parasite of tissue cells of vertebrates and also of arthropods. All species of *Chlamydia* contain antigenically similar heat-stable lipoglycoproteins. Their multiplication is inhibited by tetracycline, chloramphenicol, sulfonamide, and (as an alternative drug) erythromycin. Treatments take four to eight weeks.

Chlamydia trachomatis causes diseases of the eye (trachoma and inclusion conjunctivitis) that occur only in humans. It is the so-called TRIC agent.[1] It also causes non-ophthalmologic diseases in humans: lymphogranuloma venereum (a venereal disease), urethritis, and arthritis. *C. psittaci* is the cause of psittacosis or ornithosis in humans and various species of birds (especially parrots and pigeons) and a variety of other conditions—respiratory, arthritic, enteric, and placental—in a wide range of vertebrates other than primates.

TRACHOMA

This disease, known since antiquity, is caused by *Chlamydia trachomatis* (serotypes VR 571, VR 572, or VR 573). The World Health Organization estimates that it afflicts about 400 million persons at the present time and has blinded nearly 20 million of them. The infection is contracted in childhood and progresses toward *inflammatory chronic keratoconjunctivitis* with secondary bacterial infections and, finally, scarring of the cornea with blindness. It is spread by contact with excretions of infected eyes, dust, and probably houseflies *(Musca sorbens),* but it is not highly contagious (Fig. 34–1). No immunity to trachoma seems to develop as the result of infection, although specific antibodies have been found in eye secretions of patients. The disease is most prevalent in countries in the Mediterranean region, especially Egypt. Anyone who has ever seen the inward curvature of the eyelid and the deformation and discharges of advanced stages of this disease does not easily forget it. Control is mainly dependent on avoidance of contact, personal cleanliness, and education of the

[1]Probably "trachoma agent" would be a preferable term, to distinguish TRIC from the protozoan *Trichomonas.*

PART C BACTERIA CAUSING VENEREAL AND RELATED INFECTIONS

CHAPTER 34

public. Topical applications of tetracyclines or sulfonamides are useful in *early* cases.

INCLUSION CONJUNCTIVITIS

Inclusion conjunctivitis (IC), or blennorrhea, the result of an infection by *Chlamydia trachomatis* serotype VR 346 or VR 575 (both previously known as *Chlamydia oculogenitalis*), is, like trachoma under natural conditions, a disease of humans only. Like the gonococcus, this pathogen is an inhabitant of the genital tract that can also infect the eyes. It is transmitted during coitus and also by contaminated hands and fomites. Inclusion conjunctivitis usually develops in the newborn's eyes, unlike trachoma, which rarely occurs in neonates. Eye-to-eye infections have not been reported; however, adults may contract the ocular disease, usually from swimming pools contaminated by genital secretions.

The disease in infants resembles gonococcal ophthalmia (see Chapter 36), and it is contracted in the same way—i.e., during the birth process. Infections of the adult cervix do not cause any symptoms; in the male, the organism may be found associated with mild urethritis.

In the newborn who has been infected, a few days after birth the conjunctivae become inflamed and there is copious mucopurulent discharge. But even without therapy, the infection is self-limiting. The incubation period (five to seven days) is longer than that of gonococcal ophthalmia (two to three days). No scar or blindness develops, although the disease may take a year or longer to clear up. Diagnosis may be made by microscopic examination of stained scrapings from inflamed conjunctivae. Epithelial (not pus) cells contain diagnostically distinctive *inclusions* (organisms).

THE VENEREAL GRANULOMATOSES

Lymphogranuloma Venereum[2]

Granuloma is a pathologist's term for an abnormal, swollen tissue made up of fleshy granules, generally of an inflammatory or ulcerative character and often seen in healing tissues, especially if infected. It is sometimes colloquially called "proud flesh." Granulomas may be caused by or associated with injuries due to mechanical, chemical, or infectious agents.

Lymphogranuloma venereum (LGV) is a painful and destructive disease caused by *Chlamydia trachomatis* serotype VR 121, also known as the LGV agent.

---

[2]This disease has several names: lymphogranuloma inguinale, venereal lymphogranuloma, and so on. Do not confuse with *granuloma inguinale*, a different disease, which also has several names.

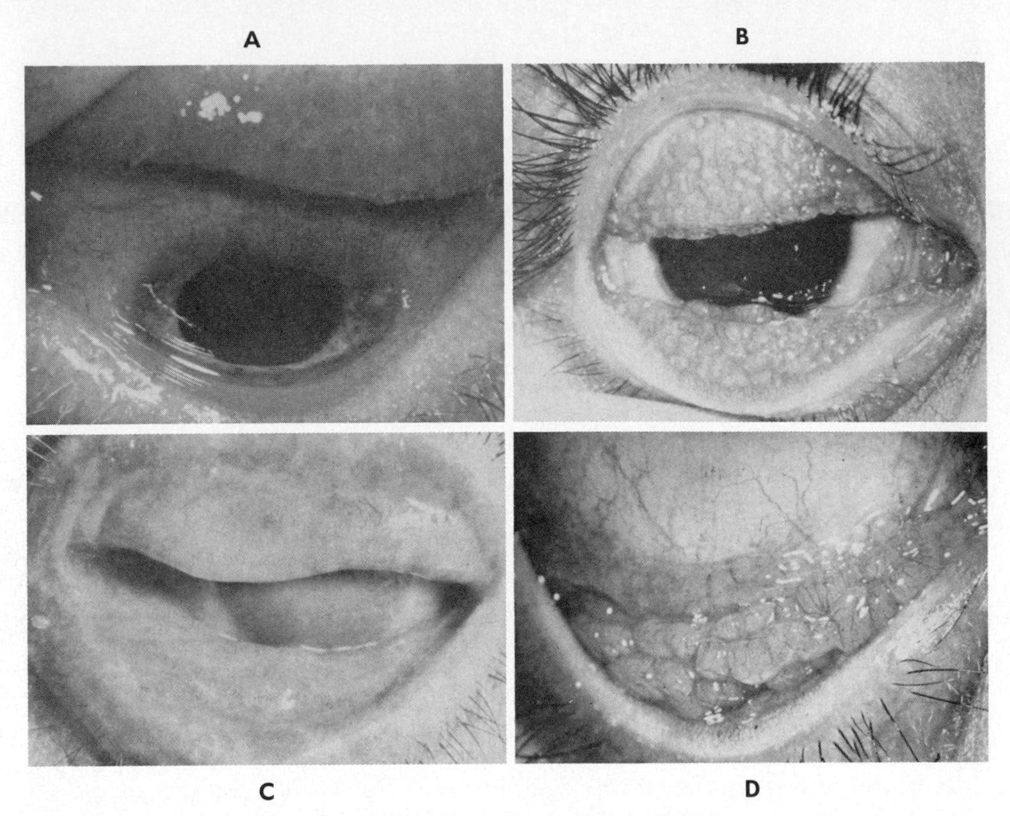

Figure 34–1 *A*, Early trachoma with follicular and papillary hypertrophy of upper tarsal conjunctiva and beginning pannus at upper limbus. *B*, Established trachoma, Stage IIa MacCallan. Note predominant follicular hypertrophy in the absence of scars. *C*, Stage IV (healed) trachoma with conjunctival and corneal scars and loss of useful vision. *D*, Inclusion conjunctivitis in young adult. Note massive follicular hypertrophy of conjunctiva of lower fornix. There was an associated urethritis. (From Thygeson, P., and Hanna, L., *In* Lennette, E. H., and Schmidt, N. J. (eds.): Diagnostic Procedures for Viral and Rickettsial Infections. 4th ed. Public Health Association. 1969.)

No immunity seems to develop toward this agent either. Another classification assigns *C. trachomatis* strains causing LGV to immunotypes L1, L2, and L3, distinct from other disease-related chlamydias. The disease is generally transmitted by sexual contact and is common in the tropics and among promiscuous persons. It begins one to 12 weeks after exposure as a small, transitory, local sore on the genitalia, which, in males, is followed by development of swollen inguinal "glands" (lymph nodes) and, perhaps weeks later, of swellings in adjacent parts. These swellings (buboes) ulcerate and discharge seropurulent fluid. In females the pelvic organs are involved, with lesions in the lower bowel. Discharges from active lesions are highly infectious and may persist for years.

Various other organs and tissues may eventually become involved and produce large swellings. Stricture of the rectum is a common complication, especially in females, and is very painful and troublesome; however, many cases pass unnoticed and serve to disseminate the disease. A good number of LGV infections give positive reactions in syphilis serology with cardiolipid antigens. For diagnosis, tests for allergy to the organism (like the Frei test) can be made. These are analogous in principle to the tuberculin test but are less reliable. More accurate for diagnosis is the isolation of the LGV strain from a bubo, the urethra, or the cervix. Alternatively, after appearance of a bubo, a complement-fixation test with a 1:32 titer is considered significant. A new microimmunofluorescent antibody test is currently being evaluated. All patients with LGV are likely to be infected with syphilis as well and should be tested for both diseases.

The tetracycline antibiotics are useful therapeutically in early and older cases, but they cannot repair damage done by infections of long standing. Sulfonamides, erythromycin, and cycloserine have also been recommended for treatment.

Granuloma Inguinale (Granuloma Venereum)

This supposedly venereal disease (it is only mildly communicable) is due to a diagnostically distinctive microorganism, *Calymmatobacterium granulomatis* (previously called *Donovania granulomatis* or Donovan bodies), which is a gram-negative, encapsulated, rod-shaped, often pleomorphic bacterium. It is readily seen within phagocytic cells in stained smears made with exudate from the lesions of the disease. It is believed by some that Donovan bodies closely resemble heavily encapsulated *Klebsiella pneumoniae*.

The disease occurs principally in the tropics. It is also not uncommon in the southern United States. Because of the similarity in names, it is sometimes confused with lymphogranuloma venereum, and it is mentioned here partly for this reason. The two diseases are similar in epidemiology and modes of transmission. They both exhibit extensive gross superficial lesions that are painful and progressive, with destruction of tissues.

Prevention of Venereal Granulomatoses. Since the causative microorganisms occur in the discharges, these, as well as dressings, clothing, and rubber gloves contaminated with them, should be handled with care and sterilized or disinfected. There is little other danger of infection. The same precautions should be used in chancroid (see Chapter 36). The organisms are easily killed by the usual disinfectants.

PSITTACOSIS (PARROT FEVER)

Although psittacosis belongs with the other respiratory diseases listed in Chapter 33, it is discussed here because of the close taxonomic relationship of its etiologic agent, *Chlamydia psittaci*, to *Chlamydia trachomatis*.

Psittacosis attacks parrots, parakeets, and related species of birds, as well as pigeons, chickens, and others, causing diarrhea, sneezing, a generally sick appearance, and often a high (and costly!) mortality rate. This condition in birds other than psittacines is usually called *ornithosis*. Psittacosis was probably first brought to the United States by infected lovebirds, and it has also appeared in European countries. It is readily transmitted to human beings by dust from the dried feces, feathers, and exudate from the nostrils and mouths of the sick birds.

In humans it causes a generalized reaction with chills, fever, headache, and vomiting. It especially attacks the lungs, causing a form of bronchopneumonia. The disease in humans is frequently fatal, and several microbiologists have lost their lives as a result of studying this pathogenic agent. Commercial handling of psittacine birds is very apt to transmit the organism to the persons engaged in the work. Purchasers of these birds not infrequently contract the disease.

Transmission and Control of Psittacosis. The nose and mouth secretions of infected human beings can transmit the causative organisms; however, most cases in humans result from contact with sick birds. The preventive precautions used for influenza and pneumonia are also applicable in human cases of psittacosis.

All parrots or other birds suspected of having the disease should be killed and incinerated, and their cages either burned or soaked in saponated solution of cresol or some other disinfectant solution.

A practical application of our knowledge of the susceptibility of *Chlamydia* to antibiotics is the feeding of appropriate antibiotics (e.g., tetracyclines) to flocks of psittacine birds to eradicate psittacosis. This is said to be very effective and is important to both the industry and the purchasers.

At least 10 strains of *Chlamydia psittaci* cause diseases in various animals. The human psittacosis strain is VR 601.

MYCOPLASMA (PPLO)

Whereas *Mycoplasma pneumoniae* (see Chapter 4) is found in patients with bronchitis and pneumonia (Chapter 33), *Mycoplasma hominis* and so-called T strains are obtained from various infections of the female upper reproductive tract. Since they are not uncommon in the normal flora of the vagina and cervix, and in arthritis and some other conditions, no etiologic role in venereal disease can be attributed with certainty to *Mycoplasma* species at this time.

SUMMARY

At present, the classification of chlamydial disease–causing microorganisms is quite simple. Only two species are recognized; these are subdivided into serotypes that have been given numbers. *Chlamydia trachomatis*, also called the TRIC agent (TRachoma Inclusion Conjunctivitis), is the cause of chronic keratoconjunctivitis, better known as *trachoma*, a blinding eye disease of 20 million people worldwide, of a total of 400 million being affected by the disease; *inclusion conjunctivitis* (IC or blennorrhea), an inflammation of the eyes of newborns infected at birth; and *lymphogranuloma venereum* (LGV or bubo), a painful and destructive disease of the genitalia transmitted by sexual contact. The tetracyclines and other antibiotics are now useful therapeutically against all of these diseases, but they cannot repair damage done by infections of long standing.

Another supposedly venereal disease that is only mildly communicable, is caused by *Calymmatobacterium granulomatis* (Donovan bodies). It is known as granuloma inguinale, or granuloma venereum, and should not be confused with lymphogranuloma venereum.

Psittacosis (parrot fever) is caused by *Chlamydia psittaci*. It is an ornithosis in birds, but the only one of these diseases also affecting man, as a type of bronchopneumonia-like condition.

REVIEW QUESTIONS

1. What were previous names of the family Chlamydiaceae?
2. Name two species of *Chlamydia* and the diseases they cause.
3. How is trachoma transmitted? What damage does it cause? What disease in the newborn is related to it? What treatments are effective against it?
4. What is LGV? How is it transmitted? How is it related to granuloma venereum?
5. How is psittacosis transmitted? What is the organism causing the disease? Who are the people most likely to be exposed to the disease? Why?
6. Why was *Mycoplasma pneumoniae* mentioned in this chapter?

Supplementary Reading

Hobson, D., and Holmes, K. K.: Nongonococcal Urethritis and Related Oculogenital Infections. Washington, D.C., American Society for Microbiology, 1977.

Jawetz, E.: Agents of trachoma and inclusion conjunctivitis. *Ann. Rev. Microbiol., 18*:301, 1964.

Lymphogranuloma venereum. *Pfizer Spectrum, 6*:484, 1958.

McComb, D. E., and Nichols, R. L.: Antibody type specificity to trachoma in eye secretions of Saudi Arab children. *Infect. Immun., 2*:65, 1970.

McDaniel, W. E.: Four lesser venereal diseases. *J. Kentucky Med. Assn., 62*:281, 1964.

Thygeson, P., and Hanna, L.: TRIC agents. *In* Lennette, E. H., and Schmidt, N. J. (eds.): Diagnostic Procedures for Viral, Rickettsial, and Chlamydial Infections. New York, American Public Health Association, 1979.

Vedros, N. A.: Species-specific antigen from trachoma and inclusion-conjunctivitis (chlamydial) agents. *J. Immunol., 99*:1183, 1967.

THE TREPONEMATOSES

THE SPIROCHAETALES

The term spirochete is commonly applied to helically coiled, flexible bacteria (see Fig. 4–4). These have been described on page 53.

Subdivisions. In the eighth edition of *Bergey's Manual of Determinative Bacteriology* the order Spirochaetales has only one family, Spirochaetaceae, which includes pathogens and saprophytic organisms of no medical significance, such as the genus *Cristispira.* Only the pathogens will be discussed here.

The pathogenic spirochetes are included in three genera of great medical importance: (1) *Borrelia,* species of which (e.g., *B. recurrentis* and several other similar species[1]) cause relapsing fever; (2) *Treponema* (of which important species are *T. pallidum,* the cause of syphilis, *T. pertenue,* the cause of the tropical scourge yaws, and *T. carateum,* the cause of pinta, a tropical skin disease); and (3) *Leptospira,* of which the species *L. interrogans* (previously known as *L. icterohaemorrhagiae)* is infamous as the cause of Weil's disease or infectious hemorrhagic jaundice, which clinically resembles yellow fever. *L. interrogans* was discussed previously with urinary and enteric pathogens (Chapter 25) because it is excreted in and transmitted by urine.

The Treponematoses. The treponemal diseases are collectively referred to as treponematoses. Generally, they are extremely destructive to tissues (Fig. 35–1). An exception is pinta, which does not produce the ulceration found in other treponemal diseases. Pinta lesions are flat, spreading, scaly, erythematous areas of skin, which are at first hyperpigmented. There is little generalized reaction. The lesions, on healing after months or years, become completely depigmented, wide, and patterned areas, sometimes showing remarkable bilateral symmetry. Transmission is by personal contact, not necessarily venereal. Serologic tests for syphilis are also positive for pinta.

Figure 35–1 Disfigurement due to yaws *(Treponema pertenue)*, characteristically restricted to skin, bones, and cartilage. Analogous destruction of tissue, both internally and externally, characterizes syphilis and bejel. These treponematoses yield readily to combined sanitation and penicillin. Once, these and similar diseases were confused with Hansen's disease (leprosy). It is easy to understand the medieval horror of lepers (and *supposed* lepers). (Courtesy of Dr. H. van Zile Hyde, Division of International Health, U.S. Public Health Service.)

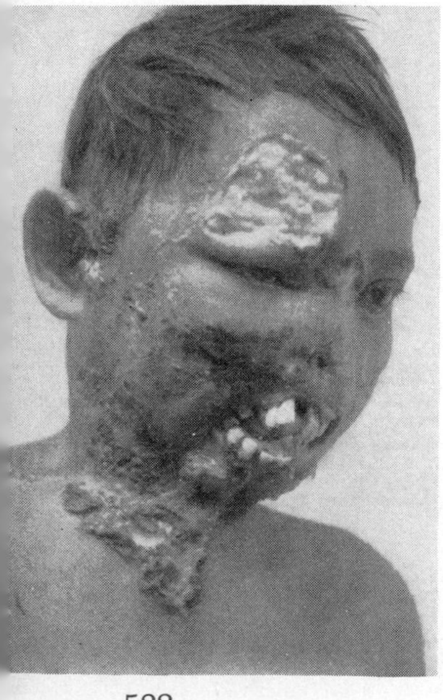

In the United States the most important species of spirochetes is *Treponema pallidum,* the cause of syphilis. In certain tropical areas, identical or very closely related species of *Treponema* cause at least two other important diseases: *yaws,* caused by *T. pertenue* and common in native tropical peoples around the world; and *bejel* or *nonvenereal syphilis,* caused by *T. pallidum* and found among some less advanced peoples in Southeastern Mediterranean and adjacent areas.

Syphilis is venereally transmitted. Yaws and bejel are spread in tropical and subtropical regions principally by nonvenereal personal or household contact, especially among children and between mothers and babies, and probably also by flies and fomites. Although they may sometimes be transmitted by sexual contact, yaws and bejel, like pinta, are not regarded as primarily venereal diseases.

SYPHILIS

Treponema pallidum was discovered in 1905 by Fritz Richard Schaudinn (1871–1906), a German scientist, in the primary sores (*chancres,* pronounced

[1] *Treponema vincentii,* possibly involved in trench mouth infections, was previously known as *Borrelia vincentii.*

CHAPTER 35

shank'ers) of persons infected with syphilis. *T. pallidum* has never been successfully cultivated in a virulent state in an artificial (nonliving) medium, although several cultures of nonvirulent spirochetes closely resembling *T. pallidum* (and probably variants of it) exist. One now widely used as a source of antigen in serologic diagnostic tests for syphilis is the Reiter treponeme; another is the Nichols treponeme. *T. pallidum* may be kept alive and motile for some time in artificial media and tissue cell cultures (2 to 2.5 days), but it does not multiply significantly, if at all. It may also be frozen with CO_2 and 15 per cent glycerol and thus remain alive for years. It is now thought (hoped) that active immunization against syphilis is feasible. However, the preparation of the vaccine depends on cultivation of the organism by methods yet to be devised.

All species of *Treponema* are morphologically alike. All are relatively small, thin, tightly coiled spirochetes (Fig. 35–2). Their eight to 14 spirals are close and regular unless protoplasmic contractions change them. The ends of the organisms are drawn out into extremely fine fibrils, which have been mistaken for flagella. Progressive motion depends presumably on the propeller-like action of the spirals when the treponemes rotate. They move slowly and flex slightly.

Demonstration of Treponema. The recognized species of *Treponema* are not easily stained, and consequently other methods are generally used to observe them microscopically:

Negative Staining. This is not really a method of staining. Appropriate material (e.g., exudate from pathologic conditions such as syphilitic ulcers) is mixed with a little India ink or nigrosin (a dense, black dye), spread thinly on a slide, and allowed to dry. The ink or nigrosin forms a dark background, and the organisms, unstained by the ink or nigrosin, appear as transparent, wavy lines since the ink and nigrosin fail to penetrate them. Many other microorganisms may be demonstrated in this way (see Fig. 9–8).

Silver Impregnation Method. In this method ammoniacal solutions of silver salts are first allowed to penetrate the cells. Metallic silver is then precipitated on the surface of the spirochetes by means of a reducing solution. The organisms appear thick and black against a yellowish background (Fig. 35–3).

Darkfield Method. Probably the quickest and easiest method of demonstrating treponemes, and indeed any bacteria in fluid material, is by means of the darkfield microscope technique. An ordinary compound microscope may easily be equipped for this work. A special substage darkfield condenser with an opaque stop in the central part is used in place of the usual type of substage condenser. This darkfield condenser is designed to prevent the entrance of any rays of light from the light source straight upward into the tube of the microscope. By means of the darkfield condenser all rays from the light source emerge from the upper surface of the slide at such an angle that they do not enter the objective lens (Fig. 35–4). The field therefore appears dark when examined with the objective lens of the microscope, hence the term *darkfield*. When a fluid containing any particles, such as dust, bacteria, or spirochetes, is placed on the slide at the focal point of these oblique rays, the oblique light rays are reflected from the surfaces of the spirochetes or other objects upward through the lenses to the eye of the observer. Each particle on the slide becomes visible as a brightly illuminated speck owing to the light reflected from its surface. The remainder of the field remains black. The principle is the same

Figure 35–2 Scrapings from a syphilitic chancre as seen with a microscope with darkfield illumination. The spirochetes are *Treponema pallidum*. The eight small, rounded objects are erythrocytes; the two larger, rounded ones are tissue cells or pus cells; the smallest irregular forms are salt crystals, cell detritus, and other bacteria (approximately ×1,000). (Courtesy of Chas. Pfizer and Co., Inc., Brooklyn, N.Y. In *J.A.M.A.*, Vol. 157, 1955.)

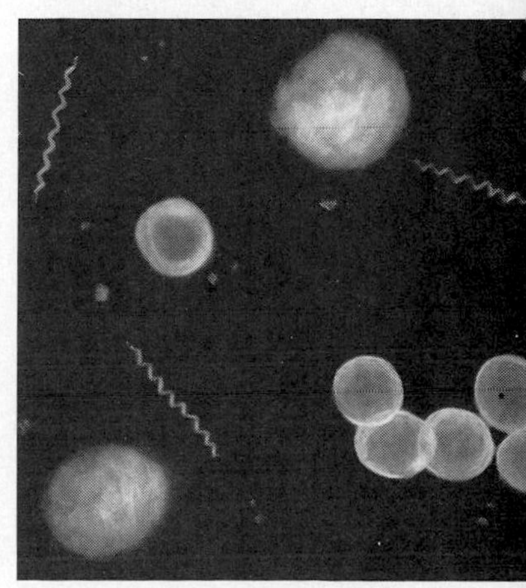

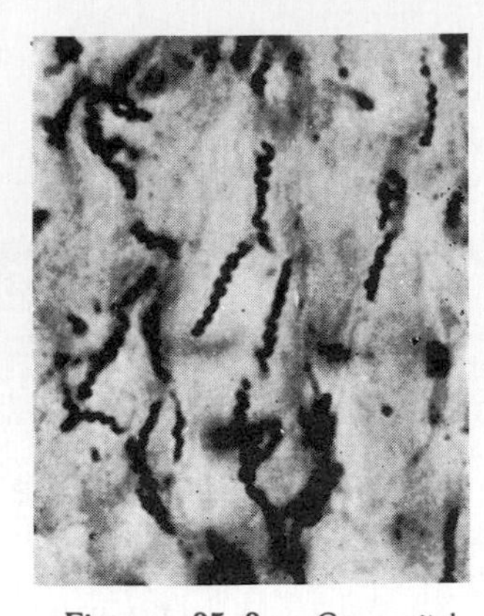

Figure 35–3 Congenital syphilis in the lung. *Treponema pallidum* demonstrated by the Levaditi silver impregnation method (approximately ×1,600). (From Smith, L. W., and Gault, E. S.: Essentials of Pathology. 2nd ed. New York, Appleton-Century-Crofts, Inc., 1942.)

as that which makes the moon and planets visible as bright objects in a dark sky, or dust particles visible in a dark room into which a single ray of sunlight falls.

Only outward form and motility are demonstrable by the darkfield, but these are two of the chief means by which treponemes and other spirochetes are distinguished.

The Phase Contrast Microscope. The use of the phase contrast condenser (especially anoptral phase) permits better contrast without having to resort to the darkfield technique.

The Fluorescence Microscope Method. This very useful method (see Chapter 20) is often used to demonstrate microorganisms in tissue sections and other materials. Its use in demonstrating pathogenic treponemes for the fluorescent antibody technique is discussed later in this chapter.

Properties of the Syphilis Organism. *Treponema pallidum* dies quickly outside the body because it is very sensitive to drying, cooling, changes in osmotic pressure, pH, and standard disinfectants. Soap and other detergents quickly destroy it. Material that has dried will not carry the disease-causing organism, but objects very recently soiled and still warm and moist with secretions containing the spirochete are possible sources of infection, especially to health personnel and laboratory workers who handle syphilitic materials and patients. The survival time of the spirochetes on ordinary objects is very short. Except as noted, the disease is very rarely (some say never) transmitted by fomites, virtually all cases being acquired by direct contact with an infected person.

The Course of Untreated Syphilis

Untreated syphilis develops through three definite stages: primary, secondary, and tertiary (Fig. 35–5).

Primary Syphilis. *Treponema pallidum* gains entrance to the body through the skin or mucous membranes. Most often it infects the genitalia during coitus, occasionally the lips during kissing. Not infrequently it is homosexually acquired. Close contact between an infected lesion and a break in the skin on the genitalia, mouth, lips, anus, face, fingers, or other part of the body is required to contract the spirochete, although it is usually coitus that spreads the disease.

The organisms begin to multiply locally in the tissue almost immediately, producing an initial lesion or *chancre* (Fig. 35–6) that appears three to four weeks after exposure. The chancre is typically a rather flat, indurated or "hard" ulcer (the so-called "hard chancre"). It is often absent, unrecognizable, or

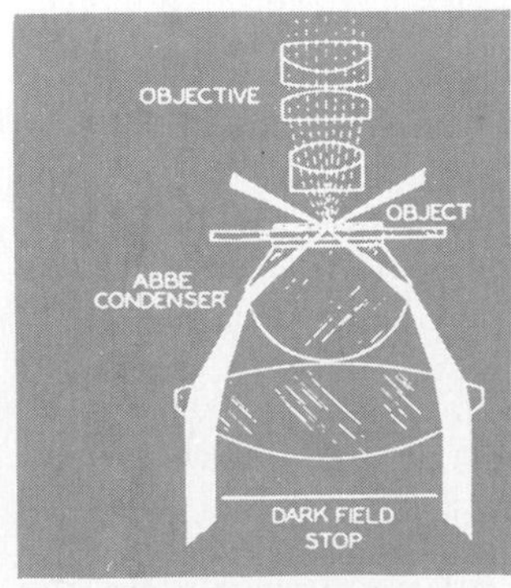

Abbé Condenser with Darkfield Stop

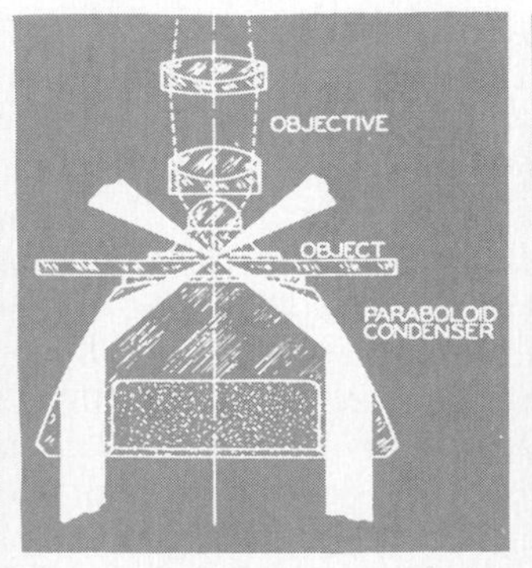

Paraboloid Condenser

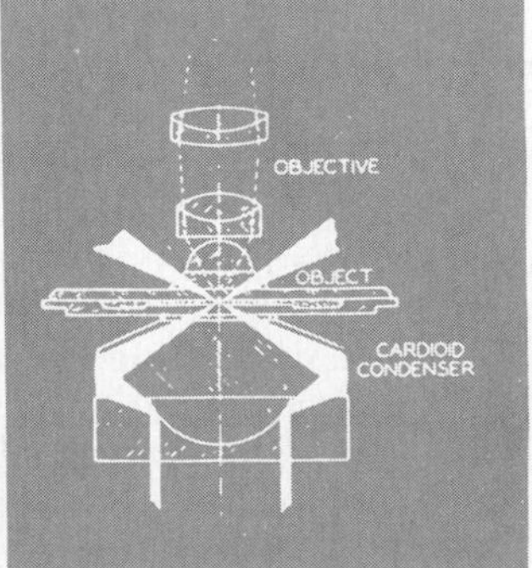

Cardioid Condenser

Figure 35–4 Various forms of condensers for oblique illumination of the darkfield. In the extreme left picture an object on the slide is reflecting light up through the objective lens. In each arrangement note that only peripheral light rays pass the condenser. (Courtesy of Bausch & Lomb Optical Co.)

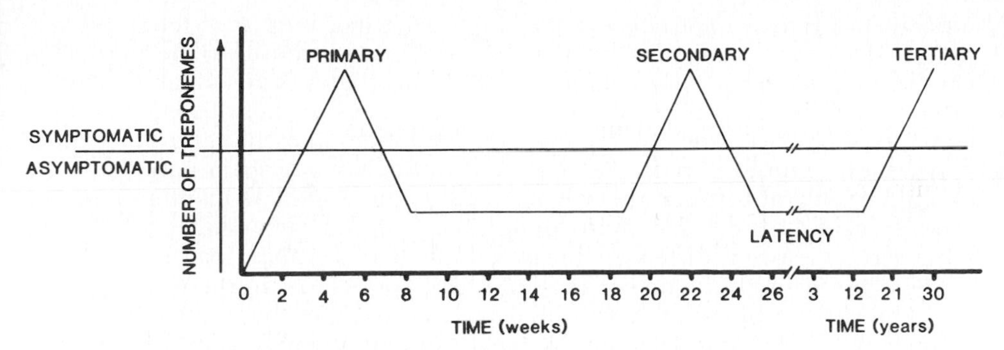

Figure 35–5. Average time intervals for the three stages of syphilis. This disease fluctuates between symptomatic and asymptomatic periods. Clinical manifestations of symptomatic syphilis result from multiplication of treponemes within various tissues. Effective host defenses then evolve that significantly decrease the number of treponemes, resulting in asymptomatic periods. (From Fitzgerald, T. J.: Pathogenesis and immunology of *Treponema pallidum.* Reproduced, with permission, from the *Annual Review of Microbiology,* Vol. *35,* p. 31. by Annual Reviews, Inc.)

inconspicuous, usually forms a temporary "scab," and finally may heal completely without treatment. However, this does not mean that the disease is cured: the organisms have simply left the initial portal of entry and invaded the body. In homosexual males primary lesions in the anal canal or perianal region may be due to syphilitic proctitis.

Within a short time, varying probably from a few hours to several days after infection, the spirochetes appear in the local lymph nodes, producing a swelling (bubo). Thence they enter the blood and are carried to every organ in the body. Within two to four weeks after infection, specific antibodies called *reagin* (a mixture of IgM and IgA antibodies) may be demonstrated in the serum. Although *T. pallidum* may be present in the blood during the incubation period of syphilis, there is little danger of transmitting the disease through blood transfusions. The spirochete does not remain viable in drawn refrigerated blood longer than 2 to 3 days, so that three-day-old blood from blood banks is safer, although any blood giving a positive serologic reaction (STS)[2] should be rejected.

Secondary Syphilis. As a general rule, although not always, a rash appears following the general invasion of the body. This seems to be of an allergic nature and indicates that the body is responding to the infection immunologically. There is nothing especially distinctive in the form or appearance of the rash. It may resemble almost any known type of rash. This is one of the reasons why syphilis, which in later stages produces such a great variety of lesions in so many organs and tissues, is often spoken of as *"The Great Imitator."* The rash may occur as late as several months after the appearance of the chancre but usually earlier, about six weeks after the chancre has healed. At about this time, sores may appear on the genitalia or in the oral cavity (mucocutaneous lesions) or both, accompanied by sore throat, slight fever, and headache. This is called the *secondary stage.* It is rarely disabling at this stage and is often ignored by the uninformed as trifling. Syphilis in this stage is, however, a highly infectious disease. The rash, ulcers, and sores swarm with spirochetes, and the disease may be spread at this stage by contact with the patient or by articles still moist with secretions; the same is true for yaws, bejel, and pinta.

Tertiary Syphilis. After a number of weeks or months these secondary symptoms also tend to recede without treatment, probably due to the development of some degree of immunity. But the patient is not cured; the living spirochetes are still boring from within, and the disease enters a *chronic stage,* called the *third* or *tertiary stage.*

Late Syphilis. Like the tubercle bacillus, *Treponema pallidum* may remain alive in the body for many years without causing any definite or acute symptoms, but it is always doing great damage during this time, and new lesions eventually appear in various parts of the body. It may attack almost any organ in the body. The lesions are called *gummas* and may be external and ulcerous (and

Figure 35–6 Chancre of the lip. (From Domonkos, A. N., et al.: Andrews' Diseases of the Skin, 7th ed. Philadelphia, W. B. Saunders Company, 1982.)

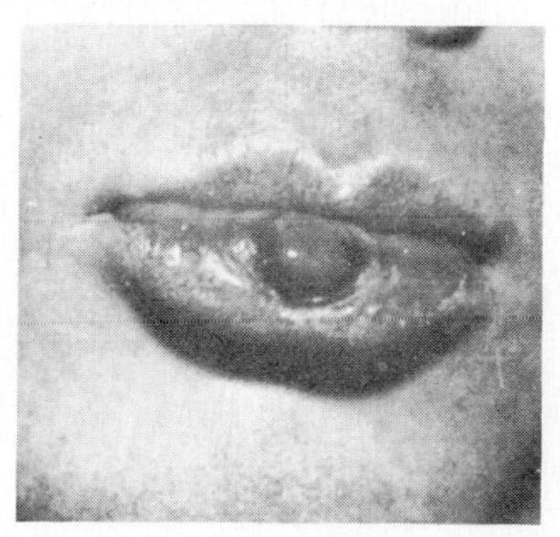

[2]STS stands for "serologic test for syphilis," and there are many different ones (see pages 527–529).

infective) or internal and abscess-like. These gummas tend to extend and also to heal at the same time, thus causing much disfigurement due to the contractile scar tissue. Healed portions rarely contain spirochetes, but exudates may be infectious. In the later stages of the disease there are long periods during which damage progresses slowly, symptoms are indefinite, and the ignorant patient may believe himself cured. The disease then breaks out again. Owing to the long time interval between the original infection and the late manifestations of syphilis, many of the symptoms and lesions of the later stages were once regarded as entirely different diseases until their syphilitic nature was fully understood through the demonstration of the spirochetes in the tissues.

The late symptoms of syphilis depend on the organ that is attacked by the spirochetes. The most frequent, serious, and disabling results of syphilitic infection are diseases of the *heart, arteries,* and *nervous system,* which often cause early death.

The spirochetes lodge particularly in the walls of the arteries, causing chronic inflammation and destruction of tissue. The vessel walls are weakened and may bulge, creating saclike dilatations called *aneurysms.* These not infrequently burst, and the patient may suddenly die of the hemorrhage. Syphilis is a principal (but not the only) cause of *aneurysm of the aorta and diseases of the aortic valves,* and it is among the several recognized causes of *arteriosclerosis* and *cerebral hemorrhage* in comparatively young persons.

Unless treatment is given early, the spirochetes lodge also in the central nervous system (brain and spinal cord), where they cause a chronic inflammation and destroy the nerve tissue. Various forms of insanity ensue. These lesions result in *paresis* (general paralysis, or dementia paralytica) and *locomotor ataxia,* or tabes. In paresis, spirochetes are present in the tissues of the brain; in tabes, in the spinal cord.

Congenital Syphilis. Syphilis is sometimes incorrectly said to be inherited. This implies transmission of the spirochete by the sperm or ovum, which probably occurs very rarely, if at all.[3] A woman may transmit syphilis to her child if she has the disease at the time of conception or acquires it during pregnancy. The spirochetes pass from the mother through the placenta into the blood of the fetus; they are one of the few organisms that can pass through the placenta. The baby is thus infected before birth and comes into the world with living spirochetes in its body. This is called "congenital syphilis." Syphilitic babies with rashes and sores (rhagades) around the mouth and anus or with "snuffles" (chronic discharges from syphilitic lesions in the nose) are infectious. There is practically no danger, however, from older congenitally syphilitic children, or from the advanced stages of syphilis in adults, if there are no open sores. "Late congenital syphilis" refers to clinical manifestations that appear only after infancy. In fact, congenital syphilis often is not suspected or is misdiagnosed, and may not be recognized until the child is 10 years old or older. Congenital syphilis generally responds very well to a full treatment with aqueous procaine penicillin (APP).

Since congenital syphilis is acquired in utero, it cannot be called a venereal disease.

Syphilis and Fetal Death. Many pregnancies in syphilitic women, unless energetically treated, end in miscarriages or stillbirths. Syphilis is a very frequent cause of fetal death unless adequate treatment is given. The child may be born alive with signs of syphilis and die very early, or these signs may develop later, depending on the status of immunity in the mother. Syphilis has been a cause of much infant mortality.

Latent Syphilis. As in the case of adults, the spirochetes may remain alive in the child's body for a number of years without causing symptoms (*latent syphilis*) and then become active. After infancy, the most serious consequences

[3]Recent studies of serum hepatitis in families suggest that this disease may be transmitted by the husband to the wife via his sperm.

of congenital syphilis are stunting of growth, diseases of the eye, deafness, epilepsy, and feeblemindedness, but these conditions are not by any means always due to syphilis.

Diagnosis of Syphilis

A very satisfactory method of diagnosis in the first stage of the disease is demonstration of the spirochetes with the darkfield microscope technique (see page 523) in fresh, unstained material obtained from the chancre or sores in the mouth or elsewhere. These contain the organisms in great numbers. Negative darkfield examinations should be repeated on three consecutive days to confirm that they are truly negative. The complement fixation, precipitin reactions, and other manifestations of immunity, such as antibodies that kill and immobilize the spirochetes, become positive only in the second or third week after the appearance of the chancre. They cannot, therefore, as a rule be used for diagnosis in the very earliest stages, although later they are reliable tests. It is very important to make the diagnosis early.

Complement Fixation (CF) Tests. Application of the CF reaction to the diagnosis of syphilis was first described by a scientist named Wassermann and usually bears his name or the name of some person who has modified the technique, such as Kolmer. Serum of the patient is required. Blood for the test, usually at least 10 ml, is drawn from a vein at the bend of the elbow.

A strongly positive CF test with cardiolipin antigen[4] is usually designated by the symbol of four pluses (+ + + +); weaker reactions are designated by three pluses (+ + +), two pluses (+ +), one plus (+), plus-minus or doubtful (±), and negative (−). There are various other notation systems designed to indicate the amount of antibody in the patient's serum. These tests require several hours to perform. Other, improved specific diagnostic complement fixation tests for syphilis (such as the TPCF test) will be mentioned later.

Precipitin Tests. These flocculation tests, as they are also called, form visible aggregates under proper conditions. Various forms of the precipitin reaction (which is closely allied to the complement fixation reaction and, in syphilis, probably dependent on the same antibody or reagin) bear the names of their inventors: Kahn, Eagle, Hinton, Kline, VDRL,[5] and Mazzini. Since performance of any of the precipitin tests requires only the patient's serum and the antigen, they are simpler, quicker, and less expensive than the complement fixation tests and are thought by many investigators to be more reliable.

The precipitin tests depend on the fact that when a concentrated alcohol-ether extract of antigen, similar to that used in the complement fixation test, is mixed in certain proportions and in a certain way with the serum of a syphilitic patient, visible particles are formed that cloud the mixture, clump together into flakes or *flocs,* and precipitate to the bottom of the tube (Fig. 35–7).

Biologic False-Positive (BFP) Reactions. None of the reactions obtained in the complement fixation or precipitin tests as just described is specific because the "antigens" used in them are not of syphilitic origin. They are called *cardiolipin* antigens. They have no relation to *T. pallidum* whatever, though for little understood reasons they react especially with serum of syphilitic patients.

Because the antigens are not specific, falsely positive reactions occur in 10 to 20 per cent of all persons who do not have syphilis. A positive reaction to tests made with such antigens does not, by itself, necessarily prove that a person

[4]Cardiolipin antigen is an alcohol-ether extract of normal beef heart lipoidal material. It serves as a *nonspecific* antigen instead of *Treponema pallidum,* which would be the specific antigen, as in the TPCF test.

[5]Venereal Disease Research Laboratories of the National Communicable Disease Center.

has syphilis. Persistently positive reactions with these antigens in healthy persons are called BFP reactions. They have no significance in relation to syphilis. BFP reactions may also be encountered with people who have malaria, relapsing fever, lymphogranuloma venereum, leprosy, yaws, or any of 18 other known diseases. BFP reactions may even occur in pregnancy. This nonspecificity, unless carefully investigated, can introduce error into the tests, sometimes with tragic results; however, in general, these serologic tests are usefully accurate in diagnosing syphilis and evaluating the effect of treatment when intelligently used. When there is doubt concerning the result of these tests recourse may be had to tests depending on specific antibodies. As mentioned previously, serologic tests for syphilis are usually designated STS. Those made with cardiolipin antigens are sometimes designated as standard serologic tests (SST).

Specific Diagnostic Tests

***Treponema pallidum* Immobilization (TPI) Test.** As the disease progresses, specific antibodies appear in the serum that immobilize and kill *T. pallidum.* These antibodies act only in the presence of complement. They are entirely distinct from the nonspecific complement-fixing and precipitin reagins previously discussed. They are highly specific, react only with *T. pallidum,*[6] and persist in the patient longer than the nonspecific complement-fixing and precipitin reagins.

The *Treponema pallidum* immobilization test is an exceedingly valuable supplement to the nonspecific complement fixation and precipitin tests in diagnostic syphilis serology. The TPI test is used mainly in large diagnostic laboratories, since it is not easy to perform, but gives excellent results when the nonspecific tests are in doubt. It is too difficult and expensive for routine use by smaller laboratories. The spirochetes used in these tests are derived from the testes of rabbits artificially infected for the purpose.

The Specific Complement Fixation (TPCF) Test. Equally specific are complement fixation tests using antigens extracted from *Treponema pallidum*

[6]The antibodies also react with a few very closely related species like *T. pertenue,* the cause of yaws, and *T. carateum,* the cause of pinta.

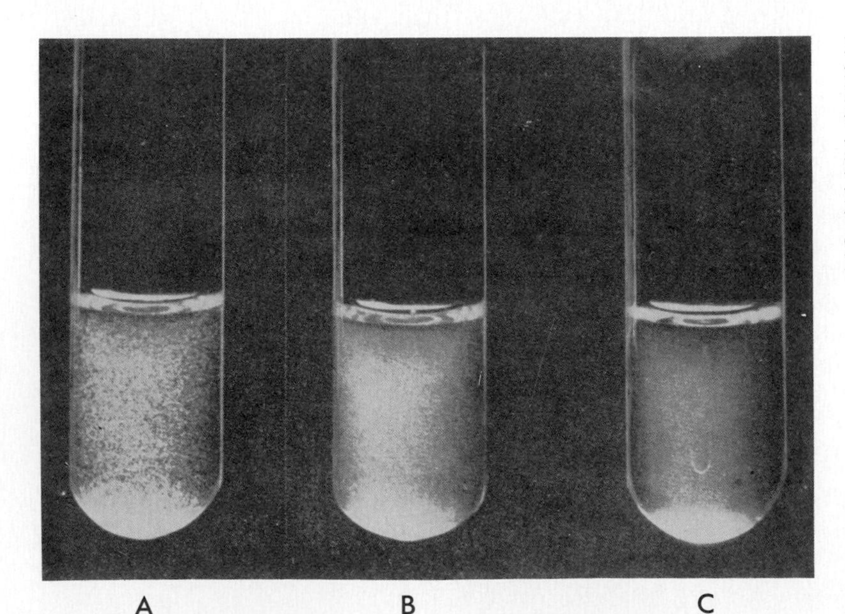

Figure 35–7 The Kahn reaction, a precipitin reaction test for syphilis. *A,* Strongly positive. Definitely visible particles are suspended in the transparent liquid. *B,* A weaker reaction. Fine particles are suspended in a somewhat turbid liquid. *C,* Negative reaction. The liquid is transparent and free from particles. (The light areas at the bottoms of the tubes are caused by reflections of light.) (From Kahn, R. L.: The Kahn Test. Baltimore, Williams & Wilkins Co., 1928.)

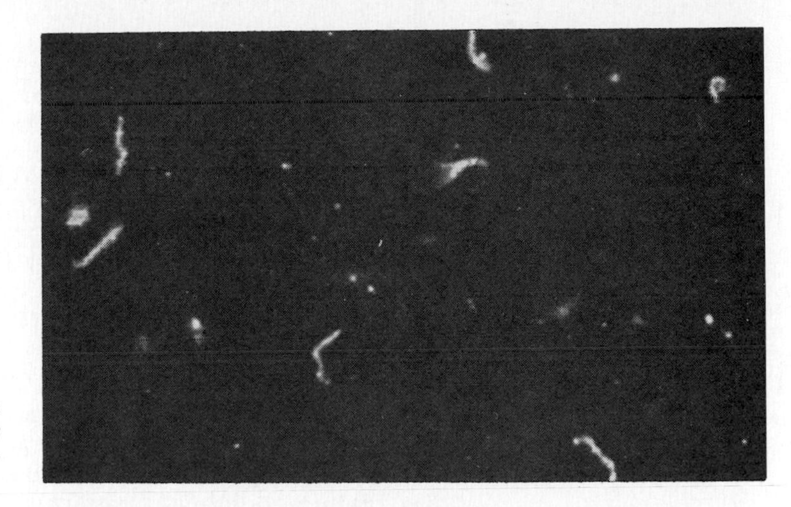

Figure 35–8 *Treponema pallidum* stained by the fluorescent-antibody technique (approximately ×1,200). (Courtesy of Baltimore Biological Laboratory, Inc.)

with a mixture of sodium desoxycholate and sodium citrate followed by acetone-ether. These tests are valuable, but preparation of such antigens is difficult and expensive. An alternative source of antigen is the cultivable Reiter spirochete.

The Reiter Protein Complement Fixation (RPCF) Test. As previously noted, although *T. pallidum* in a virulent state has not been cultivated in artificial media, very similar but avirulent spirochetes (possibly variants of *T. pallidum*) have been artificially cultivated—for example, the strain known as the *Reiter spirochetes* or Reiter treponemes *(Treponema phagedenis).* These cultured spirochetes contain antigens much like those of *T. pallidum* and are used as sources of antigen in CF tests for syphilis. These tests are thought by some to be not as specific as tests made with *T. pallidum* antigens, but they are valuable diagnostic adjuncts. Experience with them indicates that whereas cardiolipin antigens tend to give some falsely positive BFP reactions, Reiter treponeme antigens tend to err in being not sufficiently sensitive.

Fluorescent Antibody Technique. Among the most promising new developments in diagnostic techniques for infectious diseases of all kinds, including the treponematoses, are procedures based on fluorescent antibodies. The Fluorescent Treponemal Antibody Absorption (FTA-ABS) test is done by what is called the indirect method (Fig. 35–8). Details are given in the literature cited. In brief, it may be said that anti-syphilis globulins (IgM, IgA) in the patient's serum (human) are brought into contact with *Treponema pallidum* (antigen). The two combine. This combination of *T. pallidum* with anti-syphilis globulins is not visible under ordinary light. However, it is made visible by coating it with anti-human globulins conjugated with a fluorescent dye. Viewed in the microscope under ultraviolet light the antibody-coated treponemes glow with a yellow fluorescence. The same principle applies to any specific antigen-antibody combination and is widely used in diagnosis.

Other Serologic Tests. Other serologic tests are the FTA-200 (which was later modified to become the previously mentioned FTA-ABS) and the "*Treponema pallidum* immune adherence" test, known as TPIA. Some of the tests are now automated for routine use in clinical laboratories.

Other nontreponemal antigen tests are the "Unheated Serum Reagin" test (USR) and the "Rapid Plasma Reagin (RPR) (circle) Card" test. Both of these are rapid reagin tests that use a modified VDRL antigen suspension with enhanced reactivity.

Serology and Clinical Progress. A patient in the first (chancre) stage of syphilis may have a negative reaction to all serologic tests because in the very early periods of syphilis, as in most diseases, not enough antibody has appeared to give a positive reaction. The reaction usually becomes positive 10 days to two weeks after the appearance of the chancre, sometimes earlier. When treatment is instituted, the reactions usually get progressively weaker and may finally become negative. This is a favorable sign, and repeated tests are therefore made to find out how the treatment is progressing. Congenitally syphilitic children usually have a positive reaction.

If the brain or spinal cord is affected by syphilis, a positive serologic test is usually obtained with the *cerebrospinal fluid.* The reactions are practically always positive in paresis and frequently positive in locomotor ataxia.

Treatment

Penicillin in Venereal Diseases. A single treatment with 2.4 million to 4.8 million units of benzathine penicillin G will render a patient noninfectious in primary syphilis, and probably in early secondary syphilis, although the latter may require additional treatments. The same treatment will also cure early gonorrhea unless, as is increasingly frequent, the strains of *N. gonorrhoeae* are penicillin-resistant. For patients sensitive to penicillin other very effective drugs are available—tetracycline or its derivative, doxycycline (Vibramycin). Any of these drugs can have undesirable and even dangerous side effects and are to be administered only under medical supervision. It is worth noting also that other treponematoses, especially yaws, bejel, and pinta, respond equally promptly to penicillin therapy. Syphilis of more than one year's duration requires weekly injections of 2.4 million units of benzathine penicillin G for three successive weeks, or more, to acertain a cure.

When treatment is indicated in pregnancy, tetracycline is not used because of its potential toxicity to mother and child. Tetracycline should not be given to children less than 8 years of age. The risk of congenital syphilis can be greatly diminished by adequate prenatal therapy.

Unfortunately, new cases of both syphilis and gonorrhea still occur in enormous numbers annually (Fig. 35–9, Table 35–1), thanks to the ignorance, fear, indifference, and prudery of the public. Tragically, the greatest increase is among teenagers, much of it homosexually acquired. Nevertheless, the mortality from venereal disease has declined. For example, the death rate per 100,000 population from syphilis was 11.0 in 1911, 4.1 in 1950, and only about 2.0 in 1961. Yet we cannot be proud of the fact that in 1980 68,832 cases of syphilis (all stages) were reported in the United States. Of these, 27,204 were primary and secondary syphilis. This is an increase of 9.4 per cent over the number reported for 1979. In 1982, 32,553 cases were reported.

Like other communicable diseases that are a menace to public health, venereal diseases are reportable to the health department, but many infections go unreported and untreated. The decline in the death rate is the result not of a decline in new infections but of modern methods of treatment.

Importance of Prompt and Thorough Treatment. Unless all the spirochetes in the body are reached by whatever drug is used, the patient is not cured. If the drug is then discontinued, the disease will again make progress and reaction to the diagnostic serologic tests will again become positive. Drugs can arrest but not repair damage done by the spirochetes. Much depends on how soon after infection treatment is started. If the spirochetes are located in some deep, obscure, well-protected lesion in the body, drugs may not reach them for a time and repeated injections may be necessary.

The chances of permanent cure diminish in proportion to the length of time that elapses between the time of infection and the beginning of treatment. It is of the utmost importance to place every syphilitic patient under treatment as early as possible, both for his or her own sake and to protect persons with whom the patient comes into contact.

Treatment and Community Health. There are two views about the

Figure 35–9 Syphilis (primary and secondary), reported civilian case rates by year, United States, 1941–1980. (From *Morbidity and Mortality Weekly Report*, Annual Summary, 1980. U.S. Department of Health and Human Services.)

Table 35–1. Reported Cases of Venereal Disease by Age Group, United States, 1956–1970

Age Groups	1956	1960	1965	1970	1976 *Total*	Percentage Increase 1956–1976
Primary and Secondary Syphilis						
0–9	11	20 }	281 }	296 }	210	+244.2
10–14	75	139 }				
15–19	1,093	2,577	4,039	3,933	3,604	+329.7
0–19	1,179	2,736	4,320	4,229	3,814	+323.5
All Ages	6,399	16,144	23,338	21,982	23,731	+370.9
Gonorrhea						
0–9	1,222	1,619	}	7,015 }	11,867	+325.4
10–14	2,425	3,261				
15–19	44,264	53,649		144,805	261,500	+590.8
0–19	47,911	58,529		151,820	273,367	+570.6
All Ages	224,687	258,933		600,072	1,001,994	+446.0

*Adapted from a report of the U.S. Public Health Service Task Force on Eradication of Syphilis, L. Baumgartner, M.D., Chairman 1962; and National Communicable Disease Center: *Morbidity and Mortality,* 1967, Vol. 16, No. 23, U.S. Department of Health, Education, and Welfare and updated to 1976.

desirability of starting treatment during the primary stage. From the standpoint of preventive medicine, an infectious patient with an open lesion at any stage may be made noninfectious quite promptly by treatment. The patient should, on this basis, be treated immediately in order to prevent spread of the disease. Mere suppression of the infectious state, however, does not imply a cure.

This patient may actually be cured before he has time to develop any immunity, and he may promptly become infected again and be just as infectious until he is treated all over again. Some consider it useful to allow time for the disease to progress until the patient develops a good concentration of antibodies and tissue resistance, but the patient should be warned about his infectious condition, and his cooperation enlisted until the time is suitable to begin treatment. Much depends on the intelligence and cooperativeness of the patient.

Syphilis as a Family Disease. Syphilis is a family disease quite as much as tuberculosis. In the lower social levels, if the husband has the disease, it is quite probable that the wife and children have also been infected. A single ignorant or irresponsible syphilitic person can originate many new cases (Fig. 35–10).

If one person in a family is found to have syphilis, it is of the greatest importance to have the other members examined. By placing the infected members under treatment early, the late results of the disease may be avoided.

Prevention

Adequate treatment with penicillin quickly renders the syphilitic patient noninfectious, so that no special preventive procedures are necessary in contacts

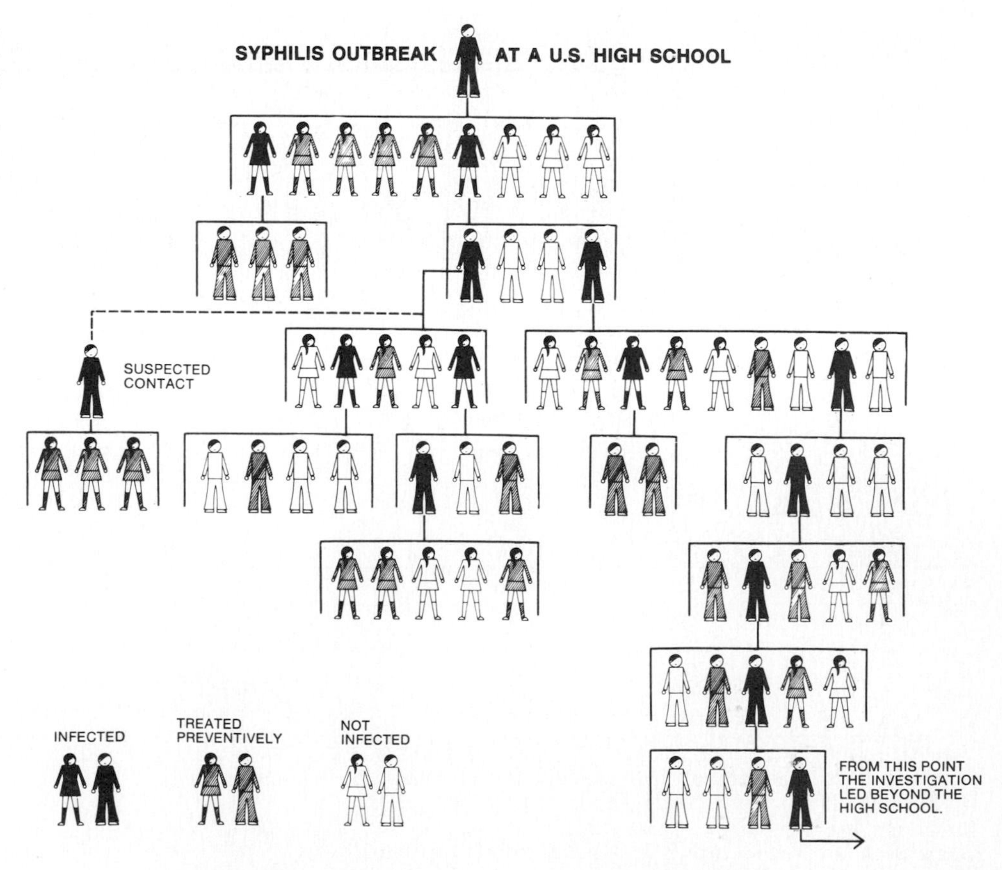

Figure 35–10 Persons involved in a high school syphilis epidemic. (From Today's VD Contact Problem—1971. Courtesy of the American Social Health Association, Committee on the Joint Statement—1970. W. L. Fleming, M. D., Chairman.)

with persons who have been so treated.[7] In dealing with untreated cases of syphilis, freshly soiled clothing, bed linen, towels, and washcloths must be regarded as potentially infectious and should be separately handled and disinfected. In the cases of persons with primary or secondary lesions of the skin, genitalia, or mouth, such things as handkerchiefs, eating utensils, and toothbrushes are also potentially infectious. Dressings of any open lesions must of course be burned or otherwise hygienically disposed of. The person faced with the problem must be guided by the nature and location of lesions, as no general rule can be laid down.

After the primary and secondary lesions have healed, and as long as no gummas or other lesions are draining to the exterior, the syphilitic patient is no longer infectious in the ordinary relations of life, although unknown numbers of spirochetes remain alive in the body. This is an important point to remember, as members of the health professions are often unnecessarily afraid of patients in the later stages of the disease. Rashes, mouth ulcers, and possibly cutaneous gummas, however, are infectious at any stage.

RABBIT SYPHILIS

This disease is caused by *Treponema paraluis-cuniculi* and is a venereal disease of rabbits. Humans are not vulnerable to this scourge; however, the close relation of this organism to *T. pallidum* has led to confusion in laboratories where rabbits are used in experimentul treponeme studies.

YAWS

As previously mentioned, yaws is not primarily a venereal disease, though its clinical progress has parallels to the venereal disease syphilis. Like syphilis, it begins with an initial lesion called a "mother yaw," accompanied by mild general symptoms. After weeks or months, very characteristic and diagnostically distinctive secondary, raspberry-like excrescences occur over the body. These are especially painful on the feet ("crab yaws"). The name frambesia, commonly given to yaws, is from the French word *framboise,* raspberry. As in syphilis, healing tends to occur but new lesions recur intermittently after long "latent" periods. There are progressive, widely destructive, and frequently revolting lesions of skin, cartilage, and bones (Fig. 35–1). Unlike syphilis, there is no involvement of the CNS, eyes, aorta, or viscera. Fatality rates are low. All serologic tests for syphilis are positive in yaws. One treatment with slowly absorbed penicillin is promptly curative. The United Nations World Health Organization (WHO) has treated millions of yaws patients with brilliant and dramatic success.

BEJEL

Bejel occurs among children in Africa, the Middle East, and Southeast Asia. It resembles yaws and produces highly infectious skin lesions. Its causative agent appears to be *Treponema pallidum,* but the early age of acquisition points to nonvenereal routes of infection combined with malnutrition. Diagnosis and treatments are like those for syphilis.

---

[7]Penicillin is also the drug of choice against bejel, yaws, and pinta.

PINTA

Treponema carateum causes pinta, a disease endemic in the Philippines, Central and South America, and the Pacific among dark-skinned people. The infection is due to direct contact or transmission by the *Hippelates* fly and causes bizarre pigmentary changes of the skin. Later the nervous system may become involved. Diagnosis and treatments are like those for syphilis.

SUMMARY

The etiologic agent of syphilis, *Treponema pallidum,* was discovered in 1905 by Schaudinn. The recognized species of the treponemes are not easily stained; consequently, other methods are generally used to observe them microscopically. These are negative staining, the silver impregnation method, the darkfield method, and use of the phase contrast microscope and the fluorescence microscope. *Treponema pallidum* does not exist long outside the body; it has not been cultured successfully in the laboratory, although the Reiter treponeme and others are being routinely reproduced.

Treponema pallidum gains entrance to the body through the skin or mucous membranes. Most often, it infects the genitalia during coitus, occasionally the lips during kissing, and possibly the anus in homosexuals. Usually, during sexual intercourse the genitalia are infected and *primary syphilis* results. A "hard chancre" appears externally, but the organism moves to every organ in the body. Antibodies (reagins) may be found in the blood two to four weeks after infection; these are useful in the *s*erologic *t*est for *s*yphilis (STS) to diagnose the disease. In secondary syphilis, a rash appears that is allergic in nature. There may be new sores in the mouth or throat or on the genitalia, with possible headache. This is a most infectious stage of "The Great Imitator," syphilis. *Tertiary syphilis* may occur weeks, months, or years later. There are gummas affecting the organs, resulting in damage to the heart, arteries, and central nervous system, leading to destruction, insanity, or death. A child may have been infected before birth with *congenital syphilis.* Many pregnancies in syphilitic women, unless energetically treated, end in miscarriages or stillbirths.

Numerous methods exist for the diagnosis of syphilis using serum to test for antibodies in the patient. Some of the most common tests are the Kolmer-Wassermann complement fixation test, the TPCF test, the TPI test, the RPCF test, some tests involving fluorescent antibodies, precipitin tests such as the Kahn, Eagle, Kline, Mazzini, and VDRL tests, and others such as the FTA-200, TPIA, and RPR tests. Benzathine penicillin G is the antibiotic of choice in the treatment of syphilis. The dosage used depends on the stage of the disease. If essential, erythromycin may be used, but other tetracyclines should be employed with caution only. It should be remembered that syphilis, like tuberculosis, is a family disease and must be so treated.

A nonvenereal *Treponema pallidum* disease is *bejel*; it resembles *yaws,* which, in turn, historically was mistaken for leprosy. *Pinta* is another treponemal disease endemic in South America, the Philippines, and other areas in the Pacific. Bejel, yaws, and pinta respond to the same diagnostic tests and treatments as syphilis does.

REVIEW QUESTIONS

1. What type of organisms are the treponemes? Name the diseases they cause and the species that cause them.
2. How may syphilis be diagnosed microscopically in the laboratory? What is the basis of the term "darkfield"?

3. Explain the following tests and describe how they differ from each other: Kolmer-Wassermann test, TPCF test, Kahn test, TPI test, and fluorescent antibody test. What other tests for syphilis can you name?
4. Define the following: hard chancre, 4+ Wassermann, "The Great Imitator," BFP, Reiter treponeme.
5. Differentiate primary, secondary, tertiary, congenital, and latent syphilis.
6. How may one become infected with syphilis?
7. What cures for syphilis exist? Which one is used most often, and why?
8. How are the following related to syphilis: rabbit syphilis, yaws, bejel, pinta?
9. What problems may exist as the result of congenital syphilis?
10. Why is syphilis considered to be a family disease?
11. Why should pregnant women have blood tests for syphilis?

Supplementary Reading

Alolerete, J. F., and Baseman, J. B.: Surface-associated host proteins or virulent *Treponema pallidum. Infect. Immun., 26*:1048, 1979.

Annotated Bibliography on Venereal Diseases for Nurses. Current ed. Washington, D.C., U.S. Department of Health, Education, and Welfare.

Baughn, R. E., and Musher, D. M.: Altered immune responsiveness associated with experimental syphilis in the rabbit: Elevated IgM and depressed IgG response to sheep erythrocytes. *J. Immunol., 120*:1691, 1978.

Caldwell, J. G.: Congenital syphilis: A nonvenereal disease. *Am. J. Nurs., 71*:1768, 1971.

Coffey, E. M., Naritomi, L. S., Ulfeldt, M. V., Bradford, L. L., and Wood, R. M.: Further evaluation of the automated fluorescent treponemal antibody test for syphilis. *Appl. Microbiol., 21*:820, 1971.

Fleming, W. L. (Chairman): Today's VD Control Problem—1971. American Social Health Association. Committee on the Joint Statement—1970.

Johnson, R. C. (ed.): The Biology of Parasitic Spirochetes. New York, Academic Press, 1976.

Manual of Tests for Syphilis. Public Health Service Publication No. 411. Washington, D.C., U.S. Government Printing Office, 1969.

Pirozzi, D.: Syphilis and the "Minor" Venereal Diseases. New York, Biomedical Information Corp., 1977.

Reed, E. L.: The rapid plasma reagin (circle) card test for syphilis as a routine screening procedure. *Pub. Health Lab., 23*:96, 1965.

U.S. Department of Health and Human Services, Public Health Service: Sexually Transmitted Diseases—Treatment Guidelines 1981. Atlanta, Georgia, Centers for Disease Control, Veneral Disease Control Services, 1982.

GONORRHEA, CHANCROID, TRICHOMONIASIS, VENEREAL HERPES, PREVENTION OF VENEREAL DISEASES

GONORRHEA

This disease, which occurs only in humans, is most commonly transmitted by coitus. When so transmitted, gonorrhea[1] is an infection primarily of the glandular tissues of the cervix (in the female) and urethra (in the male) and of adjacent reproductive organs in both males and females. It is caused by the gonococcus *Neisseria gonorrhoeae;* see Chapter 29). It begins four to ten days (rarely longer) after exposure as an acute inflammation that later (four months and longer) becomes chronic. A great deal of pus is formed in the early stages.

Gonorrhea is one of the most frequent of all the serious infectious diseases. Reliable estimates indicate that in this country over 1,600,000 cases of gonorrhea are under medical care. In addition, there are great numbers of cases, possibly another million, that remain unreported and untreated. New cases are continually occurring. In the United States the number of cases reported more than doubled between 1969 and 1980—from 498,178 to 1,004,029. Interestingly the increase was much more rapid among females than among males; from 23.75 per cent of all cases in 1969 to 40.96 per cent in 1980 (see Table 36–1 and Fig. 36–1). It has been suggested that the increase among females may be due to new freedoms (liberties?) enjoyed by women. It may also be the result of improved methods of diagnosing females, frequently a difficult process.

In 1972 (the latest comparable data on this point) the incidence of gonorrhea per 100,000 in France was 30; in Britain, 118; in Denmark, 319; in the United States, about 360; in Sweden, 514. In mainland China it was reported to be very low indeed.

In addition to gonorrhea, nongonococcal urethritis of several types is increasingly common, as are oral and anal gonococcal infections. Some of these are described later. A list of sexually transmissible diseases of humans is given in Table 36–2.

In the Female. The gonococcus causes an inflammation of the urethra, cervix uteri, vulva, various local glands in the genitalia and the uterus, fallopian tubes (salpingitis; Greek *salpinx,* tube), and ovaries. Involvement of the tubes and ovaries is quite common and necessitates many gynecologic operations. Pelvic peritonitis can cause great pain. In most gynecologic clinics it is the custom to examine routinely the cervical secretion of every patient for the gonococcus. The infection is often relatively mild and tends to chronicity, especially in adult females. In infants and prepubertal children it is severe.

The adult vaginal wall does not readily become infected with gonococci, but the vagina of preadolescent girls can easily be infected, not only with gonococci but also with other *Neisseria* (see Chapter 29), because the membrane

[1]Gonorrhea is also known as "the clap" or "GC."

CHAPTER 36

lining the vagina and covering the vulva of immature girls is thin and delicate, whereas that of the sexually mature female is thicker and more resistant to infection. Family members should be investigated for gonorrhea if preadolescent girls are infected, and child abuse should not be easily dismissed as possible cause of infection.

Proctitis, an involvement of the rectum, has been shown to be a secondary infection in about 50 per cent of females with urogenital gonorrhea. Often, it is actually *cervicitis* in the female or *urethritis* in the male (Fig. 36–2). An additional 5 to 10 per cent of infections occur at the rectal site only; numerous other infections are found in the oropharyngeal mucosa.

In the Male. The gonococcus causes inflammation of varying intensity, especially of the urethra. This is urethritis. It often spreads to adjacent portions of the genitourinary system; the epididymis, seminal vesicles, prostate, and bladder may be affected, leading to painful and disabling illness, even death. The urethra may become so scarred (often as a result of self-medication with caustic disinfectants) that urine cannot pass. This condition is commonly called *stricture* and requires surgical treatment.

Nongonococcal Urethritis. Many women with gonorrhea tend to be asymptomatic. This is certainly not true for males, among whom gonorrhea commonly causes urethritis. Gonococcal urethritis (Fig. 36–2) and nongonococcal urethritis (NGU) can be easily distinguished in most cases, unless both conditions exist simultaneously. Penicillin may be used against gonococcal urethritis but may not be effective against NGU. In multiple infections erythromycin is required to cure both conditions. Other antibiotics, useful when resistant strains of gonococci are involved, are spectinomycin, cefoxitin, tetracycline, and minocycline, depending on the complications and the tissues involved. NGU, which can cause as much damage as gonorrhea, may be due to *Chlamydia trachomatis, Ureaplasma urealyticum* (a mycoplasma), or *Trichomonas* species (protozoa) infections.

Other Consequences. Untreated gonorrhea is a frequent cause of childless marriages because it destroys parts of the genital organs. In the female the fallopian tubes are closed by scar tissue that replaces tissues destroyed by the intense local inflammation set up by the gonococci. In the male a similar process results in the occlusion of the vas deferens. These conditions prevent the ova of the female and the sperm cells of the male (respectively) from reaching the uterus, which results in sterility.

The infection of children with gonococci often takes place in families in which there is an adult case of gonorrhea. Such infection in female children may result from sleeping in the same bed with the patient or from the common use of washcloths, towels, bathtubs, and toilets. In adults, however, extracoital transmission of gonorrhea is rare.

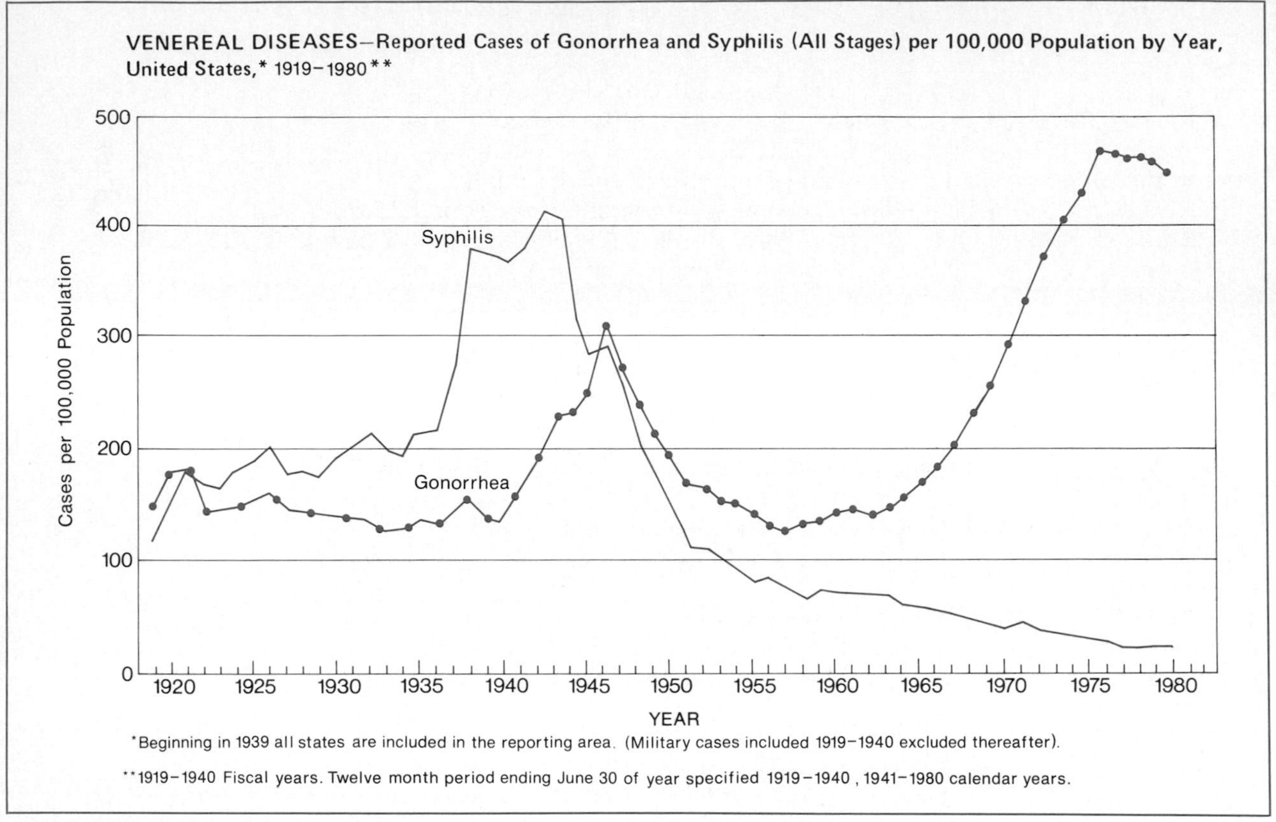

Figure 36–1 Venereal diseases—reported cases of gonorrhea and syphilis (all stages) per 100,000 population by year, United States, Fiscal Years 1919–1980. (*Morbidity and Mortality Weekly Reports,* Annual Supplement 1980.)

Ophthalmia

Ophthalmia is a severe inflammation of the eye or conjunctivae or both. It may be due to physical or chemical causes or to infection by any of several microorganisms, including staphylococci, streptococci, pneumococci, *Haemophilus* species, viruses, rickettsias, TRIC agents, and gonococci. The diagnosis should be confirmed by laboratory studies.

Gonococcal Ophthalmia. Ophthalmia caused by the gonococcus can occur in adults and in newborn babies. An adult patient with gonorrhea may infect his (or someone else's) eyes by means of his hands or fomites. Conversely, a child or adult with gonorrheal ophthalmia may infect his or her own genitalia. The organisms occur inside *pus* cells. Doctors, nurses, and medical students, if they become careless, may contract gonorrheal ophthalmia through caring for patients.

If a woman who has gonorrhea gives birth to a child, the baby's eyes may become infected during the process of birth. Unless the condition is treated promptly and vigorously, the child's eyes may be destroyed within a few days. Gonorrheal ophthalmia neonatorum was formerly the most frequent cause of blindness in infancy.

Blindness due to the gonococcus is diminishing in most countries because of the general practice of putting a drop of a silver nitrate or other prescribed antiseptic or penicillin into each eye immediately after birth (Fig. 36–3). If done properly, this is an almost sure preventive of gonococcal ophthalmia. Inclusion blennorrhea (or inclusion conjunctivitis, see Chapter 34), however, is not prevented by this means.

In most states the use of a silver preparation (silver nitrate 1 per cent) or other legally prescribed drug (antibiotics or sulfonamides or combinations) in the eyes at birth is required by law, and a report must be given to the health department within 24 hours of any inflammation of the eyes of a newborn baby. Devices for treating the eyes at birth are furnished free by most health departments. They usually consist of small wax ampules, which are crushed between the fingers so that the contents drop into the eye of the infant.

Therapeutic regimens for severe neonatal ophthalmia involve, in addition to antibiotic eye drops (not including silver nitrate), hospitalization and daily intravenous administration of aqueous crystalline penicillin for seven days.

Oral Gonorrheal Lesions

Extragenital gonorrheal lesions, either primary or secondary, are increasing in frequency. Members of the dental profession must realize that oral gonorrhea and pharyngitis are dangerous not only to the dental practitioner but also to patients who are seen after the infected individual. It is best to refer the infected patient to a physician for treatment without delay and to postpone dental procedures to a later date.

Table 36–1. Incidence of the Most Common Venereal Diseases in the United States

Disease	Year	Total New Cases Reported	Cases in Males	Cases in Females	Per Cent of Cases in Males	Per Cent of Cases in Females
Gonorrhea*	1969	478,178	364,645	113,533	76.25	23.75
	1971	670,268	—	—	—	—
	1972	767,215	504,575	262,640	65.77	34.23
	1973	842,621	509,821	332,800	60.50	39.50
	1974	906,121	534,565	364,378	59.46	40.54
	1975	999,937	593,754	406,183	59.37	40.62
	1976	1,001,994	596,613	405,381	59.54	40.45
	1977	1,022,219				
	1978	1,013,436	597,639	415,797	59.97	40.03
	1979	1,004,058	590,763	413,295	58.83	41.17
	1980	1,004,029	592,809	411,220	59.04	40.96
Syphilis*†	1974	25,385	17,903	7,482	70.52	29.48
	1975	25,561	18,429	7,132	72.09	27.90
	1976	23,731	17,301	6,430	72.90	27.09
	1977	20,399				
	1978	21,656	16,325	5,331	75.38	24.62
	1979	24,874	19,183	5,691	77.12	22.88
	1980	27,204	20,767	6,437	76.33	23.67
Genital herpes, physician consultations	1966	29,560				
	1979	60,890				
Nongonococcal urethritis (NGU)‡	1980		estimated 120,000 to 240,000			

*Data from the Center for Disease Control.

†Not including congenital syphilis. Only primary and secondary cases included.

‡An estimated 40 per cent of NGU is caused by *Chlamydia trachomatis,* and 60 per cent is due to mycoplasmas and other organisms.

Table 36–2. Sexually Transmissible Diseases of Humans

Name of Disease	Organism Causing Disease	Frequency of Disease
Gonorrhea	*Neisseria gonorrhoeae*	Currently epidemic in U.S. Classic venereal disease
Syphilis	*Treponema pallidum*	Classic venereal disease
Chancroid, soft chancre	*Haemophilus ducreyi*	Classic venereal disease
Nongonococcal urethritis	*Ureaplasma urealyticum (a mycoplasma)*	Becoming more common, epidemic in England
Lymphogranuloma venereum	*Chlamydia trachomatis*	Classic venereal disease
Granuloma inguinale	*Calymmatobacterium granulomatis,* called Donovan bodies	Classic venereal disease
Venereal herpes, genital herpes	Herpes simplex virus, HSV-?	Currently epidemic in U.S.
Trichomonas vaginitis	*Trichomonas vaginalis,* a protozoan	Curently epidemic in U.S.
Condyloma acuminatum, wartlike lesion (papilloma) on external genitalia	Papillomavirus Condylomata acuminata	Currently epidemic in U.S.
Monilial vaginitis, candidiasis	*Candida albicans,* yeast	Curently epidemic in U.S.
Pediculosis pubis, "crabs"	Lice, especially *Phthirus pubis*	Is getting more common
Scabies, also called "the itch"	*Sarcoptes scabiei,* a mite	Is getting more common
Molluscum contagiosum	A poxvirus	
Reiter's syndrome, (conjunctivitis, urethritis, polyarthritis)	May be related to the TRIC agent, *Chlamydia*	
Pinta, treponematosis	*Treponema carateum*	
Bejel (*non*-venereal syphilis)	*Treponema pallidum*	

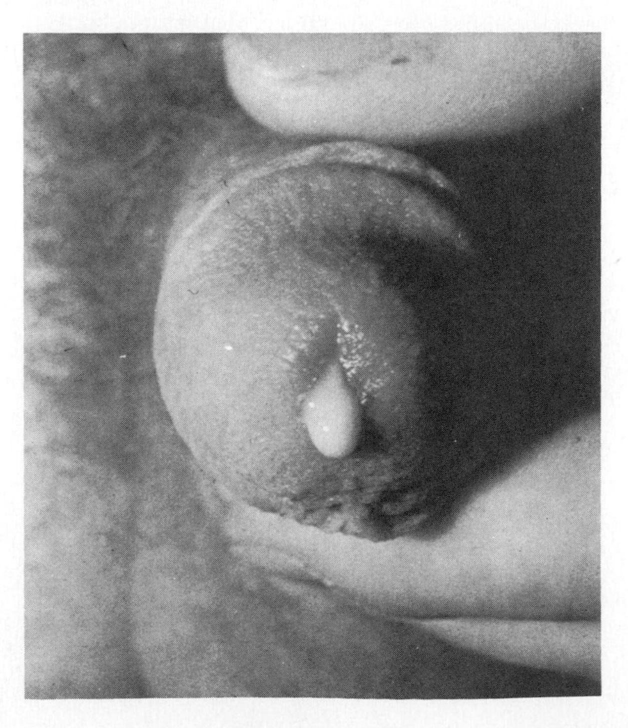

Figure 36–2 Spontaneous appearance of thick, purulent exudate at the urethral meatus is characteristic of gonococcal urethritis. (From Wisdom, A.: Color Atlas of Venerology, Chicago, Year Book Medical Publishers, Inc., 1973.)

Other Lesions

Other lesions that may result from gonococcal bacteremia include gonococcal arthritis, tenosynovitis of the knees, ankles, and wrists; proctitis; endocarditis; and meningitis. Penoanal contact with males or females may lead to anorectal gonorrhea with bloody discharges, proctitis, possible abscesses and stricture if antibiotic therapy is not forthcoming.

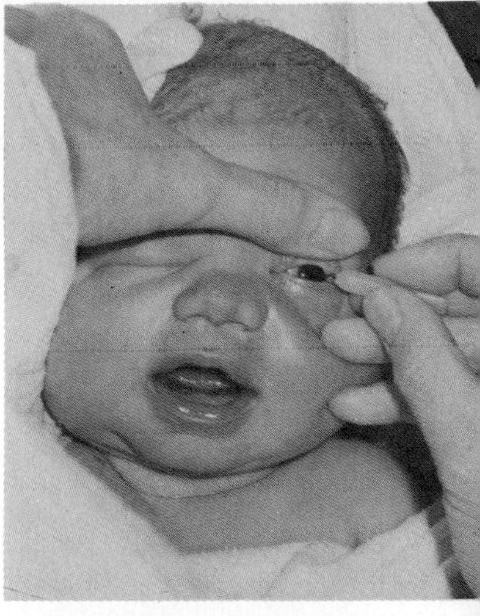

Figure 36–3 The nurse stabilizes the head of the infant with one hand and pulls down on the conjunctival sac with one finger of the other hand while dropping in the silver nitrate. (From Ingalls, A. J., and Salerno, M. C.: Maternal and Child Health Nursing. 5th ed. St. Louis: C. V. Mosby Co., 1983. Courtesy of Grossmont Hospital, La Mesa, Calif.)

Diagnosis of Gonococcal Infections

In the acute stages of gonococcal infection in a person admitting exposure, the diagnosis is usually made with the microscope by finding the distinctive gram-negative diplococci in Gram-stained specimens of pus. The cocci are characteristically enclosed within the leukocytes (Fig. 36–4). In the chronic stages gonococci are very scarce, and repeated cultural procedures may be necessary to make the diagnosis.

In female children, in persons with ophthalmia, or in adults (especially females) who effectively deny exposure, microscopic evidence is insufficient to prove that organisms resembling *Neisseria* are actually gonococci. In such instances it is necessary to inoculate "chocolate" blood-agar (blood-agar heated at 90 C for 10 minutes) containing antibiotic supplements, either vancomycin-colistin-nystatin or polymyxin-ristocetin. This medium, called Thayer-Martin medium, is prepared in Petri plates and, after inoculation, is incubated at 37 C in an atmosphere containing 10 per cent CO_2 to isolate the organisms in pure culture and make complete biochemical and serologic identifications. *Neisseria gonorrhoeae* typically ferments only glucose but without gas, and colonies of the gonococcus are catalase- and oxidase-positive. A specific fluorescent antibody staining test is also very useful in making rapid tentative identification of the organism. Complement fixation tests, although sometimes performed, are neither specific nor reliable. Error in diagnosis can be particularly embarrassing and costly for both the patient and the medical personnel.

Vaginitis or Vulvovaginitis

This condition in preadolescent girls can be due to *Neisseria* from the upper respiratory tract. The causative *Neisseria,* which are morphologically identical

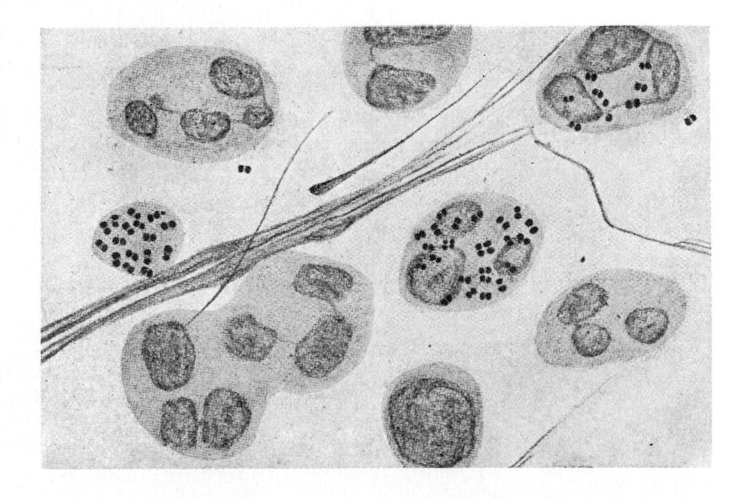

Figure 36–4 Gonococci in leukocytes in a smear of pus from a case of gonorrhea, stained with methylene blue. (Compare with Figure 29–7.) The long, fibrous objects are shreds of fibrin and mucus in the pus. Note that the pairs of gonococci are nearly within the leukocytes. (From Ford, W. W.: Bacteriology. New York, Paul B. Hoeber, 1939.)

to gonococci, are introduced into the vagina by dirty hands, clothing, towels, and so on. These organisms are easily distinguished from the gonococcus by cultural methods. Especially in children's wards and institutions, cases of suspected gonorrheal vaginitis are not infrequently innocent self-infections or fomite-borne infections with respiratory *Neisseria*. *Neisseria* common in the nose and throat and found in vaginitis are *N. sicca*, *N. subflava*, and *N. meningitidis* (the meningococcus; see Chapter 29) (Table 36–3). *Branhamella catarrhalis*, *Trichomonas vaginalis* (discussed later in this chapter), and *Candida albicans* (Chapter 38) are also causes of vulvovaginitis in both children and adults. It is sometimes difficult to cure the condition in the individual child or to prevent its transmission to other children. Some hospitals delay the admission of girls to the general children's ward until vaginal smears or cultures have been examined and found negative for *Neisseria*. Gonococcal infection of preadolescent children, especially those living with infected adults, can also occur.

Antibiotics and Neisserian Infections

A recommended schedule of treatment for uncomplicated acute gonorrhea in male and female consists of 1 g of probenecid administered orally, followed immediately by intramuscular injection of 4.8 million units of aqueous procaine penicillin G (APPG) at two or more sites in one treatment. Alternative treatments include:

1. Ampicillin, 3.5 g, plus 1 g of probenecid administered together orally.
2. Tetracycline, 0.5 g orally four times daily for ten days.
3. Spectinomycin hydrochloride, 2 g intramuscularly in one dose.
4. Cefoxitin (not for pharyngitis), 2 g intramuscularly, with 1.0 g probenecid orally.

Benzathine penicillin G, although effective for syphilis, is not recommended

Table 36–3. Taxonomy and Nomenclature of the Neisseriaceae†

Genus I.	*Neisseria*
Species:	*N. gonorrhoeae*
	N. meningitidis
	N. lactamica
	N. sicca
	N. flavescens
	N. subflava (includes *N. flava*, *N. perflava*)
	N. mucosa
Genus II.	*Branhamella*
Species:	*B. catarrhalis* (formerly *N. catarrhalis*)
	B. ovis (formerly *N. ovis*)
	B. caviae (formerly *N. caviae*)
Genus III.	*Moraxella*
Species:	*M. lacunata*
	M. nonliquefaciens
	M. bovis
	M. osloensis
	M. phenylpyruvica
	M. kingii
	*M. urethralis** (formerly *Mima polymorpha* var *oxidans*)
Genus IV.	*Acinetobacter*
Species:	*A. calcoaceticus* (formerly *Bacillus anithratum*, *Herrelea vaginicola*, or *M. lwoffi*)

*Tentatively placed in this genus.
†From Braude, A. I.: Microbiology. Philadelphia, W. B. Saunders Company, 1982.

for gonorrhea. Pregnant patients and those with complications require other therapeutic regimens.

Infections by strains of gonococci resistant to penicillin, which may eventually become more frequent, will require alternative treatment with other antibiotics. Patients with proven penicillinase-producing *Neisseria gonorrhoeae* (PPNG) should be treated with spectinomycin. Tetracycline is added for coexisting chlamydial infections.

CHANCROID

Haemophilus ducreyi was named for Ducrey, a famous Italian physician. It causes a severe ulcerative and *necrotic* (from Greek *nekros,* dead; tissue-destructive) disease of the external genitalia and adjacent parts, called *chancroid* or *soft chancre* (in contrast with the syphilitic chancre, which is hard).

Chancroid is common especially among unclean peoples. The disease is more frequent and severe in men than in women. It is transmitted primarily by sexual intercourse and very rarely by contact of the genitalia with articles (clothing, bedding, towels, toilets) freshly contaminated with the pus and serum from lesions. The causative organism is much like the influenza bacillus but is even more fragile and difficult to cultivate (Chapter 31).

After an incubation period of three days or less, a macule and then a soft, necrotic, spreading ulcer develop. The lesion spreads over the surfaces and into the deeper structures and erodes the tissues of the genitalia and adnexa. It produces painful inguinal buboes or adenitis ("swollen inguinal glands"). Diagnosis is by microscopic examination of exudate and by cultural studies of the organism. The disease may often be prevented in exposed persons by vigorous disinfection of the affected parts and by attention to personal hygiene. Tetracycline antibiotics (erythromycin) and sulfonamides (trimethoprim/sulfamethoxazole) for more than 10 days are of value in treating soft chancre.

TRICHOMONIASIS

Trichomonas vaginalis. Although differing in some morphologic details, such as number and length of flagella, size, and length of posterior spike, the protozoan *T. vaginalis* is very much like *T. hominis* (Chapter 27) and *T. tenax,* both nonpathogens. As its name implies, however, *T. vaginalis* is commonly found in the normal human vagina. It is also present in the normal human prostate and adnexa and in the urine of both males and females.

Transmission is generally by sexual contact but possibly also by common toilet seats, freshly soiled (damp) washcloths, clothing, or common vaginal douches. The protozoans are killed by a few hours of drying.

Genitourinary Trichomoniasis. Protozoa of the genus *Trichomonas* that inhabit the human genital tract ordinarily live unnoticed as parasites of humans. Under some conditions of vaginal pH, however—including, probably, infections by other microorganisms—they can multiply enormously and set up a local irritation, with itching and burning that may range from slight and transitory to prolonged and very distressing.

Diagnosis is made by finding the organisms—through microscopic examination for motile trichomonads or by means of the Romanowsky stain—in prostatic secretions, scrapings from the vagina, or urine. The organism may be cultured in trypticase-liver serum medium.

The infection may readily be cured in males by chemotherapy; in females this is more difficult. If one partner is found infected, it is advisable to examine the other. Drugs of choice are metronidazole (Flagyl), Tricofuron, and Floraquin, or estrogens when vaginal epithelium needs restoration.

VENEREAL HERPES

In numerous cases the herpes simplex virus (HSV-2) has been demonstrated in venereally (also manually) transmitted vesicular and ulcerative lesions of both male and female genitalia. Venereal herpes is unexpectedly common. The reader is referred to Chapter 33, where the herpes simplex virus is discussed in more detail. "Genital herpes" in women may become quite painful and extensive; in men the lesions remain discrete. In both sexes the disease recurs frequently and provides a portal of entry for other venereal infections.

Treatments are under study but only one can be recommended at present. On April 28, 1982, acyclovir ointment (Zovirax) was released for prescription use in the United States. Perhaps it will bring relief to the more than 20 million venereal herpes sufferers in this country.

THE PREVENTION OF VENEREAL DISEASES

The most obvious means of preventing venereal disease is, of course, to prevent contact of uninfected persons with partners of promiscuous sexual behavior and individuals known to be infected. Because this so often fails, prevention of venereal disease is one of the most urgent public health questions. One of the great difficulties is *finding* those contacts who are sources of infection. Patients need to reveal information to health personnel about contacts with persons who have venereal infections.

Unfortunately, individuals in the most infectious stages of venereal disease are usually not incapacitated and are sometimes secretive and uncooperative through fear or stupid irresponsibility. Some are indifferent, partly because they rely on effectiveness of treatment. For example, a man may become infected with *N. gonorrhoeae* by a prostitute on Sunday, infect his wife on Tuesday, and appear at the clinic for treatment on Thursday morning. Receiving a complete course of antibiotic, he returns home cured in the evening, but comes to the clinic again on Monday morning with a new acute infection contacted on Friday from his wife, whom he had infected on Tuesday. Some clinics give treatment to a person only if the sexual partner also agrees to receive treatment simultaneously, in order to avoid reinfection. Refusing to treat a patient, however, may lead to the spread of the disease.

Treatment. Syphilitic, gonorrheal, chancroid, and lymphogranuloma venereum patients can be made noninfectious in the ordinary relations of social and business life by adequate treatment with appropriate chemotherapy (see Chapter 14 and discussions of specific diseases). This is one of the most practical ways of diminishing chancroid, syphilis, and gonorrhea, and a less practical way of diminishing lymphogranuloma venereum and granuloma inguinale. Incomplete treatment, sufficient only to achieve temporary abatement of clinical symptoms, frequently results in relapse, with greater damage than ever. This is what occurs when a patient, contracting syphilis and gonorrhea at one exposure, is treated (or treats himself) for a day or two with antibiotic sufficient to cure gonorrhea but only enough to mask the early stages of syphilis, which later appears in generalized form.

Most state and many city health departments now conduct clinics where proper treatment may be obtained free of charge. Many hospitals cooperate in the campaign. They have ended the practice of requiring parental permission to treat affected preadolescents.

Personal and Household Precautions. Until the patient is adequately treated and shown to be noninfectious, the principal precautions in handling a case of any venereal disease involve preventing transfer of urethral or other discharges to other persons. Although gonococci and syphilis spirochetes cannot survive more than a few minutes of drying or an hour or so of exposure to sunlight or moist room temperature, other microorganisms are much more resistant. Therefore, all patients' towels, toilet articles, washcloths, underclothing, and bed linen should be kept apart from those belonging to others and disinfected with heat or chemicals. Bandages and dressings should be incinerated. The patient must sleep alone, and no one else should get into that bed at any time. The patient should not fondle children. Systematic washing of the hands after each voiding by the patient and by health personnel after handling infectious linens and other articles must be insisted on. Bathtubs and toilet seats must be disinfected in families in which there is an infectious patient. The use of public toilet seats is always undesirable. Paper seat covers are available in most department stores or can be improvised from a sheet of newspaper or from strips of toilet paper.

Gonorrheal and Other Ophthalmias. In any case of infectious ophthalmia, the exudate or pus is highly dangerous. Dressings soiled with exudates should be carefully handled, discarded, and incinerated. As described previously, bed linens and other items (fomites) may spread the disease. The hands are dangerous vectors of infectious materials from the eyes. If only one eye is infected, the noninfected eye should be protected with adequate bandages and the infected eye irrigated or cleaned by washing or wiping away from, instead of toward, the good eye. This may prevent transfer of the infection from one eye to the other.

Case-Finding. Finding as many of the infected persons as possible and getting them under treatment is a most important part of the prevention of venereal disease. The public health team is of great assistance here. A blood test (or other appropriate diagnostic test) should be made for every hospital, dispensary, and institutional patient, because unrecognized and far-advanced cases of syphilis are found in every large group of the general population. Most states require that all applicants for a marriage license submit to a blood test for syphilis. Surprising results are sometimes found.

Elimination of Congenital Syphilis. In order to prevent congenital syphilis, every pregnant woman should have a blood test, and all those with positive reactions should have thorough treatment. The earlier in pregnancy the mother comes under treatment, the better the chance that the child will escape injury or death. Infection of the fetus rarely occurs before the fourth month.

The children of parents known to be syphilitic should be kept under medical observation even though they show no signs or symptoms of the disease. Congenital syphilis yields to treatment much more readily in infancy than in later childhood. If congenitally syphilitic patients always received good treatment in infancy, the number of children showing the late and irreparable consequences of congenital syphilis would be much reduced.

Prophylaxis. Transmission of syphilis and gonorrhea during coitus may usually be prevented by the use of intact rubber sheaths, but this does not necessarily prevent transmission of chancroid or lymphogranuloma venereum, or of syphilis or gonorrhea if lesions are outside the protective coverage of the sheath.

Prophylaxis with 2.4 to 4.8 million units of aqueous procaine penicillin G is reportedly effective against both syphilis and gonorrhea. Allergic complications must be guarded against.

The educational, legal, and social methods of controlling venereal diseases are as important as the medical and should be briefly mentioned here. The educational campaign is being conducted by means of literature, lectures, motion pictures, sex education in the home and school, and, somewhat reluctantly, television. The aim is to teach the public about the spread, diagnosis, and treatment of these diseases and their results, using accurate, clear, and simple language and presenting the subject in an unsensational and unsentimental way. The legal attack revolves around blood and other diagnostic tests for persons intending marriage, breaking up prostitution as a business, and the protection of young people from sex racketeers, drug addiction, and the like.

SUMMARY

In the United States, the number of cases of gonorrhea doubled between 1969 and 1980. The increase was more rapid among females than among males. Nongonococcal urethritis of several types is also increasingly common.

In the female, the gonococcus *Neisseria gonorrhoeae* causes an inflammation of the urethra, cervix, uterus, vulva, various glands in the genitalia and the uterus, fallopian tubes, and ovaries. In the male, urethritis occurs, affecting the structures of the genitourinary system, such as the epididymis, seminal vesicles, prostate, and bladder. The urethra may become scarred, and stricture may result. Proctitis, involvement of the rectum, may affect as many as 50 per cent of females with urogenital gonorrhea. Many women with gonorrhea tend to be asymptomatic, and thus, diagnosis is more difficult in them than in men. Untreated gonorrhea may destroy parts of the genital organs, resulting in childless marriages.

Infections of female children with gonococci often takes place in families in which there is an adult with gonorrhea; this may be from fomites or from sleeping in the same bed. Gonorrhea in adults is not spread extracoitally. Although the pus produced in gonococcal disease is very infectious, sexual intercourse is the usual route of transmission. Several organisms may produce severe inflammation of the eye or conjunctivae or both, as in gonococcal ophthalmia. If a woman who has gonorrhea gives birth to a child, the baby's eyes may become infected during the process of birth. Gonorrheal ophthalmia neonatorum is prevented in these cases by putting a drop of silver nitrate, or other effective agent, in each eye immediately after birth.

Vulvovaginitis is an infection with *Neisseria* or *Branhamella* species in preadolescent girls; it is spread by fomites or finger contact. Like gonorrhea, it responds to treatment with ampicillin or other antibiotics.

Other infections of the genitalia are chancroid, caused by *Haemophilus ducreyi*; trichomoniasis, a protozoal disease due to *Trichomonas vaginalis*; and venereal herpes, a condition transmitted by the herpes simplex virus (HSV-2). Genital herpes may become quite painful in women; at present, there is no cure for this disease, but Zovirax ointment, recently released for prescription, may help to bring some relief.

These are all diseases that should be brought under control where prophylactic measures and proper education are provided.

REVIEW QUESTIONS

1. Why is gonorrhea on the rise in the United States' population? Are treatments available? How may the disease be prevented?
2. What other venereal disease may be transmitted together with gonorrhea?
3. How is gonorrhea diagnosed? What damage does it cause in the female? In the male?
4. Why should the sexual partner of a person with gonorrhea be treated simultaneously with the patient?
5. What problems develop with untreated gonorrhea?
6. What is vulvovaginitis? What organisms may cause it?
7. What is ophthalmia? How can it be prevented?
8. What diseases may be produced by *Haemophilus ducreyi, Trichomonas hominis,*

T. vaginalis, or HSV-2? How do these diseases differ from gonorrhea or syphilis?

9. How many different infections can you name that are considered to be venereal diseases?

Supplementary Reading

Braude, A. I.: Microbiology. Philadelphia, W. B. Saunders Company, 1982.

Center for Disease Control: Penicillinase-producing *Neisseria gonorrheae*—Worldwide. *Morbidity and Mortality Weekly Reports, 26*:153, 1977.

Janda, W. M., Bohnhoff, M., Morella, J. A., et al.: Prevalence and site pathogenic studies of *Neisseria meningitidis* and *N. gonorrhoeae* in homosexual men. *J.A.M.A., 244*:2060, 1980.

Jones, R. T., and Talley, R. S.: Simplified complete medium for the growth of *Neisseria gonorrhoeae. J. Clin. Microbiol., 5*:9, 1977.

Kellog, D. S., Jr., and Thayer, J. D.: Virulence of gonococci. *Ann. Rev. Med., 20*:323, 1969.

Lennette, E. H., et al. (Eds.): Manual of Clinical Microbiology. Washington, D.C., American Society for Microbiology, 1980.

Pirozzi, D.: Syphilis and the "Minor" Venereal Diseases. New York, Biomedical Information Corp., 1977.

Quinn, T. C., Correy, L., Chaffee, R. G., et al.: The etiology of anorectal infections in homosexual men. *Am. J. Med., 71*:395, 1981.

Raab, B., and Lorincz, A. L.: Genital herpes simplex: Concepts and treatment. *J. Am. Acad. Dermatol., 5*(3):249, 1981.

Roberts, R. B.: The Gonococcus. New York, John Wiley & Sons, 1977.

Sherris, J. C. (ed.): Laboratory Diagnosis of Gonorrhea (Cumitech 4). Washington, D.C., American Society for Microbiology, 1975.

Spectinomycin-resistant penicillinase-producing *Neisseria gonorrhoeae*—California. *Morbidity and Mortality Weekly Report, 30*:221, 1981.

Suriyanon, V., Nelson, K. E., and Ayudhya, V. C. N.: *Trichomonas vaginalis* in a perinephric abscess. *Am. J. Trop. Med. Hyg., 24*:776, 1975.

Wolner-Hanssen, P., Weström, L., and Mardh, P.-A.: Perihepatitis and chlamydial salpingitis. *Lancet, 1*:901, 1980.

BACTERIAL PATHOGENS OF THE SOIL: THE ANAEROBES AND THE AEROBES

Many species of pathogenic microorganisms occur in the soil: bacteria, viruses, fungi, protozoa, and helminths. Some injure crops or other plant life; some infect lower animals or humans. Some, e.g., *Bacillus anthracis,* occur only occasionally or accidentally in soil, as when it contains infected animal matter. Most *bacterial* pathogens of humans that are *indigenous* to fertile soil are species of the anaerobic genus *Clostridium.* One, *Clostridium botulinum,* causes a form of food poisoning that was discussed in Chapter 28. Two others (occurring also in feces) are *Cl. tetani,* the cause of lockjaw or tetanus, and *Cl. perfringens,* the principal cause of gas gangrene and one form of food poisoning. Figure 37–1 shows a simplified classification of some of these generally gram-positive rods. An alternative scheme is based primarily on motility, and another is based on characteristics of spores.

Bacillus anthracis, the cause of anthrax, is included in this chapter as a soil-borne pathogen, although it is aerobic and is usually transmitted by fomites and by the blood, fluids, and tissues of infected animals that contaminate soil. It is included here also because like all *Clostridium* species, it forms highly heat-resistant spores, an exclusive property of *Clostridium* and *Bacillus*.

THE ANAEROBES

The general discussion of anaerobiosis in Chapter 8 should be read and understood before reading the following material.

Of about 300 known species of anaerobes, only about 25 are recognized as clinically important, not excluding yet unrecognized and overlooked species (perhaps lost in isolation or cultivation). Anaerobic bacteria are the dominant microflora of the bodies of healthy individuals (Table 37–1). In the intestine they outnumber other bacteria 1,000 to 1; in the mouth and female genitalia the ratio is 10 to 1. Much remains to be learned about the pathogenicity of anaerobes.

Gas Gangrene

All species of *Clostridium* (of which there are nearly 100),[1] including those involved in gas gangrene, are obligate anaerobes and form highly heat-resistant spores. Autoclaving for at least 20 minutes at 121 C or oven baking for two hours at 165 C is necessary to kill these spores for purposes of surgical asepsis. Neither boiling nor ordinary chemical disinfectants (except certain sporicidal gases such as ethylene oxide and betapropiolactone) can be relied on to kill them. Any object open to the air and dust is likely to be contaminated with

[1]Some are *Species incertae sedis,* since they were not available for restudy to be classified in *Bergey's Manual,* Eighth Edition.

CHAPTER 37

these spores, which are found in all fertile soils. Many of them are also common inhabitants of the intestinal tract and feces of humans and animals.

Most species of *Clostridium* are harmless, and several can produce fermentations of great industrial value. Only a few need to be considered here—the species that are involved in gas gangrene, *Clostridium perfringens, Clostridium histolyticum, Clostridium paraputrificum,* and several others; and the cause of tetanus or lockjaw, *Clostridium tetani.*

Some types of *Cl. perfringens* have been implicated in outbreaks of food poisoning, especially those associated with cooked meats and soybean substitutes for meat. *Cl. botulinum* is notorious as a cause of food poisoning (see Chapter 28).

For growth, all clostridia require rich organic media—such as blood, serum, dead meat, or vegetables—and complete exclusion of atmospheric oxygen, although species vary considerably in oxygen tolerance. Special media and methods designed to exclude oxygen are generally used to cultivate them. One simple but effective apparatus is shown in Figure 8–6. (Also note Figure 37–2.) When free oxygen (air) is present, growth stops and the vegetative cells (but not the spores) die within a few minutes or hours, depending on the suspending material (e.g., saline or pus) and oxygen tolerance.

Clostridium perfringens. This organism is the most important member of the group that produces the condition called *gas gangrene.* Under proper conditions it may also cause food poisoning,[2] gastroenteritis, septicemia, pneumonia, meningitis, and burn wound sepsis.

Clostridium perfringens is a short, gram-positive, nonmotile bacillus with rounded ends. The spore is formed near the center, and, since spores resist penetration of stains, they appear as holes in the bacilli unless a special stain is used that penetrates the spore wall. When cultivated in milk, *Cl. perfringens* ferments the lactose and produces acid, which causes the milk to clot. As a result of fermentation, gas is formed, the clot is torn to shreds, and the plug may be blown from the tube. This process, called "stormy fermentation," is characteristic of *Cl. perfringens,* which is often spoken of as the "gas bacillus." Since *Cl. perfringens* is always found in the normal intestinal tract of humans and animals, it is easy to understand why its spores should be widely scattered in the soil. Deep, anaerobic wounds that are contaminated with soil, especially manured soil, may readily become infected not only with *Cl. perfringens* but also with related gas gangrene clostridia, which are similarly distributed. War wounds, especially those received on manured farmland, tragically illustrated this fact during the trench warfare of World War I.

The organisms of gas gangrene grow vigorously under proper conditions. They thrive in "cooked-meat medium," made with ground meat and broth which in fact closely resembles the necrotic tissues in a deep, contused, soil-contaminated wound likely to develop gas gangrene.

[2]Food poisoning is due to heat-resistant type A *Cl. perfringens* that usually germinates on meat. The resulting diarrhea is mild and lasts not longer than 24 hours.

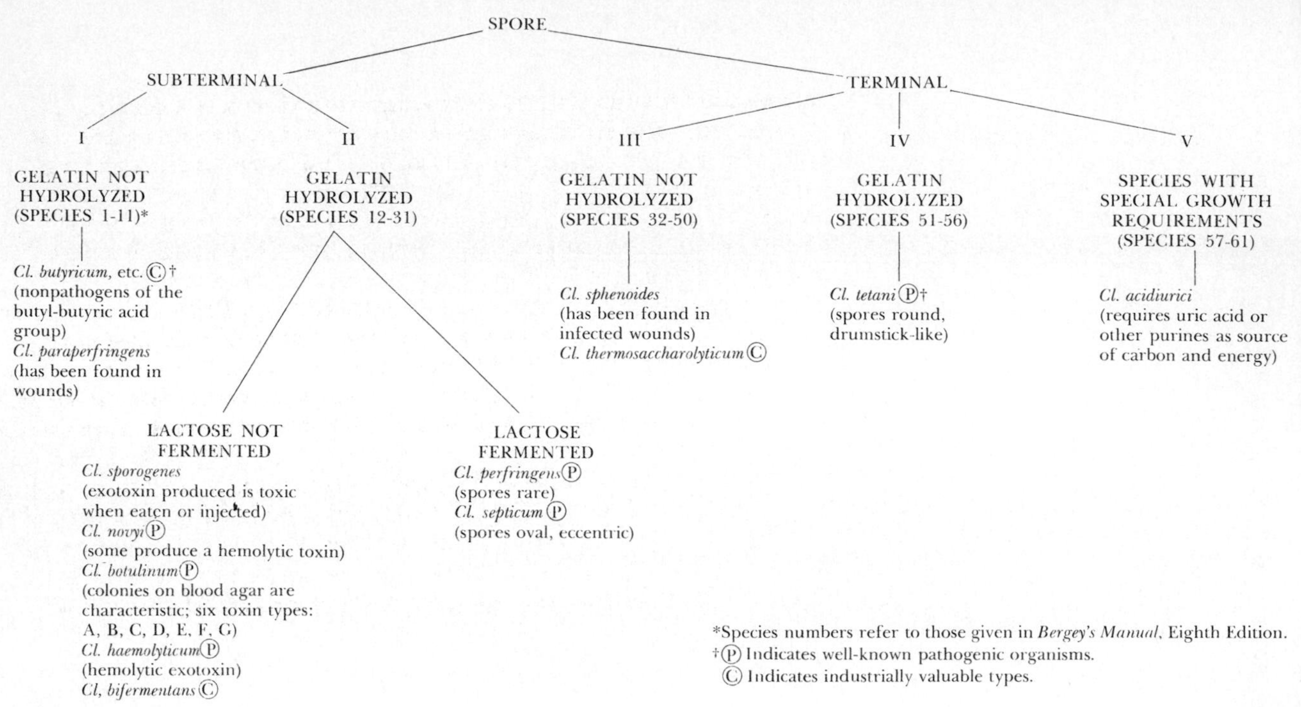

Figure 37–1 Five groups of the genus *Clostridium.* Only the more important species are listed.

Gas Gangrene. The spores of *Clostridium perfringens* and the other clostridia of gas gangrene find in a deep, dirty wound ideal conditions for germination and growth. It is warm and moist, and there is dead tissue for food, and, because the wound is deep, air is excluded and obligate anaerobes can grow. Gas is formed, the bubbles press on blood vessels, and the tissue dies, partly for lack of blood. *Cl. perfringens* produces, among several other toxins, alpha, beta, epsilon, and iota toxins and an enzyme (collagenase or kappa toxin) that helps to destroy the tissue. Other species of *Clostridium* that commonly accompany *Cl. perfringens* in soil contribute to death and destruction of the tissues.

Table 37–1. Anaerobes of Clinical Importance That Inhabit the Body

Gram Stain	Anaerobic Organism	Spore Formation	% Recovery from Human Infections	Site Usual in Humans: *Mouth*	*Female Genital Tract*	*Colon*	*Skin*
Gram-positive bacilli	*Clostridium perfringens*	+	5– 9		✓	✓	
	Cl. tetani	+					
	Cl. histolyticum	+			✓	✓	
	Cl. septicum	+			✓	✓	
	Actinomyces israelii	–		✓			
	Propionibacterium acnes	–	5–20				✓
	Eubacterium lentum	–	3– 5			✓	
Gram-negative bacilli	*Bacteroides fragilis*	–	12–23		✓	✓	
	Bacteroides melaninogenicus	–	6–10	✓	✓	✓	
	Fusobacterium nucleatum	–	2– 3	✓			
Gram-positive cocci	*Peptococcus asaccharolyticum*	–	9–10	✓	✓	✓	
	Peptostreptococcus anaerobius	–	8– 9	✓	✓	✓	
Gram-negative cocci	*Veillonella* species	–		✓			

✓ indicates that the microorganism is found in this body location.

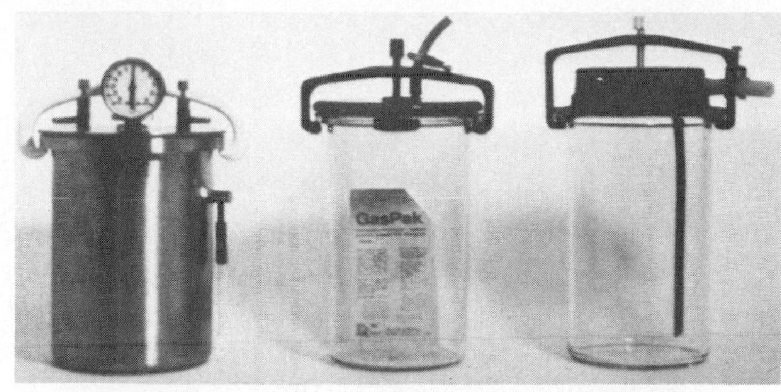

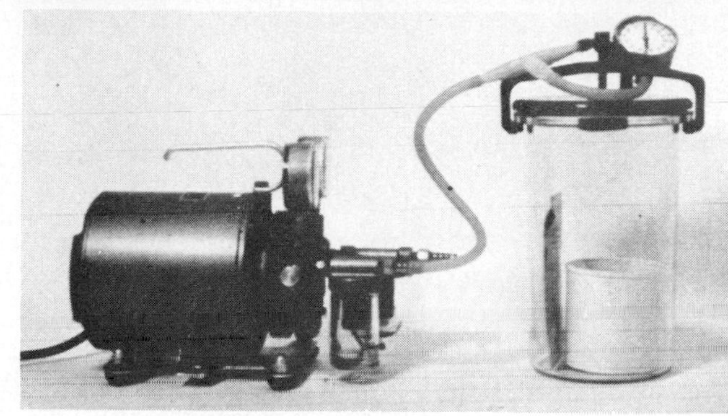

Figure 37–2 Anaerobic jars. (From Finegold, S. M., Rosenblatt, J. E., Sutter, V. L., and Attebery, H. R.: Scope Monograph on Anaerobic Infections. 2nd ed. Kalamazoo, Michigan, The Upjohn Company, 1974.)

The tissue soon becomes gangrenous and distended *(crepitant* or "bubbly") with gas, and the bacilli multiply in it and secrete more toxins. There is much foul-smelling fluid exudate. The condition spreads rapidly and may destroy a whole limb in a comparatively short time. Proteolytic species of *Clostridium,* such as *Cl. paraputrificum* and *Cl. histolyticum,* all saprophytes in the soil, contribute to the septic process by liquefying the dead tissue. There are others, more invasive, such as *Cl. septicum* and *Cl. novyi,* that cause septicemia and toxemia. Several of these organisms, including *Cl. perfringens,* produce very dangerous hemolytic exotoxins. The entire syndrome is often spoken of as a *histotoxic clostridial infection.*

Gas gangrene is not, properly speaking, a contagious disease, since it is not commonly transmitted from one person to another by close contact, as are diseases like smallpox and scarlet fever. It is *transmissible,* however. In surgical wards, operating rooms or emergency field dressing stations under war or disaster conditions, where organisms or spores from one patient's wound may be transferred to another by dirt, soiled dressings, bedding, unsterile instruments, or hands, gas gangrene may be considered a communicable or transmissible disease and controlled accordingly (Fig. 37–3). It was once a particular horror of battlefields. Depending on the site of infection, the mortality rate in gas gangrene is now 15 to 50 per cent.

Prevention. Gas gangrene can be prevented by proper and prompt treatment of wounds. Preliminary surgical cleaning (débridement) is essential to prevention. Deep, dirty wounds must be carefully cleaned, opened as much as practicable, drained, irrigated, flushed often with 100 per cent oxygen (at 3 atm) for two hours, and made to heal from the inside out. All dead tissue should be cut away. Exposure of the depths of the wound to air, as far as possible, discourages growth of the anaerobes. Polyvalent antitoxins have been employed, but their effectiveness has never been proved, and their present clinical use is minimal. Antibiotics are lifesaving, with penicillin or erythromycin being the drug of choice, in addition to clindamycin or cephalothin, but tetracycline is never used because some strains of clostridia are resistant to it.

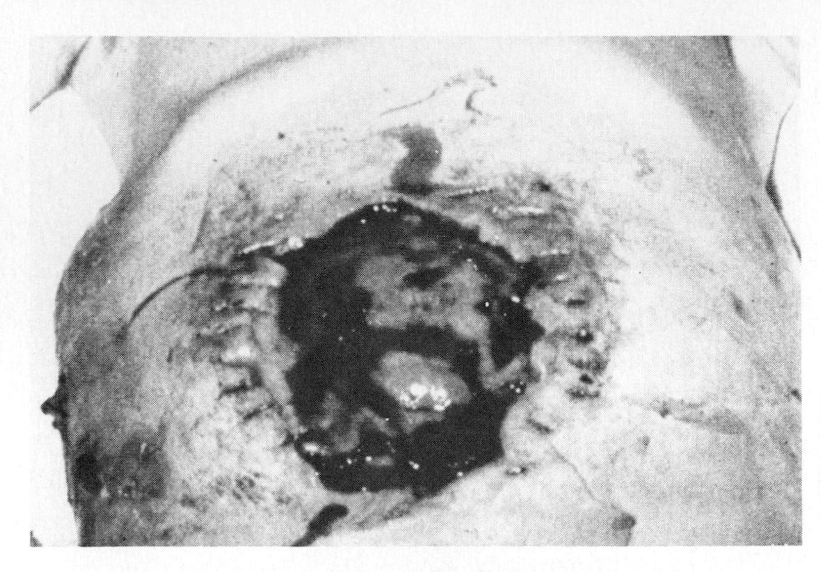

Figure 37–3 Gas gangrene of the abdominal wall following bowel surgery. (From Finegold, S. M., Rosenblatt, J. E., Sutter, V. L., and Attebery H. R.: Scope Monograph on Anaerobic Infections. 2nd ed. Kalamazoo, Michigan, The Upjohn Company, 1974.)

The prompt and early treatment of wounds in World War II and the Korean and Vietnam conflicts resulted in far fewer cases of gas gangrene than in World War I.

Several *Clostridium* antisera can be obtained commercially, suitable for animal protection, including antisera from *Cl. perfringens* types A, B, C, D, and E; *Cl. novyi* types A and B; *Cl. histolyticum; Cl. septicum; Cl. tetani;* and *Cl. botulinum* types A, B, C, D, E, and F. Bacterial strains may be typed by protecting mice with known antitoxins prior to injecting the culture.

Because gas gangrene is caused by spore-forming organisms, autoclaving, oven sterilization of infected objects, and treatment with sporicidal gases are the only safe methods of preventing transmission by fomites. Disposable objects, such as dressings, may be burned. The nurse's or surgeon's hands should be protected with sterile gloves when changing the dressings. The gloves, if not disposable, can be washed, wrapped, and autoclaved.

Tetanus

Clostridium tetani. Like *Clostridium perfringens,* this bacterium is often found in the normal intestinal tract of humans and other mammals, where it does no harm. It is very widely distributed in the soil, especially in manured farmlands. Like *Cl. perfringens,* it is dangerous to man and livestock only when it gains entrance to some deep, dirty wound where there is dead tissue. In farm accidents and war wounds, *Cl. tetani* and the gas gangrene group are often found together. Culturally, they have similar requirements. They can be differentiated in the laboratory by their microscopic appearance, specific serologic tests, and their effects on different carbohydrates, proteins, and other substances.

Newborn infants can die of tetanus *(tetanus neonatorum)* when the stump of the umbilical cord is infected with soil and/or feces. High tetanus neonatorum mortality (50 per cent or more, an estimated half million babies every year around the world) is common among primitive peoples who live in filth and have no understanding of cleanliness or sanitation, much less aseptic midwifery. But the disease is not unknown in the United States (81 cases in 1982). In fact, careless illegal abortion can produce lethal tetanus and other infections in women as well; this should not happen, but it still does, much too frequently.

The tetanus bacillus is a gram-positive, motile, spore-forming rod. The spore, situated at the end of the bacillus, gives the organism a "drumstick" appearance (Fig. 37–4). The spores, like those of other clostridia, may remain alive and virulent in soil and dust for many years.

When the bacilli grow in a wound, they produce the deadly tetanus toxin that causes the disease tetanus, also called lockjaw since it almost always affects the jaw muscles. If, however, *Cl. tetani* is taken into the intestinal tract, it is entirely harmless, a striking demonstration of the fact that many bacteria can cause an infection only when they enter by an appropriate route. Like *Cl. perfringens, Cl. tetani* cannot invade living tissue and thus lacks aggressiveness.

Tetanus Toxin. Tetanus toxin, a neurotoxin and endotoxin called *tetanospasmin,* is one of the most deadly biologic poisons known. The *injection* of 0.0008 ml of a suitable broth culture of the bacillus would probably kill a man. However, he could drink 100 times this dose with impunity, because the toxin is promptly digested in the stomach. Tetanus toxin acts on the nerves that activate muscles, producing tetanic convulsions that may lead to death (Fig. 37–5).

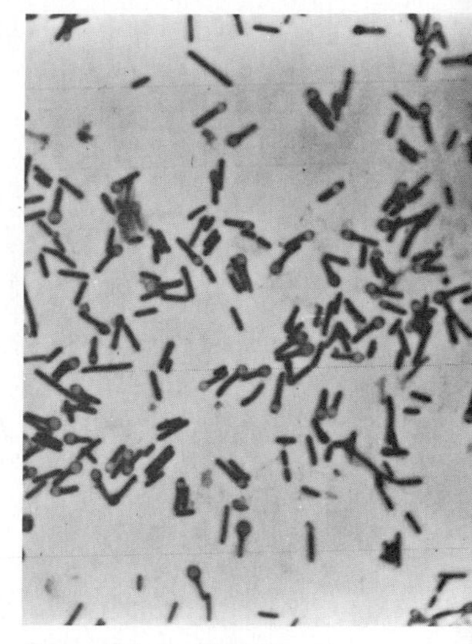

Figure 37–4 *Clostridium tetani*, a Gram-stained smear showing terminal spores that swell the rods and produce the typical drumstick appearance. Cells of *Cl. tetani* are about 0.5 to 1.0 μm in diameter and range from 2.0 to 10.0 μm in length (approximately ×1,000). (Courtesy of General Biological Supply House, Chicago, Ill.)

Tetanus Antitoxin. This is obtained in the same way as diphtheria antitoxin, that is, by drawing blood from horses that have received repeated injections of broth in which tetanus bacilli have grown and produced their toxin or toxoid. Serum is separated from the blood, the specific, tetanus-immune globulins are extracted from the serum and concentrated, and the purified concentrate is then dispensed in syringe packages for use, after proper tests for potency and sterility. The curative power of this antitoxin is limited, but is has immense value in preventing the development of tetanus. Formerly it was not uncommon to give a prophylactic or passively immunizing dose of 500 to 1,500 units to patients who had wounds in which tetanus bacilli were likely to develop, especially if they had not previously had immunizing toxoid "shots." Severe allergic reactions, however, may occur, and the patient may become permanently allergic to all therapeutic sera from horses. The necessity for the horse serum antitoxin should therefore be clearly evident before it is given to the patient.

Human tetanus–immune globulin is readily available and is being used in the United States in preference to horse serum. It has a longer half-life and is better tolerated.

After tetanus toxin has affected nerve tissue, the antitoxin cannot effectively neutralize it—hence the great importance of giving the antitoxin as soon after the injury as possible. Poor results are obtained with antitoxin given after the disease has far advanced. When symptoms of the disease are present, the

Figure 37–5 Spasms in a wounded soldier with tetanus were illustrated in this drawing by the Scottish surgeon and anatomist Sir Charles Bell in his book *The Anatomy and Philosophy of Expression*, published in 1832. The classic signs of tetanus—risus sardonicus, trismus, opisthotonus, flexion of the upper extremities, and extension of the lower extremities—are shown.

antitoxin is given in larger doses both intravenously and intraspinally, but not much can be expected of it under these circumstances. The same principles apply to the use of any antiserum (see Fig. 30–3).

Tetanus Toxoid. Persons can (and should) be actively immunized against tetanus by giving them a series of injections of tetanus toxoid. When a wound likely to cause tetanus occurs in a "toxoid-immunized" person, that person receives an immediate injection of a "booster dose" of toxoid when the wound is treated. In response, the tissues promptly form antitoxin. A prophylactic dose of antitoxin may also be given for additional protection, but only if conditions seem to warrant it.

Prevention. Unless the patient with tetanus has an open, draining wound, isolation precautions are not necessary. A patient with a draining wound requires the same isolation technique as a patient with gas gangrene. Tetanus is usually prevented entirely by proper early surgical treatment of wounds or use of toxoid and penicillin (I.V.) or tetracycline. The preferred muscle spasm relaxant is Valium.

Other Anaerobic Microorganisms

Anaerobes other than *Clostridium* are commonly responsible for brain and other abscesses. The organisms usually found are *Peptostreptococcus, Fusobacterium,* and *Bacteroides.* These occur mostly in foul-smelling mixtures of anaerobes or with aerobes or facultative organisms. Other anaerobic mixed infections may occur in the lungs or the peritoneum following injury to the bowel.

Since the gram-negative short rods *Bacteroides melaninogenicus* and *Bacteroides fragilis*[3] are normal inhabitants of the upper respiratory, genital, and intestinal tracts, and since *Bacteroides* species make up more than 95 per cent of the normal intestinal flora, they are usually also found in ulcers of the skin and mucous membranes. It is only recently that anaerobic organisms, including *Bacteroides, Peptostreptococcus anaerobius, Fusobacterium, nucleatum,* and *Peptococcus* species have been recognized as important etiologic agents of a variety of infections. These anaerobes, particularly *B. melaninogenicus* and *B. oralis,* are mentioned here to make the reader aware of their importance in the oropharynx. For every *Escherichia coli* cell in the human colon, there are about 1,000 cells of *Bacteroides fragilis,* with most of the vitamin K being synthesized by *Bacteroides* rather than by *E. coli,* as was once believed.

---

[3]*Bacteroides* species are gram-negative, nonspore-forming, strictly anaerobic rods.

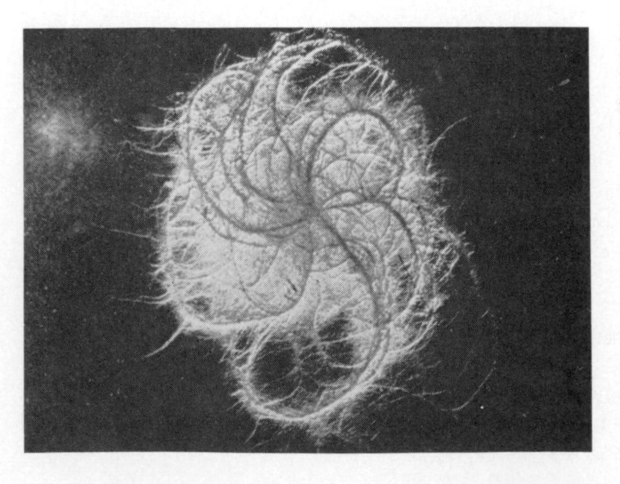

Figure 37–6 Colony of *Bacillus cereus* var. *mycoides* growing on nutrient agar. The curiously curled growth is distinctive of this organism. It is interesting to note that the growth is very frequently counterclockwise in direction. (Courtesy of Dr. F. E. Clark, U.S. Agricultural Research Service, Beltsville, Md.)

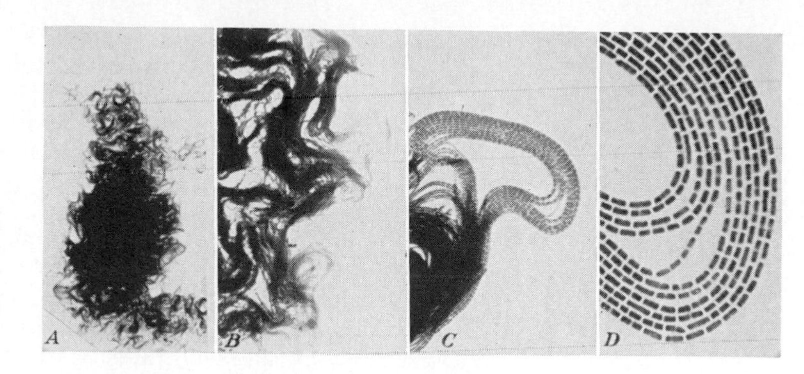

Figure 37–7 *Bacillus anthracis.* Photomicrograph of a colony on agar, stained with methylene blue. *A* shows the entire colony at an original magnification of ×45. Distinctive curled strands or "ropes" of bacilli are seen. *B, C,* and *D* show one portion of a strand at increasing magnifications (×95, ×400, and ×1,600). Note the square-ended (cigarette-shaped) bacilli in *D.* Some notion of the astronomic numbers of cells in a single small colony of bacteria may be gained from this picture. (Courtesy of Dr. C. D. Stein, Agricultural Research Service, U.S. Dept. of Agriculture, Washington, D.C.)

THE AEROBES

The Genus Bacillus

This genus includes at present 48 species, divided into two groups. Most of these bacteria are common, harmless, and useful. They are found in soil and dust virtually everywhere. The harmless species include *Bacillus subtilis* (source of the antibiotic bacitracin), *B. polymyxa* (which produces the antibiotic polymyxin), and *B. cereus* var. *mycoides* (which forms very distinctive colonies [Fig. 37–6]). These various species are found in "hay infusions" (hay soaked for a day or two in water) and are often called "hay bacilli." They are of no particular medical importance, except that their spores are very heat-resistant and often appear as embarrassing witnesses to careless sterilization by growing in supposedly sterile environments. Occasionally they cause serious infections of wounds. Several outbreaks of food poisoning have been attributed to *Bacillus cereus,* and *Bacillus subtilis* may sometimes cause human eye infections. Most of the species grow readily on simple organic media at room temperatures. They are as active in fermentation and putrefaction as *Clostridium* species. They are useful scavengers, and many produce valuable industrial products.

Spore-forming bacteria of the genus *Bacillus* are, typically, gram-positive rods that in many ways resemble those in the genus *Clostridium.* In contrast to the strictly anaerobic *Clostridium,* however, most species of the genus *Bacillus* require access to air at all times and will not grow in its absence.

Bacillus anthracis. The only species of the genus *Bacillus* that regularly causes serious infection is *Bacillus anthracis,* a large (2.5 × 10.0 μm), rod-shaped, spore-forming, gram-positive bacillus that resembles *Bacillus cereus* morphologically (Fig. 37–7). *B. anthracis* causes the disease anthrax (Greek for carbuncle: malignant boil or pustule) in farm animals and humans. The organism forms highly resistant spores that can remain alive for many years. They occur in soil and are often found in the hair of domestic animals, where they are a menace to workers in the wool, hair, hide, and leather industries. Government regulations require the proper disinfecting of hides and wool before they may be handled by workers. Brushes made from animal hair or bristles must be sterilized and so marked. Some cattle and sheep fields in France and other parts of Europe have at times been called "anthrax pastures," so permeated are they with anthrax spores.

Anthrax. Three forms of anthrax may be recognized in humans: (1) cutaneous, usually appearing on hands or forearms; (2) pulmonary, which is very severe and often fatal; and (3) gastrointestinal, which is rare and almost invariably fatal. Cutaneous anthrax, untreated, has a fatality rate of from 5 to 20 per cent. The death rate is very low if antibiotic therapy is promptly and properly administered. *Bacillus anthracis* gains entrance to the animal or human body chiefly through cuts and scratches in the skin. The bacilli multiply at the portal of entry very rapidly, forming a large black "malignant" pustule. From this site they may invade the blood, where in a few hours they become exceedingly numerous. Toxins produced by the bacilli appear to be the principal pathogenic agent in the infection. Death may be due to septicemia and usually ensues quickly. Pulmonary anthrax is associated with the wool industry and is called "wool-sorters disease"; it is due to inhalation of the spores from wool dust, especially in the presence of an existing respiratory disease (e.g., influenza). In the United States, in 1976, wool imported from Asia had to be confiscated from stores and destroyed when it was discovered to contain anthrax spores.

In anthrax, all body fluids (with the possible exception of urine) teem with the organisms. The patient or animal with anthrax is a real source of danger because the bacilli sporulate as soon as the exudates, tissues, or blood are exposed to air. The fluids and carcass of an animal dead of anthrax are clearly sources of contamination and a menace to careless handlers.

Prevention. The care of a patient with anthrax involves strict precautions and can be dangerous. It requires safe disposal of any objects or dressings contaminated with purulent materials or, in fatal cases, fluids exuding from the wounds, eyes, nose, and mouth. If feasible, these materials should be burned promptly to prevent sporulation or spread of spores. Once dried, spores may be scattered far and wide by wind and dust. It is best to use surgical gloves and a rubber or plastic apron in handling such material. Only autoclaving, oven sterilization, or exposure to sporicidal gases for a sufficient time can be relied on to disinfect articles not burned.

A precipitin test, called the *Ascoli test,* is used to diagnose anthrax in dead animals. Boiled, filtered extract of tissue of the animal suspected of having anthrax, used as antigen, is carefully layered over serum from an immunized rabbit. This is done to determine if hides removed for industrial use from such an animal are safe to process.

Pulmonary anthrax requires the same care as other highly dangerous respiratory infections (e.g., pulmonary bubonic plague), with the added precautions necessitated by the presence of disinfection-resistant spores in the exudates.

When growing in infected tissues or when cultivated in organic media containing serum and bicarbonate ions, *B. anthracis* produces a potent exotoxin, and the production of capsules is greatly enhanced. Virulence is related to the presence of both toxin and capsules. Antibodies are engendered against both. The exotoxin can be used to induce artificial immunity and is available for certain persons who are heavily exposed to anthrax. The use of antiserum containing antitoxin has been largely superseded by treatment with chemotherapy, sulfonamides, penicillin G, and tetracycline.

MELIOIDOSIS

Melioidosis, a disease first described by Whitmore in 1912, occurs in humans and some other mammals and is caused by *Pseudomonas pseudomallei.* The disease is endemic in many Southeast Asian areas and islands, but scattered cases have occurred around the globe. Melioidosis caused frequently fatal infections among the U.S. armed forces in Vietnam.

Clinically, melioidosis in humans resembles glanders in horses.[4] Typically it is a generalized infection spreading from an initial cutaneous lesion or by way of the oral or respiratory tract. It is marked by high fever, muscular pains,

[4]Glanders (also called farcy) in horses is caused by *Pseudomonas mallei*; it is transmissible to humans.

nodular abscesses, and pustules, especially in the nose, mouth, and respiratory tract, with copious mucopurulent discharges. Symptoms vary depending on which organs are affected (e.g., meningitis and pneumonia). Like many other diseases, melioidosis varies in severity from subacute to rapidly fatal (in two to four days) to chronic. In the acute cases septicemia and generalized invasions occur. General symptoms are probably due to endotoxins, exotoxins, or both.

Infection appears to be commonly derived from soil, vegetation, or water polluted by discharges from infected animals or persons, or by direct contact with discharges, blood, or tissues, all of which are highly infectious. Personnel attending cases of melioidosis must exercise the utmost care in handling patients and in disposing of fomites and all contaminated materials. Portals of entry appear not to be specific and may include cuts and scratches, mucous membranes, inhalation of infectious dust, or contact with contaminated objects. Rodents are commonly infected, arthropods may be vectors, and rat excreta may pollute water or food. The epidemiology of this condition is still uncertain.

Pseudomonas pseudomallei is a small, gram-negative, nonspore-forming rod with bipolar flagella and polar granules that are conspicuous when stained with Wayson's stain.[5] The organism grows well at 37 C on selective medium (EMB) used for Enterobacteriaceae, on Sabouraud's medium with antibiotics at pH 5.6, or on infusion agar, on which it forms smooth, viscous, honey-colored colonies. It produces acid (but no gas) from glucose, lactose, sucrose, and other carbohydrates as well as from several alcohols. It reduces nitrates and gives a positive cytochrome oxidase reaction. *Pseudomonas pseudomallei* has often been confused with *Pseudomonas mallei* (the cause of glanders), previously known as *Actinobacillus pseudomallei, Bacillus mallei,* and *Malleomyces mallei; malleus* is the Latin word for glanders.

SUMMARY

Of about 300 known species of anaerobes, only about 25 are recognized as clinically important. However, these organisms are the dominant microflora of the body in healthy individuals. Much remains to be learned about the pathogenicity of anaerobes. There are close to 100 species of *Clostridium*, most of them harmless. Clostridia are obligate anaerobes that form heat-resistant spores that are highly resistant to boiling and common disinfectants.

Clostridial pathogens are those that cause gas gangrene *(Clostridium perfringens, Cl. histolyticum, Cl. paraputrificum,* and others); the organism of food poisoning, or botulism, *Cl. botulinum*; and the tetanous (lockjaw) organism *Cl. tetani*. The last two are deadly organisms owing to their powerful toxins. In wars throughout history, *Cl. perfrigens* found its way into deep, dirty wounds and formed gas bubbles that compressed blood vessels, resulting in gas gangrene and death and destruction of tissues. War wounds were also portals of entry for *Cl. tetani*, as is the stump of the umbilical cord when soiled with dirt or feces, causing tetanus neonatorum in infants. Careless illegal abortion can also lead to lethal tetanus.

Tetanus antitoxin and tetanus toxoid are very effective in the prevention of the disease; human tetanus immune globulin is also readily available.

---

[5]Wayson's stain:

Solution *A*

Basic fuchsin	0.20 g
Methylene blue	0.75 g
Ethyl alcohol (absolute)........	20.00 ml

Solution *B*

5% phenol in distilled H_2O	200.00 ml

Add solution *A* to solution *B*. Stain for 10 to 30 seconds; wash with water.

Anaerobic, gram-negative, short rods of the genus *Bacteroides* make up more than 95 per cent of the normal intestinal flora. Many infections are most likely caused by these organisms, but they are not often isolated from the ulcer, periodontal lesion, or other lesion.

The only pathogen in the aerobic genus *Bacillus* is *Bacillus anthracis*. Although anthrax is normally a disease of livestock, *B. anthracis* also infects humans by way of cuts or scratches in the skin. A malignant pustule may form on the skin (like a boil or abscess), or pulmonary anthrax or intestinal anthrax may develop. Toxins produced by the bacilli appear to be the principal pathogenic agent in the infection. Pulmonary anthrax is also called "wool-sorters' disease." It is also an occupational hazard for workers who handle hair, hides, or leather. The use of antiserum containing antitoxin has been largely superseded by treatment with sulfonamide, penicillin G, and tetracycline.

Pseudomonas pseudomallei is the cause of melioidosis in humans, a disease that clinically resembles glanders in horses. Again, this infection appears to be commonly derived from soil or water polluted by discharges from infected animals or persons. Blood and tissues of people with melioidosis are highly infectious.

REVIEW QUESTIONS

1. What are the characteristics of the genus *Clostridium?* What diseases do clostridia cause?
2. What is gas gangrene? Under what conditions can it be found? How is it prevented?
3. What are the similarities and differences of *Cl. perfringens, Cl. tetani,* and *Cl. botulinum?*
4. What is tetanus toxin? Tetanus antitoxin? How is each prepared? What is tetanus toxoid?
5. Where is *Cl. botulinum* usually found? How does this organism cause disease in man? How can this disease be prevented?
6. What is the importance of *Bacillus anthracis* to man and animals? How is the disease anthrax transmitted to man? What precautions are important in the care of a patient with anthrax to prevent transmission of this disease? Why must these precautions be taken?
7. What other anaerobes can you name?
8. How are *Pseudomonas pseudomallei* and *Pseudomonas mallei* related? What disease does each cause?
9. What is the use of Wayson's stain?
10. What anaerobes of clinical importance inhabit the human body? How does their distribution compare with that of aerobes?
11. What is "wool-sorters' disease"?

Supplementary Reading

Angelety, L. H., and Wright, G. G.: Agar diffusion method for the differentiation of *Bacillus anthracis. Appl. Microbiol., 21*:157, 1971.

Berggren, W. L., and Berggren, G. M.: Changing incidence of fatal tetanus of the newborn. *Amer. J. Trop. Med. Hyg., 20*:491, 1971.

Bryan, F. L., and Kilpatrick, E. G.: *Clostridium perfringens* related to roast beef cooking, storage, and contamination in a fast food service restaurant. *Amer. J. Pub. Health, 61*:1869, 1971.

Finegold, S. M.: Anaerobic Bacteria in Human Disease. New York, Academic Press, 1977.

Finegold, S. M.: Anaerobic Bacteria in Human Disease. New York, Academic Press, 1977.
Finegold, S. M.: Taxonomy, enzymes, and clinical relevance of anaerobic bacteria. *Rev. Infect. Dis., 1*:248, 1979.
Gerhardt, P., Sadoff, H. L., and Costilow, R. N. (eds.): Spores. Vol. 6. Washington, D.C., American Society for Microbiology, 1975.
Heerema, M. S., Ein, M. E., Musher, D. M., et aL.: Anaerobic bacterial meningitis. *Am. J. Med., 67*:219, 1979.
Newman, M. G.: The role of *Bacteroides melaninogenicus* and other anaerobes in periodontal infections. *Rev. Infect. Dis., 1*:313, 1979.
Perlstein, M. A., Stein, M. D., and Elam, H.: Routine treatment of tetanus. *J.A.M.A., 173*:1536, 1960.
Redfearn, M. S., Palleroni, N. J., and Stanier, R. Y.: A comparative study of *Pseudomonas pseudomallei* and *Bacillus mallei. J. Gen Microbiol., 43*:293, 1966.
Smith, L., and Holdeman, L. V.: The Pathogenic Anaerobic Bacteria. Springfield, Ill., Charles C Thomas, 1968.
Smith, L. D. S.: The Pathogenic Anaerobic Bacteria. 2nd ed. Springfield, Ill., Charles C Thomas, 1975.
Sterne, M.: Pathogenic Clostridia. London, Butterworth, 1975.
Sterne, M., and van Heyningen, W. E.: The clostridia. *In* Dubos, R. J., and Hirsch, J. G.: Bacterial and Myocotic Infections of Man. 4th ed. Philadelphia, J. B. Lippincott Co., 1965.
van Heyningen, W. E.: Tetanus. *Sci. Amer., 218*:69, 1968.
Wright, G. G., Angelety, L. H., and Swanson, B.: Studies on immunity in anthrax. XII. Requirements for phosphate for elaboration of protective antigen and its partial replacement by charcoal. *Infect. Immun., 2*:772, 1970.

THE MYCOSES

Diseases caused by yeasts and molds are referred to as mycotic diseases or *mycoses*. These fungal infections may be classified into two main groups: superficial or *cutaneous* mycoses, involving the skin, hair, and nails; and *systemic* or *deep* mycoses, invading subcutaneous tissues and internal organs. Most fungi that cause *systemic* mycoses of humans live primarily in the soil as saprophytes. Most fungi that cause human *cutaneous* mycoses are not indigenous to the soil, but they are included in this chapter for convenience in discussing fungal infections in general.

THE CUTANEOUS MYCOSES

Most fungi invading the skin, hair, and nails grow as branching filaments or hyphae (Fig. 38–1). They invade only the dead outer layer of these tissues; the hyphae cannot grow significantly into the deeper or living regions. They are typical aerobes. The fungi that cause superficial mycoses are spoken of collectively as *dermatophytes,* literally, "skin plants." The diseases they cause are called dermatomycoses (singular: dermatomycosis). The dermatophytes can attack and utilize *keratin,* a tough insoluble protein, the principal component of hair, nails, horn, and epidermis.

When the *skin* is infected, toxic substances of the fungi diffuse into the living layer of dermal cells. Most people are sensitive to these materials, and their skin responds with a reddened (erythematous) area that itches. Little blisters soon appear within the erythematous area. The infecting fungus spreads radially into adjacent skin. The lay term "ringworm" for some of these conditions arose because the center of the lesion, when it becomes scaly and heals, leaves an enlarging, red, outer ring covered with blisters.

Hair can be infected within the hair follicle just above the area where the hair shaft is being formed by living cells. As these cells continue to form the hair shaft, the infected portion of the shaft is brought upward. Hyphae of some fungi may invade the internal structure of the hair shaft, which causes the hair to break readily, often within the hair follicle. This makes the infected area of the scalp appear bald. When only the surface of the hair shaft is infected by the fungal hyphae, the hair is not as fragile. It will often break, however, giving the scalp a moth-eaten appearance. Children are much more susceptible to fungal infections of the hair than are adults.

Infected *nails* become yellow and then start crumbling. Such infections are almost always secondary to fungal infection elsewhere on the body.

"Ringworm"; The Tineas

Superficial fungal infections ("ringworm") of the skin, hair, and nails are known by the medical term *tinea.* When tinea occurs on the scalp and involves the hair, the infection is called *tinea capitis.* When fungal infections occur on the foot, a condition commonly called "athlete's foot," the infection is designated

CHAPTER 38

as *tinea pedis* (Fig. 38–2). Infections on other parts of the body by various dermatophytes are also given specific names (nail, *tinea unguium;* groin, *tinea cruris;* body, *tinea corporis;* ringworm of the beard or barbers' itch, *tinea barbae).*

These infections are collectively called *dermatophytoses* or *dermatomycoses,* from the group name *dermatophyte.* The dermatophytes are closely related fungi comprising three genera and about 20 species. Some are listed in Table 38–1. Almost any species of dermatophyte may cause any form of superficial mycosis, the clinical appearance depending largely on the location of the infection.

A few of the dermatophytes live in soil.[1] Infection takes place when the fungus from the soil is rubbed onto the skin. The fungi grow as hyphae in soil or in cultures, but they also produce conidiospores. Other species occur only on the skin and hair of humans and animals. Infections by such fungi are not spread by soil but directly from one individual to another. As the infecting hyphae age in the tissues, the fungus cells may develop directly into chains of spores.

Dermatomycoses are spread easily. It is important to remember that each scale from a ringworm lesion and each stub of infected hair contains hyphae and spores. If spores are produced inside the hair shaft, the fungus is said to be of the *endothrix* type; if spores are formed on the outer surface of the shaft, the fungus is of the *ectothrix* type. Hyphae or spores grow readily if they become lodged on the skin or scalp of another susceptible individual, especially a child. The hyphae promptly grow out from the infected skin scale into the skin of the new host. Dogs, cats, cattle, and other animals are often the source of a ringworm infection. Close inspection of an infected animal will show scaly patches with a loss of hair. An infected child can transmit this infection to the family cat or puppy as well as to human playmates.

Numerous different fungi may infect the external auditory canal, causing otomycosis or myringomycosis ("hot weather ear" or "fungus ear"). Often *Candida* (a yeast), *Aspergillus, Penicillium,* or other fungi grow together with common bacterial species that may have started an infection.

Treatment of Superficial Mycoses

As has been mentioned in Chapter 14, griseofulvin (given orally with meals to enhance absorption) is quite effective if the infection is superficial. Therapeutically useful for candidiasis is topical nystatin (Mycostatin) or amphotericin B (Fungizone). Ringworm has been treated with various topical antifungal agents such as Whitefield's Ointment (to prevent scaling), and antibiotics such as haloprogin (Halotex), clotrimazole (Lotrimin), and miconazole (MicaTin). Before these antibiotics became available, treatments for tinea involved ultraviolet radiation and various other topical ointments. For candidiasis of the genitalia or tinea capitis in young girls, estrogens were often used. Although these sometimes cured the infection, they made the girls develop prematurely.

[1]Notably *Keratinomyces ajelloi.*

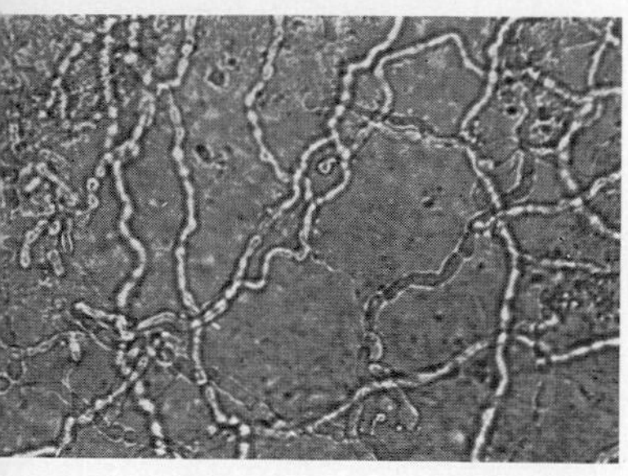

Figure 38–1 A mold (*Trichophyton*) causing "ringworm" of the feet or "athlete's foot," sometimes contracted in swimming pools and gymnasiums. This slide shows material scraped from superficial lesions on the foot and treated with 10 per cent KOH (×350). (Courtesy of Miss Rhoda W. Benham.)

THE SYSTEMIC MYCOSES

Fungi causing systemic mycoses occur naturally in soil as saprophytes. Humans and animals become infected with certain of these fungi when they inhale dust containing spores or small fragments of hyphae. When lodged in the lung, these begin to grow. If the host's defenses, such as the macrophages, are not capable of destroying these organisms, they multiply and form a lesion in the lung. The fungi may then (depending on the species) enter the lymph or blood and be carried to all parts of the body, where they may produce more lesions. In most people, however, infection is limited to a small area of the lung, and these infections soon heal.

Some soil fungi enter through wounds of the hands or feet. Millions of people are infected by pathogenic fungi every year, but only a few develop obvious disease.

Coccidioidomycosis

Coccidioidomycosis is a deep (systemic) mycosis that begins as a respiratory infection. In this stage it resembles influenza.

This disease was first recognized as occurring commonly in the San Joaquin Valley of California, and consequently the name San Joaquin Valley fever became popular. The responsible dimorphic fungus, *Coccidioides immitis,* grows as a saprophyte in the soil in many areas of the arid southwestern United States

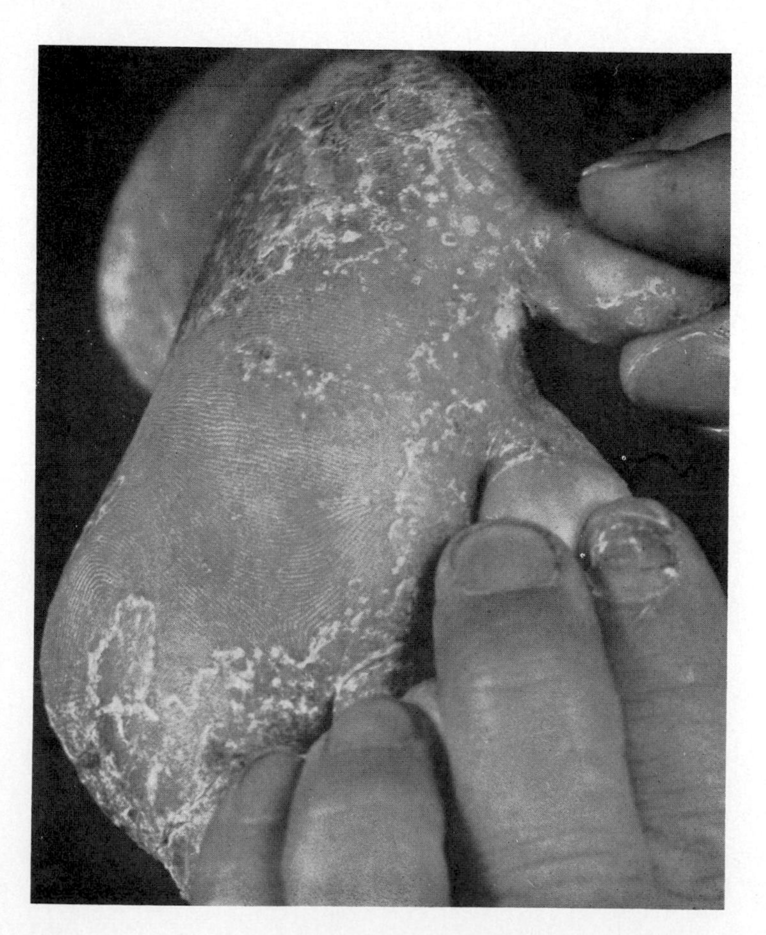

Figure 38–2 Chronic intertriginous type of tinea pedis, showing maceration and fissure between the fourth and fifth toes. (From Conant, N. F., et al.: Manual of Clinical Mycology. 3rd ed. Philadelphia, W. B. Saunders Company, 1971.)

Table 38–1. Three Genera of Dermatophytes, with Common Species of Each and Some of the Conditions They Often Cause[1]

1. *Microsporum* (invades hair and skin; rarely nails)
   M. canis (dermatomycosis in animals; ringworm; tinea capitis[2] in children)
   M. audouini (epidemic tinea capitis in children)
   M. gypseum (ringworm; tinea capitis)
2. *Trichophyton* (invades skin, hair, and nails)
   T. mentagrophytes (athlete's foot; various infections of skin and nails)
   T. rubrum (athlete's foot; tinea cruris[3]; nail infections; sycosis[4])
   T. tonsurans (tinea capitis)
   T. schoenleini (favus[5])
3. *Epidermophyton* (invades the skin and nails but not hair)
   E. floccosum (athlete's foot; tinea cruris; nail infections)

[1]With a few exceptions, almost any species may cause any of the conditions mentioned, subject to the restrictions listed for each genus regarding hair or nails; clinical differentiation is often difficult.

[2]"Ringworm" (dermatomycosis) of the scalp.

[3]"Ringworm" (dermatomycosis) of the groin.

[4]*Sycosis*—pustular inflammation of the hair follicles, especially on the face, due to dermatophytes ("barbers' itch").

[5]*Favus*—a dermatomycosis distinguished by honeycomb-like, itching, yellow crusts over the hair follicles, usually in the scalp.

and in several similar environmental zones elsewhere in the world. People living in these areas become infected by inhaling fungus-laden dust. About 60 per cent of these people experience no symptoms. Their infections are recognized only in retrospect when they are skin-tested and found to react to *coccidioidin.* This is a material derived from *C. immitis* cultures, similar to tuberculin; the skin test, called the *coccidioidin* test, is analogous in application and significance to the tuberculin test. Following injection of a small quantity of coccidioidin into the skin, a hard, red area develops in those persons who have coccidioidomycosis or who have been infected by *C. immitis* in the past. A person does not lose the ability to react to this antigen once he or she has recovered from the infection.

About 40 per cent of people infected experience symptoms—usually only a mild, influenza-like syndrome with fever. A few patients also have painful red bumps on their legs called *erythema nodosum.* In some, pulmonary symptoms are more severe, appearing as pneumonia (Fig. 38–3). In others, symptoms suggestive of tuberculosis occur. In less than 1 per cent of those infected, the fungus is disseminated to all parts of the body. Skin and bone lesions may develop, meningitis is common, and the patient is seriously ill. In a few patients there may be no dissemination, but the lung lesions may progress and giant cavities may form in the lung. Lesions of pulmonary coccidioidomycosis cast x-ray shadows that often cannot be distinguished from those of tuberculosis. The coccidioidin test is then of special value.

The fungus in infected tissue grows as a large, thick-walled, spherical cell called a *spherule,* with little spores forming inside. When the spherule ruptures, these spores are released into the tissue, and each repeats the cycle (Fig. 38–4). If pus or sputum containing these cells is placed on culture media at room temperature, the fungus grows, forming no spherules. It develops a moldlike growth identical to its growth in soil. Because of its ability either to form spherules or to grow as mold it is said to be a *diphasic* or *dimorphic* fungus. In the soil the hyphae form chains of lightweight arthrospores (Fig. 38–5), which break free and are readily blown about by the wind. Most infections occur during dust storms or when soil is disturbed, as when a tractor plows a field. Cattle, dogs, and many other animals also acquire coccidoidomycosis by inhalation of *Coccidioides* spores in the dust. Direct transmission from person to person is very rare.

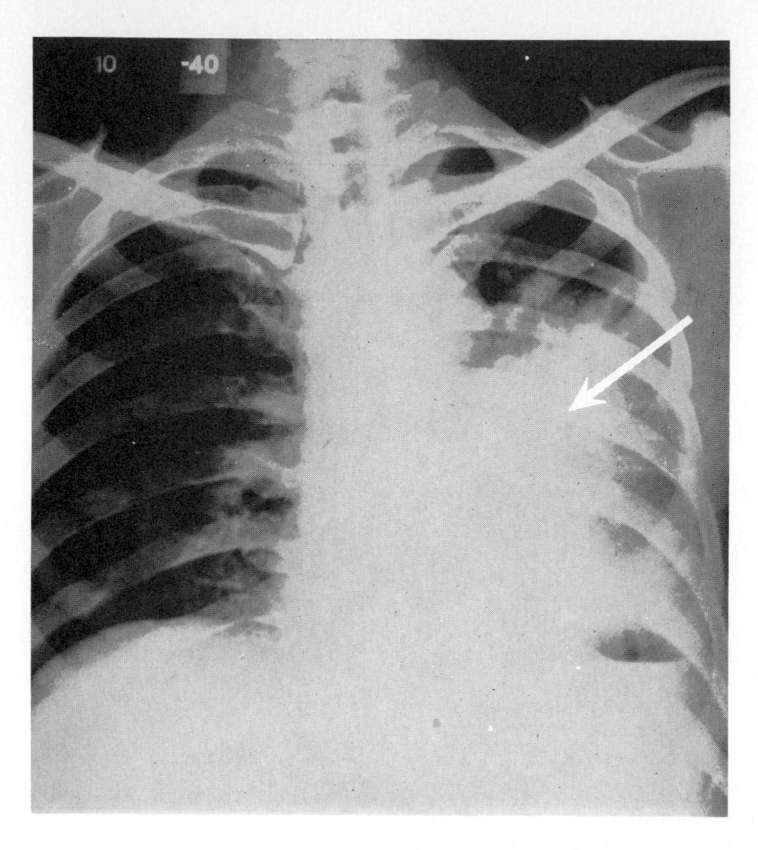

Figure 38–3 Chest x-ray of a patient with progressive coccidioidomycosis of the lungs. The light area at the right of the picture, nearly filling the patient's left chest (arrow), represents the x-ray shadows cast by the lesions in the lung. This was a fatal case. (Courtesy of Dr. C. E. Smith, *in* Conant, N. F., et al.: Manual of Clinical Mycology. 3rd ed. Philadelphia, W. B. Saunders Company, 1971.)

Primary infections of coccidioidomycosis are extremely common in endemic areas in the United States, but the disease is usually not reported to the U.S. Public Health Service. Still, in 1980, 198 cases in Arizona and 20 cases in 21 other states were optionally reported to this agency. California had 500 cases in 1976. These figures do not present a true picture, as evidenced by the fact that, in epidemic areas, 46 to 90 per cent of the population tested give a positive coccidioidin test.

The specific treatment of benefit in disseminated infections is amphotericin B (Fungizone) at a near toxic level.

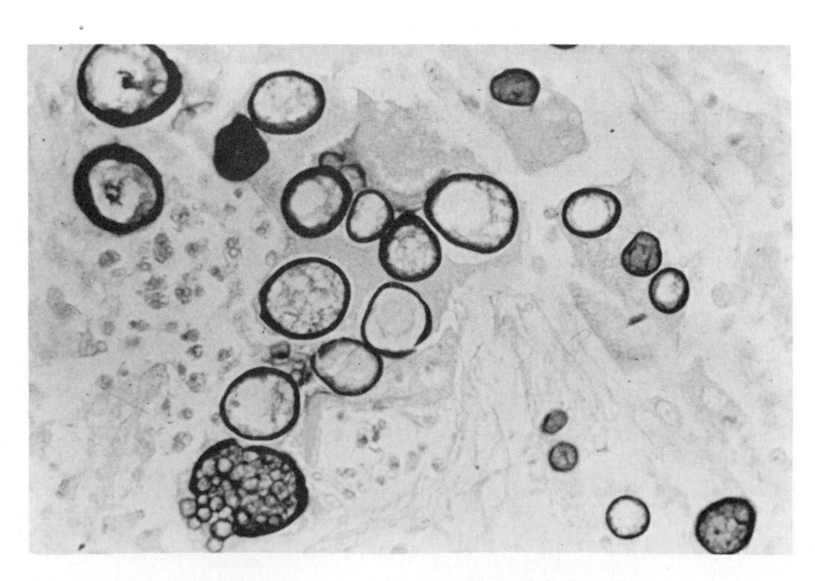

Figure 38–4 Coccidioidomycosis. Tissue section showing developing and mature spherules, some containing endospores. (From Rippon, J. W., *in* Burrows, W.: Textbook of Microbiology. Philadelphia, W. B. Saunders Company, 1973.)

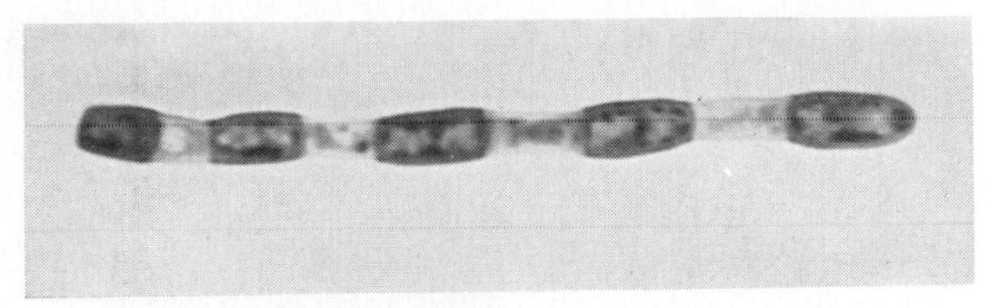

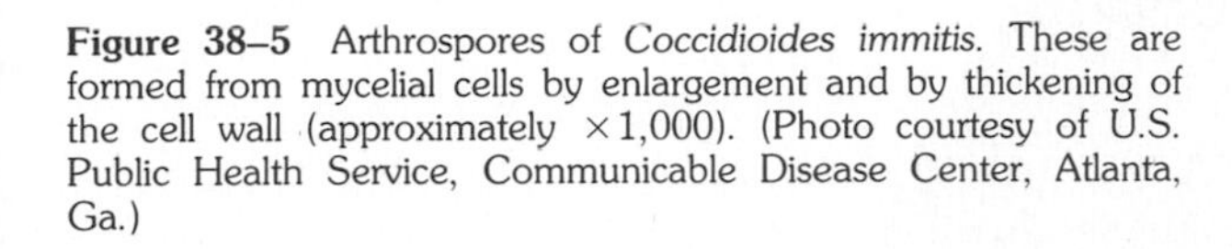

Figure 38–5 Arthrospores of *Coccidioides immitis.* These are formed from mycelial cells by enlargement and by thickening of the cell wall (approximately ×1,000). (Photo courtesy of U.S. Public Health Service, Communicable Disease Center, Atlanta, Ga.)

Histoplasmosis

Histoplasmosis (also known as Darling's disease)[2] is caused by *Histoplasma capsulatum.* This fungus grows in moist, fertile soil in many parts of the world. In the United States it has been found in soil throughout the Middle West and in the eastern half of the country. It is unusually common in soils that have been enriched by fecal material from chickens and other birds, especially starlings. As in coccidioidomycosis, infection takes place by inhalation of spores or fragments of hyphae in dust. Transmission from person to person is very rare.

Skin-testing surveys with *histoplasmin* (a material similar in principle and use to tuberculin and coccidioidin) indicate that millions of people have been infected by *H. capsulatum.* Most of them, however, do not recall having had any recognizable symptoms. Apparently, the healthy body is able to destroy the fungus before much lung tissue is involved.

X-rays made on some healthy people living in the areas where histoplasmosis is endemic often show calcified nodules resembling somewhat those of healed tuberculosis. These people react to histoplasmin skin tests but not to tuberculin unless they have had a tuberculous infection. Most of them are unaware that they have had pulmonary histoplasmosis.

In some infected individuals, large areas of the lung become involved, and these persons experience symptoms that may resemble influenza or pneumonia. Sometimes histoplasmosis may present many symptoms similar to tuberculosis, so much so that laboratory tests are necessary to distinguish which infection is present. When pulmonary symptoms are mild, the patient recovers quickly. When pulmonary disease is severe, the prognosis (outlook for recovery) may be serious.

H. capsulatum, like *C. immitis,* is a diphasic fungus. In the body it grows as small, budding, yeast-like cells that are quickly engulfed by cells of the reticuloendothelial system. The phagocytic cells are usually very active in destroying invading microorganisms, but in the serious forms of histoplasmosis this is not so. The fungus grows actively within these cells. Budding rapidly, the parasites quickly fill the phagocytes.

In symptomless or mild infections the growth of budding *Histoplasma* is controlled, but in serious infections the fungi are carried by the blood to all parts of the body and continue to invade reciculoendothelial cells. Organs such as the liver and spleen, containing large numbers of these cells, become greatly enlarged.

H. capsulatum grows in soil and on culture media at *room* temperature (about 23 C) as a mold producing characteristic conidiospores (Figs. 38–6 and 38–7). If the fungus is cultivated on special media at *body* temperature (37 C), it will grow as small, budding cells identical to those seen in infected tissue or sputum (Fig. 38–8).

Histoplasmosis also occurs in dogs, cats, rats, and other animals. For acute or chronic pulmonary cases, amphotericin B is used with caution because of its side effects.

---

[2]The disease discussed here is clinically entirely different from African histoplasmosis (often confusingly called histoplasmosis), caused by *Histoplasma capsulatum* var. *duboisii.*

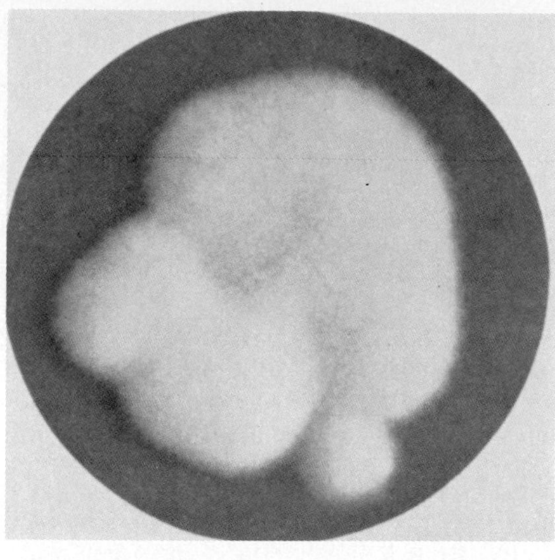

Figure 38–6 A relatively young colony of *Histoplasma capsulatum* on agar, showing the woolly, moldlike growth characteristic of this organism in aerobic cultures at about 23 C. (Courtesy of Dr. J. M. Kurung.)

North American Blastomycosis

North American blastomycosis (also known as Gilchrist's disease) and South American blastomycosis are different diseases and are caused by distinctly different dimorphic fungi.

Blastomyces dermatitidis, the cause of North American blastomycosis, is another mold indigenous to the soil. It becomes yeast-like when it grows as a pathogen in the body.

Infection begins in the lung following inhalation of a spore or hyphal fragment. Dissemination in the tissues is common and may extend to all parts of the body. Lesions occur especially frequently on the skin and in the bones (Fig. 38–9), also often in the internal male reproductive organs, involving the prostate. The disease is most common in adult males. Direct transmission from person to person is rare.

Very few people other than patients with blastomycosis react to *blastomycin*

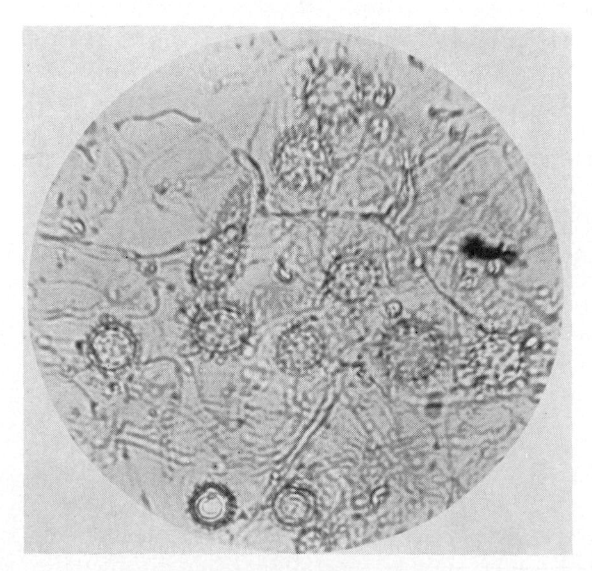

Figure 38–7 *Histoplasma capsulatum.* Moldlike phase found in cultures in the laboratory and in soil. The round bodies with projections are called tuberculate chlamydospores and are distinctive of *H. capsulatum* (approximately ×900). (Courtesy of Dr. J. M. Kurung, Ray Brook State Tuberculosis Hospital, Ray Brook, N.Y. In *Am. Rev. Tuberc.*, Vol. 66.)

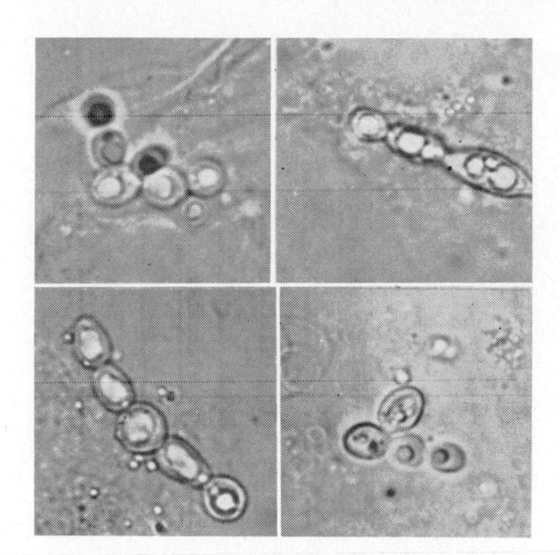

Figure 38–8 Yeast-like cells of *Histoplasma capsulatum* found in the sputum of a patient with pulmonary histoplasmosis (approximately ×1,000). (Courtesy of Dr. J. M. Kurung.)

(see histoplasmin) when skin-tested. This indicates that subclinical infections are rare and that blastomycosis is not as common as coccidioidomycosis or histoplasmosis. Blastomycosis cases due to *B. dermatitidis* have been found only in the Middle West and the eastern half of the United States and Canada; hence, this disease is often called North American blastomycosis. All infections with this fungus reported elsewhere in the world have been in people coming from this area.

No preventive measures or immunization regimens for blastomycosis are known. The treatment is amphotericin B or 2-hydroxystilbamidine as an effective alternative.

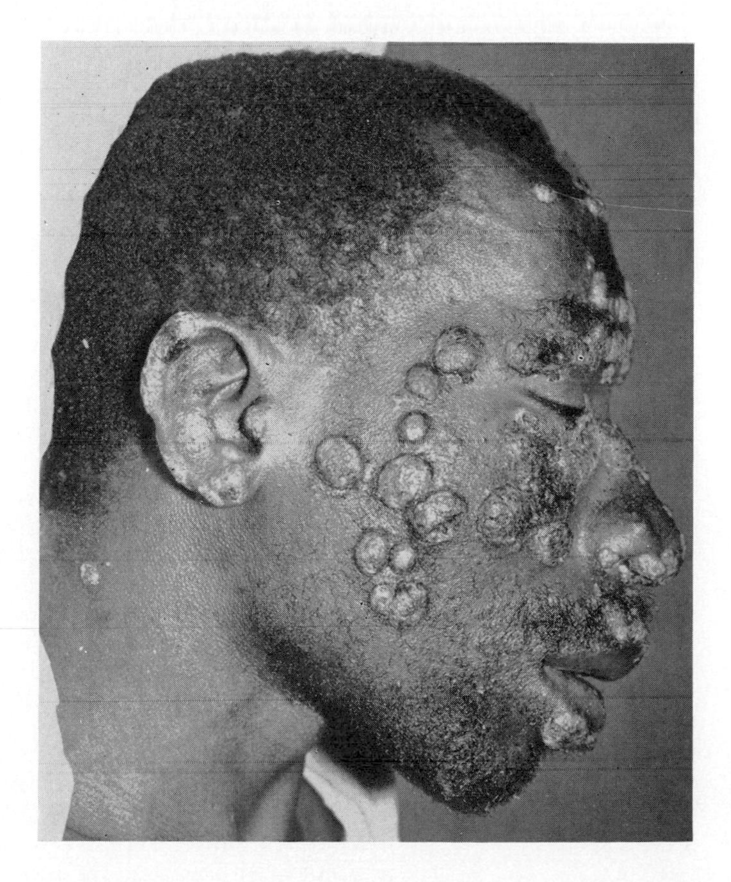

Figure 38–9 North American blastomycosis of the skin, showing multiple discrete, elevated, granulomatous lesions. (From Conant, N. F., et al.: Manual of Clinical Mycology. 3rd ed. Philadelphia, W. B. Saunders Company, 1971.)

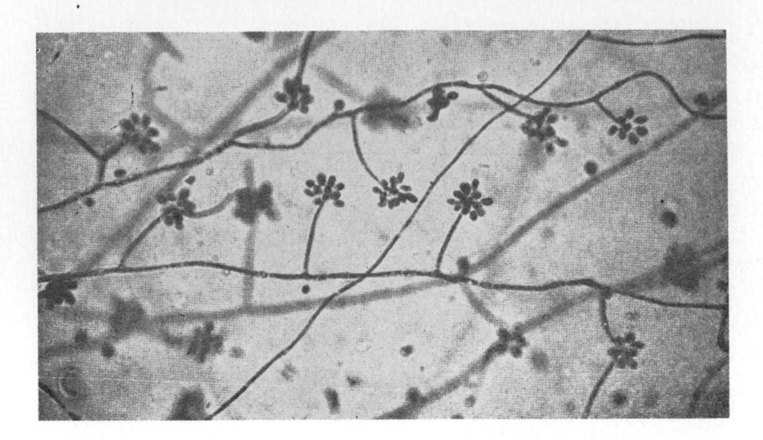

Figure 38–10 Growth of *Sporothrix schenkii*, on Sabouraud's agar, from a case of sporotrichosis. Note the distinctive arrangement of the bunches of conidia. (From Starrs, R. A., and Klotz, M. O.: North American blastomycosis (Gilchrist's disease). *Arch. Intern. Med., 82*:1, 1948.)

Paracoccidioidomycosis (South American Blastomycosis)

Paracoccidioidomycosis, also called South American blastomycosis, is a serious, often fatal chronic mycosis with lung involvement and ulcerative lesions of the skin and mucosa in the oral, nasal, or rectal areas. All viscera may be affected, especially the adrenal gland. The fungus causing this disease is *Paracoccidioides brasiliensis,* previously called *Blastomyces brasiliensis.* The organism resembles closely *Blastomyces dermatitidis,* the cause of North American blastomycosis.

The disease is endemic in South America, especially in Brazil, in rural areas. It affects adults rather than children and is perhaps 10 times as common in males as in females.

Intravenous use of amphotericin B promptly arrests the spread of lesions, although prolonged treatment with combinations of sulfonamides is also indicated to prevent relapse. Amphotericin B is very toxic.

OPPORTUNISTIC MYCOSES

Many fungi living in soil are capable of producing infection in humans under unusual conditions. The diseases they cause are of two categories.

Primary Pathogens. This category includes the diseases *sporotrichosis* (due to *Sporothrix schenckii,* Fig. 38–10), *chromoblastomycosis,* and *maduromycosis* (the last due to various fungi). In these mycoses, infection takes place only when the fungus is introduced into the skin and subcutaneous tissue by some form of trauma. A localized lesion develops, and the fungus can be found growing in the damaged tissue. Usually there is no dissemination of the organisms in the body beyond enlargement of the lesion or infection of the lymphatics draining the area. The patient does not develop widespread life-endangering disease. In each of these diseases caused by a specific *opportunistic fungus,* a more or less characteristic clinical appearance is imparted to the infected area.

Secondary Pathogens. In this category are species of saprophytic soil fungi not ordinarily pathogenic to humans. In these cases infection takes place opportunistically, i.e., only in patients who are debilitated by other diseases. For instance, a patient with leukemia or lymphoma is more susceptible to *cryptococcosis* (infection by *Cryptococcus neoformans,* a yeast-like fungus somewhat similar to *Blastomyces dermatitidis*) or to *aspergillosis* (infection by *Aspergillus fumigatus;* Chapter 3) than is someone in good health. A patient who is receiving antibiotics or cortisone for some other condition is also more likely to develop a secondary fungal infection. Most often these begin in the lung and spread to other organs of the body. In *Cryptococcus* infection, meningitis often results.

With an increasing population of older, chronically ill patients, many of whom are receiving antibiotics and steroid therapy or other drugs that tend to depress defensive functions, there will be an increasing number of opportunistic infections by fungi. The healthy individual appears to be nonsusceptible to infection, but debilitated patients cannot cope with these ordinarily harmless but potentially pathogenic fungi. Dung and dust of pigeons and starlings can be infectious.

In general the systemic mycoses are believed to be noncontagious. However, sputum, pus, dressings, and other materials that might contain organisms from the patient present a health hazard and must be disposed of properly.

Other Opportunistic Mycoses

Of the more than 70 species of *Aspergillus,* some may become associated with infections in humans. This is especially true of *Aspergillus fumigatus,* a pathogen for birds and occasionally for people. Penicilliosis is a very rare disease; only a few of the more than 140 species of *Penicillium* have been isolated from cases of this disease. Just as rare are mucormycosis and other infections caused by *Rhizopus.*

Candidiasis[3]

Candida albicans (previously known as *Monilia albicans*) is a yeast-like organism that is common to the mouth and gastrointestinal tract of many normal people. It is generally *not* indigenous to the soil. Ordinarily it is present in normal persons only in small numbers and produces no disease. Under conditions of diminished resistance to infection, local or general, this organism multiplies greatly and may even change its growth form. When it is actively invading tissue, it grows as long filaments as well as in the form of budding, yeast-like cells (Fig. 38–11).

[3]The disease is also known as candidosis or moniliasis.

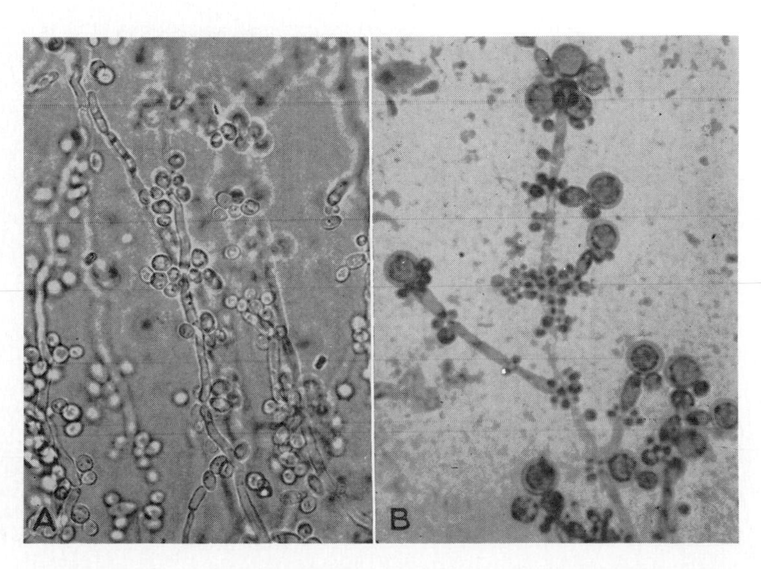

Figure 38–11 *Candida albicans. A,* Pseudohyphae and clusters of blastospores on Sabouraud's glucose agar (×650). *B,* Typical chlamydospores on corn meal agar (×750). (From Conant, N. F., et al.: Manual of Clinical Mycology. 3rd ed. Philadelphia, W. B. Saunders Company, 1971.)

Some conditions that may promote *candidiasis* (infection by *Candida albicans* and occasionally other species of *Candida*) are diabetes, general debilitation, steroid therapy, and the prolonged administration of antibiotics and certain other drugs. Pregnant women are also prone to develop *Candida* vaginitis. Some people may develop skin lesions in areas that are usually moist, warm, and somewhat acid due to skin secretions—e.g., the inframammary and perineal zones, as well as the interdigital webs and the fingernails, especially if their hands are wet for long periods of time. This is another reason why the automatic dishwasher is preferred to hand dishwashing.

Treatments often are not necessary if antibiotic therapy or certain medications for other diseases are discontinued. In gastrointestinal infections with *Candida,* nystatin (Mycostatin) is useful; amphotericin B is helpful in generalized fungal disease; nystatin or gentian violet is used topically on lesions of the skin and mucous membranes.

Thrush. *Candida* infections are commonly associated with the mouth and gastrointestinal tract, where this organism is part of the normal flora. When it invades tissues, it causes them to be inflamed, and a white *pseudomembrane* (see diphtheria) appears on the surface. On the lips and in the mouth this condition is called *thrush.* It is also known as oral moniliasis or candidiasis. Thrush is common in newborn infants whose mothers have *Candida* vaginitis. The infants acquire their infections during birth (Fig. 38–12). This infection is self-limiting, but to avoid spreading it, topical application of 1 per cent aqueous gentian violet is recommended.

Thrush is especially common in patients who have been on prolonged antibiotic treatment, which greatly reduces the gastrointestinal tract bacteria that ordinarily suppress growth of *C. albicans.* Under these conditions *C. albicans,* whose growth is not inhibited by most antibiotics, proliferates greatly and may cause diarrhea and other distressing symptoms. Perianal rashes or itching may also result.

Candida is also a common secondary or opportunistic invader, especially in the lungs. When lung tissue has been damaged by cancer, tuberculosis, or other infectious agents, this yeast can grow abundantly in the necrotic tissue and may obscure the diagnostic laboratory findings. *C. albicans* rarely enters the blood, but when it does it may create vegetations on the heart valves or lesions in the kidneys, always dangerous developments.

Candida albicans is readily transmissible by fluids and excretions from infected areas and tissues: the skin, vagina, oral mucosa, urine, feces, and so

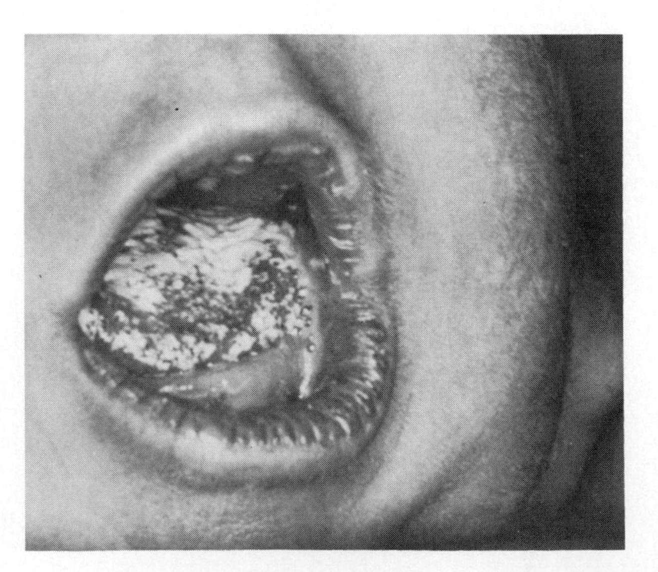

Figure 38–12 Thrush, also called stomatitis (inflammation of the mouth), caused by *Candida albicans,* is observed occasionally in the newborn or during the neonatal period. Lesions are painless, white or gray-white, slightly elevated patches resembling curdled milk. They can appear on the buccal mucosa, lips, tongue, and pharynx. They cling tenaciously, and when a patch of thrush is "wiped off" the underlying mucosa is left raw and usually shows some bleeding. (From Number Five of a combined series on variations and minor departures in infants. Courtesy of Mead Johnson Nutritional Division.)

on. Personal contact with lesions should be avoided. Fomites should be disinfected; dressings and bandages must be burned.

MYCOTOXINS

Not all fungal diseases are true infections. Some are due to ingestion of toxic metabolites of fungi, collectively called *mycotoxins.* Among the best known of these at present is the group of aflatoxins, metabolites of *Aspergillus flavus,* though a dozen or more are known. Aflatoxins and several other mycotoxins are produced when the molds grow in "spoiled" vegetable products such as hay, peanuts, grains, and so forth, commonly used as stock and poultry feeds. The role of mycotoxins in human disease has yet to be fully evaluated. Recognized cases of acute human mycotoxicosis are rare, presumably because humans do not commonly eat "moldy" or "spoiled" foods, although certain cheeses might be cited as popular exceptions. Toxins from fungal products sometimes cause severe mental derangement, as from the ingestion of muscarine from the hallucinogenic mushroom *Amanita muscaria* or of lysergic acid diethylamide (LSD).

During the Middle Ages, epidemics of St. Anthony's disease, (ergotism) caused by cereal grain infections with *Claviceps purpurea,* a fungus, ravaged much of Europe. The introduction from the New World of the potato and maize did away with *ergot* poisoning, which had caused so much cardiovascular distress and/or neurologic convulsion and confusion.

SUMMARY

Diseases caused by yeasts are referred to as mycotic diseases or mycoses. Cutaneous, or superficial, mycoses involve the skin, hair, and nails; deep, or systemic, mycoses are fungal infections of the subcutaneous tissues and internal organs. Most systemic mycoses are caused by fungi that are indigenous to the soil. Opportunistic mycoses are infections caused by normally harmless fungi; infection takes place only when the fungus is introduced into the skin and lower tissues by some form of trauma.

The most common cutaneous mycoses are the tineas, also called ringworm. Numerous species of fungi are known to cause dermatomycoses such as tinea capitis (ringworm of the scalp), tinea corporis (ringworm of the body), tinea unguium (ringworm of the nail), tinea cruris (ringworm of the groin), tinea pedis (ringworm of the foot, or athlete's foot), and others. Otomycosis, also called myringomycosis, hot-weather ear, or fungus ear, may be caused by 32 or more different species of fungi and is easily complicated by opportunistic bacterial infections.

Representative systemic mycoses are coccidioidomycosis, an infection with *Coccidioides immitis*; histoplasmosis, caused by *Histoplasma capsulatum*; and North American blastomycosis and South American blastomycosis, caused by *Blastomyces dermatitidis* and *Paracoccidioides brasiliensis*, respectively. These diseases either infect only the lungs, where calcified nodules retain the organisms, as seen with x-rays and checked by skin tests, or they spread in the body, threatening the life of the patient. Although the inhalation of spores of these fungi cause infections in millions of people in many parts of the world, the acute disease stage occurs relatively infrequently. This is further helped by treatment with amphotericin B, a very toxic but effective antifungal chemotherapeutic agent.

Candida albicans causes a yeast-like opportunistic mycosis called candidiasis, such as *Candida* vaginitis, dermatitis, and thrush.

Like bacteria, fungi, too, produce toxins, collective called mycotoxins. As an example of these mycotoxins, the reader may think of mushrooms, remembering that many are not edible, but more common are the aflatoxins of *Aspergillus* that are formed by molds when vegetable products spoil.

REVIEW QUESTIONS

1. What are cutaneous mycoses, dermatomycoses, deep mycoses, systemic mycoses, superficial mycoses, and opportunistic mycoses?
2. What is ringworm? Name different manifestations of this infection and the body regions that may be involved. What treatment exists for these conditions?
3. State other names for fungus ear.
4. Relate systemic mycoses to soil and geographic areas of the world.
5. Discuss coccidioidomycosis, histoplasmosis, North American blastomycosis, paracoccidioidomycosis, maduromycosis, and candidiasis. How do these diseases differ? How are they alike? Why?
6. State the value of the following in managing mycoses: skin tests, x-rays, amphotericin B, estrogens.
7. What is thrush? How is it treated?
8. Why should we be concerned about mycotoxins or ergot?

Supplementary Reading

Ajello, L.: Comparative ecology of respiratory mycotic disease agents. *Bacteriol. Rev., 31*:6, 1967.

Al-Doory, Y.: The Epidemiology of Human Mycotic Diseases. Springfield, Ill., Charles C Thomas, 1976.

Beneke, E. S., and Rogers, A. L.: Medical Mycology Manual. 3rd ed. Minneapolis, Burgess Publishing Co., 1971.

Berry, D. R., and Smith, J. E.: The Filamentous Fungi. New York, Halsted Press, 1975.

Braude, A. I.: Medical Microbiology and Infectious Diseases. Philadelphia, W. B. Saunders, 1981.

Burnett, J. H.: Fundamentals of Mycology. 2nd ed. New York, Crane, Russak & Co., 1977.

Conant, N. F., Smith, D. T., Baker, R. D., and Callaway, J. L.: Manual of Clinical Mycology. 3rd ed. Philadelphia, W. B. Saunders Company, 1971.

Emmons, C. W., Binford, C. H., Utz, J. P., and Kwon-Chung, K. J.: Medical Mycology. 3rd ed. Philadelphia, Lea & Febiger, 1977.

Fiese, M. J.: Coccidioidomycosis. Springfield, Ill., Charles C Thomas, 1958.

Furcolow, M. L.: Histoplasmosis. *Amer. J. Nurs., 59*:79, 1959.

Goldblatt, L. A. (ed.): Aflatoxin. New York, Academic Press, 1969.

Goodwin, R. A., Jr., and Des Prez, R. M.: Histoplasmosis. *Am. Rev. Respir. Dis., 117*:929, 1978.

Gordon, M. A., and Devine, J.: Filamentation and endogenous sporulation in *Cryptococcus neoformans. Sabouraudia, 8*:227, 1970.

Louria, D. B.: Experiences with and diagnosis of diseases due to opportunistic fungi. *Ann. N.Y. Acad. Sci., 98* (Art. 3):617, 1962.

Martinson, F. D.: Chronic phycomycosis of the upper respiratory tract. *Amer. J. Trop. Med Hyg., 20*:449, 1971.

Medoff, G., and Kobayashi, G. S.: Strategies in the treatment of systemic fungal infections. *N. Engl. J. Med., 302*:145, 1980.

Rebell, G., and Taplin, D. (eds.): Dermatophytes: Their Recognition and Identification. 2nd rev. ed. Coral Gables, Fla., University of Miami Press, 1974.
Sarosi, G. A., and Davies, S. F.: Blastomycosis. *Am. Rev. Respir. Dis., 120*:911, 1979.
Schlaegel, T. F., Jr.: Ocular Histoplasmosis. New York, Grune & Stratton, 1977.
Skinner, C. E., and Fletcher, D. W.: A review of the genus *Candida. Bacteriol. Rev., 24*:397, 1960.
Sorley, D. L., Levin, M. L., Warren, J. W., et al.: Bat-associated histoplasmosis in Maryland bridge workers. *Am. J. Med., 67*:623, 1979.
Spensley, P. C.: Mycotoxins. *Roy. Soc. Health J., 5*:248, 1970.
Walter, J. E., and Atchison, R. W.: Epidemiological and immunological studies of *Cryptococcus neoformans. J. Bacteriol., 92*:82, 1966.
Wieland, T.: Poisonous principles of mushrooms of the genus *Amanita. Science, 159*:946, 1968.

SECONDARY AND ACCIDENTAL INFECTIONS; ARTHROPOD-BORNE BACTERIAL INFECTIONS

INTRODUCTION

There are at least three ways in which microorganisms may gain entrance into the blood: as the result of other infections, or through cuts and wounds, or by way of natural vectors, such as arthropods that transmit disease-causing organisms (Table 39–1). In addition, do not forget blood transfusions!

Invasion of Blood as a Sequela to Primary Infection. As previously stated, many pathogenic organisms commonly invade the blood secondarily after they have established a primarily localized infection at the initial portal of entry (e.g., *Salmonella typhi* via the alimentary tract, *Treponema pallidum* via the genital tract, and streptococci from abscessed teeth). Infections of various parts of the body by microorganisms transmitted in blood from initial or other local sites are often called *hematogenous* infections.

Artificial or Accidental Infection of Blood

Many pathogenic agents in infected blood may be transmitted by any vector that can transfer the infectious blood into the body of another person. The vector in such infections may be artificial: unsterile hypodermic syringes, bloody dressings, any cutting or piercing instrument, unprocessed serum, or plasma. Such infections are usually accidental or irregular occurrences due to carelessness or ignorance. For example, in one dramatic instance a group of drug addicts, one of whom was malarious, all used the same needle and syringe at one sitting, and all but one (the carrier!) contracted malaria. In another, fatal instance, a pathologist pricked his finger with a knife during an autopsy on a person who had died of plague. Others have similarly acquired "blood poisoning" from human tissues carrying streptococcal or staphylococcal infection. One nurse handling bloody dressings contaminated with *Francisella tularensis* had a tiny scratch on her hand, and the microorganisms entered her blood.

Serum Hepatitis. The *homologous serum jaundice* (or hepatitis B) *virus* is found, so far as is known, only in human blood and tissues. It can infect orally or through mucous membranes and possibly through unbroken skin. It is most commonly transmitted by artificial procedures that transfer blood (or certain blood derivatives) from one person to another, such as accidents, blood transfusions, serum injections, and the use of any inadequately sterilized skin-piercing implement on a succession of persons, including hypodermic needles and syringes (before disposable ones became generally available) and needles used in obtaining blood specimens or in tattooing. These modes of transmission are entirely man-made. How such a virus comes to infect people all over the world is not clear. Apparently, chronic carriers of the virus are fairly common, and many very mild or latent infections occur. In 1980, 19,015 cases of serum hepatitis (SH) were reported in the United States. Note that infectious hepatitis (IH)—i.e., infection by hepatitis virus A—was much more common; during the

PART E PATHOGENS TRANSMITTED IN BLOOD

CHAPTER 39

same period 29,087 cases were observed. Also reported but unspecified as to type were 11,894 cases. For a discussion of both SH and IH, see pp. 412–414.

The virus is not affected by any known chemotherapeutic drugs. It differs from other known viruses (except that of infectious or *epidemic* hepatitis, which it most closely resembles) in that it is unusually resistant to heat. Ordinary 10-minute boiling cannot be relied upon to inactivate it. The only safe sterilizing procedures are autoclaving and use of the hot air oven. Disinfectant handwashing, use of sterile needles when collecting blood specimens or injecting successive persons, and handling of bloody cotton or dressings or blood during specimen-taking or surgery should be done with scrupulous care. The now almost universal use of disposable needles, syringes, and lancets (used for small blood samples) helps greatly in avoiding the transmission of this infection. Homologous serum hepatitis characteristically has a long incubation period of weeks or months. The overt disease is an acute febrile one, with general constitutional symptoms of varying severity. Hepatitis and jaundice are common but not invariable. Children are especially susceptible. Fatality rates can be high.

Australia Antigen. The virion of serum hepatitis (hepatitis B) consists of a core of DNA with a specific antigen called HB_cAg and a capsid consisting of a second specific antigen called surface antigen or HB_sAg. This antigen was formerly known as Australia antigen or AuAg because it was first described in Australia. It has also been called HBAg, HAA (Hepatitis Associated Antigen), and SHAg. HB_sAg is found in the blood or serum of patients with, convalescents from, and carriers of serum hepatitis. Its presence indicates *potential* infectiousness of the blood; it is not infectious per se.

When antiserum against the hepatitis B virus reacts with the hepatitis-associated antigen in the serum of a carrier, a precipitate forms that contains the virus. Actually, several different antigens may be seen in electron micrographs, the most infectious of which (called the Dane particle) has a core of HB_cAg.

HB_sAg has been found in mosquitoes in Senegal and in urine and feces of infected persons elsewhere. Although the virus itself has not been demonstrated in such materials, these observations indicate the possibility of transmission of the disease by vectors other than human blood and blood derivatives.

Diagnostic demonstration of this antigen in the blood by serologic means is now possible and is of great importance in the selection of blood donors. Unlike prevention of hepatitis A, immune serum globulins are of little prophylactic value against hepatitis B infection.

Natural Vectors (Sanguivorous Arthropods)

Numerous infectious agents are transmitted under natural conditions directly into the blood, mainly by sanguivorous arthropods such as malaria-transmitting mosquitoes.

Table 39–1. Blood Infections and Their Vectors

I. Sequelae to infections acquired via other portals of entry (secondary, endogenous infections).
II. Artificial or accidental vectors, such as cuts, pricks, scratches with infecting instruments; injection of infected serum or blood; careless handling of infectious blood, instruments, or dressings; careless tattooing: may transmit any pathogen present in the blood at the time, including viruses of epidemic hepatitis (hepatitis A) and serum hepatitis (hepatitis B), and those listed below.
III. Natural vectors (sanguivorous arthropods) may transmit:
  1. Bacteria (e.g., plague, tularemia, relapsing fever)
  2. Rickettsias (e.g., typhus and typhus-like fevers)[1]
  3. Viruses (e.g., yellow and dengue fevers, various encephalitides)
  4. Protozoa (e.g., malaria, African sleeping sickness, kala-azar)
  5. Helminths (e.g., various types of filariasis).

[1]We are not so well informed about the means of transmission of chlamydial diseases by blood.

The remainder of this chapter will discuss arthropods as disease vectors and explore some arthropod-borne *bacterial* diseases.

ARTHROPODS AS DISEASE VECTORS

Human-biting and animal-biting arthropods constitute one of the most important classes of disease vectors. Arthropods may conveniently be divided into two groups with respect to the means by which they transmit disease.

Mechanical Means. When flies walk on feces, boils, or sores, soiling their bodies, feet, and legs, and then walk on food, on plates, glasses, on the lips of a sleeping infant, or on the ulcers of a sick person, the implications are obvious and need no elaboration.

Transmission by Feces and Vomitus. Flies, ants, roaches, and other coprophagic (feces-eating) arthropods may harbor in their gut intestinal pathogens taken in with the feces they eat. These vermin also feed on vomitus, infectious blood, open sores, and so on. Such arthropods deposit disease agents with their fecal droppings on food and about the house, as well as on the wounds, eyes, lips, and skin of waking or sleeping human beings. Many species also regurgitate (vomit) contents of their gut when they bite. This also needs no further discussion.

Table 39–2. Some Common Arthropod-Borne Diseases and Their Vectors

Mosquitoes:
- Malaria (protozoa)
- Yellow fever (virus)
- Dengue (virus)
- Encephalitis (virus)
- Filariasis (helminths)

Ticks:
- Relapsing fever (spirochetes)
- Rocky Mountain spotted fever or tick typhus (rickettsias)
- Texas fever of cattle (protozoa)
- Tularemia or "rabbit fever" (bacteria)

Fleas of rats and other rodents:
- Endemic typhus or flea typhus (rickettsias)
- Bubonic plague (bacteria)
- Tularemia (bacteria)

Lice, ticks, fleas, and "red spider" ("chigger") larvae of mites:
- Typhus fever, tsutsugamushi or mite typhus, and related diseases (rickettsias)

Biologic Means. This generally implies passage of pathogenic agents from the gut of the insect into the tissues and body fluids of the arthropod vector, especially into the saliva, as in malaria, various forms of viral encephalitis, yellow fever, bubonic plague, and elephantiasis (filariasis). Some important diseases carried by sanguivorous arthropods, as well as the general nature of the disease agents, are tabulated in Table 39–2. The lowly bedbug has not been convicted as a disease vector, although it is often suspected of being one and probably is.

Arthropod-borne diseases prevail especially in warm regions where numerous species of sanguivorous flies, mosquitoes, and ticks abound. They occur also in temperate climates under conditions of uncleanliness and crowding in which lice, fleas, and flies flourish, and in swampy and woodland areas where mosquitoes and ticks breed.

ZOONOSES

A number of diseases can be transmitted to humans from wild animals in the forest or field, "far from the madding crowd's ignoble strife." Infectious diseases of animals transmissible to humans by any means are called *zoonoses,* a term first applied by Rudolph Virchow to the infectious diseases that human beings acquire from domestic animals. The Second Report of the Joint World Health Organization–Food and Agriculture Organization Expert Group on Zoonoses (1959) redefined zoonoses as "those diseases and infections which are naturally transmitted between vertebrate animals and man." Experimental infections from laboratory animals that do not normally harbor the infectious agents are not considered zoonoses. Over a hundred zoonoses are known today, a great increase since the turn of the century, when only a few zoonotic diseases like anthrax, cowpox, glanders, rabies, and a few zooparasitic infections were recognized.

Some Zoonoses Caused by Bacteria

Salmonelloses are common in fowl, pigs, dogs, cattle, rats, mice, primates, cats, goats, sheep, pigeons, hares, rabbits, foxes, mink, and guinea pigs, as well as other mammals, birds, reptiles, and arthropods.

Two distinct etiologic agents have been recognized as causing what had previously been considered a single pathologic entity, *rat-bite fever.* The organism causing one form of the disease, called sodoku, is *Spirillum minor.* Sodoku in humans usually results from the bite of an infected rodent. *Haverhill fever* is another form of rat-bite fever, the result of infection with the bacterium *Streptobacillus moniliformis.*

The causative agent of pseudotuberculosis in laboratory animals is *Yersinia pseudotuberculosis.* It attacks the guinea pig chiefly but can also infect mice, rabbits, and humans, but not rats.

Leptospirosis is not a serious disease among laboratory animals but may become so in the wild. One causative agent in rodents is *Leptospira interrogans,* the only recognized species in this genus. Several serotypes of *Leptospira* have been implicated in rodent disease. In many, symptomatology and pathology are varied.

Yersinia pestis[1] is named in honor of A. E. J. Yersin (1863–1943), a colleague

[1]Formerly *Pasteurella pestis.*

of Pasteur and discoverer of the bacillus of bubonic plague. *Francisella tularensis*[2] is named in honor of Thomas Francis, Jr., the American microbiologist who demonstrated the role of rodents in the transmission of tularemia.

Pasteurella multocida ("killer of many") is so named because it causes frequently fatal hemorrhagic septicemia (or hemorrhagic fever) in many species of lower vertebrates—cattle, sheep, birds, and so on—often resulting in great commercial losses. It occasionally attacks humans.

Francisella tularensis, the cause of rabbit fever (tularemia), is given the name *tularensis* for the Tulare areas near San Francisco where it was first observed causing this disease.

Yersinia pestis is the cause of bubonic plague. Since all of these diseases are primarily infections of animals transmissible to humans, they are properly classed as zoonoses.

All species of *Pasteurella, Yersinia,* and *Francisella,* are small, gram-negative, nonspore-forming rods. When observed microscopically in smears of blood, pus, or other pathologic material, they are characterized by staining (with special stains like Wayson's) more intensely at the tips than in the center (bipolar staining) (Fig. 39–1). They are easily recognized, and, with pertinent corroboratory data, the infection can be diagnosed by appearance. Cultures and animal inoculations are also useful.

These species produce bacteremias and generalized infections, which tend to localize especially in the lymph nodes. In humans, these nodes swell up and then are spoken of as *buboes,* especially when in the groin, axillae, or neck. The buboes are filled with pus and bacilli. They often rupture, and the draining pus is highly infectious. Attendants have to consider this in handling patients with any infection due to these organisms.

Hemorrhagic Fever. Since this form of bacteremia is primarily a disease of animals, it need not be discussed in detail here. The organisms *(Pasteurella multocida)* occur in blood and body discharges. They persist in the dust and dirt of animal cars and pens and are inhaled or swallowed or gain entrance through cuts and scratches in the skin. The occasional human case may be treated like tularemia (see next paragraph). Arthropod transmission of this disease is probably not important, but the organisms occur in the mouths of rats, whose bites probably infect animals and occasionally humans.

---

[2]Formerly *Pasteurella tularensis.*

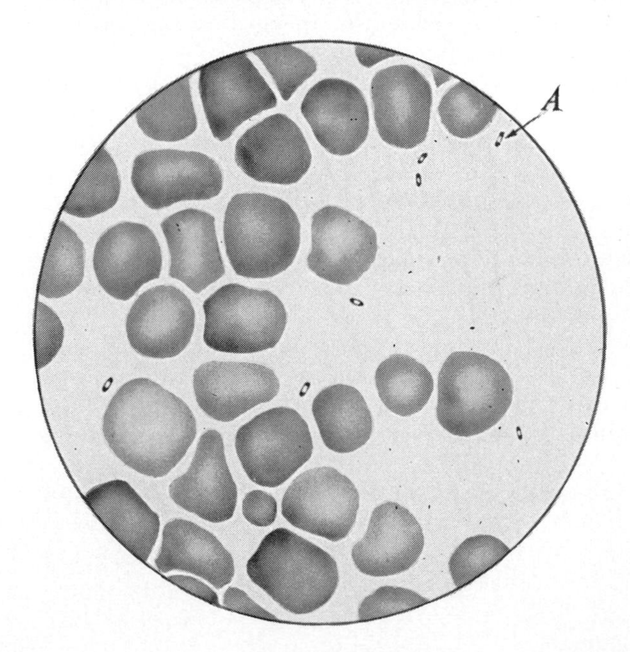

Figure 39–1 Drawing made from a microphotograph of blood of a mouse that died of hemorrhagic septicemia, a plague-like disease of animals. At *A* are seen *Pasteurella multocida,* morphologically like *Yersinia pestis,* the cause of bubonic plague. Note the rounded ends and bipolar staining distinctive of *Pasteurella* and *Yersinia.* The large rounded objects are erythrocytes distorted by shrinkage due to drying of the smear. (Courtesy of Dr. C. D. Stein, Bureau of Animal Industry, U.S. Dept. of Agriculture.)

Tularemia. This disease, often called "rabbit fever," is caused by *Francisella tularensis,* a species very similar to *Y. pestis* and *P. multocida* in many respects. It causes a similar disease among wild rodents (notably ground squirrels) and rabbits.

TRANSMISSION. Ticks, fleas, deer flies, and probably other sanguivorous arthropods found on rabbits and rodents transmit tularemia from animal to animal and to humans. A pustule forms at the site of the initial infection, and the pus is highly infectious. Tularemia often occurs in persons (hunters, market workers, housewives) who handle wild rabbits. It has a wide range of hosts, including humans; over 50 species of mammals, birds, and reptiles are susceptible to it. Why then blame it all on the poor rabbit?

Infected wild animals occasionally die on the banks of streams, making the water infectious and causing serious outbreaks of tularemia among persons drinking the untreated water—another reason for boiling or chlorinating water from unknown sources, if it must be used.

Blood, pustular sites of primary infection, and draining secondary buboes in humans are infectious. Biting arthropods must be kept away from patients as well as from other persons. Tularemia is much less fatal than plague, but it is very protracted, very debilitating, and sometimes incapacitating for long periods. In a 1971 outbreak of tularemia in Sweden the epidemiologic, bacteriologic, and clinical data indicated that the infection was generally transmitted through inhalation of dust from hay contaminated with animal feces.

Some tularemia-specific skin-test antigens have been tested and found to give a positive reaction very early in the illness (Foshay's test). in 1980, 234 cases of tularemia in humans were recorded in the United States, and in 1981, an unconfirmed 208 cases.

Bubonic Plague. This is one of the so-called "classic" diseases, a scourge of the ancient and medieval worlds and still a dangerous menace in many parts of the Eastern and Southern Hemispheres. The presence of buboes gave plague its name: bubonic. The arthropods most commonly responsible for transmission of bubonic plague to humans are fleas from rats and ground squirrels infected with *Yersinia pestis.* The disease is maintained constantly among rodents by these arthropods.

An infectious disease that is regularly transmitted among lower animals is said to be *enzootic.* If it becomes unusually prevalent among them, it is *epizootic* (compare *endemic* and *epidemic).* Plague among forest and prairie animals is called "sylvatic" or "campestral" plague (*sylvatic,* in the forest; *campestral,* in the fields). Occasional cases of plague occur in persons who handle carcasses of such animals.

Bubonic plague is primarily a disease of rats and other rodents. When these become numerous in contact with humans, bites of fleas *(Xenopsylla cheopsis,* the oriental rat flea and other species, including *Pulex irritans,* a cosmopolitan flea of humans, swine, and rodents) from infected rodents, especially rats, are likely to occur. Under such conditions plague becomes epidemic, and devastating outbreaks are likely. The *Black Death,* famous in history and legend, was bubonic plague in its acute epidemic form. The term "black death" was derived from the fact that in such epidemics the disease was marked by intense septicemia and subcutaneous hemorrhages over the whole body, giving it a dark appearance. In the United States such epidemics are now unknown because we now know that rats and ground squirrels harbor the disease-causing *Yersinia* and that their fleas transmit it, and we take appropriate measures to control both rodents and fleas. However, sporadic cases sometimes occur, especially in the southwestern United States.

Rats and ground squirrels are controlled by the United States Public Health Service in cooperation with local and private agencies. This involves poisoning, trapping, elimination of rats from ships (rats in ships from plague areas are exceedingly clever at getting ashore in spite of immigration and other officials),

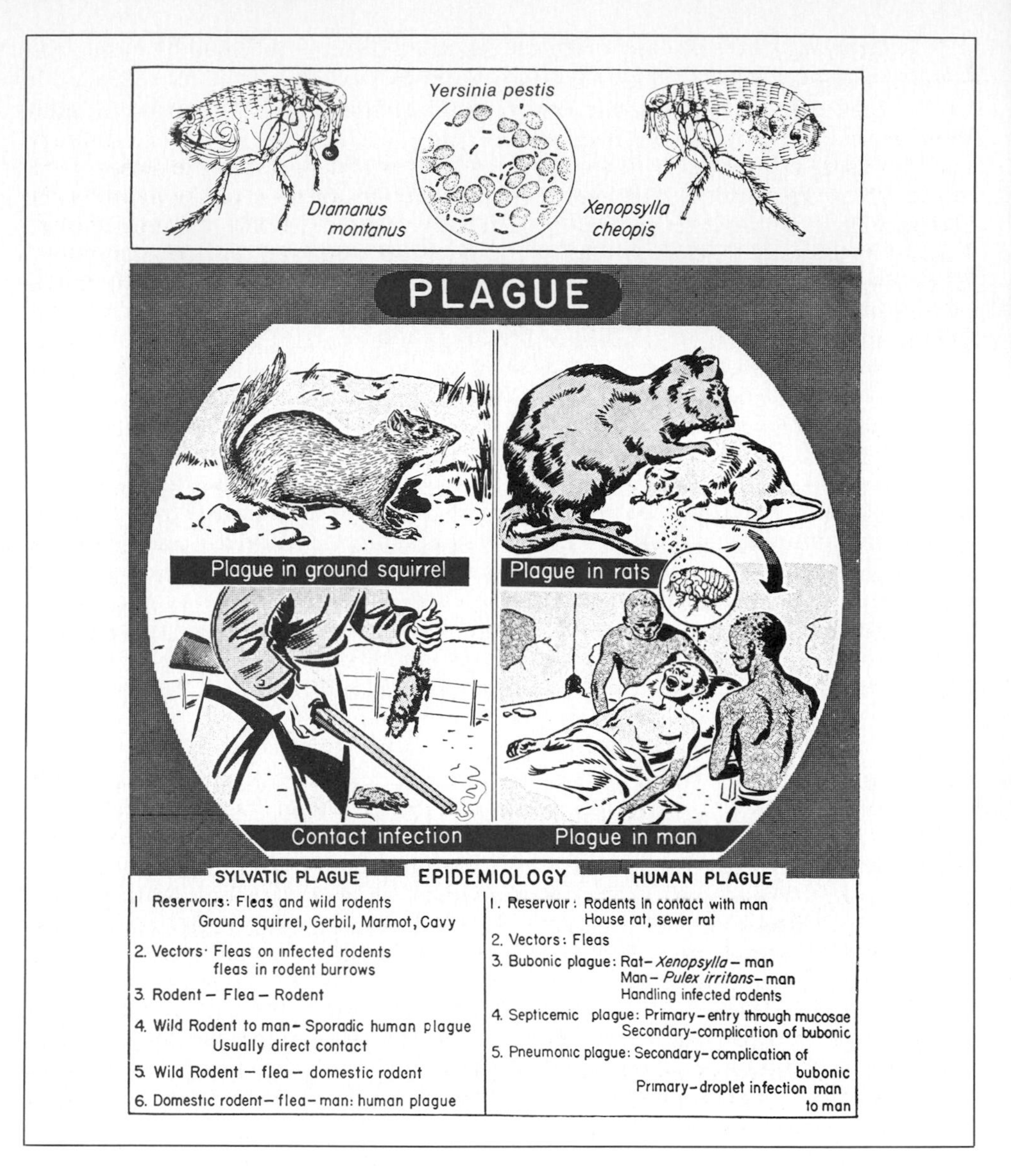

Figure 39–2 Epidemiology of plague. (From Hunter, G. W., III, Swartzwelder, J. C., and Clyde, D. F.: Tropical Medicine. 5th ed. Philadelphia, W. B. Saunders Company, 1976.)

rat-proof construction, the destruction of shelter places, and the elimination of garbage dumps, open feed bins, and similar places where rats can feed. The use of residual DDT powder and other proper insecticides (Diazinon or Gammexane where DDT is discouraged) in homes and on rat runways has been very effective in ridding the rats of their fleas, although it does not eliminate the rats.[3] When dead rats cool off, their fleas leave them and go to other warm-blooded animals. Therefore, avoid dead rats (Fig. 39–2).

The rat control bill passed in 1967 by Congress should help to reduce these potential carriers of bubonic plague. Since January 1, 1971, measures applicable to any land, sea, or air vehicles arriving from plague areas have been specified in International Health Regulations, WHO, Geneva. Medical personnel in North America will seldom see a case of plague unless they serve in the foreign service or the military in the Orient, the Mediterranean, certain South

[3]Effective rodenticides are sodium fluoroacetate (1080) and warfarin (Coumadin).

American countries, or some areas of the southwestern United States where ground squirrels and similar animals are common.

PLAGUE VACCINE. A vaccine (Haffkine's vaccine), much like antisalmonellosis vaccine, is used to immunize persons going to plague districts, e.g., military and health personnel and doctors and nurses working in plague areas. The immunity conferred is not absolute and is of short duration, probably not over six months. All contacts of a person having plague should be immunized. Persons remaining in endemic areas should receive semiannual booster doses. Early use of broad-spectrum antibiotics, especially streptomycin with tetracycline, is extremely valuable in therapy.

PNEUMONIC PLAGUE. This is really a respiratory tract disease, but it may be considered here as a development of bubonic plague. Occasionally, in a patient with flea-borne plague of the bubonic type, the organisms may invade the lungs, causing a plague pneumonia (pneumonic plague). This disease, like other diseases of the respiratory tract, is exceedingly difficult to control. It is highly contagious by means of respiratory secretions and droplets thereof, and it is 90 per cent fatal. Septicemic seeding of the lungs often leads to fulminant pneumonia. but any and every tissue of the body can be infected. Outbreaks of pneumonic plague, unless controlled expertly and promptly by state and federal health authorities, are apt to become devastating scourges. Any severe case of bubonic plague may become pneumonic and cause respiratory failure in hours. Streptomycin is the preferred antimicrobial agent applied in therapy, given in large doses. Tetracyclines and chloramphenicol are also considered effective treatments.

Prevention. The area surrounding the bite of an infected arthropod, in both plague and tularemia, is usually marked by a pustule containing highly infectious pus. In both diseases, drainage and dressings from the initial lesion, as well as other exudates, must be carefully handled and immediately disinfected, preferably burned. The clothing of the patient admitted with bubonic plague should be put into a tight bag immediately and treated to kill fleas. The patient must be absolutely isolated. Nurse, attendants, and physicians must wear complete gowns and, in cases of pneumonic plague, head coverings with celluloid masks, as well as rubber gloves. The utmost precautions must be taken with any fomites. Nothing should leave the room that has not been disinfected or enclosed. Droplet and dust infection must be most carefully prevented in pneumonic patients. Bodies are autopsied or buried only under special licenses and precautions. Rats or fleas on the premises must be eliminated. A good insect spray that really kills fleas, used daily for three successive days and repeated weekly for five or six weeks throughout the infested premises, will often eliminate fleas.

However, it is erroneous to conclude that plague simply does not occur any more in the United States. Statistics show that between the years 1965 and 1968, one person died each year in this country from this disease. Five deaths occurred among the 18 reported cases of plague in 1980, and the data shown in Figure 39–3 indicate that this pestilence is still with us.

The Genus *Borrelia*

Relapsing Fever.[4] *Borrelia*[5] is the name of a genus of spirochetes that infect the blood and tissues of humans and other animals (Fig. 39–4), and that

[4]Distinguish between *relapsing* fever, due to spirochetes of the genus *Borrelia,* and *undulant* fever, due to *Brucella abortus* and related organisms.

[5]Named for A. Borrel (1867–1936), a famous French microbiologist who made pioneering studies of these organisms.

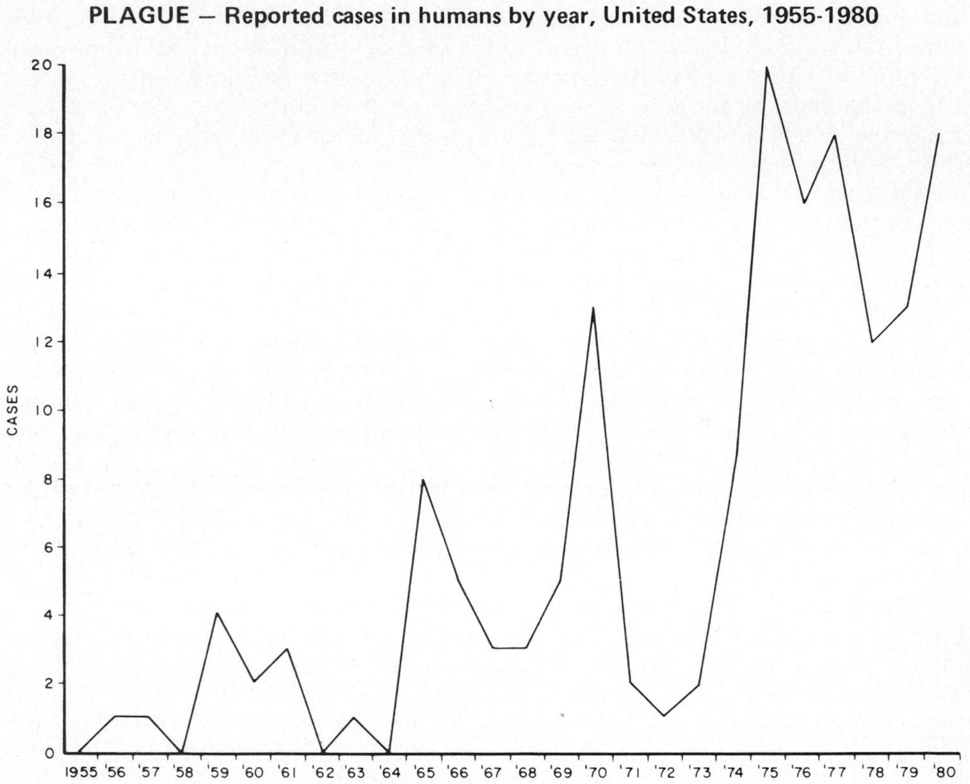

PLAGUE — Reported cases in humans by year, United States, 1955-1980

CASES

20 18 16 14 12 10 8 6 4 2 0

1955 '56 '57 '58 '59 '60 '61 '62 '63 '64 '65 '66 '67 '68 '69 '70 '71 '72 '73 '74 '75 '76 '77 '78 '79 '80

PLAGUE — Reported cases in humans by state and by age, United States, 1960-1980

Area	Total	0-9	10-19	20-29	30-39	40+
United States	**148**	**46**	**45**	**14**	**14**	**29**
Arizona	19	5	9	3	2	—
California	16	6	2	—	—	8
Colorado	10	2	3	1	1	3
Idaho	1	—	—	—	1	—
Nevada	3	1	1	—	—	1
New Mexico	90	30	27	8	10	15
Oregon	5	—	3	1	—	1
Texas	1	—	—	1	—	—
Utah	2	2	—	—	—	—
Wyoming	1	—	—	—	—	1

Five deaths occurred among the 18 reported cases of plague in 1980. Thirteen cases were acquired in New Mexico.

Figure 39–3 Plague, reported cases in humans by year, United States, 1955–1980. (From *Morbidity and Mortality Weekly Report*, Annual Summary, 1980. U.S. Department of Health and Human Services.)

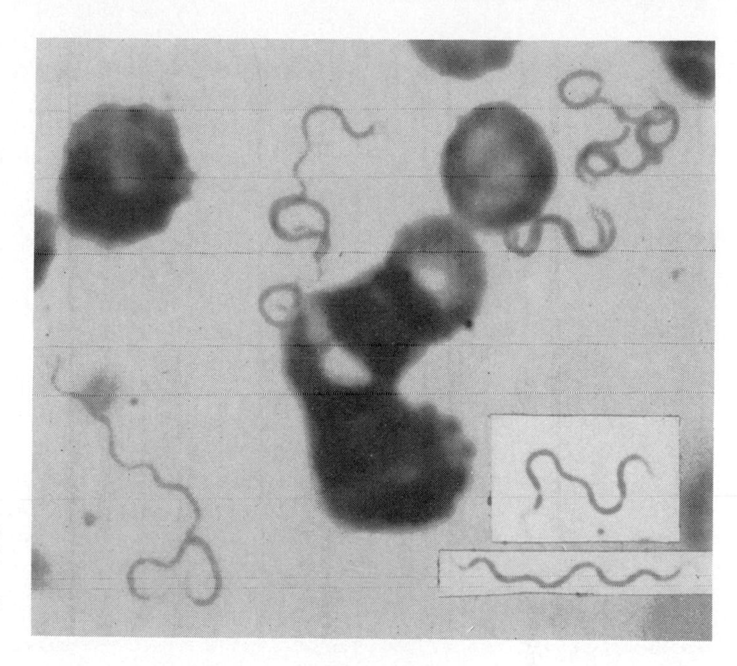

Figure 39–4 *Borrelia recurrentis* in a droplet of blood (the large, rounded, dark objects are blood cells). This is one of the species of *Borrelia* that cause relapsing fever (×2,700). (Courtesy of Dr. A. Packchanian, the University of Texas Medical Branch at Galveston.)

are transmitted by two types of arthopod: lice *(Pediculus humanus)* and ticks (of the genus *Ornithodorus)*. The spirochetes cause relapsing fever in humans. Diseases caused by *Borrelia* are often spoken of collectively as borrelioses. The borrelias are somewhat thicker and less regularly curved than treponemes, but the two have often been confused with one another. *Borrelia recurrentis,* transmitted by the body louse, is morphologically very similar to *Borrelia duttonii,* whose arthropod vector is the tick; both borrelias cause relapsing fever in humans. Several other *Borrelia* species, also transmitted by ticks, may cause the same disease in humans, although their natural hosts are a variety of wild animals.

Louse-borne Relapsing Fever. Relapsing fever due to *B. recurrentis* occurs chiefly in limited areas of Europe, North and South Africa, and Central America. The common vector is the body louse *(Pediculus humanus)*. Louse-borne diseases, including typhus fever and louse-borne relapsing fever, have been rare or nonexistent in the United States (why?). With the spread of human lice among sexually promiscuous persons, it may become more common. Since relapsing fever can occur in louse-infested populations, it often appears as an epidemic. The same populations also suffer from epidemics of typhus fever, since this is also louse-borne. Such populations generally wear clothes, i.e., they are not tropical natives who go largely naked.

Tick-borne Relapsing Fever. Cases of relapsing fever due to *B. duttonii,* seen in some of our western states, are transmitted from wild animals, especially rodents, to humans by bites of various species of soft *(Ornithodorus)* ticks. Such infectious tick bites, unlike the constant biting of lice, are rather uncommon and accidental, not epidemic, and sometimes occur during picnics in tick-infested areas.

Relapsing fever due to either species of *Borrelia* derives its name from its frequent and regular recurrences. Relapses of fever, chills, and pain recur at intervals of five to 15 days and last for several days. Usually five to eight relapses of diminishing severity occur. The spirochetes circulate in the blood during the febrile relapses and may readily be seen with the microscope in stained smears or darkfield preparations. Thompson reported in 1969 that the largest outbreak of tick-borne relapsing fever in the Western hemisphere occurred in 1968 in Washington state. It was thought that chipmunks and pine squirrels were the reservoirs of this outbreak.

Prevention and Treatment. Preventing the transmission of either louse-borne or tick-borne relapsing fever requires suitable precautions to keep the arthropod vectors from biting either patients or well persons. In military action this may be a difficult problem. Blood-engorged ticks, lice, and other anthropophilic arthropods, regardless of when or where found, should be handled with forceps and immediately incinerated.

A second set of precautions is based on the fact that the blood of patients in the febrile stage will infect through cuts or scratches on the hands or possibly through the unbroken skin. This is important to know when collecting blood specimens or giving intravenous injections.

Tick-infested houses may be dusted with lindane (BHC), as is done in Africa and certain parts of the Soviet Union. This extermination procedure is effective for about four weeks.

Most common antibiotics are effective chemotherapeutic agents against *Borrelia* that cause relapsing fever. A single fever attack may be overcome in one day with appropriate antibiotics.

SUMMARY

Microorganisms may gain entrance to the blood as a result of other infection, through cuts and wounds, by blood transfusions, or by way of arthropods that transmit disease-causing organisms when biting, stinging, or sucking. Serum hepatitis is caused by the homologous serum jaundice hepatitis B virus, which is known to occur only in human blood. Although serum hepatitis (SH) is less common than infectious hepatitis (IH), both are diseases of the liver with similar symptoms, including jaundice; in addition, neither of the viruses causing these diseases (HBV for SH and HAV for IH) is affected by any known chemotherapeutic agent, and both are highly resistant to heat. The antigen associated with the SH virus is the Australia antigen, also called HBAg, HAA, and SHAg, with the subantigens named the Dane particle, HB_sAg, and HB_cAg.

Malaria-transmitting mosquitoes are sanguivorous arthropods that transmit infectious agents directly into the blood. Other means by which flies, ants, roaches, and other insects may infect is by transmission through feces or vomitus or mechanically, for example, flies walking on feces or sores and then on food, plates, or glasses.

Diseases transmitted from wild animals to humans are called zoonoses. Some zoonoses are salmonellosis, rat-bite fever (sodoku), pseudotuberculosis, leptospirosis (Weil's disease), and tularemia (rabbit fever).

Hemorrhagic fever, a disease of animals caused by *Pasteurella multocida*, almost never affects humans; it is related to bubonic plague, a dangerous menace to all mankind throughout history. The bubonic plague is a zoonosis, a disease primarily of rats and other rodents. These animals have fleas, which can and do transmit *Yersinia pestis*, resulting in the Black Death, pestilence, the plague, or one of the many names by which this epidemic disease is known. When dead rats cool off, their fleas go to other warm-blooded hosts. The term "Black Death" was derived from the fact that in such epidemics the disease was marked by intense septicemia and subcutaneous hemorrhages over the whole body. When the organisms invade the lungs, pneumonic plague develops, a very deadly stage of the disease. Haffkine's vaccine is very effective in preventing plague, but this immunity lasts only for about six months.

Relapsing fever may result from an infection with *Borrelia recurrentis*, transmitted by the human body louse *Pediculus humanus* (remember typhus fever!), or with *Borrelia duttonii*, a tick *(Ornithodores)*-borne pathogenic spirochete.

REVIEW QUESTIONS

1. How may microorganisms gain entrance into the blood? What organisms may enter the blood in each way? What are the diseases they cause?
2. How may blood be infected accidentally? Artificially?
3. Name some diseases that cause jaundice.
4. What are the differences between SH and IH? In what ways are they similar?
5. List different names for SH.
6. What is the Australia antigen? What are sanguivorous arthropods? What are zoonoses?
7. How may arthropod vectors transmit disease-causing organisms? List diseases for each mode of transmission.
8. List the microorganisms that cause the following diseases: Haverhill fever, sodoku, rat-bite fever, tularemia, plague, hemorrhagic fever, relapsing fever.
9. List different names for the bubonic plague. How is this disease transmitted? How infectious is it? What are its symptoms? What means of prevention are known for this disease? What treatments are available?
10. Compare louse-borne to tick-borne relapsing fever. Why is it called relapsing fever?
11. What diseases may be transmitted by *Pediculus humanus?*

Supplementary Reading

Burrows, W.: Textbook of Microbiology, 20th ed. Philadelphia, W. B. Saunders Company, 1973.

Dahlstrand, S., Ringertz, O., and Zetterberg, B.: Airborne tularemia in Sweden. *Scand. J. Infect. Dis., 3*(1):7, 1971.

Emmons, R. W., Woodie, J. D., Taylor, M. S., and Nygood, G. S.: Tularemia in a pet squirrel monkey *(Saimiri sciureus). Lab. Anim. Care, 20*:1149, 1971.

Faust, E. C., Beaver, P. C., and Jung, R. C.: Animal Agents and Vectors of Human Disease. 4th ed. Philadelphia, Lea & Febiger, 1975.

Felsenfeld, O.: Borrelia, Borreliosis and Relapsing Fever. St. Louis, Warren H. Green, Inc., 1975.

Frobisher, M., Hinsdill, R. D., Crabtree, K. T., and Goodheart, C. R.: Fundamentals of Microbiology. 9th ed. Philadelphia, W. B. Saunders Company, 1974.

Goldenberg, M. I.: Laboratory diagnosis of plague infection. *Health Lab. Sci., 5*:38, 1968.

Horsfall, W. R.: Medical Entomology. New York, The Ronald Press Co., 1962.

Human bubonic plague—New Mexico. *Morbidity and Mortality, 20*:283, 1971.

Hunter, G. W., III, Swartzwelder, J. C., and Clyde, D. F.: Tropical Medicine. 5th ed. Philadelphia, W. B. Saunders Company, 1976.

Mullet, C. F.: The Bubonic Plague and England: An Essay in the History of Preventive Medicine. Lexington, The University of Kentucky Press, 1956.

Nutter, J. E.: Antigens of *Pasteurella tularensis:* Preparative procedures. *Appl. Microbiol., 22*(1):44, 1971.

Thompson, R. S., Burdorfer, W., Russell, R., and Francis, B. J.: Outbreak of tick-borne relapsing fever in Spokane County, Washington. *J.A.M.A., 210*:1045, 1969.

ARTHROPOD-BORNE RICKETTSIAL AND VIRAL INFECTIONS

RICKETTSIOSES

An infection with any of the rickettsias is properly spoken of as *rickettsiosis*. Several names of long standing, however, are commonly used to differentiate various rickettsioses transmitted by different arthropods and in different geographic areas. There are perhaps 20 such diseases, which have been placed in five major groups: *typhus fevers, spotted fevers, scrub typhus, Q fever,* and *trench fever* (Table 40–1). Of these, only two are of major importance in the United States: typhus, especially murine or rat-borne typhus, and Rocky Mountain spotted fever. Q fever is probably more widespread than is generally supposed.

The Typhus Fever Group

Louse-borne Typhus Fever

The word "typhus," freely translated from the Greek *typhos,* means stupor and refers to the stuporous, confused mental state of typhus patients. This disease, caused by *Rickettsia proważekii,* has been known since very ancient times. *Typhoid* fever (due to *Salmonella typhi*) received its name because some typhoid patients are stuporous and to this extent resemble typhus patients, but the two diseases are otherwise totally unrelated.

Epidemic or louse-borne typhus is transmitted through the feces of body lice deposited when the louse (*Pediculus humanus capitis* or *Pediculus humanus corporis*) bites.[1] The feces may then be scratched into the wound made by the bite. Louse-borne typhus occurs in Central and Eastern Europe, South and Central America, Asia, Africa, Russia, and (only rarely) in the United States and Canada. It can become epidemic when populations are thoroughly infested with lice. Troops in trenches, sailors in wooden ships, and crowded, poverty-stricken peoples during times of war or famine are especially likely to be visited by the disease. The course of history has been turned time and again by devastating epidemics of typhus fever; it has been known to decimate armies and entire populations since antiquity. It was probably in large part responsible for the eventual downfall of Napoleon's armies in Russia and possibly the defeat of the German armies in Italy and at Stalingrad in World War II.

Typhus fever is characterized by an onset of about two days' duration, during which there are nausea, headache, dizziness, and high fever. There then appears a rash, which may cover the whole trunk. It lasts for a week or more and disappears slowly. The patient is lethargic and delirious. The mortality in some epidemics is high. The blood of patients is infectious for lice, humans, and animals, although the rickettsias have not been demonstrated microscopically in the blood.

[1] In 1980, quite unexpectedly, five cases of "louse typhus" in the United States were traced to fleas from flying squirrels.

CHAPTER 40

Weil-Felix Reaction. A peculiar immunologic phenomenon, known by the name of its discoverers as the Weil-Felix reaction, occurs during typhus fever and in some other typhus-like diseases such as Rocky Mountain spotted fever and mite typhus or tsutsugamushi. It is sometimes used for diagnostic purposes. Agglutinins appear in high concentration in the blood during the first week of the diseases and later. Curiously, they are active not only against the infecting species of rickettsias but also against certain strains of *Proteus vulgaris.*[2] *Proteus* apparently has no other relationship to the rickettsial diseases.

The *Proteus* bacilli and the rickettsias both possess similar antigens, which give rise to antibodies that react with either organism. Thus, *Proteus vulgaris* provides so-called nonspecific antigens that are certain proteins that are also present in some rickettsial strains containing the specific antigens.

Some strains of *Proteus* in the nonmotile or O phase, designated OX–2, OX–19, and OX–K, are especially likely to be agglutinated by sera from persons convalescent from certain of the typhus-like rickettsial diseases (Table 40–2). These tests are useful but, like many biological tests, not wholly reliable. For example, they are often strongly positive in persons infected not with rickettsias but with strains of *Proteus.*

Complement Fixation Reaction. A more accurate method for diagnosing rickettsial infections is based on the CF reaction. The antigens used in this test are *specific,* being derived from rickettsias cultivated in the yolk sacs of living embryonated eggs. Similar material is also made into vaccines.

Indirect Fluorescent Antibody Test. These tests are group-specific but not species-specific. FA tests for rickettsial diseases are useful but serve their purpose only when used by highly skilled investigators; they are hardly of value for routine testing.

Prevention of Louse-Borne Typhus Fever. For the prevention of louse borne typhus fever, a vaccine is prepared from killed *R. prowazekii* cultivated in embryonated eggs. This has considerable value in modifying the disease. A live vaccine of strain E is now being used experimentally. The use of DDT insecticide to combat lice, along with measures of general cleanliness, was an important factor in helping American and British troops win the Italian campaign in World War II.

Flea-Borne Typhus Fever Typhus

This endemic disease is transmitted from rats to humans by the feces of infected rat fleas, which remain infectious during their lifetime. Flea typhus is also called murine or endemic typhus. Although sometimes less deadly than louse-borne typhus, it is essentially the same disease. The rash it causes is mild and may not occur at all, in contrast to the more severe and extensive rash of louse-borne typhus. Thanks to anti-rat campaigns, it has decreased in the

[2]These are gram-negative, nonspore-forming, motile rods classified with the Enterobacteriaceae. See Chapter 24.

Table 40–1. Human Rickettsial Diseases*

RICKETTSIOSIS GROUP	DISEASE NAMES	INFECTIOUS AGENT	GEOGRAPHIC REGIONS	VERTEBRATE HOSTS	ARTHROPOD TRANSMITTING AGENTS
Typhus fever group	Epidemic typhus fever, epidemic louse-borne typhus, typhus fever (classical type), typhus exanthematicus	*Rickettsia prowazekii*	In colder areas worldwide	Man	Body louse, *Pediculus humanus*
	Brill's disease (also called Brill-Zinsser disease), a recrudescence of typhus (not a new infection)	*Rickettsia prowazekii*	As typhus	Man	None
	Endemic typhus fever, endemic flea-borne typhus, murine typhus	*Rickettsia typhi* or *Rickettsia mooseri*	Worldwide	Rat, *Rattus rattus, Rattus norvegicus*	Fleas, *Xenopsylla cheopsis* or *Pulex irritans*
Spotted fever group	Rocky Mountain spotted fever, New World spotted fever, tick-borne typhus fever	*Rickettsia rickettsii* (possibly *Rickettsia canada* in North America?)	(1) Eastern U.S., Southern U.S. (2) Northwestern U.S., (3) South-western U.S., Brazil, Columbia	Rodents and other animals, dog, opossum, etc.	(1) Dog tick, *Der macentor varia-bilis* (2) Woo tick, *Dermacen tor andersoni* (3) Lone Sta tick, *Amblyomm americanum*, others
	Boutonneuse fever, Marseilles fever, Mediterranean fever, South African tick bite fever, Indian tick typhus, Kenya tick fever	*Rickettsia conorii*	(4) Mediterranean area (5) South Africa, India, Asia	Rodents and other small animals	(4) Dog tick (5 Various other ticks
	Queensland tick typhus (Note: this is not Q fever)	*Rickettsia australis*	Australia	Marsupials and rodents	Tick, *Ixodes holocyclus*
	North Asian rickettsiosis, Siberian tick typhus	*Rickettsia sibirica*	USSR, Mongolia	Rodents	Ticks of gener *Dermacentor* and *Haemaph salis*
	Rickettsial pox	*Rickettsia akari*	Cities U.S. and U.S.S.R., Equatorial and South Africa	House mouse, *Mus musculus*	Rodent mite, *Liponyssoides (Allodermanyssu sanguineus*
Scrub typhus	Scrub typhus, tsutsugamushi, rural typhus, Japanese river fever, mite typhus	*Rickettsia tsutsugamushi (Rickettsia orientalis, Rickettsia akamushi)*	Japan, Korea, China, South East Asia, Philippines, Indonesia, Australia	Rodents	Larval stage of mites *Leptotrombidium akamushi* or *Trombicula akamushi, Leptotrombidiu deliensis* or *Trombicula deliensis*
Q fever	Q fever	*Coxiella burnetii*	Worldwide, Western U.S.	Sheep, cattle, goats, birds	Various ticks
Trench fever	Trench fever, five-day fever, Voihynia fever	*Rochalimaea quintana*	Europe during World War I and II; recently in North Africa, USSR, Poland, Mexico	Man	Human louse, *Pediculus humanus*

*This table is compiled from numerous sources. No attempt is made to name *all* synonyms of organisms or all rickettsial diseases.

Table 40–2. Representative Weil-Felix Reactions of Patient's Serum

(ANTIGEN) *Proteus* STRAINS	LOUSE AND FLEA TYPHUS (CLASSIC TYPHUS) *(R. prowazekii)*	TICK TYPHUS (ROCKY MOUNTAIN SPOTTED FEVER, ETC.) *(R. rickettsii)*	MITE TYPHUS (TSUTSUGAMUSHI OR SCRUB TYPHUS) *(R. tsutsugamushi)*
OX–19	++++*	+	−
OX–2	+	+	±
OX–K	−	±	++++

*++++ = strong agglutination in relatively high dilutions of serum; +++, ++, +, and ±, lesser degrees of agglutination; − = no agglutination.

United States (Fig. 40–1). Chemotherapy has reduced the death rate to zero. Still, there were 68 cases of flea-borne typhus in the United States in 1980.

Rocky Mountain Spotted Fever

RMSF, caused by *Rickettsia rickettsii,* is like typhus fever in several respects. The infectious sheep ticks and wood ticks that transmit it probably maintain the disease among rabbits and other animals by bites. The ticks transmit the rickettsias among themselves from males to females and from females to their progeny through the eggs (i.e., transovarially) (Fig. 40–2). In patients with the disease the rash is typically confined to the arms and legs but may cover the entire body.

Rocky Mountain spotted fever is transmitted mainly by wood ticks (*Dermacentor andersoni*) in the West and mainly by dog ticks (*Dermacentor variabilis*) in the eastern part of the United States. São Paulo typhus in Brazil, transmitted by ticks (*Amblyomma cajennense*), and boutonneuse fever (fièvre boutonneuse) in Africa are closely related forms of this disease. Other rickettsioses in the *spotted fever group* resemble RMSF in many ways. These diseases differ chiefly in the names of the species of *Rickettsia* involved, in geographic distribution, and in vectors (Fig. 40–1).

An excellent preventive vaccine for RMSF is prepared from cultures of rickettsias grown in chicken embryo cell cultures in roller bottles. Hyperimmune rabbit and goat sera are also available for treatment and are probably of value if given early, but they are not much used.

To avoid tick-borne diseases, stay out of woods or meadows where ticks are known to be present. If forced to go into such areas, wear white clothing, preferably long trousers, and examine the whole body for ticks at three-hour intervals. Do not handle dogs with ticks on them, and never remove or crush engorged ticks with bare hands. Use insect repellents. In 1980, 1,163 cases of Rocky Mountain spotted fever were reported in the United States, mostly in the South Atlantic area (Fig. 40–3). Why is this disease still on the increase?

Broad-spectrum antibiotics like chlortetracycline and chloramphenicol are very valuable in treatment of rickettsial infections, especially Rocky Mountain spotted and typhus fevers. Penicillin, streptomycin, and the cephalothins are ineffective. Treatment must be started early. These antibiotics depend for their bactericidal action on interference with synthesis of cell walls, which, in ordinary aqueous environments with low osmotic pressure (near 0.85 per cent NaCl), are sufficiently strong to retain the cell contents intact. Rickettsias and chlamydias have inherently weak cell walls. They can thrive only intracellularly. Here, partly because of the relatively high osmotic pressure, neither their naturally weak cell walls nor the action of penicillin and similar antibiotics can affect them adversely. They are vulnerable, instead, to antibiotics like the tetracyclines and chloramphenicol, which interfere with protein synthesis.

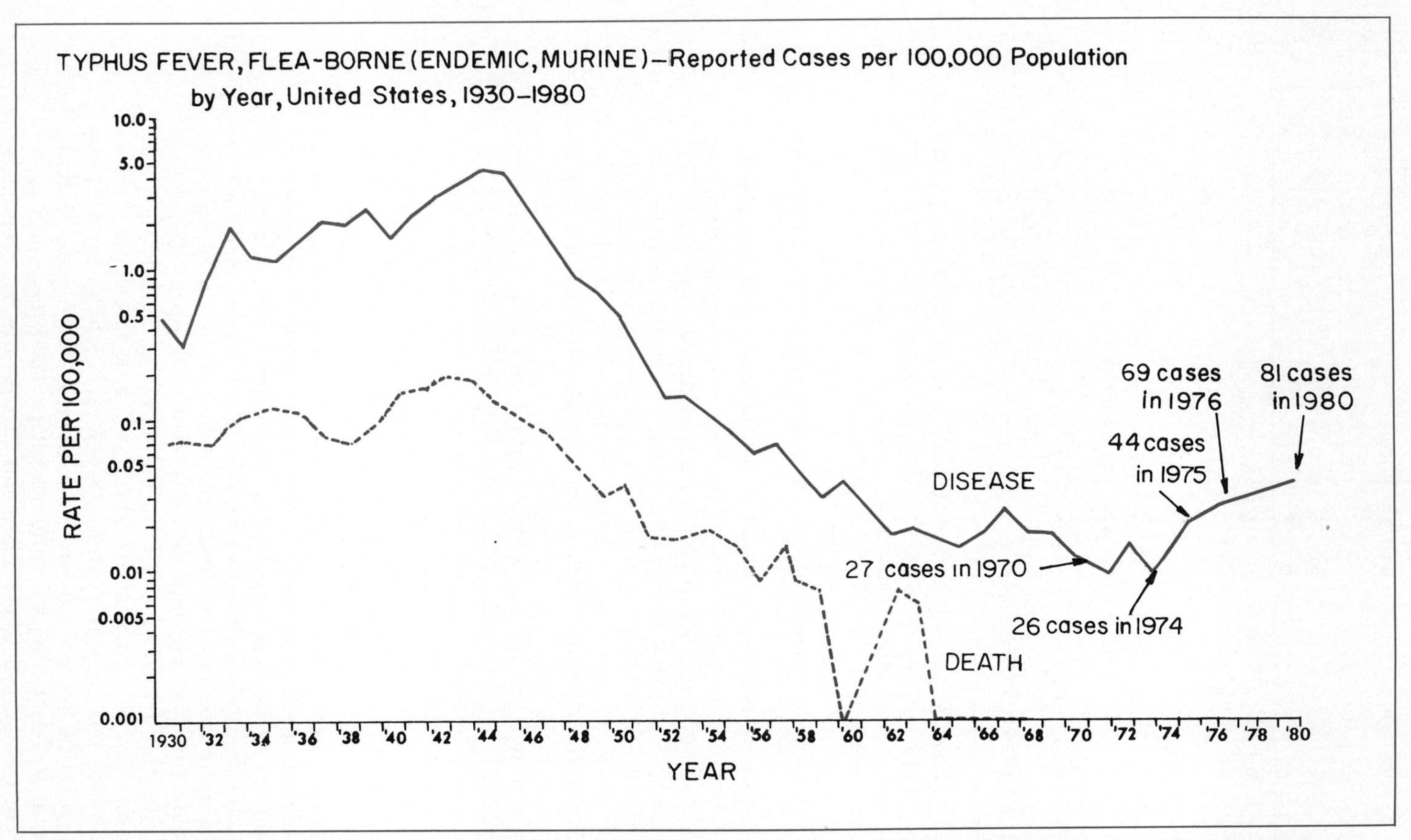

Figure 40–1 Typhus fever, flea-borne (epidemic, murine typhus). Reported cases per 100,000 population by year, 1930–1980. There was a total of 69 cases reported in 1976 and 81 cases reported in 1980. (From *Morbidity and Mortality Weekly Report,* Annual Supplements 1976 and 1980.)

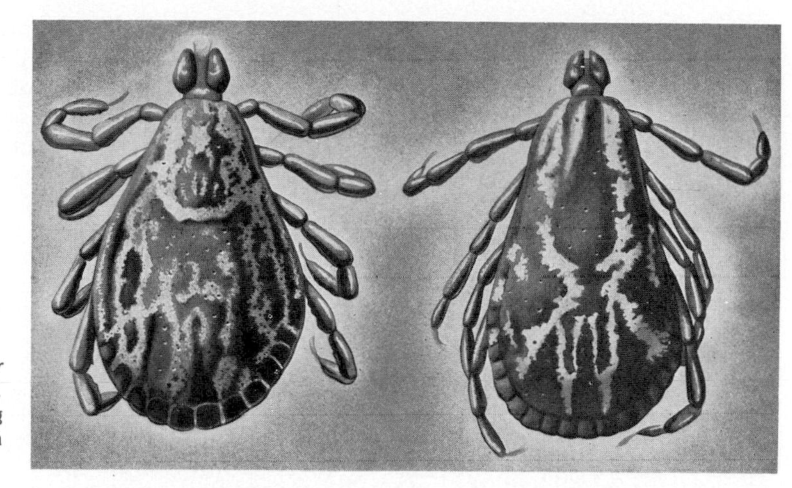

Figure 40–2 Spotted fever ticks. *Left, Dermacentor andersoni,* major vector of spotted fever in the West. *Right, Dermacentor variabilis,* common eastern dog tick, a vector of spotted fever in the East. (From Sharp & Dohme Seminar, Vol. 4, No. 2.)

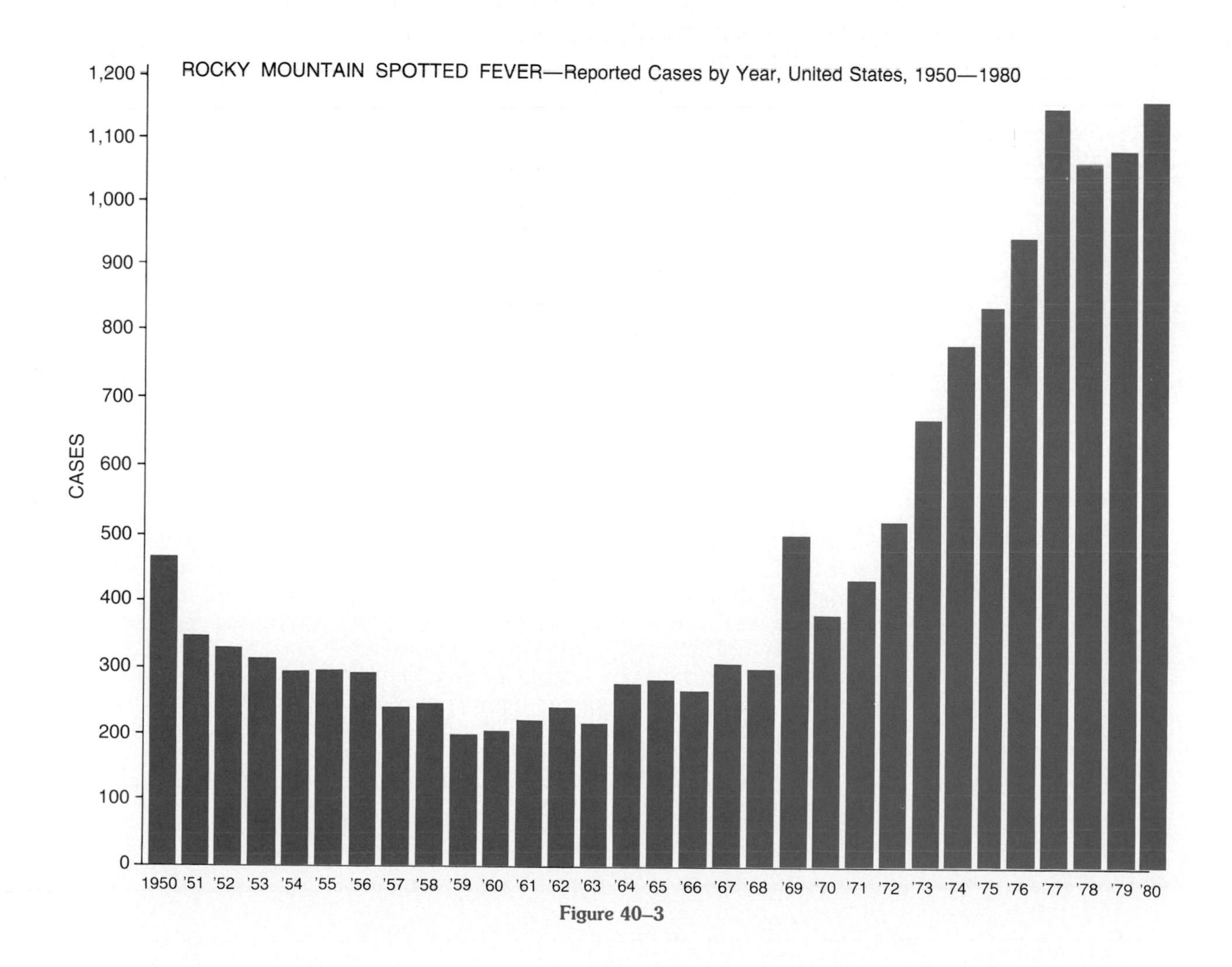

Figure 40–3

Scrub Typhus

Scrub typhus resembles typhus fever clinically except for the ulcer-like, dark scab area where the original mite bite was located, which is characteristic of scrub typhus. Prevalent in Japan, the Malay Peninsula, and the South Pacific Islands generally, scrub typhus has many names, such as mite typhus, tsutsugamushi, river fever, and swamp fever. It is caused by *Rickettsia tsutsugamushi,* and it is transmitted from rodents such as field rats and field mice to humans by the bite of the larva of a "chigger"-like mite known as *Leptotrombidium akamushi* or *Trombicula akamushi.*

Q Fever

This disease, caused by a rickettsia, *Coxiella burnetii,*[3] clinically resembles influenza or pneumonitis and in this respect differs from other known rickettsial diseases. Q fever is not one of the typhus-like rickettsioses. The disease was first observed in 1935 by Derrick and by Burnet in Australia. The designation "Q" stands for "Query," since the first observers were puzzled by the nature of the disease. It is true that this disease was discovered first in Queensland, but this did not give it the name Q fever, as is often assumed.

Q fever has since been observed in all continents and is probably quite widespread, even in areas where it has not yet been recognized. Diagnosis is based primarily on complement fixation tests and secondarily on isolation of the rickettsias, which is readily done by inoculation of animals.

The mode of transmission in nature appears to be several species of ticks that transmit *C. burnetii* to rodents. Persons working with cattle, goats, and sheep (especially during birth of young animals), as well as in stockyards, abattoirs, and dairies, and in contact with infected animal materials, seem to be especially likely to contract it. Cattle carry the infection, and drinking raw, infectious milk transmits the disease. Inhalation of dust containing the organism seems to be a common means of infection. Person-to-person transmission is rare.

Many species of ticks—some of which feed on cattle, others on sheep—are known to harbor the rickettsias. Various wild animals also are hosts to the organisms.

Q Fever and Pasteurization. Pasteurization of milk for 30 minutes at 145 F (62.8 C) or for 15 seconds at 161 F (71.7 C) is required to kill the rickettsias. They may survive for weeks in infectious butter.

Prevention. The urine in Q fever has been shown to contain *Coxiella* and should be disinfected. In addition, during the febrile stages of all rickettsial fevers the blood is always very infectious, and due care to avoid infection should be taken in drawing samples of blood for any purpose. The only acceptable method makes use of disposable needles and plastic syringes, which are subsequently incinerated in the proper manner.

Care should be taken to eliminate arthropods of all kinds as well as rats, from the premises. In epidemic typhus areas and in areas where mite typhus occurs, this may prove to be a problem of major difficulty and importance.

A vaccine of killed rickettsias is available for specially exposed workers but may cause severe reactions. Q fever vaccines are the latest rickettsial vaccines being developed.

---

[3]Named for its discoverers, H. R. Cox and F. M. Burnet.

Trench Fever

Trench fever, or five-day fever, is a disease caused by *Rochalimaea quintana* and transmitted by *Pediculus humanus,* the human body louse. This peculiar disease occurred among the armies in Europe during both World Wars. Symptoms are fever, exhaustion, headache, pain, coldness, and a rash. Despite its second name, the relapses may take place for over a year before recovery is complete. Also known as Voihynia fever, trench fever is presently occurring in North Africa, the Soviet Union, Poland, and Mexico.

ARTHROPOD-BORNE VIRAL INFECTIONS

Viruses that cause diseases in human beings and other vertebrates and that *are transmitted by* and *multiply in* sanguivorous arthropods are called "arboviruses"; *arbo* is an abbreviation of arthropod-borne.

These viruses are generally transmitted among infected birds (herons, sparrows, blackbirds, pheasants, pigeons, and many others), horses, monkeys, pigs, other mammals, and humans by mosquitoes, ticks, and sandflies, the vector and animal reservoir depending on the geographic location and the kind of virus. At least 300 different arthropod-borne viruses are recognized to cause diseases in humans, and the number is growing rapidly.

There are three major groups of arboviruses, differentiated on the basis of antigenic properties and serologic cross-reactions (complement fixation, hemagglutination-inhibition [HI], and so on). Most arboviruses are named for the place of their discovery, the animal principally infected, or the clinical effect. For example, in Group A we find Chickungunya, Semliki Forest, and Eastern and Western equine encephalitis viruses (the last two designated EEE and WEE, respectively); in Group B, dengue fever, yellow fever, St. Louis, West Nile, and Russian spring-summer encephalitis; in Group C, Marituba and Oriboca. There are also California encephalitis, Colorado tick fever, Bunyamwera, Bwamba, the Sandfly fever group (*Phlebotomus* fever) with 12 different identified viruses, and others. The student need not remember these groupings beyond the fact that they exist and that some have the mosquito as vector, others the sandfly or the tick, and some a yet unknown transmitting agent.

Prevalence of these diseases among humans implies a large reservoir of infected birds or mammals, numerous transmitting arthropods, and fairly close proximity of humans to these reservoirs and vectors. Transmission by sanguivorous arthropods produces viremia in animals during early stages of infection. Person-to-person transmission via arthropods can occur, resulting in large epidemics, as of yellow fever.

Yellow Fever

This disease has been known for centuries as a scourge of tropical Africa, tropical America, and even North Atlantic and European shipping ports, where it was introduced by sailing ships carrying infected tropical mosquitoes. It was introduced to the Western Hemisphere from Africa as early as 1500 by slave traders, probably to Central America. In 1898–1900 it nearly caused the defeat of American troops in Cuba, and along with malaria and dysentery it had earlier discouraged the French Government from completing a Panama canal. About 1900, a United States Army Commission under Major Walter Reed demonstrated that the disease is caused by a virus and that it is transmitted in populous areas principally by the bite of a mosquito called *Aedes aegypti* (Fig. 40–4), common in tropical and subtropical areas such as the U.S. Gulf Coast.

The virus circulates in the blood during the first four or five days of the initial febrile stage. Mosquitoes become infectious when they bite the patient. After about the fifth day the virus is found mainly in the viscera. Here it does extensive damage to the liver, kidneys, and blood vessels, causing jaundice, albuminuria, and hemorrhages into the gastrointestinal tract. Because it damages the viscera it is (for convenience) often spoken of as a viscerotropic virus. The histopathology of the liver in yellow fever (except Rift Valley fever) is diagnostically distinctive. The jaundice gives the disease its name. The vomitus, blackened by the blood in the stomach, gives rise to the well-known old term "black vomit," so effectively used by Captain Ahab in *Moby Dick.*

Vectors. The *Aedes aegypti*[4] mosquito is a household pest breeding mostly in artificial water deposits such as cisterns, rain gutters, flower vases, and water jars. It is not an extensive outdoor breeder. Control in urban and suburban areas therefore revolves mainly around the elimination or covering of these man-made water containers, which can eradicate the disease almost entirely.

Recent Outbreaks. In Africa, only a few sporadic cases of yellow fever were reported in 1970, in contrast to the series of outbreaks that occurred in several African countries in 1969. At that time, a total of 322 cases and 119 deaths were reported in Ghana, Mali, Nigeria, Togo, and Upper Volta. Not a single case of yellow fever has been reported in the United States since 1924.

Jungle Yellow Fever. Yellow fever has not disappeared absolutely because, as discovered in 1931, certain jungle-inhabiting mosquitoes (*Sabethes* and *Hemogogus,* and possibly other arthropods) can also transmit the disease, and certain wild animals (mainly monkeys) can act as an animal reservoir of the virus. Complete suppression of these jungle mosquitoes and monkeys is impossible under present conditions. Yellow fever contracted in the jungle or in farmlands adjacent to forests is called jungle (or sylvatic) yellow fever, but the urban and rural diseases are identical except for their vectors. Occasionally the jungle fever is introduced into a town by an infected traveler (or monkey, perchance), and then there may be an outbreak of classic urban yellow fever if the *Aedes aegypti* (or other urban mosquitoes in Africa) are prevalent in the town. Jungle yellow fever is still present in the Americas, as shown in Table 40–3.

Until not long ago it would have been very serious indeed if an infected person or mosquito were introduced into one of the Gulf or other Southern states, since there were plenty of *Aedes* there. Now the mosquitoes are being controlled. Authorities exert the utmost care to see that airplanes arriving from areas where yellow fever is known to exist contain no mosquitoes and (it is hoped) no person in the early (infective for mosquitoes) stages of yellow fever. A mosquito remains infectious for its lifetime (about two months).

Yellow Fever Vaccine. A yellow fever *vaccine* (designated as 17D) is now available and is one of the best known and safest immunizing agents. A single dose confers immunity of long, probably lifelong, duration. It is prepared in much the same way as influenza vaccine but differs in that the entire chick embryo is ground up and filtered, and the fluid constitutes the vaccine. No preservative is added. It is an active-virus vaccine. The 17D virus is the yellow fever virus that has been modified or attenuated by cultivation in tissue cultures and by animal passage. Injections of convalescent serum (antibodies) are of no demonstrated value in therapy, although they are effective in prophylaxis. Treatment is mainly supportive. Antibiotics may be of possible value in pre-

---

[4]In Ethiopia, *Aedes africanus* and *A. simpsoni.*

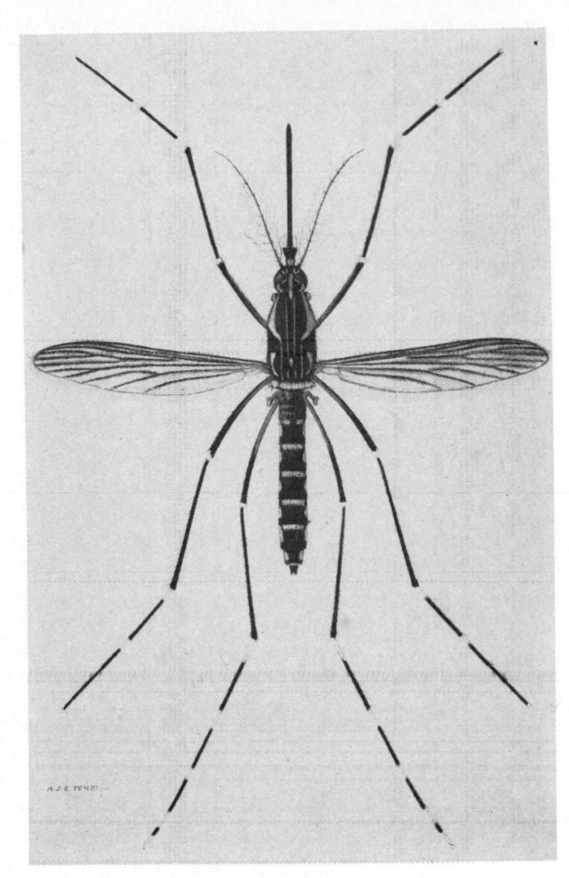

Figure 40–4 *Aedes aegypti*, the classic urban vector of yellow fever (and some other viral diseases). Note the lyre-shaped markings on the back of the thorax and the silvery bands on the legs. These handsome yellow fever–carrying mosquitoes have been common in the southeastern and U.S. Gulf states, but fortunately there is not (at present!) any virus there for them to transmit. (From Smith, K. G.: Insects and Other Arthropods of Medical Importance. New York, John Wiley and Sons, 1973.)

venting secondary bacterial infection, but they are not effective against the virus.

There are two special problems in the prevention of yellow fever. The first is that of excluding mosquitoes. The second problem is the febrile blood (first four or five days), which is highly infectious for humans, mosquitoes, and personnel of the health team.

Table 40–3. Jungle Yellow Fever—Reported Number of Cases and Deaths in South America, 1969–70*

COUNTRY	1969		1970	
	Cases	*Deaths*	*Cases*	*Deaths*
Bolivia	8	–	2	–
Brazil	4	4	2	–
Colombia	7	7	7	7
Peru	28	24	75	57
Surinam	1	1	–	–
Total	48	36	86	64

**Morbidity and Mortality*, *20*:37, 1971.

ARTHROPOD-BORNE VIRAL ENCEPHALITIDES[5]

Encephalitis is a pathologic term meaning inflammation of the brain. It may be caused by a variety of agents, both physical and chemical, as well as by infectious agents, among which are the so-called neurotropic viruses. These include poliomyelitis and rabies (previously discussed). Many of the neurotropic viruses are arboviruses, as noted earlier in this chapter.

Mosquito-Borne Viral Encephalitides

Each disease is caused by a specific virus. These viral diseases are classified in groups: e.g., group A includes Eastern equine (EEE) and Western equine encephalitis (WEE); group B, Japanese B, St. Louis (SLE), Murray Valley; LaCrosse is in the California encephalitis (CE) group. The vectors are mosquitoes like *Culiseta melanura* and several *Aedes* and *Culex* species.

By constant surveillance and examinations of mosquitoes and of sera from birds, domestic and wild animals, and humans, along with effective control of animal reservoirs and insect vectors in endemic areas by federal, state, and other health authorities, the number of cases of arboviral encephalitides occurring annually in the United States has been reduced to a very low level: in 1979, WEE, 3; SLE, 32; EEE, 3; CE, 139.

Eastern Equine Encephalitis (EEE). This is quite representative of the viral encephalitides. It occurs mainly in the eastern parts of North, Central, and South America but is not necessarily limited to these areas. It strikes mainly in the late summer and early fall, partly because it is transmitted by mosquitoes from migratory birds, reservoirs of the virus.

Mortality from this disease, as from many of the viral encephalitides, may run as high 55 per cent, or even 80 per cent among older people and children. As in most viral encephalitides, onset is sudden, with headache, fever, vomiting, drowsiness, apathy, and nervous signs and symptoms, including disturbances of reflexes, speech difficulties, rigidities of neck and back muscles, and paralyses. As in poliomyelitis and many other viral diseases, more inapparent infections than recognized cases occur.

In 1980, three EEE cases were reported in Georgia and Florida. The patients had onset of encephalitis in the second half of June. All three patients were hospitalized; two died. EEE virus infection was confirmed serologically in one of the patients. Georgia health authorities applied insecticides near the residences of the patients. An investigation of EEE activity among humans, horses, birds, and mosquitoes in the area was made.

The arthropod vectors of EEE are *Aedes, Culex, Anopheles, Culiseta melanura,* and other mosquitoes.

Diagnosis. This is most reliably done by inoculation of infected arthropods, or of febrile blood from patients, into suckling mice or other susceptible animals and observing the disease there. Retrospective diagnosis (during convalescence or after recovery) can be made by repeated complement fixation and other serologic tests to observe a rise in antibody titer over a period of a week or more.

Control. Control of the viral encephalitides depends on control or immunization of the animal hosts, destruction or avoidance of the arthropod vectors, and use of vaccines such as the inactivated-virus vaccine for Japanese

---

[5]The diseases of the central nervous system due to the so-called neurotropic viruses are collectively spoken of as the *viral encephalitides* or *encephalomyelitides* (inflammations of the brain or the brain and spinal cord).

encephalitis, made from the brains of infected mice or from infected embryonated eggs. The efficacy of such vaccines for human protection is still under evaluation.

Venezuelan Equine Encephalitis (VEE). This disease is caused by an arbovirus, immunologically distinct from others found in the United States. The 1972 outbreaks of equine encephalitis (VEE) in Mexico and Texas resulted in the deaths of many horses and a number of persons. An attenuated Venezuelan equine encephalitis virus vaccine is currently under investigation for use among high-risk individuals.

Tick-Borne Viral Encephalitides

These diseases—like Russian spring-summer encephalitis, diphasic milk fever, Central European tick-borne encephalitis, and others—occur in the USSR, Eastern and Central Europe, the British Isles, and Scandinavia, but the Powassan virus also exists in Canada and the United States. The ticks *Ixodes persulcatus* and *I. ricinus* transmit the viruses to sheep, deer, birds, and rodents and (fortunately only rarely) to humans.

SUMMARY

The rickettsioses, diseases caused by different species of *Rickettsia*, have been placed into major groups: typhus fevers, spotted fevers, scrub typhus, Q fever, and trench fever. In the typhus fever group are *louse*-borne typhus fever, best known as epidemic typhus, and *flea*-borne typhus fever, which is endemic typhus. Scrub typhus, also known as *mite* typhus, resembles typhus fever in many respects. The spotted fever group of diseases, represented by Rocky Mountain spotted fever (RMSF), or *tick* typhus, also mimics typhus fever clinically. Diseases in the spotted fever group differ chiefly in the names of the species of *Rickettsia* involved, in geographic distribution, and in the vectors that transmit the etiologic agents. The reader should best refer to Figure 10–1 for an easy classification of the rickettsioses. For diagnostic purposes, the Weil-Felix test (reaction) uses strains of *Proteus vulgaris* as the nonspecific antigen instead of the specific rickettsial strain; it tests for antibodies in the patient's serum against the *Rickettsia* causing the disease. Antibiotics are quite effective against rickettsial diseases, and vaccines, such as the one used to prevent RMSF, are excellent. It must be remembered that the vectors of rickettsioses are arthropods, lice, fleas, ticks, and mites; to destroy them is to reduce the incidence of these diseases.

Yellow fever is a viral disease of the liver that results from the bite of an infected *Aedes aegypti* mosquito. Characteristic of the infection are the jaundice and black vomit it produces. A yellow fever vaccine (designated as 17D) is one of the best-known immunizing agents.

Arthropods are also the vectors of the viral encephalitides, inflammations of the brain. Mosquito-borne viral encephalitides are caused by many specific different viruses. These viruses are classified in groups: e.g., group A includes eastern equine encephalitis (EEE) and western equine encephalitis (WEE), and group B includes Japanese B, St. Louis, and Murray Valley encephalitides. The arthropod vectors of EEE are *Aedes, Culex, Anopheles*, or *Culiseta* mosquitoes. Venezuelan equine encephalitis (VEE) is immunologically distinct from the other diseases; it results in the death of horses but has been transmitted to humans as well. Tick-borne viral encephalitides affect sheep, deer, birds, rodents, and rarely, humans.

REVIEW QUESTIONS

1. Differentiate between the major groups of rickettsioses on the basis of names of diseases, infectious agent, geographic region, vertebrate host, and arthropod vector.
2. Which diseases are clinically related to typhus fever? What differences are there?
3. What is known about the history of rickettsial diseases in war? Why?
4. Why do you think there are so many different infectious agents of the spotted fever group?
5. Explain the use and function of the Weil-Felix reaction (test). How is it similar to the Widal test? How do they differ?
6. What vaccine and what treatments are used for RMSF?
7. What are the two major vectors of arthropod-borne viral diseases? What other vector can you name?
8. What are the arboviruses? List some diseases in each of their subgroups.
9. What symptoms of yellow fever can you name? Why is it difficult to eliminate this disease? How is it transmitted? What vaccine has been developed for yellow fever?
10. Differentiate between encephalitides, encephalitis, and encephalomyelitides.
11. What is the mortality rate of eastern equine encephalitis? How is this disease diagnosed? How is it controlled?
12. Name four genera of mosquitoes.

Supplementary Reading

Burrows, W.: Textbook of Microbiology. 20th ed. Philadelphia, W. B. Saunders Company, 1973.

Cohen, S., and Sadun, E. H.: Immunology of Parasitic Infections. St. Louis, C. V. Mosby Co., 1976.

Davis, B. D.: Microbiology. New York, Harper & Row, 1980.

Donohue, J. F.: Lower respiratory tract involvement in Rocky Mountain spotted fever. *Arch. Intern. Med., 140*:223, 1980.

Follow-up on Venezuelan equine encephalitis—Texas. *Morbidity and Mortality, 20*:275, 1971.

Hahon, N., and Zimmerman, W. D.: Intracellular survival of viral and rickettsial agents at −60° C. *Appl. Microbiol., 17*:775, 1969.

Henderson, J. R., Karabatsos, N., Bourke, A. T. C., Wallis, R. C., and Taylor, R. M.: A survey for arthropod-borne viruses in south-central Florida. *Amer. J. Trop. Med. Hyg., 11*:800, 1962.

Human Venezuelan equine encephalitis—Florida. *Morbidity and Mortality, 20*:411, 1971.

Hunter, G. W., III, Swartzwelder, J. C., and Clyde, D. F.: Tropical Medicine. 5th ed. Philadelphia, W. B. Saunders Company, 1976.

Justin, O. J.: The epidemiology of murine typhus in Texas. 1969. *J.A.M.A., 214*:2011, 1970.

Klingberg, W., Klingberg, M. A., and Goldwasser, R. A.: Recrudescent typhus. *Scand. J. Infect. Dis., 2*:215, 1970.

McDade, J. E.: Determination of antibiotic susceptibility of *Rickettsia* by the plaque assay technique. *Appl. Microbiol., 18*:133, 1969.

McDade, J. E., Stakebake, J. R., and Gerone, P. J.: Plaque assay system for several species of *Rickettsia*. *J. Bacteriol., 99*:910, 1969.

Pratt, H. D.: Mites of Public Health Control and Their Importance. Atlanta, U.S. Dept. of Health, Education, and Welfare, Public Health Service, Center for Disease Control, 1976.

Vinson, J. W., Varela, G., and Molina-Pasquel, C.: Trench fever III. Induction of clinical disease in volunteers inoculated with *Rickettsia quintana* propagated on blood agar. *Am. J. Trop. Med., 18*:713, 1969.
Weinberg, E. H., Stakebake, J. R., and Gerone, P. J.: Plaque assay for *Rickettsia rickettsii. J. Bacteriol., 98*:398, 1969.
Wells, G. M., Woodward, T. E., Fiset, P., et al.: Rocky Mountain spotted fever caused by blood transfusion. *J.A.M.A., 239*:2763, 1978.
Woodman, D. R., Weiss, E., Dasch, G. A., et al.: Biological properties of *Rickettsia prowazekii* strains isolated from flying squirrels. *Infect. Immun., 16*:853, 1977.
Yellow fever in 1970—Africa and South America. *Morbidity and Mortality, 20*:332, 1971.

ARTHROPOD-BORNE PROTOZOAL AND HELMINTHIC DISEASES

PROTOZOAL DISEASES

Malaria

In the United States malaria as an epidemic or endemic disease has been virtually eliminated. Persons entering the United States from malarious areas reintroduce it, but it does not spread. In 1971 a total of 2,375 new cases of malaria were reported in the United States, most of these imported from Southeast Asia and the rest from seamen traveling in the tropics (Fig. 41–1). By 1975, the number had declined to 373, since the movement of troops from Asia had ceased. A few cases may be accounted for by blood transfusions and the use of unsterile needles by groups of drug addicts.

It is typical of our fight against pathogens that every time we think a disease is practically eliminated, we find that our vigilance still cannot be relaxed. Malaria remains widespread and death-dealing in many areas outside the United States, especially in tropical and subtropical zones around the world. Because of foreign immigration, by 1980 the number of reported malaria cases in the United States had again risen to 2,062. In 1955 its ultimate elimination from this planet became a major concern of many international health agencies such as the World Health Organization and the Pan-American Sanitary Bureau. However, in 1968 the WHO revised its scheme for global malaria eradication, considering it unattainable as long as basic medical services necessary for the detection and treatment of residual malaria remain insufficiently developed in many countries. Prior to the work of the WHO an estimated 1,692 million people were continuously endangered by malaria throughout the world; of these about two-thirds live now in regions entirely free of this disease, and other areas are making progress. Still, malaria will be with us for some time to come.

The Malarial Parasite. The disease is caused by species of the genus *Plasmodium,* a protozoan parasite belonging to the class Sporozoa. Its life history is quite complicated. It has two stages of development: the asexual stage, called *schizogony,* takes place in the human body; the sexual stage, called *gametogony* and *sporogony,* occurs in the mosquito. This is a good example of the phenomenon called *alteration of generations.* Both humans and mosquitoes are hosts of the parasite and both are necessary for its life and reproduction. Unless the parasite can pass from its human host to the mosquito and from this insect back to another person, it will die out.

Life in the Human Host. The mosquito introduces the parasites as *sporozoites* into the blood of its victim with its saliva when it bites (Fig. 41–2). The parasites undergo a period of multiplication (one to five weeks, depending on species) in cells of the liver. This is called the *exoerythrocytic* (or *preerythrocytic*) *stage.* The sporozoites have now become *merozoites.* Very soon these enter erythrocytes and grow within them. This stage of the parasite is called the *trophozoite* stage. Each parasite multiplies asexually within its red blood cell, forming a *schizont* with many small, nucleated segments (Fig. 41–3). Finally the affected red blood cell breaks up, and the segments escape into the circulating blood. Each segment is a new, active *merozoite* that is released into the plasma

CHAPTER 41

and immediately attacks another erythrocyte in order to multiply again. In this way the blood is soon teeming with the parasites *(parasitemia)*. The patient becomes anemic and weakened by the loss of so many red blood cells and probably also suffers from poisonous products, especially fever-producing agents *(pyrogens)*, formed by the parasites.

Every time successive generations of the parasite divide and burst out of the blood cells, the affected individual suffers the chills and fever so characteristic of malaria. A chill means that a fresh crop of parasites has matured and entered the circulation.

After passing through several cycles of asexual development as just described, distinctive round male and female gametes begin to appear in the blood of the patient. They are called *gametocytes*. These are larger than the asexual forms and are easily recognized in smears of the blood examined under the microscope. The gametocytes undergo no further development in the human erythrocytes, and if not taken up by a mosquito, they die. The protozoal life cycle is asexual in man and sexual in the mosquito.

Life in the Mosquito. When a *female* mosquito of the proper genus *(Anopheles)* bites a person who has malarial gametocytes in the blood, she takes these in with the blood she sucks. The gametocytes then become mature sex cells *(gametes)*, and the sexual stage begins.

After fertilization of the female gamete by the male gamete, forming a *zygote* in the stomach of the mosquito, the parasites, as *ookinetes*, invade the cells lining the mosquito's stomach and multiply there. The parasites undergo further development in a sac in the wall of the mosquito's stomach (an *oocyst*), resulting in a stage of development impossible in humans but necessary for the continued existence of the parasite. Here the fertilized parasite multiplies by fission. The oocyst ruptures, liberating numerous new young parasites *(sporozoites)*. These, after moving about for some days inside the mosquito, reach the mosquito's salivary glands and from there are injected into humans when the insect bites. The life cycle is thus complete. Because of the necessary period of sexual reproduction of the parasite, a mosquito that has bitten a malaria patient cannot transmit the disease to another person until after 10 to 21 days, depending on temperature and species. This period is spoken of as the *extrinsic incubation* period. The female mosquito, once infected, remains so for the rest of her life, which may be two months or more.

Species. There are four species of human malarial parasites. One of the most widely distributed in tropical and temperate zones is called *Plasmodium vivax*, from the vivacious activity of its trophozoite stage. It usually requires about 48 hours to complete its development within the red blood cells. The chills therefore commonly, but not invariably, occur at intervals of 48 hours (Fig. 41–4). This type of malaria is called *tertian (third) fever*.

A second, less common species, called *Plasmodium malariae*, requires about 72 hours for development, and groups of parasites mature about every fourth day. This species causes *quartan (fourth) fever*. Both species commonly cause prolonged infections with relapses unless treated.

A third form, *Plasmodium falciparum* (the word falciparum is derived from the curved or sickle-shaped [falciform] sexual cells), causes "malignant tertian" or *aestivo-autumnal fever*, and requires from 24 to 48 hours for development.

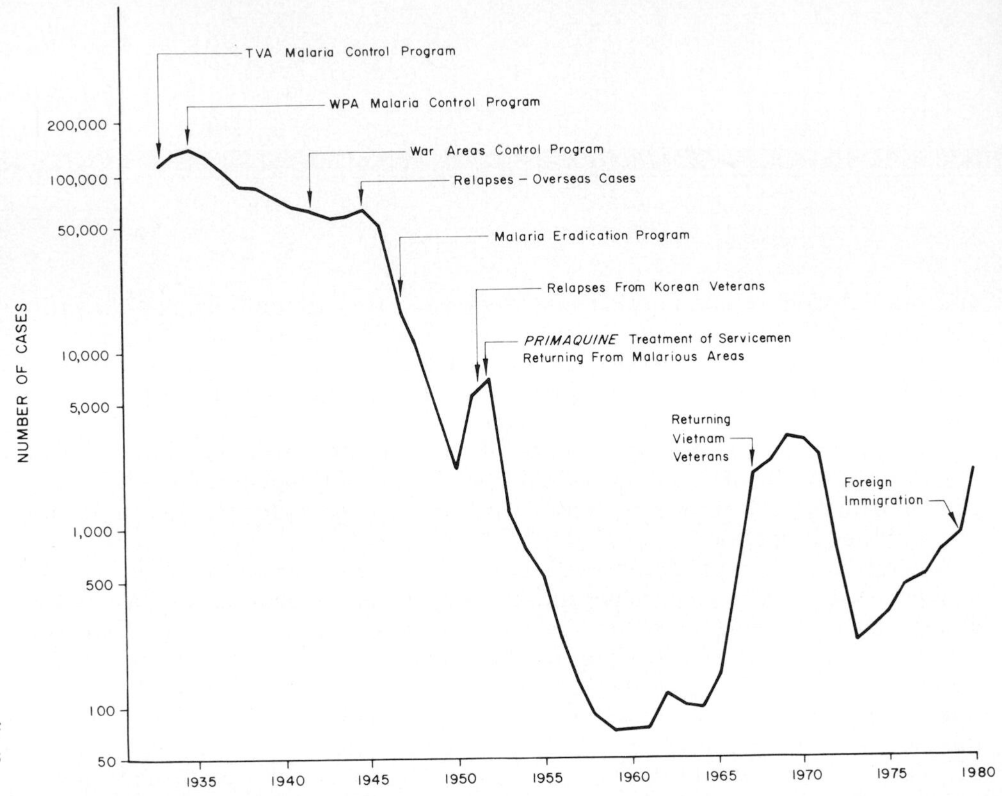

Figure 41–1 Incidence of malaria in the United States 1933–1980.

The temperature curve of this fever is irregular, and chills may occur every day or be entirely absent. It is called *aestivo-autumnal* because in temperate climates it frequently occurs in the late summer and fall. It is most prevalent in tropical zones. *Falciparum* malaria is more severe than the other forms of malaria. Most of the fatal cases of malaria are caused by *P. falciparum.*

Plasmodium ovale, a fourth species resembling *P. vivax,* causes a disease much like tertian malaria but milder.

Excepting acute falciparum malaria, malaria is usually a self-limiting disease. If the infected individual harbors the parasite in the tissues, attacks occur frequently at first, less often over the years, and finally not at all. One can never be sure that this disease is fully overcome; blood transfusions from people who have a history of malaria (however remote) must always be avoided.

Within the last decade, studies of tropical malaria have shown that species of *Plasmodium* formerly thought to be entirely confined to monkeys (simian malaria) are transmissible to humans by various jungle mosquitoes. These are typical zoonoses and are probably much more common infections of people than previously supposed.

The Mosquito Vector. There are numerous species of mosquitoes, but only a few carry the parasites of human diseases. All species that transmit malaria parasites of the four human types belong to the genus *Anopheles.* Only the female bites humans, and she does so only because blood is necessary for egg-laying. One of the most dangerous mosquitoes is *Anopheles gambiae.*

Mosquitoes, like most other two-winged insects *(Diptera),* pass through four stages of development (Fig. 41–5): the egg, the larva or "wiggler," the pupa, and the fully developed insect (imago). The first three stages develop in water.

The *Anopheles* mosquito may be recognized as she bites. First, she stands in a position like that indicated in Figure 41–5, with her hind legs raised high in the air. Second, she usually has spots of silver or gray on the wings and often

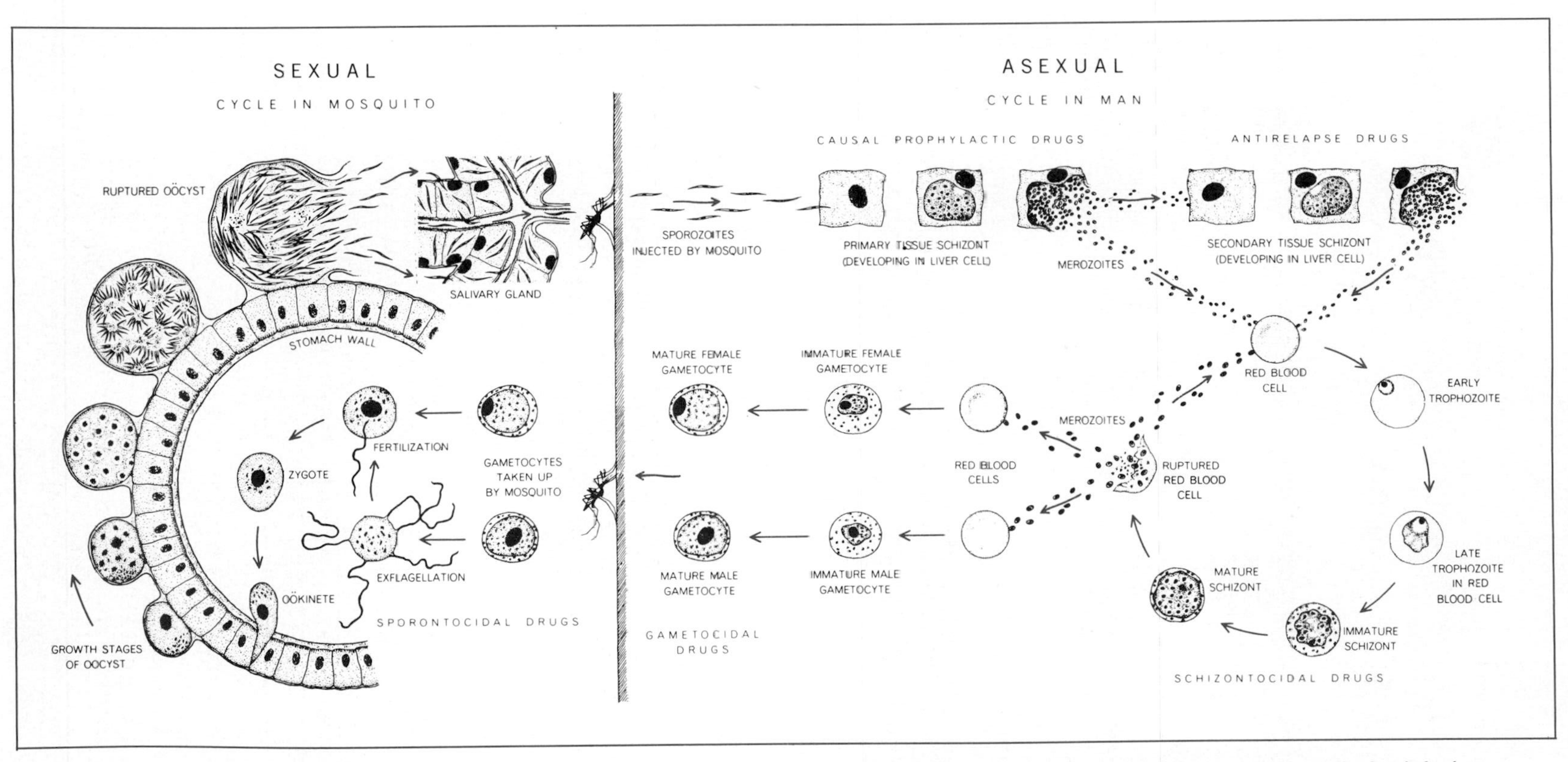

Figure 41–2 Life cycle of *Plasmodium*, the malaria parasite, in the mosquito and in man. The sporozoite from the mosquito's saliva enters the host's body and multiplies in the liver; later, as a merozoite, it enters a red blood cell, and as a schizont it destroys the cell, causing the chills and fever of malaria. When another mosquito bites, male and female gametocytes enter its stomach with blood sucked from the malarial patient. A zygote (fertilized egg) forms in the stomach wall. Once the mosquito regurgitates sporozoites into its salivary glands, it can infect another new host. Various drugs can attack the parasite at different stages as indicated. The protozoal life cycle is asexual in man and sexual in the mosquito. (From Carlos A. Alvarado and L. J. Bruce-Chwatt: Malaria. Copyright 1962 by Scientific American, Inc. All rights reserved.)

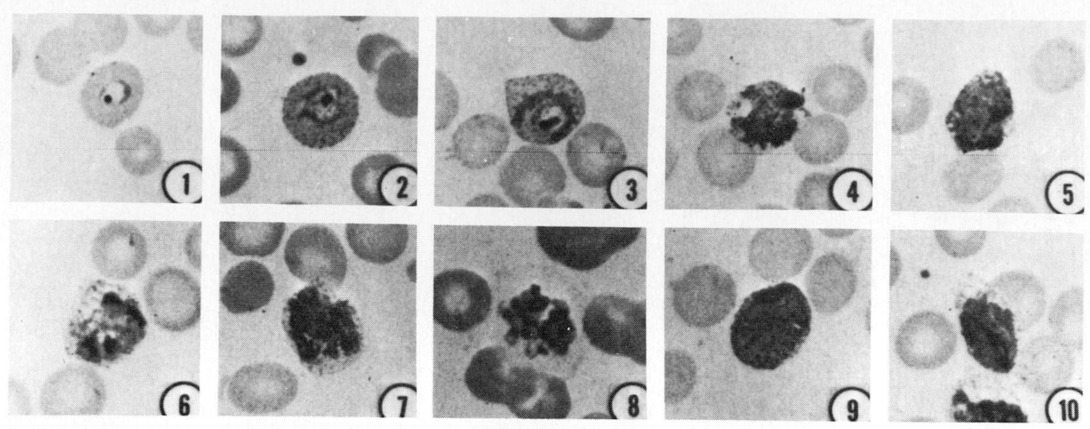

Figure 41–3 Developmental stages of *Plasmodium vivax* in erythrocytes. 1, Young trophozoite, ring stage. 2, Developing trophozoite in enlarged erythrocyte with Schüffner's dots. 3, Older trophozoite with pseudopod. 4, Mature trophozoite with irregular margin. 5, Schizont showing initial division of nucleus. 6, Older schizont with four nuclei. 7, More mature schizont. 8, Mature schizont with 16 nuclei and clumped pigment. 9, Macrogametocyte. 10, Microgametocyte. (From Wilcox, A.: Manual for the Microscopical Diagnosis of Malaria in Man. National Institute of Health Bulletin No. 180. Washington, D.C., U.S. Government Printing Office, 1942.)

gray bands on the legs. The nonmalaria-bearing varieties are usually brownish or brown-gray and stand as shown in Figure 41–5.

Diagnosis of Malaria. The laboratory diagnosis of malaria is commonly made by spreading a small drop of the patient's blood on a slide and either examining it in the fresh state or staining it with Wright's stain or any special stain used for bloodsmears. The parasites can be seen in or upon the blood cells and may have various appearances, depending on species and the stage of their development in the erythrocytes.

Control of Malaria. As with other arthropod-borne diseases, control is directed primarily against the vector arthropod. The control of mosquitoes in swamps and other breeding places is essential and often presents an engineering problem. Mosquitoes may be kept out of homes and away from sleepers by screens and insect repellents. The use of DDT, dieldrin, and other residual sprays has been effective, since they kill infected female mosquitoes in the house. Treatment of infected persons to kill the parasites in their blood is very important. Then the mosquitoes, even though they bite, do not become infected. Drugs now available include chloroquine, pyrimethamine, amodiaquine, and preferably primaquine. Quinine was mainly of historical interest until outbreaks of the falciparum form in Southeast Asia during the Vietnam conflict, where treatment with newer drugs was often ineffective against drug-resistant strains

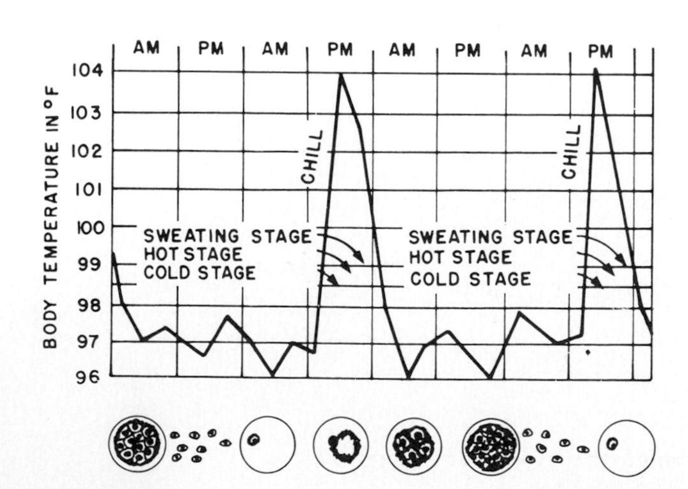

Figure 41–4 Relationship between periodicity of malarial fever and rupture of parasites from erythrocytes. (Courtesy of Encyclopaedia Britannica, 1967 ed., vol. 14, and Braude, A. I., Medical Microbiology and Infectious Disease. Philadelphia, W. B. Saunders Company, 1981, p. 1482.)

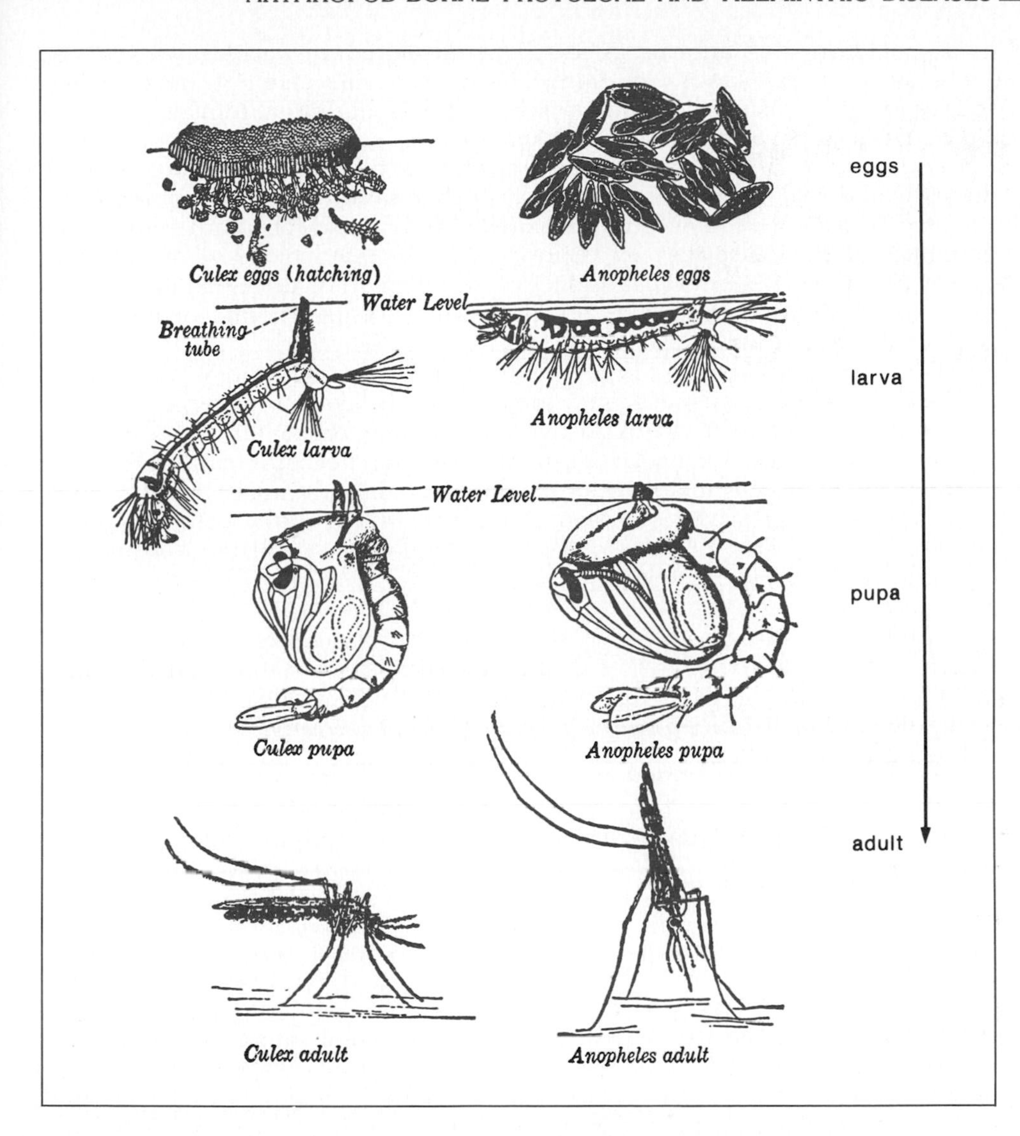

Figure 41–5 Stages in the life cycle of a common house mosquito of the genus *Culex* at the left and of a malaria mosquito, *Anopheles*, at the right. Note the distinctive biting positions of the adult mosquitoes, *Anopheles* with its hind legs up in the air. (Courtesy of U.S. Bureau of Entomology.)

of *P. falciparum* until quinine was administered together with pyrimethamine. Treatment regimens differ for the various types and stages of malaria. Some drugs are of value chiefly as prophylactics, others as curatives. Difficulties arise in control because of the development of strains of *Anopheles* that are resistant to the insecticides and strains of *Plasmodium* that are resistant to drugs.

Malaria-like Babesiosis

A malaria-like disease called babesiosis or red water fever is caused by the sporozoa *Babesia microti,* which is transmitted through the bite of blood-sucking female ticks *(Ixodes dammini)*. Erythrocytes are invaded, and the symptoms of the disease resemble those of malaria. Mice serve as the rodent reservoir.

Trypanosomiasis

The Trypanosomes. The trypanosomes, of which there are several important species, are flagellate protozoa, some of which inhabit the blood and tissues of man and animals. They are transmitted in various ways. One, *Trypanosoma equiperdum,* is transmitted by sexual contact among horses. This species causes dourine or "equine syphilis." It does not infect humans.

African Sleeping Sickness. Other species, such as *Trypanosoma brucei gambiense* and *Trypanosoma brucei rhodesiense,* are transmitted among African domestic and wild animals and to humans in the saliva of any of six species of *Glossina* (*G. palpalis,* and so on), the sanguivorous tsetse flies. Either the male or female of the tsetse fly may be infected by ingested blood of an infected person or animal. These trypanosomes cause infection of the blood and tissues of humans, often producing encephalitis, which results in the torpor called African sleeping sickness.

Developmental Stages. Trypanosomes generally pass through two or more of a series of developmental stages, some of which, like the malaria parasites, may appear only in the arthropod vector, whereas the others are found in animal hosts. As seen in Figure 41–6, at least four developmental forms are differentiated: the leishmanial, the leptomonad, the crithidial, and the adult trypanosome. Progressive elongation, change in position of the parabasal body, and longitudinal development of the flagellum and undulating membrane characterize this cycle. The first two stages do not occur in *T. b. rhodesiense* and *T. b. gambiense.*

Adult trypanosomes are wavy, spindle-shaped organisms, about 20 μm in length (without the flagellum) with pointed ends and an undulant, keel-like membrane extending from tip to tip. A flagellum, attached like an edging along the margin, extends free for perhaps 15 μm anteriorly. There are a well-defined nucleus and other functional granules typical of eukaryotic cells (Fig. 41–6).

Chagas' Disease. A species of trypanosomes in South and Central America and the southern United States, *Trypanosoma cruzi,* causes an infection of blood and tissues called Chagas' disease, in some respects similar to African sleeping

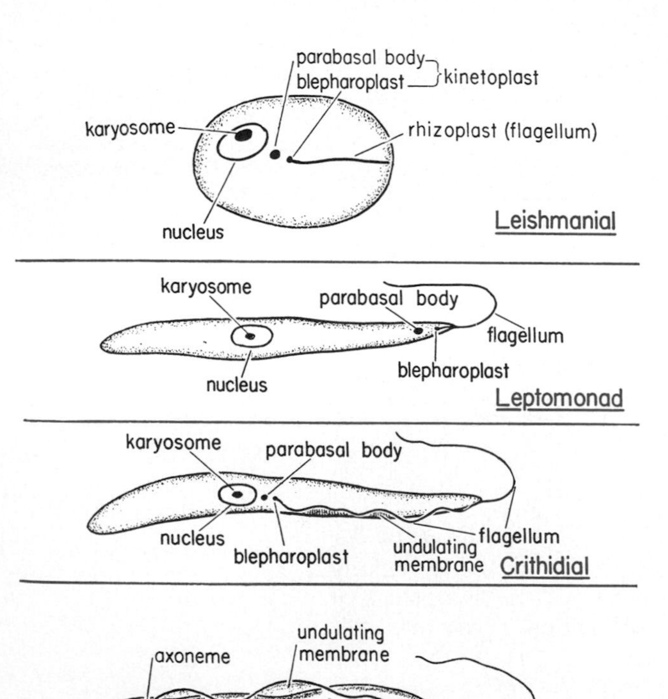

Figure 41–6 Forms of flagellate protozoa related to trypanosomes and to leishmanias. Note that from the first form (leishmanial) to the most highly developed form (trypanosomal) the flagellum becomes longer and more external and that the location of the parabasal body changes systematically. Most leishmanias lack the two lower stages. (From Hunter, G. W., III, Swartzwelder, J. C., and Clyde, D. F.: Tropical Medicine. 5th ed. Philadelphia, W. B. Saunders Company, 1976.)

sickness but without the marked brain involvement. The South American trypanosomes are transmitted in the feces of several blood-sucking bugs, often called "barberios" or "kissing bugs,"[1] represented by the species *Panstrongylus megistus* (Fig. 41–7). These bugs are distantly related to the "squash bug."

Diagnosis. Blood-infecting species, such as *Trypanosoma brucei gambiense,* when plentiful in the blood, are easily visible in blood smears stained with Wright's or another polychrome stain. This is one of the principal means of diagnosis (Fig. 41–8). In Chagas' disease, and also in African sleeping sickness, the tissues, such as muscles in the former and lymph nodes in the latter, often are infected with the parasites. In trypanosomal sleeping sickness the spinal fluid also often contains the trypanosomes, and diagnosis may be made by microscopic examination of the fluid. In Chagas' disease, the trypanosome form is present in the blood for only a short period. The parasites soon enter tissue cells and assume the leishmanial form.

In tsetse flies and in *Triatoma, Panstrongylus,* and other vectors, the trypanosomes undergo a developmental cycle, passing through the various phases shown in Figure 41–6.

Figure 41–7 *Triatoma gerstaeckeri* (♀) from Texas. This formidable bug closely resembles *Panstronglyus megistus* and *Triatoma infestans*, vectors of Chagas' disease (South American trypanosomiasis). Triatomine bugs are also known as assassin bugs, kissing bugs, wild bedbugs, or Mexican bedbugs. (Courtesy of Dr. A. Packchanian, The University of Texas Medical Branch at Galveston.)

The Leishmaniases

The organisms causing these diseases are species of protozoa of the genus *Leishmania,* named for William Leishman, discoverer of an important species. They appear to represent a somewhat less highly evolved group than the trypanosomes. They undergo a similar developmental cycle, but it proceeds only to the leptomonad stage. The leishmanial stage is found in the tissues of animal hosts, the leptomonad stage in arthropods. These protozoa characteristically cause ulcers or sores in which the leishmanias may be demonstrated microscopically and from which they may be transmitted. The four principal species in this genus are *L. donovani, L. braziliensis, L. mexicana,* and *L. tropica.*

Kala-Azar. This disease is caused by *Leishmania donovani.* It occurs in Africa, Central and South America, the Orient, India, and the Mediterranean countries, especially among children, dogs, foxes, and rodents. The liver, spleen, and bone marrow contain large numbers of the organisms (Fig. 41–9). Skin lesions are not prominent.

Because the parasite invades the internal organs, the disease may be called *visceral leishmaniasis.* It is also known as dumdum fever, black sickness, and

[1]So called because they often bite the lips of sleeping children, giving the appearance of kissing them: a true "kiss of death."

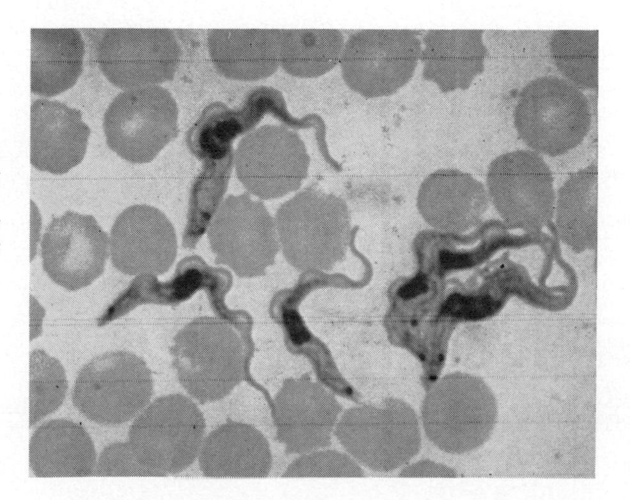

Figure 41–8 *Trypanosoma brucei gambiense* in a droplet of blood (the large, round objects are erythrocytes). This is one of the species of trypanosomes causing African trypanosomiasis (African "sleeping sickness"). Note the prominent flagellum along the edge of the wavy, keel-like membrane on each trypanosome (× 1,525). (Courtesy of Dr. A. Packchanian, The University of Texas Medical Branch at Galveston.)

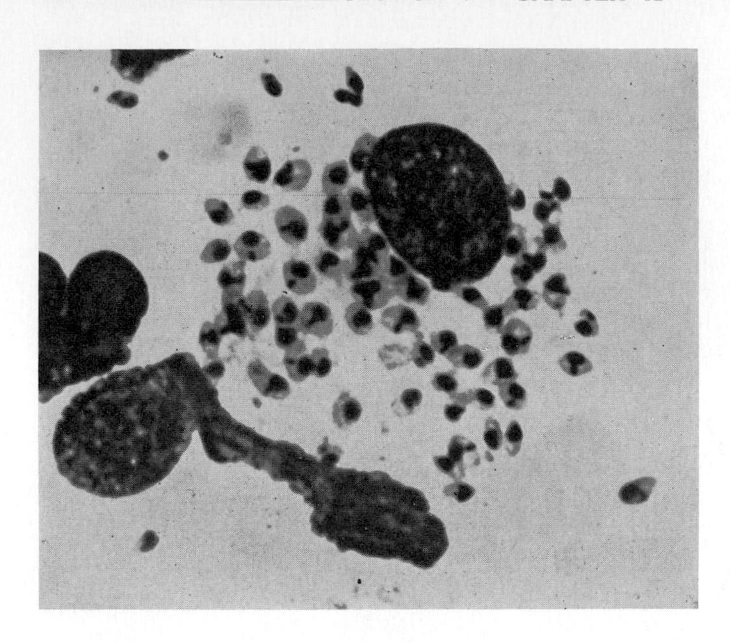

Figure 41–9 *Leishmania donovani* in stained smear from spleen puncture. Compare with the top diagram in Figure 41–6 (× 1,000). The large, deeply stained masses are cells of the spleen. (From Hunter, G. W., III, Swartzwelder, J. C., and Clyde, D. F.: Tropical Medicine. 5th ed. Philadelphia, W. B. Saunders Company, 1976.)

Table 41–1. Epidemiology of Cutaneous Leishmaniases

Cutaneous Leishmaniasis Type	Pathogenic Protozoa	Sandfly Vector	Natural Reservoir	Geographic Region	Human Disease
Old World	*Leishmania tropica minor*	*Phlebotomus*	Humans, dogs	Middle East, Mediterranean	Urban sores
	Leishmania tropica major	*Phlebotomus*	Rodents	Middle East, Mediterranean, India, China, West Africa	Rural sores
	Leishmania aethiopica	*Phlebotomus*	Hyrax (herbivorous mammals)	Ethiopia, Kenya	Simple sores, diffuse cutaneous leishmaniasis
New World	*Leishmania mexicana mexicana*	*Lutzomyia*	Forest rodents	Yucatan, Belize, Guatemala	"Chicle ulcers," diffuse cutaneous leishmaniasis
	Leishmania mexicana amazonensis	*Lutzomyia*	Rodents, marsupials, foxes	Amazonian Brazil, Venezuela	Single skin lesions, diffuse cutaneous leishmaniasis
	Leishmania braziliensis braziliensis	*Lutzomyia* and *Psychodopygus*	Forest rodents	Amazonian Brazil	Espundia (metastases)
	Leishmania braziliensis guyanensis	*Lutzomyia*	Unknown	Guyana, Surinam, Brazil, Venezuela	"Forest yaws" (metastases)
	Leishmania braziliensis panamensis	*Lutzomyia* and *Psychodopygus*	Rodents, primates, sloths	Panama, Costa Rica, Colombia	Skin ulcers (metastases)
	Leishmania braziliensis peruviana	*Lutzomyia*	Dogs	Peru, Argentina	Uta

From Braude A. I.: Medical Microbiology and Infectious Diseases, Vol. 2. Philadelphia, W. B. Saunders Company, 1981, p. 1646.

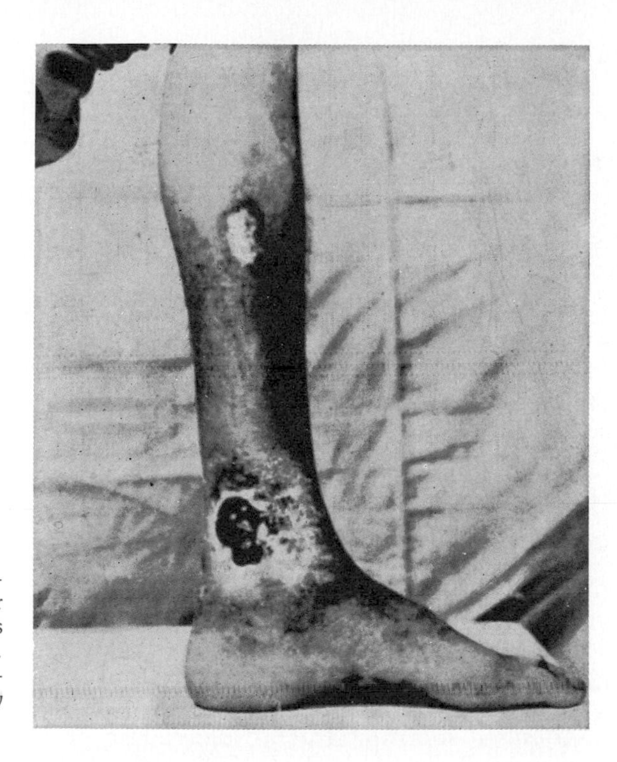

Figure 41–10 Cutaneous leishmaniasis, with ulcer near malleolus and scar from healed ulcer above. Metastases from healing ulcer are characteristic. (From Culbertson, J. T.: Medical Parasitology. New York, Columbia University Press, 1942.)

splenic anemia of infants. Kala-azar is transmitted mainly by the bites of sandflies *(Phlebotomus)*. *Leishmania* appear in nasal secretions and possibly in the urine and feces of patients. Transmission may be by these means as well as by sandflies. Kala-azar is 95 per cent fatal unless treatment is instituted.

Prevention. Since sandflies are the most important vector, they should be eliminated and avoided by means of screens, repellents, and sprays. Care should be exercised in handling articles that have been in contact with any cutaneous lesions, oral or nasal secretions, feces, or urine. The usual precautions for bedding and fomites are recommended. Treatment with stibogluconate sodium (Pentostam) or (for resistant infections from the Sudan) pentamidine is recommended. Dogs and cats in endemic zones may harbor the infection and can infect sandflies.

Cutaneous Leishmaniases. This disease (also known as Old World cutaneous leishmaniasis, oriental sore, Delhi boil, Aleppo boil, Baghdad ulcer, and so forth—see Table 41–1) is caused by *Leishmania tropica* and is found in India, the Middle East, parts of Africa, and parts of China. In South and Central America, New World cutaneous leishmaniasis is caused by *L. braziliensis* and *L. mexicana*. The lesions are large ulcers that tend to heal with scar formation (Fig. 41–10). Unlike those of kala-azar, these lesions are entirely cutaneous or in the mucous membranes, and the organisms occur in the superficial ulcers. These diseases are therefore often called *cutaneous leishmaniases*. As in kala-azar, dogs and rodents are susceptible and are sources of infection via bites of sandflies. The organisms in the ulcers may be transmitted by direct contact if the skin is scratched or abraded. Recovery and vaccines confer natural immunity.

Figure 41–11 Mucocutaneous leishmaniasis. (Courtesy of Dr. Mark R. Feldman, Louisiana State University School of Medicine, New Orleans.)

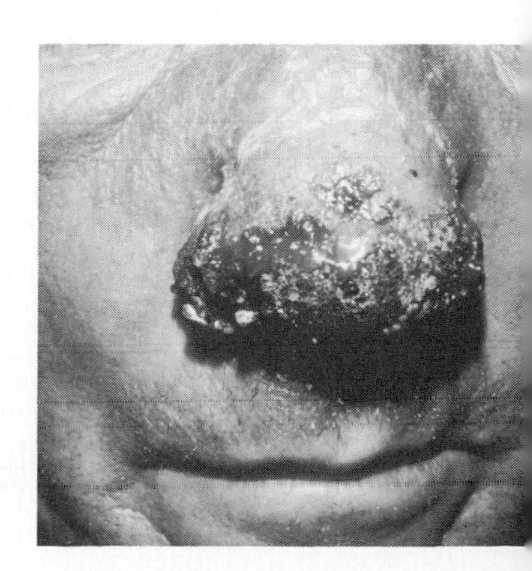

Espundia and Uta. Also called *American leishmaniasis,* two different diseases are involved in this severe and tissue-destructive (sometimes invasive) form of cutaneous leishmaniasis that invades particularly the tissues of the nasopharyngeal mucosa, nose, and adjacent tissues of the face (Fig. 41–11). The organisms occur in the ulcerous lesions and are present in the discharges from the sores. *Leishmania braziliensis peruviana* causes *uta,* resembling cutaneous leishmaniasis infection with *L. tropica*; and *Leishmania braziliensis braziliensis* is the pathogenic agent of espundia. Espundia is a more severe mucocutaneous

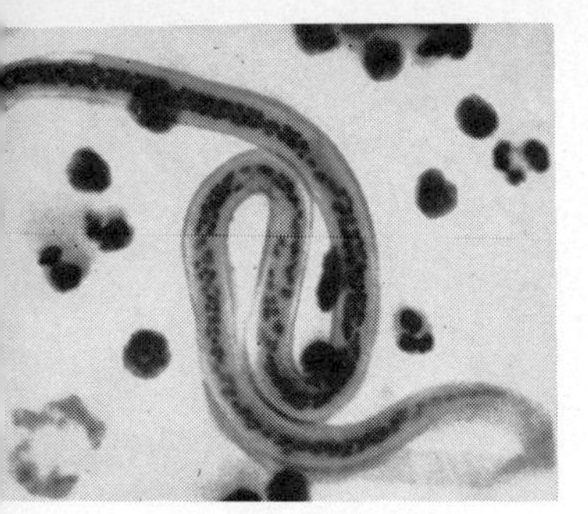

Figure 41–12 Microfilaria of *Wuchereria bancrofti* in a droplet of blood (the dark, rounded objects in the background are blood cells). *W. bancrofti* is one of the worms involved in elephantiasis (× 1,140). (Courtesy of Dr. F. Hawking, In *Sci. Amer.*, *199*:94, 1958.)

leishmaniasis that may metastasize by way of the blood stream, leading to septicemia, pneumonia, and either death or spontaneous recovery within several months. Unless the patient is treated with antimony, amphotericin B, or other, still experimental drugs (Lampit, pentamidine, or cycloguanil pamoate), the disease often progresses to a fatal termination. Dogs, cats, and rodents become infected and act as carriers. Sandflies (of the genus *Lutzomyia,* not *Phlebotomus*) transmit the organisms, but transmission may also occur as in other forms of cutaneous leishmaniasis. Preventive precautions are the same as for other cutaneous leishmaniases.

Prevention. Bandages, dressings, clothing, and bedding contaminated by the sores must be handled carefully and disinfected. As in kala-azar, transmission is mainly by sandflies, and anti-sandfly measures are indicated. Both Old and New World cutaneous leishmaniases respond to stibogluconate sodium treatment with an alternative therapy of amphotericin B available against resistant New World infections.

ARTHROPOD-BORNE HELMINTHS

Filarial Worms

These worms cause the disease *filariasis,* which is found in many tropical and some subtropical areas, depending on the distribution of certain vector arthropods. In some types of filariasis, the notorious swellings about the legs and genitalia called *elephantiasis* occur. Many infections are subclinical. Several species of filarial parasites are known, *Wuchereria bancrofti, Onchocerca volvulus, Brugia malayi,* and *Loa loa* being especially common. Filarial worms are transmitted among human beings by various biting arthropods, including mosquitoes, black flies ("coffee flies"), and midges.

Figure 41–13 Elephantiasis in a patient 63 years of age. Symptoms of swelling were first noted in both legs at age 15 and were seen in both arms at age 33. The patient also had scrotal elephantiasis. (Photograph from William A. Robinson, Papeete, Tahiti, and John F. Kessel, Medical School, University of Southern California. In Culbertson, J. T., and Cowan, C.: Living Agents of Disease. New York, G. P. Putnam's Sons, 1952.)

Filariasis. Details concerning the clinical features of the various forms of filariasis and life histories of the parasites differ among the various species, but all follow a generally similar pattern. In an infected person the adult male and female worms live in deep tissues, where they reproduce. The female produces larvae that are long and thin (around 275 μm by 7 μm), actively motile roundworms called *microfilariae* (Fig. 41–12). They move out of the deep tissues and accumulate in the peripheral blood vessels during the hours when the arthropod vector for that particular species of microfilaria bites. They seem to know when the arthropods take their blood meals, and they make a point of being on the spot to get their share![2]

In the arthropod vector they undergo a series of transformations toward maturation, and they eventually migrate to the biting parts of the arthropod. From there they enter the blood of the next person bitten. The microfilariae themselves appear to do little damage. In the human host, however, they undergo further maturation, during which they move about or congregate in masses, causing various painful swellings, the location depending on the species. Among these swellings is the notorious but not inevitable condition known as elephantiasis (Fig. 41–13), caused mainly by adults of *Wuchereria bancrofti,* which obstruct lymph channels.

Control. Chemotherapy is used to eliminate the worms from human hosts. Any person contemplating working in the tropics will do well to keep this in mind. Screening and insect repellents are used to prevent reinfection. Unless persons are constantly exposed to the infectious arthropods, the worms eventually die out in the infected person. Elimination of the sources of the arthropod vectors involves various problems in applied entomology.

SUMMARY

The only cases of malaria occurring in the United States are those that have been imported from mosquito-infested places. The disease is caused by protozoans belonging to the genus *Plasmodium,* which are introduced into the body when the female *Anopheles* mosquito bites. The protozoa circulate in the host's liver and blood, eventually entering red blood cells as merozoites, where they reproduce and rupture the cells, causing chills, fever, and other symptoms of malaria. The cycles of malaria attacks are different for the four species: *Plasmodium ovale, P. malariae, P. falciparum,* and *P. vivax. P. falciparum* infections may involve the brain and result in death. Most malaria, however, is self-limiting over the years. In the human host the life cycle of *Plasmodium* is asexual, but when a second mosquito bites it becomes infected with gametocytes that unite to form a zygote (sexual stage) in its stomach. The protozoa reach the salivary glands of the insect, and the cycle repeats itself when a new host is bitten.

Insecticide sprays, draining of swamps, and antimalarial drugs help to prevent malaria, but cures for the disease have not been found.

A malaria-like disease called babesiosis is transmitted through the bites of blood-sucking female ticks.

Flagellate protozoans of the genus *Trypanosoma* cause a series of diseases referred to as trypanosomiases. Transmitted through the bites of tsetse flies, *Trypanosoma brucei gambiense* and *Trypanosoma brucei rhodesiense* cause African sleeping sickness. *Trypanosoma cruzi,* carried by "kissing bugs," produces Chagas' disease, and *Trypanosoma equiperdum* produces a venereal disease in horses called dourine or "equine syphilis."

Transmitted through the bites of sandflies (*Phlebotomus, Lutzomyia,* or others), leishmaniases result from infections with *Leishmania donovani* (kala-

[2]Of course, no intelligence is involved. The reaction is entirely one of instinct or biochemical response and can be utterly frustrated and even reversed by various experimental procedures.

azar), *Leishmania tropica* (cutaneous sores), *Leishmania aethiopica braziliensis* (simple sores), *Leishmania mexicana* (cutaneous leishmaniasis), *Leishmania braziliensis braziliensis* (espundia), and *Leishmania braziliensis peruviana* (uta).

Filarial worms cause *filariases* and may be transmitted to humans or animals by biting arthropods such as mosquitoes, black flies, and midges. Among the filarial parasitic worms are *Wuchereria bancrofti, Onchocerca volvulus, Brugia malayi,* and *Loa loa.* Chemotherapy is a practical means of eliminating the worms from human hosts.

REVIEW QUESTIONS

1. Name the four species of organisms that cause human malaria.
2. Explain the asexual and sexual cycles of *Plasmodium.*
3. Discuss the epidemiology of malaria over the years in the United States and the rest of the world.
4. Describe methods used to control malaria. Why is this disease still with us?
5. What is meant by a self-limiting disease?
6. What is babesiosis?
7. Name some protozoans that cause trypanosomiases, and list the specific diseases they cause.
8. Specify the differences between Old World cutaneous leishmaniasis and New World cutaneous leishmaniasis.
9. Define the following: Baghdad ulcer, Delhi boil, sandflies, filariasis, helminths, tsetse flies, and elephantiasis.

Supplementary Reading

Blecka, L. J.: Concise Medical Parasitology. Reading, Mass., Addison-Wesley, 1980.

Brown, H. W.: Basic Clinical Parasitology. 4th ed. New York, Appleton-Century-Crofts, 1975.

Chin, W.: Recent developments in malaria. *South. Med. J., 71*:97, 1978.

Drugs for parasitic infections. *Med. Lett. Drugs Ther., 21*:105, 1979.

Faust, E. C., Beaver, P. C., and Jung, R. C.: Animal Agents and Vectors of Human Disease, 4th ed. Philadelphia, Lea & Febiger, 1975.

Hawking, F.: Filariasis. *Sci. Amer., 199*:94, 1958.

Horsfall, W. R.: Medical Entomology. New York, Ronald Press, 1962.

Hubbert, W. T., et al. (eds.): Diseases Transmitted from Animals to Man. 6th ed. Springfield, Ill., Charles C Thomas, 1975.

Hunter, G. W., III, Swartzwelder, J. C., and Clyde, D. F.: Tropical Medicine. 5th ed. Philadelphia, W. B. Saunders Company, 1976.

Jacoby, G. A., Hunt, J. V., Kosinski, K. S., et al.: Treatment of transfusion-induced babesiosis by exchange transfusion. *N. Engl. J. Med., 303*:1098, 1980.

Markell, E. K., and Voge, M.: Medical Parasitology. 5th ed. Philadelphia, W. B. Saunders Company, 1981.

Marsden, P. D.: Leishmaniasis. *N. Engl. J. Med., 300*:350, 1979.

Najarian, H. H. (ed.): Textbook of Medical Parasitology. Huntington, N.Y., Krieger, 1975.

Noble, E. R., and Noble, G. A.: Parasitology: The Biology of Animal Parasites. 4th ed. Philadelphia, Lea & Febiger, 1976.

Pan American Health Organization: New Approaches in American Trypanosomiasis Research. Proceedings of an International Symposium, Brazil, 1975. Washington, D.C., Pan American Health Organization, 1976.

Sassa, M. (ed.): Human Filariasis: A Global Study. Baltimore, University Park Press, 1976.

Warrell, D. A., Looareesuwan, S., Warrell, M. J., et al.: Dexamethasone proves deleterious in cerebral malaria: A double blind trial in 100 patients. *N. Engl. J. Med., 306*:313, 1982.

Williams, L. L., Jr.: Malaria eradication in the United States. *Amer. J. Public Health, 53*:17, 1963.

Young, M. D.: Malaria. *In* Hunter, G. W., III, Swartzwelder, J. C., and Clyde, D. F. (eds.): Tropical Medicine. 5th ed. Philadelphia, W. B. Saunders Company, 1976.

ALLIED HEALTH PERSONNEL, ASSISTANTS TO THE PHYSICIAN

The professional health worker is responsible for aiding the physician in the diagnosis, treatment, and prevention of all types of disease. Transmissible diseases present a challenge as great as (and sometimes greater than) other pathologic conditions. However, many of the general statements made in this chapter about transmissible diseases apply also to other illnesses.

ASSISTING THE PHYSICIAN IN DIAGNOSIS

The Collection of Specimens

It is important for anyone responsible for collecting specimens for diagnostic examination to know the correct methods and have an intelligent understanding of the nature and purpose of the procedure. The value of a specimen depends entirely on the care with which it is taken. Poorly taken specimens waste the laboratory worker's time. It may be impossible to make any diagnosis with such a specimen, or, worst of all, a wrong diagnosis may be made.

Labeling. The first consideration about a specimen is the label. Each label should have on it the patient's name and hospital number (this is more often now the Social Security number), the date, the site from which the specimen is taken (e.g., blood, throat, cervix, abscess), the physician's name, and the name of the ward or service. Pasted-on or self-adhering labels are undesirable, as they easily drop off from the smooth glass surface of a flask or tube. Adhesive tape makes very good labels for specimens. Transparent tape over a label provides added protection.

If specimens are taken from a number of patients at the same time, each one must be labeled as soon as it is taken or immediately before it is taken. Finish entirely with one specimen before going on to the next. Do not let the specimen out of the hand until the label is complete. If there is the faintest suspicion that two specimens have been mixed, both must be discarded and a new start made; or both may be tested, and if one is found "positive," both must be retested. A mix-up in throat cultures or syphilis serology tubes could have serious consequences. Petri plates should be labeled on the bottoms because the tops may be transposed.

Transmission to Laboratory. Specimens should be sent to the laboratory at the earliest possible moment. This prevents the specimen from drying and bacteria from dying, makes possible an earlier report, is a convenience to the laboratory worker in planning the work, and may be a matter of life or death to the patient; who after all is the person around whom all these activities revolve.

Swabs. Material for bacteriologic examination is often collected on *swabs.* These are put in some container (test tubes are very convenient), and the whole is sterilized before use (Fig. 42–1). Excellent swabs for all purposes are now commercially available in convenient sterile packings. An example of such a

PART F TECHNICAL RESPONSIBILITIES IN TRANSMISSIBLE DISEASES

CHAPTER 42

swab is the "Culturette," which contains Stuart's Bacterial Transport Medium with sodium thioglycolate for anaerobic bacterial growth. Following the directions on the package as given below will guarantee that the culture reaches the laboratory in a viable state:

Directions:

1. Peel open ⅓ length. Remove Culturette from package.
2. Pinch cardboard cap to grasp swab stick and remove from tube.
3. Take sample and return swab to tube, replacing cap.
4. Crush ampule by squeezing tube.
5. Push cardboard cap to bring swab into contact with moistened pledget.
6. Write patient identification on package. Replace Culturette. Send to laboratory.

Swabs are used for throat cultures, pus, and smears from the cervix, eyes, ulcers, and so on. For use in the operating room, culture tubes containing swabs are wrapped so that the outside of the tube is sterile and can be handled by the surgeon at the operating table.

Swabs made of calcium alginate (a seaweed gum) are preferable to cotton for many purposes. They are free from substances like fatty acids that may be inimical to microorganisms. Further, the alginate swab can be dissolved in a physiologic solution like Locke's with hexametaphosphate, thus releasing the entrapped organisms into the solvent.

Nurses or other health personnel may be required to use a swab to inoculate culture medium. This may be broth, or it may be agar medium slanted in a tube or flat in a Petri plate. In inoculating agar, the tip of the swab, after it has touched the desired lesion or infected place, is passed back and forth over the surface of the solid medium with gentle pressure, making a series of numerous, closely spaced zigzag paths on the surface. The important point in taking a swab culture is to touch the swab only on the spot from which the culture is desired, and nowhere else, and to send it promptly to the laboratory or use it immediately to inoculate media (Fig. 42–1).

Throat Cultures. These should be taken only with a clear view of the throat, in a good light, and using a tongue depressor. A swab is used. Material to be examined for diphtheria bacilli should be taken directly from the tonsil or from any white or inflamed spots in the throat. The same is true of cultures for scarlet fever or other infections of the throat.

Cultures from the nasopharynx are made particularly for meningococci (Fig. 42–2). These may be located high up behind the soft palate. An ordinary swab may be used by passing it back gently through the nostril (*X,* Fig. 42–2). The swab should be cultured and the culture incubated immediately because meningococci are very fragile and die quickly outside the body.

Although pharyngeal swabs most often yield positive cultures in whooping cough, it is sometimes advantageous to make "cough plates" as well. These have been mentioned previously in connection with whooping cough but are equally appropriate for other upper respiratory tract infections. A Petri plate containing medium suitable for the growth of the organism desired is held before the mouth of the patient when he coughs. The spray of sputum raised by the

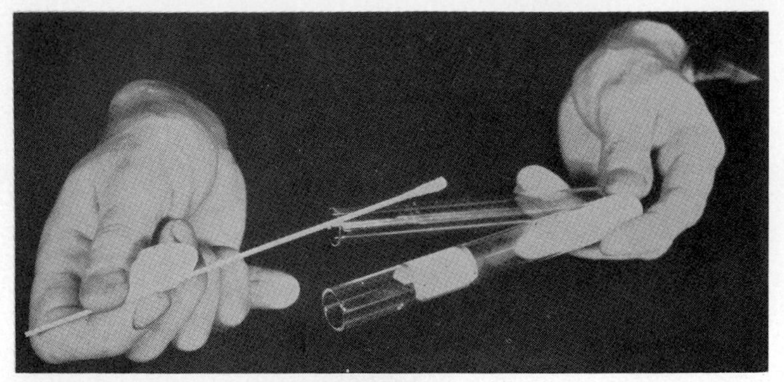

A

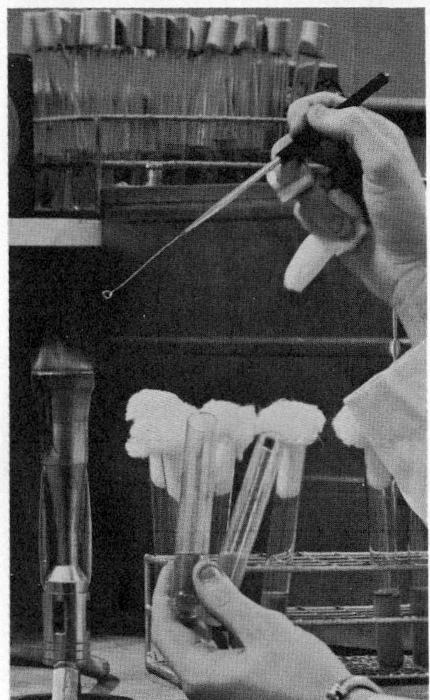

B

Figure 42–1 *A,* Method of inoculating a tube of medium by means of a swab. The plugs are held between the fingers in such a way that the ends that go into the tubes do not touch anything. The swab carrying the bacteria (e.g., a throat swab) is inserted into the tube of medium without making any surface contact. It is then wiped gently over the surface of the medium, being rotated between finger and thumb so as to bring the whole swab area into contact with the medium. It is then put back into its tube in the same careful way before replacing the plugs. The worker in the dispensary and community health field is often asked to inoculate tubes. Note that the tube of medium is labeled and is tilted downward to exclude dust. Tubes of fluid medium can be held almost horizontally without spilling.

B, Method of inoculating cultures by means of a platinum loop. Note the Bunsen burner for sterilizing the loop and singeing shreds of cotton from the tubes. In making an inoculation, the loop is first heated to redness on the burner. The sterilized loop is cooled. The cotton plugs (or metal or plastic caps) are then withdrawn and held as shown. The loop is dipped into the broth culture (or used to take up a portion of the growth on an agar slant). It is then withdrawn from the tube, and the material on the loop is quickly transferred to whatever culture is to be inoculated. The inside of either tube should never be touched by any nonsterile part of the inoculation needle. The plugs or caps are promptly replaced in or on their respective tubes. The platinum loop is then heated to redness in the flame, from the holder toward the loop, before laying it down. (*B,* courtesy of Rohm & Haas Company, Chemists, Philadelphia, Pa.)

coughing inoculates the medium. The plate is immediately covered and placed in the incubator in an inverted position. Use of the cough plate with infants is often impracticable because of the difficulty of eliciting a productive cough.

Sputum. If the examination is to be of value, the specimen must be obtained in the proper way. The sputum collected by patients in the ordinary sputum cup consists to a large extent of the material from the mouth and throat: mucus, saliva, and bits of food. To collect a sputum specimen correctly, a sterile container should be obtained from the laboratory. If the patient is able, he should brush his teeth and rinse the mouth thoroughly with water. A morning sputum sample is generally cleaner and not as contaminated as when taken during the day. Sputum may be collected in wide-mouthed ointment jars of 1 to 2 oz capacity or in other suitable watertight containers with tight covers. The sputum should be collected directly after a cough that brings secretions up from the lungs (not just from the throat) and the specimen should be sent immediately to the laboratory. The most simple techniques to make the patient cough involve either the use of ultramist or tickling the throat. Sputum is obtained from infants by swabbing the throat or, better, by gastric lavage, as infants usually swallow their sputum.

Urine. Specimens of urine for bacteriologic examination are of value only when they are collected with extreme care to *avoid* (1) extraneous contamination and (2) holding at room temperature more than one hour after voiding. There are innumerable bacteria on the skin and mucous membranes of the genitalia, and unless the mouth of the urethra and the areas around it (vulva, labia, glans penis, and adjacent areas) are cleansed thoroughly the specimen is sure to be contaminated. Normally voided urine is never sterile. In collecting the specimen, the first part of the urine is allowed to run off in order to wash out any bacteria that may be present in the urethra, and a portion of the clean-catch midstream urine is collected in a sterile container. Numbers of bacteria present may be determined by plate-count methods using 1 ml of urine diluted 1:1,000 mixed with clear nutrient agar. Quantitative loopfuls may also be spread on blood agar or another suitable medium. Overt cystitis is usually evidenced by bacteria numbering over 100,000 per ml in the urine. Pathogens may be introduced by any instrumentation or may come from vaginal or prostatic infections or kidneys (pyelonephritis) or other internal lesions. The commonest pathogens are species of *Escherichia, Klebsiella, Enterobacter, Proteus,* and *Pseudomonas*; occasionally enterococci or staphylococci may be encountered. Specimens for chemical or microscopic examination only need not be sterile but should be freshly voided.

Specimens from the kidneys are collected by catheterization of the ureters, which is a surgical procedure. A very important point about specimens from the kidneys is to be absolutely sure that the tubes from the right and left kidneys are correctly designated. A mistake may prove fatal for the patient and disastrous for the hospital, the doctor, and the nurse.

Feces. Examination of feces for living organisms is undertaken most often for pathogenic Enterobacteriaceae, cholera vibrios, the eggs of animal parasites like hookworms or tapeworms, or protozoal cysts. Stools for bacteriologic examination should be fresh. If blood or pus is present, it should be selected for the culture. A sterile tongue blade, knife, or spoon may be used to transfer a portion of stool about half the size of a walnut to the desired container. If the stool is liquid, about a teaspoonful is sufficient. Fifty-ml plastic sputum containers with screw covers are used by many hospitals, physicians, and community health agencies.

Carriers of pathogenic Enterobacteriaceae are found by making cultures from the feces. The finding of typhoid and dysentery bacilli in the stools or urine of carriers is often difficult, since the organisms are usually absent or

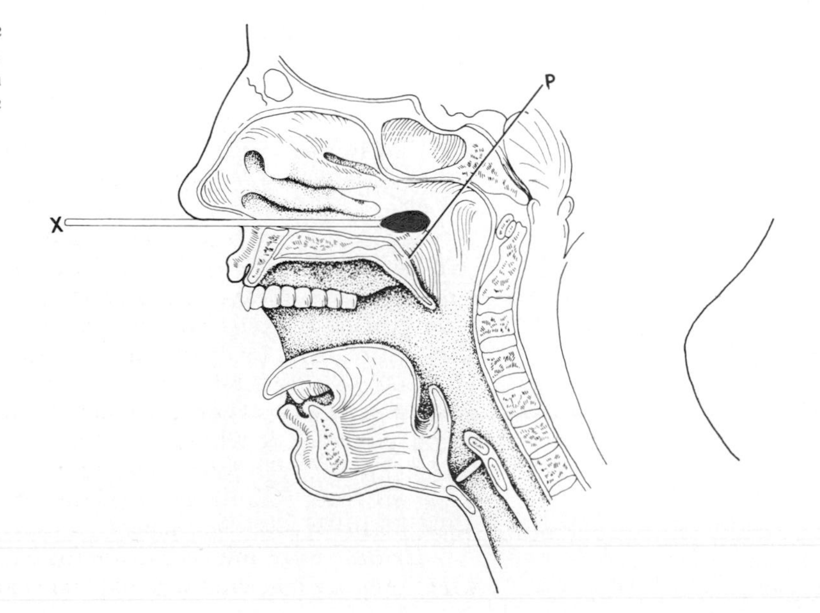

Figure 42–2 Method of taking a culture from the nasopharynx by means of a nasal swab. *P,* soft palate. The swab is passed back through the nostril as shown at *X.* It should be held very loosely and left in place if the patient moves or sneezes.

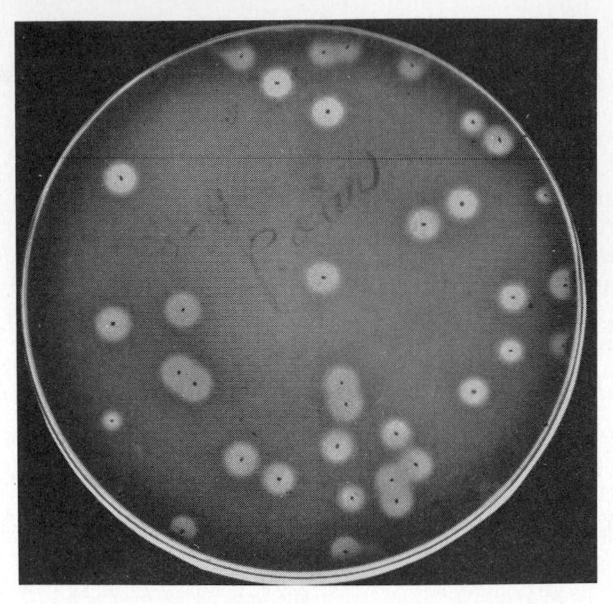

Figure 42–3 Petri plate with blood culture from patient with severe septicemia due to beta-hemolytic streptococci (3/4 life size). The use of an agar pour plate for a measured blood culture gives some idea of the number of organisms per ml of blood, thus furnishing a guide to prognosis and treatment, as well as permitting prompt isolation of the organism if tests for sensitivity to various antibiotics are desired. A broth blood culture permits the use of a larger blood sample and is therefore a more sensitive diagnostic procedure, but it is not quantitative. (Preparation by Dr. Elaine L. Updyke. Photo courtesy of U.S. Public Health Service, Communicable Disease Center, Atlanta, Ga.)

present in comparatively small numbers. Many health departments provide a special outfit for stool cultures from suspected typhoid cases or carriers. This may contain a solution of bile and an aniline dye, e.g., brilliant green. The solution is tubed in 5-ml amounts and sealed with a rubber stopper. To regulate the amount of feces added, a sterile swab is sent with each tube. The bile and brilliant green are favorable to the growth of the typhoid bacillus and unfavorable to other bacteria, so that in some cases the bacilli may be found after the tube has been in transit for two, three, and even four days when mailed to a medical laboratory, where cultures will be made. Preservative solutions, such as mixtures of sodium citrate and sodium desoxycholate, are of value when added to the specimen, but do not use too much specimen in proportion to the volume of preservative solution. A ratio of 1:5 is about right. "Preservative" solutions leave much to be desired. It is better to rush the specimen to the laboratory and culture it immediately. The stools from patients suspected of having bacillary dysentery should be cultured as soon after evacuation as possible, because dysentery bacilli die out very rapidly in feces.

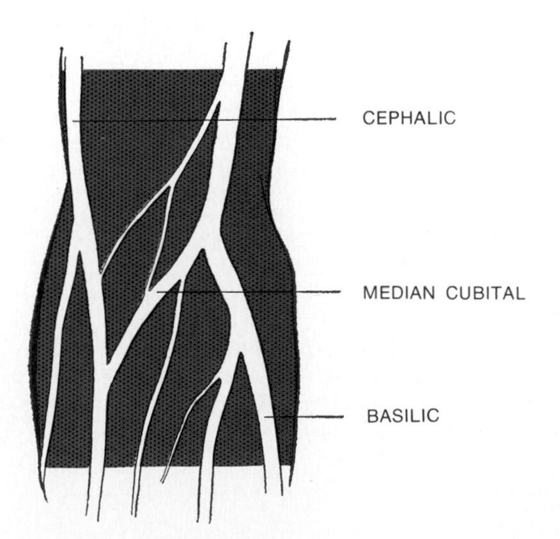

Figure 42–4 Location of the principal veins in the arm suitable for drawing blood samples. Intravenous therapy should be avoided at this location because of the bend of the elbow. (From U.S. Civil Defense Administration, Instructor's Guide, IG-11–1: Venipuncture and Intravenous Procedures. Washington, D.C.)

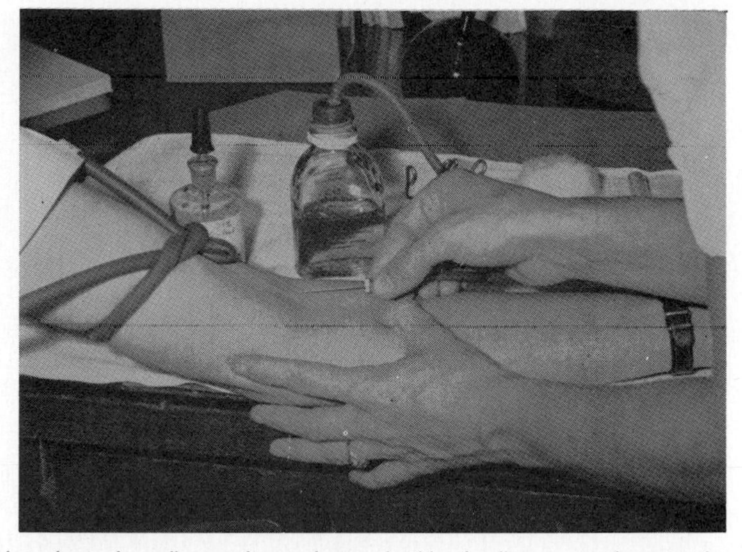

Figure 42–5 Method of taking blood for culture. A tourniquet consisting of plastic tubing is tightened moderately about the patient's upper arm. Before introducing the needle, the skin around the site of the puncture should be well washed with tincture of green soap and wiped with mild tincture of iodine, which is removed with 70 per cent alcohol. The area must not be touched with the fingers unless they are covered by sterile gloves. A sterile needle is inserted into a convenient vein in the bend of the elbow. The blood may be drawn into any sterile receptacle, such as a syringe, for transfer immediately to culture media. In this picture the needle is attached, by means of a previously prepared sterile rubber tube, to a bottle with a diaphragm-type rubber stopper and containing culture medium. The tube is clamped shut until the needle enters the vein. A sterile needle attached to the distal end of the tube pierces the rubber bottle cap; the vacuum of the specially prepared bottle draws in the blood. After the proper quantity of blood is drawn, the tourniquet is released, the tube is pinched shut, and the distal needle is withdrawn from the culture flask. Immediately afterward, the needle is withdrawn from the arm. Drops of blood remaining in the tube and needle may be used to make blood cell counts and smears for microscopic examination. A pledget of sterile cotton is pressed against the puncture in the skin until the bleeding stops. Note the clamp for closing the tube, the sterile towel, and the bottle of disinfectant. (Courtesy of U.S. Public Health Service, Communicable Disease Center, Atlanta, Ga.) (See also Figures 42–3, 42–7, and 42–8.)

Blood Cultures. Bacteria are present in the circulating blood in septicemia associated with various diseases. Cocci may be found during the severe stages of any streptococcal or staphylococcal disease (see Fig. 42–3). Typhoid bacilli are always present in the blood in the early stage of typhoid fever. Septicemia of any sort is always a serious condition. The diagnosis is made by cultivating the bacteria in the blood, usually taken from one of the large veins of the arm (Fig. 42–4).

To obtain blood for culture, the skin over the vein is prepared as detailed in Figure 42–5. Meticulous care is required to avoid (a) contamination of the specimen with microorganisms from the skin (commonly staphylococci or diphtheroids) and (b) infection of the patient. A simple instrument is shown in Figure 42–6.

Vacutainer[1] Bottles. The blood is best drawn by vacuum-container directly from the vein into two tubes or bottles from the same puncture, each tube or bottle containing 18 ml of culture medium for 2-ml samples (45 ml for 5-ml samples). The sample should fill the bottle or tube. While drawing blood the container should be held in a nearly vertical position to avoid back-flow into the vein (Fig. 42–7).

The medium commonly used is "all-purpose" broth, prereduced and presterilized, containing sodium polyanetholesulfonate or sodium amylosulfate as anticoagulant and anticomplementary agents, with penicillinase and para-aminobenzoic acid to inactivate any penicillin or sulfonamide (respectively) that might be in the blood of the patient.

After removal of the container and needle from the vein, the bleeding needle is replaced in one tube or bottle by a sterile needle, open to the air but loosely capped, for aerobic incubation (Fig. 42–8*B*); the other container, for anaerobic incubation, receives a needle similarly capped but with a valve-like device that excludes air but permits the escape of gases produced by bacterial growth (Fig. 42–8*A*). Special media may be used for organisms known to require it: e.g., fungi, *Francisella tularensis, Mycoplasma,* and *Leptospira.* One or two ml of the blood, drawn out from the container with a syringe, may be

[1]A product of BBL, a division of Bio-Quest.

Figure 42–6 The Becton Dickinson Vacutainer System consists of a combined blood collection needle, needle holder, and evacuated tube, all sterile and disposable. The system is manipulated like a syringe and yet cannot inject, since the vacuum already exists in the collecting tube. The rubber diaphragm closure of the vacuum tube is pierced by forcing the butt of the needle through it after the point has entered the vein. (Courtesy of Becton Dickinson and Company.)

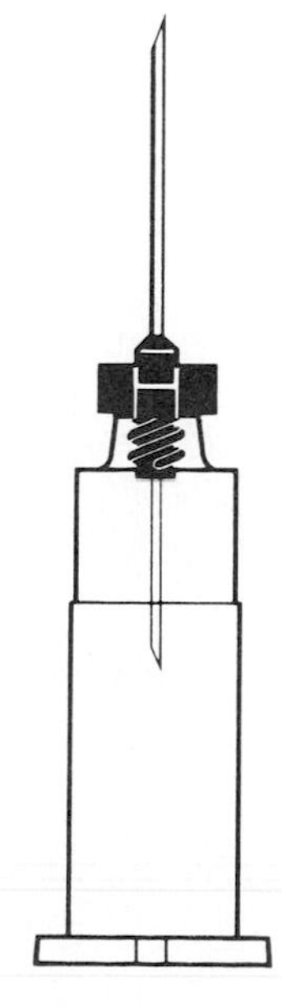

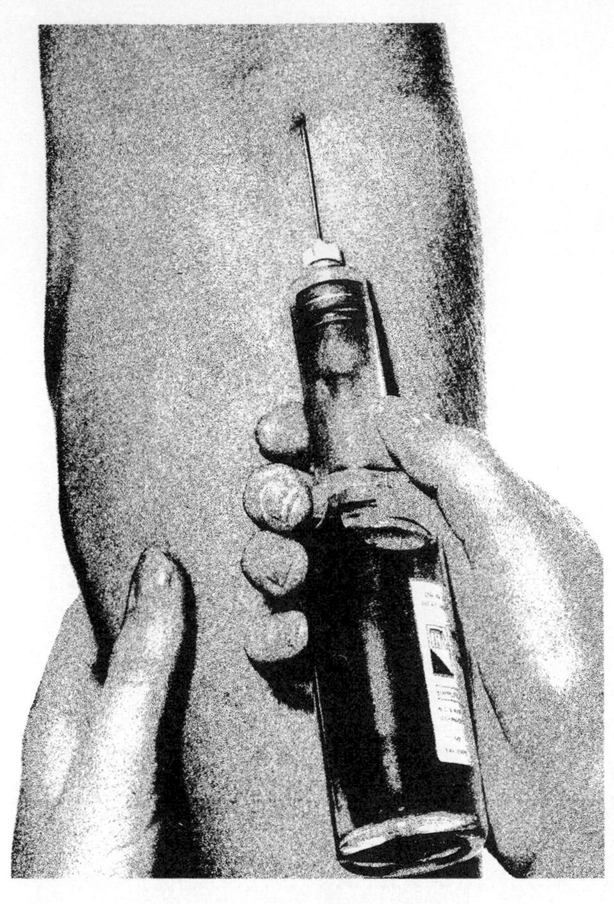

Figure 42–7 Use of bottle for the Vacutainer blood culture system. Note the near-vertical position of the bottle. (See text.) (Courtesy of Becton Dickinson and Company.)

mixed with blood agar and poured into a Petri dish. Hemolytic colonies may appear after 24 hours or more of incubation at 35 C (Fig. 42–3).

Blood for Serologic Tests. This may be obtained *after* collecting blood for culture, using the same needle and puncture but substituting empty tubes for the culture tube or bottle. Successive samples may be drawn into tubes containing anticoagulants such as sodium citrate. Blood for serologic testing is set aside to clot. It is now routine to examine the serum of every patient or person checking into a hospital with the VDRL test.

One rule should be remembered. After a needle has been used to stick a patient, it should not be inserted into a flask or tube to let the blood out of a syringe; rather, a new, sterile needle should be used to replace it. Why?

For some serologic tests, blood grouping, and leukocyte counts, only a few drops of blood are needed. Unless these are taken from the syringe or needle when blood is drawn for a serologic test or blood culture, it is necessary to puncture the finger. The ball of the index or middle finger is generally used. Before the puncture, the finger should be wiped *clean* with cotton moistened in alcohol. The needle used for making the puncture must be sterile. An excellent device is the small, sterile, individually packaged, disposable lancet available under the name of "Hemolet." Previously, more elaborate instruments, with a hidden needle activated by a spring and trigger, were used. Such needles have sometimes been very effective vectors of syphilis, various bacterial infections, homologous serum hepatitis, and other conditions when not properly heat-sterilized. It now seems strange that it took so long to recognize these once common objects as real dangers.

Blood for films to be examined microscopically is taken from the finger and spread in a thin uniform layer on an absolutely clean slide. The slides are

Figure 42–8 Venting units for Vacutainer blood culture tubes or bottles. *A*, This type excludes air but is designed to maintain a proper anaerobic atmosphere in the tube while allowing gas producing cultures to vent safely and automatically. Gas produced by the culture will cause the stopper to rise only high enough to vent, while the clip prevents the stopper from popping off. For clean and efficient subculturing, simply lift the dustcap and draw a specimen directly through the stopper. *B*, This type permits free entry of air and exit of gas (aerobic cultivation) (see text). (Courtesy of Becton Dickinson and Company.)

prepared by washing them with detergent and water, rinsing, then washing with alcohol, and drying with gauze.

It is possible to make good films of blood as shown in Figure 42–9. It is also common to make these smears on glass coverslips. A drop of blood is placed on a clean glass coverslip and covered by a second clean coverslip at an angle so that all four corners of the two coverslips protrude on all sides. The two coverslips are separated by moving them sideways as quickly as possible, and each provides a slide that can be examined microscopically.

Fluid from the Peritoneal and Pleural Cavities. This is commonly withdrawn in a syringe. The site of the needle puncture must be carefully disinfected before insertion and protected with sterile gauze after the needle is withdrawn. If tuberculosis is suspected, some of the fluid may be inoculated intraperito-

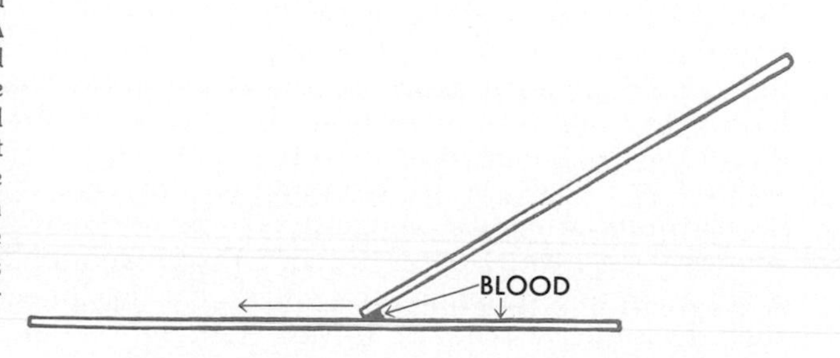

Figure 42–9 Glass slides arranged for making a film of blood for microscopic examination. A small drop of blood is first placed on the horizontal slide under the small arrow. The edge of the tilted slide is then drawn from the left into contact with the drop and the blood spreads across the slide in the angle. We are looking at the slides edgewise. The arrow shows how to move the tilted slide after the blood touches it.

neally into a guinea pig or onto suitable culture medium, or both. It is advisable also to examine a stained smear. These specimens are usually taken by the physician, but assisting personnel often prepare the sterile needles and flasks and are responsible for proper labeling and sending the specimen promptly to the laboratory.

Cerebrospinal Fluid. This is examined for the diagnosis of the different kinds of meningitis, for acute poliomyelitis, and for syphilis of the nervous system. The fluid is obtained by *lumbar puncture* and is allowed to run from the needle directly into four sterile test tubes or centrifuge tubes. The puncture is always made by a physician. The site should be carefully disinfected beforehand and covered with a sterile Band-Aid after the needle is withdrawn.

In acute meningitis caused by the meningococcus, streptococcus, or other organisms, the appearance of the fluid may range from slightly cloudy to very turbid with leukocytes or bacteria or both. Smears and cultures are made to determine what bacteria and cells are present. The tube containing such fluid must be kept warm, and the culture should be made as soon as possible (within a few minutes) after the puncture. In most such cases diagnosis is attempted with a Gram-stained or Ziehl-Neelsen–stained smear of sediment from the centrifuged specimen. In acute meningitis culture methods are usually too slow to be of immediate value for instituting therapy.

The four tubes from the spinal tap are used for (1) serologic testing, (2) protein and glucose determination, (3) cell and differential counts, and (4) bacterial cultures. In a clear or cloudy fluid the number of white cells is counted, using the same technique as in counting leukocytes in the blood. A differential count is made from a smear of the sediment. The amounts of albumin and globulin are determined chemically. Both the number of cells and the amount of protein may be much increased, although the fluid remains clear. This is of special value in differentiating paresis from other forms of neurosyphilis.

Cultures are not made on fluids from suspected cases of syphilis, but a serologic test is carried out.

Reporting the Results of Laboratory Tests

As soon as the results of laboratory tests are received, the nurse or other appropriate personnel (if the patient is in a hospital) should read them and note any discrepancies from the normal range of the particular tests. Most reports contain a listing of general standards of test values; this permits the person receiving the report to evaluate the patient's deviation from the normal range. A nurse or other assistant is, of course, expected to know, even without these charts, what constitutes expected values in healthy individuals. Because the doctor usually visits patients in a hospital only once or twice a day, the attendant persons are responsible for keeping the doctor informed of unusual events that may occur between visits. The nurse or other assistant must be able to identify these unusual events. The doctor also expects not to be called unnecessarily and that the assistants will know when a call is indicated. Reports of laboratory tests that show abnormal findings frequently provide the clues necessary for better diagnosis and immediate therapy for the patient. Failure to transmit this information to the physician may delay essential treatment, which may delay or in other ways influence the recovery of the patient. The student will rarely be held responsible for exercising the judgment necessary to determine when the physician is to be notified about abnormalities found in reports from the laboratory, but this responsibility will be expected after graduation. It is essential, therefore, that the knowledge necessary for making such decisions be acquired during training.

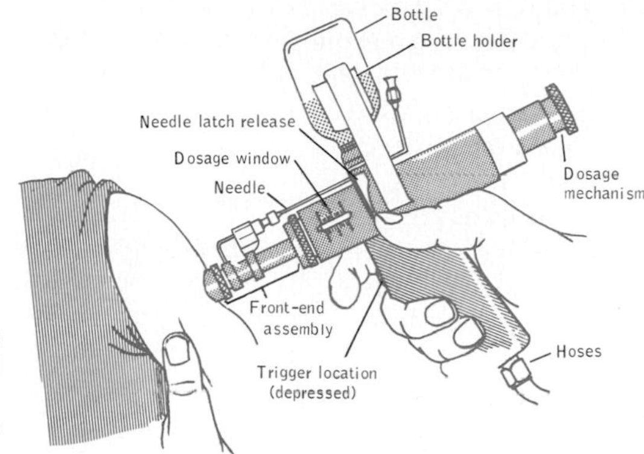

Figure 42–10 Hypospray jet injector. It is used for subcutaneous, intramuscular, and intradermal injections without a needle. (Courtesy of the Texas Department of Health Resources.)

ASSISTING THE PHYSICIAN IN TREATMENT

One of the responsibilities of every nurse or assistant is to carry out the orders that the physician gives for treatments and medications. Although it is always important to give medications on time, it is imperative in the case of antibiotics or chemotherapeutic drugs. As has been explained in Chapter 14, drug fastness and antibiotic-resistant microorganisms may develop if the blood levels of these therapeutic agents are too low. Current problems with the highly resistant *Staphylococcus aureus* may have resulted from insufficient, indiscriminate, ineffective, or belated use of an antibiotic against an infection, which reduced the blood level to a point where the more resistant individuals of that particular group of staphylococci could survive and multiply. If untoward symptoms seem to result from any treatment or medication, the nurse or assistant should report this to the physician as soon as possible.

ASSISTING IN THE PREVENTION OF TRANSMISSIBLE DISEASES

The responsibilities of the health team in the prevention of transmissible diseases fall into two categories: inhibition, destruction, and removal of pathogenic microorganisms; and assisting with programs that increase the immunity of a population. In the earlier parts of this book, the values and applications of prophylactic vaccinations were discussed in detail. One useful device for vaccinating large numbers of people (e.g., in the army) has been used very successfully; it is illustrated in Figure 42–10. Theoretically, it would be possible to prevent outbreaks of transmissible diseases if the first case in a given community could be immediately recognized and completely isolated from all other people, and if all the pathogens that leave the body of the patient were completely destroyed. Obviously this is not feasible; the early symptoms of many transmissible diseases are not easily recognized, unknown carriers for many diseases are present in the population, and not all of the methods of transfer for each disease are known or can be controlled at the present time. Nonetheless, health personnel should exert every effort to isolate patients who have transmissible diseases, particularly from any segment of the population known to have high susceptibility or high risk of complications.

Members of the health team are key figures in the control and destruction of microorganisms as they leave the body. Information on body fluids and discharges is summarized in Table 42–1. The articles designated in the right column should be thoroughly disinfected or sterilized by the procedures most effective for the causative organism and most feasible for the article in question.

Table 42–1. Body Fluids and Discharges Commonly Carrying Pathogenic Organisms, and Important Fomites Likely to Be Infectious

Body Fluids and Discharges Commonly Carrying Pathogenic Organisms	Articles in Contact with Patient to Which Health Workers Should Give Special Attention
Feces and Urine: Salmonellosis Dysentery (feces only) Cholera Undulant fever Infectious (epidemic) hepatitis Leptospirosis (urine only) Anterior poliomyelitis (feces) Hookworm and some other helminthic diseases Mumps	Bed linen, clothing, eating utensils, bed pans; anything likely to be contaminated by feces or urine. **The health worker's hands.**
Sputum, Nose and Throat Discharge: Pneumonia (due to any organism) Diphtheria Scarlet fever Septic or any other form of infectious sore throat Whooping cough Epidemic meningitis Syphilis (with open lesions of mouth or respiratory tract) Common cold, influenza, etc. Anterior poliomyelitis Measles; chickenpox; mumps Certain mold and yeast infections, such as thrush Smallpox Tuberculosis	Bed linen, clothing, handkerchiefs, towels,washcloths, toothbrushes, eating utensils, books, toys, pencils, dust from patient's room, sputum boxes or bottles, thermometers, tongue depressors, throat swabs; drinking fountains, medicine spoons, spray from coughing, sneezing, and talking; the health worker's mask, gown, shoes, gloves, apron, etc. **The health worker's hands.**
Pus or Exudate from Local Lesion: Gonorrhea Syphilis Venereal herpes Ulcers or abscesses of any kind "Pink-eye" Erysipelas Tuberculosis Blastomycosis "Trench mouth" (Vincent's angina) Any infection of eye, ear, nose, or genitalia Smallpox Plague Tularemia Undulant fever Staphylococcal and streptococcal infections	Bandages and dressings, swabs, clothing, bed linen, towels, washcloths; if the lesions are in the mouth, watch especially eating utensils, books, pencils, toys, and other articles mentioned above; toilet seats, wash water. **The health worker's hands.**
Blood: Typhoid fever (first week only) Undulant fever Malaria Leptospirosis Syphilis (not common) Generalized streptococcal and pneumococcal infections associated with "blood poisoning," scarlet fever, pneumonia and related diseases. Generalized staphylococcal infections Epidemic meningitis (early stages) Pneumonia Arthropod-borne viral diseases (early stages) Arthropod-borne bacterial diseases Rickettsial infections (early stages) Homologous serum hepatitis Infectious (epidemic) hepatitis	Any instrument used in taking blood, syringes, needles, gauze or cotton used to absorb blood; sheets or clothing or wash water contaminated with blood; obstetric instruments, dressings, etc. Arthropod vectors of the disease in question in arthropod-borne diseases. **The health worker's hands.**

SUMMARY

New methods and procedures resulting from newly manufactured devices are continuously made available for evaluation in the clinical setting. The professional health worker is expected to adopt those that are real improvements. However, certain basic principles of health care do not change. This chapter emphasizes these concepts.

The health worker must know why a specimen is taken and how to take it properly and must keep in mind that the patient's welfare depends on correct labeling and expeditious transfer of the specimen to the laboratory. For example, if it is suspected that a wound is infected by an anaerobic pathogen, the specimen must be immediately collected as anaerobically as possible and placed into an anaerobic medium, rather than just any transfer medium, to be sent to the laboratory; otherwise, the agent causing the infection will never be determined.

Certain knowledge is necessary to differentiate between methods for properly taking specimens from the throat, obtaining sputum, avoiding extraneous contamination of urine samples, obtaining fecal samples for different purposes, preparing blood cultures, and any of these with respect to the time that the specimen may be kept or transferred. The skill of taking blood for many different purposes has to be learned, of course. This chapter simply points out some of these purposes and the methods employed to meet them. Mentioned are the methods used to obtain blood to be used in serologic tests, fluid from the peritoneal and pleural cavities, and cerebrospinal fluid.

Just as important as the taking of samples is the correct reporting of the results of laboratory tests, the reporting of changes in the patient's condition to the physician, the institution and administration of proper treatment and medications, as directed, and the prevention of infections. The responsibilities of the health team in the prevention of transmissible diseases fall into two categories—inhibition, destruction, and removal of pathogenic microorganisms and assistance with programs that increase the immunity of populations.

REVIEW QUESTIONS

1. What are the functions of the medical technologist, the professional nurse, the practical nurse, the orderly, the physician assistant, and the physician?
2. Why are specimens collected? What type of specimens can you name? Cite the purpose of each.
3. What must be understood in proper labeling of a specimen? How should a specimen be transferred to a laboratory? Why? What is Stuart's Bacterial Transport Medium? How is it used?
4. What may be learned from taking specimens for culture from the throat, sputum, feces, blood, peritoneal or pleural cavities, and cerebrospinal fluid?
5. What different uses of blood taken from a patient can you name?
6. What basic rule must be remembered when a needle is used to stick a patient? What is the use of a Vacutainer bottle?
7. What are the responsibilities of every member of the health team?

Supplementary Reading

Benenson, A. S. (ed.): The Control of Communicable Diseases in Man. 12th ed. Washington, D.C., American Public Health Association, 1975.

Du Gas, R. W.: Introduction to Patient Care. 4th ed. Philadelphia, W. B. Saunders Company, 1982.

Finegold, S. M., et al.: Diagnostic Microbiology. 5th ed. St. Louis, C. V. Mosby Co., 1978.

French, R. M.: Nurse's Guide to Diagnostic Procedures. 4th ed. New York, McGraw-Hill, 1975.

Hopps, H. C.: Principles of Pathology. 2nd ed. New York, Appleton-Century-Crofts, 1964.

Isenberg, H. D., and Berkman, J. I.: Microbial diagnosis in a general hospital. *Ann. N.Y. Acad. Sci., 98*:647, 1962.

Lepper, M. H.: Collection of Specimens. *In* Lennette, E. H., et al. (eds.): Manual of Clinical Microbiology. 3rd ed. Washington, D.C., American Society for Microbiology, 1980.

Sherris, J. C. (ed.): Blood Cultures (Cumitech 1). Washington, D.C., American Society for Microbiology, 1974.

Simon, H. J.: Saprophytes in infection and the laboratory bedside interaction. *Ann. N.Y. Acad. Sci., 98*:745, 1962.

Wolff, L., et al.: Fundamentals of Nursing. 6th ed. Philadelphia, J.B. Lippincott Co., 1979.

APPENDIX A

Classification and Nomenclature of Bacteria as Given in Bergey's Manual of Determinative Bacteriology* (Eighth Edition)

KINGDOM *PROCARYOTAE*

DIVISION I. THE CYANOBACTERIA†
DIVISION II. THE BACTERIA

PART 1. PHOTOTROPHIC BACTERIA
ORDER I. RHODOSPIRILLALES
FAMILY I. RHODOSPIRILLACEAE
Genus I. *Rhodospirillum*
Genus II. *Rhodopseudomonas*
Genus III. *Rhodomicrobium*
FAMILY II. CHROMATIACEAE
Genus I. *Chromatium*
Genus II. *Thiocystis*
Genus III. *Thiosarcina*
Genus IV. *Thiospirillum*
Genus V. *Thiocapsa*
Genus VI. *Lamprocystis*
Genus VII. *Thiodictyon*
Genus VIII. *Thiopedia*
Genus IX. *Amoebobacter*
Genus X. *Ectothiorhodospira*
FAMILY III. CHLOROBIACEAE
Genus I. *Chlorobium*
Genus II. *Prosthecochloris*
Genus III. *Chloropseudomonas*
Genus IV. *Pelodictyon*
Genus V. *Clathrochloris*

PART 2. THE GLIDING BACTERIA
ORDER I. MYXOBACTERALES
FAMILY I. MYXOCOCCACEAE
Genus I. *Myxococcus*

---

*This material constitutes pages xxiii to xxvi of the eighth edition of *Bergey's Manual of Determinative Bacteriology.* © 1974, The Williams & Wilkins Co., Baltimore.

†Blue-green algae; also sometimes called Schizophyceae, Cyanophyceae, or Myxophyceae. Their classification is not detailed here.

FAMILY II. ARCHANGIACEAE
Genus I. *Archangium*
FAMILY III. CYSTOBACTERACEAE
Genus I. *Cystobacter*
Genus II. *Melittangium*
Genus III. *Stigmatella*
FAMILY IV. POLYANGIACEAE
Genus I. *Polyangium*
Genus II. *Nannocystis*
Genus III. *Chondromyces*
ORDER II. CYTOPHAGALES
FAMILY I. CYTOPHAGACEAE
Genus I. *Cytophaga*
Genus II. *Flexibacter*
Genus III. *Herpetosiphon*
Genus IV. *Flexithrix*
Genus V. *Saprospira*
Genus VI. *Sporocytophaga*
FAMILY II. BEGGIATOACEAE
Genus I. *Beggiatoa*
Genus II. *Vitreoscilla*
Genus III. *Thioploca*
FAMILY III. SIMONSIELLACEAE
Genus I. *Simonsiella*
Genus II. *Alysiella*
FAMILY IV. LEUCOTRICHACEAE
Genus I. *Leucothrix*
Genus II. *Thiothrix*

FAMILIES AND GENERA OF UNCERTAIN AFFILIATION
Genus *Toxothrix*
FAMILY ACHROMATIACEAE
Genus *Achromatium*
FAMILY PELONEMATACEAE
Genus *Pelonema*
Genus *Achroonema*
Genus *Peloploca*
Genus *Desmanthos*

PART 3. THE SHEATHED BACTERIA

Genus *Sphaerotilus*
Genus *Leptothrix*
Genus *Streptothrix*
Genus *Lieskeella*
Genus *Phragmidiothrix*
Genus *Crenothrix*
Genus *Clonothrix*

PART 4. BUDDING AND/OR APPENDAGED BACTERIA

Genus *Hyphomicrobium*
Genus *Hyphomonas*
Genus *Pedomicrobium*
Genus *Caulobacter*
Genus *Asticcacaulis*
Genus *Ancalomicrobium*
Genus *Prosthecomicrobium*
Genus *Thiodendron*
Genus *Pasteuria*
Genus *Blastobacter*
Genus *Seliberia*
Genus *Gallionella*
Genus *Nevskia*
Genus *Planctomyces*
Genus *Metallogenium*
Genus *Caulococcus*
Genus *Kusnezovia*

PART 5. **THE SPIROCHETES**
ORDER I. SPIROCHAETALES
FAMILY I. SPIROCHAETACEAE
Genus I. *Spirochaeta*
Genus II. *Cristispira*
Genus III. *Treponema*
Genus IV. *Borrelia*
Genus V. *Leptospira*

PART 6. **SPIRAL AND CURVED BACTERIA**
FAMILY I. SPIRILLACEAE
Genus I. *Spirillum*
Genus II. *Campylobacter*

GENERA OF UNCERTAIN AFFILIATION
Genus *Bdellovibrio*
Genus *Microcyclus*
Genus *Pelosigma*
Genus *Brachyarcus*

PART 7. **GRAM-NEGATIVE, AEROBIC RODS AND COCCI**
FAMILY I. PSEUDOMONADACEAE
Genus I. *Pseudomonas*
Genus II. *Xanthomonas*
Genus III. *Zoogloea*
Genus IV. *Gluconobacter*
FAMILY II. AZOTOBACTERACEAE
Genus I. *Azotobacter*
Genus II. *Azomonas*
Genus III. *Beijerinckia*
Genus IV. *Derxia*
FAMILY III. RHIZOBIACEAE
Genus I. *Rhizobium*
Genus II. *Agrobacterium*
FAMILY IV. METHYLOMONADACEAE
Genus I. *Methylomonas*
Genus II. *Methylococcus*
FAMILY V. HALOBACTERIACEAE
Genus I. *Halobacterium*
Genus II. *Halococcus*

GENERA OF UNCERTAIN AFFILIATION
Genus *Alcaligenes*
Genus *Acetobacter*
Genus *Brucella*
Genus *Bordetella*
Genus *Francisella*
Genus *Thermus*

PART 8. **GRAM-NEGATIVE, FACULTATIVELY ANAEROBIC RODS**
FAMILY I. ENTEROBACTERIACEAE
Genus I. *Escherichia*
Genus II. *Edwardsiella*
Genus III. *Citrobacter*
Genus IV. *Salmonella*
Genus V. *Shigella*
Genus VI. *Klebsiella*
Genus VII. *Enterobacter*
Genus VIII. *Hafnia*
Genus IX. *Serratia*
Genus X. *Proteus*
Genus XI. *Yersinia*
Genus XII. *Erwinia*
FAMILY II. VIBRIONACEAE
Genus I. *Vibrio*
Genus II. *Aeromonas*

Genus III. *Plesiomonas*
Genus IV. *Photobacterium*
Genus V. *Lucibacterium*

GENERA OF UNCERTAIN AFFILIATION
Genus *Zymomonas*
Genus *Chromobacterium*
Genus *Flavobacterium*
Genus *Haemophilus (H. vaginalis)*
Genus *Pasteurella*
Genus *Actinobacillus*
Genus *Cardiobacterium*
Genus *Streptobacillus*
Genus *Calymmatobacterium*
Parasites of *Paramecium*

PART 9. GRAM-NEGATIVE, ANAEROBIC BACTERIA

FAMILY I. BACTEROIDACEAE
Genus I. *Bacteroides*
Genus II. *Fusobacterium*
Genus III. *Leptotrichia*

GENERA OF UNCERTAIN AFFILIATION
Genus *Desulfovibrio*
Genus *Butyrivibrio*
Genus *Succinovibrio*
Genus *Succinomonas*
Genus *Lachnospira*
Genus *Selenomonas*

PART 10. GRAM-NEGATIVE COCCI AND COCCOBACILLI

FAMILY I. NEISSERIACEAE
Genus I. *Neisseria*
Genus II. *Branhamella*
Genus III. *Moraxella*
Genus IV. *Acinetobacter*

GENERA OF UNCERTAIN AFFILIATION
Genus *Paracoccus*
Genus *Lampropedia*

PART 11. GRAM-NEGATIVE, ANAEROBIC COCCI

FAMILY I. VEILLONELLACEAE
Genus I. *Veillonella*
Genus II. *Acidaminococcus*
Genus III. *Megasphaera*

PART 12. GRAM-NEGATIVE, CHEMOLITHOTROPHIC BACTERIA

A. ORGANISMS OXIDIZING AMMONIA OR NITRITE
FAMILY I. NITROBACTERACEAE
Genus I. *Nitrobacter*
Genus II. *Nitrospina*
Genus III. *Nitrococcus*
Genus IV. *Nitrosomonas*
Genus V. *Nitrosospira*
Genus VI. *Nitrosococcus*
Genus VII. *Nitrosolobus*

B. ORGANISMS METABOLIZING SULFUR
Genus *Thiobacillus*
Genus *Sulfolobus*
Genus *Thiobacterium*
Genus *Macromonas*
Genus *Thiovulum*
Genus *Thiospira*

C. ORGANISMS DEPOSITING IRON OR MANGANESE OXIDES

FAMILY I. SIDEROCAPSACEAE

- Genus I. *Siderocapsa*
- Genus II. *Naumanniella*
- Genus III. *Ochrobium*
- Genus IV. *Siderococcus*

PART 13. METHANE-PRODUCING BACTERIA

FAMILY I. METHANOBACTERIACEAE

- Genus I. *Methanobacterium*
- Genus II. *Methanosarcina*
- Genus III. *Methanococcus*

PART 14. GRAM-POSITIVE COCCI

A. AEROBIC AND/OR FACULTATIVELY ANAEROBIC

FAMILY I. MICROCOCCACEAE

- Genus I. *Micrococcus*
- Genus II. *Staphylococcus*
- Genus III. *Planococcus*

FAMILY II. STREPTOCOCCACEAE

- Genus I. *Streptococcus*
- Genus II. *Leuconostoc*
- Genus III. *Pediococcus*
- Genus IV. *Aerococcus*
- Genus V. *Gemella*

B. ANAEROBIC

FAMILY III. PEPTOCOCCACEAE

- Genus I. *Peptococcus*
- Genus II. *Peptostreptococcus*
- Genus III. *Ruminococcus*
- Genus IV. *Sarcina*

PART 15. ENDOSPORE-FORMING RODS AND COCCI

FAMILY I. BACILLACEAE

- Genus I. *Bacillus*
- Genus II. *Sporolactobacillus*
- Genus III. *Clostridium*
- Genus IV. *Desulfotomaculum*
- Genus V. *Sporosarcina*

GENUS OF UNCERTAIN AFFILIATION

- Genus *Oscillospira*

PART 16. GRAM-POSITIVE, ASPOROGENOUS ROD-SHAPED BACTERIA

FAMILY I. LACTOBACILLACEAE

- Genus I. *Lactobacillus*

GENERA OF UNCERTAIN AFFILIATION

- Genus *Listeria*
- Genus *Erysipelothrix*
- Genus *Caryophanon*

PART 17. ACTINOMYCES AND RELATED ORGANISMS

CORYNEFORM GROUP OF BACTERIA

- Genus I. *Corynebacterium*
  - a. Human and animal parasites and pathogens
  - b. Plant pathogenic corynebacteria
  - c. Non-pathogenic corynebacteria
- Genus II. *Arthrobacter*

GENERA INCERTAE SEDIS

- *Brevibacterium*
- *Microbacterium*

Genus III. *Cellulomonas*
Genus IV. *Kurthia*
FAMILY I. PROPIONIBACTERIACEAE
Genus I. *Propionibacterium*
Genus II. *Eubacterium*
ORDER I. ACTINOMYCETALES
FAMILY I. ACTINOMYCETACEAE
Genus I. *Actinomyces*
Genus II. *Arachnia*
Genus III. *Bifidobacterium*
Genus IV. *Bacterionema*
Genus V. *Rothia*
FAMILY II. MYCOBACTERIACEAE
Genus I. *Mycobacterium*
FAMILY III. FRANKIACEAE
Genus I. *Frankia*
FAMILY IV. ACTINOPLANACEAE
Genus I. *Actinoplanes*
Genus II. *Spirillospora*
Genus III. *Streptosporangium*
Genus IV. *Amphosporangium*
Genus V. *Ampullariella*
Genus VI. *Pilimelia*
Genus VII. *Planomonospora*
Genus VIII. *Planobispora*
Genus IX. *Dactylosporangium*
Genus X. *Kitasatoa*
FAMILY V. DERMATOPHILACEAE
Genus I. *Dermatophilus*
Genus II. *Geodermatophilus*
FAMILY VI. NOCARDIACEAE
Genus I. *Nocardia*
Genus II. *Pseudonocardia*
FAMILY VII. STREPTOMYCETACEAE
Genus I. *Streptomyces*
Genus II. *Streptoverticillium*
Genus III. *Sporichthya*
Genus IV. *Microellobosporia*
FAMILY VIII. MICROMONOSPORACEAE
Genus I. *Micromonospora*
Genus II. *Thermoactinomyces*
Genus III. *Actinobifida*
Genus IV. *Thermomonospora*
Genus V. *Microbispora*
Genus VI. *Micropolyspora*

PART 18. THE RICKETTSIAS

ORDER I. RICKETTSIALES
FAMILY I. RICKETTSIACEAE
TRIBE I. RICKETTSIEAE
Genus I. *Rickettsia*
Genus II. *Rochalimaea*
Genus III. *Coxiella*
TRIBE II. EHRLICHIEAE
Genus IV. *Ehrlichia*
Genus V. *Cowdria*
Genus VI. *Neorickettsia*
TRIBE III. WOLBACHIEAE
Genus VII. *Wolbachia*
Genus VIII. *Symbiotes*
Genus IX. *Blattabacterium*
Genus X. *Rickettsiella*

FAMILY II. BARTONELLACEAE
Genus I. *Bartonella*
Genus II. *Grahamella*
FAMILY III. ANAPLASMATACEAE
Genus I. *Anaplasma*
Genus II. *Paranaplasma*
Genus III. *Aegyptionella*
Genus IV. *Haemobartonella*
Genus V. *Eperythrozoon*
ORDER II. CHLAMYDIALES
FAMILY I. CHLAMYDIACEAE
Genus I. *Chlamydia*

PART 19. THE MYCOPLASMAS

CLASS I. MOLLICUTES
ORDER I. MYCOPLASMATALES
FAMILY I. MYCOPLASMATACEAE
Genus I. *Mycoplasma*
FAMILY II. ACHOLEPLASMATACEAE
Genus I. *Acholeplasma*

GENUS OF UNCERTAIN AFFILIATION
Genus *Thermoplasma*
Genus *Spiroplasma*

MYCOPLASMA-LIKE BODIES IN PLANTS

APPENDIX B

Sterilization Charts

Hospital Equipment and Materials Sterilizable by Ethylene Oxide*

TELESCOPIC INSTRUMENTS	PLASTIC GOODS	RUBBER GOODS	INSTRUMENTS AND EQUIPMENT	MISCELLANEOUS
Bronchoscopes	Catheters	Tubing	Cautery sets	Dilators
Cystoscopes	Nebulizers	Surgical gloves	Eye knives	Electric cords
Electrotomes	Vials	Catheters	Lamps	Hair clippers
Endoscopes	Syringes	Drain and feed sets	Needles	Miller-Abbott tube
Esophagoscopes	Gloves	Sheeting	Neurosurgical instruments	Pumps
Ophthalmoscopes	Test tubes		Scalpel blades	Motors
Otoscopes	Petri dishes		Speculae	Books
Pharyngoscopes	I.V. sets		Syringes	Toys
Proctoscopes	Infant incubators		Dental instruments	Pottery
Resectoscopes	Heart-lung machines		Oxygen tents	Blankets
Sigmoidoscopes	Heart pacemakers			Sheets
Thoracoscopes	Artificial kidney machines			Furniture
Urethroscopes				Sealed ampules
				Sutures
				Medicine droppers

*Chemical Gas Sterilization. From: Ethylene Oxide Sterilization. *J. Hosp. Res.*, 7:1, 1969. American Sterilizer Company.

Ethylene Oxide Mixtures Used in Gaseous Sterilization Procedures*

MIXTURES	MANUFACTURER
Ethylene oxide-carbon dioxide	
CARBOXIDE	
10% Ethylene Oxide 90% Carbon Dioxide	Union Carbide Corp. Linde Division, New York, N.Y.
OXYFUME STERILANT-20	
20% Ethylene Oxide 80% Carbon Dioxide	Union Carbide Corp. Linde Division, New York, N.Y.
STEROXIDE-20	
20% Ethylene Oxide 80% Carbon Dioxide	Castle Ritter Pfaudler Corp. Rochester, N.Y.
Ethylene oxide-fluorinated hydrocarbons	
CRY-OXCIDE	
11% Ethylene Oxide 79% Trichloromonofluoromethane 10% Dichlorodifluoromethane	Ben Venue Laboratories Bedford, Ohio
BENVICIDE	
11% Ethylene Oxide 54% Trichloromonofluoromethane 35% Dichlorodifluoromethane	The Matheson Co. East Rutherford, N.J.
PENNOXIDE	
12% Ethylene Oxide 88% Dichlorodifluoromethane	Pennsylvania Engineering Co. Philadelphia, Pa.
STEROXIDE-12	
12% Ethylene Oxide 88% Dichlorodifluoromethane	Castle Ritter Pfaudler Corp. Rochester, N.Y.

*Chemical Gas Sterilization. From: Ethylene Oxide Sterilization. *J. Hosp. Res.*, 7:1, 1969. American Sterilizer Company.

Sterilization of Apparatus and Supplies

ARTICLE	METHOD[1]	TIME IN MINUTES		
		at 165 C	*at 121 C*	*at 132 C*
Bronchoscopes	Autoclave		10	2–3
Bonewax	Hot Air	90		
Bougies	Chemical only			
Brushes	Autoclave		10	2–3
Catheters (Gum Elastic or Woven Silk Base)[2]	Chemical only			
Cellophane	Autoclave		30	15
Cystoscopes	Chemical only			
Diapers	Autoclave		30	15
Drains (Gutta Percha—Rubber)[2]	Autoclave		15	3
Drums—Dressing	Autoclave			
Loosely packed			30	15
Full but not compressed			45	
Electric Cords	Autoclave		10	2–3
Ether Cones	Autoclave		10	2–3
Glassware—Test Tubes, Tubing, Petri Dishes, etc.	Hot Air	120		
Glycerin	Hot Air	90		
Hard Rubber Details	Chemical only			
Instruments (In general)				
Routine	Autoclave		10	2–3
Emergency	Autoclave		5	
Extreme Emergency	Autoclave		3	
Scalpels and Scissors	Same as above			
Cataract Knives	Chemical only			
Tenotomes	Chemical only			
Urethrotomes	Chemical only			

[1] Sterilization by any approved sporicidal gas, such as ethylene oxide or beta-propiolactone, may be used in those situations in which heat and/or moisture would destroy the objects sterilized.

[2] Many supplies listed here, which were formerly assembled and sterilized as separate items, are now purchasable as complete kits or outfits, assembled for use, packaged, and properly sterilized; many are disposable.

12% Ethylene Oxide and 88% Freon-12 of the American Sterilizer Company, Erie, Pennsylvania should be used in this manner; 650 to 750 mg/liter of chamber space for 1¾ to 4 hours, at a total cycle time of 2½ to 5½ hours, temperature 125 to 135 F (52 to 57 C) and a relative humidity of 40 to 80%.

Dry heat sterilization in the oven is recommended at 340 F (171 C) for one hour, at 320 F (160 C) for two hours, or at 250 F (121 C) for six hours or longer. Common use is 165 C for two hours.

(Frobisher, M. F., Jr., Sommermeyer, L., and Goodale, R. H.: Microbiology and Pathology for Nurses. 5th ed. Philadelphia, W. B. Saunders Company, 1960.)

Sterilization of Apparatus and Supplies—Continued

ARTICLE	METHOD	TIME IN MINUTES		
		at 165 C	*at 121 C*	*at 132 C*
Intravenous Sets[2]	Autoclave		20	10
Iodoform	Do not sterilize			
Iodoform Drainage Material. Assemble previously sterilized parts under strict aseptic conditions, or purchase sterile.				
Jars—Enamelware	Autoclave		30	15
Lamb's Wool	Autoclave		30	15
Lamps, Diagnostic	Chemical only			
Maternity Packs[2]	Autoclave		30	15
Miller-Abbott Tubes	Chemical only			
Nebulizers	Chemical only			
Needles—Suture and Hypodermic	Hot Air	90		
Oils—Various	Hot Air	60		
Operating Motors—*See Manufacturer's Specifications*				
Paraffin Gauze	Hot Air	120		
Plastic Ware—*See Manufacturer's Specifications*[2]				
Proctoscopes	Chemical only			
Rubber Goods[2]	Autoclave			
Catheters			20	10
Gloves			15	3
Sheeting			20	10
Tubing			20	10
Scalpel Blades Spares—in Medicine Bottles	Autoclave		30	15
Sigmoidoscopes	Chemical only			
Solutions—Aqueous, in:	Autoclave			
2,000 ml Erlenmeyer Flask (Pyrex)[3] thin glass			20	10
2,000 ml Florence Flask (Pyrex) thin glass			20	10
1,800 ml Fenwal Flask (Pyrex) thick glass			30	15
1,000 ml Erlenmeyer Flask (Pyrex) thin glass			15	3
1,000 ml Florence Flask (Pyrex) thin glass			15	3

[3]Or Kimble products called *KIMAX*.

Table continued on following page

Sterilization of Apparatus and Supplies—Continued

ARTICLE	METHOD	TIME IN MINUTES		
		at 165 C	at 121 C	at 132 C
1,000 ml. Fenwal Flask (Pyrex)[3] thick glass	Autoclave		20	10
500 ml thin glass			12	3
250 ml thin glass			10	2–3
125 ml thin glass			8	2–3
2–4 ounce bottles, thick glass			10	2–3
Test Tubes, 150 x 18 mm.			8	2–3
Sulfa Drugs:				
Powder: At 300–315 F (150–155 C) only	Hot Air	90		
In solution	Autoclave		20	10
Surgical Packs—Major Packs[2]	Autoclave		30	15
Cotton filled Dressing Combines, Cotton Napkins, Cellulose Napkins, Gauze Sponges, etc.			20–30	10–15
Sutures				
Nonboilable Tubes	Chemical only			
Boilable Tubes	Autoclave		10	2–3
Silk, Cotton, Linen, Nylon	Autoclave		10	2–3
Syringes—Unassembled[2]	Hot Air	90		
	Autoclave		20	10
Talcum Powder	Hot Air	2 hours		
Thermometers	Chemical only			
Tongue Depressors	Autoclave		30	15
Transfusion Sets[2]	Autoclave		20	10
Trays—All kinds	Autoclave		20	10
Urethral Catheters[2]	Autoclave		10	2–3
Urethroscopes	Chemical only			
Utensils	Autoclave		15	3
Vaseline Petroleum Jelly (Petrolatum)	Hot Air	120		
Vaselinized Gauze	Hot Air	120		
Zinc Peroxide (Hold temperature at 280 F—135 C)	Hot Air	4 hours		

Packaging Materials for Articles to be Sterilized

MATERIAL	NATURE	TYPE OF PRODUCT	THICKNESS OR GRADE	SUITABLE FOR		
				Steam	***Dry Heat***	***EtO Gas***
Muslin	Textile	Wrappers	140 thread count	Yes	Yes	Yes
Jean Cloth	Textile	Wrappers	160 thread count	Yes	No	Yes
Broadcloth	Textile	Wrappers	200 thread count	Yes	No	Yes
Canvas	Textile	Wrappers	–		–Do Not Use–	
Kraft Brown	Paper	Wrappers Bags	30–40 lb	Yes	No	Yes
Kraft White	Paper	Wrappers Bags	30–40 lb	Yes	No	Yes
Glassine	Coated Paper	Envelopes Bags	30 lb	Yes	No	Yes
Parchment	Paper	Wrappers	Patapar 27–2T	Yes	No	Yes
Crepe	Paper	Wrappers	Dennison-Wrap	Yes	No	Yes
Cellophane	Cellulose Film	Tubing Bags	Weck Sterilizable	Yes	No	Yes
Polyethylene	Plastic	Bags Wrappers	1–3 mils	No	No	Yes
Polypropylene	Plastic	Film	1–3 mils	*	No	Yes
Polyvinyl Chloride	Plastic	Film Tubing	1–3 mils	No	No	Yes
Nylon	Plastic	Film Bags	1–2 mils	*	No	Yes
Polyamide	Plastic	Film Wrappers	1–2 mils	*	No	Yes
Aluminum	Foil	Wrappers	1–2 mils	No	Yes	No

*Not recommended. Difficult to eliminate air from packs.

From Perkins, J. J.: Principles and Methods of Sterilization in Health Sciences. 3rd Edition, 1980. Permission to reproduce this table has been granted by Charles C Thomas, Publisher, Springfield, Illinois.

Disinfection of Miscellaneous Objects and Substances

OBJECT OR SUBSTANCE	METHODS OF DISINFECTION	ALTERNATE METHODS
Bedpans from noninfected patient.	Wash thoroughly after each use and return to patient unit. When patient is discharged, bedpan should be cleaned thoroughly and steamed or boiled for 2 minutes.	
Bedpans from patient who has an enteric disease, brucellosis, poliomyelitis, viral hepatitis, or tuberculosis of the intestinal or urinary tract.	Disinfect in special equipment for steaming bedpans at 95 to 100 C for 30 minutes after contents have been emptied into covered pail for disinfection or otherwise disposed of in an approved manner. After steaming, wash thoroughly with soap and water and return to patient.	If bedpan steaming equipment is not available, empty contents into covered pail for disinfection, or otherwise dispose of in an approved manner. Immerse bedpan in tub of chlorinated lime (5%) or saponated solution of cresol or iodophore (5%) for 1 hour or in boiling water at 95 to 100 C for 30 minutes. Wash thoroughly and return to the patient unit.
Dishes and eating utensils from a noninfected patient.	Scrape to remove uneaten food, wash thoroughly with soap and hot water (preferably in a mechanical dishwasher so that the temperature of the water can be near boiling temperature), rinse, allow dishes to drain dry, and store in clean closed cabinets.	Where mechanical dishwashers are not available, use plenty of soap, friction and hot water. After washing, soak in clear, cool water containing at least 50 parts per million of free chlorine. Drain dry.
Dishes and eating utensils from a patient who has an infection transferable via respiratory or gastrointestinal tract.	Completely immerse and boil for 10 minutes before washing. Wash as stated above.[4]	Where feasible, use paper plates, cups and other dishes that can be burned after use.
Gowns used in care of a communicable disease.	When gown technique is indicated, gowns should be used once and discarded with contaminated linen. Reuse of gowns by the same person or others has little justification bacteriologically or esthetically.	In the home, it is usually impossible to have enough gowns for each time that a gown is needed. Care must be taken to avoid contaminating the inside of the gown when it is removed. Gowns must remain in the sickroom until removed to be disinfected every eight hours or at least daily (see *Linen*).
Gowns used in assisting with surgery.	Sterilize in the autoclave at 121 C for 30 minutes or at 132 C for 15 minutes.	
Hands of personnel caring for patients who have communicable diseases.	Wash thoroughly with soap and water, after the care of each patient, after emptying a bedpan or urinal, before preparing medications, before serving meals, before going off duty or to the dining room for their own meals. Cover each area of the hands and to the middle of the forearm twice. Be sure to pay special attention to the fingernails and areas between the fingers. Apply soothing hand lotion several times each day to prevent rough and chapped hands.	

[4]If epidemic viral hepatitis, autoclave.

Disinfection of Miscellaneous Objects and Substances—Continued

OBJECT OR SUBSTANCE	METHODS OF DISINFECTION
Hands of nurse preparing to assist with surgery.	Wash thoroughly with hexachlorophene or soap and water, covering each area of the hands and to the elbow three times. Use good friction, enough soap to make a good lather, and be sure to give special attention to the fingernails, fingertips, and areas between the fingers. Rinse hands thoroughly with water and immerse or rinse with ethyl alcohol (70%) or aqueous iodophore 1:1,000. Dry with sterile towel and put on sterile rubber gloves. Use of hexachlorophene should be limited to relatively short exposures of small areas of the body. It should be thoroughly removed.
Infectious paper, gauze or linen handkerchiefs or "wipes" from patients with respiratory diseases.	See *Infectious dressings from wounds or lesions.*
Infectious dressings from wounds or lesions.	Wrap carefully in newspaper or other wrapping so that the outside of the package is not contaminated and with enough thicknesses of wrapping so that drainage will not soak through. Burn this package completely.
Infectious feces and urine from patients who have enteric infections, brucellosis, poliomyelitis, or viral hepatitis.	In some hospitals and communities the bedpan is emptied directly into existing sewerage or hoppers. If this is not done, empty into a vessel that can be tightly covered. Break up large particles of feces with a wooden tongue blade. Do not spatter! Leave the tongue blade with the infectious material. Add an approximately equal amount of chlorinated lime (5%), laundry bleach (full strength), organic iodine disinfectant like Wescodyne, or saponated solution of cresol (5%). Do not spatter! Place cover on the vessel and secure with Scotch tape or adhesive. Allow a minimum of 1 hour contact with the disinfectant. Remove the tongue blade with a forceps, wrap in paper and burn. Boil the forceps. Do not let blade or forceps drip. Empty into the toilet or hopper.
Instruments used on infected wounds.	Boil before washing unless the infection is gas gangrene or anthrax. Wash, wrap, and resterilize by autoclaving. If the infection is gas gangrene or anthrax, handle with forceps, place in tray, and autoclave before washing. Wash and resterilize in the autoclave.

OBJECT OR SUBSTANCE	METHODS OF DISINFECTION	ALTERNATE METHODS
Linen from the bed, and bedclothing, of a patient having a communicable disease.	Collect in a bag or pillowcase, keeping the outside of the bag or pillowcase uncontaminated. Autoclave or send to laundry room if automatic washers are available. (Laundry water temperature should be close to 100 C.) If linen is from a patient who has gas gangrene or anthrax, the linen must be autoclaved.	Completely immerse in boiling water in a large container and boil for 10 minutes, or soak overnight in 5% saponated cresol solution or strong chlorine laundry bleach or iodophore (follow manufacturer's directions).

Table continued on following page

Disinfection of Miscellaneous Objects and Substances—Continued

OBJECT OR SUBSTANCE	METHODS OF DISINFECTION	ALTERNATE METHODS
Sputum and other respiratory secretions.	Collect with paper gauze or old linen handkerchiefs or in covered paper cup, wrap carefully to prevent contamination of the outside of the package, and burn. (If sputum collected in the sputum cup is copious, it is desirable to add sawdust or shredded paper to absorb the moisture.)	If burning is not feasible, soak paper handkerchiefs or container with sputum in a covered vessel with a saponated solution of cresol (5%), iodophore or chlorinated lime (5%) for a minimum of 1 hour. Discard in toilet or hopper. Covered glass containers can be reused after they have been sterilized.

OBJECT OR SUBSTANCE	METHODS OF DISINFECTION
Stethoscopes, otoscopes.	After use on a patient with a communicable disease cover the bell of a stethoscope with gauze moistened with saponated solution of cresol (5%) or ethyl alcohol (70%) for 5 minutes. The rest of the stethoscope should be washed thoroughly with soap and water and scrubbed with cresol or alcohol. Before use on a patient with a communicable disease the battery part of an otoscope can be wrapped with a small towel or strips of muslin to protect it. The otoscope tip can be boiled for 5 minutes if it is metal. If the tip is plastic, follow instructions of manufacturer. Wash off cresol with soap and water.
Thermometers—oral.	There should be sufficient supply of thermometers so that there is one available for each patient on the ward. Thermometers should be taken to patients in a clean dry container. After use: (1) Wipe clean with pledget of cotton or gauze wet with tincture of green soap mixed in equal volumes with 95 per cent ethyl alcohol. (2) Rinse thoroughly with clear water. (3) Completely immerse thermometer in 70% ethyl or rubbing alcohol, preferably containing 0.5 to 1% iodine, for 10 minutes. Rinse and dry. Store in a clean, dry, covered container. If necessary the disinfectant can be returned to the stock bottle to conserve supplies and facilitate the work of the public health worker in the field. The thermometer should be returned to its container clean and dry.
Thermometers—rectal.	Rectal thermometers should be lubricated with water-soluble lubricant. If this is not available, soap is a better lubricant bacteriologically than petrolatum or oil. Cleaning and disinfection are the same as for oral thermometers.
Toys, books, mail from a patient who has a communicable disease.	Where possible, disposable toys and books should be used for the patient who has a communicable disease. Some toys can be autoclaved, or washed with soap and water and exposed to the sun, or washed with disinfectant and sunned. Expensive books can be autoclaved if stood on end but the steam will deteriorate the bindings and covers. Outgoing mail can be autoclaved if it is written in pencil. Stamps are placed on and flaps gummed with Scotch tape after autoclaving.
Urinal from patient who has a urinary communicable disease.	See instructions for bedpan.

(Frobisher, M. F., Sommermeyer, L., and Goodale, R. H.: Microbiology and Pathology for Nurses. 5th ed. Philadelphia, W. B. Saunders Company, 1960.)

INDEX

Note: **Boldface** page numbers indicate main topic; *italic* numbers refer to illustrations; *t* indicates tables.

Abdominal wall, gas gangrene of, *552*
Abscess, 294
  amebic, *396*
  multiple, 437
  tooth, 442
  tuberculous, 480
Achromycin, 241
Acid-fast bacilli, 476
  fluorescence microscopy of, 477
Acid-fast stain, 476
Acidity, growth of microorganisms and, 135
Actidione, 241
Actinomyces, 490, 631
Actinomyces israelii, 490
Actinomycetaceae, 64
Actinomycetales, 63, *63,* 476, 632
Actinomycetes, 19, 60*t,* 63
Actinomycin A, B, C, and D, 241
Actinomycosis, 64, 490
Adenosine-5′-phosphate, *112*
Adenosine triphosphate, *112*
Adenoviruses, properties of, 93*t*
Adhesions, fibrous, 295
Adsorption, 198
Aedes aegypti, 594, *595*
Aerating sewage disposal unit, *167, 168*
Aerobes, **555**
  facultative, 133, *134*
    oxidation mechanisms of, *133*
  gram-negative, 629
  in sewage purification, 169
  obligate, 132, *134*
    oxidation mechanisms of, *133*
Aerosol sprays, as disinfectants, 227
Aestivo-autumnal fever, 601
Aflatoxins, 422, 571
African sleeping sickness, 606
Agammaglobulinemia, 336
Agar, 134
  blood, 146
    growth of hemolytic streptococci on, 439, *441*
  "chocolate," 449
  EMB, 172
  Endo, 172
  for anaerobic cultivation, 134, *135*
  for pure cultures, 142
Agar *(Continued)*
  nutrient, 145
    bacteriostatic agents with, 193
  Sabouraud's, 38
  thiosulfate citrate bile salts sucrose (TCBS), 379
  triple sugar iron (TSI), 369, *370*
Agar agar, 145
Agar dilution method, of sensitivity testing, 242, *244*
Agglutination, 320
  bacterial, *160,* 161, *161*
  flagellar, *161*
  in diagnosis of salmonellosis, 371
  of blood grouping, 273, 274, *274*
  somatic, *161*
Agglutination test, H and O, 371
Agglutinin, 161, 320, 322
  Vi, 371
Agglutinogens, ABO blood group, 273, 273*t*
  Rh, 275
Aggressiveness, of microorganism, 281
Agramonte, Aristides, 11
AIDS, 348
Air, disinfectin of, 435
  microorganisms in, 173, 179
Alastrim, 494
Alcohols, as disinfectants, 219*t,* 222
Algae
  blue-green, 19
    characteristics of, 20*t*
    structures in, *35*
  characteristics of, **34**
  eukaryotic, 34
  eukaryotic true, 19
  green, characteristics of, 20*t*
Alkalinity, growth of microorganisms and, 135
Alkylating agents, as disinfectants, 225
Allergens, 284, 340
Allergy
  cutaneous, 341
  delayed-type, cell-mediated, 340, 344
  immediate-type, antibody-mediated, 340, 341
    cytotoxic, 344
    noncytotoxic, 341
  in passive immunity, 335

Allergy *(Continued)*
  in tuberculosis, 481
  Koch's phenomenon and, 481
  types of, **340**
Alpha rays, 138
  sterilization by, 211
Alteration of generations, 600
Alum-precipitated toxoid, 463, 472
Amblyomma cajennense, 589
Amebas, **394**
  ingesting flagellate, *269*
  trophozoites and cysts of, *396*
Amebiasis, **394**
  epidemiology of, *397*
  prevention of, 398
  transmission of, 395, *397*
  treatment of, 398
Amebic abscess, *396*
  diagnosis of, 398
Amebic dysentery, **394**
  prevention of, 398
  transmission of, 395, *397*
  treatment of, 398
Ameboid movement, 394
Ammonium disinfectants, quaternary, 219*t*, 225
    formula for, *225*
Amoeba proteus, 51
Amphotericin B, 241, 242
Amphyl, 224
Anaerobes, **548**
  aerotolerant, 133
  clinically important, 550*t*
  cultivation of, 133, *134, 135*
  gram-negative, 630
    facultative, 629
  in sewage purification, 169
  obligate, 133
  strict, 133
    oxidation mechanisms of, *133*
Anaerobic culture dish, Brewer, *135*
Anaerobic jars, *551*
Anaerobiosis, 52
Anamnestic reaction, 371
Anaphylactic shock, 343
Anaphylaxis, 341
  substances involved in, 342*t*
Aneurysm, syphilitic, 526
Angina, Vincent's, 472
Animal inoculation, of tubercle bacilli, 483
  to identify microorganisms, 157, *158*
Anopheles mosquito, as vector of malaria, 602
  developmental stages of, *605*
Anoxybiontic organisms, 134
Antagonist, metabolite, 231
Anthrax, 555
  prevention of, 556
Antibacterial agents, modes of action of, 233*t*
Antibiosis, 280
Antibiotics, 230, **235.** See also *Antimicrobial agents.*
  antifungal, 241
  antineoplastic, 241
  bactericidal, 235, 236, 237*t*
  bacteriostatic, 235, 236, 237*t*
  broad-spectrum, 236, 241
  classes of, 236, 237*t*
  commonly used, 235*t*
  neisserian infections and, 542
  peptide, 241
  release cultures and, 459
Antibiotics *(Continued)*
  response of microorganisms to, 237*t*
  selective toxicity of, 232
  sensitivity testing of microorganism to, 242, *243, 244*
  source of, 236
  toxic effects of, 232
  use of, 236
    excessive or improper, effects of, 242
Antibody(ies), 159, 293, **303.** See also *Immunoglobulins.*
  ABO blood group, 273, 273*t*
  allergy mediation by, immediate-type, 340, 341
  anaphylactic, 341
  blocking, 322
  circulating, 299
  conjugated, 322, *323*
  fluorescent staining of, 322, *323*
  hapten-specific, 300, *303*
  hemagglutination-inhibiting, 322, *322*
  immobilizing, 315
  labeled, 322, *323*
  lymphocytes and, **298**
  monoclonal, 324, 337
  neutralizing, 322
  opsonic action of, 293
  rising titers of, diagnostic importance of, 321
  sensitizing, **313**
  specific, 299
    formation of, 313
Antibody-antigen reactions, 322
Antifungal antibiotics, 241
  modes of action of, 233*t*
Antigens, 159, **283**
  ABO blood group, 273, 273*t*
  Australia, 414, 575
  cardiolipin, in syphilis tests, 527
  classification of, 284, 285*t*
  endogenous, 284, 285*t*
  exogenous, 284, 285*t*
  Forssman, 284
  H, 275, 371
  infections and, **280**
  initial contact of, 283
  M, 444
  of viruses, 96
  partial, 285
  primary stimulus of, 333, *333*, 463
    slow response to, 334
  residue, 285
  secondary stimulus of, 333, *333*, 463
  sensitizing dose of, 340, 343
  sheath, 54
  toxic (shocking) dose of, 340, 343
  Vi, 371
Antihistamine, 343
Antimetabolite, 231
Antimicrobial agents, **233**
  blood levels of, maintenance of, 234
  chemotherapy and, **230**
  modes of action of, 233*t*
  selective toxicity of, 232
  sensitivity testing of microorganism to, 242, *243, 244*
Antineoplastic antibiotics, 241
Antirabies serum, equine (ARS), 328, 504
Antisepsis, 193
Antiseptic, 193
  properties of, 219*t*
Antistreptolysin, in rheumatic fever, 445

Antitoxins, 313, 322
botulism, 419
concentrated (purified), 335
delay of administration of, survival and, 462, *462*
diphtheria, therapeutic use of, 460
unit of, 461, 461*t*
tetanus, 553
"A.O.A.C. tests," 227
API-20 test, 364*t*
Apoenzyme, *105*
Appert, Nicholas, 191
Arboviruses, 593
Archaebacteria, *17*, 18
Argyrol, 222
ARS (antirabies serum, equine), 328, 504
Arteriole, 264
Arthropod, as disease vector, 576, 576*t*
sanguivorous, 575, **576**
Arthropod-borne encephalitides, viral, **596**
Arthropod-borne infection
bacterial, **576**
helminthic, **610**
protozoal, **600**
rickettsial, **586**
viral, **593**
Arthrospore, 42
of *C. immitis*, 563, *565*
Arthus reaction, 344
Ascaris lumbricoides, egg of, *400*
Ascocarp, 43
Ascoli test, 556
Ascospores, 30, *40*
of yeasts, destruction of, 195
Ascus(i), 39, *40*
Aspergillosis, 568
Aspergillus, *43*, 45
Aspergillus fumigatus, 569
Ataxia, locomotor, in syphilis, 526
Athlete's foot, 560, *562*
Atopy, 343
Aureomycin, 241
Australia antigen, 414, 575
Autoantigens, 345, 346
Autoclave
ethylene oxide sterilization in, 211, *212*
for bacterial spore destruction, 195, 197
for instrument cleaning, 209, *210*
for sterilization, **204,** *205–207*
placement of articles in, *208, 209*
indicators for, 210
Autoimmune disease, 346, 347*t*, 348
Avian complex virus, 92*t*
Axillae, microorganisms of, 180*t*
Azochloramide, as disinfectant, 220
Azotobacter chroococcum, *155*

Babesia microti, 605
Babesiosis, 605
Bacillaceae, 60*t*, 63
Bacillary dysentery. See *Shigellosis.*
Bacille Calmette-Guérin vaccine, 330
in tuberculin-positive individuals, 482
tuberculosis prevention and, 487
Bacillus(i), 50, *52*
acid-fast, 476
fluorescence microscopy of, 477
atypical acid-fast, 476
Bordet-Gengou, 470
Bacillus(i) *(Continued)*
diphtheria, 456
virulence of, 460
dysentery, 374
endospore-forming, 60*t*, 63, 631
enteric, differentiation of, 369, *370*
fusiform, *52*, 473, *473*
gas, 549
gram-negative, 52
aerobic, 60*t*, 61, 629
facultatively anaerobic, 60*t*, 62, 629
obligate anaerobic, 62
gram-positive, 50
hay, 555
Hofmann's, 465
Klebs-Löffler, 456
Koch-Weeks, 469
Morax-Axenfeld, 469
paratyphoid, 361
Schmitz, 374
Shiga's, 424
Sonne dysentery, 374
tubercle, 476
disinfection and, 479
species of, 478
types of, 50
Bacillus, 52, 555
Bacillus anthracis, 555, *555*
exotoxin of, 556
for attenuated bacterial vaccine, 330
Bacillus cereus var. *mycoides*, *554*, 555
food poisoning due to, 422
Bacillus polymyxa, 555
Bacillus subtilis, 555
spore formation in, *58*
Bacteremia, 182
Bacteria. See also *Microorganisms.*
agglutination of, *160*, 161, *161*
appendaged, 628
autotrophic, 102
biological oxidation of, 106
brownian movement of, 152
budding, 114, 628
capsule of, 53, *54*
health and, 54
virulence and, 122
cellular anatomy of, **53,** *57*
cell wall–defective variants, 72, *73*
characteristics of, 20–21*t*, **48,** 72*t*
chemolithotrophic, 630
chemosynthetic, 110, 128
classification of, 19, 48, **627**
in *Procaryotae*, 39*t*
morphologic, 50
coliform, as sanitary indicators, 377
testing for, in drinking liquids, 171, *172*
colony of. See *Colony, bacterial.*
conjugation of, *115*, 116
coryneform group of, 63, 631
culture of. See *Culture.*
drug resistance of, 120
endospores of, 55, *57, 58*
endotoxin-producing, 282
energy sources of, 108
enterotoxin-producing, 422
exotoxin-producing, 282
fimbriae of, 54
fission of, 114
flagella of, 54, *55, 56*
gene transfer in, 114
genetic recombination in, *115*, 117

Bacteria *(Continued)*
gliding, 54, 627
gram-negative, 154, 155*t,* 629, 630
gram-positive, 154, 155*t,* 631
granules of, 59, *59*
green sulfur, 19, 109
growth of, in culture, 145, *146*
helicoidal, 50, 52, *53*
heterotrophic, 102
in milk, 427
L-forms of, 72, *73*
characteristics of, 21*t*
lyophilization of, *126,* 127
major groups of, **59,** 60*t*
metabolism of, 103, 127
methanogenic, 18, 631
morphology of, 50
motility of, 54, 151*t*
multiplication of, 114
nomenclature of, 48, **627**
nonmotile, 151*t*
nonspore-forming, heat and chemical resistance of, 195, 197
nutrients of, **127**
nutrition of, 103
of soil, pathogenic, **548**
phenotype variations of, 120
photosynthetic, 109, 128
phototrophic, 627
physiology of, **102**
pigments of, 59
pili of, 54, *57*
plasmids of, 58
prosthecae of, 55
purple, 19, 109
relation of, to true fungi, 36
rod-shaped, asporogenous, 631
sheathed, 54, 628
size of, 49
spiral and curved, 629
spore-forming, heat and chemical resistance of, 195, 197
structure of, 49
subculturing of, *144*
transduction in, 115
transformation in, 116
vaccines of, 330
variable characters of, 120
virulence of, variation in, 122
zoonoses caused by, **577**
Bactericide, 192
Bacterins, 330
Bacteriochlorophyll, 32, 102, 109
Bacteriocins, 245
Bacteriolysins, 315
Bacteriophage, 76, **82.** See also *Virus.*
binal symmetry of, *81,* 83
demonstration of, 83
flexuous tailed, *78*
infection by, 83, *85, 86*
lysogeny by, 84
plaque formation by, *86,* 87
prophage state of, 84
shape of, *78,* 83, *84*
T-even, *78, 84*
Bacteriostasis, 230
definition of, 193
selective, 193, 230
Bacteroides fragilis, 554
Bacteroides melaninogenicus, 554
Bacteroides oralis, 554
Bang's disease, 387
Barberios, 607
Base layer, 243
Base plate, of bacteriophage, 83, *86*
Basic fuchsin, as bacteriostatic agent, 193
Basophil, 268
B-cells, 299. See also *B-lymphocytes.*
BCG vaccination, 330
in tuberculin-positive individuals, 482
tuberculosis prevention and, 487
Bedpan, disinfection of, 640*t*
Bedpan washer/sterilizer, 256, *256, 257*
Bedside equipment, in communicable disease, disinfection of, 255, 640–642*t*
Bejel, 522, 533, 540*t*
Benvicide, 212, 635*t*
Berkefeld filter, 213
Betadine, 221
Beta-glycoprotein, 308
Beta-propiolactone, 193, 195
sterilization by, 213
Beta rays, 138
sterilization by, 211
Binary fission, 114
Biologic false-positive reaction, in syphilis tests, 527
Biooxidation, 106, 128
mechanisms of, *133*
Biotic community, 164
Bis-phenols, as disinfectants, 224
Bittner virus, *81*
Blabarus discoidalis, 51
Black Death, 579
Blanching reaction, in scarlet fever, 446
Blastomyces dermatitidis, 566
Blastomycin, 566
Blastomycosis, North American, 566, *567*
South American, 568
Blastospores, 42
Blennorrhea, 517
Blood
clotting of, 266, *266, 272*
factors involved in, 272*t*
mechanism of, 271
composition of, **265**
cross matching of, 276
culture of, *618,* 619
in salmonellosis, 369
method of taking, *619*
veins for, location of, *618*
for serologic tests, 620
functions of, **264**
infection of, 280
artificial or accidental, 574
as sequela to primary infection, 574
vectors of, 574, 576*t*
natural, 575, **576**
microorganisms in, 182
microscopic examination of, 621, *621*
pathogens carried in, 357*t,* 624*t*
protein fractions in, 266, 266*t*
Blood agar, 146
growth of hemolytic streptococci on, 439, *441*
Blood-agar plate method, of sensitivity testing, 243
Blood bank blood, microorganisms in, 183
precautions with, 183
Blood broth, 146

Blood cells, origin of, 266, *267*
red, 267, *268*
white. See *Leukocyte(s)*.
"Blood exchange," with erythroblastosis fetalis, 276
Blood fluke, 400*t*
Blood grouping, 274, *274*
Blood platelets, *268*, 271
Blood transfusion, hepatitis risks from, 414*t*
Rh factor and, 276
Blood types, **273**
inheritance of, 273, 273*t*
B-lymphocytes, 299
from bursa of Fabricius, *302*
hapten reaction with, 300, *303*
humoral immunity and, 299
stem cell differentiation of, *298*, 299
Body
chromatin, 56
Donovan, 519
elementary, 71
Golgi, *16*
inclusion, 88
L-, 72, *73*
Negri, 503, *503*
Russell, 302
Body fluids and discharges, carrying pathogens, 624*t*
Boiling, sterilization by, 204, *204*
effectiveness of, 256
"Bombay type," 275
Booster dose, 333
Bordet-Gengou bacillus, 470
Bordetella, 470
Bordetella bronchiseptica, 470
Bordetella parapertussis, 470
Bordetella pertussis, 470, *470*
Borrelia, 581
Borrelia anserina, *53*
Borrelia duttonii, 583
Borrelia recurrentis, 583, *583*
Botulism, 419
categories of, 420
organism causing, 418
prevention of, 419
Boutonneuse fever, 588*t*, 589
Bradykinin, anaphylactic reaction to, 342*t*
"Breed count," 429
Brewer anaerobic culture dish, *135*
Bronchopneumonia, 445, 448
Broth
blood, 146
brillant green–lactose bile (BGLB), 172
glucose, carbohydrate fermentation in, *107*
infusion, 146
nutrient, 145
thioglycollate, 134, *135*
Brownian movement, 152
Bruce, Sir David, 386
Brucella, **386**
characteristics of, 387, 388*t*
species of, 386
Brucella abortus, 386
for attenuated bacterial vaccine, 330
Brucellosis, **387**
Brunhilde strain, of poliovirus, 407
Buboes, 578
Bubonic plague, 579. See also *Plague*.
Budding, bacterial, 114
by yeasts, 39, *39*
of *H. capsulatum*, 565, *567*
Burkholder, Paul R., 240
Burkitt's lymphoma, 94
Burns, of skin, 289
Bursa of Fabricius, *302*

Calcification, in tuberculosis, *480*, 481
Calcium hypochlorite, as disinfectant, 219
Calymmatobacterium granulomatis, 519
Campylobacter jejuni, 381
Campylobacteriosis, 381
Cancer, mutagenic agents and, 118
Candida albicans, 569, *569*
Candidiasis, 540*t*, 569
Candle filter, 191
Capillary vessels, 264
relation of, to tissue cells and fluids, *265*
Capneic organisms, 134
Capsid, viral, *77*, 81
Capsomer, *77*, 81, *83*
Capsular polysaccharide, 54, 282, 448
Capsule
bacterial, 53, *54*
health and, 54
virulence and, 122
of pneumococci, 447, 448
of streptococci, 442
polysaccharide, 54, 282, 448
Carbamide peroxide, 225
Carbohydrates
antigenic, 284
of beta-type streptococci, 442
decomposition of, 175
fermentation of, in glucose broth, *107*
metabolism of, 110, *111*
Carbolic, "crude," 223
Carbolic acid, 223
Carboxide, 212, 635*t*
Carbuncle, 294
Cardiolipin antigen, in syphilis tests, 527
Carrier, 173
convalescent, temporary or permanent, 366
diphtheria, 459
of enteric infection, 367
control of, 382
of epidemic meningitis, 451
of respiratory tract infections, 434
of typhoid fever, 366, 367
Carroll, James, 11
Caseation, in tuberculosis, 480
Catalysis, 104
Catheterization, sterilization for, 251
Cathode rays, sterilization by, 211
Cell, 22
abnormal, T-lymphocyte response to, 301, 302
B, 299. See also *B-lymphocytes*.
bacterial, *23*
anatomy of, **53,** *57*
blood, origin of, 266, *267*
red, 267, *268*
white. See *Leukocyte(s)*.
division of, *30*
dormant, 126
eukaryotic, 14
division of, 23, 27
giant, in tuberculosis, 480
Hfr, 116
in immune process, differentiation and localization of, *267*

Cell *(Continued)*
mast, 303
memory, 299, 312
null, 302
plasma, 299, 302
prokaryotic, 14, *23*
structure of, 23
T, 300. See also *T-lymphocytes.*
types of, **14**
vegetative, 126
wandering, 268
Cell culture, of viruses, 87, *88,* 147
Cell membrane, *16,* 24
semipermeable, 103
Cell wall, *16,* 24, *25*
Cellular immunity, **298**
T-lymphocytes and, 299, 300
Cellulitis, streptococcal, 444
Cellulose, in cell wall, 24
Celsius temperature scale, Fahrenheit equivalents of, 131*t*
Centrosome, 31
Cephalosporins, 237, 240
Cephalothin, 237, 240
Cerebrospinal fluid, collection of, 622
Cervicitis, gonococcal, 537
Cestodes, 45
eggs of, *400*
pathogenic, for humans, 400*t*
Chagas' disease, 606
Chamberland, Charles, 191
Chamberland-Pasteur filter, 213
Chancre, 522, 524, *525*
hard, 524
soft, 540*t,* 543
Chancroid, 540*t,* 543
Chemical agents. See also *Disinfectants.*
destruction of microorganisms by, **194**
properties for, 196
interaction between microorganism and, factors influencing, 199
sterilization with, 211
"Chemical nature," of disinfectant, 196
Chemolithotrophs, 128
Chemoorganotrophs, 129
Chemosynthesis, bacteria and, 110, 128
Chemotaxis, 292
Chemotherapeutic agents, **233**
blood levels of, maintenance of, 234
toxic effects of, 232
Chemotherapy, 230
antimicrobial agents and, **230**
mechanism of, 231
phagocytosis and, 231
Chemotrophs, 128
Chick embryo inoculation, 69, *70,* 87
Chickenpox, 500
characteristics of, 90*t*
Chilomastix mesnili, 396
CHINA virus, 404, 498
Chitin, in cell wall, 24
Chlamydia psittaci, 71, 511, 516, 519
Chlamydia trachomatis, 71, 516
Chlamydiaceae, 516
Chlamydiales, 68, **71**
Chlamydias, 21, **71**
characteristics of, 21*t,* 72*t*
diseases caused by, **516**
Chlamydospores, 42, *42*
Candida albicans, 569
Histoplasma capsulatum, 566
tuberculate, *566*
Chloramine-T, as disinfectant, 220, *220*
Chloramphenicol, 240
Chloride of lime, as disinfectant, 219
Chlorine, as disinfectant, 218, 219*t*
organic compounds of, 219*t,* 220, *220*
Chlorine gas, as disinfectant, 219
Chlorophyll, 108
Chlorophyta, characteristics of, 20*t*
Chloroplast, *16*
Chlorothymol, 224
Cholera, 360
Asiatic, **379**
transmission of, 380
Choleragen, 379
Chromatin body, 56
Chromatophores, 32
Chromoblastomycosis, 568
Chromosomes, 27
Cicatrization, 295
Cilia, electron micrograph of, *152*
role of, in defense against infection, 290, *290*
Cistron, 27, 118
Citric acid cycle, *113*
Climate control, in prevention of respiratory infection, 435
Clonal selection, of immunoglobulins, 304
Clone, definition of, 48
Clostridial infection, histotoxic, 551
Clostridium, 52, 548, *550*
Clostridium botulinum, 418
Clostridium perfringens, 549
food poisoning due to, 420
toxins of, 420, 421*t,* 550
Clostridium tetani, 552, *553*
Clot, blood, 266, *266, 272*
factors involved in, 272*t*
mechanism of formation of, 271
Coagulase, staphylococcal production of, 422, 436, 437
Coagulation
blood, 266, *266, 272*
factors involved in, 272*t*
mechanism of, 271
moisture and, 197
Coccidioides immitis, 562
arthrospores of, 563, *565*
Coccidioidin, 563
Coccidioidin test, 563
Coccidioidomycosis, 562, *564*
pulmonary, 563, *564*
Coccobacilli, 50, 52, *52*
gram-negative, 630
Coccus(i), 50, *52*
classification and description of, 50, 436
endospore-forming, 631
gram-negative, 60*t,* 62, 630
aerobic, 60*t,* 61, 629
anaerobic, 630
gram-positive, 60*t,* 62, 631
pathogenic, differentiation of, *440*
of respiratory tract, **436**
pyogenic, infections by, **434**
Codon, 28, 29*t*
Coenzymes, 105, *105*
used by microorganisms, 108*t*
Cohn, Ferdinand, 8
Cold, "common," 509
Cold hemagglutinin, 321
Colicins, 245
Coliforms, as sanitary indicators, 377
testing for, in drinking liquids, 171, *172*

Collagenase, 283
Colon, microorganisms, of, 180*t,* 182
Colony, bacterial, 143, *143*
appearance of, 144
counting of, 143, *145*
automatic, 144, *145*
dwarf, 122
growth curve of, 145, *146*
isolated, obtaining of, 143
mucoid, 122
rough, 122, *122*
selective, *143,* 471
smooth, 122, *122*
subculturing of, *144*
variant types of, 122, *122*
Columella, of molds, *42*
Commensalism, 280
Communicable disease
entry portals of, 354, 356–357*t*
exit portals of, 356, 356–357*t*
prevention of, 358
early history of, 8
health personnel in, 623
transmission of, factors in, **354,** *354*
vectors of, 357
Communicable disease unit, sterilization and disinfection in, 255, *255–257*
in home, 259
Community health practice, disinfection in, 258
Complement, 308, **313,** 314
activation pathway of, 314, *316–317*
fixation of, 315, 322
function of, in inflammation, 294
properdin pathway of, *318*
Complement fixation test
for rickettsial infection, 587
for syphilis, 527
Reiter protein, 529
specific, 528
Complementation analysis, 27
Concentration, necessary, of chemical disinfectant, 196
Condenser, substage, 150
Condyloma acuminatum, 540*t*
Conidia, of molds, 41, *42, 43*
destruction of, 195
of *Streptomyces,* 64, *65*
Conidiophore, *42, 43*
Conidiospore, 43, *43*
H. capsulatum, 565, *566*
Conjugation, *115,* 116
Conjunctivae, defense mechanisms of, 290
normal microorganisms of, 180*t,* 181
Conjunctivitis, acute (angular), 469
prevention and treatment of, 470
inclusion, 71, 517, *518*
Contact, of sterilizing or disinfecting agent, 199
time of, 200
Contact factor, 271
Contamination, of food, 176
sources of, 423
of pure culture, prevention of, 144
Conversion, viral, 116
Cooking, in preventing food infection and poisoning, 424
Cortisone, use of, in delayed-type allergy, 345
Corynebacterium diphtheriae, **456,** *458*
virulence of, 460
Coryneform bacteria, 63, 631
"Cough plates," 471, 615
Coxiella, heat resistance of, 425*t*
Coxiella burnetti, 68, *70,* 592
Coxsackieviruses, 410
immunological types of, *406,* 411
Cresol, adsorption of, 198
as disinfectant, *220,* 223
saponated solution of, 223
Cristispira sp., *53*
Cryoxcide, 212, 635*t*
Cryptococcosis, 568
Cryptogram, for virus designation, 76
Cultivation. See also *Culture.*
of anaerobes, 133, *134, 135*
of rickettsia, 69, *70*
of tubercle bacilli, 477, 483
of viruses, 87, *88,* **147**
selective, *143,* 471
Culture
bacterial, lyophilization of, *126,* 127
rapid methods for, 156
blood, *618,* 619
in salmonellosis, 369
method of taking, *619*
cell, of viruses, 87, *88,* 147
contaminated, 142
feces, 617
for pathogenic microorganisms, *159,* 160
in enteric infection, 369, *370*
mixed, 142
pure, 9, **142**
contamination of, prevention of, 144
growth curve of, 145, *146*
isolation of, 142
methods of, 142
of *C. diphtheriae,* 457
of yeasts, 40
release, antibiotics and, 459
throat, 615, *617*
urine, 616
in enteric infection, 371
Culture medium, 9, 145. See also *Agar; Broth.*
Bordet-Gengou, 470
defined, 147
for clostridia, 549
for tubercle bacilli, 477, 483
incubation of, 142, *144*
inoculation of, 615, *616*
Pai's, 456
surface tension reducers in, 198
synthetic, 146, 147*t*
tellurite, 457
Thayer-Martin, 541
Culturette, 615
Cyanobacteria, 19
characteristics of, 20*t*
structures in, *35*
Cyanophages, 83
Cycloheximide, 241
Cytochrome, 267
electron transport in, *114*
Cytokinesis, 23
Cytolysins, 314, 322
Cytolysis, 315
Cytomegalovirus infections, 501
Cytoplasm, 31
eukaryotic, 15, *16,* 32
prokaryotic, 16, *17, 31,* 32
Cytoplasmic streaming, 394

Dane particle, 414, 575
Darkfield microscopy, condensers for, *524*
for demonstration of treponemes, 523

Darling's disease, 565
Decay, 165
Defense mechanisms, **288,** *289*
first line of, 289
second line of, 291
species or racial, 288
Dehydrogenase, 106
Dehydrogenation, 106, 107
Deliquescent, 134
Denitrification, *178*
Denitrosification, *178*
Deoxyribonucleic acid (DNA), **25**
helices of, 25, *26*
nucleotide units in, 25, *26*
linear sequence of, 27, *28*
oncogenic viruses containing, *78,* 93*t,* 94
recombinant, 4
replication of, semiconservative, *28*
structure of, *26*
Deoxyribonucleotides, *26,* 27
Dermacentor andersoni, 589, *591*
Dermacentor variabilis, 589, *591*
Dermatomycoses, 561
Dermatophytes, 560, 563*t*
Dermatophytoses, 561
Desensitization, 343
Detergents
anionic, 225, *225*
as disinfectants, 225
as surface tension reducer, 198
biodegradable, 226
cationic, 225, *225*
nonionic, 225
Diabetes, viral theory of, 412
Diapedesis, 265, *265*
Diaphragm, iris, 150
Dick test, 446
Dientamoeba fragilis, 396
Diffusion, 103
Digestion, extraneous, 103, *104*
phagocytic, 291, *292*
Digestive tract, amebae and protozoa of, *396*
Dihydrostreptomycin, 240
Diphtheria, **456**
bacilli of, 456
virulence of, 460
immunity to, 462
immunization against, 322*t,* 464
laryngeal, 458
nasopharyngeal, 458
pathogenesis of, 457
transmission of, 458
Diphtheria toxin, 457
specificity of, 283
structure and action of, 460
toxicity of, 421*t*
toxigenicity of, 282
units of, 461*t*
Diphtheroids, 181, 464
Diplococci, 50, *52*
Disease
autoimmune, 346, 347*t,* 348
communicable. See *Communicable disease.*
infectious, etiology of, Koch's postulates of, 9
prevention of, passive immunity in, 335
sexually transmissible, 540*t*
venereal. See *Venereal disease.*
vs. infection, 281
Dishes, disinfection of, 256, 640*t*
sanitizing of, by hand, *426*
by machine, *427*
Dishwashing, proper methods of, 426, *426, 427*
Disinfectant, 192
"A.O.A.C. tests" of, 227
chemical nature of, 196
contact of, 199
time of, 200
diluted, 193
evaluation of, 227
ideal, 199
interaction between microorganism and, factors influencing, 199
necessary concentration of, 196
organic material as obstacle to, 199
pH of, 200
phenol coefficient of, 227
properties of, 219*t*
resistance to, of bacterial spores, 195
of nonspore-forming bacteria, 195
strength of, 227
surface tension of, 197
temperature of, 199
toxicity of, 198
water solubility of, 197
Disinfection, 194, **218**
agent-organism interaction in, 199
air and dust, 435
characteristics of microorganisms and, 195
definition of, 192
historical aspects of, 190
in communicable disease unit, 255, *255–257*
in community health practice, 258
in health care, **248**
in medical/surgical wards, **248**
in operating room, **252,** *253–255*
of objects and substances, 640–642*t*
properties of agent used for, 196
terminal, 252
tubercle bacilli and, 479
Disk plate method, of sensitivity testing, 242, *244*
Disk procedure, simplified, 243
DNA. See *Deoxyribonucleic acid.*
Donovan bodies, 519
Dortmund tank, *168,* 169
Dourine, 606
Dressing cart, surgical, 250, *251*
Dressings, surgical, disinfection of, 641*t*
handling procedure for, 249
Droplet infection, 174, *174*
Droplet nuclei, 174
Drug hypersensitivity, 346
Drug intolerance, 346
Drug resistance, 234
bacterial, 120
R-factors and, 241
staphylococcal, 438
DTP, 463
schedule for, 332*t*
Duck embryo vaccine (DEV), 328, 504, 505*t*
Dust, control of, in prevention of respiratory infection, 435
Dyes, as disinfectants, 226
Dysentery, amebic. See *Amebic dysentery.*
bacillary. See *Shigellosis.*
Shiga, 424
Dysentery neurotoxin, 375
toxicity of, 421*t*

Ears, microorganisms of, 180*t*
Eastern equine encephalitis, 596
Eberth, Karl Joseph, *10,* 366

ECBO virus, 404
ECF-A, anaphylactic reaction to, 342*t*
Echoviruses, 404, 411
 immunological types of, *406*
Ecology, 164, 280
 bacterial, 164
Ecosystem, 164
Edema, in inflammation, 294
Edmonston vaccine, 499
ED-pathway, 110
Edwards and Ewing classification, of Enterobacteriaceae, 360, 362*t*
Electricity, bacterial effect of, 138
Electron(s), sterilization by, 211
Electron microscope, 50, 152, *153*
 scanning, 50, *51, 150*
 shadowing of, 81, *81, 82*
Electron transport, in cytochrome system, *114*
Elementary body, 71
Elephantiasis, 610, *610*, 611
EMB agar, 172
Embden-Meyerhof-Parnas pathway, 110, *111*
Encephalitides, viral, mosquito-borne, 596
  tick-borne, 597
Encephalitis, eastern equine, 596
 Venezuelan equine, 597
Encephalomyelitis, equine, characteristics of, 91*t*
Encephalomyocarditis virus (EMC), *406*
Endo agar, 172
Endocarditis, bacterial, 442, *443*
Endo-enzymes, 104
Endolimax nana, 396
Endoplasmic reticulum, *16*
Endospores, bacterial, 55, *57, 58*. See also *Spores, bacterial.*
Endosymbionts, development of, *17*, 19
Endotoxins, 282
 bacterial, toxicity of, 421*t*
Entamebas, 394
Entamoeba coli, 396, 398
Entamoeba histolytica, 394, *397*
 cysts of, 395, *396*
 heat resistance of, 425*t*
 trophozoites of, 395, *396*
Enteric infection
 carriers of, 367
  control of, 382
 diagnosis of, 369
 prevention of, 381
 transmission of, 367
 viral, **404**
Enteric virus, in urine, 411
Enterobacteriaceae, 60*t*, 62, **360**
 biochemical tests for, 364*t*, *365*
 description of, 360
 differential characters of, 363*t*
 subdivisions of, 360, 362*t*
Enterobius vermicularis, egg of, *400*
Enteromonas hominis, 396
Enterotoxin, 283, 439
 organisms producing, 422
 staphylococcal, 421*t*, 422, 436, 439
Enterotube tests, 364*t*
Enteroviruses, **404**
 heat resistance of, 425*t*
 immunologic relationships between, *406*
Entner-Doudoroff pathway, 110
Envelope, viral, 77, 81
Enzymes, **103**
 adaptive or induced, 239
 classification of, 106
Enzymes *(Continued)*
 destruction of, by heat and chemical agents, 197
 differentiation of microorganisms by, 156
 endo-, 104
 exo-, 104
 hydrolysis by, 103, *104*
 permease, 105
 role of, in inflammation, 294
 specificity of, 105, *105*
 tissue-damaging, 283
Eosin, 268
 as bacteriostatic agent, 193
Eosinophils, 268
Epidermophyton, 563*t*
Episome, 245
Epstein-Barr virus, 94
Equine encephalitis, eastern, 596
 Venezuelan, 597
Equine syphilis, 606
Ergot poisoning, 571
E rosettes, 299
Erysipelas, 444
Erythema nodosum, 563
Erythroblastosis fetalis, 276
Erythrocytes, 267, *268*
Erythrogenic toxin, streptococcal, toxicity of, 421*t*
Escherichia coli
 chromosome-labeled, *59*
 enterotoxic, 377
 genetic map of, *117*
 genetic markers for, 117, 118*t*
 genitourinary tract infection and, 378, 378*t*
 nucleoplasm of, *58*
 pathogenic, 376
 sanitary significance of, 377
 size of, 81, *82*
 testing for, in drinking liquids, 171, *172*
 T-even phage of, *84*
Escherichieae, 376
 biochemical differentiation of, 376
 classification of, 361, 362*t*
 pathogenic, 376
 sanitary significance of, 377
Espundia, 609
Ethyl alcohol, as disinfectant, 222
Ethylene oxide, 193, 195
 as disinfectant, 219*t*, 225
 sterilization by, 211, *212*
  mixtures used in, 635*t*
  of hospital equipment and materials, 634*t*
Ethylene oxide–carbon dioxide mixtures, 635*t*
Ethylene oxide–fluorinated hydrocarbon mixtures, 635*t*
Eubacteria, *17*, 18
Eukaryon, 14
Eukaryotes, **14**
 characteristics of, 20*t*
 classification of, 19
Eumycetes, **36**
 characteristics of, 20*t*
Exanthema subitum, 500
 characteristics of, 90*t*
Excretions, of patient with communicable disease, disinfection of, 256, *256, 257*
Exfoliatin, 437
Exo-enzymes, 104
Exoerythrocytic stage, of *Plasmodium*, 600
Exotoxins, 282
 B. anthracis, 556
  toxicity of, 421*t*

Exotoxins *(Continued)*
bacterial, as immunizing agents, 331
diphtheria, 457. See also *Diphtheria toxin.*
Eye, defense mechanisms of, 290
Eyepiece, 149

Face mask, in control of respiratory infection, 435
surgical, *255*
Factor(s)
in blood coagulation, 271, 272*t*
P, 308
Rh. See *Rh factor.*
transfer, 308
V, 468
X, 468
Fahrenheit temperature scale, Celsius equivalents of, 131*t*
Falcon filter unit, *79*
Fat decomposition, 176
Feces
culture of, 617
for pathogenic microorganisms, *159*, 160
in enteric infection, 369, *370*
disinfection of, 219, 641*t*
pathogens in, 624*t*
transmission of disease-causing microorganisms from, *354*, 576
Feline complex virus, 92*t*, 94
Fermentation, 165, 175
carbohydrate, in glucose broth, *107*
chemistry of, *111*
for differentiation of Enterobacteriaceae, 364*t*
pathways of, 110
"stormy," 549
yeast, 40
Fertility factor, 116
Fetal death, syphilis and, 526
Feulgen reaction, 14
Fever blisters, 501
characteristics of, 90*t*
Fibril, axial, of spirochetes, 54, *55*
of bacteriophage, 83
Fibrin, 266
blood clot formation and, 271, *272*
defensive, 294
Fibrinogen, 266
Fibrinolysins, 283
Fibrinolysis, 271
Fibroblast, 294
Fibroblast interferon, 307
Filarial worms, 610
pathogenic, 400*t*
Filariasis, 45, 610, 611
Filter
candle, 191
membrane. See *Membrane filter.*
sterilization by, 213, *213–215*
trickling, 167, *168*
Filter bed, in water purification, 171, *171*
Filtration, sterilization by, 213
Fimbriae, bacterial, 54
Fisher Rh system, 275, 276*t*
Fission, microbial, 114
Flagella, bacterial, 54, *55*, *56*
peritrichous, 54, *55*, *56*
polar, 54
Flatworms, 45
Flea-borne typhus fever, 587, 588*t*
case rates of, *590*
Weil-Felix reaction of, 589*t*
Fleming, Sir Alexander, 238, *238*
Flexner dysentery bacillus, 374
Flies, transmission of enteric disease and, 368, *369*
tsetse, 606
Floc, in precipitation test for syphilis, 527, *528*
in water purification, 170, *171*
Flu. See *Influenza.*
Flukes, 45
blood, 400*t*
Fluorescence microscopy, 152
for demonstration of treponemes, 524, 529
of acid-fast bacilli, 477
Fluorescent antibody staining, 322, *323*
Fluorescent antibody test, for syphilis, 529
indirect, for rickettsial infection, 587
Fluorescent Treponemal Antibody Absorption (FTA-ABS) test, 529, *529*
Fluorine, 218, 221
5-Fluorocytosine, 242
Flury strain, of rabiesvirus, 503
Folic acid, 231
Fomites, *354*, 357
contamination of, 176
infectious, 624*t*
Food
as vector of microbial disease, *418*
sources of, 423
microorganisms in, 174
pathogens borne by, 420*t*, 423
heat resistance of, 424, 425*t*
transmission of disease-causing microorganisms from feces to, *354*
Food handling, contamination during, 423
sanitation in, 418, **424**
supervision of, 425
Food infection, 418, **422**
due to *Salmonella,* 372, 423, 424
prevention of, 424
vs. food poisoning, 424
Food poisoning, **418**
prevention of, 424
staphylococcal, 422, 439
vs. food infection, 424
Forceps, transfer, 250
Foreign, 284
Formaldehyde, as disinfectant, 219*t*, 225
Formula, infant, aseptic preparation of, 375
Forssman antigen, 284
Fragmentation, in septate molds, 42
Francisella tularensis, 578, 579
Freeze-drying, of bacterial culture, *126*, 127
Freezing, in prevention of food infection and poisoning, 425*t*
FTA-ABS test, 529, *529*
Fuchsin, basic, as bacteriostatic agent, 193
Fulvicin, 241
Fungal infections. See *Mycoses.*
Fungi, 19
cutaneous infection and, 560, 561, *562*
diphasic (dimorphic), 563
ectothrix type, 561
endothrix type, 561
eukaryotic, characteristics of, 36
eukaryotic true, classification of, 38*t*
fission. See *Bacteria.*
opportunistic, 568

Fungi *(Continued)*
systemic infection and, **562**
toxins of, 571
true, **36**
classification of, 38*t*
relation of, to bacteria, 36
Fungi imperfecti, 43
Fungicide, 192
Fungicidin, 241
Fungistasis, 230
Fusobacterium, 554
Fusobacterium nucleatum, 554

Gaffyka, *52*
GALT, 299
Gametocytes, malarial, 601, *603, 604*
Gametogony, of malarial parasite, 600
Gamma globulins, 266, 303. See also *Immunoglobulins.*
human, to provide passive immunity, 336
hyperimmune, 336
in measles protection, 499
in pertussis, 472
Gamma rays, 138
sterilization by, 211
Gas gangrene, 548
of abdominal wall, *552*
prevention of, 551
Gastrointestinal tract, infectious agents of, transmission of, 356*t*
Gene(s), 118
isolation of, 119
new combinations of, 119
transfer of, 120, *121*
mechanisms of, in microorganisms, 114
Generations, alteration of, 600
Genetic code, 28, 29*t*
Genetic engineering, 4, 119
techniques used in, 119
vectors for, 120, *121*
Genetic markers, for recombination, 117, 118*t*
Genetic mutation, 27
Genetic recombination, 114
frequency of, 116
in bacteria, *115*, 117
Genitalia, microorganisms of, 180*t*, 181
mucous membranes of, defense mechanism of, 291
Genitourinary tract infections, bacteria in, 378, 378*t*
Genome, duplication of, *30*
Genotype, 120
Genus, definition of, 48
Germ theory, 8
German measles. See *Rubella.*
Germicide, 192
Giant cells, in tuberculosis, 480
Giardia lamblia, 396, 399
Gilchrist's disease, 566
Gingivostomatitis, ulcerative, 472
Glanders, 556
Globulins
for postexposure rabies prophylaxis, 328
gamma. See *Gamma globulins.*
immune serum (ISG), 335
of human blood, 266, 266*t*
rabies immune (RIG), human, 328
Glucose broth, carbohydrate fermentation in, *107*
Glucuronic acid cycle, 110
Glutaraldehyde, as disinfectant, 219*t*, 225
Glycolysis, anaerobic, 110, *111*, 112
Gly-Oxide, 225
Glyoxylate cycle, 110
Golgi body, *16*
Gonorrhea, **536,** 540*t*
case rates of, 531*t*, 536, *538*
diagnosis of, 541, *541*
in female, 536
in male, 537
incidence of, 539*t*
ophthalmia due to, 538
oral lesions of, 539
prophylaxis of, 545
treatment of, 542
untreated, consequences of, 537
Gram, Hans Christian Joachim, 154, *154*
Gram stain, 154
metachromatic, 59
Granules, bacterial, 59, *59*
sulfur, of actinomycosis, *490,* 491
Granulocytes, 268
neutrophilic, 293
Granuloma, 517
Granuloma inguinale, 519, 540*t*
Granuloma venereum, 519
Granulomatoses, venereal, 517
prevention of, 519
Griseofulvin, 241
Gumma, syphilitic, 525
Gut-associated lymphoid tissue, 299

H antigen, 275
Haeckel, Ernst Heinrich, 14
Haemophilus, 468
Haemophilus aegyptius, 469
Haemophilus ducreyi, 543
Haemophilus influenzae, 468
Haffkine's vaccine, 581
Hageman factor, 271
Hair, fungal infection of, 560, 561
Halogens, as disinfectants, **218**
Halophilic organisms, 135
Hands, disinfection of, 640*t*, 641*t*
microorganisms of, 180*t*
Handwashing, in hospital care, 248
Hanging drop, 151, *151*
Hansen's disease, **488**
clinical types of, 489
lepromatous phase of, 489, *490*
tuberculosis and, 490
Hapten, 284
B-cell reaction with, 300, *303*
Hapten inhibition, 285
Haverhill fever, 577
Hay bacillus, 555
HDCV (human diploid cell rabies vaccine), 328, 504, 505*t*
Heaf test, 482
Healing process, 295
Health care, sterilization and disinfection in, **248**
Health personnel
assisting physician, **614**
in diagnosis, 614
in prevention of transmissible diseases, 623
in treatment, 623

Heart disease, rheumatic, 445
Heat
  as disinfectant, 192
  destruction of microorganisms by, **194**
    contact of, 199
      time of, 200
    intensity of, 196
    resistance to, 195, 197
  dry, sterilization by, 197, 202
  moist, coagulation and, 197
    sterilization by. See *Steam sterilization.*
Heavy metals, compounds of, for disinfection, 221
Helminths, **45**
  arthropod-borne diseases of, **610**
  infection with, preventive rules for, 402
  intestinal, 394, **399**
    eggs of, 399, *400*
  pathogenic for humans, 400–401*t*
Hemagglutination, 321
  cold, 321
  in blood grouping, 274, *274*
  indirect (passive), 321
  viral, 321
Hemagglutination-inhibiting antibodies, 322, *322*
Hemagglutinins, 321
Hematocrit, 265
Heme, 468
Hemoglobin, 267
Hemolysins, 283, 315
Hemolysis, 267, 315
  hot-cold, 441
Hemorrhagic fever, 578
Hemostasis, 271
Heparin, anaphylactic reaction to, 341, 342*t*
Hepatitis, **412**
  case rates of, *413*
  epidemic (infectious), 412
    characteristics of, 91*t*
    immunologic and biologic events of, *413*
  homologous serum, 412, 574
    characteristics of, 91*t*
  non-A, non-B, 414
  post-transfusion, 414
    risks of, 414*t*
Hepatitis virus, 412
  A, 412
    heat resistance of, 425*t*
  B, 412, 414, 574
    heat resistance of, 425*t*
  non-A, non-B, 414
Heredity, infective, 116
Herpes, venereal (genital), 540*t*, 544
    characteristics of, 90*t*
    incidence of, 539*t*
Herpes simplex, 501
  characteristics of, 90*t*
Herpes simplex virus, 501
  type 2, 94, 544
Herpes zoster, 501
  characteristics of, 90*t*
Herpesvirus hominis, 501
Herpesviruses, properties of, 93*t*, 94
Heterokaryon, 16
Hexachlorophene, *220*, 224
  for operating room use, 252
Hexose monophosphate pathway, 110
Hexylresorcinol, *220*, 224
Hfr cells, 116
HIGG (hyperimmune gamma globulin), 336
  in measles protection, 499
  in pertussis, 472
Hi-Sine, 221
Histamine, anaphylactic reaction to, 342*t*
Histiocytes, 269, 292, 293
Histoplasma capsulatum, 565, *566, 567*
Histoplasmin, 565
Histoplasmosis, 565
HMP pathway, 110
Hofmann's bacillus, 465
Holmes, Oliver Wendell, 190
Holoenzyme, *105*
Homograft reactions, 345
Homokaryon, 16
Hookworm, egg of, *400*
  pathogenic, 400*t*
Host, 164, 280
Host-parasite relationships, 280
Human beings, microorganisms in and on, **179**, 180*t*, *181*
Human diploid cell rabies vaccine (HDCV), 328, 504, 505*t*
Humoral immunity, **298**
  B-lymphocytes and, 299
Humus, 167
Huxley, Thomas, 23
Hyaluronidase, 282, 283
Hybridoma, lymphocyte, 324, 337
Hydrogen acceptors, 107
Hydrogen ion concentration, 135. See also *pH.*
Hydrogen peroxide, as antiseptic, 224
Hydrolase, 106
Hydrolysis, 103, *104*
  by compressed steam, 206
  of fat, 176
Hydrophobia, **503**. See also *Rabies.*
Hygroscopic, 134
Hymenolepis nana, egg of, *400*
Hyperemia, as sign of inflammation, 294
Hypersensitivity, 340
  drug, 346
  skin tests for, 346
Hyphae, fungal, cutaneous infection and, 560, 561, *562*
  of actinomycetes, 63, *63*
  of molds, 41, *43*
Hypoglobulinemia, 336
Hypospray jet injector, *623*

Ig immunoglobulins. See *Immunoglobulins.*
IgM–IgG switch, 305
Imhoff tanks, 167, *168*
Immobilization test, *Treponema pallidum,* 315 528
Immune-adherence phenomenon, 315
Immune globulin
  human rabies (RIG, HRIG), 328, 504
  Rh (RhIG), 277
  serum, 335
    for protection of infectious diseases, 499
Immune interferon, 307
Immune processes, cell differentiation and localization in, *267*
Immunity
  acquired, 288
  active, **312**
    artificial, **326**
      in diphtheria, 463

Immunity *(Continued)*
cellular, **298**
T-lymphocytes and, 299, 300
classification of, *289*
humoral, **298**
B-lymphocytes and, 299
natural, 288, *289*
passive, 336
to diphtheria, 462
nonspecific, **288**
passive, **334**
allergy in, 335
artificial, 334
in diphtheria, 463
therapeutic sera for, 334
in disease prevention, 335
in pertussis, 472
need for, 334
specific, **312**
to diphtheria, 462
to rabies, 503
to smallpox, 495
Immunization. See also *Vaccination.*
against diphtheria, 332*t*, 464
against influenza, 508
against measles, 499
against salmonellosis, 373
natural, in respiratory infection, 434
rabies, regimens for, 505*t*
schedule for, in children, 332*t*
Immunoelectrophoresis, 320, *320*
Immunogen, 283
Immunoglobulins, 159, **303**
classes of, 303
clonal selection of, 304
human, for passive immunity, 336
in serum, 305
molecular structure of, 303, *304, 305*
of B-cells, 299
properties of, 306*t*
Immunologic specificity, 313
Immunology, to detect viruses, 148
Immunosuppression, 346
Impetigo, streptococcal, 444
IMViC reaction, of *E. coli,* 376
Incineration, sterilization by, 202
Inclusion body, 88
Inclusion conjunctivitis, 71, 517, *518*
Inclusions, 517
Incubation, extrinsic, in malaria, 601
of culture media, 142, *144*
Infant
gonococcal ophthalmia in, prevention of, 538, *541*
parainfluenza of, 509
shigellosis in, prevention of, 375
syphilis of, 526
tetanus of, 552
"Infantile paralysis," 405, 507
characteristics of, 91*t*
Infection, **280**
accidental, 574
antigens and, **280**
arthropod-borne. See *Arthropod-borne infection.*
blood. See *Blood, infection of.*
cytomegalovirus, 501
droplet, 174, *174*
enteric. See *Enteric infection.*
enzootic, 579
Infection *(Continued)*
epizootic, 579
factors in, 281
food. See *Food infection.*
fungal. See *Mycoses.*
gonococcal, diagnosis of, 541, *541*
hematogenous, 574
histotoxic clostridial, 551
metastatic, 355
pneumococcal, 448
puerperal, 445
pyogenic cocci, **434**
resistance to, nonspecific, **288**
specific, **312**
respiratory tract. See *Respiratory tract, infection of.*
staphylococcal, 437
streptococcal, 442
prevention of, 447
subclinical, 281
of diphtheria, 459
specific active immunity from, 312
transmission of agents of, 356–357*t*
tuberculous, 485
classification of, 481
viral. See *Viral infection.*
vs. disease, 281
Infectious diseases, preventive vaccines against, 332*t*
Infective heredity, 116
Inflammation, **293**
types of, 294
Influenza, 507
characteristics of, 91*t*
complications of, 508
immunization against, active, 508
transmission of, 507
treatment of, 509
vaccines for, 329, 508, *508*
virus types of, 507
Infusion broth, 146
Ingestion, phagocytic, 291, *292*
Interferon, 307
Intestine
helminths of, 394, **399**
microorganisms of, 180*t*, 182, 381
protozoa of, **394,** *396*
Ioclide, 198, 221
Iodamoeba bütschlii, 396
Iodine, as bactericide, 219*t*, 221
organic compounds of, 221
Iodophor, 198, 219*t*, 221
Ionizing radiation, cellular effect of, 138
Iosan, 221
Iris diaphragm, 150
ISG (immune serum globulin), 335
for protection of infectious disease, 499
Isoantigens, 284
Isohemagglutination, 273, 273*t*
Isohemagglutinins, 322
Isolation technique, in home, 259
Isoniazid, 241
Isopropyl alcohol, as disinfectant, 223

Jaundice
catarrhal, 412
epidemic, 412
homologous serum, 414, 574
transfusion, 414

Jenner, Edward, *7,* 8, 326
Jungle yellow fever, 594, 595*t*

Kahn reaction, *528*
Kala-azar, 607
Kauffman-White scheme, of *Salmonella* serotypes, 371, 372*t*
Keratin, 560
Keratoconjunctivitis, inflammatory chronic, 516
Ketoconazole, 241
Kidneys, specimens from, 617
Kircher, Athanasius, 149
"Kissing bugs," 607, *607*
Klebs-Löffler bacillus, 456
Koch, Robert, 9, *9, 10,* 190
Koch-Weeks bacillus, 469
Koch's phenomenon, allergy and, 481
Koch's postulates, 10
Koplik's spots, 497
Krebs cycle, *113*

L-forms, of bacteria, 72, *73*
characteristics of, 21*t*
Laboratory study, of microorganisms, **142**
routine procedure for, 160
Laboratory tests, reporting results of, 622
Lancefield groups, 442, 443*t*
Lansing strain, of poliovirus, 407
Laryngotracheitis, 469
Laveran, Alphonse, *10,* 11
Lawn, bacterial, *86*
Lazear, Jesse, 11
Lecithinases, 282
Legionnaires' disease, 510
Legionella pneumophila, 510
Legionellosis, 511
Leishman, William, 607
Leishmania braziliensis, 609
Leishmania braziliensis braziliensis, 609
Leishmania braziliensis peruviana, 609
Leishmania donovani, 607, *608*
Leishmania mexicana, 609
Leishmania tropica, 609
Leishmaniasis, 607
American, 609
cutaneous, 609, *609*
epidemiology of, 608*t*
mucocutaneous, 609, *609*
visceral, 607
Leitz ORTHOPLAN microscope, *323*
Lens, objective, 149
ocular, 149
oil immersion, 150
Leon strain, of poliovirus, 407
Lepromin test, 490
Leprosy. See *Hansen's disease.*
Leptospira, 386, **389,** *390*
Leptospira interrogans, 53, 389, *390*
in rodent disease, 577
Leptospirosis, 386, **390**
diagnosis of, 391
in rodents, 577
rats and, 391
sanitary measures for, 392
Leptothrix buccalis, 473, *473*
Leukocidin, 283
staphylococcal production of, 436
Leukocyte(s), 267, *268*
ameboid movements of, 268, *269*
as phagocytes, 292
polymorphonuclear, 268
phagocytosis by, 269, *270*
proportions of, in blood, 269
Leukocyte count, 269
differential, 270
Leukocyte interferon, 307
Leukocytosis, 271
Leukopenia, 271
Life forms, evolutionary succession of, 5
Lip, chancre of, *525*
Lipoidal substances, antigenic, 284
Lister, Joseph, 8, *10,* 190
Listeria, 465
Listeria monocytogenes, 465
Listeriosis, 465
Liver, amebic abscess of, *396*
Lockjaw. See *Tetanus.*
Löffler's stain, 456
Loop, platinum, inoculating cultures by, *616*
Louse-borne relapsing fever, 583
prevention and treatment of, 584
Louse-borne typhus fever, 586, 588*t*
prevention of, 587
Weil-Felix reaction of 589*t*
Lumbar puncture, 622
Lungs
coccidioidomycosis of, 563, *564*
congenital syphilis in, *524*
defense mechanisms of, 290
plague pneumonia of, 581
Lymph, 264
relationship of, to cells and plasma, *265*
Lymph nodes, 264
thymic-dependent area of, *301*
tuberculous involvement of, 479, *479*
Lymphoblastoid interferon, 307
Lymphocyte(s), 268, 269, 292
antibodies and, **298**
B. See *B-lymphocytes.*
heterogeneity of, *300*
T. See *T-lymphocytes.*
Lymphocyte hybridoma, 324
Lymphogranuloma venerum, 71, 517, 540*t*
Lymphokines, cellular effect of, 344*t,* 345
Lymphoma, Burkitt's, 94
Lyophilization, *126,* 127
Lysins, 322
Lysis, by bacteriophage, 84, *85*
Lysogeny, 84
Lysol, 224
Lysozyme, in tears, 290

M antigen, 444
Macrophage, 269, 292
fixed, 292, 293
of mononuclear phagocyte system, 293
Maduromycosis, 568
Magnetism, bacterial effect of, 138
Malaria, **600**
aestivo-autumnal, 601
control of, 604
diagnosis of, 604
extrinsic incubation of, 601

Malaria *(Continued)*
fever of, periodicity of, 601, *604*
incidence of, *602*
mosquito vector of, 602
parasite of, 600. See also *Plasmodium.*
quartan, 601
tertian, 601
Malta fever, 386, 387
Mammalian cells, in culture, 87, *88*
Mantle, viral, 81
Mantoux test, 482
Mask, in control of respiratory infection, 435
surgical, *255*
Mast cell, 303
IgE-coated, 341
Measles, **497**
characteristics of, 90*t*
complications of, 498
German. See *Rubella.*
immunization for, active, 499
schedule for, 332*t*
incidence of, *329,* 497, *498*
prevention of, 498
transmission of, 497
vaccination against, 329
Medical ward, sterilization and disinfection in, **248**
Mediterranean fever, 387
Medium, culture. See *Culture medium.*
Megacins, 245
Megakaryocytes, 271
Meiosis, 27
Melioidosis, 556
Membrane
cell, *16*, 24
semipermeable, 103
limiting, of virus, 81
mucous, as defense mechanism, 290
nuclear, *16*
vacuolar, *16*
Membrane filter
for bacteriological examination of water, 172, *173*
sterilization by, 213
of fluid, 157, *157*
to identify bacteria, 157, *158*
Memory cells, 299, 312
Meningitis
cerebrospinal fluid culture in, 622
epidemic, 451
meningococcus, 451
diagnosis of, 451, *452*
prevention of, 452
therapy of, 452
Meningococcus, 450
in spinal fluid, *452*
vaccines for, 451
Mercury, bichloride of, 222
organic compounds of, as disinfectants, 219*t*, 222
Merozoites, of *Plasmodium,* 600, *603*
Mesophiles, 130, *132*
Mesosomes, 24
Metabolism, of microorganisms, 103, 127
Metabolite, 231
Metals, heavy, for disinfection, 221
Metaphyta, 22
Metazoa, 22
Metchnikoff, Elie, *10,* 291, *291*
Methyl alcohol, as disinfectant, 223
Methylene blue, as bacteriostatic agent, 193
stain, 154
Metric units, nomenclature for, 24
Microaerophiles, 134, *134*
Microbiology
health professions and, 11
history of, 4
in early disease prevention, 8
industrial, 176
study of, 4
Microbistasis, 231
Microfilaria, of *W. bancrofti, 610,* 611
Microorganisms. See also *Bacteria; Virus.*
acidity and, 135
aerobic. See *Aerobes.*
aggressiveness of, 281
alkalinity and, 135
anaerobic. See *Anaerobes.*
animal inoculation of, 157, *158*
anoxybiontic, 134
antibiotic-resistant, 234
as benefactors, 164
autotrophic, 128
capneic, 134
characteristics of, 20–21*t*
chemotrophic, 128
classification of, 14, **19**
coenzymes used by, 108*t*
colonies of. See *Colony.*
culture of, pure. See *Culture, pure.*
rapid methods for, 156
destruction of, 194
by heat and chemical agents, **194**
numbers of and, 196
properties of agent used for, 196
resistance to, 195, 197
temperatures required for, 204, *204*
differential diagnosis of, **156**
drug-fast, 234
ecological system and, **164**
electricity and, 138
entry portals of, 354, 356–357*t*
environment of, **126**
enzyme study of, 156
exit portals of, 356, 356–357*t*
foods of, 127
habitat of, 32
halophilic, 135
heterotrophic, 128
in air, 173, 179
in and on human beings, **179,** 180*t*, *181*
in blood, 182
in blood banks, 183
in food, 174
in soil, **548**
industrial use of, 176
infecting, dose of, 281
inhibition of, 194, 230
laboratory study of, **142**
lyophilization of, *126,* 127
magnetism and, 138
membrane filter identification of, 157, *157*, *158*
metabolism of, 103, 127
microaerophilic, 134, *134*
microscopy of, **148**
microwaves and, 138
mode of life of, 32
moisture needs of, 126
nutritional types of, 128

Microorganisms *(Continued)*
osmotic pressure and, 134
oxybiontic, 134
oxygen needs of, 132
parasites, 129, 280
path of, in body, 355
pathogenic. See *Pathogens.*
pathogenicity of, 281
pH relationship of, 132, 136, *137*
phototrophic, 128
radiation of, 137
removal of, 194
response of, to antibiotics, 237*t*
saprophytic, 129, 164
saprozoic, 129, 164
sensitivity testing of, 242, *243*, *244*
serologic tests for, 158
sound waves and, 138
staining of, **154**
supersonic vibrations and, 139
temperature requirements of, 130, *132*
toxigenicity of, 282
toxins of, 282
transmission of, from feces to food, *354*
unicellar, 19, **22**
vaccination with, dead, 330
living, attenuated, or harmless, 326
virulence of, 281
variations in, 122, 283
wastes toxic to, 130
Microphages, 293
Microscopes, **148**
compound, 49, 149, *149*
darkfield, condensers for, *524*
for demonstration of treponemes, 523
fluorescence, 152
for acid-fast bacilli, 477
for demonstration of treponemes, 524, 529
early observations through, 5
electron, 50, 152, *153*
scanning, 50, *51*, *150*
shadowing of, 81, *81*, *82*
hanging drop preparation for, 151, *151*
in diagnosis of tuberculosis, 483
Leitz ORTHOPLAN, *323*
optical, 149, *149*
phase contrast, 152
for demonstration of treponemes, 524
wet mount preparation for, 151
Microscopic method, direct, 429
Microsporum, 563*t*
Microwaves, bacterial effect of, 138
Milk, **427**
as vector of diphtheria, 460
bacteria in, 427
bacteriologic examination of, 429
contaminated, enteric infection and, 368
cow's, pathogens in, sources of, 429*t*, 430
disease transmitted by, 430
drawing of, "closed" methods of, 427, *428*
household care of, 430
pasteurization of, 430
supplies of, public health supervision of, 429
Milking stall, automatic, *428*
Millipore filter, 213
Milorganite, 167
Mite typhus, 592
Weil-Felix reactions of, 589*t*
Mitochondrion, *16*
Mitomycin A, B, and C, 241
Mitosis, 23, 27
Moisture, coagulation and, 197
requirements of microorganisms for, 126
Molds, 36, **40**
characteristics of, 20*t*
coenocytic, 41, *42*
common species of, 45
growth of, 44, *44*
reproduction of, 42
asexual, 42, *42*, *43*
sexual, 43, *44*
septate, 42
protoplasmic flow through, *41*, 42
Molecular filter, 213
Mollicutes, 21, 72
Molluscum contagiosum, 496, 540*t*
Monoclonal antibodies, 324
Monocytes, 269
phagocytic, 292
Mononuclear phagocyte system (MPS), 182, 269
macrophages of, 293
Mononucleosis, characteristics of, 90*t*
Morax-Axenfeld bacillus, 469
Moraxella, 469
Moraxella kingii, 469
Moraxella lacunata, 469
Mosquito, as vector, of malaria, 602
of viral encephalitis, 596
developmentul stages of, 602, *605*
Mouth, microorganisms of, 180*t*, 181
Moxam, 242
Mucopeptides, in cell wall, 24
Mucopolysaccharides, in cell wall, 24
Mucor, 45
Mucous membranes, as defense mechanism, 290
Mumps, 502
characteristics of, 90*t*
immunization schedule for, 332*t*
prevention of, 502
Murine complex virus, 92*t*
Mutagenic agents, cancer and, 118
Mutant, 118
Mutation, 118
Muton, 27, 118
Mutualism, 280
Mycelium, of actinomycetes, 63, *63*
of molds, 41
Mycobacteria, 476
differentiation of, 478, 484*t*
heat resistance of, 425*t*
properties of, 484*t*
T-strains, 520
unclassified (anonymous), 476
Mycobacteriaceae, 64
Mycobacterium leprae, 488, 489
Mycobacterium tuberculosis, 476
Mycoplasma, 21, **72,** 633
characteristics of, 21*t*, 72*t*
in venereal disease, 520
pneumonia caused by, 510
Mycoplasma hominis, 72, 520
Mycoplasma pneumoniae, 72, 510
Mycoses, **560**
cutaneous, 560
treatment of, 561
opportunistic, 568
systemic, **562**

Mycostatin, 241
Mycotoxins, 571
in food poisoning, 422
Myelocyte, 268
Myocarditis, due to diphtheria exotoxin, 458
Myringomycosis, 561

Nails, fungal infection of, 560, 561
Nasopharynx, cultures from, 615, *617*
Needles, disinfection of, 249
in community health care, 259
presterile, handling of, 249
Negative staining, 155, *155*
of treponemes, 523
Negri bodies, 503, *503*
Neisseria
infections due to, antibiotics and, 542
morphology of, 449, *450*
physiologic properties of, 449
respiratory, **449**
vaginal infections and, 542
Neisseria gonorrhoeae, 449, *450*
Neisseria meningitidis, 449, 450, *450*
Neisseriaceae, taxonomy and nomenclature of, 542*t*
Nematodes, 45
eggs of, *400*
pathogenic for humans, 400*t*
Neo-Silvol, 222
Neufeld quellung reaction, 448
Neurotoxin, dysentery, 375
toxicity of, 421*t*
Neutrons, 138
sterilization by, 211
Neutrophils, 293
polymorphonuclear, 268
Nichols treponeme, 523
Nicotinamide adenine dinucleotide, *112*
added to *Haemophilus* mejia, 468
Nicotinamide mononucleotide, *112*
Nightingale, Florence, 11
NITA virus, 404
Nitrification, 178, *178*
Nitrogen cycle, *178*
Nitrogen fixation, 176, *178*
nonsymbiotic, 177
symbiotic, 177, *177*
Nitrosification, *178*
Nocard, Edmund Etienne, *10*
Nocardiaceae, 64
Nocardiosis, 64
Nonself, 284
Nonspecific resistance, to infection, **288**
Normoblasts, 267
North American blastomycosis, 566, *567*
Nose, discharge from, pathogens in, 624*t*
microorganisms of, 179, 180*t*
Noxalactam, 242
Nuclear membrane, *16*
Nucleic acids, antigenic, 284
of viruses, 77
Nucleocapsid, *77,* 81
Nucleoid, 14, 56, *58*
replication of, 56, *59*
viral, *77,* 81
Nucleolus, *16,* 29
Nucleotides, 27
DNA, 25, *26*
RNA, 25
linear sequence of, 27, *28*
Nucleus(i), *16,* 24
bacterial, 14, 56, *58*
replication of, 56, *59*
droplet, 174
Null cells, 302
Nutrient agar, 145
bacteriostatic agents with, 193
Nutrient broth, 145
Nutrients, bacterial, **127**
Nutrition, of microorganisms, 128
osmotrophic, 103
phagotrophic, 35, 394
Nuttall, George H. F., *10*
Nyctotherus ovalis, 51
Nystatin, 241

O antigen, 371
Objective lens, 149
oil immersion, 150
Ocular lens, 149
Oidia, 42
Oncornaviruses, properties of, 92*t*
Operating room, sterilization and disinfection in, **252,** *253–255*
Ophthalmia, 538
gonococcal, 538
prevention of, 545
Opsonin, 293, 322
Opsonic action, of antibodies, 293
Oral tract, infectious agents of, transmission of, 355, *355,* 356*t*
Organelles, 15, *16,* 23
Organic material, as obstacle in destruction of microorganisms, 199
Organization, in healing process, 295
Ornithodorus, as vector, in relapsing fever, 583
Ornithosis, 519
Orthophenylphenol, 224
Osmosis, 103, 134
Osmotic pressure, growth of microorganisms and, 134
Osteomyelitis, 437
O-Syl, 224
Otomycosis, 561
Otoscope, disinfection of, 642*t*
Ovens, for sterilization, 202, *202*
Oxidase, 106
Oxidase reaction, of *Neisseria,* 449, *450*
Oxidation, biological, 106
mechanisms of, *133*
Oxidizing agents, as antiseptics, 224
Oxidoreductase, 106
Oxybiontic organisms, 134
Oxyfume, 212, 635*t*
Oxygen, requirements of microorganisms for, 132
Oxygenase, 106

"P" factor, 308
Pai's medium, 456
Pain, in inflammation, 294
Panencephalitis, subacute sclerosing, 498
Panstrongylus megistus, 607, *607*
Papovaviruses, properties of, 93*t*
Para-aminobenzoic acid, chemical structure of, 231, *232*

Paracoccidioides brasiliensis, 568
Paracoccidioidomycosis, 568
Parainfluenza, characteristics of, 91*t*
of infants and children, 509
Paralysis, in syphilis, 526
"infantile," 405, 507
characteristics of, 91*t*
Parapertussis, 470
Parasite, 129, 280
host relationship with, 280
Parasitemia, 601
Parasitism, 280
paratroph, 129
paratyphoid fever, 372
Paresis, in syphilis, 526
Parotitis, contagious. See *Mumps.*
Parrot fever, 511, 519
Pasteur, Louis, 8, *8,* 190
rabies preventive immunization of, 328
swan-neck flasks of, *9*
Pasteurella multocida, 578, *578*
Pasteurization, 191, 430
Q fever and, 592
Patch test, 483
Pathogenicity, 281
Pathogens, 129, 164
bacterial, of soil, **548**
blood-borne, 357*t,* 624*t*
enteric, path of, in body, 355
food-borne, 420*t,* 423
heat resistance of, 424, 425*t*
fungal, 568
in body fluids and discharges, 624*t*
in milk, 429*t,* 430
respiratory tract, **436**
path of, in body, 355, *355,* 356*t*
spirochetes, 522
Pediculosis pubis, 540*t*
Pediculus humanus, as vector, of relapsing fever, 583
of trench fever, 593
of typhus fever, 586
Pellicle, 144
Penicillinase, 238, *239*
Penicillins, 237, *239*
in culture medium, for selective cultivation, *143,* 471
in venereal diseases, 530
semisynthetic, effective against penicillinase, 239, 240*t*
Penicilliosis, 569
Penicillium, 43, 45
Pennoxide, 635*t*
Pentose phosphate pathway, 110
Peplomers, 82
Peptidoglycans, in cell wall, 24, *25*
Peptococcus, 554
Peptone, 146
Peptostreptococcus, 554
Peptostreptococcus anaerobius, 554
Perianal folds, microorganisms of, 180*t*
Peritoneal cavity, fluid collection from, 621
Peroxides, as antiseptics, 224
Pertussis, 471
bacillus of, 470, *470*
immunization schedule for, 332*t*
vaccine for, 472
pH, 136
determinations of, 136, *136*
effectiveness of chemical agent and, 200
pH *(Continued)*
growth of microorganisms and, 132, 136, *137*
scale of, 136, *137*
pH meter, *136*
Phage. See *Bacteriophage.*
Phagocyte(s), fixed, 269
leukocytes as, 292
mononuclear, 182, 269
Phagocytosis, as second line of defense, 291, *292*
by polymorphonuclear leukocytes, 269, *270*
chemotherapy and, 231
Phagolysosome, 291
Phase contrast microscope, 152
for demonstration of treponemes, 524
Phenocopies, 120
Phenol, *220,* 223
Phenol coefficient, 227
Phenolic disinfectants, 219*t,* 223
Phenotype, 120
fluctuation of, 120
Phosphatase test, of pasteurization, 430
Photobacteria, 19
Photolithotrophs, 128
Photoorganotrophs, 128
Photosynthesis, bacterial, 109
evolution of, major developments in, *15*
green plants and, 108
Phototrophs, 128
Picornaviruses, 404
Pigments, bacterial, 59
respiratory, 267
Pili, bacterial, 54, *57*
Pine oil, as disinfectant, 224
"Pink-eye," 469
prevention and treatment of, 470
Pinocytic vesicle, *16*
Pinta, 540*t,* 522, 534
Pinworm, egg of, *400*
pathogenic, 400*t*
Plague
bubonic, 579
campestral, 579
case rates of, 581, *582*
epidemiology of, *580*
pneumonic, 581
prevention of, 581
sylvatic, 579
vaccine for, 581
viral, 148
Plaque, bacteriophage formation of, *86,* 87
Plasma, 265
protein components of, 266, 266*t*
Plasma cell, 299, 302
Plasmid
bacterial, 58
conjugative, 116
F, 116
in staphylococci, 438
R, 116
drug resistance and, 241
Plasmodium, 600
life cycle of, 600, *603*
in human host, 600, *603*
in mosquito, 601, *603*
mosquito vector of, 602
species of, 601
Plasmodium falciparum, 601
Plasmodium malariae, 601

Plasmodium ovale, 602
Plasmodium vivax, 601
  developmental stages of, *604*
Plasmolysis, 134
Plasmoptysis, 134
"Plate count," of bacteria in milk, 429
Platelets, *268,* 271
Platinum loop, inoculating cultures by, *616*
Pleural cavity, fluid collection from, 621
Pleuropneumonia-like organisms, 72. See also *Mycoplasma.*
Pneumococci, 447
  capsule of, 447, 448
  identification of, 448
  infections by, 448
  serologic type of, 448
Pneumonia
  atypical, 510
  lobar, 448
  mycoplasmal, 510
  plague, 581
Pneumonitis, tuberculous, *479*
Poikilocytes, 267
Poisoning, food. See *Food poisoning.*
  ptomaine, 423
Poliomyelitis, **405**
  anterior, 507
    spinal cord lesions of, *406*
  characteristics of, 91*t*
  immunization schedule for, 332*t*
  paralytic, 405
    case rates of, *409*
    prevention of, Salk vaccine and, 408
  pathogenesis of, 407, *407*
  prevention of, 410
  surgery and, 410
  transmission of, 410
  vaccines for, 408
  virus of. See *Poliovirus.*
Poliovirus, *82,* 407
  immunological types of, *406,* 407
  tissue culture diagnosis of, 410
  types of, 407
Polymerase, 106
Polymorphonuclear leukocytes, 268
  phagocytosis by, 269, *270*
Polymorphonuclear neutrophils, 268
Polymyxins, 241
Polysaccharide, capsular, 54, 282, 448
Polypeptides, antigenic, 284
Potassium permanganate, as antiseptic, 224
Potassium tellurite, for selective inhibition, 457
Povidone, 221
Poxvirus, *81*
  properties of, 93*t*
PPLO, 72. See also *Mycoplasma.*
Precipitate, 318, *319*
Precipitation, 318
Precipitin, 318, 322
Precipitin reactions, 318
Precipitin test, *319*
  complete, 319*t*
  for anthrax, 556
  in gel, 319, *320*
  in syphilis, 527, *528*
Predation, 280
Pregnancy, polio vaccination in, 408
Procaryotae, 14, 102
  classifcation and nomenclature of, 19, 39*t,* **627**
Proctitis, with urogenital gonorrhea, 537
Proerythroblasts, 267
Prokaryon, 14
Prokaryotes, **14**
  characteristics of, 20–21*t*
  classification of, 19
  types of, *17,* 18
Properdin, 308
  in activation pathway of complement, *318*
Properdin system, 308
Prophage, 84
Propionibacterium acnes, 465
Prostaglandins, anaphylactic reaction to, 342*t*
Prosthecae, bacterial, 55
Protargol, 222
Proteases, 283
Protein
  allergic reaction to, in therapeutic sera, 335
  antigenic, 284
  in plasma, 266, 266*t*
    of Group A streptococci, 444
Protein conjugate, antigenic, 284
  T-cell reaction with, 300, *303*
Protein decomposition, 175
Protein synthesis, *30*
Proteus, classification of, 361, 362*t*
  genitourinary tract infection and, 378, 378*t*
Proteus vulgaris, 51
  strains of, for Weil-Felix reaction, 587, 589*t*
Protoplasm, 22
  flow of, through septate mold, *41,* 42
Protoplast, 72
Protozoa, **34**
  characteristics of, 20*t*
  classification of, 36, 37*t*
  cyst stage of, 36, *37, 396*
  diseases caused by, **600**
  intestinal, **394,** *396*
  life cycles of, 35
  multiplication of, 34
  nutrition of, 35
  structure of, 34
  trophozoite stage of, 35, *36, 396*
Pseudomembrane, in diphtheria, 457
  in thrush, 570
Pseudomonadaceae, 60*t,* 61
Pseudomonas aeruginosa, genitourinary tract infection and, 378, 378*t*
Pseudomonas pseudomallei, 556
Pseudopodia, 394
Pseudotuberculosis, in laboratory animals, 577
Psittacosis, 511, 519
  transmission and control of, 519
Psychrophiles, 130, *132*
Ptomaine poisoning, 423
Puerperal infections, 445
Pure culture. See *Culture, pure.*
Purified protein derivative, 482
Pus, 269, *270*
  formation of, in inflammation, 294
  pathogens in, 624*t*
Putrefaction, 165, 175
Pyocins, 245
Pyrogens, 601

Q fever, 68, 588*t,* 592
Quartan fever, 601

Quaternary ammonium disinfectants, 219*t*, 225
  formula for, *225*
Quellung reaction, 448

R plasmid, 116
  drug resistance and, 241
Rabbit fever, 579
Rabbit syphilis, 533
Rabies, **503**
  characteristics of, 91*t*
  control of, 504
  diagnosis of, 503
  immunity to, 503
  immunization regimens for, 505*t*
  in dogs and cats, 506
  Pasteur's preventive immunization against, 328
  prevention of, 504
  vaccines against, 328, 504, 505*t*
Rabies immune globulin, human (RIG), 328, 504
Rabiesvirus, *81*, 503
Rad, 211
Radiation, electromagnetic, sterilization with, 211
  of microorganisms, 137
Radium particles, 138
Rat-bite fever, 577
Rats, leptospirosis and, 391
R/b enteric differential system, 364*t*, *365*
Reagin, in syphilis, 525
Recon, 27, 118
Reductase test, of milk quality, 430
Reduction, 107
Reed, Walter, 11
Refractory period, in immediate-type allergy, 343
Refrigeration, in preventing food infection and poisoning, 425
Reiter treponeme, 523
  in complement fixation tests for syphilis, 529
Reiter's syndrome, 540*t*
Relapsing fever, 581
  louse-borne, 583
  prevention and treatment of, 584
  tick-borne, 583
Reoviruses, 404, 411
Replicon, 118
Reproduction, 114
Resistance. See also *Immunity*.
  acquired, 288
  drug. See *Drug resistance*.
  species or racial, 288
  to infection, nonspecific, **288**
    specific, **312**
Respiration, evolution of, *15*
Respiratory pigments, 267
Respiratory syncytial virus(RSV), 509
Respiratory tract
  infection of, **434**
    chlamydial, 511
    control of, 434
    fungous, 491
    mycoplasmal, 510
    transmission of, 355, *355*, 356*t*, 434
    viral, **494**
      transmission of, 510
  Neisseria of, **449**
  pathogens of, **436**
Reticuloendothelial cells, 269
Reticuloendothelial system (RES), 182, 269
  macrophages of, 293
Retortamonas intestinalis, 396
Reverse precautions, in sterilization, 251
Rh agglutinogen, 275
Rh factor, 275, 276*t*
  Fisher system of, 275, 276*t*
  nomenclature of, 275
  transfusion and, 276
  Wiener system of, 275, 276*t*
Rh immune globulin (RhIG), 277
Rheumatic fever, 445
Rhizobium, nitrogen-fixing by, 177, *177*
Rhizoids, *42*
Rhizopus, 45
  reproduction of, asexual, *42*
    sexual, 43, *44*
  structures of, *42*
Rhodospirillales, 19
Ribonucleic acid (RNA), **25**
  heterogeneous nuclear, 29
  messenger, 16, 27
  nuclear, 27
  nucleotide units in, 25
  oncogenic tumors containing, *78*, 92*t*, 94
  ribosomal, 16, 27
  soluble, 27
  transfer, 16, 27
Ribosome, *16*
  classes of, 18*t*
  eukaryotic, 16, *16*
  prokaryotic, 16
  proteins isolated from, *18*
Ricketts, Howard Taylor, 68
Rickettsia canada, 69
Rickettsia prowazekii, 70, 586
Rickettsia rickettsii, 70, 589
Rickettsia tsutsugamushi, 592
Rickettsia typhi, 70
Rickettsias, 19, **68**, 632
  characteristics of, 21*t*, 72*t*
  habitat of, 71
  infection with, arthropod-borne, **586**, 588*t*
  propagation of, 69, *70*
Rickettsioses, **586**, 588*t*
RIG (rabies immune globulin, human), 328, 504
"Ring test," for brucellosis, 389
Ringworm, 560
RNA. See *Ribonucleic acid*.
Rochalimaea quintana, 68, 593
Rocky Mountain spotted fever, 588*t*, 589
  case rates of, *591*
  Weil-Felix reactions of, 589*t*
Rodenticides, 580
Rods. See also *Bacillus(i)*.
  endospore-forming, 60*t*, 63, 631
  gram-negative, 52
    aerobic, 60*t*, 61, 629
    facultatively anaerobic, 60*t*, 62, 629
    obligate anaerobic, 62
  gram-positive, 50
Roseola infantum, 500
  characteristics of, 90*t*
Ross, Sir Ronald, 11
Roundworms, 45
  egg of, *400*
  pathogenic, 400*t*
Roux, Emile, *10*
RPCF test, 529

Rubella, 500
  characteristics of, 90*t*
  immunization schedule for, 332*t*
  vaccination against, 329
Rubeola. See *Measles.*
Rubratoxins, 422
Runyon, E. H., 476
Runyon groups, 484*t*
Russell bodies, 302

Sabin oral polio vaccine (OPV), 329, 408
Sabouraud's agar, 38
Salivary gland viruses, 501
Salk, Jonas, 408
Salk vaccine (IPV), 408
  use and effectiveness of, 408, *409*
Salmonella, classification of, 361, 362*t*
  food infection due to, 372, 423, 424
  food "poisoning" due to, 422, 424
  heat resistance of, 425*t*
  serotypes of, Kauffmann-White scheme of, 371, 372*t*
Salmonella paratyphi, 361
Salmonella typhi, 361, 366
  food infection due to, 423
Salmonella typhimurium, food infection due to, 423, 424
Salmonellosis, 360, **361**
  clinical management of, 373
  diagnosis of, 369
  food-borne, 372
  immunization against, 373
  laboratory diagnosis of, *159*, 160, *160, 161*
  transmission of, 367
San Joaquin Valley fever, 562
Sandflies, as vector of leishmaniasis, 609, 610
Sanitation, in food handling, 418, **424**
Sanitization, definition of, 193
Saprophytes, 129, 164
Saprozoa, 129, 164
Sarcina, *52*
Scabies, 540*t*
Scalp, microorganisms of, 180*t*
Scar formation, 295
Scarification, in tuberculin test, 482
Scarlet fever, 446
Schaudinn, Fritz Richard, 522
Schick test, 464
Schizogony, of malarial parasite, 600
Schizomycetes, 19
  characteristics of, 20*t*
  relation of, to true fungi, 36
Schizont, 600, *604*
Schmitz bacillus, 374
Scotobacteria, 19
Scrub typhus, 588*t*, 592
  Weil-Felix reactions of, 589*t*
Secretions, respiratory, of communicable disease, disposal of, *255*, 256
Sediment, of bacterial colony, 144
Sedimentation, in water purification, 170, *171*
Seed layer, 244
Seitz filter, 213
Self, 284
Seller's stain, 503
Semmelweis, Ignaz Philipp, 8, *8*, 190
Semple vaccine, 328
Sensitivity testing, methods of, 242, *243, 244*
Septa, of molds, 41
    protoplasmic flow through, *41*, 42
Septic sore throat, 444, 446
Septic tank, *166*
Septicemia, 280
  blood culture in, *618*, 619
  meningococcal, 451
  of typhoid fever, 366
Serologic tests, 158
  blood for, 620
Serotonin, anaphylactic reaction to, 342*t*
Serum, 159, 266, *266*
  immunoelectrophoresis of, *320*
  immunoglobulins in, 305
  testing of, in infection, 159
  therapeutic, preparation of, 334
Serum sickness, 343
Sewage, aerobic treatment of, *168*, 169
  anaerobic treatment of, 169
  disposal systems for, 167, *168*
    aerating, *167, 168*
    home, *166, 167*
  pressure cooking of, with Wet Air Oxidation Unit, 169, *170*
  purification of, **165**
    municipal, 167, *168*
Sexually transmissible diseases, 540*t*
Sheath antigen, 54
Shiga neurotoxin, 375
Shiga's bacillus, 424
Shigella
  biochemical differentiation of, 375*t*
  classification of, 361, 362*t*
  heat resistance of, 425*t*
  toxin production by, 374
  types of, 374
Shigella dysenteriae, food infection due to, 424
Shigellosis, 360, **373**
  diagnosis of, 374
  prevention of, 375
    in infants, 375
  transmission of, 374
Shingles, 501
  characteristics of, 90*t*
Shock, anaphylactic, 343
Silver, organic compounds of, as disinfectants, 222
Silver impregnation method, 523, *524*
Silver nitrate, as disinfectant, 222
  in prevention of gonococcal ophthalmia, 538 *541*
Silvol, 222
Sintered glass filter, 213, *213*
Skin
  as defense mechanism, 289
  burns of, 289
  cuts in, transmission of infectious agents from soil into, 357*t*
  fungal infection of, 560, 561, *562*
  microorganisms of, 179, 180*t*
  sensitivity of, following mumps, 502
  synthetic, 290
Skin tests, for hypersensitivity, 346
Sleeping sickness, African, 606
  characteristics of, 91*t*
Sludge, 167
  activated, *168*, 169
  oxidation and dewatering system for, 169, *170*

Smallpox, **494**
case rates of, 495, *496*
characteristics of, 90*t*
immunity to, 495
immunization schedule for, 332*t*
vaccination against, 326, 495
compulsory, 496
protection given by, 495, *497*
reaction to, *328*, 495
Smith, Theobald, 11
Sneezing, spread of microorganisms by, *174*
Soap, disinfectant, 224, 226
Sodium desoxycholate, as bacteriostatic agent, 193
Sodium fluoride, 218, 221
Sodium perborate, 225
Sodium hypochlorite, for disinfection, 220
Sodoku, 577
Soil, bacterial pathogens of, **548**
infectious agents in, transmission of, 357*t*
Soluble specific substance (SSS), 54, 282, 448
Solute, 134
Sonic disintegrator, *139*
Sonne dysentery bacillus, 374
Sound waves, bacterial effect of, 138
South American blastomycosis, 568
Species, definition of, 48
Specific resistance, to infection, **312**
Specificity, immunologic, 313
of toxins, 283
Specimens, collection of, **614**
labeling of, 614
transmission of, to laboratory, 614
Spherule, in coccidioidomycosis, 563, *564*
Spinal cord lesions, of anterior poliomyelitis, *406*
Spinal fluid, collection of, 622
meningococci in, *452*
Spirilla, 50, *53*
Spirillum minor, 577
Spirochaeta plicatilis, 53
Spirochaetales, 53, 522
subdivisions of, 522
Spirochetes, 53, *53*, 60*t*, 61
axial fibrils of, 54, *55*
classification of, 629
oral, 472, *473*
pathogenic, 522
Reiter, 529
relapsing fever and, 581
Spleen, thymic-dependent area of, *301*
Sporangiophores, *42*
Sporangiospores, 42, *42*
Sporangium, 42, *42*
bacterial, 56
Spores
bacterial, 55, *57, 58*
boiling of, effectiveness of, 258
dust and, 55
resistance of, to heat and chemical agents, 195, 197
sterilization and, 55
temperatures required to destroy, 204, *204*
Coccidioides, 563, *565*
fungal, 561
of molds, 41
Sporogony, of malarial parasite, 600
Sporothrix schenckii, 568, *568*
Sporotrichosis, 568
Sporozoites, of *Plasmodium*, 600, *603*
Spotted fever group, 588*t*. See also *Rocky Mountain spotted fever.*
Sputum, collection of, 616
disinfection of, 642*t*
pathogens in, 624*t*
SRS–A, anaphylactic reaction to, 342*t*
SSS (soluble specific substance), 54, 282, 448
Stain
acid-fast, 476
counter, 154
differential, 154
Gram, 154
Löffler's, 456
methylene blue, 154
Seller's, 503
Wayson's, 557
Ziehl-Neelsen, 476
Staining, **154**
bipolar, 578, *578*
fluorescent antibody, 322, *323*
negative, 155, *155*
of treponemes, 523
Staphylococci, 50, *52*, **436**
anaerobic, 447
coagulase-producing, 422, 436, 437
drug-fast, 438
enterotoxigenic, 421*t*, 422, 436, 439
Group A, serologic types of, 444
heat resistance of, 425*t*
infections with, 437
isolation of, 437
phage types of, 438
plasmids in, 438
toxins of, 283, 421*t*, 422, 436
Staphylococcus aureus, 51, 436, 437
drug-resistant, 623
food poisoning due to, 422, 439
Staphylococcus epidermidis, 436, 437
Staphylococcus saprophyticus, 436, 437
Steam sterilization, 196, **204**
compressed, 204, *205–207*
live, 204
placement of articles for, *208, 209*
Stem cells, blood cell origin from, 266
bone marrow, lymphocyte differentiation from, *298*, 299
Sterifil aseptic filtration system, 213, *214–215*
Sterilization, 190, **194**, **202**
agent-organism interaction in, factors influencing, 199
bacterial spores and, 55
by ethylene oxide, 211, *212*
mixtures used in, 635*t*
of hospital equipment and materials, 634*t*
by filtration, 213
characteristics of microorganisms and, 195
definition of, 193
dry heat, 197, 202
fractional, 204
historical aspects of, 190
in communicable disease unit, 255, *255–257*
in health care, **248**
in medical/surgical wards, **248**
in operating room, **252**, *253–255*
moist heat. See *Steam sterilization.*
of apparatus and supplies, 636–638*t*
packaging materials for, 639*t*
properties of agent used for, 196
with chemicals, 211

Sterilization *(Continued)*
with radiation, 138, 211
without heat, **210**
Sterilization charts, **634**
Sterilizer, steam-jacketed, *205*
Steroxide, 635*t*
Stethoscope, disinfection of, 642*t*
Stimuli, antigenic, primary, 333, *333*, 463
secondary, 333, *333*, 463
slow response to, 334
Stolon, of molds, *42*
Stomach, microorganisms of, 182
Stool, See *Feces*.
Strain, definition of, 48
Streptobacillus moniliformis, 577
Streptococci, 50, *52*, *54*, **439**
alpha-type hemolytic, 441, *441*, 443*t*
infections by, 442
anaerobic, 447
beta-type hemolytic, 441, *441*, 443*t*
infections by, 442
serologic groups of, 442
blood-agar types of, 439, *441*
capsules of, 442
classification of, 443*t*
double-zone beta-type, 441
fecal group, 443*t*
gamma-type nonhemolytic, 441
infections with, 442
prevention of, 447
lactic acid group, 443*t*
pyogenic, 441, 443*t*
viridans, 441, 443*t*
Streptococcus faecalis, 442
endocarditis due to, *443*
Streptococcus mitis, *441*, 442
Streptococcus pneumoniae, 447
vaccine against, 449
Streptococcus pyogenes, *51*, *441*, 442
Streptococcus salivarius, 442
Streptomyces, conidial filaments of, 64, *65*
Streptomycetaceae, 64
Streptomycin, 240
Stricture, urethral, 537
Substage condenser, 150
Substrates, 105, *105*
Sulfanilamide, 234
chemical structure of, compared with PABA, 231, *232*
Sulfonamides, 234
Sulfur granules, of actinomycosis, *490*, 491
Sunlight, bactericidal effect of, 137
Superinfection, 242
Supersonic vibrations, effect on microorganisms, 139
Surface tension, of disinfectant, 197
Surface-tension reducers, 198
Surfactant, 198
Surgery, polio and, 410
sterilized equipment for, 252, *254*
Surgical dressings, cart for, 250, *251*
disinfection of, 641*t*
handling procedure for, 249
Surgical face masks, *255*
Surgical instruments, autoclave cleaning of, 209, *210*
disinfection of, 641*t*
reserve table for, *254*
Surgical ward, sterilization and disinfection in, **248**
Svedberg unit, 16
Swabs, 614
Swan-neck flasks, Pasteur's, *9*
Swelling, in inflammation, 294
Swine flu, 507, *508*
Symbiosis, 280
Syphilis, **522**, 540*t*
as family disease, 532
case rates of, 530, *531*, 531*t*, *538*
congenital, 526
elimination of, 545
diagnosis of, 527
equine, 606
fetal death and, 526
incidence of, 539*t*
late, 525
latent, 526
oral, 473
penicillin therapy of, 530
prevention of, 532, 545
primary, 524
rabbit, 533
secondary, 525
tertiary, 525
tests for, 527
biologic false-positive reactions to, 527
clinical progress and, 529
specific, 528
treatment of, 530
untreated, course of, 524, *525*
Syringes, disinfection of, 249
in community health care, 259
presterile, handling of, 249

TAB vaccine, 373
Tabes, in syphilis, 526
Tableware, bacteriologic examination of, 427
disinfection of, 640*t*
Taenia sp., egg of, *400*
Taeniasis, epidemiology of, 401
Tapeworm, eggs of, *400*
infection with, epidemiology of, *401*
pathogenic, for humans, 400*t*
TCA cycle, *113*
TCBS agar, for vibrio, 379
T-cells, 300. See also *T-lymphocytes*.
Tears, bactericidal action of, 290
Teichoic acids, in cell wall, 24
Tellurite medium, 457
Temperature
efficiency of chemical disinfectant and, 199
growth of microorganisms and, 130, *132*
required to kill microorganisms, 204, *204*
skin, changes in, in inflammation, 294
Temperature scale, Celsius, 131*t*
Fahrenheit, 131*t*
Terminal disinfection, 252
Terramycin, 241
Tertian fever, 601
Tetanospasmin, 553
Tetanus, 552
antitoxin of, 553
immunization schedule for, 332*t*
prevention of, 554
spasms with, *553*
toxin of, 553
specificity of, 283
toxicity of, 421*t*
toxigenicity of, 282
toxoid of, 554

Tetanus neonatorum, 552
Tetracycline antibiotics, *239*, 241
Thallophytes, 37, 38*t*
Thayer-Martin medium, 541
Thermometer
  disinfection of, 248, 249, 642*t*
    in community health care, 259
  electronically registering, 248, *250*
  oral, disinfection of, 248
  rectal, disinfection of, 249
  technique of use of, 248
Thermophiles, 130, *132*
Thioglycollate broth, 134, *135*
Throat
  cultures of, 615, *617*
    of *C. diphtheriae,* 457
  discharge from, pathogens in, 624*t*
  microorganisms of, 180*t*, 181
  septic sore, 444, 446
Thrombi, 294
Thrombocytes, 271
Thrush, 570, *570*
Thymocyte, 268, 299, 301
Tick, as vector, in relapsing fever, 583
  in Rocky Mountain spotted fever, 589, *591*
  of viral encephalitis, 597
Tick-borne relapsing fever, 583
  prevention and treatment of, 584
Tick typhus, 589, 589*t*
Tine test, 482
Tinea(s), 560
Tinea capitis, 560
Tinea pedis, 561, *562*
Tissue culture, of viruses, 87, *88*, 147
Titer, rising, diagnostic importance of, 321
T-lymphocytes, 300
  allergy mediation by, delayed-type, 340, 344
  cellular immunity and, 299, 300
  lymphokines released by, cellular effect of, 344*t*, 345
  protein conjugate reaction with, 300, *303*
  stem cell differentiation of, *298*, 299
Tobacco mosaic virus, virion of, *83*
Tonsillitis, due to beta-hemolytic streptococci, 444
Tooth abscess, 442
TOPV, immunization schedule for, 332*t*
Torula, 40
Toxic shock syndrome, 438, *438*
Toxicity, of disinfectant, 198
  of waste products, to microorganisms, 130
  selective, of antibiotic, 232
Toxigenicity, 282
Toxigenicity test, in vitro, 460, *461*
Toxins, 282
  bacterial, toxicity of, 421*t*
  botulinal, 419, 421*t*
  cholera, 379
  clostridial, 420, 421*t*, 550
  diphtheria. See *Diphtheria toxin.*
  E. coli, 377
  erythrogenic, 446
  fungal, 571
  hemolytic, of group A streptococci, 444, 445
  in blood, 280
  Shigella, 374
  specificity of, 283
  staphylococcal, 283, 421*t*, 422, 436
  tetanus. See *Tetanus, toxin of.*
Toxoid(s)
  alum-precipitated, 463, 472
  as immunizing agents, 331
  cholera, 380
  fluid, 463
  tetanus, 554
  triple, 463
TPCF test, 528
TPI test, 528
Trace elements, 128
Tracheotomy, 458
Trachoma, 71, 516, *518*
Transduction, 115
  abortive, 116
Transfer factor, 308
Transfer forceps, 250
Transferase, 106
Transformation, 116
Transfusion, hepatitis risks from, 414*t*
  Rh factor and, 276
Transfusion jaundice, 414
Trematodes, 45
  pathogenic for humans, 400*t*
Trench fever, 68, 588*t*, 593
"Trench mouth," 472
Treponema carateum, 534
Treponema denticola, 53
Treponema pallidum, 53, 473, 522, *523*
  properties of, 524
  immobilization (TPI) test for, 315, 528
Treponema paraluis-cuniculi, 533
Treponema pertenue, 522, 524, 535
Treponema phagedenis, 529
  biotype, Reiter, *53*
Treponema refringens, 53
Treponema vincentii, 472, *473*, 522
Treponematoses, **522**, *522*
Treponemes
  demonstration of, 523
  Nichols, 523
  Reiter, 523
    in complement fixation test for syphilis, 529
Triatoma gerstaeckeri, 607
TRIC agent, 71, 516
Tricarboxylic acid cycle, *113*
Trichina worm, pathogenic, 400*t*
Trichomonas hominis, 396, 399
Trichomonas tenax, 396, 399
Trichomonas vaginalis, 396, 399, 543
Trichomoniasis, 543
  genitourinary, 543
  vaginal, 540*t*, 543
Trichophyton, 562, 563*t*
Trichuris trichiura, egg of, *400*
Trickling filters, 167, *168*
Triple sugar iron agar, 369, *370*
Triple toxoid, 463
Trophozoite, entameba, 395, *396*
  of *Plasmodium,* 600, *604*
  protozoal, 35, *36*, *396*
Truant fluorescence technique, 477
Trypanosoma brucei gambiense, 606, *607*
Trypanosoma brucei rhodesiense, 606
Trypanosoma cruzi, 606
Trypanosoma equiperdum, 606
Trypanosomes, 605
  developmental stages of, 606, *606*
Trypanosomiasis, 605
Tsetse flies, 606

TSI agar, 369, *370*
Tsutsugamushi, 592
  Weil-Felix reactions of, 589*t*
Tube dilution method, of sensitivity testing, 242, *243*
Tubercle, 479, 480
Tubercle bacillus, 476
  disinfection and, 479
  species of, 478
Tuberculin, old, 482
  purified protein derivative, 482
Tuberculin reaction, 345, 482
Tuberculin test, 345, 482
  methods of, 482
Tuberculosis, **479**
  allergy in, 481
  case rates of, 485, *486*
  "childhood," 480
  clinical forms of, 479
  diagnosis of, 483
  exudative lesion of, 480
  geographical distribution of, *486*
  Hansen's disease and, 490
  infections of, 485
    classification of, 481
  prevention of, BCG and, 487
  primary, 479, *479*, *480*
  primary complex, 479, *479*
    healed, *480*
  proliferative lesion of, 480
  reinfection (adult) type, 481
  socioeconomic conditions and, 487
  transmission of, 484
  vaccine for, 330
Tularemia, 579
  prevention of, 581
Turbidity, of bacterial colony, 144
Tween 80, 198
Twort-d'Herelle phenomenon, 83
Tyndall, 191
Tyndallization, 191, 204
Typhoid fever, 361
  carriers of, 366, 367
  diagnosis of, 369
  transmission of, 367
Typhoid vaccine, 330
Typhus
  endemic, 587
  epidemic, 586
  mite, 592
  São Paulo, 589
  scrub, 588*t*, 592
  tick, 589
  Weil-Felix reactions in, 589*t*
Typhus fever, 586, 588*t*
  flea-borne, 587, 588*t*
    case rates of, *590*
  louse-borne, 586, 588*t*
    prevention of, 587
  Weil-Felix reactions in, 587, 589*t*

Ultracentrifugation, of viruses, 148
Ultrafiltration, of viruses, 148
Ultrasonic energy, cleaning by, 210
Ultraviolet rays, exposure to, 137
  sterilization by, 211
Undulant fever, **387**
  prevention of, 389
  serologic diagnosis of, 388
Undulant fever *(Continued)*
  transmission of, 388
  treatment of, 389
Unicellular organisms, 19, **22**
"Universal" donor, 274
Urethritis, gonococcal, 537, *540*
  nongonococcal, 537, 540*t*
    incidence of, 539*t*
Urine
  collection of, for bacteriologic examination, 616
  culture of, in enteric infection, 371
  disinfection of, 641*t*
  enteric viruses in, 411
  examination of, in tuberculosis, 483
  pathogens in, 624*t*
Urkaryote, *17*
Uronychia sp., 51
Urticaria, 343
Uta, 609
Uterus, bacterial infection of, 182

"V" factor, 468
Vaccination. See also *Immunization; Vaccine.*
  BCG, 330
  immune reaction in, 328, *328*
  polio, in pregnancy, 408
  smallpox. See *Smallpox, vaccination against.*
  with bacterial exotoxins, 331
  with dead organisms, 330
  with living, attenuated, or harmless organisms, 326
Vaccine
  against *Streptococcus pneumoniae,* 449
  alum-precipitated, pertussis, 472
  autogenous, 331
  avianized rabies, 503
  bacille Calmette-Guérin, 330
    in tuberculin-positive individuals, 482
    tuberculosis prevention and, 487
  bacterial, 330
  booster doses of, 333
  cholera, 380
  DTP, 463
    schedule for, 332*t*
  duck embryo (DEV), 328, 504, 505*t*
  dysentery, 375
  Edmonston, 499
  for Rocky Mountain spotted fever, 589
  for typhus fever, louse-borne, 587
  Haffkine's, 581
  human diploid cell rabies (HDCV), 328, 504, 505*t*
  influenza, 329, 508, *508*
  lyophilized, 327
  measles, 329, 499
  meningococcal, 451
  mixed, 331, 472
  plague, 581
  polio, 408
  preventive, against human infectious diseases, 332*t*
  Q fever, 592
  "quad," 331
  rabies, 328, 504, 505*t*
  rubella, 329, 500
  Sabin oral polio(OPV), 329, 408
  Salk (IPV), 408
    use and effectiveness of, 408, *409*

Vaccine *(Continued)*
Semple, 328
sensitized, 331
17D, 329, 594
TAB, 373
tuberculosis, 330
typhoid, 330
viral, 329
preparation of, 327, *327*
virus variation and, 96
whooping cough, 472
yellow fever, 329, 594
Vaccinia, 495
Vaccinia virus, for smallpox vaccination, 326
Vaccinoid reaction, 495
Vacuolar membrane, *16*
Vacuole, *16*
phagocytic, 291
Vacutainer bottles, 619, *620*
venting units for, 619, *621*
Vagina, normal microorganisms of, 182
Vaginitis, 541
Candida, 540*t*, 570
due to respiratory *Neisseria,* 542
trichomonas, 540*t*, 543
Van Leeuwenhoek, Antony, 5, *6*
drawings of bacteria by, *6*
original microscope of, 5, *7*
Varicella, 500
characteristics of, 90*t*
Varicella zoster virus, 501
Variola. See *Smallpox.*
Vectors
of blood infection, 574, 576*t*
natural, 575, **576**
of infectious organisms, 357
animate, 358
inanimate, 357
of microbial disease, foods as, **418**
Venereal disease
case-finding in, 545
case rates of, 530, *531*, 531*t*, *538*
incidence of, 539*t*
penicillin in, 530
precautions in, personal and household, 544
prevention of, 544
transmission of, 356, 357*t*
treatment of, 544
Venereal granulomatoses, 517
prevention of, 519
Venereal herpes, 540*t*, 544
characteristics of, 90*t*
incidence of, 539*t*
Venezuelan equine encephalitis, 597
Ventilation, in control of respiratory infection, 435
Venule, 264
Verruca, characteristics of, 91*t*
Vi agglutinins, of typhoid, 371
Vibrios, 50, *52*, 60*t*, 62
El Tor biotype, 52, 380
Vibrio cholerae, 379, *379*
differential characters of, 363*t*
heat resistance of, 425*t*
Vibrio fetus, 379
Vibrio parahaemolyticus, 381, 422, 424
Villemin, Jean, 476
Vincent's angina, 472
Viral conversion, 116
Viral encephalitides, mosquito-borne, 596
tick-borne, 597
Viral hemagglutination, 321
Viral infection, **89**, 90–91*t*
arthropod-borne, **593**
enteric, **404**
immunity against, 346
interferon synthesis with, 307, *307*
respiratory, **494**
transmission of, 510
Viral vaccines, 329
preparation of, 327, *327*
virus variation and, 96
Viremia, 182
Virion, 27, 76
components of, *77*
forms of, *81*
of serum hepatitis, 575
Viroids, 96
Viropexis, 84
Virucide, 192
Virulence, 281
variations in, 122, 283
Virus, 22, **76**. See also *Bacteriophage.*
antigens of, 96
arbovirus, 593
Bittner, *81*
boiling of, effectiveness of, 258
capsid of, *77*, 81
characteristics of, 21*t*, 72*t*, **76**
CHINA, 404, 498
classification of, 77, 80*t*
"cold," 509
complex, *81*, 82
coxsackie, 410
immunological types of, *406*, 411
cultivation of, 87, *88*, **147**
cytomegalovirus, 501
cytopathic effect of, 88, *88*
dermotropic, 89
diseases caused by. See *Viral infection.*
disinfectant resistance of, 89
DNA, tumor-causing, *78*, 93*t*, 94
ECBO, 404
ECHO, 404, 411
immunological types of, *406*
enteric, in urine, 411
enterotropic, 94
Epstein-Barr, 94
filtration of suspensions of, *79*
heat resistance of, 88
helical, 82, *83*
hepatitis. See *Hepatitis virus.*
herpes simplex, 501
type 2, 94, 544
histologic modification of, 95
icosahedral, 82
identification of, 76
infective particle of. See *Virion.*
influenza, types of, 507
measles, 499
microscopy of, 81, *81, 82*
modifications of, 95
neurotropic, 89, 405
NITA, 404
nutrition of, 129
observation of, 148
oncogenic, 92–93*t*, 94
parainfluenza (PIV), 509
pathogenic, tissues affected by, 89, 90–91*t*
plaque formation by, 148
pneumotropic, 89
polio. See *Poliovirus.*

Virus *(Continued)*
poxvirus, *81*
propagation of, selective, 95
techniques of, 87
rabies, *81*, 503
REO, 404, 411
respiratory, transmission of, 510
respiratory syncytial (RSV), 509
RNA, tumor-causing, *78*, 92*t*, 94
salivary gland, 501
shape of, *78*, 82
size of, *78*, 81, *82*
slow, 95
tobacco mosaic, virion of, *83*
unusual, 96
vaccinia, 326
variation of, vaccines and, 96
varicella zoster, 501
vertebrate, classification of, 80*t*
viroids, 96
viscerotropic, 89
zoologic modification of, 95
Vollmer test, 482
Volutin, 59, *59*
of *C. diphtheriae,* 456, *458*
Vomit, "black," 391
transmission of infection through, arthropod-borne, 576
Von Behring, Emil, *11*
Von Pirquet test, 482
Vulvovaginitis, 541

Waksman, Selman A., *238*, 240
Wandering cells, 268
Warburg-Dickens-Lipmann pathway, 110
Warts, characteristics of, 91*t*
Wassermann test, 315
Water
boiling, sterilization by, 204, *204*
effectiveness of, 256
drinking, bacteriological examination of, 171, *172, 173*
membrane filter method of, 172, *173*
purification of, 169
fecal pollution of, coliform indicators of, 171, *172*, 377
municipal, purification of, 167, *168*
reclamation of, 170, *171*
polluted, enteric infection and, 368, *369*
purification of, **165**
solubility in, and effect of disinfectant, 197
Water reclamation plant, 170, *171*
Watkin's solution, 377
Wayson's stain, 557
Weil-Felix reaction, 587, 589*t*
Weil's disease, 391
Wescodyne, 198, 221
Wet Air Oxidation Unit, for pressure cooking of sewage, 169, *170*
Wet mount, 151
Whipworm, egg of, *400*
Whooping cough. See *Pertussis.*
Widal test, 371
Wiener Rh system, 275, 276*t*
Wool-sorter's disease, 556
Worms, filarial, 610
pathogenic, 400*t*
helminthic. See *Helminths.*
Wuchereria bancrofti, microfilaria of, *610*, 611

"X" factor, 468
X-rays, cellular effect of, 138
in diagnosis of tuberculosis, *480,* 484
sterilization by, 211

Yaws, 522, *522*, 533
Yeasts, 36, **37**
baker's, 38
brewer's, *39*
budding of, 39, *39*
cell structure of, 38
characteristics of, 20t, **37**
habitat of, 38
multiplication of, 39
asexual, 39, *39*
sexual, 39, *40*
nonpathogenic, activities of, 40
pure culture of, 40
"wild," 40
Yellow fever, 593
characteristics of, 91*t*
jungle, 594, 595*t*
vaccine for, 594
17D, 329, 594
Yersin, A.E.J., 577
Yersinia pestis, 577, 578
classification of, 361, 362*t*
Yersinia pseudotuberculosis, 577
Yolk sac injection, for rickettsia cultivation, 69, *70*

Zephiran, 198
Ziehl-Neelsen method, 476
Zinc peroxide, 225
Zoonoses, **577**
Zovirax, 501, 544
Zygospore, 43, *44*